5th International Workshop DICE2010: Space-Time-Matter – Current Issues in Quantum Mechanics and Beyond

Journal of Physics: Conference Series Volume 306

Tuscany, Italy
13 – 17 September 2010

ISBN: 978-1-61839-050-9
ISSN: 1742-6588

Printed from e-media with permission by:

Curran Associates, Inc.
57 Morehouse Lane
Red Hook, NY 12571

Some format issues inherent in the e-media version may also appear in this print version.

Printed by Curran Associates, Inc. (2011)

For permission requests, please contact the Institute of Physics
at the address below.

Institute of Physics
Dirac House, Temple Back
Bristol BS1 6BE UK

Phone: 44 1 17 929 7481
Fax: 44 1 17 920 0979

techtracking@iop.org

Additional copies of this publication are available from:

Curran Associates, Inc.
57 Morehouse Lane
Red Hook, NY 12571 USA
Phone: 845-758-0400
Fax: 845-758-2634
Email: curran@proceedings.com
Web: www.proceedings.com

5th International Workshop DICE2010: Space-Time-Matter - Current Issues in Quantum Mechanics and Beyond

Journal of Physics: Conference Series Volume 306

Tuscany, Italy
13 - 17 September 2010

ISBN: 978-1-61839-050-9
ISSN: 1742-6588

IOP Institute of Physics

TABLE OF CONTENTS

INVITED LECTURES

CONTRIBUTED PAPERS

PREFACE

These proceedings present the Invited Lectures and Contributed Papers of the *Fifth International Workshop on Decoherence, Information, Complexity and Entropy – DICE 2010*, held at **Castello Pasquini, Castiglioncello (Tuscany), 13–17 September 2010**.

These proceedings are intended to document the stimulating exchange of ideas at this conference for both the interested public and the wider scientific community, as well as for the participants. The number of participants attending this series of meetings has been growing steadily, which reflects its increasing attraction. Our intention to bring together leading researchers, advanced students, and renowned scholars from various areas in order to stimulate new ideas and their exchange across the borders of specialization seems to bear fruit.

In this way, the series of meetings has continued successfully from the beginning with DICE 2002 [1], followed by DICE 2004 [2], DICE 2006 [3], and DICE 2008 [4], uniting more than 100 participants representing almost 30 countries worldwide.

It has been a great honour and inspiration to have Professor Luc Montagnier (Nobel Prize for Medicine 2008) from the World Foundation for AIDS Research and Prevention with us, who presented the lecture *DNA waves and water* (included in this volume).

The discussions took place under the wider theme *Space-Time-Matter – current issues in quantum mechanics and beyond* in the very pleasant and inspiring atmosphere of Castello Pasquini, which – with its beautiful surroundings, overlooking the Tuscany coast – hosted the conference very successfully for the second time. The five-day program was grouped according to the following topics:

- Gravity and Quantum Mechanics
- Quantum Coherent Processes in Biology / Many-Body Systems
- From Quantum Foundations to Particle Physics
- The Deep Structure of Spacetime
- Quantum – Relativity – Cosmology

A Public Roundtable Discussion formed an integral part of the program under the theme *Sull' Onda Della Coerenza" – le nuove frontiere della scienza moderna* with the participation of E Del Giudice (INFN & Università di Milano), L Fronzoni (Università di Pisa) and G Vitiello (Università di Salerno). By now forming a tradition, this evening event drew a large audience, who participated in lively discussions until late.

The workshop was organized by L Diósi (Budapest), H-T Elze (Pisa, chair), L Fronzoni (Pisa), J Halliwell (London), E Prati (Milano) and G Vitiello (Salerno), with essential help from our conference seceretaries M Pesce-Rollins and L Baldini and from our students G Gambarotta and F Vallone, all from Pisa.

Several institutions and sponsors supported the workshop; their representatives and, in particular, the citizens of Rosignano / Castiglioncello are deeply thanked for their generous help and kind hospitality:

- *Comune di Rosignano* – A Franchi (Sindaco di Rosignano), S Scarpellini (Segreteria sindaco), L Benini (Assessore ai lavori pubblici), M Pia (Assessore all' urbanistica).
- *REA Rosignano Energia Ambiente s.p.a.* – F Ghelardini (Presidente della REA), E Salvadori (Segreteria).
- *Associazione Armunia* – M Paganelli (Direttore), G Mannari (Programmazione). Special thanks go to G Mannari and her collaborators for their advice and great help in all the practical matters that had to be dealt with in order to run the meeting at Castello Pasquini smoothly.

Funds made available by Università di Pisa, by Domus Galilaeana (Pisa), Centro Interdisciplinare per lo Studio dei Sistemi Complessi – CISSC (Pisa), Dipartmento di Matematica e Informatica (Università di Salerno), Instituto Italiano per gli Studi Filosofici – IISF (Napoli), and by the Hungarian Scientific Research Fund OTKA, are gratefully ackowledged.

Last, but not least, special thanks are due to Laura Pesce (Vitrium Galleria, San Vincenzo) for the exposition for her artwork *Dal io al cosmo* at Castello Pasquini during the conference.

The papers presented at the workshop and collected here have been edited by L Diósi, H-T Elze, L Fronzoni, J J Halliwell, E Prati, G Vitiello and J Yearsley. The proceedings essentially follow the order of presentation during the conference program, however, divided into Invited Lectures and Contributed Papers. (We regret that lectures by D Bouwmeester, G G Guerreschi, G C Ghirardi and C Kiefer could not be reproduced here, partly for copyright reasons.) In the name of all the participants, we would like to thank S Toms and G Douglas, and their collaborators at IOP Publishing (Bristol) for their friendly advice and most valuable and immediate help during the editing process and, especially, for their continuing efforts to make the *Journal of Physics: Conference Series* available to all.

Budapest, Pisa, London, Milano and Salerno, May 2011

Lajos Diósi, Hans-Thomas Elze, Leone Fronzoni, Jonathan Halliwell, Enrico Prati, Guiseppe Vitiello and James Yearsley

[1] Elze H-T (ed) 2004 Decoherence and Entropy in Complex Systems *Lecture Notes in Physics* **633** (Berlin: Springer)
[2] Elze H-T (ed) 2005 Proceedings of the Second International Workshop on Decoherence, Information, Complexity and Entropy – DICE 2004 *Braz. Journ. Phys.* **35**

2A and B pp 205-529
free access at: www.sbfisica.org.br/bjp
[3] Elze H-T, Diósi L, Fronzoni L, Halliwell J J and Vitiello (eds) 2007 Proceedings of the Third International Workshop on Decoherence, Information, Complexity and Entropy – DICE 2006 *J. Phys.: Conf. Ser.* **67**
free access at: www.iop.org/EJ/toc/1742-6596/67/1
[4] Elze H-T, Diósi L, Fronzoni L, Halliwell J J and G Vitiello (eds) 2009 Proceedings of the Fourth International Workshop on Decoherence, Information, Complexity and Entropy – DICE 2008 *J. Phys.: Conf. Ser.* **174**
free access at: www.iop.org/EJ/toc/1742-6596/67/1

5th International Workshop DICE2010 IOP Publishing
Journal of Physics: Conference Series **306** (2011) 012001 doi:10.1088/1742-6596/306/1/012001

Lessons from classical gravity about the quantum structure of spacetime

Thanu Padmanabhan
IUCAA, Pune University Campus, Ganeshkhind, Pune 411007, INDIA
E-mail: paddy@iucaa.ernet.in

Abstract. I present the theoretical evidence which suggests that gravity is an emergent phenomenon like gas dynamics or elasticity with the gravitational field equations having the same status as, say, the equations of fluid dynamics/elasticity. This paradigm views a wide class of gravitational theories — including Einstein's theory — as describing the thermodynamic limit of the statistical mechanics of 'atoms of spacetime'. Strong internal evidence in favour of such a point of view is presented using the classical features of the gravitational theories with just one quantum mechanical input, viz. the existence of Davies-Unruh temperature of horizons. I discuss several conceptual ingredients of this approach.

1. How can *classical* gravity teach us anything about *quantum* spacetime?

In one sentence, the paradigm that we will explore [1] is the following: Gravity is an emergent phenomenon like gas dynamics or elasticity with the gravitational field equations having the same status as, say, the equations of fluid dynamics/elasticity. Historically, this paradigm originated with Sakharov [2] and was interpreted in different ways by Jacobson [3], Volovik [4], Bei-Lok Hu [5] and many others. (Analogue models [6] as well as the membrane paradigm [7] for black holes have some similarities with this approach. For a small sample of more recent work reassembling different aspects of these ideas, see [8]). I will now elaborate on this theme drawing mainly from the work I was involved in. [1]

Part of this programme involves a "top-down" approach to quantum spacetime (in the sense of zooming in from the top to smaller and smaller spatial scales, like in a Google map) to learn key lessons (Sections 1–5), which are then used to provide a derivation of field equations from extremising spacetime entropy density (Section 6). Some people use the word "top-down" to mean exactly the opposite; I will use "top-down" the way I have defined it, viz. from classical to quantum domain.

One may find it surprising that such an attempt, to determine the features of the microscopic theory from knowing its properties at the macroscopic scales., is so successful. Of course, in the *strict* sense, classical theories cannot tell us anything about quantum dynamics; after all, classical physics, by definition, is independent of $\hbar$ while quantum effects do depend on $\hbar$. But there is one effect, viz., the thermodynamics of spacetime horizons [9] which brings together the principles of quantum theory and gravity. This fact, along with a judicious choice for the questions to ask, allows one to make a fairly persuasive case for the structure of quantum

[1] I will use mostly positive signature with English alphabets covering $0, 1, ...D-1$ and Greek alphabets covering the spatial indices $1, 2, ...D-1$ of a D dimensional spacetime.

spacetime. To see such a "top-down" approach in context, let me describe at least three other — more conventional — examples in which the deeper, more exact, ('bottom layer') theory leaves a tell-tale signature on the 'top layer'.

(i) *Electrons in a helium atom*: Suppose you manage to solve the Schrödinger equation for the two electrons in the helium atom and determine the energy eigenfunctions $\psi(\boldsymbol{x}_1, \boldsymbol{x}_2)$. Your experimental friend will tell you that only half of these wave functions [which are antisymmetric under the interchange $(\boldsymbol{x}_1 \Leftrightarrow \boldsymbol{x}_2)$] occur in the real world. Any amount of your staring at the Schrödinger equation for the helium atom will not tell you why nature requires this antisymmetry under pair exchange for electrons. The reason lies deep down in relativistic quantum theory but its residual effect remains as a tell-tale signature even in the $c = \infty$ limit of field theory, viz., the non-relativistic quantum mechanics.

(ii) *Boltzmann's conjecture of atoms*: Classical thermodynamics of a gas/fluid uses variables like density, pressure etc. in the continuum description. But the fact that such a fluid can store and exchange heat energy *cannot* be understood within the continuum theory. Boltzmann had the insight to suggest that thermal phenomena *demand* the existence of microscopic degrees of freedom in matter. In fact, the law of equipartition, expressed as $E/[(1/2)k_BT] = N$ relates two thermodynamic variables E, T (which are well-defined for a continuum fluid) to N, the number density of microscopic degrees of freedom, which cannot be interpreted in the continuum limit at all. The Avogadro's number, closely related to N, was determined even before we fully understood what exactly it counts and without any direct evidence for the molecular structure of matter. This is another example of our being able to say something about microscopic structure from the features of macroscopic theory.

(iii) *Equality of inertial and gravitational masses*: The most dramatic example is provided by Einstein's use of principle of equivalence which, of course, was known for centuries. Einstein realized that $m_i = m_g$ is not a trivial algebraic accident which should be taken for granted, as others before him have done, but requires an explanation. This led him to the description of gravity in terms of the geometry of spacetime. The relation $m_i = m_g$ was a signature of the deeper theory discernible in the approximate (top-layer) description.

The lesson from these three examples is obvious. For a top down approach to be useful, you need to ask the right questions! One way is to pick up features of the theory that are usually taken for granted ('algebraic accidents') — or not even noticed — and demand deeper explanations for them. This is the procedure I will follow in this programme to probe the quantum structure of spacetime from known aspects of classical gravity.

2. The conceptual background

I will describe several peculiar features of classical gravitational theories from which we can obtain a broad picture regarding the quantum microstructure of the spacetime, in the form of a series of "lessons". Most of these (starting from lesson 4!) will be specific mathematical features of the theory. But to provide the necessary backdrop, I will distill out of these mathematical features three conceptual points and describe them right at the outset.

Lesson 1: Providing a quantum description of spacetime structure is quite different from constructing a quantum theory of gravity.

In this approach, it is necessary to make a clear distinction between quantum description of spacetime structure and a theory of quantum gravity.

Classical field equations of gravity *also* happens to describe the classical dynamics of the spacetime because of the geometrical interpretation. In the emergent gravity paradigm, these field equations have a status similar to the equations of fluid mechanics or elasticity. So, if this paradigm is correct, one should *not* expect quantizing a classical theory of gravity to lead us to the quantum structure of spacetime any more than quantizing the equations of elasticity or hydrodynamics will lead us to atomic structure of matter! Quantizing the elastic vibrations of a solid will lead only to phonon physics [5] just as quantizing a classical theory of gravity will lead to graviton physics. The latter could be quite different from a description of quantum structure of spacetime just as phonon physics is quite different from the physics of the atom.

Lesson 2: The guiding principle to use for understanding the quantum microstructure of the spacetime should be the thermodynamics of horizons.

Combining the principles of GR and quantum theory is not a technical problem that could be solved just by using sufficiently powerful mathematics. It is more of a conceptual issue and decades of failure of sophisticated mathematics in delivering quantum gravity indicates that we should try a different approach. This is very much in tune with item (iii) mentioned in Sec. 1. Einstein did not create a sophisticated mathematical model for m_i and m_g and try to interpret $m_i = m_g$. He used thought experiments to arrive at a conceptual basis in which $m_i = m_g$ can be embedded naturally so that $m_i = m_g$ will cease to be an algebraic accident. Once this is done, physics itself led him to the maths that was needed.

Of course, the key issue is what could play the role of a guiding principle similar to principle of equivalence in the present context. *For this, my bet will be on the thermodynamics of horizons.*[1, 10] A successful model will have the connection between horizon thermodynamics and gravitational dynamics *at its foundation* rather than this feature appearing as a result derived in the context of certain specific solutions to the field equations. We will see evidence for its importance throughout the discussion in what follows.

Lesson 3: Think beyond Einstein gravity, black hole thermodynamics and think off-shell.

There are four technical points closely related to the above conjecture (viz., thermodynamics of horizons should play a foundational role) which needs to be recognized if this approach has yield dividends:

- One must concentrate on the general context of observer dependent, local, thermodynamics associated with the local horizons, going beyond the *black hole* thermodynamics. Black hole horizons in the classical theory are far too special, on-shell, global constructs to provide a sufficiently general back-drop to understand the quantum structure of spacetime. The preoccupation with the black hole horizons loses sight of the conceptual fact that all horizons are endowed with temperature as perceived by the appropriate class of observers. Observer dependence [11] of thermal phenomena is a feature and not a bug!
- One should also think beyond Einstein's theory and use the structure of, say, Lanczos-Lovelock models of gravity [12] in exploring the microstructure of spacetime. Previous work (starting from ref. [13]) has shown that the interpretation of gravity as an emergent phenomenon transcends Einstein's theory and remains applicable to (at least) all Lanczos-Lovelock models. Exploiting this connection will allow us to discriminate between results of general validity from those which are special to Einstein's theory in $D = 4$. Irrespective of whether Lanczos-Lovelock models are relevant to real world, they provide a good test-bed to see which concepts and results are robust and general.
- A corollary is that one should *not* think of entropy of horizons as being proportional to their area. This result, which is true in Einstein's theory, fails for all higher order Lanczos-Lovelock models [14]. But all the general thermodynamic features still continue to remain valid. Because area brings in several other closely related geometrical notions,

restricting oneself to Einstein's theory leads to an incorrect view of what entropy and quantum microstructure of spacetime could be.

- The quantum features of a theory are off-shell features. But, fortunately, classical action principles provide a window to quantum theory because of the path integral formalism. Therefore any peculiar feature of classical action principle could give us insights into the underlying quantum theory much more than the structure of field equations. This suggests that we need to look at the off-shell structure of the theory using the form of action principles rather than tie ourselves down to field equations.
  These ingredients, to a great extent, distinguish the approach I was developing from those of many others.

3. Lessons from the thermodynamics of horizons

Having outlined the broad conceptual features, I will now move on to specifics. There are four lessons one can learn from putting together well-known features of horizon thermodynamics in an appropriate manner.

Lesson 4: Temperature of horizons does not depend on the field equations of the theory and is just an indication that spacetimes, like matter, can be hot, but in a observer-dependent manner.

One can associate a temperature with any null surface that can act as horizon for a class of observers, in any spacetime (including flat spacetime). This temperature is determined by the behaviour of the metric close to the horizon and has *nothing to do with the field equations* (if any) which are obeyed by the metric.

The simplest situation is that of Rindler observers in flat spacetime with acceleration κ who will attribute a temperature $k_BT = (\hbar/c)(\kappa/2\pi)$ to the Rindler horizon — which is just a $X = T$ surface in the flat spacetime having no special significance to the inertial observers. While this result is usually proved for an eternally accelerating observer, they also hold in the (appropriately) approximate sense for an observer with variable acceleration [15]. In general, this result can be used to show that the vacuum state in a freely falling frame will appear to be a thermal state in the locally accelerated frame for high frequency modes if κ^{-1} is smaller than the local radius of (spacetime) curvature.

In the usual context of a bifurcation horizon that divides the spacetime into two causally disconnected regions R and L, the global vacuum state $|\text{vac}\rangle$ of a quantum field theory can be described by a vacuum functional $\langle\text{vac}|\phi_L,\phi_R\rangle$ in terms of the field configurations ϕ_L in L and ϕ_R in R. Using the Euclidean path integral representation for the ground state functional, one can express this functional in two different ways and obtain:

$$\langle\text{vac}|\phi_L,\phi_R\rangle \propto \int_{T_E=0;\phi=(\phi_L,\phi_R)}^{T_E=\infty;\phi=(0,0)} \mathcal{D}\phi e^{-A} \propto \int_{\kappa t_E=0;\phi=\phi_R}^{\kappa t_E=\pi;\phi=\phi_L} \mathcal{D}\phi e^{-A} \propto \langle\phi_L|e^{-(\pi/\kappa)H_R}|\phi_R\rangle \quad (1)$$

where H_R is the Hamiltonian describing the dynamics in one of the wedges [16] and κ is the acceleration. Both path integrals cover the upper-half of the Euclidean $X - T_E$ plane. The first path integral is in the global coordinate system (inertial, Kruskal ....) with time T_E running from $T_E = 0$ to $T_E = \infty$ with the boundary conditions at both limits as indicated. The second path integral is in the coordinate system adapted to region outside the horizon (Rindler, Schwarzschild .....) with the time coordinate behaving like a polar angle in the plane, going from $\kappa t_E = 0$ (the right wedge) to $\kappa t_E = \pi$ (the left wedge), with the fields taking appropriate boundary values in the two limits. Thus we get:

$$\langle\text{vac}|\phi_L,\phi_R\rangle \propto \langle\phi_L|e^{-(\pi/\kappa)H_R}|\phi_R\rangle \quad (2)$$

For describing the physics in the region outside the horizon, say in R, we will trace out the modes ϕ_L beyond the horizon. This gives a thermal density matrix for the observables in the right wedge:

$$\rho(\phi'_R, \phi_R) \propto \int \mathcal{D}\phi_L \langle \phi_L, \phi'_R|\text{vac}\rangle\langle\text{vac}|\phi_L, \phi_R\rangle \propto \langle \phi'_R|e^{-(2\pi/\kappa)H_R}|\phi_R\rangle \tag{3}$$

corresponding to the horizon temperature $T = \kappa/2\pi$. This result only depends on the near horizon geometry having the approximate form of a Rindler metric and is independent of the field equations of the theory.

Lesson 5: All thermodynamic variables are observer dependent.

An immediate consequence, not often emphasized, is that *all* thermodynamic variables must become observer dependent if vacuum acquires an observer dependent temperature. A "normal" gaseous system with "normal" thermodynamic variables (T, S, F etc.....) must be considered as a highly excited state of the inertial vacuum. It is obvious that a Rindler observer will attribute to this highly excited state different thermodynamic variables compared to what an inertial observer will attribute. Thus thermal effects in the accelerated frame brings in [11, 17] a new level of observer dependence even to *normal* thermodynamics. One need not panic if variables like entropy now acquire an observer dependence and lose their absolute nature.

Lesson 6: In sharp contrast to temperature, the entropy of horizons depends on the field equations of gravity and cannot be determined by using just QFT in a background metric.

One would have expected that if integrating out certain field modes leads to a thermal density matrix ρ, then the entropy of the system should be related to lack of information about the *same* field modes and should be given by $S = -\text{Tr}\, \rho \ln \rho$. This entropy, called entanglement entropy, (i) is proportional to area of the horizon and (ii) is divergent without a cut-off [18]. Such a divergence makes the result meaningless and thus we cannot attribute a unique entropy to horizon using just QFT in a background metric.[2] That is, while the temperature of the horizon can be obtained through the study of test-QFT in an external geometry, one cannot understand the entropy of the horizon by the same procedure.

This is because, unlike the temperature, the *entropy* associated with a horizon in the theory depends on the field equations of the theory, which we will briefly review. Given the principle of equivalence (interpreted as gravity being spacetime geometry) and principle of general covariance, one could still construct a wide class of theories of gravity. For example, if we take the action functional

$$A = \int d^D x \sqrt{-g} \left[L(R^{ab}_{cd}, g^{ab}) + L_{\text{matt}}(g^{ab}, \phi_A) \right] \tag{4}$$

where L_{matt} is the matter Lagrangian (for some matter variables denoted symbolically as ϕ_A) and vary the metric with appropriate boundary conditions, we will get the field equations (see e.g., chapter 15 of [20]):

$$\mathcal{G}_{ab} = P_a^{\ cde} R_{bcde} - 2\nabla^c\nabla^d P_{acdb} - \frac{1}{2} L g_{ab} \equiv \mathcal{R}_{ab} - \frac{1}{2} L g_{ab} = \frac{1}{2} T_{ab} \tag{5}$$

[2] In the literature, one often "regularizes" the expression for entanglement entropy by introducing a Planck scale cut-off by hand. This has no justification because of two reasons. First, a free quantum field theory in flat spacetime should not require any cut-off to give meaningful results. Second, in the conventional approach, there is no way a flat spacetime ($G = 0$) quantum field theory will know anything about Planck length. In fact, the divergence of entanglement entropy and the need for a Planck scale cut-off is an indication that there is no such thing as flat spacetime, just as there is no such thing as classical, continuum solid [19].

where $P^{abcd} \equiv (\partial L/\partial R_{abcd})$. A nice subclass of theories in which the field equations remain second order in the metric is obtained if we choose L such that $\nabla_a P^{abcd} = 0$. The most general scalar functionals $L(R^{ab}_{cd}, g^{ij})$ satisfying this condition are specific polynomials in curvature tensor which lead to the Lanczos-Lovelock models [12] with the field equations:

$$P^{de}_{ac} R^{bc}_{de} - \frac{1}{2} L \delta^b_a = \mathcal{R}^b_a - \frac{1}{2m} \mathcal{R} \delta^b_a = \frac{1}{2} T^b_a; \quad \mathcal{R}^b_a \equiv P^{de}_{ac} R^{bc}_{de}; \qquad \mathcal{R} = \mathcal{R}^a_a \tag{6}$$

The second form of the equation is valid for the $m-$th order Lanczos-Lovelock model for which $\mathcal{R} = R^{abcd}(\partial L/\partial R^{abcd}) = mL$. In the simplest context of $m = 1$ we take $L \propto R = R/16\pi$ (with conventional normalization), leading to $P^{ab}_{cd} = (32\pi)^{-1}(\delta^a_c \delta^b_d - \delta^a_d \delta^b_c)$, we get $\mathcal{R}^a_b = R^a_b/16\pi, \mathcal{G}^a_b = G^a_b/16\pi$ and one recovers Einstein's equations. The structure of the theory is essentially determined by the tensor P^{ab}_{cd} which has the algebraic symmetries of curvature tensor and is divergence-free in all indices.

In any such generally covariant theory, the infinitesimal coordinate transformation $x^a \to x^a + q^a$ leads to the conservation of a Noether current J^a (which depends on q^a) given by:

$$J^a \equiv \left(2\mathcal{G}^a_b q^b + L q^a + \delta_q v^a\right) = 2\mathcal{R}^a_b q^b + \delta_q v^a; \qquad \nabla_a J^a = 0. \tag{7}$$

where $\delta_q v^a$ represents the boundary term in the action which arises for the variation of the metric in the form $\delta g^{ab} = (\nabla^a q^b + \nabla^b q^a)$. Given $\nabla_a J^a = 0$, we can introduce an anti-symmetric tensor J^{ab} by $J^a = \nabla_b J^{ab}$. For the Lanczos-Lovelock models, one can determine $\delta_q v^a$ and show that the J^{ab} and J^a can be expressed in the form

$$J^{ab} = 2P^{abcd} \nabla_c q_d; \qquad J^a = 2P^{abcd} \nabla_b \nabla_c q_d \tag{8}$$

The field equations of Lanczos-Lovelock models, possess black hole solutions (with horizons) in asymptotically flat spacetime. Studying the physical processes occurring in such spacetimes, one can obtain an expression for the entropy of the horizon (called Wald entropy [14]) which is closely related to the Noether current J^a as follows:

$$S_{\rm Noether} \equiv \beta \int d^{D-1}\Sigma_a \, J^a = \beta \int d^{D-2}\Sigma_{ab} \, J^{ab} = \frac{1}{4} \oint_{\mathcal{H}} (32\pi \, P^{ab}_{cd}) \epsilon_{ab} \epsilon^{dc} d\sigma \tag{9}$$

where $\beta^{-1} = \kappa/2\pi$ is the horizon temperature and J^a is the Noether current $q^a = \xi^a$ where ξ^a is the Killing vector corresponding to time translation symmetry of the asymptotically static black hole solution. In the final expression the integral is over any surface with $(D-2)$ dimension which is a spacelike cross-section of the Killing horizon on which the norm of ξ^a vanishes, with ϵ_{ab} denoting the bivector normal to the bifurcation surface. Thus horizon entropy is given by an integral over the horizon surface of the P^{abcd}, which we may call the *entropy tensor* of the theory. Note that the Noether current J^a multiplied by $\beta_{loc} \equiv N\beta$, where N is the lapse function, can be thought of as the entropy current density.

In Einstein's theory, with $32\pi \, P^{ab}_{cd} = (\delta^a_c \delta^b_d - \delta^a_d \delta^b_c)$, the entropy will be one quarter of the area of the horizon. But in general, the entropy of the horizon is *not* proportional to the area and depends on the theory.[3]

This dichotomous situation as regards temperature versus entropy is the first indication that the thermodynamics of the horizon, probed by QFT in a external gravitational field, is just the tip of an iceberg. As we will see the emergent paradigm provides a better understanding of these features.

[3] This feature again shows that the entanglement entropy cannot be identified with the entropy of the Lanczos-Lovelock models without modifying the regularization procedure. In the emergent paradigm one can argue that such a modification is indeed required. Then, using a generalisation of ideas described in ref.[21], one can possibly tackle this issue. I will not this discuss here for more details, see ref. [19].

Lesson 7: The connection between horizon entropy and the conserved current arising from the diffeomorphism invariance demands deeper understanding.

Why should a current J^a, conserved due to diffeomorphism invariance of the theory have anything to do with a thermodynamical variable like entropy of horizons in the theory ?

In the conventional approach, which views $x^a \to x^a + q^a$ as a relabeling of coordinates, this question has no answer. In contrast, if we take the 'active' point of view, we notice that it shifts (virtually) the location of null surfaces and thus the information accessible to specific observers. The connection with entropy arises due to the cost of gravitational entropy involved in the virtual displacements of null horizons.

This idea can be made more precise in terms of entropy balance at local Rindler horizons [22]. Let us choose any event $\mathcal{P}$ and introduce a local inertial frame (LIF) around it with Riemann normal coordinates $X^a = (T, \mathbf{X})$ such that $\mathcal{P}$ has the coordinates $X^a = 0$ in the LIF. Let k^a be a future directed null vector at $\mathcal{P}$ and we align the coordinates of LIF such that it lies in the $X - T$ plane at $\mathcal{P}$. We next transform from the LIF to a local Rindler frame LRF with acceleration κ along the X axis. Let ξ^a be the approximate Killing vector corresponding to translation in the Rindler time such that the vanishing of $\xi^a\xi_a \equiv -N^2$ characterizes the location of the local horizon $\mathcal{H}$ in LRF. Usually, we shall do all the computation on a time-like surface infinitesimally away from $\mathcal{H}$ with $N =$ constant, called a "stretched horizon". Let the time-like unit normal to the stretched horizon be r_a.

Consider an infinitesimal displacement of a local patch of the stretched horizon in the direction of r_a, by an infinitesimal proper distance ϵ, which will change the proper volume by $dV_{prop} = \epsilon\sqrt{\sigma}d^{D-2}x$ where σ_{ab} is the metric in the transverse space. The flux of energy through the surface will be $T^a_b\xi^b r_a$ and the corresponding entropy flux can be obtained by multiplying the energy flux by $\beta_{\rm loc} = N\beta$. Hence the 'loss' of matter entropy to the outside observer because the virtual displacement of the horizon has engulfed some matter is $\delta S_m = \beta_{\rm loc}\delta E = \beta_{\rm loc}T^{aj}\xi_a r_j dV_{prop}$. Recalling from Eq. (9) that $\beta_{loc}J^a$ gives the gravitational entropy current, the change in the gravitational entropy is given by $\delta S_{\rm grav} \equiv \beta_{loc} r_a J^a dV_{prop}$ where J^a is the Noether current corresponding to the local Killing vector ξ^a given by $J^a = 2\mathcal{G}^a_b\xi^b + L\xi^a$. As the stretched horizon approaches the true horizon, it can be shown that $Nr^a \to \xi^a$ and $\beta\xi^a\xi_a L \to 0$. Hence we get, in this limit: $\delta S_{\rm grav} \equiv \beta\xi_a J^a dV_{prop} = 2\beta\mathcal{G}^{aj}\xi_a\xi_j dV_{prop}$. Comparing $\delta S_{\rm grav}$ and δS_m we see that the field equations $2\mathcal{G}^a_b = T^a_b$ can be interpreted as the entropy balance condition $\delta S_{grav} = \delta S_{matt}$ thereby providing direct thermodynamic interpretation of the field equations as local entropy balance in local Rindler frame.

In the emergent paradigm, the spacetime is analogous to a solid made of atoms and $x^a \to x^a + q^a(x)$ is analogous to the deformation of an elastic solid [23]. When such a deformation leads to changes in accessible information — like when one considers the virtual displacements of horizons — it costs some amount of gravitational entropy thereby providing a direct link between the transformation $x^a \to x^a + q^a(x)$ and spacetime entropy — a link that is lacking in the conventional approach. We will say more about this in Sec. 6.

4. Thermodynamic interpretation of field equations and action functionals

I stressed in Sec. 1 that for the top-down approach to be of use, we need to identify the 'algebraic accidents' in the top level description which are usually taken for granted without a demand for explanation. I will briefly summarize three such issues in classical gravity, which can give us clues about the microscopic theory.

Lesson 8: The gravitational field equations reduce to a thermodynamic identity on the horizon in a wide class of theories.

It can be shown that [24] the field equations in any Lanczos-Lovelock model, when evaluated on a static solution of the theory which has a horizon, can be expressed in the form of a thermodynamic identity $TdS = dE_g + PdV$. Here S is the correct Wald entropy of the horizon in the theory, E_g is a geometric expression involving an integral of the scalar curvature of the sub-manifold of the horizon and PdV represents the work function of the matter source. The differentials dS, dE_g etc. should be thought of as indicating the difference in S, E_g etc between two solutions in which the location of the horizon is infinitesimally displaced.

This equality between field equations on the horizon and the thermodynamic identity — originally obtained [25] for spherical horizons in Einstein's theory, has now been demonstrated for an impressively wide class of models [26] like stationary axisymmetric horizons and evolving spherically symmetric horizons in Einstein gravity, static spherically symmetric horizons and dynamical apparent horizons in Lanczos-Lovelock gravity, generic, static horizon in Lanczos-Lovelock gravity, three dimensional BTZ black hole horizons, FRW cosmological models in various gravity theories and even in the case Horava-Lifshitz gravity.

This result is non-trivial in the sense that the field equation on the horizon does not look very "thermodynamical" at first sight. For example, in the simplest context of spherically symmetric horizon in Einstein's theory [with $-g_{00} = g_{11}^{-1} = f(r)$ with $f(a) = 0$ determining the location of the horizon at $r = a$], the field equation on the horizon reduces to

$$\frac{c^4}{G}\left[\frac{\kappa a}{c^2} - \frac{1}{2}\right] = 4\pi P a^2 \tag{10}$$

where $\kappa = f'(a)/2$ is the surface gravity and P is the pressure of the source. As I said, this equation does not seem to have any thermodynamics in it. However, if we multiply it by da it can be re-written in the form:

$$\underbrace{\frac{\hbar}{c}\left(\frac{\kappa}{2\pi}\right)}_{k_B T}\ \underbrace{\frac{c^3}{G\hbar}d\left(\frac{1}{4}4\pi a^2\right)}_{k_B^{-1} dS}\ \underbrace{-\frac{1}{2}\frac{c^4 da}{G}}_{-dE_g} = \underbrace{Pd\left(\frac{4\pi}{3}a^3\right)}_{P\,dV} \tag{11}$$

The only extra input we needed was the expression for the horizon temperature in terms of the surface gravity which needed introducing $\hbar$ in the numerator and denominator. Similar miracle occurs in all the gravitational theories, much more general than Einstein's theory, in which entropy is no longer proportional to horizon area. As we discussed earlier, the temperature of the horizon knows *nothing* about the field equations of the theory but the entropy does. It is therefore remarkable that one obtains the correct combination TdS for a wide variety of theories showing that the information about the theory is encoded in the entropy functional, exactly as it would be for a macroscopic body.

There are significant differences between this identity $TdS = dE_g + PdV$ to which field equations reduce to and the so called Clausius relation $TdS = dE_m$ (used, for example, by Jacobson [3]) which need to be recognised:

- In addition to the obvious existence of the work term PdV, it should be stressed that E_m used in the Clausius relation $TdS = dE_m$ is related to *matter stress tensor* while E_g in the $TdS = dE_g + PdV$ is a purely geometrical construct built out of the metric. The origin of these differences can be traced to two different kinds of virtual displacements of the horizons considered in these two approaches to define the infinitesimal differences [27].

- More importantly, *while $TdS = dE_g + PdV$ holds in widely different contexts,* it has been found to be *impossible* to generalize $TdS = dE_m$ beyond Einstein's theory without introducing additional assumptions (like dissipation), the physical meaning of which remains unclear.

Incidentally, while Davies-Unruh temperature scales as $\hbar$ the entropy scales as $1/\hbar$ (coming from inverse Planck area), thereby making TdS independent of $\hbar$! This is reminiscent of the fact that in normal thermodynamics $T \propto 1/k_B, S \propto k_B$ making TdS independent of k_B. In both cases, the effects due to discrete microstructure (indicated by non-zero $\hbar$ or k_B) disappear in the continuum limit thermodynamics. Thermal phenomena require microstructure but thermodynamical laws are independent of it! Similarly we expect the thermodynamic description of spacetime to be useful and independent of exact nature of the QG description. Any ('bottom-up") model for quantum gravity which leads to horizon thermodynamics and gives Davies-Unruh temperature for QFT in the semi-classical limit, must be consistent with the ('top-down') thermodynamic description merging together in the correct limit.

Lesson 9: Holographic structure of gravitational action functionals finds a natural explanation in the thermodynamic interpretation of the field equations.

If the gravitational dynamics and horizon thermodynamics are so closely related, with field equations becoming thermodynamic identities on the horizon, then the action functionals of the theory (from which we obtain the field equations) must contain information about this connection. This clue comes in the form of another unexplained algebraic accident related to the structure of the action functional and tells us something significant about the *off-shell structure* of the theory.

Gravity is the only theory known to us for which the natural action functional preserving symmetries of the theory contain second derivatives of the dynamical variables but still leads to second order differential equations. Usually, this is achieved by separating out the terms involving the second derivatives of the metric into a surface term which is either ignored or its variation is cancelled by a suitable counter-term. However, this leads to a serious conceptual mystery in the conventional approach when we recall the following two facts together: (a) The field equations can be obtained by varying the bulk term after ignoring (or by canceling with a counter-term) the surface term. (b) But if we evaluate the surface term on the horizon of any solution to the field equations of the theory, one obtains the entropy of the horizon! *How does the surface term, which was discarded before the field equations were obtained, know about the entropy associated with a solution to those field equations?!* In the conventional approach we need to accept it as another 'algebraic accident' without any explanation and, in fact, no explanation is possible within the standard framework.

The explanation lies in the fact that the surface and bulk term of the Lagrangian are related in a specific manner thereby duplicating the information about the horizon entropy [28]. One can show that there exists a relation of the form:

$$\sqrt{-g}L_{\rm sur} = -\partial_a \left(g_{ij} \frac{\delta\sqrt{-g}L_{\rm bulk}}{\delta(\partial_a g_{ij})} \right) \tag{12}$$

All Lanczos-Lovelock action functionals have this form [13]. In fact, this relation is crucial for an action with second derivatives of the dynamical variables to still lead to field equations which are only second order — a feature shared by all the Lanczos-Lovelock models. It can be shown that this result will be true for actions that can be separated into a surface term and a bulk term with the surface term being an integral over $\partial_a(q^A\pi^a_A)$ where q^A are the dynamical variables and π^a_A are the canonical momentum. This structure allows one to interpret all these action functionals, including Einstein-Hilbert action, as providing the momentum space description (see p. 292 of [20]) of the theory.

This duplication of information also allows one relate the variation of the surface term to $\mathcal{R}^a_b$ of the theory. From Eq. (7), it follows that:

$$\begin{aligned}\int_{\partial\mathcal{V}} d^{D-1}x\sqrt{h}n_a(\delta_q v^a) &= \int_{\mathcal{V}} d^D x\sqrt{-g}\nabla_a(\delta_q v^a) \\ &= \int_{\mathcal{V}} d^D x\sqrt{-g}\nabla_a(2\mathcal{R}^a_b q^b) = \int_{\partial\mathcal{V}} d^{D-1}x\sqrt{h}n_a(2\mathcal{R}^a_b q^b)\end{aligned} \tag{13}$$

Computing the corresponding variation of matter action under the change $\delta g^{ab} = \nabla^a q^b + \nabla^b q^a$, one can construct a variational principle to obtain the field equations, purely from the surface term [29]. More importantly, since the variation of the surface term gives the change in the gravitational entropy, we see that $\mathcal{R}^{ab}$ essentially determines the gravitational entropy density of the spacetime. We will say more about this in sec. 6.

The duplication of information between surface and bulk term in Eq. (12) also allows one to obtain the full action [10] from the surface term alone using the entropic interpretation. In fact, in the the Riemann normal coordinates around any event $\mathcal{P}$ the gravitational action reduces to a pure surface term, again showing that the dynamical content is actually stored on the boundary rather than in the bulk.

Lesson 10: Gravitational actions have a surface and bulk terms because they give the entropy and energy of static spacetimes with horizons, adding up to make the action the free energy of the spacetime.

This provides yet another, direct, physical interpretation for the structure of the gravitational action functionals analyzed above. The result is most easily seen for any Lanczos-Lovelock model by writing the time component of the Noether current in Eq. (7) for the Killing vector $q^a = \xi^a = (1, \mathbf{0})$ in the form:

$$L = \frac{1}{\sqrt{-g}}\partial_\alpha\left(\sqrt{-g}\,J^{0\alpha}\right) - 2\mathcal{G}^0_0 \tag{14}$$

Only spatial derivatives contribute in the first term on the right hand side when the spacetime is static. Integrating over $L\sqrt{-g}$ to obtain the action it is is easy to see (using Eq. (9)) that the first term gives the entropy and the second term can be interpreted as energy [30].

Finally, I stress again that the real importance of these results arises from the fact that they hold for all Lanczos-Lovelock models in an identical manner.

5. The Avogadro number of the spacetime

The results described in the previous sections suggest that there is a deep connection between horizon thermodynamics and the gravitational dynamics. Because the spacetime can be heated up just like a body of gas, the Boltzmann paradigm ("If you can heat it, it has microstructure") motivates the study of the microscopic degrees of freedom of the spacetime exactly the way people studied gas dynamics *before* they understood the atomic structure of matter. There exists, fortunately, an acid test of this paradigm which it passes with flying colours.

Lesson 11: Gravitational field equations imply the law of equipartition $\Delta E = (1/2)k_B T\Delta N$ in any static spacetime, allowing the determination of density of microscopic degrees of freedom. The result again displays holographic scaling.

Boltzmann's conjecture led to the equipartition law $\Delta E = (1/2)k_B T\Delta N$ relating the number density ΔN of microscopic degrees of freedom required to store an energy ΔE at temperature

T and to the determination of Avogadro number of a gas. If our ideas are correct, we should be able to relate the E and T of a given spacetime to determine the number density of microscopic degrees of freedom of the spacetime. Remarkably enough, this can be done directly from the field equations [31]. In a hot spacetime, Einstein's equations *imply* the equipartition law

$$E = \frac{1}{2}k_B \int_{\partial\mathcal{V}} \frac{\sqrt{\sigma}\, d^2x}{L_P^2} \left\{\frac{Na^\mu n_\mu}{2\pi}\right\} \equiv \frac{1}{2}k_B \int_{\partial\mathcal{V}} dn\, T_{\rm loc} \tag{15}$$

(where $T_{\rm loc} = (Na^\mu n_\mu/2\pi)$ is the local acceleration temperature and $\Delta n = \sqrt{\sigma}\, d^2x/L_P^2$) thereby allowing us to read off the number density of microscopic degrees of freedom. We again see that gravity is holographic in the sense that the number density Δn scales as the proper area $\sqrt{\sigma}\, d^2x$ of the boundary of the region rather than the volume. (In the case of a gas, we would have got an integral over the volume of the form $dV(dn/dV)$ rather than an area integral.) We also notice that, in Einstein's theory, the number density (dn/dA) is a constant with every Planck area contributing a single degree of freedom.

The true elegance of this result again rests on the fact that it holds true for all Lanczos-Lovelock models! For a Lanczos-Lovelock model with an entropy tensor P^{ab}_{cd} one gets the result

$$E = \frac{1}{2}k_B \int_{\partial\mathcal{V}} dnT_{loc}; \qquad \frac{dn}{dA} = \frac{dn}{\sqrt{\sigma}d^{D-2}x} = 32\pi P^{ab}_{cd}\epsilon_{ab}\epsilon^{cd} \tag{16}$$

where ϵ_{ab} is the binormal on the codimension-2 cross-section. All these gravitational theories are holographic and the density of microscopic degrees of freedom encodes information about the theory through the entropy tensor. *I consider these results as the most direct evidence for the emergent paradigm of gravity.*

Lesson 12: One can obtain the Wald entropy for a general theory directly from law of equipartition.

The density of microscopic degrees of freedom obtained in Eq. (16) suggests that the entropy associated with a *general surface* in Lanczos-Lovelock models (or the entropy associated with a horizon in a more general theory) will be proportional to an integral over $P^{ab}_{cd}\epsilon_{ab}\epsilon^{cd}$. That is,[4]

$$S \propto \int_{\partial\mathcal{V}} dn \propto \int_{\partial\mathcal{V}} 32\pi P^{ab}_{cd}\epsilon_{ab}\epsilon^{cd}\sqrt{\sigma}d^{D-2}x \tag{17}$$

This is precisely the expression for Wald entropy [14] *but we have obtained it using only the equipartition law and as a local statement!*

This comes about because the field equations have a specific relationship with Noether current. Further field equations imply equipartition law while Noether current is related to Wald entropy, thereby connecting all the three. Let me indicate how this comes about by a more direct analysis. In static spacetimes, we have a Killing vector ξ^a corresponding to time translation invariance. If we take $q^a = \xi^a$, the expression for the Noether current is quite simple and we get $J^a = 2\mathcal{R}^a_b\xi^b$. Using the relations $J^a \equiv \nabla_b J^{ab}, \xi^a = Nu^a$ and the antisymmetry of J^{ab} one can easily show that:

$$D_\alpha(J^{b\alpha}u_b) = 2N\mathcal{R}_{ab}u^a u^b \tag{18}$$

This is a generalization of the relation $D_\mu(Na^\mu) = 4\pi\rho_{\rm komar}$ between the divergence of the acceleration and the Komar energy density in Einstein's theory, once again showing the role

[4] Note that in Einstein's theory, we get $\Delta n = \Delta A/L_P^2$. One usually considers this as arising due to dividing the area ΔA into Δn patches of area L_P^2. If we attribute f internal states to each patch, then the total number of microstates $\Delta\Omega$ will be $\Delta\Omega = f^{(\Delta n)}$ and $\Delta S = \ln \Delta\Omega \propto \Delta n$ which is how the extensivity $\Delta S \propto \Delta n$ arises. In a more general theory, we replace $\Delta n = \Delta A/L_P^2$ by the expression in Eq. (16).

of Noether potential J^{ab} in the dynamics. The integral version of this relation for a region $\mathcal{V}$ bounded by $\partial\mathcal{V}$ is:

$$\int_{\partial\mathcal{V}} d^{D-2}x\sqrt{\sigma}(n_i u_b J^{bi}) = \int_{\partial\mathcal{V}} d^{D-2}x\sqrt{\sigma}(N n_\alpha J^{\alpha 0}) = \int_{\mathcal{V}} 2N\mathcal{R}_{ab}u^a u^b\sqrt{h}\,d^{D-1}x \tag{19}$$

where we have used $u_a = -N\delta_a^0$ and $J^{0\alpha} = -J^{\alpha 0}$. (The middle relation shows that the result is essentially an integral over $\partial\mathcal{V}$ of $J^{bi}d\sigma_{ib}$, where $d\sigma_{ib} = (1/2)n_{[i}u_{b]}\sqrt{\sigma}d^{D-2}x$.) Now consider a static spacetime with a bifurcation horizon $\mathcal{H}$ given by the surface $N^2 \equiv -\xi^a\xi_a = 0$. The horizon temperature $T \equiv \beta^{-1} = \kappa/2\pi$ where κ is the surface gravity. Since the Wald entropy of the horizon is essentially the Noether charge (multiplied by β), we will interpret [22] the Noether charge *density* $\beta J_b u^b$ (multiplied by β) as the entropy density of the spacetime as perceived by the static observers with four velocity $u^a = \xi^a/N$, so that the total entropy is

$$S_{\rm grav}[u^i] = \beta\int_{\mathcal{V}} J_b u^b\sqrt{h}\,d^{D-1}x \tag{20}$$

Using $J^a = 2\mathcal{R}^a_b\xi^b$ and Eq. (19) and integrating the expression over a region bounded by the $N =$ constant surface, it is easy to see that

$$S = \frac{1}{2}\beta E \tag{21}$$

which is a statement of equipartition, first obtained [32] in 2004 in the form of a relation $E = 2TS$ in Einstein's theory and is generalized to all Lanczos-Lovelock models in ref.[31]. Further, if we take $\partial\mathcal{V}$ to be the horizon $\mathcal{H}$ and use $\beta T = 1$, we get the horizon entropy to be

$$S = \frac{1}{4}\int_{\mathcal{H}} dn = \frac{1}{4}\int_{\mathcal{H}} 32\pi P^{ab}_{cd}\epsilon_{ab}\epsilon^{cd}\sqrt{\sigma}d^{D-2}x \tag{22}$$

which is the standard expression for Wald entropy in a general theory thereby justifying the choice in Eq. (20). This ansatz in Eq. (20) also fixes the proportionality constant in Eq. (17) to be 1/4.

Our expressions for entropy and energy in differential form are given by $dE_{\rm komar} = (1/2)T_{\rm loc}(dn/dA)dA$, $dS = (1/4)(dn/dA)dA$. The resulting expression for TdS is essentially equivalent to what we found earlier in the case of first law, $TdS = dE_g + PdV$, applied to infinitesimal horizon displacements when the differentials appearing in the two expressions are properly related.

Lesson 13: Gravity is intrinsically quantum mechanical at all scales

The holographic nature of gravity which I have alluded to several times shows that area elements play a significant role in the microscopic description of the theory. This is directly related to the fact that the basic unit of the theory is the Planck *area* $\mathcal{A}_P \equiv (G\hbar/c^3)$. Only by taking a square root, rather artificially, one obtains the Planck *length*. Classical gravity, in fact, should be described using $\mathcal{A}_P$ rather than using G with Newton's law of gravity written in the form $F = (\mathcal{A}_P c^3/\hbar)(m_1 m_2/r^2)$. This has the crucial consequence that one cannot really take $\hbar \to 0$ limit at fixed $\mathcal{A}_P$ and call it classical gravity. Gravity is intrinsically quantum mechanical at all scales [33] because of the microstructure of spacetime.

As an aside, one may mention that, strictly speaking, normal matter is also intrinsically quantum mechanical at all scales due to the atomic structure. For example, one cannot study classical elasticity, say, by taking the strict, mathematical, limit $\hbar \to 0$ in a crystal lattice, because such a limit will also make all the electrons in the atom collapse! What we actually do is to keep $\hbar$ nonzero at subatomic scales, ensuring the atomic stability and take the $\hbar \to 0$ limit for the lattice interactions in the continuum limit to obtain the laws of elasticity. We need to do something analogous to obtain classical spacetime from quantum spacetime.

6. Entropy density of spacetime and its extremisation

So far we have been faithfully following the 'top-down' philosophy of starting from known results in classical gravity and obtaining consequences which suggests an alternative paradigm. For example, the results in the last section were obtained by starting from the field equations of the theory, rewriting them in the form of law of equipartition and thus determining the density of microscopic degrees of freedom.

Ultimately, however, we have to start from a microscopic theory and obtain the classical results as a consequence. We know that the thermodynamical behaviour of a normal system can be described by an extremum principle for a suitable potential (entropy, free energy ...) treated as a functional of appropriate variables (volume, temperature ,....). If our ideas related to gravitational theories are correct, it must be possible to obtain the field equations by extremising a suitably defined thermodynamic potential. The fact that null surfaces block information suggests that this thermodynamic potential should be closely related to null surfaces in the spacetime. This expectation turns out to be correct [34].

Lesson 14: Gravitational field equations can be obtained from an alternative, thermodynamic, extremum principle.

Recall that 'how gravity tells matter to move' can be determined by demanding the validity of special relativistic laws for *all* locally inertial observers. Similarly, 'how matter curves spacetime' can be determined by demanding that the a suitable thermodynamic potential of the microscopic degrees of freedom of the spacetime should be an extremum *all* local Rindler observers. The physical content of this potential (free energy, entropy, enthalpy ....) will depend on the context but the argument works for any one of them. The mathematics involves associating with every null vector field $n^a(x)$ in the spacetime a thermodynamic potential $\Im(n^a)$ which is quadratic in n^a and given by:

$$\Im[n^a] = \Im_{grav}[n^a] + \Im_{matt}[n^a] \equiv -\left(4P_{ab}^{cd}\nabla_c n^a \nabla_d n^b - T_{ab}n^a n^b\right) , \tag{23}$$

where P_{ab}^{cd} is a tensor having the symmetries of curvature tensor and is divergence-free in all its indices and T_{ab} is a divergence-free symmetric tensor. (Once we get the field equations we can read off T_{ab} as the matter energy-momentum tensor; the notation anticipates this result). We also know that the P^{abcd} with the assigned properties can be expressed as $P_{ab}^{cd} = \partial L/\partial R_{cd}^{ab}$ where L is the Lanczos-Lovelock Lagrangian and R_{abcd} is the curvature tensor [1]. This choice in Eq. (23) will also ensure that the equations resulting from the entropy extremisation do not contain any derivative of the metric which is of higher order than second. (More general possibilities exist which I will not discuss here.). We now demand that $\delta\Im/\delta n^a = 0$ for the variation of all null vectors n^a with the condition $n_a n^a = 0$ imposed by adding a Lagrange multiplier function $\lambda(x) g_{ab} n^a n^b$ to $\Im[n^a]$. Using

$$\frac{\partial \Im}{\partial(\nabla_c n^a)} = (-8P_{ab}^{cd}\nabla_d n^b); \quad \frac{\partial \Im}{\partial n^a} = 2[T_{ab} + \lambda(x)g_{ab}]n^b \tag{24}$$

the Euler-Lagrange equations reduce to:

$$\nabla_c\left[-8P_{ab}^{cd}\nabla_d n^b\right] = 2[T_{ab} + \lambda(x)g_{ab}]n^b \tag{25}$$

Because of the condition $\nabla_c P_{ab}^{cd} = 0$ and the antisymmetry $P_{ab}^{cd} = -P_{ab}^{dc}$ we find that all the derivatives disappear on the left hand side and an elementary calculation gives:

$$\left(2\mathcal{R}_b^a - T_b^a - \lambda\delta_b^a\right) n_a = 0 , \tag{26}$$

where $\mathcal{R}^a_b \equiv P^{jk}_{bi} R^{ai}_{jk}$. We demand that Eq. (26) should hold for all null vector fields n^a. Using the generalized Bianchi identity and the condition $\nabla_a T^a_b = 0$ we obtain [1, 34] from Eq. (26) the equations

$$\mathcal{G}^a_b = \mathcal{R}^a_b - \frac{1}{2}\delta^a_b L = \frac{1}{2}T^a_b + \Lambda\delta^a_b \tag{27}$$

where Λ is a constant. These are precisely the field equations for gravity in a theory with Lanczos-Lovelock Lagrangian L (with an undetermined cosmological constant Λ which arises as an integration constant.

The thermodynamical potential can be obtained by integrating the density $\Im[n^a]$ over a region of space or a surface etc. depending on the context. The matter part of the $\Im$ is proportional to $T_{ab}n^a n^b$ which will pick out the contribution $(\rho + p)$ for an ideal fluid, which is the enthalpy density. If multiplied by $\beta = 1/T$, this reduces to the entropy density because of Gibbs-Duhem relation. When the multiplication by β can be reinterpreted in terms of integration over $(0, \beta)$ of the time coordinate, the corresponding potential can be interpreted as entropy and the integral over space coordinates can be interpreted as rate of generation of entropy. [This was the interpretation provided in the earlier works [1, 34] but the result is independent of this interpretation as long as suitable boundary conditions can be imposed]. One can also think of $\Im[n^a]$ as an effective Lagrangian for a set of collective variables n^a describing the deformations of null surfaces.

In addition to providing a purely thermodynamic extremum principle for the field equations of gravity, the above approach also has the following attractive features.

- The extremum value of the thermodynamic potential, when computed on-shell for a solution with static horizon, leads to the Wald entropy. This is a non-trivial consistency check on the approach because it was not designed to reproduce the Wald entropy. It also shows that when the field equations hold, the total entropy of a region $\mathcal{V}$ resides on its boundary $\partial\mathcal{V}$ which is yet another illustration of the holographic nature of gravity.
- In the semi-classical limit, one can show [35] that the gravitational (Wald) entropy is quantized with $S_{\rm grav}$ [on-shell] $= 2\pi n$. In the lowest order Lanczos-Lovelock theory, the entropy is proportional to area and this result leads to area quantization. More generally, it is the gravitational entropy that is quantized. The law of equipartition for the surface degrees of freedom is closely related to this entropy quantization because both arise from the existence of discrete structures on the surfaces in question.
- The entropy functional in Eq. (23) is invariant under the shift $T_{ab} \to T_{ab} + \rho_0 g_{ab}$ which shifts the zero of the energy density. This symmetry allows any low energy cosmological constant, appearing as a parameter in the variational principle, to be gauged away thereby alleviating the cosmological constant problem to a great extent [36]. I will not discuss this issue here.

There is another way of interpreting Eq. (26) which is more in tune with the emergent perspective of gravity. Note that, while Eq. (26) holds for any vector field once the normalization condition is imposed through the Lagrange multiplier, the entropy was originally attributed to *null* vectors and hence it is natural to study Eq. (26) when $n^a = \ell^a$, the null normal of a null surface $\mathcal{S}$ in the spacetime and project Eq. (26) onto the null surface. If $\boldsymbol{\ell}$ is the normal to $\mathcal{S}$, then such a projection leads to the equations:

$$R_{mn}\ell^m q^n_a = 8\pi T_{mn}\ell^m q^n_a; \quad R_{mn}\ell^m \ell^n = 8\pi T_{mn}\ell^m \ell^n \tag{28}$$

where $q_{ab} = g_{ab} + \ell_a k_b + \ell_b k_a$ with k^a being another auxiliary null vector satisfying $\boldsymbol{\ell} \cdot \boldsymbol{k} = -1$. The metric q_{ab} with $q_{ab}\ell^b = 0 = q_{ab}k^b$ acts as a projector to $\mathcal{S}$ (see ref. [37] for details). It is possible to rewrite the first equation in Eq. (28) in the form of a Navier-Stokes equation thereby providing a hydrodynamic analogy for gravity. This equation, known in the literature as Damour-Navier-Stokes (DNS) equation [38], is usually derived by rewriting the field equations. Our analysis [37] provides an entropy extremisation principle for the DNS equation which makes the hydrodynamic analogy natural and direct.

It may also be noted that the gravitational entropy density — which is the integrand $\Im_{grav} \propto (-P^{cd}_{ab}\nabla_c\ell^a\nabla_d\ell^b)$ in Eq. (23) — obeys the relation:

$$\frac{\partial\Im_{\rm grav}}{\partial(\nabla_c\ell^a)} \propto (-P^{cd}_{ab}\nabla_d\ell^b) \propto (\nabla_a\ell^c - \delta^c_a\nabla_i\ell^i) \tag{29}$$

where the second relation is for Einstein's theory. This term is analogous to the more familiar object $t^c_a = K^c_a - \delta^c_a K$ (where K_{ab} is the extrinsic curvature) that arises in the (1+3) separation of Einstein's equations. (More precisely, the projection to 3-space leads to t^c_a.) This combination can be interpreted as a surface energy momentum tensor in the context of membrane paradigm [39] because t_{ab} couples to δh^{ab} on the boundary surface when we vary the gravitational action (see, e.g., eq.(12.109) of [20]). Equation (29) shows that the entropy density of spacetime is directly related to t^c_a and its counterpart in the case of null surface. This term also has the interpretation as the canonical momentum conjugate to the spatial metric in (1+3) context and Eq. (29) shows that the entropy density leads to a similar structure. That is, the canonical momentum conjugate to metric in the conventional approach and the momentum conjugate to ℓ^a in $S_{\rm grav}$ are essentially the same.

Further, the *functional* derivative of the gravitational entropy in Eq. (23) has the form, in any Lanczos-Lovelock model:

$$\frac{\delta\Im_{\rm grav}}{\delta\ell^a} \propto \mathcal{R}_{ab}\ell^b \propto J_a \tag{30}$$

Previous discussion has shown that the current $J_a = 2\mathcal{R}_{ab}\ell^b$ plays a crucial role in interpreting gravitational field equations as entropy balance equations. In the context of local Rindler frames, when ℓ^a arises as a limit of the time-like Killing vector in the local Rindler frame, J_a can be interpreted as the Noether (entropy) current associated with the null surface. In that case, the generalization of the two projected equations in Eq. (28) to Lanczos-Lovelock models will read as

$$J_a\ell^a = \frac{1}{2}T_{ab}\ell^a\ell^b; \quad J_a q^a_m = \frac{1}{2}T_{ab}\ell^a q^b_m \tag{31}$$

which relate the gravitational entropy density and flux to matter energy density and momentum flux. (The second equation in the above set becomes the DNS equation in the context of Einstein's theory.) All these results, including the DNS equation, will have direct generalization to Lanczos-Lovelock models which can be structured using the above concepts. We again see that all these ideas find a natural home in the emergent paradigm.

7. Concluding comments

As promised, I have presented the internal evidence hidden in the structure of classical gravitational theories which suggest that gravity is an emergent phenomenon. This evidence brings up the holographic nature of gravity in more than one way (surface density of microscopic degrees of freedom, structure of gravitational action functionals .....), provides a thermodynamic interpretation to field equations (field equations reducing to $TdS = dE_g + PdV$ on the horizons, entropy balance for virtual displacements of horizons, equipartition), allows one to explicitly determine the number density of microscopic degrees of freedomand — finally — derive the field equations from an entropy maximization procedure. The approach also clarifies several issues which have no explanation in conventional procedure and links several ideas together (like e.g. the relation between the diffeomorphism invariance and the entropy of null surfaces). All of these work in any Lanczos-Lovelock model seamlessly without we having to tinker anything.

It is worthwhile to list explicitly the questions which have natural answers in the emergent paradigm while have to be treated as algebraic accidents in the conventional approach:

(i) While the temperature of the horizon can be obtained using QFT in curved spacetime, the *corresponding* entanglement entropy is divergent and meaningless. Why?

(ii) The temperature of horizon is independent of the field equations of gravity but the entropy of the horizon depends explicitly on the field equations. What does this difference signify?

(iii) The horizon entropy can be expressed in terms of the Noether current which is conserved due to diffeomorphism invariance. Why should an infinitesimal coordinate transformation $x^a \to x^a + q^a$ have anything to do with a thermodynamic variable like entropy?

(iv) Why does gravitational field equations (which does not look very "thermodynamical"!) reduce to $TdS = dE_g + PdV$ on the horizon, picking up the correct expression for S for a wide class of theories?

(v) How come all gravitational action principle have a surface and bulk term which are related in a specific manner (see Eq. (12))? Why do the surface and the bulk terms allow the interpretation as entropy and energy in static spacetimes?

(vi) The field equations for gravity can be obtained from the bulk part of the action after discarding the surface term. But the surface term evaluated on the horizon of a solution gives the entropy of the horizon! How does the surface term — which was discarded before the field equations were obtained — know about the entropy of a solution?

(vii) Why does the gravitational field equations reduce to the equipartition form, expressible as $\Delta E = (1/2)(k_B T)\Delta n$ allowing us to determine the analog of Avogadro's number for the spacetime? And, why does the relevant microscopic degrees of freedom for a region reside on the boundary of the region?

(viii) Finally, why is it possible to derive the field equations of any diffeomorphism invariant theory of gravity by extremizing an entropy functional associated with the null surfaces in the spacetime, without treating the metric as a dynamical variable?

Obviously, any alternative perspective, including the conventional approach, need to provide the answers for the above questions if they have to be considered a viable alternative to emergent paradigm. The explanations need to work for all Lanczos-Lovelock models and not for just Einstein's theory. I think the emergent paradigm scores on all these counts and provides valuable insights into the deeper structure of the theory.

Acknowledgements
I thank the organizers of DICE 2010 conference, especially Thomas Elze, for excellent hospitality and a delightful conference.

References

[1] Padmanabhan T 2010 *Rep. Prog. Phys.* **73** 046901 (*Preprint* arXiv:0911.5004)
[2] Sakharov A D 1968 *Sov. Phys. Dokl.* **12** 1040
[3] Jacobson T 1995 *Phys. Rev. Lett.* **75** 1260
[4] Volovik G E 2003 *The universe in a helium droplet* (Oxford: Oxford University Press)
[5] Hu B L 2010 Gravity and Nonequilibrium Thermodynamics of Classical Matter *Preprint* arXiv:1010.5837; *do.* 1996 General Relativity as Geometro-Hydrodynamics *Preprint* arXiv:gr-qc/9607070
[6] For a review, see e.g., Barcelo C, Liberati S and Visser M 2005 *Living Rev. Rel.* **8** No. 12 (*Preprint* arXiv:gr-qc/0505065)
[7] Thorne K S et al. 1986 *Black Holes: The Membrane Paradigm* (Yale: Yale University Press)
[8] Yu Tian Xiao-Ning Wu 2010 *Preprint* arXiv:1012.0411; Hendi S H and Sheykhi A 2010 *Preprint* arXiv:1012.0381; Culetu H 2010 *Preprint* arXiv:1011.3343; Wei Gu, Miao Li and Rong-Xin Miao 2010 *Preprint* arXiv:1011.3419; Li-Ming Cao 2010 *Preprint* arXiv:1009.4540; Shao-Feng Wu et al. 2010

Preprint arXiv:1008.2072; Banerjee R et al. 2010 *Phys. Rev.* D **82** 124002; Yu Tian et al. 2010 *Preprint* arXiv:1007.4331; Shalyt-Margolin A E 2010 *Preprint* arXiv:1006.4979; Nicolini P 2010 *Preprint* arXiv:1005.2996; Piazza F 2010 *Preprint* arXiv:1005.5151; Fursaev D V 2010 *Phys. Rev.* D **82**:064013; Kiselev V V et al. 2010 *Mod. Phys. Lett.* A **25** 2223; Sheykhi A 2010 *Phys. Rev.* D **81** 104011; Xiao-Gang He and Bo-Qiang Ma 2010 *Preprint* arXiv:1003.1625; Jae-Weon Lee 2010 *Preprint* arXiv:1003.1878; *do.* 2010 *Preprint* arXiv:1003.4464; Banerjee R and Bibhas Ranjan Majhi 2010 *Preprint* arXiv:1003.2312; Xiao-Gang He and Bo-Qiang Ma 2010 *Preprint* arXiv:1003.2510; Yi-Fu Cai, Jie Liu and Hong Li 2010 *Preprint* arXiv:1003.4526; Ghosh S 2010 *Preprint* arXiv:1003.0285; Munkhammar J 2010 *Preprint* arXiv:1003.1262; Zhao L 2010 *Preprint* arXiv:1002.0488; Kowalski-Glikman J 2010 *Preprint* arXiv:1002.1035; Yu-Xiao Liu, Yong-Qiang Wang and Shao-Wen Wei 2010 *Preprint* arXiv:1002.1062; Rong-Gen Cai, Li-Ming Cao and Nobuyoshi Ohta 2010 *Preprint* arXiv:1002.1136; Pesci P 2010 *Preprint* arXiv:1002.1257; Yu Tian and Xiaoning Wu 2010 *Preprint* arXiv:1002.1275; Yun Soo Myung and Yong-Wan Kim 2010 *Preprint* arXiv:1002.2292; Vancea I V and Santos M A 2010 *Preprint* arXiv:1002.2454; Konoplya R A 2010 *Preprint* arXiv:1002.2818; Culetu H 2010 *Preprint* arXiv:1002.3876; Rong-Gen Cai, Li-Ming Cao and Nobuyoshi Ohta 2010 *Phys. Rev.* D **81** 061501 (*Preprint* arXiv:1001.3470); Smolin L 2010 *Preprint* arXiv:1001.3668; Fu-Wen Shu and Yungui Gong 2010 *Preprint* arXiv:1001.3237; Padmanabhan T 2010 *Preprint* arXiv:1001.3380; Makea J 2010 *Preprint* arXiv:1001.3808; Miao Li and Yi Wang 2010 *Preprint* arXiv:1001.4466; Changjun Gao 2010 *Preprint* arXiv:1001.4585; Yi Zhang, Yun-gui Gong and Zong-Hong Zhu 2010 *Preprint* arXiv:1001.4677; Culetu H 2010 *Preprint* arXiv:1001.4740; Yi Wang 2010 *Preprint* arXiv:1001.4786; Tower Wang 2010 *Preprint* arXiv:1001.4965; Shao-Wen Wei, Yu-Xiao Liu and Yong-Qiang Wang 2010 *Preprint* arXiv:1001.5238; Yi Ling and Jian-Pin Wu 2010 *Preprint* arXiv:1001.5324; Jae-Weon Lee, Hyeong-Chan Kim and Jungjai Lee 2010 *Preprint* arXiv:1001.5445; Verlinde E P 2010 *Preprint* arXiv:1001.0785

[9] Bekenstein J D 1972 *Nuovo Cim. Lett.* **4** 737–740; Hawking S W 1975 *Commun. Math. Phys.* **43** 199–220; Davies P C W 1975 *J. Phys.* A **8** 609–616; Unruh W G 1976 *Phys. Rev.* D **14** 870

[10] Padmanabhan T 2005 *Phys. Reports* **406** 49 (*Preprint* arXiv:gr-qc/0311036)

[11] Padmanabhan T 2011 *A Dialogue on the Nature of Gravity* as Chapter 2 in *Foundations of Space and Time* ed J Murugan, A Weltman and G F R Ellis (Cambridge: Cambridge University Press) (*Preprint* arXiv:0910.0839)

[12] Lanczos C 1932 *Z. Phys.* **73** 147; Lovelock D 1971 *J. Math. Phys.* **12** 498

[13] Padmanabhan T 2005 *Dark Energy: Mystery of the Millennium* Albert Einstein Century International Conference Paris *AIP Conference Proceedings* **861** 858 (*Preprint* arXiv:astro-ph/0603114)

[14] Wald R M 1993 *Phys. Rev.* D **48** 3427 (*Preprint* arXiv:gr-qc/9307038); Iyer V and Wald R M 1995 *Phys. Rev.* D **52** 4430 (*Preprint* arXiv:gr-qc/9503052)

[15] Kothawala D and Padmanabhan T 2010 *Phys. Lett.* B **690** 201-206 (*Preprint* arXiv:0911.1017); Raval A, Hu B L and Don Koks 1997 *Phys. Rev.* D **55** 4795 (*Preprint* arXiv:gr-qc/9606074)

[16] Lee T D 1986 *Nucl. Phys.* B **264** 437; Unruh W G and Weiss N 1984 *Phys. Rev.* D **29** 1656; for a textbook description, see section 14.5 of ref.[20]

[17] Marolf D, Minic D and Ross S 2004 *Phys. Rev.* D **69** 064006

[18] See, for e.g., Bombelli L et al. 1986 *Phys. Rev.* D **34** 373; Srednicki M (1993) *Phys. Rev.* D **71** 66 and for a recent review Nishioka T et al. 2009 *J. Phys.* A **42** 504008 (*Preprint* arXiv:0905.0932)

[19] Padmanabhan T 2010 *Phys. Rev.* D **82** 124025 (*Preprint* arXiv:1007.5066)

[20] Padmanabhan T 2010 *Gravitation: Foundations and Frontiers* (Cambridge, UK: Cambridge University Press)

[21] Padmanabhan T 1997 *Phys. Rev. Lett.* **78** 1854 (*Preprint* arXiv:hep-th-9608182); *do.* 1998 *Phys. Rev.* D **57** 6206 ; *do.* 1987 *Class. Qu. Grav.* **4** L107; *do.* 1985 *Annals Phys.* **165** 38-58; Srinivasan K et al. 1998 *Phys. Rev.* D **58** 044009 (*Preprint* arXiv:gr-qc-9710104)

[22] Padmanabhan T 2009 *Int. J. Mod. Phys.* D **18** 2189 (*Preprint* arXiv:0903.1254)

[23] Padmanabhan T 2004 *Int. J. Mod. Phys.* D **13** 2293-2298 (*Preprint* arXiv:gr-qc/0408051); *do.* 2005 *Brazilian J. Phys.* (Special Issue) **35**(2A) 362 (*Preprint* arXiv:gr-qc/0412068)

[24] Kothawala D and Padmanabhan T 2009 *Phys. Rev.* D **79** 104020 (*Preprint* arXiv:0904.0215)

[25] Padmanabhan T 2002 *Class. Quan. Grav.* **19** 5387 (*Preprint* arXiv:gr-qc/0204019)

[26] There is a large literature on this subject a small sample of which will be: Cai R G et al. 2008 *Phys. Rev.* D **78** 124012; Kothawala D, Sarkar S and Padmanabhan T 2007 *Phys. Lett.* B **652** 338 (*Preprint* arXiv:gr-qc/0701002); Paranjape A, Sarkar S and Padmanabhan T 2006 *Phys. Rev.* D **74** 104015 (*Preprint* arXiv:hep-th/0607240); Akbar M 2007 *Chin. Phys. Lett.* **24** 1158 (*Preprint* arXiv:hep-th/0702029); Cai R G and Kim S P 2005 *JHEP* **0502** 050 (*Preprint* arXiv:hep-th/0501055); Sheykhi A, Wang B and Cai R G 2007 *Nucl. Phys.* B **779** 1 (*Preprint* arXiv:hep-th/0701198); Sheykhi Wang A B and Cai R G 2007 *Phys. Rev.* D **76** 023515 (*Preprint* arXiv:hep-th/0701261); Shalyt-Margolin A E 2010 *Preprint* arXiv:1006.4979; Cai R G 2008 *Prog. Theor. Phys. Suppl.* **172** 100 (*Preprint* arXiv:0712.2142); Ge X H 2007 *Phys. Lett.* B

651 49 (*Preprint* arXiv:hep-th/0703253); Gong Y and Wang A 2007 *Phys. Rev. Lett.* **99** 211301 (*Preprint* arXiv:0704.0793); Wu S F, Wang B and Yang G H 2008 *Nucl. Phys.* B **799** 330 (*Preprint* arXiv:0711.1209); Wu S F et al. 2008 *Class. Quant. Grav.* **25** 235018 (*Preprint* arXiv:0801.2688); Cai R G and Ohta N 2009 *Preprint* arXiv:0910.2307

[27] Kothawala D 2010 The thermodynamic structure of Einstein tensor *Preprint* arXiv:1010.2207

[28] Mukhopadhyay A and Padmanabhan T 2006 *Phys. Rev.* D **74** 124023 (*Preprint* arXiv:hep-th/0608120)

[29] Padmanabhan T 2006 *Gen. Rel. Grav.* **38** 1547-1552; *do.* 2006 *Int. J. Mod. Phys.* D **15** 2029 (*Preprint* arXiv:gr-qc/0609012)

[30] Sanved K and Padmanabhan T 2010 *Phys. Rev.* D **82** 024036 (*Preprint* arXiv:1005.0619)

[31] Padmanabhan T 2010 *Mod. Phys. Lett.* A **25** 1129 (*Preprint* arXiv:0912.3165); *do.* 2010 *Phys. Rev.* D **81** 124040 (*Preprint* arXiv:1003.5665)

[32] Padmanabhan T 2004 *Class. Quan. Grav.* **21** 4485 (*Preprint* gr-qc/0308070)

[33] Padmanabhan T 2002 *Mod. Phys. Lett.* A **17** 1147 (*Preprint* hep-th/0205278); *do.* 2002 *Gen. Rel. Grav.* **34** 2029-2035 (*Preprint* gr-qc/0205090

[34] Padmanabhan T 2008 *Gen. Rel. Grav.* **40** 529-564 (*Preprint* arXiv:0705.2533); *ibid.* 2031-2036; Padmanabhan T and Paranjape A 2007 *Phys. Rev.* D **75** 064004 (*Preprint* gr-qc/0701003)

[35] Kothawala D, Padmanabhan T and Sarkar S 2008 *Phys. Rev.* D **78** 104018 (*Preprint* arXiv:0807.1481)

[36] Padmanabhan T 2009 *Adv. Sci. Lett.* **2** 174 (*Preprint* arXiv:0807.2356); *do.* 2003 *Phys. Rep.* **380** 235 (*Preprint* hep-th/0212290); *do.* 2005 *Class. Qu. Grav.* **22** L107-L110 (*Preprint* hep-th/0406060)

[37] Padmanabhan T 2010 Entropy density of spacetime and the Navier-Stokes fluid dynamics of null surfaces *Preprint* arXiv:1012.0119

[38] Damour T 1979 Thèse de doctorat d'État, Université Paris (available at http://www.ihes.fr/~damour/Articles/)

[39] Price R H and Thorne K S 1986 *Phys. Rev.* D **33** 915

Macroscopic quantum phenomena from the large N perspective

C H Chou[1], B L Hu[2] and Y Subaşı[2]

[1]Department of Physics, National Cheng Kung University, Tainan, Taiwan 701 and National Center for Theoretical Sciences (South), Tainan, Taiwan 701

[2] Joint Quantum Institute and Maryland Center for Fundamental Physics, University of Maryland, College Park, Maryland 20742, USA

E-mail: hubeilok@gmail.com

Abstract. Macroscopic quantum phenomena (MQP) is a relatively new research venue, with exciting ongoing experiments and bright prospects, yet with surprisingly little theoretical activity. What makes MQP intellectually stimulating is because it is counterpoised against the traditional view that macroscopic means classical. This simplistic and hitherto rarely challenged view need be scrutinized anew, perhaps with much of the conventional wisdoms repealed. In this series of papers we report on a systematic investigation into some key foundational issues of MQP, with the hope of constructing a viable theoretical framework for this new endeavour. The three major themes discussed in these three essays are the large N expansion, the correlation hierarchy and quantum entanglement for systems of 'large' sizes, with many components or degrees of freedom. In this paper we use different theories in a variety of contexts to examine the conditions or criteria whereby a macroscopic quantum system may take on classical attributes, and, more interestingly, that it keeps some of its quantum features. The theories we consider here are, the $O(N)$ quantum mechanical model, semiclassical stochastic gravity and gauge / string theories; the contexts include that of a 'quantum roll' in inflationary cosmology, entropy generation in quantum Vlasov equation for plasmas, the leading order and next-to-leading order large N behaviour, and hydrodynamic / thermodynamic limits. The criteria for classicality in our consideration include the use of uncertainty relations, the correlation between classical canonical variables, randomization of quantum phase, environment-induced decoherence, decoherent history of hydrodynamic variables, etc. All this exercise is to ask only one simple question: Is it really so surprising that quantum features can appear in macroscopic objects? By examining different representative systems where detailed theoretical analysis has been carried out, we find that there is no a priori good reason why quantum phenomena in macroscopic objects cannot exist.

1. Quantum / classical, micro / macro

There are many ways to deal with the issue of quantum-classical correspondence [1]. In the most common and traditional view the classical limit corresponds to $\hbar \to 0$, or, invoking the Bohr correspondence principle, the principal quantum number of a system $n \to \infty$, or regarding the coherent state as the 'most classical' quantum state, or the Wigner function as the 'closest to classical' distribution. Less precise criteria also abound, such as the loose concept that a system at high temperature behaves classically, or viewing the thermodynamic / hydrodynamic limits (of a quantum system) as classical. (For a description of the various criteria, see, e.g., [2]). There are holes in almost all of the above common beliefs. A more sophisticated viewpoint invokes decoherence, the process whereby a quantum system loses its coherence (measured by its quantum phase information) through interaction with its environment [3]. In this work we examine an alternative perspective, as the folklore goes, that quantum pertains to the small (mass, scale) while classical to the large (size, multiplicity). This common belief now requires a much closer scrutiny in the face of new challenges from macroscopic quantum phenomena (MQP), viz, quantum features may show up even at macroscopic scales. A common example is superconductivity where the Cooper pairs can extend to very large scales compared to interatomic distances and Bose-Einstein condensate (BEC) where a large number N of atoms occupy the same quantum state, the N-body ground state. Other examples include nanoelectromechanical devices [4] where the center of mass of a macroscopic classical object, the cantilever, obeys a quantum mechanical equation of motion. Experiments to demonstrate the quantum features such as the existence of interference between two macroscopic objects have been carried out, e.g., for C^{60} molecules passing through two slits [5] or proposed, e.g., for two mirrors [6, 7].

A most direct account of the difference between the microscopic and the macroscopic behaviours of a quantum system is by examining N, the number of physically relevant (e.g. for atomic systems, forgetting about the tighter-bound substructures) quantum particles or components in a macroscopic object. One may ask: At what number of N will it be suitable to describe the object as mesoscopic with qualitatively distinct features from microscopic and macroscopic? In the case of classical equilibrium statistical mechanics this issue is easily addressed by use of the (grand) canonical ensemble, and such inquiry is answered by understanding how the (grand) partition function behaves as a function of N and temperature T (and chemical potential μ) all the way to the thermodynamic limit. In classical nonequilibrium statistical mechanics, this issue underlies the derivation from molecular Hamiltonian dynamics the thermodynamic and kinetic properties (such as transport functions) of a gas of N molecules, and their dynamics, which possess salient dissipative and time-asymmetry features nonexistent in the microscopic dynamics. Boltzmann answered this question with great valour and magnificent success, bringing forth also the issue of time-arrow and providing a molecular dynamics basis for the Second Law. To perform a quantitative analysis of such issues one needs to work with kinetic and stochastic equations for the N particles, the Boltzmann equation -BBGKY hierarchy for (effectively) closed systems in the case where there is no distinguished party, and when there is, the Langevin equation for open systems. In the recent decade significant advances have been made in providing a molecular dynamics basis to the foundations of thermodynamics [8], relating the macroscopic thermodynamic behaviour of a gas to the chaotic dynamics of its molecular constituents. One could even calculate the range in the number of molecules where a microscopic system begins to acquire macroscopic behaviour and hence identify the approximate boundaries of mesoscopia [9].

For quantum systems one needs to deal with additional concerns of quantum coherence and entanglement which are critically important issues in quantum information processing (QIP) [10]. A fundamental issue in QIP is how the performance of a quantum information processor alters as one scales the system up. This dependence on N is known as the "scaling" problem [11]. We will defer considerations of quantum entanglement of macroscopic objects to a third paper in this series after we have a chance to explore how quantum correlations and fluctuations impact on MQP in the second paper using the n particle-irreducible (nPI) representation. There are many important and interesting issues of MQP, one subset of special interest to us is how quantum expresses itself in the macroscopic domain since usually macro conjures classicality. Thus even the simplest yet far from naive question need be reconsidered properly. For example, why is it that an ostensibly macroscopic object such as a cantilever should follow a quantum equation of motion. This center-of-mass axiom is implicitly assumed in many descriptions of MQP but rarely justified. The conditions upon which this can be justified are provided by two of us and a different third author [12] with the derivation of a master equation for N harmonic oscillators (NHO) in a finite temperature harmonic oscillator bath.

In this paper we ask the question: When will a quantum system of N particles or components begin to acquire or show classical features. We shall present an array of physical examples to illustrate how this comes about, and glean from them 1) how classicality is defined, 2) How the meaning of classicality changes in different types of theories? For 1) the criteria mentioned at the beginning and expounded further in the next section will be invoked in several concrete examples, such as, how the inflaton dynamics determines when the inflationary universe begins to behave classically, from the simplest criteria (invoking the uncertainty principle) to the more involved (correlation between canonical variables) and sophisticated (decoherence). Other criteria studied are randomization of quantum phase, entropy generation, decoherence of hydrodynamic variables due to the existence of conservation laws. For 2) we give a detailed study of a) the O(N) model: an N component quantum scalar field ϕ^4 with $\lambda\phi^4$ self interaction (in section 3) (see, e.g., [13, 14, 15]), the quantum mechanical version of which describes a system of N interacting quantum harmonic oscillators, as it is the common underlying structure of many physical examples, where some claims of macrocity (being macroscopic, for lack of a better noun) and classicality originate. For our purpose here it is sufficient to work with a zero-dimensional field theory, i.e., a quantum mechanical model – not only can one shred off the shrouded field theory technicalities but it is more directly related to atomic-optical and condensed matter set-ups for easier experimental comparisons. b) semiclassical stochastic gravity [16], and c) gauge / string theory (in section 5). We use some well-known results in quantum field theory - the large N expansion – to consider mean field behaviour and under what conditions the semiclassical limit of a quantum theory will be reached at the leading order (LO) large N expansion, and the role of quantum fluctuations at the next-to-leading-order (NLO) large N expansion.

Both the O(N) theory and semiclassical gravity show semiclassical behaviour when N becomes very large. For the former the leading order large N expansion yields a mean field theory with Vlasov equation which is time-reversal invariant. We use this model as an example to expound these issues in section 4. The next-to-leading order (NLO) expansion begins to show dissipative behaviour in the two particle irreducible (2PI) representation [17]. Using the closed-time-path (CTP) 2PI NLO effective action an H-theorem can be shown to exist for the quantum mechanical O(N) model [18]. In the latter example, perturbative quantum gravity interacting with an N component quantum scalar field emerges as the semiclassical limit when N becomes very large [19]. The equations of motion (semiclassical Einstein equation) is time-reversal invariant. The NLO expansion yields stochastic gravity [20] which has both dissipative and fluctuation features. These prior results provide a stimulus for us to inquire about the quantum-classical correspondence issue in an alternative way, i.e., instead of using the loop expansion (in orders of $1/\hbar$) where quantum features reveal at successively higher loop orders as corrections to the classical solution, we use the large N expansion, where N provides a quantitative measure of the 'magnitude' or 'complexity' of a quantum system, to examine how the semiclassical limit of a quantum theory is reached. We shall rely heavily on the results obtained in [21, 22, 23] to illustrate these crucial aspects in MQP. However, as the examples in gauge and string theory show [24, 25, 26] (section 5) the leading order large N does not necessarily correspond to a classical limit. It may correspond to a thermodynamics limit instead. These are the purposes for our invoking different theories to illustrate different criteria and see their fine and not so fine distinctions.

Perhaps it is necessary to make this remark here before we describe the contents of this paper: many ingredients in our discussions are well-known to different communities, such as the large N expansion in field theory and critical phenomena, dissipation and fluctuations in nonequilibrium statistical mechanics, stochastic equations in quantum open systems, hydrodynamic variables and quasi-classical domains in decoherent histories, etc. We see this paper's contribution is a modest one: discerning the physical essence of apparently similar technology and sharpening the often mixed-up physical meanings while focusing on the issues interspersed between the macro and the quantum. We hope to use these well-known parts to elucidate some lesser known, even misunderstood facts, and to shape a new perspective, as a first step towards constructing a theoretical foundation for MQP. [1]

Readers familiar with these theories should just skip over the technical details in those parts and concentrate on how they bring out the physical issues raised. In section 2 we display an array of criteria behind the common beliefs pertaining to quantum, classical and macro, draw the linkages between them, differentiate seemingly equivalent criteria, and issue some warnings to over-generalizations. In section 3 we present a) the $O(N)$ model, give a quick summary of the large N expansion method, leading to a set of coupled equations where one can see the variation with N in physical quantities of interest. In section 4 we use three examples to describe how a macroscopic quantum system acquires classical features: 1) The so-called 'quantum roll' model of inflationary cosmology [27] based on the $O(N)$ model, how stringent the different classicality criteria are: amongst them the use of the uncertainty relation, correlation in conjugate variables, and environment-induced decoherence. 2) Quantum Vlasov equation (QVE) as an example of a mean field theory obtained from the leading order large N (LOLN) expansion of the $O(N)$ model. Following [28] we express QVE as a coupled set of equations, one for the number of particles, the other for their correlations which contain quantum phase relations. This enables one to see how coarse-graining over the phase information brings about entropy generation which is related to uncertainty which can be used to quantify classicality. We also mention how in the decoherent history conceptual framework quasi-classical domains can appear owing to the existence of conservation laws for hydrodynamic variables, thus highlighting the role of emergence and importance of collective behaviour in MQP. In section 5 we consider the next-to-leading order large N expansion and try to identify the role of quantum fluctuations in MQP, using examples given above to make explicit the conditions a macroscopic system can shred off or retain its quantum features. We use two more illustrative theories here, b) semiclassical stochastic gravity and c) gauge / string theories to examine their LOLN and NLO dynamics and explore their classical versus quantum behaviours. We learned that there is no a priori reason why a quantum system with large N need to be classical. We discuss the differences in physical meaning between a mean field and a classical field, and under what conditions the former leads to the latter. Only for the class of Gaussian theories like the O(N) model where the mean field theory obtained from the LOLN expansion is equivalent to the classical theory would it make sense to think of deviations (from LO expansion) from the mean as providing the quantum corrections which then allows one to use large N to address the quantum- classical correspondence issue. In section 6 we summarize what we have learned in this intellectual exercise and make some general observations. The overall lesson we learned is that we need to be very careful in finding the conditions and formulating criteria for a system with large N to behave classically. To be safe, one should think of all systems as quantum intrinsically at all orders of N, and not to identify mean field as classical without knowing the structure and behavior of the theories we are talking about.

[1] Because we want to be as concrete as possible in making our points we decided to extract technical details verbatim from some representative papers on the specific topics, instead of grossing over and talking abstractly. For this we show our appreciation to the authors of these papers for their lucid presentations which facilitate our illustrative purpose. An apology to the readers goes along with our appreciation to the authors: To experts in field theory, string theory, gravity and cosmology, the themes in this paper may be too familiar to them and the inclusion of details may thus make the contents overladen. This is because we need to consider readers from the atomic, optical and condensed matter community where MQP experiments and theories are currently undertaken.

2. Large N in relation to quantum and classical

In this program we have set up for ourselves we need to clearly distinguish the relation between micro-macro / quantum-classical as well as the thermo-hydro limits. The micro-macro correspondence underlies the basic theme of statistical mechanics which has been investigated for centuries, and although mesoscopic physics has been with us for three decades many key issues bearing on quantum correlations and fluctuations remain a challenge today. The quantum-classical transition /correspondence has been studied in earnest in the last two decades and we have gained much understanding of the role environments play in bringing about a quantum system's classical behaviour, but the issue of quantum and macro remains little explored, since most common beliefs associate micro with quantum and macro with classical. MQP is a strong statement that this need not be the case. We ought to scrutinize all conventional assumptions and reexamine every single concept and criterion before reconnecting them into a new conceptual framework to meet this new challenge.

As a background for motivation and a collection of key points we list below the common beliefs for a macroscopic object to assume classical behaviour. We pay special attention to the relation of large N, quantum and classicality.

2.1. $\hbar$, loop expansion and classicality

- Why is it that $\hbar \to 0$ is often viewed as classical? This can be easily seen from the path integral formalism: $1/\hbar$ multiplies the action in the path integral, which has no further dependence on $\hbar$. The dominant contribution to the path integral comes from the path which extremizes the action, i.e., the classical trajectory. The steepest descent method becomes exact when $\hbar \to 0$. It is in this sense that taking $\hbar \to 0$ give rise to classical physics.
  Suppose the macroscopic object is made up of a space-time inhomogeneous boson condensate. Its classical behaviour manifests itself via the vacuum expectation value of a scalar field ψ:
  $$\phi(\vec{x}, t) = \langle 0|\psi(\vec{x}, t)|0\rangle$$
  where $|0\rangle$ is the vacuum with the space-time inhomogeneous boson condensate creating the macroscopic object. Note that ψ^n contains not only ϕ^n, but also the product of normal product terms. Rearranging a product of normal product terms into a normal product creates c-numbers due to the contraction of creation and annihilation operators, while each contraction creates a c-number of order $\hbar$, namely,
  $$\langle 0|\psi(\vec{x}, t)^n|0\rangle = \phi(\vec{x}, t)^n + O(\hbar).$$
  Thus only when the contractions which produce the loops in the Feynman diagrams are ignored can one replace $\langle 0|\psi^n|0\rangle$ by ϕ^n. This gives the tree level approximation which is usually termed "classical".

2.2. Large N, mean field and fluctuations

- *Fluctuations to mean*: Bohr's correspondence principle states that the classical behaviour of a quantum system with n quantum levels is established when the quantum fluctuation Δn is much smaller than the average quantum number n, i.e. $\Delta n/n \ll 1$. This condition does not depend on $\hbar$, but is controlled by the number of participating particles or quanta.
- *Large N as a mean field*: The limiting theory when $N \to \infty$ in a leading order large N expansion is usually referred to as a *mean field theory* in that contributions from fluctuations are maximally suppressed. To see this note that in every theory known to have a sensible

large N limit, the vacuum expectation of any product of operators, $\hat{A}\hat{B}$, satisfies the factorization relation

$$\langle \hat{A}\hat{B} \rangle = \langle \hat{A} \rangle \langle \hat{B} \rangle + O(1/N).$$

(This means that the disconnected Feynman graphs always dominate). Therefore the variance of any operator vanishes as $N \to \infty$,

$$\lim_{N\to\infty} (\langle \hat{A}^2 \rangle - \langle \hat{A} \rangle^2) = 0.$$

This means that the fluctuations become negligible when $N \to \infty$.

- *$\hbar \to 0$ and $N \to \infty$ are different limits.* The loop expansion and large N expansion are very different. In QCD, N determines the number of degrees of freedom ($N^2 - 1$ gluons), therefore the limit $N \to \infty$ resembles the infinite volume limit of a two-dimensional lattice system. This is different for theories involving N-component vector fields such as the $O(N)$ model described in section 3. There, the $\hbar \to 0$ limit gives the complete classical theory whereas $N \to \infty$ is the gaussian approximation of a quantum theory, for which classical and quantum dynamics are identical. So even for this class of theories these two limits are, strictly speaking, different.

2.3. *Large N, classical or thermodynamic limits?*

- *Is the mean field a classical field?* There is a common misunderstanding which equates mean field with classical field, whence $N \to \infty$ limit is viewed as a condition of classicality. But as pointed out by Habib [29, 30](and reference therein) it is only for Gaussian systems such as the O(N) QM model that this is true. It is not true, e.g., for gauge fields, as shown below.
- *Large N as thermodynamic limit?* Coleman in his 1979 Erice Lectures [31], perhaps the most commonly read introduction to large N field theory, drew a parallel with the classical limit: "There is a classical gauge field configuration, which I will call the master field, such that the large-N limits of all gauge invariant Green functions are given by their values at the master field." Along this line Witten in his 1979 Cargese lectures [25] presented an argument for the existence of a master field in the large-N limit of QCD. However, O. Haan [32] pointed out that a crucial assumption Witten made, that the expectation value of any $U(N)$ symmetric operator of the form $f(A) = \mathrm{Tr}[A(x_1)...A(x_l)]$ where A are $N \times N$ matrix fields, factorizes, i.e., $< f_1(A)...f_k(A) >= \Pi_{i=1}^k < f_i(A) > [1 + O(1/N^2)]$, does not hold. He showed that for the two matrix models this 'masterfield' does not exist. Instead of the classical limit Haan argued that the factorization of expectation values of invariant operators is a property analogous to the vanishing of fluctuations for macroscopic observables in the thermodynamic limit. We will return to this issue in section 5.3.
- *Hydrodynamics and thermodynamics limits.* See section 4.4 on this issue.

2.4. *Coherent state and classicality*

- Yaffe in [33] stated that to view Large N limits as classical mechanics requires 4 assumptions. If these assumptions are satisfied, one can generate a natural set of generalized coherent states. These coherent states may then be used to construct a classical phase space, derive a classical Hamiltonian, and the resulting classical dynamics is equivalent to the limiting form of the original quantum dynamics.
  We note that by way of a microscopic model in the open quantum system framework Zurek, Habib and Paz [34] studied environment-induced decoherence and showed that there is a tendency for a quantum system to evolve to a coherent state which is commonly viewed as

the 'most classical' quantum state. This is related to the uncertainty relation criterion for classicality as illustrated in [2, 35].

3. Quantum mechanical O(N) model: large N expansion

Large N expansion of the O(N) model in 3+1 dimensional quantum field theory has been investigated extensively in the literature. The results of immediate relevance for our purpose are from the work of Cooper, Dawson, Habib, Kluger, Mihaila, Mottola, Paz et al. For the O(N) quantum mechanical model describing the dynamics of a system of N nonlinear oscillators we follow the treatment of [21, 22], while relegating the 2PI aspects [18, 23] to our sequel paper. For this model numerical integration valid for all N is possible with the quantum roll initial condition [27] which provides a measure of how accurate the large N (LO and NLO) expansion is as a function of N. We will make use of these results to analyze the quantum-classical problem for large N quantum systems.

3.1. O(N) model

The Lagrangian for the $O(N)$ model in quantum mechanics is given by:

$$L(x, \dot{x}) = \frac{1}{2} \sum_{i=1}^{N} \dot{x}_i^2 - V(r) , \tag{3.1}$$

where $r^2 = \sum_{i=1}^{N} x_i^2$ and $V(x)$ is a potential of the form

$$V(r) = \frac{g}{8N} \left(r^2 - r_0^2\right)^2 \tag{3.2}$$

where g is the self-coupling constant (related to the λ in its field theory origin) and r_0 is where the potential goes to zero, $V(r_0) = 0$. The time-dependent Schrödinger equation for this problem is given by:

$$\mathrm{i} \frac{\partial \psi(x,t)}{\partial t} = \left\{ -\frac{1}{2} \sum_{i=1}^{N} \frac{\partial^2}{\partial x_i^2} + V(r) \right\} \psi(x,t). \tag{3.3}$$

The initial conditions for the quantum roll problem allow a numerical solution for all N, as only the radial part of the wave function enters. With the exact solution available one can examine the validity of the results from a large-N expansion.

Assuming a solution of the form

$$\psi(r,t) = r^{(1-N)/2} \phi(r,t), \tag{3.4}$$

the time dependent Schrödinger equation for $\phi(r,t)$ reduces to [36]:

$$\mathrm{i} \frac{\partial \phi(r,t)}{\partial t} = \left\{ -\frac{1}{2} \frac{\partial^2}{\partial r^2} + U(r) \right\} \phi(r,t) \tag{3.5}$$

with an effective one dimensional potential $U(r)$ given by

$$U(r) = \frac{(N-1)(N-3)}{8\, r^2} + \frac{g}{8N} \left(r^2 - r_0^2\right)^2 . \tag{3.6}$$

Upon the rescaling

$$r^2 = N y^2 , \qquad r_0^2 = N y_0^2 . \tag{3.7}$$

it becomes

$$u(y,N)=\frac{U(y)}{N}=\frac{(N-1)(N-3)}{8N^2\,y^2}+\frac{g}{8}\,(y^2-y_0^2)^2\,, \tag{3.8}$$

corresponding to the new Schrödinger equation,

$$\mathrm{i}\,\frac{\partial\phi(y,\tilde{t})}{\partial\tilde{t}}=\left\{-\frac{1}{2N^2}\frac{\partial^2}{\partial y^2}+u(y,N)\right\}\,\phi(y,\tilde{t}) \tag{3.9}$$

where $\tilde{t}=N\,t$.

To implement the large N expansion, following the original work of [37], we introduce a composite field χ and add a constraint term

$$\frac{N}{2g}\left[\chi-\frac{g}{2N}(r^2-r_0^2)\right]^2\,, \tag{3.10}$$

to the original Lagrangian which yields an equivalent Lagrangian,

$$L'(x,\dot{x},\chi)=\sum_i\frac{1}{2}\left(\dot{x}_i^2-\chi x_i^2\right)+\frac{r_0^2}{2}\chi+\frac{N}{2g}\chi^2\,. \tag{3.11}$$

One can also glean off the results from the corresponding field theory treatment in [13] (by specializing to $0+1$ dimensions, and replacing $\phi_a(t)\to x_i(t)$, $\mu^2\to-\frac{r_0^2 g}{2N}$, and $\lambda\to g$.)

3.2. *Large N expansion*

The generating function $Z[j,J]$ is given by the path integral over the background fields $x_i(t)$:

$$\begin{aligned}
Z[j,J]=\mathrm{e}^{\mathrm{i}W[j,J]}&=\int\mathrm{d}\chi\,\prod_i\mathrm{d}x_i\,\exp\Bigl\{\mathrm{i}\,S[x,\chi;j,J]\Bigr\}\,,\\
S[x,\chi;j,J]&=\int_{\mathcal{C}}\mathrm{d}t\,\Bigl\{L'+\sum_i j_i x_i+J\chi\Bigr\}\,.
\end{aligned}$$

The effective action, to order $1/N$, is obtained by integrating the path integral for the generating functional for the Lagrangian (3.11), over the x_i variables, and approximating the integral over χ by the method of steepest descent (keeping terms up to order $1/N$). A Legendre transform of the resulting generating functional then yields the effective action, which we find to be:

$$\begin{aligned}
\Gamma[q,\chi]&=\int_{\mathcal{C}}\mathrm{d}t\,\Bigl\{\frac{1}{2}\sum_i\left[\dot{q}_i^2(t)-\chi(t)\,q_i^2(t)\right]+\frac{\mathrm{i}}{2}\sum_i\ln\left[G_{ii}^{-1}(t,t)\right]\\
&\quad+\frac{r_0^2}{2}\,\chi(t)+\frac{N}{2g}\,\chi^2(t)+\frac{\mathrm{i}}{2}\ln[D^{-1}(t,t)]\Bigr\}\,,
\end{aligned} \tag{3.12}$$

where the integral is over the close time path $\mathcal{C}$, discussed in [13] and $q(t)=\langle x_i(t)\rangle$. Here $G_{ij}^{-1}(t,t')$ and $D^{-1}(t,t')$ are the lowest order in $1/N$ inverse propagators for x_i and χ, given by

$$\begin{aligned}
G_{ij}^{-1}(t,t')&=\left\{\frac{\mathrm{d}^2}{\mathrm{d}t^2}+\chi(t)\right\}\,\delta_{\mathcal{C}}(t,t')\,\delta_{ij}\equiv G^{-1}(t,t')\,\delta_{ij}\,,\\
D^{-1}(t,t')&=-\frac{N}{g}\delta_{\mathcal{C}}(t,t')-\Pi(t,t')\,,
\end{aligned}$$

where

$$\Pi(t,t') = -\frac{\mathrm{i}}{2}\sum_{i,j} G_{ij}(t,t')\,G_{ji}(t',t) + \sum_{i,j} q_i(t)\,G_{ij}(t,t')\,q_j(t')\,. \tag{3.13}$$

Here $\delta_C(t,t')$ is the closed time path delta function.

The equations of motion for the background fields $q_i(t)$, to order $1/N$, are

$$\left\{\frac{\mathrm{d}^2}{\mathrm{d}t^2} + \chi(t)\right\} q_i(t) + \mathrm{i}\sum_j \int_C \mathrm{d}t'\, G_{ij}(t,t')\,D(t,t')\,q_j(t') = 0\,, \tag{3.14}$$

with the gap equation for $\chi(t)$ given by

$$\chi(t) = -\frac{g}{2N}r_0^2 + \frac{g}{2N}\sum_i \left[q_i^2(t) + \frac{1}{\mathrm{i}}\mathcal{G}_{ii}^{(2)}(t,t)\right]. \tag{3.15}$$

The next-to-leading order x_i propagator $\mathcal{G}_{ij}^{(2)}(t,t')$ and self energy $\Sigma_{ij}(t,t')$ to order $1/N$ are

$$\begin{aligned}\mathcal{G}_{ij}^{(2)}(t,t') &= G_{ij}(t,t') - \sum_{k,l}\int_C \mathrm{d}t_1 \int_C \mathrm{d}t_2\, G_{ik}(t,t_1)\,\Sigma_{kl}(t_1,t_2)\,G_{lj}(t_2,t')\,,\\ \Sigma_{kl}(t,t') &= \mathrm{i}\,G_{kl}(t,t')\,D(t,t') - q_k(t)\,D(t,t')\,q_l(t')\,.\end{aligned} \tag{3.16}$$

(compare these equations with (2.18–2.22) of [13].)

The actual equation for $\mathcal{G}$ which follows from the effective action differs from (3.16) in that the final G in the integral equation is replaced by the full $\mathcal{G}$. This leads to a partial resummation of the 1/N corrections which guarantees positivity of $\langle x^2(t)\rangle$ (but not the full positivity for the density matrix).

In order to solve for $D(t,t')$, we first write

$$\frac{N}{g}D(t,t') = -\delta_C(t,t') + \frac{N}{g}\Delta D(t,t')\,, \tag{3.17}$$

then $\Delta D(t,t')$ satisfies the integral equation,

$$\frac{N}{g}\Delta D(t,t') = \frac{g}{N}\Pi(t,t') - \int_C \mathrm{d}t''\,\Pi(t,t'')\,\Delta D(t'',t')\,, \tag{3.18}$$

(Compare with (2.13–2.16) of Ref. [13].)

3.3. *Effective potential*

As pointed out by Mihaila et al [21] the effective potential in the large N approximation has been previously obtained by Root [38] to order $1/N$. We continue to follow their treatment. From the action 3.12, they found that [2]

$$V_{\mathrm{eff}}^{[1]}(r,\chi) = \frac{N\chi}{g}\left(\mu^2 - \frac{\chi}{2}\right) + \frac{1}{2}\chi r^2 + \frac{N}{2}\int\frac{\mathrm{d}k}{2\pi\mathrm{i}}\ln[\tilde{G}^{-1}(k)] + \frac{1}{2}\int\frac{\mathrm{d}k}{2\pi\mathrm{i}}\ln[\tilde{D}^{-1}(k)]\,, \tag{3.19}$$

[2] When x_i and χ are independent of time, one can ignore the closed time path ordering and use Fourier transforms, passing the poles by using the Feynman contour.

where χ satisfies the requirement

$$\frac{\partial}{\partial \chi} V_{\text{eff}}(r, \chi) = 0 \,. \tag{3.20}$$

Using the rescaled y variables defined in (3.7) the effective potential (3.19) becomes

$$\begin{aligned} \frac{V^{[1]}_{\text{eff}}(y, \chi)}{N} &= \frac{\chi}{2}\,(y^2 - y_0^2) - \frac{\chi^2}{2\,g} + \frac{\sqrt{\chi}}{2} \\ &\quad + \frac{1}{2\,N}\,(m_+ + m_- - 3\,\sqrt{\chi})\,. \end{aligned} \tag{3.21}$$

where $m_\pm^2 = b \pm \sqrt{b^2 - c}$, with

$$b = \frac{5}{2}\chi + \frac{g}{2}\left(y^2 + \frac{1}{2\sqrt{\chi}}\right) \tag{3.22}$$

$$c = 4\chi^2 + g\left(4y^2\chi + \frac{1}{2}\sqrt{\chi}\right) \tag{3.23}$$

The gap equation which determines χ follows from (3.20)

$$\chi = \frac{g}{2}\,(y^2 - y_0^2) + \frac{g(N-3)}{4\,N\,\sqrt{\chi}} + \frac{g}{2\,N}\,\frac{\partial(m_+ + m_-)}{\partial \chi}\,. \tag{3.24}$$

To leading order in the large N expansion, (3.21), (3.24) reduce to the parametric set [3]

$$\begin{aligned} \frac{V^{[0]}_{\text{eff}}(\chi)}{N} &= \frac{\chi^2}{2g} + \frac{\sqrt{\chi}}{4}\,, \\ y^2(\chi) &= y_0^2 + \frac{2}{g}\chi - \frac{1}{2\,\sqrt{\chi}}\,. \end{aligned} \tag{3.25}$$

These snugly placed equations are very useful for our purpose of analyzing the aforementioned issues in MQP. A result of significance is finding the lower bound on N whereby the NLO results give a good approximation to the exact. How does this bear on the issues at hand, namely, quantum behavior and classicality in macroscopic systems? We will discuss this in section 5.1.

In the next section we will focus on the leading order results which are much simpler but still provides useful insight on the macro-quantum relation.

4. Number - phase / correlation, coarse-graining / uncertainty - entropy, decoherence / classicality and hydrodynamics

In this section while staying at the leading order large N (LOLN) level we adopt another angle to analyze mean field theory and classicality. Using the uncertainty relation to find out when a quantum system begins to show classical behaviour is shown to be inadequate, as we will illustrate with an example of quantum roll in the inflationary universe. Decoherence and correlation are needed. In the second example we use a theory shown to be obtainable as the LOLN of an O(N) model which is time-reversal invariant and examine another set of criteria for classicality, that of quantum phase information and entropy generation. Although a system with no quantum phase information is usually identified as classical, yet in essence this is not the

[3] These equations (3.21) and (3.24) agree with Root , however he used the leading order expression for χ in (3.25), rather than the full χ of (3.24).

case. Fluctuations in the number density carry phase information which signifies its quantum origin. One needs to ensure that phase information at all orders vanish for all times to be able to say so, but this is not natural and not easily implementable. In the third subsection we discuss classicality from the decoherent history viewpoint, that hydrodynamic variables are more readily decohered because of the existence of conservation laws for these variables. We also mention why thermodynamic and classical limits are not identical.

4.1. Criteria of classicality: uncertainty relation, decoherence and correlation in a quantum roll

One commonly used signifier of when a quantum system begins to show classical features is the use of the uncertainty relation to distinguish these two regimes. (See [2].) A concrete example with some importance in early universe cosmology is Guth and Pi (GP)'s [27] use of a quantum mechanical model for the description of the inflaton dynamics of a scalar field (the inflaton). They solve for the evolution of the wave function in a de Sitter universe to illustrate how the inflationary transition takes place. The O(N) model has been used in Friedmann universe models by Boyanovsky [39], Cooper et al [40] and many others. The scalar field starts at the top of an inverted harmonic oscillator potential and 'rolls down'; this so-called 'quantum roll' initial condition is in fact used in the results reported in the previous section by Mihaila et al. We want to use these studies to consider the relation between large N, macroscopic quantum systems, the uncertainty relation and classicality. We first give a brief account of how GP treated this problem, then point out the deficiencies in the criterion they used for classicality. We then describe the necessary requirements for classicality as represented by the behaviour of the Wigner function, namely that it *decoheres* in the presence of an environment and it possesses *correlations* between the canonical variables as in a classical trajectory in phase space.

GP considered an eternal inflation situation with the background metric in the form

$$ds^2 = -dt^2 + a(t)^2 dx^2 \tag{4.1}$$

where the scale factor $a(t) = e^{Ht}$ for eternal inflation with the Hubble expansion rate given by $H(t) = \sqrt{8\pi G\rho_0/3}$ where ρ_0 is the vacuum energy density of the false vacuum. The inflaton field ϕ dynamics is described by the action

$$S = \int d^4x\sqrt{-g}[-\frac{1}{2}g^{\mu\nu}\partial_\mu\phi\partial_\nu\phi - V(\phi)] \tag{4.2}$$

with the potential $V(\phi)$ given by

$$V(\phi) = \frac{1}{4}\lambda\left(\phi^2 - \frac{\mu^2}{\lambda}\right)^2 \tag{4.3}$$

where μ is related to the value of the field $\phi_c = \frac{\mu}{\sqrt{\lambda}}$ at the minimum of the potential. The false vacuum energy is given by $\rho_0 = \frac{\mu^4}{4\lambda}$, giving an expansion rate $H = (\frac{2\pi G}{3\lambda})^{\frac{1}{2}}\mu^2$. From the action the inflaton field obeys the dynamical equation

$$\ddot{\phi} + 3H\dot{\phi} - a^{-2}(t)\Delta^{(3)}\phi = -\partial V/\partial\phi \tag{4.4}$$

where $\Delta^{(3)}$ denotes the 3-dim Laplace operator on the spatial hypersurface.

GP then expanded the free field (assumed to be confined within a 3-dim box of length L) in Fourier modes with σ_κ being the amplitude for the $\pm$ propagating components of the kth mode and the zero mode (treated separately because of infrared divergence considerations) and its conjugate momentum π_κ.

As initial condition GP assumed that at early times the system was in thermal equilibrium with a heat bath at temperature T_0 which rapidly drops as the universe inflates away $T = T_0 e^{-Ht}$. The expectation value of physical quantities such as σ_κ with respect to this thermal ensemble can be calculated by taking the ensemble average,

$$< \sigma_\kappa^2(t) >= \frac{1}{2}(L/2\pi)^3 |\psi(\mathbf{k}_\kappa t)|^2 \coth(\Theta_\kappa/2) \tag{4.5}$$

where ψ is given by the Hankel functions of the first kind and Θ_κ is the initial value of $\frac{\hbar\omega}{T(t)}$ at $t \to -\infty$.

To verify that the system behaves classically at late times GP calculated how the uncertainty function changes in time.

$$U_\kappa(t) = \frac{(< \sigma_\kappa^2 >< \pi_\kappa^2 >)^{\frac{1}{2}}}{\hbar/2} \tag{4.6}$$

The denominator $\hbar/2$ is the value for minimal uncertainty in the case of a Gaussian wave packet at zero temperature. Physically the κ mode amplitude can be viewed as behaving classically when its effective wavelength $\lambda_{eff} = 2\pi/k_{eff}$ (where $k_{eff} = a(t)/p(t)$, with $a(t) = e^{Ht}$, $p_\kappa = \sqrt{\mathbf{k}_\kappa^2 + \gamma^2}$, $\gamma = \sqrt{\lambda/\hbar T_0/2}$) is much larger than the horizon length.

Hu and Zhang [2] (see also Anastopoulos and Halliwell [35]) have derived an expression for the uncertainty relation at finite temperature using the Wigner function of an oscillator interacting with a thermal bath modelled by N harmonic oscillators as weighing function. They also discussed how the contributions of quantum fluctuations weighted against thermal fluctuations and the criterion of classicality and thermality.

As later investigations showed this criterion used by GP for classicality is suggestive but incomplete. Two conditions need be satisfied by the (reduced) Wigner function from the equivalent (reduced) density matrix ('reduced' refers to the situation where the system is coupled to an environment and rendered open). Wigner function has long been viewed as the quantum object 'closest to' the classical distribution function. Up to the early 90's before decoherence was considered in earnest it was viewed that if the Wigner function of a quantum system peaks along a classical trajectory, meaning that there is correlation between the coordinate x and momentum p, then one could view the system as classical. This criterion is shown to be valid only for Gaussian systems (such as a free harmonic oscillator with quadratic potential coupled bi-linearly to an environment of simple harmonic oscillators) because otherwise it may not even be positive definite [41]. In the Gaussian approximation the Wigner function is positive for all times but as was further pointed out [42] it does not describe classical correlations unless the system is coupled to an environment. Thus classical correlation and environment-induced decoherence are two necessary conditions for a quantum system to show classical behaviour.

When the wave packet spreads out further away from the top of the potential the Gaussian approximation breaks down. As shown by Lombardo et al (LMM) [43], as the coupling between the system and the environment increases, its decoherence time decreases. Due to the nonlinearities of the potential, when the coupling vanishes there is no classical limit, not even classical correlations. In fact, as reported by Antunes et al [44], in a quenched phase transition, even after classicalization has been reached, the system may display quantum behaviour again.

With these new understandings LMM reanalyzed GP's model with and without an environment, using the quantum Brownian motion [45, 46] model, especially results from the non-Markovian master equation for a general environment [47]. They found that when the system is isolated, due to the high squeezing of the initial wave packet x and p become classically correlated. The density matrix is not diagonal. The correlation time depends on the shape of the potential. Only when the particle is coupled to an environment will a *bona fide* quantum to classical transition occur. The Wigner function becomes peaked around a classical trajectory and the density matrix diagonalizes. The decoherence time depends on the diffusion coefficient in the master equation.

Now that we have analyzed the quantum mechanical model in detail how would a macroscopic system as modelled by the O(N) quantum field theory behave? Symmetry breaking in a $\phi^4 O(N)$ model has been studied in detail by many groups [14, 39]. Lombardo et al [43] carried out numerical analysis of this model and made the following observations: at late times the long-wavelength modes $k_l \ll t^{-1}$ in the wave functional become classically correlated. This is analogous to the situation described above for the inverted harmonic oscillator. For these modes the width of the Gaussian wave function increases linearly with time. Moreover, as the width of the wave function or the density matrix increases, the Wigner function becomes sharply peaked around the classical trajectory. We see that in the large N limit, and at long times, the dynamical evolution of the O(N) model shows classical correlations but not true quantum to classical transition. Just as in the inverted oscillator without environment, the density matrix does not become diagonal. The correlation time depends on the details of the potential.

In conclusion Lombardo et al [43] observe that in order to get a Wigner function that is positive definite and peaked around a classical trajectory at long times, it is necessary to have both vanishing effective mass and a Gaussian wave function. Therefore, it is quite possible that in a field theory calculation with finite N there will be no classical limit nor classical correlations unless one allows the field to interact with an environment. On this case for an analysis of how the nonequilibrium evolution of a condensate and its fluctuations vary with N, Baacke and Michalski [48] have provided analytic expressions for general N and numerical solutions for N=1,4, 10. This should be compared with the analytic expressions for the effective potential calculated up to NLO for finite N given by Mihaila et al for an O(N) model, which we will discuss in section 5.1.

4.2. Criteria of classicality: coarse-graining of quantum phase, entropy generation

We next use the entropy function derived from the number of particles and their quantum correlation in a quantum Vlasov equation to illustrate the notion of classicality. The large N aspect enters because quantum Vlasov equation is obtained as a mean field theory derivable as the leading order large N approximation from scalar QED, where N is the number of identical copies of the charged matter field placed in an electromagnetic (EM) field . This is sometimes referred to as the semiclassical limit since the matter field is fully quantized and the EM field is treated classically. Kluger, Mottola and Eisenberg [28] considered the case of a spatially homogeneous electric field, represented by the vector potential in the Coulomb gauge, $\mathbf{A} = A(t)\hat{\mathbf{z}},\ A_0 = 0$, whence the electric field is given by $\mathbf{E} = -\dot{A}\hat{\mathbf{z}} = E\hat{\mathbf{z}}$.

Assuming also that the field lives in a finite large volume V we can expand the charged scalar field operator in Fock space in Fourier modes. Since particles are physically distinct from antiparticles, we need two independent sets of destruction operators

$$\Phi(\mathbf{x},t) = \frac{1}{\sqrt{V}}\sum_{\mathbf{k}} e^{i\mathbf{k}\cdot\mathbf{x}}\varphi_{\mathbf{k}}(t) \tag{4.7}$$

$$= \frac{1}{\sqrt{V}}\sum_{\mathbf{k}} \left\{ e^{i\mathbf{k}\cdot\mathbf{x}} f_{\mathbf{k}}(t) a_{\mathbf{k}} + e^{-i\mathbf{k}\cdot\mathbf{x}} f^*_{-\mathbf{k}}(t) b^\dagger_{\mathbf{k}} \right\} . \tag{4.8}$$

Denote the time-independent annihilation operator of a particle in mode $\mathbf{k}$ by $a_{\mathbf{k}}$ and the creation of an anti-particle in mode $-\mathbf{k}$ by $b^\dagger_{\mathbf{k}}$. They obey the commutation relations

$$[a_{\mathbf{k}}, a^\dagger_{\mathbf{k}'}] = [b_{\mathbf{k}}, b^\dagger_{\mathbf{k}'}] = \delta_{\mathbf{k}\mathbf{k}'} . \tag{4.9}$$

Therefore $N_+(\mathbf{k}) \equiv \langle a^\dagger_{\mathbf{k}} a_{\mathbf{k}}\rangle, N_-(\mathbf{k}) \equiv \langle b^\dagger_{\mathbf{k}} b_{\mathbf{k}}\rangle$ are the mean numbers of particles and antiparticles respectively. Without loss of generality we can make use of the freedom in defining the initial

Journal of Physics: Conference Series **306** (2011) 012002 doi:10.1088/1742-6596/306/1/012002

phases of the mode functions to set the correlation densities $\langle a_{\mathbf{k}} a_{\mathbf{k}} \rangle = \langle b_{\mathbf{k}} b_{\mathbf{k}} \rangle = 0$. In a Hamiltonian description we can take for each mode $\mathbf{k}$

$$\varphi_{\mathbf{k}}(t) \equiv f_{\mathbf{k}}(t) a_{\mathbf{k}} + f^*_{\mathbf{k}}(t) b^\dagger_{-\mathbf{k}} \tag{4.10}$$

as the (complex) generalized coordinates of the field Φ and

$$\pi_{\mathbf{k}}(t) = \dot{\varphi}^\dagger_{\mathbf{k}}(t) = \dot{f}^*_{\mathbf{k}}(t) a^\dagger_{\mathbf{k}} + \dot{f}_{\mathbf{k}}(t) b_{-\mathbf{k}}\,, \tag{4.11}$$

as the momentum canonically conjugate to it. By virtue of the commutation relation (4.9) they obey the canonical commutation relation,

$$[\varphi_{\mathbf{k}}, \pi_{\mathbf{k}'}] = \mathrm{i}\hbar \delta_{\mathbf{k}\mathbf{k}'}\,, \tag{4.12}$$

provided that the mode functions satisfy the Wronskian condition

$$(f_k, f^*_k) = \mathrm{i}\hbar. \tag{4.13}$$

The complex amplitude function $f_{\mathbf{k}}(t)$ of the $\mathbf{k}$ th mode satisfies the wave equation

$$\frac{\mathrm{d}^2 f_{\mathbf{k}}}{\mathrm{d}t^2} + \omega^2_{\mathbf{k}}(t) f_{\mathbf{k}}(t) = 0, \tag{4.14}$$

where the time dependent frequency $\omega^2_{\mathbf{k}}(t)$ is given by

$$\omega^2_{\mathbf{k}}(t) = (\mathbf{k} - e\mathbf{A})^2 + m^2 = (k_z - eA(t))^2 + k^2_\perp + m^2\,. \tag{4.15}$$

where k_z is the constant canonical momentum in the $\hat{\mathbf{z}}$ direction while the physical (gauge-invariant) kinetic momentum is given by

$$p_z(t) = k_z - eA(t), \qquad \dot{p}_z = -e\dot{A} = eE \tag{4.16}$$

For a spatially homogeneous electric field (*i.e.*, $\nabla \cdot \mathbf{E} = 0$), by Gauss' Law, the mean charge density must vanish, i.e, $j^0(t) = 0$. and the mean current in the $\hat{\mathbf{z}}$ direction is

$$j(t) = 2e \int \mathrm{d}^3\mathbf{k}\, [k_z - eA(t)] |f_{\mathbf{k}}(t)|^2 (1 + N_+(\mathbf{k}) + N_-(-\mathbf{k})) \tag{4.17}$$

One can further restrict to the subspace of states for which

$$N_+(\mathbf{k}) = N_-(-\mathbf{k}) \equiv N_{\mathbf{k}} \tag{4.18}$$

Clearly the vacuum $N_+(\mathbf{k}) = N_-(-\mathbf{k}) = 0$ (as well as a thermal state) belongs to this class of states.

Particle pairs will be produced in a strong background field, and in turn, affect the strength and evolution of this background field. This first step is called a 'test field' approximation where the background field is assumed fixed. The second step of including the effects of created particles is called a 'backreaction problem'. It demands a self-consistent solution of the mean electric field $\mathbf{E}(t)$ coupled to the expectation value of the current $j(t)$ of the quantum charged scalar field $\varphi_{\mathbf{k}}(t)$. In a spatially homogeneous electric field, the only nontrivial Maxwell equation is simply,

$$-\dot{E}(t) = \ddot{A}(t) = j(t) \tag{4.19}$$

where the current is given by (4.17). Since the charged scalar field depends on the vector potential A to begin with, $f_{\mathbf{k}}(t)$ and $A(t)$ need be solved self-consistently from (4.14) with (4.15) and (4.19).

Journal of Physics: Conference Series **306** (2011) 012002 doi:10.1088/1742-6596/306/1/012002

4.2.1. Particle production In a time dependent background the equation for the mode functions (4.14) would in general admit time dependent solutions. This means that one set of particle states which define a Fock space at one moment of time would change at a later time into another set of particle state in a different Fock space. In particular, an initial vacuum could evolve into an n-particle state later, which means particles are being produced. This is best expressed in terms of the Bogoliubov transformation [49] between these two Fock space basis $a_{\mathbf{k}}$ and $\tilde{a}_{\mathbf{k}}$. Consider for the moment just particles, no anti-particles, and no particle interactions (see [50, 51] for that case). Let the first basis $a_{\mathbf{k}}$ be associated with modes $(f_{\mathbf{k}}, f_{\mathbf{k}}^*)$, the second basis $\tilde{a}_{\mathbf{k}}$ with modes $\left(\tilde{f}_{\mathbf{k}}, \tilde{f}_{\mathbf{k}}^*\right)$. We may expand the field operators in either base, such as

$$\varphi_{\mathbf{k}}(t) = f_{\mathbf{k}}(t)\, a_{\mathbf{k}} + f_{\mathbf{k}}^*(t)\, a_{-\mathbf{k}}^{\dagger} \tag{4.20}$$

in the first case, and

$$\varphi_{\mathbf{k}}(t) = \tilde{f}_{\mathbf{k}}(t)\, \tilde{a}_{\mathbf{k}} + \tilde{f}_{\mathbf{k}}^*(t)\, \tilde{a}_{-\mathbf{k}}^{\dagger} \tag{4.21}$$

in the second. Since both sets of solutions of the mode equations are complete, we must have

$$f_{\mathbf{k}}(t) = \alpha_{\mathbf{k}} \tilde{f}_{\mathbf{k}}(t) + \beta_k \tilde{f}_{\mathbf{k}}^*(t), \tag{4.22}$$

and its inverse

$$\tilde{f}_{\mathbf{k}}(t) = \alpha_{\mathbf{k}}^* f_{\mathbf{k}}(t) - \beta_{\mathbf{k}} f_{\mathbf{k}}^*(t). \tag{4.23}$$

The Wronskian condition $(f_{\mathbf{k}}, f_{\mathbf{k}}^*) = \left(\tilde{f}_{\mathbf{k}}, \tilde{f}_{\mathbf{k}}^*\right) = i\hbar$ imposes a condition on the Bogoliubov coefficients

$$|\alpha_{\mathbf{k}}|^2 - |\beta_{\mathbf{k}}|^2 = 1 \tag{4.24}$$

for each $\mathbf{k}$. We can thus write

$$\begin{aligned} |\alpha_{\mathbf{k}}(t)| &= \cosh r_{\mathbf{k}}(t), \\ |\beta_{\mathbf{k}}(t)| &= \sinh r_{\mathbf{k}}(t). \end{aligned} \tag{4.25}$$

where $r_{\mathbf{k}}(t)$ is called the squeeze parameter for mode $\mathbf{k}$ a terminology adopted from quantum optics. For a description of particle creation in the squeezed state language, see, e.g., Chapter 4 of [52].

The above only describes how one set of Fock states defined at one time is related to another set, but whether a vacuum can indeed be defined at any one time is a different and often more challenging problem. We won't go into the details of this important issue, but just be satisfied for our present purpose with the fact that when the background varies sufficiently slowly one can invoke the concept of an adiabatic number state [4]. This was used by [28] to calculate the charged particle production in an electric field problem described above.

The adiabatic number state $\tilde{f}_{\mathbf{k}(0)}^{+}(t)$ corresponds to the $0th$ order adiabatic vacuum [defined in (4.46) of [52] as $f_{\mathbf{k}(0)}^{+}$]

$$\tilde{f}_{\mathbf{k}}^{(0)}(t) \equiv \sqrt{\frac{\hbar}{2\omega_{\mathbf{k}}(t)}} \exp\left(-i\Theta_{\mathbf{k}}^{(0)}\right), \tag{4.26}$$

[4] This definition of a number state makes use of the fact that under adiabatic evolution, particle number is an adiabatic invariant, thus restricting its validity from the start to extremely weak or slowing varying background fields. This level of approximation will not give a good measure for on-going particle creation, as particle creation is basically a nonadiabatic process. It is however useful for quantum kinetic theory descriptions, where a quasi-particle approximation is usually introduced which amounts to incorporating only the quantum radiative corrections to the particles but not the fully field theoretical effects such as particle creation.

where $\Theta_{\mathbf{k}}^{(0)} \equiv \int^t \omega_{\mathbf{k}}(t')\mathrm{d}t'$, is the 0th order adiabatic phase. At this level of accuracy one measures particle numbers at all times with respect to the initial vacuum state at time t_0. The adiabatic particle number is defined to be [28]

$$\begin{aligned}
\tilde{N}_{\mathbf{k}}(t) &\equiv \langle \tilde{a}^{\dagger}_{\mathbf{k}}(t)\tilde{a}_{\mathbf{k}}(t)\rangle = \langle \tilde{b}^{\dagger}_{-\mathbf{k}}(t)\tilde{b}_{-\mathbf{k}}(t)\rangle = |\alpha_{\mathbf{k}}|^2\langle a^{\dagger}_{\mathbf{k}}a_{\mathbf{k}}\rangle + |\beta_{\mathbf{k}}|^2\langle b_{-\mathbf{k}}b^{\dagger}_{-\mathbf{k}}\rangle \\
&= \left(1+|\beta_{\mathbf{k}}|^2\right) N_+(\mathbf{k}) + |\beta_{\mathbf{k}}|^2\left(1+N_-(-\mathbf{k})\right) \\
&= |\beta_{\mathbf{k}}|^2 + (1+2|\beta_{\mathbf{k}}|^2)N_{\mathbf{k}} = N_{\mathbf{k}} + (1+2N_{\mathbf{k}})\,|\beta_{\mathbf{k}}(t)|^2
\end{aligned} \tag{4.27}$$

where the last line is valid only if the number of positive and negative charges are equal (cfr. (4.18)). The amount of particle production at time t in this basis is given by the expectation value of the number operator $\tilde{a}^{\dagger}\tilde{a}$ at time t with respect to the vacuum state $| >_0$ defined at t_0, (not the vacuum state $| >_t$ defined at t).

4.2.2. Number and correlation In this time-dependent particle number basis an equation can be obtained for the time rate of change of the number of particles created in each mode by differentiating (4.27),

$$\frac{\mathrm{d}}{\mathrm{d}t}\tilde{N}_{\mathbf{k}} = 2\,(1+2N_{\mathbf{k}})\,\mathrm{Re}\,(\beta^*_{\mathbf{k}}\dot{\beta}_{\mathbf{k}}). \tag{4.28}$$

We need an expression for $\dot{\beta}_{\mathbf{k}}$ in terms of $\alpha_{\mathbf{k}}, \beta_{\mathbf{k}}$ and $\Theta_{\mathbf{k}}^{(0)}(t) \equiv \int^t \omega_{\mathbf{k}}(t')\mathrm{d}t'$ (henceforth we will omit the superscript 0 on $\Theta_{\mathbf{k}}^{(0)}$). They are given by

$$\dot{\alpha}_{\mathbf{k}} = \frac{\dot{\omega}_{\mathbf{k}}}{2\omega_{\mathbf{k}}}\beta_{\mathbf{k}}\exp(2\mathrm{i}\Theta_{\mathbf{k}}), \quad \dot{\beta}_{\mathbf{k}} = \frac{\dot{\omega}_{\mathbf{k}}}{2\omega_{\mathbf{k}}}\alpha_{\mathbf{k}}\exp(-2\mathrm{i}\Theta_{\mathbf{k}})\,. \tag{4.29}$$

and thus,

$$\begin{aligned}
\frac{\mathrm{d}}{\mathrm{d}t}\tilde{N}_{\mathbf{k}} &= \frac{\dot{\omega}_{\mathbf{k}}}{\omega_{\mathbf{k}}}\,(1+2N_{\mathbf{k}})\;\mathrm{Re}\left\{\alpha_{\mathbf{k}}\beta^*_{\mathbf{k}}\exp(-2\mathrm{i}\Theta_{\mathbf{k}})\right\} \\
&= \frac{\dot{\omega}_{\mathbf{k}}}{\omega_{\mathbf{k}}}\,\mathrm{Re}\left\{\mathcal{C}_{\mathbf{k}}\exp(-2\mathrm{i}\Theta_{\mathbf{k}})\right\}\,,
\end{aligned} \tag{4.30}$$

where we have defined the time-dependent pair correlation function,

$$\mathcal{C}_{\mathbf{k}}(t) \equiv \langle \tilde{a}_{\mathbf{k}}(t)\tilde{b}_{-\mathbf{k}}(t)\rangle = (1+2N_{\mathbf{k}})\,\alpha_{\mathbf{k}}\beta^*_{\mathbf{k}}\,. \tag{4.31}$$

The pair correlation $\mathcal{C}_{\mathbf{k}}(t)$ is a very rapidly varying function, since the time dependent phases on the right side of (4.31) *add* rather than cancel. The phases, however, nearly cancel in the final combination of (4.30) to render $\tilde{N}_{\mathbf{k}}$ a slowly varying function. The time derivative of the pair correlation function is given by,

$$\begin{aligned}
\frac{\mathrm{d}}{\mathrm{d}t}\mathcal{C}_{\mathbf{k}} &= \frac{\dot{\omega}_{\mathbf{k}}}{2\omega_{\mathbf{k}}}\,(1+2N_{\mathbf{k}})\exp(2\mathrm{i}\Theta_{\mathbf{k}})\,\left(1+2|\beta_{\mathbf{k}}|^2\right) \\
&= \frac{\dot{\omega}_{\mathbf{k}}}{2\omega_{\mathbf{k}}}\left(1+2\tilde{N}_{\mathbf{k}}\right)\exp(2\mathrm{i}\Theta_{\mathbf{k}})\,.
\end{aligned} \tag{4.32}$$

4.2.3. Quantum Vlasov equation Let us return now to the two equations for the rates of change of the particle number and the quantum correlations. Solving (4.32) formally for $\mathcal{C}_{\mathbf{k}}$, assuming that $\mathcal{C}_{\mathbf{k}}$ vanishes at some $t = t_0$ which could be taken to $-\infty$, and substituting into (4.30) we obtain

$$\frac{\mathrm{d}}{\mathrm{d}t}\tilde{N}_{\mathbf{k}} = \frac{\dot{\omega}_{\mathbf{k}}}{2\omega_{\mathbf{k}}}\int_{t_0}^{t}\mathrm{d}t'\,\left\{\frac{\dot{\omega}_{\mathbf{k}}}{\omega_{\mathbf{k}}}(t')\left(1+2\tilde{N}_{\mathbf{k}}(t')\right)\cos\left[2\Theta_{\mathbf{k}}(t)-2\Theta_{\mathbf{k}}(t')\right]\right\}\,, \tag{4.33}$$

(4.33) gives the rate of particle creation in an arbitrary time-varying mean field which may be called a "quantum Vlasov equation". Note the appearance of the Bose enhancement factor $(1+2\tilde{N}_{\mathbf{k}})$ in (4.33) indicates that both spontaneous and induced particle creation are present. One important feature of (4.33) is that it is nonlocal in time, the particle creation rate depending on the entire previous history of the system. Thus particle creation in general is a *non-Markovian* process [53, 54, 55, 56]. Note that the nonlocal form of (4.33) results from solving one variable $\mathcal{C}$ in terms of the other $\tilde{N}$, each obeying a Hamiltonian equation of motion. This is a general feature of coupled subsystems.

(4.33) becomes exact in the limit in which the electric field can be treated classically, i.e the limit in which real and virtual photon emission is neglected, and there is no scattering. Inclusion of scattering processes leads to collision terms on the right side of (4.33) which are also nonlocal in general. This nonlocality is essential to the quantum description in which phase information is retained for all times. The phase oscillations in the cosine term are a result of the quantum coherence between the created pairs, which must be present in principle in any unitary evolution. However, precisely because these phase oscillations are so rapid it is clear that the integral in (4.33) receives most of its contribution from t' close to t, which suggests that some local approximation to the integral should be possible, provided that we are not interested in resolving the short time structure or measuring the phase coherence effects. The time scale for these quantum phase coherence effects to wash out is the time scale of several oscillations of the phase factor $\Theta_{\mathbf{k}}(t) - \Theta_{\mathbf{k}}(t')$, which is of order $\tau_{qu} = 2\pi/\omega_{\mathbf{k}} = 2\pi\hbar/\epsilon_{\mathbf{k}}$, where $\epsilon_{\mathbf{k}}$ is the single particle energy.

4.3. Density matrix and entropy

After the elimination of the rapid variables $\mathcal{C}_{\mathbf{k}}$ defined in (4.31) in favour of the slow variables $\mathcal{N}_{\mathbf{k}}$ one can construct the density matrix in the adiabatic number basis easily [28]. In a pure state the only nonvanishing matrix elements of ρ are in uncharged pair states with equal numbers of positive and negative charges, $\ell_{\mathbf{k}} = n_{\mathbf{k}}^{(+)} = n_{\mathbf{k}}^{(-)}$, with $\ell_{\mathbf{k}}$ the number of pairs in the mode $\mathbf{k}$, *viz.*

$$\langle 2\ell'_{\mathbf{k}}|\rho|2\ell_{\mathbf{k}}\rangle_{\rm pure} = \mathrm{e}^{\mathrm{i}(\ell_{\mathbf{k}}{}'-\ell_{\mathbf{k}})\vartheta_{\mathbf{k}}(t)}\ \mathrm{sech}^2 r_{\mathbf{k}}(t)\ \left(\mathrm{tanh} r_{\mathbf{k}}(t)\right)^{\ell_{\mathbf{k}}{}'+\ell_{\mathbf{k}}} \tag{4.34}$$

where the magnitude of the Bogoliubov transformation, $r_{\mathbf{k}}(t)$ is defined in (4.25) and its phase, $\vartheta_{\mathbf{k}}(t)$ is determined by

$$\alpha_{\mathbf{k}}\beta_{\mathbf{k}}^{*}\mathrm{e}^{-2i\Theta_{\mathbf{k}}} = -\mathrm{sinh} r_{\mathbf{k}}\ \mathrm{cosh} r_{\mathbf{k}}\ \mathrm{e}^{\mathrm{i}\vartheta_{\mathbf{k}}}\,. \tag{4.35}$$

Hence the off-diagonal matrix elements $\ell' \neq \ell$ of ρ are rapidly varying on the time scale τ_{qu} of the quantum mode functions, while the diagonal matrix elements $\ell' = \ell$ depend only on the adiabatic invariant average particle number via

$$\begin{aligned}\langle 2\ell_{\mathbf{k}}|\rho|2\ell_{\mathbf{k}}\rangle_{\rm pure} &\equiv \rho_{2\ell_{\mathbf{k}}} = \mathrm{sech}^2 r_{\mathbf{k}}\mathrm{tanh}^{2\ell_{\mathbf{k}}} r_{\mathbf{k}}\\ &= \frac{|\beta_{\mathbf{k}}|^{2\ell_{\mathbf{k}}}}{(1+|\beta_{\mathbf{k}}|^2)^{\ell_{\mathbf{k}}+1}} = \left.\frac{\mathcal{N}_{\mathbf{k}}^{\ell_{\mathbf{k}}}}{(1+\mathcal{N}_{\mathbf{k}})^{\ell_{\mathbf{k}}+1}}\right|_{\rm pure},\end{aligned} \tag{4.36}$$

and are therefore much more slowly varying functions of time. The average number of positively charged particles (or negatively charged antiparticles) in this basis is given by

$$\sum_{\ell_{\mathbf{k}}=0}^{\infty} \ell_{\mathbf{k}}\rho_{2\ell_{\mathbf{k}}} = \mathcal{N}_{\mathbf{k}}\,. \tag{4.37}$$

Thus the diagonal and off-diagonal elements of the density matrix in the adiabatic particle number basis stand in precisely the same relationship to each other and contain the same information as the particle number $\mathcal{N}_{\mathbf{k}}$ and pair correlation $\mathcal{C}_{\mathbf{k}}$ respectively.

From the (admittedly somewhat simplistic) perspective of the diagonization of the (reduced) density matrix as a signifier of classicality, this is similar to the criterion used in environment-induced decoherence alluded to in section 4.1, with one major difference: Here, quantum phase in the system is eliminated by choice of representation, while there, it is through the system's interaction with the environment. The former case needs physical justification for making such a choice, the latter needs explicit demonstration of how the pointer basis from measurement readings depend on the interaction Hamiltonian. We shall continue the first route below.

4.3.1. Classicality from ignoring the quantum phase In the density matrix (4.36) the diagonal elements $\rho_{2\ell_{\mathbf{k}}}$ may be interpreted (for a pure state) as the independent probabilities of creating $\ell_{\mathbf{k}}$ pairs of charged particles with canonical momentum $\mathbf{k}$ from the vacuum. This corresponds to disregarding the intricate quantum phase correlations between the created pairs in the unitary Hamiltonian evolution. When physics is expressed in the adiabatic particle number basis (the Fock or N representation) the phase information is ignored. The quantum density matrix in this representation produces an entropy function which reflects the entropy associated with particle creation alone but says nothing about the evolution of the quantum phase or correlation. This illustrates the crucial role played by the choice of representations in the definition of entropy associated with particle creation [57].

Results obtained from neglecting quantum phase are known to be quite accurate for long intervals of time in the back-reaction of the current on the electric field producing the pairs, because when the current is summed over all the $\mathbf{k}$ modes, the phase information in the pair correlations cancels very efficiently. Thus for practical purposes one can approximate the full Gaussian density matrix over large time intervals by its diagonal elements only, in this basis.

4.3.2. Entropy generation from particle creation Let us examine the reduced von Neumann entropy constructed from the diagonal density matrix (4.36)

$$S_{\mathcal{N}}(t) = -\sum_{\mathbf{k}} \sum_{\ell_{\mathbf{k}}=0}^{\infty} \rho_{2\ell_{\mathbf{k}}} \ln \rho_{2\ell_{\mathbf{k}}} \tag{4.38}$$

Upon substituting (4.36) into this, the sums over $\ell_{\mathbf{k}}$ are geometric series which are easily performed. The von Neumann entropy of this reduced density matrix

$$S_{\mathcal{N}}(t) = \sum_{\mathbf{k}} \{(1+\mathcal{N}_{\mathbf{k}})\ln(1+\mathcal{N}_{\mathbf{k}}) - \mathcal{N}_{\mathbf{k}} \ln \mathcal{N}_{\mathbf{k}}\} \tag{4.39}$$

is precisely equal to the Boltzmann entropy of the single particle distribution function $\mathcal{N}_{\mathbf{k}}(t)$. Hence

$$\frac{\mathrm{d}}{\mathrm{d}t} S_{\mathcal{N}} = \sum_{\mathbf{k}} \ln\left(\frac{1+\mathcal{N}_{\mathbf{k}}}{\mathcal{N}_{\mathbf{k}}}\right) \frac{\mathrm{d}}{\mathrm{d}t}\mathcal{N}_{\mathbf{k}} \tag{4.40}$$

increases if the mean particle number increases. This is always the case *on average* for bosons if one starts with vacuum initial conditions, since $|\beta_{\mathbf{k}}|^2$ is necessarily nonnegative and can only increase if it is zero initially [58]. Locally, or once particles are present in the initial state, particle number or the entropy (4.40) does not necessarily increase monotonically in time.

Hence the notion of entropy associated with particle creation, and the lore that it increases in time, is only valid for spontaneous production of bosons from an initial vacuum state. This function associated with fermions, and that associated with stimulated production of both boson and fermions, can decrease in time.

4.4. Decoherence of hydrodynamic variables, conservation laws and quasi-classical domains

To end this section on the criteria of classicality we want to describe an important formulation in how classicality emerges. This is the decoherent histories approach of Gell-Mann and Hartle [59] applied to the quantum mechanics of closed systems, which is very different from the above-mentioned environment-induced decoherence scheme applied to open quantum systems. A central theme of this program is the decoherence of hydrodynamic variables, namely, that the variables typically characterizing the quasiclassical domain of a large and possibly complex quantum system that will become classical 'habitually' are the integrals over small volumes of locally conserved densities. Examples of local densities are number density, momentum density, energy density, charge density, which are collectively called hydrodynamic variables. The question of emergent classicality then consists first of understanding why these variables enjoy this special status, and second, of deriving the familiar hydrodynamic equations for these densities from decoherent histories. The representative references are [60, 61, 62, 63]. The treatment in Brun and Halliwell [63] contains examples for ease of understanding. For deeper conceptual issues related to quasi-classical domains read Dowker and Kent [64].

We mention here those results which are directly relevant to our discussion of hydrodynamics, thermodynamics and classicality.

1) Following Gell-Mann and Hartle [60] Hartle, Laflamme and Marolf [61] showed that exactly conserved quantities are exactly decoherent, and it is conceivable that the probabilities for such histories would peak about deterministic evolution equations. Brun and Halliwell [63] derived a formula which shows the explicit connection between local conservation and approximate decoherence.

2) Halliwell [62] showed that a) for the case of the diffusion of the number density of a dilute concentration of foreign particles in a fluid, for certain physically reasonable initial states, the probabilities for the histories of the number density are strongly peaked about evolution according to the diffusion equation. b) When the initial state is a local equilibrium state, large-N statistics guarantees that the probabilities for histories of local densities are strongly peaked about a single history which describes the evolution of the mean in the local equilibrium initial state. This intuitively clear result readily connects with previous work, which shows that the mean values of local densities evolve according to hydrodynamic equations. The fact that there is essentially only one history with nonzero probability means that there is approximate decoherence.

3) Brun and Halliwell [63] considered a class of models consisting of a large number of weakly interacting components, in which the projections onto local densities may be decomposed into projections onto one of two alternatives of the individual components. One model they gave is the placement of N particles into two partitions, the other model consists of a long chain of locally coupled spins for which the Hamiltonian conserves the total spin. They computed the decoherence functional for histories of local densities, in the limit when the number of components is very large, they find that decoherence requires two conditions: a) the smearing volumes must be sufficiently large to ensure approximate conservation, and b) the local densities must be partitioned into sufficiently large ranges to ensure protection against quantum fluctuations.

The last two points are of special interest to us: namely, 2b) referring to the evolution of the mean in the local equilibrium initial state shows the relation of thermality and mean field dynamics. 3a) involving a large number of components N in a system and a large volume V smacks of thermodynamical limit, except that hydrodynamics ensures that conservation laws apply for these collective variables. Point 3b) above gives the condition for quantum fluctuations to be suppressed. We saw similar descriptions for the mean field and fluctuations in terms of leading order and NLO expansions for a large N quantum mechanical system.

Journal of Physics: Conference Series **306** (2011) 012002 doi:10.1088/1742-6596/306/1/012002

5. Next-to-leading order large N, quantum fluctuations

So far we have discussed the leading order theory of an O(N) model obtained from a large N expansion. This mean field theory and its behaviour is usually referred to as "classical". Assuming such a connection which we will examine critically in this section, we may ask: 1) Can one identify a number N_c whereby a system containing N components with $N > N_c$ will begin to acquire classical features? 2) Since a classical system is devoid of quantum fluctuations, can one extract the information related to quantum fluctuations from higher order expansions in large N? Following this logic one may hope to come up with a systematic way to identify the demarkation zone between classical and quantum features in terms of N for a quantum system with a large collection of components.

But as we forewarned in the Introduction, this equivalence of mean field with classical features is valid only for Gaussian theories. More accurately one should think about mean field rather than classical theory in the leading order large N and deviations from the mean in the NLO large N. One should therefore ask, 3) Does one see a change of qualitative behaviour in the NLO or higher order large N expansion from the mean field theory? What measures deviation from mean field? Perhaps one should not think of quantum fluctuations, but correlations, as in the BBGKY hierarchy? (The role of 2PIEA in closed (democratic) systems such as systems with N components will be discussed in our next paper.) Finally, to see the difference of behaviour from the O(N) model we have studied above one should also ask, 4) Would the large N limit of a gauge field theory behave differently, and how so? What can one say about deviation from the mean field and how the contribution of quantum fluctuations enter in these theories?

In the following three subsections we will address issues raised above: subsection 5.1 in the context of O(N) theory for question 1), subsection 5.2 in the context of stochastic gravity for queries 2) and 3), and subsection 5.3 in the context of large N gauge field theory for question 4).

5.1. How large an N will NLO give sensible results? Quantum features beyond mean field theory?

For this question let us return to the results reported in section 3 pertaining to next-to-leading order (NLO) large N expansion in an O(N) model. The authors of [21] show that the naive large N expansion violates unitarity (or more generally, positivity) leading to an instability at least for N less than some value N_T. They have provided numerical evidence that a sharp threshold at $N \sim N_T$ exists, where for $N > N_T$ they saw no instability. They trace this behaviour to be related to the nature of the effective potential at NLO: At this order, the effective potential has the property of not being defined everywhere for values of $N < N_c$, where N_c depends on the values of the parameters specifying the potential, and $N_T \sim N_c$. However, for $N > N_c$, the effective potential is defined everywhere and there is no instability.

Probing further using the effective potential for the NLO expansion these authors asked the following question: "How does the point $y_{\min}$, below which the next-to-leading order large N effective potential does not exist, changes as a function of N?" We know that at "infinite N," (leading order), $y_{\min} = 0$, but it is important to know how this limit is reached. For instance, is there a finite value of N beyond which $y_{\min} = 0$? They plotted (figure 5 of [21]) $y_{\min}$ as a function of N for the set of parameters chosen in their equation (7.15): At the next-to-leading order large N level, for $N \leq 18.6$, $y_{\min}$ is finite, but for $N \geq 18.6$, it hits the origin. Thus for $N \geq 18.6$, the authors concluded, one can associate a quantum state (though not known explicitly) with the next-to-leading order approximation.

These findings by [21] are interesting by implications. If one adheres to the conventional wisdom based on the loose concept that mean field theory obtained from LO approximation describes classical physics then one might be tempted to think of the NLO expansion as containing some quantum features. Does $N \geq N_c$ signals a transition from quantum to classical?

5.1.1. Is there a critical N_c which marks the appearance of MQP? If $N = \infty$ is used as a criterion to view an N component quantum system as reaching its classical limit, it seems logical to ask, "What is the smallest number N_c for such classical behaviour in this quantum system to disappear? The related question for our concern is, "Does the number N_c obtained by these authors carry such a meaning, as a demarkation between classical and quantum behaviors?" If so this is a very neat result because it provides a quantitative measure in how 'large' N needs to be to mark the appearance of MQP. We shall explain in what follows that unfortunately this is not the case. Life would have been much easier if it were. The key observation is that it is only for Gaussian theories that the mean field theory from LO LN approximation is equivalent to a classical theory. The critical N_c is only a measure of the validity of the NLO approximation, not a measure of the importance of quantum contributions. Quantum features are pervasive at all levels of approximation. We will explore this issue further in the context of critical phenomena in the sequel paper.

Convexity and positivity. Let us look at some technical features of the effective potential method which these authors used. The generating functional of connected Greens functions ($W \sim \ln Z$) is convex. This follows from the positivity of the two point function which is obtained by functionally differentiating W twice. Legendre transformation takes a convex function to a concave function. Paying attention to the signs in the definitions it follows that the effective potential is also convex. As a result the convexity of the effective potential is a consequence (or a guarantee, depending on which way one looks at it) of the positivity. This is the reason why the critical values N_T and N_c are about the same. Recall that N_T is the critical value of N below which an instability occurs (see figure 7 in MACDH for a plot of $< r^2 >$ as a function of t) at finite time and N_c is the critical value of N below which the effective potential ceases to exist everywhere.

Perturbative expansions of the effective potential using the background field method run into difficulties for classical potentials (tree level) that are non-convex, which is the case described in MACDH. The loop expansion develops an imaginary part because of the unstable modes, which prompt tunneling or nucleation. If one were to sum up all the loop corrections the result would be a convex effective potential (with minima at the origin). However at finite order there usually are domains where the effective potential becomes imaginary.

The NLO LN expansion is not immune to this shortcoming. It sums up graphs from all orders in loop expansion but still the dynamics is altered. The same problem as in loop expansion occurs here as well. For $N > N_c$ convexity is restored. The effective potential is convex for both classical and quantum O(N) models if we don't make any approximation. Hence the fact that convexity is restored for $N > N_c$ is a statement about the validity of the $1/N$ approximation, not about the system becoming classical or if enough quantum details have been retained.

5.1.2. Large N approximation does not capture full quantum dynamics As MACDH mention in their paper although NLO is a significant improvement over LO for sufficiently large N, at late times NLO fails to capture the nonlinear effects as well as LO does. Hence having $N > N_c$ (or $N > N_T$, since they are both of the same order) doesn't mean NLO approximation describes the dynamics accurately, it simply means the divergences and nonpositivity is avoided. At late enough times the approximation will fail (see figure 11 in MACDH). In general to make a claim about the validity of an approximation time need be specified, since any approximation will fail after some time. To inquire about the turnover from quantum dynamics to classical dynamics we are referring to the full dynamics, not just for short time transient or long time asymptote behaviours.

Thus, we see that $N > N_c$ is a criterion for the validity of the NLO approximation, as these authors meant it to be, not so much about the quantum versus classical behaviour of the N component system. As a side note: it is the NLO approximation that has this pathology not the LO. LO approximation most likely gets inaccurate and unreliable before the NLO approximation diverges. So LO approximation behaving well doesn't say anything about its accuracy. Same applies to NLO approximation for $N > N_c$.

5.2. Semiclassical and stochastic gravity as LO and NLO large N theories
We want to use these two theories as an instructional example to two important themes in our program, 1) the autocratic open quantum system (Langevin) paradigm, where gravity is treated preferentially as the system of special interest over the matter field regarded as its environment where only its coarse-grained (or averaged or integrated over) effects are accounted for. This is how the Einstein-Langevin equation (ELE) [65] was first derived. 2) the democratic large N (Boltzmann) system. We shall use an N component matter field as source to illustrate how the semiclassical Einstein (SCE) equation and the ELE can be derived as the leading order O (1) and next to leading order O(1/N) theories in the large N expansion respectively.

Semiclassical gravity refers to the theory where gravity (metric function g_{ab}) is treated classically but matter as a quantum field $\hat{\phi}_j$, the vacuum expectation value (vev) of whose (renormalized) stress energy tensor serves as the source to the semiclassical Einstein equation

$$G_{ab}[g] = \kappa \left\langle \hat{T}_{ab}[g] \right\rangle_{\text{ren}} . \tag{5.1}$$

where the subscript *ren* denotes renormalized.

Stochastic gravity [16] includes in the source of the SCE equation an additional term ξ_{ab} accounting for the fluctuations of the stress tensor operator of the quantum matter field through the Einstein-Langevin equation:

$$G_{ab}[g] = \kappa(\left\langle \hat{T}_{ab}[g] \right\rangle + \xi_{ab}[g])_{\text{ren}}. \tag{5.2}$$

The stochastic source $\xi_{ab}[g]$ at the level of Gaussian approximation is completely characterized by its correlation function in terms of the noise kernel $\mathcal{N}_{abcd}(x, y)$. The noise kernel is the symmetrized connected part of the two-point quantum correlation function of the stress tensor operator with respect to the state of the matter fields. It describes their stress-energy fluctuations via

$$\langle \xi_{ab}[g;x)\xi_{cd}[g;y) \rangle_\xi = \mathcal{N}_{abcd}(x,y) \equiv \frac{1}{2} \left\langle \left\{ \hat{t}_{ab}[g;x), \hat{t}_{cd}[g;y) \right\} \right\rangle , \tag{5.3}$$

where $\hat{t}_{ab} \equiv \hat{T}_{ab} - \langle \hat{T}_{ab} \rangle$ and $\langle \ldots \rangle$ is the usual expectation value with respect to the quantum state of the matter fields, whereas $\langle \ldots \rangle_\xi$ denotes taking the average with respect to all possible realizations of the stochastic source ξ_{ab}.

5.2.1. Open system approach: gravity interacting with quantum matter fields In the open quantum system approach one begins with a closed system, described by a density matrix $\rho(x, q; x', q', t)$, defines a system S of special interest described, say, by the x variables or fields and look for the overall coarse-grained effects of its environments E, described by the q fields, on this system. The influence functional (IF) formalism [45] offers a very elegant way of capturing these effects on a microphysics level, with full self-consistency. If we are only interested in how the state of the system is influenced by the overall effects, but not the precise states, of the environment(s), then the reduced density matrix $\rho_r(x, x', t) = \int \mathrm{d}q\ \rho(x, q; x', q, t)$ could provide the relevant information. (The subscript r stands for reduced.) It is propagated in time from t_i by the propagator $\mathcal{J}_r$:

$$\rho_r(x,x',t) = \int_{-\infty}^{+\infty} \mathrm{d}x_i \int_{-\infty}^{+\infty} \mathrm{d}x'_i \; \mathcal{J}_r(x,x',t \mid x_i,x'_i,t_i)\; \rho_r(x_i,x'_i,t_i\,) \tag{5.4}$$

Assuming that the action of the coupled system decomposes as $S = S_s[x] + S_e[q] + S_{int}[x,q]$, and that the initial density matrix factorizes (i.e., takes the tensor product form), $\rho(x,q;x',q',t_i) = \rho_s(x,x',t_i)\rho_e(q,q',t_i)$, the propagator for the reduced density matrix is given by

$$\mathcal{J}_r(x,x',t|\; x_i,x'_i,t_i) = \int_{x_i}^{x_f} Dx \int_{x'_i}^{x'_f} Dx' \; \mathrm{e}^{\mathrm{i}S_{eff}[x,x',t]}$$

where

$$S_{eff}[x,x',t] \equiv S_s[x] - S_s[x'] + S_{IF}[x,x',t] \tag{5.5}$$

is the full effective action and S_{IF} is the influence action. The influence functional $\mathcal{F}$ is defined as [5]

$$\begin{aligned}\mathcal{F}[x,x',t] &\equiv \mathrm{e}^{\mathrm{i}S_{IF}[x,x',t]} \\ &\equiv \int \mathrm{d}q_f\, \mathrm{d}q_i\, \mathrm{d}q'_i \int_{q_i}^{q_f} Dq \int_{q'_i}^{q_f} Dq' \; \mathrm{e}^{\mathrm{i}(S_e[q]+S_{int}[x,q]-S_e[q']-S_{int}[x',q'])}\rho_e(q_i,q'_i,t_i).\end{aligned} \tag{5.6}$$

S_{IF} is typically complex; its real part $\mathcal{R}$, containing the dissipation kernel **D**, contributes to the renormalization of S_s, and yields the dissipative terms in the effective equations of motion. The imaginary part $\mathcal{I}$, containing the noise kernel **N**, provides the information about the fluctuations induced on the system through its coupling to the environment.

For a system with a quadratic action such as a simple harmonic oscillator, bilinearly coupled to a Gaussian environment (made up of e.g., non-interacting harmonic oscillators) as in the quantum Brownian motion (QBM) problem treated profusely in the literature (see, e.g., [46, 47]) the influence action takes the generic form :

$$S_{\mathrm{IF}}\,[x,x'] = Z\cdot\Delta + \Delta\cdot\mathbf{H}\cdot\Sigma + \frac{\mathrm{i}}{2}\Delta\cdot\mathbf{N}\cdot\Delta, \tag{5.7}$$

where $\Delta = x'-x, \Sigma = \frac{1}{2}(x+x')$ and $\cdot \equiv \int_i^f \mathrm{d}t$. Z is a current term marking the contribution of the initial conditions, and the kernels $\mathbf{H},\mathbf{N}$ are related to the dissipation and noise kernels. In exact analogy with the above we can derive the dynamics of the gravitational field as an open system under the influence of N quantum matter fields viewed as its environment by calculating the influence action S_{IF} defined as: (details are in [67], Appendix C)

$$\mathrm{e}^{\mathrm{i}S_{\mathrm{IF}}[h,h']} = \prod_{j=1}^{N} \int \mathcal{D}\phi_j\mathcal{D}\phi'_j \mathrm{e}^{\mathrm{i}S_{\mathrm{m}}[\phi_j,h]-\mathrm{i}S_{\mathrm{m}}[\phi'_j,h']}\rho[\phi_j^{(i)},\phi_j'^{(i)}]. \tag{5.8}$$

Up to quadratic order in the metric perturbations h_{ab} it is given by [68]

$$S_{\mathrm{IF}}\,[\Sigma_{ab},\Delta_{ab}] = N\left(Z\cdot\Delta + \Delta\cdot(\mathbf{H}+\mathbf{M})\cdot\Sigma + \frac{\mathrm{i}}{8}\Delta\cdot\mathcal{N}\cdot\Delta\right), \tag{5.9}$$

[5] For more details and further discussions on how the IF is applied in statistical field theory and applications, see, e.g., [66].

where we have introduced the semisum and difference variables $\Sigma_{ab} = (h_{ab} + h'_{ab})/2$ and $\Delta_{ab} = h'_{ab} - h_{ab}$, $Z^{ab}(x) = -\frac{1}{2}\langle \hat{T}^{ab}[\hat{\phi}, g; x]\rangle$ and the kernels $\mathbf{H}$, are $\mathbf{M}$ are defined as follows:

$$H^{abcd}(x,y) = -\frac{1}{4}\mathrm{Im}\left\langle T^{*}\hat{T}^{ab}\left[\hat{\phi}, g; x\right)\hat{T}^{cd}\left[\hat{\phi}, g; y\right)\right\rangle + \frac{\mathrm{i}}{8}\left\langle\left[\hat{T}^{ab}\left[\hat{\phi}, g; x\right), \hat{T}^{cd}\left[\hat{\phi}, g; y\right)\right]\right\rangle \tag{5.10}$$

$$M^{abcd}(x,y) = -\frac{1}{2}\left(\frac{1}{\sqrt{-g(x)}}\frac{\delta\left(\left\langle \hat{T}^{ab}[\hat{\phi}, g_{ab}; x)\right\rangle\right)}{\delta g_{cd}(y)}\right), \tag{5.11}$$

where the notation T^{*} was employed to indicate that the spacetime partial derivatives appearing in the time-ordered operators also act on the theta function implementing the time ordering. The functional derivative appearing on the right-hand side of (5.11) should be understood to account only for the explicit dependence on the metric: the implicit dependence through the field operator $\hat{\phi}[g]$ is excluded.

As is known the kernels $\mathbf{H}$ and $\mathbf{M}$ exhibit divergences that are cancelled by renormalizing the gravitational coupling constant and the cosmological constant in the bare gravitational action as well as the coupling constants of the counterterms quadratic in the curvature. (We will not need terms of higher order in the metric perturbations because they give contributions to the connected part of the CTP generating functional of higher order in $1/N$.)

5.2.2. Closed system: semiclassical and stochastic gravity from large N expansion We now give a closed system treatment for the effects of an N component quantum matter field $\hat{\phi}_j$ in a background spacetime with metric g_{ab}. The semiclassical Einstein and the Einstein-Langevin equations can be identified as originating from the leading order and the next-to-leading order large N expansions. Treating perturbative quantum gravity (in terms of quantized linear metric perturbations $\hat{h}_{ab}$) interacting with N conformally invariant scalar fields $\hat{\phi}_j$, Hartle and Horowitz [19] computed the effective action to leading order in large N and showed that the semiclassical Einstein equation can be interpreted as the equation governing the evolution of the expectation value of the metric to leading order [which for this theory is actually $O(1)$]. The sources of the SCE equation are given by the expectation value of the stress tensor operator of the quantum matter fields. In the same vein Roura and Verdaguer [20] demonstrated that the next to leading order [which in this theory is $O(1/N)$] [6] contribution to the *quantum* correlation functions in a large N expansion is equivalent to the *stochastic* correlation functions obtained in the context of stochastic semiclassical gravity. Let us see how this comes about.

Consider metric perturbations around a globally hyperbolic background spacetime with metric g_{ab} and scalar curvature R interacting with N minimally coupled free scalar fields $\phi_j (j = 1, .., N)$. The action for the combined system is the sum of the gravitational action $S_{\rm g}$ plus the action for the matter fields $S_{\rm m}$. The gravitational action is given by the usual Einstein-Hilbert term, the corresponding boundary term (which should be included to have a well-defined variational problem) and the usual counterterms required to renormalize the divergences arising

[6] Note the ordering scheme used in [20, 67] is different from the customary usage. Their leading order in the large N limit refers to the lowest order in $1/N$ with a nonvanishing contribution. Hence, as we will see, the leading order for the source of the semiclassical Einstein equation, which is proportional to the expectation value of the stress tensor operator, is $1/N^0$, or $O(1)$ – this is called leading order large N in [19], whereas the leading order for the quantum two-point correlation functions is $1/N$. In the counting scheme of [19] this would be the next-to-leading order which yields stochastic gravity. In the way how it is presented here this latter expansion is to be understood in a NLO sense.

when functionally integrating the matter fields:

$$S_g = \frac{N}{2\bar{\kappa}} \int_{\mathcal{M}} d^4x \sqrt{-\tilde{g}} R(\tilde{g}) + \frac{N}{\bar{\kappa}} \int_{\mathcal{S}=\partial\mathcal{M}} d^3x \sqrt{\tilde{g}_{\mathcal{S}}} K^a_a(\tilde{g}) + \text{(counterterms)}, \tag{5.12}$$

where $\tilde{g}_{ab} = g_{ab} + h_{ab}$ is the perturbed metric, g_{ab} is the background metric and the gravitational coupling constant $\kappa = 8\pi/m_p^2$ was rescaled to $\bar{\kappa}/N$ so that the product of the rescaled gravitational constant times the number of fields remains constant in the limit $N \to \infty$. The action for the matter fields is

$$S_m = -\sum_{j=1}^{N} \int_{\mathcal{M}} d^4x \sqrt{-\tilde{g}} \frac{1}{2} \left(\tilde{g}^{ab} \nabla_a \phi_j \nabla_b \phi_j + m^2 \phi_j^2 \right), \tag{5.13}$$

where m is the mass of the scalar field. We will use the expansion in $1/N$ to systemize the contributions. At the end one can always substitute back the rescaled gravitational constant in terms of the physical one.

To make connection of the large N expansion with the theory of stochastic gravity one can use the procedure of Calzetta, Roura and Verdaguer [69] for the QBM problem to derive the symmetrized quantum correlation function for the metric perturbations (see Appendix C of [69]):

$$\begin{aligned} \frac{1}{2} \left\langle \left\{ \hat{h}_{ab}(x), \hat{h}_{cd}(x') \right\} \right\rangle &= \left\langle \Sigma^{(0)}_{ab}(x) \Sigma^{(0)}_{cd}(x') \right\rangle_{\Sigma^{(i)}_{ab}, \Pi^{cd}_{(i)}} \\ &+ \frac{\bar{\kappa}^2}{N} \left(G_{\text{ret}} \cdot \mathcal{N} \cdot (G_{\text{ret}})^T \right)_{abcd} (x, x'). \end{aligned} \tag{5.14}$$

There are two separate contributions to the two-point correlation function: the first one, called *intrinsic* fluctuations, is related to the dispersion of the initial state for the metric perturbations, whereas the second one, called *induced* fluctuations is proportional to the noise kernel and accounts for the fluctuations induced by their interaction with the quantum matter fields as its environment. Under the aforementioned conditions, the symmetrized quantum correlation function for the metric perturbations is equivalent to the stochastic correlation function obtained in stochastic semiclassical gravity by solving the Einstein-Langevin equation.

The above exposé serves to illustrate the following points:

1) In a large N expansion semiclassical gravity (SCG) corresponds to a mean field theory obtainable from the leading order $O(1)$ expansion, and stochastic gravity (STG) theory corresponds to the next-to-leading $O(1/N)$ order expansion. In this regard SCG and STG share the same properties as the $O(N)$ theory. Note however that stochastic gravity used in this more restrictive sense assumes Gaussian noise in the quantum matter field environment and weak perturbations off a background metric which is a solution of the SCE equations. In the more general sense such as used in [70] there is no such restriction and the parallel with the $O(N)$ model ends at the Gaussian level.

2) Stochastic gravity includes the fluctuations of the stress tensor in the matter sector via the noise kernel. This in turn enters into the determination of the two-point quantum correlation function for the metric perturbations. Contributions of higher order in $1/N$ enter either by including the vertices for the metric perturbations or terms from the influence functional evaluated beyond the Gaussian approximation. (e.g., in figure 1 of [67] the two diagrams shown give contributions of order $1/N^2$ to the two-point quantum correlation function for the metric perturbations.)

3) From these theories one can also see the relation between stochastic correlations and quantum fluctuations (via the loop expansion order), as well as the relation between correlation order in the Schwinger-Dyson hierarchy and the large N order. We shall dwell on the correlation aspects of MQP with the use of 2PI effective action in our sequel paper.

Note that it is by virtue of its Gaussian property that SCG as a mean field theory can be viewed as semiclassical. For STG if one does not make the Gaussian assumption for the noise, the Feynman-Vernon identity which enables one to interpret the quantum fluctuations of the matter field (as environment) as a classical stochastic source no longer holds. In what follows we give an example of field theories which are non-Gaussian. We will see that even at the rudimentary level the identification of mean field theory as classical is no longer valid.

5.3. Large N gauge field and string theory

5.3.1. 1/N expansion in gauge theories 't Hooft [24] introduced the $1/N$ expansion for nonabeliean gauge theories in order to express physical quantities as a systematic expression in powers of $1/N$. Consider the gauge group $SU(N)$. The gauge vector fields are Hermitian matrices $A^i_{\mu j}$, where both indices i, j run from 1 to N. The field strength is

$$F_{\mu\nu}{}^{i}{}_{j} = \partial_\mu A^i_{\nu j} - \partial_\nu A^i_{\mu j} + ig[A_\mu, A_\nu]^i_j, \tag{5.15}$$

where g is the gauge coupling constant. The Lagrangian density for a spinor particle of mass m_ψ with wave function ψ interacting with this gauge field is given by

$$\begin{aligned}\mathcal{L} &= -\frac{1}{4}\mathrm{Tr}(F_{\mu\nu}F^{\mu\nu}) - \bar{\psi}(i\gamma D + m_\psi)\psi \\ &= -\mathrm{Tr}\Big(\frac{1}{2}(\partial_\mu A_\nu)^2 - \frac{1}{2}\partial_\mu A_\nu \partial_\nu A_\mu + ig\partial_\mu A_\nu[A_\mu, A_\nu] - \frac{1}{4}g^2[A_\mu, A_\nu]^2\Big) \\ &\quad -\bar{\psi}(i\gamma^\mu\partial_\mu + ig\gamma^\mu A_\mu + m_\psi)\psi.\end{aligned} \tag{5.16}$$

It is convenient to introduce a double line notation to keep track of the group indices i and j. Whenever there is a close index loop in the Feynman diagram, we get a factor of N from counting the values of the index can assume. This is how the dimension N of the gauge group enters. Since all three-point vertices come with a factor of g, and all four-point vertices come with a factor of g^2, the g and N dependence of the amplitude for a general Feynman diagram is seen to be $(g^2N)^{F+2\chi-2}N^{2-2\chi}$, where F is the number of loops and χ is the genus, or number of "holes" of the polyhedron formed by the Feynman diagram. In the 't Hooft large N limit while keeping $g^2N \equiv \lambda$ fixed, we see that the N dependence of the diagram is $N^{2-2\chi}$. So the diagram without holes ($\chi = 0$) dominate. In other words, in this limit, only the planar diagrams contribute. The interactions between hadrons is expected to be a $O(1/N)$ effect. Hence in the large N limit one can consider the problem of confinement and hadron mass spectrum without the complication of residual hadron interactions. The hope was that one could solve the theory with $N = \infty$ exactly followed by an expansion in $1/N = 1/3$ for QCD.

Witten [25] referred to this limit as analogous to the classical limit for quantum mechanics, in the sense that for large N there exists one single "classical" gauge field, called "master field", which saturates the path integral and gives the dominant contribution. Such a classical field configuration was called a Witten's argument was based on the assumption that all possible invariant operators satisfy the factorization property [which holds in any $U(N)$ symmetric theory of $N \times N$ matrix fields $A_{ij}(x)$]

$$\langle f_1(A)...f_k(A)\rangle = \prod_{i=1}^{k}\langle f_i(A)\rangle[1 + O(1/N)], \tag{5.17}$$

where the $f_i(A)$ are $U(N)$ symmetric operators of the form $f(A) = \mathrm{Tr}[A(x_1)...A(x_l)]$.

However, Haan [32] pointed out that there is a gap in Witten's argument for the existence of a master field in the large N limit of QCD. He showed that the master field does not exist for the two-matrix model and the large N limit is not a classical limit. Instead, the factorization of expectation values of invariant operators is shown to be a property analogous to the vanishing of fluctuations for macroscopic observables in the thermodynamic limit.

5.3.2. Large N field theories and string theory At large N, QCD is expected to behave like a string theory. This is supported by the following observations. First, Feynman diagrams are organized in a topology expansion, just like string worldsheets. Second, the existence of Regge trajectories indicates a string-like behaviour. Third, QCD contains string-like objects, which are the electric flux tubes between quarks and antiquarks. Maldacena's AdS/CFT correspondence [26] conjectures the equivalence of $N = 4$ $SU(N)$ super Yang-Mills theory with gauge coupling g_{YM} in 4 dimensions and Type IIB superstring theory with string coupling g_s in $AdS_5 \times S^5$ space where both AdS_5 and S^5 have the same radius L. The parameters between these two theories are related by

$$g_s = g_{YM}^2, \quad L^4 = 4\pi g_s N(\alpha')^2, \tag{5.18}$$

where α' is related to the string tension. This conjecture is assumed to be valid for all values of N and of $g_s = g_{YM}^2$. Assuming the validity of Maldacena's conjecture, one can study different limits of field theory and string theory.

- First take the 't Hooft limit. This is the case where $\lambda \equiv g_{YM}^2 N = g_s N$ fixed and letting $N \to \infty$. In the Yang-Mills theory, this limit is well-defined, at least perturbatively, and corresponds to a topological expansion of the field theory's Feynman diagrams. On the AdS side, the string coupling is $g_s = \lambda/N$ and the 't Hooft limit corresponds to weak coupling string perturbation theory. The $1/N$ expansion in gauge theory corresponds to the g_s string loop expansion in the string theory side. Hence the Maldacena conjecture in this case becomes a correspondence between large N limit of gauge theories and the classical string theory. (Here classical means there is no string loop expansion in the $N \to \infty$ limit).
- The large λ limit. In this limit, one can expand the effective action in the AdS side in terms of α'. The distance scale is set by the AdS radius L which is related to the scale of Riemann tensor by $R \sim 1/L^2 = (g_s N)^{-\frac{1}{2}}/\alpha' = \lambda^{-\frac{1}{2}}/\alpha'$. Hence in the large λ limit, the classical Type IIB string theory on $AdS_5 \times S^5$ becomes classical Type IIB supergravity on $AdS_5 \times S^5$ and the α' expansion in string theory corresponds to $\lambda^{-\frac{1}{2}}$ expansion in field theory. (Here gravity is weak because the curvature scale approaches zero in this limit and one can neglect the stringy correction and approximate it by the classical supergravity action).

To summarize, let us see what all this means for the issues we wish to address in this paper: classicality and large N from the perspective of string/gauge/gravity theories.

Old school field theorists often refer to the large N limit of a Yang-Mills gauge theory as classical. Or stretching it a little, one may habitually view all large N theories as classical. The advent of string theory alters fundamentally the way we look at nature, not just pertaining to Planck scale microphysics but also in relation to gauge fields and gravity, two essential elements of nature as we see it today. To begin with, N enters in a new coupling constant called the 't Hooft coupling, in addition to the string tension, the magnitude of both and their interplay alter a theory's qualitative behaviour in a fascinating way. Another important new concept introduced by string theory is duality: between string theory and gauge theory, and between gauge theory and gravity.

The correspondence between gauge theory and string theory provides another window to examine the large N behavior of gauge theory. Under this duality relation, a Yang-Mills gauge theory ingrains also the effects of gravitational interaction: Under different regimes of 't Hooft coupling, the behavior of large N limit seen from the AdS side of the correspondence can appear either as the suppression of string loop expansion and hence a classical string theory or the suppression of the stringy effect giving rise to a weakly interacting gravity theory where the quantum effect is negligible.

We can only give an inkling of the immense body of work accumulated in the last half a century for large N, especially since 1998 when the AdS/CFT correspondence was proposed. The simple message we wish to convey here is: "exercise caution when tackling the meaning or finding the criteria of classicality": the explanation of the emergence of classical behavior or even the apparently straightforward identification of a classical regime from an interacting quantum theory is a subtle issue, as it may contain multiple structures and carry different dynamics in different interaction regimes.

6. Conclusion

Macroscopic quantum phenomena is a relatively new research venue, with exciting experiments and stimulating proposals. Surprisingly there are very little systematic theoretical investigations or even serious attempts we are aware of to construct a viable theoretical framework for this new endeavour. [7] In this paper we ask only one simple question: Is it really that surprising that quantum features can appear in macroscopic objects? Note the surprise comes from the conventional and hitherto unchallenged view that macroscopic means classical. The purpose of our series of papers is to remove this degeneracy. By examining many well-known examples where theoretical analysis has been performed in detail for different systems, we find that there is no a priori good reason why quantum phenomena in macroscopic objects cannot exist. The challenge for theorists is to identify for various systems the physical conditions MQP manifest and in what ways.

In this paper we take the large N perspective, and try to understand what the leading order and next-to-leading order large N expansions bring us. In general we can only say that the LOLN gives mean field theory and the NLO gives deviations from the mean field, both are still very much quantum in nature. Only for Gaussian models such as the $O(N)$ theory (zero-dimensional field theory, referring to the quantum mechanics of N oscillators) or in theories where the Gaussian approximation is imposed, such as semiclassical gravity and stochastic gravity, do we find that the mean field at LOLN is equivalent to classical, and the NLO incorporates deviations from the mean, which, only when one identifies the mean with classical would it be conjured as containing "quantum fluctuations". By combing through these examples we are trying to change the common attitude on a very subtle point, namely, "Don't begin with classical and try to go backwards to quantum. One should always begin with quantum and only under very special conditions and stringent criteria would a quantum system begin to show certain classical features." We want to shift the burden of proof to those who believe they live in a classical world, to find these conditions and check every criteria. If they are not completely fulfilled, well then, perhaps one should just accept that the world is fundamentally quantum and it is only under specific yet commonly encountered conditions [8] that it appears classical.

This message is of course not that foreign to researchers working in the last two decades on the quantum to classical transition or the Q-C correspondence issues. The half a dozen or so criteria we reviewed in this paper stem from these investigations. What we have done somewhat different here is to examine every criteria anew with a macroscopic system, here with the aid of large N expansions. There is an abundance of information in the particle physics literature on this subject. We brought forth some here to serve as illustrations for the one question we

[7] Maybe this is a sociological phenomenon, that exciting experiments are carried out by atomic-optical physicists while those who care or know more about the theories are mainly in the particle and gravitational physics community. If so this series of papers can be viewed as a modest attempt to bridge that divide.

[8] Specific and common are not contradictory conditions: equilibrium state is an example. It is one specific state out of a large number of states accessible to a large system, but it is also the most commonly encountered – with high probability – when the system is placed in a heat bath. Behind this is the working of central limit theorem, quantum regression theorem in counting the statistics, or the conservation laws governing the hydrodynmic variable, as far as decoherence is concerned. These theorems and laws give a deeper meaning of 'habitual'.

raised, but there is plenty more material which can be used fruitfully for performing quantitative analysis of MQP.

There are other important issues which we have left out. For example, here we use the number of components N as a measure of macroscopic, but it could be physically more meaningful to incorporate interaction strengths and dynamics, in the renormalization group sense. Thus it could be the ultraviolet versus the infrared behaviour of a (quantum) system which conveys a stronger microscopic versus macroscopic sense, as manifests in critical phenomena. It is near the critical point that the system's IR behaviour dominates, where the correlation function extends to infinity. This aspect of correlation, both classical and quantum, is the focus of our sequel paper. For interacting quantum systems we will invoke the nPI effective action and the Schwinger-Dyson equations (corresponding classically to the BBGKY correlation hierarchy) to analyze what quantum versus classical means for large/small and/or strongly/weakly correlated systems. In that context we will include critical phenomena and other statistical mechanics considerations to examine this new issue of MQP at the foundation of quantum mechanics.

Acknowledgments

We thank Charis Anastopoulos, Fred Cooper, Jonathan Halliwell, Albert Roura and Rafael Sorkin for their opinions and comments. BLH wishes to thank the organizers of DICE2010, especially Professor Thomas Elze, for their kind invitation. He also enjoyed the warm hospitality of the National Center for Theoretical Sciences (South) and the Department of Physics of National Cheng-Kung University, Tainan, Taiwan while part of this work was done. This work is supported by the National Science Council of Taiwan under grant NSC 97-2112-M-006-004-MY3 and by NSF grants PHY-0601550 and PHY-0801368 to the University of Maryland.

References

[1] Feng D H and Hu B L (eds) 1997 *Quantum classical correspondence* (Boston: International Press)
[2] Hu B L and Zhang Yuhong 1993 *Mod. Phys. Lett.* A **8** 3575; *do.* 1995 *Int. J. Mod. Phys.* A **10** 4537
[3] Paz J P and Zurek W 1999 Environment-induced decoherence and the transition from quantum to classical *Lectures at the 72nd Les Houches Summer School on "Coherent Matter Waves" Preprint* quant-ph/0010011
Zurek W H 2003 *Rev. Mod. Phys.* **75** 715-75
[4] Armour A D, Blencowe M P and Schwab K C 2002 *Phys. Rev. Lett.* **88** 148301
Blencowe M P 2004 *Phys. Rep.* **395** 159-222
[5] Arndt M, Nairz O, Vos-Andreae J, Keller C, van der Zouw G and Zeilinger A 1999 *Nature* **401** 680
[6] Marshall W, Simon C, Penrose R and Bouwmeester D 2003 *Phys. Rev. Lett.* **91** 130401
[7] Mueller-Ebhardt H, Rehbein H, Schnabel R, Danzmann K and Chen Y 2008 *Phys. Rev. Lett.* **100** 013601
[8] Dorfman R 1999 *An Introduction to Chaos in Nonequilibrium Statistical Mechanics* (Cambridge: Cambridge Univ. Press)
[9] Gaspard P 1998 *Scattering and Statistical Mechanics* (Cambridge: Cambridge Univ. Press)
[10] Hu B L and Yu T (guest eds) 2009 *Quantum Information Processing* vol 8 Special Issue on Quantum Decoherence and Entanglement (Springer)
[11] DiVincenzo D P 2000 *Preprint* quant-ph/0002077
[12] Chou C H, Hu B L and Yu Ting 2008 *Physica* A **387** 432-44
[13] Cooper F, Habib S, Kluger Y, Mottola E, Paz J P and Anderson P 1994 *Phys. Rev.* D **50** 2848
[14] Cooper F, Habib S, Kluger Y and Mottola E 1997 *Phys. Rev.* D **55** 6471
[15] Ramsey S and Hu B L 1997 *Phys. Rev.* D **56** 661
[16] Hu B L and Verdaguer E 2008 *Living Rev. Rel.* **11** 3; *do.* 2003 *Class. Quant. Grav.* **20** R1
[17] Cornwall J M, Jackiw R and Tomboulis E 1974 *Phys. Rev.* D **10** 2428
[18] Calzetta E A and Hu B L 2003 *Phys. Rev.* D **68** 065027
[19] Hartle J B and Horowitz G 1981 *Phys. Rev.* D **24** 257
[20] Roura A and Verdaguer E, unpublished. See also talk given by A. Roura in Tainan, Taiwan, Jan 2007
[21] Mihaila B, Athan T, Cooper F, Dawson J and Habib S 2000 *Phys. Rev.* D **62** 125015
[22] Mihaila B, Dawson J and Cooper F 2001 *Phys. Rev.* D **63** 096003
[23] Calzetta E and Hu B L 2002 Thermalization of an Interacting Quantum Field in the CTP-2PI Next-to-leading-order Large N Scheme *Preprint* hep-ph/0205271

[24] 't Hooft G 1974 *Nucl. Phys.* B **72** 461
[25] Witten E 1979 *Nucl. Phys.* B **160** 57
[26] Maldacena J 1998 *Ad. Theor. Math. Phys.* **2** 231
[27] Guth A H and Pi S Y 1985 *Phys. Rev.* D **32** 1899
[28] Kluger Y, Mottola E and Eisenberg J M 1998 *Phys. Rev.* D **58** 125015
[29] Habib S, Kluger Y, Mottola E and Paz J 1996 *Phys. Rev. Lett.* **76** 4660
[30] Habib S 2004 Gaussian dynamics is classical dynamics *Preprint* quant-ph/0406011
[31] Coleman S 1985 *Aspects of Symmetry* (Cambridge: Cambridge Univ. Press)
[32] Haan O 1981 *Phys. Lett.* B **106** 207
[33] Yaffe L 1982 *Rev. Mod. Phys.* **54** 407
[34] Zurek W H, Habib S and Paz J P 1993 *Phys. Rev. Lett.* **70** 1187
[35] Anastopoulos C and Halliwell J J 1995 *Phys. Rev.* D **51** 6870
[36] Blaizot J-P and Ripka G 1986 *Quantum Theory of Finite Systems* (Cambridge MA: MIT Press) p 156
[37] Coleman S, Jackiw R and Politzer H D 1974 *Phys. Rev.* D **10** 2491
[38] Root R 1974 *Phys. Rev.* D **10** 3322
[39] Boyanovsky D, de Vega H J, Holman R, Lee D -S and Singh A 1995 *Phys. Rev.* D **51** 4419
[40] Cooper F, Pi S Y and Stancioff P 1986 *Phys. Rev.* D **34** 3831
[41] Anderson A 1990 *Phys. Rev.* D **42** 585
[42] Habib S and Laflamme R 1990 *Phys. Rev.* D **42** 4056
[43] Lombardo F C, Mazzitelli F D and Monteoliva D 2000 *Phys. Rev.* D **62** 045016
[44] Antunes N D, Lombardo F C and Monteoliva D 2001 *Phys. Rev.* E **64** 066118
[45] Feynman R and Vernon F 1963 *Ann. Phys.* (N.Y.) **24** 118
[46] Caldeira A O and Leggett A J 1983 *Physica* **121A** 587
[47] Hu B L, Paz J P and Zhang Y 1992 *Phys. Rev. D* **45** 2843; *do.* 1993 *Phys. Rev. D* **47** 1576
[48] Baacke J and Michalski S 2002 *Phys. Rev.* D **65** 065019
[49] Parker L 1969 *Phys. Rev.* **183** 1057
[50] Hu B L and Kandrup H E 1987 *Phys. Rev.* D **35** 1776
[51] Elze H-T 1996 *Phys. Lett.* B **369** 295
[52] Calzetta E and Hu B L 2008 *Nonequilibrium Quantum Field Theory* (Cambridge: Cambridge Univ. Press)
[53] Birrell N D and Davies P C W 1982 *Quantum Fields in Curved Spaces* (Cambridge: Cambridge Univ. Press)
[54] Rau J 1994 *Phys. Rev.* D **50** 6911
[55] Rau J and Mueller B 1996 *Phys. Rep.* **272** 1
[56] Smolyansky S A, Roepke G, Schmidt S, Blaschke D, Toneev V D and Prozorkevich A V 1997 *Preprint* hep-ph/9712377
[57] Hu B L and Pavon D 1986 *Phys. Lett.* B **180** 329
[58] Kandrup H E 1988 *Phys. Rev.* D **37** 3505
Kandrup H E 1988 *Phys. Rev.* D **38** 1773
[59] Gell-Mann M and Hartle J B 1990 Quantum mechanics in the light of quantum cosmology *Complexity, Entropy and the Physics of Information* ed W H Zurek (Reading, MA: Addison-Wesley)
[60] Gell-Mann M and Hartle J B 1993 *Phys. Rev.* D **47** 3345
[61] Hartle J B, Laflamme R and Marolf D 1995 *Phys. Rev.* D **51** 7007
[62] Halliwell J J 1998 *Phys. Rev.* D **58** 105015
[63] Brun T and Halliwell J J 1996 *Phys. Rev.* D **54** 2899
[64] Dowker F and Kent A 1996 *J. Stat. Phys.* **82** 1575
[65] Calzetta E and Hu B L 1994 *Phys. Rev.* D **49** 6636
Hu B L and Matacz A 1995 *Phys. Rev.* D **51** 1577
Hu B L and Sinha S 1995 *Phys. Rev.* D **51** 1587
Campos A and Verdaguer E 1996 *Phys. Rev.* D **53** 1927
Lombardo F C and Mazzitelli F D 1997 *Phys. Rev.* D **55** 3889
[66] Hu B L 1994 Quantum statistical fields in gravitation and cosmology *Proc. Third International Workshop on Thermal Field Theory and Applications* ed R Kobes and G Kunstatter (Singapore: World Scientific) *Preprint* gr-qc/9403061
[67] Hu B L, Roura A and Verdaguer E 2004 *Phys. Rev.* D **70** 044002
[68] Martin R and Verdaguer E 1999 *Phys. Rev.* D **60** 084008
Martin R and Verdaguer E 1999 *Int. J. Theor. Phys.* **38** 3049
[69] Calzetta E, Roura A and Verdaguer E 2003 *Physica* A **319** 188
[70] Hu B L 2002 *Int. J. Theor. Phys.* **41** pp 2111-38

Gravitational and centrifugal quantum states of neutrons

V V Nesvizhevsky (for the GRANIT Collaboration)
Institut Laue-Langevin, 6 rue Jules Horowitz, Grenoble, France, F-38046
E-mail: nesvizhevsky@ill.eu

Abstract. Two related physical phenomena have recently been observed: quantum states of neutrons in the gravitational field above a flat mirror, and quantum states of neutrons in an effective centrifugal potential in the vicinity of a concave mirror. It is curious that the measurement of the near-surface quantum states of these two types is the first direct demonstration of the weak equivalence principle for a massive object in a quantum state. These two phenomena and their application to various domains of science will be explored using a new GRANIT spectrometer, which is under commissioning in the Institut Laue-Langevin, Grenoble, France. Most recent advances in these domains are mentioned.

1. Introduction

The present contribution is based on the talk given at the DICE-2010 conference carried out in Castiglioncello, Italy. It is devoted to two related physical phenomena: quantum states of neutrons in the gravitational field above a flat mirror (section 2), and quantum states of neutrons in an effective centrifugal potential in the vicinity of a concave mirror (section 3). These two phenomena and their application to various domains of science will be explored using a new GRANIT spectrometer (section 4), which is under commissioning at the moment. As many related topics have been extensively discussed in literature, we will present only most recent advances; otherwise we will just give proper references. All relevant aspects were analyzed in detail during our recent dedicated Workshop GRANIT-2010, carried out in Les Houches, France, 14-19 February 2010 [1].

2. Gravitational quantum states of neutrons

The phenomenon of gravitational quantum states of neutrons above a mirror, methods of their observation, related experimental results, and various applications are overviewed, for instance, in [2, 3]. A general formulation of the problem and further details are presented, for instance, in [4]-[15]. We will just explain here the nature of this phenomenon and note the state-of-the-art in the field: main results and limitations for further progress.

Imagine the following experiment, known to everyone from every-day life: If we raise a ping-pong ball to a height H_0 above a table and let it gently drop, the ball accelerates in the Earth's gravity field to the velocity $V_0 = \sqrt{2gH_0}$, where $g = 9.81\text{m/s}^2$ is the gravitational acceleration, and is then reflected from the table surface. In the case of a perfectly elastic reflection in a vacuum, the ball would return due to gravity to the surface after the time interval $\Delta\tau_0(H_0) = \sqrt{8H_0/g}$ and would then continue bouncing with the frequency $\nu_0(H_0) =$

$1/\Delta\tau_0(H_0) = \sqrt{g/(8H_0)}$. The smaller H_0 is, the larger ν_0: $\nu_0(1\text{m}) \approx 1\text{Hz}$, $\nu_0(1\text{cm}) \approx 10\text{Hz}$, $\nu_0(100\mu\text{m}) \approx 10^2\text{Hz}$, as follows from the equality of the maximum potential and kinetic energies of the ball: $Mv_0^2/2 = MgH_0$. The frequency ν_0 is independent of the ball mass M. What would happen in this experiment at very small height H_0? Would the frequency ν_0 tend to infinity? No, it would increase as long as H_0 exceeds the quantum mechanical limit H_0^{QM}, which can be estimated using the Heisenberg position-momentum uncertainty relation $H_0^{QM} M\sqrt{2gH_0^{QM}} \approx 2\pi\hbar$, where $\hbar$ is the Planck constant. For a ball in the Earths gravitational field, the height $H_0^{QM} \approx 10^{-21}\text{m}$ is too small, and quantum phenomena cannot be observed, but for an elementary particle above a mirror, for instance, for an ultracold neutron (UCN) [16, 17, 18] with the mass $m \approx 1\text{GeV}$, quantum effects can be observed at relatively large heights $H_0^{QM} \approx 10\mu\text{m}$.

In preceding experiments [13, 14, 15] wavefunction of neutrons in the lowest quantum states have been studied using so-called integral and differential methods. The integral method consists in measuring the neutron flux through a slit between the mirror on the bottom and the flat scatterer on top as a function of the slit height. The differential method consists in analyzing the spatial neutron density distribution behind the horizontal bottom mirror exit using special position-sensitive UCN detectors of ultra-high spatial resolution [15, 33]. In both cases the accuracy of measuring parameters of quantum states does not exceed and probably will never exceed a few per cent. However, much more precise experiments are motivated by eventual applications of the phenomenon of gravitation quantum states of neutrons, for instance, to searches for extra short-range forces, for constraining the neutron electric charge, or to search for physics beyond the Standard model (see, for instance [19]-[23], and numerous references in [2, 3]). In order to provide the required accuracy, we have built a new spectrometer GRANIT (see section 4), and develop new methods ([24] and section 3). The main methodical improvement will consist in measuring the frequencies of transitions between quantum states instead of measuring absolute shapes of neutron wave-functions.

3. Centrifugal quantum states of neutrons

Studies of centrifugal quantum states of neutrons [25, 26, 27, 28] are motivated by several arguments, in particular: 1) if you replace gravity in the experiment described above, with a ping-pong ball, to constant acceleration of the table then, in accordance with the weak equivalence principle, the ball would behave equivalently in the reference frame associated with the table; so we would directly demonstrate for the first time the weak equivalence principle for an object in a quantum state; 2) centrifugal quantum states of neutrons would provide a nearly ideal "quantum bouncer" analogous to gravitational quantum states of neutrons; 3) this phenomenon, under certain conditions, would be extremely sensitive to any extra forces on the characteristic scale of some tens of nanometers equal the spatial size of quantum states; 4) finally, under different conditions, we could use centrifugal quantum states of neutrons as an excellent probe for surface physics effects.

In the present talk we presented all mentioned aspects of centrifugal quantum states covered in detail in very recent references [25, 26, 27, 28]. The principle result consists in the first observation of the phenomenon itself and in extremely rich and detailed experimental data available, which are being analyzed at the moment.

4. GRANIT spectrometer

Finally, third part of presentation describes a new gravitational spectrometer GRANIT, which is under commissioning at the moment in the Institut Laue-Langevin, in Grenoble, France. As no detailed publication on the GRANIT spectrometer is available, we will present it here. The GRANIT spectrometer is a follow-up project aiming at significant improvement of all key parameters of our previous experimental installations, both measuring gravitational and centrifugal quantum states of neutrons. It would allow us both to maintain UCN in quantum states in the gravitational field for extended period of time of up to several seconds, and to measure precisely the frequency of resonant transitions between these states. It will also be a unique tool for performing a wide range of investigations not only in particle physics and fundamental interactions (neutron electric charge, spin-dependent or spin-independent short-range forces) and the foundations of quantum mechanics (interaction of a quantum system with a gravitational field, loss of quantum coherence), but also in solid-state, surface physics, and the field of experimental techniques and their possible technological applications.

The GRANIT spectrometer is installed at H172A neutron beam at the level C, ILL reactor. Compared to previous measurements (PF2, level D), it profits from 1) higher UCN density in a dedicated UCN source [29, 30] and longer measuring time due to its permanent installation; 2) lower vibration noise environment; 3) clean-room conditions allowing to work with delicate precision optical components; 4) better magnetic environment. A general scheme of installation of the GRANIT spectrometer is the following: Cold neutrons with broad wavelength distribution are delivered to the GRANIT intercalated-graphite monochromator (from the liquid-deuterium cold neutron source installed in vicinity of the active zone of the ILL high-flux reactor) through H172 neutron guide with the curvature radius of 650m, the length of 12m, and the cross-section of 8cm by 8cm; the guide walls are coated with m=3 super-mirror. Neutrons with the wavelength of 8.9A, as well as higher harmonics, are reflected in the monochromator towards the GRANIT spectrometer. They are guided by the secondary, slightly focusing, GRANIT neutron guide with the length of 4.5m; the guide walls are coated with m=2 super-mirror. 4He UCN source is installed downstream the neutron shutter at the exit of this guide. UCN are produced in the source and delivered to the GRANIT spectrometer installed inside the clean room using special UCN extraction system.

In order to reflect neutrons with the wavelength as large as 8.9E by a crystal using Bragg scattering, one has to increase the inter-planner distance. This is achieved, by intercalation of mosaic graphite crystals with potassium (K). So-called stage-2 crystals ($C_{24}K$) reflect 8.9E neutrons to the take-off angle of 61.2^o. The monochromator is an assembly of 18 such plates. The crystals are mounted pair-wise onto graphite bars which are then screwed into an indium sealed aluminum box. Using a monochromator and a secondary guide reduces the background with respect to a converter placed in the direct white beam. The price to be paid is a reduction in intensity due to imperfection of monochromator and secondary guide, and due to omission of multi-phonon process. The UCN source uses down-scattering of monochromatic cold neutrons with the wavelength of 8.9A in superfluid helium, via single-phonon excitation. The heart piece of the UCN source is the UCN converter vessel made of five rectangular tubes from BeO with inner dimension 70 by 70 mm^2 and the length 200mm each. The 1m long vessel is closed on both sides with Be windows. A hole with the diameter 25 mm in the last BeO section is used to extract UCN in the vertical direction through a chimney from stainless steel and a UCN valve made of stainless steel as well. An elbow followed by a straight section of UCN guide delivers the UCN to the spectrometer. The UCN converter vessel is situated in an aluminum vessel to contain the superfluid helium used as converter material. The vessel itself is situated within a custom-made cryostat. Primary low temperatures are provided at the two stages of a commercial 1.5W GM cold head, which cools thermal radiation screens and serves to liquefy He-4 from a gas bottle and He-3 within a closed circuit to operate secondary cooling stages. A "K-pot" serves for precooling incident He-4 to pass it through a superleak from compressed AlO powder for purification of the superfluid for the UCN converter from He-3. An evaporation stage for He-3 serves to cool the UCN converter to temperatures well below 1K, via a heat exchanger made from pure copper with electrically eroded fins on both sides.

The spectrometer consists of several neutron optical elements and UCN detectors installed on a massive granite table in an aluminum vacuum chamber with a volume of $\sim 1.5\text{m}^3$: 1) The neutron guide: it connects the spectrometer to the UCN source. A particular design of this guide provides general mechanical flexibility allowing to compensate for eventual misalignments and to decouple the spectrometer from the UCN source vibrations. Mechanical elasticity of the transport system allows decoupling GRANIT from the UCN source vibrations. An intermediate volume to store UCN is installed at the end of this guide. 2) Neutron extraction system: A semi-diffusive narrow slit [31, 32] with a height of 100-200μm allows us to extract UCN with small vertical velocity components needed for GRANIT. Other UCN will be reflected back to the source, thus reducing loss of the UCN density in the source. This configuration allows us to use a 4He UCN source in a flow-through mode. 3) The transport mirror system: to guide UCN from the exit of the semi-diffusive slit to the main mirror, we use a transport system consisting of a horizontal mirror edged by two vertical side walls, and a scatterer/absorber placed above the horizontal mirror. The transport mirror selects a few of the lowest quantum states. 4) The neutron mirror: The main optical element of the spectrometer is the trap mirror. It is a 30cm square mirror sided by four vertical walls coated with DLC (Diamond-Like Carbon). It should provide large neutron lifetimes in the gravitationally bound quantum states in the storage measuring mode. As the sensitivity of our experiment is proportional to the neutron life time in quantum states, extremely severe geometric characteristics of the mirror should be meet, such as the flatness of much better than 1μm, the roughness of 1E etc. 5) A resonant transition system [24]: For the first experiment in the flow-through mode with GRANIT we will use a resonant transition system. It consists of a multi-wire assembly installed above the transport mirror; it produces a spatially periodic gradient of the magnetic field. 6) Detectors of different type to be used: Three types of detectors will be used at the first stage of the GRANIT experiment. 3He gaseous proportional counters with extremely low background have been developed for this experiment. They will be used for integral measurements. Position-sensitive nuclear-track UCN detectors with a record resolution of $\sim 1\mu$m [15, 33] will be used to study the spatial distribution of neutrons in the quantum states. Real-time position-sensitive detectors with a resolution of a few hundred microns will measure the velocity distributions of neutrons in the quantum states. 7) Numerous spectrum-shaping and spectrum-analyzing devices as well as horizontal leveling and positioning systems will be similar to those used or discussed earlier in [13, 14, 15, 19] and [33]-[40]. The GRANIT spectrometer has to be isolated from external perturbations.

Three main systems are installed around it: 1) The anti-vibration and leveling system. The first level of protection against vibrations, also rough leveling of the installation, is provided by three pneumatic feet with controlled valves, which support the vacuum chamber. High level anti-vibration protection and leveling is provided by three high-load piezo-elements, located in the vacuum chamber under the granite table supporting all the optical elements. These piezo-elements support the granite with all optical elements on top of it. The vibrations associated with ground will be largely suppressed by the system described above. The main residual perturbation is expected to come from the mechanical connection between GRANIT and the UCN source; this effect has to be studied. One should note that the relative position of all optics elements are not affected by vibrations and slow mechanical drifts anyway, because they are all mounted rigidly to the granite table. 2) The clean room: Optical elements of GRANIT are very sensitive to dust contaminations. A precise adjustment with a typical accuracy of 1μm is required. For this reason, dust particles should be smaller than this value. Moreover, eventual dust particles present between two mirror surfaces at such a distance would damage our very delicate and expensive mirrors. Therefore the GRANIT spectrometer is installed inside a ISO 4 clean room. This clean room is equipped with a laminar air flow coming from the roof and regulated in such a way to avoid any turbulence in the central zone of the spectrometer right above optical elements. 3) The magnetic field control: Compensation of local magnetic fields and introduction of a constant field for neutron spin guiding are provided by three pairs of coils in Helmholtz configuration, installed around the chamber. The largest one is 3m large and 2.2m high. 300 turns of a 2mm diameter copper wire provides a field of 1.5mT if the current equals 6.5A. Later, for experiments requiring low values of residual magnetic fields, we will replace the coils by a μ-metal anti-magnetic screen.

The clean room is separated in three sections: the "white" section with the spectrometer inside, the "grey" section for preparation of experimental equipment, and SAS. Electronics, pumps, gas handling systems are installed below floor of the clean room and outside of the clean room. Other zones for preparation of experiments, storage of spare part of the spectrometer, as well as electronics to control experiments and the GRANIT equipment are placed outside of the clean room.

The GRANIT spectrometer will be used not only for the experimental program already previewed by the GRANIT collaboration. We are open to new ideas and proposals. During a fraction of beam-time, it will serve as "user" instrument for external experiments accepted by the ILL.

5. Conclusion

We have briefly overviewed recent advances on gravitational and centrifugal quantum states of neutrons, as well a status of the new GRANIT spectrometer. The present overview does not cover the whole announced subject, but serves rather as a reference to more detailed publications; moreover, it is focused only on activity of our GRANIT collaboration in this field.

Acknowledgments

The author is sincerely grateful to all colleagues from GRANIT collaboration, who all contributed to the results discussed here, also to the Institut Laue-Langevin and the Laboratoire de Physique Subatomique et de Cosmologie/IN2P3/UJF/INPG, Grenoble, France, and the Russian federal program "Cadres" for support of this activity.

References

[1] http://lpsc.in2p3.fr/Indico/conferenceDisplay.py?confId=371
[2] Baessler S 2009 *J. Phys.* G **36** 104005
[3] Nesvizhevsky V V 2010 *Usp. Phys. Nauk* **53**(7) 645
[4] Kogan V I and Galitskii V M 1956 *Problems in Quantum Mechanics* (Moscow: Gostekhizdat)
[5] Ter Haar D 1964 *Selected Problems in Quantum Mechanics* (New York: Academic Press)
[6] Landau L D and Lifshits E M 1964 *Quantum Mechanics: Nonrelativistic Theory* (Oxford: Pergamon Press)
[7] Langhoff P W 1971 *Am. J. Phys.* **39** 954
[8] Flügge S 1974 *Practical Quantum Mechanics I* (Berlin: Springer-Verlag)
[9] Gibbs R L 1975 *Am. J. Phys.* **43** 25
[10] Luschikov V I 1977 *Phys. Today* **30**(6) 42
[11] Luschikov V I and Frank A I 1978 *JETP Lett.* **28**559
[12] Sakurai J J 1985 *Modern Quantum Mechanics* (Menlo Park, Calif.: Benjamin/Cummings)
[13] Nesvizhevsky V V et al 2002 *Nature* **415** 297
[14] Nesvizhevsky V V et al 2003 *Phys. Rev.* D **67** 102002
[15] Nesvizhevsky V V et al 2005 *Europ. Phys. J.* C **40**479
[16] Zeldovich Ya B 1959 *Sov. Phys. JETP* **9** 1389
[17] Luschikov V I et al 1969 *JETP Lett.* **9** 23
[18] Steyerl A 1969 *Phys. Lett.* B **29** 33
[19] Nesvizhevsky V V and Protasov K V 2006 in *Trends in Quantum Gravity Research* ed D C Moore (New York: NOVA Sci. Publ.)
[20] Bertolami O and Nunes F M 2003 *Class. Quant. Grav.* **20** L61
[21] Abele H et al 2003 *Lect. Notes Phys.* **631** 355
[22] Baessler S et al 2007 *Phys. Rev.* D **75** 075006
[23] Nesvizhevsky V V et al 2008 *Phys. Rev.* D **77** 034020
[24] Kreuz M et al 2009 *NIM* A **611**326
[25] Nesvizhevsky V V et al 2008 *Phys. Rev.* A **78**(3) 033616

[26] Cubitt R et al 2009 *NIM* A **611**322
[27] Nesvizhevsky V V et al 2009 *Nature Phys.* **6**(2) 114
[28] Nesvizhevsky V V et al 2010 *New J. Phys.* **12** 113050
[29] Zimmer O et al 2007 *Phys. Rev. Lett.* **99** 104801
[30] Schmidt-Wellenburg P et al 2009 *NIM* A **611** 267
[31] Barnard J et al 2008 *NIM* A **591** 431
[32] Schmidt-Wellenburg P et al 2007 *NIM* A **577** 623
[33] Nesvizhevsky V V et al 2000 *NIM* A **440** 754
[34] Nesvizhevsky V V 2003 *Usp. Fiz. Nauk* **46** 93
[35] Nesvizhevsky V V 2004 *Usp. Fiz. Nauk* **47** 515
[36] Nesvizhevsky V V et al 2005 *J. Res. NIST* **110** 263
[37] Voronin A Yu et al 2006 *Phys. Rev.* D **73** 044029
[38] Meyerovich A E et al 2006 *Phys. Rev.* A **73** 063616
[39] Adhikari R et al 2007 *Phys. Rev.* A **75** 2007
[40] Westphal A et al 2007 *Europ. Phys. J.* C **51** 367

5th International Workshop DICE2010 IOP Publishing
Journal of Physics: Conference Series **306** (2011) 012004 doi:10.1088/1742-6596/306/1/012004

Interferometric gravitational wave detectors: state of the art and fundamental noise issues

Carlo Nicola Colacino

Dipartimento di Fisica "Enrico Fermi", Università di Pisa, Largo B. Pontecorvo 3, 56127 Pisa, Italy

E-mail: colacino@pi.infn.it

Abstract. Interferometric gravitational wave detectors, such as Virgo in Italy, LIGO in the US and GEO600 in Germany, have already completed scientific data runs. 2nd generation detectors, such as Advanced Virgo and Advanced LIGO will start operation within the next 5 years and preliminary design studies have already begun for 3rd generation detectors. It is hoped that we will soon be able to reach the holy grail of the first direct detection of gravitational radiation and start the exciting field of gravitational wave astronomy. The LIGO Scientific Collaboration (LSC) and the Virgo collaboration have published many observational results of great scientific relevance. For the future, the main challenge is to further increase the detector sensitivity by reducing the instrumental noise, in particular quantum noise, i.e. a fundamental noise term intrinsic to the measurement process itself. This can be achieved by the implementation of forefront quantum optics techniques.

1. Introduction

Several large km-scale interferometric gravitational wave detectors, like the Virgo detector in Italy, near Pisa, and the two Laser Interferometric Gravitational Observatory (LIGO) in the US, one in Hanford, WA and the other in Livingstone, LA, have taken scientific data on and off for the last three years. No detection claim has been made so far, but some scientifically important results in the form of upper limits have been obtained for the amplitude of the Stochastic Gravitational Wave Background and for the ellipticity of the pulsars. In this paper we will briefly describe their main results, then we will describe the noise terms which limit their sensitivity and possible schemes to improve the sensitivity.

2. Main scientific results obtained so far

One of the most important results obtained by the LSC and the Virgo collaboration is the result on continuous waves. According to astrophysical estimates there are roughly $\bigcirc(10^4)$ electromagnetically quiet neutron stars within a radius of 500 pc. Only a tiny fraction of them is in the band of interest for gravitational wave detectors. Pulsars in the frequency band between 50 and 1100 Hz were studied, and a new limit on the ellipticity was placed: $\epsilon \leq 10^{-6}$ [1]. This limit is more stringent than previous limits that were based on population synthesis models. The second result of great importance was published on *Nature*: it is an observational result about the Stochastic Gravitational-Wave Background [2]. The indirect but very stringent limit on the Stochastic Background coming from the Primordial Nucleosynthesis theory was surpassed, as shown in Figure 1, and even some cosmic string models were ruled out.

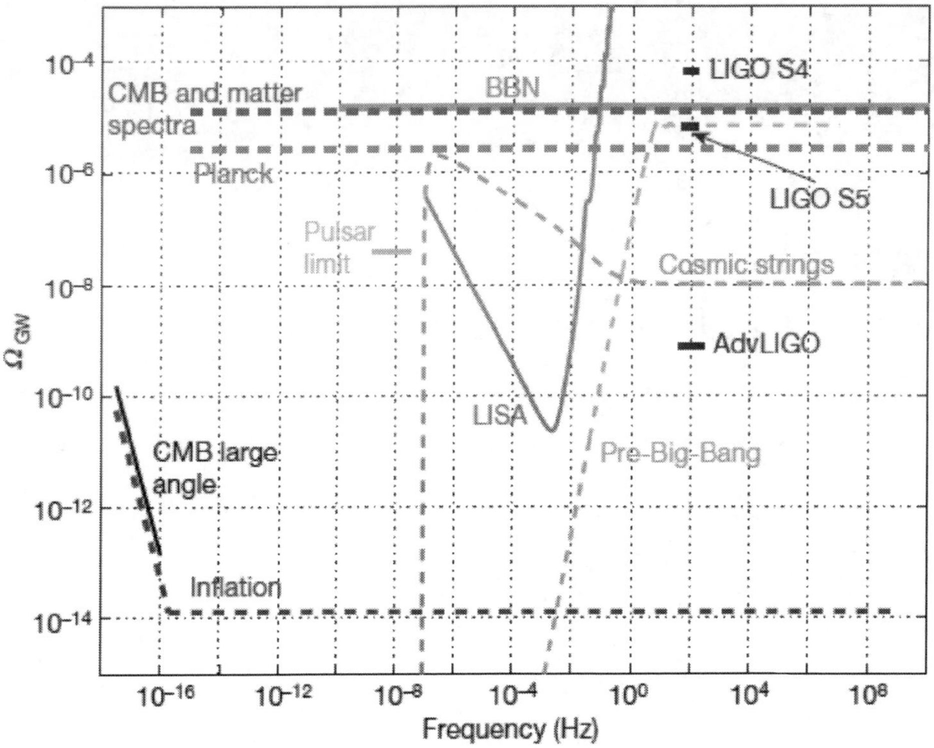

Figure 1. The BBN bound

3. Advanced detectors

2nd generation detectors will start operation around 2015. Their sensitivity is expected to be an order of magnitude better than current interferometers'. GW detectors measure the **amplitude** which scales as $1/r$, and not the power which scales as $1/r^2$. For GW detectors, the number of detectable sources grows linearly with the volume. An improvement in one order of magnitude in the sensitivity means an horizon one order of magnitude bigger and therefore an accessible volume 10^3 times bigger. For example Virgo can see a coalescence of Neutron Stars up to $d = 15$ Mpc and the event rate is $0.01 \rightarrow 0.1$ per year. A detector with a sensitivity better by an order of magnitude would thus be able to see up to $d = 150$ Mpc, with an expected event rate of $10 \rightarrow 100$ per year. Advanced detectors will hopefully mark the birth of gravitational-wave astronomy.

4. Noise issues

Test masses in interferometric GW detectors do move also because of effects that have nothing to do with a real GW signal. These effects are generically called *displacement noise*. *Seismic noise* is the noise due to human activity (cars, trains, etc) as well as to a micro-seismic background (sea tides). Seismic isolation systems make this noise relevant only below a certain frequency threshold, also called seismic cutoff; currently, the lowest cutoff, at 3 Hz, is obtained by Virgo thanks to its superattenuator system [3]. *Newtonian noise* is given by the fluctuations of the static gravitational field. It is neglibigle in the whole Virgo-LIGO band, 1 Hz – 10 kHz but represents a *fundamental limit for ground-based detectors* [4]. No ground-based detector can go below 1 Hz, although noise reduction schemes based on accelerometers have been proposed [5].

Thermal noise induces vibrations both in the suspension wires and in the mirrors. Its effects can be understood in terms of the Fluctuation-Dissipation Theorem. The main contribution between 3 and 30 Hz comes from the pendulum thermal noise (horizontal displacement of test masses). Violin modes, i.e. fluctuations of the normal modes of the wire, give some spikes from 300 Hz onwards. Test-mass thermal noise is largely dominated by the Brownian motion of the mirrors, which is the limiting noise source between 30 and 250 Hz. Another important contribution is the thermo-refractive noise (refraction index of the coating depends on temperature). To these noise terms we must add what we call "accidental noise", i. e. noise such as the residual pressure of the vacuum, micro-roughness of the mirrors, fluctuations of the laser power as well as technical noise (longitudinal control noise) and spurious coupling between seismic noise and the electromagnetic fields.

5. Quantum noise and the Standard Quantum Limit

Quantum noise is intrinsic to the measurement process. Quantum theory introduces an intrinsic uncertainty into the amplitude and phase. There are two major contributions: *Radiation-pressure noise* and *Shot noise* [5]. Radiation-pressure noise is given by uncertainties in the mirror position due to quantum fluctuations exerting fluctuating radiation pressure on the mirrors. Shot-noise uncertainties are due to quantum mechanical Poissonian fluctuations in the number of photons at the interferometer output. Radiation-pressure noise and shot-noise are associated with two orthogonal quadratures of the electromagnetic field [6]. An important concept is that of the *Standard Quantum Limit* (SQL). For gravitational-wave detectors, the SQL is obtained by exactly balancing radiation-pressure and shot-noise. It is the intrinsic limit on the accuracy with which any position-sensing device can determine the position of a free mass. The SQL in a GW detector with input power I_0 can be written as $S_x S_p \geq 1/4$, where $S_x \propto 1/I_0$ is the power spectral density of position fluctuations due to the uncertainty in the number of photons and $S_p \propto I_0/f^2$ is due to the back-action (radiation-pressure). The two noise terms dominate in different frequency regions. For current detectors, whose laser input power is low, radiation-pressure is completely negligible. For Advanced detectors radiation-pressure will also be important. Optical noise sources implement the SQL only if they are *uncorrelated*. In Figure 4 we show the contributions of the two quantum noise sources.

There exists an optimal quantum power that minimises the total quantum noise. Quantum noise is due to vacuum fluctuations entering the unused port at the beam-splitter. Shot noise can be reduced by rendering asymmetric the noise entering that port. A **squeezed** state is asymmetric by definition and can thus reduce shot noise. Injecting squeezed vacuum (non-linear crystals) into the unused port can reduce the dominant noise source by a factor $S_h \propto \exp(-2R)$ where $\exp(-2R)$ is the power squeeze factor. This configuration is equivalent to an unsqueezed interferometer with an increase to the laser power by the squeeze factor. Squeezing can thus help reduce the circulating power without losing in sensitivity. Injected squeezed vacua is not the only possibility to use squeezed states: if carrier light with side bands reflects off a mirror, the side bands are **ponderomotively squeezed**. Correlations are created between the two quadratures. If the position signal is measured in the phase quadrature, we have $b = \kappa(f)\triangle A + \triangle\phi + h$. These correlations are not useful if direct photon detection is used, because in this case we would reduce the noise in one quadrature, but increase it in the other. If we manage to measure an admixture of quadratures with **frequency dependent** homodyne detection scheme (angle) that is function of κ, then it is possible to completely remove radiation-pressure noise from the measurement [7]. There are some experimental requirements that have to be met, in order for squeezing to be useful: Squeezing in the *whole detection band*10 Hz $\rightarrow$ 10 kHz: this is experimentally feasible, as shown by R. Schnabel's group in Hanover and by P K Lam's group in Canberra. Other requirements are a high level of squeezing, ideally 10 dB or higher (and the current record is currently 13 dB), long term stability, which, like the previous item, is

Journal of Physics: Conference Series **306** (2011) 012004 doi:10.1088/1742-6596/306/1/012004

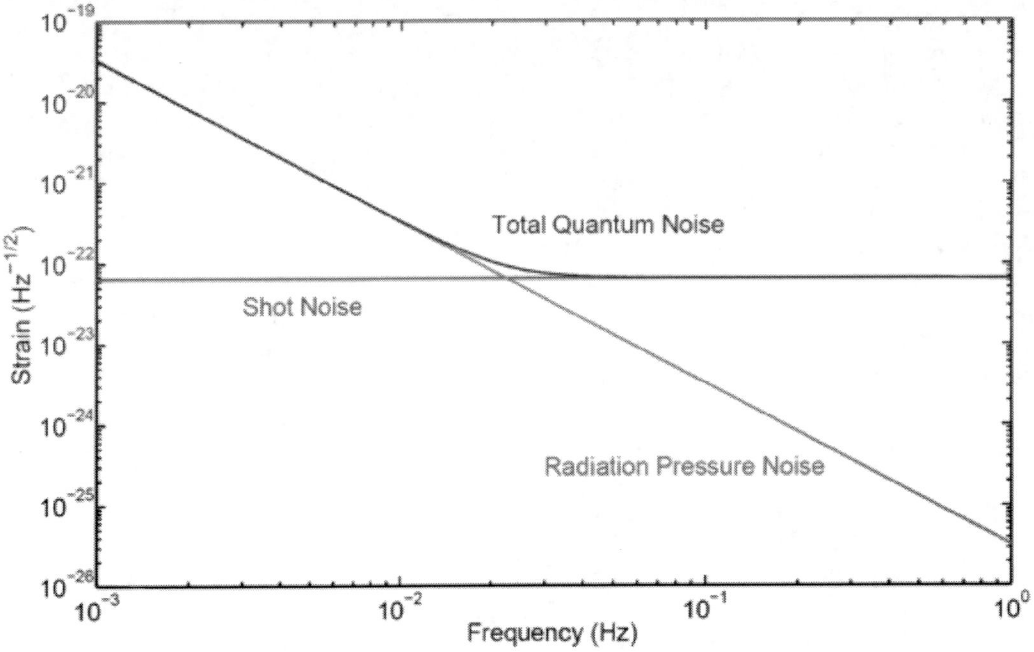

Figure 2. The total quantum noise

mainly a technical problem and a frequency-dependent squeezing angle, which is perhaps the most challenging issue. 3rd generation detectors, i.e. detectors that will follow Advanced LIGO and Advanced Virgo, such as the Einstein Telescope (ET) are likely to be quantum noise limited over the whole frequency range and therefore will certainly implement squeezing [8], as well as other techniques of non-classical interferometry [9].

6. Holographic noise

Quantum theory implies a minimum time, the Planck time $t_{\rm Pl} = \sqrt{hG_{\rm N}/c^5} = 5 \times 10^{-44}$ s and a minimum length, the Planck length $l_{\rm Pl} = \sqrt{hG_{\rm N}/c^3} = 1.616 \times 10^{-35}$ m. According to Craig Hogan, the minimum interval of time might be studied directly with GW interferometers [10]: the discretisation of the space-time manifold would give rise to transverse uncertainties in position. This phenomenon is called **holographic noise** and it is a precise, parameter-free prediction. Interferometers could in principle detect quantum indeterminacy of holographic geometry because the indeterminacy in the beam splitter position would insert holographic noise into the signal. At a frequency $f = c/2L$ the shear fluctuations have a power spectral density $S^2_{\rm H} = L\triangle\theta^2 = t_{\rm Pl}$. Planck's time is the only relevant quantity in this expression. The power spectral density of shear perturbations is flat $h_{\rm H} = \sqrt{t_{\rm Pl}} = 2.3 \times 10^{-22}$ Hz$^{-1/2}$. This is a general property of holographic quantum geometry. GEO600 meausred an unaccounted for excess of noise: by adding the holographic contribution, the agreement is better (although not perfect), as can be seen from Figure 5, a slide taken from a talk by Craig Hogan himself.

Subsequent measurements however have not confirmed the detection of the holographic noise. GEO 600 was very sensitive to transverse displacements because it was the only detector to use the Signal-Recycling mirror. Advanced detectors are likely to have a signal-recycling mirror too so hopefully the issue will be clarified.

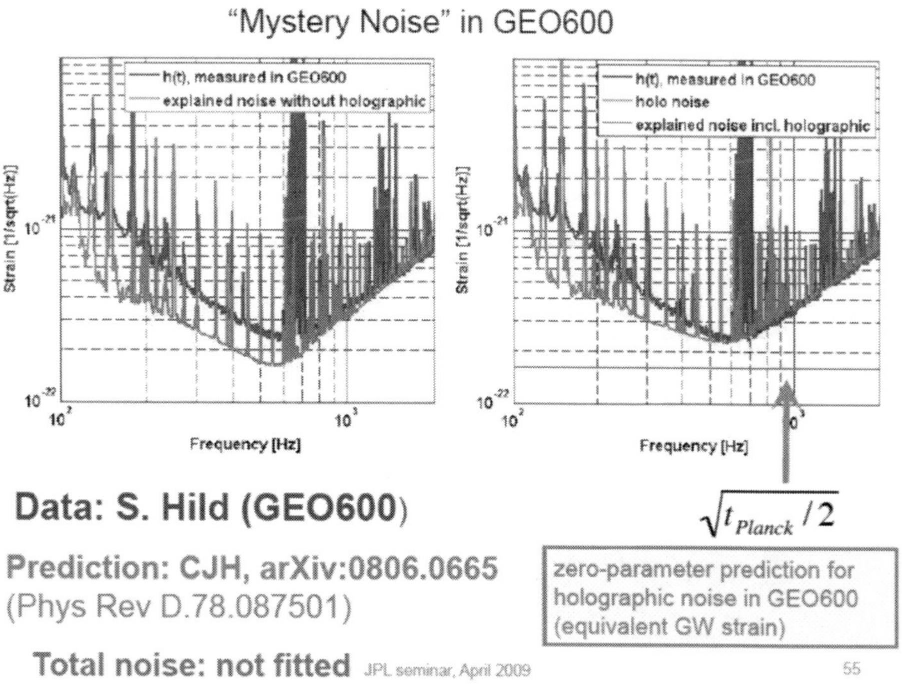

Figure 3. GEO600 with and without holographic noise

7. Summary

Interferometric GW detectors such as LIGO and Virgo have not yet detected gravitational radiation but the data they have collected have led to very interesting scientific results, such as an improvement on the BBN bound for the stochastic gravitational radiation of cosmological origin. Advanced detectors, which will start operation in 5 years' time, will almost surely mark the first direct detection of gravitational radiation and hopefully also give birth to a new and exciting branch of astronomy, gravitational-wave astronomy. Such detectors, and their successors, will be quantum limited and will make use of non-classical light states, thus relating a purely classical theory such as Einstein's General Relativity to frontier research in Quantum Optics. As a side-effect, GW-detectors might also give us some hints towards a quantum theory of gravity.

References

[1] B Abbott et al 2009 *Phys. Rev. Lett.* **102** 111102
[2] The LIGO Scientific Collaboration & The Virgo Collaboration 2009 *Nature* **460** 990
[3] The Virgo Collaboration 2010 *Astroparticle Physics* **33** 182-189
[4] Beker M, Cella G, DeSalvo R, Doets M, Grote H, Harms J, Hennes E, Mandic V, Rabeling D, van den Brand J and van Leeuwen C 2010 *General Relativity and Gravitation* 1-34
[5] Maggiore M 2007 *Gravitational Waves. Volume 1: Theory and Experiment* (Oxford: Oxford University Press)
[6] Fox M 2003 *Quantum optics* (Oxford: Oxford University Press)
[7] Corbitt T, Ottaway D, Innerhofer E, Pelc J and Mavalvala N 2006 *Phys. Rev.* A **74** 021802
[8] Khalaidovski A 2010 private communication at *ET 3rd Annual Meeting*, Budapest, November 2010
[9] Tamburini F, de Matos C J, Perdigues Armengol J M and Colacino C N 2009 *Il Nuovo Cimento* B **124**(05) 511-523
[10] Hogan C 2008 *Phys. Rev.* D **78** 087501

Journal of Physics: Conference Series **306** (2011) 012005 doi:10.1088/1742-6596/306/1/012005

A realist interpretation of quantum mechanics based on undecidability due to gravity

Rodolfo Gambini[1], **Luis Pedro García-Pintos**[1] **and Jorge Pullin**[2]

[1] Instituto de Física, Facultad de Ciencias, Universidad de la República, Iguá 4225, CP 11400 Montevideo, Uruguay

[2] Department of Physics and Astronomy, Louisiana State University, Baton Rouge, LA 70803-4001

E-mail: `pullin@lsu.edu`

Abstract. We summarize several recent developments suggesting that solving the problem of time in quantum gravity leads to a solution of the measurement problem in quantum mechanics. This approach has been informally called "the Montevideo interpretation". In particular we discuss why definitions in this approach are not *for all practical purposes* (fapp) and how the problem of outcomes is resolved.

1. Introduction

The problem of measurement in quantum mechanics arises in standard treatments as the requirement of a reduction process when a measurement takes place. Such process is not contained within the unitary evolution of the quantum theory but has to be postulated externally and is not unitary. It is usually justified through the interaction with a large, classical measuring device.

Two main objections have been levied onto two aspects of the solution of the problem of measurement through decoherence.

1) Since the evolution of the system plus environment is unitary, the coherence could potentially be regained.

2) The second criticism has to do with the fact that in a picture where evolution is unitary "nothing ever occurs". This is Bell's "and/or" problem. The final reduced density matrix of the system plus the measurement device will describe a set of coexistent options and not alternative options with definite probabilities.

We shall argue that a solution to the problem of time in quantum gravity leads to a modification of quantum mechanics such that the two above objections can be overcome. The result is an objective description of quantum mechanics that is compatible with unitarity and does not require a reduction postulate.

The plan of the article is as follows. In the next section we briefly describe the proposed solution to the problem of time in quantum gravity. In section 3. we discuss how such solution leads to a modification of Schrödinger's equation. In section 4. we discuss the solution to the objections to environmental decoherence. In section 5. we argue that fundamental undecidability leads to the solution of the problem of outcomes and macro-objectification. We end with a discussion.

2. The problem of time in quantum gravity

Quantum gravity is expected to be most relevant in situations where one cannot assume that one has a clock external to the system under study (for instance when one considers the universe as a whole as is done in quantum cosmology). As a consequence of that one has to choose as clock one of the variables of the system under study. Since the variables of the system under study are all represented by quantum operators, time is not going to be a classical parameter as in ordinary quantum mechanics.

In fact, one can choose to describe ordinary quantum mechanics in terms of a "real clock", i.e. a variable represented by a quantum operator subject to quantum fluctuations, etc. There usually is an objection that no monotonous variable can be conjugate to the Hamiltonian and this is seen as an obstruction to time being represented by a physical variable. But this is easily solved: one picks variables that work well as clocks for a while. If one needs a longer lasting clock one adds another variable when the first stopped being monotonous, pretty much like the hour hand on a clock does when the minute hand goes around a complete circle. The way to describe quantum mechanics with a real clock is relationally. You pick an observable, let us call it $T(t)$ and then pick a set of observable quantities we want to study that commute with $T(t)$, i.e. $O^1(t), \ldots, O^N(t)$. Here we denote by t the ideal classical time that appears in the ordinary Schrödinger equation. One then computes the conditional probability,

$$P\left(O^i = O_0^i | T = T_0\right) = \lim_{\tau \to \infty} \frac{\int_{-\tau}^{\tau} dt \, \mathrm{Tr}\left(P_{O_0^i}(t) P_{T_0}(t) \rho P_{T_0}(t)\right)}{\int_{-\tau}^{\tau} dt \, \mathrm{Tr}\left(P_{T_0}(t)\rho\right)}, \tag{1}$$

where $P_{O_0^i}(t)$ is the projector on the eigenspace associated with the eigenvalue Q_0^i at time t and similarly for $P_{T_0}(t)$. These conditional probabilities are positive and add to one. If one chooses operators with continuum spectrum such probability of giving a definite value vanishes, one has to ask about the probability of being in an interval around the eigenvalue and the above expression is unchanged.

There is a technical problem in carrying out the above construction in theories like general relativity that are generally covariant and it is the construction of observables to use in the construction. This was for many years a stumbling block for carrying out this program. Recently we proposed to use evolving constants of the motion as the observables and this construction has appeared satisfactory in model systems that could not be treated before. Since we are orienting this article towards readers not primarily interested in quantum gravity, we omit the details here and refer to our paper on the subject [1].

3. A modified Schrödinger equation

An important question is how the conditional probabilities we introduced in the previous section evolve. We have been able to show that if one makes some judicious assumptions, namely, that the clock does not interact with the system, that the clock is in a highly classical state (a coherent state where the "hand" of the clock is sharply peaked in space and moves in a monotonous way), then one can define a density matrix labeled by the eigenvalues of T,

$$\rho(T) \equiv \int_{-\infty}^{\infty} dt \, U_{\text{sys}}(t) \rho_{\text{sys}} U_{\text{sys}}(t)^\dagger \mathcal{P}_t(T), \tag{2}$$

where the "sys" subscript indicates that it is the density matrix and the evolution operator corresponding to the system under study, it does not include the clock. We are assuming the density matrix of the whole universe is of the form, $\rho = \rho_{\text{sys}} \otimes \rho_{\text{clock}}$. The probability $\mathcal{P}_t(T)$ is the probability that the clock reads when the ideal Schrödinger time is t. Such probabilities are not directly observable, that is the reason the variable t always appears integrated. The explicit expression for the probability is,

$$\mathcal{P}_t(T) = \frac{\mathrm{Tr}\left(P_T(t)\rho\right)}{\int_{-\infty}^{\infty} dt' \ \mathrm{Tr}\left(P_T(t')\rho\right)}. \tag{3}$$

If one assumes one has a clock that follows the ideal Schrödinger time perfectly, then $\mathcal{P}_t(T) = \delta(t-T)$. In reality there will be departures. If we assume the departures are very small,

$$\mathcal{P}_t(T) = \delta(t-T) = \delta(T-t) + a(T)\delta'(T-t) + b(T)\delta''(T-t) + \ldots, \tag{4}$$

one can show that the density matrix satisfies an approximate Schrödinger equation (for a pedagogical discussion of the derivation, see [2]),

$$\frac{\partial\rho(T)}{\partial T} = i[\rho(T), H] + \sigma(T)[H,[H,\rho(T)]]. \tag{5}$$

where $\sigma(T) = \partial b(T)/\partial T$. This type of equation is a particular case of the Lindblad equation considered in decoherence, but in this particular form it has the virtue that it conserves energy. The effect of the extra term is to make the off diagonal elements of the density matrix in the energy basis go to zero exponentially. That is, the evolution in terms of the real clock is not unitary. The underlying evolution in terms of the idealized time t is, but the "real clock" T cannot keep track of it accurately enough to keep a unitary evolution. After some time, for a given T, one will have a distribution of possible t's associated with it and, even if one started with a pure state, one ends with a superposition of pure states for a given value of T. In what follows each time we refer to unitarity we will be speaking of the evolution in the underlying unobservable parameter. The observable evolution depends on the particular implementation of the clock we are taking and it is not unitary.

At this point it is worthwhile asking: can one make the effect arbitrarily small by choosing better clocks? That requires to estimate the minimum uncertainty one can have in a clock. A full discussion would require more understanding of quantum gravity that we have right now. There have been several heuristic estimates in the literature [3] and some controversy surrounding them. We prefer to just consider generically that the uncertainty in the measurement of time has to grow with the observed time and if the effect is due to gravity, Planck's time $t_{\rm Planck} = 10^{-44}s$ should be involved. This leads to postulate an uncertainty in the measurement of a period of time T of the form,

$$\delta T = t_{\rm Planck}^{1-a} T^a. \tag{6}$$

The heuristic arguments we mentioned above yield $a = 1/3$ or $a = 1/2$, but in practice this makes little difference. As long as the clock error increases with the time measured, there will be a fundamental loss of coherence due to the use of real clocks.

A similar effect is introduced by the use of "real rods" to measure space. Such a discussion requires quantum field theory and had not been developed fully, so we refer the reader to Ref. [4].

4. Solution to the objections to environmental decoherence

The presence of a fundamental source of loss of coherence helps deal with the objections levied against the solution of the problem of measurement through environmental decoherence. The two main objections can be characterized as "the information is still there" and the "and/or" problem or "problem of outcomes". We discuss them in the following two subsections.

4.1. Impossibility of recovering the information

This objection is as follows: although a quantum system interacting with an environment with many degrees of freedom will very likely give the appearance that the initial quantum coherence of the system is lost, the information about the original superposition could be recovered, for instance, by carrying out a measurement that includes the environment. The fact that such measurements are hard to carry out in practice does not prevent the issue from existing as a conceptual problem.

To analyze this point in some detail, as is customary in discussions of decoherence, let us consider a variation of a model proposed by Zurek [5] in which the quantum system, the environment, and the measuring apparatus are under control. The system consists of a cavity with a uniform magnetic field in the z (vertical) direction. Inside the cavity is a spin S that represents the "needle" of the measuring device. A steady flux of spins is pumped into the cavity in a horizontal stream and each spin interacts with S a finite amount of time τ. The stream constitutes "the environment". We also assume the spins of the environment are separated enough to avoid considering interactions among them. The interaction Hamiltonian is given by,

$$\hat{H}_k^{\rm int} = f_k\left(\hat{S}_x\hat{S}_x^k + \hat{S}_y\hat{S}_y^k + \hat{S}_z\hat{S}_z^k\right), \tag{7}$$

with f_k the coupling constants. The Hamiltonian due to the presence of the magnetic field when the k-th particle is in the cavity is,

$$\hat{H}_k^B = \gamma_1 B\hat{S}_z \otimes \hat{I}_k + \gamma_2 B\hat{I} \otimes S_z^k, \tag{8}$$

where $\hat{I}$ is the identity matrix acting on the Hilbert space of the needle and I_k is the identity in the Hilbert space of the k-th particle. The introduction of a constant magnetic field pointing in a given direction, is in order to have a definite pointer basis.

We have discussed this model in detail in [6], so we will not repeat all details here. We outline the main result. d'Espagnat [7] suggested that for systems like this one can define a global observable,

$$\hat{M} \equiv \hat{S}_x \otimes \prod_k^N \hat{S}_x^k. \tag{9}$$

This observable has the property that if one starts from a normalized initial state given by,

$$|\Psi(0)\rangle = (a|+\rangle + b|-\rangle)\prod_{k=1}^N \otimes\left[\alpha_k|+\rangle_k + \beta_k|-\rangle_k\right], \tag{10}$$

where $|\pm\rangle$ are eigenstates of σ_z, and computes the expectation value of the observable in a state that has evolved unitarily one gets,

$$\langle\psi|M|\psi\rangle = ab^*\prod_k^N\left[\alpha_k\beta_k^* + \alpha_k^*\beta_k\right]e^{-2i\Omega_k\tau} + a^*b\prod_k^N\left[\alpha_k\beta_k^* + \alpha_k^*\beta_k\right]e^{2i\Omega_k\tau}, \tag{11}$$

where $\Omega_k = \sqrt{4f_k^2 + B^2(\gamma_1-\gamma_2)^2}$ with γ_1 the magnetic moment of the spins of the environment and γ_2 that of the "needle". On the other hand, if one assumes that a collapse of the wavefunction has occurred, one obtains:

$$\langle\psi|M|\psi\rangle = 0. \tag{12}$$

If one now studies the same observable but assuming the modified Schrödinger evolution we discussed in the previous section one gets,

$$\begin{aligned}\langle \hat{M} \rangle &= ab^* e^{-i2N\Omega T} e^{-4NB^2(\gamma_1-\gamma_2)^2\theta} \prod_k^N \left[\alpha_k \beta_k^* e^{-16B^2\gamma_1\gamma_2\theta} + \alpha_k^* \beta_k\right] \\ &+ba^* e^{i2N\Omega T} e^{-4NB^2(\gamma_1-\gamma_2)^2\theta} \prod_k^N \left[\alpha_k \beta_k^* + \alpha_k^* \beta_k e^{-16B^2\gamma_1\gamma_2\theta}\right] \end{aligned} \quad (13)$$

where $\Omega = B(\gamma_1 - \gamma_2)$. We see the expectation value is exponentially damped with an exponent that is large. Actually putting in realistic numbers one concludes that if one has an "environment" with more than 10^7 particles the expectation value decays too fast to be measured within the accuracies of the experiment, as we will discuss in the next subsection.

Due to the fundamental decoherence induced by quantum clocks the expectation value of the observables exponentially decreases and is more and more difficult to distinguish from the vanishing value resulting from collapse. A similar analysis allows to show that "revivals" [8] of coherence of the wavefunction of the system plus the measurement device (but excluding the environment) are also prevented by the modified evolution. When the multi-periodic functions in the coherences tend to take again the original value, after a Poincaré time of recurrence, the exponential decay for sufficiently large systems completely hides the revival under the noise amplitude.

Thus, the difficulties found in testing macroscopic superpositions in a measurement process are enhanced by the corrections resulting from the use of physical clocks.

4.2. Fundamental undecidability and the problem of outcomes

To address the second objection to the decoherence solution to the measurement problem, the "and/or" problem, or as others put it "nothing ever happens", we will introduce a criterion to determine when an event takes place in a theory where evolution is always unitary.

When one takes into account the way that time enters in general covariant systems including the quantum fluctuations of the clock, the unitary evolution of the total system (system plus apparatus plus environment) becomes indistinguishable from a statistical mixture of all the alternative states that correspond to each of the possible outcomes of a measuring device. By indistinguishable we mean that all physical predictions of two states cannot be distinguished due to fundamental limitations in measurement. We call such situation "undecidability". We will claim that an event takes places when undecidability has been reached. Therefore undecidability is truly a comparison between the quasi-unitary evolution in physical time described by the Lindblad equation and the statistical mixture of the states associated with the different outcomes of the measuring device.

The fact that in the definition of undecidability we had to invoke fundamental limitations in measurement raises the question, if the definition is "for all practical purposes" only or if it is fundamental in nature. That is, it is not due to a technological limitation, but due to limitations imposed by the laws of physics. We are going to show that undecidability is not only for all practical purposes (FAPP) but fundamental (see [9] for details). From the previous discussion one can gather that as one considers environments with a larger number of degrees of freedom and as longer time measurements are considered, distinguishing between collapse and unitary evolution becomes harder. But is this enough to be a fundamental claim?

Given the expectation value we discussed,

$$< \hat{M} >\sim \exp\left(-6NB^2(\gamma_1-\gamma_2)^2 T_{\rm Planck}^{4/3} \tau^{2/3}\right), \quad (14)$$

is it possible for a big enough ensemble to distinguish it from zero?

Brukner and Kofler [10] have recently proven that from a very general quantum mechanical analysis together with bounds from special and general relativity there is a fundamental uncertainty in the measurements of angles associated with quantum mechanical spin, even if one uses a measuring device of the size of the observable Universe. The bound is given by $\Delta\theta \geq \ell_{\rm Planck}/R$, with $\ell_{\rm Planck}$ the Planck length and R the radius of the universe. This uncertainty in the measurement of angles in our example implies there is an uncertainty in the measurement of $< M >$. If that uncertainty is bigger than the value of $< M >$ one can measure, there is no way of distinguishing a collapse from a unitary evolution, even in principle. For the model in question this occurs for $N \sim 10^7$ spins. So one sees that even if one considers ensembles of growing size, this does not help in distinguishing the expectation value in the case of collapse or unitary evolution since there are fundamental limitations to measurement. It is like attempting to measure a distance with an accuracy greater than the smallest spacing on a ruler. No matter how many measurements one makes, the error will be at least as large as the smallest spacing on the ruler.

The problem of macro-objectification of properties may be described according with Ghirardi [11] as follows: "how, when, and under what conditions do definite macroscopic properties emerge (in accordance with our daily experience) for systems that, when all is said and done, we have no good reasons for thinking they are fundamentally different from the micro-systems of which they are composed?" We think that undecidability provides an answer to this problem. If a system suffers an interaction with an environment such that the density matrix is such that all the physical predictions of such a state cannot be distinguished from those of a statistical mixture of the states corresponding to different outcomes of the collapsed system, we will claim that an event took place. A fundamental ingredient is that in order to distinguish the density matrices one will be limited by fundamental limitations on the measurements of physical quantities and preparations of physical states. A detailed implementation will depend on the system under study. An example is the limitation on the measurements of spins of Brukner and Kofler we discussed in this section. Notice that for a quantum micro-system in isolation, events would not occur. However for a quantum system interacting with an environment, events will be plentiful. The same goes for a system being measured by a macroscopic measuring device.

5. Discussion

Gravity fundamentally limits how accurate our measurements of space and time can be. This requires reformulating quantum mechanics in terms of real clocks and rods, that have errors in their measurements. The resulting picture of quantum theory is one where there is a fundamental loss of quantum coherence: pure states evolve into mixed states as our clocks and rods cannot keep track of the unitary evolution. This eliminates the problem of revivals and global observables in the solution to the measurement problem through environmental decoherence: just waiting longer does not improve things as more quantum coherence is lost. One also has a definition for when an event takes place: when the fundamental loss of coherence is such that one cannot distinguish the unitary evolution from a reduction (in the sense previously explained), an event has taken place. We call this situation "undecidability" between reduction and unitary evolution. One therefore has a complete, objective formulation of quantum mechanics with a unitary evolution and a notion of event that does not imply the collapse of the wavefunction. We call it the Montevideo interpretation of quantum mechanics. There even exists an axiomatic formulation of the theory [12]. We refer the reader to the references for further details.

Acknowledgments

Part of this work was done in collaboration with Rafael Porto and Sebastián Torterolo. This work was supported in part by grant NSF-PHY-0650715, funds of the Hearne Institute for Theoretical Physics, CCT-LSU, Pedeciba and ANII PDT63/076.

References:

[1] Gambini R, Porto R A, Pullin J and Torterolo S 2009 *Phys. Rev.* D **79** 041501 (*Preprint* arXiv:0809.4235)
[2] Gambini R, Porto R and Pullin J 2007 *Gen. Rel. Grav.* **39** 1143 (*Preprint* arXiv:gr-qc/0603090)
[3] Károlhyázy F, Frenkel A and Lukács B 1986 in "Quantum concepts in space and time" ed R Penrose and C Isham (Oxford: Oxford University);
Ng Y J and van Dam H 1995 *Annals N. Y. Acad. Sci.* **755** 579 (*Preprint* arXiv:hep-th/9406110); *do.* 1994 *Mod. Phys. Lett.* A **9** 335;
Amelino-Camelia G 1994 *Mod. Phys. Lett.* A **9** 3415 (*Preprint* arXiv:gr-qc/9603014);
Lloyd S and Ng J 2004 *Scientific American* November Issue
[4] Gambini R, Porto R A and Pullin J 2008 *Phys. Lett.* A **372** 1213-1218 (*Preprint* arXiv:0708.2935)
[5] Zurek W 1982 *Phys. Rev.* D **26** 1862
[6] Gambini R, Pintos L P G and Pullin J 2010 *Found. Phys.* **40** 93-115 (*Preprint* arXiv:0905.4222)
[7] d'Espagnat B 1995 *Veiled reality* (New York: Addison Wesley)
[8] Gambini R and Pullin J 2007 *Found. Phys.* **37** 1074-1092 (*Preprint* quant-ph/0608243)
[9] Gambini R, Garcia-Pintos L P and Pullin J 2010 *Preprint* arXiv:1009.3817
[10] Brukner C and Kofler J 2010 Are there fundamental limits for observing quantum phenomena from within quantum theory? *Preprint* arXiv:1009.2654
[11] G. Ghirardi 2007 *Sneaking a look at God's cards* (Princeton, NJ: Princeton University Press)
[12] Gambini R, Garcia-Pintos L P and Pullin J 2010 Complete quantum mechanics: An Axiomatic formulation of the Montevideo interpretation *Preprint* arXiv:1002.4209

5th International Workshop DICE2010 IOP Publishing
Journal of Physics: Conference Series **306** (2011) 012006 doi:10.1088/1742-6596/306/1/012006

The gravity-related decoherence master equation from hybrid dynamics

Lajos Diósi

Research Institute for Particle and Nuclear Physics, H-1525 Budapest 114, P.O.Box 49, Hungary

E-mail: `diosi@rmki.kfki.hu`

Abstract. Canonical coupling between classical and quantum systems cannot result in reversible equations, rather it leads to irreversible master equations. Coupling of quantized non-relativistic matter to gravity is illustrated by a simplistic example. The heuristic derivation yields the theory of gravity-related decoherence proposed longtime ago by Penrose and the author.

1. Introduction

There is at least one classical dynamic system whose quantization is still problematic, that is gravity. Experimental evidences up to now have not indicated that gravitation should be quantized. Methods of quantizing the classical equations of gravity have failed to become generally accepted. So, gravitation could happen to be classical. Classical dynamics of gravitation would couple to quantum dynamics of other fields. This mathematical problem is not at all trivial.

Hybrid quantum-classical systems are important not only for gravity but for molecular and nuclear systems as well where we might need to go beyond the mean-field approximation. Hybrid dynamics has obtained certain theoretical importance in foundations, in cosmology, in measurement problem. A very incomplete list of related works [1, 2, 3, 4, 5, 6, 7, 8, 9, 10, 11, 12] shows the diversity of motivations.

The particular goal of the present work is to re-derive a former heuristic gravity-related decoherence model. According to it, the Newton gravitational field ϕ possesses a universal uncertainty $\delta\phi$ which is modelled by a zero-mean white-noise with spatial correlation [13, 14]

$$\langle\delta\phi(r,t')\delta\phi(s,t)\rangle_{noise} = \frac{G\hbar}{|r-s|}\delta(t'-t)\ , \tag{1}$$

where G is the Newton constant. This noise causes a certain universal decoherence between 'macroscopically' different mass distributions. The following quantum master equation can be derived [14, 15] for the density matrix $\hat\rho_Q$ of the quantized system:

$$\frac{d}{dt}\hat\rho_Q = -\frac{i}{\hbar}[\hat H_Q+\hat H_G,\hat\rho_Q] - \frac{1}{2}\int_{r,s}\frac{G/\hbar}{|r-s|}[\hat f(r),[\hat f(s),\hat\rho_Q]\ , \tag{2}$$

where $\hat{H}_Q$ is the Hamiltonian, $\hat{H}_G$ is the standard Newton pair-potential, $\hat{f}(r)$ is the operator of spatial mass density. (We have omitted the integral volume elements dr, ds and we keep this shorthand notation throughout our work.) The decoherence mechanism encoded in Eqs. (1,2) coincides (upto a missing factor 1/2) with Penrose's proposal [16, 17]

$$\frac{1}{4}\int_{r,s}\frac{G/\hbar}{|r-s|}[f(r)-f'(r)][f(s)-f'(s)] \tag{3}$$

for the rate of decoherence between two different mass distributions f and f'.

Sec. 2 introduces the mathematical model of hybrid dynamics which will be applied to the matter-gravity quantum-classical hybrid system in Sec. 3 including a re-derivation of Eqs. (1,2).

2. Hybrid dynamics

We are going to discuss a possible mathematical model for hybrid composite systems consisting of one quantum and one classical subsystem.

2.1. Hybrid state, expectation values

The generic state of a hybrid system is the hybrid density

$$\hat{\rho} \equiv \hat{\rho}(q,p) \tag{4}$$

which is a non-negative matrix in the Hilbert space of the quantum subsystem, depending on the phase space point (q,p) of the classical subsystem. Its quantum marginal

$$\hat{\rho}_Q \equiv \int \hat{\rho}(q,p)dqdp \tag{5}$$

is the usual normalized density matrix of the quantum subsystem, while its classical marginal

$$\rho_C(q,p) \equiv \mathrm{tr}\hat{\rho}(q,p) \tag{6}$$

is the usual normalized Liouville density of the classical subsystem. (We mention the notion of conditional state $\hat{\rho}_Q|_{qp} \equiv \hat{\rho}(q,p)/\rho_C(q,p)$ of the quantum subsystem; conditional state of the classical subsystem wouldn't make sense in general.)

The interpretation of the hybrid state follows from the interpretation of density matrices and Liouville densities, and goes like this. Consider an arbitrary phase-point-dependent Hermitian matrix $\hat{A}(q,p)$, i.e.: a hybrid observable. Its expectation value in the hybrid state $\hat{\rho}$ is defined by

$$\langle \hat{A}(q,p)\rangle_{\hat{\rho}} \equiv \mathrm{tr}\int \hat{A}(q,p)\hat{\rho}(q,p)dqdp\,. \tag{7}$$

The expectation value has the usual statistical interpretation: if we measure the value of the hybrid observable $\hat{A}(q,p)$ repeatedly on a large ensemble of identically prepared hybrid systems characterized by the hybrid state $\hat{\rho}$, the average value of the measured outcomes will tend to the above calculated expectation value.

Once we have specified the general features of hybrid states, we can go on and learn about their dynamics.

2.2. Dirac, Poisson, Aleksandrov brackets

Let us begin with two independent systems, one is quantum and the other is classical. Our quantum system's state is described by the density matrix $\hat{\rho}_Q$. The evolution is governed by the von Neumann equation with Hamiltonian $\hat{H}_Q$:

$$\frac{d}{dt}\hat{\rho}_Q = -\frac{i}{\hbar}[\hat{H}_Q, \hat{\rho}_Q] \equiv -\frac{i}{\hbar}\left(\hat{H}_Q\hat{\rho}_Q - \hat{\rho}_Q\hat{H}_Q\right) . \tag{8}$$

For our classical system, we have chosen the classical canonical Liouville theory because of its best formal match with the quantum theory. The state of the classical system is described by the normalized phase space density $\rho_C = \rho_C(q,p)$. The evolution is governed by the Liouville equation with Hamilton function $H_C = H_C(q,p)$:

$$\frac{d}{dt}\rho_C = \{H_C, \rho_C\}_P \equiv \sum_n \left(\frac{\partial H_C}{\partial q_n}\frac{\partial \rho_C}{\partial p_n} - \frac{\partial \rho_C}{\partial q_n}\frac{\partial H_C}{\partial p_n}\right) , \tag{9}$$

where $q = (q_1, q_2, \ldots, q_n)$ and $p = (p_1, p_2, \ldots, p_n)$ stand for the canonical pairs of coordinates and momenta. The classical structure $\{\ ,\ \}_P$ is called Poisson bracket, the quantum structure $-(i/\hbar)[\ ,\ ]$ is called Dirac bracket.

Now we can take the above quantum and classical systems to construct a single composite hybrid system. We can easily form the hybrid composite state:

$$\hat{\rho} = \hat{\rho}(q,p) \equiv \hat{\rho}_Q \rho_C(q,p) , \tag{10}$$

as well as the hybrid Hamiltonian:

$$\hat{H} = \hat{H}(q,p) \equiv \hat{H}_Q + H_C(q,p) . \tag{11}$$

We write the hybrid evolution equation in the form

$$\frac{d}{dt}\hat{\rho} = -\frac{i}{\hbar}[\hat{H}, \hat{\rho}] + \{\hat{H}, \hat{\rho}\}_P , \tag{12}$$

which is equivalent with the previous evolution equations (8) and (9) for $\hat{\rho}_Q$ and ρ_C separately.

Of course, we would like to introduce interaction between the quantum and classical subsystems, i.e., we introduce the hybrid interaction Hamiltonian $\hat{H}_{QC} = \hat{H}_{QC}(q,p)$ which is a certain Hermitian matrix that also depends on the canonical coordinates of the classical subsystem. Now our total hybrid Hamiltonian reads

$$\hat{H} = \hat{H}(q,p) \equiv \hat{H}_Q + H_C(q,p) + \hat{H}_{QC}(q,p) , \tag{13}$$

and Aleksandrov [1] proposed the following evolution equation:

$$\frac{d}{dt}\hat{\rho} = -\frac{i}{\hbar}[\hat{H}, \hat{\rho}] + \mathrm{Herm}\{\hat{H}, \hat{\rho}\}_P . \tag{14}$$

The structure on the r.h.s. is called the Aleksandrov bracket, being the simplest generalization of the naive interaction-free structure (12).

The Aleksandrov equation (14) does not preserve the factorized form (10) of the hybrid state $\hat{\rho}(q,p)$, which is the legitimate consequence of the interaction. But a serious drawback arises. As pointed out by Boucher and Traschen [2], the positivity of $\hat{\rho}(q,p)$ may not be preserved, the Aleksandrov equation is fundamentally defective whereas it can be a useful approximation beyond the mean-field model.

2.3. Blurring Dirac+Poisson

However disappointing it is, the algebraic unification of classical and quantum dynamics runs into difficulties [2, 3, 4, 5]. Perhaps the deepest reason is that the notion of deterministic classical trajectories of one subsystem becomes lost under the influence of the other subsystem which is subject to quantum uncertainties. I proposed a possible remedy long time ago [3], another approach was shown together with Gisin and Strunz [6]; further mathematical structures of hybrid dynamics appear from time to time [7, 8, 9, 10, 11, 12]. A comparative analysis is missing. The present work follows the method of Ref. [3].

In the hybrid Hamiltonian (13), the interaction can always be written in the form

$$\hat{H}_{QC} = \sum_r \hat{J}_Q^r J_C^r \,, \tag{15}$$

where Ref. [3] called $\hat{J}_Q^r$ and J_C^r as 'currents' that build up the hybrid coupling. This time, with one eye on the forthcoming application in Sec. 3, we write the hybrid interaction this way:

$$\hat{H}_{QC} = \sum_r \hat{f}^r \phi^r \,. \tag{16}$$

Let us blur the interacting 'currents' by auxiliary classical noises δf^r and $\delta\phi^r$:

$$\hat{H}_{QC}^{noise} = \sum_r \left(\hat{f}^r + \delta f^r(t)\right)\left(\phi^r + \delta\phi^r(t)\right) \,. \tag{17}$$

Accordingly, we replace the hybrid Hamiltonian (13) by the noisy one:

$$\hat{H}^{noise} = \hat{H}_Q + H_C + \hat{H}_{QC}^{noise} \,, \tag{18}$$

and we replace the Aleksandrov equation (14) by its noise-averaged (blurred, coarse-grained) version:

$$\frac{d}{dt}\hat{\rho} = \left\langle -\frac{i}{\hbar}[\hat{H}^{noise}, \hat{\rho}] + \mathrm{Herm}\{\hat{H}^{noise}, \hat{\rho}\}_P \right\rangle_{noise} \,. \tag{19}$$

Suppose the noises δf and $\delta\phi$ are independent white-noises of zero mean, with the following auto-correlations:

$$\langle \delta f^r(t')\delta f^s(t)\rangle_{noise} = D_Q^{rs}\delta(t'-t) \,, \tag{20}$$

$$\langle \delta\phi^r(t')\delta\phi^s(t)\rangle_{noise} = D_C^{rs}\delta(t'-t) \,. \tag{21}$$

The two correlation matrices D_Q^{rs} and D_C^{rs} can be chosen in such a way that the noise-averaged Aleksandrov equation (19) guarantee the positivity of the coarse-grained hybrid density $\hat{\rho}$. The price to pay is the loss of reversibility: the mathematically correct hybrid equation will be a certain irreversible master (kinetic) equation.

2.4. Hybrid master equation

By hand, we added a noisy part (17) to the total hybrid Hamiltonian (13):

$$\hat{H}^{noise} = \hat{H} + \sum_r (\hat{f}^r \delta\phi^r + \phi^r \delta f^r) \,. \tag{22}$$

Compared to Eqs. (17,18), the term $\delta f \delta\phi$ has been omitted because it cancels from both the Dirac and Poisson brackets. In the evolution equation (19), the 1st order contribution of the noisy part is vanishing on average. In 2nd order, the term $\hat{f}\delta\phi$ adds the structure $-[\hat{f}, [\hat{f}, \hat{\rho}]]$ and the term $\phi\delta f$ adds the structure $\{\phi, \{\phi, \hat\rho\}_P\}_P$ to the r.h.s. of the naive Aleksandrov equation (14):

$$\frac{d}{dt}\hat\rho = -\frac{i}{\hbar}[\hat H, \hat\rho] + \mathrm{Herm}\{\hat H, \hat\rho\}_P - \frac{1}{2\hbar^2}\sum_{r,s} D_C^{rs}[\hat f^r, [\hat f^s, \hat\rho]] + \frac{1}{2}\sum_{r,s} D_Q^{rs}\{\phi^r, \{\phi^s, \hat\rho\}_P\}_P\,. \tag{23}$$

This hybrid master equation, i.e.: the noise-averaged Aleksandrov equation (19), preserves the positivity of $\hat\rho$ if the correlation matrices satisfy

$$D_Q D_C \geq \frac{\hbar^2}{4}\,. \tag{24}$$

The proof was given in Ref. [3] (it has remained largely unnoticed).

The first irreversible term on the r.h.s. of the hybrid master equation (23) is a typical decoherence term dephasing the superpositions between different configurations of the 'currents' f. The kernel D_C plays the role of decoherence matrix. The second irreversible term concerns the classical subsystem, yielding diffusion in its phase space.

3. Gravity-related decoherence

Hybrid dynamics will be applied to the interaction of quantized massive particles with classical gravitational field.

3.1. Hybrid master equation for matter plus gravity

The Hamiltonian $\hat H_Q$ will stand for the quantized non-relativistic matter. We'll need its mass density field operator $\hat f$, which should be the non-relativistic limit of $\hat T_{00}/c^2$, the 00 component of the Einstein energy-momentum tensor divided by c^2 (c is the speed of light). For point-like particles of mass m and position operator $\hat x$, the mass density would be

$$\hat f(r) = \sum m\delta(r - \hat x)\,, \tag{25}$$

where summation extends for all massive particles. In fact, a certain short-length cutoff is always understood, for details of this issue see Ref. [18, 19, 20] and references therein.

As to the classical gravitational field, we start with a scalar relativistic model and we take the non-relativistic limit $c \to \infty$ afterwards. We define the Newton potential $\phi \equiv \frac{1}{2}c^2(g_{00} - 1)$ where g_{00} is the 00 component of the Einstein metric tensor. The canonical coordinates q of Sec. 2 will be played by the Newton field $\phi(r)$, the conjugate canonical momenta p will be denoted by the field $\xi(r)$. The Hamilton function(al) can be chosen as

$$H_C(\phi, \xi) = \int_r \left(2\pi G c^2\, \xi^2 + \frac{|\nabla\phi|^2}{8\pi G}\right)\,. \tag{26}$$

To construct the hybrid interaction, we couple the quantized mass density $\hat f$ to the Newton field:

$$\hat H_{QC}(\phi) = \int_r \hat f(r)\phi(r)\,. \tag{27}$$

Note, in parentheses, that $\hat H_{QC}$ does not depend on the canonical momenta ξ. Looking back to notations in Sec. 2, we recognize the identity of this interaction with (16) provided we replace $\sum_r, \hat f^r, \phi^r$ by $\int_r, \hat f(r), \phi(r)$, respectively. Therefore the equations of Sec. 2 apply here conveniently. Let us form the total hybrid Hamiltonian (13):

$$\hat H(\phi, \xi) = \hat H_Q + H_C(\phi, \xi) + \hat H_{QC}(\phi)\,. \tag{28}$$

Write down the hybrid master equation (23) for the state $\hat\rho(\phi,\xi)$ of our hybrid system:

$$\frac{d}{dt}\hat\rho = -\frac{i}{\hbar}[\hat H,\hat\rho] + \mathrm{Herm}\{\hat H,\hat\rho\}_P \\ -\frac{1}{2\hbar^2}\int_{r,s} D_C(r,s)[\hat f(r),[\hat f(s),\hat\rho]] + \frac{1}{2}\int_{r,s} D_Q(r,s)\{\phi(r),\{\phi(s),\hat\rho\}_P\}_P\,. \tag{29}$$

For preserving the positivity of the state $\hat\rho$, the decoherence and diffusion kernels should satisfy the condition (24). Below, we propose a heuristic resolution of the remaining ambiguity as to the concrete form of D_Q and D_C.

3.2. *The gravity-related decoherence matrix*

In the Newtonian non-relativistic limit $c\to\infty$, the mean-field Poisson equation is valid. In our notations (7):

$$\Delta\left\langle\phi(r)\right\rangle_{\hat\rho} = 4\pi G\left\langle\hat f(r)\right\rangle_{\hat\rho}\,. \tag{30}$$

Let us recall the definitions (20) and (21) of the correlation matrices (kernels) D_Q and D_C, respectively. If we imposed the Poisson equation $\Delta\delta\phi = 4\pi G\delta f$ on the fluctuations $\delta f,\delta\phi$ in Eqs. (20,21), respectively, then the above two correlation kernels would become related by

$$\Delta\Delta' D_C(r,r') = (4\pi G)^2 D_Q(r,r')\,. \tag{31}$$

(We are aware of a conflict with the derivation of the hybrid master equation where we considered $\delta\phi$ and δf uncorrelated. This is conceptual problem, it remains harmless for our resulting hybrid master equation but it may become relevant for later developments.) From the inequality (24), the minimum blurring corresponds to $D_C D_Q = \hbar^2/4$. Hence the unique translation invariant solution reads

$$D_C(r,s) = \frac{G\hbar}{2}\frac{1}{|r-s|}\,. \tag{32}$$

This is the correlation (1) of universal gravitational fluctuations [13, 14] apart from the numeric factor 1/2.

3.3. *Reduced quantum master equation*

Let us invoke the expansion (9) of the Poisson bracket and calculate the full expansion of the r.h.s. of the hybrid master equation (29):

$$\frac{d}{dt}\hat\rho = -\frac{i}{\hbar}[\hat H_Q,\hat\rho] - \frac{i}{\hbar}\int_r \phi(r)[\hat f(r),\hat\rho] \\ -\int_r\left(4\pi Gc^2\xi(r)\frac{\delta\hat\rho}{\delta\phi(r)} + \frac{1}{4\pi G}\Delta\phi(r)\frac{\delta\hat\rho}{\delta\xi(r)}\right) + \mathrm{Herm}\int_r \hat f(r)\frac{\delta\hat\rho}{\delta\xi(r)} \\ -\frac{1}{2\hbar^2}\int_{r,s} D_C(r,s)[\hat f(r),[\hat f(s),\hat\rho]\,] + \frac{1}{2}\int_{r,s} D_Q(r,s)\frac{\delta^2\hat\rho}{\delta\xi(r)\delta\xi(s)}\,. \tag{33}$$

We are going to derive the reduced dynamics of the quantized matter. Its density matrix is the marginal (5) of the hybrid state:

$$\hat\rho_Q = \int\hat\rho(\phi,\xi)\mathcal{D}\phi\mathcal{D}\xi\,. \tag{34}$$

Let us try to obtain a closed master equation for $\hat\rho_Q$. Integrate both sides of (33) by $\mathcal{D}\phi$ and $\mathcal{D}\xi$. Most terms on the r.h.s. cancel and we are left with

$$\frac{d}{dt}\hat\rho_Q = -\frac{i}{\hbar}[\hat H_Q,\hat\rho_Q] - \frac{i}{\hbar}\int_r \phi(r)[\hat f(r),\hat\rho(\phi)]\mathcal{D}\phi - \frac{1}{2\hbar^2}\int_{r,s} D_C(r,s)[\hat f(r),[\hat f(s),\hat\rho]]\,, \tag{35}$$

where $\hat{\rho}(\phi) = \int \hat{\rho}(\phi, \xi)\mathcal{D}\xi$. Suppose the post-mean-field Ansatz:

$$\int \Delta\phi(r)\hat{\rho}(\phi)\mathcal{D}\phi = 4\pi G \,\mathrm{Herm}\hat{f}(r)\hat{\rho}_Q \,, \tag{36}$$

which is stronger than the standard mean-field equation (30). From this Ansatz, we express the functional integral on the r.h.s. of (35):

$$\int \phi(r)\hat{\rho}(\phi)\mathcal{D}\phi = -G\mathrm{Herm} \int_s \frac{\hat{f}(s)}{|r-s|}\hat{\rho}_Q \,. \tag{37}$$

Subsequently, the complete term in question becomes Hamiltonian:

$$-\frac{i}{\hbar}\int_r \phi(r)[\hat{f}(r), \hat{\rho}(\phi)]\mathcal{D}\phi = -\frac{i}{\hbar}[\hat{H}_G, \hat{\rho}_Q] \,, \tag{38}$$

where $\hat{H}_G$ is the well-known Newtonian pair-potential:

$$\hat{H}_G = -\frac{G}{2}\int_{r,s} \frac{\hat{f}(r)\hat{f}(s)}{|r-s|} \,. \tag{39}$$

The Eq. (35) has thus led to the following closed master equation for the evolution of the quantized matter:

$$\frac{d}{dt}\hat{\rho}_Q = -\frac{i}{\hbar}[\hat{H}_Q + \hat{H}_G, \hat{\rho}_Q] - \frac{1}{2\hbar^2}\int_{r,s} D_C(r,s)[\hat{f}(r), [\hat{f}(s), \hat{\rho}_Q] \,, \tag{40}$$

which, invoking D_C from (32), is the master equation (2) [14, 15] apart from an additional numeric factor 1/2 (cf. also Ref. [21]) in front of the decoherence term .

4. Concluding remarks

We have constructed a possible consistent hybrid dynamics coupling classical gravity to quantized non-relativistic matter. We were able to re-derive the former heuristic equations of gravity-related universal decoherence.

We admit a certain elusiveness of our procedure. In order to borrow self-dynamics to the Newtonian field ϕ, we attributed the status of relativistic scalar field to ϕ, with the non-relativistic limit to be taken on the hybrid master equation. On the other hand, the proposed hybrid master equation assumes the Markovian approximation since it contains classical white-noise fields. Coexistence of relativistic and white-noise fields may be an issue in itself. Therefore our hybrid master equation is heuristic rather than exact, and it remains so until, e.g., we construct some explicit solutions at least. The non-relativistic limit $c \to \infty$ was treated formally, without check for its existence. This is why we keep talking about the post-mean-field Ansatz rather than the post-mean-field equation despite the heuristic proof presented in the Appendix.

In any case, the formal re-derivation of the gravity-related decoherence master equation from a certain closed set of equations is reassuring and indicates the possible relevance of hybrid dynamics in foundations [22, 23].

Acknowledgments

This work was supported by the Hungarian Scientific Research Fund under Grant No. 75129.

Appendix

We ought to show that the post-mean-field Ansatz (36) is consistent with the non-relativistic limit $c \to \infty$ of the hybrid master equation (29). We begin with the heuristic derivation of the standard mean-field equation (30).

From the hybrid master equation (33) and the definition of expectation values (7), calculate the time-derivatives of the classical mean fields $\langle\phi\rangle_{\hat\rho}$ and $\langle\xi\rangle_{\hat\rho}$:

$$\frac{d}{dt}\langle\phi(r)\rangle_{\hat\rho} = 4\pi Gc^2\langle\xi(r)\rangle_{\hat\rho}\,, \tag{.1}$$

$$\frac{d}{dt}\langle\xi(r)\rangle_{\hat\rho} = -\frac{1}{4\pi G}\langle\Delta\phi(r) - 4\pi G\hat f(r)\rangle_{\hat\rho}\,. \tag{.2}$$

From the first equation wee see that $\langle\xi\rangle_{\hat\rho} \to 0$ in the non-relativistic limit. It also means that the l.h.s. of the second equation vanishes, hence we have:

$$\langle\Delta\phi(r) - 4\pi G\hat f(r)\rangle_{\hat\rho} = 0\,. \tag{.3}$$

This is just the standard mean-field equation (30).

Let us upgrade it into the post-mean-field Ansatz (36). To this end, we calculate the time-derivative of the expectation value of the simple hybrid quantity $\hat A\phi$ where $\hat A$ itself is an arbitrary constant Hermitian matrix:

$$\begin{aligned}\frac{d}{dt}\langle\hat A\phi(r)\rangle_{\hat\rho} &= 4\pi Gc^2\langle\hat A\xi(r)\rangle_{\hat\rho} + \frac{i}{\hbar}\Big\langle[\hat H_Q + \int_s\phi(s)\hat f(s), \hat A\phi(r)] - \frac{1}{2\hbar^2}\int_{r,s}D_C(u,s)[\hat f(u),[\hat f(s),\hat A\phi(r)]]\Big\rangle_{\hat\rho}\\ &= 4\pi Gc^2\langle\hat A\xi(r)\rangle_{\hat\rho} + \mathcal{O}(1)\,.\end{aligned} \tag{.4}$$

Here $\mathcal{O}(1)$ stands for terms that does not contain the diverging factor c^2. In the non-relativistic limit, the quantity $\langle\hat A\xi(r)\rangle_{\hat\rho}$ must vanish. Since $\hat A$ can be any Hermitian operator, the 'partial' expectation value, i.e.: the expression (7) without the trace and $\hat A$, does already vanish:

$$\int\xi(r)\hat\rho(\phi,\xi)\mathcal{D}\phi\mathcal{D}\xi = 0\,. \tag{.5}$$

In general, the hybrid dynamics (33) yields

$$\frac{d}{dt}\langle\hat A\phi(s)F(\xi)\rangle_{\hat\rho} = 4\pi Gc^2\langle\hat A\xi(r)F(\xi)\rangle_{\hat\rho} + \mathcal{O}(1)\,, \tag{.6}$$

where $F(\xi)$ can be a generic functional of ξ. Repeating the previous arguments, we conclude that the 'partial' expectation value of $\xi(r)F(\xi)$ must vanish for $c \to \infty$. Since $F[\xi]$ is an arbitrary functional, we come to the conclusion that, in the non-relativistic limit, $\xi(r)$ is identically zero. Now we derive the time-derivative of $\langle\hat A\xi(r)\rangle_{\hat\rho}$ from (33):

$$\begin{aligned}\frac{d}{dt}\langle\hat A\xi(r)\rangle_{\hat\rho} = &-\frac{1}{4\pi G}\langle\hat A\Delta\phi(r) - 4\pi G\mathrm{Herm}\hat A\hat f(r)\rangle_{\hat\rho}\\ &+\frac{i}{\hbar}\Big\langle[\hat H_Q + \int_s\phi(s)\hat f(s), \hat A\xi(r)] - \frac{1}{2}\int_{r,s}D_C(u,s)[\hat f(u),[\hat f(s),\hat A\xi(r)]]\Big\rangle_{\hat\rho}\,.\end{aligned} \tag{.7}$$

As we said, in the limit $c \to \infty$ we can set ξ to zero identically for all time and position therefore the above equation reduces to:

$$\langle\hat A\Delta\phi(r) - 4\pi G\mathrm{Herm}\hat A\hat f(r)\rangle_{\hat\rho} = 0\,. \tag{.8}$$

The l.h.s. vanishes for all $\hat A$, hence we can remove the trace and $\hat A$ from the hybrid expectation value, yielding:

$$\int\Big(\Delta\phi(r) - 4\pi G\mathrm{Herm}\hat f(r)\Big)\,\hat\rho(\phi,\xi)\mathcal{D}\phi\mathcal{D}\xi = 0\,. \tag{.9}$$

This is exactly the post-mean-field Ansatz (36) if we invoke the definition (5) of $\hat\rho_Q$.

References

[1] Aleksandrov I V 1981 *Z. Naturforsch.* **A 36** 902
[2] Boucher W and Traschen J 1988 *Phys. Rev.* **D 37** 3522
[3] Diósi L 1995 Quantum dynamics with two Planck constants and the semiclassical limit *Preprint* quant-ph/9503023
[4] Salcedo L L 1996 *Phys. Rev.* **A 54** 3657
[5] Caro J and Salcedo L L 1999 *Phys. Rev.* **A 60** 842
[6] Diósi L, Gisin N and Strunz W T 2000 *Phys. Rev.* **A 61** 22108
[7] Prezhdo O V and Kisil V V 1997 *Phys. Rev.* **A 56** 162
[8] Dias N C 2001 *J. Phys.: Math. Gen.* **A 34** 771
[9] Peres A and Terno D 2001 *Phys. Rev.* **A 63** 022101
[10] Sahoo D 2004 *J. Phys.: Math. Gen.***A 37** 997
[11] Kisil V V 2005 *Europhys. Lett.* **72** 873
[12] Reginatto M and Hall J W 2009 *J. Phys.: Conf. Ser.* **174** 012038
[13] Diósi L and Lukács B 1987 *Annln. Phys.* **44** 488
[14] Diósi L 1987 *Phys. Lett.* **A 120** 377
[15] Diósi L 1989 *Phys. Rev.* **A 40** 1165
[16] Penrose R 1994 *Shadows of the mind* (Oxford: Oxford University Press)
[17] Penrose R 1996 *Gen. Rel. Grav.* **28** 581
[18] Diósi L 2005 *Braz. J. Phys.* **35** 260
[19] Diósi L 2007 *J. Phys.: Math. Gen.* **A 40** 2989–2995
[20] Diósi L 2007 *J. Phys.: Conf. Ser.* **67** 012024
[21] Diósi L 2009 *J. Phys.: Conf. Ser.* **174** 012002
[22] Diósi L 2000 *Lect. Notes Phys.* **538** 243 (*Preprint* quant-ph/9902087)
[23] Diósi L 2000 *New Trends in Quantum Mechanics* ed H.D. Doebner, S.T. Ali, M. Keyl and R. Werner (Singapore: World Scientific) p 78 (*Preprint* quant-ph/9903088)

DNA waves and water

L Montagnier[1,2], J Aissa[2], E Del Giudice[3], C Lavallee[2], A Tedeschi[4], and G Vitiello[5]

[1] World Foundation for AIDS research and Prevention (UNESCO), Paris, France
[2] Nanectis Biotechnologies, S.A. 98 rue Albert Calmette, F78350 Jouy-en-Josas, France
[3] IIB, International Institute for Biophotonics, Neuss, Germany
[4] WHITE HB, Milano, Italy
[5] Dipartimento di Matematica e Informatica, Università di Salerno and
INFN, Gruppo Collegato Salerno, I-84100 Salerno, Italy

E-mail: vitiello@sa.infn.it

Abstract. Some bacterial and viral DNA sequences have been found to induce low frequency electromagnetic waves in high aqueous dilutions. This phenomenon appears to be triggered by the ambient electromagnetic background of very low frequency. We discuss this phenomenon in the framework of quantum field theory. A scheme able to account for the observations is proposed. The reported phenomenon could allow to develop highly sensitive detection systems for chronic bacterial and viral infections.

1. Introduction

Over the last 60 years, the development of basic knowledge in biology as well as many medical applications owes much to the discoveries made in DNA. Here is a partial list emphasizing the main advances in DNA discovery:

1944 Transformation of bacteria by DNA (O. Avery, C. McLeod, and M. McCarty)
1953 Double helix structure elucidated (J. Watson, F. Crick, M. Wilkins, R. Franklin)
1956 DNA polymerase (A. Kornberg)
1968 Restriction enzymes (W. Arber)
1969 Reverse transcription of retroviruses (H. Temin, D. Baltimore)
1976 DNA sequencing (A. Maxam, W. Gilbert, F. Sanger)
1986-1988 Polymerase chain reaction (K. Mullis) Taq polymerase (R.K. Saiki)
2001 First Human Genome Sequence
2004-2010 High-Throughput DNA Sequencing.

On the other hand, in the same times evidence has been accumulated on the influence of electromagnetic (em) fields on living organisms. The frequencies of the involved em fields cover different intervals corresponding to the different scales present in the organisms. In the present paper, by referring to recently published experimental results [1, 2, 3], we discuss the appearance of a new property of DNA correlated with the induction of extremely low frequency (ELF) em fields. These fields can be induced by suitable procedures in water dilutions which become able to propagate the information contained in the DNA of the original organisms to other ones.

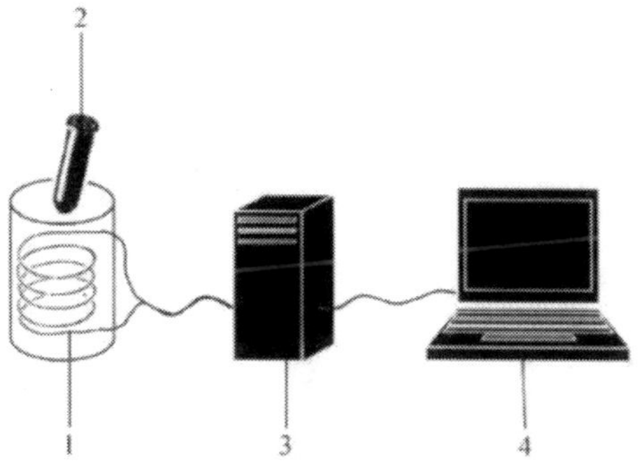

Figure 1. Device for the capture and analysis of em signals. (1) Coil made up of copper wire, impedance 300 Ohms. (2) Plastic stoppered tube containing 1 ml of the solution to be analyzed. (3) Amplifier. (4) Computer. From Ref. [1]

The paper includes three Sections: the new facts, a theoretical scheme where to discuss them and the medical applications.

2. The new facts: a new property of DNA and the induction of electromagnetic waves in water dilutions

The story started ten years ago when one of us (L.M.) studied the strange behaviour of a small bacterium, a frequent companion of HIV, *Mycoplasma pirum*, and like HIV a lover of human lymphocytes. L.M. was trying to separate the bacterium, which is about 300 *nm* in size, from viral particles whose size is about 120 *nm* by filtration using filters of 100 *nm* and 20 *nm*. Starting with pure cultures of the bacterium on lymphocytes, the filtrates were indeed sterile for the bacterium when cultured on a rich cellular medium, SP4. Polymerase chain reaction (PCR) and nested PCR, based on primers derived from a gene of *M. pirum* which had been previously cloned and sequenced, adhesin, were negative in the filtrate. However, when the filtrate was incubated with human lymphocytes, (previously controlled for not being infected with the mycoplasma) the mycoplasma with all its characteristics was regularly recovered! Then the question was raised: what kind of information was transmitted in the aqueous filtrate? It was the beginning of a long lasting investigation bearing on the physical properties of DNA in water. Indeed, a new property of *M. pirum* DNA was found: the emission of low frequency waves in some water dilutions of the filtrate, soon extended to other bacterial and viral DNAs.

The apparatus used to detect the electromagnetic signals comprises a solenoid capturing the magnetic component of the waves produced by the DNA solution in a plastic tube converting the signals into electric current. This current is then amplified and finally analyzed in a laptop computer using specific software (Fig. 1).

Here is a brief summary of the laboratory observations, which are described in more detail in [1, 2, 3]:

1) Ultra Low Frequency Electromagnetic Waves (ULF 500−3000 Hz) were detected in certain dilutions of filtrates (100 *nm*, 20 *nm*) from cultures of micro-organisms (virus, bacteria) or from the plasma of humans infected with the same agents (Fig. 2). Same results are obtained from their extracted DNA.

2) The electromagnetic signals (EMS) are not linearly correlated with the initial number of bacterial cells before their filtration. In one experiment the EMS were similar in a suspension of *E. coli* cells varying from 10^9 down to 10. It is an all or none phenomenon.

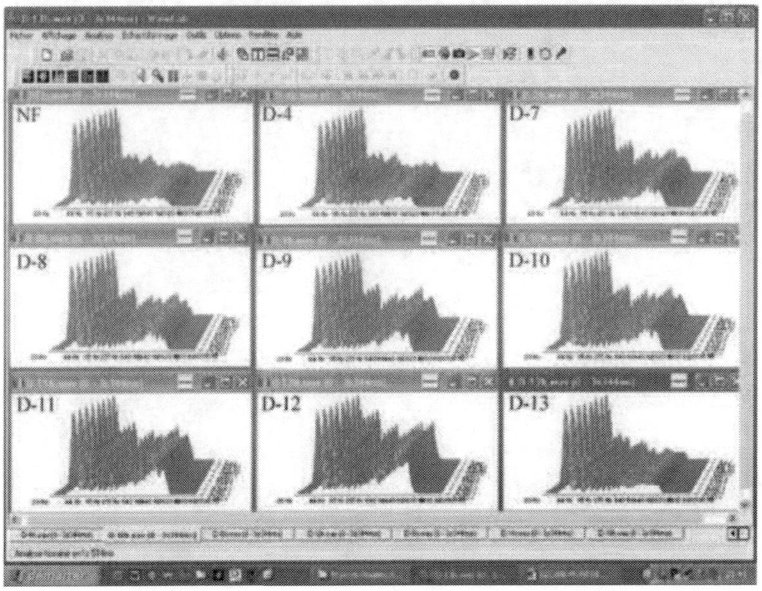

Figure 2. Typical signals from aqueous dilutions of *M. pirum* (Matlab software). Note the positive signals from D-7 to D-12 dilutions. From Ref. [1]

3) EMS are observed only in some high water dilutions of the filtrates. For example, from 10^{-9} to 10^{-18} dilutions in some preparations of *E. coli* filtrates.

4) In the case of *M. pirum*, an isolated single gene (adhesin, previously cloned and sequenced) could induce the EMS. As the gene was cloned in two fragments, each of the isolated fragments was able to generate EMS, suggesting that a short DNA sequence was sufficient to induce the signals. Similarly, a short HIV DNA sequence (104 base pair) is found to be sufficient to produce the EMS.

5) Some bacteria are not producing EMS: this is the case of probiotic bacteria such as *Lactobacillus* and also of some laboratory strains of *E. coli* used as cloning vector.

6) These studies have been extended to viruses, although not all the viral families have been explored. Similar EMS were detected from some exogenous retroviruses (HIV, FeLV), hepatitis viruses (HBV, HCV), and influenza A (in vitro cultures). In general, EMS are produced by 20 nm filtrates of viral suspensions or from the extracted DNA: a question remains for RNA viruses (HCV, influenza) as to whether the RNA from the mature viral particles is a source of EMS, or not. In the case of HIV, EMS are not produced by the RNA of viral particles, but rather are produced by the proviral DNA present in infected cells. In the case of bacteria, EMS are produced by 100 nm filtrates and not by 20 nm filtrates, indicating that the size of the structures producing EMS is ranging between 20 and 100 nm. This justifies the name of nanostructures. These studies are highly suggestive that one is dealing with nanostructures made of water. Highly purified water samples are used, although one cannot exclude the role of minimal traces of impurities. The EMS production by the nanostructures is resistant to: Rnase treatment, Dnase (while this will destroy the DNA at the origin of EMS), Protease (proteinase K), Detergent (SDS). However, they are sensitive to heat (over 70 $°C$) and freezing (-80 $°C$). This sensitivity is reduced when dealing with purified short DNA sequences. The technical conditions for EMS induction is summarized by the following list:

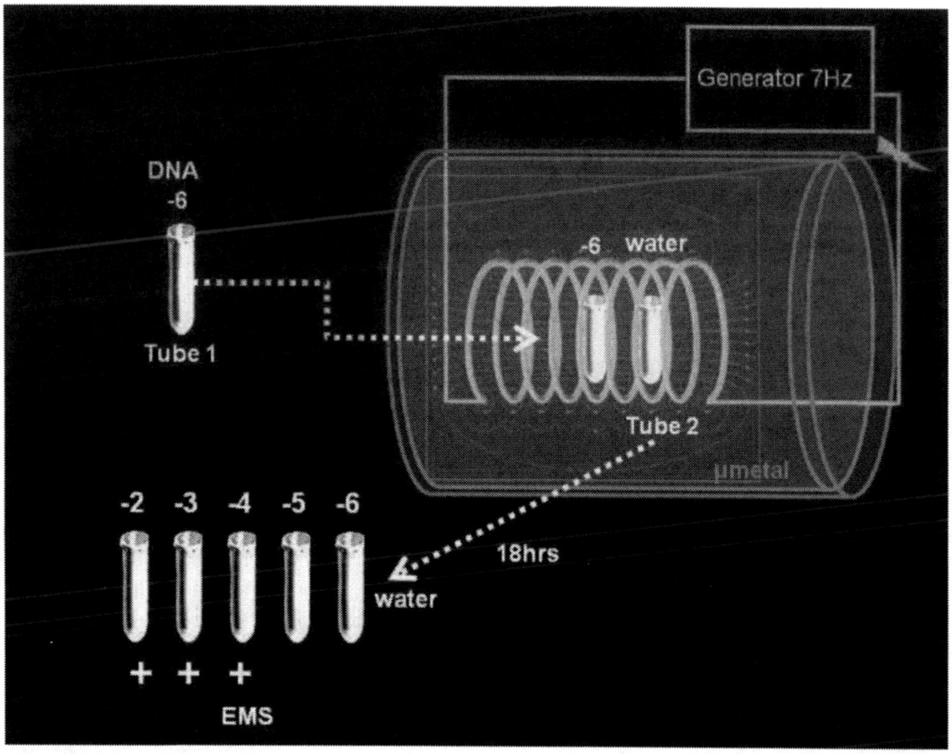

Figure 3. Transmission of DNA genetic information into water through electromagnetic waves. From Ref. [3]

- Filtration: $\frac{450}{100}$ nm for bacterial DNA, $\frac{450}{20}$ nm for viral DNA
- High dilutions in water
- Mechanical agitation (Vortex) between each dilution
- Excitation by the electromagnetic background of extremely low frequency (ELF), starting very low at 7 Hz. The excitation is not induced when the system is shielded by a mu-metal cage.

The stimulation by the electromagnetic background of very low frequency is essential. The background is either produced from natural sources (the Schumann resonances [4] which start at 7.83 Hz) or from artificial sources.

2.1. Transmission of DNA sequence through waves and water

In further experiments a fragment of HIV DNA taken from its long terminal repeat (LTR) has been used as the DNA source. This fragment was amplified by PCR (487 base pairs) and nested PCR (104 base pairs) using specific primers. In a first step, 10 fold serial DNA dilutions were made starting from a 2 ng/ml solution; the production of EMS under the ambient electromagnetic background was detected in dilutions 10^{-4} to 10^{-7}. Then the following steps were taken. As shown in Fig. 3, one of the positive dilutions (say 10^{-6}) was placed in a container shielded by 1 mm thick layer of mu-metal (an alloy absorbing ultralow frequency waves). In its vicinity another tube containing pure water was placed. The water content of each tube was filtered through 450 nm and 20 nm filters and diluted from 10^{-2} to 10^{-15}. A copper solenoid is placed around them and receives a low intensity electric current oscillating at 7 Hz, produced by an external generator. The produced magnetic field is maintained for 18 hours at room temperature. EMS are then recorded from each tube. Now also the tube containing water emits EMS, at the dilutions corresponding to those positive for EMS in the original DNA tube. This result shows that, upon 7 Hz excitation, the transmission into pure water of the oscillation of the nanostructures initially originated from DNA has been achieved. The following controls were found to suppress the EMS transmission in the water tube:

- Time of exposure of the two tubes less than $16 - 18$ hrs
- No coil
- Generator of magnetic field turned off
- Frequency of excitation $< 7\ Hz$
- Absence of DNA in tube 1.

At this point the most critical step was undertaken, namely to investigate the specificity of the induced water nanostructures by recreating from them the DNA sequence. For this all the ingredients to synthesize the DNA by polymerase chain reaction (nucleotides, primers, polymerase) were added to the tube of signalized water. The amplification was performed under classical conditions (35 cycles) in a thermocycler. The DNA produced was then submitted to electrophoresis in an agarose gel. The result was that a DNA band of the expected size of the original LTR fragment was detected. It was further verified that this DNA had a sequence identical or close to identical to the original DNA sequence of the LTR. In fact, it was 98 % identical (2 nucleotide difference) out of 104. This experiment was found to be highly reproducible (12 out of 12) and was also repeated with another DNA sequence from a bacterium, *Borrelia burgdorferi*, the agent of Lyme disease. It was shown clearly that the water nanostructures and their electromagnetic resonance can faithfully perpetuate DNA information. These elements give support to a provocative explanation of our *Mycoplasma pirum* filtration experiment (Fig. 1): the nanostructures induced by *M. pirum* DNA in the filtered water represent different segments of its genomic DNA. Each nanostructure when in contact with the human lymphocytes is retro-transcribed in the corresponding DNA by some cellular DNA polymerases. Then there is a certain probability (even very low) that each piece of DNA recombines within the same cell to other pieces for reconstructing the whole DNA genome. We have to assume that in presence of the eukaryotic cells the synthesis of the mycoplasma components (membrane lipids, ribosomes) can be also instructed by the mycoplasma DNA. One single complete mycoplasma cell is then sufficient to generate the whole infection of lymphocytes. Recent experiments of the C. Venter group have shown [5] that a synthetic genomic DNA is sufficient to maintain all the characteristics of a mycoplasma. All the steps assumed in the regeneration from water can be analyzed and open to verification.

3. The theoretical framework

The above experimental observations fit into the physical view which addresses biological dynamics as an interplay of chemical processes and em interactions; in other words, as an array of em assisted biochemical reactions.

We will try to interpret the above experimental results in the framework of a recently proposed theory of liquid water based on Quantum Field Theory (QFT) [6]-[11]. This theory is intrinsically non-linear and it provides the suitable tools to describe a complex ensemble of processes which are also non-linear.

Let us first summarize the main points of such a theory, whose details can be found in the quoted references.

The starting point is the realization that liquid water molecules cannot be assumed to be bound by purely static interactions (H-bonds, electric dipole-dipole interaction). Their binding is actually induced by the time-dependent radiative long range em field. Short range static bonds, such as H-bonds, then set in as a consequence of the molecule condensation induced by such long range radiative fields.

The main results of the theory are [6]-[11]:

a) An ensemble of molecules interacting with the radiative em field acquires, above a density threshold and below a critical temperature, a new non-trivial minimum energy state, different from the usual one where the oscillations of the molecules are uncorrelated and the em field is vanishing. The new minimum energy state implies a configuration of the system where all molecules enclosed within an extended region, denominated Coherence Domain (CD), oscillate in unison in tune with an em field trapped within the CD. The size of this extended region is just the wavelength λ of the trapped em field. The collective coherent oscillation of the molecules component the CD occurs between the individual molecule ground state and an excited state whose volume, according to atomic physics, is wider than the ground state volume. The wavelength λ of the trapped em field depends on the excitation energy E_{exc} through the equation:

$$\lambda = \frac{hc}{E_{exc}} \tag{1}$$

The CD is a self-produced cavity for the em field because of the well known Anderson–Higgs–Kibble mechanism [9] which implies that the photon of the trapped em field acquires an imaginary mass, becoming therefore unable to leave the CD. It is just this self-trapping of the em field that guarantees that the CD energy has a finite lower bound. Because of this self-trapping the frequency of the CD em field becomes much smaller than the frequency of the free field having the same wavelength. The above results apply to all liquids. The peculiarity of water is that the coherent oscillation occurs between the ground state and an excited state lying at 12.06 eV just below the ionization threshold (12.60 eV). In the case of liquid water, the CD (whose size is 100 nm according to Eq. (1)) includes an ensemble of almost free electrons which are able to accept externally supplied energy and transform it into coherent excitations (vortices) whose entropy is much lower than the entropy of the incoming energy. Consequently, water CDs could become dissipative structures in the sense of the thermodynamics of irreversible processes [12]-[14].

b) Coherence among molecules is counteracted at any non-vanishing temperature T by thermal collisions which could put molecules out of tune, like in the Landau picture of liquid Helium [15]. The competition between electrodynamic attraction and thermal noise produces a permanent crossover of molecules between a coherent regime and a non-coherent one. For a given value of T, the total number of coherent and non-coherent molecules are constant, but each molecule oscillates between the two regimes producing a continuous change of the space distribution of the coherent and non-coherent fractions of the molecules. It is just this flickering landscape of the two phases of liquid water to produce, in the case of experiments whose resolution time is long enough, the appearance of water as a homogeneous liquid. However, the above property holds only for bulk water. Near a surface, the attraction between water molecules and the surface could protect the coherent structure from the thermal noise, giving rise to a stabilization of the coherent structure. This is in particular the case of living organisms where water molecules are bound to membranes or biomolecule backbones. In this case CDs live long enough to exhibit the peculiar properties of coherence.

c) CDs store externally supplied energy in form of coherent vortices. These vortices are long lasting because of coherence, so that a permanent inflow of energy produces a pile up of vortices; they sum up to give rise to a unique vortex whose energy is the sum of the partial energies of the excitations which have been summed up. In this way water CDs can store a sizeable amount of energy in a unique coherent excitation able to activate molecular electron degrees of freedom; this high-grade energy is the sum of many small contributions, whose initial entropy was high.

d) CDs oscillate on a frequency common to the em field and the water molecules and this frequency changes when energy is stored in the CD. When the oscillation frequency of the CD matches the oscillation frequency of some non aqueous molecular species present on the CD boundaries, these "guest" molecules become members of the CD and are able to catch the whole stored energy, which becomes activation energy of the guest molecules; consequently, the CD gets discharged and a new cycle of oscillation could start. The above mechanism fits the intuition of Albert Szent-Gyorgyi [16] who proposed half a century ago that water surrounding biomolecules should be at the origin of the excitations of molecule electron levels responsible for chemical reactions. Moreover, should the ensemble of frequencies able to attract the component monomers of a polymer be excited in the water CD, the polymer would be created by the attraction of the monomers on the CD, provided that they be present in the solution. In this way it is possible to induce the polymerization of monomers by supplying to the water CDs of the monomer solution the em fields having the relevant frequencies (electromagnetic information).

e) A collective performance of water CDs, which could give rise to a biochemical activity synchronous in a mesoscopic region, should demand a uniform rate of energy loading for all the involved CDs. This requirement is satisfied by a mechanism which includes electrolyte ions, whose essential role in the biological dynamics is widely recognized. Ions close to water CDs are attracted by the em field trapped in the domains; so they are kept orbiting around the domain moving at a circular speed proportional to the so called cyclotron frequency ν_c :

$$\nu_c = \frac{1}{2\pi}\frac{q}{m}B \tag{2}$$

where q and m are the electric charge and the mass of the ion, respectively, and B is the magnetic field. Since DNA and also proteins are polyelectrolytes, they are surrounded by a cloud of positive counter-ions; ions having a cyclotron frequency in the interval between 1 and 100 Hz play an important role. It has been experimentally detected by M. Zhadin [17] and later by Zhadin and Giuliani [18] that by applying a magnetic field, having a frequency which matches the ion cyclotron frequency, on a system where ions are present, these ions are extracted from their orbits. This mechanism has been theoretically clarified in [19]. Due to the conservation of angular momentum, the extraction of ions from the cyclotron orbits produces a rotational motion of the quasi free electrons of the water CDs, which therefore become energetically excited [20]. In the case where the ion concentration could be assumed to be uniform in a mesoscopic region and the externally applied magnetic field has also a mesoscopic size, the amount of energy excitation could be assumed uniform in a region including a large number of water CDs which correspondingly are excited in a uniform way, thus ensuring the coherence among them. The persistence in time of such extremely low frequency magnetic fields guarantees a steady excitation of the water CDs and correspondingly of the biochemical activity catalyzed by them.

Let us analyze now the experimental results reported in Section 2 within the theoretical framework summarized above.

The role played by the background of low em frequency is understood by observing that in order to load energy in the water CDs, we need a resonant alternating magnetic field. In higher organisms, such as the humans, this field can be produced by the nervous system. Elementary organisms, such as bacteria, should use environmental fields. Good candidates are the Schumann modes of the geomagnetic field [4]. These modes are the stationary modes produced by the magnetic activity (lightnings or else) occurring in the shell whose boundaries are the surface of the Earth and the conductive ionosphere, which acts as a mirror wall for the wavelengths higher than several hundreds of meters. These stationary modes should have a frequency ν_s which in the ideal case is [4]

$$\nu_s(n) = \frac{c}{2\pi R}\sqrt{n(n+1)}\ , \tag{3}$$

where R is the radius of the Earth. The real Earth-ionosphere cavity is not an ideal one, so that real frequencies are a little bit lower than the values given by Eq. (3). Actually, peaks around 7.83, 14.3, 20.8, 27.3 and 33.8 Hz are experimentally found.

Consequently, in order to produce the energy loading of CDs, the biological system should select ion species having a $\frac{q}{m}$ ratio such that, given the local value of B in the organism, the value given by Eq. (2) fits with one of the Schumann resonances. The local value of B is expected to be not much different from the value of the Earth's magnetic field, which is in the order of 50 microtesla.

As explained in the point e above, ions are extracted from their orbits when a magnetic field B is applied so to match the ion cyclotron frequency. This is done indeed by the Schumann mode of 7.83 Hz of the geomagnetic field. Such an extraction of ions from their cyclotron orbits produces in turn, because of conservation of angular momentum, a counter-rotation of the plasma of quasi-free electrons in the CDs, whose frequency depends, of course, on the number of involved ions, namely on their concentration, which therefore is the only relevant variable. The phenomenon occurs in the same way on all the CDs of the system, whose number is irrelevant for this purpose, in agreement with the item 2 in Section 2.

As a result, the so induced rotation of the plasma of the quasi-free electrons in the CDs produces the observed EMS signal.

A confirmation that the observed frequencies are produced by magnetic activity is given by the circumstance that the excitation is prevented by mu-metal absorption.

Summarizing, the above analysis fits with the role that the frequency at 7.83 Hz plays in the observations (see Section 2).

It is interesting to observe that, should the cyclotron orbits around the water shell be saturated by an ion species which does not match the Schumann resonances, the activity of the biological system would be inhibited. This prediction is in agreement with facts since we know that there are ions promoting biological activity and ions inhibiting it. The above conclusion holds, of course, if the only em background is the natural one (Schumann modes) or an artificial em background with frequencies similar to the Schumann ones; should an artificial em background with a different frequency distribution be present, a reshuffling of the favorable and unfavorable ion species would occur. This feature could provide a rationale for the observed impact of ELF fields on the physiological activity.

We observe also that the dependence on the concentration is at the origin of the dependence of the signal frequencies on the aqueous dilution. Since angular momentum is quantized, a threshold in the dilution (ion concentration) value is naturally expected, as indeed observed. In more detail, suppose one is able to extract n ions per CD, then the plasma of N quasi-free electrons starts to rotate with a frequency much higher than the ion cyclotron frequency since electron mass is much smaller than the ion mass. A corresponding co-resonating field appears in the surroundings of the rotating CDs and could be at the origin of an extended coherence among CDs.

The CD frequency is decreased by the increase of aqueous dilution. The existence of a window of dilutions (see point 3 in Section 2) could be understood by presuming that the signal produced by the lower dilutions could have a frequency higher than the interval of the values detectable by the used instruments. Higher dilutions, on the contrary, could produce no signal because the ion concentration is decreased below the threshold able to excite the CDs.

We note that the coherence among the CDs in a region of mesoscopic extension, produced in the presence of the excitation energy in such a region, may sustain the non-dispersive propagation of the em radiative field. Here a peculiar role is played by the specific DNA structure and by the coherent water surrounding it. This is a point to be further analyzed on the basis of the specific chemical structure of the DNA fragments used in the experiments. Such an important investigation is in our future plans.

The field associated to the observed EMS, generated through the highly non-linear mechanism described above, may also trigger the coherence among the CDs in a second tube containing pure water where it is allowed to penetrate under the conditions set in the experiments described in subsection 2.1. This phenomenon is similar to the proximity effect observed in two superconducting samples or in the arrays of Josephson junctions, by which the samples or the junctions fall into a phase-locking regime. In the case under study, such a phase-locking regime manifests itself in the observed transmission of the excited em field of the water microstructures which surrounded the DNA in the first tube. Once the water CDs of the second tube are excited by the em field coming from the first tube, the DNA is fabricated according to the process suggested in the point d of the present section. The phase-locking at the specific frequency of the em field propagating in the original DNA tube is clearly reflected in the specificity of the induced water microstructures out of which the original DNA sequence (98 % identical) is recreated. On the other hand, the observed high reproducibility of this experiment finds its explanation in the high stability of the coherent structures (CDs and coherent clusters of CDs) into play. A series of experiments aimed at verifying the correctness of the described theoretical scheme have recently been undertaken.

Finally, we observe that, at the present stage of the theoretical analysis, our discussion can only lead us to qualitative agreement with the features observed in the experiments. For a quantitative fitting we need to introduce specific models in the frame of the general theoretical scheme depicted above. We leave this to a future work.

4. Conclusions and medical applications

In this paper we have described the experiments showing a new property of DNA and the induction of electromagnetic waves in water dilutions. We have briefly depicted the theoretical scheme which can explain qualitatively the features observed in these experiments.

We remark that it is possible to detect the same EMS from the plasma of patients suffering from various infections and chronic diseases. The plasma has to be kept fresh and unfrozen. If the plasma is frozen at $-70\ ^{\circ}C$ then one must extract the DNA in order to recover the signals. The DNA can be also extracted from tissue biopsies. The list of diseases in which EMS have been found (such as Alzheimer, Parkinson, Multiple Sclerosis, various neuropathies, chronic Lyme syndrome, rheumatoid arthritis) indicate clearly that their presence is not limited to diseases known to be of infectious origin: The fact that EMS have been found in diseases not known to be of infectious origin is intriguing, and leads us to seek bacterial or viral factors in these diseases. A special case is that of HIV. Signals have been regularly detected coming from HIV DNA sequences in the blood of patients treated by antiretroviral therapy and responding well to that treatment by the disappearance of viral RNA copies in the circulating blood. This would indicate that such DNA comes from a reservoir not accessible to the classical treatment, and not from viral particles circulating in the blood. Moreover, not only the plasma of the patients, but also the red blood cell fraction contains DNA emitting signals. This is intriguing, as the red cells do not contain any cellular DNA, and as the virus does not bind to erythrocyte membrane. The possibility that a third element is involved is under investigation. The hypothesis [2] has been put forward that the antiretroviral treatment, which includes reverse transcriptase inhibitors, is itself selecting for a new way of viral DNA replication involving one or several cellular DNA polymerases. As for the *M. pirum* DNA, it is suggested that the HIV DNA fragments and their nanostructures present in the blood may not originate from cell lysis but, on the contrary, represent pieces of definite size able to recombine in the appropriate recipient cells (lymphocytes) to form whole genomic DNA and finally regenerate infectious virus. Whatever is the origin of this DNA, its easy detection by electromagnetic signals may render it a unique biomarker for attacking the viral reservoir. Its existence in the blood by more classical PCR technology has also been confirmed [2]. One thus has a powerful tool to test new types of treatment aimed at eradicating HIV infection, which has never yet been achieved, by the treatments actually in use. This is particularly important for patients in countries in which HIV prevalence is very high, ($5 - 10$ % in a large part of sub-Saharan Africa). The process of launching clinical trials in West and South Africa to test new therapeutics is planned. Their efficacy will be monitored by this new test, together with the improvement of more classical parameters, evaluating the full restoration of the immune system. The objective here is eradication of HIV infection, so that it will not be necessary for patients to be treated for life by a combination of toxic and expensive drugs. Our work is interdisciplinary, involving biologists, physicists, and medical doctors. There are of course many unresolved questions raised by our findings, which deserve more work and more interactions. DNA signalling is stimulated by 7 Hz naturally occurring waves on earth. Waves produced by the human brain are also in the range of 7 Hz. These are interesting questions to be asked and possibly answered.

References

[1] Montagnier L, Aïssa J, Ferris S, Montagnier J-L and Lavallee C 2009 *Interdiscip. Sci. Comput. Life Sci.* **1** 81
[2] Montagnier L, Aïssa J, Lavallee C, Mbamy M, Varon J and Chenal H 2009 *Interdiscip. Sci. Comput. Life Sci.* **1** 245
[3] Montagnier L 2010 *Lindau Nobel Conference* www.lucmontagnierfoundation.org
[4] Nickolaenko A P and Hayakawa M 2002 *Resonances in the Earth-ionosphere Cavity* (Dordrecht-Boston-London: Kluwer Academic Publishers)
[5] Gibson D G *et al* 2010 *Science* **329** 52
[6] Preparata G 1995 QED *Coherence in Matter* (Singapore: World Scientific)
[7] Arani R, Bono I, Del Giudice E, Preparata G 1995 *Int. J. Mod. Phys. B* **9** 1813
[8] Del Giudice E, Preparata G and Vitiello G 1988 *Phys. Rev. Lett.* **61** 1085
[9] Del Giudice E and Vitiello G 2006 *Phys. Rev. A* **74** 022105
[10] Del Giudice E and Tedeschi A 2009 *Electr. Biol. Med.* **26** 48
[11] Del Giudice E, Spinetti P R and Tedeschi A 2010 *Water* **2** 566
[12] Del Giudice E, Pulselli R.M and Tiezzi E 2009 *Ecol. Model* **220** 1874
[13] Marchettini N, Del Giudice E, Voeikov V L and Tiezzi E 2010 *J. Theo. Bio.* **265** 511
[14] Voeikov V.L, Del Giudice E 2009 *Water Journal* **1** 52
[15] Landau L.D 1941 *J. Physics USSR (Moscow)* **5** 71
[16] Szent-Gyorgyi A 1957 *Bioenergetics* (New York, NY: Academic Press Inc)
[17] Zhadin M. N, Novikov V V, Barnes F S and Pergola N F 1998 *Bioelectromagn.* **19**(1) 41
[18] Zhadin M and Giuliani L 2006 *Electr. Biol. Med.* **25**(4) 269
[19] Del Giudice E, Fleischmann M, Preparata G and Talpo G 2002 *Bioelectromagnetics* **27** 522
[20] Del Giudice E and Giuliani L 2010 *Eur. J. of Oncol. Library* **5** 7

Journal of Physics: Conference Series **306** (2011) 012008 doi:10.1088/1742-6596/306/1/012008

Quantum mechanical aspects of cell microtubules: science fiction or realistic possibility?

Nick E Mavromatos

CERN - Theory Division, CH-1211 Geneva 23, Switzerland

On leave from: King's College London, Physics Department, Strand, London WC2R 2LS, UK

E-mail: nikolaos.mavromatos@kcl.ac.uk

Abstract. Recent experimental research with marine algae points towards quantum entanglement at ambient temperature, with correlations between essential biological units separated by distances as long as 20 Angströms. The associated decoherence times, due to environmental influences, are found to be of order 400 fs. This prompted some authors to connect such findings with the possibility of some kind of quantum computation taking place in these biological entities: within the decoherence time scales, the cell "quantum calculates" the optimal "path" along which energy and signal would be transported more efficiently. Prompted by these experimental results, in this talk I remind the audience of a related topic proposed several years ago in connection with the possible rôle of quantum mechanics and/or field theory on dissipation-free energy transfer in microtubules (MT), which constitute fundamental cell substructures. The basic assumption was to view the cell MT as quantum electrodynamical cavities, providing sufficient isolation *in vivo* to enable the formation of electric-dipole quantum coherent solitonic states across the tubulin dimer walls. Crucial to this, were argued to be the electromagnetic interactions of the dipole moments of the tubulin dimers with the dipole quanta in the ordered water interiors of the MT, that play the rôle of quantum coherent cavity modes. Quantum entanglement between tubulin dimers was argued to be possible, provided there exists sufficient isolation from other environmental cell effects. The model was based on certain ferroelectric aspects of MT. Subsequent experiments *in vitro* could not confirm ferroelectricity at room temperatures, however they provided experimental measurements of the induced electric dipole moments of the MT under the influence of external electric fields. Nevertheless, this does *not* demonstrate that *in vivo* MT are not ferroelectric materials. More refined experiments should be done. In the talk I review the model and the associated experimental tests so far and discuss future directions, especially in view of the algae photo-experiments.

1. Quantum Mechanics and Biology: fiction or fact?

It is a common perception that Quantum Mechanics (QM) pertains to the *small* (*micro*scopic) and *cold*, whilst classical physics affects the large (*macro*scopic) and complex systems, usually embedded in relatively *hot environments*.

Elementary particles are from the above point of view the best arena for studying quantum effects, and this has lead to important discoveries regarding the structure of our Universe at microscopic scales, of length size less than 10^{-18} m [1]. However, there are important examples from condensed matter physics where quantum effects manifest themselves at relatively *large distances* and/or *high temperatures*. A famous example is the superconductivity phenomenon [2], where formation of electron pairs exhibiting quantum coherence at macroscopic scales of order of a few thousand Angströms (*i.e.* at distances three orders of magnitude larger than the atomic scale) is responsible for electric-current transport virtually dissipation free. Subsequently, high-temperature superconductors, with critical temperatures for the onset of superconductivity up to 140 K (albeit with coherence length of order of a few Angströms), have also been discovered [3]. In ref. [5] the authors describe the *macroscopic* entanglement of two samples of Cs atoms (containing more than 10^{12} atoms) at *room temperature*. Quite recently, it has also been demonstrated experimentally that in certain polymer chains, and under certain circumstances, one may observe quantum phenomena associated with intrachain (but not interchain) coherent electronic energy transport at *room temperature* [4].

A natural question, therefore, which comes to one's mind is whether similar quantum phenomena may occur in biological systems, which are certainly complex, relatively large (compare to atomic physics scales) entities, living at room temperatures. In fact this is an old question, dating back to Schroedinger [6], who in his famous 1944 Book entitled "*What is life*", attempted to argue that certain aspects of living organisms, such as mutations (changes in the DNA sequence of a cell's genom or a virus), might not be explainable by classical physics but required quantum concepts, for instance quantum leaps.

Several years later, H. Fröhlich [7] have suggested that macroscopic quantum coherent phenomena may be responsible for dissipation-free energy and signal transfer in biological systems through coherent excitations in the microwave region of the spectrum due to nonlinear couplings of biomolecular dipoles. The frequency with which such coherent modes are 'pumped' in biological systems was conjectured to be of order

$$t_{\text{coherence Froehlich}} \sim 10^{-11} - 10^{-12}\ \text{s}\ . \tag{1}$$

which is known as Fröhlich's frequency.

Soon after, A.S. Davydov [8], proposed that *solitonic excitation states* may be responsible for dissipation-free energy transfer along the α-helix self-trapped amide in a fashion similar to superconductivity: there are two kinds of excitations in the α-helix: deformational oscillations in the α-helix lattice, giving rise to quantized excitations ("phonons"), and internal amide excitations. The resulting non-linear coupling between these two types of excitations is a Davydov soliton, which traps the vibrational energy of the α-helix and thus prevents its distortion (solitons are classical field theory configurations with finite energy).

In a rather different approach, F. Popp [9] suggested that studies of the statistics of counts of photons in ultra-weak bioluminescence in the visible region of the spectrum point towards the existence of a coherent component linked to the living state. There have also been attempts [10] to link the mechanisms for quantum coherence in Biology suggested by Fröhlich and Popp.

In the 1990's a suggestion on the rôle of quantum effects on brain functioning, and in particular on conscious perception, has been put forward by R. Penrose and S. Hameroff [11], who concentrated on the microtubules (MT) [12] of the brain cells. In particular, they noted that one may view the tubulin protein dimer units of the MT as a quantum two-state system, in coherent superposition. The model of [11] assumes, without proof, that sufficient environmental isolation occurs, so that the *in vivo* system of MT in the brain undergoes *self-collapse*, as a result of sufficient growth that allowed it to reach a particular threshold, namely a critical mass/energy, related to *quantum gravity* (orchestrated reduction method). This type of collapse should be distinguished from the standard *environmental decoherence* that physical quantum systems are subjected to [13]. In this way, the authors of [11] argue that decoherence times of order O(1 s), which is a typical time for conscious perception, may be achieved, thereby deducing that consciousness is associated with *quantum computations* in the mind.

Unfortunately, in my opinion, environmental decoherence, even for *in vivo* MT, cannot be ignored. In a series of works, which I will review below [14, 15], we have developed a *quantum electrodynamics cavity model* for MT, in which electromagnetic interactions between the electric dipole moments of the tubulin protein dimer units and the corresponding dipole quanta in the (thermally isolated) water interiors of the *in vivo* MT, are argued to be the dominant forces, inducing environmental entanglement and eventual decoherence [13] in at most $O(10^{-6}-10^{-7})$ s. Such times are much shorter than the required time scale for conscious perception, but have been argued to be sufficient for *dissipation-less* energy transfer and signal transduction along moderately long MT of length sizes of order $\mu m = 10^{-6}$ m. As I will discuss below, the basic underlying mechanism is the formation of appropriate *solitonic* dipole states along the protein dimer walls of the MT, which are reminiscent of the quantum coherent states in the Fröhlich-Davydov approach. We have also speculated [15] that under sufficient environmental isolation, which however is not clear if it can be achieved in *in vivo* MT systems, these coherent states may provide the basis for an operation of the MT as quantum logic and information teleporting gates. At any rate, our main concern in the above works was the search for, and modeling of, possible quantum effects in cell MT which may not be necessarily associated with conscious perception. In fact in this talk I will disentangle the latter from dissipation-free energy and signal transfer in biological matter, which I will restrict my attention to.

All the above are so far mere speculations. Until recently, there was no strong experimental evidence (if at all) to suggest that macroscopic quantum coherent phenomena might have something to do with the living matter. There were of course consistency checks with such assumptions, as is, for instance, the work of [16], which, by performing experiments on the brain of *Drosophila*, provided consistency checks for the cavity model of MT [14, 15], especially on its possible rôle for brain memory function. Nevertheless, despite the interesting and quite delicate nature of such experiments, one could not extract conclusive experimental evidence for quantum aspects of the brain.

To such skepticism, I would also like to add the fact that some theoretical estimates on the environmental decoherence time in brain microtubules, performed by Tegmark [17], in a model-independent way (which, however, as I will discuss below, is probably misleading), place the relevant decoherence MT time scales in the range

$$t_{\rm decoh\ MT\ estimate} \in 10^{-20} - 10^{-13}\ {\rm s}\ , \qquad (2)$$

depending on the specific environmental source. Such estimates made the author of [17] to suggest that there is no rôle of quantum physics in the functioning of the brain. For the purposes of our talk below, I notice that it is the upper limit in (2) that has been proposed in [17] as a conservative estimate on the characteristic decoherence time for MT, the main source of decoherence being assumed to be the Ca^{2+} ions in each of the 13 microtubular protofilaments. Although I would disagree with such estimates, for reasons to be explained below, nevertheless, I will point out later in the talk that, even if a MT decoherence time scale of order 10^{-13} s is realized in Nature, this still allows for quantum effects to play a significant rôle associated with optimization of efficient energy and signal transfer.

At any rate, because of such a distinct lack of experimental confirmation so far, many scientists believed that any claim on a significant rôle of quantum physics in biology constituted science *fiction*.

2. Recent experimental evidence for biological quantum entanglement?

The situation concerning the experimental demonstration of a concrete rôle of quantum physics on basic functions of living matter started changing in 2007, when research work on photosynthesis in plants [18] has presented rather convincing experimental evidence that light-absorbing molecules in some photosynthetic proteins capture and transfer energy according to *quantum-mechanical probability laws* instead of classical laws at temperatures up to 180 K.

Even more excitingly, in the beginning of this year, compelling experimental evidence on quantum effects on living matter at ambient temperatures was provided in ref. [19]. Using photo

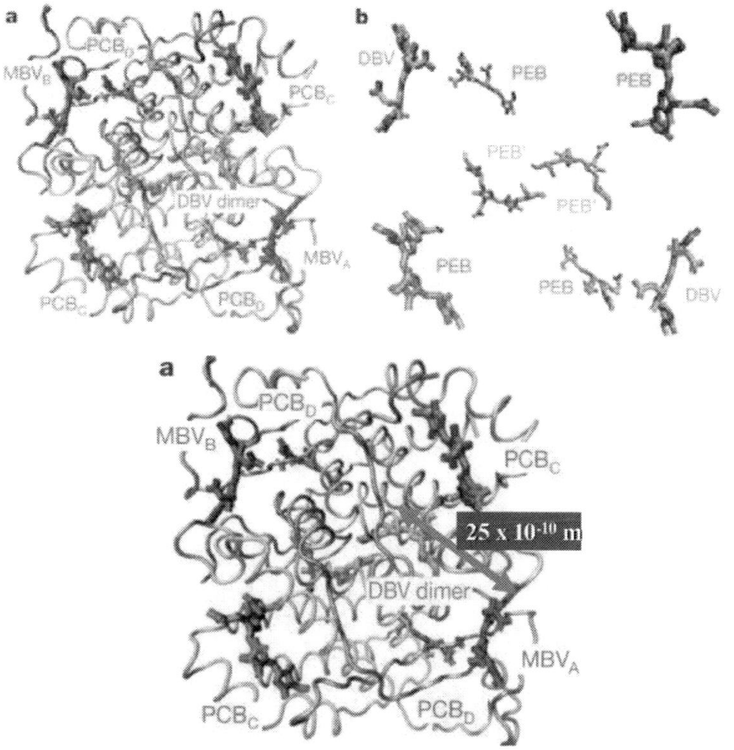

Figure 1. ***Top figures:*** **a.** Structural model of one type of *Cryptophytae Marine Algae* (CMA) protein antenna, PC645 [19]. The eight bilin molecules (Chromophores) responsible for light harvesting are indicated in various colours. **b.** The Chromophores from the structural model of the second type of CMA protein antenna studied in [19], PE545. ***Bottom figure:*** the same as in **a.** above, but with the alleged quantum-entanglement (coherent-wiring) distance of about 25 Angström between bilin molecules indicated by a red double arrow.

echo spectroscopy methods on two kinds of light-harvesting proteins, isolated appropriately from *cryptophyte marine algae*, the authors of [19] have demonstrated that there exist long-lasting electronic oscillation excitations with (quantum) correlations across the 5 nm long proteins, even at *room temperatures* of order 294 K.

More specifically, there are eight light-harvesting molecules (pigments-*Chromophores*, *i.e.* substances capable of changing colour when hit by light as a result of selective wavelength absorption) inside the protein antennae of marine algae (see fig. 1). The authors of [19] studied the electronic absorption spectrum of this complex system, and the results are indicated in fig. 2.

In the experiments, a laser pulse (indicated by a dashed line in fig. 2) of about 25 fs duration is applied to the biological entities, exciting a coherent superposition (in the form of a wave packet) of the protein antenna's vibrational-electronic eigenstates (the relevant absorption bands are indicated by coloured bars in fig. 2). The relevant theory, pertaining to the quantum evolution of a system of coupled bilin molecules with such initial conditions, predicts that the relevant excitation subsequently oscillates in time between the positions at which the excitation is localized, with distinct correlations and anti-correlations in phase and amplitude (*c.f.* fig. 3). Such coherent oscillations last until the natural eigenstates are restored due to *decoherence*, as a consequence of environmental entanglement [13]. The experimental results of [19] confirmed such a behaviour (*c.f.* fig. 4), thereby indicating a quantum superposition of the electronic structure

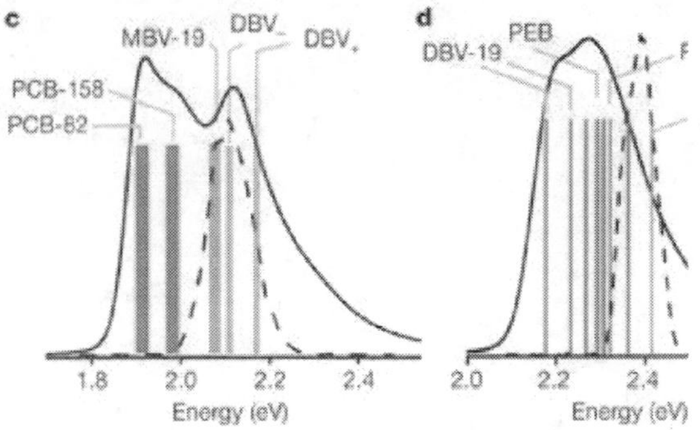

Figure 2. c. The approximate electronic absorption energies of the bilin molecules indicated in fig. 1 for the PC645 protein in aqueous buffer at ambient temperatures (294 K). **d.** The same but for the protein PE545, with the same external conditions (pictures taken from [19]). The externally applied laser pulse that excites the system is indicated by a dashed line. Coloured bars denote the absorption band positions.

of the bilin molecule dimer DBV (dihydrobiliverdin) for some time, which in the experiments was found to be relatively long, for a room temperature system, of order

$$t_{\rm decoh} = 400 \text{ fs} = 4 \cdot 10^{-13} \text{ s} . \tag{3}$$

The quantum oscillations of the DVB molecules were transmitted to the other bilin molecules in nthe complex, at distances 20 Angströms apart, as if these molecules were connected by springs. The authors of [19], therefore, suggested that distant molecules within the photosynthetic proteins are '*entangled*' together by *quantum coherence* ("coherently wired" is the used terminology) for more efficient light harvesting in marine algae. In other words, by exploiting such correlations, the biological cell '*quantum calculates*' – within the decoherence time scale (3) – which is the most efficient way and path to transport energy across macroscopically large distances of order of a few nm (*path optimization*). Some authors would interpret this behaviour as a prototype of 'quantum computation', although personally I believe we are rather far from rigorously demonstrating this.

There is an interesting question as to what guaranteed sufficient environmental isolation of the bilin molecules in the *cryptophytae* antenna studied in [19] so as to have such relatively long decoherence times (3) at room temperatures. The authors speculate that this might be due to the fact that, unlike most of photosynthetic pigments in Nature, the eight bilin molecules in this case are covalently bound to the proteins of the *cryptophytae* antenna complex.

These are, in my opinion, quite exciting experimental results that, for the first time, provide concrete evidence for quantum entanglement over relatively large distances in living matter at ambient temperature, and suggest a rather non-trivial rôle of quantum physics in path optimization for energy and information transport. Given that I have worked in the past on such issues regarding Cell Microtubules (MT), it is rather natural to revisit the pertinent theoretical models, in light of the findings of ref. [19]. It is the point of this talk, therefore, to first review the theoretical models of [14, 15] suggesting quantum coherent properties of MT, discuss the associated predictions as far as the scale of quantum entanglement and decoherence times are concerned, and attempt a comparison with the data of [19].

Caution should be exercised here. Algae light-harvesting antennae and MT are entirely different biological entities. Nevertheless, they are both highly complex protein bio-structures,

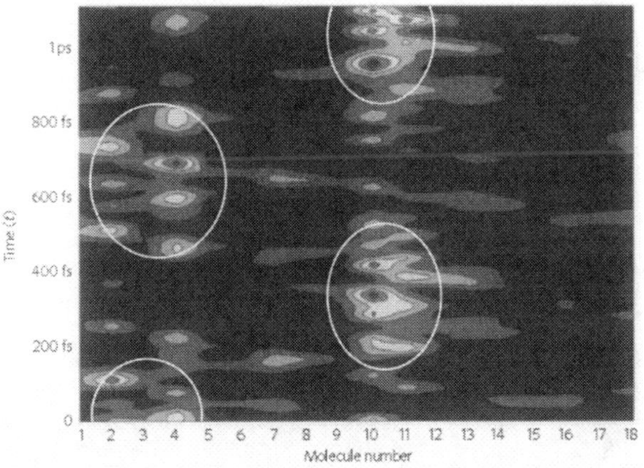

Figure 3. Theoretical Calculations of the electronic excitation dynamics in photosynthetic light-harvesting biological complexes [20]. The example is for the light-harvesting complex of the bacterium *Rhodopseudomonas acidophila*, which contains eighteen pigment molecules, arranged in a circle. Upon an external photo-stimulus, such as a laser pulse, the pigments enter excited electronic states by the absorption of photons, with coherent quantum correlations among the various pigments. The picture shows the probability of such an excitation to reside in a certain position in the complex (the dark-blue regions correspond to zero such probability, while the red-coloured regions indicate maximal probability). The vertical axis is time (in fs), while the horizontal line refers to the pigment molecules numbered 1 to 18. The white circles indicate the (small) number of molecules among which coherent oscillations of the excitation occur. In the model of [20] such oscillations have a half period of 350 fs. Similar oscillatory dynamics is observed in the algae complexes in [19] (*c.f.* fig. 4).

and the fact that quantum effects may play an essential rôle in energy transfer in algae at room temperature, as seems to be indicated by the recent experiments, might be a strong indication that similar coherent effects also characterize energy and signal transfer in MT *in vivo*, as conjectured in [14]. At this point, to avoid possible misunderstandings, I would like to stress once more that even if this turns out to be true, it may have no implications for conscious perception or in general brain functioning [21], although, of course, such exciting possibilities cannot be excluded.

The structure of the remainder of the talk is as follows: in the next section 3, I review the basic features of the MT cavity model and discuss its predictions, especially in the light of the recent findings of [19]. In section 4, I discuss some experimental tests of the model, especially as far as ferroelectric properties are concerned, which unfortunately are not conclusive. I also discuss avenues for future experiments that could confirm some other aspects of the model, including a possible extension of the photo-experiments of [19] to MT complexes. Finally, section 5 contains our conclusions.

3. Cavity Model for Microtubules (MT) revisited: quantum coherence and dissipation-free energy transfer in biological cells

Microtubules (MT) [12] are paracrystalline *cytoskeletal* structures that constitute the fundamental scaffolding of the cell. They play a fundamental rôle in the cell *mitosis* and are also believed to play an important rôle in the transfer of electric signals and, more general, of energy in the cell. They are cylindrical structures (*c.f.* fig. 5) with external cross section

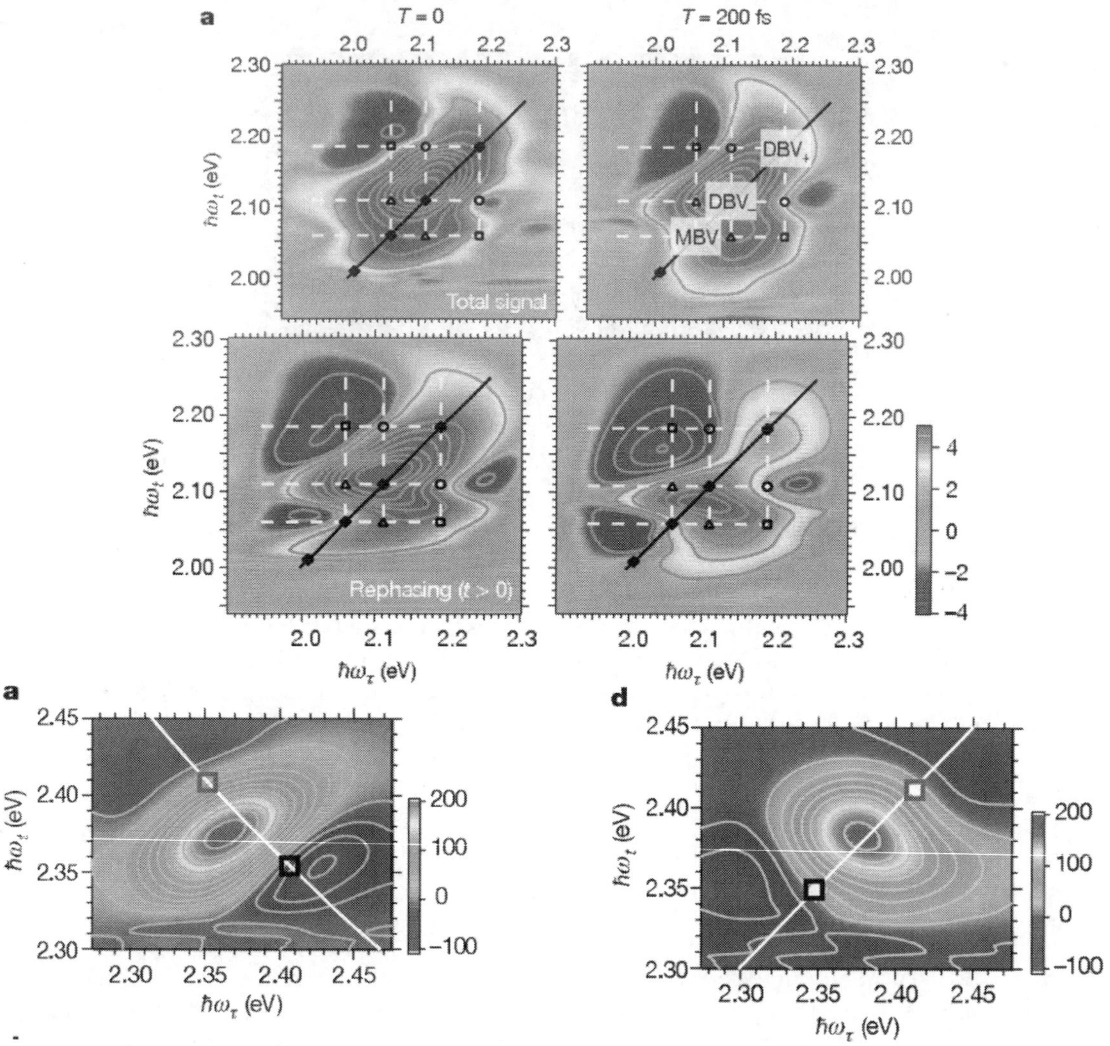

Figure 4. The two-dimensional photon echo (2DPE) data for PC645 protein of photosynthetic marine algae (upper picture) and for PE545 protein (lower picture). In the upper picture **a.**, the right column shows the data for a time of 200 fs since the excitation by the external pulse. The 2DPE data show the signal intensity in an arcsinh scale plotted as a function of the coherence frequency ω_τ and emission frequency ω_t. The lower picture shows data for times 100 fs, left picture **a.** involves re-phasing of the real signal, while in picture **d.** such a re-phasing has not been performed (pictures taken from [19]).

diameter of about 25 nm and internal diameter 15 nm. A moderately long MT may have a length of the order of a few μm $= 10^{-6}$ m. Their exterior walls consist of tubulin protein units (*c.f.* fig. 6). The tubulin protein dipers are characterized by two hydrophobic pockets, of length 4 nm $= 4 \cdot 10^{-9}$ m each (the total length of a dimer being about 8 nm), and they come in two conformations, $\alpha-$ and $\beta-$ tubulin, depending on the position of the unpaired charge of 18 e relative to the pockets.

The tubulin also has an electric dipole moment. A complete electron microscope chartography of the tubulin protein dimer is available today at 3.5 Angström resolution [23]. This allows for

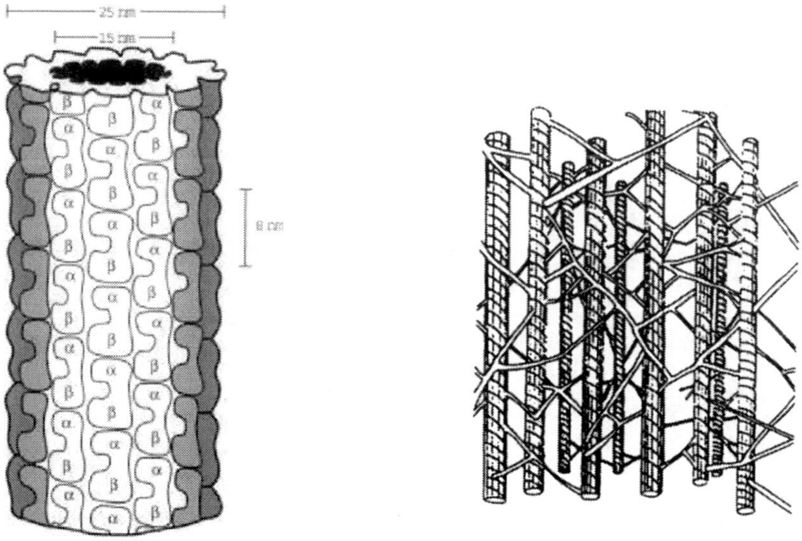

Figure 5. ***Left Picture***: A Microtubule arrangement in the cell. The exterior walls consist of tubulin protein dimers which are arranged in 13 protofilaments (vertical chain-like structures, parallel to the main axis of MT). The interiors are full of ordered water molecules. ***Right Picture:*** A network of Microtubules in a cell. The subtance connecting the various MTs together are the Microtubule Associated Proteins (MAP) (figures from ref. [22]).

theoretical modelling and computer calculations of the electric dipole moment of the dimers as well as of the entire MT [24, 25]. Current simulations have shown that the bulk of the tubulin's electric dipole moment lies on an axis *perpendicular* to the protofilament axis of the MT and only a fifth of the total tubulin dipole moment lies parallel to it (see also discussion in section 4). However, for the purposes of constructing a (rather simplified) model of MT dynamics [22, 14] that captures the essential features of dissipation-free energy and information transfer, it suffices to observe that the two conformations of the tubulin dimer differ by a relative angle (of about 29^0) relative to the protofilament axis in the monomers orientation. This will have implications on the electric dipole moment of the monomer, as indicated in fig. 6. In this simplified picture, one ignores the components of the electric dipole perpendicular to the protofilament's axis, and concentrates rather on a description of the array of the dipole oscillators along the MT protofilaments by a single effective degree of freedom, namely the projection, u_n, on the MT cylinder's axis of the displacement of the n-th tubulin monomer from its equilibrium position. The strong uniaxial dielectric anisotropy of the MT supports this picture, which enables one to view the MT as one-space dimensional crystals.

As we shall discuss below, this rather simplified geometry captures essential features of the MT, insofar as soliton formation and dissipation-free energy transfer are concerned. It is understood, though, that microscopic detailed simulations of the complete MT, which recently started becoming available [26], should eventually be used in order to improve the theoretical modelling of MT dynamics [22, 14] and allow for more accurate studies of their possible quantum entanglement aspects.

3.1. Classical solitons in MT and dissipation-free energy transfer

Based on such ingredients, the authors of [22] have attempted to discuss a *classical physics* model for dissipationless energy transfer across a MT, by conjecturing ferroelectric properties

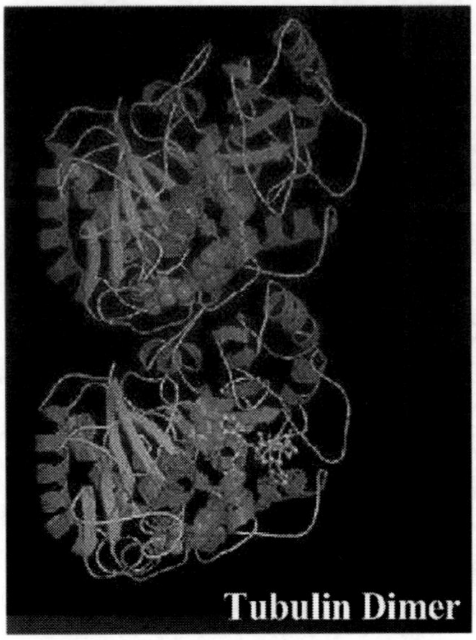

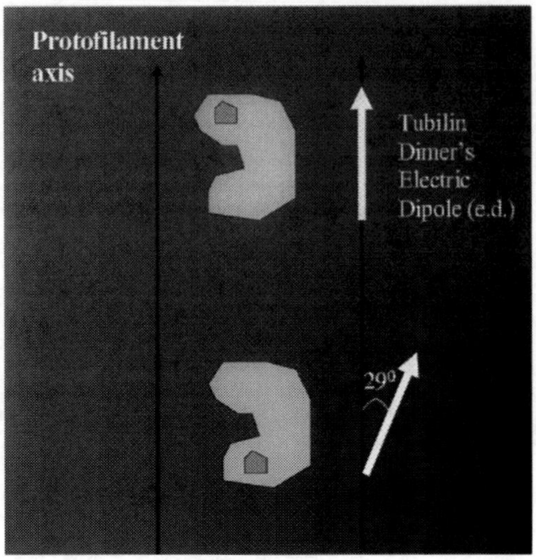

Figure 6. ***Left Picture***: Electron Microscopy chartography of tubulin dimer at 3.5 Angström resolution [23]. ***Right Picture:*** Schematic view of the two conformations α, β and the position (relative to the MT axis) of the electric dipole moment. The two conformations arise from the position of the unpaired electric charge (red polygon) relative to the two hydrophobic pockets of the tubulin dimer [12]. In simplified models, the two monomer conformations differ by a relative displacement of the monomer's electric-dipole orientation by an angle of about 29^0 relative to the protofilament axis. More complicated geometries for the permanent electric dipole moment of the tubulin dimer may occur in realistic systems, according to detailed simulations [24, 25], in which the bulk of it lies in a direction perpendicular to the main symmetry axis of a MT.

for MTs at room temperatures, and thus describing the essential dynamics by means of a lattice ferroelectric-ferrodistortive one-spatial dimensional chain model [27].

In the approach of [22], as mentioned previously, the relevant degree of freedom was the displacement vector $\vec{u}$ arising from the projections of the electric dipole moments of the $\alpha-$ and $\beta-$ tubulin conformations onto the protofilament axis. As a result of the inter and intra-chain interactions, this vector may well be approximated by a *continuously* interpolating variable, at a position x along a MT protofilament, which at time t has a value

$$u(x,t). \tag{4}$$

The time dependence is associated with the dipole oscillations in the dimers.

The relevant continuum Hamiltonian, obtained from the appropriate Lattice model [27] reads then [22]:

$$H = kR_0^2(\partial_x u)^2 - M(\partial_t u)^2 - \frac{A}{2}u^2 + \frac{B}{4}u^4 + qEu\ , \quad A = -|\text{const}|\,(T - T_c)\ , \tag{5}$$

where the critical temperature T_c for the on-set of ferroelectric order is assumed at room temperatures. k is a stiffness parameter, R_0 is the equilibrium lattice spacing between adjacent dimers and the term linear in u is due to the influence of an external electric field of intensity $\vec{E}$, assumed parallel to the protofilament axis (x axis for concreteness in the model we are discussing); q = 36e is the mobile charge (the reader should recall that in MT there is an

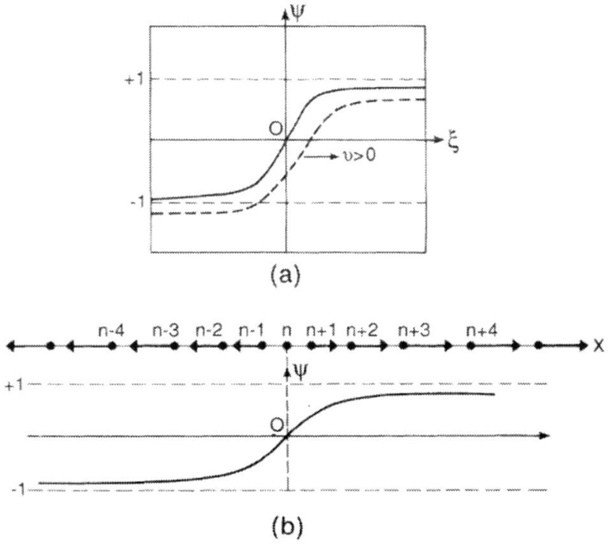

Figure 7. The kink soliton solution (8) of friction equation (9), along a protofilament (x-) axis of a MT, which is a travelling wave with velocity v along the x direction. The points n in the lower picture denote dimer positions in an one-dimensional lattice chain, representing the protofilament tubulin dimers (from ref. [22]).

unpaired charge 18 e in each dimer conformation) and M is the characteristic mass of the tubulin dimers. The simple double-well u^4 non-linear potential terms in (5) have been assumed in [22] to describe inter-protofilament interactions. In [14, 15] we have generalized such potential terms to arbitrary polynomial terms $V(u)$ of a certain degree.

The presence of ordered water in the MT interior (fig. 5) is approximated in this approach by the *addition* to the *equations of motion* derived from the Hamiltonian (5) of a friction term, linear in the time derivative of u, with a phenomenological coefficient γ:

$$M\frac{\partial^2 u(x,t)}{\partial t^2} - kR_0^2\frac{\partial^2 u(x,t)}{\partial x^2} - Au + Bu^3 + \gamma\frac{\partial u(x.t)}{\partial t} - qE = 0. \tag{6}$$

As is well known from mechanics, when friction is present, the standard lagrangian formalism breaks down, unless one enhances the degrees of freedom of the system to include the dynamics of the environment, in order to provide a microscopic dynamical description of the friction. We shall come to this point later on. For the moment, we take into account the presence of the ordered water environment and its effects on the dimers merely by the above-mentioned friction term in eq. (6). This friction should be viewed as an environmental effect, which however does not lead to energy dissipation, as a result of the formation of non-trivial solitonic structures in the ground-state of the system and the non-zero constant force due to the electric field. This is a well known result, directly relevant to energy transfer in biological systems [28]. Indeed, equation (6) admits as a unique bounded solution a *kink soliton*, acquiring the form of a travelling wave [22] (see fig. 7) [1]:

$$\psi(\xi) = a + \frac{b-a}{1+e^{\frac{b-a}{\sqrt{2}}\xi}}, \tag{8}$$

$$\psi \equiv \frac{u(\xi)}{\sqrt{A/B}}, \quad \xi = \alpha(x-ut), \quad \alpha \equiv \sqrt{\frac{|A|}{M(v_0^2-v^2)}},$$

with $\psi(\xi)$ satisfying the equation (primes denote derivatives with respect to ξ):

$$\psi'' + \rho\psi' - \psi^3 + \psi + \sigma = 0, \qquad \rho \equiv \gamma v[M|A|(v_0^2-v^2)]^{-\frac{1}{2}}, \qquad \sigma = q\sqrt{B}|A|^{-3/2}E, \tag{9}$$

and the parameters b, a and d are defined as:

$$(\psi-a)(\psi-b)(\psi-d) = \psi^3 - \psi - \left(\frac{q\sqrt{B}}{|A|^{3/2}}E\right). \tag{10}$$

The quantity

$$v_0 \equiv \sqrt{k/MR_0} \tag{11}$$

is the "sound" velocity, which is of order 1 km/s for the system at hand [22].

The kink (8) propagates along the protofilament axis with fixed velocity

$$v = v_0[1 + \frac{2\gamma^2}{9d^2M|A|}]^{-\frac{1}{2}}. \tag{12}$$

This velocity depends on the strength of the electric field E through the dependence of d on E via (10). Notice that, due to *friction*, $v \neq v_0$, and this is essential for a non-trivial second derivative term in (9), necessary for wave propagation. For realistic biological systems $v \simeq 2$ m/s (although under certain circumstances it may even be of order 20 m/s) [22]).

With a velocity of this order, the travelling waves of kink-like excitations of the displacement field $u(\xi)$ transfer energy along a moderately long microtubule of length $L = 10^{-6}m$ in about

$$t_T = 5 \times 10^{-7}\text{s}. \tag{13}$$

The total energy of the kink solution (8) is easily calculated to be:

$$E = \frac{1}{R_0}\int_{-\infty}^{+\infty} dx H = \frac{2\sqrt{2}}{3}\frac{A^2}{B} + \frac{\sqrt{2}}{3}k\frac{A}{B} + \frac{1}{2}M^*v^2 \equiv \Delta + \frac{1}{2}M^*v^2, \tag{14}$$

and is *conserved* in time. The 'effective' mass M^* of the kink is given by

$$M^* = \frac{4}{3\sqrt{2}}\frac{MA\alpha}{R_0B}. \tag{15}$$

[1] In [14, 15] we have generalized the double-well u^4 potential terms to arbitrary polynomial terms $V(u)$ of a certain degree. In the mathematical literature [29] there has been a classification of solutions of friction equations (9) with generalized potentials $V(u)$. For certain forms of the potential [14, 15] the solutions include *kink solitons* with the more general structure:

$$u(x,t) \sim c_1\left(\tanh[c_2(x-vt)] + c_3\right), \tag{7}$$

where c_1, c_2, c_3 are constants depending on the parameters of the dimer lattice model. For the double-well potential of [22], of course, the soliton solutions reduce to (8).

The first term of equation (14) expresses the binding energy of the kink and the second the resonant transfer energy. In realistic biological models the sum of these two terms, of order of 1 eV, dominate over the third term [22]. On the other hand, the effective mass in (15) is of order $5 \times 10^{-27} kg$, which is about the proton mass ($1GeV$) (!).

This amount of energy (14) is then transferred across a moderately long MT in a time scale (13), virtually *free* from *dissipation*.

The above *classical* kink-like excitations (8) have been discussed in [22] in connection with physical mechanisms associated with the hydrolysis of GTP (Guanosine-ThreePhosphate) tubulin dimers to GDP (Guanosine-DiPhosphate) ones.

However, because the two forms of tubulins correspond to different conformations α and β, it is conceivable to speculate that the *quantum mechanical oscillations* between these two forms of tubulin dimers might be associated with a *quantum version* of kink-like excitations in the MT network, in which the solution (8) is viewed as a *macroscopic coherent state*. In such a picture, this state would be itself the result of *incomplete environmental decoherence* [13]. Indeed, it is known that certain quantum systems, with specific Hamiltonian interactions with the environment, undergo *incomplete* localization of their quantum states, in the sense that localization stops before it is complete. In such cases, a quantum coherent ('*pointer*') state is formed as a result of decoherence [30]. This is argued to be the case in the so-called cavity model of MT proposed in [14], which we now come to review.

3.2. Quantization of the soliton solutions

Mathematically speaking, a semiclassical quantization of solitonic states of the form (8) (or more generally (7)), has been considered in [14] in a way independent of any microscopic model (for the water-induced friction).

To this end, one assumes the existence of a canonical second quantized formalism for the (1+1)-dimensional scalar field $u(x,t)$, based on creation and annihilation operators $a_k^\dagger$, a_k. One then constructs a squeezed vacuum state [31]

$$|\Psi(t)>= N(t)e^{T(t)}|0> \ ; \qquad T(t) = \frac{1}{2}\int\int dxdy u(x)\Omega(x,y,t)u(y) \ , \tag{16}$$

where $|0>$ is the ordinary vacuum state annihilated by a_k, and $N(t)$ is a normalization factor to be determined. $\Omega(x,y,t)$ is a complex function, which can be split in real and imaginary parts as

$$\begin{aligned} \Omega(x,y,t) &= \frac{1}{2}[G_0^{-1}(x,y) - G^{-1}(x,y,t)] + 2i\Pi(x,y,t) \\ G_0(x,y) &= <0|u(x)u(y)|0> \ . \end{aligned} \tag{17}$$

The squeezed coherent state for this system can be then defined as [31]

$$|\Phi(t)>\equiv e^{iS(t)}|\Psi(t)> \ ; \qquad S(t) = \int_{-\infty}^{+\infty} dx[D(x,t)u(x) - C(x,t)\pi(x)] \ , \tag{18}$$

with $\pi(x)$ the momentum conjugate to $u(x)$, and $D(x,t)$, $C(x,t)$ real functions. With respect to this state, $\Pi(x,t)$ can be considered as a momentum canonically conjugate to $G(x,y,t)$ in the following sense

$$<\Phi(t)| - i\frac{\delta}{\delta\Pi(x,y,t)}|\Phi(t)>= -G(x,y,t) \ . \tag{19}$$

The quantity $G(x,y,t)$ represents the *modified boson field* around the soliton. From the expression (18), one may also identify $C(x,t)$ with the dynamical field representing solitonic excitations, which in our case are the quantum-corrected solitons of the dipole configurations of the tubulin dimers in the MT wall,

$$u_q(x,t) = C(x,t) \ . \tag{20}$$

To determine the functions C, D and Ω one applies the *Time-Dependent Variational Approach* (TDVA) [31] according to which

$$\delta \int_{t_1}^{t_2} dt < \Phi(t)|(i\partial_t - H)|\Phi(t) >= 0 \ , \tag{21}$$

where H is the canonical Hamiltonian of the system. This leads to a canonical set of (quantum) Hamilton equations

$$\begin{aligned} \dot{D}(x,t) &= -\frac{\delta \mathcal{H}}{\delta C(x,t)} \ , \qquad \dot{C}(x,t) = \frac{\delta \mathcal{H}}{\delta D(x,t)} \ , \\ \dot{G}(x,y,t) &= \frac{\delta \mathcal{H}}{\delta \Pi(x,y,t)} \ , \qquad \dot{\Pi}(x,y,t) = \frac{\delta \mathcal{H}}{\delta G(x,y,t)} \ , \end{aligned} \tag{22}$$

where the quantum energy functional $\mathcal{H}$ is given by[31]

$$\mathcal{H} \equiv < \Phi(t)|H|\Phi(t) >= \int_{-\infty}^{\infty} dx \mathcal{E}(x) \ , \tag{23}$$

with

$$\begin{aligned} \mathcal{E}(x) = \frac{1}{2}D^2(x,t) &+ \frac{1}{2}(\partial_x C(x,t))^2 + \mathcal{M}^{(0)}[C(x,t)] + \\ +\frac{1}{8} < x|G^{-1}(t)|y > &+ 2 < x|\Pi(t)G(t)\Pi(t)|y > + \frac{1}{2} lim_{x \to y} \nabla_x \nabla_y < x|G(t)|y > - \\ -\frac{1}{8} < x|G_0^{-1}|y > &- \frac{1}{2} lim_{x \to y} \nabla_x \nabla_y < x|G_0(t)|y > \ . \end{aligned} \tag{24}$$

Above we used the following operator notation in coordinate representation $A(x,y,t) \equiv < x|A(t)|y >$, and

$$M^{(n)} = e^{\frac{1}{2}(G(x,x,t)-G_0(x,x))\frac{\partial^2}{\partial z^2}} U^{(n)}(z)|_{z=C(x,t)} \ ; \qquad U^{(n)} \equiv d^n U/dz^n \ . \tag{25}$$

The function U denotes the potential of the original soliton Hamiltonian, H. Notice that the quantum energy functional is conserved in time, despite the various time dependencies of the quantum fluctuations. This is a consequence of the canonical form (22) of the Hamilton equations.

Performing the functional derivations in (22) one obtains

$$\begin{aligned} \dot{D}(x,t) &= \frac{\partial^2}{\partial x^2}C(x,t) - \mathcal{M}^{(1)}[C(x,t)] \\ \dot{C}(x,t) &= D(x,t) \ , \end{aligned} \tag{26}$$

which after elimination of $D(x,t)$, yields a modified soliton equation for the (*quantum corrected*) field $u_q(x,t) = C(x,t)$ (*c.f.* (20)) [31]

$$\partial_t^2 u_q(x.t) - \partial_x^2 u_q(x,t) + \mathcal{M}^{(1)}[u_q(x,t)] = 0 \ , \tag{27}$$

with the notation

$$M^{(n)} = e^{\frac{1}{2}(G(x,x,t)-G_0(x,x))\frac{\partial^2}{\partial z^2}} U^{(n)}(z)|_{z=u_q(x,t)} \ , \quad \text{and} \quad U^{(n)} \equiv d^n U/dz^n \ .$$

The quantities $M^{(n)}$ carry information about the quantum corrections. For the kink soliton (7) the quantum corrections (27) have been calculated explicitly in ref. [31], thereby providing us with a *concrete example* of a large-scale quantum coherent *squeezed* state. The whole scheme may be thought of as a *mean-field-approach* to quantum corrections to the soliton solutions.

Having established these facts, it is interesting to attempt to formulate a microscopic model for MT, leading to the above-described solitonic coherent states. Such a model has been proposed in [14] and elaborated further in [15], where its quantum information processing aspects have been discussed. The basic ingredient of the model consists of viewing the MT as *quantum electrodynamical cavities*. We next proceed to review briefly the model's basic features.

3.3. The Quantum Cavity Model of MT

According to the Quantum Cavity Model of MT, the two conformations α and β, which, as mentioned above, differ by the positions of the unpaired electrons relative to the two hydrophobic pockets of the dimer (fig. 6), are viewed as two different states of a two-state quantum system, excitable by an external stimulus. For instance, applying an external pulse, or an electric field, excites a quantum superposition of these two states, and there are quantum oscillations between the two dimer configurations, which damp out after a finite time, as a consequence of environmental decoherence. This aspect is shared of course with the starting point of the Hemeroff-Penrose model for quantum MT [11]. However the mechanisms underlying the formation of the quantum coherent states and their eventual environmental decoherence are entirely different, as we now proceed to discuss.

In this scenario, Tubulin is viewed as a typical *two-state* quantum mechanical system, where the dimers couple to conformational changes with $10^{-9} - 10^{-11}$s transitions, corresponding to an angular frequency $\omega \sim \mathcal{O}(10^{10}) - \mathcal{O}(10^{12})$ Hz. In [14] we assumed the upper bound of this frequency range to represent (in order of magnitude) the characteristic frequency of the dimers, viewed as a two-state quantum-mechanical system:

$$\omega_0 \sim \mathcal{O}(10^{12}) \text{ Hz} \ . \tag{28}$$

In the quantum Cavity Model, the MT are viewed as *quantum electrodynamical cavities*. In fact the *thermally isolated cavity regions* are thin interior regions of thickness of order of a few Angströms *near* the *dimer* walls, in which the electric dipole-dipole interactions between the dimers and the ordered water molecules overcome thermal losses, and provide the necessary conditions for quantum coherent states to be formed along the tubulin dimer walls of the MT [2].

Indeed, if we consider two electric dipole vectors $\vec{d_i}$, $\vec{d_j}$, at locations i and j at a relative distance r_{ij}, one pertaining to a water molecule, and the other to a protein dimer in the MT chain, then the dipole-dipole interaction has the form:

$$E_{dd} \sim -\frac{1}{4\pi\varepsilon}\frac{3(\hat{\eta}.\vec{d_i})(\hat{\eta}.\vec{d_j}) - \vec{d_i}.\vec{d_j}}{|r_{ij}|^3} \ , \tag{29}$$

[2] At this point we would like to point out the following. It is known experimentally [32], that in a thin exterior neighborhood of MT there are areas of atomic thickness, consisting of charged ions, which isolate the MT from thermal losses. This means that the electrostatic interactions overcome thermal agitations. It seems theoretically plausible, albeit yet unverified, that such thermally isolated exterior areas can also operate as *cavity regions*, in a manner similar to the areas interior to MT. At this point it is unclear whether there exist the necessary coherent dipole quanta in the ionic areas. Further experimental and theoretical (simulational) work needs to be done regarding this issue.

where $\hat{\eta}$ is a unit vector in the direction of $\vec{r}_{ij}$, and ε is the dielectric constant of the medium.

First of all we note that, since each dimer has a mobile charge [12]: $q = 18 \times 2e$, e the electron charge, one may *estimate* the electric dipole moment of the dimer roughly as

$$d_{\rm dimer} \sim 36 \times \frac{\varepsilon_0}{\varepsilon} \times 1.6 \times 10^{-19} \times 4 \times 10^{-9}\ {\rm C\,m} \sim 3 \times 10^{-18}\ {\rm C\,m} = 90\ {\rm Debye}\ . \qquad (30)$$

where we used the fact that a typical distance for the estimate of the electric dipole moment for the 'atomic' transition between the α, β conformations is of $\mathcal{O}(4\ \rm nm)$, *i.e.* of order of the distance between the two hydrophobic dimer pockets. We also took account of the fact that, as a result of the water environment, the electric charge of the dimers appears to be screened by the relative dielectric constant of the water, $\varepsilon/\varepsilon_0 \sim 80$. We note, however, that the biological environment of the unpaired electric charges in the dimer may lead to further suppression of d_{dimer} in (30) [3].

Assuming that the medium between the ordered-water molecules in the layer and the MT dimers corresponds exclusively to the tubulin protein, with a typical value of dielectric constant $\varepsilon \sim 10$ [33], and taking into account the generic (conjectural) values of the electric dipole moments for tubulin dimers (30) and water molecules [14] (see discussion below eq. (31)), we easily conclude that the dipole-dipole interactions (29) may overcome thermal losses at room temperatures, $\sim k_B T$, for distances $|r_{ij}|$ of up to a few tenths of an Angström. Notice that for such distances the respective order of the energies is $\mathcal{O}(10^{-2}\ \rm eV)$. Such thin cavities may not be sufficient to sustain quantum coherent modes for sufficiently long times necessary for dissipation-less energy transfer along the MT.

However, isolation from thermal losses could be assisted enormously by the existence of a *ferroelectric* transition below some critical temperature T_c, for the system of protein dimers in MT. In the models of [22] and ours [14] this is precisely what happens, with the critical temperature for the onset of ferroelectricity near ambient temperature

Ferroelectricity implies an effective dielectric "constant" $\varepsilon(\omega) < 1$ in (29); in such a case, these interactions can overcome thermal losses at room temperatures for up to *a few* Angströms. An additional possibility would be that for this range of frequencies a *negative* (dynamical) dielectric constant arises. This would mean that the dimer walls become *opaque* for the modes in a certain range of frequencies [14] below some critical value, thereby providing concrete support to the idea of MT behaving as isolated 'cavities', by trapping such modes. As explained in [14], such frequencies occur naturally within the range of frequencies of our model.

Inside such thin interior cavity regions, there are *quantum coherent modes*, which in [14] have been argued to be the *dipole quanta*, conjectured to exist in water in ref. [34], and discussed in this conference. These coherent modes arise from the interaction of the electric dipole moments of the water molecules with the quantized radiation of the electromagnetic field [34], which may be self-generated in the case of MT arrangements [22, 14]. The corresponding Hamiltonian interaction terms are of the form [34]

$$H_{ow} = \sum_{j=1}^{M} [\frac{1}{2I} L_j^2 + \vec{A} \cdot \vec{d}_{ej}]\ , \qquad (31)$$

where $\vec{A}$ is the quantized electromagnetic field in the radiation gauge, M is the number of water molecules, L_j is the total angular momentum vector of a single molecule, I is the associated (average) moment of inertia, and d_{ej} is the electric dipole vector of a single molecule,

[3] The reader is referred to section 4 below for further discussion on detailed simulations of the permanent electric dipole moment of the tubulin dimer. The order of magnitude of our estimates below, however, is not affected by such detailed considerations.

$|d_{ej}| \sim 2e \otimes d_e$, with $d_e \sim 0.2$ Angström. As a result of the dipole-radiation interaction in (31) coherent modes emerge, which in [34] have been interpreted as arising from the quantization of the Goldstone modes responsible for the *spontaneous breaking* of the electric dipole (rotational) symmetry. Such modes are termed '*electric dipole quanta*' (EDQ). It is our view that such modes do not characterize ordinary water, but may well arise in the ordered water in the MT interior, which is a different phase of water. Such quanta play a rôle similar to the coherent modes of quantized electromagnetic radiation in ordinary cavities.

In our cavity model for MT [14] such coherent modes play the role of '*cavity modes*' in the quantum optics terminology [35]. These in turn interact with the dimer structures, mainly through the unpaired electrons of the dimers, leading to the formation of a quantum coherent solitonic state that may extend over the entire MT, and under sufficient isolation even over the entire MT network. As mentioned above, such states may be identified [14] with semi-classical solutions of the friction equations (27). In the model of [14], such coherent, *almost classical*, states are viewed as the result of *incomplete decoherence* of the dimer system due to its interaction/coupling with the water environment [13]. Incomplete decoherence may characterize some systems, in the sense that environmentally induced decoherence stops before complete localization of the quantum state. If this happens, then such partial decoherence time scales could be identified with the time taken for the dimer solitons to form.

In [14] estimates for the *formation time* of such solitons have been given, using conformal field theory methods for the description of the dynamics of the dipoles along the (one-dimensional) protofilaments, represented as Ising spin chains. Admittedly, these are crude approximations, and thus the so-obtained formation time estimates may not be accurate. For completebess we mention that such formation times of the solitons along the dimer walls may be of order 10^{-10} s, although smaller times cannot be excluded. This is the time scale over which solitonic coherent pointer states in the MT dimer system are formed ('*pumped*'), according to our scenario, which is not far from the originally assumed Fröhlich's coherence time scale (1). Eventually, these soliton coherent states will decohere to purely classical configurations. The decoherence scales involved in this second stage can be estimated by applying standard cavity quantum optics considerations to the MT system (see fig. 8).

Indeed, the above-mentioned dimer/water coupling leads to a situation analogous to that of atoms in interaction with coherent modes of the electromagnetic radiation in *quantum optical cavities*. An important phenomenon characterizes such interactions, namely the so-called *Vacuum-Field Rabi Splitting* (VFRS) effect [36]. VFRS appears in both the emission and absorption spectra of atoms [37]. For our purposes below, we shall review the phenomenon by restricting ourselves for definiteness to the absorption spectra case.

Consider a collection of N atoms of characteristic frequency ω_0 inside an electromagnetic cavity. Injecting a pulse of frequency Ω into the cavity causes a doublet structure (splitting) in the absorption spectrum of the atom-cavity system with peaks at [36]:

$$\Omega = \omega_0 - \Delta/2 \pm \frac{1}{2}(\Delta^2 + 4N\lambda^2)^{1/2} , \tag{32}$$

where $\Delta = \omega_c - \omega_0$ is the detuning of the cavity mode, of frequency ω_c, compared to the atomic frequency. For resonant cavities the splitting occurs with equal weights

$$\Omega = \omega_0 \pm \lambda\sqrt{N} . \tag{33}$$

Notice here the *enhancement* of the effect for multi-atom systems $N >> 1$. The quantity $2\lambda\sqrt{N}$ is called the 'Rabi frequency' [36]. From the emission-spectrum analysis an estimate of λ can be inferred which involves the matrix element, $\vec{d}$, of atomic electric dipole between the energy states of the two-level atom [36]:

$$\lambda = \frac{E_c \vec{d}.\vec{\epsilon}}{\hbar} , \tag{34}$$

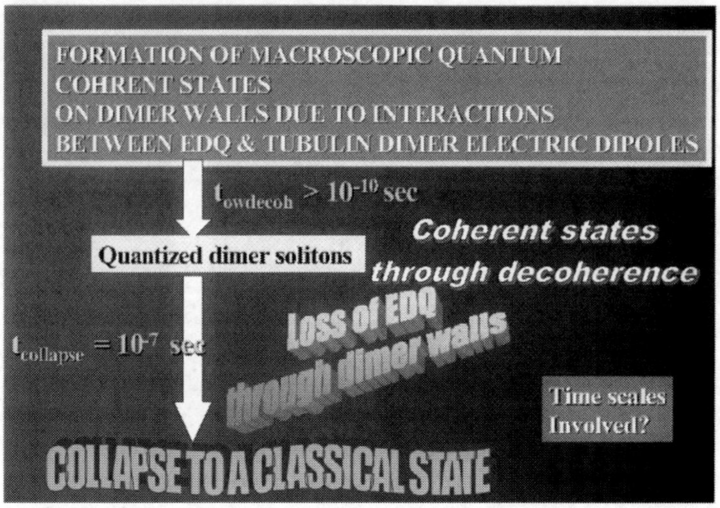

Figure 8. In the Cavity model for MT of [14], there are two stages of quantum decoherence: one incomplete one, arising from the rôle of ordered water as an environment to the dimer system, which results in the formation of the soliton coherent state, and the second one, much longer, which results in the complete collapse of the soliton coherent state, as a result of the the loss of dipole quanta and energy through the imperfect dimer walls. Several uncertainties as to the precise origin of the incomplete decoherence exists in the model, which may shorten significantly the first time scale, below the $O(10^{-10})$ s. For dissipationless energy transfer in MT, it is the second (complete) decoherence time scale that it is important.

where $\vec{\epsilon}$ is the cavity (radiation) mode polarisation, and

$$E_c \sim \left(\frac{2\pi\hbar\omega_c}{\varepsilon V}\right)^{1/2} \tag{35}$$

is the r.m.s. vacuum (electric) field amplitude at the center of a cavity of volume V, and of frequency ω_c, with ε the dielectric constant of the medium inside the volume V. In atomic physics the VFRS effect has been confirmed by experiments involving beams of Rydberg atoms resonantly coupled to superconducting cavities [38].

In the analogy between the thin cavity regions near the dimer walls of MT with electromagnetic cavities, the role of atoms in this case is played by the unpaired two-state electrons of the tubulin dimers [14] oscillating with a frequency (28). To estimate the Rabi coupling between cavity modes and dimer oscillations, one should use (34) for the MT case.

In [14] we have used some simplified models for the ordered-water molecules, which yield a frequency of the coherent dipole quanta ('cavity' modes) of order [14]:

$$\omega_c \sim 6 \times 10^{12}\ \mathrm{s}^{-1}\ . \tag{36}$$

Notably this is of the same order of magnitude as the characteristic frequency of the dimers (28), implying that the dominant cavity mode and the dimer system are almost in resonance. Note that this is a feature shared by atomic physics systems in cavities, and thus we can apply the pertinent formalism to our system. Assuming a relative dielectric constant of water w.r.t to that of vacuum ε_0, $\varepsilon/\varepsilon_0 \sim 80$, one obtains from (35) for the case of MT cavities:

$$E_c \sim 10^4\ \mathrm{V/m}\ . \tag{37}$$

Electric fields of such a magnitude can be provided by the electromagnetic interactions of the MT dimer chains, the latter viewed as giant electric dipoles [22]. This suggests that the coherent modes ω_c, which in our scenario interact with the unpaired electric charges of the dimers and produce the kink solitons (8), (27) along the chains, owe their existence to the (quantized) electromagnetic interactions of the dimers themselves.

The Rabi coupling for an MT with $\mathcal{N}$ dimers, then, is estimated from (34) to be of order:

$$\text{Rabi coupling for MT} \equiv \lambda_{MT} = \sqrt{\mathcal{N}}\lambda \sim 3 \times 10^{11}\ \text{s}^{-1}\ , \tag{38}$$

which is, on average, an order of magnitude smaller than the characteristic frequency of the dimers (28).

In the above analysis, we have assumed that the system of tubulin dimers interacts with a *single* dipole-quantum coherent mode of the ordered water and hence we ignored dimer-dimer interactions. More complicated cases, involving interactions either among the dimers or of the dimers with more than one radiation quanta, which undoubtedly occur *in vivo*, may affect these estimates. Moreover, the use of more sophisticated models for the description of water, are in need. We hope to be able to come back to such improved analyses in the near future, especially in view of the rapid recent developments on the experimental detection of quantum coherent effects in biological systems [18, 19].

For the time being, we note that the presence of a Rabi coupling between water molecules and dimers provides a *microscopic* description for the *friction* that leads to quantum coherent solitonic states (8), (27) of the electric dipole quanta on the tubulin dimer walls. To estimate the decoherence time we remark that the main source of dissipation (environmental entanglement) comes from the imperfect walls of the cavities, which allow *leakage of coherent modes and energy*. The time scale, T_r, over which a cavity-MT dissipates its energy, can be identified in our model with the average life-time t_L of a coherent-dipole quantum state, which has been found to be [14]: $T_r \sim t_L \sim 10^{-4}$ s. This leads to a first-order-approximation estimate of the quality factor for the MT cavities, $Q_{MT} \sim \omega_c T_r \sim \mathcal{O}(10^8)$. We note, for comparison, that high-quality cavities encountered in Rydberg atom experiments dissipate energy in time scales of $\mathcal{O}(10^{-3})-\mathcal{O}(10^{-4})$ s, and have Q's which are comparable to Q_{MT} above.

Applying standard quantum mechanics of quantum electrodynamical cavities [35, 36, 38], we may express the pertinent decoherence time in terms of T_r, the Rabi coupling λ and the number of oscillation quanta n included in a coherent mode of dipole quanta in the ordered water. The result is [14]:

$$t_{collapse} = \frac{T_r}{2n\mathcal{N}\sin^2\left(\frac{\mathcal{N}n\lambda^2 t}{\Delta}\right)}\ . \tag{39}$$

The time t represents the 'time' of interaction of the dimer system with the dipole quanta. A reasonable estimate [14] of this time scale can be obtained by equating it with the average life-time of a coherent dipole-quantum state, which, in the super-radiance model of [39], used in our work, can be estimated as [4]

$$t \sim \frac{c\hbar^2 V}{4\pi d_{ej}^2 \epsilon N_w L}\ , \tag{40}$$

with d_{ej} the electric dipole moment of a water molecule, L the length of the MT, and N_w the number of water molecules in the volume V of the MT. For moderately long MT, with length $L \sim 10^{-6}$ m and $N_w \sim 10^8$, a typical value of t is:

$$t \sim 10^{-4}\ \text{s}\ . \tag{41}$$

[4] In this model the entire interior volume of the MT is taken into account. In our cavity model, however, it is only a thin cylindrical region near the dimer walls that plays the rôle of the cavity. Nevertheless, this does not affect our estimates, since the density of the relevant water molecules N_w/V is assumed uniform.

On the assumption that a typical coherent mode of dipole quanta contains an average of $n = \mathcal{O}(1) - \mathcal{O}(10)$ oscillator quanta, and for detuning between the cavity mode and the characteristic frequency of the dimer quanta of order $\Delta/\lambda \sim \mathcal{O}(10) - \mathcal{O}(100)$, the analysis of [14] yields the following estimate for the collapse time of the kink coherent state of the MT dimers due to cavity dissipation:

$$t_{collapse} \sim \mathcal{O}(10^{-7}) - \mathcal{O}(10^{-6})\ \mathrm{s}\ . \tag{42}$$

This is of the same order as the time scale (13) required for energy transport across the MT by an average kink soliton in the models of [22]. The result (42), then, implies that quantum physics is relevant as far as *dissipationless* energy transfer across the MT is concerned.

In view of this specific model, we are therefore in stark *disagreement* with the conclusions of Tegmark in [17], *i.e.* that *only* classical physics is relevant for studying the energy and signal transfer in biological matter. Tegmark's conclusions did not take proper account of the possible isolation against environmental interactions, which seems to occur inside certain regions of MT with appropriate geometry and properties. The latter can screen decoherening effects in the immediate environment of an MT, for instance those due to neighboring ions used in [17] to arrive at the estimate (2) for the decoherence time of MT.

However, I must stress once again that the above estimates are very crude. We do not have a good microscopic model for the ordered water in the MT interiors, nor we understand the detailed structure of the water-dimer couplings. Above, we have ignored interactions among dimers, when we considered the dipole quanta-dimer coupling, assuming that there is a single coupling between a dipole quantum and a dimer oscillator. Moreover, there is no fundamental reason why the detuning Δ should have the order assumed. All these complicate accurate estimates of the decoherence time, and the actual situation may be very different. One needs, first of all, to develop theoretical models for the ordered water, and use sophisticated numerical simulations in order to estimate the various quantum optical characteristics of the model, such as vacuum Rabi splittings, dipole electric moments *etc.* This is not done as yet. We hope that such detailed modelling will be available in the near future. Thus, although we disagree with Tegmark's conclusion that quantum effects play no essential role in energy and information process in biological systems, such as the brain, one cannot exclude the possibility that decoherence time estimates as short as (2) may, after all, characterize MT. As I will argue below, though, this is still a long enough time interval to allow for an essential rôle of quantum physics to be played in the functioning of MT.

3.4. MT cavity model parameters revisited in light of algae experiments

In this subsection I would like to re-analyze the characteristic features of the cavity model of MT in light of the recent experimental evidence on quantum coherent effects in protein antennae of marine cryptophyte algae [19]. This is essential, in the sense that the recent experiments gave us a feeling on the scale and the order of magnitude of the quantum coherent effects that appear to characterize biological complex systems at room temperatures.

First of all, let us remind the reader of the analogies between the cases of the photosynthetic algae and the MT, as far as their quantum aspects are concerned. In algae, one applies an external stimulus in the form of an external laser pulse, which excites quantum superpositions of electronic states in the protein-antennae pigments. There are coherent quantum oscillations and entanglement between pigments spatially separated at distances as long as 25 Angströms, at *room temperatures.* The eventual environmental decoherence, induced by the rest of the algae complex, damps out such oscillations. The collapse of the pertinent wavefunctions occurs within 400 fs time intervals, which sets the time scale (3) as the characteristic time scale for decoherence of this system at ambient temperatures.

In photosynthetic systems like algae, the (quantum) oscillation periods are of order of a few hundreds of fs [20]. One may naively think that such short decoherence time scales have no relevance for any interesting biophysical process. However, the authors of [19], and I tend to agree with them, speculated that such a time period is sufficient for the bio-complex of algae to *quantum compute* the most efficient, *optimal*, way for energy and information transport across the pigments.

The experiments of ref. [19], as well as the relevant simulations of such bio systems [20], constitute strong (experimental) evidence that quantum effects at ambient temperatures, playing a rôle in the system's functioning, are *facts*, which, however, characterize *some* but *not all* photosynthetic biosystems. The cryptophyte algae have the pigments co-valently bound with the rest of the complex, and this was argued in [19] to be the main reason why quantum coherence at ambient temperatures can occur, entangling parts of the complex at distances over 20 Angströms.

Tubulin dimers, on the other hand, have been theorized to be characterized by quantum superporsitions of their two conformations α and β. In this picture, the dimers constitute two-state quantum systems, oscillating between their two energy eigenstates with a frequency (28), corresponding to a period of $\sim$ 6000 fs. For comparison, we mention that the characteristic period of quantum oscillations among the photosynthetic pigments in the calculated dynamics of *Rhodopseudomonas acidophila* [20] is about 500 fs (*c.f.* fig. 3), while in the photosynthetic algae studied in [19], the period of the quantum oscillations is of order 60 fs. Presumably such quantum superpositions in tubulin can also be excited by external stimuli, such as applied (or self-generated [22, 14]) electric and magnetic fields, photon laser pulses *etc.* When the quantum oscillating dimers find themselves bound in the walls of a MT, coherence has been argued to extend over macroscopic scales at distances of order μm, across the entire MT. The cavity model of MT has been invoked in [14, 15] as a plausible microscopic model underlying such properties. The unpaired charge of the tubulin dimer constitutes the two-state quantum system, and the rest of the tubulin protein complex, as well as the ordered water environment of the MT and its C-termini appendices [12], constitute the environment in this case.

If sufficient isolation from other electrostatic effects, such as Ions in MT [17], occurs in MT, then decoherence times as long as $10^{-6} - 10^{-7}$ s can be achieved, (42). As we have seen above, for this to happen [14] one needs strong electric dipole-dipole interactions, between tubulin protein dimers and ordered water molecules inside the thin cavity regions near the dimer walls. This would provide an analogue to the strong co-valence binding of the pigments into the protein antennae of the cryptophyte algae [19].

In view of the fact that *not all* photosynthetic systems exhibit such a strong binding, and that the latter has been linked [19] to the persistence of quantum coherence for relatively long times at ambient temperature, one is tempted to conjecture that such strong dipole-dipole interactions and cavity region realization may *not* characterize *all* MT in *all* biosystems. For instance, it would be interesting from the point of view of conscious perception, if only brain MT exhibited such strong cavity effects. Unfortunately the current experimental evidence towards this is nonexistent, and it is generally believed that there is a universal structure of MT, at least as far as the ordered water environment and the basic chemistry of tubulin dimers are concerned. But one cannot exclude surprises. Moreover, one should not dismiss the possibility that potentially important differences between *in vitro* and *in vivo* properties of biological matter may occur. The latter point of view is supported by the fact that living matter constitutes a non-equilibrium system, as we have also heard in the talk by E. del Giudice in this conference, and hence many concepts and properties we are accustomed to in physics, that pertain to equilibrium situations, may not characterize systems out of equilibrium. Nevertheless, at present such non universal behavior of MT constitutes a mere speculation, lacking any experimental confirmation or evidence.

In view of the *experimental findings* of [19] on quantum decoherence times of order 400 fs, it would be interesting to examine what range of the parameters of the MT cavity model of [14] could lead to such short decoherence time scales. Of course, MT and photosynthetic algae are completely different biological entities, and there is no *a priori* reason why quantum effects, if any observed in MT, should be of the same order as in cryptophyte algae. Nevertheless, the question has a physical meaning, given in particular the fact that such short decoherence scales are of the same order as those induced by the neighboring ions in MT, (2) [17].

In our model of [14], the main source of dissipation is assumed to come from the loss of dipole quanta through the imperfect dimer walls. From (39), (40) we observe that an immediate six-order of magnitude reduction in the decoherence time could be achieved by a corresponding reduction in the time T_r over which the cavity MT dissipates its energy, which in turn would reflect on a much shorter life time of the alleged dipole quanta inside the cavity regions (40), which T_r is identified with. Such a reduction appears not natural within the super-radiance model of [39], used in our approach. Indeed, the Rabi coupling λ that could conceivably change, even by several orders of magnitude, given the rough way its magnitude was estimated in our work so far, appears in (39) inside a sinusoidal function, which is always less than unity. Hence, a change in the order of magnitude of λ cannot reduce the decoherence time significantly.

The only parameters of the model that could conceivably change and reduce (39) by six orders of magnitude would be: (i) the density of ordered water molecules near the dimer walls N_w/V, which in our model had so far been assumed uniform inside the MT interior regions, and (ii) the number of coherent dipole quanta. However, we find that changes of an unnatural magnitude in these two quantities would be required in order for (39) to be reduced to the order of O(400) fs. It is hard to imagine, for instance, how we can arrange for $n \sim 10^6$ dipole quanta to be contained in a coherent cavity mode in the model. The cavity model of [14], makes use of mesoscopic cavities, with n=O(10) at most, which is a natural number for dipole quanta inside the conjectured thin interior cavity regions of MT.

On the other hand, such short decoherence times in the quantum oscillations of the dimers, if observed, would indeed point towards the fact that the cavity model of MT is in trouble, although quantum mechanics could still play an important rôle. For instance, as we have already discussed, short decoherence times of order (2) could be provided by the electrostatic effects of the Ca^{2+} and other ions [17] in the C-termini and other neighborhoods of the dimers. In such a case, although the cavity model of MT would be probably invalidated, nonetheless we would still disagree with the conclusions of [17] that quantum effects play no rôle in efficient energy transfer in the (brain) cells. Following the speculations in [19], we may conjecture that such short times may be sufficient for the MT to *quantum calculate* (through the quantum entanglement of (part of) its tubulin subunits) the *optimal direction* along which energy and information would be most efficiently transported. This would be a pretty important function. In this scenario, although energy would still be transported according to classical physics, nevertheless its optimum path would have been decided by a *quantum computation*, lasting at most 10^{-13} s.

Another important aspect of the algae photosynthetic systems was the fact that they involved only a relatively small number of fundamental 'units' (8) that undergo quantum superpositions and are entangled at room temperatures. On the other hand, the MT involve a rather large number of tubulin dimers, more than 100 in each protofilament. It may actually be that in realistic situations *not all*, but only a *small subset* of relatively nearby tubulin dimers in a MT, separated by distances of up to 40 Angström, are actually "coherently wired". This could still be sufficient for the MT to *quantum calculate* the *optimal* direction for energy and signal transfer, as in algae protein antennae [19].

However, it may also be the case that the much longer –as compared to algae – decoherence time scales in the cavity model of MT, can actually be explained, in a rather counterintuitive way, by the fact that a much larger number of fundamental "units" are coherently entangled in

MT. Indeed, there are situations in physics where such a phenomenon occurs [15]. As mentioned in the introduction of the talk, in ref. [5] the authors describe the macroscopic entanglement of two samples of Cs atoms at room temperature. The entangling mechanism is a pulsed laser beam and although the atoms are far from cold or isolated from the environment, partial entanglement of bulk spin is unambiguously demonstrated for 10^{12} atoms for $\sim 0.5ms$. The system's resilience to decoherence is in fact *facilitated* by the existence of a *large number of atoms*: even though atoms lose the proper spin orientation continuously, the *bulk* entanglement is not immediately lost. Quantum informatics, the science that deals with ways to encode, store and retrieve information written in qubits, has to offer an alternative way of interpreting the surprising resilience of the Cs atoms by using the idea of "*redundancy*". Simply stated, information can be stored in such a way that the logical (qu)bits correspond to many physical (qu)bits and thus are resistant to corruption of content.

Yet another way of looking at this is given in the work of ref. [40], where the authors have demonstrated experimentally a *decoherence-free quantum memory of one qubit* by encoding the qubit into the "decoherence-free subspace" (DFS) of a pair of trapped Berrilium $^9Be^+$ ions. They achieved this by exploiting a "safe-from-noise-area" of the Hilbert space for a *superposition* of two basis states for the ions, thus encoding the qubit in the superposition rather than one of the basis states. By doing this they achieved decoherence times on average an order of magnitude longer.

Both of the above works show that it is possible to use DFS, error correction and high redundancy to both store information and to keep superpositions and entanglements alive for biologically relevant times in macroscopic systems at high temperature. Thus it nay not be entirely inappropriate to imagine that in biological *in vivo* regimes, one has, under certain circumstances, such as the ones specified above, similar entanglement of tubulin/MT arrangements [15].

I believe that very interesting future experiments can be done with MT, which could shed light on the above aspects of MT as quantum devices, which presently belong to the realm of *science fiction*. Let me now discuss briefly some of such experiments.

4. Experimental tests of the MT cavity model: past and future

I proceed now to discuss some experimental tests of the cavity model of MT presented in the previous section. I will concentrate on direct physical tests of the model, rather than expanding on its physiological or quantum information aspects. As I have already mentioned in the beginning of the talk, there are very interesting experiments, for instance, showing consistency of the model as far as its predictions on the memory function of the brain are concerned [16, 21], but I will not touch upon such aspects here. I am also not qualified to discuss experiments to study physiological/biological properties of MT and/or the tubulin dimers in general.

Some of the experiments I will describe below, require for their interpretation detailed theoretical knowledge on the structure of the tubulin dimers and MT, especially as far as their electrical properties are concerned. Computer molecular simulations for the permanent electric dipole moments of the tubulin, to be discussed briefly below, can be found in [24, 25] and are based on the structure of the tubulin dimer provided by electron-microscope studies at 3.5 Angström resolution in ref. [23]. For detailed recent molecular dynamics simulations on the structure of the MT, which may also be useful in future studies and tests of properties of MT, including our cavity model, I refer the reader to the interesting works of ref. [26]. However, for the past experiments I will discuss below, we used the simulations and tubulin structure data available up to the year of the experiments (2005).

4.1. Measurement of electric dipole moments and ferroelectric properties tests

An essential aspect of the model is its ferroelectric features at room temperatures, namely the induction of a permanent electric dipole moment that remains after switching off an externally applied electric field. We have performed such experiments with porcine brain MT in 2005, in collaboration with the molecular biology group of E. Unger [41]. We have examined the effects

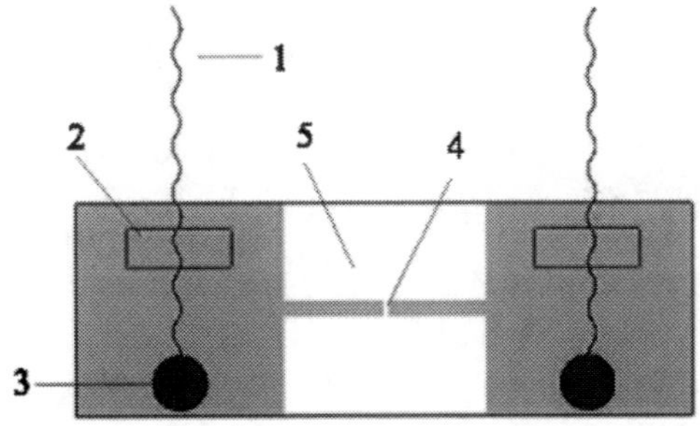

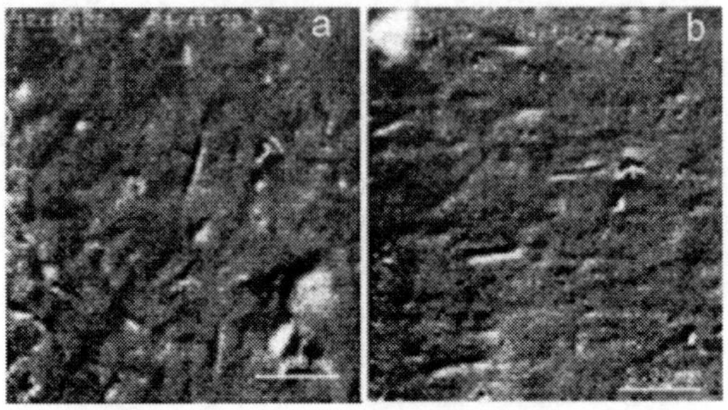

Figure 9. ***Left Picture***: Experimental arrangement for measuring the electron dipole moment of MT in suspension. The picture shows the electrode system used for the measurements in alternating electric fields. The gray areas denote gold layers. 1: thin, flexible wires for the application of the voltage; 2: strips of adhesive tape as pull relief; 3: drops of conductive silver for contacting the wires with the gold; 4: electrode gap filled with the sample of porcine brain MT; 5: gold-free glass areas. ***Middle and Right Pictures:*** **a.(Middle:)** Random positions of MT in the absence of electric field. This state is again realized after switching off the field. **b. (Right):** Alignment of MT in the direction of the externally applied alternating electric field 210,000 V/m and 2 MHz frequency. The visualization of MT has been achieved by video-enhanced DIC Microscopy (Pictures taken from ref. [41]).

of both constant (up to 2 × 103 V/m) and high-frequency alternating fields (up to 2.1 × 105 V/m, with frequencies from 200 kHz to 2 MHz) on suspended porcine microtubules. At pH 6.8 and 120 mM ionic strength, constant fields cause a motion of microtubules toward the anode (*c.f.* fig. 9).

The electrophoretic mobility amounts to 2.6×10^{-4} cm^2/V s, reflecting a negative net charge of approximately 0.2 elementary charges per tubulin dimer. The moving microtubules are randomly space oriented. Alternating high-frequency fields induce electric dipoles and align the microtubules parallel to the field direction. By determining the angular velocity of the turning microtubules, we estimate a dipole moment for the MT roughly

$$p_{\rm MT} = 34,000 \text{ Debye} \tag{43}$$

at 2.1 × 105 V/m and 2MHz frequency. By comparing the potential energy of the dipole in the applied field with the thermal energy of microtubules, we obtained a minimum value of 6,000 Debye, necessary for efficient alignment.

Unfortunately *no evidence* for permanent electric dipole has been found at ambient temperatures, where the experiment has been performed. Thus ferroelectricity has *not* been confirmed as yet. In what follows, I will attempt a comparison of these results with theoretical estimates and seek possible explanations for our inability to observe a permanent electric dipole moment, other than the straightforward dismissal of any ferroelectric properties of MT.

Computer molecular simulations of the permanent electric dipole moment of the tubulin dimers [24], taking into account their detailed structure [23], have shown that the bulk of it is directed in a direction perpendicular to the protofilament axis (x-axis) of the MT, and only about a fifth of the total electric dipole moment is along the x-axis:

$$p_x = 337 \text{ Debye} , \qquad p_y = -1669 \text{ Debye} , \qquad p_z = 198 \text{ Debye} . \tag{44}$$

Taking into account the screening effect of water (whose dielectric constant may be as high as 80) on the dimer charges, yields the following suppressed estimate for the (permanent) dimer dipole moment [25]

$$p_{\rm dimer} = 90 \text{ Debye} . \tag{45}$$

The unpaired charges,that is, the net charges inside the hydrophobic pockets of the dimers, that appear isolated from their environment, may lead to even further suppression , that is, one may take as an order of magnitude [41]

$$p_{\rm dimer} = 15 \text{ Debye} . \tag{46}$$

One may speculate that the other components perpendicular to the microtubule axis will be neutralized in the cylindrical microtubular geometry and screened by the environment, thereby leaving the p_x component as the dominant contribution to the total dipole moment of the microtubule. The reader is reminded at this point that such an assumption also characterizes the simplified ferroelectric one-dimensional lattice model for MT, discussed in [22] and adopted in our studies in [14].

If this is the case, then taking into account that in moderately long microtubules of an average length L = 3.5 μm with 12 protofilaments each (as used in the experiments of [41]), there are about $\mathcal{N} = 5280$ tubulin dimers of average length 8 nm each, one would arrive at the most optimistic estimate for the total dipole moment (all dimer dipoles contributing equally to the x-direction) in the range

$$p_{\rm total} = \mathcal{N} p_{\rm dimer} = 79,200 \text{ Debye} . \tag{47}$$

This is likely to be further suppressed if details on the water and other environments and geometry are properly taken into account.

Such a suppression of the total dipole moment may provide an explanation for our inability to observe an alignment of microtubules in constant electric fields up to 2×103 V/m. As discussed in [41], estimates of the interaction energy of the supposed permanent dipoles in the electric fields show that this energy is too small to overcome the influence of the thermal energy. Safety requirements prevented the use of higher intensity constant fields in our experiments. On the other hand, as we already mentioned, the application of alternating fields with intensities as high as 2×105 V/m, at 2 MHz frequency, induce alignment of the MT along the direction of the field (*c.f.* fig. 9), from which we estimated the dipole moment (43).

The results of this experiment, therefore, probably imply that a possible permanent part of the dipole moment does not play a role for the orientation at high frequencies because the dipole cannot follow the changes of the field. Only the induced part of the dipole is responsible for the orientation, because the torque does not depend on the field direction. Comparing the potential energy of the dipole in the applied field with the thermal energy of microtubules, shows, as mentioned above, that a minimum value of the dipole moment is necessary for a successful orientation.

Thus the non-observation of ferroelectric properties in our experiment does not falsify the ferroelectric-ferrodistortive models of MT [22] on which the cavity approach is based [14, 15]. Further, refined, experiments in this direction are certainly due.

4.2. Quantum-optics-inspired experiments for the cavity model of MT: are they feasible?

One of the most important ingredients in the approach of [14] was the rôle of the coherent cavity modes of the electric dipole quanta. Their (Rabi) dipole coupling with the dipoles of the tubulin dimers provides the necessary "friction" environment, as we discussed above, which is responsible for the soliton (coherent) states that transfer energy across the MT in a dissipation-free manner. One of the decisive tests of the MT cavity model, therefore, would be to observe such coherent modes/Rabi couplings, adapting properly the pertinent quantum optics/cavity electrodynamics experiments [38, 35] in the MT situation. For completeness, below I will discuss these experiments, with a view to see whether there is a possibility of applying them to the biological systems at hand.

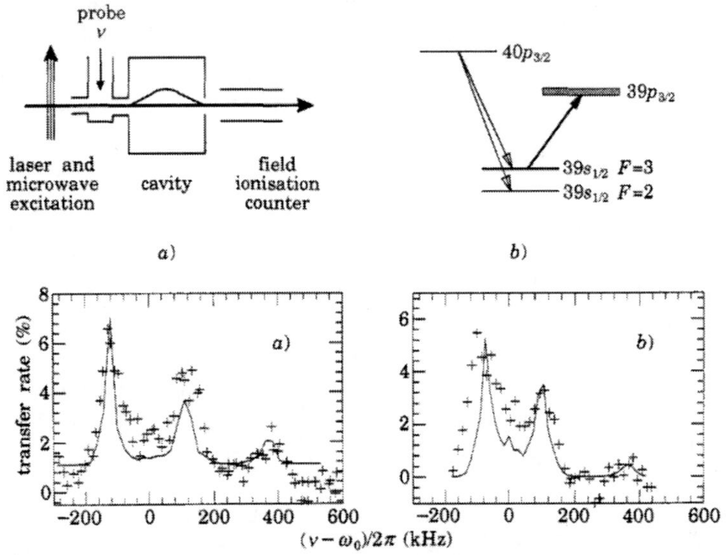

Figure 10. ***Upper Figures***: **a)** Experimental set up for the measurement of the Vacuum Field Rabi splitting in atomic physics. The atomic beam is represented by a horizontal arrow. The atoms are prepared in some initial configurations ($39s_{1/2}$ with hyperfine two-state splitting $F = 2, 3$, indicated by arrows pointing downwards) by laser and microwave excitation. The probing electric field (with a frequency ν and sine arch time variation of its amplitude, indicated in the figure by a continuous curve inside the cavity region) is injected into the cavity through the same wave guides as the atoms. **b.)** The cavity mode couples the $39s_{1/2}$ F=3 hyperfine level to the $39p_{3/2}$ levels (their hyperfine structure remains unresolved in the conditions of the experiment). This is indicated by an arrow pointing upwards. ***Lower Figures:*** The observed Rabi spectra, corresponding to the following number of atoms in the cavity: **a.)** N = 10, **b.)** N = 5. Crosses indicated experimental points, continuous lines are theoretical simulations. $\omega_0/2\pi = 68.38$ GHz is the frequency of the $39s_{1/2}$ F=3 $\rightarrow$ $39p_{3/2}$ transition. The transition $39s_{1/2}$ F=2 $\rightarrow$ $39p_{3/2}$ has a shifted frequency $(\omega_0 + \Delta)/2\pi$, $\Delta/2\pi = 320$ KHz (Pictures and results taken from ref. [38]).

In cavity quantum electrodynamics, the easiest way to observe the Rabi splitting of atoms [38], discussed briefly in subsection 3.3 above, (32), (33), is through the experimental set up indicated in fig. 10. Some two-state Rydberg atoms (or, more generally (*c.f.* fig. 10), atoms properly prepared so as to resemble two-state systems) are injected through a wave guide into a cylindrical superconducting cavity, which in the experiments of [38] is cooled down to 1.7 K. Before entering the cavity the atoms are prepared in some configuration of the hyperfine splitted $39s_{1/2}$ F=2, and F=3 states. The cavity mode couples the $39s_{1/2}$ F= 3 state to the $39p_{3/2}$ state. As a result of this coupling, upon the application of the external electric field of frequency ν, the cavity mode does not resonate at the bare atom frequency but exhibits instead two peaks in the corresponding absorption spectrum, as indicated in the lower part of fig. 10. From these peaks one clearly can compute the Rabi coupling, using (32) or (33).

However, experimentally (*c.f.* fig. 10) there are more features and structures than the two symmetric absorption peaks theory predicts: the two observed Rabi main peaks appear asymmetric and there is a much weaker third peak at frequencies around $(\omega_0 + \Delta)/2\pi$. The latter feature is clearly associated with the non-resonant coupling of the $39s_{1/2}$ F=2 atoms to the cavity modes, since $(\omega_0 + \Delta)/2\pi$ is the frequency of the $39s_{1/2}$ F=3 $\rightarrow 39p_{3/2}$ transition (*c.f.* fig. 10). The former are associated with the motion of the atoms, as well as the fluctuations in the atom number inside the cavity. Numerical simulations of such effects confirm the experimental results and thus provide convincing explanations of the Rabi-splitting phenomenon in realistic systems [38].

Such measurements are then used for the classification of atoms that are employed in quantum-optics experimental demonstrations of environmentally-induced decoherence effects in atomic physics [42]. In these experiments, two-state atoms in a quantum superporition are sent through isolated cavities filled by microwaves. The corresponding Rabi couplings of the atoms to the cavity coherent modes cause a shift in the phase of the microwave field, by different amounts. The experiment involves Rydberg atoms that interact one at a time with the few photon coherent modes ($O(1-10)$), trapped in the cavity. In this way, the field in the cavity is also put in a superposition of two states. The exchange of energy of the field with its environment, and the loss of photon coherent cavity modes through the (imperfect) cavity walls, imply decoherence and therefore the eventual collapse of the field superposition into a single definite state. The authors of [42] have observed experimentally this decoherence, while it unfolded, via the study of correlations between the energy levels of pairs of atoms sent through the cavity with various time delays between the atoms.

In the case of the cavity model of MT, there are formal similarities between the quantum superpositions of the two energy levels of the two-state atoms used in the above-described quantum optics experiments and those of the two conformations of the tubulin dimers. However, from the technical point of view, the situation is much more complicated. Unlike atoms, the MT complexes contain many entities in their environment, exhibiting motion, vibrations (due to (room) temperature effects) *etc*, which complicate simple tests in search of Rabi-couplings between the dimer excitations, playing the rôle of the atoms, and the electric-dipole quanta inside the cavity regions of the MT, playing the rôle of the quantized cavity field modes in the atomic physics experiments. Such couplings would lead [14] to frequency splittings, with the characteristic peaks (*c.f.* fig. 10) (32), (33), in the appropriate absorption spectra of MT (see discussion in subsection 3.3).

Unfortunately, I am not qualified to discuss in detail the feasibility of such experiments. I can only make speculations which, from an experimental point of view, may not be realizable in practice. Nevertheless, as this is a talk, I am free to speculate, so here are my thoughts/suggestions on such experiments: one needs, first of all, to have isolated microtubules, probably in suspension. One should apply laser fields through them and study the absorption spectrum. Since the conformational changes of the tubulin dimers in MT are expected theoretically [14] to be in the thousand GHz region (28), one should arrange for the externally applied fields to have frequencies near such values. According to the model of cavity MT, reviewed above, the characteristic coherent modes of the cavity regions of the MT, excited by the field, are almost in resonance with the dimer oscillations, and hence standard Rabi-splitting phenomenology (33) should be expected, if the model correctly describes nature.

The induced Rabi splitting in the frequencies of the absorption spectra of MT, if observed, would then constitute compelling evidence for the existence of *both* isolated cavity regions in MT interiors and coherent modes inside such regions, which in the model of [14] would be the dipole quanta of the water molecules, suggested in [34]. It goes without saying, that in view of the complicated MT structure, and the associated motions and vibrations of the various biological entities in the tubulin protein or the C-termini appendices, there would be non resonant couplings of atoms and ions in the dimers to the cavity electron dipole quanta, in addition to the simple dipole Rabi couplings examined so far. These would complicate the resulting spectra, however one should still expect to see pronounced Rabi peaks, in analogy with the atomic physics experiments of fig. 10.

Unfortunately, observations of such effects alone (even if they are realized in nature) would not constitute a proof of the quantized nature of the dimer excitations of the MT. Although the latter is a plausibility, and according to such interpretations, the (conjectural) thin interior cavity regions of MT would entail vacuum fluctuations of the ordered-water dipole quanta that would split the resonance line of the dimers by an amount proportional to the collective dimer-water-cavity coupling (a sort of induced dynamical Stark Effect on the quantum superpositions of the dimer states), alternative explanations of the Rabi splitting phenomenon [38] exist. The latter point towards an interpretation of the phenomenon as a consequence of the fact that the MT dimer *medium* behaves as a *refractive* one with a *classical complex* (*i.e* containing imaginary parts) index of refraction that splits the cavity mode into two components.

To demonstrate, therefore, unambiguously the existence of quantum coherent effects in MT one needs to explicitly observe experimentally the quantum oscillations of the dimers and measure their environmentally induced decoherence. A straightforward extension of the atomic physics experiments, observing quantum decoherence in electrodynamical cavities [42] may, unfortunately, not be feasible in the case of MT. Indeed, in the case of the atomic physics experiments of ref. [42], a Rydberg atom beam in a quantum superposition is sent through a cavity, *exits* from it and is eventually counted in one or the other Rydberg state by appropriate ionizing detectors, so that only decoherence of the atom-cavity system ("atom plus measuring apparatus") is measured in the experiment. In contrast, in the case of MT, the "cavity" regions are attached to the "atoms", being part of the MT structure. One cannot separate the quantum oscillations of the dimers from the rest of the MT and its ordered water interiors. Thus, the methods of [42], that could in principle measure quantum decoherence of the field-dimer (and hence of the entire MT) system, as the latter unfolds, appear inapplicable.

On the other hand, it seems to me that Photon echo absorption data, like the ones in *Cryptophyte Algae* [19], when appropriately adapted to the much more complex case of MT, might be a way forward, in order to observe the decoherence of the coupled "field-dimer" system inside the MT. Presently I do not know whether such measurements are feasible in the near future, nevertheless, I find the prospect of performing such experiments very exciting, and I am sure there are ways one can proceed along these lines in the near future.

5. Conclusions and outlook

The exciting experimental developments in the light-harvesting *Cryptophyte Algae* provided compelling evidence on an important rôle of quantum effects in biological systems at room temperature. Specifically, on exciting by photon pulses certain dimer pigments (DVB) of the photosynthetic protein antennae of the algae into a quantum superposition of appropriate electronic states, one observes experimentally, by means of two-dimensional photon echo (absorption) data [19], the quantum oscillations between the two electronic states of the DVB dimers, as well the quantum entanglement of the DVB molecules with the other pigments, at distances of order 20 Angstrom away. In this way, the entanglement is responsible for "coherently wiring" pigment molecules across the entire protein antennae. The experiment has been performed at ambient temperatures (294 K).

The situation is reminiscent of coupled oscillators through extended springs. Such action at a distance is the result of quantum correlations between the quantum states of the pigments. The eventual decoherence of the relevant oscillations, induced by the complex environment of the protein antennae, has been observed to last for about 400 fs. Although this time scale is relatively short, nevertheless, the authors of ref. [19] have argued that it may be sufficiently long for the protein antenna to *quantum calculate* in which direction energy and information would be transported more efficiently. Thus the observed coherent 'wiring' across the entire antenna complex is thereby linked with energy transfer optimization in photosynthetic algae.

Not all photosynthetic proteins exhibit such a behavior, which in [19] has been attributed to the fact that the pigments are covalently bound in the protein complex, unlike the case of other light-harvesting biosystems. This strong bounding is probably responsible for the rather long (from the point of view of information transfer optimization) decoherence time scales at room temperatures.

Such exciting developments brought back to life, at least in the mind of the speaker (!), a rather old model of biological microtubules (MT) viewed as quantum electrodynamical cavities, developed in [14] and elaborated further in [15]. In this model, quantum coherent oscillators in the dimer walls was the result of environmental entanglement of the tubulin protein dimers (viewed as a two state quantum system, oscillating between the two conformational states, α and β - tubulin) with quantum dipole quanta inside thin interior regions of the MT, extending from the dimer walls inwards, for a few Angströms, which are full of ordered water. These regions have been argued to play the rôle of thermally isolated cavity regions, as a result of strong dipole-dipole interactions between the electric dipoles of the dimers and those of the dipole quanta in the ordered water. The latter arise as a result of the interaction of water molecules with quantized electromagnetic fields inside the cavity regions.

Electromagnetic fields in MT can be either externally applied or self generated. In this talk, it has been argued that strong experimental evidence in favor of the existence of dipole quanta in the interior of MT would be the observation of characteristic Rabi splitting in the photon absorption frequency spectra of tubulin dimers, under the influence of externally applied electromagnetic fields of definite frequency.

The dipole-dipole water/dimer coupling is held responsible [14] for inducing macroscopic quantum coherence across the entire MT, of some μm$=10^{-6}$ m long, which would manifest itself as a dimer-dipole kink soliton, responsible for dissipation-free energy and signal transfer. It has been estimated that, within some natural range of the parameters of the model of [14], the decoherence time scale, within which such solitonic states collapse to a classical state, could be as long as

$$t_{\rm MT\ decoh}\mathcal{O}\left(10^{-6}-10^{-7}\right)\ {\rm s}\ . \tag{48}$$

under the assumption that the dominant source of decoherence is the loss of dipole quanta through the imperfect cavity walls. Such long decoherence times may be achieved if there is ferroelectric behaviour of MT complexes at room temperature. They are argued to be sufficiently long for dissipation-free energy and signal transfer by the kink soliton along the MT protofilaments. Our mechanism for the emergence of coherent states in MT arrangements therefore inside the cell can be summarized by the 'basic cycle' shown in figure 11.

Much shorter decoherence times of the quantum excitations of the dimers are expected if the ferroelectric behaviour and the strong dipole-dipole water/dimer coupling do not exist in Nature. In particuler, as explained in [17], decoherence times of order of a few hundreds of fs may be the result of Ca^{2+}-ion induced decoherence in MT. Although, such generic estimates have been argued in [17] to eliminate any essential rôle of quantum physics in biological MT, in particular brain MT, nevertheless, in view of the findings of [19], such conclusions should be revised.

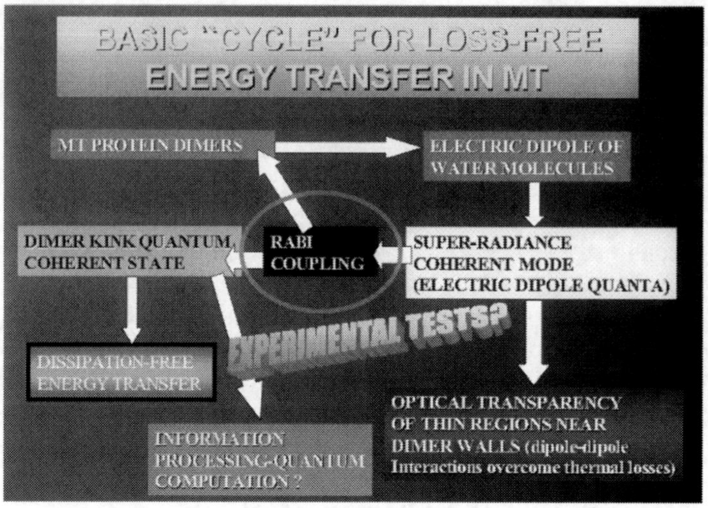

Figure 11. Suggested basic cycle, summarizing functioning and properties of the quantum cavity model for Microtubules [14], and its rôle in dissipation-free energy and information transfer.

Decoherence times, of order of a few hundreds of fs, *may be* sufficiently long for optimization of the direction along which energy and information transfer in MT would be more efficient, even if the latter occurs by means of classical physics (mechanics) laws of transport rather than quantum. Thus, even if the cavity model of MT turns out *not* to provide the correct description of natural MT, some sort of quantum computation in energy and signal optimization could still take place inside the biological cells, in the way conjectured above.

All such features would leave the realm of *science fiction* and become *realistic possibilities* or even realities, once some experimental evidence on the basic properties of models entailing quantum dynamics of MT emerges. The latter can come in many ways, from a direct observation of quantum oscillations of dimers in an MT, using photon echo data as in [19], assuming that such a technique is feasible in MT, and indirect verification of the existence of cavity regions in their interior, through the observation of Rabi-splitting peaks in appropriate absorption spectra, to an explicit demonstration of ferroelectric properties of MT at room temperatures, through the measurement of permanent dipole moments and other phenomena characteristic of ferroelectricity [43]. Unfortunately, at present there is no such experimental evidence. However, the situation can easily change in the near future.

I would like to close this talk by stating once more something that physicists tend to forget, when experimenting with living matter. *In vitro* (Laboratory) situations may be entirely different from actual *in vivo* situations. Living organisms are non equilibrium systems, as emphasized in this conference by E. Del Giudice, hence, their behaviour may be entirely different from dead *in vitro* samples in suspension. It is then possible that some quantum aspects of *in vivo* Microtubules are not easily (if at all) revealed in the laboratory. More sophisticated than usual techniques might be needed to bring forward such aspects.

I sincerely believe that the emergence of quantum phenomena in light-harvesting algae will open new and exciting avenues for research in this direction, and, if such coherent phenomena prove to characterize also the fundamental building blocks of the cell, the Microtubules, this may eventually lead to establishing a possibly important rôle of quantum physics on basic functions of the cell. Time will probably tell whether all these constitute *science fiction* or *facts...*

Acknowledgements

I thank the organizers of the DICE 2010 Workshop (Castello Pasquini, Castiglioncello (Italy), September 11-18 2010) for the invitation and the opportunity to talk in this interesting and thought stimulating conference. I also acknowledge informative discussions on molecular modelling with C. Molteni. The work of N.E.M. is partially supported by the European Union through the FP6 Marie Curie Research and Training Network *UniverseNet* (MRTN-CT-2006-035863).

References

[1] See, *e.g.*: Amsler C *et al.* [Particle Data Group] 2008, Review of particle physics *Phys. Lett.* B **667** 1
[2] Onnes H K 1911 *Commun. Phys. Lab. Univ. Leiden* **12** 120; Meissner W and Ochsenfeld R 1933 *Naturwiss.* **21** (44) 787 [doi:10.1007/BF01504252]; London F and London H 1935 *Proc. R. Soc. London* A **149** (866) 71 [doi:10.1098/rspa.1935.0048]; Ginzburg V L and Landau L D 1950 *Zh. Eksp. Teor. Fiz.* **20** 1064; Bardeen J, Cooper L N and Schrieffer J R 1957 *Phys. Rev.* **106** (1) 162; *ibid.* 1957 **108** (5) 1175
[3] Bednorz J G and Mueller K A 1986 *Z. Phys.* B **64** (2) 189; for a review see: Leggett A 2006 *Nature Physics* **2** 134 [doi:10.1038/nphys254]
[4] Collini E and Scholes G D 2009 *Science* **323** 369
[5] Julsgaard B, Kozhekin A and Polzik E 2001 *Nature* **413** 400
[6] Schrödinger E 1944 *What is Life* (Cambridge: Cambridge University Press, reprinted 1992 [ISBN 0521427088])
[7] Frohlich H 1968 *Int. J. Quantum Chem.* **2** 641; *do.* 1970 *Nature* **228** 1093; *do.* 1972 *Phys. Lett.* A **39** (2) 153; *do.* 1968 in *Bioelectrochemistry* ed F Guttman and H Keyzer (New York: Plenum)
[8] Davydov A S 1977 *J. Theoret. Biology* **66** (2) 379
[9] Popp F A 1979 in *Electromagnetic Bio- Information* ed F A Popp, G Becker, H L Konig and W Peschka (Munich-Vienna-Baltimore: Urban & Schwarzenberg)
[10] see: Swain J 2006 *Preprint* arXiv:physics/0603137, and references therein
[11] Hameroff S and Penrose R 1995 *J. Consciousness Studies* **2** 98; *ibid.* 1996 **3** 36; *ibid.* 1996 *Mathematics and Computers in Simulation* **40** 453; *do.* 1996 in *Towards a science of Consciousness, The First Tucson Discussions and Debates* ed S Hameroff *et al.* (Cambridge MA: MIT Press) 507. See also: Penrose R 1989 *The Emperor's New Mind* (Oxford: Oxford Univ. Press); *do.* 1994 *Shadows of the Mind* (Oxford: Oxford Univ. Press)
[12] See, *e.g*: Dustin P 1984 *MicroTubules* (Berlin: Springer); Engleborghs Y 1992 *Nanobiology* **1** 97
[13] Zurek W H 2003 *Rev. Mod. Phys.* **75** 715, and references therein
[14] Mavromatos N E and Nanopoulos D V 1997 *Int. J. Mod. Phys.* B **11** 851 (*Preprint* arXiv:quant-ph/9512021); *do.* 1998 *Int. J. Mod. Phys.* B **12** 517 (*Preprint* arXiv:quant-ph/9708003); *do.* 1998 *Adv. Struct. Biol.* **5** 283 (*Preprint* arXiv:quant-ph/9802063)
[15] Mavromatos N E 1999 *Bioelectrochemistry and Bioenergetics* **48** 273; Mavromatos N E, Mershin A and Nanopoulos D V 2002 *Int. J. Mod. Phys.* B **16** 3623-3642 (*Preprint* arXiv:quant-ph/0204021)
[16] Mershin A, Nanopoulos D V and Skoulakis E M C 2000 *Preprint* arXiv:quant-ph/0007088; Mershin A, Pavlopoulos E, Fitch O, Braden B C, Nanopoulos D V and Skoulakis E M 2004 *Learn Mem.* **11** (3) 277; see also: Mershin A *et al.* in [21] below, and references therein
[17] Tegmark M 2000 *Phys. Rev.* E **61** 4194 (*Preprint* arXiv:quant-ph/9907009)
[18] Engel G S *et al.* 2007 *Nature* **446** 782; Lee H, Cheng Y C and Fleming G R 2007 *Science* **316** 1462; Mercer I P *et al.* 2009 *Phys. Rev. Lett.* **102** 057402
[19] Collini E *et al.* 2010 *Nature* **463** 644
[20] Novoderezhkin V I, Rutkauskas D and van Grondele R 2006 *Biophys. J.* **90** 2890; van Grondele R and Novoderezhkin V I 2006 *Phys. Chem. Chem. Phys.* **8** 793; also in: 2010 *Nature* **463** 614, *news and views* on the results of ref. [19]
[21] For potential implications of quantum effects in brain functioning and the associated experimental tests, the reader may be referred to the collection of relevant articles in: Tuszynski J A (ed) 2006 *The Emerging Science of Consciousness*, The Frontiers Collection (Berlin, Heidelberg: Springer-Verlag Berlin, Heidelberg [ISBN 3-540-23890-5]), and references therein
[22] Satarić M V, Tuszyński J A and Zakula R B 1993 *Phys. Rev.* E **48** 589. For an updated analysis see: Satarić M V and Tuszyński J A 2005 *J. Biol. Phys.* **31** 487
[23] Löwe J, Li H, Downing K H and Nogales E 2001 *J. Mol. Biol.* **313** 1045. See also: Nogales E, Wolf S G and Downing K H 1998 *Nature* **391** 199

[24] See: Tuszynski J A, Brown J A, Crawford E, Carpenter E J, Nip M L A, Dixon J M and Sataric M V 2005 *Mathematical and Computer Modelling* **41** 1055, and references therein; also: Tuszynski J A, Luchko T, Portet S and Dixon J M 2005 *European J. Phys. E: Soft Condensed Matter* **17** (1) 29
[25] Mershin A, Kolomenski A A, Schuessler H A and Nanopoulos D V 2004 *Biosystems* **77** 73
[26] Wells D B and Aksimentiev A 2010 *Biophys. J.* **99** 629
[27] Collins M A, Blumen A, Currie J F and Ross J 1979 *Phys. Rev.* B **19** 3630; *ibid.* 3645
[28] Lal P 1985 *Phys. Lett.* A **111** 389
[29] Otwinowski M, Paul R and Laidlaw W G 1988 *Phys. Lett.* A **128** 483
[30] Zurek W H, Habib S and Paz J P 1993 *Phys. Rev. Lett.* **70** 1187
[31] Tsue Y and Fujiwara Y 1991 *Progr. Theor. Phys.* **86** 469
[32] Sackett D L 1995 Structure and Function in the "Tubulin Dimer and the Role of the Acidic Carboxyl Terminus", in *Subcellular Biochem.* **24**, *Proteins: Structure, Function and Engineering* ed B B Biswas and S Roy (New York: Plenum Press)
[33] Brown J A and Tuszynski J A 1997 *Phys. Rev.* E **56** 5834
[34] Del Giudice E, Preparata G and Vitiello G 1988 *Phys. Rev. Lett.* **61** 1085. For a discussion on the rôle of non-equilibrium field theory in living matter, in a more general setting, see: Del Giudice E, Doglia S, Milani M and Vitiello G 1985 *Nucl. Phys.* B **251** (FS 13) 375; *do.* 1986 *Nucl. Phys.* B **275** (FS 17) 185
[35] Haroche S and Raimond J M 1994 *Cavity Quantum Electrodynamics* ed P Berman (New York: Academic Press) 123, and references therein
[36] Sanchez-Mondragon J J, Narozhny N B and Eberly J H 1983 *Phys. Rev. Lett.* **51** 550
[37] Agarwal G S 1984 *Phys. Rev. Lett.* **53** 1732
[38] Bernardot F *et al.* 1992 *Europhysics Lett.* **17** 33
[39] Jibu M, Hagan S, Hameroff S R, Pribram K and Yasue K 1994 *Biosystems* **32** 195
[40] Kielpinski D, Meyer V, Rowe M A, Sackett C A, Itano W M, Monroe C and Wineland D J 2001 *Science* **291** 1013
[41] Böhm K, Mavromatos N E, Michette A, Stracke R and Unger E 2005 *Electromagnet. Biol. and Medicine* **24** 297
[42] Brune M *et al.* 1996 *Phys. Rev. Lett.* **77** 4887
[43] Mavromatos N E, Nanopoulos D V, Samaras I and Zioutas K 1998 *Adv. Struct. Biol.* **5** 127 (*Preprint* arXiv:quant-ph/9803005); Tuszynski J A, Craddock T J A and Carpenter E J 2008 *J. Computational and Theoret. Nanoscience* **5** (10) 2022

To make a nanomechanical Schrödinger-cat mew

Tamás Geszti
Department of Physics of Complex Systems, Eötvös University, Budapest, Hungary
E-mail: geszti@elte.hu

Abstract. By an explicite calculation of Michelson interferometric output intensities in the optomechanical scheme proposed by Marshall *et al.* [1], an oscillatory factor is obtained that may go down to zero just at the time a visibility revival ought to be observed. Including a properly tuned phase shifter offers a simple amendment to the situation. By using a Pockels phase shifter with fast time-dependent modulation in one arm, one may obtain further possibilities to enrich the quantum state preparation and reconstruction abilities of the original scheme, thereby improving the chances to reliably detect genuine quantum behaviour of a nanomechanical oscillator.

Detecting genuine quantum effects in nanomechanical systems is a highly desirable scope, expected to be reached in the near future. The major challenges are: efficient cooling close to the ground state, sufficiently strong coupling to a well-identified quantum system, reliable preparation and identification of non-classical states of the nanomechanical component. Most of the experimental studies are done on mechanical oscillators, considered harmonic for the small displacements characteristic for the quantum domain. For those systems ground-state cooling means $k_B T < \hbar\omega$ which is relatively easy to fulfil for hard (high-frequency) oscillators [2]. For them, however, preparing and analyzing quantum states is overly demanding [3]. The efforts to reconcile these conflicting requirements have seen rapid and most competitive advance with the participation of a number of experimental groups. A still more ambitious, so far elusive prospect of related studies would be to detect deviations from standard quantum mechanics, predicted by several theoretical studies [4].

An original and promising project, using a soft (low-frequency) nanomechanical oscillator, coupled by a mirror to one of the paths of a Michelson interferometer, and using Fabry-Perot cavities (on both paths, to preserve interference) to strengthen the optomechanical coupling, has been started by Marshall *et al.* [1], and following extensive development of an optical feedback cooling technique [5], is now approaching the level of seeing quantum effects.

According to the argument of Ref. [1], for a single-photon source of frequency ω_c, and a vibrating mirror of effective mass M and vibration frequency ω_m, acting as one of the mirrors delimiting a Fabry-Perot resonator cavity of length L, the dimensionless coupling constant is $\kappa = (\omega_c/\omega_m)(\sqrt{(\hbar/2M\omega_m)}/L)$. With the vibrating-mirror cavity on interferometer arm A, and the cavity with two rigid mirrors on arm B, starting with a mirror in its quantum mechanical ground state $|0\rangle_m$, coupling during time t results in an entangled photon-mirror state

$$|\Psi(t)\rangle = \frac{e^{-i\omega_c t}}{\sqrt{2}} \left(e^{i\varphi(t)} |A\rangle |\alpha(t)\rangle_m + |B\rangle |0\rangle_m \right), \tag{1}$$

where $|A\rangle = |1\rangle_A |0\rangle_B$ and $|B\rangle = |0\rangle_A |1\rangle_B$ denote one-photon states with the photon in arm A

and in arm B resp., whereas, $|\alpha(t)\rangle_m$ and $|0\rangle_m$ are two coherent states of respective complex amplitudes $\alpha(t)$ and 0 of the vibrating mirror, appearing as a pair of Schrödinger cat states entangled to the respective orthogonal photon states $|A\rangle$ and $|B\rangle$. Finally,

$$\varphi(t) = \kappa^2(\omega_m t - sin\omega_m t); \quad \alpha(t) = \kappa(1 - e^{-i\omega_m t}). \tag{2}$$

The density matrix $\hat{R}(t) = |\Psi(t)\rangle\langle\Psi(t)|$ corresponding to the pure state $|\Psi\rangle$ can be expanded on the photon state basis $|A\rangle = \begin{pmatrix} 1 \\ 0 \end{pmatrix}$, $|B\rangle = \begin{pmatrix} 0 \\ 1 \end{pmatrix}$ in the form of a two-by-two matrix

$$\hat{R}(t) = \begin{pmatrix} \hat{\rho}_{AA}(t) & \hat{\rho}_{AB}(t) \\ \hat{\rho}_{BA}(t) & \hat{\rho}_{BB}(t) \end{pmatrix}, \tag{3}$$

where the matrix elements are operators acting on variables of the vibrating mirror. It has been pointed out [1, 6] that the crucial quantity characterizing the interference pattern to be observed is the nondiagonal element on that basis, $\hat{\rho}_{AB}(t) = \langle A|\Psi(t)\rangle\langle\Psi(t)|B\rangle$, which carries extra information about the motion of the vibrating mirror, missing from the more usual reduced density matrix $\hat{\rho}_m(t) = \hat{\rho}_{AA}(t) + \hat{\rho}_{BB}(t)$, obtained by tracing out $\hat{R}(t)$ over photon variables.

Interference visibility, as expressed by $\hat{\rho}_{AB}(t)$, carries a signature of the motion of the vibrating mirror, but the usefulness of that signature is seriously limited by interactions of the mirror with the environment. Neglecting direct environmental effects on the photon subsystem, the above expansion remains valid, allowing decoherence and friction of the vibrating mirror to be included in the dynamics of $\hat{\rho}_{AB}(t)$ and analyzed by standard tools [6, 7].

Returning for a moment to the decoherence-free case covered by Eq. (1), let us quote the result of Ref. [1]:

$$\mathrm{Tr}_m\, \hat{\rho}_{AB}(t) = \frac{e^{i\varphi(t)}}{2}\, \mathrm{Tr}_m \left[|\alpha(t)\rangle_m\, {}_m\langle 0|\right] = \frac{1}{2} e^{i\varphi(t) - |\alpha(t)|^2/2}, \tag{4}$$

where Tr_m means trace over mirror variables. From here one concludes that the crucial quantity to keep under control is $|\alpha(t)|$, and - using Eq. (2) - the observation should be carried out at multiples of time $t = 2\pi/\omega_m$, where $|\alpha(t)|$ returns to 0 and the interference signal can be large enough to be observed, under the extremely stringent condition that decoherence could be suppressed by sufficient cooling and isolation from mechanical supports.

That treatment catches the essential point of the phenomenon but fails to give a correct description of the underlying oscillatory behaviour, which is relevant for a reliable identification of superposition states of the mirror. To show that, we proceed by noticing that Eq. (1) is not the final state observed by the photon detectors, since the photons pass once more through the beam splitter, before reaching output ports C (facing A) and D (facing B). On the time scale of whatever happens to the vibrating mirror, this is a fast and coherence-preserving unitary operation on the photon subsystem, transforming the intermediate state (1) into the final one

$$\frac{1}{2} e^{-i\omega_c t} \left[|C\rangle \left(|0\rangle_m + i e^{i\varphi(t)} |\alpha(t)\rangle_m \right) + i\, |D\rangle \left(|0\rangle_m - i e^{i\varphi(t)} |\alpha(t)\rangle_m \right) \right], \tag{5}$$

with $|C\rangle = |1\rangle_C|0\rangle_D = (|A\rangle + i|B\rangle)/\sqrt{2}$ and $|D\rangle = |0\rangle_C|1\rangle_D = (i|A\rangle + |B\rangle)/\sqrt{2}$. Carrying out the above unitary transformation on the full density matrix, now we can trace over mirror (and eventually, for a full analysis, environmental) variables to obtain directly the density matrix of the photons to be detected, in the form

$$\mathbf{R}^{phot} = \begin{pmatrix} \mathrm{Tr}_m\, \hat{\rho}_{CC} & \mathrm{Tr}_m\, \hat{\rho}_C \\ \mathrm{Tr}_m\, \hat{\rho}_{DC} & \mathrm{Tr}_m\, \hat{\rho}_{DD} \end{pmatrix}, \tag{6}$$

with the diagonal elements

$$\left.\begin{array}{l} I_C \\ I_D \end{array}\right\} = \left.\begin{array}{l} \mathrm{Tr}_m\,\hat{\rho}_{CC} \\ \mathrm{Tr}_m\,\hat{\rho}_{DD} \end{array}\right\} = \frac{1}{2} \mp \mathrm{Im}\left(\mathrm{Tr}_m\,\hat{\rho}_{AB}\right), \tag{7}$$

describing the output intensities on the two respective ports, and the off-diagonal elements

$$\mathrm{Tr}_m\,\hat{\rho}_{CD} = \left[\mathrm{Tr}_m\,\hat{\rho}_{DC}\right]^* = \frac{1}{2}\mathrm{Re}\left(\mathrm{Tr}_m\,\hat{\rho}_{AB}\right) + i\left(\frac{1}{2} - \mathrm{Tr}_m\,\hat{\rho}_{AA}\right), \tag{8}$$

describing cross-correlations between the two detectors.

Of particular interest are the output intensities, and we see from Eq. (7) that, instead of the modulus of $\mathrm{Tr}_m\,\hat{\rho}_{AB}(t)$, they display the *imaginary* part of the same quantity. Evaluating, like before, the decoherence-free case, we obtain

$$\left.\begin{array}{l} I_C \\ I_D \end{array}\right\} = \frac{1}{2}\left(1 \mp e^{-|\alpha(t)|^2/2}\sin\left[\varphi(t) + \chi(t)\right]\right) \tag{9}$$

$$= \frac{1}{2}\left(1 \mp e^{-\kappa^2(1-\cos\omega_m t)}\sin\left[\kappa^2(\omega_m t - \sin\omega_m t) + \chi(t)\right]\right), \tag{10}$$

where we have used Equation (2), and for the discussion to follow, included an additional phase shift $\chi(t)$ created by a phase shifter device on arm A of the Michelson interferometer, not present in the original setup [1].

Having to measure the imaginary part of $\mathrm{Tr}_m\,\hat{\rho}_{AB}(t)$ makes a huge difference. Without the phase shifter, that factor oscillates as $\sin\kappa^2(\omega_m t - \sin\omega_m t)$, which - depending on the actual value of the optomechanical coupling constant κ - may go down close to zero just about time $(2\pi/\omega_m)n$, when the nth visibility revival is expected. There is a simple amendment though: one should include a constant phase shifter and tune it to

$$\chi_n = 2\pi\left(\frac{2n+1}{4} - \kappa^2\right), \tag{11}$$

to match the nth maximum of the oscillatory factor with the expected revival time. With that, revivals become observable in the interferometric intensity patterns, as soon as decoherence effects are suppressed to the desired level [8]; of course, to see at least $n = 1$ would be the minimum requirement. Alternatively, if noise level of data allows, scanning with χ to locate the maximum signal can be a way to measure the coupling constant κ.

The above reasoning connects the Marshall *et al.* scheme to a simple token introduced by Yurke and Stoler [9], used in various experiments [10, 11] to overcome some of the difficulties associated with the entanglement present e.g. in Eq. (1), and create a visible interference pattern of the two Schrödinger cat components, as apparent in Eq. (5), observable through straightforward intensity measurement of a chosen component of the associated two-state quantum system. In the Marshall *et al.* scheme it is the last passage through the Michelson beamsplitter that - acting as a $\pi/2$ quasi-spin rotation of the two-state photon basis - offers a kind of gratis implementation of the Yurke and Stoler procedure; this was the original motivation for the present work.

Alas, just seing the visibility revival would not prove quantumness [7]. Therefore to reach that final scope it may be necessary to add a time-dependent phase shifter $\chi(t)$ to the photon subsystem. That can strongly enhance the possibilities of preparing and analysing quantum states of an oscillator, as demonstrated by the successful application of phase controlling protocols in ion-trap Schrödinger cat experiments [10]. Optical phase shifters based on the Pockels effect are commercially available, and admit electric modulation at frequencies up to a few GHz [12], with the potentiality to make the Marshall *et al.* nanomechanical Schrödinger cat mew more clearly about his/her quantum behaviour. Some of the possibilities offered by that token will be analyzed in a forthcoming paper.

Acknowledgments
Support by the Hungarian Scientific Research Fund OTKA under Grant No. T75129 is acknowledged. I thank Lajos Diósi for helpful discussions.

References
[1] Marshall W, Simon C, Penrose R and Bouwmeester D 2003 *Phys. Rev. Lett.* **91** 130401
[2] O'Connell A D *et al.* 2010 *Nature* **464** 697
[3] For a recent discussion, see Vanner M R *et al.* 2010 *Preprint* arXiv:1011.0879v1
[4] Ghirardi G C, Rimini A and Weber T 1986 *Phys. Rev.* D **34** 470; Ghirardi G C, Pearle P and Rimini A 1990 *Phys. Rev.* A **42** 78; Diósi L 1987 *Phys. Lett.* A **120** 377; Diósi L 1989 *Phys. Rev.* A **40** 1165; Penrose R 1996 *Gen. Rel. Grav.* **28** 581; see also Geszti T 2004 *Phys. Rev.* A **69** 032110
[5] Kleckner D and Bouwmeester D 2006 *Nature* **444** 75
[6] Adler S L, Bassi A and Ippoliti E 2005 *J. Phys.* A **38** 2715; Bassi A, Ippoliti E and Adler S L 2005 *Phys. Rev. Lett.* **94** 030401
[7] Bernád J Z, Diósi L and Geszti T 2007 *Phys. Rev. Lett.* **97** 250404
[8] Another possibility is to measure cross-correlations between the two output ports, the oscillations of which are shifted in phase by $\pi/4$ with respect to the one-port intensities.
[9] Yurke B and Stoler D 1986 *Phys. Rev. Lett.* **57** 13
[10] Monroe C, Meekhof D M, King B E and Wineland D J 1996 *Science* **272** 1131
[11] Brune M, Hagley E, Dreyer J, Maître X, Maali A, Wunderlich C, Raimond J M and Haroche S 1996 *Phys. Rev. Lett.* **77** 4887
[12] Saleh B E A and Teich M C 2007 *Fundamentals of Photonics* 2nd ed (Hoboken, NJ: Wiley) Chapter 20

General linear dynamics – quantum, classical or hybrid

H-T Elze, G Gambarotta and F Vallone
Dipartimento di Fisica "Enrico Fermi", Largo Pontecorvo 3, I-56127 Pisa, Italia
E-mail: elze@df.unipi.it

Abstract. We describe our recent proposal of a path integral formulation of classical Hamiltonian dynamics. Which leads us here to a new attempt at hybrid dynamics, which concerns the direct coupling of classical and quantum mechanical degrees of freedom. This is of practical as well as of foundational interest and no fully satisfactory solution of this problem has been established to date. Related aspects will be observed in a general linear ensemble theory, which comprises classical and quantum dynamics in the form of Liouville and von Neumann equations, respectively, as special cases. Considering the simplest object characterized by a two-dimensional state-space, we illustrate how quantum mechanics is special in several respects among possible linear generalizations.

1. The notorious linearity of the evolution equations

This article is motivated by discussions of the origin of quantum mechanics, as presented, for example, in Refs. [1, 2]. These suggest that quantum mechanics could possibly *emerge* by coarse-graining and sufficiently far from the Planck scale. Thus, the quantum phenomena would arise from dynamics beneath.

Numerous indications exist which support this view. Nevertheless, the general principles governing such hypothetical dynamics are under debate [3, 4, 5, 6, 7, 8, 9, 10, 11, 12, 13, 14] and, perhaps, have not even been touched upon, so far.

In this situation, it may be worth while to focus on aspects of quantum and classical mechanics which bring them as closely together as possible, yet sharpen remaining differences. The *linearity* of the respective evolution equations must certainly be counted among such aspects.

Indeed, it is remarkable that dynamical evolution in physics, to a large extent, can be encoded in one or the other of three linear equations:

$$\partial_t \rho = \{H, \rho\} \ , \qquad (1)$$

$$i\partial_t \Psi = \hat{H}\Psi \ , \qquad (2)$$

$$i\partial_t \hat{\rho} = [\hat{H}, \hat{\rho}] \ , \qquad (3)$$

which, respectively, are the *Liouville equation* (1), governing the evolution of a classical phase space density ρ in terms of the Poisson bracket $\{\ ,\ \}$ and the Hamilton function H of the object under study, the *Schrödinger equation* (2) for the quantum mechanical wave function(al) in terms of the appropriate Hamilton operator $\hat{H}$, and the *von Neumann equation* (3) for the density operator $\hat{\rho}$ in terms of the commutator $[\ ,\ ]$ with $\hat{H}$.

We emphasize that all three equations pertain to (or can be interpreted as) ensemble theories. They have the generic structure:

$$i\partial_t \text{ ``state''} = \hat{\mathcal{L}} \text{ ``state''} \quad , \tag{4}$$

where the "state" appears only *linearly* in all cases. This first-order in time differential equation, of course, assumes widely different forms depending on the forces or dynamics under study. That is, the "Liouville superoperator" $\hat{\mathcal{L}}$ on the right-hand side allows to incorporate physics ranging from classical non-relativistic single particles all the way to relativistic quantum fields, employing a phase space ensemble of initial conditions in classical statistical mechanics and the functional Schrödinger picture for interacting quantum fields.

Since the Schrödinger equation describes special cases (pure states) of situations (pure and mixed states) covered by the von Neumann equation, we consider Eqs. (1) and (3), eventually in the form of Eq. (4), in the following.

In passing, we recall that *nonlinear* modifications of quantum mechanics have been proposed and severely constrained in various ways, see, for example, Refs. [15, 16, 17, 18].

In distinction, modifications of the linear Eq. (3), in particular, have been derived for "open systems" or quantum mechanical objects interacting with an environment (such as a heat bath of oscillators, etc.). – Here, one is clearly *not* interested in modifying quantum mechanics but in applying it to complex situations. – The generic form of the resulting *consistent* dynamics, replacing Eq. (3) while preserving the defining properties of the density operator, is known as the *Lindblad equation* [19, 20]. Its generator provides an important example of a *completely positive map*. Some aspects of this will be discussed in more detail in Section 4, in order to contrast it with modifications of the linear dynamics introduced there.

A hypothetical fundamental coupling of quantum mechanical and classical degrees of freedom – *"hybrid dynamics"* – instead, presents a departure from quantum mechanics. While this does not touch the linearity, in the sense of Eq. (4), consistency of the resulting theory has to be carefully examined.

We remark that there is also a practical interest in certain forms of hybrid dynamics. The Born-Oppenheimer approximation, for example, is based on a separation of interacting slow and fast degrees of freedom of a compound object. The former are treated as approximately classical while the latter as of quantum mechanical nature. [1] It must be emphasized, however, that in these cases, hybrid dynamics is considered as relevant *approximate description* of an intrinsically quantum mechanical object. This has recently been employed, for example, in a re-derivation of geometric forces and Berry's phase [22].

Such considerations are and will become increasingly important for precise manipulations of quantum mechanical objects by apparently, for all practical purposes classical means, especially in (sub-)nanotechnological devices.

However, this may be also relevant for studies of the *measurement problem*: Namely, quantum theory, endowed with the Copenhagen interpretation, *assumes* a coupling of the quantum mechanical object to the classical measuring (or manipulating) apparatus. However, it remains profoundly silent about *how* this is realized and how the ensuing state reduction ("collapse of the wave function") proceeds. – Truly hybrid dynamics, which does not arise from an approximation, augmented by effects of environment induced decoherence and anharmonic forces, must have something to say about measurement situations, if it can be formulated consistently.

Presently, we are motivated by interest in the *back-reaction* effect of quantum fluctuations on classical degrees of freedom, in particular if they are physically distinct. We recall here

[1] Mean field theory, based on the expansion of quantum mechanical variables into a classical part plus quantum fluctuations, leads to another approximation scheme and another form of hybrid dynamics. This is discussed more generally for macroscopic quantum phenomena in Ref. [21].

especially discussions of the "semiclassical" Einstein equation coupling the classical metric of spacetime to the expectation value of the energy-momentum tensor of quantized matter fields. Can this be made into a consistent hybrid theory leaving gravity unquantized? This has recently been re-examined, for example, in Refs. [23] and [24, 25, 26]; earlier related work includes Refs. [27, 28, 29, 30, 31, 32, 33, 34].

Concerning the origin of quantum mechanics from a coarse-grained deterministic dynamics [1, 2], to which we alluded above, the back-reaction problem can be more provocatively stated as the problem of the interplay of fluctuations among underlying deterministic and emergent quantum mechanical degrees of freedom. Or, in short: *"Can quantum mechanics be seeded?"*

In this paper, we present a few ingredients which may be helpful in further studies of these questions. – In the following Section 2, we begin with a brief recapitulation of our path integral formulation of classical Hamiltonian dynamics. Some simple observations following from this lead us to take up the subject of hybrid dynamics in Section 3. In Section 4, we continue the discussion from a different angle, namely with some remarks on problems encountered when trying to embed classical or quantum mechanics in a more general linear dynamics.

2. A path integral for classical Hamiltonian dynamics

Our aim here is to solve the Liouville equation (1) with the help of a suitable propagator which, in turn, is represented as a path integral. We summarize essential steps, while more details of the derivation can be found in Ref. [35].

The following considerations are rather independent of the number of degrees of freedom and apply to matrix or Grassmann valued variables as well; field theories require a classical functional formalism, which has been considered elsewhere [8, 36]. We consider a one-dimensional system, for simplicity.

2.1. The quantum-like version of the Liouville equation

We assume conservative forces acting on the classical object under study and that Hamilton's equations are determined by the generic Hamiltonian function:

$$H(x,p) := \frac{1}{2}p^2 + V(x) \ , \tag{5}$$

where x and p denote generalized coordinate and momentum, respectively, and where $V(x)$ stands for an external potential (a mass parameter will be inserted later).

An ensemble of such objects can be described by a probability distribution function ρ depending on the x,p-coordinates of phase space and time. Such a distribution evolves according to the *Liouville equation*:

$$-\partial_t\rho = \frac{\partial H}{\partial p}\cdot\frac{\partial\rho}{\partial_x} - \frac{\partial H}{\partial x}\cdot\frac{\partial\rho}{\partial_p} = \{p\partial_x - V'(x)\partial_p\}\rho \ , \tag{6}$$

with $V'(x) := \mathrm{d}V(x)/\mathrm{d}x$. – The relative minus sign in the Poisson bracket, or between terms here, reflects the symplectic phase space symmetry. It will give rise to a commutator structure, which reminds one of quantum mechanics, as we shall see momentarily.

Following a Fourier transformation, $\rho(x,p;t) = \int \mathrm{d}y\ \mathrm{e}^{-ipy}\rho(x,y;t)$, and a transformation of the effectively *doubled number of spacelike coordinates*,

$$Q := x + y/2 \ , \quad q := x - y/2 \ , \tag{7}$$

the Liouville equation becomes:

$$i\partial_t\rho \ = \ \{\hat{H}_Q - \hat{H}_q + \mathcal{E}(Q,q)\}\rho \ , \tag{8}$$

$$\hat{H}_\chi := -\frac{1}{2}\partial_\chi^2 + V(\chi) \;, \quad \text{for } \chi = Q, q \;, \tag{9}$$

$$\mathcal{E}(Q,q) := (Q-q)V'(\frac{Q+q}{2}) - V(Q) + V(q) = -\mathcal{E}(q,Q) \;. \tag{10}$$

Thus, it bears strong resemblance to the *von Neumann equation*, cf. Eq. (3), considering $\rho(Q,q;t)$ as matrix elements of a density operator $\hat{\rho}(t)$.

We automatically recover the Hamiltonian operator $\hat{H}$ related to the Hamiltonian function, Eq. (5), as in quantum theory. Furthermore, reality and normalization of the phase space probability distribution $\rho(x,p;t)$ translate into hermiticity and trace normalization of the density operator $\hat{\rho}(t)$ [35].

However, an essential dynamical feature here consists in the interaction $\mathcal{E}$ between *bra*- and *ket*-states. Thus, generally, the Hilbert space and its dual are coupled by a genuine *superoperator*, a concept to be defined in the following subsection. [2]

It is remarkable that the interaction between bra- and ket-states vanishes under certain circumstances:

$$\mathcal{E} \equiv 0 \iff \text{potential } V(x) \text{ is constant, linear, or harmonic} \;, \tag{11}$$

rendering a Liouville superoperator of quantum mechanical form, i.e., as in Eq. (3). [3] This has been discussed in Ref. [2] under the perspective of having quantum phenomena emerge due to discrete spacetime structure.

In the following, we will study in more detail the classical dynamics described by Eq. (8), or by appropriate generalizations, and pay particular attention to the presence of the superoperator $\mathcal{E}$, when comparing with the von Neumann equation and its solution by a propagator.

2.2. The superspace

The dynamics of density operators in the general form of Eq. (4) can be conveniently rewritten by introducing the concept of *superspace*, also called *Liouville space* [35, 37]. [4]

Let $\hat{H}$ denote a Hamiltonian operator, as in quantum theory, and let $\{|j\rangle\}$, $j = 1,\dots,N$, present a complete orthonormal set of basis states, assuming here that the relevant Hilbert space is N-dimensional. Then, the matrix elements of the von Neumann equation (3) read:

$$i\partial_t\rho_{jk} = [(\hat{H}\rho)_{jk} - (\rho\hat{H})_{jk}] \;, \quad j,k = 1,2,\dots,N \;, \tag{12}$$

with a density matrix ρ of N^2 elements. Or, written as in Eq. (4):

$$i\partial_t\rho_{jk} = \sum_{l,m}\mathcal{L}_{jk,lm}\rho_{lm} \equiv (\hat{\mathcal{L}}\rho)_{jk} \;, \tag{13}$$

where the *Liouville superoperator* $\hat{\mathcal{L}}$ is now defined by its matrix elements:

$$\mathcal{L}_{jk,lm} := H_{jl}\delta_{mk} - \delta_{jl}H_{mk} \;. \tag{14}$$

[2] The interaction $\mathcal{E}$ is antisymmetric under $Q \leftrightarrow q$. It follows that the complete Liouville superoperator on the right-hand side of Eq. (8), to be compared with Eq. (4), has a symmetric spectrum with respect to zero and, in general, will not be bounded below. Related observations were discussed, for example, in Refs. [6, 8, 9, 10].

[3] Analogously, the vanishing of $\mathcal{E}$ in a field theory amounts to having massive or massless free fields, with or without external sources, and with or without bilinear couplings. In these cases, anharmonic forces or interactions are absent.

[4] No relation with the case of supersymmetry is implied.

This simple rewriting may suggest to introduce a space in which the density operator is a vector. This is the role of the Liouville space (or superspace).

The dynamics of density operators can then be described in parallel for classical and quantum mechanics. In general, they will differ, of course, by the precise form of the superoperator, examples of which we have seen here and in the previous subsection.

Given the Hilbert space, as above, the density operator can be expanded as:

$$\hat{\rho} = \sum_{j,k} \rho_{jk} |j\rangle\langle k| \ . \tag{15}$$

The family of N^2 operators $|j\rangle\langle k|$, with $j, k = 1, \cdots, N$, can be interpreted as a complete set of matrices, or vectors, such that the density operator becomes:

$$|\rho\rangle = \sum_{j,k} \rho_{jk} |jk\rangle\rangle \ , \tag{16}$$

where the "ket" $|jk\rangle\rangle$ denotes the Liouville space *vector* representing the Hilbert space *operator* $|j\rangle\langle k|$. Similarly, we introduce a "bra"-vector $\langle\langle jk|$ as the Hermitian conjugate of $|jk\rangle\rangle$.

Consequently, any operator $\hat{A}$ is represented by a vector, denoted by $|A\rangle\rangle$, and can be expanded as:

$$|A\rangle\rangle = \sum_{j,k} A_{jk} |jk\rangle\rangle \ , \tag{17}$$

where A_{jk} are the usual matrix elements $\langle j|\hat{A}|k\rangle$. – Furthermore, with a bra-vector $\langle\langle B|$ representing $\hat{B}^\dagger$, the scalar product of two operators is defined by:

$$\langle\langle B|A\rangle\rangle := \mathrm{Tr}(\hat{B}^\dagger \hat{A}) \ . \tag{18}$$

This implies the orthonormality condition:

$$\langle\langle jk|mn\rangle\rangle = \mathrm{Tr}(|k\rangle\langle j|m\rangle\langle n|) = \delta_{kn}\delta_{jm} \ , \tag{19}$$

in analogy to and based on $\langle j|k\rangle = \delta_{jk}$. – Finally, consider the scalar product:

$$\langle\langle jk|A\rangle\rangle = \mathrm{Tr}(|k\rangle\langle j|\hat{A}) = \sum_{l} \langle l|k\rangle\langle j|\hat{A}|l\rangle = \langle j|\hat{A}|k\rangle \equiv A_{jk} \ . \tag{20}$$

Upon substitution in Eq. (17), this yields:

$$|A\rangle\rangle = \sum_{j,k} |jk\rangle\rangle\langle\langle jk|A\rangle\rangle \ . \tag{21}$$

This is consistent with the following completeness relation in Liouville space:

$$\sum_{j,k} |jk\rangle\rangle\langle\langle jk| = \mathbf{1} \ . \tag{22}$$

It is easy to see now that the Liouville space is a linear space and that the density operator $\hat{\rho}$, in particular, is a vector in this space. – To conclude these formal considerations, we define a linear operator in terms of its matrix elements:

$$\hat{\mathcal{F}} := \sum_{j,k,m,n} |jk\rangle\rangle\langle\langle jk|\hat{\mathcal{F}}|mn\rangle\rangle\langle\langle mn| \equiv \sum_{j,k,m,n} \mathcal{F}_{jk,mn} |jk\rangle\rangle\langle\langle mn| \ , \tag{23}$$

an example of which is given in Eq. (14).

The importance of Liouville space for classical *and* quantum dynamics resides in that Liouville and von Neumann equations, both, assume the form of Eq. (4), incorporating an appropriate superoperator $\hat{\mathcal{L}}$, cf. Eqs. (8)–(10) and (12)–(14), respectively. Thus, we find a close formal similarity between the structure of these equations and the Schrödinger equation (2).

Therefore, it is plausible that formal derivations or techniques concerning the solution of the Schrödinger equation can be transferred to the cases of Liouville or von Neumann equations with the help of Liouville space notions. This regards perturbation theory as much as nonperturbative methods, the path integral approach in particular, to which we turn next.

2.3. The Liouville propagator and path integral

The technical ingredients of the Feynman path integral approach, for the derivation of quantum mechanical propagators in particular, are very well known [38, 39]. We will make use of these ideas, in order to derive a path integral for the propagator of density matrices based on the Liouville space formulation of the preceding subsection.

Our derivation relies on the close formal similarity between the classical Liouville equation and the von Neumann equation on one hand side and the Schrödinger equation on the other, in the appropriate representation that we discussed. [5]

In particular, the formal solutions of the classical Liouville equation and of the quantum mechanical von Neumann equation, both, can be written in the form:

$$|\rho(t)\rangle\rangle = \mathrm{e}^{-i\hat{\mathcal{L}}(t-t_0)/\hbar}|\rho(t_0)\rangle\rangle \ , \tag{24}$$

where $\hat{\mathcal{L}}$ is the relevant Liouville superoperator. Here, we have:

$$\langle\langle Q,q|\hat{\mathcal{L}}|Q',q'\rangle\rangle = \delta(Q-Q')\delta(q-q')(\hat{H}(Q)-\hat{H}(q)+\mathcal{E}(Q,q)) \ , \tag{25}$$

where $\hat{H}$ denotes the appropriate Hamilton operator in coordinate representation, as indicated, which alone is relevant for the von Neumann equation, while $\mathcal{E}$ represents the additionally present superoperator term for classical dynamics, cf. Subsection 2.1.

In order to solve the problem of time evolution in the present case, we need to know the (super)matrix elements entering the propagation equation:

$$\langle\langle Q,q|\rho(t)\rangle\rangle = \int \mathrm{d}Q'\mathrm{d}q' \ \langle\langle Q,q|\mathrm{e}^{-i\hat{\mathcal{L}}(t-t_0)/\hbar}|Q',q'\rangle\rangle\langle\langle Q',q'|\rho(t_0)\rangle\rangle \ , \tag{26}$$

in analogy to the case of a state vector evolving according the Schrödinger equation. Not surprisingly, we can now follow the usual steps [38, 39], in order to construct the path integral representation of our propagator [35].

In this derivation, one has to pay attention to a suitable generalization of the Trotter product formula. This works out in a straightforward way; the relevant definitions and details of the proof are given in Ref. [40].

Rewriting the Eq. (26) as:

$$\rho(Q,q;t) = \int \mathrm{d}Q'\mathrm{d}q' \ \mathcal{G}(Q,q;t|Q',q';t_0)\rho(Q',q';t_0) \ , \tag{27}$$

our interest is to know the *superpropagator* $\mathcal{G}$.

[5] In this subsection, we reinstate $\hbar$ explicitly.

Following the recipe to arrive at a path integral representation, we make use here of suitably inserted complete sets of superspace vectors, such as:

$$\int \mathrm{d}Q\mathrm{d}q\ |Q,q\rangle\rangle\langle\langle Q,q| = \mathbf{1}\ , \tag{28}$$

and, correspondingly, for momentum space, cf. Eq. (22). Using the plane wave relation between coordinate and momentum eigenfunctions, we also employ:

$$\langle\langle P,p| = \frac{1}{2\pi\hbar}\int \mathrm{d}Q\mathrm{d}q\ \exp\big(-\frac{i}{\hbar}(PQ-pq)\big)\langle\langle Q,q|\ . \tag{29}$$

Furthermore, the orthogonality relation $\langle\langle Q,q|Q',q'\rangle\rangle = \delta(Q-Q')\delta(q-q')$, cf. Eq. (19), implies:

$$\langle\langle P,p|Q,q\rangle\rangle = \frac{1}{2\pi\hbar}\exp\big(-\frac{i}{\hbar}(PQ-pq)\big)\ . \tag{30}$$

Then, with all following steps of the derivation in parallel with the usual ones, it is straightforward to obtain the *Liouville path integral* [35, 40]:

$$\mathcal{G}(Q_f,q_f;t|Q_i,q_i;t_0) = \int \mathcal{D}Q\mathcal{D}q\ \exp\big(\frac{i}{\hbar}\mathcal{S}[\dot{Q},Q;\dot{q},q]\big)\ , \tag{31}$$

with the boundary conditions $Q(t_i) = Q_i$, $q(t_i) = q_i$, $Q(t_f) = Q_f$, and $q(t_f) = q_f$, otherwise unrestricted paths, and with the *superaction* $\mathcal{S}$ defined by:

$$\mathcal{S} := \int_{t_0}^{t}\mathrm{d}\tau\ \Big(\tfrac{m}{2}\dot{Q}^2 - V(Q) - (\tfrac{m}{2}\dot{q}^2 - V(q)) - \mathcal{E}(Q,q)\Big)\ , \tag{32}$$

for a particle of mass m. We recall that $\mathcal{E} \equiv 0$ corresponds to evolution according to the von Neumann equation, whereas $\mathcal{E} \neq 0$ represents the *only* modification due to classical dynamics, in accordance with the Liouville equation, cf. Eqs. (8)–(10).

However simple this result may seem, the Eqs. (31)–(32) describe time evolution of the full density matrix. The particular new feature is that formally *classical dynamics is treated on the same footing as quantum mechanics*, differing only in the action entering the phase in the integrand of the path integral.

These considerations are limited neither by one dimension nor by single-particle physics, but can be extended all the way to relativistic field theories. This offers new calculational tools, new approximation methods in particular, and may be of interest for applications in classical statistical mechanics. In the following, however, we turn to the quantum-classical divide.

2.4. Simple properties of the Liouville path integral

First of all, it seems worth while to record a few simple properties of the Liouvile path integral.

We observe that in the absence of forces, $V \equiv 0$, the classical and quantum mechanical propagators *coincide*, according to Eqs. (31)–(32). Which implies that classical and quantum mechanical behaviour can differ at most in the initial states that are being propagated, in this case. – This holds true even for $V \neq 0$, if $\mathcal{E} \equiv 0$, i.e., for harmonic forces, cf. (11). – Thus, the cherished "textbook effect" of the *spreading of wave packets* is not a peculiar (kinematical) quantum effect, but rather an effect of the particular states considered!

A simple calculation, taking into account our above transformations from x,p- to Q,q-coordinates (and back), shows that a *classical point particle* with initial phase space distribution $\propto \delta(x-x_0)\delta(p-p_0)$ is propagated to the distribution $\propto \delta(x-x_0-p_0t/m))\delta(p-p_0)$, i.e. along a straight line path, as expected. – Instead, a free Gaussian wave packet in Q,q-space spreads in the way described in textbooks on quantum mechanics.

Finally, we can convince ourselves that the classical dynamics is properly represented by the Liouville propagator, in general, by undoing the coordinate transformations (7) directly in the path integral representation of the superpropagator:

$$y := Q - q \ , \quad x := \frac{Q+q}{2} \ , \tag{33}$$

a transformation with unit Jacobian. Following a partial integration of the superaction, the path integral over the y-coordinate simply yields a functional δ-"function" and results in:

$$\mathcal{G}(x_f, y_f; t | x_i, y_i; t_0) = \int \mathcal{D}x \ \exp(im\dot{x}_f y_f) \ \delta[m\ddot{x} + V'(x)] \ \exp(-im\dot{x}_i y_i) \ , \tag{34}$$

with the boundary conditions $x(t_{i,f}) = x_{i,f}$, similarly for the velocities, and where $y_{i,f}$ denote the initial and final values of the y-variable, respectively. The phase factors stem from the partial integration of the superaction.

Thus, we find the expected result that only paths following solutions of the *classical* equations of motion contribute to the propagator. The role of the phase factors is easily understood by recalling Eq. (27): After the inverse coordinate transformations (33), one of the two ordinary integrations there becomes here, including the relevant (initial) phase factor:

$$\int \mathrm{d}y_i \ \mathrm{e}^{-im\dot{x}_i y_i} \rho(x_i, y_i; t_0) = \rho(x_i, m\dot{x}_i \equiv p_i; t_0) \ , \tag{35}$$

i.e., this incorporates the inverse Fourier transformation, back to the momentum variable of phase space. The second (final) phase factor, then, is necessary for the propagator to fulfil the important semi-group property [38].

3. Hybrid dynamics for two interacting objects

We have seen that the path integral for the propagator of the classical Liouville equation shows significant similarity with the propagator of the quantum mechanical von Neumann equation. This suggests a new perspective on *hybrid dynamics* – the hypothetical direct coupling of quantum and classical degrees of freedom that we discussed in Section 1.

In particular, we are interested here in the propagator for the density matrix of a bi-partite system, composed of a classical and a quantum mechanical particle of masses m, m' in external potentials V, v, respectively. Correspondingly, the relevant coordinates will be denoted by Q, q and Q', q', respectively. Our previous considerations lead us to propose the following *hybrid superpropagator* (cf. Eq. (31)):

$$\mathcal{G}_{\mathrm{h}}(Q_f, q_f; Q'_f, q'_f; t | Q_i, q_i; Q'_i, q'_i; t_0) = \int \mathcal{D}Q\mathcal{D}q\mathcal{D}Q'\mathcal{D}q' \ \exp\big(\frac{i}{\hbar}\mathcal{S}[\dot{Q}, Q; \dot{q}, q; \dot{Q}', Q'; \dot{q}', q']\big) \ , \tag{36}$$

where the superaction naturally consists of three contributions, $\mathcal{S} \equiv \mathcal{S}_{\mathrm{cl}} + \mathcal{S}_{\mathrm{qm}} + \mathcal{S}_{\mathrm{h}}$:

$$\mathcal{S}_{\mathrm{cl}} := \int_{t_0}^{t} \mathrm{d}\tau \ \Big(\tfrac{m}{2}(\dot{Q}^2 - \dot{q}^2) - (Q-q)V'(\frac{Q+q}{2})\Big) \ , \tag{37}$$

$$\mathcal{S}_{\mathrm{qm}} := \int_{t_0}^{t} \mathrm{d}\tau \ \Big(\tfrac{m}{2}(\dot{Q}'^2 - \dot{q}'^2) - (v(Q') - v(q'))\Big) \ , \tag{38}$$

$$\mathcal{S}_{\mathrm{h}} := -\int_{t_0}^{t} \mathrm{d}\tau \ \mathcal{V}(Q, q; Q', q') \ . \tag{39}$$

Several remarks are in order here. The contribution $\mathcal{S}_{\rm cl}$ is the classical superaction arrived at in Eqs. (31)–(32); we just inserted $\mathcal{E}$, Eq. (10), explicitly and collected terms. Correspondingly, the contribution $\mathcal{S}_{\rm qm}$ is the action for the propagator of the von Neumann equation, i.e., it describes the quantum mechanical particle in the usual way. Thus, in the absence of the interaction term $\mathcal{S}_{\rm h}$, the hybrid superpropagator $\mathcal{G}_{\rm h}$ factorizes into the corresponding ones for Liouville and von Neumann equations. This would consistently describe a composite system of two independent, classical and quantum mechanical particles.

The interaction term $\mathcal{S}_{\rm h}$ introduces the coupling responsible for *hybrid dynamics*. – For our present purposes, we restrict this coupling by two consistency requirements: [6]
(**A**) $\mathcal{S}_{\rm cl}+\mathcal{S}_{\rm h}$ describes a classical particle interacting with another particle (coordinates Q',q'), as if the latter were *classical*;
(**B**) $\mathcal{S}_{\rm qm}+\mathcal{S}_{\rm h}$describes a quantum mechanical particle interacting with another particle (coordinates Q,q), as if the latter were *quantum mechanical*.

In view of the general form of the external potential terms in Eqs. (37) and (38), these conditions can only be fulfilled, if $\mathcal{V}(Q,q;Q',q')$ is *harmonic*, i.e., is a polynomial of degree less than or equal to two in all variables. Thus, the hybrid coupling is of a form that admits a classical or quantum mechanical interpretation, depending on whether it is viewed from the classical or quantum mechanical subsystem.

An example is provided by a distance dependent oscillator potential, $\lambda^{-1}\mathcal{V}:=(x-x')^2$. Following the derivation of the classical superpropagator, this becomes, in the above coordinates:

$$\begin{aligned}\lambda^{-1}\mathcal{V}(Q,q;Q',q') &= \Big((Q-q)\partial_{(Q+q)/2}+(Q'-q')\partial_{(Q'+q')/2}\Big)\big((Q+q)/2-(Q'+q')/2\big)^2 \\ &= (Q-q)(Q+q)+(Q'-q')(Q'+q')^2-2(QQ'-qq')\;, \qquad (40)\end{aligned}$$

i.e., it can be seen as coupling between two classical particles, fulfilling condition (**A**). On the other hand, the separable oscillator terms here also equal $Q^2-q^2+Q'^2-q'^2$, i.e., are of quantum mechanical form, in accordance with (11), and the bilinear coupling can as well be seen as quantum mechanical, fulfilling condition (**B**).

Interesting consequences of these conditions, determining a *harmonic hybrid interaction*, are summarized in the following Table:

V	v	$\mathcal{V}$	resulting dynamics
h	h	h	CL or QM
anh	h	h	CL
h	anh	h	QM
anh	anh	h	???

Indicated are the nature of the potentials acting on the classical and quantum mechanical particles and their interaction, V, v, and $\mathcal{V}$, respectively – h: harmonic; anh: anharmonic – and the character of the resulting hybrid dynamics – CL: classical; QM: quantum mechanical.

For example, if the classical and quantum mechanical particles are governed by a harmonic potential and an anharmonic potential, respectively, then the resulting dynamics of the composite system can be considered as that of two bilinearly coupled quantum mechanical particles, one of which moves in a harmonic potential. – The case of both potentials being anharmonic is most interesting. However, no general statements can be made in this case without further study. The quantum mechanical or classical behaviour of the composite system might very well depend on *where* (concerning its variables) one is looking at this object! [7]

[6] The question of consistency of the map induced by the superpropagator which evolves the quantum-classical hybrid density will be discussed in more detail elsewhere.
[7] These findings might help with the problem posed in Section 1: *"Can quantum mechanics be seeded?"*

3.1. Intra- and interspace entanglement

There is a *qualitative difference between CL and QM*, contained in the full path integral for the superpropagator, Eqs. (36)–(39). Following standard arguments which lead back from a path integral to an equivalent equation of motion for the (hybrid) density operator $\hat{\rho}$, several related observations may be interesting.

The QM evolution is generated by a commutator of the Hamiltonian with the density operator. Generally, this superposes and, for multi-partite systems, in particular, entangles underlying bra- and ket-states separately, $\propto H_{ij}\rho_{jk} - \rho_{ij}H_{jk}$ (using a convenient notation with discrete indices). For a bi-partite system, it is revealing to write the relevant interaction terms explicitly:

$$[\hat{H}_{int}, \hat{\rho}] = \hat{H}_1\hat{\rho}_1 \otimes \hat{H}_2\hat{\rho}_2 - \hat{\rho}_1\hat{H}_1 \otimes \hat{\rho}_2\hat{H}_2 \ , \tag{41}$$

for an interaction $\hat{H}_{int} := \hat{H}_1 \otimes \hat{H}_2$, with the factors acting on subsystems "1" and "2", respectively, and where $\hat{\rho} = \hat{\rho}_1 \otimes \hat{\rho}_2$, for a separable initial state. This has been called *dynamically assisted entanglement generation*, see, for example, Refs. [41, 42, 43].

It may come as a surprise that the CL evolution does this just as well, due to the structure of the superoperator. For polynomial interactions, in particular, the superoperator *always* contains a contribution proportional to the usual QM terms.

However, the CL evolution, generally, produces additional correlations in $\hat{\rho}$, due to terms contained in $\mathcal{L}_{ij;kl}\rho_{kl}$ which *entangle bra- and ket-states.* – In comparison with Eq. (41), such terms can have the unfamiliar structure:

$$\hat{H}_1'\hat{\rho}_1 \otimes \hat{\rho}_2\hat{H}_2' - \hat{\rho}_1\hat{H}_1' \otimes \hat{H}_2'\rho_2 \ , \tag{42}$$

which differs decidedly from a commutator. – This leads us to distinguish *intra-* (i.e., within given tensor product Hilbert space of subsystems "1" and "2") and *inter-space entanglement* (i.e., between said Hilbert space and its dual).

Consider, for example, the anharmonic potential $V(x_1 - x_2) := \lambda(x_1 - x_2)^4$ for a bi-partite system consisting of particles "1" and "2". Similarly as before, this leads here to the interaction:

$$\mathcal{V}(Q_1, q_1; Q_2, q_2) = \frac{1}{2}\lambda(Q_1 - q_1 - (Q_2 - q_2))(Q_1 + q_1 - (Q_2 + q_2))^3 \ , \tag{43}$$

in terms of, by now, familiar variables, taking into account subsystems "1" and "2" with Q's and q's refering to bra- and ket-states, respectively. Besides separable terms, $\propto (Q_a - q_a)(Q_a + q_a)^3$, $a = 1, 2$, there are terms which mix (and entangle) variables of both subsystems, as usual in QM. However, there are also additional terms that refer to Hilbert space and its dual simultaneously (and entangle corresponding states), for example, $\propto Q_aQ_bq_b^2$, $b \neq a$.

In retrospect, somehow, such difference between CL and QM evolution had to be expected: instead with superstates $|Q, q\rangle\rangle$, we could have worked with superstates $|x, p\rangle\rangle$, relating to coordinates and momenta of the classical theory. There, coordinates and momenta end up tightly correlated, due to Hamilton's equations, and produce inter-space entanglement in an interacting bi-partite system.

Thus, the confrontation of CL with QM, as in our side-by-side study, is quite revealing. In particular, we speculate that this opens new views on generating entanglement in multipartite systems, perhaps, by evolving through quasiclassical stages or by making use of decohered intermediary states. [8]

[8] Previous considerations of the semiclassical regime, such as in Refs. [41, 42], were motivated as suitable approximations of the quantum mechanical evolution, in particular, for studies of the different decoherence properties between classically regular and chaotic systems. Our results seem to show that crossing the quantum-classical divide may offer an additional resource for entanglement generation and related "truly quantum" phenomena. This might be related to a common "underlying reality" of CL and QM physics, assumed to consist, for example, only in statistical correlations in Refs. [44, 45].

Concerning the quantum-classical divide, the present analysis shows that there is an appealing, if not puzzling formal similarity between CL and QM. However, this demonstrates one more time that what has been discussed in various ways as CL limit of QM – and which is similarly relevant for "emergent QM" – deserves further study.

While our work has been concerned mainly with the evolution of CL or QM objects, we recall that V.I. Man'ko and collaborators have pointed out that classical states may differ widely from what could be obtained as the "$\hbar \to 0$" limit of quantum mechanical ones. They show that all states can be classified by their 'tomograms' as *either* CL *or* QM, CL *and* QM, and *neither* CL *nor* QM [46].

The classical limit can be considered a limit "*For All Practical Purposes*", gradually approached with decoherence as an essential but insufficient ingredient or, formally, following the mnemonic "$\hbar \to 0$" rule. However, in order to truly bridge the qualitative difference between intra- and inter-space entanglement that we find, and explain the "Man'ko classes of states", some unknown dynamics seems missing. [9]

The problems discussed here lead us to the question: "Does the $\hbar \to 0$ deformation of QM provide the only *interesting linear dynamics* besides QM itself?"

4. How special is quantum mechanics?

We are interested here once more in the structure of the linear evolution equations that we have discussed side by side, namely the classical Liouville equation and the von Neumann equation of quantum mechanics. However, we consider this with respect to a hypothetical most *general linear dynamics* that can be represented in the generic form of Eq. (4):

$$\partial_t \rho_{ij} = \frac{1}{i} \sum_{k,l} \mathcal{L}_{ij,kl} \rho_{kl} \quad , \tag{44}$$

where the indices $(i, j, k, l = 1, \ldots, N)$ refer to a discrete, finite dimensional Hilbert space, which we assume for simplicity. Thus, generally, there are $N^2 \times N^2$ complex coefficients to be specified. Imposing the *constraint* that the 'density matrix' is and remains Hermitian, which requires that $\mathcal{L}_{ij,kl} = -\mathcal{L}^*_{ji,lk}$, the set of coefficients can be specified by N^4 *real parameters.*

This should be compared with the von Neumann equation – which we discussed, in order to introduce the concept of a superoperator, cf. Eqs. (12)–(14) in Section 2.2. In terms of a Hermitian Hamiltonian, $\hat{H}$, the matrix elements of which are specified by N^2 *real parameters,* the corresponding superoperator has been obtained as: $\mathcal{L}_{ij,kl} \equiv H_{ik}\delta_{lj} - \delta_{ik}H_{lj}$.

Therefore, we can write a symbolic relation, concerning the number of real parameters which determine the respective dynamical equation:

$$\text{“ QM} \sim \sqrt{\text{GL}}\text{ ”} \quad , \tag{45}$$

i.e., the number of real parameters entering the quantum mechanical evolution law (QM) scales with the Hilbert space dimension like the square root of the corresponding number for the most general linear dynamics (GL) which preserves hermiticity. By choosing the eigenstates of the Hamiltonian as basis of the Hilbert space, the number of relevant parameters in QM could be further reduced to N, the number of real eigenvalues of $\hat{H}$.

We may wonder whether the commutator in Eq. (12), $[\hat{H}, .]$, presents the *minimal structure* preserving hermiticity, normalization, and positivity of the density matrix. – Conversely, can the different numbers of real parameters between a more general linear evolution law and QM be attributed to an *attractor mechanism* or *information loss* if and when QM is emergent [1, 2]?

[9] A simple attractor model, motivated by assumptions about effects of fundamental spacetime discreteness [2], has been discussed in Ref. [3].

As a first step to investigate these issues, we consider here whether QM admits a more general linear evolution. – We recall the following result for the QM of *open systems*, see Ref. [47] and references there:

- If and only if the Liouville superoperator $\hat{\mathcal{L}}$ entering the right-hand side of Eq. (44) generates a *Hermitian, trace preserving, completely positive map*, $\hat{\mathcal{M}}(t)\hat{\rho}(0) := \exp(-i\hat{\mathcal{L}}t)\hat{\rho}(0)$, then it can be written in canonical *Kraus form*:

$$\hat{\mathcal{M}}(t)\hat{\rho}(0) = \sum_k \hat{M}_k(t)\hat{\rho}(0)\hat{M}_k^{\dagger}(t) \quad , \tag{46}$$

  with $\sum_k \hat{M}_k^{\dagger}\hat{M}_k = \mathbf{1}_{N\times N}$.
- Such a map $\hat{\mathcal{M}}$ applied to a Hermitian, positive semidefinite, normalized density operator yields another one and linear maps beween density operators can be written in Kraus form.

An evolution equation corresponding to such a map, in general, contains the von Neumann equation as a special case. – In fact, the solution of the von Neumann equation (12) is provided by a unitary transformation, $\hat{\rho}(t) = \hat{U}(t)\hat{\rho}(0)\hat{U}(t)^{\dagger}$, which presents the simplest case with only one unitary Kraus operator, $\hat{M}(t) \equiv \hat{U}(t) := \exp(-i\hat{H}t)$.

Therefore, any generalization which maintains the defining properties of density operators, generally, will need additional Kraus operators specified by additonal parameters. – In this sense, the commutator defining the von Neumann equation presents the *minimal structure* indeed.

Furthermore, it has been shown that the dynamics generated by $\hat{\mathcal{M}}$ of the Kraus form, in general, is equivalent to a non-unitary reduced dynamics of a unitary dynamics on a bigger (tensor product) Hilbert space "$\mathcal{S}$ystem$\otimes\mathcal{E}$nvironment" [47]:

$$\hat{\mathcal{M}}\hat{\rho}_S = \mathrm{Tr}_E\left(\hat{U}_{big}\hat{\rho}_S \otimes \hat{\rho}_E\hat{U}_{big}^{\dagger}\right) \quad , \tag{47}$$

with $\hat{U}_{big}\hat{U}_{big}^{\dagger} = \hat{U}_{big}^{\dagger}\hat{U}_{big} = 1$. Hence, we learn:

- Hermitian, trace preserving, completely positive maps, while generalizing the von Neumann dynamics, are *not* general enough to "leave QM". They describe the QM of *open systems.*

However, this kind of maps does *not* exhaust the larger set of maps generated by all possible Liouville superoperators.

We conclude that one cannot "leave QM" by invoking *more general linear dynamics than that generated by trace preserving, completely positive maps* without affecting properties and interpretation of the states represented by density matrices. [10]

It is useful to recall here (the motivation behind) the assumptions made concerning the properties and interpretation of density operators. – A density operator $\hat{\rho}$ is required to be **i)** *Hermitian*, i.e., to have real eigenvalues, in order to qualify as an observable. This is needed, in turn, if one requires $\hat{\rho}$ to be **ii)** *positive-semidefinite* and **iii)** *normalized.* All three properties, together, are necessary assumptions for the standard probability interpretation of the eigenvalues of a density matrix, according to the Born rule.

A recurrent theme, when comparing quantum with classical states, for example, with the help of the Wigner (function) transform of the density matrix, is the appearance of negative eigenvalues or *"negative probabilities"*. Numerous attempts have been made to give a satisfactory physical interpretation and mathematically consistent definition to these, see Refs. [45, 48, 49] and further references there.

Instead of entering this discussion, we look at the arguably simplest example of *general linear dynamics* in the following. This is obtained by giving up the requirement **ii)** above, i.e., by abandoning positivity, and by considering a two-dimensional Hilbert space.

[10] It will be interesting to see how (any form of) hybrid dynamics, cf. Section 3., fares in this respect.

4.1. General linear 2 × 2 dynamics: a model

A convenient parametrization of the most general Liouville superoperator $\hat{\mathcal{L}}$ for a two-dimensional state space can be written with the help of the Pauli matrices $\vec{\sigma}$. Thus, the (super)matrix elements $\mathcal{L}_{ij,kl}$, $i,j,k,l = 1,2$, are defined by:

$$\mathcal{L}_{ij,kl} := i\mathcal{G}_{\mu\nu}\langle i|a_\mu + \vec{b}_\mu \cdot \vec{\sigma}|k\rangle\langle l|a_\nu + \vec{b}_\nu \cdot \vec{\sigma}|j\rangle \ , \tag{48}$$

where a summation over repeated indices is understood; in this notation we have $(\hat{\mathcal{L}}\hat{\rho})_{ij} = \mathcal{L}_{ij,kl}\rho_{kl}$. Hermiticity requires $\mathcal{G}_{\mu\nu} = \mathcal{G}^*_{\nu\mu}$. Generally, three vectors $\vec{b}_{\mu=0,1,2} \in \mathbf{R}^3$ are needed. Furthermore, we define:

$$\mathcal{G} := \begin{pmatrix} 1 & 0 & 0 \\ 0 & g_{11} & g_{12} \\ 0 & g^*_{12} & g_{22} \end{pmatrix} \tag{49}$$

i.e., where g is a Hermitian 2×2 matrix. This Ansatz saturates the expected number of $N^4 = 2^4 = 16$ real parameters, since a_μ, $\vec{b}_\mu$, and g contribute 3, 9, and 4 parameters, respectively. In order that the map generated by $\hat{\mathcal{L}}$ be trace preserving, we must have $0 \stackrel{!}{=} i\partial_t \mathrm{Tr}\hat{\rho} = \mathrm{Tr}\hat{\mathcal{L}}\hat{\rho}$, for all $\hat{\rho}$, which leads to $\mathcal{L}_{ii,kl} \stackrel{!}{=} 0$. This yields four real constraints,

$$a \cdot \mathcal{G} \cdot a + \vec{b} \cdot \mathcal{G} \cdot \vec{b} \ \stackrel{!}{=} \ 0 \ , \tag{50}$$

$$2a \cdot \mathrm{Re}\mathcal{G} \cdot \vec{b} + ig_{ij}\vec{b}_i \times \vec{b}_j \ \stackrel{!}{=} \ 0 \ , \tag{51}$$

which reduce the number of available real parameters to twelve.

Next, we conveniently choose the vectors $\vec{b}_{0,1,2}$ to form a right-handed orthogonal system; with respect to suitable coordinates, this sets six vector components to zero. Thus, we are left with six real parameters, three of which could characterize the QM evolution of a two-state object, while three pertain to the generalization we are concerned with.

Finally, we assume for simplicity that the 'QM part' of $\hat{\mathcal{L}}$ is diagonal, corresponding to a spin-1/2 particle in a constant external magnetic field parallel to the quantization axis, for example. Setting one of the remaining parameters to zero, we obtain a simplified model with altogether three real parameters, $\alpha, \beta, \gamma \in \mathbf{R}$:

$$\begin{aligned} \mathcal{L}_{ij,kl} \ := \ & \alpha\Big\{\langle i|\sigma_z|k\rangle\delta_{lj} - \delta_{ik}\langle l|\sigma_z|j\rangle\Big\} \\ & +\beta\gamma\Big\{\langle i|\sigma_z|k\rangle\langle l|\sigma_x|j\rangle - \langle i|\sigma_x|k\rangle\langle l|\sigma_z|j\rangle\Big\} + i\beta\gamma\Big\{\langle i|\sigma_y|k\rangle\delta_{lj} + \delta_{ik}\langle l|\sigma_y|j\rangle\Big\} \\ & +i\beta^2\delta_{ik}\delta_{lj} + i\gamma^2\langle i|\sigma_y|k\rangle\langle l|\sigma_y|j\rangle - i(\beta^2+\gamma^2)\langle i|\sigma_z|k\rangle\langle l|\sigma_z|j\rangle \ , \end{aligned} \tag{52}$$

where QM terms $\propto \alpha\sigma_z$ feature in the first line. All other contributions on the right-hand side have no counterpart in QM; this will become obvious by explicitly solving the model. The first two terms in the second line and the two last ones in the third couple bra- and ket-states, which we discussed as a consequence of classical evolution in Sections 2.1 and 3.1. It is easy to verify that the generator $-i\hat{\mathcal{L}}$ is Hermitian and trace preserving, as it should.

Taking hermiticity and trace normalization of the density matrix into account by:

$$\hat{\rho} \equiv \begin{pmatrix} \rho_{11} & \rho_{12} \\ \rho^*_{12} & 1-\rho_{11} \end{pmatrix} \ , \tag{53}$$

we write the resulting evolution equation explicitly:

$$i\partial_t\hat{\rho} \ = \ \hat{\mathcal{L}}\hat{\rho} \ = \tag{54}$$

$$\begin{pmatrix} i\gamma^2(1-2\rho_{11}) & 2\beta\gamma + (2\alpha + i[2\beta^2+\gamma^2])\rho_{12} - i\gamma^2\rho^*_{12} \\ -2\beta\gamma - (2\alpha - i[2\beta^2+\gamma^2])\rho^*_{12} - i\gamma^2\rho_{12} & -i\gamma^2(1-2\rho_{11}) \end{pmatrix} .$$

Note that for $\beta,\gamma\rightarrow 0$, we recover the von Neumann equation, e.g., for a spin-1/2 particle in a constant magnetic field. In this case, besides the Hamiltonian, $\hat{H}:=\alpha\sigma_z$, also the total spin-squared $S^2:=\vec{\sigma}^2/4\propto\mathbf{1}$ is a constant of motion.

4.2. Constants of motion

We remark that if a constant operator, $\hat{C}$, obeys $\mathrm{Tr}(\hat{C}\hat{\rho}(0))=\mathrm{Tr}(\hat{C}\hat{\rho}(t))$, for all solutions $\hat{\rho}$ of the general linear evolution equation (44), then $C_{ij}\mathcal{L}_{ji,kl}=0$, which generalizes the vanishing of the commutator $[\hat{C},\hat{H}]_{lk}$ in QM, for constants of motion. This is fulfilled for any $\hat{C}\equiv c\mathbf{1}$, with constant c, due to the preservation of the trace normalization of the density matrix, incorporated by $\hat{\mathcal{L}}_{ii,kl}\stackrel{!}{=}0$. However, this raises also the interesting question, whether there exist conserved superoperators, $\hat{Q}$, which can be defined by the vanishing supercommutator:

$$[\hat{Q},\hat{\mathcal{L}}]_{ij,mn}:=Q_{ij,kl}\mathcal{L}_{kl,mn}-\mathcal{L}_{ij,kl}Q_{kl,mn}\stackrel{!}{=}0\ . \tag{55}$$

In the present case, this amounts to a 4×4 matrix equation, to be studied.

4.3. Solution of the 2×2 model

We now turn to the explicit solution of Eq. (54), which can be represented as follows. The matrix elements are:

$$\rho_{11}(t)=1-\rho_{22}(t) = \frac{1}{2}+\Big(\rho_{11}(0)-\frac{1}{2}\Big)\mathrm{e}^{-2\gamma^2t}\ , \tag{56}$$

$$\rho_{12}(t)=\rho_{21}^*(t) = r_+\mathrm{e}^{\Omega_+t}+r_-\mathrm{e}^{\Omega_-t}+r_c\ , \tag{57}$$

where $r_\pm$ are determined by initial conditions. [11] The remaining constants are defined by:

$$r_c := -\beta\gamma\frac{\alpha-i\beta^2}{\alpha^2+\beta^2(\beta^2+\gamma^2)}\ , \tag{58}$$

$$\Omega_\pm := 2\beta^2+\gamma^2\pm 2i\alpha(1-\gamma^4/4\alpha^2)^{1/2}\ , \tag{59}$$

and we may choose $\alpha\geq 0$. Diagonalizing the resulting normalized Hermitian density matrix, we obtain the eigenvalues:

$$\rho_\pm(t)=\frac{1}{2}\pm\Big((\rho_{11}(0)-1/2)^2\mathrm{e}^{-4\gamma^2t}+|\rho_{12}(t)|^2\Big)^{1/2}\ , \tag{60}$$

with $|\rho_{12}|^2$ given by Eqs. (57)–(59).

Several features of the solutions of our model seem unexpected. – To begin with, for $\beta,\gamma\rightarrow 0$, we obtain:

$$\rho_{11}(t)=1-\rho_{22}(t) = \rho_{11}(0)\ , \tag{61}$$

$$\rho_{12}(t)=\rho_{21}^*(t) = r_+\mathrm{e}^{2i\alpha t}+r_-\mathrm{e}^{-2i\alpha t}\ , \tag{62}$$

which amounts to the usual QM solution only, if we choose $r_+\equiv 0$. From a QM perspective, the *larger set of solutions*, parametrized by a larger set of initial conditions, is quite surprising.

[11] Interestingly, the underlying equation for the off-diagonal matrix elements, for $\gamma\neq 0$ in particular, is of second order; it decouples into two first order equations in the limit $\gamma\rightarrow 0$, independently of the value of β.

It leads to time dependent 'probabilities', if we try to interpret the eigenvalues of $\hat{\rho}$ in the usual way:

$$\rho_\pm(t) = \frac{1}{2} \pm \left((\rho_{11}(0) - 1/2)^2 + |r_+|^2 + |r_-|^2 + 2\mathrm{Re}(r_+ r_-^* \mathrm{e}^{4i\alpha t}) \right)^{1/2} \; . \tag{63}$$

While these 'probabilities' are real and normalized, by construction of our model, they might temporarily fall outside the interval $[0,1]$, depending on the parameters $r_\pm$.

This remark applies also for the general case, with $\beta, \gamma \neq 0$, where amplitudes grow $\propto \exp(2\beta^2 + \gamma^2)t$, see Eqs. (57),(59).

The effect of such anomalous 'probabilities' is clearly seen in the following mean square deviations sensitive to fluctuations. With $\langle \hat{O} \rangle := \mathrm{Tr}(\hat{O}\hat{\rho})$, $\sigma^2_{x,y,z} = \mathbf{1}$, and $\langle \sigma_x \rangle = 2\mathrm{Re}(\rho_{12})$, $\langle \sigma_y \rangle = -2\mathrm{Im}(\rho_{12})$, $\langle \sigma_z \rangle = 2\rho_{11} - 1$, we obtain:

$$\Delta^2_x := \langle \sigma^2_x \rangle - \langle \sigma_x \rangle^2 \;=\; 1 - \left(2\mathrm{Re}(\rho_{12}) \right)^2 \; , \tag{64}$$

$$\Delta^2_y := \langle \sigma^2_y \rangle - \langle \sigma_y \rangle^2 \;=\; 1 - \left(2\mathrm{Im}(\rho_{12}) \right)^2 \; , \tag{65}$$

$$\Delta^2_z := \langle \sigma^2_z \rangle - \langle \sigma_z \rangle^2 \;=\; 4\rho_{11}(1 - \rho_{11}) \; . \tag{66}$$

While $\Delta^2_z \stackrel{t\to\infty}{\longrightarrow} 1$, for $\gamma \neq 0$, which would correspond to a completely mixed state in QM, $\Delta^2_{x,y}$ can have oscillatory contributions with growing amplitude $\propto \exp \Omega_\pm t$, see Eqs. (56)–(59). – Also note that the would-be QM energy expectation varies in time, $\alpha\langle \sigma_z \rangle = \alpha(2\rho_{11}(0) - 1)\exp(-\gamma^2 t)$, approaching the stationary mixed-state value zero.

To summarize: *Modifying a QM two-state model by applying a general linear perturbation, however small – which preserves hermiticity and trace normalization of the density matrix – opens the possibility of a larger state space, reflected in a doubling of the number of degrees of freedom.* [12]

We emphasize that the QM evolution becomes exponentially unstable when such perturbation is introduced. Generically, this spoils the standard interpretation of the eigenvalues of the density matrix as probabilities and can result in interesting oscillatory effects, as we have seen. [13]

It is an interesting question, whether there is a "classical" formulation of general linear dynamics, suggested by a doubled number of certain degrees freedom and "anomalous probabilities" that seem invariably to appear – and which remind of similar phenomena encountered when relating quantum and classical mechanics, e.g., via the Wigner function.

5. Conclusions

In this article we have summarized our earlier derivation of a *path integral* for classical Hamiltonian systems based on an ensemble description [35, 40].

This leads us to point out the correlation properties of classical dynamics in parallel to the quantum mechanical ones and to identify characteristic similarities and differences. In particular, it seems useful to distinguish *intra- and interspace entanglement.*

The former has been held characteristic of quantum mechanics and a feature of superpositions of tensor product states. The latter concerns classical mechanics only; it correlates Hilbert space states and their duals. However, surprisingly, for anharmonic potentials or interactions, classical mechanics additionally shows intraspace entanglement, as in quantum mechanics.

As a first application, we propose a new formulation of *hybrid dynamics*, i.e., based on a hypothetical direct coupling between quantum and classical objects. This may be of practical as well as foundational interest.

[12] I.e., a doubling of the initial conditions for the off-diagonal matrix elements here; it will be interesting to see, whether this extends to all degrees of freedom in a more general model.

[13] One might speculate on the relevance for flavour oscillations.

Finally, we study a generalization of quantum evolution, *general linear dynamics*, where the evolution is generated by a superoperator that preserves hermiticity and trace normalization of density matrices. We argue that one cannot "leave QM" without giving up one of the three defining properties of density matrices, to be Hermitian, normalized, and positive-semidefinite.

In the most simple case of a two-state system, we solve such dynamics explicitly. We show that the corresponding von Neumann dynamics becomes *exponentially unstable* under the influence of a general linear perturbation. Most interestingly, it leads to the appearance of "anomalous probabilities" and an enlargement of the state space, possibly pointing towards a sort of prequantum dynamics. – We intend to study more complex objects consisting of such two-state systems as building blocks which interact. This may be useful in trying to understand quantum mechanics as an emergent phenomenon [1].

Acknowledgments
We thank M.J. Everitt, F. Finster, A. Khrennikov, V.I. Man'ko, and T. Padmanabhan for discussions and M.J.W. Hall for correspondence.

References

[1] 't Hooft G 2010 Classical cellular automata and quantum field theory *Int. J. Mod. Phys.* A **25** No. 23 4385-4396; *do.* 2009 *Preprint* arXiv:0908.3408
[2] Elze H-T 2009 Does quantum mechanics tell an atomistic spacetime? *J. Phys.: Conf. Ser.* **174** 012009 (*Preprint* arXiv:0906.1101)
[3] Elze H-T 2009 *Int. J. Qu. Inf. (IJQI)* **7** 83-96 (*Preprint* arXiv:0806.3408)
[4] Elze H-T 2008 *J. Phys. A.: Math. Theor.* **41** 304020 (*Preprint* arXiv:0710.2765)
[5] 't Hooft G 2007 *AIP Conf. Proc.* **957** 154-163 (*Preprint* arXiv:0707.4568)
[6] 't Hooft G 2007 *J. Phys.: Conf. Ser.* **67** 012015 (*Preprint* arXiv:quant-ph/0604008)
[7] 't Hooft G 2003 *Int. J. Theor. Phys.* **42** 355-361; *do.* 1999 *Class. Quant. Grav.* **16** 3263-3279
[8] Elze H-T 2006 *J. Phys.: Conf. Ser.* **33** 399-404 (*Preprint* arXiv:gr-qc/0512016); *do.* 2005 *Braz. J. Phys.* **35** 343-350; *do.* 2003 *Phys. Lett.* A **310** 110-118
[9] Blasone M, Jizba P and Kleinert H 2005 *Ann. Phys.* **320** 468-486 (*Preprint* arXiv:quant-ph/0504200); *do.* 2005 *Braz. J. Phys.* **35** 497-502; *do.* 2005 *Phys. Rev.* A **71** 052507
[10] Blasone M, Jizba P and Vitiello G 2001 *Phys. Lett.* A **287** 205-210 (*Preprint* arXiv:hep-th/0007138)
[11] Markopoulou F and Smolin L 2004 *Phys. Rev.* D **70** 124029 (*Preprint* arXiv:gr-qc/0311059)
[12] Adler S L 2005 *Quantum Mechanics as an Emergent Phenomenon* (Cambridge, UK: Cambridge U. Press)
[13] Wetterich C 2008 *J. Phys.: Conf. Ser.* **174** 012008 (*Preprint* arXiv:0811.0927); *do.* 2008 Probabilistic observables, conditional correlations, and quantum physics *Preprint* arXiv:0810.0985; *do.* 2008 Quantum entanglement and interference from classical statistics *Preprint* arXiv:0809.2671
[14] Isidro J M, Santander J L G and Fernandez de Cordoba P 2008 A note on the quantum-mechanical Ricci flow *Preprint* arXiv:0808.2717; *do.* 2008 Ricci flow, quantum mechanics and gravity *Preprint* arXiv:0808.2351
[15] Bialynicki-Birula I and Mycielski J 1976 *Ann. Phys. (N.Y.)* **100** 62
[16] Kibble T W B 1978 *Commun. Math. Phys.* **64** 73
[17] Kibble T W B and Randjbar-Daemi S 1980 *J. Phys.* A **13** 141
[18] Weinberg S 1989 *Phys. Rev. Lett.* **62** 485; *do.* 1989 *Ann. Phys. (N.Y.)* **194** 336
[19] Lindblad G 1976 *Commun. Math. Phys.* **48** 119-130
[20] Gorini V, Kossakowski A and Sudarshan E C G 1976 *J. Math. Phys.* **17** 821-825
[21] Chou C H, Hu B-L and Subaşi Y 2011 Macroscopic quantum phenomena from the large N perspective *Preprint*, in this proceedings volume
[22] Zhang Q and Wu B 2006 *Phys. Rev. Lett.* **97** 190401
[23] Diósi L 2010 The gravity-related decoherence master equation from hybrid dynamics *Preprint* arXiv:1101.0672, in this proceedings volume
[24] Hall M J W and Reginatto M 2005 *Phys. Rev.* A **72** 062109
[25] Hall M J W 2008 *Phys. Rev.* A **78** 042104
[26] Reginatto M and Hall M J W 2009 *J. Phys.: Conf. Ser.* **174** 012038 (*Preprint* arXiv:0905.2948)
[27] Diósi L 1984 *Phys. Lett.* A **105** 199
[28] Diósi L 2005 *Braz. J. Phys.* **35** 260-265 (*Preprint* arXiv:quant-ph/0412154v2)
[29] Penrose R 1998 *Phil. Trans. R. Soc.* **356** 1927
[30] Adler S L 2003 *Stud. Hist. Philos. Mod. Phys.* **34** 135

[31] Ellis J, Mavromatos N E and Nanopoulos D V 1992 *Phys. Lett.* B **293** 37-48 (*Preprint* arXiv:hep-th/9207103v2)
[32] Gambini R, Porto R A and Pullin J 2008 *Phys. Lett.* A **372** 1213-1218 (*Preprint* arXiv:0708.2935v2)
[33] Hu B-L 2009 *J. Phys.: Conf. Ser.* **174** 012015 (*Preprint* arXiv:0903.0878)
[34] Diósi L 2009 *J. Phys.: Conf. Ser.* **174** (*Preprint* arXiv:0902.1464) 012002
[35] Elze H-T, Gambarotta G and Vallone F 2011 *Int. J. Qu. Inf. (IJQI)* **9** Suppl. **1** 203-224 (*Preprint* arXiv:1006.1569)
[36] Elze H-T 2007 *Int. J. Theor. Phys.* **46** No. 8 2063-2081 (*Preprint* arXiv:hep-th/0510267)
[37] Mukamel S 1995 *Principles of nonlinear optical spectroscopy* (Oxford: Oxford Univ. Press)
[38] Schulman L S 1981 *Techniques and applications of path integration* (New York: Wiley)
[39] Kleinert H 2009 *Path integrals in quantum mechanics, statistics, polymer physics, and financial markets* 5th edition (Singapore: World Scientific)
[40] Vallone F 2010 (Master thesis, Universitá di Pisa), unpublished
[41] Jacquod Ph 2004 *Phys. Rev. Lett.* **92** 150403 (*Preprint* arXiv:quant-ph/0308099)
[42] Jacquod Ph and Petitjean C 2009 *Adv. in Phys.* **58** vol. 2, 67 (*Preprint* arXiv:0806.0987)
[43] Busse M and Hornberger K 2010 *J. Phys. A: Math. Theor.* **43** 015303 (*Preprint* arXiv:0910.1062)
[44] Wetterich C 2009 Zwitters: particles between quantum and classical *Preprint* arXiv:0911.1261
[45] Khrennikov A 2003 *"Interpretations of Probability"* (Utrecht and Boston: VSP)
[46] Man'ko O V and Man'ko V I 2004 *J. Russ. Laser Res.* **25**(5) 477 (*Preprint* arXiv:quant-ph/0407183)
[47] Diósi L 2007 *A Short Course in Quantum Information Theory, Lecture Notes in Physics* **713** (Berlin: Springer)
[48] Mückenheim W 1986 *Phys. Rep.* **133**(6) 337
[49] Burgin M 2010 Interpretations of Negative Probabilities *Preprint* arXiv:1008.1287

Recent experimental progresses in testing Quantum Mechanics

Giorgio Brida, Marco Genovese and Fabrizio Piacentini
INRIM, strada delle Cacce 91, 10135 Torino, Italy
E-mail: m.genovese@inrim.it

Abstract. In this paper we will show some recent experiments performed in the "Carlo Novero" labs of the Optics Division of the National Institute of Metrological Research (INRIM, Torino, Italy), in order to discriminate between Standard Quantum Mechanics and some specific, restricted class of Hidden Variable Theories (HVTs).
The first experiment will be on the testing of two restricted Local Realistic Theories (LRTs) [1, 2], properly built to describe polarization entangled photons experiments, whose inequalities are not affected by the detection loophole. The second and third experiments instead will perform the Alicki - Van Ryn non-classicality test [3] on single particles, by means of two different setups based on heralded single photon sources.

1. Introduction

Quantum Mechanics represents nowadays one of the pillars of modern physics: so far a huge amount of theoretical predictions deriving from this theory has been confirmed by very accurate experimental data, while the theory is at the basis of a large spectrum of researches ranging from solid state physics to cosmology, from bio-physics to particle physics. Furthermore, in the last years the possibility of manipulating single quantum states has fostered the development of promising quantum technologies as quantum information (calculus, communication, etc.), quantum metrology, quantum imaging, ...

Nevertheless, even after a pluri-decennial debate many problems related to the foundations of this theory persist [4], like non-local effects of entangled states, wave function reduction and the concept of measurement in Quantum Mechanics, the transition from a microscopic probabilistic world to a macroscopic deterministic world described by classical mechanics (macro-objectivation) and so on. Problems that, beyond their fundamental interest in basic science, now also concern the impact of these developing technologies.

In this proceeding we review some recent experimental works in the field: in particular, some specific experiments realized at INRIM will be described in some detail (concerning tests of specific local realistic models, wave particle duality). The first experiment to be described regards the testing of two specific, restricted local realistic models [1, 2] , properly built for experiments with entangled photons; the interesting feature of these models is the fact that they are free from the detection loophole, thus they don't rely on the fair sampling assumption (the quantum efficiency of the system plays a role in the inequalities built for these models). The experimental data collected violated these inequalities, showing instead a perfect agreement with Quantum Mechanics predictions [5]. The second one is a non-classicality test at the single particle level, in two different experimental configurations [6, 7].

2. Tests on specific loophole-free LR models

Though a complete, loophole-free test of Bell inequalities needs some technological improvements to be performed, it is possible to theorize and check particular restricted families of LHVTs, giving birth to Bell-like inequalities free from detection loophole and, as a consequence, not relying anymore on the Fair Sampling assumption (an additional hypothesis saying that the sub-sample of hidden variables collected in the experiments can be considered a faithful representation of the whole HV distribution). In the first part of this proceeding, starting from generic considerations on experiments based on polarization entangled photons, we present an experiment [5] addressed to test two specific LHVT models, on the basis of two inequalities (not affected by the detection loophole, since they depend also on the system's quantum efficiency η).

2.1. The restricted LHV models

Let's begin with the first LHVT family, the simplest one: if we consider the standard setup for polarization entangled photons experiments, with a Parametric Down-Conversion (PDC) [8, 9] source, with two spatially separated fiber collectors sending each of the down-converted pair of photons to a polarizer and a detector on each optical path (with a tunable angle ϕ_i with respect to the horizontal axis), the single counts R_i, the coincidences R_{12} and the whole pair production R_0 can be written as (for a Ψ^+ Bell state):

$$R_1 = R_2 = \frac{1}{2}\eta R_0 \quad R_{12} = \frac{1}{4}\eta^2 R_0[1 + V\cos(2\phi)] \tag{1}$$

where $\phi = |\phi_1 - \phi_2|$ is the angle between the two polarization axes of the polarizers, V the *visibility* of the state and η the detection efficiency (assumed to be identical for the two arms). A LHV model mimicking Quantum Mechanics results, in this case, could be characterized by the equations [1]:

$$\rho(|\lambda_1 - \lambda_2|) \geq 0 \qquad 0 \leq P_j(|\lambda_j - \phi_j|) \leq 1 \qquad j = 1, 2$$

$$\int \rho(|\lambda_1 - \lambda_2|) d\lambda_1 d\lambda_2 = 1 \tag{2}$$

where λ_1, λ_2 are angular hidden variables and ϕ_1, ϕ_2 the angles of the polarization planes with respect to an axis orthogonal to pump direction and polarization; P_j is the probability of a photon of passing through the polarizer at ϕ_j.
The predictions for the single count and coincidence rates given by this LHVT are:

$$\frac{R_j}{R_0} = \int \rho(|\lambda_1 - \lambda_2|) P_j(|\lambda_j - \phi_j|) d\lambda_1 d\lambda_2 \tag{3}$$

$$\frac{R_{12}}{R_0} = \int \rho(|\lambda_1 - \lambda_2|) \prod_{j=1}^{2} P_j(|\lambda_j - \phi_j|) d\lambda_1 d\lambda_2 \tag{4}$$

In the limit of low η ($\leq 30\%$) and high visibility of the entangled state ($V \geq 95\%$), by introducing the visibility parameters:

$$V_A\big|_{SQM} = \frac{R_{12}(0) - R_{12}(\pi/2)}{R_{12}(0) + R_{12}(\pi/2)} \qquad V_B\big|_{SQM} = \sqrt{2} \cdot \frac{R_{12}(\pi/8) - R_{12}(3\pi/8)}{R_{12}(\pi/8) + R_{12}(3\pi/8)} \tag{5}$$

one can derive the inequality [1]:

$$\frac{V_B}{V_A} \geq F = 1 + \cos^2\left(\frac{\pi\eta}{2}\right)\left[V_B - \frac{\sin^2(\pi\eta/2)}{(\pi\eta/2)^2}\right] \tag{6}$$

that can be violated in Standard Quantum Mechanics (where, in the ideal case, $V_B/V_A = 1$), as we will se in the following.
Now, let us "extend" this particular local realistic model to a larger LHVT family, again free from detection loophole: keeping in mind the same considerations that led to eq. (1), one can take into account a set of four hidden variables $\lambda \equiv \{\chi_1, \chi_2, \mu_1, \mu_2\}$ (of two different types). The equations defining this LHVT are [2]:

$$\rho(\lambda) \equiv \rho_\chi(\chi_1, \chi_2) \cdot g_1(\mu_1) \cdot g_2(\mu_2) \qquad \rho_\chi(\chi_1, \chi_2),\ g_1(\mu_1),\ g_2(\mu_2) \geq 0$$
$$\int \rho_\chi(\chi_1, \chi_2) d\chi_1 d\chi_2 = \int g_j(\mu_j) d\mu_j = 1$$
$$0 \leq P_j(\chi_j, \phi_j) = \int Q_j(\mu_j, \chi_j, \phi_j) g_j(\mu_j) d\mu_j \leq 1 \qquad \left(0 \leq Q_j \leq 1\right) \tag{7}$$

In a system with rotational invariance, in which $R_{12}(\phi_1, \phi_2) = R_{12}(|\phi_1 - \phi_2|)$, by putting $P_1 = P_2 = P$ (perfect symmetry between the apparatuses on the two PDC correlated branches), and one can replace equations (7) with:

$$\rho(\lambda) \equiv \rho_\chi(|\chi_1 - \chi_2|) \cdot g_1(\mu_1) \cdot g_2(\mu_2) \qquad \rho_\chi(|\chi_1 - \chi_2|),\ g_1(\mu_1),\ g_2(\mu_2) \geq 0$$
$$\int \rho_\chi(|\chi_1 - \chi_2|) d\chi_1 d\chi_2 = \int g_j(\mu_j) d\mu_j = 1$$
$$0 \leq P(|\chi_j - \phi_j|) = \int Q_j(\mu_j, |\chi_j - \phi_j|) g_j(\mu_j) d\mu_j \leq 1 \qquad \left(0 \leq Q_j \leq 1\right) \tag{8}$$

giving, after integrating over the μ_j variables and fixing the function periodicity to π, the single counts and coincidence rates:

$$\frac{R_j}{R_0} = \int \rho_\chi(|\chi_1 - \chi_2|) P(|\chi_j - \phi_j|) d\chi_1 d\chi_2 \tag{9}$$
$$\frac{R_{12}}{R_0} = \int \rho_\chi(|\chi_1 - \chi_2|) \prod_{j=1}^{2} P(|\chi_j - \phi_j|) d\chi_1 d\chi_2 \tag{10}$$

where the integration interval is $\left[-\frac{\pi}{2}, \frac{\pi}{2}\right]$. Let us also add the conditions:

$$\frac{d\rho_\chi(|\chi_1 - \chi_2|)}{d(|\chi_1 - \chi_2|)} \leq 0 \qquad \frac{dP(|\chi_1 - \chi_2|)}{d(|\chi_1 - \chi_2|)} \leq 0 \tag{11}$$

the first one indicating that the pairs with $\chi_1 = \chi_2$ are most likely produced in the EPR photons source and the second giving rise to a greater detection probability for incoming photons with polarization close to the polarization plane of the polarizer.
After some manipulations, we can extract from this model the parameter $D(\eta)$:

$$D(\eta) \equiv \Delta = \frac{4}{3\pi}\sqrt{\frac{2}{3\eta} - \frac{1}{2} - \frac{\sin^4(\pi\eta/2)}{(\pi\eta/2)^4}}\left(V - \frac{\sin^2(\pi\eta/2)}{(\pi\eta/2)^2}\right)_+^{\frac{3}{2}} \tag{12}$$

giving the deviation from SQM of the best model of our LHV family. At this point, every experiment, if responding to this LHVT laws, would give a SQM-LHVT discrepancy equal or greater than $D(\eta)$: by considering a finite set of values for the variable ϕ ($\phi_j = \frac{(j-1)\pi}{n}$ $j = 1, ..., n$),spanning in the interval $[0, \pi]$, we can write:

$$\Delta_{exp}^{LHVT} = \left\langle \left[\frac{R_{12}(\phi)}{\langle R_{12}(\phi) \rangle} - 1 - V\cos(2\phi) \right]^2 \right\rangle^{\frac{1}{2}} =$$

$$= \left\{ \frac{1}{n} \sum_{j=1}^{n} \left[\frac{R_{12}(\phi_j)}{\left(\frac{1}{n} \sum_{j=1}^{n} R_{12}(\phi_j) \right)} - 1 - V\cos(2\phi_j) \right]^2 \right\}^{\frac{1}{2}} \geq D(\eta) \tag{13}$$

while in SQM Δ_{exp} should be zero. The minimum for Δ_{exp}^{LHVT} is reached by putting:

$$V = 2\frac{\sum_{j=1}^{n} R_{12}(\phi_j)\cos(2\phi_j)}{\sum_{j=1}^{n} R_{12}(\phi_j)} \tag{14}$$

Substituting this V into eq. (13), one obtains the final inequality [2]:

$$\Delta_{min} = \left[\frac{n\sum_{j=1}^{n} R_{12}^2(\phi_j)}{\left(\sum_{j=1}^{n} R_{12}(\phi_j) \right)^2} - \frac{2\left(\sum_{j=1}^{n} R_{12}(\phi_j)\cos(2\phi_j) \right)^2}{\left(\sum_{j=1}^{n} R_{12}(\phi_j) \right)^2} - 1 \right]^{\frac{1}{2}} \geq D(\eta) \tag{15}$$

where also the V appearing in $D(\eta)$ (eq. (12)) takes the form (14). This inequality, satisfied by our extended LHVT, can be violated by Standard Quantum Mechanics predictions ($\Delta_{min}^{SQM} = 0$). In the next part we will try to exclude these theories by means of the experimental violation of the obtained Bell-like inequalities (6) and (15).

2.2. Loophole-free LHV theories experimental test

The test on these inequalities is performed with the experimental setup reported in Fig. 1.

It consists of a 0.2 W, 398 nm pulsed (with 150 fs pulses) laser beam, obtained by second harmonic generation from a Ti:Sapphire beam at 796 nm, pumping a type-II BBO $5 \times 5 \times 0,5$ mm crystal. The produced collinear degenerate PDC is to a non-polarizing 50%-50% beam splitter (BS): through the BS, the quantum state becomes:

$$|HV\rangle \rightarrow \frac{1}{2}\Big(|H_r V_r\rangle + |H_t V_t\rangle + |V_r H_t\rangle + |H_r V_t\rangle \Big) \tag{16}$$

where r and t indicate the reflection and transmission directions.

By discarding the events with both the EPR photons transmitted or reflected, one obtains the Bell state:

$$|\psi_+\rangle = \frac{1}{\sqrt{2}}\Big(|V_r H_t\rangle + |H_r V_t\rangle \Big) \tag{17}$$

On the (t) arm we place a Glan-Thomson polarizer, a pinhole, a narrow band interference filter (with FWHM 10 nm, λ_{peak} 795 nm) and a fiber coupler connected with a multi-mode fiber to the detector, a Perkin-Elmer SPCM-AQR-14-FC-9491 Si-SPAD. The (r) arm instead hosts a Glan-Thomson polarizer (identical to the one on the t arm), a large band interference filter (FWHM 40 nm, λ_{peak} 800 nm) and again a fiber coupler connected with a multi-mode fiber to a

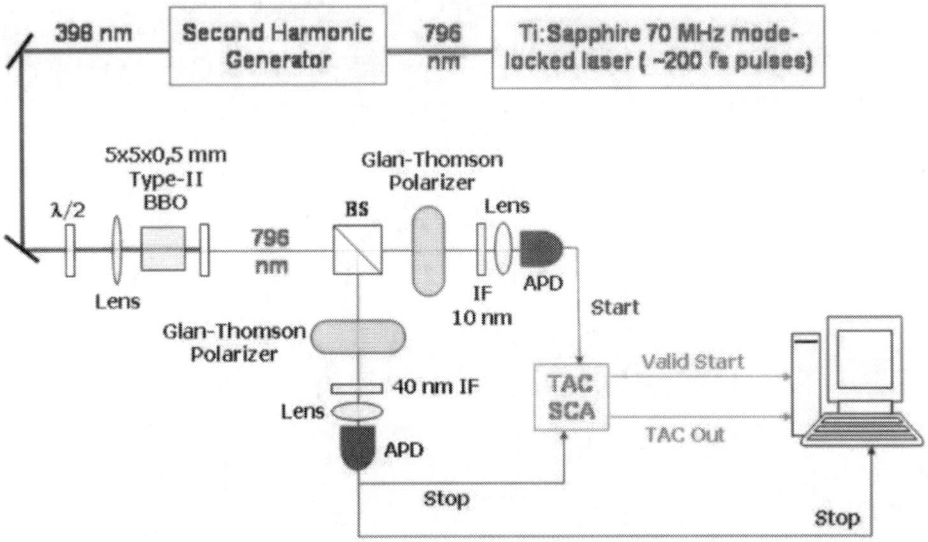

Figure 1. Scheme of the setup for the LR experiment. Entanglement here is generated by the non-polarizing BS, that with a 50% probability splits the photons belonging to the same correlated pair, sending them to the two polarization analyzers.

CH violation	σ_{CH}	violation
0,1026	0,0021	47,8 σ

Table 1. CH inequality violation.

Perkin-Elmer SPCM-AQR-14-FC-9491 Si-SPAD. The counts of the detector on the t arm were were the Start for a Time-To-Amplitude converter (TAC), triggering a collection window of 20 ns, while the arm r detector output was the Stop. The 20 ns coincidence window is set properly to exclude spurious coincidences with PDC photons of the following pulse.
To violate the inequalities of the two LHVTs, we fix the polarization axis of both the Glan-Thomson polarizers at an angle of $\frac{\pi}{4}$ with the horizontal plane, and then we perform a scan of the coincidence rates $R(\phi)$ (we remind that ϕ is the difference in angle between the polarization planes) by rotating the r polarizer by $\frac{\pi}{8}$ for each acquisition step, until we cover the whole interval $\phi \in [0, \pi[$. To evaluate the total coincidence number R_{tot} we perform a measurement after removing the polarizers, and in the end the background is evaluated and subtracted.
The obtained results [5], reported in fig. 2, show a very good agreement between theoretical fit and experimental data (correlation factor $r = 0,99994$).

A first test for our entangled state is the check (within the fair sampling assumption) of the Clauser-Horne inequality, written in the form:

$$\frac{\left|R(\frac{\pi}{8}) - R(\frac{3\pi}{8})\right|}{R_{tot}} \leq \frac{1}{4} \tag{18}$$

The collected data lead to the violation reported in Table 1, that certifies the "goodness" of our $|\psi\rangle_+$ Bell state, since the CH inequality is clearly violated by it.
After that, we try to test the first LHVT model inequality (6), and for this purpose we evaluate

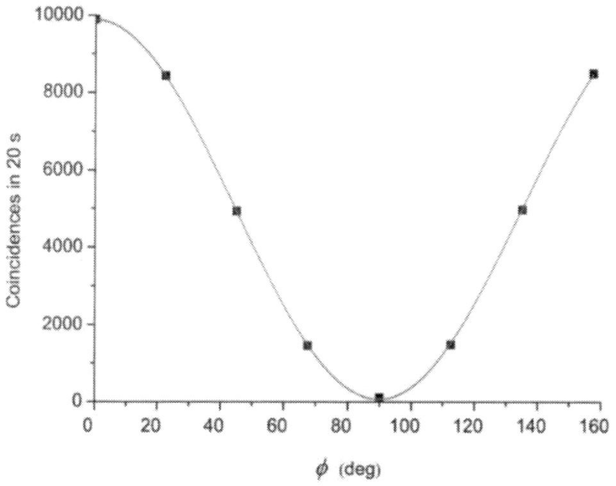

Figure 2. Coincidence rates $R_{12}(\phi)$ versus ϕ (angle between the t and r polarizers axes): the matching between the experimental data and the theoretical curve is optimal (correlation coefficient $r \simeq 1$).

the quantities V_B, V_A and F, obtaining the violation reported in Table 2.

$F - V_B/V_A$	$\sigma_{(F-V_B/V_A)}$	violation
0,0671	0,0057	11,7 σ

Table 2. Inequality (6) violation.

$D(\eta) - \Delta_{min}$	$\sigma_{(D(\eta)-\Delta_{min})}$	violation
0,0073	0,0022	3,3 σ

Table 3. Inequality (15) violation.

The final test, made upon the second LHV model presented in the previous section (leading to inequality (15), is performed by computing the values $D(\eta)$ and Δ_{min}, achieving the results shown in Table 3.

Tables 2 and 3 tell us that, in our experiment, a violation of the predictions made upon the LHV models presented in [1] and [2] occurred, while SQM predictions were confirmed [5].
In conclusion, our experimental results clearly demonstrate (without the detection loophole issue) that these two restricted local realistic models fail to reproduce all the quantum mechanical predictions, failing to be a valid candidate to replace Standard Quantum Mechanics.

3. The Alicki - Van Ryn test of quantumness at the single particle level

3.1. Theoretical framework

Some years ago, a test of quantumness at the single qubit level was proposed [3]. This test is very appealing both because of its simplicity and ability to show that some quantum states in a two dimensional Hilbert space cannot be classical. Since this is a test of single particle states, there is no reference made to the locality problem, rather it is a proper test of nonclassicality with respect to the possibility of an underlying hidden variable theory. However, it allows to perform tests on specific classes of states, although like the Bell test, it is subject to a number of loopholes depending on its experimental implementation.
Given any pair of positive real functions $\mathcal{A}, \mathcal{B}$ obeying the relation:

$$\langle 0 \leq \mathcal{A}(x) \leq \mathcal{B}(x) \rangle, \tag{19}$$

then for any probability distribution $\rho(x)$ it must be true that

$$\langle \mathcal{A}^2 \rangle \equiv \int \mathcal{A}^2(x)\rho(x)dx \leq \int \mathcal{B}^2(x)\rho(x)dx \equiv \langle \mathcal{B}^2 \rangle. \tag{20}$$

In a quantum mechanical framework, one can find pairs of observables $\widehat{A}, \widehat{B}$ such that the minimum eigenvalue of $\widehat{B} - \widehat{A}$ is greater than zero, which we refer to as the inequality:

$$0 \leq \widehat{A} \leq \widehat{B}. \tag{21}$$

The commutation relations of the classical approach prescribe that for all systems described by the density matrix $\widehat{\rho}$ one has:

$$\langle \widehat{A}^2 \rangle \leq \langle \widehat{B}^2 \rangle, \tag{22}$$

where $\langle \widehat{O} \rangle \equiv \text{Tr}[\widehat{O}\, \widehat{\rho}]$, while quantum theory predicts that for determined states we have:

$$\langle \widehat{A}^2 \rangle > \langle \widehat{B}^2 \rangle. \tag{23}$$

This discrepancy between classical (in the sense discussed above) and quantum predictions can be tested experimentally on an ensemble of single particles. In the following we present two experimental implementations of this test by using the polarization degree of freedom of single photons [6].

We underline that the condition given in Eq. (19) defines the class of hidden variable theories that can be tested by violation of the inequality (20).
We performed two different experiments to verify the proposed test, both of them with horizontally polarized photons ($|H\rangle$). The chosen observables are:

$$\widehat{A} = a_0 \widehat{P}_\alpha \tag{24}$$

$$\widehat{B} = b_0 \left[p_1 \widehat{P}_\beta + (1 - p_1) \widehat{P}_{\beta+\pi/2} \right], \tag{25}$$

being a_0 and b_0 numerical parameters, $\widehat{P}_\theta$ the projector on the state $|s(\theta)\rangle = \cos\theta|H\rangle + \sin\theta|V\rangle$ (and $\widehat{P}_{\theta+\pi/2}$ the projector on the orthogonal state $\sin\theta|H\rangle - \cos\theta|V\rangle$), and $0 \leq p_1 \leq 1$. The expectation value $\langle \widehat{A} \rangle$ can be obtained experimentally by projecting heralded photons onto the state $|s(\alpha)\rangle$, while $\langle \widehat{B} \rangle$ is realized with an experimental setup that projects heralded photons onto the state $|s(\beta)\rangle$ with probability p_1, and onto the state $|s(\beta+\pi/2)\rangle$ with probability $(1-p_1)$.
In our first experimental scheme [6] the probabilistic projection is achieved by means of a variable

beam-splitter whose splitting ratio is set to p_1.
The experimental measurement of both $\langle\widehat{A}\rangle$ and $\langle\widehat{A}^2\rangle$ is achieved by projecting the photon onto the state $|s(\alpha)\rangle$. To measure $\langle\widehat{B}^2\rangle$, where $\widehat{B}^2 = b_0^2\left[p_1^2\widehat{P}_\beta + (1-p_1)^2\widehat{P}_{\beta+\pi/2}\right]$, however, it is necessary to change the beam splitter ratio to $p_2 = \frac{p_1^2}{p_1^2+(1-p_1)^2}$. Thus the operator $\widehat{B}^2$ becomes:

$$\widehat{B}^2 = b_0^2\frac{1-2\sqrt{(1-p_2)p_2}}{(1-2p_2)^2}\left[p_2\widehat{P}_\beta + (1-p_2)\widehat{P}_{\beta+\pi/2}\right]. \tag{26}$$

We assume that the beam splitter randomly and fairly splits the incoming photons, with probabilities equivalent to the splitting ratio of "classical" waves. The second experimental scheme followed a similar approach for modeling the operators involved but with the difference that the probabilities p_1 and p_2 are then considered as part of the spectrum of the operators and they do not need to be experimentally evaluated.
By setting the parameters: $a_0 = 0.74$, $b_0 = 1.2987$, $p_1 = 4/5$, $p_2 = 16/17$, $\alpha = 11/36\ \pi$, and $\beta = 5/12\ \pi$, the results predicted by quantum mechanics are $\langle\widehat{B}^2\rangle - \langle\widehat{A}^2\rangle = -0.0449$, and $\langle\widehat{B}\rangle - \langle\widehat{A}\rangle = 0.0685$, while the minimum eigenvalue of $\widehat{B} - \widehat{A}$ is $d_- = 0.0189$, where

$$d_- \equiv \frac{1}{2}\left\{b_0 - a_0 - \sqrt{a_0^2 + b_0^2\ (1-2p_1)^2 + 2\ a_0\ b_0\ (1-2p_1)\ \cos[2(\alpha-\beta)]}\right\}. \tag{27}$$

As mentioned before, the test proposed in Ref. [3] concerns the class of HVTs satisfying Eq. (19) only: the boundary of the class of HVTs identified by this condition, as well as the possibility of enlarging the class by modifying this method, is a very important question, but is beyond the aim of this paper.

3.2. First experiment: 1550 nm heralded single photons

In our experimental setup (Fig. 3), single photons are produced by heralding the presence of a photon by the detection of another one produced in pair in Parametric Down-Conversion (PDC). Our PDC source is a 5 mm long periodically-poled MgO-doped lithium niobate (PPLN) crystal, pumped by a CW laser at 532 nm, that produces pairs of correlated photons at 810 nm and 1550 nm [6]. A cutoff filter blocks the pump laser light at the crystal's output and a dichroic mirror separates the 810 nm and 1550 nm photons. Interference filters at 810 and 1550 nm with a FWHM of respectively 10 nm and 30 nm further suppress fluorescence from the PPLN crystal, reducing background counts.
A proper characterization of this single-photon source is performed to ensure that the correlation between the signal photons at 810 and 1550 nm dominates over the background of accidental coincidences, and that multi-photon emission is negligible (i.e. when conditioned on a photon detection at 810 nm, the probability to observe two photons in a 1550 path is negligible).
For the experiment, photons in the heralding arm are routed by a single-mode fiber (SMF) to a Si-SPAD operating in Geiger mode, while photons in the heralded arm, coupled into a second SMF, are sent to an all-fiber variable beam splitter, implementing the probabilistic projections according to the parameter values determined above. This variable beam-splitter is made from an optical switch that can route heralded photons with an adjustable splitting ratio into two different optical paths. The input polarization state in each optical path after the beam splitter is controlled using a three paddle single-mode fiber polarization rotator followed by a rotating polarizer (POL). This scheme allows us to set the polarizers to perform projections on $\widehat{P}_\alpha$, $\widehat{P}_\beta$, and $\widehat{P}_{\beta+\pi/2}$, and set the splitting probability of the beam splitter to p_1 or p_2 to make necessary measurements of the observables. After passing through the polarizers that performed the

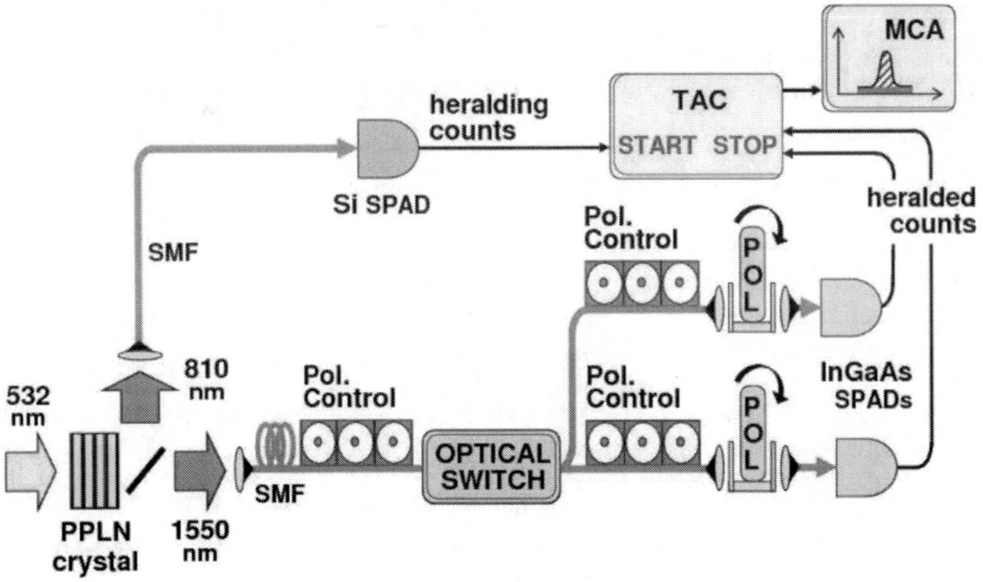

Figure 3. First experimental setup. 1550 nm heralded photons are generated by PDC in a PPLN crystal pumped by a 532 nm laser. The heralded photons are sent to the measurement apparatus in order to evaluate the observables $\langle\widehat{A}\rangle$, $\langle\widehat{A}^2\rangle$, $\langle\widehat{B}\rangle$, and $\langle\widehat{B}^2\rangle$.

projections, the heralded photons were finally sent to InGaAs-SPADs gated by the Si-SPAD heralding counts.
We determine the true coincidence probability for each gate, rather than using the raw measured counts, in order to eliminate the contribution of accidental coincidences, detector deadtimes, and drifts. The probability for each measurement i was evaluated according to:

$$\eta_i(\theta, p) = \frac{N_i(\theta, p)}{M_{g,i}}, \tag{28}$$

where $N_i(\theta, p)$ is the number of coincidences sent with probability p towards the detection system with the polarizer projecting photons onto the state $|s(\theta)\rangle$, and $M_{g,i}$ is the number of heralding gate counts. Thus, $\langle\widehat{P}(\theta)\rangle$ was estimated as:

$$\mathcal{E}[\langle\widehat{P}(\theta)\rangle] = \frac{\sum_i \eta_i(\theta, p)}{\sum_i[\eta_i(\theta, p) + \eta_i(\theta + \pi/2, p)]}, \tag{29}$$

while the probability of sending a photon towards a detection system was estimated as:

$$\mathcal{E}[p] = \frac{\sum_i[\eta_i(\theta, p) + \eta_i(\theta + \pi/2, p)]}{\sum_i \begin{bmatrix} \eta_i(\theta, p) + \eta_i(\theta, 1-p) + \\ \eta_i(\theta + \pi/2, p) + \eta_i(\theta + \pi/2, 1-p) \end{bmatrix}}. \tag{30}$$

Using Eqs. (29) and (30) we computed the experimental values of $\langle\widehat{A}\rangle$, $\langle\widehat{A}^2\rangle$, $\langle\widehat{B}\rangle$, and $\langle\widehat{B}^2\rangle$ (Table 4). From the same experimental results we obtained an indirect evaluation of the minimum eigenvalue of $\widehat{B} - \widehat{A}$ as (0.0101 ± 0.0065), showing that we have met requirement (3) (we point out that the high relative uncertainty of this evaluation is due to its indirect determination). From the value of $\mathcal{E}[\langle\widehat{B}^2\rangle - \langle\widehat{A}^2\rangle]$ we show a clear violation of the classical limit (Eq. (22)).

Quantity	Measurement	QM theory
$\mathcal{E}[\langle\widehat{B}\rangle - \langle\widehat{A}\rangle]$	0.0581 ± 0.0049 (± 0.0112)	0.0685
$\mathcal{E}[\langle\widehat{B}^2\rangle - \langle\widehat{A}^2\rangle]$	-0.0403 ± 0.0043 (± 0.0066)	-0.0449
		(> 0 HVTs)
$\mathcal{E}[p_1]$	0.80 ± 0.01	0.800
$\mathcal{E}[p_2]$	0.94 ± 0.01	0.941

Table 4. Experimental results obtained with the first experimental scheme. The total uncertainties (in parentheses) account for both statistical and systematic effects.

In Table 4 can be found both the statistical and total uncertainties. Statistical uncertainties include those due to Poisson counting statistics as well as those due to random misalignment of the polarizers (we estimate an angular uncertainty of 2.5°), while the total uncertainties also include systematic effects such as the uncertainty in setting the optical switch voltage bias used to obtain the required splitting ratio.
Furthermore, we analyzed how the non-ideal (multi-photon) behavior of our single-photon source might have affected the experimental results, and we found its effects to be negligible, being more than an order magnitude below the listed uncertainties.

3.3. Second Experimental test: 800 nm heralded single photons

This second experimental setup (Fig. 4) hosts a heralded single-photon source based on 800 nm photon pairs produced by non-collinear degenerate PDC in a 5 mm long $LiIO_3$ bulk crystal, pumped by a 400 nm pulsed beam obtained by duplicating the frequency of the output of a MIRA laser pumped by a 5W VERDI. On the photon pair channels interference filters at 800 nm with a FWHM of 20 nm suppress fluorescence from the $LiIO_3$ crystal reducing background counts. Microscope objectives (20X) collect photons on multi-mode fibers addressing them to Si-SPADs operating in Geiger mode. One channel is the heralding arm collecting the start counts for the coincidence apparatus, while the other channel is the heralded arm sending the stop counts for the coincidence apparatus. We perform the projections, according to the parameter values previously determined by means of a wave plate and a polarizing beam-splitter.
This scheme allows us to set the angle of the wave plate to perform projections on $\widehat{P}_\alpha$, $\widehat{P}_\beta$, and $\widehat{P}_{\beta+\pi/2}$, making all the necessary measurements of the physical observables. To determine the true coincidence probability for each pulse, we use the same approach exploited for the first experiment measurements. The probability for each measurement i was evaluated following an approach analogous to Eq. (28) where $N_i(\theta)$ is the number of coincidences with the wave plate and the polarizing beam splitter projecting photons onto the state $|s(\theta)\rangle$, and $M_{g,i}$ is the number of heralding counts. Thus, in each experimental configuration, $\langle\widehat{P}(\theta)\rangle$ was estimated as Eq. (29). We computed the experimental values of $\langle\widehat{A}\rangle$, $\langle\widehat{A}^2\rangle$, $\langle\widehat{B}\rangle$, and $\langle\widehat{B}^2\rangle$ as seen in Table 5. From the value of $\mathcal{E}[\langle\widehat{B}^2\rangle - \langle\widehat{A}^2\rangle]$ we show a 46 standard deviations violation of the classical limit (Eq. (22)).

As well as with the first source, also this heralded single-photon source is independently characterized to ensure that the signal photons dominates over the background of accidental coincidences and that multi-photon emission is negligible (see last subsection).

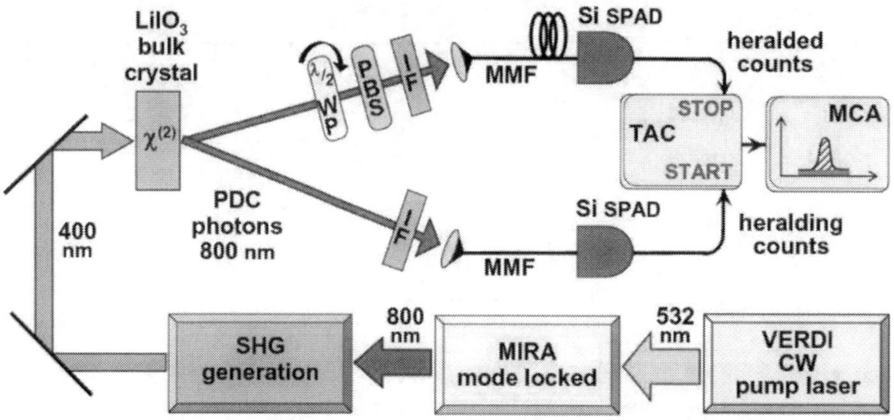

Figure 4. Second experimental setup. A PDC heralded single-photon source generates pairs of photons at 800 nm: the heralded single photons are sent to the measurement apparatus designed to evaluate the observables $\langle\widehat{A}\rangle$, $\langle\widehat{A}^2\rangle$, $\langle\widehat{B}\rangle$, and $\langle\widehat{B}^2\rangle$.

Quantity	Measurement	QM theory
$\mathcal{E}[\langle\widehat{B}\rangle - \langle\widehat{A}\rangle]$	0.0701 ± 0.0015	0.0685
$\mathcal{E}[\langle\widehat{B}^2\rangle - \langle\widehat{A}^2\rangle]$	-0.0461 ± 0.0010	-0.0449
		(> 0 HVTs)

Table 5. Experimental results obtained with the second setup, with statistical and total uncertainties and quantum mechanics theoretical predictions.

3.3.1. Single-photon sources characterization A necessary requirement for a convincingly realizing the Alicki - Van Ryn proposal [3] is a demonstration that our sources in fact produces single-photon states: because of that, we verify that the possibility of having more than one photon in the heralded arm after detecting the heralding photon is very low.
For the analysis of the source of the first experiment we use the main setup (Fig. 3), but with the polarizers removed and the splitting ratio of the optical switch set to $p = (0.50 \pm 0.01)$. In the case of the second setup we need to substitute the multi-mode fiber on the heralded arm with a passive multi-mode fiber beam-splitter 50:50 sending photons to two Si-SPADs.
The efficiency of a single-photon source can be described by means of the two parameters $\Gamma_1 = Q(1)/Q(0)$ and $\Gamma_2 = Q(2)/Q(1)$, where $Q(0)$ is the probability that for each heralding count neither the detector in the heralded arm fires, $Q(1)$ is the probability of detecting just one count for each herald, and $Q(2)$ is the probability of observing a coincidence for each heralding count from simultaneous firings by the two SPADs on the heralded channel.
The probability of a specific detector firing due to a heralded optical pulse containing n photons is:

$$\begin{aligned} Q(1|n) &= \sum_{m=0}^{n}[1-(1-\tau)^m]B(m|n;p) = \\ &= 1-(1-p\,\tau)^n, \end{aligned} \tag{31}$$

where p is the splitting ratio of the optical switch in the first experiment or the splitting ratio of the passive beam-splitter in the second one, $B(m|n;p) = n![m!\,(n-m)!]^{-1}p^m(1-p)^{n-m}$ is

Source type	Γ_1	Γ_2	Γ_2/Γ_1
Single-photon	$\tau/[2(1-\tau/2)]$	0	0
Poisson	$e^{\tau\mu/2}-1$	$1-e^{-\tau\mu/2}$	≈ 1 (when $\tau\mu \ll 1$)
I Exp	$(4.14 \pm 0.06)\cdot 10^{-3}$	$(0.66 \pm 0.06)\cdot 10^{-3}$	0.16 ± 0.01
I Exp (bkg subtr.)	$(4.02 \pm 0.06)\cdot 10^{-3}$	$(0.37 \pm 0.36)\cdot 10^{-3}$	0.09 ± 0.09
II Exp	$(2.82 \pm 0.01)\cdot 10^{-2}$	$(0.26 \pm 0.04)\cdot 10^{-2}$	0.09 ± 0.02
II Exp (bkg subtr.)	$(2.49 \pm 0.01)\cdot 10^{-2}$	$(0.15 \pm 0.05)\cdot 10^{-2}$	0.07 ± 0.02

Table 6. Two-photon characterization of an ideal single-photon source, a Poisson source and our two single photon sources (with and without background subtraction).

the binomial distribution representing the splitting of n photons towards the two SPADs on the heralded channel , and τ is the detection efficiency of each SPAD (that also accounts for all collection and optical losses in the channel). In the same way, the probability of observing a coincidence between the two SPADs due to a heralded optical pulse with n photons is:

$$\begin{aligned} Q(2|n) &= \sum_{m=0}^{n}[1-(1-\tau)^m][1-(1-\tau)^{n-m}]B(m|n;p) \\ &= 1-(1-p\,\tau)^n-[1-(1-p)\tau]^n+(1-\tau)^n. \end{aligned} \tag{32}$$

Thus we get $Q(1)=\sum_n Q(1|n)\mathcal{P}(n)$, and $Q(2)=\sum_n Q(2|n)\mathcal{P}(n)$ for $\mathcal{P}(n)$ being the general probability distribution of the number of photons in a heralded optical pulse. Setting $p=0.5$, in the case of an ideal single-photon source ($\mathcal{P}(n)=\delta_{n,1}$) we obtain $Q(1)=\tau/2$, and $Q(2)=0$, corresponding to $\Gamma_2=0$ and $\Gamma_1=\tau/[2(1-\tau/2)]$; while for a Poissonian source ($\mathcal{P}(n)=\mu^n e^{-\mu}/n!$) we obtain $Q(1)=1-\exp(-\tau\mu/2)$, and $Q(2)=[1-\exp(-\tau\mu/2)]^2$, corresponding to $\Gamma_2=1-\exp(-\tau\mu/2)$ and $\Gamma_1=\exp(\tau\mu/2)-1$ (meaning $\Gamma_2\simeq\Gamma_1=\tau\mu/2$ when $\tau\mu\ll 1$).
See Table 6 for comparison between the ideal sources above and our two implementations.

From our experimental data we obtained, with background subtraction, results for Γ_2 that are in both experiments compatible with 0 as for ideal single-photon sources. We also note that $\mathcal{E}[\Gamma_1]$ for the first experiment, is in agreement with the estimated optical losses and a previous detector calibration [10].

An alternative characterization metric for single-photon sources, was proposed by Grangier *et al.* [11]. They introduced an "anticorrelation criterion" based on the parameter $\alpha = Q(2)/[Q^{(I)}(1)\;Q^{(II)}(1)]$ ((I), (II) indicate the two detectors after the variable beam splitter). For an ideal single-photon source $\alpha=0$, while $\alpha\geq 1$ corresponds to classical sources. From our experimental data we obtain in the case of IR region heralded photons $\mathcal{E}[\alpha]=(0.18\pm0.02)$ and $\mathcal{E}[\alpha]=(0.11\pm0.11)$ with and without background subtraction, respectively, while for the visible heralded photons we obtain $\mathcal{E}[\alpha]=(0.10\pm0.02)$ and $\mathcal{E}[\alpha]=(0.06\pm0.02)$. Thus also the control based on the anticorrelation criterion ensures that conditional single-photon output dominates for both our sources.

In conclusion, we have investigated the theoretical proposal for a non-classicality test on single particles [3]. While the utility of this test is open to question, as it does not apply to every conceivable HVT like Bell's inequalities but only the class of HVTs satisfying Eq. (19), we have nonetheless experimentally implemented it as proposed. Following the test's protocol, our two set of measurement results are clearly incompatible with the sub-class of HVTs defined by Eq. (19), while being well predicted by QM. This effort represents a first step in the direction of providing a sharp difference between QM and HVTs at the single qubit level.

References

[1] Santos E 2004 *Phys. Lett.* A **327** 33
[2] Santos E 2007 *Europ. Phys. Journ.* D **42** 501
[3] Alicki R and van Ryn N 2008 *J. Phys.* A **41** 062001
[4] Genovese M 2010 *ASL* **3** 100501
[5] Brida G, Genovese M and Piacentini F 2007 *EPJD* **44** 547
[6] Brida G et al 2008 *Opt. Exp.* **16** 11750
[7] Brida G et al 2010 *PRA* **79** 044102
[8] Klyshko D N 1988 *Photons and Nonlinear Optics* (Gordon and Breach Science Publisher)
[9] Mandel L and Wolf E 1985 *Optical Coherence and Quantum Optics* (Cambridge, UK: Cambridge Univ. Press)
[10] Schettini V, Polyakov S V, Degiovanni I P, Brida G, Castelletto S and Migdall A L 2007 *IEEE J. Sel. Top. Quantum Electron.* **13** 978
[11] Grangier P, Roger G and Aspect A 1986 *Europhys. Lett.* **11** 173

Exploring the foundations of physics in solid state systems at ultra-low energy

Enrico Prati

Laboratorio Materiali e Dispositivi per la Microelettronica, Consiglio Nazionale delle Ricerche - IMM, Via Olivetti 2, I-20041 Agrate Brianza, Italy

E-mail: enrico.prati@cnr.it

Abstract. The emergence of quantum field theory as a coarse grained effective model grounded on a more fundamental level is often conjectured. The hypothesis is based both on the lack of a unified theory capable to consistently include gravity and on the existence of several known classes of low energy effects in which effective properties emerge on a coarse grained level. Solid state physics presents major examples of emergence. Effective properties of electrons and light in solid state media mimic, through both collective and coarse grained properties, some fundamental characteristics attributed to particles and spacetime. Here, three low energy mechanisms of the solid state physics connected to those properties are considered, with a particular attention to the relevant energy scale: zero mass fermions in graphene, asymptotic freedom in semiconductor quantum dots, and effective curvature of space in negative refractive index metamaterials.

1. Introduction

There is an increasing interest in the concept of emergence in complex systems, and a particular importance is attributed to the emergence in physics [1, 2]. Discoveries like those of metamaterials [3, 4], graphene [5], and the development of solid state quantum dots [6, 7] have contributed to the large number of either predicted or unexpected effects which occur when some properties of a substrate are non trivially exploited. Metamaterials provide a clear example of emergence. They are composite materials obtained by repeating a cell made of conductive and non conductive regions, and their effective response to an incident electromagnetic field depends on the shape of the unit cell and not on the constituent materials [8]. In other words, the effective properties at a sufficiently myopic scale depend on both the organization of the elementary constituents and the structure of the substrate itself, instead of the specific physical properties of its constituents. This mechanism leads towards several effects like effective electromagnetic media. There are several examples in other branches of solid state physics which provide a ground for emergence. On the other hand, the physics of elementary particles, devoted to explain the property of the ultimate constituents and forces of natural phenomena, achieves several results from the hypothesis of a space-time background and unobserved fields (Higgs, superparticles, gravitons) [9, 10]. It is conjectured that apparently elementary particles could be obtained as the manifestation at low energy of a more fundamental structure. In order to bridge the emergence of mechanism in solid state physics with properties relevant in particle physics, in this paper I will review the most important examples contributing to the topic with a particular attention to the energy scale. In Section II the bending of the light in metamaterials is reviewed. Section III is devoted to the connection between the massless relativistic Dirac field and the electrons in graphene, whose discovery has been recently awarded with the Nobel Prize. Finally, the analogy between the Kondo effect in semiconductor quantum dots and the asymptotic freedom (another discovery awarded with the Nobel Prize) of heavy quark will be discussed.

2. Light bending in space: transformation optics in metamaterials

Electromagnetic metamaterials [3, 8] are artificially structured materials designed to interact with electromagnetic waves. A metamaterial is a composite medium in which the crystal cells are generally realized by structures of dielectrics and conductors. For wavelengths of the radiation much longer than the lattice spacing, the response is averaged on the cells and the effective dielectric permittivity $\epsilon(\omega)$, magnetic permeability $\mu(\omega)$, and refractive index $n(\omega)$ may realize values not obtainable in nature. Such materials can be designed to behave at a predictable frequency with peculiar characteristics, such as negative effective permeability [3] and permittivity [8], so the consequences of wave propagation, calculated in 1968 by Veselago [11], are dramatically different with respect to those in ordinary media. A periodic structure can be treated as photonic crystal for all ratios of the wavelength relative to the scale of inhomogeneity, but when the wavelength is very large relative to the inhomogeneity, then an effective medium description is also valid. In the transitional regime between the effective medium and the photonic crystal descriptions of a periodic structure the definitions of ϵ and μ are approximate but yet have practical utility. Metamaterials fall in this regime. In partial parallel with the curvature of space-time in general relativity, the light bending in space can be obtained by considering gradients in the constitutive tensor which describes ϵ and μ in their most general covariant form. Covariant Maxwell equations are expressed as:

$$F_{\alpha\beta,\mu} + F_{\beta\mu,\alpha} + F_{\mu\alpha,\beta} = 0 \tag{1}$$

$$G^{\alpha\beta}_{,\alpha} = J^{\beta} \tag{2}$$

where

$$F_{\alpha\beta} = \begin{pmatrix} 0 & E_1 & E_2 & E_3 \\ -E_1 & 0 & -cB_2 & cB_2 \\ -E_2 & cB_3 & 0 & -cB_1 \\ -E_3 & -cB_2 & cB_1 & 0 \end{pmatrix}, G_{\alpha\beta} = \begin{pmatrix} 0 & -cD_1 & -cD_2 & -cD_3 \\ cD_1 & 0 & -H_2 & H_2 \\ cD_2 & H_3 & 0 & -H_1 \\ cD_3 & -H_2 & H_1 & 0 \end{pmatrix} \tag{3}$$

while $J^{\beta} = (c\rho, \mathbf{J})$ and the constitutive relations are given by $G^{\alpha\beta} = \frac{1}{2}C^{\alpha\beta\mu\nu}F_{\mu\nu}$ [12]. The constitutive relations reflect the space curvature in the sense that they generate a coordinate transformation Λ such that:

$$C^{\alpha'\beta'\mu'\nu'} = \left|det\left(\Lambda^{\alpha'}_{alpha}\right)\right|^{-1} \Lambda^{\alpha'}_{\alpha}\Lambda^{\beta'}_{\beta}\Lambda^{\mu'}_{\mu}\Lambda^{\nu'}_{\nu}C^{\alpha\beta\mu\nu} \tag{4}$$

where the Jacobian is $\Lambda^{\alpha'}_{\alpha} = \frac{\partial x^{\alpha'}}{\partial x^{\alpha}}$ so one defines the metric tensor $g^{i'j'} = \Lambda^{i'}_{j}\Lambda^{j'}_{k}\delta^{jk}$ which relates to the ϵ and μ tensors:

$$\epsilon_{i'j'} = \left|det\left(g^{i'j'}\right)\right|^{-1/2} g^{i'j'}\epsilon \quad \mu_{i'j'} = \left|det\left(g^{i'j'}\right)\right|^{-1/2} g^{i'j'}\mu \tag{5}$$

Metamaterials are particularly suitable to create gradients of the constitutive parameters in space by grading the artificial density [13]. Regions with inverted sign parameters in a frequency band are possible. Cloaking of objects can be achieved in some specific frequency range by means of graded density metamaterials with negative parameters [14]. The energy scale of metamaterials is that associated to photons. The operability is generally obtained from the X band (10 GHz) to infrared (100-200 THz), but also for visible light (up to 800 THz). In terms of energy the range is between $4 \cdot 10^{-5}$ eV to 3.3 eV. The wavelength fixes the size of the cell of factor of at least 1/6-1/10 of $3 \cdot 10^{-2}$ m to $3.7 \cdot 10^{-7}$ m respectively. To conclude, transformation optics allows arbitrary light bending in coarse grained metamaterials at low energy and at a wanted wavelength, provided that the latter is long if compared to the unit cell size.

3. Massless relativistic Dirac field: electrons in graphene

The first paper on the band structure of the graphene has been published in 1946 by P.R. Wallace [15]. Initially believed of impossible realization, the scientific community had to wait 2004 before the first realization of graphene conceived by K. S. Novoselov and A. A. Geim to experimentally exploit its peculiar properties [5]. The structure of the electronic energy bands and Brillouin zones for graphite was developed using the tight binding approximation, for which the total Hamiltonian H is approximated by a single atomic orbital Hamiltonian in a restricted Hilbert space. Graphite consists of a semimetal with zero activation energy, for which there are no free electrons at zero temperature, but they are created at non zero temperature by excitation to a band contiguous to the highest one. Graphite has the very well know exagonal structure of carbon atoms and possesses four valence electrons. Three of them form tight bonds with neighboring atoms in the plane. Their wave functions are of the form

$$\frac{1}{\sqrt{3}}\left(\psi(2s)+\sqrt{2}(\sigma,2p)\right)\,(i=1,2,3) \tag{6}$$

If $X(r)$ is the normalized orbital $2p_z$ wave function for an isolated atom, then the wave function in the tight binding approximation has the form:

$$\Psi = \phi_1 + \lambda\phi_2 \tag{7}$$

where

$$\phi_1 = e^{2\pi i\mathbf{k}\cdot\mathbf{r}_A}X\left(\mathbf{r}-\mathbf{r}_A\right)\,\phi_2 = e^{2\pi i\mathbf{k}\cdot\mathbf{r}_B}X\left(\mathbf{r}-\mathbf{r}_B\right) \tag{8}$$

The wavefunction Ψ can be considered a Dirac bi-spinor. The lattice index ranging on the two sites A and B is called *pseudospin* in obvious analogy with the spin of standard bi-spinors. Indeed, Ψ is a two component wavefunction made of the two separate wavefunctions of the $2p_z$ electron of each cell. According to the analogy, Ψ, which is distributed along thousend of lattice cells is called *quasi-particle* and describes an almost free electron. The plane wave behaviour of the electrons, one can turn to the Dirac formalism and express the Hamiltonian as:

$$H = \hbar v_F = \begin{pmatrix} 0 & k_x - ik_y \\ k_x + ik_y & 0 \end{pmatrix} = \hbar v_F \sigma\cdot\mathbf{k} \tag{9}$$

where k is the momentum of the quasiparticle, σ is the 2-dimensional Pauli matrix vector, v_F is the k-independent speed of electrons, which corresponds to the limiting speed of light for the system. Experimentally one finds that $c/v_F \approx 300$. From the band theory point of view, graphene has a linear dispersion relation up to 1 eV (a property referred as Dirac cones, for which the cones are with respect to the k_x and k_y coordinates) and it is a gap-less semimetal. One may define the chirality, which is the projection of σ along $\mathbf{k}$. The chirality is positive (+) for electrons and negative for holes. Chirality manifests itself by observing the (chiral) quantum Hall effect, a modified version of the ordinary Hall effect, for which

$$\sigma_{xy} = \pm 4e^2/h\left(N+\frac{1}{2}\right) \tag{10}$$

where the factor 4 accounts the spin and the valley degeneracy. In other words, two massive leptons in a two dimensional lattice behave as a charged almost massless particle. The formal analogy with the known almost massless particle (ν_e neutrino) is broken by the non zero electric charge of the quasi particles in graphene.

4. Asymptotic freedom: Kondo effect in quantum dots

In this section the analogy between asymptotic freedom of heavy quarks and screened impurities in semiconductor quantum dot is reviewed. There is a fundamental difference between Quantum ElectroDynamics (QED) and Quantum ChromoDynamics (QCD) [16], arising from the fact that the first is governed by a force between charged particles, while the latter between colored particles. Indeed, while the photons, which carry the force between charged particles are electrically neutral, the gluons - carrying the force between colored particles, are colored. Consequently, a quark is surrounded by a sea of virtual gluons that arise from quantum fluctuations, and the color of the virtual gluons enhance the color of the quarks. A probe coming closer and closer to the quark is influenced less and less by the virtual gluons, so that the effective color charge of the quark seems to weaken. This mechanism is called asymptotic freedom. Surprisingly, the asymptotic freedom has been observed first in another field of physics. Fifteen years ago, it was experimentally discovered that the resistivity of some metals containing impurities, contrarily to the expected decreasing trend by lowering its temperature, have a minimum of resistivity at some temperature. Lowering further the temperature, an increase of resistivity was observed. This effect was first explained by Juno Kondo with the screening of an impurity spin by a sea of conduction electrons. The direct consequence was a free spin behaviour at high energy. The Kondo effect first observed in metals, has been later observed also in semiconductor quantum dots [17, 18]. Today, it is a very hot topic of solid state physics and nanotechnology the control of transparency of tunnel barriers in quantum dots, in order to create the Kondo effect in semiconductor quantum dots. Let us first introduce the subject with the case of the metal, for historical reasons. In the case of a metal, the Kondo Hamiltonian of the conduction electrons and an impurity with spin $S = 1/2$ are expressed with (single channel spin 1/2 Kondo effect 1CK [21]:

$$H = \sigma_{k\sigma}\epsilon_k c^{+}_{k\sigma} c_{k\sigma} + J\boldsymbol{S} \cdot s(\boldsymbol{0}) \tag{11}$$

where $c^{+}_{k\sigma}$ is the creation operator of a conduction electron with momentum k and spin σ, $s(\boldsymbol{0})$ is the total spin of the conductive electrons in the site of the impurity in the position $\boldsymbol{0}$, while J is the antiferromagnetic coupling factor. The resistivity is expressed as:

$$\rho(T) = \rho_0 + aT^2 + bT^5 + c[J^2 + J^3 D(E_F) ln(E_{BW}/T)] \tag{12}$$

which sets a Kondo temperature corresponding to $E_K = kT_K = E_{BW}e^{-1/JD(E_F)}$ where E_{BW} is the conduction electron band width. The effective coupling $J(E)$ scales inverse logarithmically with energy: this scaling form is identical to the scaling of g^2 in the $3 + 1$ dimensional Yang-Mills gauge theory in the high energy limit (short distance), being g the coupling constant for the Yang-Mills non-Abelian gauge fields. The logarithm accounts the increase of the resistivity at very low temperature. The formalism can be idealized as a weak coupling at high temperature between the impurity spin and the conduction electron spins, which collapse to a strong coupling at low temperature, where the screening creates a non magnetic state (a singlet between the impurity and the pseudofermion generated by the electrons). Solid state quantum dots are nanometric boxes with size of tens of nanometers, where localized electrons are confined. Thermal effects are negligible below 1 K [19, 20] and electrons can be maintained and manipulated one by one [7]. Let us now move to the interesing elementary example of a 1 impurity Kondo quantum dot (1IK), which paves the way to the relevant case of the 2 impurities Kondo effect (2IK). The

Hamiltonian of the case of 1IK, where the impurity is connected with two leads (left lead (L) and right lead (R)):

$$H = H_L + H_R + H_{QD} + H_I \tag{13}$$

where H_L,H_R and H_{QD} are the second quantization Hamiltonians of the electrons contained in the L and R leads, and the quantum dot respectively. More interestingly, the interaction is expressed by:

$$H_I = \sigma_{k\sigma} V_{Lk\sigma} \left[c^+_{Lk\sigma} d_\sigma + h.c.\right] + V_{Rk\sigma} \left[c^+_{Rk\sigma} d_\sigma + h.c.\right] \tag{14}$$

By introducing a rotation $c^+_{+k\sigma} = cos\theta c^+_{Lk\sigma} + sin\theta c^+_{Rk\sigma}$ and $c^+_{-k\sigma} = -sin\theta c^+_{Lk\sigma} + cos\theta c^+_{Rk\sigma}$ the interaction is re-written as:

$$H_I = \sigma_{k\sigma} \sqrt{V^2_{Lk\sigma} + V^2_{Rk\sigma}} \left[c^+_{+k\sigma} d_\sigma + h.c.\right] \tag{15}$$

This example is introductory to a more interesting case, directly connected with the physics of heavy quarks. Let us now consider the case of 2 impurities in series connected to the left and the right leads (2IK). In this case, the physics is expressed by the interaction Hamiltonian, which reads, in a compact expression:

$$H_I = J_0[s(\mathbf{r}_1) \cdot \mathbf{S}_1 + s(\mathbf{r}_2) \cdot \mathbf{S}_2] + J_1 \mathbf{S}_1 \cdot \mathbf{S}_2 \tag{16}$$

with obvious notation, where 1 and 2 correspond to the two impurities. The full expression goes beyond the scope of this review. This interaction provides the competition between two possible ground state configurations: when the antiferromagnetic coupling J_1 is weak, in comparison to the Kondo effect between spin and leads J_0, two separate Kondo systems are formed. In the opposite case of a strong antiferromagnetic coupling J_1, the two impurities lock in a singlet and no Kondo screening occurs. To resume, the single impurity (1IK) show asymptotic freedom as a low energy bound state and high energy free state, a problem which can be mapped in the 1+1 dimensional fermionic systems which quartic interaction, which is fundamentally different from the Yang-Mills non-abelian gauge theories. On the contrary, the double impurity (2IK) mimics a two heavy quark system. The formation of the spin singlet at low energies inspires similar situation in SU(3) of an ultra-heavy quark and anti-quark pair in a light quark sea. Indeed, the tunable heavy-quark/anti-heavy-quark interaction (a linear attractive potential in distance) is similar to J_1 in the Kondo problem: for a weak $q\bar{q}$ interaction, the two are screened by the light quark sea, resulting in two separate bound states. For strong $q\bar{q}$ interaction, they form a bound state much like the two-impurity Kondo situation. The energy scale of Kondo physics is set by the Kondo temperature. Experimentally one observes the Kondo temperature between 1 K and 4 K [22] which corresponds to few hundreds of μeV.

5. Conclusion

To conclude, the emergence of phenomena typical of particle physics at very low energy in solid state physics have been reviewed. Major examples in three different branches of solid state physics have been reported, namely metamaterials, graphene and semiconductor quantum dots. Low energy physics may reveal mechanisms capable to recreate the emergence of similar phenomena in particle physics and quantum field theory, at a very different length and energy scales. It provides a new perspective to consider the role of the properties of the elementary constituents from the point of view of the structure and of the organization.

References

[1] Wayne A 2009 *Emergence in Physics*, Philosophy Compass (Wiley Online Library)
[2] Licata I and Sakaji A 2008 *Physics of Emergence and Organization* (Singapore: World Scientific Publishing)

[3] Pendry J B , Holden A J and Stewart W J 1999 Magnetism from conductors and enhanced nonlinear phenomena *IEEE Trans. Microw. Th. Tech.* **47** 2075-2084
[4] Prati E 2004 Crossover between the cell size and the wavelength of the incident radiation in metamaterials *Microw. Opt. Technol. Lett.* **40** (4) 272
[5] Novoselov K S, Geim A K, Morozov S V, Jiang D, Katsnelson M I, Grigorieva I V, Dubonos S V and Firsov A A 2005 Two-dimensional gas of massless Dirac fermions in graphene *Nature* **438** 197-200
[6] Tarucha S, Austing D G, Honda T, van der Hage R J and Kouwenhoven L P 1996 Shell Filling and Spin Effects in a Few Electron Quantum Dot *Phys. Rev. Lett.* **77** 3613-3616
[7] Prati E et al. 2011 Adiabatic charge control in Silicon single donor quantum dots *Appl. Phys. Lett.* **98** 053109
[8] Smith D R, Padilla W J, Vier D C, Nemat-Nasser S C and Schultz S 2000 Composite medium with simultaneously negative permittivity and permeability *Phys. Rev. Lett.* **84** 4184-4187
[9] Bernardi G, Carena M and Junk T 2007 Higgs bosons: theory and searches *Reviews of Particle Data Group: Hypothetical particles and Concepts*
[10] De Gouvea A, Moroi T and Murayama H 1997 Cosmology of Supersymmetric Models with Low-energy Gauge Mediation *Phys. Rev.* D **56** 1281-1299
[11] Veselago V G 1968 The electrodynamics of substances with simultaneously negative values of epsilon and mu *Sov. Phys. USPEKHI* **10** 509-514
[12] Schurig D et al. 2006 *Optics Express* **14** (21) 9794
[13] Prati E and Amabile C 2010 Lattice Density Effect on the Negative Refractive Index of a Uniaxial Metamaterial, in *Metamaterials – From Modeling and Fabrication to Application* ed J L-W Li, N Engheta, R Pachter and M Tanielian *Mater. Res. Soc. Symp. Proc.* Vol. **1223**E (Warrendale, PA) 1223-EE02-05
[14] Pendry J 2009 *Physics* **2** 95
[15] Wallace P R 1947 The Band Theory of Graphite *Phys. Rev.* **71** 622-634
[16] Glashow S L, Iliopoulos J and Maiani L 1970 Weak Interactions with Lepton-Hadron Symmetry *Phys. Rev.* D **2** 1285-1292
[17] Kouwenhoven L and Glazman L 2001 Revival of the Kondo effect *Physics World* **14** (1) 33-38
[18] Cronenwett S M , Oosterkamp T H and Kouwenhoven L P 1998 A Tunable Kondo Effect in Quantum Dots *Science* **281** 5376 540-544
[19] Prati E 2010 Finite Quantum Grand Canonical Ensemble and Temperature from Single Electron Statistics in a Mesoscopic Device *Journ. Statist. Mechanics* P01003
[20] Prati E et al. 2010 Measuring the Temperature of a Mesoscopic Electron System by means of Single Electron Statistics *Appl. Phys. Lett.* **96** 113109
[21] Chang A M and Chen J C 2009 The Kondo effect in coupled-quantum dots *Rep. Prog. Phys.* **72** 096501
[22] Lansbergen G P, Tettamanzi G C, Verduijn J, Collaert N, Biesemans S, Blaauboer M and Rogge S 2010 *Nano Letters* **10** 455

Generalized clocks in timeless canonical formalism

Enrico Prati

Laboratorio Materiali e Dispositivi per la Microelettronica, Consiglio Nazionale delle Ricerche - IMM, Via Olivetti 2, I-20041 Agrate Brianza, Italy

E-mail: `enrico.prati@cnr.it`

Abstract. Hamiltonian dynamics is recast in a timeless formalism in which parameter time σ is derived from the generalized coordinates, the Hamiltonian invariance on trajectories, and the Maupertuis principle. In order to define a time variable T in macroscopic systems, the cyclicity in the phase space replaces the self consistent assumption of time periodicity generally adopted for real clocks. Generalized clocks are defined in physical systems of sufficient complexity. Under suitable assumptions, physical systems can be separated in a subsystem to be dynamically described, and another cyclic subsystem which has the role of generalized clock. The latter provides a discrete approximation of the parameter time, called metric time. The stability prescription of generalized clocks guarantees that dynamics is expressed by the same equations of motion parametrized by the parameter time, in terms of metric time at the desired approximation. The timeless Hamiltonian framework, together with the definition of generalized clock, provide a ground to account the fundamental timelessness of nature, and the experimental evidence of time evolution in macroscopic systems experienced by the observers.

1. Introduction

After the pioneering work of De Witt [1], timeless approach to fundamental physics [2] has recently obtained a renewed interest[3, 4, 5]. The odd nature of time arises from a large number of reasons, first of all the discrepancy between the satisfactory description of dynamics in terms of time evolution, and the fundamental timelessness of general relativity and canonical quantum gravity [1, 4, 5, 6, 7, 8, 9]. Furthermore, several contradictions emerge when different theories are compared, like classical and quantum mechanics, for which time is an external parameter to generate a strongly continuous unitary group of transformations $U(1)$ [10], special and general relativity, quantum field theory, where time is the negative metric signature coordinate of a 4-dimensional differential manifold [11, 12] and canonical quantum gravity. On the other hand, in metrology clock time is a metric time operatively defined at rest. Time is also sometimes associated to the concept of irreversibility and entropy [13, 14, 15]. Such variety of properties and domains of applicability reveals the lack of a satisfactory and universal well posed definition of time.

When considering the problem of the nature of time, there are implicitly two distinct problems to address. The first is about whether or not time must be included in the list of fundamental quantities of Nature, well defined at all the possible energy and length scales. The second is the description of the emergence of time metrology based on operatively defined clocks, and the explanation of time evolution experienced by an observer. The present work strongly supports the view that time is not a fundamental quantity of Nature. The starting point is to observe that time is used both to indicate the parameter used in dynamic equations to describe change in the phase space, and the discretized quantity measured by some macroscopic instruments treated as reference clocks. In most of the scientific literature, the two are implicitly considered the same quantity. Hamiltonian mechanics, which governs the dynamics of generalized coordinates or quantum fields, can be rigorously well defined without the concept of time. As a consequence, it becomes apparently even more difficult to solve the second problem, because time disappears from the list of observable quantities, even if there is a field of metrology entirely devoted to time and frequency measurements conceptually based on the common experience of time.

A good theoretical model capable to be predictive and satisfactory without time answers only partly to the problem of the nature of time: it says what time is not [7]. In the present work the answer to the first problem is addressed, with a particular attention in the definition of the time parameter in a Hamiltonian system in terms of other quantities. Next, also the second part of the problem is addressed, in order to account the experimentally measured and experienced clock time. The connection between the experimental clock time and the theoretical parameter time is the main goal of the present contribute.

As already pointed out by Newton itself, the fact that time is not a measurable quantity [1, 16, 17, 18] can be clarified as follows. One observes that a clock measures with some uncertainty an hypothetical 'true' external time t as a classical quantity $T_i(t)$ where the index i spans the clocks. The other observable quantities $O_j(t)$ are detected as $O_j(T_1)$, $O_j(T_2)$,.... where j spans the observables. However, the clock used to label the dynamical quantities of the system is in turn object of a measurement which establishes its value, accuracy and stability, by means of another clock. Consequently, being the second clock subject to the same check by the first, a two-clock time measurement is required to determine the fractional frequency stability from the Allan variance of both. The reference standard is expressed by $T_2(T_1)$ and $T_1(T_2)$,[7, 19] without any explicit use of t. In other words, a clock is not capable to measure such hypothetical external parameter time, but only self consistent quantities assisted by the recursive definition of *period*.

In the following we addresses the problem of explaining the macroscopic correspondence of clock time with parameter time of dynamics and we provide a universal definition of time for a Hamiltonian system in terms of generalized coordinates change in the phase space.

I will concentrate on a variational approach which enables the introduction of time in a physical theory in two steps. The method provides a parameter, called parameter time, which does not correspond to a specific observable quantity. However it can be put in correspondence with measurable quantities via cyclic phenomena. This is achieved by dividing a system in opportune subsystems. The present approach partially recalls the distinction presented in Ref.[4] between parametric (proper) time and discrete physical time. Differently from there, here no compactified extra-dimensions are required to introduce a detector operator, neither a lapse function or other parameters to appear in the Lagrange function.

In Section II the definition of parameter time in the framework of timeless Hamiltonian theories is presented. Time emerges as the natural parameter after one imposes a variational principle on a timeless action. The approach is applied in the subsection 2.1 to classical mechanics, and extended to quantum field theory in the subsection 2.2. Section III is devoted to connect the parametric time with clock metric time measured by means of realistic devices. In Section IV the conclusions are briefly discussed.

2. Parameter time in a Hamiltonian timeless scenario

The Maupertuis [20, 21, 22] action principle generates the dynamics without explicitly using time in the Hamilton. The interest is restricted to closed systems, so parameter independent Hamiltonians are considered. The variational principle, the Hamiltonian and the generalized coordinates are consequently expressed in a timeless framework. In the following I will show that the imposition of both the variational principle and the stationarity of the Hamiltonian individuate a special parametrization among all the possible parametrizations, which is the one commonly used to describe dynamics. In the following the corresponding parameter is indicated by σ and corresponds to the parameter τ of Ref. [20], and to parameter time t of ix_0 in ordinary Hamiltonian theory. The main difference from the latter is given by its derivation in a timeless framework. The capability of defining Hamiltonian mechanics without the concept of time will require consequently that some extra hypothesis are assumed in order to provide a definition of clock time. Its correspondence with the parameter σ is defined and discussed in the next section.

2.1. *Parameter time in timeless classical mechanics*

The Hamilton equations are expressed in timeless formalism from a variational principle on asynchronous varied trajectories. The time independent Hamiltonian $H(\mathbf{p},\mathbf{q})$ is a function of the generalized three dimensional coordinates $\mathbf{p}$ and $\mathbf{q}$. The independence of H from time reduces the degrees of freedom to $2n-1$. It is necessary to assume that it exists a set of trajectories in the coordinates space μ for which H is constant.

In order to determine the parametrization imposed by the stationarity of the action, a generic parametrization of the points of the trajectories is first assumed. Such arbitrary parametrization λ gives $q_i = q_i(\lambda)$ and $p_i = p_i(\lambda)$ where all the functions belong to C^2 on the interval $[\lambda_A, \lambda_B] \in \mathcal{R}$. The Hamiltonian $H(\mathbf{p},\mathbf{q})$ does not depend explicitly on λ. In order to impose a variational principle on the trajectory it is now considered a variation that is normally used to impose asynchronous varied trajectories in canonical formalism to derive Hamilton equation from the Maupertuis principle. A new parametrization σ of the generalized coordinates and of λ is now defined, under the condition that $\frac{d\lambda}{d\sigma} \neq 0$ on $[\sigma_A, \sigma_B]$.

Such distinction between λ and σ represents a subtle principle and technical difference from the approach of Ref.[23, 24]. The stationarity of the action is imposed:

$$A = \int p_i dq_i \tag{1}$$

where the Einstein summation on the repeated indexes is adopted and $i = 1, 2, 3$. The Maupertuis variational principle reads

$$\delta A = \delta \int p_i dq_i = 0 \tag{2}$$

The imposition of the stationarity of the action is given by the variation of the trajectories. Neglecting as usual second order perturbations and integrating by parts where necessary, one has:

$$d\sigma = \left(\frac{\partial H}{\partial p_i}\right)^{-1} dq_i = -\left(\frac{\partial H}{\partial q_i}\right)^{-1} dp_i \tag{3}$$

under the hypothesis that $\left(\frac{\partial H}{\partial p_i}\right) \neq 0$ and $\left(\frac{\partial H}{\partial q_i}\right) \neq 0$. They differ from the Hamilton equations since σ does not represent the macroscopic metric time. On the contrary, it only represents the natural parameterization of the system imposed by the energy conservation. *σ has nothing to do with the quantity measured by clocks.*

2.2. *Parameter time in quantum field theory*

The most convenient formalism to extend the action principle to general relativity and to quantum mechanics is the extended presymplectic approach [7]. There, dynamics is expressed on the unparameterized curve γ in the relativistic configuration space $C = \mathcal{R} \times C_0$, where C_0 is the m-dimensional space of coordinates q^i, which extremizes the integral

$$A[\gamma] = \int_{\gamma} \theta \tag{4}$$

where

$$\theta = p_i dq^i + p_t dt \tag{5}$$

is the natural one-form defined on the cotangent space T^*C and the constraint

$$H(q^i, t, p_i, p_t) = 0 \tag{6}$$

where H is the relativistic Hamiltonian. In the extended presymplectic formalism, the variational principle reads:

$$\delta A[\gamma] = \delta \int_{\gamma} \theta = 0 \tag{7}$$

Such principle allows a quantum extension, which goes beyond the scopes of the present section. Both the lagrangian and the extended presymplectic formalism consider time as a part of the manifold where physics is defined. Time t or x_0 assumes a role comparable to that of space, even when starting with an unparameterized curve as happens in presymplectic approach. Technically, since the action admits invariance under reparameterization of time (spacetime in relativistic domain), it does not represent a problem. Here, in order to avoid the use of the concept of time, the configuration space is only C_0 instead of $C = \mathcal{R} \times C_0$ and the extended configuration space will only include fields and their conjugate momenta (generalized fields).

A Hamiltonian operator $H = \int d^3x\mathcal{H}$ is given, where $\mathcal{H}$ is the Hamiltonian density. The Hamiltonian operator H acts as a constraint for quantum field dynamics. The action, in terms of a quantum fields $\psi_i(x)$ and the conjugate coordinates $\pi_i(x)$, can be re-expressed as:

$$A = \int d^3x \int d\psi_i \pi_i \tag{8}$$

where the Einstein summation on the repeated indexes is adopted. The roman index spans on the space dimensions 1, 2, and 3. To define time as the natural parameterization of change in the generalized coordinate space μ_Q, the points of the trajectories $f(q_i, p_i) = 0$ are replaced in QFT by space configurations of the generalized field $Q = (\psi_i(\mathbf{x}), \pi_i(\mathbf{x}))$ in μ_Q . In the classical case neighboring position and momentum states are associated to the parameter σ, while in QFT σ labels the generalized field with support in $\mathcal{R}^3$. Two arrays of fields variate the quantum fields and their conjugate fields respectively. As in the previous case, the extremality of the action is obtained under the condition that:

$$d\sigma = \left(\frac{\delta\mathcal{H}}{\delta\pi_i}\right)^{-1}_{\psi_i} d\psi_i(\mathbf{x}) = -\left(\frac{\delta\mathcal{H}}{\delta\psi_i}\right)^{-1}_{\pi_i} d\pi_i(\mathbf{x}) \tag{9}$$

The parameter σ belongs to $\mathcal{R}$ by construction. The parameterization of the field distribution is locally achieved by tagging neighboring configurations with the parameter σ.

3. Definition of generalized clock time

σ has the property of providing a special parameterization suitable for describing dynamics, but it is not an observable quantity. In order to explain the macroscopic experience of time in complex systems, an observable quantity T is built. T realizes an experimentally measurable discrete approximation of σ. Since (metric) time is operatively defined by clock standards based on the period of an oscillator, it is only defined in such systems complex enough to contain a subsystem acting as such a clock. Unfortunately the definition of periodicity implicitly assumes that an external time is available in order to compare a period with the next one, which is meaningless in a timeless framework. Consequently, the concept of *period* is relaxed to the concept of *cycle* in the phase space μ or in μ_Q. Defining the *clock time* T, measured for example by atomic clocks, corresponds to label simultaneous occurences in the phase space of two or more subsystems where one is identified as the clock. The clock corresponds to the cyclic subsystem, as defined below. The dynamics of the i-th observable O_i will consequently be expressed by the simple law involving σ:

$$O_i(T) \cong O_i(\sigma) \tag{10}$$

Let's consider a Hamiltonian system S separable in two independent subsystems S_1 and S_2, so that all the states are represented by factorized (eigen)states of their respective Hamiltonian $\psi_1 \otimes \psi_2 \in H_1 \otimes H_2$ where H_1 and H_2 are the Hilbert spaces of the subsystems 1 and 2 respectively. From the previous analysis, the system S owns a unique natural parameter time σ which is well defined also separately for the two subsystems by construction. We now define the properties required by the system S_1 to act as a clock in S in order to describe dynamics in S_2. For a given $\bar{\sigma}$, a state $\psi \in H_1 \otimes H_2$ consists of the tensor product of the state $\psi_1(\bar{\sigma}) \in H_1$ and the state $\psi_2(\bar{\sigma}) \in H_2$. We say that $\bar{\psi}_1$ has multiplicity κ_{AB} on the interval (σ_A, σ_B) if there are κ_{AB} values of $\tilde{\sigma}_i \in (\sigma_A, \sigma_B)$ such that $\psi_1(\tilde{\sigma}_i) = \bar{\psi}_1$ where $i \in (0, \kappa_{AB})$. We say that the subsystem S_1 is cyclic in the phase space if

(i) its path in the phase space is closed,
(ii) its velocity $|dQ/d\sigma| \neq 0$ and it is smooth,
(iii) the multiplicity κ_{AB} of a state vector in the System 1 monotonically grows with the interval (σ_A, σ_B) and it tends to infinity when $\sigma_A \to -\infty \wedge \sigma_B \to +\infty$.

The second requirement grants that the realizations of two contiguous states occur along the σ axis by respecting the order of the parameter σ. The third requirement that the clock never stops and its velocity in the phase space is enough to grant that the number of cycles is not finite.

Given the interval (σ_A, σ_B), it is now defined the set $\Omega(\sigma_A, \sigma_B) \subset H_2$:

$$\Omega(\sigma_A, \sigma_B) = \{\psi_2(\sigma) \in H_2 | \sigma \in (\sigma_A, \sigma_B)\} \tag{11}$$

An arbitrary origin σ_0 is fixed for the parameter time. We associate to such origin the arbitrary initial states $\bar{\psi}_1 = \psi_1(\sigma_0)$ and $\bar{\psi}_2 = \psi_2(\sigma_0)$. Macroscopic time duration $T^{(S_1)}$ of the interval (σ_A, σ_B) measured by the cyclic subsystem S_1 is given by the number k_{AB} of states $\psi_2(\sigma) \in \Omega$ so that $\psi_1(\sigma) = \bar{\psi}_1$. More explicitly, one has

$$T_{AB}^{(S_1)} \equiv k_{AB} \tag{12}$$

A good clock has the property of being *stable* (small standard deviation) and *accurate* (high Q factor of the resonance associated to the clock) [19, 25]. Since the accuracy refers to the arbitrary resonance frequency of the time standard (for example the Cesium resonance frequency), the present analysis considers only the requirement of stability. Given a target standard deviation Σ required in an experiment performed on the subsystem S_2 in the interval (σ_A, σ_B), for an integration time τ, the *generalized clock* has to fulfill the following prescription:

$$\epsilon \equiv E^2 \left[T_{i,i+1}^{(S_1)}\right] < \Sigma \tag{13}$$

where E^2 is the standard deviation and

$$\sigma_{i+1} = \sigma_i + \tau \tag{14}$$

where $i = 0...N_{AB}$ with $N_{AB} = (\sigma_B - \sigma_A)/\tau$. Consequently, the definition of clock metric time loses of validity for time intervals $T^{(S_1)}$ comparable with the clock period, and for shorter time intervals. Under such hyphotesis, dynamics of observables in the interval (σ_A, σ_B) is approximated by the discrete valued equations:

$$x_\rho(T_i) \cong x_\rho(\sigma_i \pm \epsilon) = x_\rho(\sigma_i) \pm O_{x\rho}[\sigma_i, \epsilon] \tag{15}$$

$$p_\rho(T_i) \cong p_\rho(\sigma_i \pm \epsilon) = p_\rho(\sigma_i) \pm O_{p\rho}[\sigma_i, \epsilon] \tag{16}$$

where $\rho = 1, 2, 3$, and $O_{x\rho}[\sigma, \epsilon]$ and $O_{p\rho}[\sigma, \epsilon]$ are higher order quantities in ϵ. Such equations provide the bridge between parameter time of Hamiltonian timeless formalism, and the experimentally defined clock time experienced by observers.

4. Conclusion

I've presented an approach to provide the correspondence between timeless physics in the microscopic domain, and macroscopic time metrology. Consistently with the discussion, some considerations on the use of the concept of time in theoretical physics is implied. Since clock time is by definition fundamentally discrete and it depends on the specific fabrication of the clock, a (macroscopic) measurement of time below one cycle (period) of the time standard is meaningless. At the present time the most advanced available clock technology is given by single ion atomic clocks based on Al^+/Hg^+ with a fractional uncertainty of about $1 - 2 \times 10^{17}$ [26]. The presented approch implies for example that Planck time scale is an extrapolation, an extension of the concept of clock time beyond its field of definition. To conclude an explicit Hamiltonian framework, entirely developed without the concept of time, has been defined. The time and frequency metrology has been mapped in the equations of motion expressed as a function of the time parameter obtained in the timeless framework, by defining cyclic subsystems capable to account the (discrete) definition of clock time. The present work provides a framework capable to account the timelessness of nature at a fundamental level, and to explain how clock time can be defined in metrology and experiments, consistently with the dynamics of relations between variables and parameter time evolution itself.

Acknowledgments

I gratefully acknowledge E. Minguzzi, Y. Itin, P. Zenczykowski, and H.-T. Elze for the helpful suggestions to improve the first and the second version of the manuscript, and to J. Barbour for the useful criticisms.

References

[1] De Witt B S 1967 Quantum theory of gravity. I *Phys. Rev.* **160** 1113
[2] Rovelli C 1990 Quantum mechanics without time: A model *Phys. Rev.* D **42** 2638-2646
[3] Callender C 2010 Is time an illusion? *Scientific American* **6**
[4] Elze H-T 2004 Quantum mechanics and discrete time from "timeless" classical dynamics *Springer Lect. Notes Phys.* LNP **633** 196
[5] Butterfield J and Isham C 2006 The arguments of time, emergence of time in quantum gravity (Oxford: Oxford University Press)
[6] Rovelli C and Smolin L 1994 The Physical Hamiltonian in nonperturbative quantum gravity *Phys. Rev. Lett.* **72** 446
[7] Rovelli C 2006 *Quantum Gravity* (Cambridge: Cambridge University Press)
[8] Anderson E 2007 Emergent semiclassical time in quantum gravity: I. Mechanical models *Class. Qu. Grav.* **24** 2935-2977
[9] Albrecht A and Iglesias A 2008 The clock ambiguity and the emergence of physical laws *Phys. Rev.* D **77** 063506
[10] Reed M and Simon B 1972 *Methods of Modern Mathematical Physics, I: Functional Analysis* (New York: Academic Press) p. 262

[11] Itin Y and Hehl F 2004 Is the Lorentz signature of the metric of spacetime electromagnetic in origin? *Annals Phys.* **312** 60-83
[12] Minguzzi E and Sanchez M 2008 The Causal hierarchy of spacetimes, in *Recent developments in pseudo-Riemannian geometry* ed D V Alekseevskii and H Baum (European Mathematical Society)
[13] Lebowitz J L 1993 Boltzmann's entropy and time's arrow *Physics Today* **46**(9) 32-38
[14] Lebowitz J L Macroscopic laws, microscopic dynamics, time's arrow and Boltzmann's entropy 1993 *Physica* A **194**(1-4) 1-27
[15] Rovelli C 1993 Statistical mechanics of gravity and the thermodynamical origin of time *Class. Quantum Grav.* **10** 1549-1566
[16] Banks T 1985 *Nucl. Phys.* B **249** 322
[17] Halliwell J J and Hawking S W 1985 *Phys. Rev.* D **31** 1777
[18] Brout R and Venturi G 1989 *Phys. Rev.* D **39** 2436
[19] Levine 1999 *Rev. Sci. Instrum.* **70**(6) 2567
[20] Arnold V I 1989 *Mathematical Methods Of Classical Mechanics* 2nd ed. (Berlin: Springer) Chap. IX
[21] Landau L D and Lifshitz E M *Vol. 1: Mechanics*
[22] Gray C G, Karl G and Novikov V A 1999 From Maupertuis to Schroedinger. Quantization of classical variational priciples *Am. J. Phys.* **67**(11) 959
[23] Barbour J B 1994 The timelessness of quantum gravity: I. The evidence from the classical theory *Class. Qu. Grav.* **11** 2875
[24] Barbour J B 1994 The timelessness of quantum gravity: II. The appearance of dynamics in static configurations *Class. Qu. Grav.* **11** 2898
[25] Heavner P, Jefferts S R, Donley E A, Shirley J H and Parker T E 2005 NIST-F1: Recent improvements and accuracy evaluations *Metrologia* **42**(5) 411
[26] Oskay W H et al. 2006 Single-atom optical clock with high accuracy *Phys. Rev. Lett.* **97** 020801

Fundamental physics in space with the Fermi Gamma-ray Space Telescope

Luca Baldini (for the Fermi LAT Collaboration)
INFN-Sezione di Pisa, Largo B. Pontecorvo 3, 56127, Pisa, Italy
E-mail: `luca.baldini@pi.infn.it`

Abstract. Successfully launched in June 2008, the Fermi Gamma-ray Space Telescope, formerly named GLAST, has been observing the high-energy gamma-ray sky with unprecedented sensitivity for more than two years, opening a new window on a wide variety of exotic astrophysical objects. This paper is a short overview of the main science highlights, aimed at non-specialists, with emphasis on those which are more directly connected with the study of fundamental physics—particularly the search for signals of new physics in the diffuse gamma-ray emission and in the cosmic radiation and the study of Gamma-Ray Burst as laboratories for testing possible violations of the Lorentz invariance.

1. Introduction

Designed to survey the gamma-ray sky in the broad energy range from 20 MeV to more than 300 GeV, with the additional capability of studying transient phenomena at lower energies, Fermi is *de facto* the reference space-based gamma-ray observatory of this decade. It largely surpasses the previous generations of gamma-ray telescopes in terms of effective area, energy range, instrumental dead time, angular resolution and field of view. It has the ability to observe 20% of the sky at any time which, in the nominal scanning mode of operation, allows to view the entire sky every three hours.

The Fermi observatory carries two instruments on-board: the Gamma-ray Burst Monitor (GBM) [2] and the Large Area Telescope (LAT) [3]. The GBM, sensitive in the energy range between 8 keV and 40 MeV, is designed to observe the full sky not occulted by the Earth with rough directional capabilities (at the level of one to a few degrees) for transient point sources, particularly Gamma-Ray Bursts (GRBs).

The LAT is a pair conversion telescope for photons above 20 MeV (up to hundreds of GeV). Though owing most of the basic design to its predecessors—particularly the Energetic Gamma Ray Experiment Telescope (EGRET) [4] on-board the CGRO mission—it exploits the state of the art in terms of detector technology, which allows for a leap in sensitivity of a factor of ten or more. The design, construction and operation of such a complex detector is a fascinating subject on its own and the interested readers can refer to [3] and references therein.

Based on the talk that I presented at the fifth international workshop DICE (held in Castiglioncello, Italy, in September 2010), this is a very *biased* and unconventional review of the Fermi science highlights of the first two years of operation. In fact a significant part of the material that would naturally belong to a review paper is deliberately omitted, here, in favor of a few items of discussion that I think will be of interest for the conference audience. For a more orthodox and comprehensive overview, the reader can refer to [1] and references therein.

1.1. *The gamma-ray sky*

Figure 1 shows the full sky above 200 MeV as seen by the Large Area Telescope; it is a simple map in Galactic coordinates of the event rate corrected for the exposure, with no further processing. The most prominent feature is the plane of our Galaxy, which is indeed very bright in gamma-rays.

Figure 1. Sky map of the gamma-ray candidates above 200 MeV collected by the LAT in the first year of operation. The diffuse emission along the Galactic plane (and especially around the Galactic center), as well as the brightest point sources are clearly visible.

Quite a few point sources can be seen, both along the Galactic plane and at high celestial latitudes (most of the latter being of extragalactic origin). In fact a detailed analysis of the LAT data reveals many more than those evident at first glance: the first Fermi LAT catalog [5], based on one year's worth of data, contains 1451 sources—to be compared with the 271 detected by the EGRET telescope over its entire mission lifetime [6]. Though one of the main science objectives of the Fermi mission, the study of gamma-ray point sources is beyond the scope of this review.

The vast majority of what remains in the sky-map once the contribution from point sources is subtracted is produced by *diffuse* processes taking place in our Galaxy—primarily the interaction of charged Cosmic Rays (CR) with the Galactic interstellar gas and radiation fields. This component (concentrated onto the Galactic plane but not limited to it) goes under the name of Diffuse Galactic Emission (DGE) and accounts for some 90% of the total gamma-ray intensity in the LAT energy range.

The presence of a much fainter isotropic emission, commonly known as the Extragalactic Gamma-ray Background (EGB) is a firmly established experimental fact [17]. The search for a possible truly diffuse component (on top of contributions from unresolved faint sources and misidentified CR background) is one of the primary scientific interests associated to this type of radiation. The DGE and EGB and their relevance in the search for hints of new physics will be discussed in more detail in section 3.

2. Search for photon lines

The quest for a possible narrow line in the diffuse gamma-ray background arises naturally in the context of indirect Dark Matter (DM) searches, where one can look for photons produced in two-body DM particle annihilations $\chi\chi \to \gamma X$ or decays $\chi \to \gamma X$. Since in most scenarios dark matter is electrically neutral (and therefore does not couple directly to photons) such processes only occur at higher orders and the corresponding expected branching ratios are strongly suppressed. Despite this, a photon line, if present, is comparatively easy to identify and distinguish from the standard astrophysical sources of gamma rays—whose flux is dominant in most situations. Unlike most of the cases that we shall examine in the following sections, this channel is peculiar in that it features a distinctive experimental signature that, if observed, would incontrovertibly indicate new physics at work.

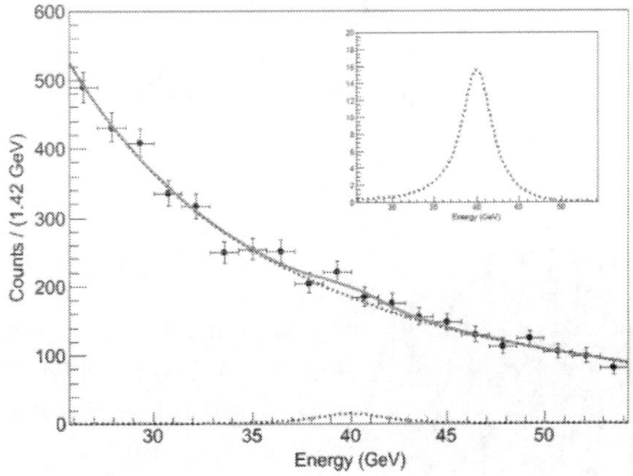

Figure 2. Binned representation of the fit procedure (here centered at 40 GeV) used to derive the upper limit on the flux of a possible photon line contribution in the all-sky (except for part of the Galactic plane) spectrum (from [8]). The two dotted lines represent the background (modeled with a power law) and the signal from the fit, respectively, while the red line is their sum. The insert shows a close-up of the instrument response to a monochromatic line at 40 GeV, which is used to model the signal.

Sure enough, the detector response to a monochromatic line is *not* a monochromatic line and the effect of the finite energy resolution can hardly be ignored. That said, this response can be modeled by means of Monte Carlo simulations (see the insert in Figure 2) and verified with tests at accelerators so that it can be effectively folded into the procedure used to asses the statistical significance of a possible line component in the measured count spectrum.

No significant evidence of gamma-ray line(s) has been found, based on the first 11 months of data, between 30 and 200 GeV [8] and work is ongoing to extend the energy range of the analysis and include more data. The detailed discussion of the upper limits obtained and their relevance in the context of specific models is beyond the scope of this brief overview.

3. Cosmic rays and diffuse gamma-ray radiation

Galactic cosmic rays and gamma rays are closely related to each other. As mentioned in section 1.1, the diffuse gamma-ray emission is mainly produced by cosmic-ray protons and He nuclei interacting with the interstellar gas (through subsequent π^0 decay) and by Cosmic-Ray Electrons (CRE) interacting with the gas itself (via bremsstrahlung), with the ambient low-energy photons (via inverse Compton scattering) and with the Galactic magnetic field (via synchrotron emission). As a consequence, there is a conspicuous number of different ingredients going into the modeling of the Galactic diffuse emission: the energy spectra of the various cosmic-ray species, the interaction cross sections, the distribution of the atomic and molecular gas in the interstellar medium and the details of the interstellar radiation field [9]. Most of these ingredients, indeed, are known to a sufficient level of accuracy to allow realistic numerical simulations of the high-energy Galaxy in which the CR and gamma-ray observables can be self-consistently modeled (see for instance [10]). Still, there are significant uncertainties connected with the predictions of those models. And it goes without saying that such uncertainties are, by their nature, of primary importance to any search for signatures of new physics.

Over the last few years there has been a tremendous interest in the possibility of exotic sources of cosmic rays and/or gamma rays. This excitement has been indeed fostered by a number of measurements supposedly indicating unexpected *features* in the basic observables. Terms like "GeV excess", "ATIC peak" or "PAMELA positron fraction" (which we shall discuss in detail in a moment) have become part of the common jargon of the community, with a profusion of scientific papers aimed at interpreting the apparent "anomalies" and the "excesses" in the data.

Overall, if we look back at the beginning of 2009 (i.e. a few months after the launch of Fermi), it is fair to say that several independent pieces of experimental evidence were (to different extents) difficult to reconcile with our basic understanding of the cosmic-ray propagation in the Galaxy. As we shall see in this section, the situation has largely changed over the last two years.

3.1. Galactic diffuse radiation

The Galactic diffuse gamma-ray emission is not only important in itself for the strong scientific interest it bears; it is also important as it constitutes a bright foreground for the analysis of point sources (influencing the determination of their position and flux) and for the much fainter isotropic diffuse component.

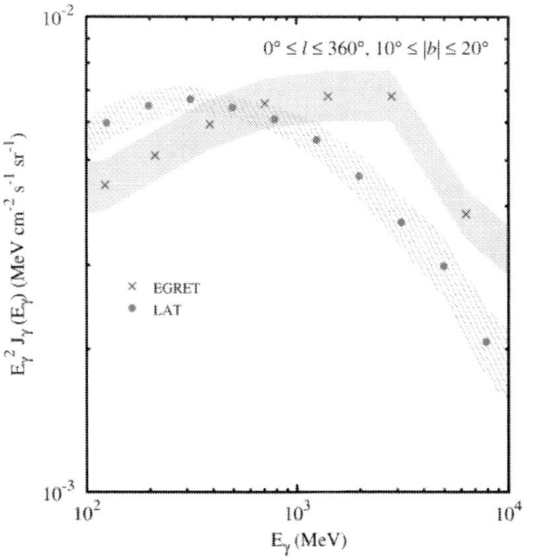

Figure 3. Average diffuse gamma-ray emission intensity at intermediate galactic latitudes $10° \leq |b| \leq 20°$ (from [13]). The shaded/hatched bands superimposed to the data points represent the systematic uncertainties for the Fermi LAT (red) and the EGRET (blue) instruments, respectively. The EGRET "GeV excess" is not confirmed by the LAT.

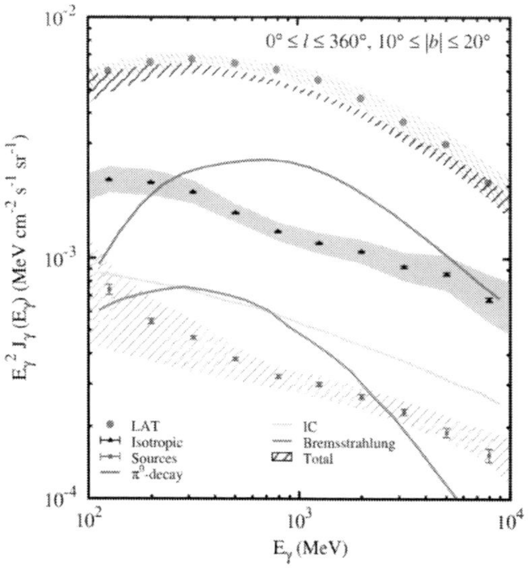

Figure 4. Comparison of the LAT spectrum shown in figure 3 with an *a priori* diffuse Galactic emission model (black hatched band). The model components (π^0 decay, Inverse Compton (IC) and bremsstrahlung) are included for completeness, along with the contributions of point sources and isotropic background, derived from the fit to the LAT data.

The Fermi LAT collaboration has measured the spectrum of the DGE at intermediate latitudes, between 100 MeV and 10 GeV, based on the first 5 months of operations [13]. The spectrum is significantly softer than that measured by EGRET (Figure 3) and it is consistent with an a-priori—i.e. reflecting the state of the art in terms of local cosmic-ray data but not tuned to reproduce the gamma-ray data—model (Figure 4). Particularly, the excess emission above 1 GeV measured by EGRET, which in the literature is often informally referred to simply as the "GeV excess", is not confirmed by the LAT.

The region of interest $10° \leq |b| \leq 20°$ (or about 15° from the Galactic plane) was chosen to maximize the signal to noise ratio and, at the same time, minimize the uncertainties in the modeling (in order to model the diffuse emission on the Galactic plane it is necessary to trace the cosmic rays through the whole Galaxy). Work is ongoing in the Collaboration for a more thorough analysis, not limited to the region of interest and the energy range in [13]. It is already clear, however, that most of the scientific exegesis of the original EGRET claim is to be radically reviewed in the light of the new Fermi results.

3.2. Isotropic diffuse radiation

First detected by the second Small Astronomy Satellite (SAS-2) mission [14] and later confirmed by EGRET [15], the extragalactic diffuse gamma-ray background is generally considered to be the superposition of contributions from unresolved extragalactic sources and truly diffuse emission processes—the latter potentially including imprints of new physics.

The measurement of the EGB is, to many respects, more challenging than the others being reviewed in this paper. Not only because it is intrinsically faint, but also because, being isotropic and stationary, it is not possible to exploit any spatial and/or temporal signature to enhance the signal-to-noise ratio. In fact the estimation of the residual contamination of misidentified charged cosmic rays (whose flux is also isotropic) is one of the critical aspects of the analysis.

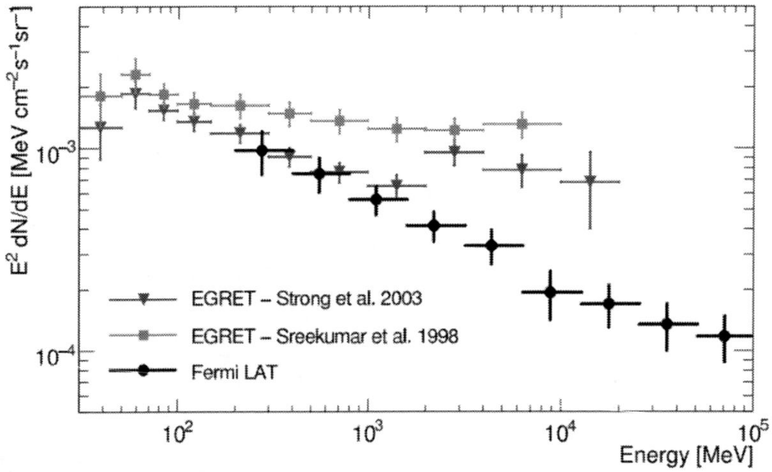

Figure 5. EGB intensity measured by Fermi (from [13]), compared with the one derived from the EGRET data in [15] and the re-analysis [16], based on a different model for the Galactic diffuse emission.

Some 5 years after the original EGRET publication [15], a re-analysis of the EGRET data [16], based on a different model of the diffuse Galactic emission, revealed a distinctive peak in the spectrum around 3 GeV, which triggered the interest of the community in what might have been its origin. To this respect, it is interesting to note how the blue and red data points in figure 5 are based on the *same* experimental data. The substantial difference between the two spectra is only attributable to a difference in the underlying model for the DGE (which, as we mentioned before, is a strong foreground for the measurement of the EGB). This, in turn, is a good illustration of how often putative signs of new physics are inextricably mixed with genuine astrophysical processes, and the uncertainties on the latter can have significant effects. For what it's worth, it is the opinion of the author that this aspect does not always receive the attention it deserves.

The spectrum of the extragalactic gamma-ray background measured by the Fermi LAT [17], shown in Figure 5, is significantly softer than the EGRET spectrum. It is compatible with a featureless power law with spectral index $\Gamma = 2.37 \pm 0.05$ and does not confirm the spectral feature evident in the EGRET re-analysis. A detailed study of the Fermi source count distributions [18] shows that unidentified Blazars can only account for 40% of the EGB flux at most. The nature of the remaining fraction is of the highest scientific interest and will be investigated as more data are collected.

3.3. Leptons in cosmic rays

Charged cosmic rays have been extensively studied since the mid 1960s with balloon-borne and satellite experiments—and, at comparatively higher energies, by means of ground arrays. We have, to date, a large, well-established body of knowledge which includes spectra, relative abundances, composition and time variation. Among the different species, energetic cosmic-ray electrons and positrons are peculiar in that they rapidly loose energy through synchrotron radiation in the Galactic magnetic field and inverse Compton scattering on the interstellar radiation field. Therefore, at high energy, they effectively probe the nearby Galactic space.

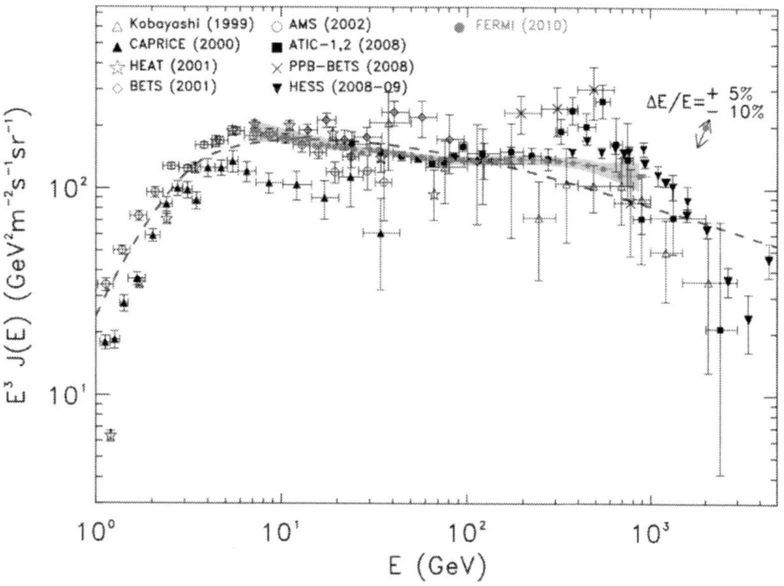

Figure 6. Cosmic-ray electron spectrum measured by the Fermi LAT (red points, from [35]). The gray shaded band indicates the systematic uncertainties associated with the flux values, while the blue dashed line represents the prediction of a diffusive propagation model tuned on the pre-Fermi data [19]. The other recent measurements included are from [20, 21, 22, 23, 24, 25, 26, 27, 28].

In 2008 the balloon-borne Advanced Thin Ionization Calorimeter (ATIC) experiment reported evidence for a prominent, sharp spectral feature between 300 and 800 GeV in the inclusive $e^+ + e^-$ spectrum [25]—which has become widely known as the "ATIC peak". One of the possible interpretations put forward by the authors was the existence of a Kaluza-Klein particle at the TeV scale annihilating directly into electron-positron pairs. This measurement, selected by the American Institute of Physics as one of the 10 Top Physics Stories of 2008, has stirred tremendous interest in the high-energy astrophysics community, with the ATIC peak often discussed in conjunction with another measurement, published shortly after by the PAMELA collaboration—namely the evidence that the fraction of positron in the $e^+ + e^-$ spectrum rises with energy [29].

In 2009 the Fermi LAT collaboration has published the first systematic-limited measurement of the spectrum of cosmic-ray electrons and positrons [31] between 20 GeV and 1 TeV. Though harder than previously thought, the Fermi spectrum does not show evidence of any prominent feature. This has been later confirmed by the H.E.S.S. experiment [28], as well as by a dedicated analysis of the Fermi data aimed at the highest end of the energy range [35].

Figure 6 shows the Fermi LAT spectrum, compared with the predictions of *a* pre-Fermi model [19]. Whether the difference between the two is due to the limits in our understanding of the cosmic-ray propagation in the Galaxy or to the presence of additional (astrophysical or exotic) sources of high-energy electrons is a question with no definitive answer, at this point [32].

3.4. Further comments and remarks

The results from the first two years of operation of the Fermi LAT have radically changed our understanding of the diffuse gamma-ray radiation and cosmic-ray propagation. Most of the claims of prominent departures from the accepted scientific paradigms, in fact, have not been confirmed and might be, at least partially, just instrumental artifacts. It is probably fair to say that the positron fraction measured by the PAMELA experiment is still at odds with the standard picture of secondary cosmic-ray production. The Fermi LAT and the forthcoming Alpha Magnetic Spectrometer (AMS-02) space experiment will help to shed light on the subject [30]. In the meantime papers like [33] and [34] might be telling us that what we really learn from the "lepton excesses" is that we don't know enough astrophysics, yet.

That all said, the overall tone of this section should not be perceived as too dismissive. Diffuse gamma-ray emission and cosmic-rays, when interpreted as different pieces of a single puzzle, still constitute a very exciting perspective for the quest of new physics in space—particularly in the context of indirect dark matter searches. Along with the energy spectra additional observables (i.e. the anisotropies in the arrival directions [36]) will be of great interest, as new, more accurate measurements become available and our understanding of the astrophysical environment improves.

4. Gamma-Ray Bursts and tests of Lorentz invariance violation

Gamma-Ray Burst (GRB) are the most violent explosions in the Universe. The isotropic energy emission, that can be estimated based on the measured fluence, can be as high as 10^{54} erg for the brightest bursts, with a huge energy release in keV–MeV gamma-rays. This naturally suggests a narrow beaming of the emission—still, a beaming factor of the order of 10^{-3} implies a considerable energy budget of 10^{51} erg.

GRBs have been extensively studied at relatively low energies over the last two decades, the most notable observed features being:

- the cosmological origin (they are isotropically distributed in the sky and have been observed up to a redshift $z \approx 8$);
- the bimodal duration distribution (with *short* bursts lasting for ≈ 1 s and *long* ones lasting for tens of s);
- the rapid variability of the light curves, down to time scales of the order of the ms.

The Fermi LAT has effectively opened a new observational window, enabling the systematic study of the prompt emission above 100 MeV on a large sample of bursts—with supporting observations by the GBM at lower energy being crucial in putting the new information in the context of what is already known (see Figure 7). The GBM detects about 250 GRB per year, roughly half of them being in the LAT field of view; some 10% of the latter (or ≈ 10 per year) are bright enough at high energy to be detected by the Large Area Telescope. As of the end of 2010, the Fermi LAT collaboration has reported the detection of 22 GRBs through the Gamma-ray Burst Coordination Network (GCN) [37].

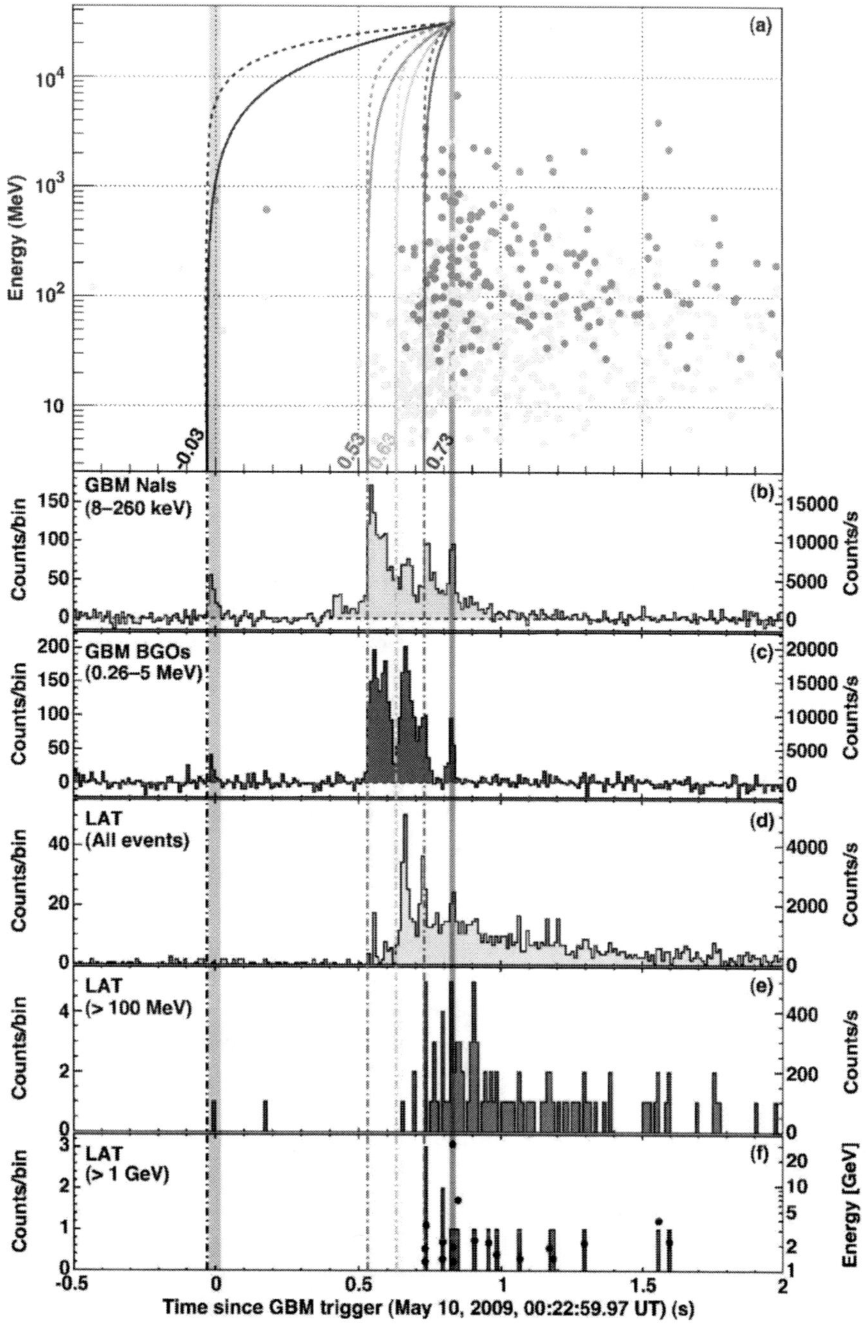

Figure 7. Light curves of the short GRB 090510 in different energy bands (from [42]). The top panel shows the photon energy vs. the photon arrival time for two different event selections (see [42] for more details). The remaining plots are the GBM and LAT light curves from lowest to highest energy (top to bottom). The overlaid curves, which are normalized to pass through the highest energy (31 GeV) photon, represent the relation between the energy and the arrival time for linear (solid line) and quadratic (dashed line) LIV—assuming that the photon is emitted in correspondence of the GBM onset (30 ms before the trigger time).

From the standpoint of this brief overview, gamma-ray bursts are mainly interesting as laboratories to test possible violations of the Lorentz Invariance (LIV). There are indeed several theoretical frameworks—particularly in the context of some of the Quantum Gravity (QG) formalisms currently being investigated—that predict (or can accommodate) a modification of the standard photon dispersion relation [38, 39, 40]. This happens in the form of an energy-dependent term ceasing to be negligible at some mass scale $m_{\rm QG}$, possibly of the order of the Planck mass $m_{\rm P} \approx 1.22 \times 10^{19}$ GeV:

$$E^2 = p^2c^2 + \Delta_{\rm QG}(E, p^2, m_{\rm QG}) \qquad (1)$$

One can therefore imagine to expand the additional term in series of $E/m_{\rm QG}$ up to the first non-vanishing order n, which yields a non trivial expression for the speed of light in vacuum as a function of the photon energy:

$$v(E) = c\left[1 \pm \left(E/m_{\rm QG}c^2\right)^n\right] \qquad (2)$$

where the $\pm$ sign distinguishes the *subluminal* and the *superluminal* cases. The experimental signature of such a scenario would be the fact that two photons of different energies, emitted simultaneously, will travel with different velocities and arrive to the observer with some relative time delay. Even a tiny variation in the speed of light, when accumulated over cosmological distances, can in principle be revealed at energies much lower than the characteristics scale $m_{\rm QG}c^2$. If, for the sake of simplicity, we limit ourselves to linear LIV (i.e. $n = 1$) the relevant expression for the time delay of two photons emitted by a distant astrophysical source at redshift z is:

$$\Delta t = \pm\frac{1}{H_0}\frac{\Delta E}{m_{\rm QG}}\int_0^z \frac{(1+\xi)}{\sqrt{\Omega_\Lambda + \Omega_m(1+\xi)^3}}d\xi \qquad (3)$$

The short duration, rapid variability and cosmological distance make GRBs for perfect candidates to constraint possible Lorentz invariance violations. Though the Fermi collaboration has previously set stringent lower limits on the mass scale $m_{\rm QG}$ with the long GRB 080916C [41], here we shall limit the discussion to the short GRB 090510 [42]. As shown in Figure 7, this burst features a 31 GeV photon emitted 0.829 s after the GBM trigger time. Thanks to the synergy with other x-ray and optical instruments, it was possible to measure the redshift of the source, which turned out to be $z = 0.903 \pm 0.003$. Sure enough, we don't know *exactly* when this particular photon was emitted, relative to the much more abundant low-energy gamma-rays, so that this arrival time cannot be readily interpreted as a *time delay*. However, if we assume that the photon itself was not emitted before the beginning of the precursor of the burst (30 ms before the GBM trigger), that gives us all the necessary ingredients to set a robust lower limit $m_{\rm QG} > 1.19 m_{\rm P}$ in the linear ($n = 1$) subluminal case. In fact this is the most stringent limit of its kind available so far, and, effectively, it strongly disfavors any scenario characterized by a linear, subluminal LIV. The assumption on which it is based is very reasonable and somehow supported by the fact that in none of the bursts detected by the Fermi LAT high-energy photons were detected before the onset of the low-energy emission—on the contrary, in many cases there was evidence for a spectral evolution with a consequent delay of the high-energy emission onset. By using more sophisticated approaches (such as the one described in [43]) or relaxing the basic assumption about the emission time of the highest energy photon it is possible to obtain comparable or more stringent limits, in both the subluminal and the superluminal case; the reader is referred to [42] for further details.

It is worth stressing that this is another case in which the effect we are trying to measure (in this case the delay accumulated by the photon during the propagation) *competes* with another physical effect (i.e. the intrinsic time delay at the source) which is of the same order or magnitude or bigger and which is essentially impossible to infer with enough accuracy. A definitive evidence for a Lorentz invariance violation might come in the form of the observation of a redshift-dependent effect on a large sample of bursts, which Fermi might be able to achieve before the end of the mission.

5. More discussion items and future prospects

We conclude this overview by briefly introducing a few more items of interest, some of which might be the object of future Fermi observations.

5.1. Pulsar timing array detection of gravitational waves

Pulsars constitute one of the primary scientific targets for Fermi—in fact the Large Area Telescope has already increased the number of known gamma-ray pulsars by roughly an order of magnitude from the 6 detected by EGRET. Though most of the scientific interest associated with these discoveries is of purely astrophysical nature, Milli-Second Pulsars (MSP)—old neutron stars spinning at a rate of hundreds of revolutions per seconds—are effectively accurate cosmic clocks suitable for tests of fundamental physics.

It was suggested already in the late 1970s [44, 45] that gravitational waves could induce small perturbations in the arrival times of the radio pulses and that those perturbations could be potentially detectable. It actually did not take too long to realize that monitoring the time residuals of an *array* of pulsars had the potential for a significant increase in sensitivity [46]. Based on this idea, *Pulsar timing arrays* are meant to observe, over periods of years, the small, correlated effects that gravitational waves have on the arrival times of the radio pulses at Earth. The whole subject recently experienced a renewed interest as, by studying unassociated high-energy sources detected by the Fermi LAT in the first months of operation, radio astronomers discovered 17 new millisecond pulsars (corresponding to a 30% increase in the overall number of known MSPs). This is really the product of a synergic cooperation between the Fermi Collaboration and the largest radio telescopes in the world: in many cases the gamma-ray data are too sparse to detect pulsation in the unidentified sources, whereas radio astronomers can take advantage of the positions of potential pulsar candidates to complement time-consuming sky surveys and improve the discovery reach.

Pulsar timing already provides the most compelling experimental evidence for the existence of gravitational radiation through the orbital decay of the binary system PSR 1913+16 [47]. However, this is generally regarded as an *indirect* evidence, in that what is observed is the effect of the emission on the emitter. On the other hand, pulsar timing arrays really act as gravitational wave detectors. As more millisecond pulsars are identified (and monitored by radio telescopes), they could provide the first *direct* evidence for the existence of gravitational radiation in the ultra low-frequency regime, where the ground based interferometers are not sensitive.

5.2. Gravitational lensing

The light curves of distant, gravitationally lensed Active Galactic Nuclei (AGN) measured by the Large Area Telescope can potentially be used to test general relativity. Multiple images of strongly lensed sources can be resolved in the optical band. The study of the time lag between such images (which can be as large as weeks, due to the different propagation times along the different light paths) is a standard observational technique, crucial for the modeling of this kind of systems. This is not directly possible in gamma-rays, as gamma-ray telescopes do not feature the necessary angular resolution. Nonetheless, if a flare is detected from such a source, the high-energy light curve might show evidence for the gravitational *echoes*. A comparison between the optical and the gamma-ray light curves can test whether photons differing by eight orders of magnitude in energy are deflected in the same way, as general relativity predicts.

5.3. Search for axion-like particles

Axion-Like Particles (ALP) are hypothetical particles featuring a direct photon coupling in the presence of an external magnetic field. Phenomenologically, this gives origin to a neutrino-like oscillation behavior between ALPs and photons that can possibly lead to observable effects on the energy spectra of distant gamma-ray sources [48]. The photon-axion mixing can take place both in the gamma ray-source, if this is strongly magnetized, and in the magnetic fields encountered along the path to the observer. In the latter case the propagation over cosmological distances compensates for the weakness of the intergalactic magnetic fields (which is believed to be of the order of nG). The prospect of detecting such modifications to the spectra of gamma-ray sources is complementary to ground-based experiments such as CAST [49] or PVLAS [50].

6. Conclusions

We briefly reviewed some of the science highlights of the Fermi gamma-ray space telescope after the first two years in orbit. The main message that this overview is aimed to convey is that the study of fundamental physics with space-based gamma-ray astronomy is indeed an exciting opportunity. Gamma-rays directly probe some of the most violent phenomena in the Universe; they give access to physical systems in the strong gravitational and/or magnetic field regime, along with photon propagation over cosmological distances—in brief, many of the conditions that would be impossible to set up in a laboratory on the Earth. On the other hand, life in space is to many respects more difficult than that in a controlled experimental environment; the physical effects that one is trying to measure are often interleaved with genuine astrophysical phenomena which (i) cannot be controlled and (ii) are extremely hard to understand and model to the needed degree of accuracy.

Fermi turned two years in orbit on June, 2010, and it is definitely living up to its expectations in terms of scientific results delivered to the community. The mission is planned to continue at least three more years (possibly eight) with many remaining opportunities for discoveries. All the Fermi gamma-ray data (along with the science analysis software tools developed and used by the Collaboration) are publicly available through the Fermi Science Support Center [51].

Acknowledgments

The work reported here is a result of the efforts of all members of the Fermi LAT and GBM Collaborations.

The *Fermi* LAT Collaboration acknowledges generous ongoing support from a number of agencies and institutes that have supported both the development and the operation of the LAT as well as scientific data analysis. These include the National Aeronautics and Space Administration and the Department of Energy in the United States, the Commissariat à l'Energie Atomique and the Centre National de la Recherche Scientifique / Institut National de Physique Nucléaire et de Physique des Particules in France, the Agenzia Spaziale Italiana and the Istituto Nazionale di Fisica Nucleare in Italy, the Ministry of Education, Culture, Sports, Science and Technology (MEXT), High Energy Accelerator Research Organization (KEK) and Japan Aerospace Exploration Agency (JAXA) in Japan, and the K. A. Wallenberg Foundation, the Swedish Research Council and the Swedish National Space Board in Sweden.

Additional support for science analysis during the operations phase is gratefully acknowledged from the Istituto Nazionale di Astrofisica in Italy.

References

[1] Michelson P F, Atwood W B and Ritz S 2010 *Rep. Prog. Phys.* **73** 074901
[2] Meegan C et al 2007 *AIP Conf. Proc.* **921** 13
[3] Atwood W B et al [Fermi LAT collaboration] 2009 *Astrophys. J.* **697** 1071
[4] Thompson D J et al 1993 *Astrophys. J. Suppl. Series* **86** 629

[5] Abdo A A et al [Fermi LAT collaboration] 2010 *Astrophys. J. Suppl. Series* **188** 405
[6] Hartman R C et al 1999 *Astrophys. J. Suppl. Series* **123** 79
[7] Martinez M 2009 *J. Phys.: Conf. Ser.* **171** 012013
[8] Abdo A A et al [Fermi LAT collaboration] 2010 *Phys. Rev. Lett.* **104** 091302
[9] Strong A W, Moskalenko I V and Ptuskin V S 2007 *Ann. Rev. Nucl. Part. Sci.* **57** 285
[10] Strong A W and Moskalenko I V 1998 *Astrophys. J.* **509** 212
[11] Hunter S D et al 1997 *Astrophys. J.* **481** 205
[12] de Boer W et al 2005 *Astron. Astrophys.* **444** 51
[13] Abdo A A et al [Fermi LAT collaboration] 2009 *Phys. Rev. Lett.* **103** 251102
[14] Fichtel C E at al 1978 *Astrophys. J.* **222** 833
[15] Sreekumar P et al 1998 *Astrophys. J.* **494** 523
[16] Strong A W et al 2004 *Astrophys. J.* **613** 956
[17] Abdo A A et al [Fermi LAT collaboration] 2010 *Phys. Rev. Lett.* **104** 101101
[18] Abdo A A et al 2010 *Astrophys. J.* **720** 435
[19] Strong A W, Moskalenko I V and Reimer O 2004 *Astrophys. J.* **613** 962
[20] Kobayashi T et al 1999 *Proc. of the 26th International Cosmic-Ray Conf.* **3** 61
[21] Boezio M et al [CAPRICE collaboration] 2000 *Astrophys. J.* **532** 653
[22] Du Vernois M A et al [HEAT collaboration] 2001 *Astrophys. J.* **559** 296
[23] Torii S et al [BETS collaboration] 2001 *Astrophys. J.* **559** 973
[24] Aguilar M et al [AMS collaboration] 2002 *Phys. Rep.* **366** 331
[25] Chang J et al [ATIC collaboration] 2008 *Nature* **456** 362
[26] Yoshida K et al [PPB-BETS collaboration] 2008 *Advan. Space Research* **42** 1670
[27] Aharonian F et al [H.E.S.S collaboration] 2008 *Phys. Rev. Lett.* **101** 261104
[28] Aharonian F at al [H.E.S.S. collaboration] 2009 *Astron. Astrophys.* **508** 561
[29] Adriani O et al [PAMELA collaboration] 2009 *Nature* **458** 607
[30] Pato M, Lattanzi M and Bertone G 2010 *Journ. Cosm. Astropar. Phys.* **12** 020
[31] Abdo A A et al [Fermi LAT collaboration] 2009 *Phys. Rev. Lett.* **102** 181101
[32] Grasso D at al 2009 *Astropart. Phys.* **32** 140
[33] Blasi P 2009 *Phys. Rev. Lett.* **103** 051104
[34] Katz B et al 2010 *Mon. Not. R. Astron. Soc.* **405** 1458
[35] Ackermann M et al [Fermi LAT collaboration] 2010 *Phys. Rev.* D **82** 092004
[36] Ackermann M et al [Fermi LAT collaboration] 2010 *Phys. Rev.* D **82** 092003
[37] Gamma-ray Burst Coordination Network, `http://gcn.gsfc.nasa.gov/`
[38] Amelino-Camelia G, Ellis J, Mavromatos N E, Nanopoulos D V and Sarkar S 1998 *Nature* **393** 763
[39] Mattingly D 2005 *Living Rev. Relativity* **8** 5
[40] Kostelecky V A and Mewes M 2008 *Astrophys. J. Lett.* **689** L1
[41] Abdo A A et al [Fermi LAT collaboration] 2009 *Nature* **323** 1688
[42] Abdo A A et al [Fermi LAT collaboration] 2009 *Nature* **462** 331
[43] Scargle J D, Norris J P and Bonnel J T 2008 *Astrophys. J.* **673** 972
[44] Sazhin M V 1978 *Soviet Astr.* **22** 36
[45] Detweiler S 1979 *Astrophys. J.* **234** 1100
[46] Foster R S and Backer D C 1990 *Astrophys. J.* **361** 300
[47] Taylor J H and Weisberg J M 1982 *Astrophys. J.* **253** 908
[48] Sánchez-Conde M A et al 2009 *Phys. Rev.* D **79** 123511
[49] Andriamonje S et al 2007 *J. Cosm. Astropart. Phys.* **4** 10
[50] Zavattini E et al [PVLAS collaboration] 2006 *Phys. Rev. Lett.* **96** 110406
[51] Fermi Science Support Center, `http://fermi.gsfc.nasa.gov/ssc/`

The No-Scale Multiverse at the LHC

Tianjun Li[1,2], **James A. Maxin**[1], **Dimitri V. Nanopoulos**[1,3,4,a] **and Joel W. Walker**[5]

[1]George P. and Cynthia W. Mitchell Institute for Fundamental Physics, Texas A&M University, College Station, TX 77843, USA
[2]Key Laboratory of Frontiers in Theoretical Physics, Institute of Theoretical Physics, Chinese Academy of Sciences, Beijing 100190, P. R. China
[3]Astroparticle Physics Group, Houston Advanced Research Center (HARC), Mitchell Campus, Woodlands, TX 77381, USA
[4]Academy of Athens, Division of Natural Sciences, 28 Panepistimiou Avenue, Athens 10679, Greece
[5]Department of Physics, Sam Houston State University, Huntsville, TX 77341, USA

E-mail: [a]`dimitri@physics.tamu.edu`

Abstract. We present a contemporary perspective on the String Landscape and the Multiverse of plausible string, M- and F-theory vacua, seeking to demonstrate a non-zero probability for the existence of a universe matching our own observed physics within the solution ensemble. We argue for the importance of No-Scale Supergravity as an essential common underpinning for the spontaneous emergence of a cosmologically flat universe from the quantum "nothingness". Our context is a highly detailed phenomenological probe of No-Scale $\mathcal{F}$-$SU(5)$, a model representing the intersection of the $\mathcal{F}$-lipped $SU(5) \times U(1)_X$ Grand Unified Theory (GUT) with extra TeV-Scale vector-like multiplets derived out of $\mathcal{F}$-theory, and the dynamics of No-Scale Supergravity. The latter in turn imply a very restricted set of high energy boundary conditions.

We present a highly constrained "golden point" located near $M_{1/2} = 455$ GeV and $\tan\beta = 15$ in the $\tan\beta - M_{1/2}$ plane, and a highly non-trivial "golden strip" with $\tan\beta \simeq 15$, $m_t = 173.0$-174.4 GeV, $M_{1/2} = 455$-481 GeV, and $M_V = 691$-1020 GeV, which simultaneously satisfies all the known experimental constraints, featuring also an imminently observable proton decay rate. We supplement this bottom-up phenomenological perspective with a top-down theoretical analysis of the one-loop effective Higgs potential. A striking consonance is achieved via the dynamic determination of $\tan\beta$ and $M_{1/2}$ for fixed Z-boson mass at the local *minimum minimorum* of the potential, that being the secondary minimization of the spontaneously broken electroweak Higgs vacuum $V_{\rm min}$. By also indirectly determining the electroweak scale, we suggest that this constitutes a complete resolution of the Standard Model gauge hierarchy problem.

Finally, we present the distinctive collider level signatures of No-Scale $\mathcal{F}$-$SU(5)$ for the $\sqrt{s} = 7$ TeV LHC, with 1 fb^{-1} of integrated luminosity. The characteristic feature is a light stop and gluino, both sparticles lighter than all other squarks, generating a surplus of ultra-high multiplicity (≥ 9) hadronic jet events. We propose modest alterations to the canonical background selection cut strategy which are expected to yield significantly enhanced resolution of the characteristic ultra-high jet multiplicity $\mathcal{F}$-$SU(5)$ events, while readily suppressing the contribution of all Standard Model processes, and allowing moreover a clear differentiation from competing models of new physics, most notably minimal supergravity. Detection by the LHC of the ultra-high jet signal would constitute a suggestive evocation of the intimately linked stringy origins of $\mathcal{F}$-$SU(5)$, and could possibly provide a glimpse into the underlying structure of the fundamental string moduli, possibly even opening a darkened glass upon the hidden workings of the No-Scale Multiverse.

Journal of Physics: Conference Series **306** (2011) 012015 doi:10.1088/1742-6596/306/1/012015

1. Introduction

The number of consistent, meta-stable vacua of string, M- or (predominantly) F-theory flux compactifications which exhibit broadly plausible phenomenology, including moduli stabilization and broken supersymmetry [1, 2, 3, 4, 5, 6], is popularly estimated [7, 8] to be of order 10^{500}. It is moreover currently in vogue to suggest that degeneracy of common features across these many "universes" might statistically isolate the physically realistic universe from the vast "landscape", much as the entropy function coaxes the singular order of macroscopic thermodynamics from the chaotic duplicity of the entangled quantum microstate. We argue here though the counter point that we are not obliged *a priori* to live in the likeliest of all universes, but only in one which is possible. The existence merely of a non-zero probability for our existence is sufficient.

We indulge for this effort the fanciful imagination that the "Multiverse" of string vacua might exhibit some literal realization beyond our own physical sphere. A single electron may be said to wander all histories through interfering apertures, though its arrival is ultimately registered at a localized point on the target. The journey to that destination is steered by the full dynamics of the theory, although the isolated spontaneous solution reflects only faintly the richness of the solution ensemble. Whether the Multiverse be reverie or reality, the conceptual superset of our own physics which it embodies must certainly represent the interference of all navigable universal histories.

Surely many times afore has mankind's notion of the heavens expanded - the Earth dispatched from its central pedestal in our solar system and the Sun rendered one among some hundred billion stars of the Milky Way, itself reduced to one among some hundred billion galaxies. Finally perhaps, we come to the completion of our Odyssey, by realizing that our Universe is one of at least 10^{500} so possible, thus rendering the anthropic view of our position in the Universe (environmental coincidences explained away by the availability of $10^{11} \times 10^{11}$ solar systems) functionally equivalent to the anthropic view of the origin of the Universe (coincidences in the form and content of physical laws explained away by the availability, through dynamical phase transitions, of 10^{500} universes). Nature's bounty has anyway invariably trumped our wildest anticipations, and though frugal and equanimous in law, she has spared no extravagance or whimsy in its manifestation.

Our perspective should not be misconstrued, however, as complacent retreat into the tautology of the weak anthropic principle. It is indeed unassailable truism that an observed universe must afford and sustain the life of the observer, including requisite constraints, for example, on the cosmological constant [9] and gauge hierarchy. Our point of view, though, is sharply different; We should be able to resolve the cosmological constant and gauge hierarchy problems through investigation of the fundamental laws of our (or any single) Universe, its accidental and specific properties notwithstanding, without resorting to the existence of observers. In our view, the observer is the output of, not the *raison d'être* of, our Universe. Thus, our attention is advance from this base camp of our own physics, as unlikely an appointment as it may be, to the summit goal of the master theory and symmetries which govern all possible universes. In so seeking, our first halting forage must be that of a concrete string model which can describe Nature locally.

2. The Ensemble Multiverse

The greatest mystery of Nature is the origin of the Universe itself. Modern cosmology is relatively clear regarding the occurrence of a hot big bang, and subsequent Planck, grand unification, cosmic inflation, lepto- and baryogenesis, and electroweak epochs, followed by nucleosynthesis, radiation decoupling, and large scale structure formation. In particular, cosmic inflation can address the flatness and monopole problems, explain homogeneity, and generate the fractional anisotropy of the cosmic background radiation by quantum fluctuation of the inflaton field [10, 11, 12, 13, 14]. A key question though, is from whence the energy of the Universe arose.

Interestingly, the gravitational field in an inflationary scenario can supply the required positive mass-kinetic energy, since its potential energy becomes negative without bound, allowing that the total energy could be exactly zero.

Perhaps the most striking revelation of the post-WMAP [15, 16, 17] era is the decisive determination that our Universe is indeed globally flat, *i.e.* , with the net energy contributions from baryonic matter $\simeq 4\%$, dark matter $\simeq 23\%$, and the cosmological constant (dark energy) $\simeq 73\%$ finely balanced against the gravitational potential. Not long ago, it was possible to imagine the Universe, with all of its physics intact, hosting any arbitrary mass-energy density, such that "$k = +1$" would represent a super-critical cosmology of positive curvature, and "$k = -1$" the sub-critical case of negative curvature. In hindsight, this may come to seem as naïve as the notion of an empty infinite Cartesian space. The observed energy balance is highly suggestive of a fundamental symmetry which protects the "$k = 0$" critical solution, such that the physical constants of our Universe may not be divorced from its net content.

This null energy condition licenses the speculative connection *ex nihilo* of our present universe back to the primordial quantum fluctuation of an external system. Indeed, there is nothing which quantum mechanics abhors more than nothingness. This being the case, an extra universe here or there might rightly be considered no extra trouble at all! Specifically, it has been suggested [10, 11, 12, 18, 19] that the fluctuations of a dynamically evolved expanding universe might spontaneously produce tunneling from a false vacuum into an adjacent (likely also false) meta-stable vacuum of lower energy, driving a local inflationary phase, much as a crystal of ice or a bubble of steam may nucleate and expand in a super-cooled or super-heated fluid during first order transition. In this "eternal inflation" scenario, such patches of space will volumetrically dominate by virtue of their exponential expansion, recursively generating an infinite fractal array of causally disconnected "Russian doll" universes, nesting each within another, and each featuring its own unique physical parameters and physical laws.

From just the specific location on the solution "target" where our own Universe landed, it may be impossible to directly reconstruct the full theory. Fundamentally, it may be impossible even in principle to specify why our particular Universe is precisely as it is. However, superstring theory and its generalizations may yet present to us a loftier prize - the theory of the ensemble Multiverse.

3. The Invariance of Flatness

More important than any differences between various possible vacua are the properties which might be invariant, protected by basic symmetries of the underlying mechanics. We suppose that one such basic property must be cosmological flatness, so that the seedling universe may transition dynamically across the boundary of its own creation, maintaining a zero balance of some suitably defined energy function. In practice, this implies that gravity must be ubiquitous, its negative potential energy allowing for positive mass and kinetic energy. Within such a universe, quantum fluctuations may not again cause isolated material objects to spring into existence, as their net energy must necessarily be positive. For the example of a particle with mass m on the surface of the Earth, the ratio of gravitational to mass energy is more than nine orders of magnitude too small

$$\left| -\frac{G_N M_E m}{R_E} \right| \div mc^2 \simeq 7 \times 10^{-10} \; , \tag{1}$$

where G_N is the gravitational constant, c is the speed of light, and M_E and R_E are the mass and radius of the Earth, respectively. Even in the limiting case of a Schwarzschild black hole of mass M_{BH}, a particle of mass m at the horizon $R_S = 2G_N M_{BH}/c^2$ has a gravitational potential which is only half of that required.

$$\left| -\frac{G_N M_{BH} m}{R_S} \right| = \frac{1}{2} mc^2 \qquad (2)$$

It is important to note that while the energy density for the gravitational field is surely negative in Newtonian mechanics, the global gravitational field energy is not well defined in general relativity. Unique prescriptions for a stress-energy-momentum pseudotensor can be formulated though, notably that of Landau and Lifshitz. Any such stress-energy can, however, be made to vanish locally by general coordinate transformation, and it is not even entirely clear that the pseudotensor so applied is an appropriate general relativistic object. Given though that Newtonian gravity is the classical limit of general relativity, it is reasonable to suspect that the properly defined field energy density will be likewise also negative, and that inflation is indeed consistent with a correctly generalized notion of constant, zero total energy.

A universe would then be in this sense closed, an island unto itself, from the moment of its inception from the quantum froth; Only a universe *in toto* might so originate, emerging as a critically bound structure possessing profound density and minute proportion, each as accorded against intrinsically defined scales (the analogous Newton and Planck parameters and the propagation speed of massless fields), and expanding or inflating henceforth and eternally.

4. The Invariance of No-Scale SUGRA

Inflation, driven by the scalar inflaton field is itself inherently a quantum field theoretic subject. However, there is tension between quantum mechanics and general relativity. Currently, superstring theory is the best candidate for quantum gravity. The five consistent ten dimensional superstring theories, namely heterotic $E_8 \times E_8$, heterotic $SO(32)$, Type I, Type IIA, Type IIB, can be unified by various duality transformations under an eleven-dimensional M-theory [20], and the twelve-dimensional F-theory can be considered as the strongly coupled formulation of the Type IIB string theory with a varying axion-dilaton field [21]. Self consistency of the string (or M-, F-) algebra implies a ten (or eleven, twelve) dimensional master spacetime, some elements of which – six (or seven, eight) to match our observed four large dimensions – may be compactified on a manifold (typically Calabi-Yau manifolds or G_2 manifolds) which conserves a requisite portion of supersymmetric charges.

The structure of the curvature within the extra dimensions dictates in no small measure the particular phenomenology of the unfolded dimensions, secreting away the "closet space" to encode the symmetries of all gauged interactions. The physical volume of the internal spatial manifold is directly related to the effective Planck scale and basic gauge coupling strengths in the external space. The compactification is in turn described by fundamental moduli fields which must be stabilized, *i.e.* , given suitable vacuum expectation values (VEVs).

The famous example of Kaluza and Klein prototypes the manner in which general covariance in five dimensions is transformed to gravity plus Maxwell theory in four dimensions when the transverse fifth dimension is cycled around a circle. The connection of geometry to particle physics is perhaps nowhere more intuitively clear than in the context of model building with $D6$-branes, where the gauge structure and family replication are related directly to the brane stacking and intersection multiplicities. The Yukawa couplings and Higgs structure are in like manners also specified, leading after radiative symmetry breaking of the chiral gauge sector to low energy masses for the chiral fermions and broken gauge generators, each massless in the symmetric limit.

From a top-down view, Supergravity (SUGRA) is an ubiquitous infrared limit of string theory, and forms the starting point of any two-dimensional world sheet or D-dimensional target space action. The mandatory localization of the Supersymmetry (SUSY) algebra, and thus the momentum-energy (space-time translation) operators, leads to general coordinate invariance of the action and an Einstein field theory limit. Any available flavor of Supergravity will not however suffice. In general, extraneous fine tuning is required to avoid a cosmological constant which scales like a dimensionally suitable power of the Planck mass. Neglecting even the question of whether such a universe might be permitted to appear spontaneously, it would then be doomed to curl upon itself and collapse within the order of the Planck time, for comparison about 10^{-43} seconds in our Universe. Expansion and inflation appear to uniquely require properties which arise naturally only in the No-Scale SUGRA formulation [22, 23, 24, 25, 26].

SUSY is in this case broken while the vacuum energy density vanishes automatically at tree level due to a suitable choice of the Kähler potential, the function which specifies the metric on superspace. At the minimum of the null scalar potential, there are flat directions which leave the compactification moduli VEVs undetermined by the classical equations of motion. We thus receive without additional effort an answer to the deep question of how these moduli are stabilized; They have been transformed into dynamical variables which are to be determined by minimizing corrections to the scalar potential at loop order. In particular, the high energy gravitino mass $M_{3/2}$, and also the proportionally equivalent universal gaugino mass $M_{1/2}$, will be established in this way. Subsequently, all gauge mediated SUSY breaking soft-terms will be dynamically evolved down from this boundary under the renormalization group [27], establishing in large measure the low energy phenomenology, and solving also the Flavour Changing Neutral Current (FCNC) problem. Since the moduli are fixed at a false local minimum, phase transitions by quantum tunneling will naturally occur between discrete vacua.

The specific Kähler potential which we favor has been independently derived in both weakly coupled heterotic string theory [28] and the leading order compactification of M-theory on S^1/Z_2 [29], and might be realized in F-theory models as well [30, 31, 32, 33]. We conjecture, for the reasons given prior, that the No-Scale SUGRA construction could pervade all universes in the String Landscape with reasonable flux vacua. This being the case, intelligent creatures elsewhere in the Multiverse, though separated from us by a bridge too far, might reasonably so concur after parallel examination of their own physics. Moreover, they might leverage via this insight a deeper knowledge of the underlying Multiverse-invariant master theory, of which our known string, M-, and F-theories may compose some coherently overlapping patch of the garment edge. Perhaps we yet share appreciation, across the cords which bind our 13.7 billion years to their corresponding blink of history, for the common timeless principles under which we are but two isolated condensations upon two particular vacuum solutions among the physical ensemble.

5. An Archetype Model Universe

Though we engage in this work lofty and speculative questions of natural philosophy, we balance abstraction against the measured material underpinnings of concrete phenomenological models with direct and specific connection to tested and testable particle physics. If the suggestion is correct that eternal inflation and No-Scale SUGRA models with string origins together describe what is in fact our Multiverse, then we must as a prerequisite settle the issue of whether our own phenomenology can be produced out of such a construction.

In the context of Type II intersecting D-brane models, we have indeed found one realistic Pati-Salam model which might describe Nature as we observe it [34, 35, 36]. If only the F-terms of three complex structure moduli are non-zero, we also automatically have vanishing vacuum energy, and obtain a generalized No-Scale SUGRA. It seems to us that the string derived Grand Unified Theories (GUTs), and particularly the Flipped $SU(5) \times U(1)_X$ models [37, 38, 39], are also candidate realistic string models with promising predictions that can be tested at the Large Hadron Collider (LHC), the Tevatron, and other future experiments.

Let us briefly review the minimal flipped $SU(5) \times U(1)_X$ model [37, 38, 39]. The gauge group of the flipped $SU(5)$ model is $SU(5) \times U(1)_X$, which can be embedded into $SO(10)$. We define the generator $U(1)_{Y'}$ in $SU(5)$ as

$$T_{\mathrm{U(1)_{Y'}}} = \mathrm{diag}\left(-\frac{1}{3}, -\frac{1}{3}, -\frac{1}{3}, \frac{1}{2}, \frac{1}{2}\right). \tag{3}$$

The hypercharge is given by

$$Q_Y = \frac{1}{5}\left(Q_X - Q_{Y'}\right). \tag{4}$$

In addition, there are three families of SM fermions whose quantum numbers under the $SU(5) \times U(1)_X$ gauge group are

$$F_i = (\mathbf{10}, \mathbf{1}),\ \bar{f}_i = (\bar{\mathbf{5}}, \mathbf{-3}),\ \bar{l}_i = (\mathbf{1}, \mathbf{5}), \tag{5}$$

where $i = 1, 2, 3$.

To break the GUT and electroweak gauge symmetries, we introduce two pairs of Higgs fields

$$H = (\mathbf{10}, \mathbf{1}),\ \overline{H} = (\overline{\mathbf{10}}, \mathbf{-1}),\ h = (\mathbf{5}, \mathbf{-2}),\ \overline{h} = (\bar{\mathbf{5}}, \mathbf{2}). \tag{6}$$

Interestingly, we can naturally solve the doublet-triplet splitting problem via the missing partner mechanism [39], and then the dimension five proton decay from the colored Higgsino exchange can be highly suppressed [39]. The flipped $SU(5) \times U(1)_X$ models have been constructed systematically in the free fermionic string constructions at Kac-Moody level one previously [39, 40, 41, 42, 43], and in the F-theory model building recently [30, 31, 32, 33, 44, 45].

In the flipped $SU(5) \times U(1)_X$ models, there are two unification scales: the $SU(3)_C \times SU(2)_L$ unification scale M_{32} and the $SU(5) \times U(1)_X$ unification scale $M_{\mathcal{F}}$. To separate the M_{32} and $M_{\mathcal{F}}$ scales and obtain true string-scale gauge coupling unification in free fermionic string models [46, 43] or the decoupling scenario in F-theory models [44, 45], we introduce vector-like particles which form complete flipped $SU(5) \times U(1)_X$ multiplets. In order to avoid the Landau pole problem for the strong coupling constant, we can only introduce the following two sets of vector-like particles around the TeV scale [46]

$$Z1 : XF = (\mathbf{10}, \mathbf{1})\ ,\ \overline{XF} = (\overline{\mathbf{10}}, \mathbf{-1})\ ; \tag{7}$$

$$Z2 : XF\ ,\ \overline{XF}\ ,\ Xl = (\mathbf{1}, \mathbf{-5})\ ,\ \overline{Xl} = (\mathbf{1}, \mathbf{5})\ , \tag{8}$$

where

$$XF \equiv (XQ, XD^c, XN^c)\ ,\quad \overline{Xl}_{(\mathbf{1},\mathbf{5})} \equiv XE^c\ . \tag{9}$$

In the prior, XQ, XD^c, XE^c, XN^c have the same quantum numbers as the quark doublet, the right-handed down-type quark, charged lepton, and neutrino, respectively. Such kind of the models have been constructed systematically in the F-theory model building locally and dubbed $\mathcal{F} - SU(5)$ within that context [44, 45]. In this paper, we only consider the flipped $SU(5) \times U(1)_X$ models with $Z2$ set of vector-like particles. The discussions for the models with $Z1$ set and heavy threshold corrections [44, 45] are similar.

These models are moreover quite interesting from a phenomenological point of view [44, 45]. The predicted vector-like particles can be observed at the Large Hadron Collider, and the partial lifetime for proton decay in the leading $(e|\mu)^+\pi^0$ channels falls around 5×10^{34} years, testable at the future Hyper-Kamiokande [47] and Deep Underground Science and Engineering Laboratory (DUSEL) [48] experiments [49, 50, 51]. The lightest CP-even Higgs boson mass can be increased [52], hybrid inflation can be naturally realized, and the correct cosmic primordial density fluctuations can be generated [53].

6. No-Scale Supergravity

In the traditional framework, supersymmetry is broken in the hidden sector, and then its breaking effects are mediated to the observable sector via gravity or gauge interactions. In GUTs with gravity mediated supersymmetry breaking, also known as the minimal supergravity (mSUGRA) model, the supersymmetry breaking soft terms can be parameterized by four universal parameters: the gaugino mass $M_{1/2}$, scalar mass M_0, trilinear soft term A, and the ratio of Higgs VEVs $\tan\beta$ at low energy, plus the sign of the Higgs bilinear mass term μ. The μ term and its bilinear soft term B_μ are determined by the Z-boson mass M_Z and $\tan\beta$ after the electroweak (EW) symmetry breaking.

To solve the cosmological constant problem, No-Scale supergravity was proposed [22, 23, 24, 25, 26]. No-scale supergravity is defined as the subset of supergravity models which satisfy the following three constraints [22, 23, 24, 25, 26]: (i) The vacuum energy vanishes automatically due to the suitable Kähler potential; (ii) At the minimum of the scalar potential, there are flat directions which leave the gravitino mass $M_{3/2}$ undetermined; (iii) The super-trace quantity $\text{Str}\mathcal{M}^2$ is zero at the minimum. Without this, the large one-loop corrections would force $M_{3/2}$ to be either zero or of Planck scale. A simple Kähler potential which satisfies the first two conditions is

$$K = -3\ln(T + \overline{T} - \sum_i \overline{\Phi}_i \Phi_i) , \qquad (10)$$

where T is a modulus field and Φ_i are matter fields. The third condition is model dependent and can always be satisfied in principle [54].

The scalar fields in the above Kähler potential parameterize the coset space $SU(N_C + 1, 1)/(SU(N_C + 1) \times U(1))$, where N_C is the number of matter fields. Analogous structures appear in the $N \geq 5$ extended supergravity theories [55], for example, $N_C = 4$ for $N = 5$, which can be realized in the compactifications of string theory [28, 29]. The non-compact structure of the symmetry implies that the potential is not only constant but actually identical to zero. In fact, one can easily check that the scalar potential is automatically positive semi-definite, and has a flat direction along the T field. Interestingly, for the simple Kähler potential in Equation (10) we obtain the simplest No-Scale boundary condition $M_0 = A = B_\mu = 0$, while $M_{1/2}$ may be non-zero at the unification scale, allowing for low energy SUSY breaking.

It is important to note that there exist several methods of generalizing No-Scale supergravity, for instance, the previously mentioned Type II intersecting D-brane models [34, 35, 36], and the compactifications of M-theory on S^1/Z_2 with next-to-leading order corrections, where we have obtained a generalization employing modulus dominated SUSY breaking [56, 57, 58, 59, 60]. Similarly, mirage mediation for the flux compactifications can be considered as another form of generalized No-Scale supergravity [61, 62]. In this paper we concentrate on the simplest No-Scale supergravity, reserving any such generalizations for future study.

7. The Golden Point

First, we would like to review the golden point of No-Scale and no-parameter $\mathcal{F}$-$SU(5)$ [63]. In the No-Scale context, we impose $M_0 = A = B_\mu = 0$ at the unification scale $M_{\mathcal{F}}$, and allow distinct inputs for the single parameter $M_{1/2}(M_{\mathcal{F}})$ to translate under the RGEs to distinct low scale outputs of B_μ and the Higgs mass-squares $M^2_{H_u}$ and $M^2_{H_d}$. This continues until the point of spontaneous breakdown of the electroweak symmetry at $M^2_{H_u} + \mu^2 = 0$, at which point minimization of the broken potential establishes the physical low energy values of μ and $\tan\beta$. In practice however, this procedure is at odds with the existing `SuSpect 2.34` code [64] base from which our primary routines have been adapted. In order to impose the minimal possible refactoring, we have instead opted for an inversion wherein $M_{1/2}$ and $\tan\beta$ float as two effective degrees of freedom. Thus, we do not fix $B_\mu(M_{\mathcal{F}})$. We take $\mu > 0$ as suggested by the results of $g_\mu - 2$ for the muon, and assume as in prior work that the masses for the vector-like particles are universal at 1 TeV.

The relic LSP neutralino density, WIMP-nucleon direct detection cross sections and photon-photon annihilation cross sections are computed with `MicrOMEGAs 2.1` [65] wherein the revised `SuSpect` RGEs have also implemented. We use a top quark mass of $m_t = 173.1$ GeV [66] and employ the following experimental constraints: (1) The WMAP 7-year measurements of the cold dark matter density [15, 16, 17], $0.1088 \leq \Omega_\chi \leq 0.1158$. We allow Ω_χ to be larger than the upper bound due to a possible $\mathcal{O}(10)$ dilution factor [67] and to be smaller than the lower bound due to multicomponent dark matter. (2) The experimental limits on the FCNC process, $b \to s\gamma$. We use the limits $2.86\times10^{-4} \leq Br(b \to s\gamma) \leq 4.18\times10^{-4}$ [68, 69]. (3) The anomalous magnetic moment of the muon, $g_\mu - 2$. We use the 2σ level boundaries, $11\times10^{-10} < \Delta a_\mu < 44\times10^{-10}$ [70]. (4) The process $B_s^0 \to \mu^+\mu^-$ where we take the upper bound to be $Br(B_s^0 \to \mu^+\mu^-) < 5.8 \times 10^{-8}$ [71]. (5) The LEP limit on the lightest CP-even Higgs boson mass, $m_h \geq 114$ GeV [72, 73].

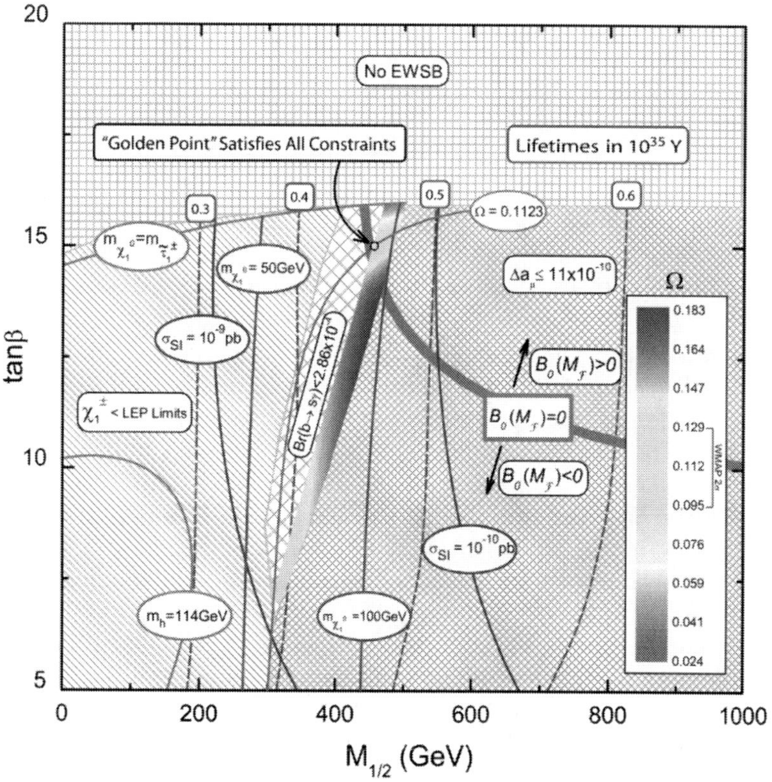

Figure 1. Viable parameter space in the $\tan\beta - M_{1/2}$ plane. The "Golden Point" is annotated. The thin, dark green line denotes the WMAP 7-year central value $\Omega_\chi = 0.1123$. The dashed purple contours label $p \to (e|\mu)^+\pi^0$ proton lifetime predictions, in units of 10^{35} years.

In the $\tan\beta$ – $M_{1/2}$ plane, $B_\mu(M_{\mathcal{F}})$ is then calculated along with the low energy supersymmetric particle spectrum and checks on various experimental constraints. The subspace corresponding to a No-Scale model is clearly then a one dimensional slice of this manifold, as demonstrated in Figure 1. It is quite remarkable that the $B_\mu(M_{\mathcal{F}}) = 0$ contour so established runs sufficiently perpendicular to the WMAP strip that the point of intersection effectively absorbs our final degree of freedom, creating what we have labeled as a No-Parameter Model.

It is truly extraordinary however that this intersection occurs exactly at the centrally preferred relic density, that being our strongest experimental constraint. We emphasize again that there did not have to be an experimentally viable $B_\mu(M_{\mathcal{F}}) = 0$ solution, and that the consistent realization of this scenario depended crucially on several uniquely identifying characteristics of the underlying proposal. Specifically again, it appears that the No-Scale condition comes into its own only when applied at near the Planck mass, and that this is naturally identified as the point of the final $\mathcal{F}$-$SU(5)$ unification, which is naturally extended and decoupled from the primary GUT scale only via the modification to the RGEs from the TeV scale $\mathcal{F}$-theory vector-like multiplet content. The union of our top-down model based constraints with the bottom-up experimental data exhausts the available freedom of parameterization in a uniquely consistent and predictive manner, phenomenologically defining a truly Golden Point near $M_{1/2} = 455$ GeV and $\tan\beta = 15$ GeV.

Table 1. Spectrum (in GeV) for the Golden Point of Figure 1. Here, $\Omega_\chi = 0.1123$, $\sigma_{SI} = 1.9 \times 10^{-10}$ pb, and $\langle\sigma v\rangle_{\gamma\gamma} = 1.7 \times 10^{-28}\ cm^3/s$. The central prediction for the $p \to (e|\mu)^+\pi^0$ proton lifetime is 4.6×10^{34} years.

$\widetilde{\chi}_1^0$	95	$\widetilde{\chi}_1^\pm$	185	$\widetilde{e}_R$	150	$\widetilde{t}_1$	489	$\widetilde{u}_R$	951	m_h	120.1
$\widetilde{\chi}_2^0$	185	$\widetilde{\chi}_2^\pm$	826	$\widetilde{e}_L$	507	$\widetilde{t}_2$	909	$\widetilde{u}_L$	1036	$m_{A,H}$	920
$\widetilde{\chi}_3^0$	821	$\widetilde{\nu}_{e/\mu}$	501	$\widetilde{\tau}_1$	104	$\widetilde{b}_1$	859	$\widetilde{d}_R$	992	$m_{H^\pm}$	925
$\widetilde{\chi}_4^0$	824	$\widetilde{\nu}_\tau$	493	$\widetilde{\tau}_2$	501	$\widetilde{b}_2$	967	$\widetilde{d}_L$	1039	$\widetilde{g}$	620

Our Golden point features $M_{1/2} = 455.3$ GeV, $\tan\beta = 15.02$, and is in full compliance with the WMAP 7-year results, with $\Omega_\chi = 0.1123$. It also satisfies the CDMS II [74], Xenon 100 [75], and FERMI-LAT space telescope constraints [76], with $\sigma_{SI} = 1.9 \times 10^{-10}$ pb and $\langle\sigma v\rangle_{\gamma\gamma} = 1.7 \times 10^{-28}\ cm^3/s$. The proton lifetime is about 4.6×10^{34} years, which is well within reach of the upcoming Hyper-Kamiokande [47] and DUSEL [48] experiments. Inspecting the supersymmetric particle and Higgs spectrum for the Golden Point of Table 1 reveals that the additional contribution of the 1 TeV vector-like particles lowers the gluino mass quite dramatically. The gluino mass M_3 runs flat from the M_{32} unification scale to 1 TeV as shown in Figure 2, though, due to supersymmetric radiative corrections, the physical gluino mass at the EW scale is larger than M_3 at the M_{32} scale. This is true for the full parameter space. For our data point, the LSP neutralino is 99.8% Bino. Similarly to the mSUGRA picture, our this point is in the stau-neutralino coannihilation region, but the gluino is lighter than the squarks in our models.

We plot gauge coupling and gaugino mass unification for the Golden Point in Figure 2. The figure explicitly demonstrates the two-step unification of flipped $SU(5) \times U(1)_X$. In addition, we present the RGE running for the μ term, the SUSY breaking scalar masses, trilinear A-terms, and bilinear B_μ term in Figure 3. Note in particular that the EW symmetry breaking occurs when $H_u^2 + \mu^2$ goes negative.

8. The Golden Strip

Second, we shall review the golden strip of correlated top quark, gaugino, and vectorlike mass in No-Scale, no-parameter $\mathcal{F}$-$SU(5)$ [77]. From the above discussions, only a small portion of viable parameter space is consistent with the $B_\mu(M_{\mathcal{F}}) = 0$ condition, which thus constitutes a strong constraint. Since the boundary value of the universal gaugino mass $M_{1/2}$, and even the unification scale $M_{\mathcal{F}} \simeq 7.5 \times 10^{17}$ GeV itself, are established by the low energy experiments via RGE running, we are not left with any surviving scale parameters in the present model. The floor of the "valley gorge" in Figure 4 represents accord with the $B_\mu = 0$ target for variations

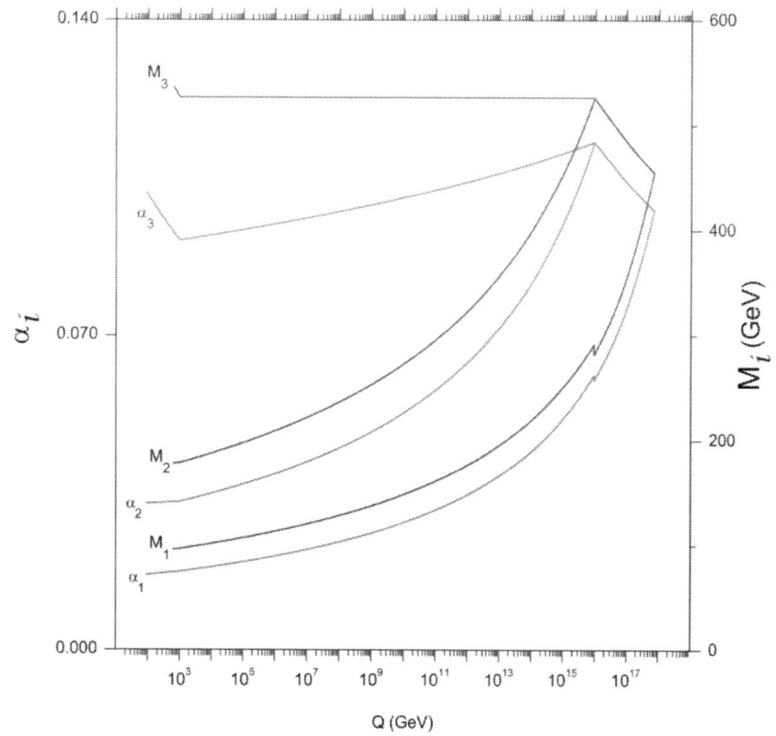

Figure 2. RGE Running of the SM gauge couplings and gaugino masses from the EW scale to the unification scale $M_{\mathcal{F}}$. There is a discontinuity in the running at M_{32} due to remixing of the hypercharge with the residual Abelian phase from the breaking of SU(5)

in $(M_{1/2}, M_{\rm V})$. We fix $\tan\beta = 15$, as appears to be rather generically required in No-Scale $\mathcal{F}$-$SU(5)$ to realize radiative EWSB and match the observed CDM density.

We have allowed for uncertainty in the most sensitive experimental input, the top quark mass, by effectively redefining $m_{\rm t}$ as an independent free parameter. Lesser sensitivities to uncertainty in $(\alpha_{\rm s}, M_{\rm Z})$ are included in the ± 1 GeV deviation from strict adherence to $B_\mu = 0$. We have established that there is a two dimensional sheet (of some marginal thickness to recognize the mentioned uncertainty) defining $|B_\mu(M_{\mathcal{F}})| \leq 1$ for each point in the three dimensional ($M_{1/2}$, $M_{\rm V}$, $m_{\rm t}$) volume, as shown in Figures (5,6). This sheet is inclined in the region of interest at the very shallow angle of 0.2° to the ($M_{1/2}$,$M_{\rm V}$) plane, such that $m_{\rm t}$ is largely decoupled from variation in the plane.

The $(g-2)_\mu$ and $b \to s\gamma$ constraints vary most strongly with $M_{1/2}$. The two considered effects are each at their lower limits at the boundary, but they exert pressure in opposing directions on $M_{1/2}$ due to the fact that the leading gaugino and squark contributions to $Br(b \to s\gamma)$ enter with an opposing sign to the SM term and Higgs contribution. For the non-SM contribution to Δa_μ, the effect is additive, and establishes an upper mass limit on $M_{1/2}$. Incidentally, the same experiment forms the central rationale for the adoption of sign$(\mu) > 0$, such that appropriate interference terms between SM and SUSY contributions are realized. Conversely, the requirement that SUSY contributions to $Br(b \to s\gamma)$ not be overly large, undoing the SM effect, requires a sufficiently large, *i.e.* lower bounded, $M_{1/2}$. The WMAP-7 CDM measurement, by contrast, exhibits a fairly strong correlation with both $(M_{1/2}, M_{\rm V})$, cross-cutting the $M_{1/2}$

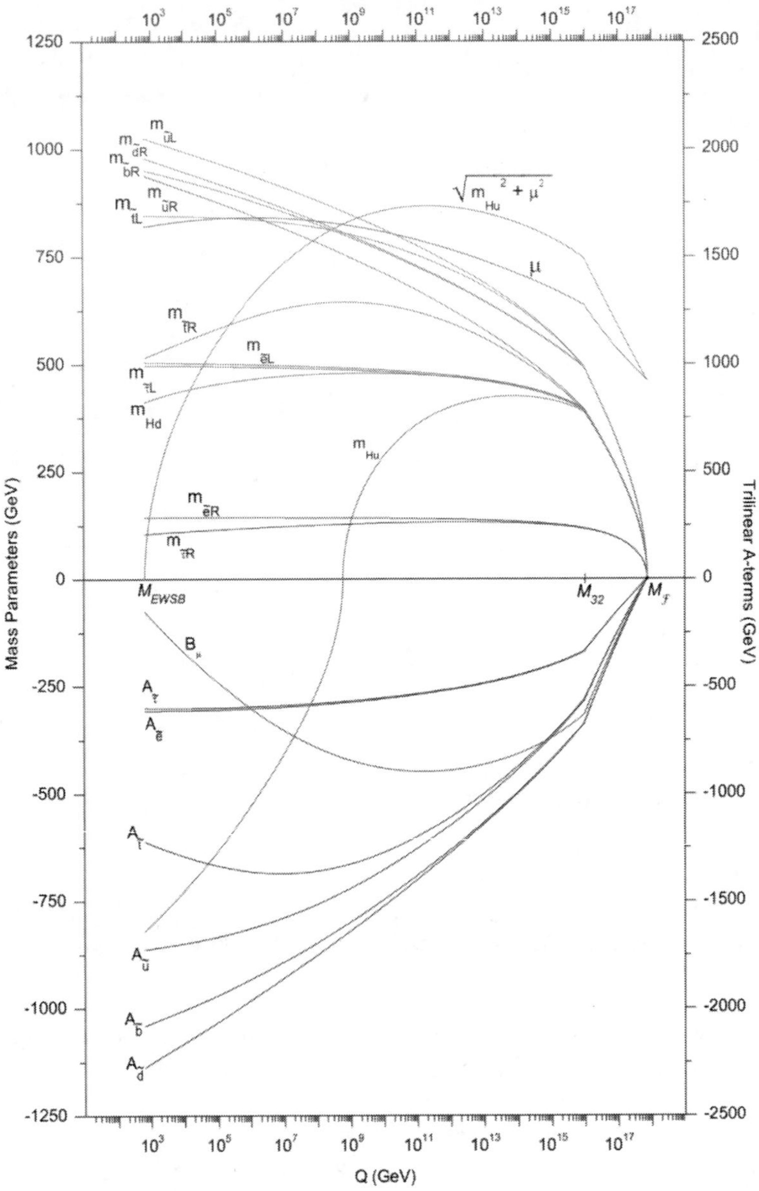

Figure 3. RGE Running of the μ term and SUSY breaking soft terms from the EW scale to the unification scale $M_{\mathcal{F}}$.

bound, and confining the vectorlike mass to 691-1020 GeV. We note that the mixing of the SM fermions and vector particles may give additional contributions to $Br(b \to s\gamma)$ and Δa_μ, but we do not consider them here.

The intersection of these three key constraints with the $|B_\mu(M_{\mathcal{F}})| \leq 1$ surface, as depicted in Figures (5,6), defines the "golden strip" of No-Scale $\mathcal{F}$-$SU(5)$. All of the prior is accomplished with no reference to the experimental top quark mass, redefined here as a *free* input. However, the extremely shallow angle of inclination (0.2°) of the $|B_\mu(M_{\mathcal{F}})| \leq 1$ sheet above the $(M_{1/2}, M_{\rm V})$ plane and into the $m_{\rm t}$ axis causes the golden strip to *imply* an exceedingly narrow range of compatibility for $m_{\rm t}$, between 173.0-174.4 GeV, in perfect alignment with the physically observed

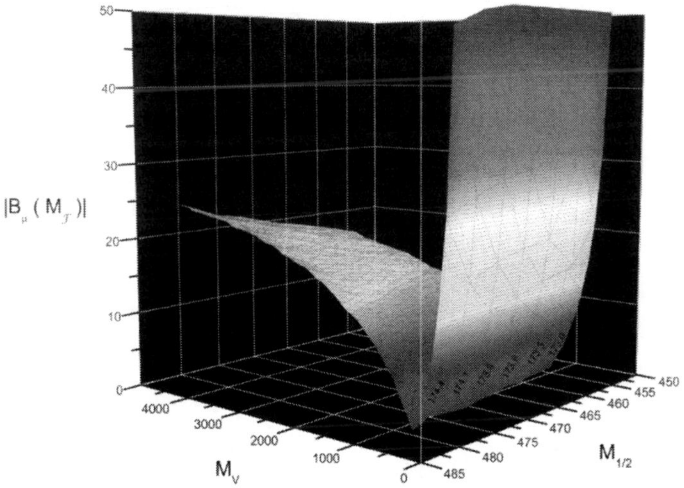

Figure 4. The $B_\mu = 0$ target for variations in $(M_{1/2}, M_{\rm V})$, with $\tan\beta = 15$. The specific $m_{\rm t}$ which is required to minimize $|B_\mu(M_{\mathcal{F}})|$ is annotated along the solution string.

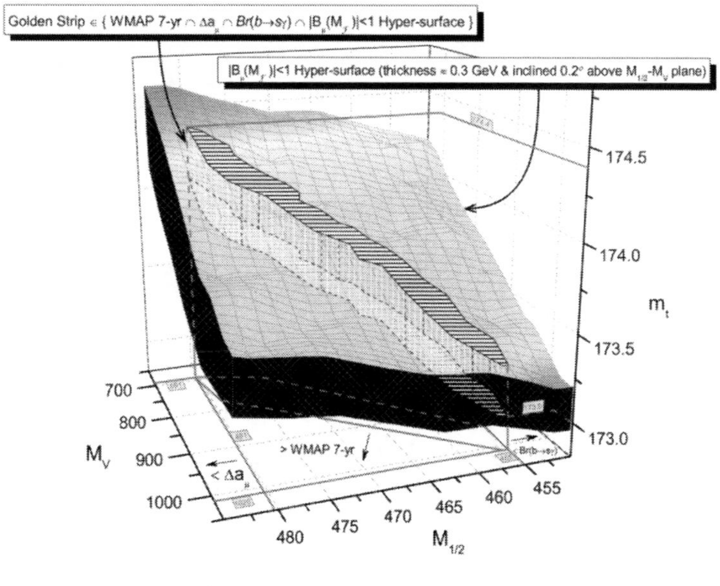

Figure 5. With $\tan\beta \simeq 15$ fixed by WMAP-7, the residual parameter volume is three dimensional in $(M_{1/2}, M_{\rm V}, m_{\rm t})$, with the $|B_\mu(M_{\mathcal{F}})| \leq 1$ (slightly thickened) surface forming a shallow $(0.2°)$ incline above the $(M_{1/2}, M_{\rm V})$ plane.

value of $m_{\rm t} = 173.1 \pm 1.3$ GeV [66].

Within the golden strip, we select the benchmark point of Table 2. The golden strip is further consistent with the CDMS II [74] and Xenon 100 [75] upper limits, with the spin-independent

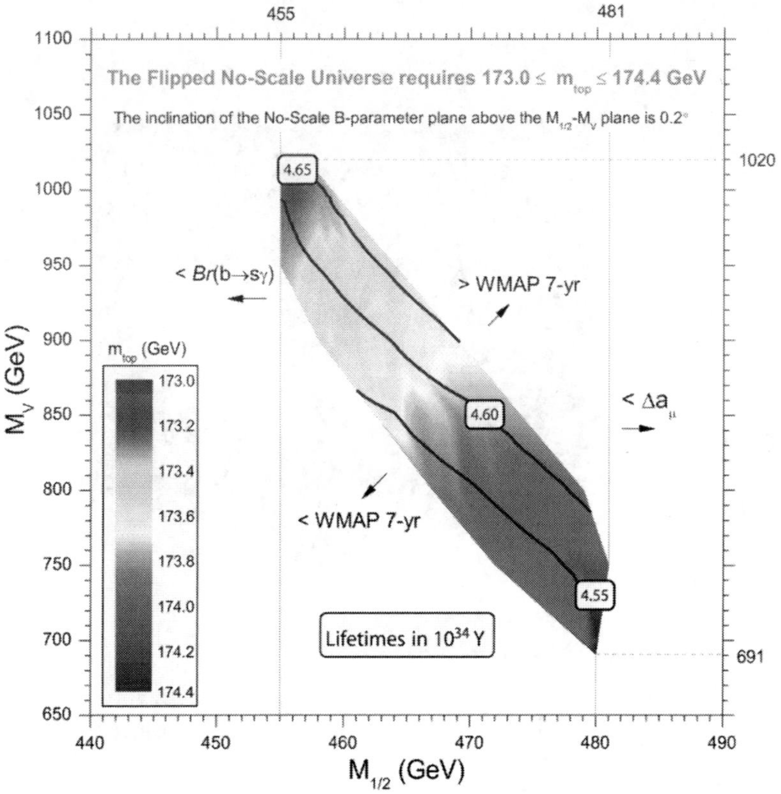

Figure 6. A flattened presentation of the the $(M_{12}, M_{\rm V})$ plane depicted in Figure 5. The overlayed blue contours mark the $p \to (e|\mu)^+\pi^0$ proton lifetime prediction, in units of 10^{34} years.

cross section extending from $\sigma_{SI} = 1.3\text{-}1.9 \times 10^{-10}$ pb. Likewise, the allowed region satisfies the Fermi-LAT space telescope constraints [76], with the photon-photon annihilation cross section $\langle\sigma v\rangle_{\gamma\gamma}$ ranging from $\langle\sigma v\rangle_{\gamma\gamma} = 1.5\text{-}1.7 \times 10^{-28}\ cm^3/s$.

Table 2. Spectrum (in GeV) for the benchmark point. Here, $M_{1/2} = 464$ GeV, $M_V = 850$ GeV, $m_t = 173.6$ GeV, $\Omega_\chi = 0.112$, $\sigma_{SI} = 1.7 \times 10^{-10}$ pb, and $\langle\sigma v\rangle_{\gamma\gamma} = 1.7 \times 10^{-28}\ cm^3/s$. The central prediction for the $p \to (e|\mu)^+\pi^0$ proton lifetime is 4.6×10^{34} years. The lightest neutralino is 99.8% Bino.

$\widetilde{\chi}_1^0$	96	$\widetilde{\chi}_1^\pm$	187	$\widetilde{e}_R$	153	$\widetilde{t}_1$	499	$\widetilde{u}_R$	975	m_h	120.6
$\widetilde{\chi}_2^0$	187	$\widetilde{\chi}_2^\pm$	849	$\widetilde{e}_L$	519	$\widetilde{t}_2$	929	$\widetilde{u}_L$	1062	$m_{A,H}$	946
$\widetilde{\chi}_3^0$	845	$\widetilde{\nu}_{e/\mu}$	513	$\widetilde{\tau}_1$	105	$\widetilde{b}_1$	880	$\widetilde{d}_R$	1018	$m_{H\pm}$	948
$\widetilde{\chi}_4^0$	848	$\widetilde{\nu}_\tau$	506	$\widetilde{\tau}_2$	514	$\widetilde{b}_2$	992	$\widetilde{d}_L$	1065	$\widetilde{g}$	629

9. The Super No-Scale Mechanism

In the following sections, we would like to review our study on Super No-Scale $\mathcal{F}\text{-}SU(5)$ [78, 79]. The single relevant modulus field in the simplest string No-Scale supergravity is the Kähler modulus T, a characteristic of the Calabi-Yau manifold, the dilaton coupling being irrelevant. The F-term of T generates the gravitino mass $M_{3/2}$, which is proportionally equivalent to $M_{1/2}$. Exploiting the simplest No-Scale boundary condition at $M_{\mathcal{F}}$ and running from high energy to low energy under the RGEs, there can be a secondary minimization, or *minimum minimorum*, of the minimum of the Higgs potential $V_{\rm min}$ for the EWSB vacuum. Since $V_{\rm min}$ depends on $M_{1/2}$, the gaugino mass $M_{1/2}$ is consequently dynamically determined by the equation $dV_{\rm min}/dM_{1/2} = 0$, aptly referred to as the "Super No-Scale" mechanism [78, 79].

It could easily have been that in consideration of the above technique, there were: A) too few undetermined parameters, with the $B_\mu = 0$ condition forming an incompatible over-constraint, and thus demonstrably false, or B) so many undetermined parameters that the dynamic determination possessed many distinct solutions, or was so far separated from experiment that it could not possibly be demonstrated to be true. The actual state of affairs is much more propitious, being specifically as follows. The three parameters M_0, A, B_μ are once again identically zero at the boundary because of the defining Kähler potential, and are thus known at all other scales as well by the RGEs. The minimization of the Higgs scalar potential with respect to the neutral elements of both SUSY Higgs doublets gives two conditions, the first of which fixes the magnitude of μ. The second condition, which would traditionally be used to fix B_μ, instead here enforces a consistency relationship on the remaining parameters, being that B_μ is already constrained.

In general, the $B_\mu = 0$ condition gives a hypersurface of solutions cut out from a very large parameter space. If we lock all but one parameter, it will give the final value. If we take a slice of two dimensional space, as has been described, it will give a relation between two parameters for all others fixed. In a three-dimensional view with B_μ on the vertical axis, this curve is the "flat direction" line along the bottom of the trench of $B_\mu = 0$ solutions. In general, we must vary at least two parameters rather than just one in isolation, in order that their mutual compensation may transport the solution along this curve. The most natural first choice is in some sense the pair of prominent unknown inputs $M_{1/2}$ and $\tan\beta$, as was demonstrated in Ref. [78, 79].

Having come to this point, it is by no means guaranteed that the potential will form a stable minimum. It must be emphasized that the $B_\mu = 0$ No-Scale boundary condition is the central agent affording this determination, as it is the extraction of the parameterized parabolic curve of solutions in the two compensating variables which allows for a localized, bound nadir point to be isolated by the Super No-Scale condition, dynamically determining *both* parameters. The background surface of $V_{\rm min}$ for the full parameter space outside the viable $B_\mu = 0$ subset is, in contrast, a steadily inclined and uninteresting function. Although we have remarked that $M_{1/2}$ and $\tan\beta$ have no *directly* established experimental values, they are severely indirectly constrained by phenomenology in the context of this model [63, 77]. It is highly non-trivial that there should be accord between the top-down and bottom-up perspectives, but this is indeed precisely what has been observed [78, 79].

10. The Minimum Minimorum of the Higgs Potential: Fixing M_Z

We employ an effective Higgs potential in the 't Hooft-Landau gauge and the $\overline{\rm DR}$ scheme, given summing the following neutral tree (V_0) and one loop (V_1) terms.

$$\begin{aligned} V_0 &= (\mu^2 + m_{H_u}^2)(H_u^0)^2 + (\mu^2 + m_{H_d}^2)(H_d^0)^2 \\ &\quad -2\mu B_\mu H_u^0 H_d^0 + \frac{g_2^2 + g_Y^2}{8}\left[(H_u^0)^2 - (H_d^0)^2\right]^2 \\ V_1 &= \sum_i \frac{n_i}{64\pi^2} m_i^4(\phi)\left(\ln\frac{m_i^2(\phi)}{Q^2} - \frac{3}{2}\right) \end{aligned} \tag{11}$$

Above, $m_{H_u}^2$ and $m_{H_d}^2$ are the SUSY breaking soft masses of the Higgs fields H_u and H_d, g_2 and g_Y are the gauge couplings of $SU(2)_L$ and $U(1)_Y$, n_i and $m_i^2(\phi)$ are the degree of freedom and mass for ϕ_i, and Q is the renormalization scale. We include the complete Minimal Supersymmetric

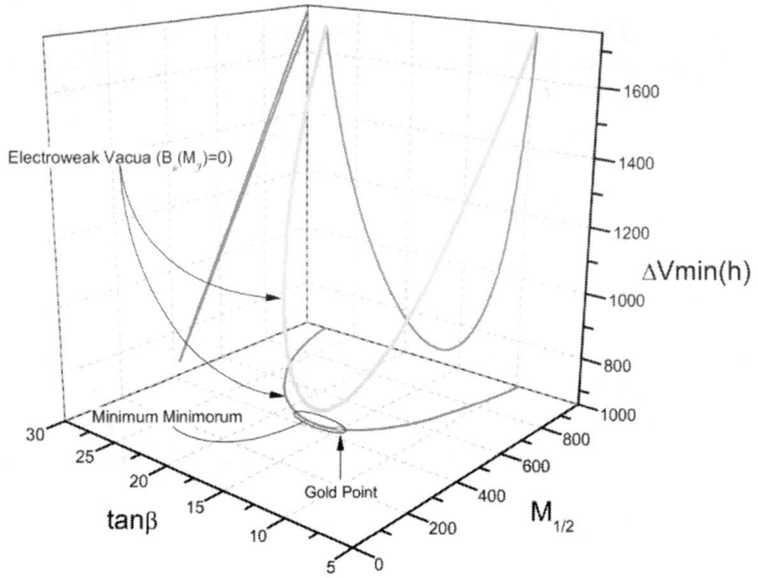

Figure 7. The *minimum* $V_{\rm min}$ of the Higgs effective potential (green curve, GeV) is plotted as a function of $M_{1/2}$ (GeV) and $\tan\beta$, emphasizing proximity of the "golden point" of Ref. [63] to the dynamic region of the $V_{\rm min}$ *minimorum.*

Standard Model (MSSM) contributions to one loop, following Ref. [80], although the result is phenomenologically identical accounting only the leading top and partner stop terms. Since the minimum of the electroweak (EW) Higgs potential $V_{\rm min}$ depends implicitly on $M_{3/2}$, the gravitino mass is determined by $dV_{\rm min}/dM_{3/2} = 0$. Being that $M_{1/2}$ is proportional to $M_{3/2}$, it is equivalent to employ $M_{1/2}$ directly as our modulus parameter, from which all other SUSY breaking soft terms here derive. In our numerical results in the figures, we shall designate differences in the fourth-root of the effective Higgs potential as $\Delta V_{min}(h) \equiv V_{\rm eff}^{1/4}$, measured in units of GeV relative to an arbitrary overall zero-offset.

Factors explicit within the potential are obtained from our customized extension of the `SuSpect 2.34` [64] codebase, including a self-consistency assessment [63] on $B_\mu = 0$. We apply two-loop RGE running for the SM gauge couplings, and one-loop running for the SM fermion Yukawa couplings, μ term and soft terms.

Studying $V_{\rm min}$ generically in the $(M_{1/2}, \tan\beta)$ plane, no point of secondary minimization is readily apparent in the strong, roughly linear, downward trend with respect to $M_{1/2}$ over the region of interest. However, the majority of the plane is not in physical communication with our model, disrespecting the fundamental $B_\mu = 0$ condition. Isolating only the compliant contour within this surface, *mirabile dictu,* a parabola is traced with nadir alighting gentle upon our original golden point, as in Figure (7). Restoring parameterization freedom to $(M_{\rm V}, m_{\rm t})$, we may scan across the corresponding golden point of each nearby universe variant, reconstructing in their union the previously advertised golden strip, as in Figure (8). Notably, the theoretical restriction on $\tan\beta$ remains stable against variation in these parameters, exactly as its experimental counterpart. We find it quite extraordinary that the phenomenologically preferred region rests precisely at the curve's locus of symmetric inflection. Note in particular that it is the selection of the parabolic $B_\mu = 0$ contour out of the otherwise uninteresting $V_{\rm min}(M_{1/2}, \tan\beta)$ inclined surface which allows a clear *minimum minimorum* to be established. We reiterate that consistency of the dynamically positioned $M_{1/2}$ and $\tan\beta$ with the golden

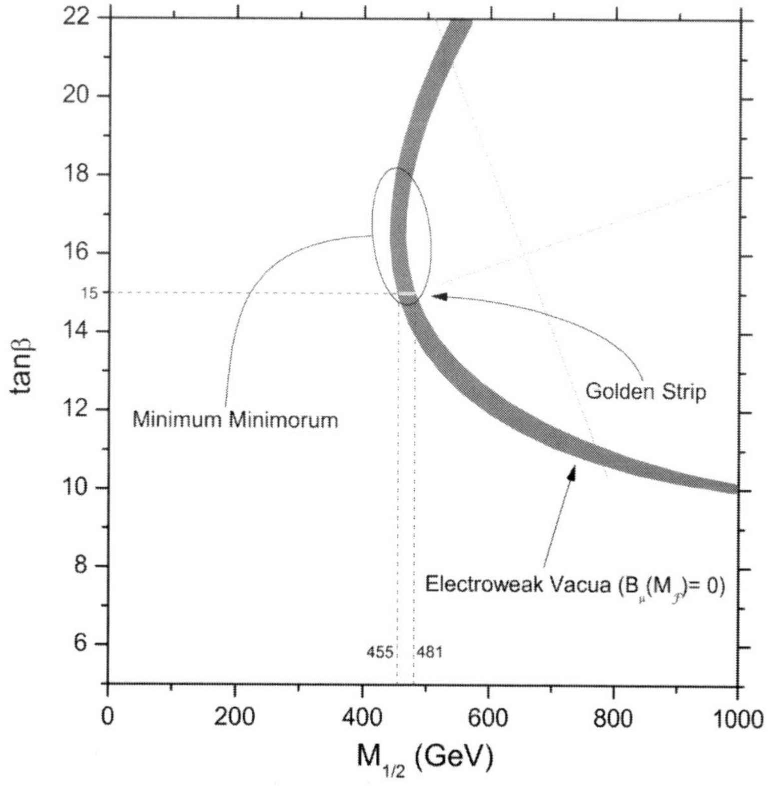

Figure 8. The projection onto the ($M_{1/2}$,tanβ) plane of Figure 7 is further detailed, expanding to span the boundary cases of the Ref. [77] "golden strip". The symmetry axis of the $B_\mu = 0$ parabola is rotated slightly above the $M_{1/2}$ axis.

strip, implies consistency with all current experimental data.

A strongly linear relationship is observed between the SUSY and EWSB scales with $M_{\mathrm{EWSB}} \simeq 1.44\ M_{1/2}$, such that a corresponding parabolic curve may be visualized. There is a charged stau LSP for $\tan\beta$ from 16 to 22, and we connect points with correct EWSB smoothly on the plot in this region. If $\tan\beta$ is larger than 22, the stau is moreover tachyonic, so properly we must restrict all analysis to $\tan\beta \leq 22$.

11. The Gauge Hierarchy Problem

Not only must we explain stabilization of the electroweak scale against quantum corrections, but we must also explain why the electroweak scale and TeV-sized SUSY breaking soft-terms are "initially" positioned so far below the Planck mass. These latter components of the "gauge hierarchy" problem are the more subtle. In their theoretical pursuit, we do not though feign ignorance of established experimental boundaries, taking the phenomenologist's perspective that pieces fit already to the puzzle stipulate a partial contour of those yet to be placed. Indeed, careful knowledge of precision electroweak scale physics, including the strong and electromagnetic couplings, the Weinberg angle and the Z-mass are required even to run the one loop RGEs. In the second loop one requires also *minimally* the leading top quark Yukawa coupling, as deduced from m_t, and the Higgs VEV $v \equiv \sqrt{\langle H_u \rangle^2 + \langle H_d \rangle^2} \simeq 174$ GeV, established in turn from measurement of the effective Fermi coupling, or from M_Z and the electroweak couplings.

Reading the RGEs up from M_Z, we take unification of the gauge couplings as evidence of a GUT. Reading them in reverse from a point of high energy unification, we take the heaviness of the top quark, via its large Yukawa coupling, to dynamically drive the term $M^2_{H_u} + \mu^2$ negative, triggering spontaneous collapse of the tachyonic vacuum, *i.e.* radiative electroweak symmetry breaking. Minimization of this potential with respect to the neutral components of H_u and H_d yields two conditions, which may be solved for $\mu(M_Z)$ and $B_\mu(M_Z)$ in terms of the constrained Higgs VEVs, which are in turn functions of M_Z (considered experimentally fixed) and $\tan\beta \equiv \langle H_u \rangle / \langle H_d \rangle$ (considered a free parameter).

Restricting to just the solution subset for which $B_\mu(M_Z)$ given by EWSB stitches cleanly onto that run down under the RGEs from $B_\mu(M_{\mathcal{F}}) = 0$, $\tan\beta$, or equivalently μ, becomes an implicit function of the single moduli $M_{1/2}(M_{\mathcal{F}})$. The pinnacle of this construction is the Super No-Scale condition $dV_{\rm min}/dM_{1/2} = 0$, wherein $M_{1/2}$, and thus also $\tan\beta$, are dynamically established at the local *minimum minimorum*. By comparison, the standard MSSM construction seems a hoax, requiring horrendous fine tuning to stabilize if viewed as a low energy supergravity limit, and moreover achieving TeV scale EW and SUSY physics as a simple shell game by manual selection of TeV scale boundaries for the soft terms $M_{1/2}$, M_0, and A.

Strictly speaking, having effectively exchanged input of the Z-mass for a constraint on $\mu(M_{\mathcal{F}})$, we dynamically establish the SUSY breaking soft term $M_{1/2}$ and $\tan\beta$ *within* the electroweak symmetry breaking vacua, *i.e.* with fixed $v \simeq 174$ GeV. However, having predicted $M_{\mathcal{F}}$ as an output scale near the reduced Planck mass, we are licensed to invert the solution, taking $M_{\mathcal{F}}$ as a high scale *input* and dynamically addressing the gauge hierarchy through the standard story of radiative electroweak symmetry breaking. This proximity to the elemental high scale of (consistently decoupled) gravitational physics, arises because of the dual flipped unification and the perturbing effect of the TeV multiplets, and is not motivated in standard GUTs. Operating the machinery of the RGEs in reverse, we may transmute the low scale M_Z for the high scale $M_{\mathcal{F}}$, emphasizing that the fundamental dynamic *correlation* is that of the *ratio* $M_Z/M_{\mathcal{F}}$, taking either as our input yardstick according to taste. For fixed $M_{\mathcal{F}} \simeq 7 \times 10^{17}$ GeV, in a single breath we receive the order of the electroweak scale, the Z-mass, the Higgs bilinear coupling μ, the Higgs VEVs, and all other dependent dimensional quantities, including predictions for the full superparticle mass spectrum. It is in this sense that we claim a complete resolution of the gauge hierarchy problem, within the context of the Super No-Scale $\mathcal{F}$-$SU(5)$ model.

12. The GUT Higgs Modulus

An alternate pair of parameters for which one may attempt to isolate a $B_\mu = 0$ curve, which we consider for the first time in this work, is that of $M_{1/2}$ and the GUT scale M_{32}, at which the $SU(3)_C$ and $SU(2)_L$ couplings initially meet. Fundamentally, the latter corresponds to the modulus which sets the total magnitude of the GUT Higgs field's VEVs. M_{32} could of course in some sense be considered a "known" quantity, taking the low energy couplings as input. Indeed, starting from the measured SM gauge couplings and fermion Yukawa couplings at the standard 91.187 GeV electroweak scale, we may calculate both M_{32} and the final unification scale $M_{\mathcal{F}}$, and subsequently the unified gauge coupling and SM fermion Yukawa couplings at $M_{\mathcal{F}}$, via running of the RGEs. However, since the VEVs of the GUT Higgs fields H and $\overline{H}$ are considered here as free parameters, the GUT scale M_{32} must not be fixed either. As a consequence, the low energy SM gauge couplings, and in particular the $SU(2)_L$ gauge coupling g_2, will also run freely via this feedback from M_{32}.

We consider this conceptual release of a known quantity, in order to establish the nature of the model's dependence upon it, to be a valid and valuable technique, and have employed it previously with specific regards to "postdiction" of the top quark mass value [77]. Indeed, forcing the theoretical *output* of such a parameter is only possible in a model with highly constrained physics, and it may be expected to meet success only by intervention of either grand coincidence or grand conspiracy of Nature.

For this study, we choose a vector-like particle mass $M_V = 1000$ GeV, and use the experimental top quark mass input $m_t = 173.1$ GeV. We emphasize that the choice of $M_V = 1000$ GeV is not an arbitrary one, since a prior analysis [77] has shown that a 1 TeV vector-like mass is in compliance with all current experimental data and the No-Scale B_μ=0 requirement.

In actual practice, the variation of M_{32} is achieved in the reverse by programmatic variation of the Weinberg angle, holding the strong and electromagnetic couplings at their physically measured values. Figure 9 demonstrates the scaling between $\sin^2(\theta_W)$, M_{32} (logarithmic axis), and the Z-boson mass. The variation of M_Z is attributed primarily to the motion of the electroweak couplings, the magnitude of the Higgs VEV being held essentially constant. We ensure also that the unified gauge coupling, SM fermion Yukawa couplings, and specifically also the Higgs bilinear term $\mu \simeq 460$ GeV, are each held stable at the scale $M_{\mathcal{F}}$ to correctly mimic the previously described procedure.

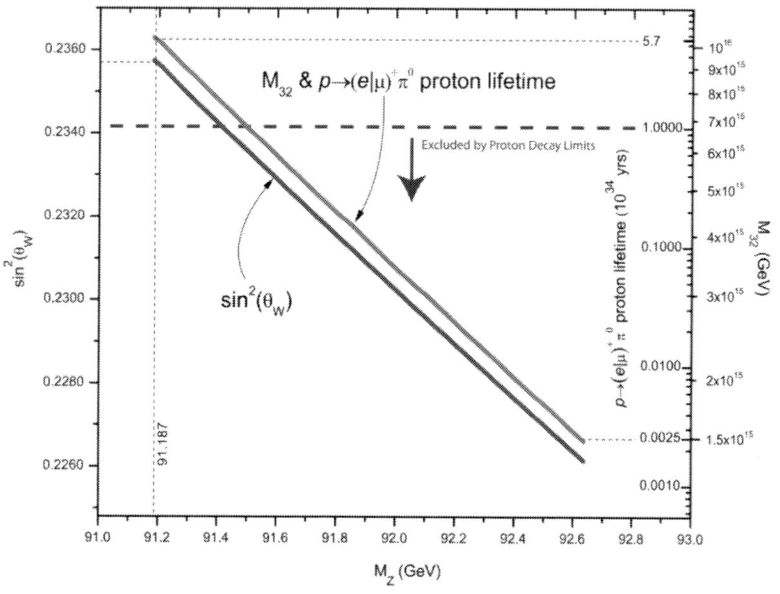

Figure 9. The interrelated variation of $\sin^2(\theta_W)$, the GUT scale M_{32} (logarithmic axis), and the Z-boson mass M_Z is demonstrated for the parameter strips which preserve $B_\mu = 0$ and $\mu = 460$ GeV at $M_{\mathcal{F}}$. The variation in M_Z is linked dominantly to motion of the EW couplings via $\sin^2(\theta_W)$. Also shown is the corresponding predicted proton lifetime in the leading $(e|\mu)^+\pi^0$ channels, in units of 10^{34} years, with the current lower bound of 1.0×10^{34} years indicated by the dashed horizontal purple line.

The parameter ranges for the variation depicted in Figure 9 are $M_Z = 91.18 - 92.64$, $\sin^2(\theta_W) = 0.2262 - 0.2357$, and $M_{32} = 1.5 \times 10^{15} - 1.04 \times 10^{16}$ GeV, and likewise also the same for Figures (10-16), which will feature subsequently. The *minimum minimorum* falls at the boundary of the prior list, dynamically fixing $M_{32} \simeq 1.0 \times 10^{16}$ GeV and placing $M_{1/2}$ again in the vicinity of 450 GeV. The low energy SM gauge couplings are simultaneously constrained by means of the associated Weinberg angle, with $\sin^2(\theta_W) \simeq 0.236$, in excellent agreement with experiment. The corresponding range of predicted proton lifetimes in the leading $(e|\mu)^+\pi^0$ modes is $2.5 \times 10^{31} - 5.7 \times 10^{34}$ years [49, 50]. If the GUT scale M_{32} becomes excessively light, below about 7×10^{15} GeV, then proton decay would be more rapid than allowed by the recently

Journal of Physics: Conference Series **306** (2011) 012015 doi:10.1088/1742-6596/306/1/012015

updated lower bound of 1.0×10^{34} years from Super-Kamiokande [81].

We are cautious against making a claim in precisely the same vein for the dynamic determination of $M_Z \simeq 91.2$ GeV, since again the crucial electroweak Higgs VEV is not a substantial element of the variation. However, in *conjunction* with the radiative electroweak symmetry breaking [82, 83] numerically implemented within the `SuSpect 2.34` code base [64], the fixing of the Higgs VEV and the determination of the electroweak scale may also plausibly be considered legitimate dynamic output, *if* one posits the M_F scale input to be available *a priori*.

The present minimization, referencing $M_{1/2}$, M_{32} and $\tan\beta$, is again dependent upon M_V and m_t, while the previously described [78] determination of $\tan\beta$ was, by contrast, M_V and m_t invariant. Recognizing that a minimization with all three parameters simultaneously active is required to declare all three parameters to have been simultaneously dynamically determined, we emphasize the mutual consistency of the results. We again stress that the new *minimum minimorum* is also consistent with the previously advertised golden strip, satisfying all presently known experimental constraints to our available resolution. It moreover also addresses the problems of the SUSY breaking scale and gauge hierarchy [78], insomuch as $M_{1/2}$ is determined dynamically.

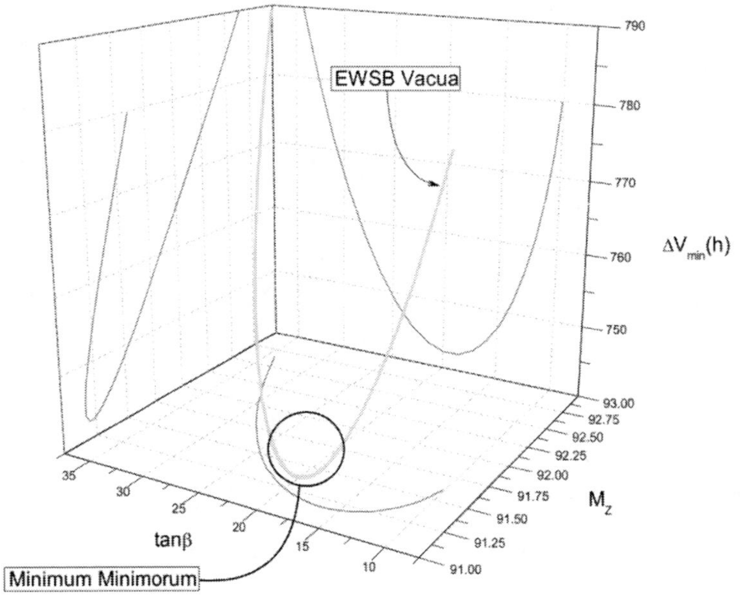

Figure 10. Three-dimensional graph of $(M_Z, \tan\beta, \Delta V_{min}(h))$ space (green curve). The projections onto the three mutually perpendicular planes (red curves) are likewise shown. M_Z and $\Delta V_{min}(h)$ are in units of GeV. The dynamically preferred region, allowing for plausible variation, is circled and tipped in gold.

13. The Minimum Minimorum of the Higgs Potential: Fixing Yukawa Couplings and μ at $M_{\mathcal{F}}$

We have revised the `SuSpect 2.34` code base [64] to incorporate our specialized No-Scale $\mathcal{F}$-$SU(5)$ with vector-like mass algorithm, and accordingly employ two-loop RGE running for the SM gauge couplings, and one-loop RGE running for the SM fermion Yukawa couplings, μ term, and SUSY breaking soft terms. For our choice of $M_V = 1000$ GeV, $m_t = 173.1$ GeV, and

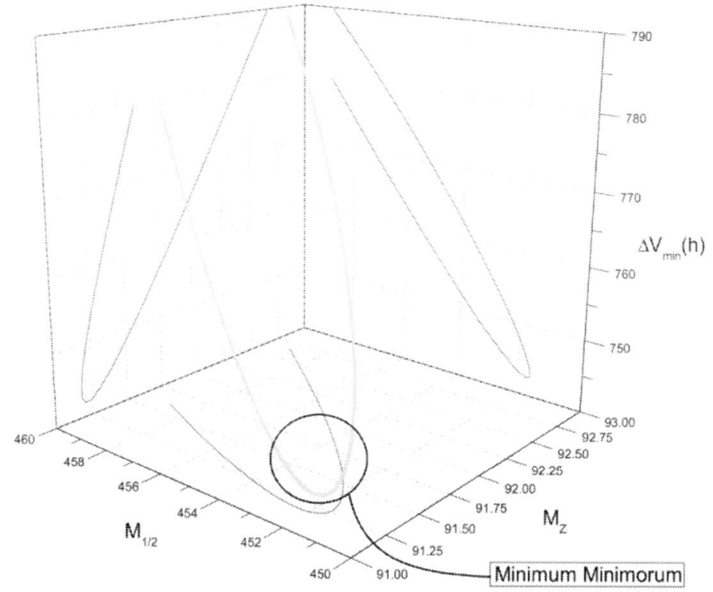

Figure 11. Three-dimensional graph of $(M_Z, M_{1/2}, \Delta V_{min}(h))$ space (green curve). The projections onto the three mutually perpendicular planes (red curves) are likewise shown. M_Z, $M_{1/2}$, and $\Delta V_{min}(h)$ are in units of GeV. The dynamically preferred region, allowing for plausible variation, is circled and tipped in gold.

$\mu(M_{\mathcal{F}}) \simeq 460$ GeV, we present the one-loop effective Higgs potential $\Delta V_{min}(h)$ in terms of M_Z and $\tan\beta$ in Figure 10, in terms of M_Z and $M_{1/2}$ in Figure 11, in terms of $M_{1/2}$ and $\tan\beta$ in Figure 12, and in terms of v and $M_{1/2}$ in Figure 13, where $v = \sqrt{v_u^2 + v_d^2}$, $v_u = \langle H_u^0 \rangle$, and $v_d = \langle H_d^0 \rangle$. These figures clearly demonstrate the localization of the *minimum minimorum* of the Higgs potential, corroborating the dynamical determination of $\tan\beta \simeq 15 - 20$ and $M_{1/2} \simeq 450$ GeV in [78].

Additionally, we exhibit the $(M_Z, M_{1/2}, M_{32})$ space in Figure 14, the $(M_Z, M_{1/2}, v)$ space in Figure 15, and the $(M_Z, M_{1/2}, g)$ space in Figure 16, where $g = \sqrt{g_2^2 + g_Y^2}$. Figure 14 demonstrates that $M_{32} \simeq 1.0 \times 10^{16}$ GeV at the *minimum minimorum*, which correlates to $M_Z \simeq 91.2$ GeV, or more directly, $\sin^2(\theta_W) \simeq 0.236$. Together, the alternate perspectives of Figures 14, 15, and 16 complete the view given in Figures 10, 11, 12, and 13 to visually tell the story of the dynamic interrelation between the M_Z, $M_{1/2}$, and M_{32} scales, as well as the electroweak gauge couplings, and the Higgs VEVs. The curves in each of these figures represent only those points that satisfy the $B_\mu = 0$ requirement, as dictated by No-Scale supergravity, serving as a crucial constraint on the dynamically determined parameter space. Ultimately, it is the significance of the $B_\mu = 0$ requirement that separates the No-Scale $\mathcal{F}$-$SU(5)$ with vector-like particles from the entire compilation of prospective string theory derived models. By means of the $B_\mu = 0$ vehicle, No-Scale $\mathcal{F}$-$SU(5)$ has surmounted the paramount challenge of phenomenology, that of dynamically determining the electroweak scale, the scale of fundamental prominence in particle physics.

We wish to note that recent progress has been made in incorporating more precise numerical calculations into our baseline algorithm for No-Scale $\mathcal{F}$-$SU(5)$ with vector-like particles. Initially, when we commenced the task of fully developing the phenomenology of this model, the extreme

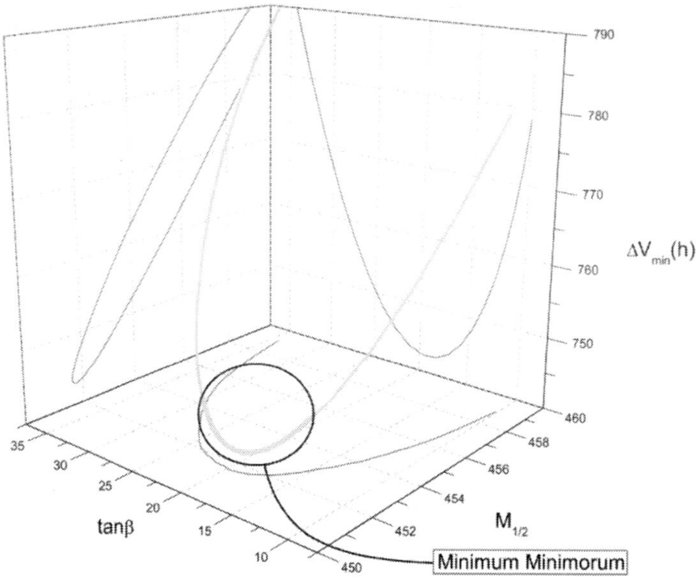

Figure 12. Three-dimensional graph of $(M_{1/2}, \tan\beta, \Delta V_{min}(h))$ space (green curve). The projections onto the three mutually perpendicular planes (red curves) are likewise shown. $M_{1/2}$ and $\Delta V_{min}(h)$ are in units of GeV. The dynamically preferred region, allowing for plausible variation, is circled and tipped in gold.

complexity of properly numerically implementing No-Scale $\mathcal{F}$-$SU(5)$ with vector-like particles compelled a gradual strategy for construction and persistent enhancement of the algorithm. Preliminary findings of a precision improved algorithm indicate that compliance with the 7-year WMAP relic density constraints requires a slight upward shift to $\tan\beta \simeq 19 - 20$ from the value computed in Ref. [63], suggesting a potential convergence to even finer resolution of the dynamical determination of $\tan\beta$ given by the Super No-Scale mechanism, and the value demanded by the experimental relic density measurements. We shall furnish a comprehensive analysis of the precision improved algorithm at a later date.

14. Probing The Blueprints of the No-Scale Multiverse at the Colliders

We offer here a brief summary of direct collider, detector, and telescope level tests which may probe the blueprints of the No-Scale Multiverse which we have laid out. As to the deep question of whether the ensemble be literal in manifestation, or merely the conceptual superset of unrealized possibilities of a single island Universe, we pretend no definitive answer. However, we have argued that the emergence *ex nihilo* of seedling universes which fuel an eternal chaotic inflation scenario is particularly plausible, and even natural, within No-Scale Supergravity, and our goal of probing the specific features of our own Universe which might implicate its origins in this construction are immediately realizable and practicable.

The unified gaugino $M_{1/2}$ at the unification scale $M_{\mathcal{F}}$ can be reconstructed from impending LHC events by determining the gauginos M_1, M_2, and M_3 at the electroweak scale, which will in turn require knowledge of the masses for the neutralinos, charginos, and the gluino. Likewise, $\tan\beta$ can be ascertained in principle from a distinctive experimental observable, as was accomplished for mSUGRA in [84]. We will not undertake a comprehensive analysis here of the reconstruction of $M_{1/2}$ and $\tan\beta$, but will offer for now a cursory examination of typical events

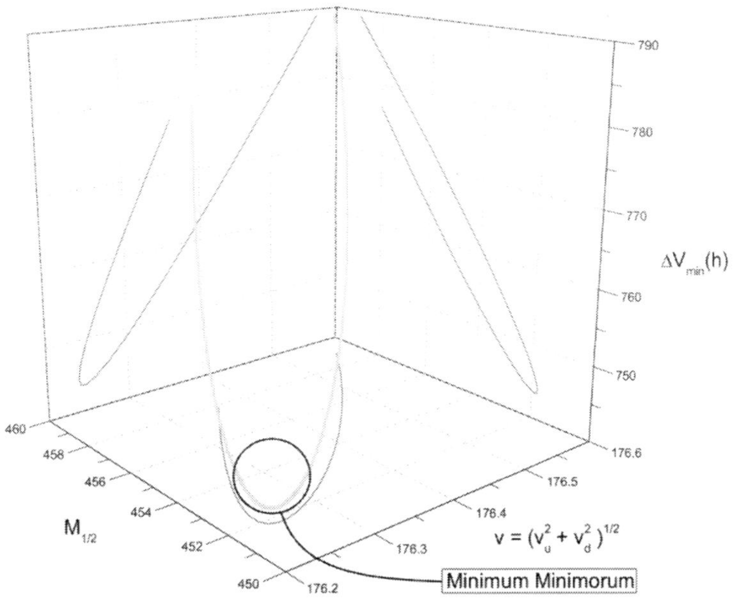

Figure 13. Three-dimensional graph of $(v, M_{1/2}, \Delta V_{min}(h))$ space (green curve). The projections onto the three mutually perpendicular planes (red curves) are likewise shown. $M_{1/2}$, v, and $\Delta V_{min}(h)$ are in units of GeV. The dynamically preferred region, allowing for plausible variation, is circled and tipped in gold.

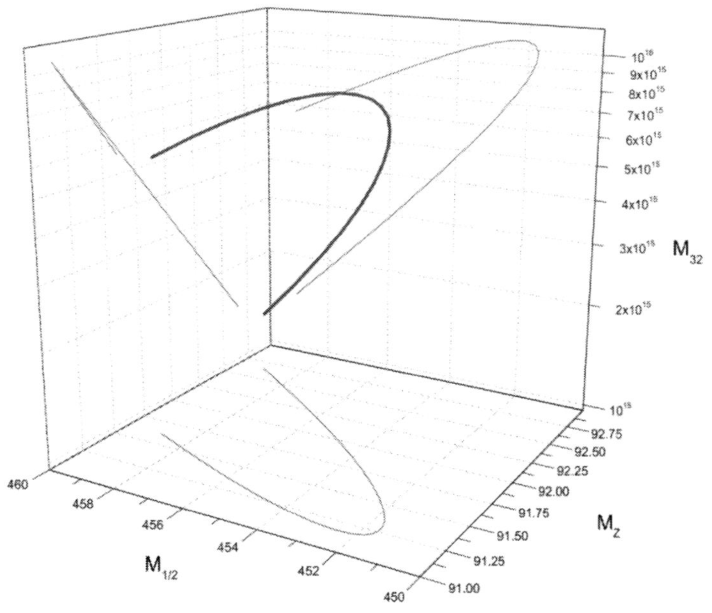

Figure 14. Three-dimensional graph of $(M_Z, M_{1/2}, M_{32})$ space (blue curve). The projections onto the three mutually perpendicular planes (red curves) are likewise shown. M_Z, $M_{1/2}$, and M_{32} are in units of GeV.

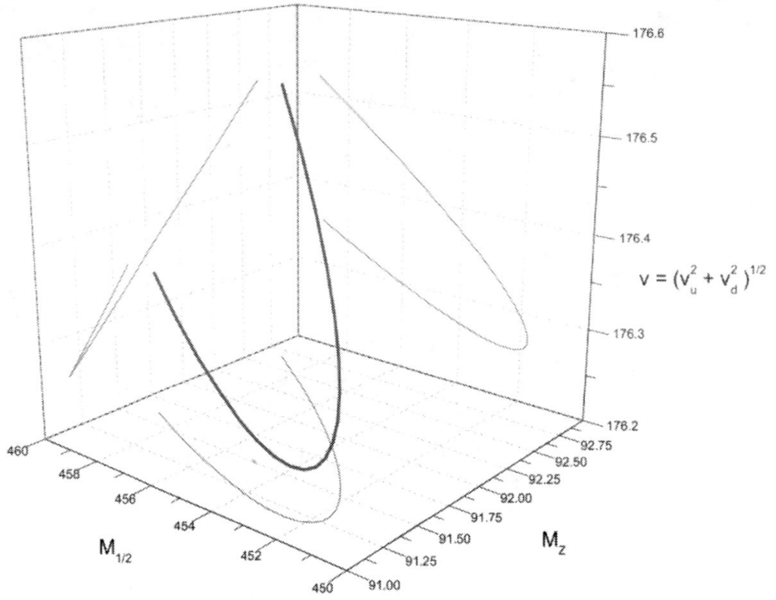

Figure 15. Three-dimensional graph of $(M_Z, M_{1/2}, v)$ space (purple curve). The projections onto the three mutually perpendicular planes (red curves) are likewise shown. M_Z, $M_{1/2}$, and v are in units of GeV.

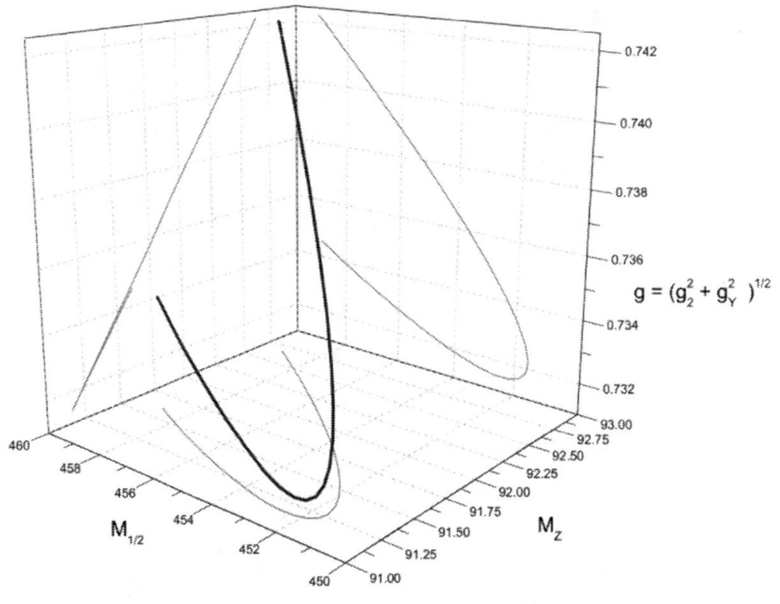

Figure 16. Three-dimensional graph of $(M_Z, M_{1/2}, g)$ space (royal blue curve). The projections onto the three mutually perpendicular planes (red curves) are likewise shown. M_Z and $M_{1/2}$ are in units of GeV.

expected at the LHC. We will present a detailed compilation of the experimental observables necessary for validation of the No-Scale $\mathcal{F}$-$SU(5)$ at the LHC in the following section.

For the benchmark SUSY spectrum presented in Table 3, we have adopted the specific values $M_{1/2} = 453$, $\tan\beta = 15$ and $M_Z = 91.187$. We expect that higher order corrections will shift the precise location of the *minimum minimorum* a little bit, for example, within the encircled gold-tipped regions of the diagrams in the prior section. We have selected a ratio for $\tan\beta$ at the lower end of this range for consistency with our previous study [78], and to avoid stau dark matter.

Table 3. Spectrum (in GeV) for the benchmark point. Here, $M_{1/2} = 453$ GeV, $M_V = 1000$ GeV, $m_t = 173.1$ GeV, $M_Z = 91.187$ GeV, $\mu(M_{\mathcal{F}}) = 460.3$ GeV, $\Delta V_{min}(h) = 748$ GeV, $\Omega_\chi = 0.113$, $\sigma_{SI} = 2 \times 10^{-10}$ pb, and $\langle\sigma v\rangle_{\gamma\gamma} = 1.8 \times 10^{-28}\ cm^3/s$. The central prediction for the $p \to (e|\mu)^+\pi^0$ proton lifetime is around 4.9×10^{34} years. The lightest neutralino is 99.8% Bino.

$\widetilde{\chi}_1^0$	94	$\widetilde{\chi}_1^\pm$	184	$\widetilde{e}_R$	150	$\widetilde{t}_1$	486	$\widetilde{u}_R$	947	m_h	120.1
$\widetilde{\chi}_2^0$	184	$\widetilde{\chi}_2^\pm$	822	$\widetilde{e}_L$	504	$\widetilde{t}_2$	906	$\widetilde{u}_L$	1032	$m_{A,H}$	916
$\widetilde{\chi}_3^0$	817	$\widetilde{\nu}_{e/\mu}$	498	$\widetilde{\tau}_1$	104	$\widetilde{b}_1$	855	$\widetilde{d}_R$	988	$m_{H^\pm}$	921
$\widetilde{\chi}_4^0$	821	$\widetilde{\nu}_\tau$	491	$\widetilde{\tau}_2$	499	$\widetilde{b}_2$	963	$\widetilde{d}_L$	1035	$\widetilde{g}$	617

At the benchmark point, we calculate $\Omega_\chi = 0.113$ for the cold dark matter relic density. The phenomenology is moreover consistent with the LEP limit on the lightest CP-even Higgs boson mass, $m_h \geq 114$ GeV [72, 73], the CDMS II [74] and Xenon 100 [75] upper limits on the spin-independent cross section σ_{SI}, and the Fermi-LAT space telescope constraints [76] on the photon-photon annihilation cross section $\langle\sigma v\rangle_{\gamma\gamma}$. The differential cross-sections and branching ratios have been calculated with `PGS4` [85] executing a call to `PYTHIA 6.411` [86], using our specialized No-Scale algorithm integrated into the `SuSpect 2.34` code for initial computation of the sparticle masses.

The benchmark point resides in the region of the experimentally allowed parameter space that generates the relic density through stau-neutralino coannihilation. Hence, the five lightest sparticles for this benchmark point are $\widetilde{\chi}_1^0 < \widetilde{\tau}_1^\pm < \widetilde{e}_R < \widetilde{\chi}_2^0 \sim \widetilde{\chi}_1^\pm$. Here, the gluino is lighter than all the squarks with the exception of the lighter stop, so that all squarks will predominantly decay to a gluino and hadronic jet, with a small percentage of squarks producing a jet and either a $\widetilde{\chi}_1^\pm$ or $\widetilde{\chi}_2^0$. The gluinos will decay via virtual (off-shell) squarks to neutralinos or charginos plus quarks, which will further cascade in their decay. The result is a low-energy tau through the processes $\widetilde{\chi}_2^0 \to \widetilde{\tau}_1^\mp\tau^\pm \to \tau^\mp\tau^\pm\widetilde{\chi}_1^0$ and $\widetilde{\chi}_1^\pm \to \widetilde{\tau}_1^\pm\nu_\tau \to \tau^\pm\nu_\tau\widetilde{\chi}_1^0$.

The LHC final states of low-energy tau in the $\mathcal{F}$-$SU(5)$ stau-neutralino coannihilation region are similar to those same low-energy LHC final states in mSUGRA, however, in the stau-neutralino coannihilation region of mSUGRA, the gluino is typically heavier than the squarks. The LHC final low-energy tau states in the stau-neutralino coannihilation regions of $\mathcal{F}$-$SU(5)$ and mSUGRA will thus differ in that in $\mathcal{F}$-$SU(5)$, the low-energy tau states will result largely from neutralinos and charginos produced by gluinos, as opposed to the low-energy tau states in mSUGRA resulting primarily from neutralinos and charginos produced from squarks.

Also notably, the TeV-scale vector-like multiplets are well targeted for observation by the LHC. We have argued [77] that the eminently feasible near-term detectability of these hypothetical fields in collider experiments, coupled with the distinctive flipped charge assignments within the multiplet structure, represents a smoking gun signature for Flipped $SU(5)$, and have thus coined the term *flippons* to collectively describe them. Immediately, our curiosity is piqued by the recent announcement [87] of the DØ collaboration that vector-like quarks have been excluded up to a bound of 693 GeV, corresponding to the immediate lower edge of our anticipated range for their discovery [77].

15. The Ultra-High Jet Signal of No-Scale $\mathcal{F}$-$SU(5)$ at the $\sqrt{s} = 7$ TeV LHC

The Large Hadron Collider (LHC) at CERN has been accumulating data from $\sqrt{s} = 7$ TeV proton-proton collisions since March 2010. It is expected to reach an integrated luminosity of 1 fb^{-1} by the end of 2011, all in search of new physics beyond the SM. SUSY, which provides a natural solution to the gauge hierarchy problem, is the most promising extension of the SM. Data corresponding to a limited 35 pb^{-1} has already established new constraints on the viable parameter space [88, 89, 90] due to the unprecedented center of mass collision energy. The search strategy for SUSY signals in early LHC data has been actively and eagerly studied by quite a few groups [91, 92, 93, 94, 95], with particular focus on the parameter space featuring a traditional mass relationship between squarks and the gluino, such as a gluino heavier than all squarks or a gluino lighter than all squarks.

A question of great interest is whether there exist SUSY models which are well motivated by a fundamental theory such as string theory, which can be tested in the initial LHC run, permitting a probe of the UV physics close to the Planck scale. In this Section we present such a model. It is well known that the supersymmetric flipped $SU(5) \times U(1)_X$ models can solve the doublet-triplet splitting problem elegantly via the missing partner mechanism [40, 41, 42]. To realize the string scale gauge coupling unification, two of us (TL and DVN) with Jiang proposed the testable flipped $SU(5) \times U(1)_X$ models with TeV-scale vector-like particles [46], where such models can be realized in the $\mathcal{F}$-ree $\mathcal{F}$-ermionic string constructions [43] and $\mathcal{F}$-theory model building [44, 45], dubbed $\mathcal{F}$-$SU(5)$. In particular, we find the generic phenomenological consequences are quite interesting [44, 45, 51].

In the simplest No-Scale supergravity, all the SUSY breaking soft terms arise from a single parameter $M_{1/2}$. The spectra in the entire Golden Strip are therefore very similar up to a small rescaling on $M_{1/2}$, with equivalent sparticle branching ratios. This leaves invariant most of the "internal" physical properties, whereas this rescaling ability on $M_{1/2}$ is not apparent in alternative SUSY models. For our analysis here, we use a vector-like particle mass of $M_V \sim 1000$ GeV, which exists in the viable parameter space; We emphasize that this is not an arbitrary choice [77], though these vector-like particles could be to heavy for observation in the early LHC run. The gluino mass is about 550 GeV, which should by contrast allow for direct testing of No-Scale $\mathcal{F}$-$SU(5)$ at the early LHC run.

To represent our model for this phase of analysis, we select the No-Scale $\mathcal{F}$-$SU(5)$ benchmark point of Table 4. The optimized signatures presented here offer an alluring testing vehicle for the stringy origin of $\mathcal{F}$-$SU(5)$. This point is again representative of the entire highly constrained $\mathcal{F}$-$SU(5)$ viable parameter space. The SUSY breaking parameters for this point slightly differ from previous $\mathcal{F}$-$SU(5)$ studies [63, 77, 79] insomuch as more precise numerical calculations have been incorporated into our baseline algorithm. The masses shift a few GeV from the spectra given in previous work, but where different, we believe this to be the more accurate representation. The branching ratios and decay modes of the spectrum in Table 4 and the spectra in [63, 77, 79] are identical, so the physical properties are consistent before and after code improvements. Thus, the signatures studied here will be common to the spectra provided previously.

Table 4. Spectrum (in GeV) for $M_{1/2} = 410$ GeV, $M_V = 1$ TeV, $m_t = 174.2$ GeV, $\tan\beta = 19.5$. Here, $\Omega_\chi = 0.11$ and the lightest neutralino is 99.8% bino.

$\widetilde{\chi}_1^0$	76	$\widetilde{\chi}_1^\pm$	165	$\widetilde{e}_R$	157	$\widetilde{t}_1$	423	$\widetilde{u}_R$	865	m_h	120.4
$\widetilde{\chi}_2^0$	165	$\widetilde{\chi}_2^\pm$	756	$\widetilde{e}_L$	469	$\widetilde{t}_2$	821	$\widetilde{u}_L$	939	$m_{A,H}$	814
$\widetilde{\chi}_3^0$	752	$\widetilde{\nu}_{e/\mu}$	462	$\widetilde{\tau}_1$	85	$\widetilde{b}_1$	761	$\widetilde{d}_R$	900	$m_{H^\pm}$	820
$\widetilde{\chi}_4^0$	755	$\widetilde{\nu}_\tau$	452	$\widetilde{\tau}_2$	462	$\widetilde{b}_2$	864	$\widetilde{d}_L$	942	$\widetilde{g}$	561

For the initial phase of generation of the low order Feynman diagrams which may link the incoming beam to the desired range of hard scattering intermediate states, we have used the program `MadGraph 4.4` [96]. These diagrams were subsequently fed into `MadEvent` [96] for appropriate kinematic scaling to yield batches of Monte Carlo simulated parton level scattering events. The cascaded fragmentation and hadronization of these events into final state showers of photons, leptons, and mixed jets has been handled by `PYTHIA` [86], with `PGS4` [85] simulating the physical detector environment. We implement MLM matching to preclude double counting of final states, and use the CTEQ6L1 parton distribution functions to generate the SM background. All 2-body SUSY processes are simulated. The b-jet tagging algorithm in `PGS4` is adjusted to update the b-tagging efficiency to ~60%. We veto an event if any of the following conditions are met: $p_T < 100$ GeV for the two leading jets; $p_T < 350$ GeV for all jets; pseudorapidity $|\eta| > 2$ for the leading jet; missing energy $\not{E}_T < 150$ GeV; isolated photon with $p_T > 25$ GeV; or isolated electron or muon with $p_T > 10$ GeV. Likewise, we discard any single jet with $|\eta| > 3$. These cuts are quite standard, but alone they are insufficient to reveal the ultra-high multiplicity jet event signature; We must also investigate the event cut on the number of jets and the p_T cut on a single jet to preserve ultra-high jet events.

The detector simulations use the spectrum for the $\mathcal{F}$-$SU(5)$ point in Table 4. The most significant asset of the spectrum for our analysis is the relationship between the stop, gluino, and other squarks. The distinctive mass pattern of $m_{\tilde{t}_1} < m_{\tilde{g}} < m_{\tilde{q}}$ is the smoking gun signature and possibly a unique characteristic of only $\mathcal{F}$-$SU(5)$. To gain a comparison of the model studied here with more standardized SUSY models, we examine the ten "Snowmass Points and Slopes" (SPS) benchmark points [97] for suitable samples. We find that none of the ten SPS benchmarks support the $m_{\tilde{t}_1} < m_{\tilde{g}} < m_{\tilde{q}}$ mass pattern. This critical element is indicative of how unique the $\mathcal{F}$-$SU(5)$ signal could be. Previous minimal supersymmetric SM studies focused on signals from a low-multiplicity of jets, whereas the aforementioned mass pattern is expected to show a very high-multiplicity of jets. For the SPS benchmarks, we only consider those spectra not light enough to have been excluded by the initial phase of LHC data, or those not too heavy for early LHC production. A few points satisfy these criteria, though we select only one since we anticipate the corollary points to exhibit analogous characteristics. For our analysis here, we use the SPS SP3 benchmark.

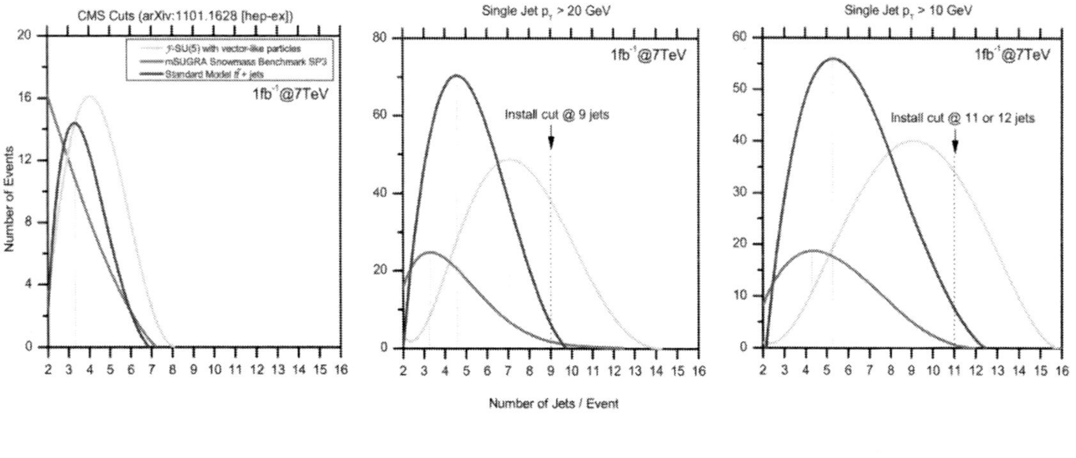

Figure 17. Distribution of events per number of jets. For clarity of the peaks, polynomials have been fitted over the histograms.

Considering the large number of hadronic jets we are examining for our signatures, there is little intrusion from SM background processes after post-processing cuts. We examine the background processes studied in [91, 92, 93, 94, 95, 98] and our conclusion is that only the $t\bar{t}$+$jets$

possesses the requisite minimum cross-section and sufficient number of jets to intrude upon the $\mathcal{F}$-$SU(5)$ signatures. Processes with a higher multiplicity of top quarks can generate events with a large number of jets, however, the cross-sections are sufficiently suppressed to be negligible, bearing in mind the large number of ultra-high jet events which our model will generate. The same is true for those more complicated background processes involving combinations of top quarks, jets, and one or more vector bosons, where the production counts for 1 fb^{-1} of luminosity are again sufficiently small. Furthermore, we neglect the QCD 2,3,4 jets, one or more vector bosons, and $b\bar{b}$ processes since none of these can sufficiently produce events with 9 or more jets after post-processing cuts have been applied.

The $\mathcal{F}$-$SU(5)$ with vector-like particles mass pattern produces events with a high multiplicity of virtual stops, which concludes in events with a very large number of jets through the dominant chains $\widetilde{g} \to \widetilde{t}_1\bar{t} \to t\bar{t}\widetilde{\chi}_1^0 \to W^+W^-b\bar{b}\widetilde{\chi}_1^0$ and $\widetilde{g} \to \widetilde{t}_1\bar{t} \to b\bar{t}\widetilde{\chi}_1^+ \to W^-b\bar{b}\widetilde{\tau}_1^+\nu_\tau \to W^-b\bar{b}\tau^+\nu_\tau\widetilde{\chi}_1^0$, as well as the conjugate processes $\widetilde{g} \to \widetilde{\bar{t}}_1 t \to t\bar{t}\widetilde{\chi}_1^0$ and $\widetilde{g} \to \widetilde{\bar{t}}_1 t \to \bar{b}t\widetilde{\chi}_1^-$, where the W bosons will produce mostly hadronic jets and some leptons. Additionally, the heavy squarks will produce gluinos by means of $\widetilde{q} \to q\widetilde{g}$. In Figure 17 we plot the number of jets per event versus the number of events for three distinct scenarios. We suppress the noise on the histogram contour to admit a more lucid distinction of the peaks in the number of jets, and fit polynomials over the data points and conceal the histograms. This allows us to gauge an appropriate selection cut for the number of jets to maximize our signal to background ratio, while assessing the impact of the selection cuts implemented by the CMS Collaboration in [88, 99]. As depicted in Figure 17, the first pane displays a comparison of the number of jets when employing the prior CMS cuts, while the remaining two panes present the results for the post-processing selection cuts defined in this paper, discriminating between two explicit cuts of the minimum p_T for a single jet. Figure 17 demonstrates that the CMS cuts of [88, 99] discard all the high-multiplicity jets, converting the events with at least 9 jets to events with few jets, thus, all information on these events with a large number of jets is lost. To retain the events with a high multiplicity of jets, we explore alternative cuts by shifting the minimum p_T for a single jet lower to the two cases of 10 GeV and 20 GeV. A minimum jet p_T of 20 GeV is secure from interfering with jet fragmentation, which typically occurs in the realm below 10 GeV, indicating that 10 GeV is certainly fringe. We see in Figure 17 that both the 10 GeV and 20 GeV jet p_T cuts preserve the high number of jets, permitting an obvious choice for location of the cut on the minimum number of jets. We thus adopt a revised cut of single jet $p_T > 20$ GeV and total number of jets greater than 9. To assess the discovery potential, we plot the number of events per 200 GeV versus H_T, where $H_T = \sum_{i=1}^{N_{jet}} E_T^{j_i}$. Figure 18 delineates the convincing separation between the $\mathcal{F}$-$SU(5)$ signal and the SM $t\bar{t} + jets$ and the SP3 point. The total number of events are summarized in Table 5. We also include one measure of discovery threshold that compares the number of signal events S to the number of background events B, where we require $\frac{S}{\sqrt{B}} > 5$. Notice that $\mathcal{F}$-$SU(5)$ comfortably surpasses this requirement.

Table 5. Total number of events for 1 fb^{-1} and $\sqrt{s} = 7$ TeV. Minimum p_T for a single jet is $p_T > 20$ GeV.

	$\mathcal{F}$-$SU(5)$	$SP3$	$t\bar{t} + jets$
Events	93.2	2.4	10
$\frac{S}{\sqrt{B}}$	29.5	0.76	

The spectrum of Table 4 exceeds the LEP constraints on the lightest neutralino $\widetilde{\chi}_1^0$ and lightest stau $\widetilde{\tau}_1$, and even more tantalizing, the close proximity of the stau mass beyond the LEP reach suggests imminent discovery at LHC. The stau presence can be reconstructed, for instance, from the dominant $\mathcal{F}$-$SU(5)$ process $\widetilde{g} \to \widetilde{t}_1\bar{t} \to b\bar{t}\widetilde{\chi}_1^+ \to W^-b\bar{b}\widetilde{\tau}_1^+\nu_\tau \to W^-b\bar{b}\tau^+\nu_\tau\widetilde{\chi}_1^0$. The inference

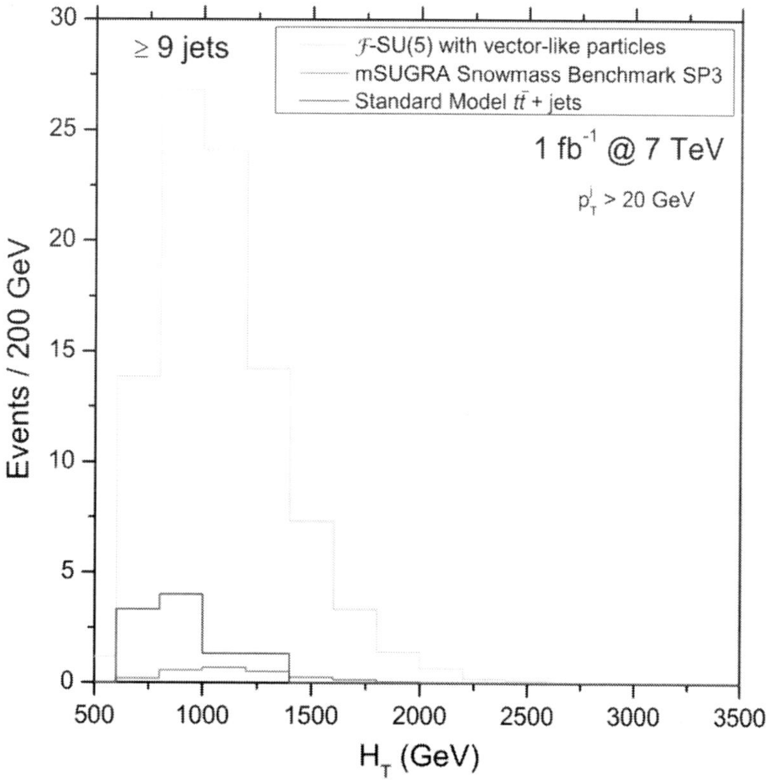

Figure 18. Counts for events with $\geq$ 9 jets.

of the short-lived stau in the $\mathcal{F}$-$SU(5)$ SUSY breaking scenario from tau production assumes fruition of the expected much improved tau detection efficiency at LHC.

16. Conclusion

The advancement of human scientific knowledge and technology is replete with instances of science fiction transitioning to scientific theory and eventually scientific fact. The conceptual notion of a "Multiverse" has long fascinated the human imagination, though this speculation has been largely devoid of a substantive underpinning in physical theory. The modern perspective presented here offers a tangible foundation upon which legitimate discussion and theoretical advancement of the Multiverse may commence, including the prescription of specific experimental tests which could either falsify or enhance the viability of our proposal. Our perspective diverges from the common appeals to statistics and the anthropic principle, suggesting instead that we may seek to establish the character of the master theory, of which our Universe is an isolated vacuum condensation, based on specific observed properties of our own physics which might be reasonably inferred to represent invariant common characteristics of all possible universes. We have focused on the discovery of a model universe consonant with our observable phenomenology, presenting it as confirmation of a non-zero probability of our own Universe transpiring within the larger String Landscape.

The archetype model universe which we advance in this work implicates No-Scale supergravity as the ubiquitous supporting structure which pervades the vacua of the Multiverse, being the crucial ingredient in the emanation of a cosmologically flat universe from the quantum "nothingness". In particular, the model dubbed No-Scale $\mathcal{F}$-$SU(5)$ has demonstrated remarkable consistency between parameters determined dynamically (the top-down approach) and parameters determined through the application of current experimental constraints (the bottom-up approach). This enticing convergence of theory with experiment elevates No-Scale $\mathcal{F}$-$SU(5)$, in our estimation, to a position as the current leading GUT candidate. The longer term viability of this suggestion is likely to be greatly clarified in the next few years, based upon the wealth of forthcoming experimental data.

We have presented a highly constrained "golden point" located near $M_{1/2} = 455$ GeV and $\tan\beta = 15$ in the $\tan\beta - M_{1/2}$ plane, and a highly non-trivial "golden strip" with $\tan\beta \simeq 15$, $m_t = 173.0$-174.4 GeV, $M_{1/2} = 455$-481 GeV, and $M_V = 691$-1020 GeV, which simultaneously satisfies all the known experimental constraints, featuring moreover an imminently observable proton decay rate. In addition, we have studied the one-loop effective Higgs potential, and considered the "Super-No-Scale" condition. With a fixed Z-boson mass, we dynamically determined $\tan\beta$ and $M_{1/2}$ at the local *minimum minimorum* of the Higgs potential, while simultaneously indirectly determining the electroweak scale, thus suggesting a complete resolution of the gauge hierarchy problem in the Standard Model (SM). Furthermore, fixing the SM fermion Yukawa couplings and μ term at the $SU(5)\times U(1)_X$ unification scale, we dynamically determine the ratio $\tan\beta \simeq 15 - 20$, the universal gaugino boundary mass $M_{1/2} \simeq 450$ GeV, and consequently also the total magnitude of the GUT-scale Higgs VEVs, while constraining the low energy SM gauge couplings. In particular, these local *minima minimorum* lie within the previously described "golden strip", satisfying all current experimental constraints.

The LHC era has long been anticipated for the expected revelations of physics beyond the Standard Model, as the quest for experimental evidence and insight into the structure of the underlying theory at high energies is enticingly close at hand. Consequently, the field of prospective supersymmetry models has grown as fingerprints of these models at LHC are studied. Nevertheless, our exploration of recently published signatures for supersymmetry discovery reveals a common focus toward low-multiplicity jet events. However, we showed here that manipulation of LHC data skewed toward these low jet events could mask an authentic supersymmetry signal. We have offered a clear and convincing ultra-high jet multiplicity signal for events with at least nine jets, unmistakable for the Standard Model or minimal supergravity. Notably, the optimized post-processing selection cuts outlined here are essential for discovery of supersymmetry if $\mathcal{F}$-$SU(5)$ is indeed proximal to the physical model. Our revised cuts are not drastic, with the two chief adjustments being lowering the minimum p_T for a single jet to 20 GeV, and raising the minimum number of jets in an event to nine. Recognition of such a signal of stringy origin at the LHC could not only reveal the flipped nature of the high-energy theory, but might also shed light on the geometry of the hidden compactified six-dimensional manifold in the string derived models, and even possibly on the hidden structure of the No-Scale Multiverse.

The blueprints which we have outlined here, integrating precision phenomenology with prevailing experimental data and a fresh interpretation of the Multiverse and the Landscape of String vacua, offer a logically connected point of view from which additional investigation may be mounted. As we anticipate the impending stream of new experimental data which is likely to be revealed in ensuing years, we look forward to serious discussion and investigation of the perspective presented in this work. Though the mind boggles to contemplate the implications of this speculation, so it must also reel at even the undisputed realities of the Universe, these acknowledged facts alone being manifestly sufficient to humble our provincial notions of longevity, extent, and largess. The stakes could not be higher or the potential revelations more profound.

Acknowledgments
D.V.N. would like to thank Thomas Elze for his kind invitation to DICE 2010. This research

was supported in part by the DOE grant DE-FG03-95-Er-40917 (TL and DVN), by the Natural Science Foundation of China under grant numbers 10821504 and 11075194 (TL), and by the Mitchell-Heep Chair in High Energy Physics (JM).

References

[1] Bousso R and Polchinski J 2000 Quantization of four-form fluxes and dynamical neutralization of the cosmological constant *JHEP* 06:006
[2] Giddings S B, Kachru S and Polchinski J 2002 Hierarchies from fluxes in string compactifications *Phys. Rev.* D **66** 106006
[3] Kachru S, Kallosh R, Linde A D and Trivedi S P 2003 De Sitter vacua in string theory *Phys. Rev.* D **68** 046005
[4] Susskind L 2003 The Anthropic landscape of string theory
[5] Denef F and Douglas M R 2004 Distributions of flux vacua *JHEP* 0405:072
[6] Denef F and Douglas M R 2005 Distributions of nonsupersymmetric flux vacua *JHEP* 0503:061
[7] Denef F, Douglas M R and Florea B 2004 Building a better racetrack *JHEP* 0406:034
[8] Denef F, Douglas M R and Kachru S 2007 Physics of String Flux Compactifications *Ann. Rev. Nucl. Part. Sci.* **57** 119–144
[9] Weinberg S 1987 Anthropic Bound on the Cosmological Constant *Phys. Rev. Lett.* **59** 2607
[10] Guth A H 1981 The Inflationary Universe: A Possible Solution to the Horizon and Flatness Problems *Phys. Rev.* D **23** 347–356
[11] Linde A D 1982 A New Inflationary Universe Scenario: A Possible Solution of the Horizon, Flatness, Homogeneity, Isotropy and Primordial Monopole Problems *Phys. Lett.* B **108** 389–393
[12] Albrecht A and Steinhardt P J 1982 Cosmology for Grand Unified Theories with Radiatively Induced Symmetry Breaking *Phys. Rev. Lett.* **48** 1220–1223
[13] Ellis J R, Nanopoulos D V, Olive K A and Tamvakis K 1983 Primordial Supersymmetric Inflation *Nucl. Phys.* B **221** 524
[14] Nanopoulos D V, Olive K A, Srednicki M and Tamvakis K 1983 Primordial Inflation in Simple Supergravity *Phys. Lett.* B **123** 41
[15] Spergel D N et al. 2003 First year Wilkinson Microwave Anisotropy Probe (WMAP) observations: Determination of cosmological parameters *Astrophys. J. Suppl.* **148** 175–194, 2003
[16] Spergel D N et al. 2007 Wilkinson Microwave Anisotropy Probe (WMAP) three year results: implications for cosmology *Astrophys. J. Suppl.* **170** 377
[17] Komatsu E et al. 2010 Seven-Year Wilkinson Microwave Anisotropy Probe (WMAP) Observations: Cosmological Interpretation
[18] Steinhardt P J 1984 The Very Early Universe. Proc. Nuffield Workshop, Cambridge, UK, June 21 - July 9, 1982 (Cambridge, UK: Cambridge Univ. Press) pages 251–266
[19] Vilenkin A 1983 The Birth of Inflationary Universes *Phys. Rev.* D **27** 2848
[20] Witten E 1995 String theory dynamics in various dimensions *Nucl. Phys.* B **443** 85–126
[21] Vafa C 1996 Evidence for F theory *Nucl. Phys.* B **469** 403–418
[22] Cremmer E, Ferrara S, Kounnas C and Nanopoulos D V 1983 Naturally Vanishing Cosmological Constant in $N = 1$ Supergravity *Phys. Lett.* B **133** 61
[23] Ellis J R, Lahanas A B, Nanopoulos D V and Tamvakis K 1984 No-Scale Supersymmetric Standard Model *Phys. Lett.* B **134** 429
[24] Ellis J R, Kounnas C and Nanopoulos D V 1984 Phenomenological $SU(1,1)$ Supergravity *Nucl. Phys.* B **241** 406
[25] Ellis J R, Kounnas C and Nanopoulos D V 1984 No Scale Supersymmetric Guts *Nucl. Phys.* B **247** 373–395
[26] Lahanas A B and Nanopoulos D V 1987 The Road to No Scale Supergravity *Phys. Rep.* **145** 1
[27] Giudice G F and Rattazzi R 1999 Theories with gauge mediated supersymmetry breaking *Phys. Rep.* **322** 419–499
[28] Witten E 1985 Dimensional Reduction of Superstring Models *Phys. Lett.* B **155** 151
[29] Tianjun Li, Lopez J L and Nanopoulos D V 1997 Compactifications of M-theory and their phenomenological consequences *Phys. Rev.* D **56** 2602–2606
[30] Beasley C, Heckman J J and Vafa C 2009 GUTs and Exceptional Branes in F-theory - I *JHEP* 01:058
[31] Beasley C, Heckman J J and Vafa C 2009 GUTs and Exceptional Branes in F-theory - II: Experimental Predictions *JHEP* 01:059
[32] Donagi R and Wijnholt M 2008 Model Building with F-Theory
[33] Donagi R and Wijnholt M 2008 Breaking GUT Groups in F-Theory

[34] Cvetic M, Tianjun Li and Tao Liu 2004 Supersymmetric patiSalam models from intersecting D6-branes: A Road to the standard model *Nucl. Phys.* B **698** 163–201
[35] Chen C-M, Tianjun Li, Mayes V E and Nanopoulos D V 2008 A Realistic world from intersecting D6-branes *Phys. Lett.* B **665** 267–270
[36] Chen C-M, Tianjun Li, Mayes V E and Nanopoulos D V 2008 Towards realistic supersymmetric spectra and Yukawa textures from intersecting branes *Phys. Rev.* D **77** 125023
[37] Barr S M 1982 A New Symmetry Breaking Pattern for $SO(10)$ and Proton Decay *Phys. Lett.* B **112** 219
[38] Derendinger J P, Kim J E and Nanopoulos D V 1984 Anti-$SU(5)$ *Phys. Lett.* B **139** 170
[39] Antoniadis I, Ellis J R, Hagelin J S and Nanopoulos D V 1987 Supersymmetric Flipped $SU(5)$ Revitalized *Phys. Lett.* B **194** 231
[40] Antoniadis I, Ellis J R, Hagelin J S and Nanopoulos D V 1988 GUT Model Building with Fermionic Four-Dimensional Strings *Phys. Lett.* B **205** 459
[41] Antoniadis I, Ellis J R, Hagelin J S and Nanopoulos D V 1988 An Improved SU(5) x U(1) Model from Four-Dimensional String *Phys. Lett.* B **208** 209–215
[42] Antoniadis I, Ellis J R, Hagelin J S and Nanopoulos D V 1989 The Flipped SU(5) x U(1) String Model Revamped *Phys. Lett.* B **231** 65
[43] Lopez J L, Nanopoulos D V and Yuan K-J 1993 The Search for a realistic flipped $SU(5)$ string model *Nucl. Phys.* B **399** 654–690
[44] Jing Jiang, Tianjun Li, Nanopoulos D V and Xie D 2009 F-$SU(5)$ *Phys. Lett.* B **677** 322–325
[45] Jing Jiang, Tianjun Li, Nanopoulos D V and Xie D 2010 Flipped $SU(5) \times U(1)_X$ Models from F-Theory *Nucl. Phys.* B **830** 195–220
[46] Jing Jiang, Tianjun Li and Nanopoulos D V 2007 Testable Flipped $SU(5) \times U(1)_X$ Models *Nucl. Phys.* B **772** 49–66
[47] Nakamura K 2003 Hyper-Kamiokande: A next generation water Cherenkov detector *Int. J. Mod. Phys.* A **18** 4053–4063
[48] Raby S et al. 2008 DUSEL Theory White Paper
[49] Tianjun Li, Nanopoulos D V and Walker J W 2011 Elements of F-ast Proton Decay *Nucl. Phys.* B **846** 43–99
[50] Tianjun Li, Maxin J A, Nanopoulos D V and Walker J W 2011 Dark Matter, Proton Decay and Other Phenomenological Constraints in $\mathcal{F}$-$SU(5)$ *Nucl. Phys.* B, in press
[51] Tianjun Li, Nanopoulos J W and Walker J W 2010 Fast Proton Decay *Phys. Lett.* B **693** 580
[52] Huo Y-J, Tianjun Li, Tong C-L and Nanopoulos D V 2011 in preparation
[53] Kyae B and Shafi Q 2006 Flipped SU(5) predicts delta(T)/T *Phys. Lett.* B **635** 247–252
[54] Ferrara S, Kounnas C and Zwirner F 1994 Mass formulae and natural hierarchy in string effective supergravities *Nucl. Phys.* B **429** 589–625
[55] Cremmer E and Julia B 1979 The SO(8) Supergravity *Nucl. Phys.* B **159** 141
[56] Nilles H-P, Olechowski M and Yamaguchi M 1997 Supersymmetry breaking and soft terms in M theory *Phys. Lett.* B **415** 24–30
[57] Nilles H-P, Olechowski M and Yamaguchi M 1998 Supersymmetry breakdown at a hidden wall *Nucl. Phys.* B **530** 43–72
[58] Lukas A, Ovrut B A and Waldram D 1998 On the four-dimensional effective action of strongly coupled heterotic string theory *Nucl. Phys.* B **532** 43–82
[59] Lukas A, Ovrut B A, Stelle K S and Waldram D 1999 The Universe as a domain wall *Phys. Rev.* D **59** 086001
[60] Tianjun Li 1999 Soft terms in M theory *Phys. Rev.* D **59** 107902
[61] Choi K, Falkowski A, Nilles H-P, Olechowski M and Pokorski S 2004 Stability of flux compactifications and the pattern of supersymmetry breaking *JHEP* 0411:076
[62] Choi K, Falkowski A, Nilles H-P and Olechowski M 2005 Soft supersymmetry breaking in KKLT flux compactification *Nucl. Phys.* B **718** 113–133
[63] Tianjun Li, Maxin J A, Nanopoulos D V and Walker J W 2011 The Golden Point of No-Scale and No-Parameter $\mathcal{F}$-$SU(5)$ *Phys. Rev.* D, in press
[64] Djouadi A, Kneur J-L and Moultaka G 2007 SuSpect: A Fortran code for the supersymmetric and Higgs particle spectrum in the MSSM *Comput. Phys. Commun.* **176** 426–455
[65] Belanger G, Boudjema F, Pukhov A and Semenov A 2009 Dark matter direct detection rate in a generic model with micrOMEGAs2.1 *Comput. Phys. Commun.* **180** 747–767
[66] Combination of CDF and DØResults on the Mass of the Top Quark 2009.
[67] Mavromatos N E and Nanopoulos D V 2009 The MAGIC of SSC and how it affects LHC *AIP Conf. Proc.* **1166** 26–32
[68] Barberio E et al. 2007 Averages of $b-$hadron properties at the end of 2006

[69] Misiak M et al. 2007 The first estimate of Br($\overline{B} \to X_s\gamma$) at $\mathcal{O}(\alpha_s^2)$ *Phys. Rev. Lett.* **98** 022002
[70] Bennett G W et al. 2004 Measurement of the negative muon anomalous magnetic moment to 0.7-ppm *Phys. Rev. Lett.* **92** 161802
[71] Aaltonen T et al. 2008 Search for $B_s^0 \to \mu^+\mu^-$ and $B_d^0 \to \mu^+\mu^-$ decays with $2fb^{-1}$ of $p\bar{p}$ collisions *Phys. Rev. Lett.* **100** 101802
[72] Barate R et al. 2003 Search for the standard model Higgs boson at LEP *Phys. Lett.* B **565** 61–75
[73] Yao W M et al. 2006 Review of particle physics *J. Phys.* G **33** 1–1232
[74] Ahmed Z et al. 2009 Search for Weakly Interacting Massive Particles with the First Five-Tower Data from the Cryogenic Dark Matter Search at the Soudan Underground Laboratory *Phys. Rev. Lett.* **102** 011301
[75] Aprile E et al. 2010 First Dark Matter Results from the XENON100 Experiment *Phys. Rev. Lett.* **105** 131302
[76] Abdo A A et al. 2010 Constraints on Cosmological Dark Matter Annihilation from the Fermi-LAT Isotropic Diffuse Gamma-Ray Measurement *JCAP* 1004:014
[77] Tianjun Li, Maxin J A, Nanopoulos D V and Walker J W 2010 The Golden Strip of Correlated Top Quark, Gaugino, and Vectorlike Mass In No-Scale, No-Parameter F-SU(5)
[78] Tianjun Li, Maxin J A, Nanopoulos D V and Walker J W 2010 Super No-Scale $\mathcal{F}$-$SU(5)$: Resolving the Gauge Hierarchy Problem by Dynamic Determination of $M_{1/2}$ and $\tan\beta$
[79] Tianjun Li, Maxin J A, Nanopoulos D V and Walker J W 2011 Blueprints of the No-Scale Multiverse at the LHC
[80] Martin S P 2002 Two-loop effective potential for the minimal supersymmetric standard model *Phys. Rev.* D **66** 096001
[81] Nishino H et al. 2009 Search for Proton Decay via $p \to e^+\pi_0$ and $p \to \mu^+\pi_0$ in a Large Water Cherenkov Detector *Phys. Rev. Lett.* **102** 141801
[82] Ellis J R, Nanopoulos D V and Tamvakis K 1983 Grand Unification in Simple Supergravity *Phys. Lett.* B **121** 123
[83] Ellis J R, Hagelin J S, Nanopoulos D V and Tamvakis K 1983 Weak Symmetry Breaking by Radiative Corrections in Broken Supergravity *Phys. Lett.* B **125** 275
[84] Arnowitt R L, Dutta B, Gurrola A, Kamon T, Krislock A et al. 2008 Determining the Dark Matter Relic Density in the mSUGRA (X0(1))- tau Co-Annhiliation Region at the LHC *Phys. Rev. Lett.* **100** 231802
[85] Conway J et al. 2009 Pgs4: Pretty good (detector) simulation 2009
[86] Sjostrand T, Mrenna S and Skands P Z 2006 PYTHIA 6.4 Physics and Manual *JHEP* 05:026
[87] Abazov V M et al. 2010 Search for single vector-like quarks in $p\bar{p}$ collisions at $\sqrt{s} = 1.96$ TeV *Phys. Rev. Lett.*, long author list - awaiting processing
[88] Khachatryan V et al. 2011 Search for Supersymmetry in pp Collisions at 7 TeV in Events with Jets and Missing Transverse Energy
[89] Barreiro Guimaraes da Costa J et al. 2011 Search for supersymmetry using final states with one lepton, jets, and missing transverse momentum with the ATLAS detector in sqrts = 7 TeV pp
[90] Barreiro Guimaraes da Costa J et al. 2011 Search for squarks and gluinos using final states with jets and missing transverse momentum with the ATLAS detector in sqrt(s) = 7 TeV proton-proton collisions
[91] Baer H, Barger V, Lessa A and Tata X 2010 Capability of LHC to discover supersymmetry with $\sqrt{s} = 7$ TeV and 1 fb^{-1} *JHEP* 06:102
[92] Kane G L, Kuflik E, Ran Lu and Wang L-T 2011 Top Channel for Early SUSY Discovery at the LHC
[93] Feldman D, Freese K, Nath P, Nelson B D and Peim G 2011 Predictive Signatures of Supersymmetry: Measuring the Dark Matter Mass and Gluino Mass with Early LHC data
[94] Buchmueller O et al. 2011 Implications of Initial LHC Searches for Supersymmetry
[95] Guchait M and Sengupta D 2011 Searches for Supersymmetry at the LHC with 7 TeV energy
[96] Alwall J et al. 2007 MadGraph/MadEvent v4: The New Web Generation *JHEP* 09:028
[97] Allanach B C et al. 2002 The Snowmass points and slopes: Benchmarks for SUSY searches *Eur. Phys. J.* C **25** 113–123
[98] Altunkaynak B, Holmes M, Nath P, Nelson B D and Peim G 2010 SUSY Discovery Potential and Benchmarks for Early Runs at $\sqrt{s} = 7$ TeV at the LHC *Phys. Rev.* D **82** 115001
[99] Search strategy for exclusive multi-jet events from supersymmetry at CMS. 2009. CMS PAS SUS-09-001

Spacetime discreteness, Lorentz invariance and locality

Fay Dowker
Blackett Lab, Imperial College, London SW7 2AZ, UK
E-mail: f.dowker@imperial.ac.uk

Abstract. If spacetime is fundamentally discrete, as many workers on quantum gravity suspect, then continuum symmetries such as Lorentz invariance are imperilled. Lorentz symmetry can nevertheless be compatible with spacetime atomicity but the price is locality. Indeed, causal sets are discrete and Lorentz invariant but highly nonlocal. The question becomes: can the nonlocality be tamed far enough to allow known local physics to emerge at large distances? We give evidence that the answer to this question is yes.

1. Introduction

From the earliest days of the causal set approach to the problem of quantum gravity [1], it has been recognised that the discreteness and Lorentz invariance of the theory give rise to a radical nonlocality [2, 3]. The type of nonlocality that arises can be argued to occur in any fundamentally discrete approach to quantum gravity in which Lorentz symmetry is not broken. If spacetime is made of atoms, the argument goes, then the number of atoms one Planck time to the past of any given one, P, must be infinite since – in the approximating infinite Minkowski spacetime which these atoms comprise – they would be distributed roughly along the hyperboloid which is one Planck unit to the past of P. This is exactly the form of nonlocality that causal sets exhibit and for a while it blocked our understanding of how local physics could arise from causal set theory. Even understanding how local physics could be compatible with a discrete spacetime background such as a causal set was difficult since our familiar physical laws are expressed in local terms as differential equations and a discrete differential operator is a difference operator which compares the values of a physical variable at neighbouring points. When there are infinitely many nearest neighbours to a given node, it appears that one is going to struggle to reproduce any differential evolution laws on a causal set. However, we now know that we can define discrete operators that approximate the scalar D'Alembertian in any spacetime dimension.

2. Causal Sets

Recall that a *causal set* (*causet* for short) is a locally finite partial order, *i.e.* it is a pair $(\mathcal{C}, \preceq)$ where $\mathcal{C}$ is a set and $\preceq$ is a relation on $\mathcal{C}$ which is reflexive ($x \preceq x$), acyclic ($y \preceq x \preceq y \Rightarrow y = x$) and transitive ($z \preceq y \preceq x \Rightarrow z \preceq x$). Local finiteness is the condition that the cardinality of any *order interval* is finite, where the (inclusive) order interval between a pair of elements $y \preceq x$ is defined to be $I(x,y) := \{z \in \mathcal{C} \,|\, y \preceq z \preceq x\}$. We call x the *top element* and y the *bottom element* of $I(x,y)$. We write $y \prec x$ when $y \preceq x$ and $y \neq x$. We define $n(x,y) := |I(x,y)|$, the cardinality of the inclusive order interval, and call a relation $y \prec x$ a *link* if $n(x,y) = 2$. A *chain* is a totally ordered subset of $\mathcal{C}$.

Sprinkling is a random process that produces a causet which is a discretisation of a d-dimensional, causal, Lorentzian manifold $(\mathcal{M}, g)$. It is a Poisson process of selecting points in $(\mathcal{M}, g)$, independently at random, with density ρ so that the expected number of points sprinkled in a region of spacetime volume V is ρV. In quantum gravity we expect that the density is Planckian so that $\rho = l^{-d}$ where l is of order the Planck length, but in this paper we treat ρ as a parameter to be varied. This process generates a causet whose elements are (identified with) the sprinkled points and whose order is that induced by the manifold's causal order restricted to the sprinkled points. If $(\mathcal{M}, g)$ is of finite volume, the causet generated is almost surely finite and so the process defines a probability distribution on the set of finite causets (*aka* the set of finite partial orders). Henceforth, for ease of notation, we will drop the explicit reference to the metric g and refer, for example, to a spacetime as $\mathcal{M}$. Sprinkling is not a physical process. It plays a purely kinematical role and expresses the discrete-continuum correspondence: a causet $\mathcal{C}$ is well approximated by a Lorentzian manifold $\mathcal{M}$ if it could have been generated, with relatively high probability, by sprinkling into $\mathcal{M}$. In other words, $\mathcal{C}$ is well approximated by a manifold $\mathcal{M}$ if there exists an embedding $i : \mathcal{C} \hookrightarrow \mathcal{M}$ such that (i) $x, y \in \mathcal{C}$, $y \preceq x$ iff $i(y) \in J^-(i(x))$ and (ii) the number of elements embedded in any sufficiently nice, large region of volume V is approximately ρV. Strictly, this is a *conjecture*, the "Hauptvermutung" of the causal set approach, but it is supported by much evidence including the result that a distinguishing Lorentzian geometry is fully determined by its causal structure and spacetime volume measure [4, 5, 6].

The causal set hypothesis is that in the deep quantum regime of very small scales, spacetime is no longer described by a metric on a differentiable manifold, but by a causal set. Just as ordinary matter appears smooth and continuous to us, but is really made of atoms and molecules, so, it is proposed, spacetime appears continuous at large scales but is really a causal set and the continuum spacetime of our experience is just an approximation to the discrete order. The order of the causal set gives rise to the spacetime causal order, the number of elements gives rise to the spacetime volume and together these contain enough information to give the spacetime metric in the continuum approximation. These correspondences can be summarised in Sorkin's slogan: "Number + Order = Geometry."

3. Particle dynamics

Putting matter on a causal set even kinematically is not a trivial question. The two things we are able to do easily are to define a massive point particle trajectory as a chain and to define a scalar field on a causal set as a map from the causet to some space of scalars. In both these cases we now have quasilocal dynamical models which approximate the continuum dynamics. The latter will be the subject of the next section. In this section we describe how a dynamical rule for particle motion can be given that is quasilocal despite the radical nonlocality of causal sets.

First we note that on a causet which is a sprinkling into Minkowski spacetime, there exists a very good discrete notion of timelike geodesic between two elements: a longest chain. We know that as the sprinkling density tends to infinity, longest chains tend to straight line geodesics. However, using this as a rule for how a particle should move on a causal set is extremely nonlocal since to determine its path, the particle must know in advance where it will end up. We show that there is a way for a particle at a given element of the causal set to make a quasilocal decision about which element to move to next [7].

Consider a point-particle of mass m on a causet $\mathcal{C}$ that is a sprinkling into Minkowski space $\mathbb{M}^4$. We take its trajectory to be a chain in $\mathcal{C}$. So, given an initial portion of the trajectory up to some element e_n of $\mathcal{C}$, how is its continuation to be determined? We assume that the trajectory's past determines its future, but that only a certain proper time τ_f into the past of the trajectory is allowed to be relevant. The following approximately Markovian, Lorentz invariant rule is a simple discrete analogue of geodesic motion.

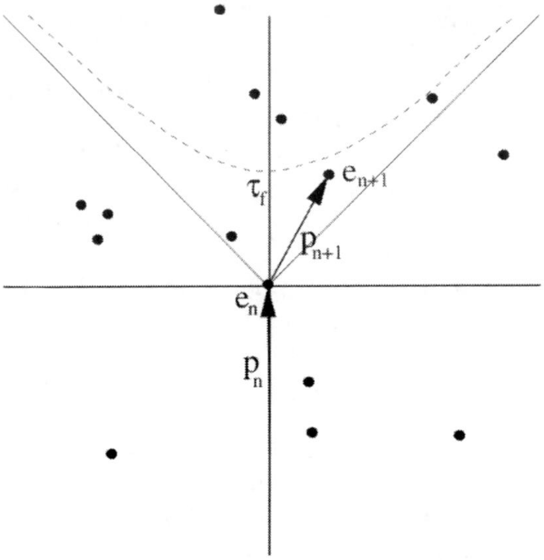

Figure 1. A portion of $1+1$ Minkowski space where the dots represent elements of a causet sprinkled into it. The trajectory of the particle has reached e_n with momentum p_n (the frame having been chosen so that the three-momentum is zero). The dotted line is the hyperbola of points a proper time τ_f to the future of e_n. The element within proper time τ_f of e_n that best preserves the momentum is e_{n+1}. The ratio of τ_f^{-2} to the density of sprinkling has been exaggerated here to emphasize the momentum change in one step. In a more realistic model τ_f would be larger.

The future trajectory is constructed inductively, as illustrated in fig. 1. Starting from an element e_n and a momentum p_n, the next element in the trajectory, e_{n+1}, must be chosen. For convenience we have drawn the diagram in the rest frame at e_n. The new momentum p_{n+1} is defined to be proportional to the vector between e_n and e_{n+1}. The selection of e_{n+1} is made such that e_{n+1} is in the causal future of e_n and is within proper time τ_f of e_n and so that $|p_{n+1} - p_n|$ is minimized. In the figure, this means that e_{n+1} is within the future light cone of e_n, below the dotted hyperbola, and such that the vector from e_n to e_{n+1} is as close to the vertical as possible. This realizes the ideas that the trajectory should be as close to a straight line as possible and that the dynamics should be approximately Markovian if the "forgetting time" τ_f is small compared to observation scales (which is possible even at the same time as being large compared to the discreteness scale if the latter is Planckian).

Our model implies random fluctuations in the momentum of the particle. Since the hypothesized effect is supposed to occur on very small scales, it should be possible to approximate it over macroscopic distances by a diffusion equation (hydrodynamic limit).

Consider the scalar (not scalar density) probability distribution ρ on $H^3 \times \mathbb{M}^4$ where H^3 is the mass shell, hyperbolic 3-space. ρ is a function of momentum, p^ν, spacetime position, x^μ, and proper time τ. With the condition that the process be Lorentz invariant, the following equation for ρ can be derived by following the prescription set out in [8] for stochastic evolution on a manifold of states:

$$\frac{\partial \rho}{\partial \tau} = k \nabla_p^2 \rho - \frac{1}{mc^2} p^\mu \frac{\partial}{\partial x^\mu} \rho \tag{1}$$

where ∇_p^2 is the Laplacian on H^3, m is the mass of the particle, and k is a constant (that will depend on the parameters of the discrete process, such as the forgetting time τ_f).

This model of particle swerves is rather unphysical. It treats the particle as if it were of zero size whereas we know that protons, for example, are composite particles with structure and will have a complicated way relationship with any discrete spacetime structure. Also the model is classical, indeed deterministic, rather than quantum. Most seriously, it could not possibly be fundamental, since the law of motion of the trajectory is not formulated only in terms of the causet, but refers also to the approximating Minkowski spacetime. However, it is easy to construct models whose rules are intrinsic to the causal set [9, 10] and whose continuum limit will be the same diffusion equation (up to the free parameter of the diffusion constant) presented above. This is because the diffusion equation is the unique Markovian, Poincaré invariant, relativistically causal, diffusion law in vacuum that preserves $\rho \geq 0$. Given this uniqueness property, the equation should be insensitive to variations in the microscopic model that underlies it, so long as the latter is Lorentz invariant, causal and (approximately) Markovian. So although the above discrete swerves model is almost certainly unphysical in its details, the macroscopic phenomenology of (1) is so robust that it transcends the limitations of any particular microscopic model.

4. Scalar D'Alembertians

We turn now to the other "easy case" of matter on a causal set: scalar field theory. To define a classical scalar field dynamics on a causal set sprinkled into Minkowski spacetime we require a discrete operator that is well-approximated by the continuum D'Alembertian. For a sprinkling into 2 dimensional Minkowski spacetime this was constructed by Henson and Sorkin [11, 12]. We also have analogous operators in 4 dimensions [13] and in various other dimension (see the contribution by Glaser in this volume) but for the purposes of this article we focus on 2 dimensions.

Consider the following definition [12] of a discrete D'Alembertian, B, on a causet $\mathcal{C}$ that is a sprinkling, at density $\rho = l^{-2}$, into 2-d Minkowski space $\mathbb{M}^2$. Let $\phi : \mathcal{C} \to \mathbb{R}$ be a real scalar field, then

$$B\phi(x) := \frac{4}{l^2}\Big[-\frac{1}{2}\phi(x) + \Big(\sum_{y\in L_1} -2\sum_{y\in L_2} + \sum_{y\in L_3}\Big)\phi(y)\Big]\,, \tag{2}$$

where the sums run over 3 *layers* L_i, $i = 1, \ldots, 3$,

$$L_i := \{y \in \mathcal{C} :\ y \prec x \text{ and } n(x, y) = i + 1\} \tag{3}$$

and $n(x, y) := |I(x, y)|$. So, for example, layer L_1 is the set of all elements y that are linked to x and as described above, they will be distributed close to a hyperboloid that asymptotes to the past light cone of x and is proper time l away from x.

How does $B\phi$ on $\mathcal{C}$ compare to $\Box\phi$ on $\mathbb{M}^2$? To answer this, we need to relate the fields on $\mathbb{M}^2$ and on $\mathcal{C}$. Let ϕ be a scalar field of compact support on $\mathbb{M}^2$. The above definition of the discrete operator on a scalar field on a causal set defines a random variable $\mathbf{B}(\phi, x, \rho)$ for every point x of $\mathbb{M}^2$ and every density ρ in the following way. Sprinkle into $\mathbb{M}^2$ at density ρ and add a marked element at x to the resulting causal set to form causet $\mathcal{C}$. Consider the field ϕ restricted to the sprinkled causal set $\mathcal{C}$. The value of the random variable $\mathbf{B}(\phi, x, \rho)$ for that realisation of the process is $B\phi$ evaluated at the marked element of the sprinkled causal set $\mathcal{C}$.

The mean of that random variable can be calculated using the Poisson distribution

$$\langle \mathbf{B}(\phi, x)\rangle = -2\rho\phi(x) + 4\rho^2 \int_{J^-(x)} d^2y \left(1 - 2\rho V_{xy} + \frac{1}{2}(\rho V_{xy})^2\right) e^{-\rho V_{xy}}\phi(y) \tag{4}$$

where V_{xy} is the volume (area) of the spacetime causal interval, $[x, y] := J^+(y) \cap J^-(x)$, between x and y. We have dropped the explicit mention of the density ρ – it remains understood that the random variable and its mean depend on ρ. Equation (4) can be understood thus: $\rho\, d^2y$ is the probability that an element is sprinkled in an elemental volume at y; $e^{-\rho V_{xy}}$, $\rho V_{xy} e^{-\rho V_{xy}}$ or $\frac{1}{2}(\rho V_{xy})^2 e^{-\rho V_{xy}}$ is the probability that there is no element, one element or two elements, respectively, sprinkled in $[x, y]$.

The form of (4) is of an integral operator acting on the continuum field ϕ and we call that operator $\bar{B}$:

$$\bar{B}\phi(x) := \langle \mathbf{B}(\phi, x) \rangle \tag{5}$$

It can be shown that

$$\bar{B}\phi(x) \to \Box\phi(x) \quad \text{as} \quad \rho \to \infty\,. \tag{6}$$

The limit is approached when ρ is large enough that the field ϕ hardly varies on the length scale l in a neighbourhood of x. This happens due to the judicious choice of the coefficients $1, -2, 1$ in equation (2) which mean that the contributions from the three layers that would diverge in the limit as $\rho \to \infty$ actually cancel out. In 4 dimensions, another layer is needed in the expression for the operator B, with appropriate coefficients [13], and in other dimensions other choices can be made (see Glaser's article in this volume).

So, the expected value of $B\phi$ is close to $\Box\phi$ but what about the typical value? Simulations show that the fluctuations around the mean actually grow as the sprinkling density tends to infinity and so the typical value will be far from the continuum result [12]. In order to ensure that the typical value will be a good approximation to the continuum value, it is necessary to smear the operator B over an intermediate length scale, $l_k >> l$, to form operator B_k whose mean is [12]:

$$\bar{B}_k\phi(x) = \frac{4}{l_k^2}\Big[-\frac{1}{2}\phi(x) + \frac{1}{l_k^2}\int_{y\in J^-(x)} d^2y\; \phi(y)\, e^{-\xi}(1 - 2\xi + \xi^2)\Big], \tag{7}$$

where $\xi := l_k^{-2} V_{xy}$. Working back, one can show that the discrete operator, B_k, with this mean is

$$B_k\phi(x) = \frac{4}{l_k^2}\Bigg[-\frac{1}{2}\phi(x) + \epsilon\sum_{y\prec x} f(n(x,y) - 2, \epsilon)\phi(y)\Bigg], \tag{8}$$

where $\epsilon = (l/l_k)^4$ and

$$f(n,\epsilon) = (1-\epsilon)^n \left[1 - \frac{2\epsilon n}{1-\epsilon} + \frac{\epsilon^2 n!}{(n-2)!(1-\epsilon)^2}\right]. \tag{9}$$

B_k reduces to B when $\epsilon = 1$. B_k effectively samples ϕ over elements in 3 broad bands with a characteristic depth l_k, the bands' contributions being weighted with the same set of alternating sign coefficients as in B. Since (7) is just (4) with ρ replaced by l_k^{-2}, the mean of $B_k\phi(x)$ is close to $\Box\phi(x)$ when the characteristic scale over which ϕ varies is large compared to l_k.

For B_k and a fixed nonlocality scale l_k simulations show that the fluctuations die away as the sprinkling density tends to infinity, *i.e* as ϵ tends to zero [12]. Thus, by choosing the nonlocality scale appropriately, we have our discrete operator, B_k, which is well-approximated by the continuum D'Alembertian when applied to scalar fields that vary slowly on the nonlocality scale.

5. Causal set action

The operators B and B_k are defined in terms of the order relation on $\mathcal{C}$ alone and so can be applied to a scalar field on *any* causet. If, therefore, $(\mathcal{M}, g)$ is a 2-d curved spacetime and ϕ

is a scalar field on $\mathcal{M}$, we can compute $B_k\phi(x)$ on a sprinkling into $\mathcal{M}$ and calculate its mean $\bar{B}_k\phi(x)$. In the presence of curvature,

$$\bar{B}_k\phi(x) = \frac{2}{l_k^2}\Big[-\phi(x) + \frac{2}{l_k^2}\int_{y\in J^-(x)} d^2y\sqrt{-g}\, e^{-\xi}(1 - 2\xi + \frac{1}{2}\xi^2)\phi(y)\Big] \tag{10}$$

where $\xi = l_k^{-2}V_{xy}$.

This can be evaluated using Riemann normal coordinates and we find [13]

$$\lim_{l_k\to 0}\bar{B}_k^{(i)}\phi(x) = \left(\Box - \frac{1}{2}R(x)\right)\phi(x) \tag{11}$$

where R is the Ricci scalar curvature. The limit is a good approximation to the mean when the field ϕ varies slowly over length scales l_k *and* the radius of curvature $R^{-\frac{1}{2}} >> l_k$.

If the damping of fluctuations found in simulations in flat space are indicative of what happens in curved space then, for a fixed large enough IR cutoff, L, the non-locality length scale l_k can be chosen such that $l << l_k << L$ and the value of $B_k\phi$ for a single sprinkling will be close to the mean. If B_k is applied to the constant field $\phi = -2$, we therefore obtain an expression that is close to the scalar curvature of the approximating spacetime.

We can now define a one parameter family of candidate actions, $S_k[\mathcal{C}]$, for a causal set, $\mathcal{C}$, by summing $B_k(-1)$ over the elements of $\mathcal{C}$, times $\hbar l^2$ to get the units right, times a number of order one. When the nonlocality length l_k equals the discreteness length l, $B_k = B$ and the action, $S[\mathcal{C}]$ takes a particularly simple form as an alternating sum of numbers of small order intervals in $\mathcal{C}$. Up to factors of order one, we have:

$$\frac{1}{\hbar}S^{(2)}[\mathcal{C}] = N - 2N_1 + 4N_2 - 2N_3 \tag{12}$$

where N is the number of elements in $\mathcal{C}$ and N_i is the number of $(i+1)$-element inclusive order intervals in $\mathcal{C}$.

There are also candidate actions in 4 dimensions [13] and in other dimensions (Glaser's article in this volume). This proposed causal set action brings with it a host of questions, including whether or not it shows anything of the the topological character of the continuum 2-d action (see Benincasa's article in this volume).

6. Discussion

There are various lessons that can be learnt from the particle swerves model and the discrete D'Alembertians described here. One main one, referred to in the title, is that Lorentz invariance, spacetime discreteness and approximate locality can co-exist despite the tension that exists between the three. Another is that the concrete kinematics of causal set theory allow us to do quantum gravity phenomenology even in the absence of a full fundamental theory of quantum causal sets. In particular it allows us to model Planck scale spacetime uncertainty in a way that respects Lorentz symmetry. This is something that is notoriously hard to do as evidenced by the fact that most quantum gravity phenomenology models violate Lorentz invariance. Even if one does not believe in causal sets as fundamental, the approach is therefore useful as a tool for modelling the effects of a spacetime which has no structure at smaller than Planckian scales.

Another lesson is that the nonlocality of causal sets cannot be completely tamed. It cannot be relegated to the Planck scale. Instead it rears its head at an intermediate mesoscale between the observation/cosmological scale and the fundamental scale. It is interesting to note that there is a similar "three scale" structure in other approaches to quantum gravity. In string/M theory, there is the 11 dimensional Planck scale and the cosmological scale, and between them there is the string scale at which stringy effects manifest themselves but full quantum gravity effects are not yet apparent. In non-commutative geometry there is the discreteness scale and the cosmological scale and an intermediate scale at which non-commutative effects show up but the full fundamental theory is not needed. In a similar way, in causal set theory we see the emergence of a nonlocality scale between the discreteness scale and the cosmological scale at which novel phenomena occur. In all these theories, the fact that this scale is potentially much larger than the relatively inaccessible fundamental scale is both a challenge – we have to show our theories are compatible with known observations – and an opportunity. Quantum gravity phenomena may already be within our observational scope.

Acknowledgments

I thank my collaborators Dionigi Benincasa, Lisa Glaser, Joe Henson, Bernhard Schmitzer and Rafael Sorkin for helpful discussions. This work was supported in part by the Royal Society Grant IJP 2006/R2.

References

[1] Bombelli L, Lee J H, Meyer D and Sorkin R 1987 *Phys. Rev. Lett.* **59** 521
[2] Moore C 1988 *Phys. Rev. Lett.* **60** 655–655
[3] Bombelli L, Lee J, Meyer D and Sorkin R D 1988 *Phys. Rev. Lett.* **60** 656
[4] Hawking S W, King A R and McCarthy P J 1976 *J. Math. Phys.* **17** 174–181
[5] Malament D B 1977 *J. Math. Phys.* **18** 1399–1404
[6] Levichev A V 1987 *Soviet Math. Dokl.* **35** 452–455
[7] Dowker F, Henson J and Sorkin R D 2004 *Mod. Phys. Lett.* A **19** 1829–1840 (*Preprint* arXiv:gr-qc/0311055)
[8] Sorkin R 1986 *Ann. Phys.* **168** 119–147
[9] Philpott L, Dowker F and Sorkin R D 2009 *Phys. Rev.* D **79** 124047 (*Preprint* arXiv:0810.5591)
[10] Philpott L 2010 *Class. Qu. Grav.* **27** 042001 (*Preprint* arXiv:0911.5595)
[11] Henson J 2006 *Approaches to Quantum Gravity: Towards a New Understanding of Space and Time* ed Oriti D (Cambridge, UK: Cambridge University Press) (*Preprint* arXiv:gr-qc/0601121)
[12] Sorkin R D 2006 *Approaches to Quantum Gravity: Towards a New Understanding of Space and Time* ed D Oriti (Cambridge, UK: Cambridge University Press) (*Preprint* arXiv:gr-qc/0703099)
[13] Benincasa D M T and Dowker F 2010 *Phys. Rev. Lett.* **104** 181301 (*Preprint* arXiv:1001.2725)

Scalar Field Theory on a Causal Set in Histories form

Rafael D Sorkin

Perimeter Institute, 31 Caroline Street North, Waterloo ON, N2L 2Y5 Canada
Department of Physics, Syracuse University, Syracuse, NY 13244-1130, U.S.A.

E-mail: `rsorkin@perimeterinstitute.ca`

Abstract. We recast into histories-based form a quantum field theory defined earlier in operator language for a free scalar field on a background causal set. The resulting decoherence functional resembles that of the continuum theory but the counterpart of the d'Alembertian operator $\Box$ is nonlocal and a generalized inverse of the discrete retarded Green function. We comment on the significance of this and we also suggest how to include interactions.

1. Introduction

At this stage in the development of fundamental physics one does not know whether anything like a quantum field will turn out to exist at the Planck scale, or whether on the other hand quantum field theory will come to be seen as a relatively low energy description of completely different, more fundamental "degrees of freedom". Once one abandons the continuum, a spinor or a vector is no longer a particularly natural object, and in the context of causal set theory further difficulties spring from the radical nonlocality inherent in a discrete Lorentzian causal structure. Such difficulties suggest that the type of quantum field theory embodied in the so called standard model will not survive at a more fundamental level.

Nevertheless, any attempt to push quantum field theory down to a more fundamental level seems bound to be illuminating, and as it happens, we now possess a theory that describes a free scalar field propagating in a fixed, background causal set. [1]

On the basis of this theory, one can hope to understand better the influence of discreteness on such phenomena as Hawking radiation or the propagation of wave packets over great distances. As formulated so far, however, the theory in question speaks the language of operators and state-vectors. The field is described by attaching to each element x of the causet C an operator $\phi(x)$ acting in a Hilbert space built up from a "vacuum state" by the action of creation operators a^*. Although all this is very familiar, one might wish also for a formulation based on "histories" (a history in this case being a function $\phi : C \to \mathbf{R}$). [2]

Indeed, some of us regard histories-based formulations of quantum theories as more basic and more satisfactory than operator formulations, both for the purposes of quantum gravity and for the sake of philosophical understanding. How, for example, might one expect a dynamical theory of causal sets to look? The only way open to a "law of motion" for causets has so far seemed to pass through something like a path integral (or more precisely a decoherence functional),

[1] For background on causal sets see [1], [2], [3], [4], [5], [6].

[2] Sometimes people say "classical history" to emphasize that $\phi(x)$ is a c-number and not an operator. But I will avoid that usage here, in order to avoid any implication that ϕ must obey the classical equations of motion.

and one might say the same about quantum gravity in general. Even with the causet limited to the role of "arena", the knowledge of how to write down a path integral for a scalar field on a background causet, would allow the concept of "anhomomorphic coevent" to brought to bear on the question of what sort of reality quantum field theory is in fact describing [7]. In a more practical vein, one encounters the key question of how to generalize from a free field to an interacting one, something which so far has not been possible with the operator formulation. As we will see, a path integral formulation can suggest a very definite answer to this question.

Of equal practical importance is the question of the classical limit. Is it possible in this limit to express the equations of motion of a scalar field in such a manner that solutions could be built up iteratively given the values of the field in a sufficiently great initial portion of the causet? If it were, then computer simulations of wave packet propagation through the causet would become possible. On the other hand, if the relevant field equation were to involve reference to the distant future, such simulations would become much more difficult, or even impossible. In fact an equation of motion of exactly the required, retarded sort has been proposed [8], but in what sense can we claim that it represents the classical limit of our discrete quantum field theory?

In this connection, the following difficulty appears to arise [9]. Suppose that the dynamics of the scalar field is to be defined through an action-integral (or sum in the case of the causet), and that the integrand at point x depends on the field in the entire past of x. When one varies $\phi(x)$ in such an expression, the effect will be felt everywhere in the future of x. The resulting Euler-Lagrange equations will thus relate $\phi(x)$ not only to the past but to the future. How then could retarded equations of motion result? We will return to this issue in sections 4 and 5.

2. Review of the continuum theory in operator form

Nothing could be more familiar than a free scalar field in Minkowski spacetime together with its vacuum-state and the concomitant Fock representation of the field operators. But the way in which these things are ordinarily introduced, starting from the Klein-Gordon field equations and the equal-time commutation relations, is not quite suitable for transposition to the context of a discontinuous structure like the causal set which not only does away with continuous time, but also resists any notion of "Cauchy data at a moment of time". [3] Nor does the normally crucial distinction between positive and negative frequency make much sense in a causal set, which insofar as it resembles a spacetime at all, resembles much more a spacetime with curvature than one without it. For such reasons, it has proven more fruitful to rest the derivation of the scalar quantum field theory as much as possible on the commutation relations alone, or more precisely on the retarded Green function, in a way that I will review here [10].

By way of preparation, let us first review the theory of a "gaussian" noninteracting (massive or massless) real scalar field ϕ in flat spacetime. [4]

2.1. *the usual story*

As this subject is commonly introduced, the input to the theory comprises first of all the operator equations of motion in the form of the Klein-Gordon equation, $(\Box - m^2)\phi(x) = 0$. Supplementing these equations (plural since there is one for each x) are the canonical commutation relations,

[3] A causal set does admit the notion of *slice* as maximal antichain Σ, but such a subset unfortunately is not equipped to fill the role of an equal-time hypersurface, or more generally of a Cauchy surface in a globally hyperbolic spacetime, the most important difference being that the great majority of causal *links* will "pass thru" Σ without meeting it at all; that is, there will exist pairs of causet elements x, y such that x is to the past of Σ, y to its future, and yet neither Σ nor any other part of the causet contains an element causally intermediate between x and y. In this sense "information can pass through Σ without registering on it", the impossibility of which is precisely what characterizes a Cauchy surface in the continuum.

[4] with a metric η_{ab} of signature $(- + ++)$ and with $\Box = \eta^{ab}\partial_a\partial_b$.

expressed at a fixed moment of time as $[\phi(x), \partial\phi(y)/\partial t] = i\delta^{(3)}(x-y)$. One then expands ϕ in terms of (four-dimensional) plane waves and identifies the (suitably normalized) coefficients of the positive-frequency waves with annihilation operators $a(k)$. Introducing, finally, the vacuum state $|0\rangle$ via the conditions $a(k)|0\rangle = 0$, one obtains a concrete family of field operators $\phi(x)$ acting in a Hilbert space spanned by states of the form $a^*(k_1)^{n_1} a^*(k_2)^{n_2} a^*(k_3)^{n_3} \cdots |0\rangle$. From an algebraic standpoint, the equations of motion together with the commutation relations define an abstract *-algebra generated by symbols $\phi(x)$, while the choice of vacuum induces an irreducible representation of this algebra in a particular sort of Hilbert space. The full input to the theory is thus the field equations, the commutation relations, and the choice of vacuum (although it is partly a matter of opinion whether or not a change of representation or of vacuum should be regarded as changing the theory per se).

For our purposes, the most important thing to notice is the need for consistency between the field equations and the commutation relations. In the development we have just called to mind, this is ensured by the fact that the latter are invariant under the time-evolution generated by the former, but we can appreciate better what is going on if we express the commutation relations in a manifestly covariant form, the so called Peierls form. (The Lorentz invariance of the theory then follows trivially.) To that end, let $G(x,y) = G^{ret}(x,y)$ be the retarded Green function belonging to the wave-operator $\Box - m^2$:

$$(\Box - m^2)G^{ret}(x,y) = \delta^{(4)}(x-y) ,$$

where G is required to vanish unless $x \succ y$, meaning that y is within or on the past lightcone of x. Further define

$$\Delta(x,y) = G(x,y) - G(y,x) , \tag{1}$$

which is just the difference between the retarded and advanced Green functions, since (as in fact holds in any globally hyperbolic spacetime) the advanced Green function coincides with the "transpose" of the retarded one. The commutation relations then assume the remarkably simple form,

$$[\phi(x), \phi(y)] = i\Delta(x,y) .$$

If to this equation one applies the wave operator $(\Box - m^2)$ (acting on x), one trivially obtains zero on the left hand side. The required consistency then follows from the fact that the right hand side also vanishes because Δ is the difference of two Green functions for $(\Box - m^2)$. Notice here that the last three equations all make perfect sense, and are valid, in any globally hyperbolic spacetime. The resulting definition of a quantum field theory accordingly goes through as well, except that something is needed to replace the notion of positive frequency if one wishes to define a vacuum state.

Now that we have a field theory in operator form, the next question is how to render it into histories form. For this there is a standard procedure that I won't review here, because we will meet it again in detail when we come to the case of the causal set. Instead, let me just quote the resulting decoherence functional, which by definition attaches a complex amplitude to every *pair* of spacetime histories of the scalar field. (A pair of histories can be called a "Schwinger history" in honor of the so called Schwinger-Kel'dysh version of the path integral, which operates precisely with such pairs.) To avert confusion between ϕ as a ("classical") history and ϕ as an operator-valued field, let me for now use the symbols ξ and $\bar{\xi}$ to represent histories. The decoherence functional $D(\xi, \bar{\xi})$ is then given formally by the expression

$$\exp\{iS(\xi)\} \, \exp\{iS(\bar{\xi})\}^* \, \delta(\xi|\Sigma_T, \bar{\xi}|\Sigma_T) \tag{2}$$

where $S(\xi) = \frac{1}{2}\int(-(\nabla\xi)^2 - m^2\xi^2)$ plus an implicit further term that incorporates the effect of the "initial state", in this case the vacuum. Also Σ_T is a spacelike future boundary at which the spacetime has been truncated and the δ-function involving it has the effect of forcing the otherwise independent histories ξ and $\bar{\xi}$ to share the same restriction to this final boundary. From this, the decoherence functional extended to sets of histories and the corresponding quantum measure can be obtained in the usual way [11].

For future reference let us also define here the so called Wightman or "two-point" function given by the expectation of $\phi(x)\phi(y)$ operators in the vacuum state:

$$W(x,y) = \langle \phi(x)\phi(y) \rangle \tag{3}$$

and let us note also that in the vacuum

$$W(x) := \langle \phi(x) \rangle = 0\ . \tag{4}$$

Because the vacuum is a "gaussian state" the entire field theory is effectively encapsulated in these two equations, a fact we will use heavily in our construction of the decoherence functional in the causet case.

2.2. a different starting point

The logic of the "usual story" that we have just rehearsed is simplest if we present the commutation relations in their Peierls form. We might then summarize in part the steps we followed as: wave operator $\to G^{ret} \to \Delta \to [\phi, \phi]$. But consistency between the wave-operator, $\Box - m^2$, and the consequent commutation relations required, as we saw, that the transpose of the retarded Green function also be a Green function for the same wave operator. Since no such relation is known in the causet case, it turns out that one can get farther by changing the starting point slightly from the equations of motion to the retarded Green function G^{ret}. From it we can derive the commutators in just the same way. Indeed, one can pass freely from G^{ret} to Δ and back by means of equation (1) and the inverse relation $G^{ret}(x,y) = \Delta(x,y)$, which holds when $x \succ y$. (In equation form, $G^{ret}(x,y) = \theta(x,y)\Delta(x,y)$ where θ is the "covariant Heaviside function".)

But how to continue without the aid of any equations of motion? The surprising answer discovered implicitly in [10] is that one can pass directly from the commutator function Δ to the two-point function $W(x,y)$ by thinking of $\Delta(x,y)$ as a real skew matrix Δ^{xy} and forming the positive part of the corresponding Hermitian operator $i\Delta$. Our new logic will therefore go as follows: $G^{ret} \to \Delta \to W$. It would be difficult to imagine anything much simpler!

Notice in this connection that W always is, in fact, a positive (semidefinite) operator, as follows immediately from the fact that $v^* W v \geq 0$ for any complex vector v. Indeed, positivity of the Hilbert space inner product yields for any such v_x , $\sum (v_x)^* W^{xy} v_y = \sum (v_x)^* \langle \phi^x \phi^y \rangle v_y = \langle X^\dagger X \rangle = ||X|0\rangle||^2 \geq 0$, where $X = \sum v_x \phi^x$. It's equally true that $\overline{W}$, the complex conjugate matrix of W, is positive semidefinite and that $i\Delta = W - \overline{W}$, as follows from: $(W - \overline{W})^{xy} = \langle \phi^x \phi^y \rangle - \overline{\langle \phi^x \phi^y \rangle} = \langle \phi^x \phi^y \rangle - \langle \phi^y \phi^x \rangle = \langle [\phi^x, \phi^y] \rangle = \langle i\Delta^{xy} \rangle = i\Delta^{xy}$. Thus, $i\Delta$ is always the difference of two positive matrices W and $\overline{W}$. What our prescription for W is adding to this fact is that these operators have orthogonal support:

$$W\overline{W} = \overline{W}W = 0\ . \tag{5}$$

One might view this as a kind of "ground-state condition" imposed on W beyond what follows automatically from the fact that it is the two-point function of a selfadjoint operator ϕ^x. From (5) follows immediately a simple matrix equation giving W in terms of Δ (and therefore in terms of G):

$$W = 1/2(i\Delta + \sqrt{-\Delta^2})\ . \tag{6}$$

In practice one would often compute $\sqrt{-\Delta^2}$ by diagonalizing Δ. Once this was done, one would obtain W just by expanding $i\Delta$ in terms of its eigenvectors and retaining the terms with positive eigenvalues, and one could regard this as the practical meaning of our prescription for obtaining W from Δ.

For future reference let me introduce the notation $R = \frac{1}{2}\sqrt{-\Delta^2}$, a manifestly real and positive matrix. Comparison with (6) then shows that R is just the real part of W, whose decomposition into real and imaginary parts is thus given by

$$W = R + i\Delta/2 \; . \tag{7}$$

It's worth noticing here that the definitions we have given yield a free field theory with a distinguished "ground state" for any globally hyperbolic spacetime or region of spacetime. (Modulo, however, certain technical questions about convergence that arise when the region's volume is infinite.) Notice also that nowhere do these definitions refer to any notion of positive frequency. In Minkowski spacetime, the new prescription can easily be seen to reproduce the usual vacuum (it more or less had to by Lorentz invariance). In this sense it generalizes the Minkowski vacuum to the case of arbitrary curvature.

Perhaps also it's also worth mentioning a possible further generalization that would tie W less closely to G. In order to reproduce correctly the commutation relations, it is necessary only that the imaginary part of W coincide with $\Delta/2$. It is also necessary for mathematical consistency that W be positive, and perhaps other conditions are needed as well, but it is clearly not necessary that W simply be the positive part of $i\Delta$. One could thus consider relaxing condition (5) as well as the assumption that $\langle\phi\rangle = 0$. As will be seen, neither generalization would require major changes to our discussion below of the decoherence functional on a causal set.

3. Scalar field theory on a causet in operator form

Let us now leave the continuum behind and turn to the case of a causal set. Henceforth our discussion will be in the context of a fixed, finite causal set C, and it will assume further that a fixed "retarded Green function" has been adopted, which I'll write as either $G(x, y)$ or G_{xy} , where x and y vary over the elements of C. In some of the most important cases a convincing candidate for G is known and has been tested in practice to a greater or lesser extent. These cases include zero mass in a causet intended to approximate an arbitrary 2-dimensional spacetime, as well as arbitrary mass in a causet intended to approximate a flat spacetime of dimension either 2 or 4. (See [12].) In addition, the "retarded d'Alembertians" of [8] and [13] could be inverted to produce candidate Green functions in more general causets, but this type of prescription has not yet been put to a test.

Starting from G and following the same steps as delineated above yields then a scalar field theory for the given causet, complete with operators $\phi(x)$, vacuum-state $|0\rangle$, and Wightman function $W(x, y) = \langle\phi(x)\phi(y)\rangle$, where all expectations are taken with respect to the state $|0\rangle$.

As we have already noted, the fact that this theory-cum-state is in some sense gaussian means that all its consequences can be recovered starting from W alone. It might be interesting to tease out better in exactly what sense the word "gaussian" can be applied to our causet theory. Having no Cauchy surfaces, it also has no straightforward Schrödinger representation, and therefore no vacuum wave function whose exponential form could be held up as the meaning of the word. On the other hand, the fact that the ϕ^x are linear combinations of raising and lowering operators implies that Wick's theorem will work in the expected manner, and this is all that we will need. Indeed, we will need only the following consequence of Wick's theorem, where Φ stands for any linear combination of the field operators $\phi(x)$.

$$\langle \exp\{\Phi\}\rangle = \exp\left\{\frac{\langle\Phi\Phi\rangle}{2}\right\} \tag{8}$$

In view of equation (20) of the appendix, the following calculation demonstrates this identity.

$$\langle \exp\{\Phi\}\rangle = \sum_n \frac{\langle \Phi^n \rangle}{n!} = \sum_n \frac{\langle \Phi^{2n} \rangle}{(2n)!} = \sum_n \frac{(2n-1)!!\langle \Phi\Phi \rangle^n}{(2n)!} = \sum_n \frac{1}{n!}\frac{\langle \Phi\Phi \rangle}{2}^n$$

This seems a good place to mention a way of thinking about W that is more geometrical in nature than the characterizations we have encountered so far and that involves only real, as opposed to complex, vector spaces. Namely W can be represented geometrically as a family of mutually orthogonal, "axially weighted 2-planes" in the real vector space $\mathbf{R}^N$, where N is the cardinality of the underlying causet, these being a family of pairs (ϖ, λ) where ϖ is an oriented 2-plane in $\mathbf{R}^N$ and λ is a strictly positive real number corresponding to one of the eigenvalues of $i\Delta$. This one can see by splitting the eigenvectors of $i\Delta$ into their real and imaginary parts, or equivalently by decomposing Δ as a sum of terms $a \wedge b$ with $a, b \in \mathbf{R}^N$ being orthogonal vectors. The latter vectors then provide the basis of a singular-value decomposition of the real matrix Δ. Thanks to these relationships, one can construct W directly from Δ by forming the singular value decomposition of the latter, thereby economizing — in computer simulations — on memory and CPU-time, since only real numbers need be involved.

We are almost ready to turn our efforts to producing a histories version of our theory, but first, perhaps a simple example of the scheme $G \to \Delta \to W$ would be in order. Consider then a causet of only two elements, $e_0 \prec e_1$, making up a 2-chain. (Given the interpretation of W as a collection of weighted 2-planes, we could in this very simple case pass immediately to a unique W without bothering with G and Δ, but that would not illustrate the general situation.) That G is retarded means precisely that $G^{01} = 0$, but since only off-diagonal terms survive the antisymmetrization in (1), we might as well suppose that the diagonal of G vanishes as well. Up to sign and normalization, G is then unique:

$$G = \begin{pmatrix} 0 & 0 \\ 1 & 0 \end{pmatrix},$$

from which there result immediately $\Delta = \begin{pmatrix} 0 & -1 \\ 1 & 0 \end{pmatrix}$ and

$$i\Delta = \begin{pmatrix} 0 & -i \\ i & 0 \end{pmatrix} = \sigma_2$$

(the Pauli matrix). Substituting this into (6) then furnishes W as

$$W = \frac{1}{2}(1 + \sigma_2) = \frac{1}{2}\begin{pmatrix} 1 & -i \\ i & 1 \end{pmatrix},$$

i.e. $2R$ in (7) is simply the identity matrix. We could also have arrived at the same result by diagonalizing $i\Delta$ explicitly as $\sigma_2 = uu^\dagger - \overline{u}\overline{u}^\dagger$, where $u = 1/\sqrt{2}\begin{pmatrix} 1 \\ i \end{pmatrix}$, and then discarding the negative term. Notice finally that W correctly satisfies the "ground-state condition" (5). Given that W is hermitian, (5) is just the assertion that $W\widetilde{W} = 0$, tilde denoting the matrix transpose. Alternatively expressed, this says that the dot-product (the bilinear one, not the hermitian one) of any row (resp. column) with itself or with any other row must vanish, and this is true by inspection.

4. Scalar field theory on a causet in histories form

Henceforth it will be convenient to redefine $G = G^{jk}$ to be the retarded Green function *with its diagonal set to zero.* Since Δ is G minus its transpose, it is unaffected by this modification, and since W was defined through Δ, it is not affected either. Equivalently, we could take W as our starting point and then simply define G to be twice the retarded half of the imaginary part of W. Either way its clear that for purposes of this paper, there's no harm in dropping diag(G). But in a larger context this change would need to be borne in mind. For example, if we had obtained G by inverting some retarded D'alembertian operator for the causet, then we could not recover the latter from G without restoring its diagonal.

By a histories-based formulation of a field theory, one means (for a real scalar field) a formulation that works directly with field configurations $\phi : C \to \mathbf{R}$ ("histories"), and avoids reference to field operators or state-vectors, except possibly as auxiliary technical devices [14]. In practical terms, this means a path-integral formulation, although one might hope that more general frameworks will one day be available. [5] In order that a framework based on path-integrals be self-contained, moreover, it seems necessary to assign amplitudes to Schwinger histories rather than to individual histories, as the more familiar path-integral does. That is to say that one needs to express the dynamics in terms of a *decoherence functional* D, from which the corresponding quantal measure [15] can then be computed. Since the definitions of these things have been presented many times in the literature, I'll not repeat them here. Rather, I'll begin by writing down the expression (an expectation of a product of projectors) that one needs to evaluate in order to recover the decoherence functional from a theory expressed in the operator language.

In line with the definitions laid down above, let ξ and $\overline{\xi}$ be two (completely independent) histories, each being specified by a list of real numbers ξ^x, one for each element x of the causet. The complex number $D(\xi, \overline{\xi})$ is then given by the equation

$$D(\xi,\overline{\xi}) = \left\langle \delta(\phi^1 - \overline{\xi}^1)\delta(\phi^2 - \overline{\xi}^2)\cdots\delta(\phi^N - \overline{\xi}^N)\delta(\phi^N - \xi^N)\cdots\delta(\phi^2 - \xi^2)\delta(\phi^1 - \xi^1)\right\rangle \quad (9)$$

Here $\phi^j = \phi(x_j)$, $\xi^j = \xi(x_j)$, etc, where $j = 1\cdots N$ is any *natural labeling* of C. In other words the elements $x \in C$ of the causet must be labeled so that no element with a smaller label ever temporally follows an element with a higher label: $x_j \prec x_k \Rightarrow j < k$. Such a labeling always exists, and any two natural labelings are guaranteed to produce the same result for $D(\xi, \overline{\xi})$ because $\Delta(x, y)$, and hence $[\phi(x), \phi(y)]$, automatically vanishes when x and y are causally unrelated ("spacelike"). Notice that D is normalized such that $\int d^N\xi d^N\overline{\xi} D(\xi, \overline{\xi}) = 1$.

Re-expressing the δ-functions as integrals puts (9) into the form

$$\int_{-\infty}^{\infty} \frac{d\lambda_1}{2\pi}\frac{d\lambda_2}{2\pi}\cdots\frac{d\overline{\lambda}_2}{2\pi}\frac{d\overline{\lambda}_1}{2\pi} e^{-i\overline{\lambda}_1\overline{\xi}^1 - i\overline{\lambda}_2\overline{\xi}^2 \cdots - i\lambda_2\xi^2 - i\lambda_1\xi^1}\left\langle e^{i\overline{\lambda}_1\phi^1}e^{i\overline{\lambda}_2\phi^2}\cdots e^{i\lambda_2\phi^2}e^{i\lambda_1\phi^1}\right\rangle, \quad (10)$$

which represents $D(\xi, \overline{\xi})$ as the Fourier transform of what one might call its "non-commutative characteristic function" in analogy with the concept of characteristic function that figures in ordinary probability theory. (Here again the real numbers λ_j and $\overline{\lambda}_j$ are independent parameters of the Fourier transform. The bar is not being used to denote complex conjugation.)

Before plunging into the evaluation of this expression for the decoherence functional, let us evaluate the simpler expression which results from integrating out the $\overline{\xi}$ and all but one of

[5] perhaps obtained by freeing the concept of preclusion from its dependence on the notion of quantum measure and measure-zero, or perhaps even by freeing anhomomorphic inference from its dependence on preclusion, cf. [7].

the variables $\xi = \xi(x)$ in $D(\xi, \overline{\xi})$, which mathematically is analogous to a marginal probability density for the remaining variable $\phi(x)$. After doing the integrals, we are left simply with

$$\int \frac{d\lambda}{2\pi} \exp\{-i\lambda\xi\} \langle \exp\{i\lambda\phi(x)\}\rangle ,$$

which in light of the identity (8) turns into

$$\int \frac{d\lambda}{2\pi} \exp\left\{-w^2\lambda^2/2 - i\lambda\xi\right\} = \frac{1}{\sqrt{2\pi w}} e^{\frac{-\xi^2}{2w}} ,$$

where I have written ξ for $\xi(x)$ and w for $\langle\phi(x)\phi(x)\rangle$. Unsurprisingly for a free field, we just obtain a gaussian.

The next simplest case, which already illustrates most of the complications, results from integrating out all but two of the field-values, and yields an integral of the form

$$\int \frac{d\alpha}{2\pi}\frac{d\beta}{2\pi} \exp\{-i\alpha\xi - i\beta\eta\} \langle \exp\{i\alpha\phi(x)\} \exp\{i\beta\phi(y)\}\rangle .$$

We can again avail ourselves of (8), but only after bringing to bear the well known identity

$$e^A e^B = e^{A+B} e^{\frac{1}{2}[A,B]} , \tag{11}$$

which holds whenever the commutator $[A, B]$ is a c-number. In virtue of (11) we find

$$\exp\{i\alpha\phi(x)\}\exp\{i\beta\phi(y)\} = \exp\left\{i\alpha\phi(x) + i\beta\phi(y) - \frac{1}{2}\alpha\beta[\phi(x),\phi(y)]\right\}$$

which can also be written as $\exp\left\{i\alpha\phi(x) + i\beta\phi(y) - \frac{1}{2}\alpha\beta(W^{xy} - W^{yx})\right\}$ because

$$[\phi(x),\phi(y)] = \phi(x)\phi(y) - \phi(y)\phi(x) = \langle\phi(x)\phi(y) - \phi(y)\phi(x)\rangle = W^{xy} - W^{yx} ,$$

the original commutator being a c-number. Hence,

$$\begin{aligned}
&\langle \exp\{i\alpha\phi(x)\}\exp\{i\beta\phi(y)\}\rangle \\
&= \exp\left\{-\frac{1}{2}\left(\langle(\alpha\phi(x) + \beta\phi(y))^2\rangle + \alpha\beta(W^{xy} - W^{yx})\right)\right\} \\
&= \exp\left\{-\frac{1}{2}\left(\alpha^2 W^{xx} + \beta^2 W^{yy} + \alpha\beta W^{xy} + \alpha\beta W^{yx} + \alpha\beta W^{xy} - \alpha\beta W^{yx}\right)\right\} \\
&= \exp\left\{-\frac{1}{2}\left(\alpha^2 W^{xx} + \beta^2 W^{yy} + 2\alpha\beta W^{xy}\right)\right\} \\
&= \exp\left\{-\frac{1}{2}\left(\begin{pmatrix}\alpha & \beta\end{pmatrix}\begin{pmatrix}W^{xx} & W^{xy}\\ W^{xy} & W^{yy}\end{pmatrix}\begin{pmatrix}\alpha\\ \beta\end{pmatrix}\right)\right\}
\end{aligned}$$

The important thing to notice here is that the order of the indices in W^{xy} copies the order of x and y in the original factors.

Following this pattern, it is straightforward to evaluate the quantity within angle brackets in (10). The result is

$$\exp\left\{-\frac{1}{2}\left(\overrightarrow{W}^{jk}\overline{\lambda}_j\overline{\lambda}_k + \overleftarrow{W}^{jk}\lambda_j\lambda_k + 2W^{jk}\overline{\lambda}_j\lambda_k\right)\right\} \tag{12}$$

where $\overrightarrow{W}^{jk}$ denotes W^{jk} if $j \leq k$ and W^{kj} if $k \leq j$, for example $\overrightarrow{W}^{23} = \overrightarrow{W}^{32} = W^{23}$; and $\overleftarrow{W}$ follows the reversed convention. (That the arrows in the first and second terms point in opposite directions merely reflects the opposite ordering of the causet elements in the two halves of the Schwinger history.) Notice in connection with (12) that the matrices $\overleftarrow{W}$ and $\overrightarrow{W}$ are complex conjugates of each other, and recall also that W itself is positive semidefinite.

Let us observe in passing that integrating out the $\overline{\xi}$ variables from $D(\xi, \overline{\xi})$ would have the effect of setting $\overline{\lambda} = 0$ in (12), leaving in the exponent only $-\frac{1}{2}\overleftarrow{W}^{jk}\lambda_j\lambda_k$, which involves only the time-ordered Wightman function, or equivalently the "Feynman Green function".

Now since $D(\xi, \overline{\xi})$ is nothing more than the Fourier transform of (12), its evaluation would be relatively carefree, were it not for the fact that the quadratic form in (12) is not invertible. This will prevent $D(\xi, \overline{\xi})$ from assuming the pure gaussian form that it would otherwise have had, and that one might naively have expected from a free field theory. Instead, it will be the product of a gaussian with a δ-function that will enforce certain "constraints" on its arguments.

The further manipulations to be done will become more transparent if at this point we switch to a matrix notation, and I'll continue to use a tilde to denote transpose. Then for example, we can re-express (12) as $\exp\left\{-\frac{1}{2}Q\right\}$ where

$$Q = \lambda \overleftarrow{W} \lambda + \overline{\lambda} \overrightarrow{W} \overline{\lambda} + 2\overline{\lambda} W \lambda \ .$$

Let us now express Q directly in terms of G and R using that, according to its definition, $W = R + i\Delta/2 = R + (i/2)(G - \widetilde{G})$, and noting also that $\overleftarrow{R} = \overrightarrow{R} = R$ and $\overleftarrow{\Delta} = -\overrightarrow{\Delta} = G + \widetilde{G}$, whence

$$\overleftarrow{W} = \overleftarrow{R + i\Delta/2} = R + i\overleftarrow{\Delta}/2 = R + (i/2)(G + \widetilde{G})$$

whence similarly $\overrightarrow{W} = R - (i/2)(G + \widetilde{G})$. Expanding $\overleftarrow{W}$ this way yields

$$\lambda \overleftarrow{W} \lambda = \lambda(R + (i/2)(G + \widetilde{G}))\lambda = \lambda R \lambda + i\lambda G \lambda$$

with similar expansions for $\overline{\lambda}\overrightarrow{W}\overline{\lambda}$ and $2\overline{\lambda}W\lambda$. Adding these together then reveals that Q is expressed most simply in terms of sum and difference variables for λ and $\overline{\lambda}$.

$$Q = (\lambda + \overline{\lambda})R(\lambda + \overline{\lambda}) + i(\lambda + \overline{\lambda})G(\lambda - \overline{\lambda}) \ .$$

Its fourier transform can then be expressed in terms of the corresponding sum and difference variables for ξ and $\overline{\xi}$. To that end, let us define

$$K = \lambda + \overline{\lambda} \ , \ k = \frac{1}{2}(\lambda - \overline{\lambda}) \ , \quad \phi = \frac{1}{2}(\xi + \overline{\xi}) \ , \ \varphi = \xi - \overline{\xi}$$

Here ϕ, the mean of the two halves of the Schwinger history, is what is sometimes called "the classical field" because $\xi = \overline{\xi}$ in the classical limit, while φ represents in some sense the deviation from classicality. [6]

With these definitions, $\lambda\xi + \overline{\lambda\xi} = K\phi + k\varphi$, and the fourier transform yielding $D(\xi, \overline{\xi})$ turns into

$$\int d^N K \, d^N k \exp\left\{-iK\phi - ik\varphi - \frac{1}{2}KRK - iKGk\right\} \tag{13}$$

where I've left out an overall normalization which needn't be carried along because it can be recovered at the end from the condition $\int d^N\xi \, d^N\overline{\xi} \, D(\xi, \overline{\xi}) = D(\Omega, \Omega) = 1$.

Evidently the conjugate variable k occurs only linearly in the exponent of (13), in the combination $k\varphi + KGk = k(\widetilde{G}K + \varphi)$; the dk integral therefore produces the δ-function I

[6] Please don't confuse ϕ as defined here with the same symbol used earlier for the field *operator*.

mentioned earlier, which constrains φ to be in the image of $\widetilde{G}$, the "advanced Green function". Equivalently, it constrains φ to be orthogonal to $\ker(G) = \{K \mid GK = 0\}$. We now have

$$D(\xi, \overline{\xi}) = (const) \int d^N K \; \delta(\widetilde{G}K + \varphi) \; \exp\left\{-\frac{1}{2}KRK - iK\phi\right\} . \tag{14}$$

When φ is in the image of $\widetilde{G}$, as required for D to be nonzero, (14) has the form of a gaussian integral over an affine subspace of $\mathbf{R}^N$, namely the space of solutions to the equation $\widetilde{G}K + \varphi = 0$. To convert (14) into an integral over a full vector space, let $I : V \to \mathbf{R}^N$ be the inclusion map of $V = \ker \widetilde{G}$ into $\mathbf{R}^N$ and let K_0 be any solution of the equation $\widetilde{G}K_0 = -\varphi$, so that the space over which we have to integrate is $K_0 + V$. (Concretely we can represent I by a matrix $I_j^{\&\alpha}$, where the vectors $I_j^{\&\alpha}$ for fixed α furnish a basis for V.) The substitution of $K_0 + K$ for K then yields our integral in the form

$$\begin{aligned}
&\int_{K \in K_0 + V} d^{N'} K \; \exp\left\{-\frac{1}{2}KRK - iK\phi\right\} \\
&= \int_{K \in V} d^{N'} K \; \exp\left\{-\frac{1}{2}(K + K_0)R(K + K_0) - i(K + K_0)\phi\right\} \\
&= \exp\left\{-\frac{1}{2}K_0RK_0 - iK_0\phi\right\} \int_{K \in V} d^{N'} K \; \exp\left\{-\frac{1}{2}KRK - K(RK_0 + i\phi)\right\}
\end{aligned}$$

where $N' = \dim \ker \widetilde{G}$ is the dimensionality of V. Then with the aid of the matrix I we obtain an integral over $\mathbf{R}^{N'}$ by putting Iv in place of K :

$$\exp\left\{-\frac{1}{2}K_0RK_0 - iK_0\phi\right\} \int d^{N'} v \; \exp\left\{-\frac{1}{2}v\widetilde{I}RIv - (K_0R + i\phi)Iv\right\} . \tag{15}$$

The result is now an ordinary gaussian integral over $\mathbf{R}^{N'}$ of the form $\int \exp\left\{-\frac{1}{2}vAv + Bv\right\}$ with $A = \widetilde{I}RI$ and $B = (K_0R + i\phi)I$. Under the assumption that A is invertible, this integral evaluates to (a constant multiple of) $\exp\left\{-\frac{1}{2}B(1/A)B\right\}$. When expanded out, the exponent here consists of three terms:

$$\phi I(\widetilde{I}RI)^{-1}\widetilde{I}\phi \; , \tag{16 – 1}$$

$$-iK_0\,(1 - RI(\widetilde{I}RI)^{-1}\widetilde{I})\,\phi \; , \tag{16 – 2}$$

$$-\frac{1}{2}K_0\,(R - RI(\widetilde{I}RI)^{-1}\widetilde{I}R)\,K_0 \; . \tag{16 – 3}$$

We now have the decoherence functional in hand, except that (16) is expressed in terms of the ambiguous constant K_0 rather than φ. For consistency, this apparent dependence on K_0 must be illusory, i.e. (16) must be independent of K_0 for fixed φ.

In considering (16), let us observe first of all that one would expect I to be "small" since one would expect $\dim \ker \widetilde{G}$ to be small compared to N, the number of elements in our causet. It can't vanish entirely, given that G is strictly retarded — now that we've removed its diagonal — but an analogy with the continuum might suggest that it would receive contributions only from the "boundary region" of the causet. To the extent that I actually can be neglected, (14) could have been evaluated without doing any integrations at all, as the δ-function therein sets K to $-\widetilde{G}^{-1}\varphi$, yielding $D = \exp\left\{-\frac{1}{2}\varphi(G^{-1}R\widetilde{G}^{-1})\varphi + i\varphi G^{-1}\phi\right\}$. Now a natural symbol to represent G^{-1} would be $\Box$, since in the continuum G is in some sense the inverse of $\Box$. In the present setting we can also appreciate that G^{-1} (if it actually existed) would be retarded too because G itself is retarded. One might then refer to $\Box = G^{-1}$ as a "retarded d'Alembertian" and its transpose $\widetilde{\Box}$ as an "advanced d'Alembertian". With this nomenclature we would have $D = \exp\left\{-\frac{1}{2}\varphi(\Box R\widetilde{\Box})\varphi + i\varphi\Box\phi\right\}$. Moreover, in the continuum, $\Box R = \Re(\Box W) = 0$, therefore one might also expect $-\frac{1}{2}\varphi(\Box R\widetilde{\Box})\varphi$ to be small in the causal set. In that case the whole of the decoherence functional would simplify to $\exp\{i\varphi\Box\phi\}$. Remarkably, this is exactly the decoherence functional of the continuum theory, as an integration by parts reveals, given that $\varphi = 0$ on the future boundary. More precisely, the continuum decoherence functional is this expression together with the δ-function in (2) that mandates the just-mentioned equality between ξ and $\overline{\xi}$ on the future boundary, cf. [11]. Such a δ-function is present implicitly in (14) and explicitly in (9), but I don't know how to extract it as such from a well-defined continuum limit of the discrete theory. We'll return to the constraints on φ shortly.

It is useful at this point to distinguish conceptually between the space $\mathbf{R}^N$ of field-configurations and the space $\mathbf{R}^N$ of configurations of the dual variables K, λ, etc. Since we have already employed the symbol Ω for the former, we might as well denote the latter by Ω^* , using a star to denote dual space. I'll also write V for $\mathbf{R}^{N'}$ considered as $\ker\widetilde{G}$ considered as a separate vector space. Then $I : V \to \Omega^*$, $\widetilde{I} : \Omega \to V^*$, $G : \Omega^* \to \Omega$, $\widetilde{G} : \Omega^* \to \Omega$, etc.

Let us return now to the full decoherence functional as given by (16). Its second member can be written as $-iK_0(1-\Pi)\phi$ if we define Π to be the operator $\Pi = RI(\widetilde{I}RI)^{-1}\widetilde{I}$. Here, of course, we continue to assume that $(\widetilde{I}RI)^{-1}$ exists. From its definition it is trivial to check that Π, though not hermitian, is a projection in the sense that $\Pi^2 = \Pi$; and also that

$$\widetilde{I}\,\Pi = \widetilde{I}\ , \quad \widetilde{\Pi}\,I = I\ . \tag{17}$$

Now the term by which $K_0 = -\widetilde{G}^{-1}\varphi$ is ambiguous is by definition Iv for arbitrary $v \in V$. Hence the ambiguity in $-K_0(1-\Pi)\phi$ is precisely $(Iv)(1-\Pi)\phi = v\widetilde{I}(1-\Pi)\phi = v(\widetilde{I}-\widetilde{I}\Pi)\phi$, which vanishes by (17). Consequently, it is justified to write (16-2) simply as $i(\widetilde{G}^{-1}\varphi)(1-\Pi)\phi = i\varphi G^{-1}(1-\Pi)\phi$. In view of these considerations, it is natural to define

$$\Box = G^{-1}(1-\Pi)\ , \tag{18}$$

meaning that $\Box$ solves the equation $G\Box = 1-\Pi$. We have just seen that $\Box\phi$ is unambiguous when contracted with any $\varphi \in \mathrm{im}(\widetilde{G})$. Thus $\Box$ itself is uniquely defined if we construe it as a map from Ω to $(\mathrm{im}\,\widetilde{G})^*$, the dual vectorspace to $(\mathrm{im}\,\widetilde{G})$: in symbols, $\Box : \Omega \to (\mathrm{im}\,\widetilde{G})^* \subseteq \Omega^*$. Regarded however as mapping Ω to all of Ω^*, $\Box$ is ambiguous by the addition of any linear operator with image in $(\mathrm{im}\,\widetilde{G})^0 = \ker G$. A nice idea at first sight would be to take advantage of this ambiguity to render $\Box$ fully retarded, but unfortunately that is not possible. Reasoning similar to the above also proves that $G\Box G = G$, that $\mathrm{im}\,\widetilde{\Pi} = \mathrm{im}\,I = \ker\widetilde{G}$, that $\ker\Pi = \mathrm{im}\,G$, and therefore that $\mathrm{im}(1-\Pi) = \mathrm{im}\,G$. Also that $\Pi R = R\widetilde{\Pi} = RI(\widetilde{I}RI)^{-1}\widetilde{I}R$.

So far, we have been occupied primarily with the analysis of (16-2), but we also need to deal briefly with (16-1) and (16-3). The former is complete as it stands, but the latter requires to be expressed in terms of φ, just as we did for (16-2). To that end, rewrite (16-3) as $-\frac{1}{2}K_0(1-\Pi)RK_0 = -\frac{1}{2}K_0(1-\Pi)^2RK_0 = -\frac{1}{2}K_0(1-\Pi)R(1-\widetilde{\Pi})K_0$ and observe that the ambiguity of adding a vector Iv to K_0 drops out just as before because $(1-\widetilde{\Pi})I = 0$. Continuing on, and replacing K_0 by $-\widetilde{G}^{-1}\varphi$, we obtain with the aid of equation (18), $-\frac{1}{2}\varphi G^{-1}(1-\Pi)R(1-\widetilde{\Pi})\widetilde{G}^{-1}\varphi = -\frac{1}{2}\varphi\Box(1-\Pi)R\widetilde{\Box}\varphi$, which can also be written as $-\frac{1}{2}(\widetilde{\Box}\varphi)R(\widetilde{\Box}\varphi)$. Collecting our three terms then yields in total

$$D(\xi,\overline{\xi}) = (constant)\,\exp\left\{i\varphi\Box\phi - \frac{1}{2}(\widetilde{I}\phi)(\widetilde{I}RI)^{-1}(\widetilde{I}\phi) - \frac{1}{2}(\widetilde{\Box}\varphi)R(\widetilde{\Box}\varphi)\right\} \tag{19}$$

In view of the rather intricate derivation we've just been through it seems worth recording here the much simpler expression that results from integrating out the $\bar{\xi}$ variables in $D(\xi,\bar{\xi})$ or simply from returning to (9) and deleting the factors involving $\bar{\xi}$. One sees immediately from (12) that — provided that $\overleftarrow{W}$ is invertible — the result will be

$$D(\xi,\Omega) = \int d^N\bar{\xi}\, D(\xi,\bar{\xi}) = (const)\,\exp\left\{\frac{i}{2}\xi\Box_F\xi\right\}$$

where I've written $\Box_F$ for $i\overleftarrow{W}^{-1}$, a notation that seems natural because $-i\overleftarrow{W}$ corresponds to the Feynman Green function of the continuum theory.

5. Consequences and questions

Perhaps the most important conclusion from the above work is that the equations of motion for the "classical field" ϕ are essentially a retarded non-local version of the continuum equations $\Box\phi = 0$. Equations of precisely this sort were proposed in [8], but until now there has seemed to be no sound reason for preferring retarded equations over, for example, the time-reverse or some combination of the two. Granted that retardation greatly facilitates computer solution, but that would be more an opportunistic reason than one of principle. Retardation also agrees better with the notion that physics in the classical regime, even if it were to be non-local, would still be "causal", but here we have reached a similar conclusion in a far more convincing manner!

But why should one refer to $\Box\phi = 0$ as the classical equations of motion? Expressed through the decoherence functional, classicality amounts to the vanishing of φ, i.e. to the equality of ξ with $\bar{\xi}$ (in which case both coincide with ϕ). But if we vary φ in (19) and then set $\varphi = 0$, the result is precisely $\Box\phi = 0$. Further support for this point of view comes from the fact, already alluded to, that the second and third terms in (19) seem likely to be relatively negligible, having no further significance than either initial-conditions or small corrections. In the case of the second term specifically (the term in ϕ), its character as a kind of initial condition or "state" can be perceived fairly clearly, because it depends on ϕ only through $\widetilde{I}\phi$, which is a combination of contractions of ϕ with vectors in $\ker\widetilde{G}$. But since $\widetilde{G}$ is "future-looking" it will annihilate any function with support on the minimal level of C. Plausibly then, its entire kernel is comprised of functions supported near this bottom level, whence $\widetilde{I}\phi$ would depend only on such initial values of ϕ.

Of course if we vary all the arguments in (19), we obtain more equations, and taken together they tell us more than just $\Box\phi = 0$. In fact they constrain ϕ itself to be zero (along with φ). This however only reflects the fact that our action (or rather decoherence functional) implicitly includes a specification of "initial conditions", something like an "initial state wave-function" in the continuum theory. In fact exactly the same thing happens in that case: a unique "classical" solution is picked out. In the present context it's unremarkable that the corresponding solution is $\phi = 0$, because our construction of the decoherence functional assumed early on that (4) was true.

We might also ask how an apparently time-symmetric scheme has given rise to a very time-asymmetric set of classical equations. To the extent that we can regard $\Box$ as retarded, the combination $i\varphi\Box\phi$ is entirely retarded in the "classical component" ϕ and advanced in the "non-classical component" φ. Just how this has come about seems mysterious, but logically the asymmetry had to originate in the order chosen for the factors in (9). Perhaps a careful tracing through of the steps leading from there to (19) would shed more light on the question.

Of course the operator $\Box$ is not fully retarded but only approximately so (if our plausibility arguments are valid). Should this worry us? I'm not sure, but one reason for equanimity at this point springs from the observation that a dynamics taking place in a frozen causal set can only be a partial reflection of full quantum gravity. Really the causet itself needs to be included in the decoherence functional. Not having done this we have restricted its future development in an unphysical manner, and thereby we have imposed a sort of future boundary condition. It would be no surprise if such a condition were to be reflected to some extent in the dynamics of the variables (namely the scalar field) that we have left free. (Whether we can be equally happy with an anti-causal equation for φ is another question however.)

A question might be raised here about our assumption that the matrix $\widetilde{I}RI$ is invertible. How well does this hold up in practice? Quite well if the few numerical experiments that I've done are indicative, but so far they have been limited in scope. An interesting exception, however, springs from the existence of so called non-Hegelian elements or pairs within the causal set C, viz. pairs of incomparable elements x and y which share the same relation to every third element $z \in C$. Such elements cannot be distinguished from each other by the causal structure, and in consequence the "difference function" defined by $f(x) = +1$, $f(y) = -1$, and $f(z) = 0$ otherwise, will be in the kernel of G, Δ, etc. In particular it will be in the kernel of $\widetilde{I}RI$, rendering the latter non-invertible. But the problem this might seem to raise solves itself. Indeed x and y are in some sense not two distinct elements at all, and the theory seems to know this, as one can see by returning to equation (13) or (14) and noticing that f is a degenerate direction for the quadratic form involved. Consequently the resulting decoherence functional will vanish unless $\xi(x) = \xi(y)$ and $\overline{\xi}(x) = \overline{\xi}(y)$. The theory thus lives in effect on the quotient causet C' obtained by identifying x with y, and one might just as well eliminate y (say) at the outset and continue in this manner until no further non-Hegelian elements remain.

Many more questions could be raised at this point, but let me just conclude with two of a more general nature, the first being this. All our work herein has presupposed a fixed, background causal set C with a finite number of elements. But in reality, one expects the causet to be growing and infinite, or at least potentially infinite toward the future. We cannot actually handle dynamical growth without setting up a full quantum gravity theory, but we can certainly imagine C being infinite toward the future. In that case our theory herein would correspond to only some initial portion of C (a so called partial stem) and this partial theory should be obtained from the full one by restriction. Conversely, if the theory is formulated initially on the stems, then the need arises for consistency between the partial theories belonging to the different stems of C. (From a histories point of view, almost all quantum theories are built up this way from partial theories which provide the measures of so-called cylinder events. See for example [16] or [17].) It is therefore an important question whether the theory herein can be, or can be adapted to be, a component of a coherent theory defined consistently on an untruncated causal set.

The other general question concerns interactions. The theory we have been working with herein is that of the causet counterpart of a free field, as manifested in the quadratic nature of (19). In its original operator form, however, the theory does not readily suggest a generalization to include interactions. A potential advantage of the histories formulation, then, is that it does suggest such a generalization. To incorporate a ϕ^4 interaction, for example, one need only include a multiple of $i\xi^4 - i\overline{\xi}^4$ into the exponent of (19). For the sake of consistency, one should check that the decoherence functional, thus modified, remains positive semidefinite, but this can be done. It would be interesting to compare the consequences of this generalized theory with those of its continuum counterpart.

Appendix. Proof of an identity used in the text

In proving equation (8) we used the following identity (20), which is a special case of Wick's theorem. Indeed, the coefficient $(n-1)!! = (n-1)(n-3)\cdots 5\cdot 3\cdot 1$ in (20) is easily seen to count the number of ways of pairing up the $2n$ factors of Φ in the expression $\langle\Phi^n\rangle = \langle\Phi\Phi\Phi\cdots\Phi\rangle$.

Here we prove this special case directly by a straightforward method.

$$\langle \Phi^{2n} \rangle = (2n-1)!!\ \langle \Phi\Phi \rangle^n \tag{20}$$

Suppose first that $\Phi = a + a^*$ with a normalized as usual so that $[a, a^*] = 1$. Also let $|0\rangle$ be the "vacuum" relative to a with $\langle \cdot \rangle = \langle 0| \cdot |0\rangle$. Noticing that $a|0\rangle = 0$ and $\langle 0|a^* = 0$, and that $[a, \Phi] = [a, a + a^*] = [a, a^*] = 1$, we have then

$$\langle \Phi^n \rangle = \langle (a + a^*)\Phi^{n-1} \rangle = \langle a\Phi^{n-1} \rangle = \langle [a, \Phi^{n-1}] \rangle = \langle (n-1)\Phi^{n-2} \rangle = (n-1)\langle \Phi^{n-2} \rangle \ ,$$

where we used the fact that $[a, \cdot]$ is a derivation (it obeys the Leibniz rule). Since $\langle \Phi^0 \rangle = 1$, it follows by induction that for even n, $\langle \Phi^n \rangle = (n-1)!!$. When n is odd $\langle \Phi^n \rangle$ of course vanishes, since an odd number of applications of a or a^* to the vacuum can never bring one back to the vacuum.

Finally, consider the more general situation where Φ is some linear combination of field operators $\phi(x)$. Because $\Phi = \Phi^*$ we can always write it in the form $\phi = A + A^*$, where A contains only the terms with lowering operators a. The proof then proceeds as before except that in place of $[a, a^*] = 1$ we have instead $[A, A^*] = \langle \Phi\Phi \rangle$, as we see from $\langle \Phi\Phi \rangle = \langle (A + A^*)(A + A^*) \rangle = \langle AA^* \rangle = \langle [A, A^*] \rangle = [A, A^*]$.

Acknowledgments
For very helpful discussion of, and feedback on, the histories formulation presented herein, I would like to thank Niayesh Afshordi and in addition the participants of the meeting on causal sets held December, 2009 at the Dublin Institute of Advanced Studies. Research at Perimeter Institute for Theoretical Physics is supported in part by the Government of Canada through NSERC and by the Province of Ontario through MRI.

References
[1] Luca Bombelli, Joohan Lee, David Meyer and Rafael D. Sorkin 1987 "Spacetime as a Causal Set" *Phys. Rev. Lett.* **59** 521-524

[2] Rafael D. Sorkin 2005 "Causal Sets: Discrete Gravity (Notes for the Valdivia Summer School)", in *Lectures on Quantum Gravity* (Series of the Centro De Estudios Científicos), proceedings of the Valdivia Summer School, held January 2002 in Valdivia, Chile, ed Andrés Gomberoff and Don Marolf (New York: Plenum, 2005) (*Preprint* gr-qc/0309009, http://www.pitp.ca/personal/rsorkin/some.papers/)

[3] Joe Henson 2006 "The causal set approach to quantum gravity" *Preprint* gr-qc/0601121. This is an extended version of a review published in in *Approaches to Quantum Gravity – Towards a new understanding of space and time*, ed Daniele Oriti (Cambridge: Cambridge University Press, 2009) (ISBN: 978-0-521-86045-1), pages 26-43

[4] Fay Dowker 2005 "Causal sets and the deep structure of Spacetime", in *100 Years of Relativity - Space-time Structure: Einstein and Beyond* ed Abhay Ashtekar (Singapore: World Scientific), to appear, *Preprint* gr-qc/0508109

[5] Petros Wallden 2010 "Causal Sets: Quantum gravity from a fundamentally discrete spacetime" *J. Phys.: Conf. Ser.* **222** 012053 (*Preprint* arXiv:1001.4041v1 [gr-qc])

[6] Sumati Surya 2011 "Directions in Causal Set Quantum Gravity" *Preprint* arXiv:1103.6272v1 [gr-qc]

[7] Rafael D. Sorkin 2010 "Logic is to the quantum as geometry is to gravity" in G.F.R. Ellis, J. Murugan and A. Weltman (eds) *Foundations of Space and Time* (Cambridge: Cambridge University Press), to appear, *Preprint* arXiv:1004.1226 [quant-ph], http://www.pitp.ca/personal/rsorkin/some.papers/
Rafael D. Sorkin 2007 "An exercise in 'anhomomorphic logic' " *J. Phys.: Conf. Ser. (JPCS)* **67** 012018, a special volume edited by L. Diosi, H.-T. Elze and G. Vitiello, and devoted to the Proceedings of the DICE-2006 meeting, held September 2006, in Piombino, Italia, (*Preprint* quant-ph/0703276, http://www.pitp.ca/personal/rsorkin/some.papers/)

[8] Rafael D. Sorkin 2007 "Does Locality Fail at Intermediate Length-Scales?" in *Approaches to Quantum Gravity – Towards a new understanding of space and time*, ed Daniele Oriti (Cambridge: Cambridge University Press, 2009) (ISBN: 978-0-521-86045-1), pages 26-43 (*Preprint* gr-qc/0703099, http://www.pitp.ca/personal/rsorkin/some.papers/)

[9] Don Marolf, private communication
[10] Steven Johnston 2009 "Feynman Propagator for a Free Scalar Field on a Causal Set" *Phys. Rev. Lett* **103** 180401 (*Preprint* arXiv:0909.0944 [hep-th])
[11] Xavier Martin, Denjoe O'Connor and Rafael D. Sorkin 2005 "The Random Walk in Generalized Quantum Theory" *Phys. Rev.* D **71** 024029 (*Preprint* gr-qc/0403085, http://www.pitp.ca/personal/rsorkin/some.papers/)
[12] Steven Johnston 2008 "Particle propagators on discrete spacetime" *Class. Quant. Grav.* **25** 202001 (*Preprint* arXiv:0806.3083 [gr-qc])
[13] Dionigi M. T. Benincasa and Fay Dowker 2010 "The Scalar Curvature of a Causal Set" *Phys. Rev. Lett.* **104** 181301 (*Preprint* arXiv:1001.2725)
[14] Rafael D. Sorkin 2007 "Quantum dynamics without the wave function" *J. Phys. A: Math. Theor.* **40** 3207-3221 (http://stacks.iop.org/1751-8121/40/3207) (*Preprint* quant-ph/0610204, http://www.pitp.ca/personal/rsorkin/some.papers/)
[15] Rafael D. Sorkin 1994 "Quantum Mechanics as Quantum Measure Theory" *Mod. Phys. Lett.* A **9** (No. 33) 3119-3127 (*Preprint* gr-qc/9401003, http://www.pitp.ca/personal/rsorkin/some.papers/)
[16] S. Gudder and Rafael D. Sorkin 2011 "Two-site quantum random walk" *General Relativity and Gravitation*, to appear, *Preprint* arXiv:1105.0705, http://www.pitp.ca/personal/rsorkin/some.papers/)
[17] Rafael D. Sorkin 2011 "Toward a 'fundamental theorem of quantal measure theory'" *Mathematical Structures in Computer Science*, to appear, *Preprint* arXiv:1104.0997, http://www.pitp.ca/personal/rsorkin/some.papers/141.fthqmt.pdf

In Search Of A Covariant Quantum Measure

Sumati Surya
Raman Research Institute, Sadashivanagar, Bangalore, 560080, India &
Department of Physics, McGill University, Montreal, Canada
E-mail: ssurya@rri.res.in

Abstract. In the quantum measure approach, quantum theory is viewed as a generalisation of classical stochastic theory rather than of classical deterministic theory. An observable or "beable" is thus a *quantum measurable* set in an appropriate event algebra. Thus, the observables in quantum cosmology are covariant quantum measurable sets. In a quantum system with a strongly positive quantum measure or decoherence functional one can construct a vector valued measure whose domain is the algebra of events or physical questions. Using results in the literature, this allows us to specify when a vector valued measure on finite time events can be extended to a measure over infinite time, or equivalently, covariant events. For the class of complex percolation dynamics we show that there is a generic obstruction to this process and that covariant observables can be constructed only for a very restricted class of dynamics.

The Copenhagen interpretation of quantum theory imposes, at the most fundamental level, a dichotomy between the classical measuring device and the observed quantum system. This dichotomy is drawn into sharp and uncomfortable relief in quantum cosmology where there is no external measuring device that can collapse the wave function of the universe. Unlike most areas of physics where this problem can be shelved as a conceptual or philosophical issue, it is an unavoidable stumbling block in constructing a *physical* theory of quantum cosmology. Quantum cosmology, in addition, requires covariant observables, and is hence best suited for a sum-over-histories or path integral formulation, rather than "moment-of-time" formulations using operators and Hilbert spaces.

A typical example of the class of questions of interest to quantum cosmology is the "bounce-question": how likely is it for the universe to have undergone a cosmic bounce? In a sum-over-histories framework, the "bounce event" is the set of all possible spacetimes α_{bounce} which undergo a bounce. Is this class of spacetimes typical? For this we need to be able to put something like a measure on α_{bounce}. A classical or probability measure is required to satisfy the classical Kolmogorov sum rule

$$P(\gamma_1 \cup \gamma_2) = P(\gamma_1) + P(\gamma_2), \tag{1}$$

for distinct pairs of histories. However, we know that this rule does not work in quantum theory because of quantum interference. Instead one requires a generalised *quantum measure* on the set of covariantly defined events.

Of course, obtaining the probability of events analogous to α_{bounce} in a *classical* stochastic system is fairly straightforward. For example, the "return event" α_{return}: "the random walker who begins her walk from the origin eventually returns to the origin" is the set of paths that return to the origin, and can be assigned a probability. This return event bears more than a fleeting resemblance to the bounce event in quantum cosmology and suggests a deep parallel between quantum theory and classical stochastic theory. This is the viewpoint adopted in the quantum measure approach, where quantum theory is taken to be a generalisation of classical stochastic theory, rather than classical deterministic theory [1, 2, 3, 4, 5, 6]. Classical stochastic theory has a precise mathematical framework which we can generalise to quantum theory. In this framework the analogy between the return event and the bounce event manifests itself clearly.

Classical stochastic theory is most elegantly described in the language of of measure theory. It is defined by the triple $(\Omega, \mathcal{A}, \mu)$, where Ω is a *sample space* of histories or spacetime configurations, $\mathcal{A}$ is an *event algebra* or set of propositions about the system, and μ is a probability measure $\mu : \mathcal{A} \to [0, 1]$ satisfying Eqn (1). The event algebra $\mathcal{A}$ is a set of subsets of Ω closed under the finite set operations of union, intersection and complementation. $\mathcal{A}$ is said to be a *sigma algebra* if it is also closed under countable set operations.

A quantum measure space is, in analogy, defined by the triple $(\Omega, \mathcal{A}, \mu)$, where Ω and $\mathcal{A}$ are defined as above, and the *quantum measure* $\mu : A \to \mathbb{R}^+$ is required to satisfy, instead of Eqn (1), the weaker Quantum Sum Rule (QSR) [1]

$$\mu(\alpha \cup \beta \cup \gamma) = \mu(\alpha \cup \beta) + \mu(\alpha \cup \gamma) + \mu(\beta \cup \gamma) - \mu(\alpha) - \mu(\beta) - \mu(\gamma) \tag{2}$$

for mutually disjoint sets $\alpha, \beta, \gamma \in \mathcal{A}$. How is one to interpret μ? To begin with if $\mu(\alpha) = 0$ for some $\alpha \in \mathcal{A}$ then it is possible to say with certainty that the event α *does not happen*, independent of whether the system undergoes a measurement or not. In order to make more definite post or pre-dictions, however, a more full fledged interpretation is required, and this is provided by the Anhomomorphic logic or Coevent interpretation [7, 8][1]. We will not discuss this interpretation here, since we are concerned with the more primitive task of constructing a covariant quantum measure space.

Let us return to the example of the classical random walk. Since we wish to ask questions that are not limited to a certain final time, the sample space Ω is the set of all future-infinite paths, starting out at the origin, say, at time $t = 0$. The dynamics is constructed by assigning transition probabilities from some initial state (x_i, t_i) to a final state (x_f, t_f). A transition probability is obtained by simply summing up the probability $P(\gamma_i)$ of each path γ_i with the same initial and final states. This gives a "classical" sum over histories for the transition probability. For the return event, however, one needs to consider paths of indefinite time. This event is obtained by taking the union of all paths γ_i^t that return to the origin for the first time at some time t, $\Gamma^t = \cup_{i \in I(t)} \gamma_i^t$ (where $I(t)$ indexes the set of returning paths at time t), and then taking the union $\cup_{t=1}^{\infty} \Gamma^t$, i.e., over *all* possible times t. Is such a set measurable?

In order to be able to frame this question precisely, one needs to first define the event algebra more carefully. The measure we have described is defined on finite time paths which are not subsets of Ω. However for each finite time or "truncated" path γ^t we can assign a *cylinder set* $\mathrm{cyl}(\gamma^t) \equiv \{\gamma \in \Omega | \gamma(t') = \gamma^t(t') \forall 0 \leq t' \leq t\}$ which is a subset of Ω. The set of cylinder sets generates an event algebra $\mathcal{A}$ over which the probability measure μ is defined, i.e., $\mu(\mathrm{cyl}(\gamma^t)) = P(\gamma^t)$. However, since $\mathcal{A}$ is constructed purely out of finite set operations, the return event does not belong to $\mathcal{A}$ but rather to $S_{\mathcal{A}}$, which is the (unique) "sigma-algebra completion" of $\mathcal{A}$ [10, 11]. Indeed, all events which are independent of truncation time belong to $S_{\mathcal{A}}$ but not to $\mathcal{A}$. This is an important point, since independence from truncation time is one of the main ingredients required for constructing covariant observables in quantum cosmology. In order to construct observables from $S_{\mathcal{A}}$, however, μ must extend from A to a measure on $S_{\mathcal{A}}$, and in order for this extension to make sense, it should be unique. Thus, we can ask the precise question: does the rule Eqn (1) still hold for countable unions? For classical measures, the

[1] It is useful to point out that in a certain approach to classical proability theory based on the so-called "Cournot principle" [9], events with vanishing probabilities are used to determine all events of interest.

Caratheodary-Hahn theorem ensures that a finitely additive probability measure $\mu : \mathcal{A} \to [0,1]$ extends to a *unique* countably additive probability measure $\overline{\mu} : S_{\mathcal{A}} \to [0,1]$, thus making it possible to give an unambiguous answer to the return question.

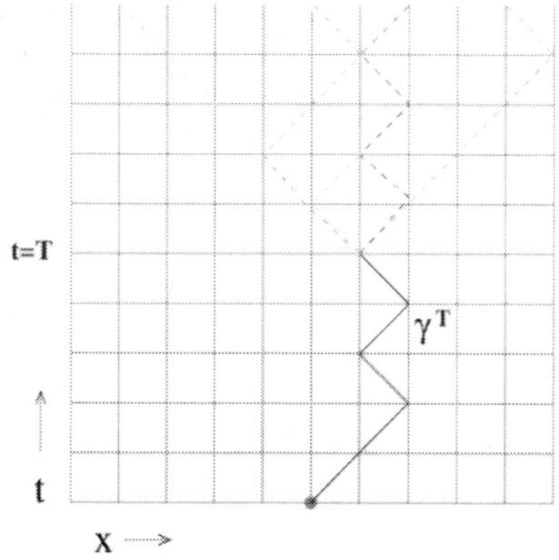

Figure 1. The cylinder set cyl(γ^T) constructed from a truncated history γ^T is a subset in the space of all future infinite paths. We show an example of this set for the classical random walk.

In a theory of quantum cosmology, any infinite time event like the bounce event has no time-label attached to it and is hence a candidate for a covariant observable, so long as the measure itself is invariant. It is therefore of importance to know whether analogs of the Caratheodary-Hahn extension theorem exist for quantum measures. In order to be able to address this in a precise manner, we find it useful to consider only those quantum measures which are derived from a *decoherence functional* $D : \mathcal{A} \times \mathcal{A} \to \mathbb{C}$. Here, D is (i) Hermitian (ii) finitely bi-additive and (iii) strongly positive [4]. This last condition is the statement that the eigenvalues of D constructed as a matrix over a set of histories $\{\alpha_i\}$ are non-negative. The quantum measure for any $\alpha \in \mathcal{A}$ is then given by $\mu(\alpha) \equiv D(\alpha, \alpha) \geq 0$. Strong positivity of D means that a histories Hilbert space $\mathcal{H}$ can be constructed from $\mathcal{A}$, via the GNS procedure, with the inner product given by the decoherence functional [4, 12]. As we now show, the quantum measure can be expressed as a finitely additive *vector* measure over $\mathcal{H}$. This is an important technical step since it brings the quantum measure back into the folds of standard measure theory.

We briefly review the construction of $\mathcal{H}$ from [12]. Let V be the space of complex valued functions on A which are non-zero only on a finite number of elements of A. V is the free vector space over $\mathcal{A}$, with inner product

$$\langle u, v \rangle_V \equiv \sum_{\alpha \in \mathcal{A}} \sum_{\beta \in \mathcal{A}} u^*(\alpha) v(\beta) D(\alpha, \beta). \tag{3}$$

The histories Hilbert space $\mathcal{H}$ is then constructed by taking the set of Cauchy sequences $\{u_i\}$ in V and quotienting by the equivalence relation

$$\{u_i\} \sim \{v_i\} \quad \text{if} \quad \lim_{i \to \infty} \parallel u_i - v_i \parallel_V = 0, \tag{4}$$

where the norm is given by the inner product. Thus,

$$\begin{aligned} [\{u_i\}] + [\{v_i\}] &\equiv [\{u_i + v_i\}] \\ \lambda[\{u_i\}] &\equiv [\{\lambda u_i\}], \\ \langle [\{u_i\}], [\{v_i\}] \rangle &\equiv \lim_{i \to \infty} \langle u_i, v_i \rangle_V, \end{aligned} \tag{5}$$

for all $[\{u_i\}], [\{v_i\}] \in \mathcal{H}$, $\lambda \in \mathbb{C}$. In [12] it was shown that for a large class of examples, this Hilbert space is canonically isomorphic to the standard Hilbert space.

The *quantum vector measure* $\mu_{\mathsf{v}} : A \to \mathcal{H}$ is given by

$$\mu_{\mathsf{v}}(\alpha) \equiv [\chi_\alpha] \in H, \tag{6}$$

where $[\cdot]$ denotes the equivalence class under (4) and χ_α denotes the constant Cauchy sequence $\{\chi_\alpha\}$ for the indicator function $\chi_\alpha : A \to \{0, 1\}$

$$\chi_\alpha(\beta) = \begin{cases} 1 & \text{if } \beta = \alpha, \\ 0 & \text{if } \beta \neq \alpha. \end{cases} \tag{7}$$

Thus,

$$\langle \mu_{\mathsf{v}}(\alpha), \mu_{\mathsf{v}}(\beta) \rangle = D(\alpha, \beta), \tag{8}$$

with the inner product taken in $\mathcal{H}$. The finite bi-additivity of D implies that μ_{v} is finitely additive

$$\mu_{\mathsf{v}}\left(\bigcup_{i=1}^{n} \alpha_i\right) = \sum_{i=1}^{n} \mu_{\mathsf{v}}(\alpha_i), \tag{9}$$

for n mutually disjoint sets $\alpha_i \in A$. Hence μ_{v} is a vector measure on A.

The pre-fix "quantum" for the above vector measure refers only to the range H, which is the histories Hilbert space. This vector measure does indeed satisfy additivity, i.e., Eqn (1) and is thus a "bonafide" measure.

We can thus express the quantum measure space as the triple $(\Omega, \mathcal{A}, \mu_{\mathsf{v}})$, and our quest for infinite time or covariant events reduces to the question – is there a unique extension of the measure to $(\Omega, S_{\mathcal{A}}, \overline{\mu_{\mathsf{v}}})$? The answer is, in general, no. However, if μ_{v} satisfies sufficiently stringent convergence properties, then uniqueness is guaranteed by the Caratheodary-Hahn-Kluvanek theorem (see [13] for a complete statement). For a finite dimensional $\mathcal{H}$, one of these conditions refers to the total variation:

$$|\mu_{\mathsf{v}}(\alpha)| = \sup_{\pi(\alpha)} \sum_{\rho} \parallel \mu_{\mathsf{v}}(\alpha_\rho) \parallel, \tag{10}$$

where the supremum is over all finite partitions $\pi(\alpha) = \{\alpha_\rho\}$ of $\alpha \in \mathcal{A}$. $|\mu_{\mathsf{v}}|$ is itself a non-negative finitely additive pre-measure on $\mathcal{A}$ and is countably additive iff μ_{v} is (Prop. 9, Chapter 1.1, [13]). For finite $\mathcal{H}$, a necessary condition for an extension is that μ_{v} be of *bounded variation*, i.e., $|\mu_{\mathsf{v}}(\alpha)| < \infty$ for all $\alpha \in \mathcal{A}$.

We will now focus on a concrete example of quantum cosmology, namely a complex percolation dynamics for causal sets. In the causal set approach to quantum gravity, the spacetime continuum is replaced by a locally finite partially ordered set, the causal set [14]. In the continuum approximation of the theory, the order relation corresponds to the causal ordering and on an average, the number of discrete events in any causally convex region of the causal set corresponds to the continuum volume. We refer the reader to the literature for a more detailed introduction to causal sets [14, 15]. For the purpose of our analysis all that we will require is a measure theoretic characterisation of the dynamics.

We give a brief description of the classical sequential growth models of [16, 17]. Sequential growth means that a labelled causal set is grown element by element starting from a single element. Every new element is added with some probability either to the future of an existing element or left unrelated to it. This process generates labelled causal sets, but the probabilities can be chosen in a label independent way. In classical sequential growth, the probabilities are required to be Markovian, label independent and also satisfy a local causality condition

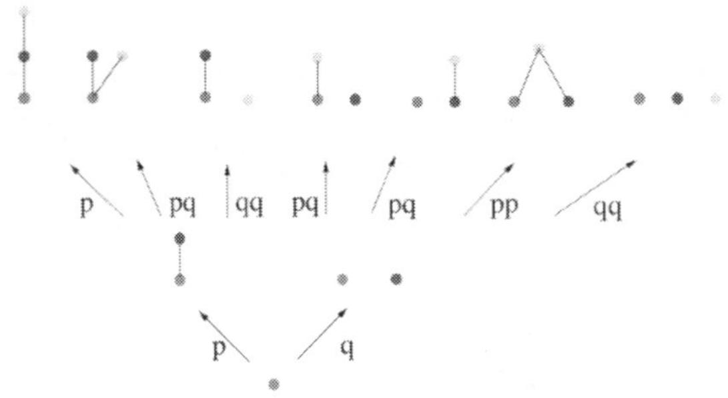

Figure 2. Transitive percolation dynamics for causal sets.

called Bell causality. Figure 2 shows an important example of sequential growth, the transitive percolation model, which will be our starting point for a quantum percolation model. Here, $p \in [0,1]$ is the probability that the new element is to the immediate future of an existing element and $q = 1 - p$ is then the probability that it is unrelated. If the growth continues for infinite time, it generates causal sets that are past finite.

Since the growth process generates causal sets that are labelled by the sequence in which each element is born, the sample space Ω is the set of past finite labelled causal sets. As in the classical random walk, associated with every finite labelled causal set C^n is a cylinder set $\text{cyl}(C^n)$ which is a subset of Ω containing labelled past finite causal sets whose first n elements form the sub-causal set C^n. These cylinder sets form an event algebra $\mathcal{A}$, with measure given by taking the product of probabilities in constructing C^n via the above sequential growth process. The covariant observables of the classical dynamics are, however measurable sets in the space Ω' of *unlabelled* past finite causal sets – in other words, the time foliation or labelling that is given by a specific growth process should not matter. One way to obtain these covariant observables is then to extend the measure from $\mathcal{A}$ to $S_{\mathcal{A}}$ on Ω and then consider the quotient sigma algebra $S'_{\mathcal{A}}$ over relabellings [19]. Thus, in order to get covariant observables from a a labelled growth dynamics, the extension from $(\Omega, \mathcal{A}, \mu)$ to $(\Omega, S_{\mathcal{A}}, \overline{\mu})$ is crucial before one can take the quotient $(\Omega', S'_{\mathcal{A}}, \overline{\mu}')$ over relabellings. While the Caratheodary-Hahn theorem guarantees this for the classical stochastic growth, as we will see, it is not always guaranteed for a quantum growth process.

In direct analogy with classical transitive percolation, we define a complex percolation model as follows. We allow $p \in \mathbb{C}$ which then gives an *amplitude* for transition, rather than a probability. In addition, we assume that

$$D(\text{cyl}(C^n), \text{cyl}(C'^n)) = \psi^*(C^n)\psi(C'^n) \tag{11}$$

where $\psi(C^n)$ is the amplitude for the transition from the empty set to the n-element causal set C^n. ψ is thus a complex measure on $\mathcal{A}$, and hence also a vector measure. The associated histories Hilbert space $\mathcal{H}$ can be shown to also be one-dimensional, so that the quantum vector measure μ_{v} on $\mathcal{A}$ is simply a complex measure. In this case, standard results in measure theory [18] imply that in order for μ_{v} to extend uniquely to a μ'_{v} on $S_{\mathcal{A}}$, it must have a bounded *total variation*. This corresponds to an unconditional convergence of the vector measure over all partitions.

Using simple properties of the sequential growth dynamics it is possible to show that $p+q = 1$ and that for p, q not real, $p+q = 1+\zeta$ where $\zeta > 0$ [20]. The Markov sum rule is then used to show that the total variation is bounded only for p, q real. These are the, somewhat oddly labelled, "real" complex-percolation models. In these models, even though the quantum measure is not additive, the observables of the theory are identical to those of classical transitive percolation. In particular, the observables can be entirely characterised by "stem sets". A *stem* is the causal set analog of a "past set" and a *stem set* is a subset of the set of unlabelled causal sets containing all causal sets which have the same stem [19]. Thus, for example, it is possible to assign a measure to the event, "the causal set is originary", i.e., the set of all causal sets in Ω with an "initial" element to the past of all other elements.

The lack of an extension for generic complex percolation models however suggests that these models cannot take on covariant robes in a straightforward way. In classical sequential growth models, the extension to the full sigma algebra is guaranteed, and one can then use this to define a quotient measure space that is fully label invariant. Without an extension, it is not clear that there is a well defined or canonical procedure using which covariant questions can be extracted. One lesson that can be learned from this example is that breaking covariance by introducing a time foliation may not always allow a fully quantum covariant theory to be constructed. Of course, if one could begin with a quantum measure on an event algebra over unlabelled causal sets, then even if some classical events are not quantum measurable, the set of covariant observables would be non-trivial. However, to find a simple set of manipulable rules with which to define such a measure seems intrinsically harder without first using the set of labelled causal sets as an intermediary. Admittedly, this is a poorly investigated area, and it may be that an appropriate choice of label invariant event algebras (for example the stem sets of [19] described above) will reveal potentially new and interesting dynamical rules with which to construct covariant observables.

Acknowledgments

This work was done in collaboration with Fay Dowker and Steven Johnston and was supported in part by the Royal Society International Joint Project 2006-R2.

References

[1] Sorkin R D 1994 *Mod. Phys. Lett.* A **9** 3119 (*Preprint* arXiv:gr-qc/9401003)
[2] Sorkin R D 1994 *Proceedings 4TH Drexel Symposium on Quantum Nonintegrability* 8-11 Sep (Philadelphia PA) (*Preprint* arXiv:gr-qc/9507057)
[3] Salgado R B 2002 *Mod. Phys. Lett.* A **17** 711 (*Preprint* arXiv:gr-qc/9903015)
[4] Martin X, O'Connor D and Sorkin R D 2005 *Phys. Rev.* D **71** 024029 (*Preprint* arXiv:gr-qc/0403085)
[5] Surya S and Wallden P 2010 *Found. Phys.* **40** 585 (*Preprint* arXiv:0809.1951)
[6] Gudder S 2009 *J. Math. Phys.* **50** 123509 (*Preprint* arXiv:0909.2203)
[7] Sorkin R D 2007 *J. Phys. Conf. Ser.* **67** 012018 (*Preprint* arXiv:quant-ph/0703276)
[8] Dowker F and Ghazi-Tabatabai Y 2008 *J. Phys.* A **41** 105301 (*Preprint* arXiv:0711.0894 [quant-ph])
[9] Ghazi-Tabatabai Y and Wallden P 2009 *J. Phys. : Conf. Ser.* **174** 012054 (*Preprint* arXiv:0907.0754v1[quant-ph])
[10] Kac M 1957 *Probability and related topics in physical sciences, Lectures in applied mathematics* (Interscience Publishers Inc)
[11] Halmos P R 1978 *Measure Theory (Graduate Texts in Mathematics) (v. 18)* (Springer)
[12] Dowker H F, Johnston S and Sorkin R 2010 *J. Phys.* A **43** 275302 (*Preprint* arXiv:1002.0589)
[13] Diestel J and Uhl J J 1977 *Vector Measures, Mathematical Surveys, Number 15* (American Mathematical Society)
[14] Bombelli L, Lee J H, Meyer D and Sorkin R 1987 *Phys. Rev. Lett.* **59** 521
[15] Sorkin R D *Preprint* gr-qc/0309009.
[16] Rideout D P and Sorkin R D 2000 *Phys. Rev.* D **61**, 024002 (*Preprint* arXiv:gr-qc/9904062)
[17] Rideout D P 2001 *Dynamics of causal sets, PhD. Thesis* (Syracuse University) (*Preprint* arXiv:gr-qc/0212064)
[18] Rudin W 1987 *Real and Complex Analysis* (McGraw-Hill International Editions)
[19] Brightwell G, Dowker H F, Garcia R S, Henson J and Sorkin R D 2003 *Phys. Rev.* D **67** 084031 (*Preprint* arXiv:gr-qc/0210061)
[20] Dowker F, Johnston S and Surya S 2010 *J. Phys.* A **A43** 505305 (*Preprint* arXiv:1007.2725 [gr-qc])

Noncommutative geometry, symmetries and quantum structure of space-time

T R Govindarajan[1], Kumar S Gupta[2], E Harikumar[3] and S Meljanac[4]

[1] Institute of Mathematical Sciences, CIT Campus, Taramani, Chennai 600113, India
[2] Theory Division, Saha Institute of Nuclear Physics, 1/AF Bidhannagar, Kolkata 700064, India
[3] School of Physics, University of Hyderabad, Hyderabad 500046, India
[4]Rudjer Bošković Institute, Bijenička c.54, HR-10002 Zagreb, Croatia

E-mail: `trg@imsc.res.in`, `kumars.gupta@saha.ac.in`, `harisp@uohyd.ernet.in`, `meljanac@irb.hr`

Abstract. We discuss how space-time noncommutativity affects the symmetry groups and particle statistics. Assuming that statistics is superselected under a symmetry transformation, we argue that the corresponding flip operator must be twisted. It is argued that the twisted statistics naturally leads to a deformed oscillator algebra for scalar fields in such a background.

1. Introduction

Describing physics at the Planck scale is one of the most challenging issues at the present time. To probe physics at Planck Scale l_p, the Compton wavelength $\frac{1}{M}$ of the probe must be less than l_p, hence $M > \frac{1}{l_p}$, i.e. Planck mass. This requires an energy of the order of $\frac{1}{l_p}$ to be concentrated for some time in the region where the measurement is taking place. Such a process generates a gravitational field and when this field becomes so strong as to prevent light or other signals from leaving the region in question, the process of measurement itself becomes fuzzy. Measurements with very short time scales also require very high energies, leading to a similar picture. Based on these arguments, Doplicher et al (DFR) [1, 2] obtained uncertainty relations between the coordinates, which they showed could be deduced from a commutation relation of the type

$$[q_\mu, q_\nu] = iQ_{\mu\nu}, \tag{1}$$

where q_μ are self-adjoint coordinate operators, μ, ν run over space-time coordinates and $Q_{\mu\nu}$ is an antisymmetric tensor. Thus noncommutative geometry emerges as a plausible paradigms for describing the quantum structure of space-time at the Planck scale.

The simplest possibility for $Q_{\mu\nu}$ is that it commutes with the coordinate operators, which corresponds to the Moyal algebra. However, there are much more general possibilities for $Q_{\mu\nu}$ allowed by the DFR analysis. In this talk we shall be concerned with one such possibility known as the κ-Minkowski algebra [3, 4, 5, 6, 7, 8, 9, 10, 11, 12, 13]. The κ-Minkowski space-time arises naturally in the description of certain noncommutative black holes and FRW cosmologies [14, 15, 16, 17]. In the context of noncommutative BTZ black hole, the κ-Minkowski algebra describes a noncommutative cylinder, where the spectrum of the time operator is quantized

[14, 15, 18, 19, 20]. The κ-Minkowski algebra is also interesting in the context of doubly special relativity and in putting bounds on the quantum gravity scale using the astrophysical data from the gamma ray bursts [21, 22, 23, 24, 25].

Hopf algebra plays an important role in analyzing symmetries of quantum field theories on noncommutative spaces [26]. In the Hopf algebra approach, the underlying symmetry algebra of the noncommutative theory acts on multi-particle states through the twisted coproducts of the symmetry generators. Such a generalized concept of symmetry leads to the notion of twisted statistics [27, 28, 29, 30, 31, 32, 33, 34]. The notion of twisted statistics in κ-Minkowski space-time leads to a generalized oscillator algebra, which plays an important role in the analysis of the corresponding quantum field theories [35, 36, 37, 38, 39, 40, 41, 42, 43].

In this talk we shall first discus the κ-Minkowski algebra and discuss some of its realizations. This will be followed by the analysis of twisted statistics in κ-Minkowski space-time. Finally we shall discuss scalar field theory and deformed oscillator algebras in this geometry.

2. κ-Minkowski algebra

The κ-Minkowski algebra is defined by

$$[\hat{x}_i, \hat{x}_j] = 0, \qquad [\hat{x}_0, \hat{x}_i] = ia\hat{x}_i, \tag{2}$$

where $a = \frac{1}{\kappa}$ and $i = 1, 2, ...n-1$. In terms of the Minkowski metric $\eta_{\mu\nu} = \text{diag}(-1, 1, 1, 1......, 1)$, $\mu = 0, 1, 2, ...n-1$ we can define $x^\mu = \eta^{\mu\alpha}x_\alpha$ and $\partial^\mu = \frac{\partial}{\partial x_\mu} = \eta^{\mu\alpha}\partial_\alpha$ which satisfy the relations

$$[x_\mu, x_\nu] = 0, \qquad [\partial_\mu, \partial_\nu] = 0, \qquad [\partial^\mu, x_\nu] = \eta^\mu_\nu, \qquad [\partial_\mu, x_\nu] = \eta_{\mu\nu}. \tag{3}$$

For later use, we also define $p_\mu = -i\partial_\mu$, so that $[p_\mu, x_\nu] = -i\eta_{\mu\nu}$.

We seek realizations of the noncommutative coordinates in terms of the commuting coordinates x_μ and corresponding derivatives ∂_μ as a power series. A class of such realizations is given by

$$\hat{x}_\mu = x^\alpha \Phi_{\alpha\mu}(\partial). \tag{4}$$

It is easy to see that these coordinates obey $[\partial_\mu, \hat{x}_\nu] = \Phi_{\mu\nu}(\partial)$. In this paper we work with a specific class of realization satisfying $[\partial_\mu, \hat{x}_\nu] = \Phi_{\mu\nu}(\partial)$ given by

$$[\partial_i, \hat{x}_j] = \delta_{ij}\varphi(A), \quad [\partial_i, \hat{x}_0] = ia\partial_i\gamma(A), \tag{5}$$

$$[\partial_0, \hat{x}_i] = 0, \qquad [\partial_0, \hat{x}_0] = \eta_{00} = -1, \tag{6}$$

where $A = -ia\partial_0$. An explicit solution of this realization is

$$\begin{aligned} \hat{x}_i &= x_i\varphi(A) \\ \hat{x}_0 &= x_0\psi(A) + iax_i\partial_i\gamma(A). \end{aligned} \tag{7}$$

Using the above realization in Eqn. (2), we get

$$\frac{\varphi'}{\varphi}\psi = \gamma - 1, \tag{8}$$

where φ' is the derivative of φ with respect to its argument $ia\partial_0$ and these functions satisfy the boundary conditions $\varphi(0) = 1, \psi(0) = 1$ and $\gamma(0) = \varphi'(0) + 1$ is finite and all are positive functions. Further demanding that the commutators of the Lorentz generators with the κ-space coordinates be linear in $\hat{x}_\mu$ as well as in the generators and have smooth commutative limit as the deformation parameter $a \to 0$ imposes further requirements on these functions and one can

Journal of Physics: Conference Series **306** (2011) 012019 doi:10.1088/1742-6596/306/1/012019

easily see that there are only two class of realizations possible, viz: one where $\psi = 1$ and a second one where $\psi = 1 + 2A$. We restrict ourselves to the case $\psi = 1$.

It may be noted that for $\varphi_S(A) = \frac{A}{e^A - 1}$ there exists a covariant realization corresponding to Weyl symmetric ordering. This realization is given by

$$\hat{x}_\mu = x_\mu \varphi_S(A) + i a_\mu x_\alpha \partial^\alpha \gamma_S(A), \qquad x_\alpha \partial^\alpha = \eta^{\alpha\beta} x_\alpha \partial_\beta, \tag{9}$$

where $A = i a_0 \partial^0 = -i a \partial_0$. Here we choose $a_\mu = (a, 0, ..., 0)$ to be time-like, which can be chosen to be is space-like or light-like as well.

The coproducts Δ_φ of the derivative operators in the realization given in Eqn. (7) are

$$\Delta_\varphi(\partial_0) = \partial_0 \otimes I + I \otimes \partial_0 \equiv \partial_0^x + \partial_0^y, \tag{10}$$

$$\Delta_\varphi(\partial_i) = \varphi(A \otimes I + I \otimes A) \left[\frac{\partial_i}{\varphi(A)} \otimes I + e^A \otimes \frac{\partial_i}{\varphi(A)} \right]. \tag{11}$$

The mapping between the functions on κ-space to that of commutative space also defines a star-product, which naturally depends on the realization φ. For any realization φ, the star-product can be expressed in terms of the twist element $\mathcal{F}_\varphi$ as

$$f \star_\varphi g = m_0(\mathcal{F}_\varphi f \otimes g) = m_\varphi(f \otimes g), \tag{12}$$

where f and g are functions of the commutative coordinates and m_0 is the usual pointwise multiplication map in the commutative algebra of smooth functions and this can be re-expressed as

$$(f *_\varphi g)(x) = m_0 \left(e^{x_i (\Delta_\varphi - \Delta_0) \partial_i} f(u) g(t) \right) |_{u=t=x_i}, \tag{13}$$

where Δ_φ is the twisted coproduct given in Eqn. (11) and the undeformed coproducts is given by $\Delta_0 = \partial \otimes I + I \otimes \partial$. Comparing Eqn. (12) and Eqn. (13), we find the twist element as

$$\mathcal{F}_\varphi = e^{x_i (\Delta_\varphi - \Delta_0) \partial_i}. \tag{14}$$

Using Eqn. (11) and Eqn. (13), we find that the $*$-product can be written as

$$(f \star_\varphi g)(x) = e^{x_i \partial_i^u \left(\frac{\varphi(A_u + A_t)}{\varphi(A_u)} - 1 \right) + x_i \partial_i^t \left(\frac{\varphi(A_u + A_t)}{\varphi(A_t)} e^{A_u} - 1 \right)} f(u) g(t) |_{u=t=x_i}. \tag{15}$$

The explicit form of the corresponding twist element is now given by

$$\mathcal{F}_\varphi = e^{N_x ln \frac{\varphi(A_x + A_y)}{\varphi(A_x)} + N_y \left(A_x + ln \frac{\varphi(A_x + A_y)}{\varphi(A_y)} \right)}, \tag{16}$$

where $N_x = x_i \frac{\partial}{\partial x_i}$ [7, 9, 10].

One can now get the modified momentum addition rules for κ-space from the coproducts given in Eqn. (10) and Eqn. (11) also. Thus going to momentum space we find

$$[K_\varphi(p, q)]_\mu = -i \Delta_\varphi(\partial_\mu),$$

$$K_\varphi(p, q) x = -(p_0 + q_0) x_0 + \varphi(-a p_0 - a q_0) \left[\frac{p_i x_i}{\varphi(-a p_0)} + \frac{e^{-a p_0}}{\varphi(-a q_0)} q_i x_i \right]. \tag{17}$$

Similarly, we can also obtain the twist element in the momentum space, denoted by $\mathcal{F}$, which tells how the star-product acts on the momentum space. Starting from

$$\mathcal{F} f(x) g(y) \equiv \mathcal{F} \int d^4k d^4q e^{ikx} \tilde{f}(k) e^{iqy} \tilde{g}(q) \tag{18}$$

and using the action of $\mathcal{F}$ on plane waves, we can easily get

$$\mathcal{F} \tilde{f}(k) \otimes \tilde{g}(q) = \mathcal{F}(i \frac{\partial}{\partial k}, k, i \frac{\partial}{\partial q}, q) \tilde{f}(k) \tilde{g}(q). \tag{19}$$

The above result will be of use to obtain the twisted commutation relations between the Fourier coefficients, necessary to discuss the twisted oscillators.

3. Twisted statistics

It is known that for the QFT's defined on Moyal plane, the twisted coproduct rules affect the statistics [28, 29, 30, 31, 32, 33, 38, 43]. This is natural as the physical theory has to be invariant under the action of the underlying symmetry group of the space (or spacetime) and the definition of statistics should also be invariant under this group action. This ensures that the statistics is superselected. Such a superselection rule is implemented by demanding that the flip operator commutes with the coproduct. As the coproduct rule is now changed, we do expect a corresponding change in the definition of flip operator also.

In the commutative case, the flip operator is defined through its action on multi-particle states. Without loss of generality, let us consider a two-particle state $f \otimes g \in \mathcal{A}_0 \otimes \mathcal{A}_0$. The action of flip operator on this (tensor product) state is given by

$$\tau_0(f \otimes g) = g \otimes f. \tag{20}$$

It is easy to see that $(\tau_0)^2 = I$. Symmetric and antisymmetric states of the physical Hilbert space are projected from the tensor product state as

$$\frac{1}{2}(1 \pm \tau_0)(f \otimes g) = \frac{1}{2}(f \otimes g \pm g \otimes f), \tag{21}$$

respectively. Since this definition of (anti)symmetric states should remain invariant under the action of the underlying symmetry, its clear that the flip operator must commute with the symmetry generator. Since Λ, a typical element of the symmetry group acts on the tensor-product state through some representation D as

$$\Lambda : f \otimes g = (D \otimes D)\Delta(\Lambda) f \otimes g, \tag{22}$$

this requirement implies that the coproduct $\Delta(\Lambda)$ commutes with the flip operator τ_0. Thus in the commutative space the flip operator τ_0 is superselected so as to have vanishing commutators with all observables. In this case of noncommutative theories, as we have seen, the coproducts get twisted and the twisted coproducts do not satisfy

$$[\Delta_\varphi, \tau_0] \neq 0. \tag{23}$$

Thus, the meaning of (anti)symmetric states defined using τ_0 are no longer invariant. We are, thus forced to define a new twisted flip operator which commutes with the coproduct action. Since $\Delta_\varphi = \mathcal{F}_\varphi^{-1} \Delta_0 \mathcal{F}_\varphi$, where Δ_0 is the coproducts of the undeformed Poincaré algebra, we are immediately led to the twisted flip-operator

$$\tau_\varphi = \mathcal{F}_\varphi^{-1} \tau_0 \mathcal{F}_\varphi, \tag{24}$$

which satisfies

$$[\Delta_\varphi, \tau_\varphi] = 0. \tag{25}$$

Using this twisted flip operator we can define an invariant definition of symmetric and antisymmetric states as $\frac{1}{2}(1 \pm \tau_\varphi)(f \otimes g)$, respectively. The twisted flip operator for a generic φ realization can be easily obtained using Eqn. (16) in Eqn. (24) as

$$\tau_\varphi = e^{i(x_i p_i \otimes A - A \otimes x_i p_i)} \tau_0, \tag{26}$$

where $A = -ia\partial_0$. In the limit $a \to 0$, we get back the familiar commutative flip-operator, smoothly. It is interesting to note that the τ_φ given above is independent of φ.

4. Deformed oscillator algebra

Having defined the twisted flip operator τ_φ, we are now in a position to define (anti) symmetric states of a theory defined in the κ-Minkowski space. We start by defining the deformed bosonic state as

$$f \star_\varphi g = m_\varphi(f \otimes g) = m_\varphi \tau_\varphi(f \otimes g). \tag{27}$$

Using the definitions of m_φ and τ_φ(see Eqn. (12) and Eqn. (24)) in the above, we get

$$f \otimes g = \tau_\varphi(f \otimes g), \tag{28}$$

or, equivalently, we can write

$$\mathcal{F}_\varphi(f \otimes g) = \tilde{\mathcal{F}}_\varphi(f \otimes g), \tag{29}$$

where we have used the mirror twist operator $\tilde{\mathcal{F}}_\varphi = \tau_0 \mathcal{F}_\varphi \tau_0$. Now defining the twisted tensor product $f \otimes_\varphi g$ as $\mathcal{F}_\varphi(f \otimes g)$, from the above, we get

$$f \otimes_\varphi g = \tau_0(f \otimes_\varphi g). \tag{30}$$

For the product of two bosonic fields $\phi(x)$ and $\phi(y)$ under interchange, now we pick up an additional factor compared to the commutative case. This can be calculated using Eqn. (28) and one gets

$$\phi(x) \otimes \phi(y) - e^{-(A \otimes N - N \otimes A)} \phi(y) \otimes \phi(x) = 0. \tag{31}$$

Expressing ϕ in the above equation using Fourier transforms and using the twisted flip operator in momentum space, we are led to the deformed commutation relations between the annihilation operators as

$$\tilde{\phi}(k)\tilde{\phi}(p) = e^{-ia\left[k_0(\partial_{p_i} p_i) - p_0(\partial_{k_i} k_i)\right]} \tilde{\phi}(p)\tilde{\phi}(k). \tag{32}$$

The $\Phi(x)$ appearing in the generalized Klein-Gordon equation can be expressed as

$$\Phi(x) = \int d^4p \delta(p_0^2 - \omega^2) \tilde{A}(p) e^{-ip\cdot x}, \tag{33}$$

where $\omega = \sqrt{p_i^2 + m^2}$. Using the mode decomposition

$$\Phi(x) = \int \frac{d^3p}{\sqrt{p_i^2 + m^2}} \left[A(\omega, \vec{p}) e^{-ip\cdot x} + A^\dagger(\omega, \vec{p}) e^{ip\cdot x}\right], \tag{34}$$

from Eqn.(32) we get

$$A^\dagger(p_0, \vec{p}) A(q_0, \vec{q}) - e^{-a(q_0 \partial_{p_i} p_i + \partial_{q_i} q_i p_0)} A(q_0, \vec{q}) A^\dagger(p_0, \vec{p}) = -\delta^3(p - q), \tag{35}$$

$$A^\dagger(p_0, \vec{p}) A^\dagger(q_0, \vec{q}) - e^{-a(-q_0 \partial_{p_i} p_i + \partial_{q_i} q_i p_0)} A^\dagger(q_0, \vec{q}) A^\dagger(p_0, \vec{p}) = 0, \tag{36}$$

$$A(p_0, \vec{p}) A(q_0, \vec{q}) - e^{-a(q_0 \partial_{p_i} p_i - \partial_{q_i} q_i p_0)} A(q_0, \vec{q}) A(p_0, \vec{p}) = 0. \tag{37}$$

For the choice $\varphi = e^{-\frac{A}{2}} = e^{-\frac{ia\partial_0}{2}}$, the generalized Klein-Gordon equation is

$$\left[\partial_i^2 + \frac{4}{a^2} sinh^2(\frac{ia\partial_0}{2}) - m^2\right] \Phi = 0. \tag{38}$$

We can decompose this field in positive and negative frequency modes and thus

$$\Phi(x) = \int \frac{d^4p}{2\Omega_k(p)} \left[A(\omega_k, \vec{p}) e^{-ip\cdot x} + A^\dagger(\omega_k, \vec{p}) e^{ip\cdot x}\right], \tag{39}$$

where $A^\dagger(\pm\omega_k, \vec{p}) = A^\dagger(\mp\omega_k, \vec{p})$. In the above, we have used

$$p_0^\pm = \pm\omega_k(p) = \pm\frac{2}{a}sinh^{-1}(\frac{a}{2}\sqrt{p_i^2+m^2}), \tag{40}$$

$$\Omega_k(p) = \frac{1}{a}sinh(a\omega_k(p)). \tag{41}$$

Using this in Eqn.(32), for the field satisfying the deformed generalized Klein-Gordon equation given above (Eqn.(38)), we obtain various twisted commutation relations between creation and annihilation operators. They are

$$A^\dagger(p_0,\vec{p})A(q_0,\vec{q}) - e^{-a(q_0\partial_{p_i}p_i+\partial_{q_i}q_ip_0)}A(q_0,\vec{q})A^\dagger(p_0,\vec{p}) = -\delta^3(p-q), \tag{42}$$

$$A^\dagger(p_0,\vec{p})A^\dagger(q_0,\vec{q}) - e^{-a(-q_0\partial_{p_i}p_i+\partial_{q_i}q_ip_0)}A^\dagger(q_0,\vec{q})A^\dagger(p_0,\vec{p}) = 0, \tag{43}$$

$$A(p_0,\vec{p})A(q_0,\vec{q}) - e^{-a(q_0\partial_{p_i}p_i-\partial_{q_i}q_ip_0)}A(q_0,\vec{q})A(p_0,\vec{p}) = 0. \tag{44}$$

Note that p_0 and q_0 are as given in Eqn. (40). From this, one can easily derive the following relations

$$A^\dagger(p_0, e^{-\frac{aq_0}{2}}\vec{q})A^\dagger(q_0, e^{\frac{ap_0}{2}}\vec{q}) - \mathcal{F}(q,p)A^\dagger(q_0, e^{-\frac{ap_0}{2}}\vec{q})A^\dagger(p_0, e^{\frac{aq_0}{2}}\vec{q}) = 0, \tag{45}$$

$$A(p_0, e^{\frac{aq_0}{2}}\vec{p})A(q_0, e^{-\frac{ap_0}{2}}\vec{q}) - \mathcal{F}(-q,-p)A(q_0, e^{\frac{ap_0}{2}}\vec{q})A(p_0, e^{-\frac{aq_0}{2}}\vec{p}) = 0, \tag{46}$$

$$A^\dagger(p_0, e^{\frac{aq_0}{2}}\vec{p})A(q_0, e^{\frac{ap_0}{2}}\vec{q}) - \mathcal{F}(-q,p)A(q_0, e^{-\frac{ap_0}{2}}\vec{q})A^\dagger(p_0, e^{-\frac{aq_0}{2}}\vec{p}) = -\delta^3(p-q). \tag{47}$$

where $\mathcal{F}(q,p) = e^{3a(q_0-p_0)}$. The above relations (42-47) define the new deformed oscillator algebra for a scalar field in the κ-Minkowski space-time, which would be relevant for future calculations in this background.

5. Conclusion

In this talk we have discussed the κ-Minkowski algebra and its twisted coproducts. Assuming that particle statistics is superselected, i.e. it is invariant under symmetry transformations, this approach leads to a twisted flip operator and twisted statistics. Applying to bosonic fields, we have calculated the deformed oscillator algebra in the κ-Minkowski background. Such a deformed oscillator algebra would be useful in evaluating the amplitudes for physical processes involving scalar fields in this background.

References

[1] Doplicher S, Fredenhagen K and Roberts J E 1994 *Phys. Lett.* B **331** 39
[2] Doplicher S, Fredenhagen K and Roberts J E 1995 *Commun. Math. Phys.* **172** 187
[3] Lukierski J, Nowicki A, Ruegg H and Tolstoy V N 1991 *Phys. Lett.* B **264** 331
[4] Lukierski J, Nowicki A and Ruegg H 1992 *Phys. Lett.* B **293** 344
[5] Lukierski J and Ruegg H 1994 *Phys. Lett.* B **329** 189
[6] Lukierski J, Ruegg H and Zakrzewski W J 1995 *Ann. Phys.* **243** 90
[7] Meljanac S and Stojic M 2006 *Eur. Phys. J.* C **47** 531
[8] Durov N, Meljanac S, Samsarov A and Škoda Z 2007 *J. Algebra* **309** 318
[9] Meljanac S, Kresic-Juric S and Stojic M 2007 *Eur. Phys. J.* C **51** 229
[10] Meljanac S, Samsarov A, Stojic M and Gupta K S 2008 *Eur. Phys. J.* C **53** 295
[11] Meljanac S and Kresic-Juric S 2009 *J. Phys.* A **42** 365204
[12] Meljanac S and Kresic-Juric S 2010 *Preprint* arXiv:1004.4647
[13] Meljanac S and Samsarov A 2010 *Preprint* arXiv:1007.3943
[14] Dolan B, Gupta K S and Stern A 2007 *Class. Quantum Grav.* **24** 1647
[15] Dolan B, Gupta K S and Stern A 2009 *J. Phys.: Conf. Ser.* **174** 012023
[16] Schupp P and Solodukhin S 2009 *Preprint* arXiv:0906.2724
[17] Ohl T and Schenkel A 2009 *JHEP* **0910** 052

[18] Chaichian M, Demichev A, Presnajder P and Tureanu A 2001 *Eur. Phys. J.* C **20** 767
[19] Chaichian M, Demichev A, Presnajder P and Tureanu A 2001 *Phys. Lett.* B **515** 426
[20] Balachandran A P, Govindarajan T R, Martins A G and Teotonio-Sobrinho P 2004 *JHEP* **0411** 068
[21] Abdo A et al 2009 *Nature* **323** 1688
[22] Abdo A et al 2009 *Nature* **462** 331
[23] Amelino-Camelia G 2009 *Nature* **462** 291
[24] Amelino-Camelia G and Smolin L 2009 *Phys. Rev.* D **80** 084017
[25] Borowiec A, Gupta K S, Meljanac A and Pachol A 2010 *Europhys. Lett.* **92** 20006
[26] Majid S 1988 *Class. Quant. Grav.* **5** 1587
[27] Grosse H 1979 *Phys. Lett.* B **86** 267
[28] Oeckl R 2000 *Nucl. Phys.* B **581** 559
[29] Chaichian M, Nishijima K and Tureanu A 2003 *Phys. Lett.* B **568** 146
[30] Balachandran A P, Mangano G, Pinzul A and Vaidya S 2006 *Int. J. Mod. Phys.* A **21** 3111
[31] Balachandran A P, Govindarajan T R, Mangano G, Pinzul A, Qureshi B A and Vaidya S 2007 *Phys. Rev.* D **75** 045009
[32] Balachandran A P, Pinzul A, Qureshi B A and Vaidya S 2007 *Phys. Rev.* D **76** 105025
[33] Aschieri P, Lizzi F and Vitale P 2008 *Phys. Rev.* D **77** 025037
[34] Piacitelli G 2009 *Preprint* arXiv:0901.3109
[35] Arzano M and Marciano A 2007 *Phys. Rev.* D **76** 125005
[36] Young C A S and Zegers R 2008 *Nucl. Phys.* B **797** 537
[37] Daszkiewicz M, Lukierski J and Woronowicz M 2008 *Phys. Rev.* D **77** 105007
[38] Govindarajan T R, Gupta K S, Harikumar E, Meljanac S and Meljanac D 2008 *Phys. Rev.* D **77** 105010
[39] Young C A S and Zegers R 2008 *Nucl. Phys.* B **804** 342
[40] Arzano M and Benedetti D 2009 *Int. J. Mod. Phys.* A **24** 4623
[41] Daszkiewicz M, Lukierski J and Woronowicz M 2007 *Preprint* arXiv:0712.0350
[42] Daszkiewicz M, Lukierski J and Woronowicz M 2009 *J. Phys.* A **42** 355201
[43] Govindarajan T R, Gupta K S, Harikumar E, Meljanac S and Meljanac D 2009 *Phys. Rev.* D **80** 025014

Torsion, a multivalued gauge degree of freedom in Einstein's gravity

H Kleinert

Institut für Theoretische Physik, Freie Universität Berlin, Arnimallee 14, D-14195 Berlin, Germany
and
ICRANet, Piazzale della Republica 10, I-65122, Pescara, Italy

E-mail: h.k@fu-berlin.de

Abstract. We show that there is an entire gauge symmetry of a novel kind which interpolates between an infinity of formulations of the laws of gravity, ranging from Einsteins pure curvature formulation to a pure torsion formulation in a teleparallel geometry. As a consequence, torsion and curvature are not independent and that torsion is an alternative description of curvature in gravity.

1. It is well known that the solutions of Einstein equations can equally well be obtained in a Riemann-Cartan spacetime with torsion, in which the Cartan curvature vanishes identically. This *teleparallel formulation* of gravity has been proposed by Einstein as an alternative to his famous theory of relativity in Riemannian spacetime, and its properties have been studied in many publications [1, 2]. In this formulation, the Einstein-Hilbert action is equal to a combination of scalars formed from torsion tensors [2]. In this contribution to the special volume on analog models of gravitational theories I would like to point out that there is an entire family of theories in which a combination of curvature and torsion provide us with an alternative description of the same gravitational forces [3, 4, 5]. This result, obtained first in Ref. [5], was reached on the basis of the close analogy of spaces with gravity and torsion to crystals with defects [6, 7].

2. A nontrivial extension of Einsteins theory in curved spacetime to Riemann-Cartan spacetime was advanced since 1959 [7, 8, 9]. It has the appealing feature that it can be rewritten as a gauge theory that is invariant under local Poincaré transformations, i.e., under both local translations and local Lorentz transformations. This brings the geometric theory of gravity to a similar form as the gauge theories of weak, electromagnetic, and strong interactions. In the gauge formulation, torsion is treated as an independent field which couples only to the intrinsic spin of the elementary particles.

The extension has, however, several unsatisfactory features. First, it has the same problem as Einstein's theory that it cannot be quantized and is necessarily classical. But spin carries a power of $\hbar$, and this is zero in the classical limit. So there exists, strictly speaking, no classical source of torsion. Moreover, since torsion is assumed to couple to spin with the universal gravitational coupling strength, its smallness implies that spin cannot play any sizable role in the forces between celestial bodies. For example, even if the earth consisted only of polarized atoms, its total intrinsic spin would be 10^{15} times smaller than the rotational spin around the axis.

A second argument against a gravitational theory with torsion was brought forward by Weinberg [1]. Since torsion is a tensor there is really no need for it to ensure general coordinate invariance.

A third argument is that the coupling of spinning particles to torsion would make the photon massive. To avoid this, the authors of [8] forbid the direct coupling to photons. They ignore the fact that by the allowing spin-1 ρ mesons to couple to torsion, their electromagnetic coupling will make photons massive after all.

Until the recent results of the satellite experiment Gravity Probe B [11] some people had hoped that the observed Lense-Thirring effect would deviate from Einstein's prediction. But it did not. Thus a simple possibility of discovering torsion experimentally has faded [12].

These problems make it doubtful that the generalization of gravity to Riemann-Cartan spacetime proposed in the papers [7, 8, 9] has a chance of being true. In this note we want to give a symmetry argument for this. It is inspired by a simple analog model of gravity, a *world crystal* with defects [9, 13] [2] whose lattice constant is of the order of a Planck length. Some consequences of such a world crystal were pointed out in a recent study of black holes in such a scenario [19].

In the world crystal, there exists a new type of extra gauge symmetry in which zero torsion is merely a special gauge. A completely equivalent gauge is the absence of Cartan curvature, which is found in Einstein's teleparallel universe. And there exists an infinite number of intermediate gauges.

3. To prepare the grounds for the argument, recall that a crystal can have two different types of topological line-like defects [7, 9], which in a four-dimensional world crystal are world surfaces (which may be the objects of string theory).

First, there are translational defects called *dislocations* (Fig. 1). These are produced by a

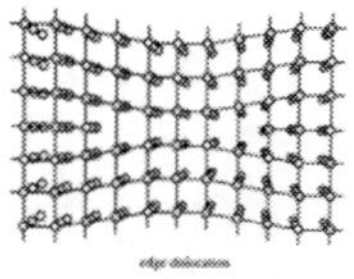

Figure 1. Formation of a dislocation line (of the edge type) by a Volterra process. The Burgers vector **b** characterizes the missing layer. There exist two more types where **b** points in orthogonal directions.

cutting process due to Volterra: a single-atom layer is removed from the crystal, allowing the remaining atoms to relax to equilibrium under the elastic forces. A second type of topological defects is of the rotation type, the so-called *disclinations* (Fig. 2). They arise by removing an entire wedge from the crystal and re-gluing the free surfaces.

The defects cause a failure of derivatives to commute in front of the displacement field $u_i(\mathbf{x})$. In three dimensions, the dislocation density is given by the tensor

$$\alpha_{ij}(\mathbf{x}) = \epsilon_{ikl}\nabla_k\nabla_l u_j(\mathbf{x}). \tag{1}$$

If $\omega_i \equiv \frac{1}{2}\epsilon_{ijk}[\nabla_j u_k(\mathbf{x}) - \nabla_k u_j(\mathbf{x})]$ denotes the local rotation field, the disclination density is defined by

[1] See the letter exchanges in the 2006 issue *Physics Today* **60**, 10, 16 following Weinberg's article [10].

[2] The gravitational action in this model is obtained by integrating out the lattice vibrations. It is thus a consequence of their *entropy*. This mechanism has recently been emphasized by Verlinde [14]. The same mechanism had been used a long time ago to generate stiffness of strings [15, 16] This is of course just reformulation of good-old Sakharov's idea [17, 18].

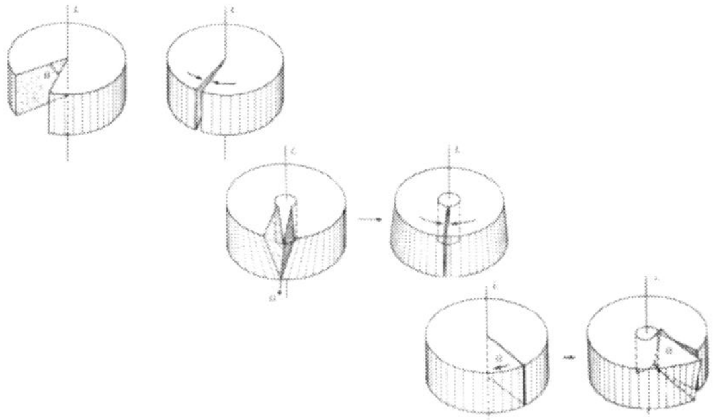

Figure 2. Three different possibilities of constructing disclinations: wedge, splay, and twist disclinations. They are characterized by the Frank vector $\mathbf{\Omega}$.

$$\theta_{ij}(\mathbf{x}) = \epsilon_{ikl}\nabla_k\nabla_l\omega_j(\mathbf{x}). \tag{2}$$

The defect densities satisfy the conservation laws

$$\nabla_i\theta_{ij} = 0, \quad \nabla_i\alpha_{ij} = -\epsilon_{jkl}\theta_{kl}. \tag{3}$$

These are fulfilled as Bianchi identities if we express $\theta_{ij}(\mathbf{x})$, $\alpha_{ij}(\mathbf{x})$ in terms of plastic gauge fields $\beta^p_{kl}, \phi^p_{lj}$, setting $\theta_{ij} = \epsilon_{ikl}\nabla_k\phi^p_{lj}$, $\alpha_{il} = \epsilon_{ijk}\nabla_j\beta^p_{kl} + \delta_{il}\phi^p_{kk} - \phi^p_{li}$. The defect densities are invariant under the gauge transformations $\beta^p_{kl} \to \beta^p_{kl} + \nabla_k u^p_l - \epsilon_{klr}\omega^p_r$, $\phi^p_{li} \to \phi^p_{li} + \partial_l\omega^p_i$, where $\omega^p_i \equiv \frac{1}{2}\epsilon_{ijk}\nabla_j u^p_k$. Thus $h_{ij} \equiv \beta^p_{ij} + \epsilon_{ijk}\omega^p_k$ and $A_{ijk} \equiv \phi^p_{ij}\epsilon_{jkl}$ are *translational* and *rotational* defect gauge fields in the crystal.

4. In order to appreciate the strategy of the paper, let us discuss a simple analog field theoretic model. Instead of Einstein's theory, we consider a model of a simple real field ρ with an Euclidean Lagrangian $\mathcal{L} = (\partial_\mu\rho)^2 - \rho^2 + \rho^4$ and a partition function $Z = \int\mathcal{D}\rho\, e^{-\int dx\mathcal{L}}$. The field ρ plays the role of the real metric tensor field $g_{\mu\nu}$. We may extend the above model trivially by an extra gauge field A_μ that has no physical consequences. This is done by re-expressing the Lagrangian in terms of a complex field $\psi = e^{i\theta}\rho$ and the gauge field A_μ as $\bar{\mathcal{L}} = |(\partial_\mu - iA_\mu)\psi|^2 - |\psi|^2 + |\psi|^4$. Now we form the partition function $\bar{Z} = \int\mathcal{D}\psi\mathcal{D}\psi^*\mathcal{D}A_\mu\Phi\, e^{-\int dx\bar{\mathcal{L}}}$, where Φ is an arbitrary gauge-fixing functional multiplied by the associated Faddeev-Popov determinant.

It is easy to verify that the classical field equations following from the new $\bar{Z}$ are exactly the same as those of the original Z.

The same thing holds for the flcutuating theory. The gauge field A_μ possesses two different spin contents. One is of spin zero, a pure gauge field $A_\mu = \partial_\mu\Lambda$ which possesses no electromagnetic field strength and leaves the physical content of the field system completely invariant. The second content has spin 1, where A^μ is orthogonal to the spin-zero part. This does possess an electromagnetic field $F_{\mu\nu} = \partial_\mu A_\nu - \partial_\nu A_\mu$. . Since the action is independent of $F_{\mu\nu}$ this spin-1 content does not contribute to the partition function $\bar{Z}$ either. If an action term of the Maxwell type $F_{\mu\nu}F^{\mu\nu}$ were present in the exponent of $\bar{Z}$, the theory would be only invariant under ordinary gauge transformations with single-valued scalar fields $\Lambda(x)$. Without such a Maxwell term, it is invariant under *all* gauge transformations with single and multi-valued [9] scalar fields $\Lambda(x)$, and thus to fluctuation of electromagnetic fields.

There is, obviously, no way of observing A_μ. In the gravitational theory to be discussed, the partition function Z of the ρ field plays the role of Einstein's theory formulated in terms of the metric field $g_{\mu\nu}$. The reformulation $\bar{Z}$ in terms of a gauge field will be the gauge field reformulation of Einstein's theory. The decomposition

$$\rho = \psi^*\psi = (\rho e^{-i\theta})(e^{i\theta}\rho) \tag{4}$$

will be the analog of the analog of this decomposition made for the metric rather than the real field ρ. The local phase θ may be any arbitrary function, single- as well as multivalued.

5. The Volterra processes can be represented mathematically by multivalued transformations from an Euclidean crystal with coordinates $\bar{x}^a$ to a crystal with defects and coordinates x^μ, as illustrated in Figs. 3 and 4 for two-dimensional crystals.

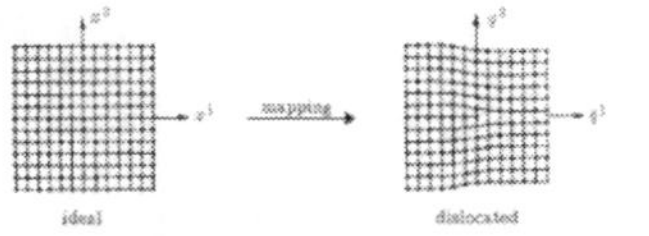

Figure 3. Multivalued mapping of the perfect crystal to an edge dislocation with a Burgers vector **b** pointing in the 2-direction.

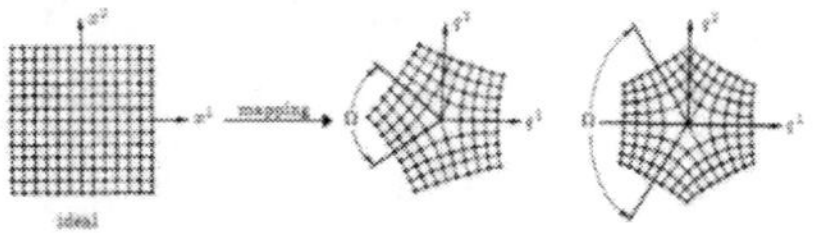

Figure 4. Multivalued mapping of the perfect crystal to a wedge disclination of Frank vector Ω in the third direction.

For an edge dislocation the mapping is $\bar{x}^1 = x^1$, $\bar{x}^2 = x^2 + (b/2\pi)\phi(x)$, where $\phi(x) \equiv (1/2\pi)\arctan(x^2/x^1)$. Initially, this function has a cut from the origin towards left infinity. In a second step, the cut is removed and the multivalued version of the arctan is taken. This makes $\phi(\mathbf{x})$ the Green function of the commutator $[\partial_1, \partial_2]$: $(\partial_1\partial_2 - \partial_2\partial_1)\phi(x) = \delta^{(2)}(\mathbf{x})$. For a wedge disclination, the mapping is $d\bar{x}^i = \delta^i{}_\mu\,[x^\mu + (\Omega/2\pi)\varepsilon^\mu{}_\nu x^\nu \phi(x)]$.

A combination of the two

$$\eta_{ij}(\mathbf{x}) \equiv \theta_{ij}(\mathbf{x}) - \tfrac{1}{2}\nabla_m[\epsilon_{min}\alpha_{jn}(\mathbf{x}) + \{ij\} + \epsilon_{ijn}\alpha_{mn}] \tag{5}$$

forms the *defect tensor*

$$\eta_{ij}(\mathbf{x}) \equiv \epsilon_{ikl}\epsilon_{jmn}\nabla_k\nabla_m u^p_{ln}(\mathbf{x}), \quad u^p_{ln} \equiv \tfrac{1}{2}(\beta^p_{ln} + \beta^p_{nl}). \tag{6}$$

It is a symmetric tensor due to the conservation laws (3), and represents the Einstein tensor $\bar{G}_{ij} \equiv \bar{R}_{ij} - \frac{1}{2}g_{ij}\bar{R}_k{}^k$ of the geometry of the world crystal [3].

The expressions can easily be defined on a simple-cubic world crystal if we replace ∇_i by lattice derivatives, as shown in [7, 9]. There it is also shown that, in three spacetime dimensions, the disclination density $\theta_{ij}(\mathbf{x})$ represents the Einstein tensor $G^{\rm C}_{ij}$ associated with the *Cartan curvature tensor* $R^{\rm C}_{ijk}{}^l$ of the Riemann-Cartan geometry of the world crystal. The relation is

$$G^{\rm C}_{ji}(\mathbf{x}) = \epsilon_{ikl}\nabla_k\nabla_l\omega_j(\mathbf{x}) = \theta_{ij}(\mathbf{x}). \tag{7}$$

The dislocation density $\alpha_{ij}(\mathbf{x})$ represents the torsion $S_{lkj} = \frac{1}{2}(\Gamma_{lkj} - \Gamma_{klj})$ of the Riemann-Cartan geometry. Here the relation is

$$\alpha_{ij} = \epsilon_{ikl}S_{lkj}. \tag{8}$$

6. The standard form of a defect with Burgers vector b_l and Frank vector Ω_q has a displacement field

$$u_l(\mathbf{x}) = -\delta(\mathbf{x}, V)[b_l + \epsilon_{lqr}\Omega_q(x_r - \bar{x}_r)], \tag{9}$$

[3] Geometric objects formed from Christoffel symbols are denoted by a bar. Otherwise the full affine connection is used.

where ϵ_{lqr} is the antisymmetric unit tensor, $\bar{x}_r$ the axis of rotation of the disclination part, and $\delta(\mathbf{x}; V)$ is the delta function on the volume V, i.e., in three dimensions:

$$\delta(\mathbf{x}; V) = \int_V d^3x' \, \delta^{(3)}(\mathbf{x} - \mathbf{x}'). \tag{10}$$

Its derivative is the delta function on the Volterra surface S of V:

$$-\nabla\delta(\mathbf{x}; V) = \delta(\mathbf{x}; S) = \int_S d\mathbf{S}' \, \delta^{(3)}(\mathbf{x} - \mathbf{x}'). \tag{11}$$

For the new gauge symmetry, the crucial observation is that as a simple consequence of (9), a dislocation line in the world crystal can either be obtained by a Volterra process of cutting out a thin slice of material of thickness $\mathbf{b}$, or alternatively by cutting out a wedge of Frank vector $\mathbf{\Omega}$, and reinserting it at distance $\mathbf{b}$ from the cut. Thus the dislocation line is indistinguishable from a pair of disclination lines with opposite Frank vector Ω whose axes of rotation are separated by a distance $\mathbf{b}$ (Fig. 5a). Conversely, a disclination line is equivalent to a stack of dislocation lines with fixed Burgers vector $\mathbf{b}$ (Fig. 5b).

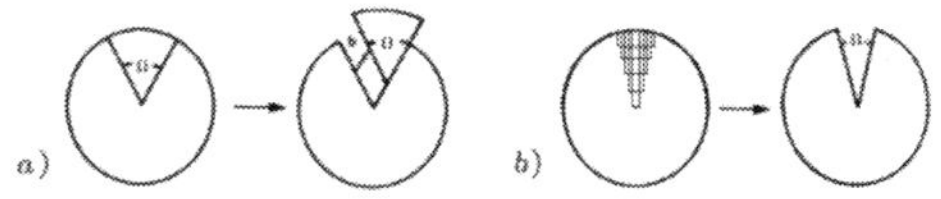

Figure 5. Equivalence between a) dislocation and pair of disclination lines, b) disclination and stack of dislocation lines.

Analytically, this is most easily seen in the two-dimensional version of the relation (5):

$$\eta_{33} = \theta_{33} + \epsilon_{3mn}\nabla_m\alpha_{3n}. \tag{12}$$

Each term is invariant under the plastic gauge transformations $\beta^p_{kl} \to \beta^p_{kl} + \nabla_k u^p_l - \epsilon_{kl}\omega^p_3$, $\phi^p_l \to \phi^p_l + \partial_l\omega^p_3$. The general defect has a displacement field

$$u_l = -\delta(V_2)[b_l - \Omega\epsilon_{3lr}(x_r - \bar{x}_r)]. \tag{13}$$

The first term is a dislocation, the second term a disclination. According to Fig. 5, the latter can be read as a superposition of dislocations with the same Burgers vector $\tilde{b}_l = -\int_{\bar{x}}^{x} dx'_r \Omega\epsilon_{3lr}$. The former may be viewed as a dipole of disclinations: $-\bar{\nabla}_l[-\frac{1}{2}b_m\epsilon_{3km}]\epsilon_{3kr}(x_r - \bar{x}_r)$.

7. Generalizing the defect relations (5) and (12) to $D \geq 4$ spacetime dimensions and allowing for large deviations from Euclidean space, we find the defect relation

$$\bar{G}_{\mu\nu} = G_{\mu\nu} - \tfrac{1}{2}D^{*\lambda}\left(S_{\mu\nu,\lambda} - S_{\nu\lambda,\mu} + S_{\lambda\mu,\nu}\right) \tag{14}$$

where $\bar{G}_{\mu\nu}$ is the Einstein tensor and $G_{\mu\nu}$ its Cartan version, while $S_{\mu\kappa}{}^{,\tau}$ is the Palatini tensor related to the torsion field $S_{\mu\kappa}{}^{\tau}$ by

$$\tfrac{1}{2}S_{\mu\kappa}{}^{,\tau} \equiv S_{\mu\kappa}{}^{\tau} + \delta_\mu{}^\tau S_{\kappa\lambda}{}^\lambda - \delta_\kappa{}^\tau S_{\mu\lambda}{}^\lambda. \tag{15}$$

The symbol D_μ denotes the covariant derivative defined by $D_\mu v_\nu \equiv \partial_\mu v_\nu - \Gamma_{\mu\nu}{}^\lambda v_\lambda$, $D_\mu v^\lambda \equiv \partial_\mu v^\lambda + \Gamma_{\mu\nu}{}^\lambda v^\nu$, and $D^*_\mu \equiv D_\mu + 2S_{\mu\kappa}{}^\kappa$. The defect conservation laws (3) read

$$D^*_\mu G_\lambda{}^\mu + 2S^{\nu\lambda\kappa}G_{\kappa\nu} - \tfrac{1}{2}S^{\nu\kappa,\mu}R_{\lambda\mu\nu\kappa} = 0, \tag{16}$$

$$D^{*\mu}S_{\lambda\kappa,\mu} = G_{\lambda\kappa} - G_{\kappa\lambda}. \tag{17}$$

They are Bianchi identities ensuring the single-valuedness of observables, connection $\Gamma_{\mu\nu}{}^\lambda$ and metric $g_{\mu\nu}$, via the integrability conditions $[\partial_\sigma, \partial_\tau]\Gamma_{\mu\nu}{}^\lambda = 0$ and $[\partial_\sigma, \partial_\tau]g_{\mu\nu} = 0$.

In a four-dimensional Riemann-Cartan spacetime, the geometry is described by the direct generalizations of *translational* and *rotational* defect gauge fields h_{ij} and A_{ijk}, which are here the vierbein field $h^\alpha{}_\mu$, and the spin connection $A_{\mu\alpha}{}^\beta$. The square of the former is the metric $g_{\mu\nu} = h^\alpha{}_\mu h_{\alpha\nu}$. The latter is defined by the covariant derivative $D_\lambda h_\beta{}^\mu = \partial_\lambda h_\beta{}^\mu - A_{\lambda\beta}{}^\gamma h_\gamma{}^\mu + \Gamma_{\lambda\nu}{}^\mu h_\beta{}^\nu \equiv D^L_\lambda h_\beta{}^\mu + \Gamma_{\lambda\nu}{}^\mu h_\beta{}^\nu$. The field strength of $A_{\mu\alpha}{}^\beta \equiv (A_\mu)_\alpha{}^\beta$

$$F_{\mu\nu\beta}{}^{\gamma} \equiv \{\partial_\mu A_\nu - \partial_\nu A_\nu - [A_\mu, A_\nu]\}_\lambda{}^\kappa, \tag{18}$$

determines the Cartan curvature $R_{\mu\nu\lambda}{}^{\kappa} \equiv h^\beta{}_\lambda h_\gamma{}^\kappa F_{\mu\nu\beta}{}^\gamma$. The field strength of $h^\gamma{}_\nu$ is the torsion:

$$S_{\alpha\beta}{}^{\gamma} \equiv \tfrac{1}{2} h_\alpha{}^\mu h_\beta{}^\nu [D^L_\mu h^\gamma{}_\nu - (\mu \leftrightarrow \nu)]. \tag{19}$$

The relations (14), (16), and (17) follow from this.

8. The theory is gauge invariant under local Lorentz transformations as a direct consequence of the fact that the metric can alternatively be written as [compare (4)]

$$g_{\mu\nu} = h^\gamma{}_\mu \Lambda^a{}_\gamma \Lambda_a{}^\beta h_{\beta\nu}, \tag{20}$$

where $\Lambda_a{}^\beta$ is an *arbitrary local* Lorentz transformation (single- as well as multi-valued), and that the Hilbert-Einstein Lagrangian $\mathcal{L}_{\rm EH} = -(1/2\kappa)R$ is *independent* of $\Lambda^a{}_\alpha$. The extra $\Lambda_a{}^\beta$ transforms the gauge field $A_{\mu\alpha}{}^\beta$ as

$$A_{\mu\alpha}{}^\beta \to A_{\mu\alpha}{}^\beta + \Delta A_{\mu\alpha}{}^\beta, \quad \Delta A_{\mu\alpha}{}^\beta \equiv \Lambda_a{}^\beta \partial_\mu \Lambda^a{}_\alpha. \tag{21}$$

At this point are ready to introduce the *new* gauge invariance announced in the title: we allow $\Lambda_a{}^\beta$ in Eq. (20) to be a *multivalued Lorentz transformation*. This is *not* integrable, so that $\Delta A_{\mu\alpha}{}^\beta$ is a *nontrivial* gauge field. Indeed, the rotational field strength $F_{\mu\nu\alpha}{}^\gamma$ can be expressed as $F_{\mu\nu\alpha}{}^\gamma \equiv \Lambda_a{}^\gamma[\partial_\mu, \partial_\nu]\Lambda^a{}_\alpha \neq 0$ and yields a nonzero Cartan curvature $R_{\mu\nu\lambda}{}^\kappa \neq 0$. The important observation is that a *multivalued* $\Lambda^a{}_\alpha$ is able to *change the geometry* [4]. The right-hand side of (14) is *independent* of the vector field $A_{\mu\alpha}{}^\beta$. This allows us to move torsion into Cartan curvature and back, fully or partially, by complete analogy with the defect transformations in two-dimensional crystals in Fig. 21. We can choose for $A_{\mu\alpha}{}^\beta$ any function we like. For example we may choose it to make the torsion vanish, and $A_{\mu\alpha}{}^\beta$ reduces to the usual spin connection of Einstein's gravity, the well-known combination of the objects of anholonomity $\Omega_{\mu\nu}{}^\lambda = \frac{1}{2}[h_\alpha{}^\lambda \partial_\mu h^\alpha{}_\nu - (\mu \leftrightarrow \nu)]$. In the opposite extreme $A_{\mu\alpha}{}^\beta = 0$, the Cartan curvature is zero, spacetime is teleparallel, and the Lagrangian is equal to the combination of torsion tensors: $\mathcal{L}_S = -(1/2\kappa)(-4D_\mu S^\mu + S_{\mu\nu\lambda}S^{\mu\nu\lambda} + 2S_{\mu\nu\lambda}S^{\mu\lambda\nu} - 4S^\mu S_\mu)$, where $S_\mu \equiv S_{\mu\nu}{}^\nu$. In any of the new gauges, the field equations follow from the Hilbert-Einstein Lagrangian

$$\mathcal{L}_{\rm EH} = -(1/2\kappa)R + \mathcal{L}_S, \tag{22}$$

where $A_{\mu\alpha}{}^\beta$ is fixed by any some convenient gauge-fixing functional. It does not follow any field equation.

9. Adding matter fields of masses m to the Einstein Lagrangian, and varying with respect to $h^\alpha{}_\mu$, we find in the zero-torsion gauge the Einstein equation

$$G_{\mu\nu} = \kappa T_{\mu\nu}, \tag{23}$$

where $T_{\mu\nu}$ is the sum over the symmetric energy-momentum tensors of all matter fields. Each contains the canonical energy-momentum tensor ${}^{\rm m}\Theta_{\mu\nu}$ and the spin current densities ${}^{\rm m}\Sigma_{\mu\nu}{}^\lambda$ in the combination due to Belinfante [20] [5],

$${}^{\rm m}T_{\kappa\nu} = {}^{\rm m}\Theta_{\kappa\nu} - \tfrac{1}{2}D^{*\mu}\left({}^{\rm m}\Sigma_{\kappa\nu,\mu} - {}^{\rm m}\Sigma_{\nu\mu,\kappa} + {}^{\rm m}\Sigma_{\mu\kappa,\,\nu}\right), \tag{24}$$

which is the matter analog of the defect relation (14).

[4] The new freedom brought about by multivalued gauge transformations in many areas of physics is explained in the textbook [9]. For instance, we can *derive* the physical laws in a magnetic field from those in field-free space, thus finding the minimal coupling rule of the magnetic vector potential. Similarly, we can *derive* the physical laws in curved space from those in flat space.

[5] For more details see Sect. 17.7 in the textbook [9].

The new gauge invariance of (14) has the physical consequence that the external gravitational field in the far-zone of a celestial body does not care whether angular momentum comes from rotation of matter or from internal spins. The off-diagonal elements of the metric in the far-zone, and thus the Lense-Thirring effect measured in [11], depend only on the total angular momentum $J^{\lambda\mu} = \int d^3x(x^\lambda T^{\mu 0} - x^\mu T^{\lambda 0})$, which by the Belinfante relation (24) is the sum of orbital angular momentum $L^{\lambda\mu} = \int d^3x(x^\lambda \Theta^{\mu 0} - x^\mu \Theta^{\lambda 0})$ and spin $S^{\lambda\mu} = \int d^3x \Sigma^{\lambda\mu,0}$. A star consisting of polarized matter has the same external gravitational field in the far-zone as a star rotating with the corresponding orbital angular momentum. This is the *universality of orbital momentum and intrinsic angular momentum* in gravitational physics observed in Ref. [21].

Since torsion is merely a *new-gauge degree* of freedom in describing a gravitational field, it cannot be detected experimentally, not even by spinning particles. A field with arbitrary spin may be coupled to gravity via the covariant derivative $D_\mu \equiv \partial_\mu \mathbf{1} + \frac{i}{2} A_{\mu\alpha}{}^\beta \Sigma^\alpha{}_\beta$, where $\Sigma^\alpha{}_\beta$ are the generators of the Lorentz group, in the Dirac case $\Sigma_{\alpha\beta} = \frac{i}{4}[\gamma^\alpha, \gamma_\beta]$. But since the torsion is a tensor, we may equally well use an infinity of alternative covariant derivatives $D^q_\mu \equiv \partial_\mu \mathbf{1} + \frac{i}{2} A^q_{\mu\alpha}{}^\beta \Sigma^\alpha{}_\beta$, where $A^q_{\mu\alpha}{}^\beta \equiv A_{\mu\alpha}{}^\beta - qK_{\mu\alpha}{}^\beta$, and $K_{\mu\alpha\beta} = h_\alpha{}^\nu h_\beta{}^\lambda K_{\mu\nu\lambda} \equiv h_\alpha{}^\nu h_\beta{}^\lambda (S_{\mu\nu\lambda} - S_{\nu\lambda\mu} + S_{\lambda\mu\nu})$. Any coupling constant q is permitted by covariance. In order to see which q is physically correct we come back to the above-discussed photon mass problem, and consider the covariant electromagnetic field tensor $F^q_{\mu\nu} \equiv D^q_\mu A_\nu - D^q_\nu A_\mu$. Working out the covariant derivative we find $\partial_\mu A_\nu - \partial_\nu A_\mu - 2(1-q) S_{\mu\nu}{}^\lambda A_\lambda$, which shows that Maxwell Lagrangian $-\frac{1}{4} F^q_{\mu\nu} F^{q\,\mu\nu}$ acquires a mass term, unless we fix the coupling strength to the value $q = 1$.

For this value of q, a little algebra [7, 9] shows that the torsion drops out from the gauge field $A^q_{\mu\alpha}{}^\beta$. This reduces to the good-old Fock-Ivanenko spin connection that has been used in Einstein gravity without torsion:

$$A^1_{\mu\alpha}{}^\beta = \bar{A}_{\mu\alpha}{}^\beta = h_\alpha{}^\nu h^{\beta\lambda}(\Omega_{\mu\nu\lambda} - \Omega_{\nu\lambda\mu} + \Omega_{\lambda\mu\nu}). \tag{25}$$

Having ensured that the photon does not couple to torsion, this choice also prevents all all other spinning baryonic matter to do so, thus avoiding that a photon mass arises via virtual processes.

10. In summary, we have shown that if the Hilbert-Einstein Lagrangian is expressed in terms of the translational and rotational gauge fields $h^\alpha{}_\mu$ and $A^1_{\mu\alpha}{}^\beta$, the Cartan curvature can be converted to torsion and back, totally or partially, by a *new type of multivalued gauge transformation* in Riemann-Cartan spacetime. In this general formulation, Einstein's original theory is obtained by going to the zero-torsion gauge, while his teleparallel theory is in the gauge in which the Cartan curvature tensor vanishes. But any intermediate choice of the field $A_{\mu\alpha}{}^\beta$ is allowed.

Higher gradient terms in the elastic energy of the world crystal are capable of generating an action of the gauge field [6] $A_{\mu\alpha}{}^\beta$. These would break the above gauge symmetry and give the gauge field a life of its own. The coupling to spin, however, must go only via the torsion-free soin connection $A^1_{\mu\alpha}{}^\beta$ to avoid a photon mass.

References

[1] Debever R 1979 (ed) *Elie Cartan–Albert Einstein, Letters on Absolute Parallelism 1929-1932* (Princeton: Princeton University Press)

[2] Hayashi K and Shirafuji T 1979 *Phys. Rev.* D **19** 3524

[3] Arcos H I and Pereira J G 2004 *Preprint* arXiv:gr-qc/0408096v2; *do.* 2005 *Int. J. Mod. Phys.* D **13** 2193 (*Preprint* arXiv:gr-qc/0501017v1); De Andrade V C, Arcos H I and Pereira J G 2004 *Preprint* arXiv:gr-qc/0412034

[4] De Andrade V C, Pereira J G 1999 *Int. J. Mod. Phys.* D **8** 14 (*Preprint* arXiv:gr-qc/9708051)

[6] See p. 1453 in Ref. [7] (**physik.fu-berlin.de/~kleinert/b1/gifs/v1-1453s.html**).

[5] Kleinert H 2010 *Preprint* physik/fu-berlin.de/ kleinert/385/385.pdf
[6] Bilby B A, Bullough R and Smith E 1955 *Proc. Roy. Soc. London* A **231** 263;
Kondo K 1962 in *Proceedings of the II Japan National Congress on Applied Mechanics, Tokyo, 1952* publ. in *RAAG Memoirs of the Unified Study of Basic Problems in Engineering and Science by Means of Geometry* Vol. **3** ed K Kondo (Gakujutsu Bunken Fukyu-Kai) p. 148
[7] Kleinert H 1989 Gauge Fields in Condensed Matter, Vol. II – Stresses and Defects (Singapore: World Scientific) pp. 744-1443 (www.physik.fu-berlin.de/~kleinert/b2)
[8] Utiyama R 1956 *Phys. Rev.* **101** 1597;
Kibble T W B 1961 *J. Math. Phys.* **2** 212;
Sciama D W 1964 *Rev. Mod. Phys.* **36** 463;
Hehl F W, Von der Heyde P, Kerlick G D and Nester J M 1976 *Rev. Mod. Phys.* **48** 393;
Hehl F W, McCrea J D, Mielke E W and Neemann Y 1995 *Phys. Rep.* **258** 1;
Hammond R T 2002 *Rep. Prog. Phys.* **65** 599;
Ni Wei-Tou 2010 *Rep. Prog. Phys.* **73** 056901 (*Preprint* arXiv:0912.5057)
[9] Kleinert H 2008 *Multivalued Fields in Condensed Matter, Electromagnetism, and Gravitation* (Singapore: World Scientific) pp. 1-497 (www.physik.fu-berlin.de/~kleinert/b11)
[10] Weinberg S 2005 *Physics Today* **58** 31
[11] See einstein.stanford.edu
[12] Mao Yi, Tegmark M, Guth A, Cabi S 2007 *Phys. Rev.* D **76** 104029
Flanagan É É and Rosenthal E 2007 *Phys. Rev.* D **75** 124016
[13] Kleinert H 1987, *Ann. d. Physik* **44** 117 (http://physik.fu-berlin.de/~kleinert/172/172.pdf)
[14] Verlinde E P 2001 *Preprint* arXiv:1001.0785
[15] Kleinert 1988 Dynamical generation of string tension and stiffness *Phys. Lett.* B **211** 151
[16] Kleinert H 1989 Membrane stiffness from van der Waals forces *Phys. Lett.* A **136** 253
[17] Sakharov A D 1976 *Dokl. Akad. Nauk SSSR* **170** 70 [*Soviet Physics-Doklady* **12** 1040 (1968)]
[18] Schmidt H J 2000 *Gen. Rel. Grav.* **32** 361 (www.springerlink.com/content/t51570769p123410/fulltext.pdf)
[19] Jizba P, Kleinert H and Scardigli F Uncertainty relation on world crystal and its applications to micro black holes *Preprint* arXiv:0912.2253
[20] Belinfante F J 1039 *Physica* **6** 887
[21] Kleinert H 2000 *Gen. Rel. Grav.* **32** 1271 (physik.fu-berlin.de/~kleinert/271/271j.pdf)

Prequantum classical statistical field theory: background field as a source of everything?

Andrei Khrennikov
International Center for Mathematical Modelling
in Physics and Cognitive Sciences
Linnaeus University, Växjö, Sweden

E-mail: andrei.khrennikov@lnu.se

Abstract. Prequantum classical statistical field theory (PCSFT) is a new attempt to consider quantum mechanics (QM) as an emergent phenomenon, cf. with De Broglie's "double solution" approach, Bohmian mechanics, stochastic electrodynamics (SED), Nelson's stochastic QM and its generalization by Davidson, 't Hooft's models and their development by Elze. PCSFT is a comeback to a purely wave viewpoint on QM, cf. with early Schrödinger. There is no quantum particles at all, only waves. In particular, photons are simply wave-pulses of the classical electromagnetic field, cf. SED. Moreover, even massive particles are special "prequantum fields": the electron field, the neutron field, and so on. PCSFT claims that (sooner or later) people will be able to measure components of these fields: components of the "photonic field" (the classical electromagnetic field of low intensity), electronic field, neutronic field, and so on. At the moment we are able to produce quantum correlations as correlations of classical Gaussian random fields. In this paper we are interested in mathematical and physical reasons of usage of Gaussian fields. We consider prequantum signals (corresponding to quantum systems) as composed of a huge number of wave-pulses (on very fine prequantum time scale). We speculate that the prequantum background field (the field of "vacuum fluctuations") might play the role of a source of such pulses, i.e., the source of everything.

1. Introduction

At the very beginning of QM the idea that quantum mechanics is simply a special model of *wave mechanics* was quite popular. It was supported by the discovery of "quantum wave equation" by Schrödinger and by the association with each particle its wave-length, the De Broglie wave-length. However, in spite of increasing success in "technical applications" (concrete spectra were found by using Schrödinger's representation), quantum wave mechanics was ideologically inconsistent. Both Schrödinger and De Broglie should give up in front of the difficulties of the interpretation of "quantum waves" as real physical waves. Although both, Schrödinger and De Broglie, did not like the Copenhagen interpretation, they were not able to present a consistent "physical waves approach" to QM. Finally, the interpretation of quantum waves as waves of probability (proposed by Born) became commonly accepted.

One of the main problems of the "physical wave interpretation" was the impossibility to describe a composite system by waves defined on the physical space, $X = \mathbf{R}^3$. The wave function of a composite system is defined on the space $X_m = \mathbf{R}^{3m}$, where m is the number of subsystems. Pauli wrote that one may consider physical waves, but they will be defined on nonphysical space. In particular, this problem was the main reason why Schrödinger accepted the probabilistic interpretation of the wave function.

Nevertheless, after 80 years of the dominance of the Copenhagen interpretation, nowadays various wave models are very popular in attempts to go beyond QM. Stochastic electrodynamics (SED), e.g., [1], is one of the most successful models. Some ideas of SED are quite similar to those which will be discussed in this paper. A crucial point of SED is the assumption on the presence of the so called background field (zero-point field, field of quantum fluctuations). This idea is very physical. Everybody agrees that the "totally empty space" is only a mathematical idealization. By SED there are no "free quantum particles". It is impossible to isolate, e.g., an electron from the zero-point field. Dynamics of quantum systems is a motion in the sea of quantum fluctuations. The latter produce new dynamical and statistical effects which are known as quantum effects. By SED these effects are purely classical. The mystery of these effects is in the use of incomplete information, cf. with Einstein's views.

My model PCSFT [2]–[7], cf. SED [1], 't Hooft's approach and its development by Elze [8], [9], reproduces all probabilistic predictions of QM. Moreover, PCSFT provides a possibility to go beyond QM. However, in this paper we "simply" reproduce predictions of QM. In this approach quantum systems are described by *classical random fields*, i.e., classical fields depending on a random parameter. Operation with random fields is equivalent to operation with ensembles of fields. Thus our theory is classical statistical mechanics of fields (not particles).

Quantum averages can be interpreted as approximations of averages of functionals of classical ("prequantum") fields with respect to fields' fluctuations. By restricting the class of classical variables to *quadratic functionals of fields* we reproduce quantum averages precisely. In [10] I generalized PCSFT to composite quantum systems. The main discovery was that, opposite to a rather common opinion, a composite system can be described by *Cartesian product of state spaces* (like in classical physics) and not by the tensor product of them (like in QM). The natural interpretation of a pure quantum state for a composite system is proposed: *it is a symbolic (vector) representation of the nondiagonal block in the covariance matrix of the random field describing a composite system.* The interpretation of a pure state due to Dirac and von Neumann seems to be an *artifact of the conventional mathematical description* of micro-systems.

At the moment we are able to produce quantum correlations as correlations of classical *Gaussian* random fields. In this paper we are interested in mathematical and physical reasons of usage of Gaussian fields. We consider prequantum signals (corresponding to quantum systems) as composed of a huge number of wave-pulses (on very fine prequantum time scale). We speculate that the prequantum background field (the field of "vacuum fluctuations") might play the role of a source of such pulses, i.e., the source of everything.

To simplify considerations, we will study quantum systems with finite dimensional Hilbert spaces. Moreover, we consider a toy-model with the real Hilbert state space.

2. PCSFT-representation of QM-averages

Here we briefly present essentials of PCSFT for ensembles composed of single particles, see [2]–[7]. We start with the remark that both classical statistical mechanics (CSM) and QM are statistical theories which predict probabilities for various variables with respect to large (in the limit infinitely large) ensembles of physical systems. Both theories have the same aim: to predict averages of these variables. At the first sight these theories differ crucially. In particular, they are based on the use of (at the first sight) totally different mathematical tools.

In CSM averages are given by integrals (over the space of parameters characterizing systems); variables are given by functions of parameters, $\phi \to f(\phi)$, where $\phi \in M$ and M is space of states. Denote by μ the probability distribution of random parameters, it is a probability measure on

M. Average of a variable $f(\phi)$ with respect to the statistical state μ is given by

$$\langle f\rangle_\mu = \int_M f(\phi)d\mu(\phi). \tag{1}$$

By using the language of probability theory we can say that there is given a random vector ϕ with values in M. Then $\langle f\rangle_\phi = Ef(\phi) = \langle f\rangle_\mu$. Here and everywhere below E denotes classical mathematical expectation (average).

To finish this short remark on CSM, we emphasize that if systems $S_i, i = 1,2,...,k$, have state spaces M_i, respectively, then the composite system $S = (S_i)_{i=1}^k$ has state space $M = M_1 \times \ldots \times M_k$, Cartesian product of state spaces M_i. Ensembles of S-systems are described by random vectors in $M : \omega \to \phi(\omega) = (\phi_1(\omega), \ldots, \phi_k(\omega))$. A trivial, but important, remark is that in general components of ϕ are not independent. There are nontrivial correlations between them. The best way to describe these correlations is to use the *covariance operator* (it will be defined a little bit later).

In QM-formalism observables are given by self-adjoint operators and states by density operators. These operators act in the Hilbert space. A special class of states, so called pure states, are represented by normalized vectors of Hilbert space. Average of an observable $\widehat{A}$ with respect to a state ρ is defined as

$$\langle \widehat{A}\rangle_\rho = \mathrm{Tr}\rho\widehat{\mathrm{A}}. \tag{2}$$

If quantum systems $S_i, i = 1,2,...,k$, have state spaces H_i, respectively, then the system $S = (S_i)_{i=1}^k$ has state space $H = H_1 \otimes \ldots \otimes H_k$, the tensor product of state spaces H_i.

Take Hilbert space M as space of classical states. To simplify the presentation, we assume that M is a real Hilbert space with the scalar product $(\cdot,\cdot)$. Consider a probability distribution μ on M having zero average (it means that $\int_M (y,\phi)d\mu(\phi) = 0$ for any $y \in M$.) and covariance operator D which is defined by a symmetric positively defined bilinear form:

$$(Dy_1, y_2) = \int_M (y_1,\phi)(y_2,\phi)d\mu(\phi), y_1, y_2 \in M, \tag{3}$$

By scaling we obtain operator

$$\rho = D/\mathrm{TrD}, \tag{4}$$

with $\mathrm{Tr}\rho = 1$. It can be considered as (Landau-von Neumann) density operator.

By PCSFT a quantum state, a density operator, is simply the symbolic representation of the covariance operator of the corresponding prequantum (classical) probability distribution.

We emphasize that at the moment we consider PCSFT $\to$ QM correspondence for a single system. It will be essentially different and more complicated for composite systems.

In general a probability distribution is not determined by its covariance operator. Thus the correspondence PCSFT $\to$ QM is not one-to-one. However, if the class of prequantum probability distributions is restricted to Gaussian, then this correspondence becomes one-to-one. In PCSFT classical variables are defined as functions from (Hilbert) state space M to real numbers,

$$f = f(\phi).$$

By PCSFT a quantum observable, a self-adjoint operator, is simply a symbolic representation of f by means of its second derivative (Hessian),

$$f \to \widehat{A} = \frac{1}{2}f''(0).$$

This correspondence is neither one-to-one. However, by restricting the class of classical variables to quadratic forms on Hilbert state space M,

$$f_A(\phi) = (\widehat{A}\phi, \phi),$$

we make correspondence PCSFT $\rightarrow$ QM one-to-one. And finally, we present the basic equality

$$\langle f_A \rangle_\mu (\equiv E f_A(\phi)) = \int_M f_A(\phi) d\mu(\phi) = \mathrm{Tr} D \widehat{\mathrm{A}} = (\mathrm{Tr D})\ \langle \widehat{\mathrm{A}} \rangle_\rho. \tag{5}$$

Thus quantum average $\langle \widehat{A} \rangle_\rho$ can be obtained as just scaling of the classical average. We remark that scaling parameter TrD is, in fact, the dispersion of the probability distribution μ :

$$\sigma^2 \equiv E||\phi||^2 = \int_M ||\phi||^2 d\mu(\phi) = \mathrm{Tr}\ D. \tag{6}$$

It describes the strength of deviations of a random vector ϕ from its average. The latter is zero in our model. If $M = L_2$ then we can consider ϕ as a random field. Its average is the field which is equal to zero everywhere. Suppose that σ^2 is very small, so deviations from zero are small. Since

$$\langle \widehat{A} \rangle_\rho = \frac{1}{\sigma^2} \langle f_A \rangle_\mu, \tag{7}$$

the quantum average can be considered as the average of amplified prequantum fluctuations.

This CSM-model for averages over ensembles consisting of single quantum particles was presented in [2]- [7]. To make it consistent with a coming CSM-model for ensembles of quantum composite systems, suppose that prequantum fields (representing quantum particles) "live" not in vacuum ("empty space"), but that, like in SED, space is filled by a background random field of the white noise type. At the moment we forget about scaling with respect to the dispersion of the prequantum random field. It is important if one proceeds beyond QM, but in this paper we just reproduce statistical predictions of QM.

Let ρ be a density operator. Set $D = \rho$. This operator determines uniquely the Gaussian prequantum random vector. (We repeat once again that we consider the case of a single system!) Set $\tilde{D} = D + \alpha I$, where I is the unit operator. Consider corresponding Gaussian random vector ϕ. Denote by μ its probability distribution. We have

$$E f_A(\phi) = \langle f_A \rangle_\mu = \mathrm{Tr} \tilde{D} \widehat{\mathrm{A}} = \mathrm{Tr} \rho \widehat{\mathrm{A}} + \alpha \mathrm{Tr} \widehat{\mathrm{A}}.$$

In this model (a modification of PCSFT created in [2]- [7]) the quantum average can be obtained as a shift of classical average:

$$\langle \widehat{A} \rangle_\rho = \langle f_A \rangle_\mu - \alpha \mathrm{Tr} \widehat{\mathrm{A}}. \tag{8}$$

3. Operator considerations

Let W be a real Hilbert space. We denote the space of self-adjoint operators acting in W by the symbol $\mathcal{L}_s(W)$. Since in this paper we consider only the finite dimensional real case, this space can be realized as the space of all symmetric matrices.

Let H_1 and H_2 be two real (finite dimensional) Hilbert spaces. We put $H = H_1 \otimes H_2$. Any vector $\Psi \in H$ can be represented in the form

$$\Psi = \sum_{j=1}^m \psi_j \otimes \chi_j,\ \psi_j \in H_1, \chi_j \in H_2, \tag{9}$$

and it determines a linear operator from H_2 to H_1

$$\widehat{\Psi} \phi = \sum_{j=1}^m (\phi, \chi_j) \psi_j,\ \phi \in H_2. \tag{10}$$

We remark that this operator is uniquely determined by Ψ, i.e., it does not depend on representation (9). Its adjoint operator Ψ^* acts from H_1 to H_2 :

$$\widehat{\Psi}^* \psi = \sum_{j=1}^{m} (\psi, \psi_j) \chi_j, \psi \in H_1. \tag{11}$$

Of course,

$$\widehat{\Psi}\widehat{\Psi}^* : H_1 \to H_1$$

and

$$\widehat{\Psi}^*\widehat{\Psi} : H_2 \to H_2$$

and these operators are self-adjoint and positively defined.

Consider operator

$$\rho = \Psi \otimes \Psi : H_1 \otimes H_2 \to H_1 \otimes H_2$$

and the operators

$$\rho^{(1)} \equiv \mathrm{Tr}_{\mathrm{H}_2} \rho$$

and

$$\rho^{(2)} \equiv \mathrm{Tr}_{\mathrm{H}_1} \rho.$$

If the vector Ψ is normalized by 1, then ρ is the density operator corresponding to the pure state Ψ and the operators $\rho^{(i)} \equiv \rho^{S_i}, i = 1, 2$, are the reduced density operators. These density operators describe quantum states of subsystems $S_i, i = 1, 2$, of a composite quantum system $S = (S_1, S_2)$.

For any $\Psi \in H_1 \otimes H_2$, the following equalities hold:

$$\rho^{(1)} = \widehat{\Psi}\widehat{\Psi}^*, \ \rho^{(2)} = \widehat{\Psi}^*\widehat{\Psi}. \tag{12}$$

For any pair of operators $\widehat{A}_j \in \mathcal{L}_s(H_j), j = 1, 2$,

$$\mathrm{Tr}\widehat{\Psi}\widehat{\mathrm{A}}_2\widehat{\Psi}^*\widehat{\mathrm{A}}_1 = \langle \widehat{\mathrm{A}}_1 \otimes \widehat{\mathrm{A}}_2 \rangle_\Psi \equiv (\widehat{\mathrm{A}}_1 \otimes \widehat{\mathrm{A}}_2 \Psi, \Psi). \tag{13}$$

Let the state vectors of systems S_1 and S_2 belong to Hilbert spaces H_1 and H_2, respectively. Then by QM the state vector Ψ of the composite system $S = (S_1, S_2)$ belongs to $H = H_1 \otimes H_2$. We remark that the interpretation of the state vector $\Psi \in H$ of a composite system is not as straightforward as for a single system. It is known that, in general, the pure state Ψ of a composite system does not determine pure states for its components. This viewpoint matches well with the PCSFT-approach. We interpret a normalized vector $\Psi \in H$ not as the state vector of a concrete composite system $S = (S_1, S_2)$, but as one of blocks of the covariance operator for the prequantum random field $\omega \to \phi(\omega) = (\phi_1(\omega), \phi_2(\omega))$ describing $S = (S_1, S_2)$.

4. PCSFT-representation of QM-correlations for pure quantum states

Let ϕ_1 and ϕ_2 be two random vectors, in Hilbert spaces H_1 and H_2, respectively. Consider Cartesian product of these Hilbert spaces:

$$\mathbf{H} = H_1 \times H_2$$

(don't mix with $H = H_1 \otimes H_2$) and the random vector

$$\omega \to \phi(\omega) = (\phi_1(\omega), \phi_2(\omega)) \in \mathbf{H}$$

such that:

a) its expectation $E\phi = 0$;
b) its dispersion $\sigma^2(\phi) = E||\phi||^2 < \infty$.
Take its covariance operator D which is determined by the symmetric (positive) bilinear form:

$$(Du, v) = E(u, \phi)(v, \phi),$$

where vectors $u, v \in \mathbf{H}$. This operator has the block structure

$$D = \begin{pmatrix} D_{11} & D_{12} \\ D_{21} & D_{22} \end{pmatrix},$$

where $D_{ii} : H_i \to H_i, D_{ij} : H_j \to H_i$. The operator is self-adjoint. Hence $D_{ii}^* = D_{ii}$, and $D_{12}^* = D_{21}$. Here:

$$(D_{ij}u_j, v_i) = E(u_j, \phi_j)(v_i, \phi_i), u_i \in H_i, v_j \in H_j.$$

Let $\widehat{A}_i \in \mathcal{L}_s(H_i), i = 1, 2$. It determines the quadratic function on the Hilbert space H_i :

$$f_{A_i}(\phi_i) = (\widehat{A}_i \phi_i, \phi_i).$$

We recall that we proceed in the finite dimensional real case. Thus $H_1 = \mathbf{R}^m$ and $H_2 = \mathbf{R}^n$, and $\mathbf{H} = \mathbf{R}^{n+m}$;

$$\phi_1(\omega) = (\phi_{11}(\omega), \phi_{12}(\omega), ..., \phi_{1m}(\omega))$$

and

$$\phi_2(\omega) = (\phi_{21}(\omega), \phi_{22}(\omega), ..., \phi_{2n}(\omega)).$$

Here D is represented by $(n + m) \times (n + m)$ matrix, D_{11} by $m \times m$ matrix and D_{22} by $n \times n$ matrix; all these matrices are positively defined; D_{12} is represented by $m \times n$ matrix.

Consider a Gaussian random vector. It is determined by its probability distribution on the space $\mathbf{H}$. In the case $D > 0$ it has the density:

$$p_D(u) = \frac{1}{\sqrt{(2\pi)^n \mathrm{detD}}} e^{\frac{-1}{2}(Ku,u)}, \; K = D^{-1}.$$

The inverse matrix K of the covariance matrix D also has the block structure. For the variable $u = (x, y), x \in H_1, y \in H_2$, the Gaussian density can be written as

$$p_D(x, y) = \frac{1}{\sqrt{(2\pi)^n \mathrm{detD}}} e^{\frac{-1}{2}[(K_{11}x,x)+2(K_{12}y.x)+(K_{22}y,y)]}.$$

Finally, we remark that in the finite dimensional case quadratic functions have the form:

$$f_{A_1}(x) = \sum_{i,j=1}^{m} A_{1ij}x_i x_j, x \in H_1, \;\; f_{A_2}(y) = \sum_{i,j=1}^{m} A_{2ij}y_i y_j, y \in H_2.$$

For any Gaussian random vector $\phi = (\phi_1, \phi_2)$ having zero average and any pair of operators $\widehat{A}_i \in \mathcal{L}_s(H_i), i = 1, 2$, the following equality takes place:

$$\langle f_{A_1}, f_{A_2} \rangle_\phi \equiv E f_{A_1}(\phi_1) f_{A_2}(\phi_2)$$

$$= (\mathrm{Tr}D_{11}\widehat{A}_1)(\mathrm{Tr}D_{22}\widehat{A}_2) + 2\mathrm{Tr}D_{12}\widehat{A}_2 D_{21}\widehat{A}_1. \tag{14}$$

To prove this, we calculate the integral

$$\int_{\mathbf{R}^{n+m}} (\widehat{A}_1 x, x)(\widehat{A}_2 y, y) p_D(x, y) dx dy,$$

see [10], where this integral is calculated for an arbitrary real Hilbert space.

We remark that

$$\mathrm{Tr} D_{ii} \widehat{A}_i = E f_{A_i}(\phi_i), i = 1, 2. \tag{15}$$

Thus we have

$$E f_{A_1} f_{A_2} = E f_{A_1} E f_{A_2} + 2 \mathrm{Tr} D_{12} \widehat{A}_2 D_{21} \widehat{A}_1. \tag{16}$$

Now take an arbitrary pure state of a composite system $S = (S_1, S_2)$, a normalized vector $\Psi \in H = H_1 \otimes H_2$.

Consider a Gaussian vector random field such that $D_{12} = \widehat{\Psi}$. By operator equality (13) the last summand in the right-hand side of (16) is equal to the QM-average. Hence, we obtain:

$$\frac{1}{2} E(f_{A_1} - E f_{A_1})(f_{A_2} - E f_{A_2}) = (\widehat{A}_1 \otimes \widehat{A}_2 \Psi, \Psi) \equiv \langle \widehat{A}_1 \otimes \widehat{A}_2 \rangle_\Psi, \tag{17}$$

or, for the covariance of two classical random vectors f_{A_1}, f_{A_2}, we have:

$$\frac{1}{2} \mathrm{cov}\, (f_{A_1}, f_{A_2}) = \langle \widehat{A}_1 \otimes \widehat{A}_2 \rangle_\Psi. \tag{18}$$

Operators D_{ii} are responsible for averages of random variables $\omega \to f(\phi_i(\omega))$, i.e., depending only on one of components of the vector random field ϕ. In particular, $E f_{A_i}(\phi_i) = \mathrm{Tr} D_{ii} \widehat{A}_i$.

We will construct such a random field that these averages will match those given by QM. It is natural to take, see [10] for discussion, the covariance operator:

$$D_\Psi = \begin{pmatrix} \widehat{\Psi}\widehat{\Psi}^* + I & \widehat{\Psi} \\ \widehat{\Psi}^* & \widehat{\Psi}^* \Psi + I \end{pmatrix}.$$

Suppose now that ϕ is a random vector with the covariance operator D_Ψ. Then $E f_{A_1}(\phi_1) = \mathrm{Tr}[\widehat{\Psi}\widehat{\Psi}^* + I]\widehat{A}_1 = (\widehat{A}_1 \otimes I \Psi, \Psi) + \mathrm{Tr} \widehat{A}_1 = \langle \widehat{A}_1 \rangle_\Psi + \mathrm{Tr} \widehat{A}_1$, where $\langle \widehat{A}_1 \rangle_\Psi$ is the quantum average of the observable $\widehat{A}_1$. Thus:

$$\langle \widehat{A}_1 \rangle_\Psi = E f_{A_1}(\phi_1) - \mathrm{Tr} \widehat{A}_1. \tag{19}$$

In the same way we obtain

$$\langle \widehat{A}_2 \rangle_\Psi = E f_{A_2}(\phi_2) - \mathrm{Tr} \widehat{A}_2. \tag{20}$$

These relations for averages combined with relation (18) for correlations provide coupling between PCSFT and QM.

5. Why Gaussian?

Although we try to proceed as far as possible with prequantum random fields having arbitrary probability distributions, the assumption that these fields are Gaussian essentially simplifies the theory. In particular, only for Gaussian fields we were able to establish the correspondence between classical (prequantum) and quantum correlations. Only for Gaussian fields the correspondence between quantum states (density operators) and prequantum random fields is one-to-one. The latter is very important for objectification of quantum states. In the absence of the one-to-one correspondence quantum states are merely the labels for huge classes of prequantum random fields. In this case preparation procedures of the modern quantum physics are not good enough to generate the same type of random signals in all runs of the experiment.

What can be physical reasons for usage of Gaussian probability distributions for prequantum random fields?

To find an answer to this question, we shall explore the analogy between PCSFT and the classical signal theory.

5.1. Pulse processes

An important class of random signals is given by pulse processes:

$$\xi(s) = \sum_m a_m F(s - s_m), \tag{21}$$

where s_m is the instant of generation of the mth pulse, a_m is its amplitude and $F(s)$ describes the form of a pulse. As always, it is assumed that $|F(s)| \to 0, |s| \to \infty$, sufficiently quickly. For example, $\xi(s) = \sum_m a_m \delta(s - s_m)$, the sum of δ-pulses.

The random parameters a_m and s_m of the random process $\xi(s)$ satisfy the following natural assumptions:

a). All amplitudes $\{a_m\}$ and instances $\{s_m\}$ of generation of pulses are independent; amplitudes are equally distributed and instances of time are equally distributed as well

$$w(a_1 \ldots, a_m, \ldots; s_1, \ldots, s_m, \ldots) da_1 \ldots da_m \ldots ds_1 \ldots ds_m \ldots$$

$$= \Pi_m w_a(a_m) da_m w_s(s_m) ds_m.$$

b). The probability of appearance of a pulse in the interval $[s, s + ds]$ does not depend on s and it is proportional ds, i.e.,

$$w_s(s) ds = n_1 ds, n_1 = \text{const}. \tag{22}$$

The latter implies that the probability of appearance of n pulses in the interval T is given by the Poisson distribution:

$$P(n) = \frac{\bar{n}^n e^{-\bar{n}}}{n!}, \bar{n} = n_1 T. \tag{23}$$

Set

$$\bar{a} = \bar{a}_m = \int_{-\infty}^{+\infty} a_m w_a(a_m) da_m,$$

the amplude mean value (we assumed that amplitudes of all pulses are equally distributed, so the average does not depend on m) and set

$$\overline{a^2} = \overline{a_m^2} = \int_{-\infty}^{+\infty} a_m^2 w_a(a_m) da_m.$$

Then the dispersion of $\xi(s)$ is given by

$$\sigma_\xi^2 \equiv \sigma_{\xi(s)}^2 = n_1 \overline{a^2} \int_{-\infty}^{+\infty} F^2(\theta) d\theta.$$

It does not depend on s (a feature of stationary of $\xi(s)$). We consider the case of symmetric fluctuations of amplitudes, i.e., $\bar{a} = 0$. In this case $\bar{\xi}(s) = 0$ and $\sigma_{\xi(s)}^2 = E\xi^2(s)$.

It can be proved that the density of the probability distribution $w(\xi)$ of the process $\xi(s)$ can be approximated (under some conditions) by the density of the Gaussian distribution with zero mean value and the dispersion which is equal to the dispersion of $\xi(s)$:

$$w(\xi) \approx \frac{1}{\sqrt{2\pi\sigma_\xi^2}} e^{-\frac{\xi^2}{2\sigma_\xi^2}} + \ldots$$

(As a consequence of stationarity of $\xi(s)$, its probability distribution does not depend on s.) If $\int_{-\infty}^{+\infty} F(\theta) d\theta = 1$ and the pulse duration is of the magnitude $\delta > 0$, then the approximation is good under the condition

$$n_1 \delta >> 1 \tag{24}$$

Journal of Physics: Conference Series **306** (2011) 012021 doi:10.1088/1742-6596/306/1/012021

This means that the number of pulses during the interval δ must be very large. In other words, we can say: at the moment s,the process $\xi(s)$ is composed of pulses which appeared in the interval $[s-\delta, s]$, because earlier pulses died to the moment s. Thus the $n_1\delta$ is the mean value of summands $x_m = a_m F(s-s_m)$ composing $\xi(s)$ at each fixed s. The condition (24 can be interpreted in the following: the distribution $w(\xi)$ of the pulse process $\xi(s)$ is close to the Gaussian distribution if a huge number of pulses is mixed in $\xi(s)$ at each moment s. If the frequency of generation of pulses is low then the distribution of $\xi(s)$ would depend essentially on the form of an individual pulse and the Gaussian approximation would not be useful.

Thus the Gaussian law can be used to describe (of course, approximately) random pulse processes for sources of pulses which produce pulses with very high frequency (comparing with the average duration of pulses).

5.2. *Wave-pulses*

Consider now a source of wave-pulses, in (21) the coefficients belong not to the field of complex numbers **C**, but to the space of square integrable fields $L_2(\mathbf{R}^3)$

$$\phi(s) = \sum_m \phi_m F(s-s_m), \tag{25}$$

where $\phi_m \equiv \phi_m(x,\omega)$ are random fields and $s_m \equiv s_m(\omega)$ are random variables. At each instant of time s_m, a spatial wave $\phi_m(x)$ is produced. In the abstract formalism the coefficients ϕ_m of randomly generated pulses belong to complex Hilbert space H. We can repeat previous considerations for such random fields. We again assume that all random fields ϕ_m have the same probability distribution on H, say μ. The only restriction is that its mean value equals to zero. The covariation operator of ν is denoted by D_ν. Consider now the probability distribution, say w_ϕ of the wave-pulses process $\phi(s)$. It is possible to prove that $\phi(s)$ is a stationary process and its probability distribution can be approximated by the Gaussian probability distribution ν_{D_ϕ}, where D_ϕ is the covariance operator of $\phi(s)$.

$$D_\phi = \Big(n_1 \int_{-\infty}^{+\infty} F^2(\theta)d\theta\Big) D_\mu \tag{26}$$

For the pulse-function $F(s) = 1/\delta$, if $s \in [0,\delta]$, and 0, if $s \notin [0,\delta]$, we obtain $\int_{-\infty}^{+\infty} F(\theta)d\theta = 1$ and $\int_{-\infty}^{+\infty} F^2(\theta)d\theta = 1/\delta$. Hence,

$$D_\phi = \frac{n-1}{\delta} D_\mu \tag{27}$$

If the condition (24) holds, i.e., if the wave-pulses are produced densily on the duration interval δ, then

$$\sigma^2_{w_\phi} = \sigma^2_{\mu_{D_\phi}} = \frac{n_1^2 \mathrm{Tr} D_\mu}{n_1\delta} << 1 \tag{28}$$

for

$$\mathrm{Tr} D_\mu \sim \frac{1}{n_1^2} \tag{29}$$

Thus if

$$\sigma_\mu \sim \frac{1}{n_1}, \tag{30}$$

then, instead of the wave-pulses process $\phi(s)$, we can consider a Gaussian random field $\sim N(0, D_\phi)$, where D_ϕ is given by (27). If the prequantum random field is really produced (at the prequantum time scale) as a stationary process of the wave-pulses type, then it is *mathematically* justified to operate everywhere with Gaussian random field. Of course, the same

Gaussian random field approximates a huge variety wave-pulses processes. From this viewpoint, the QM approximation of the prequantum randomness can be embedded in the following scheme:

[prequantum RF] → [Gaussian RF] → [covariance operator] → [density operator],

where RF is the abreviation for random field. The first step is important from the mathematical viewpoint, it simplifies essentially the description of coupling of prequantum and quantum structure, e.g., for composite systems. However, from the physical viewpoint this step can be omitted. We state again that we proceed under the assumption that the prequantum random field is a wave-pulses process. (This assumption is also a question for the debate. Under this assumption the dream of one-to-one correspondence between prequantum random fields and quantum states does not come true! The quantum language was elaborated to put a kind of order on prequantum randomness. Instead of describing a huge class of random fields fluctuating on very fine prequantum time scale and composed of a huge number of wave-pulses, QM invented the notion of a quantum state. The latter is mathematically presented by a density operator ρ which is nothing else than the normalization by dispersion of the covariance operator of the prequantum (stationary) random field $\phi(s)$.

5.3. Where are the sources of prequantum pulses?

The picture of prequantum random fields as stationary (on the prequantum time scale) stochastic processes composed of wave-pulses is very attractive. In particular, it provides a coupling between classical statistical randio-physics and electro-engineering. The main physical problem in the justification of this viewpoint to prequantum random fields is the identification of physical sources of wave-pulses. Two fundamentally different approaches to this problem can be discussed.

A). The operational approach teaches us that the QM-formalism describes preparation and measurement procedures. By extending this approach beyond QM we might identify sources of wave-pulses as ordinary sources of quantum signals (quantum particles). We just consider a fine time scale (the prequantum scale) and speculate that each quantum particle is, in fact, a huge train of wave-pulses produced on this time scale. The duration of a prequantum pulse is so short that it cannot be observed at the time scale of measurement (the quantum scale, i.e., the scale described by QM).[1]

Thus individual summands in (26) are unobservable; only the integral result of summation of a huge number of such wave-pulses is observable in the form of a quantum particle. This approach works well in standard experiments where we can identify sources of ensembles of quantum particles and, hence, by this interpretations trains of wave-pulses. Such a source produces a stationary process (on the prequantum scale) and a detector (which is always of a threshold type) divides this process into portions. The latter are called quantum particles. This model works especially well for experiments of quantum optics where electromagnetic signals (including entangled bisignals) are identified (in the quantum formalism) with ensembles of photons. The question whether the electromagnetic field is quantized in the empty space or only in the process of interaction with matter is in the very heart of QM. It is still the subject of intensive debates.

However, this approach to sources of prequantum Poissonian trains of wave-pulses (as the operational approach in general) induces the impression that the whole story is about *macroscopic* preparation and measurement devices. Even in the PCSFT-framework the presented operational-like approach ignores individual quantum systems. As in the quantum operational model, there are sources and measurement devices. The only difference is that in our model quantum sources produce trains of wave-pulses of extremely short duration. But,

[1] For example, the Schrödinger dynamics is performed on this time scale.

what about an individual electron which exists somewhere in space-time (without any relation to our experimental activity!)?

In principle, one can say that in quantum mechanics we always work with huge statistical ensembles and we cannot say anything about features of an individual quantum system. And this is a correct statement. However, this is a delicate point and we discuss it carefully. Of course, the modern experimental physics demonstrates great achievements. One-particle experiments are performed in, e.g., quantum optics or neutron-interferometry. However, we (first of all) should not forget that one-particle experiments are still statistical by their nature. Thus, e.g., one-photon sources are only statistically (with a high probability) one-photon emission devices.

However, it is even more important to understand that all quantum experiments are based on time-stationary sources and all quantum averages are, in fact, time-averages. One cannot exclude that properties of quantum systems emitted by sources reflect the time-structure of processes in sources. Thus, even by staying on the conventional quantum position, we cannot exclude that quantum systems emitted by a source carry the prints of the time structure of processes in sources.

In this situation it is natural to advertize the viewpoint that the QM (and, hence, prequantum formalism) is not about isolated systems, but correlations between sources and detectors. The additional PCSFT contribution is the description of transmission of information from a source to a detector with the aid of a classical wave signal. However, we may also explore another possibility to identify a source of prequantum random field, a train of wave-pulses.

B). The background field can play the role of a source of stationary prequantum fields which are considered as quantum systems in QM. By this interpretation each isolated quantum system is nothing else as a source of a special stationary random field. The background field in a special domain of space produces a train of wave-pulses which are dissipated on this field. The balance between generation and dissipation produce a stationary random field. The background field has to be considered as a fundamental field which is able to produce random fields corresponding to all possible sorts of quantum systems. In the future we can hope to see a fantastic picture of radiation of wave-pulses of, e.g., electronic random field from nothing and the corresponding stationary signal. Different types of quantum particles correspond to different structures of time-series (26): combination of wave-amplitudes, intensity of pulses n_1 and the duration of δ.

However, at the moment we prefer to keep to the operational-like interpretation, see A), since B) sounds too speculative (even in the PSCFT-approach). Moreover PCSFT is not rigidly coupled to the model based on wave-pulses. Thus one can proceed without coupling the prequantum random fields with wave-pulses processes.[2]

Acknowledgments

This paper was completed during my visit to University of Rome-2, Torvergata, September 2011, I would like to thank Luigi Accardi for fruitful discussions and hospitality.

References

[1] De la Pena L and Cetto A M 1996 *The Quantum Dice: An Introduction to Stochastic Electrodynamics* (Dordrecht: Kluwer)
[2] Khrennikov A 2005 *J. Phys. A: Math. Gen.* **38** 9051
[3] Khrennikov A 2005 *Found. Phys. Letters* **18** 637
[4] Khrennikov A 2006 Physics Letters A **357** 171
[5] Khrennikov A 2006 *Found. Phys. Lett.* **19** 299
[6] Khrennikov A 2006 *Nuovo Cimento* B **121** 505
[7] Khrennikov A 2008 *Physics Letters* A **372** 6588
[8] Elze H-T 2009 *J. Phys.: Conf. Ser.* **174** 012009

[2] Of course, possibility to explain usage of Gaussian random fields is one of very attractive features of this model.

[9] Elze H-T 2007 *J. Phys.: Conf. Ser.* **67** 012016
[10] Khrennikov A 2009 *J. Russian Laser Research* **30** 472

Shortcuts to quantum adiabatic processes

J G Muga[1], Xi Chen[1,2], E Torrontegui[1], S Ibáñez[1], I. Lizuain[1] and A Ruschhaupt[3]

[1] Departamento de Química Física, Universidad del País Vasco - Euskal Herriko Unibertsitatea, Apdo. 644, Bilbao, Spain

[2] Department of Physics, Shanghai University, 200444 Shanghai, P. R. China

[3] Institut für Theoretische Physik, Leibniz Universität Hannover, Appelstraße 2, 30167 Hannover, Germany

E-mail: jg.muga@ehu.es

Abstract. We review different ways to accelerate adiabatic processes in cold atom physics and atomic state preparation. Trap expansions or contractions and atomic transport may be accelerated by an invariant-based inverse engineering approach. Berry's inverse engineering method is also applied to produce fast versions of adiabatic passage methods.

1. Introduction

In quantum mechanics an "adiabatic process" is a slow change of Hamiltonian parameters that keeps the populations of the instantaneous eigenstates constant. These processes are frequently chosen to drive or prepare states in a robust and controllable way and have also been proposed to solve complicated computational problems. The main drawback is that they are, by definition, slow. It is thus quite natural to look for "shortcuts to adiabaticity" cutting down the time to arrive at the same final state, possibly up to phase factors. This may be done by designing optimal adiabatic pathways, or by admitting transient excitations in the "instantaneous basis" that diagonalizes the Hamiltonian. Several works have recently proposed ways to achieve this goal for general or specific cases. One of the early applications was particle transport without heating, see [1] and references therein. Another important application is frictionless harmonic trap compressions or expansions for state preparation. They may be accelerated with "bang-bang" (piecewise constant frequency) methods [2], or designing, by inverse engineering techniques, a time dependent frequency for which the expanding modes associated with Lewis-Riesenfeld invariants [3] take the state from the initial to the final potential configuration [4, 5]. This inverse engineering approach has been implemented experimentally to decompress ^{87}Rb cold atoms in a harmonic magnetic trap [6]. The extension to Bose-Einstein condensates may be carried out with a variational ansatz [7] and has been also realized experimentally [8]. The invariants-based method has been proposed to cool mechanical resonators as well [9]. In this paper we review first invariant-based inverse engineering: Section 2 sets the general formalism, Sections 3 and 4 deal with expansions, and Section 5 is devoted to transport.

Berry has proposed a different approach [10], based on a Hamiltonian $\mathcal{H}(t)$ such that the adiabatic approximation of the state evolution under a time-dependent reference Hamiltonian $H_0(t)$ becomes the exact dynamics with $\mathcal{H}(t)$. This has been applied, at least formally, to spins in magnetic fields [10], harmonic oscillators [5], or to speed up adiabatic state-preparation methods such as Rapid Adiabatic Passage (RAP), Stimulated Rapid Adiabatic Passage (STIRAP) and its variants [11]. This work is summarized in Section 6.

2. Invariant-based inverse engineering methods

Lewis and Riesenfeld provided the relation between the solutions of the Schrödinger equation of a system with time-dependent Hamiltonian and the eigenstates of the corresponding dynamical invariants [3]. They paid special attention to the time-dependent harmonic oscillator and its invariants quadratic in position and momentum. Later on Lewis and Leach found, in the framework of classical mechanics, the general form of the Hamiltonian compatible with invariants quadratic in momentum [12], including non harmonic potentials. This, together with the corresponding quantum results by Dhara and Lawande [13] constitutes the basis of the concepts and equations presented in this section.

A one-dimensional Hamiltonian with an invariant quadratic in momentum must have the form $H = p^2/2m + V(q,t)$, (q and p denote operators or numbers, and the context should clarify their meaning.) with the potential [12, 13], see the Appendix,

$$V(q,t) = -F(t)q + \frac{m}{2}\omega^2(t)q^2 + \frac{1}{\rho(t)^2}U\left[\frac{q-\alpha(t)}{\rho(t)}\right]. \tag{1}$$

ρ, α, ω, and F are arbitrary functions of time that satisfy the two auxiliary equations

$$\ddot{\rho} + \omega^2(t)\rho = \frac{\omega_0^2}{\rho^3}, \tag{2}$$

$$\ddot{\alpha} + \omega^2(t)\alpha = F(t)/m, \tag{3}$$

where ω_0 is a constant. Their physical interpretation will be explained below. The quadratic dynamical invariants are given, up to a constant factor, by

$$I = \frac{1}{2m}[\rho(p - m\dot{\alpha}) - m\dot{\rho}(q-\alpha)]^2 + \frac{1}{2}m\omega_0^2\left(\frac{q-\alpha}{\rho}\right)^2 + U\left(\frac{q-\alpha}{\rho}\right), \tag{4}$$

and verify

$$\frac{\mathrm{d}I}{\mathrm{d}t} \equiv \frac{\partial I(t)}{\partial t} + \frac{1}{i\hbar}[I(t), H(t)] = 0, \tag{5}$$

so that $\frac{d}{dt}\langle\psi(t)|I(t)|\psi(t)\rangle = 0$ for any wave function $\psi(t)$ that evolves with H. These invariants are quite useful. For example, $\psi(t)$ may be expanded in terms of constant coefficients c_n and eigenvectors ψ_n of I,

$$\psi(q,t) = \sum_n c_n e^{i\alpha_n}\psi_n(q,t), \tag{6}$$

$$I(t)\psi_n(q,t) = \lambda_n\psi_n(q,t), \tag{7}$$

where the λ_n are time-independent eigenvalues. The spectrum of I may be discrete and/or continuous, so ψ_n may have continuum normalization or be normalized to one as in the applications discussed here. The α_n are phases that satisfy [3, 13]

$$\hbar\frac{d\alpha_n}{dt} = \left\langle\psi_n\left|i\hbar\frac{\partial}{\partial t} - H\right|\psi_n\right\rangle, \tag{8}$$

$$\alpha_n = -\frac{1}{\hbar}\int_0^t \mathrm{d}t \quad \frac{\quad}{\rho^2} + \frac{m(\dot{\alpha}\rho - \alpha\dot{\rho})^2}{2\rho^2}\Bigg). \tag{9}$$

The functions ψ_n are in practice obtained as [13]

$$\psi_n(q,t) = e^{\frac{im}{\hbar}\left[\dot\rho q^2/2\rho + (\dot\alpha\rho - \alpha\dot\rho)q/\rho\right]} \frac{1}{\rho^{1/2}} \phi_n\Bigg(\underbrace{\frac{q-\alpha}{\rho}}_{=:\sigma}\Bigg) \tag{10}$$

from the solutions $\phi_n(\sigma)$ (normalized in σ-space) of the auxiliary Schrödinger equation

$$\left[-\frac{\hbar^2}{2m}\frac{\partial^2}{\partial\sigma^2} + \frac{1}{2}m\omega_0^2\sigma^2 + U(\sigma)\right]\phi_n = \lambda_n\phi_n. \tag{11}$$

The strategy of invariant-based inverse engineering methods is to design the auxiliary functions ρ and α first to achieve the desired objectives, and deduce the Hamiltonian afterwards. In most applications so far the important point is to control the boundary conditions of the auxiliary functions and their time derivatives at initial and final times. In particular, they may are set so that the eigenvectors of H and I coincide at initial and final times, for an exception see [1], so that the process produces no final excitation. It is however not an adiabatic process in the usual sense since, at intermediate times, excitations in the instantaneous basis are allowed.

3. Realizing fast trap expansions without final excitation

Performing fast expansions of trapped atoms without loosing or exciting them is important, for example to reduce velocity dispersion and collisional shifts in spectroscopy and atomic clocks, to reach very low temperatures unaccessible by standard cooling techniques or, in experiments with optical lattices, to extend the atomic cloud before turning on the lattice. Trap compressions are also quite common to prepare the state for further experiments. For a harmonic oscillator, we may address these expansion or contraction processes with the invariants theory by setting $\alpha = U = F = 0$. This means that (3) does not play any role and the important auxiliary equation is the "Ermakov equation" (2). Moreover ρ becomes proportional to the standard deviation of the "expanding (or contracting) modes" $e^{i\alpha_n}\psi_n$, which are exact solutions of the time dependent Schrödinger equation.

Here we shall discuss the expansion from $\omega(0) = \omega_0$ to $\omega(t_f) = \omega_f$, the treatment for compressions is very similar. Choosing $\rho(0) = 1$, $\dot\rho(0) = 0$, $H(0)$ and $I(0)$ commute, they actually become equal, and have common eigenfunctions at $t = 0$. Consistent with the Ermakov equation, $\ddot\rho(0) = 0$ holds as well. At t_f we impose $\rho(t_f) = \gamma = (\omega_0/\omega_f)^{1/2}$, $\dot\rho(t_f) = 0$, $\ddot\rho(t_f) = 0$. In this manner the expanding mode is an instantaneous eigenvector of H at $t = 0$ and t_f, regardless of the exact form of $\rho(t)$. To fix $\rho(t)$, one chooses a functional form to interpolate between these two times, flexible enough to satisfy the boundary conditions. A simple possibility is the polynomial ansatz

$$\rho(t) = \sum_{j=0}^{5} a_j t^j, \tag{12}$$

where the coefficients are determined by solving the equations imposed by the boundary conditions,

$$\rho(t) = 6\,(\gamma-1)\,s^5 - 15\,(\gamma-1)\,s^4 + 10\,(\gamma-1)\,s^3 + 1. \tag{13}$$

Here $s = t/t_f$.

The next step is to solve for $\omega(t)$ in (2). This procedure poses no fundamental lower limit to t_f, which could be in principle arbitrarily small. There are nevertheless practical limitations and/or prices to pay. For short enough t_f, $\omega(t)$ may become purely imaginary at some t [4] and the potential becomes a parabolic repeller. Another potential difficulty is that the transient energy required may be too high, as discussed in the following section.

4. Transient energy excitation for the time dependent harmonic oscillator

We shall examine now the energy "cost" of shortcut processes; more precisely, the transient excitation energies [18], centering the analysis on the expansion (or compression) of a harmonic oscillator. We may expect on intuitive grounds that the transient energy and the time of the process are somehow "conjugate", i.e., that the former increases when decreasing the later, but the details of this relation, and the role played by other parameters that define the process, such as initial and final frequencies, have to be clarified for fundamental reasons and for the applications. In particular, the energy excitation will set limits to the possible speed-up. Since actual traps are only approximately harmonic, large transient energies will imply perturbing effects of anharmonicities and thus undesired excitations of the final state, or atom loss.

Knowing the transient excitation energy is also important to quantify the principle of unattainability of zero temperature, which was first enunciated by Nernst. This principle is usually formulated as the impossibility to reduce the temperature of any system to the absolute zero in a finite number of operations, and identified with the third law of thermodynamics. Kosloff and coworkers [2] have restated the unattainability principle in quantum refrigerators as the vanishing of the cooling rate when the temperature of the cold bath approaches zero, and quantify it by the scaling law that relates cooling rate and cold bath temperature. We shall examine here the consequences of the transient energy excitation on the unattainability principle in two ways: for a single, isolated expansion, and considering the expansion as one of the branches of a quantum refrigerator cycle.

A lower bound $\mathcal{B}_n$ for the time-averaged energy of the n-th expanding mode (time averages from 0 to t_f will be denoted by a bar) is found by applying calculus of variations [18], so that $\overline{E_n} \geq \mathcal{B}_n$. If the final frequency ω_f is small enough to satisfy $t_f \ll 1/\sqrt{\omega_0\omega_f}$, and $\gamma \gg 1$, the lower bound has the asymptotic form

$$\mathcal{B}_n \approx \frac{(2n+1)\hbar}{2\omega_f t_f^2}. \tag{14}$$

A consequence is that

$$t_f \geq \sqrt{\frac{(2n+1)\hbar}{2\omega_f \overline{E_n}}}. \tag{15}$$

In (15) $\overline{E_n}$ may be limited by some maximal value, because of anharmonicities or a finite trap depth. Then the scaling is fundamentally the same as for bang-bang methods [2], and leads to a cooling rate $R \propto T_c^{3/2}$ in an inverse quantum Otto cycle, although the proportionality factor may be improved by increasing the allowed $\overline{E_n}$. This dependence had been previously conjectured to be a universal one characterizing the unattainability principle for any cooling cycle [19]. Our results provide strong support for the validity of this conjecture within the set of processes defined by time-dependent frequencies and call for further testing and study.

Independently of the participation of the harmonic trap expansion as a branch in a refrigerator cycle, we may apply the previous analysis also to a single expansion, assuming that the initial and final states are canonical density operators characterized by temperatures T_0 and T_f. These are related by $T_f = (\omega_f/\omega_0)T_0$ for a population-preserving process. In a parabolic potential expansion, the unattainability of a zero temperature can be thus reformulated as follows: The transient excitation energy becomes infinite for any population-preserving and finite-time process when the final temperature is zero (which requires $\omega_f = 0$). The excitation energy has to be provided by an external device, so there remains a fundamental obstruction to reach $T_f = 0$ in a finite time, in the form of the need for a source of infinite power.

With respect to the standard deviation [18], the dominant dependences of the time averages found numerically scale on ω_f and t_f in the same way as the average energy. These dependences are different from the ones in the Anandan-Aharonov relation [20]

$$\overline{\Delta H}\, t_f \geq \frac{h}{4}, \tag{16}$$

where

$$\overline{\Delta H} = \frac{\int_0^{t_f} \Delta H(t) dt}{t_f}, \tag{17}$$

The AA bound is correct but not tight.

We have considered so far simple processes in which the only external manipulation consists in shaping $\omega(t)$. As shown in [18], one might design even faster processes by adding terms to the harmonic oscillator Hamiltonian [10], but their physical implementation remains challenging.

5. Fast and efficient transport

A key operation for manipulating cold neutral atoms and ions is their fast efficient transport by means of a moving trap. There is a broad set of conditions, heating mechanisms, transport distances (from microns to tens of centimeters), transport times, and accelerations involved in transport processes, but there are also common elements that allow for a generic theoretical treatment as the one presented in [1] and summarized here. Ideally, transport should be fast, lossless, and lead to a final state as close as possible ("faithful") to the initial one, up to global phase factors, in the transporting trap frame.

As we have done for expansions, we may use the dynamical invariants associated with the Hamiltonian of an atom in a one dimensional moving trap to engineer the trap motion and perform fast atomic transport without final vibrational heating. The atom is driven through a non-adiabatic shortcut to the result of adiabatic, slow trap motion. For harmonic potentials this requires designing appropriate trap trajectories, whereas, in anharmonic traps, perfect transport may be achieved by applying an extra field to compensate the forces in the rest frame of the trap. The results can be extended as well to atom stopping or launching. Limitations due to geometrical constraints, energies and accelerations involved are analyzed in [1], also the relation to previous approaches (based on classical trajectories or "fast-forward" [21] and "bang-bang" methods) which can be integrated in the invariant-based framework.

Trap expansions and contractions imply a time dependent ρ function [4], but a large family of transport problems may be described by taking

$$\rho(t) = 1,\ \omega^2(t) = \omega_0^2 \tag{18}$$

so (2) plays no role and (3) is the only relevant auxiliary equation. We shall assume that the conditions (18) hold, and consider two main scenarios.

5.1. Main scenarios

(*i*) *Rigid harmonic oscillator driven by the "transport function"* $q_0(t)$ ("harmonic transport" for short). Suppose that a harmonic trap is moved from $q_0(0)$ at time $t = 0$ to $d = q_0(t_f)$ at a time t_f. In (1) this implies

$$F = m\omega_0^2 q_0(t),\ \omega(t) = \omega_0,\ U = 0. \tag{19}$$

Adding to V the global term $m\omega_0^2 q_0^2/2$, the trap potential can be written as a moving harmonic oscillator $m\omega_0^2[q - q_0(t)]^2/2$,

$$H = p^2/2m + m\omega_0^2[q - q_0(t)]^2/2, \tag{20}$$

and α can be identified as a classical trajectory q_c. Equation (3) becomes

$$\ddot{q}_c + \omega_0^2(q_c - q_0) = 0. \tag{21}$$

Now $\lambda_n = E_n = (n+1/2)\hbar\omega_0$, and the "transport mode" $e^{i\alpha_n}\psi_n$ takes the form

$$e^{i\alpha_n}\psi_n = e^{-\frac{i}{\hbar}\left[E_n t + \int_0^t \frac{m\dot{q}_c^2}{2}dt'\right]} e^{im\dot{q}_c q/\hbar}\phi_n(q-q_c). \tag{22}$$

Efficient transport is engineered by designing first an appropriate classical trajectory $q_c(t)$, from which the trap motion trajectory $q_0(t)$ is deduced via (21).

A variant is vertical transport with a gravity force, so that $F = m\omega_0^2 q_0 - mg$ and (21) becomes

$$\ddot{q}_c + \omega_0^2(q_c - q_0) = -g. \tag{23}$$

Apart from pure rest-to-rest transport, it is also possible to design stopping or launching processes without losses or heating [1].

A major concern in practice for all these applications is to keep the harmonic approximation valid. This may require an analysis of the actual potential and of the excitations taking place along the non-adiabatic transport process. Without performing such detailed analysis, the feasibility of the approach for a given transport objective set by the pair d, t_f can be estimated by comparing lower excitation bounds. These are obtained using calculus of variations as we have discussed before for expansions. Writing the average potential energy for a transport mode as $\langle V(t)\rangle = \frac{\hbar\omega_0}{2}(n+1/2) + E_P$, we find [1]

$$\overline{E_P} \geq \frac{6md^2}{t_f^4\omega_0^2}. \tag{24}$$

This bound describes the relevant dependences, as shown by numerical comparisons with actual time-averaged energies for polynomial trajectories, and sets a t_f^{-4} scaling, compare this with the milder dependence on t_f^{-2} of the time-averaged transient energy in invariant-based inverse-engineered expansion processes [18].

(ii) Arbitrary-trap driven transport with compensating force ("compensating force approach" for short). Now, in (1),

$$\omega = \omega_0 = 0, \tag{25}$$
$$F = m\ddot{q}_0. \tag{26}$$

In this scenario the trap potential $U[q - q_0(t)]$ is arbitrary, in particular it could be harmonic, and it is rigidly displaced along $q_0(t)$, so that α in (3) may be identified with the transport function q_0. Inverse engineering in this case is based on choosing the boundary conditions for q_0 [1]. In addition to U, there is a compensating linear potential term $-mq\ddot{q}_0$ in H,

$$H = p^2/2m - mq\ddot{q}_0 + U(q - q_0). \tag{27}$$

The corresponding force compensates the inertial force due to the trap motion in the rest frame of the trap, in such a way that the wave function in that frame is not modified up to a time dependent global phase factor. This Hamiltonian has been proposed by Mashuda and Nakamura using a very different "fast-forward" scaling technique [21].

6. Accelerating adiabatic processes in two and three level atoms

Two major routes for manipulating the state of a quantum system with interacting fields are resonant pulses, and adiabatic methods such as "Rapid" Adiabatic Passage (RAP), Stimulated Raman Adiabatic Passage (STIRAP), and their variants. Simple fixed-area resonant pulses may be fast if intense enough, but they are also unstable with respect to errors or fluctuations of the parameters, whereas adiabatic passage is slow but robust. For many applications we would like a fast and robust process, two requirements that are particularly demanding if quantum computing is to become feasible. It is possible to design pulse sequences which are more stable than single pulses, but their use is limited by several technical reasons [11]. In NMR, composite pulses are being superseded by adiabatic passage methods, which have also been very successful in laser cooling, chemical reaction dynamics, metrology, atom optics, interferometry, or cavity quantum electrodynamics. When robustness is the primary concern, they are quite sufficient, and have become common, basic operations for quantum information processing. However, if speed is also important the limitations may be severe. Given the difficulties of composite pulses, we may look for robustness and high operation rates by shortening the duration of adiabatic methods. A shortcut to adiabatic passage [11] ("SHAPE"), which is fast and stable, may be found using the "transitionless quantum driving" algorithm proposed by Berry [10]. We shall discuss here speeded-up version of (2-level) Rapid Adiabatic Passage (RAP). 3-level systems and STIRAP are examined in [11].

In the transitionless quantum driving algorithm [10], a Hamiltonian $\mathcal{H}(t)$ is designed so that the adiabatic approximation for the time-dependent wavefunction evolving with a reference Hamiltonian $H_0(t)$ becomes exact. The simplest Hamiltonian, $H_1(t)$, steers the dynamics along the instantaneous eigenstates $|\lambda_n(t)\rangle$ of $H_0(t)$ without inducing transitions among them and without phase factors, formally in an arbitrary time,

$$H_1(t) = i\hbar \sum_n |\partial_t \lambda_n\rangle\langle\lambda_n|. \tag{28}$$

The addition of H_0 is possible, but not necessary; it affects the phases but not the populations. H_1 may thus supplement H_0, when $\mathcal{H} = H_0 + H_1$, or substitute it, if $\mathcal{H} = H_1$. The physical realizability of H_1 has to be studied in each system. For example, when H_0 describes a particle in a time-dependent harmonic potential, H_1 becomes a non-local interaction [5], whereas for a particle with spin in a time dependent magnetic field, H_1 becomes a time-dependent magnetic field [10]. For the atomic two- and three-level systems, H_1 involves auxiliary laser or microwave interactions. The optional addition of H_0 implies different physical implementations [11].

The "Rapid Adiabatic Passage" technique inverts the population of two-levels, $|1\rangle$ and $|2\rangle$, by sweeping the radiation slowly through resonance. This method originated in Nuclear Magnetic Resonance but is used in virtually all fields where 2-level systems may be controlled by external interactions. The term "rapid" here means that the frequency sweep is shorter than the life-time of spontaneous emission and other relaxation times.

We shall formulate H_0, H_1, or $\mathcal{H}$ in an interaction picture so they will carry an I superscript, to distinguish them from Schrödinger picture Hamiltonians with S subscript. These subscripts also apply to wave functions. To set H_0^I, let us assume a semiclassical treatment of the (electric dipole) interaction with the electric field $\mathbf{E}$. Within the rotating wave approximation (RWA), the Hamiltonian for a laser interaction with linear polarization and time dependent phase $\Phi(t)$ is

$$H^S(t) = \frac{\hbar}{2}[|2\rangle\langle 1|\Omega_R(t)e^{-i\Phi(t)} + |1\rangle\langle 2|\Omega_R(t)e^{i\Phi(t)} + \omega_0(t)(|2\rangle\langle 2| - |1\rangle\langle 1|)] \tag{29}$$

with $\Omega_R(t)$ real, and a transition frequency $\omega_0(t)$ which may be time dependent. We assume that $\Phi(0) = 0$. For a laser with constant angular frequency ω_L, $\Phi(t) = \omega_L t$. In a more general, time dependent case, this notation could be maintained as in [11], but $\omega_L(t)$ would not be the instantaneous frequency. Here, instead, we shall use the symbol $\omega_L(t)$ for the instantaneous frequency, namely,

$$\omega_L(t) = d\Phi/dt. \tag{30}$$

In a laser-adapted interaction picture based on the zeroth order Hamiltonian

$$h_L(t) = \frac{\hbar\omega_L(t)}{2}(|2\rangle\langle 2| - |1\rangle\langle 1|), \tag{31}$$

and its unitary evolution operator[1]

$$U_0(t,0) = e^{-\frac{i}{\hbar}\int_0^t h_L(t')dt'} = e^{-i\Phi(t)/2}|2\rangle\langle 2| + e^{i\Phi(t)/2}|1\rangle\langle 1|, \tag{32}$$

the dynamics of the interaction-picture wave function $\psi^I(t) = U_0^\dagger(t,0)\psi^S(t)$ is governed by the Hamiltonian

$$H_0^I(t) = U_0^\dagger(t,0)[H^S(t) - h_L(t)]U_0(t,0). \tag{33}$$

Using the convention $|1\rangle = \binom{1}{0}$, $|2\rangle = \binom{0}{1}$,

$$H_0^I(t) = \frac{\hbar}{2}\begin{pmatrix} -\Delta(t) & \Omega_R(t) \\ \Omega_R(t) & \Delta(t) \end{pmatrix}, \tag{34}$$

where $\Delta(t) = \omega_0(t) - \omega_L(t)$ is the effective detuning, controlled by a change in the carrier frequency or an alteration of the Bohr frequency by Zeeman or Stark shifts. The instantaneous eigenvectors

$$\begin{aligned} |\lambda_-(t)\rangle &= \cos[\theta(t)/2]|2\rangle - \sin[\theta(t)/2]|1\rangle, \\ |\lambda_+(t)\rangle &= \sin[\theta(t)/2]|2\rangle + \cos[\theta(t)/2]|1\rangle, \end{aligned} \tag{35}$$

are given in terms of the mixing angle $\theta(t) \equiv \arccos[-\Delta(t)/\Omega(t)]$ and eigenvalues $E_\mp(t) = \mp\hbar\Omega(t)/2$, where $\Omega = \sqrt{\Delta^2(t) + \Omega_R^2(t)}$. If the adiabaticity condition $\frac{1}{2}|\Omega_a(t)| \ll |\Omega(t)|$, is satisfied, where $\Omega_a(t) \equiv \dot{\theta} = [\Omega_R(t)\dot{\Delta}(t) - \dot{\Omega}_R(t)\Delta(t)]/\Omega^2(t)$, the state that evolves from $|\psi^I(t_i)\rangle = |\lambda_\pm(t_i)\rangle$ and follows the adiabatic approximation

$$|\psi_\pm^I(t)\rangle = \exp\left\{-\frac{i}{\hbar}\int_{t_i}^t dt' E_\pm(t')\right\}|\lambda_\pm(t)\rangle, \tag{36}$$

whereas transitions will occur otherwise. Different adiabatic passage schemes correspond to pairs of functions $\Omega_R(t)$ and $\Delta(t)$ for which $\psi_\pm^I$ goes from one bare state to the other. The simplest one is the Landau-Zener scheme, where Ω_R is constant and $\Delta(t)$ is linear in time. Independently of the scheme, $H_1^I(t)$ takes here the form, using Eqs. (28) and (35),

$$H_1^I(t) = \frac{\hbar}{2}\begin{pmatrix} 0 & -i\Omega_a(t) \\ i\Omega_a(t) & 0 \end{pmatrix}, \tag{37}$$

where, up to a phase factor, Ω_a plays the role of the Rabi frequency for a fast-driving field. In principle H_1^I drives the system along the H_0^I-adiabatic path in an arbitrarily short time, but there are practical limitations such as the laser power available. In this respect a comparison with H_0^I-dynamics is only fair if $|\Omega_a|$ is smaller or approximately equal to the peak Rabi frequency in the original laser setup. In a range of interaction times that break down the adiabaticity condition, the dynamics can be driven along the H_0^I-adiabatic path while fulfilling the inequalities $|\Omega_a| \leq |\Omega| \leq |\Omega_0|$.

The physical meaning of the fast-driving term is better seen in the Schrödinger picture. For $\mathcal{H}^I = H_0^I + H_1^I$, then $\mathcal{H}^S(t) = \frac{\hbar}{2}\{[\Omega_R(t) + i\Omega_a(t)]|2\rangle\langle 1|e^{-i\Phi(t)} + [\Omega_R(t) - i\Omega_a(t)]|1\rangle\langle 2|e^{i\Phi(t)} + \omega_0(t)(|2\rangle\langle 2| - |1\rangle\langle 1|)\}$. This implies two lasers with the same frequency, orthogonal polarization, and differently shaped time-dependent intensities. Instead, when $\mathcal{H}^I = H_1^I$, $\mathcal{H}^S(t) = \frac{\hbar}{2}[i\Omega_a(t)|2\rangle\langle 1|e^{-i\Phi(t)} - i\Omega_a(t)|1\rangle\langle 2|e^{i\Phi(t)} + \omega_L(t)(|2\rangle\langle 2| - |1\rangle\langle 1|)]$, which requires only one interaction but also level shift engineering to implement $\omega_L(t)$ as the transition frequency. The $\pm i$ factors in (37) can be dropped, which amounts to redefine the states with constant phase factors or to perform an axis rotation, without altering the population transfer.

[1] Because of the notation in [11] the term "quasi interaction picture" was suggested. Using ω_L as the instantaneous frequency makes clear that the transformation is in fact a canonical interaction picture with a time-dependent zeroth order Hamiltonian.

7. Conclusions

The acceleration of quantum adiabatic processes is important both for fundamental and applied physics. The techniques reviewed here may be applied to more complicated systems, in particular to condensates [7, 8] by means of scaling relations, or to other interacting systems such as the Tonks-Girardeau gas. We are currently working on further aspects of this promising field, such as stability versus noise, and effects of higher dimensions.

Acknowledgments

We are grateful to our collaborators D. Guery-Odelin, S. Schmidt and A. del Campo who have worked out with us the original material reviewed here. E. T. and S. I. acknowledge financial support from the Basque Government (Grants No. BFI08.151 and BFI09.39). Funding by the Basque Government (Grant IT472-10), the Ministerio de Ciencia e Innovación (FIS2009-12773-C02-01), the National Natural Science Foundation of China (No. 60806041), the Shanghai Rising-Star Program (No. 08QA14030), the Shanghai Leading Academic Discipline (No. S30105), and Juan de la Cierva Programme is acknowledged.

Appendix A: Quadratic-in-p invariants

We shall now find quadratic invariants in p for the time-dependent Hamiltonian [14, 15, 3, 16, 12]

$$H(q,p,t) = \frac{p^2}{2m} + V(q,t).$$

Classically an invariant satisfies

$$\frac{dI}{dt} = \frac{\partial I}{\partial t} + [I,H] = \frac{\partial I}{\partial t} + \frac{\partial I}{\partial q}\frac{\partial H}{\partial p} - \frac{\partial I}{\partial p}\frac{\partial H}{\partial q} = 0. \tag{A.1}$$

We assume the following form

$$I(q,p,t) = p^2 f_2(q,t) + p f_1(q,t) + f_0(q,t), \tag{A.2}$$

and replace it in (A.1). Equating the coefficients of each power in p to zero gives a system of partial differential equations,

$$0 = \frac{\partial f_2}{\partial q}, \tag{A.3}$$

$$0 = \frac{\partial f_2}{\partial t} + \frac{1}{m}\frac{\partial f_1}{\partial q}, \tag{A.4}$$

$$0 = \frac{1}{m}\frac{\partial f_0}{\partial q} - 2f_2\frac{\partial V}{\partial q} + \frac{\partial f_1}{\partial t}, \tag{A.5}$$

$$0 = \frac{\partial f_0}{\partial t} - f_1\frac{\partial V}{\partial q}. \tag{A.6}$$

From (A.3), $f_2 = 2a(t)$. Replacing this into (A.4), $f_1 = -2m\dot{a}(t)q + b(t)$, where the dot denotes the derivative with respect to time t, and $b(t)$ is an arbitrary function. Replacing f_1 and f_2 into (A.5) and integrating with respect to q, one finds

$$V(q,t) = \frac{1}{4a}\left[\frac{1}{m}f_0(q,t) + \dot{b}q - m\ddot{a}q^2\right] + g(t), \tag{A.7}$$

where $g(t)$ is another arbitrary function. Finally, with this expression of $V(q,t)$, (A.6) reads

$$\frac{1}{m}(2m\dot{a}q - b)\frac{\partial f_0}{\partial q} + 4a\frac{\partial f_0}{\partial t} = (2m\dot{a}q - b)(2m\ddot{a}q - \dot{b}), \tag{A.8}$$

which is a linear first-order partial differential equation for f_0 whose characteristics are

$$\frac{dq}{dt} = \frac{1}{4am}(2m\dot{a}q - b),$$
$$\frac{df_0}{dt} = \frac{1}{4a}(2m\dot{a}q - b)(2m\ddot{a}q - \dot{b}) = \frac{d}{dt}\left[\frac{1}{2a}\left(m\dot{a}q - \frac{b}{2}\right)^2\right].$$

Integrating these two equations we find that

$$v = f_0(q,t) - \frac{1}{2a}\left(m\dot{a}q - \frac{b}{2}\right)^2,$$
$$u = \frac{q}{a^{1/2}} + \frac{1}{4m}\int_0^t dt'\,\frac{b(t')}{a^{3/2}(t')},$$

where v and u are arbitrary constants along the characteristics. Therefore the general solution to (A.8) is

$$f_0(q,t) = \frac{1}{2a}\left(m\dot{a}q - \frac{b}{2}\right)^2 + G\left(\frac{q}{a^{1/2}} + \frac{1}{4m}\int_0^t dt'\,\frac{b(t')}{a^{3/2}(t')}\right),$$

where G is an arbitrary function. With this form for f_0 and defining

$$\rho = 2a^{1/2},$$
$$-\frac{\alpha}{\rho} = \frac{1}{8m}\int_0^t dt'\,\frac{b(t')}{a^{3/2}(t')},$$
$$\tilde{G}(x) = G(2x),$$

(A.7) implies that an invariant quadratic in p exists only if the potential has the form

$$V(q,t) = m\left(\frac{\ddot{\rho}\alpha}{\rho} - \ddot{\alpha}\right)q - \frac{m}{2}\frac{\ddot{\rho}}{\rho}q^2 + \frac{1}{m\rho^2}\tilde{G}\left(\frac{q-\rho}{\alpha}\right), \qquad \text{(A.9)}$$

where we have chosen $g(t) = -b^2/8a$ such that no additive functions appear in the potential. This classical invariant, (A.2), or equivalently

$$I(q,p,t) = \frac{1}{2m}[\rho(m\dot{\alpha} - p) + m\dot{\rho}(q-\alpha)]^2 + \frac{1}{m}\tilde{G}\left(\frac{q-\alpha}{\rho}\right), \qquad \text{(A.10)}$$

becomes the quantum invariant when the canonical momentum p is replaced by the quantum-mechanical operator $-i\hbar\partial/\partial q$, ρ being a c-number [16, 17]. We have divided the original invariant (A.2) by the mass m to have units of energy. This makes it equal to the Hamiltonian at time $t = 0$ in the expansions of Section 3.

The previous result can be written in terms of the arbitrary functions $\omega(t)$, $F(t)$, $U[(q-\alpha)/\rho]$ and the arbitrary constant ω_0 using (2), (3), and

$$U\left(\frac{q-\alpha}{\rho}\right) = \frac{1}{m}\tilde{G}\left(\frac{q-\alpha}{\rho}\right) - \frac{1}{2}m\omega_0^2\left(\frac{q-\alpha}{\rho}\right)^2. \qquad \text{(A.11)}$$

The invariant takes the form (4) and the corresponding potential becomes (1).

References

[1] Torrontegui E, Ibáñez S, Chen X, Ruschhaupt A, Guéry-Odelin D and Muga J G 2010 Fast atomic transport without vibrational heating *Preprint* arXiv:1010.3271v1

[2] Salamon P, Hoffmann K H, Rezek Y and Kosloff R 2009 *Phys. Chem. Chem. Phys.* **11** 1027

[3] Lewis H R and Riesenfeld W B 1969 *J. Math. Phys.* **10** 1458

[4] Chen X, Ruschhaupt A, Schmidt S, del Campo A, Guéry-Odelin D and Muga J G 2010 *Phys. Rev. Lett.* **104** 063002

[5] Muga J G, Chen X, Ibáñez S, Lizuain I and Ruschhaupt A 2010 *J. Phys. B: At. Mol. Opt. Phys.* **43** 085509

[6] Schaff J F, Song X L, Vignolo P and Labeyrie G 2010 *Phys. Rev.* A **82** 033430

[7] Muga J G, Chen X, Ruschhaupt A and Guéry-Odelin D 2009 *J. Phys. B: At. Mol. Opt. Phys.* **42** 241001

[8] Schaff J F, Song X L, Capuzzi P, Vignolo P and Labeyrie G 2010 Shortcut to adiabaticity for a Bose-Einstein condensate *Preprint* arXiv:1009.5868v1

[9] Li Y, Wu L A and Wang Z D 2010 Fast cooling of mechanical resonator with time-controllable optical cavities *Preprint* arXiv 1008.1630v2

[10] Berry M V 2009 *J. Phys. A: Math. Theor.* **42** 365303

[11] Chen X, Lizuain I, Ruschhaupt A, Guéry-Odelin D and Muga J G 2010 *Phys. Rev. Lett.* **105** 123003

[12] Lewis H R and Leach P G 1982 *J. Math. Phys.* **23** 2371

[13] Dhara A K and Lawande S W 1984 *J. Phys.* A **17** 2324

[14] Lewis H R 1967 *Phys. Rev. Lett* **18** 510

[15] Lewis H R 1968 *J. Math. Phys.* **9** 1976

[16] Hartley J G and Ray J R 1981 *Phys. Rev.* A **24** 2873

[17] J. R. Ray and J. L. Reid 1979 *Phys. Lett.* A **71** 317

[18] Chen X and Muga J G 2010 *Phys. Rev.* A **82** 053403

[19] Rezek Y, Salamon P, Hoffmann K H and Kosloff R 2009 *Europhys. Lett.* **85** 30008

[20] Anandan J and Aharonov Y 1990 *Phys. Rev. Lett.* **65** 1697

[21] Masuda S and Nakamura K 2010 *Proc. R. Soc.* A **466** 1135

Decoherent histories analysis of minisuperspace quantum cosmology

J J Halliwell

Blackett Laboratory
Imperial College
London SW7 2BZ
UK

E-mail: `j.halliwell@imperial.ac.uk`

Abstract. Recent results on the decoherent histories quantization of simple cosmological models (minisuperspace models) are described. The most important issue is the construction, from the wave function, of a probability distribution answering various questions of physical interest, such as the probability of the system entering a given region of configuration space at any stage in its entire history. A standard but heuristic procedure is to use the flux of (components of) the wave function in a WKB approximation as the probability. This gives sensible semiclassical results but lacks an underlying operator formalism. Here, we supply the underlying formalism by deriving probability distributions linked to the Wheeler-DeWitt equation using the decoherent histories approach to quantum theory, building on the generalized quantum mechanics formalism developed by Hartle. The key step is the construction of class operators characterizing questions of physical interest. Taking advantage of a recent decoherent histories analysis of the arrival time problem in non-relativistic quantum mechanics, we show that the appropriate class operators in quantum cosmology are readily constructed using a complex potential. The class operator for not entering a region of configuration space is given by the S-matrix for scattering off a complex potential localized in that region. We thus derive the class operators for entering one or more regions in configuration space. The class operators commute with the Hamiltonian, have a sensible classical limit and are closely related to an intersection number operator. The corresponding probabilities coincide, in a semiclassical approximation, with standard heuristic procedures.

1. Introduction

1.1. Opening remarks

The problem of finding a sensible quantization of the Wheeler-DeWitt equation of minisuperspace quantum cosmology,

$$H\Psi = 0 \tag{1}$$

continues to attract considerable interest [1, 2, 3, 4, 5]. Although the setting of this problem is simple cosmological models with just a handful of homogeneous parameters, the techniques employed in answering this question may be relevant to general approaches to quantum gravity, such as the loop variables approach or causal set approach. This is because the central difficulty in consistently quantizing and interpreting the Wheeler-DeWitt equation is the absence of a variable to play the role of time and all approaches to quantum gravity must confront this issue at some stage [6].

A frequently studied example consists of a closed FRW cosmology with scale factor $a = e^\alpha$ and a homogeneous scalar field ϕ with (inflationary) potential $V(\phi)$ [7]. The Wheeler-DeWitt equation for this model is

$$\left(\frac{\partial^2}{\partial\alpha^2} - \frac{\partial^2}{\partial\phi^2} + e^{6\alpha}V(\phi) - e^{4\alpha} \right) \Psi(\alpha,\phi) = 0 \tag{2}$$

Given suitable boundary conditions, one can solve this equation for the wave function $\Psi(\alpha,\phi)$ and attempt to use it to answer a number of interesting cosmological questions. There are many such questions: Is there a regime in which the universe behaves approximately classically? What is the probability that the universe expands beyond a given size a_0? What is the probability that the universe has a certain energy density at a given value of the scale factor? What is the probability that the universe's history passes through a given region Δ of configuration space, characterized by certain ranges of a and ϕ?

Such questions, of necessity, do not involve the specification of an external time. Classically, the system's trajectories in minisuperspace may be written as paths $(\alpha(t), \phi(t))$, but here t is a convenient but unphysical parameter that labels the points along the paths – it does not correspond to the physical time measured by an external clock. The absence of a physical time is reflected in the quantum theory by the fact that the quantum state obeys the Wheeler-DeWitt equation Eq.(1), not a Schrödinger equation, and it is this difference that presents such a challenge to conventional quantization methods.

Although very plausible heuristic semiclassical methods exist for formulating and answering the above questions (in particular, the WKB interpretation [1]), it is of interest to see whether or not there exists a precise and well-defined quantum-mechanical scheme underlying these heuristic methods. We are not looking for high standards of mathematical rigour – just a standard quantum-mechanical framework of operators, inner product structures etc, obeying some reasonable requirements. A broad framework along these lines is the generalized quantum mechanics of Hartle [8]. The purpose of this paper is to carry out a specific implementation of Hartle's general framework, building also on earlier related attempts [9, 10, 11, 12, 13, 14, 15]. A much more detailed account of this work is given in Ref.[16].

1.2. Inner products and operators for the Wheeler-DeWitt equation

In the Wheeler-DeWitt equation (1), H is the Hamiltonian operator of a minisuperspace model with n coordinates q^a, and is typically of the form

$$H = -\nabla^2 + U(q) \tag{3}$$

where ∇ is the Laplacian in the minisuperspace metric f_{ab}, which has signature $(- + + + \cdots)$. It is naturally linked to the current,

$$J_a = i\left(\Psi^* \overrightarrow{\partial}_a \Psi - \Psi^* \overleftarrow{\partial}_a \Psi\right) \tag{4}$$

which is conserved

$$\nabla \cdot J = 0 \tag{5}$$

Closely associated is the Klein-Gordon inner product defined on a surface Σ

$$(\Psi,\Phi)_{KG} = i\int_\Sigma d\sigma^a \left(\Psi^* \overrightarrow{\partial}_a \Phi - \Psi^* \overleftarrow{\partial}_a \Phi\right) \tag{6}$$

where $d\sigma^a$ is a surface element normal to Σ. In flat space with a constant potential, the Wheeler-DeWitt equation is just the Klein-Gordon equation. Its solutions may be sorted out into positive and negative frequency in the usual way. With a little attention to sign, it is then possible to use components of the current to define probabilities.

However, in general, it is not possible to sort the solutions to the Wheeler-DeWitt equation into positive and negative frequency. This is one manifestation of the problem of time and more elaborate methods are required to associate probabilities with the Wheeler-DeWitt equation. There are two main issues: finding an inner product, and then finding suitable operators.

The issue of finding a suitable positive inner product is reasonably straightforward and goes by the name of Rieffel induction, or the induced (or physical) inner product [9, 17]. We consider first the usual Schrödinger inner product,

$$\langle \Psi_1 | \Psi_2 \rangle = \int d^n q \; \Psi_1^*(q) \Psi_2(q) \tag{7}$$

We then consider eigenvalues of the Wheeler-DeWitt operator

$$H | \Psi_{\lambda k} \rangle = \lambda | \Psi_{\lambda k} \rangle \tag{8}$$

where k labels the degeneracy. These eigenstates will satisfy

$$\langle \Psi_{\lambda' k'} | \Psi_{\lambda k} \rangle = \delta(\lambda - \lambda') \delta(k - k') \tag{9}$$

from which it is clear that this inner product diverges when $\lambda = \lambda'$. The induced inner product on a set of eigenstates of fixed λ is defined, loosely speaking, by discarding the δ-function $\delta(\lambda - \lambda')$. That is, the induced (or physical) inner product is then defined by

$$\langle \Psi_{\lambda k'} | \Psi_{\lambda k} \rangle_{phys} = \delta(k - k') \tag{10}$$

This procedure can be defined quite rigorously, and has been discussed at some length in Refs.[17, 9]. It is readily shown that the induced inner product coincides with the Klein-Gordon inner product when a division into positive and negative frequencies is possible, with the signs adjusted to make it positive. (This is described in Appendix A of Ref.[16]).

Turning now to the construction of interesting operators, the interesting dynamical variables associated with the Wheeler-DeWitt equation are those that commute with H. This is because the constraint equation is related to reparametrization invariance – which is reflected in the absence of a physical time variable – and we seek operators which are invariant. A wide class of operators commuting with H are of the form

$$A = \int_{-\infty}^{\infty} dt \; B(t) \tag{11}$$

which clearly commute with H since

$$\begin{aligned} e^{iHs} A e^{-iHs} &= \int_{-\infty}^{\infty} dt \; B(t+s) \\ &= A \end{aligned} \tag{12}$$

Many examples are given in Refs.[5, 17, 11]. However, another way of constructing such operators involves taking products,

$$A = \prod_{t=-\infty}^{\infty} B(t) \tag{13}$$

which may be shown to commute with H using essentially the same argument [12], but clearly further mathematical detail is required to give meaning to the infinite product. (Here t is the unphysical parameter time).

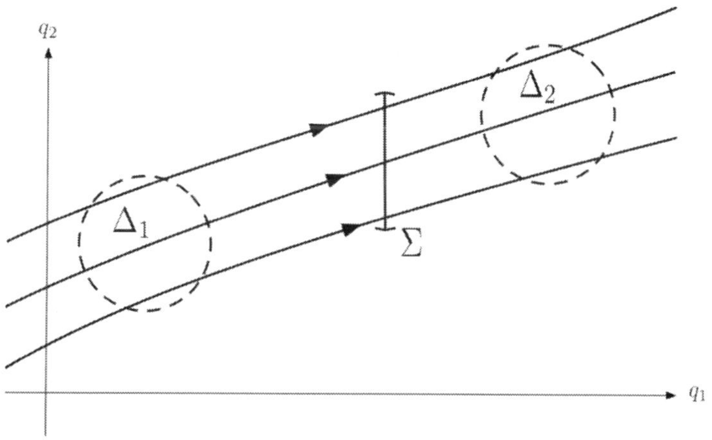

Figure 1. Given a solution Ψ to the Wheeler-DeWitt equation, what is the probability that the system enters a series of regions $\Delta_1, \Delta_2 \cdots$ in configuration space, or crosses a surface Σ, at any stage in the system's entire history?

Given these prescriptions for inner products and operators, one may then attempt to construct operators and probabilities implementing some of the questions mentioned above. We will focus on the following general question: Given a solution Ψ to the Wheeler-DeWitt equation, what is the probability of finding the system in a region Δ of configuration space, or of crossing a surface Σ, at any stage in the system's history? The question is depicted in Figure 1.

Questions involving surface crossings are not unlike more familiar questions in non-relativistic quantum mechanics, where there is a physical time parameter, but the key difference in quantum cosmology is that even classically, a given trajectory will typically cross a fixed surface more than once. It is precisely these sorts of issues that need to be carefully phrased in the quantum theory. Whilst the operator formalism briefly outlined above has been used to address such questions [5], the problems of characterizing properties of trajectories and surface crossings in quantum theory is naturally accommodated in the decoherent histories approach to quantum theory.

1.3. The decoherent histories approach

A more general approach to quantizing and interpreting the Wheeler-DeWitt equation is provided by the decoherent histories approach to quantum theory, and this approach will be the focus of this paper [18, 19, 20, 21, 22, 23, 24, 25, 26]. In this approach, probabilities are assigned to histories using the formula

$$p(\alpha) = \mathrm{Tr}\left(C_\alpha \rho C_\alpha^\dagger\right) \tag{14}$$

Here, ρ is the initial state (in our case a pure state) and C_α is a class operator characterizing the histories α of interest. In non-relativistic quantum mechanics these class operators are given by time-ordered strings of projection operators,

$$C_\alpha = P_{\alpha_n}(t_n) \cdots P_{\alpha_2}(t_2) P_{\alpha_1}(t_1) \tag{15}$$

(or by sums of such strings). The class operators always satisfies the condition

$$\sum_\alpha C_\alpha = 1 \tag{16}$$

For the reparametrization invariant systems considered here, the definition of the class operators is more subtle, and we return to this below.

Because of interference between pairs of histories, probabilities cannot always be assigned. To check for interference we therefore consider the decoherence functional,

$$D(\alpha, \alpha') = \mathrm{Tr}\left(C_\alpha \rho C_{\alpha'}^\dagger\right) \tag{17}$$

When

$$D(\alpha, \alpha') = 0 \tag{18}$$

for all pairs of histories in the set with $\alpha \neq \alpha'$, we say that there is decoherence of the set of histories and probabilities may be assigned Eq.(14). When there is decoherence, it is easily seen from Eq.(16) that the probabilities Eq.(14) are also given by

$$p(\alpha) = \mathrm{Tr}\left(C_\alpha \rho\right) = \mathrm{Tr}\left(C_\alpha^\dagger \rho\right) \tag{19}$$

Decoherence guarantees that this expression is real and positive, even though the class operators are not positive or hermitian operators in general.

The structure of the decoherent histories approach is very general and may be applied to a wide variety of situations, given an initial state, class operators, and a suitable inner product structure with which to construct the decoherence functional. A useful formulation of the decoherent histories approach encapsulating this generality is Hartle's generalized quantum mechanics [8]. This in essence defines a class of quantum theories through a decoherence functional obeying some simple requirements but does not rely on the specific form Eq.(15) of the class operators used in non-relativistic quantum mechanics. That framework is the background of what we do here, but it differs from the specific implementation of the framework in Ref.[8] (and elsewhere) through our choice of inner products and class operators.

For the application to the Wheeler-DeWitt system considered here, the initial state is take to be a solution to the Wheeler-DeWitt equation and the inner product is the induced inner product described above. The most crucial element is the specification of the class operators C_α. As indicated, in generalized quantum mechanics they can be more general than the non-relativistic version, Eq.(15). The product in the string can be taken to be continuous time [26]. Also, it is often valuable, sometimes essential, to allow the projectors to be replaced by more general operators, such as POVMs. The class operators must also properly characterize the histories one is interested in. It is not always obvious how to do this but useful clues often come from looking at the classical analogue of the class operator (where all the projectors commute).

Here, we are interested in histories which enter a region Δ of configuration space, or which cross a surface Σ, but without regard to time. This absence of a physical time variable seems particularly challenging, given that time seems to be central to the definition of non-relativistic analogue, Eq.(15). Closely related to this is the role of the constraint equation, Eq.(1). As noted already, these two features are directly related to the underlying symmetry of the theory – reparametrization invariance – and this symmetry is respected if the class operators commute with the constraint,

$$[H, C_\alpha] = 0 \tag{20}$$

Eq.(20) is in keeping with standard procedures of Dirac quantization [5]) and also ensures that the class operators have a sensible classical limit [12].

Some partially successful attempts to construct such class operators have been given previously [9, 10, 11, 12], but ran into various problems to do with the Zeno effect and with compatibility with the constraint equation. The main aim of this paper is to give fully satisfactory definitions of class operators for quantum cosmological models and explore their decoherence properties and probabilities.

Journal of Physics: Conference Series **306** (2011) 012023 doi:10.1088/1742-6596/306/1/012023

1.4. Some properties of the Wheeler-DeWitt equation and the WKB interpretation

To prepare the way for the full decoherent histories analysis of quantum cosmology, it is important to discuss some properties of the Wheeler-DeWitt equation and review the commonly used heuristic semiclassical interpretation of the wave function, since a proper quantization must recover this structure in some limit.

The Wheeler-DeWitt equation Eq.(3) is a Klein-Gordon equation in a curved configuration space with indefinite metric $f_{ab}(q)$ and potential $U(q)$ which can be positive or negative. The curvature effects of the metric are not significant in relation to the issues addressed in this paper, so we will assume for simplicity that the metric is flat.

The classical constraint equation corresponding to the Wheeler-DeWitt equation Eq.(3) is

$$\frac{1}{4} f_{ab} \dot{q}^a \dot{q}^b + U = 0 \tag{21}$$

from which one can see that the classical trajectories are timelike in the region $U > 0$ and spacelike in $U < 0$. (The timelike direction is that of increasing scale factor and the spacelike directions correspond to matter and anisotropic modes). The quantum case has analogous features. In simple models such as Eq.(2), the character of the solutions to the Wheeler-DeWitt equation depends on the sign of U. For large scale factors, $U > 0$ and the wave function is oscillatory, corresponding, very loosely, to a quasiclassical regime, and for small scale factors, $U < 0$, and the wave function is exponential, corresponding to a classical forbidden regime. However, there are certain types of models (such as those with an exponential potential for the scalar field), in which the identification of the oscillatory and exponential regions depends also on whether the constant U surfaces are spacelike or timelike [27]. We will not address this here.

One can also associate a Feynman propagator G_F with the Wheeler-DeWitt operator,

$$\begin{aligned} G_F &= \int_0^\infty dt\ e^{-iHt-\epsilon t} \\ &= \frac{-i}{H - i\epsilon} \end{aligned} \tag{22}$$

where $\epsilon \to 0+$. (Numerous propagator-like objects of this type have been considered in quantum cosmology [28]). Locally, on scales smaller than the scale on which the potential U significantly varies, the propagator $G_F(q, q')$ will be essentially identical in its properties with the analogous object for the relativistic particle for flat space. Therefore, in the region $U > 0$, for points q and q' which are timelike separated, it will propagate positive frequency solutions to the future and negative frequency solutions to the past. It will be exponentially suppressed for initial and final points q and q' which are spacelike separated (this is the familiar "propagation outside the lightcone" effect). For $U < 0$, similar statements hold but with timelike and spacelike reversed. Similar features hold on larger scales in a semiclassical approximation. These properties are important to understand the class operator constructed below.

Very plausible but heuristic answers to questions concerning crossing surfaces and entering regions are readily found using the WKB approximate solutions to the Wheeler-DeWitt equation and the Klein-Gordon current [1]. In the oscillatory regime, the Wheeler-DeWitt equation may be solved using the WKB ansatz

$$\Psi = R e^{iS} \tag{23}$$

where the rapidly varying phase S obeys the Hamilton-Jacobi equation

$$(\nabla S)^2 + U = 0 \tag{24}$$

and the slowly varying prefactor R obeys

$$\nabla \cdot (|R|^2 \nabla S) = 0 \tag{25}$$

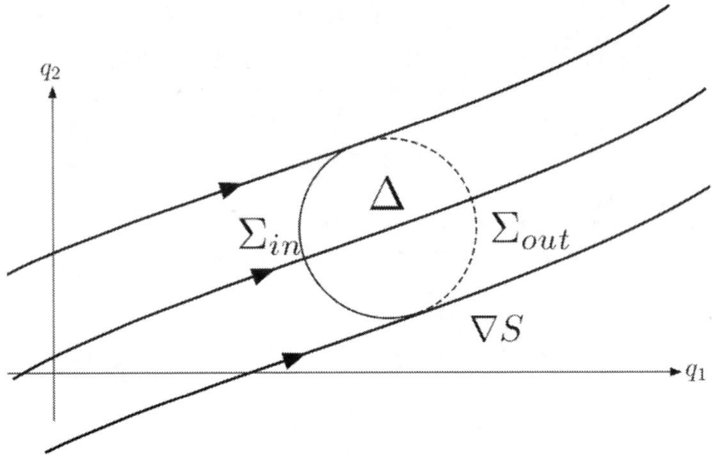

Figure 2. A WKB wave function with phase S corresponds to a set of classical trajectories with tangent vector ∇S. The probability for entering a region Δ is the amount flux of the wave function intersecting Δ. It may be expressed either in terms of the ingoing flux across Σ_{in} or equally, in terms of the outgoing flux across Σ_{out}.

The latter equation is just current conservation for the WKB current,

$$J = |R|^2 \nabla S \tag{26}$$

Wave functions of the WKB form Eq.(23) indicate a correlation between position and momenta of the form

$$p = \nabla S \tag{27}$$

and this suggests that the wave function Eq.(23) corresponds to an ensemble of classical trajectories satisfying Eq.(27). The current J may then be used to define a measure on this set of trajectories. For example, we consider a surface Σ and choose the normal n^a to the surface so that $n \cdot \nabla S > 0$. Then the probability of the system crossing the surface is taken to be

$$p(\Sigma) = \int_\Sigma d^{n-1}q \; n^a J_a \tag{28}$$

Of particular interest here is the probability of entering a region Δ. This is clearly related to the flux at the boundary of the region. The current will typically intersect the boundary of Δ *twice*. However, we can split the boundary Σ of Δ into two sections: Σ_{in} at which the current is ingoing and Σ_{out} at which the current is outgoing. The probability of entering Δ is may then be expressed in the two forms

$$\begin{aligned} p(\Delta) &= -\int_{\Sigma_{in}} d^{n-1}q \; n^a J_a \\ &= \int_{\Sigma_{out}} d^{n-1}q \; n^a J_a \end{aligned} \tag{29}$$

where here we have defined the normal n^a to point outwards. See Figure 2. These two forms are equivalent since the current is locally conserved.

Typically, little is said of the regions in which the wave function is exponential, except that they are similar to tunneling regions in non-relativistic quantum mechanics, so are classically forbidden in some sense. However, the wave function is not necessarily small in these regions, so there is surely more to it than this. In Ref.[29], it was noted that, unlike the oscillatory regions, the exponential regions do not indicate a correlation between position and momenta of the form Eq.(27), and it seems that this should be significant in some way.

Whilst the WKB interpretation is very plausible and adequate for most situations of interest, it leaves many questions unanswered. The main issue is to understand the operator origin of these probabilities, in terms of the language of operators commuting with H developed above. Furthermore, what can one say about superpositions of WKB states? Can interference between them be neglected? Also, what can one say about the exponential, classically forbidden regions? Is there a more precise way of saying that they are non-classical?

1.5. This paper

The purpose of this paper is to present a decoherent histories quantization of the Wheeler-DeWitt equation, and in particular to exhibit exactly defined class operators which characterize histories entering regions of configuration space or crossing surfaces, without reference to an external time. The key idea is to use a complex potential to define the class operator for not entering a region Δ in configuration space. In particular, we take the class operator for not entering to be the S matrix describing scattering off a complex potential localized in Δ. This turns out to have all the right properties – it avoids the Zeno effect, is compatible with the constraint and has a sensible classical limit.

In Section 2, to explain and motivate the use of a complex potential we review the use of such potentials in the decoherent histories analysis of the arrival time problem in non-relativistic quantum mechanics. In Section 3, we describe the construction of class operators for the Wheeler-DeWitt equation using a complex potential. The properties of the class operator for entering a region Δ are described in Section 4. The class operator has a sensible classical limit which, crucially, registers just one intersection of a trajectory with a surface, even when the trajectory intersects twice or more. In the quantum case, the class operator is an operator describing the ingoing flux across the boundary of the region, an expected result on semiclassical grounds, and is closely related to an intersection number operator.

In Section 5 we consider the WKB regime and show that the decoherent histories analysis reproduces the expected heuristic interpretation of the wave function. We summarize and conclude in Section 6.

2. The arrival time problem in non-relativistic quantum theory

In this section we summarize some of the key features of the arrival time problem in non-relativistic quantum mechanics [30]. These details are very relevant to the quantum cosmology case and in particular motivate the use of complex potentials in the definition of class operators.

In the one-dimensional statement of the arrival time problem, one considers an initial wave function $|\psi\rangle$ concentrated in the region $x > 0$ and consisting entirely of negative momenta. The question is then to find the probability $\Pi(\tau)d\tau$ that the particle crosses $x = 0$ between time τ and $\tau + d\tau$. See Figure 3.

The canonical answer is the current density

$$J(\tau) = \frac{(-1)}{2m} \langle\psi_\tau| \left(\hat{p}\delta(\hat{x}) + \delta(\hat{x})\hat{p}\right) |\psi_\tau\rangle \tag{30}$$

where $|\psi_\tau\rangle$ is the freely evolved state. This has the correct semiclassical limit, but can be negative for certain types of states consisting of superpositions of different momenta (backflow states). It is of interest to explore whether this simple semiclassical result emerges from more elaborate measurement-based or axiomatic schemes. Many such schemes naturally lead to the

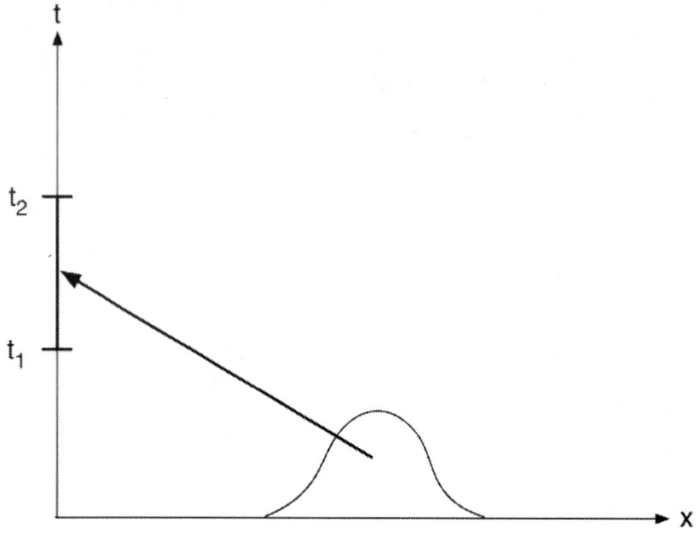

Figure 3. The quantum arrival time problem in non-relativistic quantum mechanics. Given an initial state localized entirely in $x > 0$ and consisting entirely of negative momenta, what is the probability that the particle crosses the origin during the time interval $[t_1, t_2]$?

use of a complex potential $-iV(x)$ in the Schrödinger equation [31], where

$$V(x) = V_0\theta(-x) \tag{31}$$

With such a potential, the state at time τ is

$$|\psi(\tau)\rangle = \exp\left(-iH\tau - V_0\theta(-x)\tau\right)|\psi\rangle \tag{32}$$

where H is the free Hamiltonian. The idea here is that the part of the wave packet that reaches the origin during the time interval $[0, \tau]$ should be absorbed, so that $\langle\psi(\tau)|\psi(\tau)\rangle$ is the probability of not crossing $x = 0$ during the time interval $[0, \tau]$. The probability of crossing between τ and $\tau + d\tau$ is then

$$\Pi(\tau) = -\frac{d}{d\tau}\langle\psi(\tau)|\psi(\tau)\rangle \tag{33}$$

This may be approximately evaluated in the limit $V_0 \ll E$ (where E is the typical energy scale), which is the limit of negligible reflection off the complex potential. The result is

$$\Pi(\tau) = 2V_0 \int_0^\tau dt\ e^{-2V_0(\tau - t)}\ J(t) \tag{34}$$

The probability for crossing during a finite interval $[\tau_1, \tau_2]$ is then given by

$$p(\tau_2, \tau_1) = \int_{\tau_1}^{\tau_2} d\tau\ \Pi(\tau) \tag{35}$$

and if this time interval is sufficiently large compared to $1/V_0$, we have

$$p(\tau_2, \tau_1) \approx \int_{\tau_1}^{\tau_2} dt\ J(t) \tag{36}$$

which means that the dependence on the potential drops out entirely at sufficiently coarse grained scales, and there is agreement with semiclassical expectations involving the current $J(t)$.

In the decoherent histories analysis of the arrival time problem [32, 33], we first consider the construction of the class operator C_{nc} for not crossing the origin during the finite time interval $[0,\tau]$. We split the time interval into N parts of size ϵ, and the class operator is provisionally defined by

$$C_{nc} = Pe^{-iH\epsilon}P\cdots e^{-iH\epsilon}P \tag{37}$$

where there are $N+1$ projections $P=\theta(\hat{x})$ onto the positive x-axis and N unitary evolution operators in between. One might be tempted to take the limit $N\to\infty$ and $\epsilon\to 0$, but this yields physically unreasonable results. This limit actually yields the restricted propagator in $x>0$,

$$\begin{aligned} C_{nc} &= g_r(\tau,0) \\ &= Pe^{-iPHP\tau} \end{aligned} \tag{38}$$

This object is also given by the path integral expression

$$\langle x_1|g_r(\tau,0)|x_0\rangle = \int_r \mathcal{D}x \exp(iS) \tag{39}$$

where the integral is over all paths from $x(0)=x_0$ to $x(\tau)=x_1$ that always remain in $x(t)>0$. However, the class operator defined by Eq.(38) has a problem with the Zeno effect – it consists of continual projections onto the region $x>0$ and as a result the wave function never leaves the region. This is reflected in the fact that the restricted propagator g_r is unitary in the Hilbert space of states with support only in $x>0$. This is a serious difficulty which has plagued a number of earlier works in this area [34, 35, 36].

To avoid the Zeno effect, the key is to keep ϵ non-zero. The class operator Eq.(37) is not easy to work with for finite ϵ, but fortunately a result first suggested by Echanobe et al. comes to the rescue [37, 38]. This is that the string of operators in Eq.(37) is in fact approximately equivalent to evolution in the presence of the complex potential $-iV$ introduced above. That is,

$$Pe^{-iH\epsilon}P\cdots e^{-iH\epsilon}P \approx \exp\left(-iH_0\tau - V_0\theta(-\hat{x})\tau\right) \tag{40}$$

This connection is valid as long as $\epsilon \ll 1/(\Delta H)$ and ϵ and V_0 are related by $V_0\epsilon\approx 4/3$ [38]. It strongly suggests that the class operator C_{nc}, normally defined by a string of projection operators, is justifiably defined instead using a complex potential. That is, we define

$$C_{nc} = \exp\left(-iH_0\tau - V_0\theta(-\hat{x})\tau\right) \tag{41}$$

The subsequent decoherent histories analysis was described in detail in Ref.[32]. The corresponding class operator for crossing during a time interval $[\tau_1,\tau_2]$ was shown to be

$$C_c(\tau_2,\tau_1) = \int_{\tau_1}^{\tau_2} dt\ e^{-iH_0(\tau-t)}Ve^{-iH_0t-Vt} \tag{42}$$

and again in the approximation $V_0 \ll E$ and for time intervals greater than $1/V_0$ this may be shown to have the simple and appealing form

$$C_c(\tau_2,\tau_1) \approx \int_{\tau_1}^{\tau_2} dt\ e^{-iH\tau}\frac{(-1)}{2m}\ \left(\hat{p}\ \delta(\hat{x}_t)\ +\ \delta(\hat{x}_t)\ \hat{p}\right) \tag{43}$$

which is now independent of the V_0. (These definitions of class operators differ by a factor of $\exp(-iH\tau)$ from those defined in Section 1C, but this does not make any difference in the decoherence functional and probabilities). With this class operator, one can show that there is decoherence of histories for a variety of interesting initial states, such as wave packets (but not for superposition states with backflow), and for such states, the general result Eq.(19) implies that the probability for crossing is simply

$$p(\tau_2, \tau_1) = \langle \Psi | C_c(\tau_2, \tau_1) | \Psi \rangle \tag{44}$$

which agrees precisely with the semiclassically expected result, Eq.(36).

In summary, the decoherent histories analysis of the arrival time problem in non-relativistic quantum mechanics indicates that it is reasonable to define class operators for not entering a space time region using a complex potential, as in Eq.(41), and that such a definition gives sensible a semiclassical limit, independent of the potential, at sufficiently coarse grained scales.

3. Construction of the class operators using a complex potential

We now come to the central issue concerning this paper, which is the construction of class operators for the decoherent histories analysis of the Wheeler-DeWitt equation. We seek class operators for the system Eq.(3) describing histories which enter or do not enter the region Δ, without specification of the time at which they enter. It is easiest to first focus on the class operator $\bar{C}_\Delta$ for not entering and the class operator for entering is then given by

$$C_\Delta = 1 - \bar{C}_\Delta \tag{45}$$

The earliest attempts to define class operators for the Wheeler-DeWitt equation involved defining $\bar{C}_\Delta$ as a sort of propagator obtained by integrating restricted propagators of the form Eq.(39) over an infinite range t (now regarded as the unphysical parameter time) [9, 10, 11]. However, in addition to having problems with the Zeno effect (which was not in fact appreciated in these earlier works), such constructions are difficult to reconcile with the constraint equation and some ad hoc modifications of the basic construction were required to give sensible answers.

A rather different approach to constructing $\bar{C}_\Delta$ was given in Ref.[12]. This was again problematic, but we review the construction here, since it is readily modified to yield a successful definition of the class operators. We denote by P the projector onto Δ and $\bar{P}$ the projector onto the outside of Δ. Our provisional proposal for the class operator for trajectories not entering Δ is the time-ordered infinite product,

$$\bar{C}_\Delta = \prod_{t=-\infty}^{\infty} \bar{P}(t) \tag{46}$$

where t is the unphysical parameter time. Subject to a more precise definition, given shortly, this object has the required properties. It is a string of projectors. Classically, it is equal to one for trajectories which remain outside Δ at every moment of parameter time. Also, it commutes with H, at least formally.

To define this more precisely, we first consider the product of projectors at a discrete set of times, t_1, $t_1 + \epsilon$, $t_1 + 2\epsilon, \cdots, t_1 + n\epsilon = t_2$. We define the intermediate quantity, $\bar{C}_\Delta(t_2, t_1)$ as the continuum limit of the product of projectors,

$$\bar{C}_\Delta(t_2, t_1) = \lim_{\epsilon \to 0} \bar{P}(t_2)\bar{P}(t_2 - \epsilon) \dots \bar{P}(t_1 + \epsilon)\bar{P}(t_1) \tag{47}$$

where the limit is $n \to \infty$, $\epsilon \to 0$ with $t_2 - t_1$ fixed. The desired class operator is then

$$\bar{C}_\Delta = \lim_{t_2 \to \infty, t_1 \to -\infty} \bar{C}_\Delta(t_2, t_1) \tag{48}$$

The class operator operator is clearly closely related to the restricted propagator g_r (the generalization of Eq.(39)) in the region outside Δ, since we have

$$\bar{C}_\Delta(t_2,t_1) = e^{iHt_2}\ g_r(t_2,t_1)\ e^{-iHt_1} \tag{49}$$

and therefore

$$\bar{C}_\Delta = \lim_{t_2\to\infty,t_1\to-\infty} e^{iHt_2}\ g_r(t_2,t_1)\ e^{-iHt_1} \tag{50}$$

The class operator commutes with H. This is because, from Eq.(49)

$$e^{iHs}\bar{C}_\Delta(t_2,t_1)e^{-iHs} = e^{iH(t_2+s)}\ g_r(t_2,t_1)\ e^{-iH(t_1+s)} \tag{51}$$

This becomes independent of s as $t_2\to\infty$, $t_1\to-\infty$, hence

$$[H,\bar{C}_\Delta] = 0 \tag{52}$$

The problem with this definition, however, is that it suffers from the Zeno effect, exactly like the analogous expression Eq.(37) (in the limit $\epsilon\to 0$) in the non-relativistic arrival time problem. But the key idea here is that we may also get around the problem in the same way, using a complex potential. That is, we "soften" the restricted propagator and make the replacement

$$g_r(t_2,t_1) \to \exp\left(-iH(t_2-t_1) - V(t_2-t_1)\right) \tag{53}$$

where $V(q) = V_0 f_\Delta(q)$. Here, $V_0 > 0$ is a constant and $f_\Delta(q)$ is the characteristic function of Δ, so is 1 in Δ and 0 outside it. Or we may equivalently write $V = V_0 P$, where recall, P is the projector onto Δ. This means that our new definition for the class operator is

$$\bar{C}_\Delta = \lim_{t_2\to\infty,t_1\to-\infty} e^{iHt_2}\ \exp\left(-i(H-iV)(t_2-t_1)\right)\ e^{-iHt_1} \tag{54}$$

The class operator thus defined has an appealing form: it is the S-matrix for system with Hamiltonian Eq.(3) scattering off the complex potential V. As required, it commutes with H. This is the most important definition of the paper.

Now an important observation. The class operator derived above has been defined as a time-ordered product of operators in which the direction of parameter time increases from right to left. However, since parameter time is unphysical, there is absolutely no reason why the parametrization should not run in the opposite direction. This produces an operator which is the hermitian conjugate of Eq.(54). As one can see from Eq.(19), this makes no difference in the final expressions for probabilities. In fact, it is most natural to define the class operator in such a way that it is invariant under reversing the direction of parametrization. We thus define a modified class operator which is hermitian:

$$\bar{C}'_\Delta = \frac{1}{2}\left(\bar{C}_\Delta + \bar{C}^\dagger_\Delta\right) \tag{55}$$

In what follows, for simplicity, we will primarily work with the non-hermitian class operator Eq.(54) and revert to the hermitian one, Eq.(55) where appropriate. The difference between them will turn out to be significant only for the class operator for two or more regions.

We now cast the above class operator in a more useable form. The following identities are readily derived:

$$e^{-i(H-iV)(t_2-t_1)} = e^{-iH(t_2-t_1)} - \int_{t_1}^{t_2} dt\ e^{-iH(t_2-t)}Ve^{-i(H-iV)(t-t_1)} \tag{56}$$

$$= e^{-iH(t_2-t_1)} - \int_{t_1}^{t_2} dt\ e^{-i(H-iV)(t_2-t)}Ve^{-iH(t-t_1)} \tag{57}$$

Inserting the second expression in the first, we obtain

$$\begin{aligned} e^{-i(H-iV)(t_2-t_1)} &= e^{-iH(t_2-t_1)} - \int_{t_1}^{t_2} dt\; e^{-iH(t_2-t)} V e^{-iH(t-t_1)} \\ &+ \int_{t_1}^{t_2} dt \int_{t_1}^{t} ds\; e^{-iH(t_2-t)} V e^{-i(H-iV)(t-s)} V e^{-iH(s-t_1)} \end{aligned} \tag{58}$$

Inserting in the expression for the class operator (54) and taking the limit, we obtain

$$\begin{aligned} \bar{C}_\Delta &= 1 - \int_{-\infty}^{\infty} dt\; V(t) \\ &+ \int_{-\infty}^{\infty} dt \int_{-\infty}^{t} ds\; e^{iHt} V e^{-i(H-iV)(t-s)} V e^{-iHs} \end{aligned} \tag{59}$$

The class operator for entering the region is therefore given by

$$C_\Delta = \int_{-\infty}^{\infty} dt\; V(t) - \int_{-\infty}^{\infty} dt \int_{-\infty}^{t} ds\; e^{iHt} V e^{-i(H-iV)(t-s)} V e^{-iHs} \tag{60}$$

This is an exact and useful form for the class operator for entering Δ, and it is easily confirmed that it is of the form Eq.(11) so commutes with H.

Now we use a simple but useful semiclassical approximation. Noting that $V = V_0 P$, where P recall projects into Δ, note that the expression

$$V e^{-i(H-iV)(t-s)} V \tag{61}$$

describes propagation with the complex Hamiltonian $H - iV$ between two points that lie inside Δ. This can easily be represented by a path integral in which it seems plausible that the dominant paths between this two end-points will lie entirely inside Δ, as long as the boundary is reasonable smooth and the semiclassical paths are not too irregular. If this is true, we may replace $V = V_0 P$ by the constant complex potential $V = V_0$, that is,

$$V e^{-i(H-iV)(t-s)} V \approx V e^{-i(H-iV_0)(t-s)} V \tag{62}$$

Propagation with a complex potential in Eq.(61) will also involve reflection off the boundary of the region, in which case the semiclassical approximation Eq.(62) may fail. However, reflection is small for sufficiently small V_0 [32, 31], and we will see this in more detail in Section 5. Hence we expect the semiclassical approximation Eq.(62) to hold for small V_0.

With this useful approximation (which does not affect the fact that the class operator commutes with H), we have

$$C_\Delta = \int_{-\infty}^{\infty} dt\; V(t) - \int_{-\infty}^{\infty} dt \int_{-\infty}^{t} ds\; V(t) V(s) e^{-V_0(t-s)} \tag{63}$$

Again using $V = V_0 P$, this is easily rewritten

$$C_\Delta = \int_{-\infty}^{\infty} dt\; P(t) \int_{-\infty}^{t} ds\; V_0 e^{-V_0(t-s)} \dot{P}(s) \tag{64}$$

This is the main result of the paper: a class operator commuting with H, describing histories which enter the region Δ. It is valid in the approximation Eq.(62), which is sufficient to cover the key case of histories which, classically, intersect the boundary of Δ twice.

For systems whose classical paths intersect the surfaces of interest more than two times, the semiclassical approximation Eq.(62) will not be valid, but the exact result Eq.(60) may still be used. It may also be of interest to explore higher order semiclassical approximations obtained by iterations of the basic result Eq.(58). This will not be explored here.

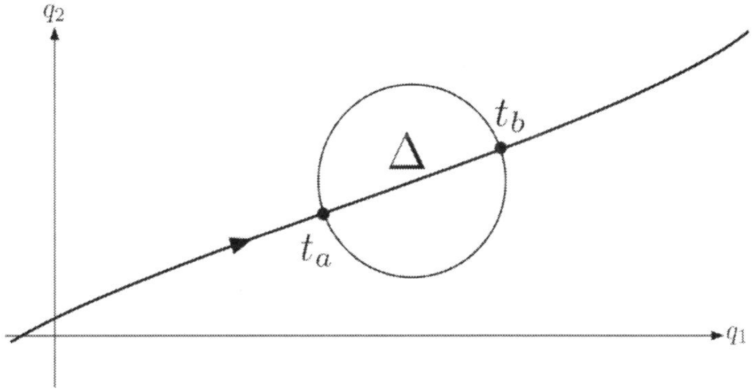

Figure 4. A classical trajectory $q^a(t)$ intersecting Δ enters at parameter time t_a and leaves at parameter time t_b.

4. Properties of the class operators

It is easy to show that the class operator for a single region Eq.(64) has a sensible classical limit. Classically, $P(t)$ is a function on classical trajectories with $P(t) = 1$ when the classical trajectory is in Δ and is zero otherwise. Suppose a given trajectory $q^a(t)$ enters Δ at some stage in its history so intersects the boundary twice (recalling we have essentially assumed no more than two intersections in the semiclassical approximation Eq.(62)). For a given choice of parametrization of the trajectory, it enters Δ at parameter time t_a and leaves at time $t_b > t_a$ (see Figure 4). (We may parameterize it in the opposite direction, with the same ultimate result).

The derivative of $P(s)$ is

$$\dot{P}(s) = \delta(s - t_a) - \delta(s - t_b) \tag{65}$$

and we have

$$C_\Delta = \int_{t_a}^{t_b} dt \int_{t_a}^{t} ds \ V_0 e^{-V_0(t-s)} \left[\delta(s - t_a) - \delta(s - t_b)\right] \tag{66}$$

Since $s \leq t \leq t_b$, the second δ-function in Eq.(66) makes no contribution to the integral. This is exactly the desired property – the expression for C_Δ registers only the first intersection of the trajectory with the boundary, but not the second intersection. The integral is easily evaluated with the result,

$$C_\Delta = 1 - e^{-V_0(t_b - t_a)} \tag{67}$$

This is approximately 1, as required, as long as

$$V_0(t_b - t_a) \gg 1 \tag{68}$$

We now give the broad picture in the quantum case and confirm some of the details in some simple models in the next section. Suppose we operate with C_Δ on an eigenstate $|\Psi_\lambda\rangle$ of H. The time integrals in Eq.(64) are easily carried out and the result is

$$C_\Delta|\Psi_\lambda\rangle = 2\pi V_0 \delta(H - \lambda) \ P G_V \dot{P} |\Psi_\lambda\rangle \tag{69}$$

where

$$G_V = \int_0^\infty dt \ e^{-i(H-\lambda)t - V_0 t} \tag{70}$$

$$= \frac{-i}{H - \lambda - iV_0} \tag{71}$$

and we have used

$$\delta(H-\lambda) = \frac{1}{2\pi}\int_{-\infty}^{\infty} dt\ e^{-i(H-\lambda)t} \tag{72}$$

The $\dot{P}$ term is a current operator on the boundary Σ of Δ and may be written

$$\begin{aligned}\dot{P} &= i[H, P] \\ &= -\hat{p}_n \delta_\Sigma(\hat{q}) - \delta_\Sigma(\hat{q})\hat{p}_n\end{aligned} \tag{73}$$

where

$$\delta_\Sigma(\hat{q}) = \int_\Sigma d^{n-1}q\ |q\rangle\langle q| \tag{74}$$

is a δ-function operator in the surface Σ and $\hat{p}_n = n^a\hat{p}_a$ is the component of the momentum operator normal to Σ with n^a the outward pointing normal. The normal n^a will depend on q in general so there may be an operator ordering issue in making $\hat{p}_n$ hermitian. The difference between different ordering will involve a term of the form $\nabla_a n^a$, essentially the extrinsic curvature of Σ, and this will be small if Δ is sufficiently large and its boundary reasonably smooth. Note that it is sometimes also convenient to write $\dot{P}$ as

$$\dot{P} = i\int_\Sigma d^{n-1}q\ |q\rangle(\overrightarrow{\partial_n} - \overleftarrow{\partial_n})\langle q| \tag{75}$$

It is very useful to separate the current operator at the boundary into ingoing and outgoing parts according to the sign of $\hat{p}_n$ at the boundary

$$\dot{P} = (\dot{P})_{in} - (\dot{P})_{out} \tag{76}$$

where $(\dot{P})_{in}$ consists of the components of $\dot{P}$ with incoming momentum

$$(\dot{P})_{in} = -\hat{p}_n\theta(-\hat{p}_n)\delta_\Sigma(\hat{q}) - \delta_\Sigma(\hat{q})\hat{p}_n\theta(-\hat{p}_n) \tag{77}$$

and similarly for $(\dot{P})_{out}$. Examples of these expressions in particular models will be given in the following sections. The restriction to positive or negative p_n means that it is generally difficult to express $(\dot{P})_{in}$ and $(\dot{P})_{out}$ in the form Eq.(75), involving the derivative $\hat{p}_n = -i\partial_n$, unless operating on a state (such as a WKB state) with simplifying properties. Note that these definitions require only that the local flux operator on a *given* surface can be split into ingoing and outgoing parts. To do so globally on a family of surfaces is generally impossible, essentially due to the problem of time, but fortunately this is not required here.

The quantity G_V has the form of a Feynman propagator, Eq.(22), with V_0 playing the role of the "$i\epsilon$ prescription", as long as V_0 is sufficiently small (in comparison to an appropriate energy scale contained in H, such as p_0 in the case of the Klein-Gordon equation). We therefore expect it to have properties similar to G_F, as described in Section 1D.

However, V_0 is not set to zero exactly and this in fact means that G_V has two properties not possessed by G_F. First, the non-zero V_0 produces a suppression for widely separated initial and final points. In a path integral representation of $G_V(q,q')$, the sum will be dominated by classical paths from q to q', to which one may associate the total parameter time τ of the path. From the integral representation Eq.(70) it can be seen that G_V will have an overall exponential suppression factor of the form $\exp(-V_0\tau)$. Thus, propagation with G_V will suppress configurations q and q' connected by classical trajectories of parameter time duration of greater than $1/V_0$. Second, recalling that the class operator is closely related to the S-matrix for scattering off a complex potential there will be some reflection involved and this appears in properites of G_V – it will not exactly possess the Feynman properties of propagating positive frequencies to the future negative frequencies to the past, but may, for example, propagate some positive frequencies to the past. However, this "non-Feynman" propagation will be small for sufficiently small V_0, as we will see in the next section, so this possibility will be ignored.

These properties of G_V are crucial to understanding the properties of C_Δ. First note that

$$\langle q|PG_V\dot{P}|\Psi_\lambda\rangle = -i\int_\Sigma d^{n-1}q'\ f_\Delta(q)G_V(q,q')(\overrightarrow{\partial_n} - \overleftarrow{\partial_n})\Psi_\lambda(q') \tag{78}$$

In this expression the propagator $G_V(q,q')$ propagates from the boundary Σ to a point q on the interior. The fact that G_V is, approximately, a propagator of the Feynman type means two things. First, there will be initial and final points q, q' for which the propagator is very small (due to "propagation outside the lightcone", discussed in Section 1D). Secondly, when it is not small, one would expect it to involve only *ingoing* modes at the boundary (to the extent that reflection is ignored). Hence the PG_V terms effectively restrict the current operator $\dot{P}$ at the boundary to ingoing modes only – exactly the result we are looking for. We therefore replace $\dot{P}$, with the current operator $(\dot{P})_{in}$ involving ingoing modes only

$$C_\Delta|\Psi_\lambda\rangle = 2\pi V_0\delta(H-\lambda)\ PG_V(\dot{P})_{in}|\Psi_\lambda\rangle \tag{79}$$

Now note that since the PG_V terms have done their job of selecting the ingoing modes, they may be eliminated. More precisely, we write $P = 1-\bar{P}$, where $\bar{P}$ is the projector onto the region outside Δ, and noting that that

$$V_0\delta(H-\lambda)G_V = \delta(H-\lambda) \tag{80}$$

we find

$$\begin{aligned} C_\Delta|\Psi_\lambda\rangle &= 2\pi\delta(H-\lambda)(\dot{P})_{in}|\Psi_\lambda\rangle \\ &- 2\pi\delta(H-\lambda)\bar{P}G_V(\dot{P})_{in}|\Psi_\lambda\rangle \end{aligned} \tag{81}$$

Consider the second term on the right-hand side. It consists of ingoing modes at the boundary of Δ propagated with G_V to a final point which is *outside* Δ. Semiclassically, this corresponds to paths which have to traverse the entire width of Δ, so if Δ is sufficiently large, this will take a long parameter time τ and, as discussed above, G_V will contain a suppression factor of $\exp(-V_0\tau)$ in comparison to the first term in Eq.(81). This is exactly analogous to the suppression of the second term in the classical case, Eq.(67). We thus deduce that

$$C_\Delta|\Psi_\lambda\rangle \approx 2\pi\delta(H-\lambda)(\dot{P})_{in}|\Psi_\lambda\rangle \tag{82}$$

Note that the approximation leading to dropping the second term in Eq.(81) also ensures that the result is independent of V_0, like the classical case in the appropriate limit. Eq.(82) is the main result of this section.

Eq.(82) may also be expressed in terms of $(\dot{P})_{out}$ defined in Eq.(76). To see this, note that, since $\dot{P} = i[H,P]$, we have

$$\delta(H-\lambda')\dot{P}|\Psi_\lambda\rangle = i(\lambda'-\lambda)\delta(H-\lambda')P|\Psi_\lambda\rangle \tag{83}$$

This is zero for $\lambda = \lambda'$, as long as $\langle\lambda|P|\lambda\rangle$ is well-defined, where $|\lambda\rangle$ are eigenstates of H. (It may not be well-defined if P projects onto an infinite region – see Appendix A). Eq.(76) then implies

$$\delta(H-\lambda)(\dot{P})_{in}|\Psi_\lambda\rangle = \delta(H-\lambda)(\dot{P})_{out}|\Psi_\lambda\rangle \tag{84}$$

so Eq.(82) may be expressed in terms of either the ingoing or outgoing flux or both.

It is also useful to note that the right-hand side of Eq.(82) may be written

$$2\pi\delta(H-\lambda)(\dot{P})_{in}|\Psi_\lambda\rangle = I_\Sigma|\Psi_\lambda\rangle \tag{85}$$

where

$$I_\Sigma = \int_{-\infty}^{\infty} dt\ e^{iHt}(\dot{P})_{in}e^{-iHt} \tag{86}$$

This is an intersection number operator for ingoing flux at the boundary Σ of Δ. Hence the class operator is essentially the intersection number I_Σ, and clearly commutes with H. It is classically equal to 1 for trajectories with ingoing flux at Σ (with no more than two intersections of the boundary, in the approximation we are using), and zero for trajectories not intersecting Σ. Of course, one may have guessed this approximate formula for the class operator, but a class operator is fundamentally defined as a product of projectors (or quasi-projectors) and the derivation given here makes it clear how this obvious guess arises from the fundamental definition. Note also the the class operator is hermitian in this case, so there is no need to consider the modified propagator Eq.(55).

The form Eq.(82) may be used to check for decoherence of histories in specific models and we will see this later. When there is decoherence of histories, the probabilities are given by the average of a single class operator, Eq.(19), which in this case reads

$$\langle\Psi_{\lambda'}|C_\Delta|\Psi_\lambda\rangle = 2\pi\langle\Psi_\lambda|(\dot{P})_{in}|\Psi_\lambda\rangle\ \delta(\lambda-\lambda') \tag{87}$$

Following the induced inner product prescription, we drop the δ-function on the right and then set $\lambda=\lambda'=0$, so the probability in terms of a solution $|\Psi\rangle$ of the Wheeler-DeWitt equation is

$$\langle\Psi|C_\Delta|\Psi\rangle_{phys} = 2\pi\langle\Psi|(\dot{P})_{in}|\Psi\rangle \tag{88}$$

This is essentially the ingoing Klein-Gordon flux on the boundary of Δ, as expected. (The factor of 2π relates to the induced inner product as described in Appendix A of Ref.[16]).

The above derivation is verified in more detail in Ref.[16] using specific examples. Furthermore, it is also shown there that the form of the class operator for n regions is,

$$C_n = \frac{1}{n!}\left(I_{\Sigma_1}I_{\Sigma_2}\cdots I_{\Sigma_n} + \ permutations\right) \tag{89}$$

where "permutations" means add all possible permutations of the n regions, to give a total of $n!$ terms. This final result has a particularly natural form which also suggests that it may have simple path integral representations. This will be explored elsewhere. (See also Ref.[42] for similar expressions, derived using a detector model for quantum cosmology).

5. WKB regime

The most important case in which to check the ideas developed above is in the WKB regime. In the oscillatory regime, the solutions to the Wheeler-DeWitt equation have the form

$$\Psi = Re^{iS} \tag{90}$$

where R and S obey Eqs.(24), (25) as described earlier. More generally, the wave function is a superposition of WKB wave functions but we consider first the case of a single term. The heuristic interpretation of such states was described in Section 1D. Our aim is to show that the decoherent histories analysis reproduces the heuristic scheme. WKB states are locally plane wave states of the type considered in the previous Section, so we will appeal to that analysis to understand the properties WKB states.

We consider the action of the class operator for a single region Δ on a WKB state. We first consider the class operator Eq.(82) acting on WKB state (regularized by making it an eigenstate of H with eigenvalue λ)

$$\begin{aligned} \langle q|C_\Delta|\Psi_\lambda\rangle &= 2\pi\langle q|\delta(H-\lambda)(\dot{P})_{in}|\Psi_\lambda\rangle \\ &= 2\pi i\int_{\Sigma_{in}} d^{n-1}q' \ \langle q|\delta(H-\lambda)|q'\rangle(\overrightarrow{\partial_n}-\overleftarrow{\partial_n})\Psi_\lambda(q') \end{aligned} \tag{91}$$

Here Σ_{in} denotes the sections of the boundary where the flux is ingoing,

$$n\cdot\nabla S<0 \tag{92}$$

where n^a is the outward pointing normal. The key property of the WKB wave function $\Psi_\lambda(q')$ is that it has momentum $p=\nabla S(q')$ at each point q' on the boundary. When the operator $\delta(H-\lambda)$ is applied, from the representation Eq.(72), we see that it's effect is to evolve the state $\Psi_\lambda(q')$ (restricted to Σ_{in}) forwards and backwards in parameter time, and then integrates over all times. Semiclassically, the evolution of the state $\Psi_\lambda(q')$ will be concentrated along the classical trajectories defined by initial positions q' in Σ_{in} and momenta $p=\nabla S(q')$, with some spreading of the wave packet, but this will be small if Δ is reasonably large. (This is analogous to the model of Section 5B).

We thus see the following: the wave function $\langle q|C_\Delta|\Psi_{\lambda},\rangle$ in Eq.(91) is spatially localized around the tube of classical trajectories passing through Δ with momenta $p=\nabla S$ (depicted in Figure 2). This may be approximately written in the alternative form

$$\langle q^a|C_\Delta|\Psi_\lambda\rangle\approx\theta(\tau_\Delta-\epsilon)\ Re^{iS} \tag{93}$$

where $\epsilon>0$ is a small parameter to regularize the θ-function at zero argument. Here, $\tau_\Delta(q)$ is the parameter time spent by the classical trajectory $q_{cl}(t)$ (with initial value q and momentum $p=\nabla S(q)$) in the region Δ and may be written

$$\tau_\Delta(q)=\int_{-\infty}^{\infty}dt\ f_\Delta(q_{cl}(t)) \tag{94}$$

This has the property that

$$\nabla S\cdot\nabla\tau_\Delta=0 \tag{95}$$

since $\nabla S\cdot\nabla$ simply translates along the classical trajectories. It follows from Eq.(25) that Eq.(93) is in fact a WKB solution to the Wheeler-DeWitt equation, since the θ-function essentially modifies the prefactor R but in a way that it still satisfies Eq.(25). (Of course, this is expected because C_Δ commutes with H).

Eq.(93) is a very useful result and allows us to check for decoherence very easily. The action of the class operator for not entering Δ is clearly

$$\langle q^a|\bar{C}_\Delta|\Psi_\lambda\rangle\approx\theta(\epsilon-\tau_\Delta)\ Re^{iS} \tag{96}$$

which is a WKB state localized on the set of trajectories not entering Δ. It immediately follows that

$$\langle\Psi_{\lambda'}|\bar{C}_\Delta C_\Delta|\Psi_\lambda\rangle\approx 0 \tag{97}$$

since the two states Eqs.(93), (96) are localized about complementary regions. There is therefore approximate decoherence of histories for a single WKB packet and for histories entering or not entering a single region Δ, as long as Δ is sufficiently large.

The key reason for the decoherence with a single WKB packet is related to the approximate determinism of the WKB wave functions: fixing values of position to lie on Σ_{in} also fixes the momenta, since $p = \nabla S$, so that Eq.(91) is concentrated along a tube of classical trajectories.

More generally, the initial state will be a superposition of WKB wave packets,

$$\Psi = \sum_k R_k e^{iS_k} \tag{98}$$

The component states in this sum are typically approximately orthogonal to each other as long as the phases S_k are sufficiently different. (This will depend on the detailed dynamics of the model). Following the analogous example in the simple models of Section 5B, we would expect that the class operators will not disturb the approximate orthogonality of these states as long as the region Δ is sufficiently large – we expect the cross terms in the decoherence functional to average to zero because of oscillations. Therefore, superpositions may in practice be treated as mixtures at sufficiently coarse-grained scales. Note that this statement also applies the special state,

$$\Psi = R\left(e^{iS} + e^{-iS}\right) \tag{99}$$

which arises from the Hartle-Hawking "no boundary" proposal [45]. That is, the interference between the two terms may be neglected.

Given decoherence, the probabilities are given by the general expression Eq.(88). It then follows that the probabilities for entering Δ coincide with Eq.(29), the sought-after result.

Now consider the case of probabilities for histories entering two regions, as described by the class operator Eq.(89). Since the two-region class operator is a sum of products of one-region class operators, its effect on the WKB wave functions is easy to see. The action of a single class operator gives Eq.(93). But since this is still a wave function of the WKB type, the action of a second class operator yields

$$\langle q^a | C_{\Delta_1 \Delta_2} | \Psi_\lambda \rangle \approx \theta(\tau_{\Delta_1} - \epsilon)\theta(\tau_{\Delta_2} - \epsilon)\ Re^{iS} \tag{100}$$

That is, it is a WKB wave function but restricted in such as way that it's flux passes through both regions. See Figure 5. It is again easy to see that there is decoherence of histories and the probability is given by an expression of the form Eq.(29), but with the integral over the subset of incoming flux at Δ_1 which goes on to intersect Δ_2.

From the above observations one can also see why the exponential WKB wave functions will *not* lead to decoherence of histories. The exponential wave functions have the form

$$\Psi = R\ e^{-I} \tag{101}$$

where I is real. The key difference between states of this type and the oscillatory type Eq.(90) is that they do not have a correlation between positions and momenta [29]. One would therefore expect the evolution of the state Eq.(91) to be spread all over the configuration space and not concentrated around a particular region. That is, these states do not have the approximate determinism of the oscillatory states. The states $C_\Delta|\Psi\rangle$ and $\bar{C}_\Delta|\Psi\rangle$ would then not be approximately orthogonal so there will be no decoherence of histories.

The decoherence of histories described here has arisen because of the approximate determinism of the oscillatory WKB states, together with the approximate orthogonality properties that arise when the regions Δ are sufficiently large. At finer grained scales, decoherence of histories may only be possible in more complicated models in which there is an environment of some sort. Models along these lines, in more basic approaches to quantum cosmology, have been considered previously [46], and one might expect that they may be adapted to the decoherent histories approach to quantum cosmology. (See also Ref.[11]).

In summary, we have derived from the decoherent histories approach the probabilities normally used in the heuristic WKB interpretation. To address these issues in more detail will require more specific quantum cosmological models. This will be considered elsewhere.

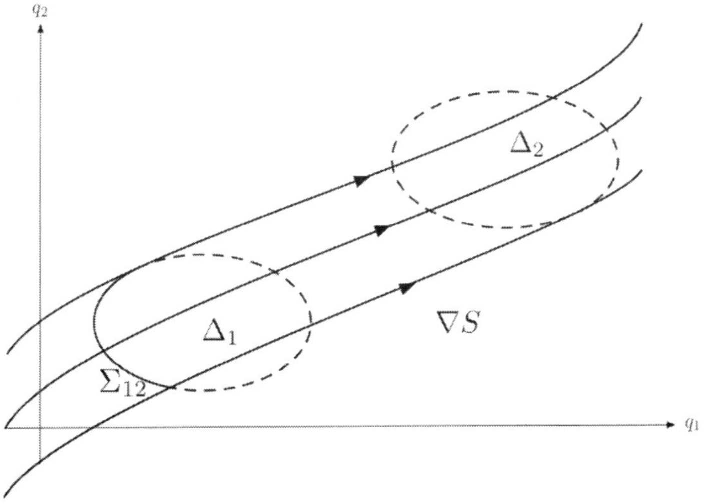

Figure 5. The class operator $C_{\Delta_1\Delta_2}$ for two regions operating on a WKB wave function produces another WKB state localized around the flux passing through both regions. This is the same as the flux entering Δ_1 across the surface Σ_{12}.

6. Discussion and further issues

We have presented a properly defined quantization procedure for quantum cosmology using the decoherent histories approach to quantum theory and derived from this the frequently used but heuristic WKB interpretation, involving fluxes of the WKB wave function.

The key idea was to use a complex potential to define the class operators for not entering a region of configuration space. This method is adequately justified by its successful use in the arrival time problem in non-relativistic quantum theory. We showed that the class operators defined in this way have all the desired properties – they have the correct classical limit, are compatible with the constraint equation, and do not have difficulties with the Zeno effect. In a semiclassical approximation, they have an appealing form in terms of intersection number operators. They give sensible results in simple models and there is approximate decoherence of histories for certain types of initial state at sufficiently coarse grained scales.

Future papers will address the more detailed application of this approach to specific models and will also undertake a comparison of the decoherent histories approach described here to other approaches [5].

Acknowledgments

I am very grateful to Thomas Elze for organizing such a stimulating meeting. I am also very grateful to Jim Hartle, Don Marolf and James Yearsley for useful conversations. I would also like to thank James Yearsley for his help in preparing the figures.

References

[1] Halliwell J J 1991 in *Quantum Cosmology and Baby Universes* ed S Coleman, J B Hartle, T Piran and S Weinberg (Singapore: World Scientific), also available as *e-print* gr-qc/0909.2566. These introductory lectures on quantum cosmology contain a detailed account of the WKB interpretation.
[2] Kiefer C and Sandhoefer B 2008 *Preprint* gr-qc/0804.0672; Wiltshire D L 2001 *Preprint* gr-qc/0101003
[3] Hartle J B, Hawking S W and Hertog T 2008 *Phys. Rev.* D **77** 123537; *do.* 2008 *Phys. Rev. Lett.* **100** 201301
[4] DeWitt B 1962 in *Gravitation: An Introduction to Current Research* ed L Witten (New York: John Wiley and Sons)

[5] Rovelli C 1991 *Class. Quant. Grav.* **8** 297; *ibid.* p 317; *do.* 1990 *Phys. Rev.* **42** 2638; *do.* 1991 *Phys. Rev.* **43** 442; Marolf D 1995 *Class. Quant. Grav.* **12** 1199; . p 1441; Hajicek P 1995 *J. Math. Phys.* **36** 4612; Montesinos M, Rovelli C and Thiemann T 1999 *Phys. Rev.* D **60** 044009; Montesinos M 2001 *Gen. Relativ. Gravit.* **33** 1; Gambini R and Porto R A 2001 *Phys. Rev.* D **63** 105014

[6] Kuchar K 1992 in *Proceedings of the 4th Canadian Conference on General Relativity and Relativistic Astrophysics* ed G Kunstatter, D Vincent and J Williams (Singapore: World Scientific); *do.* 1993 *Preprint* gr-qc/9304012; Isham C 1992 *Preprint* gr-qc/9210011

[7] Hawking S W 1984 *Nucl. Phys.* B **239** 257

[8] Hartle J B 1995 in *Proceedings of the 1992 Les Houches School, Gravity and its Quantizations* ed B Julia and J Zinn-Justin (Elsevier Science B.V.)

[9] Hartle J B and Marolf D 1997 *Phys. Rev.* D **56** 6247

[10] Halliwell J J and Thorwart J 2001 *Phys. Rev.* D **64** 124018

[11] Halliwell J J and Thorwart J 2002 *Phys. Rev.* D **65** 104009

[12] Halliwell J J and Wallden P 2006 *Phys. Rev.* D **73** 024011

[13] Craig D and Hartle J B 2004 *Phys. Rev.* D **69** 123525

[14] Whelan J 1994 *Phys. Rev.* D **50** 6344

[15] Anastopoulos C and Savvidou K 2000 *Class. Quant. Grav.* **17** 2463; *do.* 2005 *Class. Quant. Grav.* **22** 1841

[16] Halliwell J J 2009 *Phys. Rev.* D **80** 124032

[17] Ashtekar A, Lewandowski J, Marolf D, Mourao J and Thiemann T 1995 *J. Math. Phys.* **36** 6456; Higuchi A 1991 *Class. Quant. Grav.* **8** 1983; Giulini D and Marolf D 1999 *Class. Quant. Grav.* **16** 2489; *ibid.* 2479; Embacher F 1998 *Hadronic J.* **21** 337; Landsmann N 1995 *J. Geom. Phys.* **15** 285; Marolf D 2000 *preprint* gr-qc/0011112

[18] Gell-Mann M and Hartle J B 1990 in *Complexity, Entropy and the Physics of Information, SFI Studies in the Sciences of Complexity* Vol. **VIII** ed W Zurek (Reading, Mass.: Addison Wesley); *do.* 1990 in *Proceedings of the Third International Symposium on the Foundations of Quantum Mechanics in the Light of New Technology* ed S Kobayashi, H Ezawa, Y Murayama and S Nomura (Tokyo: Physical Society of Japan)

[19] Gell-Mann M and Hartle J B 1993 *Phys. Rev.* D **47** 3345

[20] Griffiths R B 1984 *J. Stat. Phys.* **36** 219; *do.* 1993 *Phys. Rev. Lett.* **70** 2201; *do.* 1996 *Phys. Rev.* A **54** 2759; *do.* 1998 *Phys. Rev.* A **57** 1604

[21] Omnès R 1988 *J. Stat. Phys.* **53** 893; *ibid.* p 933; *ibid.* p 957; *do.* 1989 *J. Stat. Phys.* **57** 357; *do.* 1990 *Ann. Phys.* **201** 354; *do.* 1992 *Rev. Mod. Phys.* **64** 339

[22] Omnès R 1997 *J. Math. Phys.* **38** 697

[23] Halliwell J J 1994 in *Fundamental Problems in Quantum Theory* ed D Greenberger and A Zeilinger *Annals of the New York Academy of Sciences* **775** 726

[24] Halliwell J J 2004 in *Decoherence and Entropy in Complex Systems – Springer Lecture Notes in Physics* LNP **633** ed H-T Elze (Berlin: Springer)

[25] Halliwell J J 1999 *Phys. Rev.* D **60** 105031

[26] Isham C 1994 *J. Math. Phys.* **23** 2157; Isham C and Linden N 1994 *J. Math. Phys.* **35** 5452; *do.* 1995 *J. Math. Phys.* **36** 5392 (1995)

[27] Halliwell J J 1986 *Nucl. Phys.* B **266** 228; *do.* 1987 *Nucl. Phys.* B **286** 729

[28] Halliwell J J 1988 *Phys. Rev.* D **38** 2468; Halliwell J J and Hartle J B 1990 *Phys. Rev.* D **41** 1815; *do.* 1991 *Phys. Rev.* D **43** 1170 (1991); Teitelboim C 1982 *Phys. Rev.* D **25** 3159; *do.* 1983 *Phys. Rev.* D **28** 297

[29] Halliwell J J 1987 *Phys. Rev.* D **36** 3626

[30] Muga J G, Sala Mayato R and Egusquiza I L (eds) 2002 *Time in Quantum Mechanics* (Berlin: Springer); Muga J G and Leavens C R 2000 *Phys. Rep.* **338** 353

[31] Muga J G, Brouard S and Macias D 1995 *Ann. Phys. (NY)* **240** 351; Blanchard Ph and Jadczyk A 1996 *Helv. Phys. Acta* **69** 613; Palao J P, Muga J G, Brouard S and Jadczyk A 1997 *Phys. Lett.* A **233** 227; Ruschhaupt A 2002 *J. Phys.* A **35** 10429; Muga J G, Palao J P, Navarro B and Egusquiza I L 2004 *Phys. Rep.* **395** 357; Halliwell J J 1999 *Prog. Theor. Phys.* **102** 707

[32] Halliwell J J and Yearsley J M 2009 *Phys. Rev.* A **79** 062101

[33] Halliwell J J and Yearsley J M 2009 *Phys. Lett.* A **374** 154

[34] Yamada N and Takagi S 1991 *Prog. Theor. Phys.* **85** 985; *ibid.* **86** 599; *do.* 1992 *Prog. Theor. Phys.* **87** 77; Yamada N 1992 *Sci. Rep. Tôhoku Uni., Series 8* **12** 177; *do.* 1996 *Phys. Rev.* A **54** 182

[35] Halliwell J J and Zafiris E 1998 *Phys. Rev.* D **57** 3351-3364

[36] Hartle J B 1991 *Phys. Rev.* D **44** 3173

[37] Echanobe J, del Campo A and Muga J G 2008 *Phys. Rev.* A **77** 032112

[38] Halliwell J J and Yearsley J M 2010 *J. Phys.* A **43** 445303

[39] Auerbach A and Kivelson S 1985 *Nucl. Phys.* B **257** 799; van Baal P 1993 in *Lectures on Path Integration* ed H A Cerdeira *et al.* (Singapore: World Scientific) [also available from the website

http://www.lorentz.leidenuniv.nl/vanbaal/HOME/publ.html]

[40] Halliwell J J and Ortiz M E 1993 *Phys. Rev.* D **48** 748

[41] Halliwell J J 1995 *Phys. Lett.* A **207** 23

[42] Halliwell J J 2001 *Phys. Rev.* D **64** 044008

[43] Schulman L S 1981 *Techniques and Applications of Path Integration* (New York: Wiley and Sons) ch. 7

[44] Mott N F 1929 *Proc. Roy. Soc.* A **124** 375 reprinted in Wheeler J and Zurek W (eds) 1983 *Quantum Theory and Measurement* (Princeton, New Jersey: Princeton University Press). For further discussions of the Mott calculation see also Bell J S 1987 *Speakable and Unspeakable in Quantum Mechanics* (Cambridge: Cambridge University Press); Broyles A A 1993 *Phys. Rev.* A **48** 1055; Castagnino M and Laura R 2000 *Int. J. Theor. Phys.* **39** 1737; Barbour J 1994 *Class. Quant. Grav.* **11** 2853; *ibid.* p 2875

[45] Hartle J B and Hawking S W 1983 *Phys. Rev.* D **28** 2960

[46] Halliwell J J 1989 *Phys. Rev.* D **39** 2912; Kiefer C 1987 *Class. Quant. Grav.* **4** 1369; Padmanabhan T 1989 *Phys. Rev.* D **39** 2924

The Fermionic Projector, entanglement and the collapse of the wave function

Felix Finster
Fakultät für Mathematik, Universität Regensburg, 93040 Regensburg, Germany
E-mail: `Felix.Finster@mathematik.uni-r.de`

Abstract. After a brief introduction to the fermionic projector approach, we review how entanglement and second quantized bosonic and fermionic fields can be described in this framework. The constructions are discussed with regard to decoherence phenomena and the measurement problem. We propose a mechanism leading to the collapse of the wave function in the quantum mechanical measurement process.

1. An introduction to the Fermionic Projector approach

In this section we give a brief introduction to the framework of the fermionic projector, outlining a few ideas, methods and results (for details see [1, 2] and the references in the review papers [3, 4]).

1.1. The idea of spontaneous structure formation

Due to the ultraviolet divergences of quantum field theory and the difficulties in quantizing gravity, it is generally believed that our notions of space and time should be modified on a microscopic scale, usually thought of as the Planck scale. However, there is no agreement on what the right framework for physics on the Planck scale should be. A major difficulty is that if one modifies the microscopic structure of Minkowski space or of a Lorentzian manifold in a naive way, the gauge invariance and the diffeomorphism invariance of the resulting physical theory are not preserved. For example, a simple ultraviolet regularization is to replace Minkowski space by a four-dimensional lattice. If one now works instead of partial derivatives with difference quotients, then the gauge invariance is lost. A better method is to work with the integrals of the gauge potentials along the links of the lattice, which restores gauge invariance and leads to the framework of lattice gauge theories. However, lattice gauge theory is not compatible with the equivalence principle, because after performing a diffeomorphism, the lattice points are no longer in a regular configuration. Taking a somewhat different point of view, the problem of a lattice is that it uses structures (like the nearest neighbor relation or the lattice spacing) which refer to an ambient space-time, although we wanted to avoid this background space-time for the formulation of the physical model. More generally, one would like to have a framework which is background independent and respects the fundamental physical principles.

The fermionic projector approach provides a framework which meets the above requirements. We begin with a finite set of space-time points $M = \{1, \ldots, m\}$, without presupposing any relations between the space-time points. On this finite set of points, we then define an "ensemble of wave functions" $\psi_1, \ldots, \psi_f$. We introduce an action $\mathcal{S}$ which to every configuration of wave functions associates a positive number. By minimizing $\mathcal{S}$, the wave functions tend to get into a configuration where the action is smaller. This process can be understood as a self-organization of the wave functions, driven by our action principle. In the resulting minimizing configuration, the wave functions are typically delocalized in the sense that they are non-zero on many space-time points. Thus by considering the correlations of the wave functions at different space-time points, one gets relations between these points. With this mechanism, referred to as *spontaneous structure formation*, additional structures and objects arise in space-time. The idea is that all the structures needed for the formulation of physics, like the causal structure, gauge potentials and fields, the Lorentzian metric, etc., are generated in this way. Finally, evaluating the Euler-Lagrange equations corresponding to our action principle in terms of all these structures should give the physical equations.

Before we can make this idea mathematically precise, we need to specify in which space the wave functions $\psi_i(x)$ should take their value. To this end, we associate to every space-time point $x \in M$ a complex vector space S_x, referred to as the *spin space*. We endow this vector space with an indefinite inner product $\prec . | . \succ_x$, the *spin scalar product*. Then a wave functions at x takes values in the corresponding spin space

$$\psi_i \,:\, x \mapsto \psi_i(x) \in S_x \,.$$

In physical applications, a wave function should correspond to a four-component Dirac wave function with the usual inner product $\prec \psi | \phi \succ_x = \overline{\psi}(x)\phi(x)$, where $\overline{\psi} = \psi^\dagger \gamma^0$ is the adjoint spinor (this will be made precise in Section 1.4 below). With this in mind, we assume that S_x is four-dimensional, and that the spin scalar product is indefinite of signature $(2, 2)$. Taking the direct sum of all the spin spaces,

$$H = \bigoplus_{x \in M} S_x \,, \qquad <\psi|\phi> := \sum_{x \in M} \prec \psi(x) | \phi(x) \succ_x \,,$$

we obtain an inner product space $(H, <.|.>)$ of dimension $4m$ and signature $(2m, 2m)$. We can then describe the evaluation of a wave function at a space-time point x by a projection operator E_x,

$$\psi(x) = E_x \, \psi \qquad \text{with} \qquad E_x \,:\, H \to S_x \,:\, \psi \mapsto \psi(x) \,.$$

Moreover, it is convenient to describe the ensemble of wave functions $\psi_1, \ldots, \psi_f \in H$ by a projector P on the subspace of H spanned by these wave functions. This motivates the framework of fermion systems in discrete space-time which we now introduce.

1.2. An action principle for fermion systems in discrete space-time

We let $(H, <.|.>)$ be a finite-dimensional complex vector space endowed with an indefinite inner product (thus $<.|.>$ is non-degenerate, but not positive definite). Next, we let $M = \{1, \ldots, m\}$ be a finite set. To every point $x \in M$ we associate a projector E_x (a projector in H is defined just as in Hilbert spaces as a linear operator which is idempotent and symmetric). We assume that these projectors are orthogonal and complete in the sense that

$$E_x \, E_y \;=\; \delta_{xy} \, E_x \qquad \text{and} \qquad \sum_{x \in M} E_x \;=\; \mathbb{1} \,. \tag{1}$$

Furthermore, we assume that the images $E_x(H) \subset H$ of these projectors are non-degenerate subspaces of H, which all have the same signature $(2, 2)$. The points $x \in M$ are called *discrete space-time points*, and the corresponding projectors E_x are the *space-time projectors.* The structure $(H, <.|.>, (E_x)_{x \in M})$ is called *discrete space-time.* The particles of our system are described by one more projector P in H, the so-called *fermionic projector*, which has the property that its image $P(H)$ is a negative definite subspace of H. The resulting system $(H, <.|.>, (E_x)_{x \in M}, P)$ is referred to as a *fermion system in discrete space-time.*

Let us briefly discuss these definitions and introduce a convenient notation. The vectors in the image of P have the interpretation as the occupied fermionic states of our system, and thus the rank of P gives the *number of particles* $f := \dim P(H)$. The space-time projectors E_x can be used to project vectors of H to the subspace $E_x(H) \subset H$. Using a more graphic notion, we also refer to this projection as the *localization* at the space-time point x. For the localization of a vector $\psi \in \mathcal{H}$ we use the short notation

$$\psi(x) := E_x \psi \tag{2}$$

and refer to $\psi(x)$ as the corresponding *wave function.* The wave functions at x span the vector space $E_x(H)$, referred to as the *spin space* at x and denoted by S_x. It is endowed with the inner product $<.|E_x.>$ of signature $(2,2)$, which we also denote by $\prec.|.\succ$ and refer to as the *spin scalar product.* Using the relations (1), we can then write

$$<\psi|\phi> = \sum_{x \in M} \prec \psi(x)|\phi(x) \succ . \tag{3}$$

The localization of the fermionic projector is denoted by $P(x,y) := E_x \, P \, E_y$. This operator maps $S_y \subset H$ to S_x, and we usually regard it as a mapping between these subspaces,

$$P(x,y) = E_x \, P \, E_y \; : \; S_y \to S_x \, .$$

Again using (1), we can write the wave function corresponding to $P\psi$ as follows,

$$(P\psi)(x) \;=\; E_x \, P\psi \;=\; \sum_{y \in M} E_x \, P \, E_y \, \psi \;=\; \sum_{y \in M} (E_x \, P \, E_y) \, (E_y \, \psi) \, ,$$

and thus

$$(P\psi)(x) \;=\; \sum_{y \in M} P(x,y) \, \psi(y) \, . \tag{4}$$

This relation resembles the representation of an operator with an integral kernel, and therefore we call $P(x,y)$ the *discrete kernel* of the fermionic projector.

In order to introduce our action principle, for any $x, y \in M$ we define the *closed chain* A_{xy} by

$$A_{xy} \;=\; P(x,y) \, P(y,x) \; : \; S_x \to S_x \, . \tag{5}$$

We denote the eigenvalues of A_{xy} counted with algebraic multiplicities by $\lambda_1^{xy}, \ldots, \lambda_4^{xy}$. We define the Lagrangian $\mathcal{L}$ by

$$\mathcal{L}[A_{xy}] = \sum_{i,j=1}^{4} \Big(|\lambda_i^{xy}| - |\lambda_j^{xy}| \Big)^2 \tag{6}$$

(for simplicity we here leave out the constraints and consider only the so-called critical case of the auxiliary Lagrangian). Our action principle is to

$$\text{minimize the action} \quad \mathcal{S}[P] = \frac{1}{8} \sum_{x,y \in M} \mathcal{L}[A_{xy}] \, , \tag{7}$$

where we consider variations of the fermionic projector, keeping the number of particles as well as discrete space-time fixed. For a more general mathematical exposition of this variational principle and the existence theory we refer to [5]. For a discussion of the underlying physical principles see [6, Section 2]. We finally remark that generalizing the above setting leads to the framework of *causal fermion systems* (see [7] and the references therein).

1.3. Effects of spontaneous structure formation

We now discuss two effects of spontaneous structure formation: the spontaneous breaking of the permutation symmetry and the generation of a discrete causal structure.

We first point out that in the above definitions, we did not distinguish an ordering of the space-time points; all our notions are symmetric under permutations of the points of M. This suggests that a minimizing fermionic projector should also be permutation symmetric. Surprisingly, this is not the case. Indeed, in [8] it is shown under general assumptions that that the fermionic projector cannot be permutation symmetric. This shows rigorously that the effect of spontaneous structure formation really occurs, giving rise to a non-trivial symmetry group acting on the discrete space-time points. Before stating the main result, we need to specify what we mean by a symmetry of our fermion system. The group of all permutations of the space-time points is the symmetric group, denoted by S_m.

Definition 1.1 *A subgroup* $\mathcal{O} \subset S_m$ *is called* **outer symmetry group** *of the fermion system in discrete space-time if for every* $\sigma \in \mathcal{O}$ *there is a unitary transformation* U *such that*

$$UPU^{-1} = P \qquad \text{and} \qquad UE_xU^{-1} = E_{\sigma(x)} \quad \text{for all } x \in M\,. \tag{8}$$

Theorem 1.2 (spontaneous breaking of the permutation symmetry) *Suppose that* $(H, <.|.>, (E_x)_{x \in M}, P)$ *is a fermion system in discrete space-time. Assume that the number of space-time points and the number of particles* f *are in the range*

$$m > 10 \qquad \text{and} \qquad 2 < f < m-1\,.$$

Then the fermion system cannot have the outer symmetry group $\mathcal{O} = S_m$.

In particular, this theorem applies in the physically interesting case of many particles and even more space-time points.

More detailed information on the space-time relations generated by the spontaneous structure formation is obtained by analyzing the eigenvalues of the closed chain. Note that in an indefinite inner product space, the eigenvalues of a self-adjoint operator A need not be real, but alternatively they can form complex conjugate pairs (see [9] or [5, Section 3]). This fact can be used to introduce a notion of causality.

Definition 1.3 (causal structure) *Two space-time points* $x, y \in M$ *are called* **timelike** *separated if the spectrum of* A_{xy} *is real. The points are* **spacelike** *separated if the spectrum of* A_{xy} *forms two complex conjugate pairs having the same absolute value. In all other cases, the two points are* **lightlike** *separated.*

This notion is compatible with our action principle in the following sense. Suppose that two space-time points x and y are spacelike separated. Then the eigenvalues λ_i^{xy} all have the same absolute value, and thus the Lagrangian (6) vanishes. A short calculation shows that the first variations of the action also vanish, so that A_{xy} does not enter the corresponding Euler-Lagrange equations. This can be seen in analogy to the usual notion of causality that points with spacelike separation cannot influence each other.

We point out that by analyzing the kernel of the fermionic projector in more detail, one can even distinguish a direction of time, define a parallel transport of spinors and introduce curvature, thus giving a proposal for a "Lorentzian quantum geometry." The interested reader is referred to [10] and the survey article [7].

1.4. The correspondence to Minkowski space

We now describe how to get a connection between discrete space-time and Minkowski space. The simplest method for getting a correspondence to relativistic quantum mechanics in Minkowski space is to replace the discrete space-time points M by the space-time continuum $\mathbb{R}^4$ and the sums over M by space-time integrals. For a vector $\psi \in H$, the corresponding localization $E_x\psi$ should be a 4-component Dirac wave function, and the spin scalar product $\prec\psi|\phi\succ$ on $E_x(H)$ should correspond to the usual Lorentz invariant scalar product on Dirac spinors $\overline{\psi}\phi$ with $\overline{\psi} = \psi^\dagger\gamma^0$ the adjoint spinor. In view of (4), the discrete kernel should go over to the integral kernel of an operator P on the Dirac wave functions,

$$(P\psi)(x) \;=\; \int_M P(x,y)\,\psi(y)\,d^4y\,.$$

The image of P should be spanned by the occupied fermionic states. We take Dirac's concept literally that in the vacuum all negative-energy states are occupied by fermions forming the so-called *Dirac sea*. Thus we are led to describe the vacuum by the integral over the lower mass shell

$$P^{\text{sea}}(x,y) \;=\; \int \frac{d^4k}{(2\pi)^4}\,(k_j\gamma^j + m)\,\delta(k^2-m^2)\,\Theta(-k^0)\,e^{-ik(x-y)}\,. \tag{9}$$

(Here Θ is the Heaviside function, and in the terms k^2 and $k(x-y)$ we use the Minkowski inner product, using the signature convention $(+ - - -)$). Computing the Fourier integral, one sees that $P(x,y)$ is a smooth function, except on the light cone $\{(y-x)^2 = 0\}$, where it has poles and singularities.

Let us compare Definition 1.3 with the usual notion of causality in Minkowski space. Even without computing the Fourier integral (9), it is clear from Lorentz symmetry that for every x and y for which the Fourier integral exists, $P(x,y)$ can be written as

$$P(x,y) \;=\; \alpha\,(y-x)_j\gamma^j + \beta\,\mathbb{1} \tag{10}$$

with two complex coefficients α and β. Taking the conjugate, we see that

$$P(y,x) \;=\; \overline{\alpha}\,(y-x)_j\gamma^j + \overline{\beta}\,\mathbb{1}\,.$$

As a consequence,

$$A_{xy} \;=\; P(x,y)\,P(y,x) \;=\; a\,(y-x)_j\gamma^j + b\,\mathbb{1} \tag{11}$$

with real parameters a and b given by

$$a \;=\; \alpha\overline{\beta} + \beta\overline{\alpha}\,, \qquad b \;=\; |\alpha|^2\,(y-x)^2 + |\beta|^2\,. \tag{12}$$

Applying the formula $(A_{xy} - b\mathbb{1})^2 = a^2\,(y-x)^2$, one can easily compute the roots of the characteristic polynomial of A_{xy},

$$\lambda_1 \;=\; \lambda_2 \;=\; b + \sqrt{a^2\,(y-x)^2}\,, \qquad \lambda_3 \;=\; \lambda_4 \;=\; b - \sqrt{a^2\,(y-x)^2}\,. \tag{13}$$

If the vector $(y-x)$ is timelike, we see from the inequality $(y-x)^2 > 0$ that the λ_j are all real. Conversely, if the vector $(y-x)$ is spacelike, the term $(y-x)^2 < 0$ is negative. As a consequence, the λ_j form a complex conjugate pair. This shows that for Dirac spinors in Minkowski space, Definition 1.3 is indeed consistent with the usual notion of causality. Moreover, from the fermionic projector (9) one can deduce the Minkowski metric as well as all all the familiar objects of Dirac theory. For example, the non-negative quantity $\prec\psi|\gamma^0\psi\succ$ has the interpretation as the probability density of the Dirac particle. Polarizing and integrating over space yields the scalar product

$$(\psi|\phi) = \int_{t=\text{const}} \prec\!\psi(t,\vec{x})\,|\,\gamma^0\phi(t,\vec{x})\!\succ\, d\vec{x}\,. \tag{14}$$

For solutions of the Dirac equation, current conservation implies that this scalar product is time independent. In this way, we can describe Dirac theory in Minkowski space in the setting of fermionic projector.

A more difficult question is if and in which sense the distribution (9) is a minimizer of our action principle (7). We refer the interested reader to the survey article [3] and the references therein.

1.5. A formulation of quantum field theory

The above correspondence to vacuum Dirac sea structures can also be used to analyze our action principle for interacting systems in the so-called *continuum limit.* We now outline a few ideas and constructions (for details see [1, Chapters 4], [2] and the survey article [4]). First, it is helpful to observe that the vacuum fermionic projector (9) is a solution of the Dirac equation $(i\gamma^j\partial_j - m)P^{\text{sea}}(x,y) = 0$. To introduce the interaction, we replace the free Dirac operator by a more general Dirac operator, for example involving gauge potentials or a gravitational field. Thus, restricting attention to the simplest case of an electromagnetic potential A, we demand that

$$\big(i\gamma^j(\partial_j - ieA_j) - m\big)\,P(x,y) = 0\,. \tag{15}$$

Moreover, we introduce particles and anti-particles by occupying (suitably normalized) positive-energy states and removing states of the sea,

$$P(x,y) = P^{\text{sea}}(x,y) - \frac{1}{2\pi}\sum_{k=1}^{n_f} |\psi_k(x)\!\succ\prec\!\psi_k(y)| + \frac{1}{2\pi}\sum_{l=1}^{n_a} |\phi_l(x)\!\succ\prec\!\phi_l(y)|\,. \tag{16}$$

Using the so-called causal perturbation expansion and light-cone expansion, the fermionic projector can be introduced uniquely from (15) and (16).

It is important that our setting so far does not involve the field equations; in particular, the electromagnetic potential in the Dirac equation (15) does not need to satisfy the Maxwell equations. Instead, the field equations should be derived from our action principle (7). Indeed, analyzing the corresponding Euler-Lagrange equations, one finds that they are satisfied only if the potentials in the Dirac equation satisfy certain constraints. Some of these constraints are partial differential equations involving the potentials as well as the wave functions of the particles and anti-particles in (16). In [2], such field equations are analyzed in detail for a system involving an axial field. In order to keep the setting as simple as possible, we here consider the analogous field equation for the electromagnetic field

$$\partial_{jk}A^k - \Box A_j = e\sum_{k=1}^{n_f} \prec\!\psi_k|\gamma_j\psi_k\!\succ - e\sum_{l=1}^{n_a} \prec\!\phi_l|\gamma_j\phi_l\!\succ\,. \tag{17}$$

With (15) and (17), the interaction as described by the action principle (7) reduces in the continuum limit to the *coupled Dirac-Maxwell equations.* The many-fermion state is again described by the fermionic projector, which is built up of *one-particle wave functions.* The electromagnetic field merely is a *classical bosonic field.* Nevertheless, regarding (15) and (17) as a nonlinear hyperbolic system of partial differential equations and treating it perturbatively, one obtains all the Feynman diagrams which do not involve fermion loops. Taking into account that by exciting sea states we can describe pair creation and annihilation processes, we also get all diagrams involving fermion loops. In this way, we obtain agreement with perturbative quantum field theory (for details see [2, §8.4]).

2. Microscopic mixing of decoherent subsystems

In the recent paper [11], a method was developed for describing entanglement and second quantized fermionic and bosonic fields in the framework of the fermionic projector. We now outline a few constructions and results from this paper.

2.1. Description of entangled second quantized fermionic fields

Recall that in the setting of Section 1.2 we described the fermion system by a projector P on an f-dimensional, negative definite subspace of H. It is now useful to choose an orthonormal basis $\psi_1, \ldots, \psi_f$ of $P(H)$ (meaning that $<\psi_i|\psi_j> = -\delta_{ij}$). Then the discrete kernel of the fermionic projector can be written in bra/ket-notation as

$$P(x,y) = -\sum_{j=1}^{f} |\psi_j(x) \succ \prec \psi_j(y)| \,. \tag{18}$$

In order to get a connection to the Fock space formalism, we take the wedge product of the one-particle wave functions,

$$\Psi := \psi_1 \wedge \cdots \wedge \psi_f \,, \tag{19}$$

to obtain a vector in the fermionic Fock space. This shows that the fermions can indeed be described by a second-quantized field. However, as Ψ is a Hartree-Fock state, it is impossible to describe entanglement (see for example [11, Example 3.1]).

This shortcoming is removed by introducing the concept of *microscopic mixing*. It is based on the assumption is that P has a non-trivial microstructure. This assumption seems natural in view of setting in discrete space-time (see Section 1.2), where space-time is not smooth on the microscopic scale. "Averaging" this microstructure over macroscopic regions of space-time gives rise to an effective kernel $P(x,y)$ of a more general form, making it possible to describe entanglement.

To explain the construction, suppose that we want to describe the fermionic state $\Psi = \Psi^{(1)} + \Psi^{(2)}$, being a superposition of two states, each of which is an f-particle Hartree-Fock state of the form

$$\Psi^{(a)} := \phi_1^{(a)} \wedge \cdots \wedge \phi_f^{(a)} \qquad (a = 1, 2) \tag{20}$$

(the generalization to a finite number of subsystems is straightforward). We now decompose Minkowski space M into two disjoint subsets

$$M = M_1 \cup M_2 \qquad \text{with} \qquad M_1 \cap M_2 = \varnothing \,, \tag{21}$$

which should be fine-grained in the sense that every macroscopic region of space-time intersects both subsets (as indicated in Figure 1). We introduce one-particle wave functions ψ_j which on M_a coincide with $\phi_j^{(a)}$,

$$\psi_j = \psi_j^{(1)} + \psi_j^{(2)} \qquad \text{with} \qquad \psi_j^{(a)} = \chi_{M_a} \, \phi_j^{(a)} \tag{22}$$

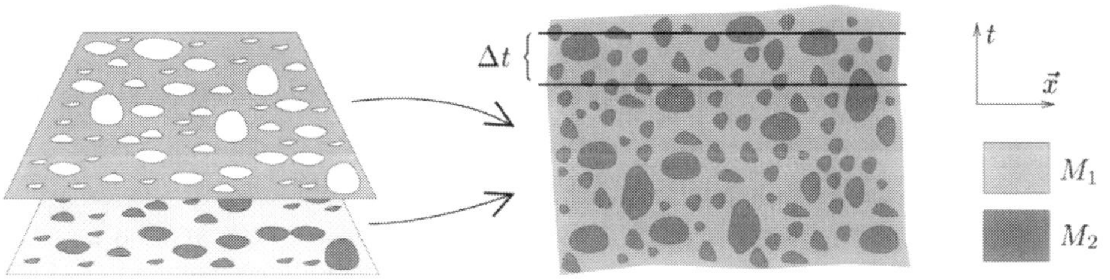

Figure 1. Example of a microscopic mixing of two space-time regions.

(here χ_{M_a} denotes the characteristic function defined by $\chi_{M_a}(x) = 1$ if $x \in M_a$ and $\chi_{M_a}(x) = 0$ otherwise). The form of the resulting fermionic projector (18) depends on whether its arguments x and y are in M_1 or M_2,

$$P(x,y) = -\sum_{j=1}^{f} |\psi_j^{(a)}(x)\succ\prec\psi_j^{(b)}(y)| \quad \text{if } x \in M_a \text{ and } y \in M_b\,.$$

If both arguments are in the same subsystem, then the fermionic projector is composed of the wave functions of this subsystem. But if the arguments are in different subsystems, then the fermionic projector involves both wave functions $\phi_j^{(1)}$ and $\phi_j^{(2)}$, thus giving correlations between the two subsystems. The idea is to remove all these correlation terms by bringing the wave functions of the subsystems "out of phase." In technical terms, we transform the wave functions of the second subsystem by a unitary matrix U,

$$\psi_j^{(1)} \to \psi_j^{(1)}\,, \qquad \psi_j^{(2)} \to \tilde{\psi}_j^{(2)} := \sum_{k=1}^{f} U_{jk}\,\psi_k^{(2)} \quad \text{with} \quad U \in \mathrm{SU}(f)\,. \tag{23}$$

This transformation does not change the fermionic projector, if its two arguments are in the same subsystem, because in the case $x, y \in M_2$, the unitarity of U yields that

$$\begin{aligned} P(x,y) \to -\sum_{j=1}^{f} |\tilde{\psi}_j^{(2)}(x)\succ\prec\tilde{\psi}_j^{(2)}(y)| &= -\sum_{j,k=1}^{f} (UU^\dagger)_{jk}\, |\psi_j^{(2)}(x)\succ\prec\psi_k^{(2)}(y)| \\ &= -\sum_{j}^{f} |\psi_j^{(2)}(x)\succ\prec\psi_j^{(2)}(y)| = P(x,y)\,. \end{aligned}$$

However, if the two arguments of the fermionic projector are in different space-time regions, the operator U does not drop out. For example, if $x \in M_2$ and $y \in M_1$,

$$P(x,y) \to -\sum_{j=1}^{f} |\tilde{\psi}_j^{(2)}(x)\succ\prec\psi_j^{(1)}(y)| = -\sum_{j,k=1}^{f} U_{jk}\, |\psi_j^{(2)}(x)\succ\prec\psi_k^{(1)}(y)|\,. \tag{24}$$

In the special case when U is a diagonal matrix whose entries are phase factors,

$$U = \mathrm{diag}(e^{i\varphi_1}, \ldots, e^{i\varphi_f}) \qquad \text{with} \qquad \sum_{j=1}^{f} \varphi_j = 0 \mod 2\pi\,,$$

the summations in (24) reduce to one sum involving the phase factors,

$$P(x,y) \to -\sum_{j=1}^{f} e^{i\varphi_j} \, |\psi_j^{(2)}(x)\succ\prec\psi_j^{(1)}(y)| \, .$$

If the angles φ_j are chosen stochastically, the phases of the summands are random. As a consequence, there will be cancellations in the sum. At this point, it is important to remember that the number f of fermions of our system is very large, because we also count the states of the Dirac sea (see Section 1.5). As a consequence, $P(x,y)$ will be very small. More generally, we find that if U is a random matrix, $P(x,y)$ is multiplied by a factor $f^{-\frac{1}{2}}$ if x and y lie in different subsystems.

Next we need to homogenize the wave functions by "taking averages" over macroscopic regions of space-time. Our concept is that this homogenization should take place in the quantum mechanical measurement process, because the measurement device is also composed of wave functions which are "smeared out" on the microscopic scale. In order to implement this idea, we introduce the so-called *measurement scalar product*, where we replace the spatial integral in (14) by integrals over a strip of width Δt in space-time (see Figure 1 and the more detailed treatment in [11, Section 4.2 and 4.3]). As a consequence of the homogenization process, the system is described effectively by the sum of the fermionic projectors corresponding to the two subsystems.

In order to clarify the above construction in the Fock space formalism, we next consider the many-particle wave function (19). Using (22), we obtain

$$\Psi = \psi_1 \wedge \cdots \wedge \psi_f = (\psi_1^{(1)} + \psi_1^{(2)}) \wedge \cdots \wedge (\psi_f + \psi_f) \, . \tag{25}$$

Multiplying out, we obtain many summands. Two of these summand correspond to the Hartree-Fock states of the two subsystems (20), but we obtain many other contributions. These additional contributions are again removed by the transformation (23). Namely, this transformation leaves the many-particle wave functions of the subsystems remain unchanged, because

$$\Psi^{(2)} \to \tilde{\Psi}^{(2)} = \tilde{\psi}_1^{(2)} \wedge \cdots \wedge \tilde{\psi}_f^{(2)} = \det U \; \psi_1^{(2)} \wedge \cdots \wedge \psi_f^{(2)} = \Psi^{(2)} \, . \tag{26}$$

But all the summands obtained by multiplying out (25) which involve factors $\psi_j^{(1)}$ as well as $\psi_k^{(2)}$ contain matrix elements of U. These matrix elements again become small if f gets large. As a consequence, the system is described effectively by the sum of the Hartree-Fock states of the subsystems. The homogenization process can again be described by the measurement scalar product. It is most convenient to introduce on the many-particle wave functions the scalar product induced by the measurement process, giving rise to the effective fermionic Fock space $\mathfrak{F}^{\text{eff}}$.

The above consideration can be understood using the notion of decoherence. If the one-particle wave functions $\psi_j^{(1)}$ and $\psi_j^{(2)}$ are coherent or "in phase", then the fermionic projector $P(x,y)$ has the usual form, no matter whether x and y are in the same subsystem or not. If however the wave functions in the subregions are decoherent or "out of phase", then the fermionic projector $P(x,y)$ becomes very small if x and y are in different subregion. We refer to this effect as the *decoherence between space-time regions.* It should be carefully distinguished from the decoherence of the many-particle wave function (see for example [12]). Namely, as we saw in (26), in our case the many-particle wave functions remain unchanged. Thus they remain coherent, no decoherence in the sense of [12] appears. But the one-particle wave functions are decoherent (23), implying that correlations between the two subsystems become small.

We finally remark that the mixing of subsystems can be generalized to the so-called *holographic mixing*, where the subsystems need no longer be localized in disjoint regions of space-time. We refer the interested reader to [11, Section 5.3].

2.2. Description of second quantized bosonic fields

As outlined in Section 1.5, in the continuum limit the interaction is described by the Dirac equation coupled to classical field equations. If our system involves a microscopic mixing of several subsystems, we can take the continuum limit of each subsystem separately. Labeling the subsystems by an index $a = 1, \ldots, L$ and again considering for simplicity the electromagnetic interaction, the interaction of the ath subsystem is described in analogy to (15) and (17) by the coupled Dirac-Maxwell equations

$$\begin{aligned} \Big(i\gamma^j(\partial_j - ieA_j^{(a)}) - m\Big) P^{(a)}(x,y) &= 0 \\ \partial_j{}^k A_k^{(a)} - \Box A_j^{(a)} &= e \sum_{k=1}^{n_f} \prec \psi_k^{(a)} | \gamma_j \psi_k^{(a)} \succ - e \sum_{l=1}^{n_a} \prec \phi_l^{(a)} | \gamma_j \phi_l^{(a)} \succ . \end{aligned} \tag{27}$$

In particular, the subsystems have an independent dynamics as described by these equations. To clarify the physical picture, we again point out that the underlying space-time is not smooth on the microscopic scale, but it is decomposed into subsystems M_a which are microscopically mixed. If x and y are in different subsystems, the fermionic projector $P(x,y)$ is very small due to decoherence, implying that the causal relations of Definition 1.3 no longer agree with the causal structure in Minkowski space. In other words, the usual causal relations are not valid between points in different subsystems. This means that the subsystems should be regarded as having a simultaneous and independent existence in our space-time (as indicated in Figure 1 by the two "space-time sheets").

The bosonic field in (27) is purely classical. The key observation for passing from a classical to a second-quantized bosonic field is that the dynamics of a free second-quantized bosonic field can be described equivalently by a superposition of solutions of the classical field equations (for details see [11, Section 5.1]). Thus to describe a free second-quantized bosonic field, it suffices to consider the microscopic mixing of subsystems (27) (each with a classical bosonic field $A^{(a)}$ satisfying the classical field equations), and to associate to every subsystem a complex number $\phi(a)$. The resulting function $\phi(a)$ can be interpreted as a wave function on the space of classical field configurations. For clarity, we first construct $\phi(a)$ in the case when no fermions are present. In this case, we take the wedge product of all the states $\psi_j^{(a)}$ which form the fermionic projector of the subsystem a to obtain a many-particle wave function $\Psi^{(a)}$. We then compare this wave function with the corresponding wave function $\check{\Psi}$ of a reference system (i.e. of a Dirac sea configuration in the presence of the external field $A^{(a)}$). These two wave functions coincide up to a proportionality factor, and thus we can define $\phi(a)$ by

$$\Psi^{(a)} = \phi(a)\, \check{\Psi} \qquad \text{with} \qquad \phi(a) \in \mathbb{C} . \tag{28}$$

The last construction immediately generalizes to the situation when fermionic particles and/or anti-particles are present. In this case, one decomposes the many-particle state into a component involving the particles and anti-particle states in (16) and the sea sates,

$$\Psi^{(a)} = \Big(\psi_1^{(a)} \wedge \cdots \wedge \psi_{n_f}^{(a)} \wedge \phi_1^{(a)} \wedge \cdots \wedge \phi_{n_a}^{(a)}\Big) \wedge \Big[\psi_{n+1} \wedge \cdots \wedge \psi_f\Big] . \tag{29}$$

Then after homogenizing, the particles and anti-particles give rise to a vector of the fermionic Fock space $\mathcal{F}^{\text{eff}}$, whereas the sea states can be used to define in analogy to (28) the second-quantized bosonic wave function ϕ.

We finally point that in the above constructions, we worked with *free* bosonic fields. To get the connection to interacting quantum fields, we recall that in the continuum limit (see Section 1.5 above), one gets all the Feynman diagrams. We thus obtain agreement also with the quantitative predictions of perturbative quantum field theory. Nevertheless, the fermionic projector approach is not equivalent to a local interacting quantum field theory (for example in the canonical formulation on a Fock space). Clarifying the connections and differences of the two frameworks is work in progress.

3. Physical interpretation

We now discuss the previous constructions and results from a conceptual point of view. For clarity, we try to explain the physical picture in simple examples, without aiming for mathematical rigor nor maximal generality.

3.1. The superposition principle

In the standard formulation of quantum physics, it is a general principle that superpositions of quantum states can be formed (for a good exposition see [12, Section 2.1]). The superposition principle holds for the one-particle wave functions in quantum mechanics just as well as for the quantum state describing the whole physical system. In the framework of the fermionic projector, however, where the whole system is described by a projector in an indefinite inner product space, the validity of the superposition principle is not obvious. We now explain this point in detail.

In the framework of the fermionic projector, the superposition principle arises on different levels. On the fundamental level of discrete space-time, the wave functions are vectors in the indefinite inner product space $(\mathcal{H}, <.|.>)$, so that superpositions of one-particle wave functions can be formed. In the continuum limit, where the interaction is described by the Dirac equation (15) coupled to a classical field (17), we thus obtain the superposition principle for the Dirac wave functions. Moreover, since the Maxwell equations are linear, the superposition principle also holds for classical electromagnetic waves.

For many-particle states, however, the superposition principle does not hold on the fundamental level. In particular, it is impossible to form the naive superposition of two physical systems, simply because the linear combination of two projectors in general is no longer a projector. But the superposition principle does hold for the effective fermionic many-particle wave function obtained by decomposing the system into decoherent subsystems and homogenizing on the microscopic scale (see Section 2.1). For the free second-quantized electromagnetic field as described in Section 2.2, the superposition principle corresponds to taking linear combinations of the complex-valued wave function ϕ defined on the classical field configurations. Since the function ϕ is constructed out of the fermionic many-particle wave function (see (28)), linear combinations are again justified exactly as for the fermionic many-particle.

We conclude that in the framework of the fermionic projector, the superposition principle again holds. It is possible to form linear combinations of macroscopic systems (like a dead and a living cat). Nevertheless, the framework of the fermionic projector differs from standard quantum theory in that for many-particle wave functions, the superposition principle does not hold on the fundamental level, but it is merely a consequence of the microscopic mixing of decoherent subsystems. This means that the superposition principle is overruled in situations when our action principle in discrete space-time (7) needs to analyzed beyond the continuum limit. We will come back to this point in Section 3.4 in the context of collapse phenomena.

3.2. The measurement problem and decoherence

One of the most controversial and difficult points in the understanding of quantum physics is the so-called measurement problem. In simple terms, it can be understood from the dilemma that on one side, the dynamics of a quantum system is described by a linear evolution equation in a Hilbert space (for example the Schrödinger equation), and considering the measurement

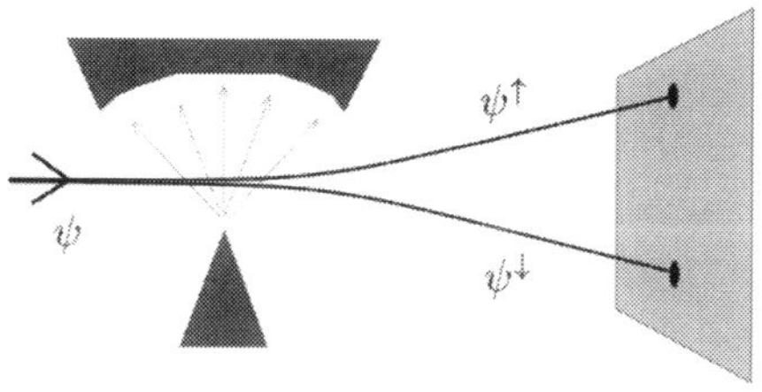

Figure 2. The Stern-Gerlach experiment.

apparatus as part of the system, one would expect that this linear and deterministic quantum evolution alone should give a complete description of physics. But on the other side, the Copenhagen interpretation requires an external observer, who by making a measurement triggers a "collapse" or "reduction" of the wave function to an eigenstate of the observable. It is not obvious how the external observer can be described within the linear quantum evolution. Also, the statistical interpretation of the expectation value in the measurement process does not seem to correspond to the deterministic nature of the quantum evolution.

The measurement problem has been studied extensively in the literature, and many different solutions have been proposed (see for example [12, 13, 14, 15, 16, 17]). Here we shall not try to enter an exhaustive discussion or comparison of the different interpretations of quantum mechanics. We only explain how our concepts fit into the picture and give one possible interpretation which corresponds to the personal preference of the author. But it is well possible that the framework of the fermionic projector can be adapted to other interpretations as well.

We begin by considering the *Stern-Gerlach experiment*. Thus a beam of atoms passes through an inhomogeneous magnetic field. Decomposing the wave function ψ into the components with spin up and down,

$$\psi = \psi^\uparrow + \psi^\downarrow \,, \tag{30}$$

these two components feel opposite magnetic forces. As a consequence, the beam splits up into two beams, leading to two exposed dots on the photographic material (see Figure 2). If the intensity of the beam is so low that only one atom passes through the magnetic field, then either the upper or the lower dot will be exposed, both with probability one half. It is impossible to predict whether the electron will fly up or down; only probabilistic statements can be made.

Let us try to describe the Stern-Gerlach experiment in the framework of the fermionic projector. For simplicity, we replace the atom by an electron (disregarding the Lorentz force due to the electron's electric charge). Then at the beginning, the system is described by a classical external magnetic field and a Dirac wave function $\psi(t, \vec{x})$ which at time $t = 0$ has the form of a wave packet moving towards the magnetic field. This situation is modeled by the fermionic projector in the continuum limit (16) for one particle; the dynamics is described by the Dirac equation (15) in the given external field. Solving the Dirac equation, the wave function splits into two components (30), which are deflected upwards and downwards, respectively. Writing the contribution of the wave function to the fermionic projector (16) as

$$-\frac{1}{2\pi}\, |\psi(x) \succ \prec \psi(y)| = -\frac{1}{2\pi}\Big(|\psi^\uparrow(x) \succ \prec \psi^\uparrow(y)| + |\psi^\downarrow(x) \succ \prec \psi^\downarrow(y)| \tag{31}$$

$$+ |\psi^\uparrow(x) \succ \prec \psi^\downarrow(y)| + |\psi^\downarrow(x) \succ \prec \psi^\uparrow(y)| \Big), \tag{32}$$

one gets contributions of different type. Namely, the two summands in (31) are localized at the upper and lower electron beam, respectively. The two summands in (32), however, are delocalized and give correlations between the two beams. As observed in [2, Chapter 10], the Euler-Lagrange equations in the continuum limit cannot be satisfied if general delocalized contributions to the fermionic projector are present. This means that there should be a mechanism which tries to avoid nonlocal correlations as in (32). A possible method for removing the nonlocal correlations is to divide the system into two decoherent subsystems (as shown in Figure 1, although at this stage they do not necessarily need to be microscopically mixed), in such a way that $\psi^\uparrow$ belongs to the first and $\psi^\downarrow$ to the the second subsystem. Then the continuum limit is to be taken separately in the two subsystems. The contribution of the wave function in the two subsystems simply is

$$-\frac{1}{2\pi}\,|\psi^\uparrow(x)\succ\prec\psi^\uparrow(y)| \qquad \text{and} \qquad -\frac{1}{2\pi}\,|\psi^\downarrow(x)\succ\prec\psi^\downarrow(y)|\,,$$

respectively. Thus the delocalized terms (32) no longer occur, so that the problem of solving the Euler-Lagrange equations observed in [2, Chapter 10] has disappeared. This consideration gives a possible mechanism for the *generation of subsystems*.

Let us carefully discuss different notions of *decoherence*. First of all, the space-time points of the two subsystems should be decoherent, in the sense that the states of the fermionic projector are unitarily transformed in the second subsystem (23). This decoherence has the effect that the fermionic projector $P(x, y)$ becomes very small if x and y are in different subsystems (see (24)), implying that the two subsystems have an independent dynamics in the continuum limit. However, the many-particle wave function of the system is not affected by the decoherence between the space-time points (see (26)). Rewriting it similar to (29) as the wedge product of the one-particle wave function ψ with the sea states, one sees that the quantum mechanical wave functions of the two subsystems are still coherent. In particular, if the two beams interfered with each other (for example after redirecting them with additional Stern-Gerlach magnets), they could be superposed quantum-mechanically, giving rise to the usual interference effects of the double slit experiment. We also point out that the dynamics of each subsystem is still described by the Dirac equation (15) in the external magnetic field. Since the Dirac equation is linear, solving it for ψ is the same as solving it separately for the two components $\psi^\uparrow$ and $\psi^\downarrow$. Thus at this point, the dynamics is not affected by the decomposition into subsystems; we still have the linear deterministic dynamics as described by the Dirac equation.

As just explained, at this stage the generation of subsystems has no effect on the dynamics of the system. This suggests that it should not be observable whether the subsystems have formed or not. This motivates us to demand that expectation values taken with respect to the measurement scalar product (as introduced in Section 2.1) should not be affected by the generation of subsystems. Keeping in mind that, after a suitable homogenization process, the measurement scalar product coincides with the integral (14), we can say alternatively that the process of *generation of subsystems should preserve the probability densities.* This condition also ensures that when we get the connection to the statistical description of the measurement process, the probabilities are indeed given by the spatial integrals of the absolute square of the wave functions. Since the probability density is the zero component of the probability current $\prec\psi|\gamma^j|\psi\succ$, we can say equivalently that the generation of subsystems should respect *current conservation.* This assumption seems reasonable, because current conservation holds in each subsystem as a consequence of the Dirac equation, and we merely extend this conservation law to the situation when the number of subsystems changes.

The dynamics becomes more complicated when the wave function approaches the screen, because the interaction with the electrons of the photographic material can no longer be described by an external field. Instead, one must consider the coupled Dirac-Maxwell equations (15) and (17) for a many-electron system. Since the interaction is no longer linear, it now makes a difference that the two subsystems have an independent dynamics. More specifically, in the first subsystem the wave function in the upper beam interacts with the electrons near the upper impact point, whereas the second subsystem describes the interaction of the lower beam. The whole system is a superposition of these two systems, described mathematically by a vector in the Fock space $\mathfrak{F}^{\text{eff}}$. Exactly as explained in [12, Chapter 3], the different interaction with the environment leads to a decoherence of the many-particle wave functions of the two subsystems. Thus now it is no longer possible to form quantum mechanical superpositions of the wave functions of the two subsystems. The whole system behaves like a statistical ensemble of the two subsystems. Following the resolution of the measurement problem as proposed in [12], one should regard the human observer as being part of the system. Thus the observer is also decomposed into two observers, one in each subsystem. The two observers measure different outcomes of the experiment. Due to the decoherence of their wave functions, the two observers cannot communicate with each other and do not even experience the existence of their counterparts. From the point of view of the observer, the outcome of the experiment can only be described statistically: the electron moves either up or down, both with probability one half.

Other experiments like the *spin correlation experiment* can be understood similarly. One only needs to keep in mind that if entanglement is present, then the subsystems must be microscopically mixed. The quantum state of each subsystem is not entangled. But homogenizing on the microscopic scale leads to an effective description of the system by an entangled state in the Fock space $\mathfrak{F}^{\text{eff}}$.

We finally remark that the mechanism for the generation of subsystems proposed above could be made mathematically precise by analyzing the action principle (7) in the discrete setting, going beyond the approximation of the continuum limit. One should keep in mind that on this level, our action principle *violates causality*. Thus it is conceivable that the formation of subsystems depends on later measurements or that subsystems tend to form eigenstates of the subsequent measurement device. However, such effects can hardly be verified or falsified in experiments, and therefore we will not consider them here (for a discussion of a measurable effect of causality violation see [2, Section 8.2]).

3.3. The wave-particle duality

Following the above arguments, one can also understand the wave-particle duality in a way where the wave function is the basic physical object, whereas the particle character is a consequence of the interaction as described by the action principle (7). To explain the idea, we return to the Stern-Gerlach experiment of Figure 2. In the previous section, we justified that if one electron flies through the magnetic field, the atom will expose either the upper or the lower dot on the screen, much in contrast to the behavior of a classical wave, which would be observable at both dots of the screen at the same time. Repeating the arguments of the previous section on the scale of the atoms of the photographic material, we conclude that more and more subsystems will form, which become decoherent as explained in [12, Chapter 3]. For an observer in one of the subsystems, the electron will not expose the whole dot on the screen uniformly, but it will only excite one atom of the photographic material. As a consequence, the electron appears like a point particle. Again, the outcome of the measurement can only be described statistically.

3.4. The collapse of the wave function

The resolution of the measurement problem in Section 3.2 is conceptually convincing and explains the experimental observations. It goes back to Everett's "relative state interpretation" and has found many different variations (see [13] and [15]). But no matter which interpretation one prefers, there always remains the counter-intuitive effect that all the possible outcomes of experiments are realized as components of the quantum state of the system. Thus when time evolves, the quantum state disintegrates into more and more decoherent components, which should all describe a physical reality. This phenomenon, which is often subsumed under the catchy but oversimplified title "many-worlds interpretation," is difficult to imagine and hard to accept. Another criticism is that decoherence leads to an effective description by a density operator, which however does not uniquely determine the Fock states of the decoherent components (for details see [18, Section 4] or [16, Section 6]). In order to avoid these problems, it has been proposed to introduce a mechanism which leads to a "collapse" or "reduction" of the wave function. Different mechanism for a collapse have been discussed, in particular models where the collapse occurs at discrete time steps [19] or is a consequence of a stochastic process [20]. These models have in common that the superposition principle is overruled by a nonlinear component in the dynamics. The nonlinearity is chosen to be so weak that it does not contradict the experimental evidence for a linear dynamics. In other words, the nonlinear term is so small that it cannot be detected experimentally. But nevertheless, this term can be arranged to prevent superpositions of macroscopically different wave functions.

In the framework of the fermionic projector, the dynamics as described by the action principle (7) is nonlinear. This indeed provides a *new collapse mechanism*, as we now explain. Suppose that our system is described by many decoherent subsystems. Since perfect decoherence seems impossible to arrange, the kernel of the fermionic projector $P(x, y)$ will in general not vanish identically if x and y are in two different subsystems. This gives rise to a contribution to the action (7) which, although being small due to decoherence, is strictly positive. These contributions, which "mix" different subsystems, grow quadratically with the number of subsystems, thus penalizing a very large number of subsystems. This shows that there is a mechanism which tries to reduce the number of subsystems, violating the superposition principle and the independent dynamics of the subsystems. It seems very plausible that this mechanism leads to a collapse of the wave function (although a derivation from the action principle (7) or a quantitative analysis has not yet been given). We also point out that, just as in other collapse theories, the collapse itself seems very difficult to observe. Namely, as systems whose quantum states are decoherent no longer interact with each other, it is impossible to decide whether there are other worlds beside the one observed by us, or whether the whole system has collapsed into the state which we observe. For this reason, despite its importance for the interpretation of quantum theory, the issue of the collapse of the wave function is often regarded as being speculative. But at least, the framework of fermion systems in discrete space-time outlined in Section 1.2 provides a well-defined mathematical setting for studying collapse phenomena. Analyzing this framework in more detail might help to overcome the open problems discussed in [17], thus leading to a fully satisfying quantum theory.

To summarize the physical interpretation, the framework of the fermionic projector seems in agreement with the superposition principle and the decoherence phenomena which explain the appearance of our classical world as well as the wave-particle duality. Our description is more concrete than the usual Fock space formulation because the decoherent components of the quantum state should all be realized in space-time by the states of the fermionic projector. Moreover, we saw qualitatively that our action principle (7) provides a mechanism for a collapse of the wave function and a reduction of the number of decoherent components. In view of the fact that this collapse seems very difficult to observe in experiments, the remaining question of how many "different worlds" are realized in our space-time seems of more philosophical nature. The personal view of the author is that the fermionic projector should only realize one macroscopic world, but at present this is mere speculation.

References

[1] Finster F 2006 *The Principle of the Fermionic Projector* AMS/IP Studies in Advanced Mathematics **35** (Providence, RI: American Mathematical Society) (*Preprints* arXiv:hep-th/0001048, arXiv:hep-th/0202059, arXiv:hep-th/0210121)
[2] Finster F 2009 *Preprint* arXiv:0908.1542

[3] Finster F 2009 in *Quantum Field Theory* ed B Fauser, J Tolksdorf and E Zeidler E (Birkhäuser Verlag) pp 235–259 (*Preprint* arXiv:0712.0685)
[4] Finster F 2009 *Preprint* arXiv:0911.2102
[5] Finster F 2007 *Calc. Var. Partial Differential Equations* **29** 431–453 (*Preprint* arXiv:math-ph/0503069)
[6] Finster F 2006 *Quantum gravity* ed Fauser B, J Tolksdorf and E Zeidler (Basel: Birkhäuser) pp 263–281 (*Preprint* arXiv:gr-qc/0601128)
[7] Finster F, Grotz A and Schiefeneder D 2010 *in preparation*
[8] Finster F 2007 *Adv. Theor. Math. Phys.* **11** 91–146 (*Preprint* arXiv:math-ph/0601039)
[9] Gohberg I, Lancaster P and Rodman L 2005 *Indefinite linear algebra and applications* (Basel: Birkhäuser Verlag)
[10] Finster F and Grotz A 2010 *in preparation*
[11] Finster F 2010 *J. Phys. A: Math. Theor.* **43** 395302 (*Preprint* arXiv:0911.0076)
[12] Joos E, Zeh H, Kiefer C, Giulini D, Kupsch J and Stamatescu I O 2003 *Decoherence and the appearance of a classical world in quantum theory* 2nd ed (Berlin: Springer-Verlag)
[13] Wheeler J and Zurek W (eds) 1983 *Quantum theory and measurement* Princeton Series in Physics (Princeton, NJ: Princeton University Press)
[14] Dürr D and Teufel S 2009 *Bohmian Mechanics* (Dordrecht, Heidelberg, London, New York: Springer-Verlag)
[15] Cini M and Lévy-Leblond J M (eds) 1990 *Quantum theory without reduction* (Bristol: Adam Hilger Ltd.)
[16] Albert D 1992 *Quantum mechanics and experience* (Cambridge, MA: Harvard University Press)
[17] Pearle P 2007 *J. Phys.* A **40** 3189–3204
[18] Ghirardi G 2009 *Preprint* arXiv:0904.0958v1
[19] Ghirardi G, Rimini A and Weber T 1986 *Phys. Rev.* D **34** (3) 470–491
[20] Pearle P 1989 *Phys. Rev.* A **39** 2277–2289

Journal of Physics: Conference Series **306** (2011) 012025 doi:10.1088/1742-6596/306/1/012025

Divergence on the horizon

Judy Kupferman

Ben Gurion University, Beer Sheva, Israel

E-mail: `Judithku@bgu.ac.il`

Abstract. Black hole entropy has been shown by 't Hooft to diverge at the horizon, whereas entanglement entropy in general does not. We show that because the region near the horizon is a thermal state, entropy is linear to energy, and energy at a barrier is inversely proportional to barrier slope, and diverges at an infinitely sharp barrier as a result of position/momentum uncertainty. We show that 't Hooft's divergence at the black hole is also an example of momentum/position uncertainty, as seen by the fact that the "brick wall" which corrects it in fact smooths the sharp boundary into a more gradual slope. This removes a major obstacle to identification of black hole entropy with entanglement entropy.

1. Introduction

Two of the possible characterizations of black hole entropy are both proportional to area but they apparently do not coincide. One is entanglement entropy, arising from correlations between that within the black hole and that without [1][2][3], and the other is the statistical mechanics definition related to the number of states of a system in the vicinity of a black hole [4][5]. Both types of entropy are expressed by calculating $-Tr(\rho ln\rho)$ of fields exterior to the black hole and the question naturally arises: do they both refer to the same thing? One problem in uniting these descriptions is the question of divergence at the black hole horizon. Entanglement entropy may have UV divergence which will need to be renormalized, but it does not diverge at a particular location. However, in the statistical mechanics calculation first carried out by 't Hooft, the integrand diverges at the black hole boundary. We will show that the problem of divergence at the boundary can be resolved. It is not a unique black hole characteristic but rather a result of quantum uncertainty, and the correct expression must involve smearing out the boundary. In this way the two expressions may coincide, and we may consider whether black hole entropy is simply entanglement entropy. This talk is based on [6].

For a massive black hole in the region just outside the horizon, it has been shown that entanglement entropy coincides with the thermodynamic definition of entropy [7][8]. Thermodynamic entropy is linear to energy. The expectation value for energy is also linear to momentum uncertainty, $\langle E\rangle \sim \langle p^2\rangle = \Delta p^2$. Therefore whenever energy is calculated at a precisely defined location we expect it to diverge due to the uncertainty relationship between x and p . We show that when any system is cut off by a barrier, the increase in energy is proportional to the narrowness of the barrier, and diverges as the barrier becomes infinitely sharp. Thus black hole entropy, which is proportional to energy, will diverge on the horizon because the horizon is a sharply defined barrier. We also show that this is consistent with 't Hooft's "brick wall" cutoff.

First we review these two characterizations of black hole entropy, entanglement and statistical, focusing on behavior at the horizon. Then we show in detail the relationship between energy and entropy in a massive black hole. We examine behavior of energy at a boundary between two subsystems, first for the non-relativistic and then for the relativistic case, and show that in both cases energy diverges as the boundary becomes sharp. Finally, we examine 't Hooft's calculation leading to divergence at the boundary, and find that his relocation of the boundary to avoid divergence is equivalent to smearing out the boundary. Therefore here too the divergence is related to sharpness of the boundary, and thus it reflects behavior of any thermal system at a boundary, and is not unique to a black hole. This conclusion lends weight to the possibility that black hole entropy is in fact entanglement entropy, because the peculiar divergence which seemed to single it out from other entangled systems is just a result of quantum x/p uncertainty.

2. Entropy at the horizon

The area law was first formulated by Bekenstein, beginning from the observation that black hole area must increase as does entropy [9], and strengthened by thermodynamic calculations [4]. Bekenstein did not call this entanglement entropy but he did describe it as a measure of the inaccessibility of information to an exterior observer as to which internal configuration of the black hole is realized in a given case. In fact black holes have entanglement entropy by definition. Entanglement entropy quantifies the extent to which a state is mixed. If the universe is described as a pure state, the black hole horizon divides it into that within and that without the hole, each of which is a mixed state. Therefore black holes have entanglement entropy, and the question is whether black hole entropy is anything more, or whether entanglement entropy saturates the definition.

't Hooft calculated thermodynamic characteristics of a black hole, among them entropy, and in doing so found a divergence of the integrand at the horizon [5]. He overcame the problem by adjusting the limits of integration to a "brick wall" a finite infinitesimal distance from the horizon, and various other solutions have been adopted since [10]. Entanglement entropy, on the other hand, has ultraviolet divergence, and a UV cutoff must be employed, but it does not diverge at any particular location. Srednicki in his landmark paper [1] calculated entropy by tracing over the degrees of freedom of part of the system, and found it proportional to area as well. He numerically obtained an expression for entropy $S = 0.30M^2R^2$,where $R = (n + \frac{1}{2})a$, a is lattice spacing and n the number of discrete oscillators. He also defines M as the inverse lattice spacing a^{-1} so that the actual expression is $S = 0.30(n+\frac{1}{2})^2$ which diverges for an infinite number of oscillators, but not at a particular location. Other treatments of entanglement entropy in general [11][12] also find UV divergence but not at any particular location. Therefore horizon divergence would seem be a special characteristic of black hole entropy, and would preclude its being a manifestation of entanglement entropy as such. We show below that this is not the case.

For a massive BH in equilibrium the space just outside the hole near the horizon can be treated as a thermal state in Rindler space [7][8]. In this case entanglement entropy coincides with thermal entropy, as follows. To find entanglement entropy we take the trace of part of the system. If that part of the system is a thermal state, the partial trace is a thermal density matrix.

$$
\begin{aligned}
S &= -Tr\left(\rho_{part} ln \rho_{part}\right) \\
\rho_{part} &= \frac{1}{Z}\sum_i e^{-\beta E_i}|E_i><E_i| \\
\langle E\rangle &= \frac{1}{Z}\sum_i E_i e^{-\beta E_i} \\
S &= (-\frac{1}{Z}\sum_i e^{-\beta E_i}) \times (-\beta \sum_i E_i - lnZ) \\
&= \beta < E > + lnZ.
\end{aligned} \tag{1}
$$

For a scalar field with finite temperature ln Z is a constant, so the entropy is linear to the expectation value of the energy. Therefore in the case of a massive black hole the entanglement entropy behaves as does the energy. So instead of calculating entropy, which is a complicated non local quantity, we can calculate the reduced density matrix of a subsystem and look at the behavior of its energy, which is a simpler local quantity.

3. Nonrelativistic treatment

We first write down the expectation value for energy in a nonrelativistic many body system. In order to look at behavior at a boundary, we insert a window function into the system. The window function mimics the horizon by "truncating" part of space for the particles. However it does not affect boundary values of the system. An example of such a window function would be the Heaviside step function. Here we provide our window function with a varying slope to see how sharp localization affects the energy divergence.

In a nonrelativistic many body system, taking spinless particles for simplicity, the field state and the kinetic energy operator for a free particle are as follows;

$$
\left|\Psi^{\dagger}(\vec{r})\right\rangle = \Psi^{\dagger}(\vec{r})\,|0\rangle = \sum_{\vec{p}} \frac{e^{-i\vec{p}\vec{r}}}{\sqrt{V}} a^{\dagger}_{\vec{p}}\,|0\rangle
$$

The energy operator is $T = \sum_{\vec{p}} \frac{p^2}{2m} a^{\dagger}_{\vec{p}} a_{\vec{p}}$, and its expectation value is $\langle E\rangle = \left\langle 0|\Psi T \Psi^{\dagger}|0\right\rangle$. In configuration space

$$
\Psi T \Psi^{\dagger} = \int_{-\infty}^{\infty} d^3r \frac{1}{2m} \nabla_r \Psi\left(\vec{r}\right) \nabla_r \Psi^{\dagger}\left(\vec{r}\right) \tag{2}
$$

With a window function $f(\vec{r})$, the field operator becomes

$$
\Psi_W = \int d^3r\, f\left(\vec{r}\right) \Psi^{\dagger}\left(\vec{r}\right) = \int d^3r\, f\left(\vec{r}\right) \sum_{\vec{p}} \frac{e^{-i\vec{p}\vec{r}}}{\sqrt{V}} a^{\dagger}_{\vec{p}}
$$

The ladder operators on the vacuum give delta functions, resulting in

$$
\langle E\rangle = \frac{1}{2m} \int_{-\infty}^{\infty} d^3r \nabla_r f^{*}\left(\vec{r}\right) \nabla_r f(\vec{r}) \tag{3}
$$

For details see [6]. This represents the contribution of the window to the expectation value of the energy.[1]

[1] In addition the state would have a wave function, e.g. $g(\vec{r})$, which would give us the energy of the particle regardless of window, but here we have taken $g(\vec{r}) = 1$ for simplicity.

We take the window function to represent a boundary with a varying slope. Technically we achieve this by combining a Gaussian with the Heaviside step function, and bound the step for purposes of normalization (see figure 1). In the figure, the amplitude to see the particle through the window falls off gradually with the slope, so that on the left side (from −1 to 0 in the graph) it is constant, and then it slopes down to zero with width L. The sharp drop on the left (at −1 in the graph) is for purposes of normalization, but does not affect the result; taking the cutoff at other values gives essentially the same result: energy which is inversely proportional to L . This is graphed in one dimension for clarity of graphing, but three dimensions give a symmetrical version of the same results.

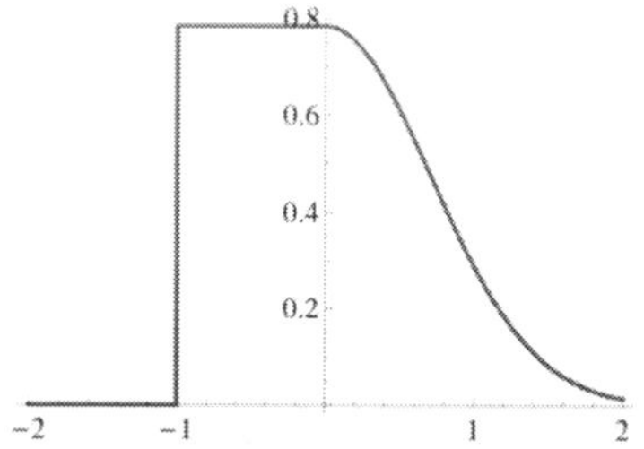

Figure 1. Half-Gaussian window, bounding the wave function, where we look at the part on the left of the slope. The horizon varies from a sharp step at x=0 to a gradual slope as in the drawing. The step at $x = -1$ is just for normalization; in fact space on the left continues to $-\infty$.

We calculate the contribution to the energy expectation value for a system with this window function and obtain

$$\langle E \rangle = \frac{1}{2m\left(L^2 + 2L\sqrt{\frac{2}{\pi}}\right)} \tag{4}$$

where L refers to width of the slope and m is particle mass. Obviously this diverges when the slope becomes perfectly sharp, L=0, as seen in figure 2 where we have taken $m = 1$.

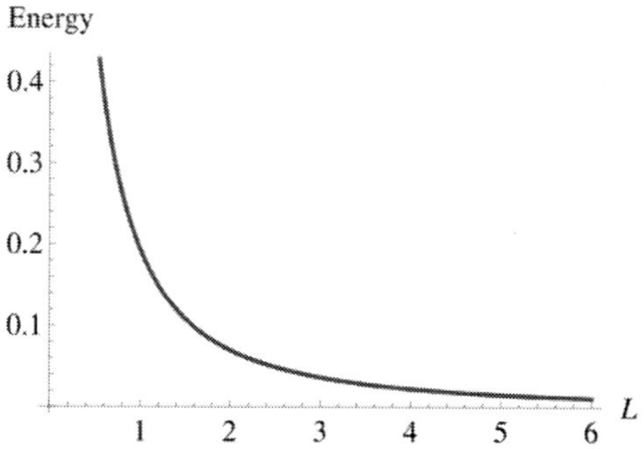

Figure 2. $\langle E \rangle$ for sloping window, as a function of slope width.

We have obtained the same result using completely different functions for the slope. For example, taking the window function as arctangent gives a slope which can be varied by using $Arctan[nx]$; larger n gives steeper slope. We took $\sqrt{Arctan[nx] + \frac{\pi}{2}}$ so that all values would be positive and to simplify the numerical calculations, and again the energy was seen to be inversely proportional to width of slope. Thus we have seen that the energy increases as the barrier grows sharper. That is, the more sharply the position of the dividing barrier is specified, the closer the energy gets to divergence. We now note that $E = \frac{1}{2m} \langle\psi| p^2 |\psi\rangle$. Since $\langle\psi| p |\psi\rangle = 0$, $\langle\psi| p^2 |\psi\rangle = (\Delta p)^2$, so that as energy diverges so do fluctuations in momentum. This divergence occurs as the width is taken to zero, and may be seen as an example of position/momentum uncertainty.

4. Relativistic treatment

In a relativistic system the energy operator is taken from the energy momentum tensor: $\hat{H} = T^{00} = \int \frac{d^3k}{(2\pi)^3} k_0 a_k^\dagger a_k$ taking $c = \hbar = 1$. A state with window function, as before, is $\int d^3r\, f(\vec{r})\, \Psi^\dagger(\vec{r})\, |0\rangle$ where the field operator here is the relativistic one.

In order to look for the various expectation values we need the relativistic scalar product:

$$\langle\varphi|\phi\rangle = \frac{1}{2} \int d^3x \left(\varphi i\partial_0 \phi - (i\partial_0 \varphi)\, \phi\right) \tag{5}$$

where the expression is sandwiched between the vacuum.[2] Thus for energy we write

$$\left\langle \varphi \left| \hat{H} \right| \varphi \right\rangle = \frac{1}{2} \int d^3x \left(\varphi i\partial_0 \left(\hat{H}\varphi \right) - (i\partial_0\, \varphi)\, \hat{H}\varphi \right). \tag{6}$$

Next we plug in the state with the window function. Since in the relativistic case $E^2 \sim p^2$ and we are interested in momentum fluctuations, we calculate $\left\langle \varphi \left| \hat{H}^2 \right| \varphi \right\rangle$ and obtain

$$\frac{1}{2} \int_{-\infty}^{\infty} d^3r \nabla_r f^*(\vec{r})\, \nabla_r f(\vec{r}). \tag{7}$$

For details see [6]. This clearly has the same behavior as the non relativistic case.

We have seen that in just as in the nonrelativistic treatment, energy is seen to diverge the more sharply position is specified. For the non relativistic case we had $E \sim p^2$ whereas in the relativistic case $E^2 \sim p^2$ but in both cases the divergence may be seen as an example of position/momentum uncertainty.

5. Statistical mechanics calculation

't Hooft treats the black hole as a quantum system, and solves the wave equation in the Schwartzschild metric:

$$\left(1 - \frac{2M}{r}\right)^{-1} E^2\varphi + \frac{1}{r^2}\partial_r \left(r\left(r - 2M\right)\partial_r\right)\varphi - \left(\frac{l\left(l+1\right)}{r^2} + m^2\right)\varphi = 0. \tag{8}$$

He identifies k, the wave number, by isolating the radial derivative on one side of the equation and the rest of the expression on the other. Thus he obtains an expression for $k^2(r)\varphi$, where $k^2(r)$ is the eigenvalue of the second radial derivative. Labelling the metric coordinates by indices for short, he obtains

$$k^2 = g_{rr}\left(g^{tt}E^2 - \left(\frac{l\left(l+1\right)}{r^2} + m^2\right)\right)$$

[2] If this use of scalar product is unfamiliar, note that $\left\langle \hat{H} \right\rangle = \langle\varphi\, |i\partial_0|\, \varphi\rangle$ with this formula gives the same expression as $\left\langle \int d^3r T^{00} \right\rangle$.

Journal of Physics: Conference Series **306** (2011) 012025 doi:10.1088/1742-6596/306/1/012025

Using a WKB approximation he obtains the number of modes by integrating the wave number over the region outside the black hole horizon:

$$\begin{aligned}\pi N &= \int_{2M}^{L} dr\, k\,(r,l,m) \qquad (9)\\ &= \int_{2M}^{L} dr\left(1-\frac{2M}{r}\right)^{-1}\int(2l+1)dl\sqrt{E^2-\left(1-\frac{2M}{r}\right)\left(\frac{l\,(l+1)}{r^2}+m^2\right)}\end{aligned}$$

where L is an infrared cutoff. However the redshift leads this to diverge at the lower limit, that is, the horizon itself. To avoid this 't Hooft takes the lower limit a slight distance away from the horizon, his well known "brick wall," so that the lower limit becomes $2M+h$. From this expression he obtains the energy and entropy, which diverge as $h \to 0$.

't Hooft's adjustment of the lower bound of the integral from $2M$ to $2M+h$ is equivalent to a change of variable which leaves the bound at $2M$ but adds a factor into the redshift:

$$\int_{2M+h}^{L} dr\left(1-\frac{2M}{r}\right)^{-1} = \int_{2M}^{L} d\tilde{r}\left(1-\frac{2M}{\tilde{r}+h}\right)^{-1}$$

We examine the change in the redshift. In the $\tilde{r}$ system the redshift is

$$\left(1-\frac{2M}{\tilde{r}}\right)^{-1}.$$

So the altered redshift is the same as taking the redshift and multiplying it by another function, thus:

$$\begin{aligned}\left(1-\frac{2M}{\tilde{r}+h}\right)^{-1} &= \left(1-\frac{2M}{\tilde{r}}\right)^{-1} f(r,h)\\ f\,(r,h) &= \frac{(\tilde{r}+h)\,(\tilde{r}-2M)}{\tilde{r}\,(\tilde{r}-2M+h)}\\ \int_{0}^{L} dr g_{rr}(r) f(r,h)\Theta(r-2M) &= \int_{0}^{L} dr g_{rr}(r)\tilde{\Theta}(r-2M) \qquad (10)\end{aligned}$$

Thus we see that 't Hooft's changed lower limit is exactly equivalent to modification of the step function. The new step function is no longer a sharp step, but rather has a width of h (for $r=2M+h$, $\tilde{\Theta}=1/2$). Figure 3 shows a graph of this softened step function $\tilde{\Theta}(r-2M)$ for various values of h, ranging from a sharp step at $h=0$ to a wider and wider slope. For $h=0$ 't Hooft had a divergence and when he changed the lower limit, equivalent to widening the horizon slope, the divergence vanished.

This is in fact position/momentum uncertainty. A closer look at eq.9 and at the effect of the modifying function $f(r,h)$ makes this clear. The integrand of eq.9 is $k(r)$. Just as in the previous sections $\langle k^2\rangle = \Delta k^2$, and its root is just the integrand of eq. 9. Multiplying this integrand by the modifying function $f(r,h)$ is equivalent to multiplying it by Δx since as we have shown, it widens the slope. Eq. 9 with the "brick wall" expressed as a softening function

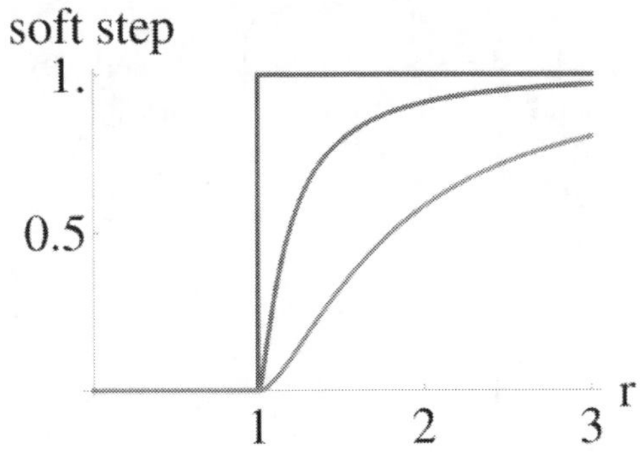

Figure 3. $\tilde{\Theta}(r-2M)$ at the horizon as function of $r/2M$. Curves have $h=0$ (sharp step), $h=0.1$ and $h=0.9$ (lowest)

becomes $\int_{2M}^{L} dr \Delta k \Delta x$. Without the brick wall, $\Delta x = 0$ and we had $\Delta k \to \infty$. Adding the brick wall is directly equivalent to widening Δx and as a result the expression becomes finite.

Another way of looking at it is that the "brick wall" is equivalent to a modification of the redshift: $g_{rr} \to \tilde{g}_{rr} = g_{rr} f(r,h)$. This may become more intuitive from the following illustration: I hover outside a black hole and send an astronaut in towards it radially, and require of the astronaut to send me back progress reports every kilometer; we all know the progress reports will come more and more slowly, so that it seems to me that he has slowed down, and come to a full stop at the horizon. I send another astronaut after him and the same thing happens. So I send a whole fleet of astronauts, tied together by a rope at regular intervals, as in dangerous mountain climbing. It will seem to me that as they near the horizon they all slow down, the rope slackens so they grow closer and closer, and at the horizon eventually they will all crowd together and I'll have an infinite density of astronauts there. This is the effect of the redshift. However, if I smooth out the horizon - stretch out the redshift, so to speak - they will not crowd up at one point and the density won't be infinite. This is, so to speak, quantum uncertainty delta(x) delta(astronaut).

't Hooft's displacement of the boundary prevents divergence at the black hole horizon by distorting the horizon so that it is no longer a sharply located step function but rather a more gradual transition. By doing so he actually shrinks the momentum uncertainty. Therefore we see that this too is an example of x/p quantum uncertainty, and not a unique characteristic of a black hole.

6. Discussion

Energy has been shown to diverge as the boundary between two subsystems becomes sharp. The divergence is due to the fact that the energy is a simple function of momentum. For the nonrelativistic case it is easy to see that $\langle E \rangle = \frac{1}{2m} \langle p^2 \rangle = \frac{1}{2m} \Delta p^2$and so divergence at a sharp boundary is just due to the quantum $\Delta p \Delta x$ uncertainty. The relativistic expression for $\langle E^2 \rangle$has exactly the same form as the nonrelativistic result and so in both cases energy divergence at an infinitely sharp boundary is a consequence of x/p uncertainty.

The region near the boundary of a black hole is a thermal state, where the entropy is linear to energy. Therefore black hole entropy will diverge at the boundary as well. We have not proven that there is no other cause of the divergence, unique to a black hole. But we have shown that regardless of any other cause, there would be divergence at the boundary as a result of the uncertainty principle. We have also shown that 't Hooft's divergence at the black hole is also an example of momentum/position uncertainty, as seen by the fact that the "brick wall" which corrects it in fact smooths the sharp boundary into a more gradual slope.

We may now consider whether the entanglement and statistical mechanics definitions of black hole entropy might refer to the same thing. Both are proportional to area. The UV divergence may be renormalized with a cutoff, and the boundary divergence by smearing out the boundary, so that these no longer preclude unification of the two expressions. The argument can then be made that black hole entropy is due to entanglement, that is, to quantum correlations between the two parts of the system inside and outside the black hole, and that counting the number of states is tantamount to counting the correlations.

However there is a third definition with which these two must now be reconciled. Black hole entropy has also been shown - from thermodynamic considerations [9] as well as explicit calculations in string theory [13] to equal one fourth of the horizon area. An open problem is to obtain the factor of 1/4 in these two cases as well.

This work has been supported by grant 239/10 of the Israel Science Foundation. We would like to thank Merav Hadad for many useful discussions.

References

[1] Srednicki M 1993 *Phys. Rev. Lett.* **71** 666.
[2] Callan C G, Wilczek F 1994 *Phys. Lett.*B **333** 55.
[3] Susskind L, Uglum J 1994 *Phys Rev* D **50**, 2700.
[4] Gibbons G W, Hawking S W 1977 *Phys Rev* D **15**, 2752.
[5] 't Hooft G 1985 *Nucl. Phys. B* **256** 727.
[6] Brustein R, Kupferman J 2010 arXiv:hep-th/1010.4157v1.
[7] Kabat D N, Strassler M J 1994 *Phys. Lett. B* **329**, 46.
[8] Brustein R, Einhorn M B, Yarom A 2006 *JHEP* **0601** 098.
[9] Bekenstein Jacob D 1994 arXiv:gr-qc/9409015v2.
[10] Demers J G, Lafrance R, Myers R C (1995) *Phys Rev* D **52**, 2245.
[11] Bombelli L, Koul R K, Lee J, Sorkin R D 1986 *Phys Rev* D **34**, 373.
[12] Plenio M B, Eisert J, Dreissig J, Cramer M 2005 *Phys. Rev. Lett.* **94** 060503.
[13] Strominger A, Vafa C 1996 *Phys. Lett. B* **379**, 99.

Quantum Mechanics on a Planck Lattice and Black Hole Physics

Petr Jizba[1], **Hagen Kleinert**[2], **Fabio Scardigli**[3]

[1] FNSPE, Czech Technical University in Prague, Břehová 7, 115 19 Praha 1, Czech Republic
[2] ITP, Freie Universität Berlin, Arnimallee 14 D-14195 Berlin, Germany
[3] Leung Center for Cosmology and Particle Astrophysics (LeCosPA), Department of Physics, National Taiwan University, Taipei 106, Taiwan

E-mail: p.jizba@fjfi.cvut.cz

E-mail: h.k@fu-berlin.de

E-mail: fabio@phys.ntu.edu.tw

Abstract. We study uncertainty relations as formulated in a crystal-like universe, whose lattice spacing is of order of Planck length. For Planck energies, the uncertainty relation for position and momenta has a lower bound equal to zero. Connections of this result with double special relativity, and with 't Hooft's deterministic quantization proposal, are briefly pointed out. We then apply our formulae to micro black holes, and we derive a new mass-temperature relation for Schwarzschild (micro) black holes. In contrast to standard results based on Heisenberg and stringy uncertainty relations, we obtain both a finite Hawking's temperature and a zero rest-mass remnant at the end of the micro black hole evaporation.

1. Introduction

The idea of a discrete structure of the space-time can be traced back to the mid 50's when John Wheeler presented his *space-time foam* concept [1]. Recent advances in physics indicate that the notion of space-time as a continuum is indeed likely to be superseded at the Planck scale, where gravitation and strong-electro-weak interactions become comparable in strength. In fact, discrete structures of space-time appear, as a rule, in many quantum-gravity models. Loop quantum gravity, non-commutative geometry or cosmic cellular automata may serve as examples. Perhaps the simplest toy-model for planck physics is a *discrete lattice*.

Recently one of us (HK) proposed a universe with a peculiar "foam" structure, that can be modeled by a simple discrete space-time Planck Lattice, whose lattice constant is of order Planck length. This discrete, crystal-like universe was termed "world crystal". There, the geometry of Einstein and Einstein-Cartan spaces can be considered as being a manifestation of the defect structure of the crystal [2]. Curvature is due to rotational defects, torsion due to translational defects. On the coarse-grained scale, the memory of the crystalline structure is smeared out, leading thus to ordinary Einstein's and Einstein-Cartan's theory.

In this paper we study the generalized uncertainty principle [3, 4, 5] (GUP) associated with quantum mechanics formulated on the world crystal, and derive physical consequences for the corresponding micro black holes. As micro black holes could be potentially created at LHC already in the TeV range (as certain theories with extra-dimensions indicate), our results could contribute into the ongoing discussion about observable signatures of micro black holes.

2. Differential calculus on a lattice

On a 1D lattice, the nodes are at $x_n = n\epsilon$, with $n \in \mathcal{Z}$. One can then define two fundamental derivatives, so called forward and backward derivatives, namely

$$(\nabla f)(x) = \frac{1}{\epsilon}\left[f(x+\epsilon) - f(x)\right] \quad \text{and} \quad (\bar{\nabla} f)(x) = \frac{1}{\epsilon}\left[f(x) - f(x-\epsilon)\right]. \tag{1}$$

The corresponding integration is performed as a summation:

$$\int dx\, f(x) \;\equiv\; \epsilon \sum_x f(x), \tag{2}$$

where x runs over all x_n. For periodic functions on the lattice or for functions vanishing at the boundary of the world crystal, the lattice derivatives can be subjected to the lattice version of the integration by parts rule:

$$\sum_x f(x)\nabla g(x) \;=\; -\sum_x g(x)\bar{\nabla} f(x), \tag{3}$$

$$\sum_x f(x)\bar{\nabla} g(x) \;=\; -\sum_x g(x)\nabla f(x). \tag{4}$$

One can also define the lattice Laplacian as

$$\nabla\bar{\nabla} f(x) = \bar{\nabla}\nabla f(x) = \frac{1}{\epsilon^2}\left[f(x+\epsilon) - 2f(x) + f(x-\epsilon)\right], \tag{5}$$

which reduces in the continuum limit to an ordinary Laplace operator ∂_x^2. The above calculus can be naturally extended to any number D of dimensions.

3. Position and momentum operator on a 1D lattice

To build Quantum Mechanics on a 1D lattice we define a *scalar product* as

$$\langle \psi_1 | \psi_2 \rangle \;=\; \epsilon \sum_x \psi_1^*(x)\psi_2(x). \tag{6}$$

It follows from Eq. (3) that

$$\langle f | \nabla g \rangle \;=\; -\langle \bar{\nabla} f | g \rangle, \tag{7}$$

so that $(i\nabla)^\dagger = i\bar{\nabla}$, and neither $i\nabla$ nor $i\bar{\nabla}$ are hermitian operators. The lattice Laplacian (5), however, is hermitian under the scalar product (6).

The position operator $\hat{X}_\epsilon$ acting on wave functions of x is defined by a multiplication with x

$$(\hat{X}_\epsilon f)(x) \;=\; x f(x). \tag{8}$$

To ensure hermiticity of the lattice momentum operator $\hat{P}_\epsilon$, we relate it to the symmetric lattice derivative. Using (1) we have

$$(\hat{P}_\epsilon f)(x) \;=\; \frac{\hbar}{2i}\left[(\nabla f)(x) + (\bar{\nabla} f)(x)\right] \;=\; \frac{\hbar}{2i\epsilon}\left[f(x+\epsilon) - f(x-\epsilon)\right]. \tag{9}$$

For small ϵ, this reduces to the ordinary momentum operator $\hat{p} \equiv -i\hbar\partial_x$:

$$\hat{P}_\epsilon \;=\; \hat{p} + \mathcal{O}(\epsilon^2)\,. \tag{10}$$

Thus, the "canonical" commutator between $\hat{X}_\epsilon$ and $\hat{P}_\epsilon$ on the lattice reads

$$\left([\hat{X}_\epsilon, \hat{P}_\epsilon] f\right)(x) \;=\; \frac{i\hbar}{2}[f(x+\epsilon) + f(x-\epsilon)] \;\equiv\; i\hbar(\hat{I}_\epsilon f)(x)\,. \tag{11}$$

The operators $\hat{X}_\epsilon$, $\hat{P}_\epsilon$, and $\hat{I}_\epsilon$ are hermitian under the scalar product (6). They form a $E(2)$ Lie algebra, which reduces to the standard Weyl-Heisenberg algebra in the limit $\epsilon \to 0$: $\hat{X}_\epsilon \to \hat{x}$, $\hat{P}_\epsilon \to \hat{p}$, $\hat{I}_\epsilon \to \mathbf{1}$. Ordinary QM is thus obtained from lattice QM by a contraction of the $E(2)$ algebra via the limit $\epsilon \to 0$.

Reminding that $\exp(\epsilon\partial_x) f(x) = f(x+\epsilon)$, we can rewrite the commutation relation (11) as

$$\left([\hat{X}_\epsilon, \hat{P}_\epsilon] f\right)(x) \;=\; i\hbar\cos\left(\epsilon\hat{p}/\hbar\right) f(x)\,. \tag{12}$$

The latter allows to identify the lattice unit operator $\hat{I}_\epsilon$ with $\cos\left(\epsilon\hat{p}/\hbar\right)$.

4. Uncertainty relations on lattice

Let us define the uncertainty of an observable A in a state ψ as the standard deviation, i.e.,

$$(\Delta A)_\psi \;\equiv\; [\langle\psi|(\hat{A} - \langle\psi|\hat{A}|\psi\rangle)^2|\psi\rangle]^{1/2}\,. \tag{13}$$

Via the Schwarz inequality we get the Uncertainty Relation on a lattice in the form

$$\Delta X_\epsilon \Delta P_\epsilon \;\geq\; \frac{1}{2}\left|\langle\psi|[\hat{X}_\epsilon, \hat{P}_\epsilon]|\psi\rangle\right| \;=\; \frac{\hbar}{2}\left|\langle\psi|\hat{I}_\epsilon|\psi\rangle\right| = \frac{\hbar}{2}\left|\langle\psi|\cos\left(\epsilon\hat{p}/\hbar\right)|\psi\rangle\right|\,. \tag{14}$$

We denote $\langle\psi|\cdots|\psi\rangle$ as $\langle\cdots\rangle_\psi$ and study two critical energy regimes of (14):

- I) The long-wavelengths regime, where $\langle\hat{p}\rangle_\psi \to 0$. In this regime the GUP (14) can be written for Planck lattices as (cf. (10))

$$\Delta X_\epsilon \Delta P_\epsilon \;\geq\; \frac{\hbar}{2}\left(1 - \frac{\epsilon^2}{2\hbar^2}\,(\Delta P_\epsilon)^2\right)\,. \tag{15}$$

- II) The regime near the boundary of the first Brillouin zone, where the averaged momentum takes its maximum value $\langle\hat{p}\rangle_\psi \to \pi\hbar/2\epsilon$. Here, using again (10), we can write

$$\Delta X_\epsilon \Delta P_\epsilon \;\geq\; \frac{\hbar}{2}\left(\frac{\pi}{2} - \frac{\epsilon}{\hbar}\,\langle\hat{P}_\epsilon\rangle_\psi\right)\,. \tag{16}$$

We see that, as the momentum reaches the boundary of the Brillouin zone, the right-hand side of (16) vanishes, so that lattice quantum mechanics at short wavelengths *can* exhibit classical behavior — there is no irreducible lower bound for uncertainties of two complementary observables! This result is analogous to the one found, on a different ground, by Magueijo and Smolin in *deformed special relativity* [6]. So we see that the world crystal physics can become "classical" for energies close to the Brillouin zone, i.e., for Planck energies. This is also a recurrent theme in 't Hooft's "deterministic" quantum mechanics[7], where a deterministic theory at the Planck scale, supplemented with a dissipation mechanism, can give rise to a genuine quantum mechanical behavior at larger scales.

5. World-crystal micro black hole physics

Following a strategy outlined in Ref. [8] we can now use (15) and (16) to obtain the mass-temperature relation for the world crystal (micro) black holes. Heisenberg's microscope argument implies that the smallest resolvable spatial detail δx goes roughly as the wavelength of the probing photon. If E is the (average) energy of photons used in the microscope, then standard Heisenberg principle dictates

$$\delta x = \frac{\hbar c}{2E} \,. \tag{17}$$

From (15), the lattice version of the Heisenberg formula (17) reads

$$\delta X_\epsilon \simeq \frac{\hbar c}{2E_\epsilon}\left[1 - \frac{\epsilon^2}{2\hbar^2 c^2}(E_\epsilon)^2\right]. \tag{18}$$

This links the (average) wavelength of a photon to E_ϵ. The position uncertainty δX_ϵ of unpolarized photons of Hawking radiation just outside the event horizon is of the order of the Schwarzschild radius R_S. This follows directly from the geometry, reminding that the average wavelength of the Hawking radiation is of the order of the geometrical size of the hole. So, we can write

$$\delta X_\epsilon \simeq 2\mu R_S = 2\mu \ell_p m \,, \tag{19}$$

with $R_S = \ell_p m$, where $m = M/M_p$ is the black hole mass in Planck units ($M_p = \mathcal{E}_p/c^2$, $\mathcal{E}_p = \hbar c/2\ell_p$), and μ is a free parameter to be specified. According to the equipartition law, the average energy E_ϵ of unpolarized photons is $E_\epsilon \simeq k_B T$. For a lattice spacing $\epsilon = a\ell_p$ the constant μ can be fixed by the condition that in the continuum limit $\epsilon \to 0$ the formula (18) should predict the standard semiclassical Hawking temperature $T = \hbar c/(4\pi k_B R_S)$. This gives $\mu = \pi$. Again using Planck units for the temperature $\mathcal{E}_p = k_B T_p/2$, $\Theta = T/T_p$, we may finally rewrite (18) as

$$2m = \frac{1}{2\pi\Theta} - \zeta^2\, 2\pi\Theta \,, \tag{20}$$

where we define the deformation parameter $\zeta = a/2\sqrt{2}\pi$.

The result coming from ordinary Heisenberg uncertainty principle, to which, in the continuum limit $\epsilon \to 0$, Eq.(18) reduces, is

$$m \;=\; \frac{1}{4\pi\Theta} \,, \tag{21}$$

which is the dimensionless version of Hawking's formula for macroscopic black holes. The result implied by the stringy uncertainty relation [4] is

$$2m \;=\; \frac{1}{2\pi\Theta} + \zeta^2\, 2\pi\Theta \,. \tag{22}$$

The phenomenological consequences of these three relations are quite different. In Fig. 1 we compare the three results, including also the curve for the ordinary Hawking relation (21). Considering m and Θ as functions of time, we can follow the evolution of a micro black hole from the curves plotted there.

For the stringy GUP, we have a maximum temperature $\Theta_{\max} = 1/(2\pi\zeta)$, and a minimum rest mass $m_{\min} = \zeta$. The end of the evaporation process is reached after a finite time, the

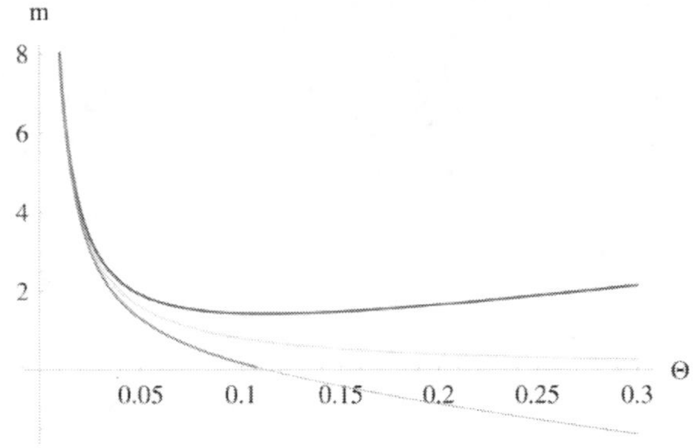

Figure 1. Diagrams for the three *mass-temperature* relations: ours (red), Hawking's (green), and stringy GUP result (blue), with $\zeta = \sqrt{2}$, as an example. As a consequence of the lattice uncertainty principle the evaporation ends at a *finite* temperature with a *zero* rest-mass remnant.

final temperature is finite, and there is a remnant of a finite rest mass. Should we have used the standard Heisenberg uncertainty principle we would have find the usual Hawking formula. This predicts that the evaporation process ends, after a finite time, with a zero mass and a worrisome infinite temperature. In contrast to these results, our lattice GUP predicts a finite final temperature $\Theta_{\max} = 1/(2\pi\zeta)$, with a *zero-mass remnant.* Our result (20) coming from the lattice GUP formula (15) avoids at once the difficulties of an infinite final temperature and the existence of finite-mass black hole remnants in the universe. The analysis with the GUP short wave limit (16) fully confirms the results previously obtained.

6. Entropy

From the first law of black hole thermodynamics [9] we know that the differential of the thermodynamical entropy of a Schwarzschild black hole reads

$$dS = \frac{dE}{T_H}, \tag{23}$$

where dE is the amount of energy swallowed by a black hole with Hawking temperature T_H. In Eq. (23) the increase in the internal energy is equal to the added heat because a black hole makes no mechanical work when its entropy/surface changes (its expanding surface does not exert any pressure). Rewriting Eq. (23) with the dimensionless variables m and Θ we get $dS = k_B\, dm/(2\Theta)$. Inserting here formula (20) we find

$$dS = \frac{k_B}{2}\frac{dm}{\Theta} = -\frac{k_B}{4}\left(\frac{1}{2\pi\Theta^3} + \frac{2\pi\zeta^2}{\Theta}\right) d\Theta\,. \tag{24}$$

By integrating dS we obtain $S = S(\Theta)$. Just as formula (20), the relation (24) can be trusted only for $\Theta \ll \Theta_{\max} = 1/2\pi\zeta$. Thus, when integrating (24), we should do this only up to a cutoff $\tilde{\Theta}_{\max} \ll \Theta_{\max}$. The additive constant in S can then be fixed by requiring that $S = 0$ when $\Theta \to \tilde{\Theta}_{\max}$. This is equivalent to what is usually done when calculating the Hawking temperature for a Schwarzschild black hole. There, one fixes the additive constant in the entropy integral to be zero for $m = 0$, so that $S(m = 0) = S(\Theta \to \infty) = 0$ (the minimum mass attainable in the standard Hawking effect is $m = 0$). The integral yields

$$S(\Theta) = \frac{k_B}{16\pi}\left(\frac{1}{\Theta^2} - \frac{1}{\tilde{\Theta}^2_{\max}} + 8\pi^2\zeta^2 \log\frac{\tilde{\Theta}_{\max}}{\Theta}\right). \tag{25}$$

The entropy is always positive, and $S \to 0$ for $\Theta \to \tilde{\Theta}_{\max}$.

7. Heat Capacity

With the entropy formulae (24) and (25) at hand we can now compute the heat capacity of a (micro) black hole in the world-crystal. This will give us important insights on the final stage of the evaporation process. Again, we shall obtain formulae valid only for $\Theta \ll \Theta_{\max} = 1/(2\pi\zeta)$. The heat capacity C of a black hole is defined via the relation $dQ = dE = CdT$. The pressure exerted on the environment by the expanding black hole surface is zero. Hence we do not need to specify which C is meant. Therefore we have $C = T\left(\frac{dS}{dT}\right) = \Theta\left(\frac{dS}{d\Theta}\right)$, which yields

$$C = -\frac{\pi k_B}{2}\left[\zeta^2 + \frac{1}{(2\pi\Theta)^2}\right]. \tag{26}$$

From this clearly follows that C is always negative. Most condensed-matter systems have $C > 0$. However, because of instabilities induced by gravity this is generally not the case in astrophysics [10], especially in black hole physics. A Schwarzschild black hole has $C < 0$ which indicates that the black hole becomes hotter by radiating. The result (26) implies that this scenario holds also for micro black holes in the world-crystal. In case of stringy GUP, we have to use Eq. (22) as the mass-temperature formula. The expression for the heat capacity then reads

$$C = \frac{\pi k_B}{2}\left[\zeta^2 - \frac{1}{(2\pi\Theta)^2}\right]. \tag{27}$$

Since also here $0 < \Theta < \Theta_{\max} = 1/2\pi\zeta$, black holes have negative specific heat also according to the stringy GUP. However, stringy GUP displays a striking difference with respect to lattice GUP. In fact, since in principle we can trust Eq. (22) also when $\Theta \simeq \Theta_{\max} = 1/2\pi\zeta$, then from (27) we have, in such limit, $C = 0$. This means that for the stringy GUP the specific heat vanishes at the end point of the evaporation process in a finite time, so that the black hole at the end of its evolution cannot exchange energy with the surrounding space. In other words, the black hole stops to interact thermodynamically with the environment. The final stage of the Hawking evaporation, according to the stringy GUP scenario, contains a Planck-size remnant with a maximal temperature $\Theta = \Theta_{\max}$, but thermodynamically inert. The remnant behaves like an elementary particle — there are no internal degrees of freedom to excite in order to produce a heat absorption or emission.

8. Connection with double special relativity

A different application relates to the idea of double (or doubly, or deformed) special relativity (DSR) (see, e.g. Ref. [6]). The general idea is that if the Planck length is a truly universal quantity, then it should look the same to any inertial observer. This demands a modification (deformation) of the Lorenz transformations, to accommodate an invariant length scale. In Ref. [6] the nonlinearity of the deformed Lorenz transformations lead the authors to novel commutators between space-time coordinates and momenta, depending on the energy

$$\left[\hat{x}^i, \hat{p}_j\right] = i\hbar\left(1 - \frac{E}{\mathcal{E}_p}\right)\delta^i_j\,, \tag{28}$$

where E is the energy scale of the particle to which the deformed Lorenz boost is to be applied, while $\mathcal{E}_p$ is the Planck energy. This suggests that they have an energy-dependent Planck "constant" $\hbar(E) = \hbar(1 - E/\mathcal{E}_p)$. Their model also implies that $\hbar(E) \to 0$ for $E \to \mathcal{E}_p$. For energies much below that Planck regime, the usual Heisenberg commutators are recovered, but when $E \simeq \mathcal{E}_p$ one has $\hbar(\mathcal{E}_p) \simeq 0$. So the Planck energy is not only an invariant in this model, but the world looks also apparently classical at the Planck scale, similarly as in 't Hooft's proposal.

The connections of the DSR model with our proposal are at this point self evident. Our GUP (14) implies that, at the boundary of the Brillouin zone, when $\langle \hat{p} \rangle_\psi \to \hbar\pi/2\epsilon$, i.e. for Planck energies $E_\epsilon \simeq (2\sqrt{2}/a)\mathcal{E}_p$, the fundamental commutator vanishes $[\hat{X}_\epsilon, \hat{P}_\epsilon] \simeq 0$, and since $\Delta X_\epsilon \Delta P_\epsilon \geq 0$, lattice quantum mechanics at short wavelengths allows for classical behavior, that is uncertainties of two complementary observables *can* be *simultaneously* zero. However, if we express the fundamental commutator (12) of our model in terms of energy, using the exact dispersion relation for photons inferred from the lattice Laplacian (5), we find (for $\epsilon = a\ell_p$)

$$\left[\hat{X}_\epsilon, \hat{P}_\epsilon\right] f(x) \quad = \quad i\hbar \left(1 - \frac{a^2}{8} \frac{\hat{E}^2}{\mathcal{E}_p^2}\right) f(x) \,. \tag{29}$$

This means that the deforming term in our model is quadratic in the energy, instead of being linear in the energy as the DSR model (28).

9. Discussion and Summary

It should be noted that the present lattice generalization of the uncertainty principle is not an approximate description, but it is an exact formula necessarily implied by our model of lattice space time. The great majority of the GUP research has always borrowed the deformed commutator $[\hat{x}, \hat{p}] = i\hbar(1 + \kappa\hat{p}^2)$ either from string theory [4], or from heuristic arguments about black holes [5]. To be precise, even in string theory the formula expressing the GUP is not derived from the basic features of the model, but instead it appears as an economic description of the high-energy gedanken experiments of string scatterings. In contrast to this, we have *derived* all the results from a simple lattice model of space-time, and from the analytic structure of the basic commutator (12).

We have calculated the uncertainties on a crystal-like universe whose lattice spacing is of the order of Planck length — the so-called world crystal. When the energies lie near the border of the Brillouin zone, i.e., for Planck energies, the uncertainty relations for position and momenta pose a zero lower bound. Hence the world crystal universe can become "deterministic" at Planck energies. In this high-energy regime, our lattice commutator resembles the double special relativity result of Magueijo and Smolin. The scenario in which the universe at Planck energies is deterministic and classical, rather than being dominated by quantum fluctuations, is also a starting point in 't Hooft's "deterministic" quantum mechanics.

Using this new generalized uncertainty relation we derived a new mass-temperature relation for Schwarzschild (micro) black holes. In contrast to standard results based on Heisenberg or stringy uncertainty relations, our mass-temperature formula predicts both finite Hawking's temperature and a zero rest-mass remnant at the end of the evaporation process. Especially the absence of remnants is a welcome bonus which allows to avoid such conceptual difficulties as entropy/information problem or why we do not experimentally observe the remnants that must have been prodigiously produced in the early stage of the universe.

Finally, the connections between DSR and our generalized uncertainty relation have been illustrated. The fundamental commutator in DSR as well as in our lattice GUP goes to zero at Planck energy. The world should therefore be manifestly "classical" in the Planck regime — a feature very different from the common believe.

Acknowledgments

We are grateful to Prof. Thomas Elze and the organizers of the Fifth International Workshop DICE2010 (Castiglioncello, Sept. 13-17, 2010) for cordial hospitality. P.J. was supported by the Ministry of Education of the Czech Republic through Grant No. MSM 6840770039.

References

[1] see e.g. Wheeler J A and Ford K W 2000 *Geons, Black Holes, and Quantum Foam* (New York: W W Norton & Co Inc)
[2] Kleinert H 2008 *Multivalued Fields* (Singapore: World Scientific)
[3] Yang C N 1947 *Phys. Rev.* **72** 874
[4] Veneziano G 1986 *Europhys. Lett.* **2** 199; Gross D J, Mende P F 1987 *Phys. Lett.* B **197** 129; Konishi K, Paffuti G, Provero P 1990 *Phys. Lett.* B **234** 276
[5] Scardigli F 1999 *Phys. Lett.* B **452** 39
[6] Magueijo J, Smolin L 2003 *Phys. Rev.* D **67** 044017
[7] 't Hooft G 1999 *Class. Quant. Grav.* **16** 3263; Blasone M, Jizba P, Vitiello G 2001 *Phys. Lett.* A **287** 205
[8] Adler R J, Chen P, Santiago D I 2001 *Gen. Rel. Grav.* **33** 2101
[9] Bardeen J M, Carter B, Hawking S W 1973 *Commun. Math. Phys.* **31** 161
[10] Thirring W 1970 *Z. Physik* **239** 339

Journal of Physics: Conference Series **306** (2011) 012027 doi:10.1088/1742-6596/306/1/012027

Noisy cooperative intermittent processes: From blinking quantum dots to human consciousness

Paolo Allegrini,[1,2] Paolo Paradisi,[1,3] Danilo Menicucci,[1,3] Remo Bedini,[1,2] Angelo Gemignani,[1,4] Leone Fronzoni[5]

[1]Istituto di Fisiologia Clinica (IFC-CNR), Via Moruzzi 1, 56124 Pisa, Italy
[2]Centro EXTREME, Scuola Superiore S. Anna, P.zza Martiri della Libertà 7, 56127 Pisa, Italy
[3]Istituto di Scienza e Tecnologie dell'Informazione "A. Faedo" (ISTI-CNR), Via Moruzzi 1, 56124 Pisa, Italy
[4]Dipartmento di Scienze Fisiologiche, Università di Pisa, Via San Zeno 31, 56127 Pisa, Italy
[5]Dipartimento di Fisica "E. Fermi"-Università di Pisa and INFM CRS-SOFT, Largo Pontecorvo 3, 56127 Pisa, Italy

E-mail: allegrini@ifc.cnr.it

Abstract. We study the superposition of a non-Poisson renewal process with the presence of a superimposed Poisson noise. The non-Poisson renewals mark the passage between meta-stable states in system with self-organization. We propose methods to measure the amount of information due to the two independent processes independently, and we see that a superficial study based on the survival probabilities yield stretched-exponential relaxations. Our method is in fact able to unravel the inverse-power law relaxation of the isolated non-Poisson processes, even when noise is present. We provide examples of this behavior in system of diverse nature, from blinking nano-crystals to weak turbulence. Finally we focus our discussion on events extracted from human electroencephalograms, and we discuss their connection with emerging properties of integrated neural dynamics, i.e. consciousness.

1. Introduction

Intermittence is an ubiquitous phenomenon in physics, and is an indication of the sporadic decrease in entropy production, due to formation of collective behaviors. In the special case when the classical Kolmogorv-Sinai Entropy (entropy-production rate) is null, a non-trivially evolving system is commonly referred to as as complex, and inverse-power-law (or fractal) statistics is found for both spatial and temporal correlations. The paradigms for these behaviors are found in turbulence. The classic theory of Kolmogorov [1] provides an analytical treatment, based on Gaussian statistics and self-similarity, for fractal behavior of the statistics of velocity fluctuations in fully developed turbulence. These fluctuations, through a branching process of energy transport, are due to bursts of activity at different scales. Deviations to the theory, due to extreme velocity excursions that happen more often than expected on the basis of Gaussian statistics are referred to as "intermittency" [2, 3]. Time is also intermittent, since bursts of activity are separated by temporal epochs of quiescence, termed "laminar regions" whose durations are in turn distributed as inverse-power laws [4, 5].

Large interest has been recently devoted to the fluorescent intermittence of quantum dots [6]. Quantum jumps have been first revealed by Dehmelt [7]. However, the Dehmelt's experiment was on a single atom in a trap, and a Poisson intermittence [8] was observed. The blinking quantum dot phenomenon is in fact produced by a collection of atoms in a nanostructure, and the intermittent effect is non-Poissonian. Although a definitive theoretical explanation of the phenomenon is still missing, it seems obvious that collective fluctuations in the nano-crystal structure is ultimately responsible for the non-Poisson blinking [6]. The time distance between two consecutive transitions from the light-on (off) to the light-off (on) condition is given by an inverse power law distribution density with a power index $\mu < 2$. Recently, a random-walk approach (in the nanostructure configuration space) [9] theoretically yields $\mu = 1.5$, a typical index for return times, i.e. the Sparre-Andersen theorem [10]. A careful analysis of the experimental data reveals that μ can be different form 1.5 and that the process is renewal [11].

Moreover, the authors of [12] developed a theoretical model of synchronizing units generating intermittence similar to that produced by blinking quantum dots. Remarkably, the intermittent phenomenon correspond to the global properties produced by the cooperation among many units, each of which, in the absence of cooperation, may be described very well by using the Poisson statistics of quantum jumps. It is only when the cooperation parameter reaches the critical value, corresponding to a phase transition from the uncorrelated to the correlated condition, that the signal generated by the sum of all individual contributions exhibits the typical intermittent behavior of blinking quantum dots. A theoretical explanation, again based on returns of a random walker in a shallow potential, again yields $\mu = 1.5$ [13]. The proposed model can be looked at as a paradigm of decision making, namely the emergence of intelligent behavior out of non-intelligent but cooperating agents, and has been proposed as a model for a critical brain.

The hypothesis of a critical brain, on the other hand, is becoming increasingly popular. Since the seminal ideas of Turing [14] more experimental data have been reported supporting this view, a notable example being the discovery of intermittent avalanches by Beggs and Plenz [15], who invoke self-organized criticality (SOC) [16] to explain their findings [17]. In this case the model consists in a branching process of information transport through layers of neurons, which, in terms of a parameter describing the probability of successfully transmitting a signal at each step, can be subcritical (information quenching, yielding null information transfer), critical (optimal information exchange) and supercritical (information explosion yielding saturation). This particular model predicts, at the critical point, neural avalanches whose size and durations are inverse-power-law distributed, with indexes respectively 1.5 and 2.0, in agreement with the experimental data of Beggs and Plenz.

Criticality in human brain has been recently reviewed by Ref. [18], focusing again on inverse-power laws and the presence of avalanches (as in SOC), and also by the scale-free-network representation of cross correlations between pairs of units, superimposable to what can be found in a dynamical Monte-Carlo of a critical Ising model. Another illuminating review is the work of Werner [19], showing that inverse-power laws and intermittency are ubiquitous in the brain, from single cells to complex behavior. At one of the highest levels, i.e. at the level of consciousness, it has been recently hypothesized that the dynamical pattern of "thoughts" is a serial conscious output which integrates non-conscious units which are in fact working in a highly parallel way, and that the serial, operational time is in fact characterized by the system visiting meta-stable states, with information only increasing at the transitions between a metastable state and the next [20]. This hypothesis may be linked to the heteroclinic orbits of Ref. [21], where it was proven that reproducible metastable states correspond to neural response to stimuli (e.g. odors) and can be associated to segmentation in electroencephalogram (EEG), namely the fact that EEG signals frequently looks like juxtapositions of epochs with fixed frequency and amplitude, with abrupt changes, called rapid transition processes (RTP), from epoch to epoch [22].

We point out that an aspect which is frequently overlooked is the "renewal" character of intermittency, namely that the time durations of the laminar regions, or of the operational epochs, are mutually independent. This means that long-range autocorrelations in the signals are dominated by the waiting time distribution $\psi(\tau)$, i.e. by the distribution density of the temporal durations of the metastable states. From an entropic point of view, this means that entropy increases during transitions, while is kept frozen when cooperative states are activated. It has been recently shown that the fluctuations of the order parameter in a critical system undergoes similar dynamics [23], so that there is a strong connection between critical systems and temporal intermittency, while space intermittency was seen at criticality a few decades ago [24], to model massless-to-mass transition (intermittency in hadron formation [25]).

Having discussed the ubiquity of intermittency in cooperative structures in this Section, we focus, respectively in Section 2 and 3, on two particular aspects, namely on formally deriving a new form of linear response theory for intermittent systems, as it allows a new kind of control, and, second, at the level of signal analysis, on how to assess intermittency in the presence of superimposed noise.

2. Linear Response for intermittent processes

The fluctuation-dissipation of first kind [26] plays a quite important role in physics, as the response to external perturbation, which can be experimentally detected, brings information on the systems dynamics, thereby helping the foundation of proper models for the processes under study. The most general form of linear response theory is given by

$$\Pi(t) = \epsilon \int_0^t \chi(t,t')\xi_P(t')dt', \tag{1}$$

where $\Pi(t)$ is the susceptibility, $\xi_P(t)$ is a function describing the perturbation time dependence and ϵ is the perturbation strength. The Fluctuation-Dissipation Theorem (FDT) is a theoretical proposal for the two-times function $\chi(t,t')$, the response function. In the literature of non-Poisson systems (e.g. glasses) [27], the form ordinarily adopted for the response $\chi(t,t')$ is

$$\chi(t,t') = \frac{d}{dt'}\langle A(t)B(t')\rangle, \tag{2}$$

where A is the variable of which we study the response to perturbation and B is the variable through which the system is coupled to the external perturbation. This simple formula is very attractive. First of all, it does not imply that the correlation function depends on $t-t'$, thereby going beyond the stationary restriction of the conventional linear response theory [26]. Then, it fits the physical intuition insofar as the derivative with respect to $t' < t$ is claimed [27] to be equivalent to enforcing the causality principle. The ultimate theoretical foundation of this theory is given [27] by the Onsager perspective [28], based on variable probabilities.

In this Section we propose a new approach focusing on the events responsible for the variables time evolution, rather than on the variables probabilities. For simplicity's sake we make the assumption $A = B = \xi_S$ and we assume the variable ξ_S dichotomous. This restriction does not prevent us from applying our approach to real physical systems. In fact, our approach is inspired by the phenomenon of molecular intermittency, and the dichotomous blinking of quantum dots [6] has been proved to be non-ergodic and renewal [11]. We prove that, at variance with (2),

$$\chi(t,t') = -\frac{d}{dt}\Psi(t,t'), \tag{3}$$

where $\Psi(t,t') \equiv \langle\xi_S(t)\xi_S(t')\rangle$, and the system is prepared at $t = 0$.

A renewal system is characterized by a waiting-time distribution $\psi(\tau)$, meaning a succession of waiting times $\{\tau_i\}$; next we define a dichotomous variable $\xi(t)$

$$\xi(t) = \begin{cases} \text{Rand}[+1;-1] & \text{if } t = t_n \\ \xi(t-dt) & \text{otherwise,} \end{cases} \tag{4}$$

with $t_n = \sum_{i=0}^{n} \tau_i$, meaning that during the quiescent regions of lengths $\{\tau_i\}$ no change of sign can happen. Only at the end of these regions the variable ξ may change or keep the sign, with probability 1/2, according to a fair coin-tossing procedure, indicated above with the function Rand[+1;−1]. We now dynamically generate $\{\tau_i\}$ with an inverse-power law distribution as follows. We consider the following equation of motion for a real positive variable $y \in (0,1)$

$$\dot{y} = ay^z, \tag{5}$$

and consider the time τ it takes for y to reach the border $y = +1$ starting form an initial condition y_0. We integrate (5) by quadrature and obtain

$$\int_{y_0}^{+1} \frac{dy}{y^z} = a\tau, \tag{6}$$

yielding a relation $y_0(\tau)$. If now we impose that every time the border is reached the new initial condition is taken randomly, namely with a constant density $p(y_0) = 1$, we find

$$\psi(\tau) = \left| \frac{dy_0}{d\tau} \right| = \frac{(\mu-1)T^{\mu-1}}{(T+\tau)^{\mu-1}}, \tag{7}$$

where $T = 1/[a(z-1)]$ and $\mu = z/(z-1)$. The renewal condition is ensured by the randomly selected re-injection in y_0. This model, a continuous-time version of the map [4], depends on two parameters, T and μ. T is the lapse of time necessary for the expression in the r.h.s of (7) to become identical to an inverse-power law. μ, on the other hand, is a fractional index that marks the presence and measures the degree of complexity.

Here let us focus our attention to linear response to small external perturbations. It is arguable that the parameter to be perturbed in (7) is T, rather than μ, since the time scale of a perturbation is normally finite. This is true, for instance, both in the case of looking at the relaxation after a kick, and in the case of a periodic perturbation. In both cases the perturbation should not be able to change the scaling μ of the system, since this is the result of infinitely long scales. We therefore imagine that a linear perturbation to a complex system can be simply modeled as a perturbation on T, or equivalently, on a. The dynamical equation (5) becomes

$$\dot{y} = a[1 \pm \epsilon\xi_P(t)]y^z, \tag{8}$$

where ϵ is a small parameter $(0 < \epsilon \ll 1)$, and $\xi_P(t)$ is the external forcing. Notice that this forcing does not change the weak repeller topology in the dynamical system. In other words, the trajectory is still laminar near the origin and its long-range properties are still dominated by a Pomeau-Manneville type of intermittency (this holds true also in multidimensional systems).

Integrating (8) yields the analogous of (6)

$$\int_{y_0^\pm}^{+1} \frac{dy}{y^z} = a\int_{t'}^{t'+\tau_\pm} [1 \pm \epsilon\xi_P(t'')]dt'', \tag{9}$$

where $y_0^\pm$ denotes the reinjection at t' corresponding to a choice of the state "plus" or "minus" for the variable ξ. If re-injection is kept flat, $y_0^\pm = y_0$ and the l.h.s's of (9) and of (6) coincide. Equating the r.h.s's results in an iterative relation for the perturbed waiting times $\tau_\pm$

$$\tau_\pm = \tau \mp \epsilon[F(t'+\tau_\pm) - F(t')], \tag{10}$$

where the function F is the integral function of ξ_P. $t = t' + \tau$ denotes the end of the quiescent region, or, according to the model (8), the instant when $y = +1$.

Let us rephrase the model and the coin-tossing procedure using the density approach. Let us define a vector $\mathbf{p} = (p_1, p_2)$. We impose equiprobability after the coin tossing. If we toss the coin at any discrete time we have

$$\mathbf{p}_{m+1} = [I - A]\mathbf{p}_m, \text{ where } A = \frac{1}{2}\begin{pmatrix} 1 & -1 \\ -1 & 1 \end{pmatrix} \tag{11}$$

is the transition matrix and I is the identity matrix. The sporadic tossing of the coin is easily modeled by imagining a scalar characteristic function $r(m)$ which assumes the value 1 when events occur (at t_n) and 0 elsewhere, multiplying A in (11). This means

$$\mathbf{p}_{m+1} = [I - r(m)A]\mathbf{p}_m, \text{ with } r(m) = \sum_n \delta_{m,t_n}, \tag{12}$$

where t_n are the transition times. In the continuous limit the Kroeneker delta's of (12) become Dirac delta's. Their intensity must so large that after the first coin tossing $p_1 = p_2$, thus

$$\mathbf{p}(t + dt) - \mathbf{p}(t) = \lim_{M \to 0} \delta(t - t_n)\frac{dt}{M}\mathbf{A}\mathbf{p}(t) \tag{13}$$

and hence

$$\dot{\mathbf{p}}(t) = -r(t)A\mathbf{p}(t), r(t) = \lim_{M \to 0}\left[\sum_n \frac{\delta(t - t_n)}{M}\right]. \tag{14}$$

This means that $r(t)$ is a sum of $1/M$-normalized density functions whose amplitudes tend to infinity in the continuous limit. This is correct since the formal solution of (13) tells us that $p_1(t)$ and $p_2(t)$ become 1/2 after the first coin tossing and never move afterwards. Let us define $\sigma \equiv p_1 - p_2$ the equation for σ, $\dot{\sigma} = -r(t)\sigma$, is readily solved

$$\sigma(t) = \exp\left[-\int_0^t r(t')dt'\right]\sigma(0) \Rightarrow \langle\sigma(t)\rangle = \left\langle \exp\left[-\int_0^t r(t'dt')\right]\right\rangle \langle\sigma(0)\rangle, \tag{15}$$

where we used the fact that the system prepared at $t = t_p = 0$, so we can factorize the average over the initial condition. The first term in the r.h.s of (15) is a product of step function since

$$\exp\left[-\int_0^t r(t')dt'\right] = \lim_{M \to 0}\prod_n \exp\left[-\int_0^t \frac{\delta(t' - t_n)}{M}dt'\right], \tag{16}$$

and each exponential is unity for $t \leq t_n$, dropping to the vanishing value $\exp(-1/M)$ for $t > t_n$. Only the first coin tossing, at time t_1 is important, so we can formally write (15) in the form

$$\left\langle \exp\left[-\int_0^t r(t'dt')\right]\right\rangle = \Pr\left[t \leq t_1\right]. \tag{17}$$

The r.h.s. in (17) is in fact a conditional probability, the condition being a coin tossing occurring at $t_p = 0$, and can be identified with the survival probability

$$\left\langle \exp\left[\int_0^t r(t'dt')\right]\right\rangle = \Psi(t) \equiv 1 - \int_0^t \psi(t')dt', \tag{18}$$

and therefore (15) is a zero-aged Onsager relation $\langle\sigma(t)\rangle = \Psi(t)\langle\sigma(0)\rangle$.

Imagine now that observation starts at $t' > 0$. In this case (17) must be modified to account for a further condition in the probability, namely that no observation was made in the interval $[0, t')$. This probability is the t'-aged survival probability $\psi(t, t')$, which obeys the relation

$$\Psi(t, t') \equiv \int_t^\infty \psi(t'', t')dt'' \Rightarrow \psi(t, t') = -\frac{d\Psi(t, t')}{dt}, \tag{19}$$

the density $\psi(t,t')$ being the t'-aged waiting time distribution, defined as the probability density of observing a coin tossing in the interval $(t, t+dt)$, with $t > t'$, given a coin tossing at $t = 0$. Eq. (18) is thus generalized as follows

$$\left\langle \exp\left[\int_{t'}^{t} r(t''dt'')\right]\right\rangle = \Psi(t,t'). \tag{20}$$

Applying a perturbation means changing the transition matrix. In general we can write

$$\dot{\mathbf{p}}(t) = -R(t)\mathbf{p}(t), \text{ with } R(t) = \frac{1}{2}\begin{pmatrix} r_+(t) & -r_-(t) \\ -r_+(t) & r_-(t) \end{pmatrix}. \tag{21}$$

It is straightforward to evaluate the equation for σ. Eq (21) becomes

$$\dot{p_1} = \frac{-r_+p_1 + r_-p_2}{2}, \qquad \dot{p_2} = \frac{r_+p_1 - r_-p_2}{2}, \tag{22}$$

and taking the difference of the two equations yields $\dot{\sigma} = -S(t)\sigma - D(t)$, where

$$S(t) = \frac{r_+ + r_-}{2}, \qquad D(t) = \frac{r_+ - r_-}{2}, \tag{23}$$

whose formal solution is

$$\sigma(t) = -\int_0^t dt' \exp\left[-\int_{t'}^t dt'' S(t'')\right] D(t') + \exp\left[-\int_0^t dt' S(t')\right]\sigma(0). \tag{24}$$

Let us go back to (10). The bias is given by the fact that the duration of the quiescent regions become shorter (longer) according to the system being in a status with the same (opposite) sign of the perturbation, after which a fair coin is tossed. It is clear that the overall bias cannot be due to the cumulative effect of previous drawings happened in the past, but one should focus his attention on the unperturbed coin-tossing times t, which will be turned into the perturbed coin-tossing time t'. In other words, the probability of a bias at time t will depend on the probability of having an unperturbed event at that same time t. We can use Eq. (10) to write the perturbed time t' in terms of the unperturbed t To simplify our treatment let us focus on the first event time, $t = t_1$. According to this notation

$$r(t) = \lim_{M\to 0}\frac{1}{M}\delta(t-\tau), \text{and } r_\pm(t') = \lim_{M\to 0}\frac{1}{M}\delta(t'-\tau_\pm), \tag{25}$$

and Eq. (10) yields $t = t' \pm \epsilon\int_0^{t'} dt''\xi_P(t'')$. Using the Jacobian methods for the Dirac's deltas,

$$r_\pm(t') = r[t(t')]\left|\frac{dt}{dt'}\right| \Rightarrow r_\pm(t') = r(t)\left[1 \pm \epsilon\xi_P(t')\right], \tag{26}$$

therefore, plugging (26) into the formal solution (24), we have

$$\sigma(t) = -\int_0^t dt' \exp\left[-\int_{t'}^t dt'' S(t'')\right]\frac{r_+(t') - r_-(t')}{2} + \exp\left[-\int_0^t dt' S(t')\right]\sigma(0). \tag{27}$$

We can now assume that $S \equiv (r_+ + r_-)/2 = r$, this approximation being exact for vanishing ϵ, since the two symmetric deltas converge to the unperturbed one. Making use of (26) we write

$$\sigma(t) = -\epsilon\int_0^t dt' \exp\left[-\int_{t'}^t dt'' r(t'')\right] r(t)\xi_P(t') + \exp\left[-\int_0^t dt' r(t')\right]\sigma(0). \tag{28}$$

Making averages and taking $\langle\sigma(0)\rangle = 0$ Eq. (28) becomes identical to (1) with a response function given by (3), which is the result we are seeking: making use of (20),

$$\chi(t,t') = \left\langle \exp\left[-\int_{t'}^t dt'' r(t'')\right] r(t)\right\rangle = \left\langle \frac{d\exp\left[-\int_{t'}^t dt'' r(t'')\right]}{dt}\right\rangle = -\frac{d\Psi(t,t')}{dt}. \tag{29}$$

3. Superposition of Poisson noise to non-Poisson signals

To study the effect of noise on signals we study a model adopted in [29] for human language and in bioinformatics [30], with the name Copying Mistake Map (CMM). The CMM model $\{\xi(t)\}$ consists of the superposition of two sequences, a non-Poisson renewal Sequence n, or $\{\xi_n(t)\}$, and a Poisson Sequence p, or $\{\xi_p(t)\}$, with $\psi_n(\tau) \propto \tau^{-\mu}$ for $\tau \to \infty$, and $\psi_p(\tau) = \gamma \exp(-\gamma\tau)$. The sequences of event times are

$$t_i^{(n)} = \sum_{k=1}^{i} \tau_i^{(n)} k \quad \text{and} \quad t_i^{(p)} = \sum_{k=1}^{i} \tau_k^{(p)}. \tag{30}$$

The signal is unity in these points, and zero otherwise, namely

$$\xi_n(t) = \delta_{t,t_i^{(n)}}, \qquad \xi_p(t) = \delta_{t,t_i^{(p)}}, \tag{31}$$

where $\delta_{i,j}$ is a Kroeneker delta. The CMM model is

$$\xi(t) = \xi_p \vee \xi_n \approx \xi_p + \xi_n \tag{32}$$

where $\vee$ is the boolean OR, and we used the fact that Prob$[\xi_p = \xi_n = 1]$ is vanishingly small.

The survival probability $\Psi(\tau)$ for the combined sequence (32) reads

$$\Psi(\tau) = P(n)e^{-\gamma\tau}\Psi_n(\tau) + P(p)e^{-\gamma\tau}\Psi_n^{\infty}(\tau), \tag{33}$$

where $\Psi_n^{\infty}(t)$ is the stationary correlation function of the non-Poisson process. In (33) $P(p)$ denotes the probability for an event to belong to the Poisson series, while $P(n)$ to the non-Poisson one. In the stationary regime $\mu > 2$ The probabilities $P(p)$ and $P(n)$ are inversely proportional to the respective average times:

$$P(n) \propto \frac{1}{\langle\tau_n\rangle} \quad ; \quad P(p) \propto \frac{1}{\langle\tau_p\rangle}, \tag{34}$$

where $\langle\tau_{p,n}\rangle = \int_0^{\infty} \tau\psi_{p,n}(\tau)d\tau$, namely

$$\langle\tau_p\rangle = \frac{1}{\gamma} \quad ; \quad \langle\tau_n\rangle = \frac{T}{\mu - 2}. \tag{35}$$

After normalization, and using (33) to evaluate the global mean time $\langle\tau\rangle$

$$P(n) = \frac{\langle\tau_p\rangle}{\langle\tau_p\rangle + \langle\tau_n\rangle} = (1-\gamma)\langle\tau\rangle; \quad P(p) = \frac{\langle\tau_n\rangle}{\langle\tau_p\rangle + \langle\tau_n\rangle} = \gamma\langle\tau\rangle. \tag{36}$$

$\Psi_n^{\infty}(t) \equiv \int_t^{\infty} \psi_n^{\infty}(\tau)d\tau$ denotes the infinitely aged survival probability, where $\psi_n^{t_a}(t)$ is the infinitely aged waiting-time distribution, with a change notation with respect of Section 2 to allow infinite age. It is calculated using $R(t)$, the probability of having an event at t, as [31]

$$\psi_n^{t_a}(\tau) \equiv \psi(\tau + t_a, t_a) = \int_{-t_a}^{0} R(t' + t_a)\psi_n(\tau - t')dt', \text{ where}$$
$$R \equiv \sum_{i=0}^{\infty} \psi_n^i(t), \quad \psi_n^i(t) = \int_0^{\infty} dt' \psi_n^{i-1}(t - t')\psi_n(t'), \tag{37}$$

and $\psi_n^0(t) = \delta(t)$ is a Dirac δ. For $t_a = \infty$ $R = 1/\langle\tau_n\rangle$ [31], so

$$\psi_n^{\infty}(\tau) = \frac{1}{\langle\tau_n\rangle}\int_{-\infty}^{0} \psi_n(\tau - t')dt' = \frac{\Psi_n(\tau)}{\langle\tau_n\rangle}. \tag{38}$$

Notice that before the exponential cutoff in (33) we have the joint action of two inverse-power laws, with indexes $\mu - 1$ and $\mu - 2$. If γ were vanishingly small, the asymptotic behavior would be dictated by the index $\mu - 1$. We recall that the treatment herein holds when $\mu > 2$, but it can be generalized for $\mu < 2$, where $\Psi_n^\infty(t)$ tends to a non-vanishing constant, while $\langle\tau\rangle \to \gamma^{-1}$ as $t \to \infty$. Numerically, the presence of two inverse-power decays and the exponential cut-off can make the decay of the survival probability indistinguishable from a stretched exponential.

This is apparent in Fig. 1, showing $\Psi(t)$ for typical cascading events (RPT in more than one channel) stemming from multi-channel EEG. Here we show Subject 5 of the experiment discussed in Ref. [32]. We show two sets of data points, one stemming from EEG data (open squares), and another with a Monte Carlo simulation of the model (open circles) with ψ_n stemming from (7), with $T = 6.0$s and $\mu = 2.05$, $\gamma = 0.006$Hz, and the same length of the data analyzed (5 minutes, sampling ratio 2ms). The curves stemming from the two point sets are indistinguishable. Notably, the dotted line, indicating a best stretched-exponential fit for $\Psi(t)$ seem to accurately describe both data sets, as earlier stated. It is in fact indistinguishable from data for $t > 30$ms.

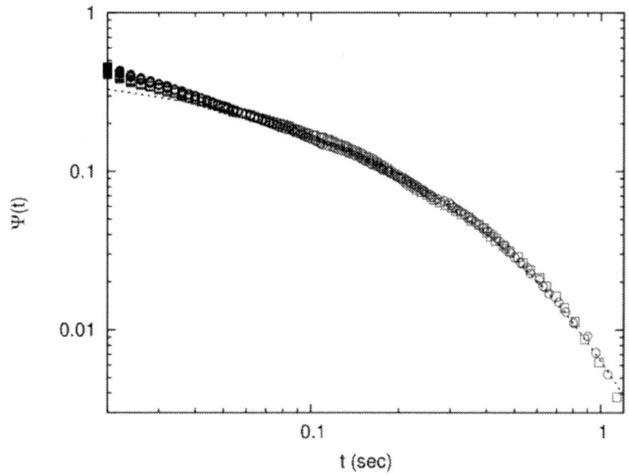

Figure 1. Survival probability. Open squares stem from EEG RTPs data. Open circles stem from a numerical simulation of the CMM with $r = 0.006$ Hz, $T = 6.0$s and same statistics of the data. Dotted line is a stretched exponential $A\exp[(t/B)^\alpha]$, with $A = 0.5$.

Let us use a more sensible correlational analysis on the CMM model. Following [32] we build a diffusing trajectory from $\xi(t)$ using two different rules. We make use of DFA to unravel long-range correlations. The correlation function of the signal $\xi(t)$ is proportional to the second derivative of $\sigma^2(t) \equiv \langle x^2(t)\rangle$, where $\dot{x} = \xi$ or, equivalently $x(t) = \int_{t_0}^{t_0+t} dt'\xi(t')$, and average $\langle\cdots\rangle$ is over all possible values of t_0 (moving windows). Therefore, if $\langle\xi(t_0)\xi(t_0+t)\rangle$ is not integrable, namely it decays as $t^{-\beta}$ with $0 < \beta < 1$, then asymptotically $\sigma^2(t) \sim t^{2H}$, with $H = 1 - \beta/2$.

In Rule #1 $\xi(t)$ is simply (32). As earlier stated, $\xi(t) = \xi_p(t) + \xi_n(t)$ and since the two sequences p and n are mutually independent, it is possible to separate their second moment contributions x_p and x_n and we have that

$$\sigma^2(t) = \langle x_p^2(t)\rangle_p + \langle x_n^2(t)\rangle_n, \tag{39}$$

where the suffix p (n) on the angular brackets means the average on the p (n) sequence alone. The Poisson process yields standard diffusion $\langle x_p^2\rangle_p \propto t$, while for renewal non-Poisson processes

$\langle x_n^2(t)\rangle_n \propto t^{4-\mu}$. The non-Poisson process dominates in the asymptotic limit. However, the presence of the Poisson Process yields a transient $H = 0.5$ in the short-time limit. This behavior is illustrated in Fig. 2, bottom curves, where we compare the analyses same EEG data of Fig. 1 with the thoeretical prediction, with $\mu = 2.05$.

Rule #2 consists of making a walker assume a constant velocity ± 1 within laminar regions, chosen with a coin-tossing procedure for every region. In this case (39) does not hold: While for this rule we again have $\langle x_p^2\rangle_p \propto t$ and $\langle x_n^2(t)\rangle_n \propto t^{4-\mu}$, the asymptotics are dominated by the Poisson process, while the non-Poisson scaling is confined in the short-time limit. Fig. 2 illustrates this behavior both for the model and the experimental data.

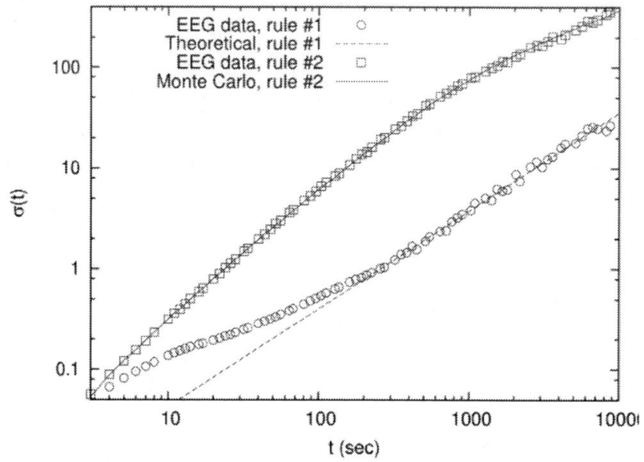

Figure 2. DFA Analysis: Rule #1 for EEG data (open squares) compared with theoretical prediction $\sigma(t) \propto t^{2-\mu/2}$ (dashed line); Rule #2 for EEG data (open circles) compared with same rule on the same Monte Carlo simulation of Fig. 1 (solid line).

In conclusion, studying concurrent (multi-channel) abrupt transitions in EEGs we conclude that a single complex renewal process (plus Poisson noise) underlies non sensory-costrained mental activity. Since (results will be reported elsewhere) single-channel transitions, although with similar $\psi(\tau)$, do not obey this simple dynamics, we claim that a serial (i.e. computationally non parallel) binding activity among neural groups is at the basis of the holistic output called consciousness. It is important to acknowledge that classifying the thought-related high level activity of the brain as a non-Poisson renewal process is of a great importance, since, as discussed in detail is Section 2, these processes obey a new form of Linear Response theory,of which we provide a detailed derivation in Section 2. The proof exposed here is mathematically simpler than what has been published elsewhere and it does not suffer from limitations due to discrete time as in [33] or to perturbations which are analytical in time as in [34]. It can be generalized to other forms of $\psi(\tau)$, as in the treatment of Ref. [35]. We think that the new Linear Response is important both for its implications and for its generality, from blinking quantum dots to human consciousness. The authors of Ref. [36] theoretically predict a response to abrupt and persistent stimuli, which is extremely slow in all the range $1 \leq \mu \leq 3$. These predictions have been experimentally confirmed in physical systems (nematic liquid crystals) [37] studying slow relaxation upon an abrupt changes of μ; at the level of cognitive neuroscience, it is worth noting that this slow response was conjectured to dominate the phenomenon of memory loosing, as discussed in [38]. As far as periodical perturbations are concerned, we refer the reader to the experimental observations of Refs [39] and [40], respectively again on liquid crystals and neuroscience. Finally, non-Poisson renewal systems effectively respond to perturbations only if the stimulus is of the same class of complexity, with a maximal information transport for $\mu \approx 2$, as recently theoretically assessed by Aquino *et al.* [41]. This phenomenon is termed "complexity matching". While an experimental assessment of these latter predictions is still missing, we note that this maximal information exchange can constitute a basis for explaining the surprising coincidence of $\mu \approx 2.1$ both in EEGs and in human language [29]. In general, the complexity matching effect can provide the basis for studying the interactions between complex systems or networks, for instance between the central nervous system and the peripheral autonomic system.

References

[1] Kolmogorov A 1941 *Dokl.Akad. Nauk.* **26** 115
[2] Kolmogorov A 1962 *J. Fluid Mech.* **13** 82
[3] Kraichnan R H 1994 *Phys. Rev. Lett.* **72** 1016
[4] Pomeau Y and Manneville P 1980 *Coommun. Math. Phys.* **74** 189; Gaspard P and Wang X -J 1988 *Proc. Natl. Acad. Sci. USA* **85** 4591
[5] Paradisi P, Cesari R, Contini D, Donateo A and Palatella L 2009 *Eur. Phys. J. Special Topics* bf 174 207
[6] Frantsuzov P, Kuno M, Jankó B and Marcus R A 2008 *Nature Physics* **4** 519
[7] Dehmelt H G 1975 *Bull. Am. Phys. Soc.* **20** 60
[8] Cook J and Kimble H J 1985 *Phys. Rev. Lett.* **54** 1023
[9] Frantsunov P and Marcus R A 2005 *Phys. Rev.* B **72** 155321
[10] Sparre Andersen E 1953 *Math. Scand.* **2** 263
[11] Bianco S, Grigolini P and Paradisi P 2005 *J. Chem. Phys.* **123** 174704
[12] Bianco S, Geneston E, Grigolini P and Ignaccolo M 2008 *Physica* A **387** 1387
[13] Turalska M, Lukovic M, West B J and Grigolini P 2009 *Phys Rev.* E **80** 021110
[14] Turing A M 1957 *Mind* **59** 236
[15] Beggs J M and Plenz D 2003 *J Neurosci.* **23** 11167
[16] Bak P, Tang C and Wiesenfeld K 1987 *Phys. Rev. Lett.* **59** 381
[17] Zapperi S, Baekgaard L K and Stanley H E 1995 *Phys. Rev. Lett.* **75** 4071; De Arcangelis L, Perrone-Capano C and Herrmann,H.J. 2006 *Phys. Rev. Lett.* **96** 028107; Pellegrini G L , De Arcangelis L, Herrmann H J and Perrone-Capano C 2007 *Phys. Rev.* E **76** 016107.
[18] Chialvo D R 2010 *Nature Physics* **6** 744
[19] Werner G 2010 Front. Physio. **1** 15
[20] Fingelkurts A A, Fingelkurts A A and Neves C H F *Physics of Life Reviews* **7** 195
[21] Afraimovich V, Zhigulin V and Rabinovich M 2004 *Chaos* **14** 1123
[22] Brodsky B E, Darkhovsky B S, Kaplan A Ya and Shishkin S L 1999 *Comput. Meth. Programs Biomed.*, **60** 93
[23] Contoyiannis Y F and Diakonos F K 2000 *Phys. Lett.* A **268** 286
[24] Wosiek J 1988 *Acta Phys. Pol.* B **19** 863
[25] Bialas A and Peschanski R 1986 *Nucl. Phys.* B **273** 703
[26] Kubo R 1957 *J. Phys. Soc. Japan* **12** 570
[27] Crisanti A and Ritort F 2003 *J. Phys.* A **36** R-181
[28] Onsager L 1931 *Phys. Rev.* **37** 405 (1931); Onsager L 1931 *Phys. Rev.* **38** 2265
[29] Allegrini P, Grigolini P and Palatella L 2004 *Chaos, Solitons and Fractals* **20** 95
[30] Allegrini P, Grigolini P and West B J 1996 *Phys. Lett.* A **211** 217
[31] Allegrini P, Aquino G, Grigolini P, Palatella L and Rosa A 2003 *Phys. Rev.* E **68** 056123; Allegrini P, Grigolini P, Palatella L and West B J 2004 *Phys Rev.* E **70** 046118
[32] Allegrini P, Menicucci D, Bedini R, Fronzoni L, Gemignani A, Grigolini P, West B J and Paradisi P 2009 *Phys. Rev.* E **80** 061914
[33] West B J, Geneston E L and P. Grigolini P 2008 *Phys. Rep.* **468**1
[34] Aquino G, Grigolini P and West B J 2007 *Europhys. Lett.* **80** 10002
[35] Budini A A and Grigolini P 2009 *Phys. Rev.* A **80** 022103
[36] Allegrini P, Bologna M, Grigolini P and West B J 2007 *Phys. Rev. Lett.* **99** 010603
[37] Silvestri L, Fronzoni L, Grigolini P and Allegrini P 2009 *Phys. Rev. Lett.* **102** 014502
[38] West B J and Grigolini P 2010 *Biol. Cybern* **103** 167
[39] Allegrini P, Bologna M, Fronzoni L, Grigolini P and Silvestri L 2009 *Phys. Rev. Lett.* **103** 030602
[40] West B J and Grigolini P 2010 *Physica* A **389** 5718
[41] Aquino G, Bologna M, Grigolini P and West B J 2010 *Phys. Rev. Lett.* **105** 040601

Entanglement, decoherence and thermal relaxation in exactly solvable models

Oleg Lychkovskiy[1]

[1] Institute for Theoretical and Experimental Physics, 117218, B. Cheremushkinskaya 25, Moscow, Russia

E-mail: lychkovskiy@itep.ru

Abstract. Exactly solvable models provide an opportunity to study different aspects of reduced quantum dynamics in detail. We consider the reduced dynamics of a single spin in finite XX and XY spin 1/2 chains. First we introduce a general expression describing the evolution of the reduced density matrix. This expression proves to be tractable when the combined closed system (i.e. open system plus environment) is integrable. Then we focus on comparing decoherence and thermalization timescales in the XX chain. We find that for a single spin these timescales are comparable, in contrast to what should be expected for a macroscopic body. This indicates that the process of quantum relaxation of a system with few accessible states can not be separated in two distinct stages – decoherence and thermalization. Finally, we turn to finite-size effects in the time evolution of a single spin in the XY chain. We observe three consecutive stages of the evolution: regular evolution, partial revivals, irregular (apparently chaotic) evolution. The duration of the regular stage is proportional to the number of spins in the chain. We observe a "quiet and cold period" in the end of the regular stage, which breaks up abruptly at some threshold time.

1. Introduction

In the present contribution several results and observations are presented which concern the dynamics of exactly solvable models and especially one-dimensional spin chains. The integrability allows to obtain rigorous results, however we believe that the applicability of general conclusions based on these results is not restricted by integrable models only. The paper is organized as follows. In the next section the quantum Liouville equation is rewritten in the form which may be tractable when the model is exactly solvable. In section 3 a sketch of the XY model is given. In section 4 a comparison of decoherence and thermalization timescales in the isotropic XY (i.e. XX) model is carried out. Section 5 is devoted to the dynamical regular-to-chaotic transition in the XY chain. In section 6 we focus on the quiet and cold period in evolution of a spin in the XY chain which takes place in the end of the regular evolution.

2. Reduced dynamics of a spin: general expression

Consider a quantum system (described by a Hilbert space $\mathcal{S}$), which interacts with a quantum environment or bath (described by a Hilbert space $\mathcal{B}$). The composite system with Hilbert space $\mathcal{H} = \mathcal{S} \otimes \mathcal{B}$ is considered to be closed. Its density matrix evolves according to

$$\rho(t) = \exp(-\mathrm{i}Ht)\rho(0)\exp(\mathrm{i}Ht), \quad (1)$$

The Hamiltonian may be decomposed as

$$H = H^{\mathcal{S}} + H^{\mathcal{B}} + H^{\mathcal{SB}}, \tag{2}$$

where $H^{\mathcal{S}}$ and $H^{\mathcal{B}}$ are self-Hamiltonians of the system and the bath correspondingly, and $H^{\mathcal{SB}}$ is an interaction Hamiltonian. Here and in what follows the usage of superscripts and subscripts $\mathcal{S}$, $\mathcal{B}$ is believed to be self-explanatory.

The spectral decomposition of the total Hamiltonian reads

$$H = \sum_{n=1}^{d} E_n |\Psi_n\rangle\langle\Psi_n|, \tag{3}$$

where Ψ_n are the eigenvectors and $d \equiv \dim \mathcal{H} = d_{\mathcal{S}} d_{\mathcal{B}}$.

The time evolution of $\rho^{\mathcal{S}}(t)$ reads

$$\rho^{\mathcal{S}}(t) = \sum_{n=1}^{d}\sum_{m=1}^{d} \langle\Psi_n|\rho(0)|\Psi_m\rangle e^{-\mathrm{i}(E_n-E_m)t} \rho^{\mathcal{S}}_{nm}, \tag{4}$$

where the $d_{\mathcal{S}} \times d_{\mathcal{S}}$ matrices

$$\rho^{\mathcal{S}}_{nm} \equiv \mathrm{tr}_{\mathcal{B}} \, |\Psi_n\rangle\langle\Psi_m| \tag{5}$$

are introduced (not to be confused with matrix elements!). Evidently, these matrices encode the dynamics of the open system $\mathcal{S}$, while the matrix elements $\langle\Psi_n|\rho(0)|\Psi_m\rangle$ describe the initial conditions.

In the case when the system $\mathcal{S}$ is represented by a single spin, its reduced density matrix $\rho^{\mathcal{S}}(t)$ and matrices $\rho^{\mathcal{S}}_{nm}$ may be parameterized by polarization vectors:

$$\rho^{\mathcal{S}}(t) = (1 + \mathbf{p}(t)\boldsymbol{\sigma})/2, \quad \rho^{\mathcal{S}}_{nm} = (\delta_{nm} + \mathbf{p}_{nm}\boldsymbol{\sigma})/2. \tag{6}$$

The polarization vector of a spin $\mathbf{p}(t)$ belongs to a unit sphere which is known as the Bloch sphere. The length of the polarization vector equals 1 for a pure state and is less than 1 for a mixed state. Its time evolution is given by

$$\mathbf{p}(t) = \sum_{n=1}^{d}\sum_{m=1}^{d} \langle\Psi_n|\rho(0)|\Psi_m\rangle e^{-\mathrm{i}(E_n-E_m)t} \mathbf{p}_{nm}. \tag{7}$$

As a rule this expression is not very useful in practical calculations (however, it is used to prove a necessary condition for thermalization of a spin coupled to a bath, see ref. [1]). However, when the model is integrable, one has a chance to calculate $\mathbf{p}_{nm}$ explicitly. If this is done and if the initial condition is of some simple form (e.g., the bath is in a thermal state), then eq.(7) may appear to be tractable and allow to study the reduced evolution of a spin in detail. We will demonstrate this is the case of integrable one-dimensional spin chain – the XY chain. Before we turn to this specific model let us give an example of initial conditions that simplify the problem greatly. These conditions correspond to the uncorrelated system and environment, when the system is in arbitrary pure state, and the environment is in thermal state with infinite temperature:

$$\rho(0) = 2^{-N}(1 + \mathbf{p}_0\boldsymbol{\sigma}) \otimes 1^{\mathcal{B}}. \tag{8}$$

In this case eq.(7) reads

$$\mathbf{p}(t) = 2^{-N} \sum_{n=1}^{d}\sum_{m=1}^{d} (\mathbf{p}_0, \mathbf{p}_{mn}) e^{-\mathrm{i}(E_n-E_m)t} \mathbf{p}_{nm}. \tag{9}$$

3. XY model on a circle

Consider a chain of N coupled spins 1/2 with the following Hamiltonian:

$$H = \kappa \sum_{j=1}^{N} ((1+\gamma)\sigma_j^x \sigma_{j+1}^x + (1-\gamma)\sigma_j^y \sigma_{j+1}^y) + h \sum_{j=1}^{N} \sigma_j^z. \tag{10}$$

Here the index $N+1$ is identified with 1, and N is supposed to be even. The following parameters enter the Hamiltonian:

- κ – coupling constant,
- γ – anisotropy parameter,
- h – magnetic field.

We consider the case of weak and moderate coupling: $h > \kappa$.

For the purposes of the present study we regard the first spin as the system $\mathcal{S}$, and other $(N-1)$ spins – as the environment $\mathcal{B}$.

Finite XY model is "almost diagonalizable" through the sequential Jordan-Wigner, Fourier and Bogolyubov transformations (see e.g. [2][3]). An important property of the XY Hamiltonian is that it commutes with the parity operator $\Pi \equiv \prod_{j=1}^{N} \sigma^z$:

$$[\Pi, H] = 0. \tag{11}$$

The diagonalization leads to the "almost free fermion form" of the Hamiltonian:

$$H = P^{\text{odd}} \sum_{q=-N/2+1}^{N/2} E_q c^{\dagger}_{\text{odd }q} c_{\text{odd }q} + P^{\text{even}} \sum_{q=-N/2+1/2}^{N/2-1/2} E_q c^{\dagger}_{\text{even }q} c_{\text{even }q}, \tag{12}$$

where $\{c_{\text{odd }q}\}$ and $\{c_{\text{even }q}\}$ are two sets of fermion operators (note, however, that the operators from different sets do not satisfy anticommutation relations!), P^{odd} and P^{even} are parity projectors,

$$P^{\text{even}} \equiv (1+\Pi)/2, \qquad P^{\text{odd}} \equiv (1-\Pi)/2 \tag{13}$$

and fermion energy is defined as

$$E_q = \sqrt{(h - \kappa\cos(2\pi q/N))^2 + (\gamma\kappa\sin(2\pi q/N))^2}. \tag{14}$$

4. Decoherence versus thermalization timescales

Decoherence is a process of the dynamical vanishing of non-diagonal entries of the reduced density matrix $\rho^{\mathcal{S}}(t)$ written in the pointer basis. Thermalization is a process of the energy exchange between the system and the environment leading to thermal equilibrium. What is the relation between the timescales of these processes? Evidently, for macroscopic bodies the decoherence timescale is much less than the thermalization timescale. Namely, decoherence timescale should be *microscopic*, in order to ensure the fast quantum-to-classical transition as it is required by our everyday experience. In contrast, the thermalization timescale is usually *macroscopic*. In fact, this hierarchy of the timescales is the basis for the very distinction between the decoherence and the thermalization: the former emerges as a first short stage of quantum relaxation, and the latter – as a succeeding long stage.

Is the above reasoning applicable to small quantum systems, in particular, to a single spin 1/2? Very often one implicitly assumes that it is. However, below we present a counterexample: the decoherence and thermalization timescales of a spin in the XY spin chain appear to be of the same order.

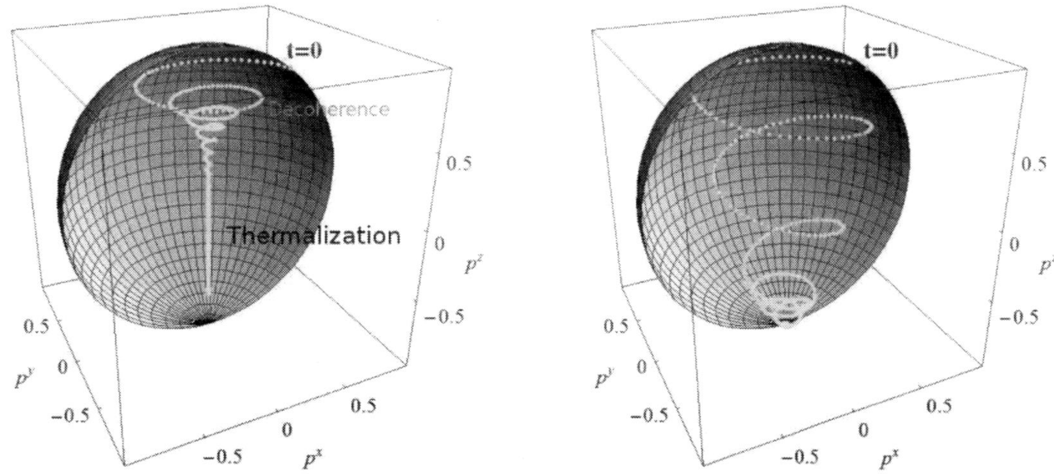

Figure 1. Patterns of the spin polarization evolution $\mathbf{p}(t)$. Left: what may be expected for the relaxation of a spin in the analogy to the relaxation of macroscopic bodies ("artist's view"). Right: real relaxation of a spin in the XX chain (plot). The points correspond to the values of $\mathbf{p}(t)$ from $t = 0$ to $t = 4\kappa^{-1}$ with the step $0.01\kappa^{-1}$. The initial value reads $\mathbf{p}(0) = (0.6, 0, 0.8)$. See the text for other details.

The easiest way to show this is to consider a special case with zero anisotropy ($\gamma = 0$, XX chain) and small coupling ($\kappa \ll h$). The latter condition ensures that the pointer basis is the eigen basis of the self-Hamiltonian $h\sigma_1^z$ [4] in the sense that $\rho^S(t)$ is approximately diagonal in this basis for almost all times [5]. Let us take the initial state to be

$$\rho(0) = (1 + \mathbf{p}_0\boldsymbol{\sigma}_1)/2 \otimes |\downarrow_2\downarrow_3 \dots \downarrow_N\rangle\langle\downarrow_2\downarrow_3 \dots \downarrow_N|, \tag{15}$$

which corresponds to the environment at zero temperature and arbitrary polarization $\mathbf{p}_0$ of the first spin (which constitutes the system). We are interested in the evolution of the central spin polarization $\mathbf{p}(t)$. In the analogy to the relaxation in macroworld one could expect the following relaxation pattern (see the left picture on Fig. 1). First p_x and p_y would vanish abruptly (or, equivalently, $\rho^S(t)$ would become approximately diagonal in the pointer basis $\{|\downarrow\rangle, |\uparrow\rangle\}$), which would constitute decoherence. In contrast, p_z would decrease gradually revealing the energy flow from the central spin to cold environment. This pattern allows to unambiguously separate decoherence from thermalization, as is evident from Fig. 1, left.

However the described above pattern is not realized in reality. Instead decoherence and thermalization proceed simultaneously, see Fig. 1, right. To be more exact, the process of relaxation can not be separated in time in two different stages – decoherence and thermalization. This can be seen from the approximate analytical solution which is valid for $0 \le t < N/\kappa$ [6]:

$$\begin{aligned} p^x &= J_0(\kappa t)(p_0^x \cos ht - p_0^y \sin ht), \\ p^y &= J_0(\kappa t)(p_0^x \sin ht + p_0^y \cos ht), \\ p^z &= -1 + (1 + p_0^z) J_0^2(\kappa t), \end{aligned} \tag{16}$$

where J_0 is Bessel function.

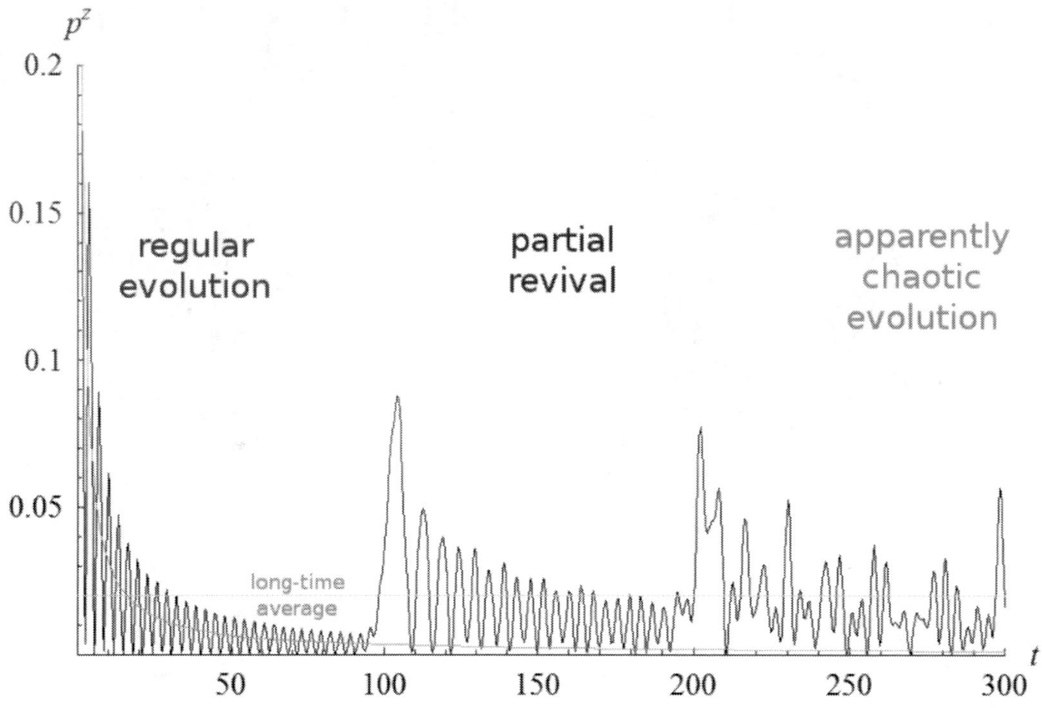

Figure 2. Evolution of $p^z(t)$ for $\gamma = 0,\ h = 5,\ \kappa = 1,\ N = 100$.

One can see that the relaxation of all components of $\mathbf{p}(t)$ occurs at a common timescale $\sim \kappa^{-1}$. In a system which consists of a single spin there is no large dimensionless parameter which could be responsible for the hierarchy of timescales. This reasoning suggests that the absence of such hierarchy is a general property of quantum systems with small number of accessible states rather than a peculiar feature attributed to integrability of the XY chain. However further work is necessary to show this with more rigor.

5. Finite-size effects in XY chain

Integrable spin chains are often studied in the $N \to \infty$ approximation. Although this allows to use the powerful machinery of Quantum Field Theory, some interesting features of the evolution may disappear in this approximation. The present section is devoted to one of such features – the regular-to-chaotic transition during the evolution of $\mathbf{p}(t)$. A similar effect in a somewhat different setting is observed (but not discussed in detail) in [7].

In what follows we take the initial state (8) which corresponds to the environment at $T = \infty$. As was already discussed such choice of the initial conditions simplifies the problem. However one should be cautious when interpreting the infinite temperature state $2^{-N+1}1^{\mathcal{B}}$. Indeed, consider a thermal state with some high temperature T. It is well approximated by $2^{-N+1}1^{\mathcal{B}}$ whenever $T \gg Nh$ but not just $T \gg h$. Evidently the former condition (in contrast to the latter one) becomes more stringent with the growth of N and appears to be impracticable in the limit $N \to \infty$. However this subtlety is not important in our discussion: the qualitative picture of the regular-to-chaotic transition does not depend on the temperature of the environment.

We focus our attention on the evolution of the z-component of polarization. Using the explicit expressions for the eigenstates of the XY model one gets from eq.(9) the following *exact* formula (the details of calculation will be given elsewhere):

$$p^z(t) = \frac{1}{2} p_0^z (A_{\rm odd}^2 + A_{\rm even}^2 + B_{\rm odd}^2 + B_{\rm even}^2), \tag{17}$$

where

$$\begin{aligned} A_{\rm even}(t) &= N^{-1} \sum_{q=-N/2+1/2}^{N/2-1/2} \cos E_q t, & A_{\rm odd}(t) &= N^{-1} \sum_{q=-N/2+1}^{N/2} \cos E_q t, \\ B_{\rm even}(t) &= N^{-1} \sum_{q=-N/2+1/2}^{N/2-1/2} \cos\theta_q \sin E_q t, & B_{\rm odd}(t) &= N^{-1} \sum_{q=-N/2+1}^{N/2} \cos\theta_q \sin E_q t, \end{aligned} \tag{18}$$

$$\cos\theta_q = \frac{h - \kappa\cos(2\pi q/N)}{E_q}. \tag{19}$$

The plot of $p^z(t)$ for $\gamma = 0$, $h = 5$, $\kappa = 1$, $N = 100$ is presented on Fig. 2. One can distinguish three successive stages of evolution:

(i) $0 \le \kappa t \lesssim N$: regular evolution. At this stage $p^z(t)$ is well described by the approximate expression [6]

$$p^z(t) \simeq p_0^z J_0^2(\kappa t). \tag{20}$$

(ii) $N \lesssim \kappa t \lesssim$ few N : partial revivals.

(iii) $\kappa t \gtrsim$ few N : apparently chaotic evolution.

As the duration of the regular stage is proportional to N, it is the only stage which may be catched by the $N \to \infty$ approximation. Evidently this stage ends up when the perturbation initially localized in the first site of the chain propagates through the circle and returns back to the first site. However the abruptness of the deviation of $p^z(t)$ from the regular expression (20) at some threshold time calls for additional explanation, which will be the purpose of the future work.

The second stage has a well-defined starting point, but its termination point is not well-defined. One can clearly see one partial revival at $N \lesssim \kappa t \lesssim 2N$ in Fig. 2, but the succeeding evolution can not be characterized that unambiguously. In fact the second stage gradually changes to the third one – apparently chaotic. Note that due to the Poincare recurrence theorem the revivals should interrupt the chaotic evolution at long times.

On Fig. 3 one can see the plots for $p^z(t)$ for various anisotropies and magnetic fields. Evidently large anisotropy suppresses the oscillations, while large magnetic field amplifies them.

6. Quiet and cold period

The asymptotic behavior of the Bessel function is given by

$$J_0(x) \simeq \sqrt{2/\pi x}\cos(x - \pi/4). \tag{21}$$

Therefore, according to eq.(20), during the regular stage the smoothed behavior of $p^z(t)$ in the XX chain is described by

$$\tilde{p}^z(t) = \frac{p_0^z}{\pi\kappa t}. \tag{22}$$

At the same time the long-time average of $p^z(t)$ is given by

$$\overline{p}^z \simeq 2p_0^z/N. \tag{23}$$

Note that at times $N(2\pi\kappa)^{-1} \lesssim t \lesssim N\kappa^{-1}$ the smoothed $p^z(t)$ is *less* than the average value $\overline{p}^z$. This means that during this time interval the energy of the central spin is less than its long-time average energy. Also the oscillations of $p^z(t)$ are suppressed at this time. Thus we observe the quiet and cold period which precedes the onset of chaotic evolution. Such "calm before the storm" may be of practical importance in the context of quantum information processing.

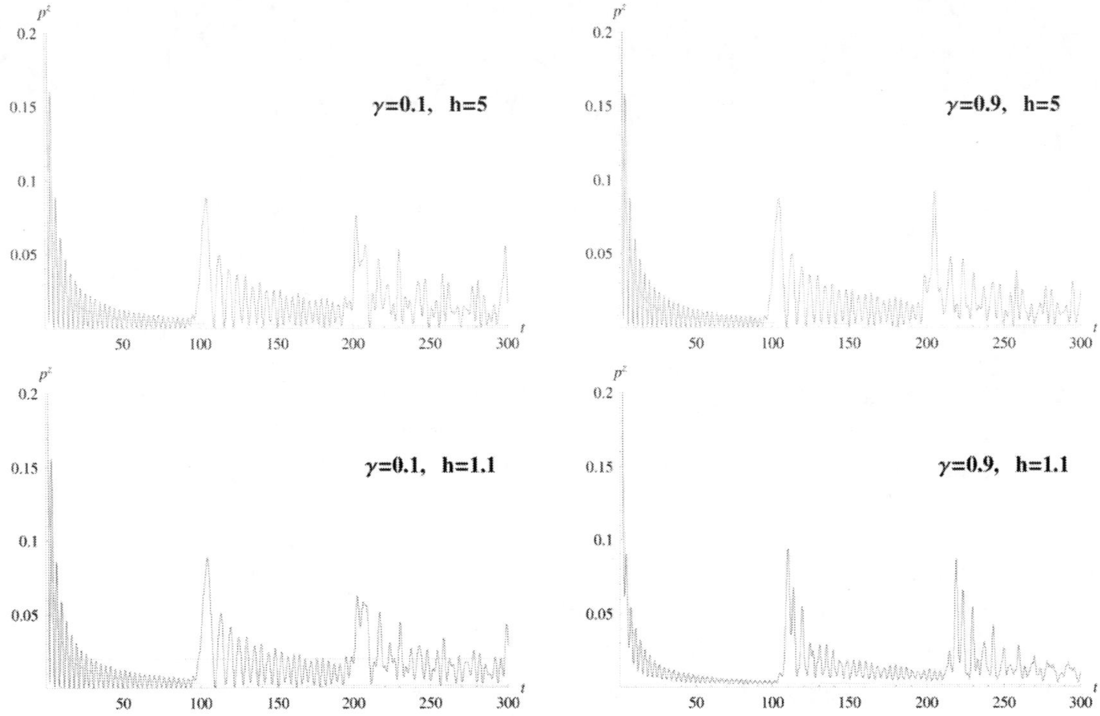

Figure 3. Evolution of $p^z(t)$ for various anisotropies and magnetic fields. $\kappa = 1,\ N = 100$.

Acknowledgments

The author acknowledges the partial support from grants NSh-4172.2010.2, RFBR-11-02-00778, RFBR-10-02-01398 and from the Ministry of Education and Science of the Russian Federation under contracts №№ 02.740.11.5158, 02.740.11.0239.

References

[1] Lychkovskiy O 2010 *Phys. Rev. E* **82** 011123
[2] Izergin A G, Kapitonov V S and Kitanin N A 2000 *J. Math. Sci.* **100** 2120
[3] De Pasquale A and Facchi P 2009 *Phys. Rev. A* **80** 032102
[4] Paz J P and Zurek W H 1999 *Phys. Rev. Lett.* **82** 5181
[5] Gogolin C 2010 *Phys. Rev. E* **81** 051127
[6] Niemeijer Th 1967 *Physica* **36** 377
[7] Mossel J and Caux J-S 2010 *New J. Phys.* **12** 055028

Journal of Physics: Conference Series **306** (2011) 012029 doi:10.1088/1742-6596/306/1/012029

Entanglement evolution for quantum trajectories

S. Vogelsberger[1] and D. Spehner[1,2]

[1] Institut Fourier, Université Joseph Fourier and CNRS, B.P. 74, F-38402 Saint Martin d'Hères, France

[2] Laboratoire de Physique et Modélisation des Milieux Condensés, Université Joseph Fourier and CNRS, B.P. 166, F-38042 Grenoble, France

E-mail: sylvain.vogelsberger@ujf-grenoble.fr

Abstract. Entanglement is a key resource in quantum information. It can be destroyed or sometimes created by interactions with a reservoir. In recent years, much attention has been devoted to the phenomena of entanglement sudden death and sudden birth, i.e., the sudden disappearance or revival of entanglement at finite times resulting from a coupling of the quantum system to its environment [1, 2, 3]. We investigate the evolution of the entanglement of noninteracting qubits coupled to reservoirs under monitoring of the reservoirs by means of continuous measurements. Because of these measurements, the qubits remain at all times in a pure state, which evolves randomly. To each measurement result (or "realization") corresponds a quantum trajectory in the Hilbert space of the qubits. We show that for two qubits coupled to independent baths subjected to local measurements, the average of the qubits' concurrence over all quantum trajectories is either constant or decays exponentially. The corresponding decay rate depends on the measurement scheme only. This result contrasts with the entanglement sudden death phenomenon exhibited by the qubits' density matrix in the absence of measurements. Our analysis applies to arbitrary quantum jump dynamics (photon counting) as well as to quantum state diffusion (homodyne or heterodyne detections) in the Markov limit. We discuss the best measurement schemes to protect the entanglement of the qubits. We also analyze the case of two qubits coupled to a common bath. Then, the average concurrence can vanish at discrete times and may coincide with the concurrence of the density matrix. The results explained in this article have been presented during the "Fifth International Workshop DICE2010" by the first author and have been the subject of a prior publication [4].

1. Introduction

As already noted by Erwin Schrödinger in 1935 "Entanglement is not one but rather the characteristic trait of Quantum Mechanics" [5]. Nowadays, entanglement is considered as a key resource in quantum information. However, it can be destroyed by interactions with reservoirs. When the two parts of a bipartite system are coupled to independent baths, entanglement typically disappears after a finite time [1, 3, 6]. This phenomenon is called "entanglement sudden death" (ESD). When the two parts of the system are coupled to a common bath, sudden revivals of entanglement may also take place after the state has become separable [7, 8, 9].

In this article, we investigate the evolution of the entanglement of two qubits coupled to two or one bath when the baths are monitored by continuous measurements. Because of these measurements, the qubits remain at all times in a pure state $|\psi(t)\rangle$, which evolves randomly. To each measurement result (or "realization") corresponds a quantum trajectory $t \in \mathbb{R}_+ \mapsto |\psi(t)\rangle$ in the Hilbert space $\mathbb{C}^4$ of the qubits. In the Markovian regime, the dynamics is given by the quantum jump (QJ) model [10, 11] or, in the case of homodyne and heterodyne detections, by the so-called quantum state diffusion (QSD) models [11, 12, 13]. We study how the entanglement of the qubits evolves in time by calculating the average $\overline{C_{\psi(t)}}$ of the Wootters concurrence over all quantum trajectories (QTs). $\overline{C_{\psi(t)}}$ differs in general from the concurrence $C_{\rho(t)}$ of the density matrix $\rho(t) = \overline{|\psi(t)\rangle\langle\psi(t)|}$ (here and in what follows the overline denotes the average over all QTs) [14, 15]. For two qubits coupled to independent baths subjected to local measurements, we find that

$$\overline{C_{\psi(t)}} = C_0\, e^{-\kappa t} \tag{1}$$

where $C_0 = C_{\psi(0)}$ is the initial concurrence and $\kappa \geq 0$ depends on the measurement scheme but not on the initial state $|\psi(0)\rangle$. In particular, if $C_0 > 0$ and $t_{\rm ESD} \in [0, \infty]$ is the time at which entanglement disappears in the density matrix (assuming that this time is finite), then $C_{\rho(t)} = 0$ at times $t \geq t_{\rm ESD}$ whereas $\overline{C_{\psi(t)}}$ can only vanish asymptotically. The continuous measurements on the two baths thus protect on average the qubits from ESD. Of course, this does not mean that *all* random wavefunctions $|\psi(t)\rangle$ remain entangled at all times. But in some cases, such as for pure dephasing and for infinite temperature baths, one can find measurement schemes such that $\kappa = 0$; then, for all QTs, if the qubits are maximally entangled at $t = 0$ (i.e., $C_0 = 1$) they remain maximally entangled at all times.

We investigate the best measurement schemes to protect entanglement. Related strategies using quantum Zeno effect [16], entanglement distillation [17], quantum feedback [18], and encoding in qutrits [19] have been proposed. It is assumed in this work that the measurements on the baths are performed by perfect detectors. The impact of detection errors has been studied in [19]. When the qubits are coupled to a common bath, we find that $\overline{C_{\psi(t)}}$ has a more complex time behavior than in (1) and may vanish at finite discrete times and be equal to the density matrix concurrence for some initial states. In [18, 20], the authors study the problem to find measurement schemes depending on the initial state such that $\overline{C_{\psi(t)}} = C_{\rho(t)}$.

This article is organized as follows. We begin to recall the definition of the concurrence and explain why it is a good measure of entanglement (Sec. 2). We review the quantum jump model in Sec. 3. In Sec. 4, we prove the formula (1) in the case of two two-level atoms coupled to the electromagnetic field initially in the vacuum for photon counting detection. We generalize this result in Sec. 5 to general quantum jump dynamics. In Secs. 6 and 7 we treat the case of homodyne and heterodyne detections and we compare the decay of $\overline{C_{\psi(t)}}$ with that found for photon counting. Finally, Sec. 8 is devoted to the evolution of the entanglement of two qubits coupled to a common bath at zero temperature monitored by continuous measurements.

2. Entanglement measures

The entanglement of formation [21] of a bipartite quantum system $\mathcal{S}$ in a pure state $|\psi\rangle$ is defined by means of the von Neumann entropy $E_\psi = -\operatorname{tr}(\rho_A \ln \rho_A) = -\operatorname{tr}(\rho_B \ln \rho_B)$ of the density matrices $\rho_A = \operatorname{tr}_B(|\psi\rangle\langle\psi|)$ and $\rho_B = \operatorname{tr}_A(|\psi\rangle\langle\psi|)$ of the two subsystems A and B composing $\mathcal{S}$. If $\mathcal{S}$ is in a mixed state, E_ρ is the minimum of $\sum p_k E_{\psi_k}$ over all convex decompositions $\rho = \sum_k p_k |\psi_k\rangle\langle\psi_k|$ of its density matrix (with $p_k \geq 0$ and $\|\psi_k\| = 1$). When A and B have two-dimensional Hilbert spaces, $E_\rho = f(C_\rho)$ is related to the concurrence [22] C_ρ by a convex increasing function $f : [0,1] \to [0, \ln(2)]$. A state ρ is separable if and only if $C_\rho = 0$, i.e., $E_\rho = 0$. For a pure state

$$C_\psi = |\langle \sigma_y \otimes \sigma_y T \rangle_\psi| \tag{2}$$

where $\sigma_y = {\rm i}(|{\downarrow}\rangle\langle{\uparrow}| - |{\uparrow}\rangle\langle{\downarrow}|)$ is the y-Pauli matrix, $T : |\psi\rangle = \sum_{s,s'} c_{ss'}|s,s'\rangle \mapsto \sum_{s,s'} c_{ss'}^{*}|s,s'\rangle$ the anti-unitary operator of complex conjugation in the canonical basis $\{|s,s'\rangle = |s\rangle \otimes |s'\rangle;\, s, s' = \uparrow, \downarrow\}$ of $\mathbb{C}^2 \otimes \mathbb{C}^2$, and $\langle \cdot \rangle_\psi = \langle\psi| \cdot |\psi\rangle$ the quantum expectation in state $|\psi\rangle$.

For quantum trajectories, one has always $\overline{E_{\psi(t)}} \geq E_{\rho(t)}$, this inequality being strict excepted if the decomposition

$$\rho(t) = \overline{|\psi(t)\rangle\langle\psi(t)|} = \int \mathrm{d}p[\psi]\, |\psi(t)\rangle\langle\psi(t)| \tag{3}$$

realizes the infinum defining $E_{\rho(t)}$. Thanks to the convexity of f, $\overline{E_{\psi(t)}} \geq f(\overline{C_{\psi(t)}})$. Thus Eq. (1) shows that for independent baths and if $C_0 > 0$ then $\overline{E_{\psi(t)}} \geq f(C_0 e^{-\kappa t}) > 0$ whatever the measurement scheme.

It is legitimate to ask which entanglement measure should be averaged. The concurrence is a natural candidate as it corresponds for pure states to the maximum over all self-adjoint local observables J_A and J_B with norms $\|J_A\|, \|J_B\| \leq 1$ of the covariance $|\langle J_A \otimes J_B\rangle_{\psi(t)} - \langle J_A \otimes 1_B\rangle_{\psi(t)}\langle 1_A \otimes J_B\rangle_{\psi(t)}|$. Moreover, as we will see below, $\overline{C_{\psi(t)}}$ is easy to calculate in the Markovian regime and it gives a lower bound on $\overline{E_{\psi(t)}}$.

3. Quantum jump model

Let us briefly recall the QJ dynamics [10, 23]. As a result of a measurement on a particle (e.g. a photon) of the bath scattered by the qubits, the qubits' wavefunction suffers a quantum jump

$$|\psi(t)\rangle \longrightarrow |\psi_{\mathrm{jump}}^{(m,i)}\rangle = \frac{J_m^i|\psi(t)\rangle}{\|J_m^i|\psi(t)\rangle\|} \tag{4}$$

where the jump operator J_m^i is related to the particle-qubits coupling and the indices m, i label all possible measurement results save for the most likely one, which we refer as a "no detection". In the small coupling limit, the probability that a measurement in the small time interval $[t, t+\mathrm{d}t]$ gives the result (m, i) is very small and equal to $\mathrm{d}p_m^i(t) = \gamma_m^i \|J_m^i|\psi(t)\rangle\|^2 \mathrm{d}t$. The jump rate γ_m^i does not depend on $|\psi(t)\rangle$ and is proportional to the square of the particle-qubits coupling constant. In the no-detection case the wavefunction of the qubits evolves as

$$|\psi(t+\mathrm{d}t)\rangle = \frac{e^{-\mathrm{i}H_{\mathrm{eff}}\mathrm{d}t}|\psi(t)\rangle}{\|e^{-\mathrm{i}H_{\mathrm{eff}}\mathrm{d}t}|\psi(t)\rangle\|} \quad , \quad H_{\mathrm{eff}} = H_0 - \frac{\mathrm{i}}{2}\sum_{m,i} \gamma_m^i J_m^{i\dagger} J_m^i \tag{5}$$

where H_0 is the Hamiltonian of the qubits. The probability that no jump occurs in the time interval $[t_0, t]$ is $p_{\mathrm{nj}}(t_0, t) = \|e^{-\mathrm{i}H_{\mathrm{eff}}(t-t_0)}|\psi(t_0)\rangle\|^2$. It is not difficult to show [10] that the density matrix $\rho(t) = \overline{|\psi(t)\rangle\langle\psi(t)|}$ satisfies the Lindblad equation

$$\frac{\mathrm{d}\rho}{\mathrm{d}t} = -\mathrm{i}[H_0, \rho] + \sum_{m,i} \gamma_m^i \Big(J_m^i \rho J_m^{i\dagger} - \frac{1}{2}\{J_m^{i\dagger} J_m^i, \rho\}\Big) \tag{6}$$

where $\{.,.\}$ denotes the anticommutator. It is known that many distinct QJ dynamics unravel the same master equation (6).

For two qubits coupled to independent reservoirs R_A and R_B, the jump operators are *local*, i.e., they have the form

$$J_m^A \otimes 1_B \quad , \quad 1_A \otimes J_m^B \tag{7}$$

depending on whether the measurements are performed on R_A or R_B. Here J_m^i are 2×2 matrices.

The absence of ESD, mentioned in the introduction, for the mean concurrence of two qubits coupled to independent baths can be traced back to the existence of trajectories for which $|\psi(t)\rangle$ remains entangled at all times. Actually, for a trajectory without jump, $|\psi_{\mathrm{nj}}(t)\rangle \propto e^{-\mathrm{i}tH_{\mathrm{eff}}}|\psi(0)\rangle$ up to a normalization factor, see (5). By (7) and for noninteracting qubits, the effective Hamiltonian H_{eff} is the sum of two (commuting) local operators acting on each qubit. If $|\psi_{\mathrm{nj}}(t)\rangle$ would be separable at a given time t then, by reversing the dynamics (i.e., by applying $e^{\mathrm{i}tH_{\mathrm{eff}}}$

Journal of Physics: Conference Series **306** (2011) 012029 doi:10.1088/1742-6596/306/1/012029

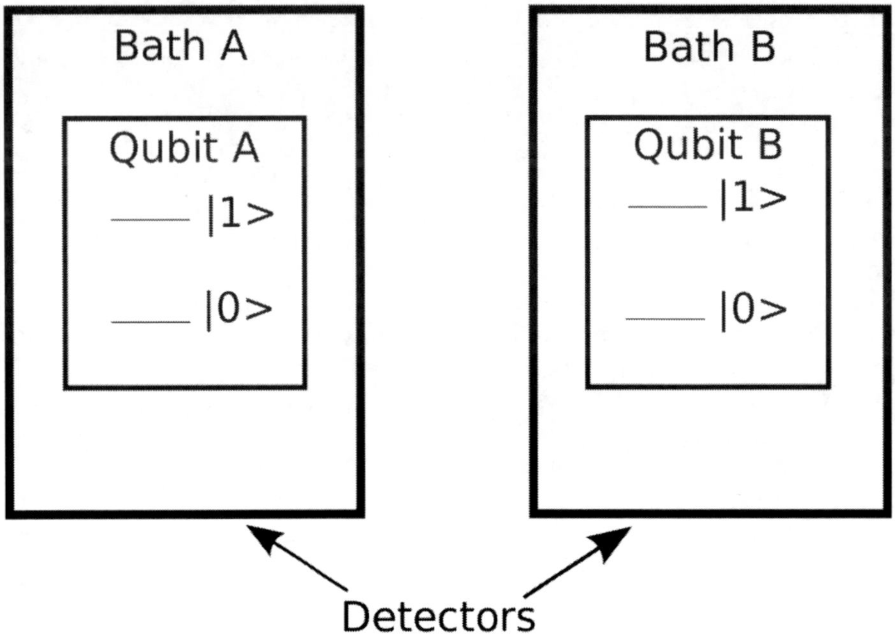

Figure 1. Two qubits coupled to two baths surrounded by perfect photon counters.

to $|\psi_{\rm nj}(t)\rangle$) one would deduce that $|\psi(0)\rangle$ is separable. Hence, $C_{\psi_{\rm nj}(t)} > 0$ if $C_0 > 0$. But the no-detection probability between times 0 and t is nonzero and thus $\overline{C_{\psi(t)}} > 0$ at all times. Note that this argument does not apply if nonlocal observables of the two baths are measured or if the two qubits are coupled to a common bath, due to the non-locality of J_m.

4. Photon counting

Let us illustrate the random dynamics described previously on a simple and experimentally relevant example [24]. Each qubit is a two-level atom coupled resonantly to the electromagnetic field initially in the vacuum (zero-temperature photon bath). The two atoms are far from each other and thus interact with independent baths. Two perfect photon counters D_i make a click when a photon is emitted by qubit i ($i = A, B$), whatever the direction of the emitted photon (see figure 1). Doing the rotating wave approximation, the jump operators are $J^i_- = \sigma^i_- = |\downarrow\rangle\langle\uparrow|$. For simplicity we take $H_0 = 0$.

If no photon is detected in the time interval $[0, t]$, the qubits wavefunction is, by (5),

$$|\psi(t)\rangle = \mathcal{N}(t)^{-1} \sum_{s,s'=\uparrow,\downarrow} c_{ss'}\, e^{-\gamma_{ss'}t/2}|s, s'\rangle \tag{8}$$

with :

(i) $c_{ss'} = \langle s, s'|\psi(0)\rangle$

(ii) $\mathcal{N}(t)^2 = \sum_{s,s'} |c_{ss'}|^2 e^{-\gamma_{ss'}t}$

(iii) $\gamma_{\uparrow\uparrow} = \gamma_A + \gamma_B$, $\gamma_{\uparrow\downarrow} = \gamma_A$, $\gamma_{\downarrow\uparrow} = \gamma_B$ and $\gamma_{\downarrow\downarrow} = 0$ (γ_i being the jump rate for detector D_i).

The concurrence (2) of the state $|\psi(t)\rangle$ is $C(t) = C_0\mathcal{N}(t)^{-2}e^{-(\gamma_A+\gamma_B)t/2}$ with $C_0 = 2|c_{\uparrow\uparrow}c_{\downarrow\downarrow} - c_{\uparrow\downarrow}c_{\downarrow\uparrow}|$.

If a photon is detected at time t_j by, say, the photon counter D_A, the qubits are just after the jump (4) in the separable state $|\psi(t_j+)\rangle \propto |\downarrow\rangle \otimes (c_{\uparrow\uparrow}e^{-\gamma_{\uparrow\uparrow}t_j/2}|\uparrow\rangle + c_{\uparrow\downarrow}e^{-\gamma_{\uparrow\downarrow}t_j/2}|\downarrow\rangle)$. Since neither a jump nor the inter-jump dynamics can create entanglement (the jump operators (7)

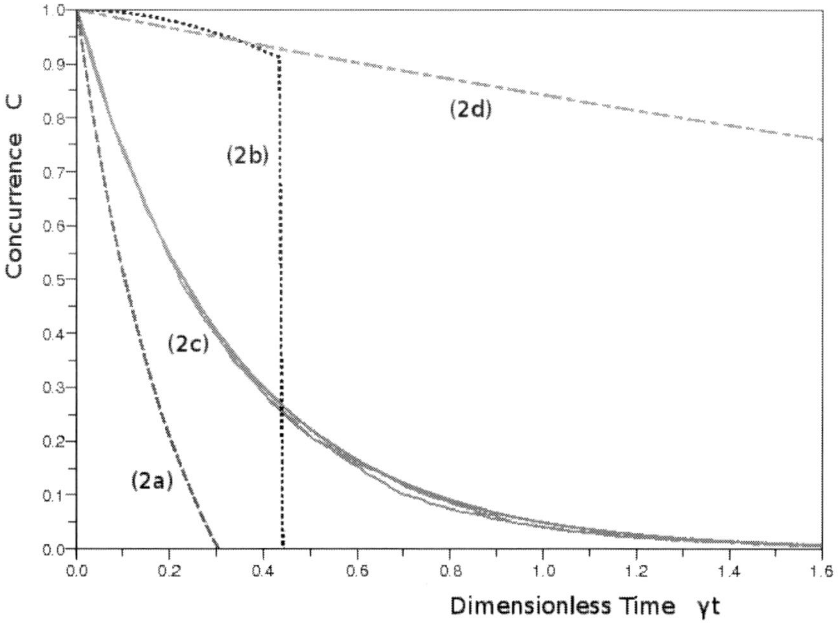

Figure 2. Concurrences of two qubits coupled to independent baths at positive temperature as a function of γt for $\gamma_+^i = \gamma_-^i/2 = \gamma$ with $|\Psi(0)\rangle = \frac{1}{\sqrt{2}}(|\uparrow\uparrow\rangle - \mathrm{i}|\downarrow\downarrow\rangle)$.
(2a) $C_{\rho(t)}$ for the density matrix (blue dashed lines).
(2b) $C_{\psi(t)}$ for a single trajectory (black dotted line).
(2c) $\overline{C_{\psi(t)}}$ averaged over 1500 trajectories (red plain line) and from Eq. (10) (red solid lines).
(2d) $\overline{C_{\psi(t)}}$ for the best measurement scheme (green dotted-dashed line).

being local), the qubits state remains separable at all times $t \geq t_j$, even if more photons are subsequently detected. Thus $C(t) = 0$ if at least one photon is detected in the time interval $[0, t]$.

Averaging over all realizations of the QTs and using the probability $p_{\mathrm{nj}}(0, t) = \mathcal{N}(t)^2$ that no photon is detected in $[0, t]$, one finds

$$\overline{C(t)} = C_0 e^{-(\gamma_A + \gamma_B)t/2}. \tag{9}$$

This argument is easily extended to baths at positive temperatures T_i. One must then add two jump operators $J_+^i = \sigma_+^i$ with rates $\gamma_+^i = e^{-\frac{\hbar\omega^i}{kT_i}}\gamma_-^i$ such that $\hbar\omega^i$ is the energy difference between the two levels of qubit i and k is the Boltzmann constant. Then the mean concurrence is given by

$$\overline{C(t)} = C_0 e^{-(\gamma_+^A + \gamma_-^A + \gamma_+^B + \gamma_-^B)t/2}. \tag{10}$$

It is compared in Fig. 2 with the concurrence for the density matrix obtained by solving the master equation (6) which shows ESD for all initial states.

5. General quantum jump dynamics

We now consider a general QJ dynamics with jump operators given by (7). The Hamiltonian of the qubits has the form $H_0 = H_A \otimes 1_B + 1_A \otimes H_B$. Let $K = K_A \otimes 1_B + 1_A \otimes K_B$ with

$$K_i = \frac{1}{2}\sum_m \gamma_m^i \, J_m^{i\dagger} J_m^i, \tag{11}$$

γ_m^i being the jump rate for the detector D_i ($i = A, B$). We first assume that no jump occurs between t and $t + \mathrm{d}t$. By expanding the exponential in (5) one gets

$$\begin{aligned} C(t+\mathrm{d}t) &= p_{\mathrm{nj}}(t, t+\mathrm{d}t)^{-1} \big| \langle \sigma_y \otimes \sigma_y T \rangle_{\psi(t)} \\ &+ \mathrm{i}\mathrm{d}t \langle H_{\mathrm{eff}}^\dagger \sigma_y \otimes \sigma_y T + \sigma_y \otimes \sigma_y T H_{\mathrm{eff}} \rangle_{\psi(t)} + \mathcal{O}(\mathrm{d}t)^2 \big| \end{aligned} \tag{12}$$

where $p_{\mathrm{nj}}(t, t+\mathrm{d}t) = \langle 1 - 2K\mathrm{d}t + \mathcal{O}(\mathrm{d}t)^2 \rangle_{\psi(t)}$ is the probability that no jump occurs between t and $t + \mathrm{d}t$. Now, for any local operator O_i acting on qubit i, one has

$$\langle O_i \sigma_y \otimes \sigma_y T \rangle_{\psi(t)} = \langle \sigma_y \otimes \sigma_y T O_i^\dagger \rangle_{\psi(t)} = \frac{\mathcal{C}(t)}{2} \mathrm{tr}_{\mathbb{C}^2}(O_i) \tag{13}$$

with

$$\mathcal{C}(t) = \langle \sigma_y \otimes \sigma_y T \rangle_{\psi(t)} = 2\big(c_{\uparrow\downarrow}(t) c_{\downarrow\uparrow}(t) - c_{\uparrow\uparrow}(t) c_{\downarrow\downarrow}(t)\big)^* \tag{14}$$

and $c_{ss'}(t) = \langle s, s' | \psi(t) \rangle$. Since $C(t) = |\mathcal{C}(t)|$, this gives

$$C(t+\mathrm{d}t)\, p_{\mathrm{nj}}(t, t+\mathrm{d}t) = C(t)\Big(1 - \mathrm{tr}_{\mathbb{C}^4}(K)\frac{\mathrm{d}t}{2} + \mathcal{O}(\mathrm{d}t^2)\Big). \tag{15}$$

If detector D_i gives the result m in the time interval $[t, t+\mathrm{d}t]$, we have by virtue of (4)

$$C_{\mathrm{jump}}^{(m,i)}(t+\mathrm{d}t) = \frac{\gamma_m^i \, \mathrm{d}t}{\mathrm{d}p_m^i(t)} C(t) \big| \mathrm{det}_{\mathbb{C}^2}(J_m^i) \big| \tag{16}$$

where we have used the identity

$$\langle O_i^\dagger \sigma_y \otimes \sigma_y O_i \rangle_{\psi(t)} = \mathcal{C}(t) \mathrm{det}_{\mathbb{C}^2}(O_i^\dagger). \tag{17}$$

Collecting the previous formulas and using the Markov property of the jump process, one gets $\overline{C(t+\mathrm{d}t)} = \overline{C(t)}(1 - \kappa_{\mathrm{QJ}}\, \mathrm{d}t + \mathcal{O}(\mathrm{d}t^2))$ with [4]

$$\begin{aligned} \kappa_{\mathrm{QJ}} &= \frac{1}{2}\mathrm{tr}_{\mathbb{C}^4}(K) - \sum_{m,i} \gamma_m^i |\mathrm{det}_{\mathbb{C}^2}(J_m^i)| \\ &= \sum_{m,i} \frac{\gamma_m^i}{2} \big(|\langle \uparrow | \tilde{J}_m^i | \uparrow \rangle - \langle \downarrow | \tilde{J}_m^{i\dagger} | \downarrow \rangle|^2 + |\langle \uparrow | 2\Re \tilde{J}_m^i | \downarrow \rangle|^2 \big) \geq 0 \end{aligned} \tag{18}$$

where we have introduced $\tilde{J}_m^i = e^{-\mathrm{i}\theta_m^i} J_m^i$ and $2\Re \tilde{J}_m^i = \tilde{J}_m^i + \tilde{J}_m^{i\dagger}$, $2\theta_m^i$ being the argument of $\mathrm{det}_{\mathbb{C}^2}(J_m^i)$. Letting $\mathrm{d}t$ go to zero, one obtains $\mathrm{d}\overline{C(t)}/\mathrm{d}t = -\kappa_{\mathrm{QJ}}\, \overline{C(t)}$. The solution of this differential equation is given by (1). Note that $\kappa_{\mathrm{QJ}} = 0$ if all matrices J_m^i are self-adjoint and traceless. This is the case e.g. for pure dephasing (jump operators $J^i = \vec{v}_i \cdot \vec{\sigma}^i$ with $\vec{v} \in \mathbb{R}^3$ and $\vec{\sigma}$ the vector formed by the three Pauli matrices σ_x, σ_y and σ_z).

We can now give the optimal measurement scheme to protect the entanglement of two qubits coupled to independent baths at positive temperatures. Let us replace the photon-counting jump operators $J^i_\pm = \sigma^i_\pm$ by $I^i_\mu = \sum_{m=\pm} (\gamma^i_m/\gamma^i_\mu)^{\frac{1}{2}} u^i_{\mu,m} \sigma^i_m$ where $U_i = (u^i_{\mu,m})^{m=\pm}_{\mu=1,\cdots,N}$ are unitary $2\times N$ matrices. This corresponds to a rotation of the measurement basis and gives another unraveling of the master equation (6). Let us stress that the new jump operators I^i_μ still act locally on each qubits. Using (18), $\sum_\mu |u^i_{\mu,\pm}|^2 = 1$, $\sum_\mu u_{\mu,+}{}^* u_{\mu,-} = 0$ and optimizing over all unitaries U_i, one finds that the smallest disentanglement rate arises for e.g. $u^i_{1,\pm} = \pm u^i_{2,\pm} = 1/\sqrt{2}$ ($N=2$) and is given by

$$\kappa^{\rm opt}_{\rm QJ} = \frac{1}{2} \sum_{i=A,B} \left(\sqrt{\gamma^i_-} - \sqrt{\gamma^i_+} \right)^2 . \tag{19}$$

The decay of $\overline{C_{\psi(t)}}$ for this optimal measurement scheme is shown in Fig. 2. Note that $\kappa^{\rm opt}_{\rm QJ} = \kappa_{\rm QJ}$ at zero temperature and $\kappa^{\rm opt}_{\rm QJ} = 0$ (perfect protection) at infinite temperature.

6. Homodyne detection

Let us come back to our example of two atoms coupled to the electromagnetic field initially in the vacuum. If homodyne photo-detection is used instead of photon counting, the jump operators become $J^i_{\pm\alpha} = \sigma^i_- \pm \alpha_i$ with the rates $\gamma^i_{\pm\alpha} = \gamma_i/2$, α_i being the amplitudes of the classical laser fields (there are now four jump operators since each homodyne detection involves two photon counters) [25]. By applying (18), we find that the disentanglement rate for homodyne detection is the same as for photon counting, $\kappa_{\rm QJ}(\alpha) = (\gamma_A + \gamma_B)/2$.

Let us now consider a general QJ model with jump operators J^i_m. A new unraveling of the master equation (6) is obtained from the new QJ dynamics with jump operators $J^i_{m,\pm\alpha} = J^i_m \pm \alpha^i_m$ and rates $\gamma^i_{m,\pm\alpha} = \gamma^i_m/2$. For large positive laser amplitudes $\alpha^i_m \gg 1$, this dynamics converges after an appropriate coarse graining in time to the quantum state diffusion (QSD) model described by the stochastic differential equation [12]

$$\begin{aligned} |{\rm d}\psi\rangle = & \Bigg[(-{\rm i}H_0 - K){\rm d}t + \sum_{m,i} \gamma^i_m \left(\Re\langle J^i_m\rangle_\psi \, J^i_m - \frac{(\Re\langle J^i_m\rangle_\psi)^2}{2} \right) {\rm d}t \\ & + \sum_{m,i} \sqrt{\gamma^i_m} (J^i_m - \Re\langle J^i_m\rangle_\psi) {\rm d}w^i_m \Bigg] |\psi\rangle \end{aligned} \tag{20}$$

where ${\rm d}w^i_m$ are the Itô differentials for independent real Wiener processes satisfying the Itô rules ${\rm d}w^i_m {\rm d}w^j_n = \delta_{ij}\delta_{mn}{\rm d}t$.

It is shown in [4] by taking the limit $\alpha^i_m \gg 1$ of the mean concurrence for the quantum jump dynamics with jump operators $J^i_{m,\pm\alpha}$ that the concurrence for the QSD model is still given by Eq. (1) but with a new rate

$$\kappa_{\rm ho} = \frac{{\rm tr}_{\mathbb{C}^4}(K)}{2} - \sum_{m,i} \gamma^i_m \left(\Re \det_{\mathbb{C}^2}(J^i_m) + \frac{(\Im \, {\rm tr}_{\mathbb{C}^2}(J^i_m))^2}{2} \right) . \tag{21}$$

Unlike $\kappa_{\rm QJ}$, $\kappa_{\rm ho}$ changes when the operators J^i_m in (20) acquire a phase factor $e^{-{\rm i}\theta^i_m}$. This arises for homodyne detection with complex laser amplitudes $\alpha^i_m = |\alpha^i_m| e^{{\rm i}\theta^i_m}$ as $|\alpha^i_m| \gg 1$. Minimizing over all laser phases θ^i_m yields

$$\kappa^{\rm opt}_{\rm ho} = \frac{1}{2}{\rm tr}_{\mathbb{C}^4}(K) - \sum_{m,i} \gamma^i_m \left(\left| \det_{\mathbb{C}^2}(J^i_m) - \frac{1}{4}[{\rm tr}_{\mathbb{C}^2}(J^i_m)]^2 \right| + \frac{1}{4} \left| {\rm tr}_{\mathbb{C}^2}(J^i_m) \right|^2 \right) . \tag{22}$$

Using the inequality $|a-b^2|+|b|^2 \geq |a|$ for $a, b \in \mathbb{C}$, we have

$$\kappa_{\rm ho}^{\rm opt} \leq \kappa_{\rm QJ} \,. \tag{23}$$

Thus optimal homodyne detection protects entanglement better than or as well as photon counting.

7. Heterodyne detection

This case is very similar to the previous one. The corresponding jump operators $J^i_{m,\pm\alpha}(t_q) = J^i_m \pm \alpha^i_m e^{i\Omega^i_m t_q}$ depend on the time t_q of occurrence of the qth jump [25] and the associated rates are $\gamma^i_{m,\pm\alpha} = \gamma^i_m/2$.

For $\alpha^i_m > 0$, in the limit $(\alpha^i_m)^2 >> \Omega^i_m/\gamma^i_m >> 1$ of large laser intensities and rapidly oscillating laser amplitudes, and with an appropriate coarse graining in time, the QJ dynamics with jump operators $J^i_{m,\pm\alpha}(t_q)$ converges to the QSD model given by the stochastic differential equation [12, 23]

$$\begin{aligned}|{\rm d}\psi\rangle = \; & \Bigg[(-{\rm i}H_0 - K){\rm d}t + \frac{1}{2}\sum_{m,i} \gamma^i_m \left(\langle J^i_m \rangle^*_\psi J^i_m - \frac{|\langle J^i_m\rangle_\psi|^2}{2} \right) {\rm d}t \\ & + \sum_{m,i} \sqrt{\gamma^i_m} \left((J^i_m - \frac{1}{2}\langle J^i_m\rangle_\psi){\rm d}\xi^i_m - \frac{1}{2}\langle J^i_m\rangle^*_\psi {\rm d}\xi^{i\,*}_m \right) \Bigg] |\psi\rangle\end{aligned} \tag{24}$$

where ${\rm d}\xi^i_m$ are the Itô differential of independent complex Wiener processes satisfying the Itô rules ${\rm d}\xi^i_m {\rm d}\xi^j_n = 0$ and ${\rm d}\xi^i_m {\rm d}\xi^{j\,*}_n = \delta_{ij}\delta_{mn}dt$.

It is proven in [4] that the mean concurrence is still given by Eq. (1) with the rate

$$\kappa_{\rm het} = \frac{{\rm tr}_{\mathbb{C}^4}(K)}{2} - \frac{1}{4}\sum_{m,i} \gamma^i_m |\,{\rm tr}_{\mathbb{C}^2}(J^i_m)|^2 \,. \tag{25}$$

We note that $\kappa_{\rm het} \geq \kappa^{\rm opt}_{\rm ho}$, so that the best measurement scheme to protect the qubits against disentanglement is given by homodyne detection with optimally chosen laser phases.

8. Qubits coupled to a common bath

If the qubits are closed from each other then the situation is different from that studied in Sec. 4 : the two qubits with equal frequencies are now coupled resonantly to the same modes of the electromagnetic field. We assume that the field is initially in the vacuum. A photon counter D makes a click when a photon is emitted by qubit A or B. The jump operator $J = \sigma_- \otimes 1_B + 1_A \otimes \sigma_-$ is now nonlocal. We take $H_0 = 0$. Proceeding as for independent baths, the contribution to the mean concurrence of QTs with no jump between 0 and t is $p_{\rm nj}(0,t)C_{\rm n.j.}(t) = |\langle e^{-Kt}\sigma_y \otimes \sigma_y T e^{-Kt}\rangle_{\psi(0)}|$ and can be calculated with the help of (14).

QTs having one jump between 0 and t gives a contribution which can be found by multiplying $C_{1{\rm j},t_j}(t) = 2\mathcal{N}^{-2}_{1{\rm j},t_j}(t)e^{-2\gamma t}|c_{\uparrow\uparrow}|^2$ by the probability density that the jump occurs at time $t_j \in [0,t]$, given by $\gamma p_{\rm nj}(t_j,t)\|J|\psi(t_j-)\rangle\|^2 p_{\rm nj}(0,t_j) = \gamma\mathcal{N}_{1{\rm j},t_j}(t)^2 = \gamma\,\|e^{-K(t-t_j)}Je^{-Kt_j}|\psi(0)\rangle\|^2$, and integrating over t_j from 0 to t.

After two clicks, $|\psi(t)\rangle = |\downarrow\downarrow\rangle$ is in an invariant separable state. Therefore, trajectories with more than one jump do not contribute to the mean concurrence.

Setting $c_\pm = c_{\uparrow\downarrow} \pm c_{\downarrow\uparrow}$, one gets [4]

$$\overline{C(t)} = \frac{1}{2}|c_-^2 - c_+^2 e^{-2\gamma t} + 4c_{\uparrow\uparrow}c_{\downarrow\downarrow}e^{-\gamma t}| + 2|c_{\uparrow\uparrow}|^2\gamma t e^{-2\gamma t}. \tag{26}$$

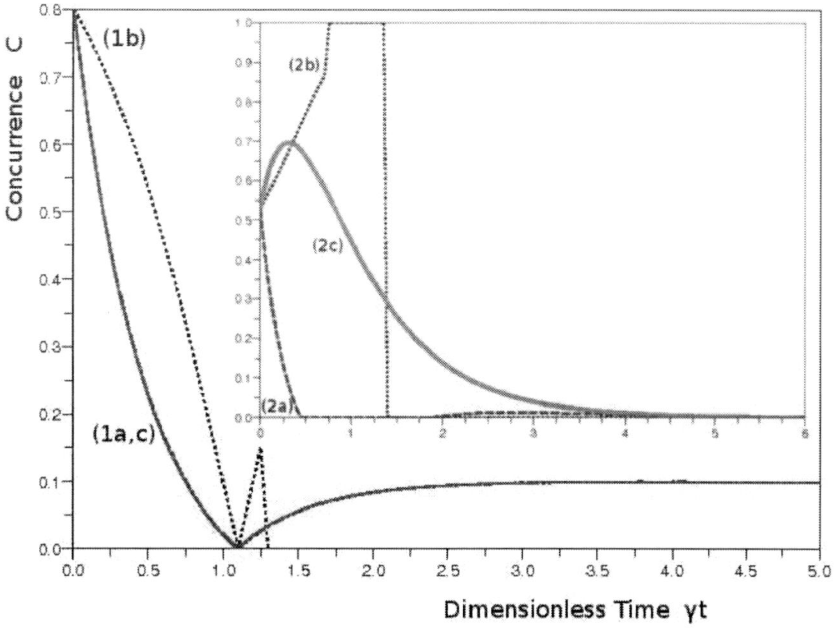

Figure 3. Concurrence of two qubits coupled to a common bath at zero temperature as a function of γt with $|\Psi(0)\rangle = \frac{2}{\sqrt{5}}|\uparrow\uparrow\rangle + \frac{1}{\sqrt{5}}|\downarrow\downarrow\rangle$.
(1a) $C_{\rho(t)}$ (blue dashed lines).
(1b) $C_{\psi(t)}$ for a single trajectory (black dotted line).
(1c) $\overline{C_{\psi(t)}}$ given by (26) (red plain line superposed on the blue line).
Inset (2) is the same for $|\Psi(0)\rangle = \frac{7i}{\sqrt{53}}|\uparrow\uparrow\rangle + \frac{2i}{\sqrt{53}}|\downarrow\downarrow\rangle$.

The time behavior of the concurrence (26) depends strongly on the initial state. For $c_{\uparrow\uparrow} = 0$ and $\arg(c_{\uparrow\downarrow}) = \arg(c_{\downarrow\uparrow})$, $\overline{C(t)} = 0$ if and only if $t = \gamma^{-1}\ln(c_+/c_-)$. Thus entanglement is lost after a finite time but this loss is immediately followed by a revival.

For any initial state such that $c_{\uparrow\uparrow} = 0$, $\overline{C(t)} = \frac{1}{2}|c_-^2 - c_+^2 e^{-2\gamma t}|$ and it can be shown that $\overline{C(t)}$ coincides with $C_{\rho(t)}$. In contrast, if $c_{\uparrow\uparrow} \neq 0$ then $C_{\rho(t)}$ and $\overline{C(t)}$ have very different behaviors as we can see in Fig. 3. Finally, $\overline{C(t)}$ converges at large times to the same limit $C_\infty = |c_-|^2/2$ as the concurrence $C_{\rho(t)}$ [16, 26].

9. Conclusion

We have found explicit formulas for the mean concurrence of quantum trajectories and have shown that the measurements on the baths may be used to protect the entanglement of two qubits. These results shed new light on the phenomenon of ESD. It should be possible to check our findings experimentally by employing similar optical devices as in Ref. [6].

Note: after the completion of this work we learned that related results have been obtained in [27].

Acknowledgments

We thank P. Degiovanni and A. Joye for interesting discussions. We acknowledge financial support from the Agence Nationale de la Recherche (Grant No. ANR-09-BLAN-0098-01).

References

[1] Diósi L 2003 *Irreversible Quantum Dynamics* (Lectures Notes in Physics vol 622) ed F Benatti and R Floreanini (Berlin: Springer) p 157
[2] Dodd P J and Halliwell J J 2004 *Phys. Rev.* A **69** 052105
[3] Yu T and Eberly J H 2004 *Phys. Rev. Lett.* **93** 140404
[4] Vogelsbeger S and Spehner D 2010 *Phys. Rev.* A **82** 052327
[5] Schrödinger E 1935 *Proceedings of the Cambridge Philosophical Society* **31** 555
[6] Almeida M P *et al* 2007 *Science* **316** 579
[7] Braun D 2002 *Phys. Rev. Lett.* **89** 277901
[8] Ficek Z and Tanás R 2006 *Phys. Rev.* A **74** 024304
[9] Mazzola L *et al* 2009 *Phys. Rev.* A **79** 042302
[10] Dalibard J, Castin Y and Mølmer K 1992 *Phys. Rev. Lett.* **68** 580
[11] Carmichael H J 1993 *An Open System Approach to Quantum Optics* (Lecture Notes in Physics m18) (Berlin: Springer)
[12] Wiseman H M and Milburn G H 1993 *Phys. Rev.* A **47** 642
[13] Gisin N and Percival I C 1992 *J. Phys A: Math. Gen.* **25** 5677
[14] Nha H and Carmichael H J 2004 *Phys. Rev. Lett.* **93** 120408
[15] Carvalho A R R *et al* 2007 *Phys. Rev. Lett.* **98** 190501
[16] Maniscalco S *et al* 2008 *Phys. Rev. Lett.* **100** 090503
[17] Mundarain D M and Orszag M 2009 *Phys. Rev.* A **79** 052333
[18] Carvalho A R R and Hope J J 2007 *Phys. Rev.* A **76** 010301(R)
[19] Mascarenhas E *et al* 2010 *Phys. Rev.* A **81** 032310
[20] Viviescas C, Guevara I, Carvalho A R R, Busse M and Buchleitner A 2010 *Phys. Rev. Lett.* **105** 210502
[21] Bennett C H *et al* 1996 *Phys. Rev.* A **54** 3824
[22] Wootters W K 1998 *Phys. Rev. Lett.* **80** 2245
[23] Breuer H-P and Petruccione F 2002 *The Theory of Open Quantum Systems* (Oxford: Oxford University Press)
[24] Haroche S and Raimond J-M 2006 *Exploring the Quantum: Atoms, Cavities and Photons* (Oxford: Oxford University Press)
[25] Plenio M B and Knight P L 1998 *Rev. Mod. Phys.* **70** 101
[26] Orszag M and Hernandez M 2010 *Adv. Opt. Photon* **2** 229
[27] Mascarenhas E, Cavalcanti D, Vedral V and França Santos M 2010 *Preprint* arXiv:1006.1233v1 [quant-ph]

Quantum discord in quantum computation

Aharon Brodutch[1], Alexei Gilchrist[1], Daniel R. Terno[1], and Christopher J. Wood[2]

[1]Centre for Quantum Computer Technology, Department of Physics and Astronomy, Macquarie University, Sydney NSW 2113, Australia

[2] Institute for Quantum Computing, University of Waterloo, 200 University Av. West Waterloo Ontario N2L 3G1, Canada

E-mail: aharon.brodutch@mq.edu.au

Abstract. Quantum discord is a measure of the quantumness of correlations. After reviewing its different versions and properties, we apply it to the questions of quantum information processing. First we show that changes in discord in the processed unentangled states indicate the need for entanglement in the distributed implementation of quantum gates. On the other hand, it was shown that zero system-environment discord is a necessary and sufficient condition for applicability of the standard completely positive description of the system's evolution. We demonstrate that this result does not translate into useful quantum process tomography. Depending on the details of the preparation procedure only absence of any initial correlations may guarantees consistency of the process tomography.

1. Introduction

What makes correlations quantum? What makes quantum computer tick? There is no definite answer to these questions, even if entanglement is a primary suspect. It is crucial in performing tasks that cannot be implemented by local operations and classical communications (LOCC), its presence in the system is a natural telltale that something non-local is aground, and one entangling gate is a part of universal sets of gates [1, 2].

Nevertheless, the story is not so simple. The "sausage" or "four and a half tatami teahouse" states [3] are nine orthogonal pure product states that are shared between the two parties — A(lice) and B(ob). Despite absence of entanglement Alice and Bob cannot distinguish between them by LOCC. Similarly the DQC1 algorithm demonstrates speed up without entanglement [4].

Quantum discord [5, 6] is one of the measures of quantum correlations. We review its three versions and their properties in the next Section. Since its introduction it was found to be related to the efficiency of various Maxwell's demons [7, 8], quantum phase transitions [9, 10], and quantum state merging [11, 12]. It was also argued that zero discord is a necessary and sufficient condition for the evolution of a system to be of a completely positive type [14, 15], thus being amenable to a relatively simple characterization, in quantum tomography.

We present two of its uses. First, ability to create entanglement is necessary for execution of two-qubit quantum gates even when they are applied to unentangled states and create no entanglement. Changes in discord indicate the failure of a local implementation. Second, having a zero system-environment discord does not guarantee that the resulting tomographic description is an adequate representation of the relevant gate action. Hence zero discord is not directly related to a "simple" evolution of the system.

2. Three types of discord

2.1. Correlations and entropies

The Shannon entropy [16] of a classical discrete probability distribution over a random variable A with values a, $p(a) \equiv p_a$, is defined by $H(A) = -\sum_a p_a \log p_a$. Entropy of the joint probability distribution over AB, $H(AB)$, is defined analogously. The Bayes theorem relates joint and conditional probabilities,

$$p(a,b) = p(a|b)p(b) = p(b|a)p(a), \tag{1}$$

where $p(a|b)$ is a conditional probability of $A = a$ given that $B = b$. The conditional entropy of A given B,

$$H(A|B) = \sum_b p_b H(A|b) = -\sum_{a,b} p(a,b) \log p(a|b) \tag{2}$$

is a weighted average of the entropies of A given a particular outcome of B.

Correlations between two probability distributions are measured by the symmetric mutual information. It has two equivalent expressions [16],

$$I(A:B) = H(A) + H(B) - H(A,B)\;;J(A:B) = H(A) - H(A|B) = H(B) - H(B|A). \tag{3}$$

Quantum-mechanical (von Neumann) entropy [1, 2, 17] of a system with a density operator ρ is defined as $S(\rho) = -\mathrm{tr}\,\rho \log \rho$. It minimizes the Shannon entropy of probability distributions that result from rank-1 positive operator-valued measures (POVMs) that are applied to the state ρ. The minimum is reached on a probability distribution A_ρ^Π that results from a projective measurement $\Pi = \{\Pi_a, a = 1, \ldots d\}$, $\sum_a \Pi_a = \mathbb{1}$, $\Pi_a \Pi_b = \delta_{ab}\Pi_a$, which is constructed from the eigenstates of ρ,

$$S(\rho) = \min_\Pi H(A_\rho^\Pi), \tag{4}$$

i. e.,

$$S(\rho) = H(A_\rho^{\Pi^*}), \qquad \rho = \sum_a p_a \Pi_a^*, \qquad p_a \geq 0, \quad \sum_a p_a = 1. \tag{5}$$

The first expression for mutual information has an obvious quantum generalization,

$$I(\rho_{AB}) := S(\rho_A) + S(\rho_B) - S(\rho_{AB}), \tag{6}$$

and represents the total amount of quantum and classical correlations [18].

To obtain a quantum version of $J(A:B)$, it is necessary to determine a conditional state of the subsystem B [8]. Given a complete projective measurement Π on A, a quantum definition of J follows its interpretation as the information gained about the system B from the measurement on A [5],

$$J^{\Pi^A}(\rho_{AB}) := S(\rho_B) - S(\rho_B|\Pi^A), \tag{7}$$

where the conditional entropy is now given by $S(\rho_B|\Pi^A) := \sum_a p_a S(\rho_{B|\Pi_a})$.

The post-measurement state of B that corresponds to the outcome $A = a$ is

$$\rho_{B|\Pi_a} = (\Pi_a \otimes \mathbb{1}_B \rho_{AB} \Pi_a \otimes \mathbb{1}_B)/p_a, \qquad p_a = \mathrm{tr}\,\rho_A \Pi_a., \tag{8}$$

while the state of B remains unchanged, $\rho_B = \mathrm{tr}_A \rho_{AB} = \sum_a p_a \rho_{B|\Pi_a}$.

2.2. Discords D_1 and D_2

Unlike their classical counterparts, the quantum expressions are generally inequivalent and $I(\rho_{AB}) \geq J^{\Pi^A}(\rho_{AB})$ [5, 6]. The difference between these two quantities is

$$D_1^{\Pi^A}(\rho_{AB}) := S(\rho_A) + S(\rho_B|\Pi^A) - S(\rho_{AB}). \quad (9)$$

Its dependence on the measurement procedure is removed by minimizing the result over all possible sets of Π, resulting in the quantum discord [5]

$$D_1^A(\rho_{AB}) := \min_{\Pi^A} D_1^{\Pi^A}(\rho_{AB}). \quad (10)$$

Similarly,

$$J_1^A(\rho_{AB}) := \max_{\Pi^A} J^{\Pi^A}(\rho_{AB}). \quad (11)$$

It is possible to define the discord when the difference is minimized over all possible POVM Λ^A [6]. However, unless stated otherwise we restrict ourselves to rank 1 projective measurements.

An explicit form of a post-measurement state will be useful in the following text. We denote this state as $\rho'_X \equiv \rho_X^{\Pi^A}$, where the subscript X stands for A, B, or AB, and use the former expression if it does not lead to confusion. After a projective measurement Π^A, the state of the system becomes

$$\rho'_{AB} = \sum_a p_a \Pi_a \otimes \rho_B^a, \quad (12)$$

where p_a and $\rho_B^a \equiv \rho_{B|\Pi_a}$ are given by Eq. (8).

The discord of the state ρ_{AB} is zero if and only if it is a mixture of products of arbitrary states of B and projectors on A [5],

$$\rho_{AB} = \sum_a p_a \Pi_a \otimes \rho_B^a, \qquad p_a \geq 0, \quad \sum_a p_a = 1. \quad (13)$$

By using this decomposition and properties of the entropy of block-diagonal matrices [20] we can identify $J^{\Pi^A}(\rho_{AB}) \equiv I(\rho'_{AB})$, because $S(\rho'_A) = H(A_\rho^\Pi)$ and

$$S(\rho'_{AB}) = H(A_\rho^\Pi) + S(\rho_B|\Pi^A). \quad (14)$$

The discord is not a symmetric quantity: it is possible to have states with $D_1^A(\rho_{AB}) \neq D_1^B(\rho_{AB})$.

Another possibility is to set

$$J_2^{\Pi^A} := S(\rho_A) + S(\rho_B) - [H(A_\rho^\Pi) + S(\rho_B|\Pi^A)] = S(\rho_A) + S(\rho_B) - S(\rho'_{AB}), \quad (15)$$

arriving to the quantum discord as defined in [7]

$$D_2^A(\rho_{AB}) := \min_{\Pi}[H(A_\rho^\Pi) + S(\rho_B|\Pi^A)] - S(\rho_{AB}), \quad (16)$$

where the quantity to be optimized is a sum of post-measurement entropies of A and B. Using Eq. (4) we see that $D_1 \leq D_2$. It is also easy to see that $D_1 = 0 \Leftrightarrow D_2 = 0$. Using Eqs. (12) and (14) we obtain a different expression for D_2:

$$D_2^{\Pi^A}(\rho_{AB}) = S(\rho_{AB}^{\Pi^A}) - S(\rho_{AB}). \quad (17)$$

Since the definition of the discord(s) involves optimization, the analytic expressions are known only in some particular cases. Moreover, typically it is important to know whether the discord is zero or not, while the numerical value itself is less significant.

2.3. Zero discord and D_3

It follows from Eq. (13) that if the spectrum of a reduced state $\rho_A = \sum_a p_a \Pi_a$ is non-degenerate, then its eigenbasis gives a unique family of projectors Π that results in the zero discord for ρ_{AB}. Hence, a recipe for testing states for zero discord and for finding the optimal basis is to trace out a subsystem that is left alone (B), to diagonalize ρ_A and to calculate the discord in the resulting eigenbasis.

If the state ρ_A is degenerate, a full diagonalization should be used. For the state with the form of Eq. (13), each of the reduced states ρ_B^a can be diagonalized as

$$\rho_B^a = \sum_b r_b^a P_a^b, \qquad P_a^b P_a^{b'} = \delta^{bb'} P_a^b. \tag{18}$$

The eigendecomposition of the state ρ_{AB} then easily follows. Writing it as

$$\rho_{AB} = \sum_{a,b} w_a r_b^a \Pi_a \otimes P_a^b, \tag{19}$$

it is immediate to see that its eigenprojectors are given by $\Pi_a \otimes P_a^b$. Hence, if ρ_B has a degenerate spectrum, but ρ_{AB} does not, the structure of its eigenvectors reveals if it is of a zero or nonzero discord. Hence we established [8]

Property 1. The eigenvectors of a zero discord state $D_1^A(\rho_{AB}) = 0$ satisfy

$$\rho_{AB}|ab\rangle = r_{ab}|ab\rangle, \quad \Rightarrow \quad |ab\rangle\langle ab| = \Pi_a \otimes P_a^b. \tag{20}$$

□

This consideration leads to the simplest necessary condition for zero discord [19]:

Property 2. If $D_1^A(\rho_{AB}) = 0$, then $[\rho_A \otimes \mathbb{1}_B, \rho_{AB}] = 0$. Hence a non-zero commutator implies $D_1^A(\rho_{AB}) > 0$. □

Naturally, if the state has zero discord, and the eigenbasis is only partially degenerated, we can use it to reduce the optimization space. On the other hand, the diagonalizing basis Π_* is not necessarily the optimal basis $\hat{\Pi}$ or $\check{\Pi}$ that enters the definition of D_1 or D_2, respectively. Consider, for example, a two-qubit state

$$\rho_{AB} = \tfrac{1}{4}(\mathbb{1}_{AB} + b\sigma_A^z \otimes \mathbb{1}_B + c\sigma_A^x \otimes \sigma_B^x), \tag{21}$$

where σ_X^a are Pauli matrices on the relevant spaces, $X = A, B$, and the constants b and c are restricted only by the requirements that ρ_{AB} is a valid density matrix. For this state $\rho_B = \mathbb{1}/2$ and $\rho_A = \mathrm{diag}(1+b, 1-b)/2$. After the measurement in the diagonalizing basis $\Pi^z = ((\mathbb{1}+\sigma^z)/2, (1-\sigma^z)/2)$ the conditional state of B becomes

$$\rho_{B|\Pi_\pm^z} = \mathbb{1}/2, \tag{22}$$

and the conditional entropy is maximal, $S(\rho_B|\Pi^z) = \log 2$.

On the other hand, in the basis $\Pi^x = ((\mathbb{1}+\sigma^x)/2, (1-\sigma^x)/2)$ the probabilities of the outcomes are equal, $p_+ = p_- = 1/2$, but the post-measurement states of B are different from the maximally mixed one,

$$\rho_{B|\Pi_\pm^x} = \tfrac{1}{2}(\mathbb{1} \pm c\sigma^x), \tag{23}$$

so the entropy $S(\rho_B|\Pi^z) \geq S(\rho_B|\Pi^x)$.

This discrepancy motivates us to define a new version of the discord [8]:

$$D_3^A(\rho_{AB}) := S(\rho_A) - S(\rho_{AB}) + S(\rho_B|\Pi_*^A), \tag{24}$$

where Π_*^A is the set of eigenprojectors of ρ_A. For the degenerate case it can be introduced using the continuity of entropy in finite-dimensional systems [20]. By applying Eq. (4) to the subsystem A we find that D_3 simultaneously holds the analog of Eq. (2.2),

$$J_3(\rho_{AB}) \equiv I(\rho_{AB}^{\Pi_*^A}), \tag{25}$$

and of Eq. (17),

$$D_3(\rho_{AB}) \equiv S(\rho_{AB}^{\Pi_*^A}) - S(\rho_{AB}). \tag{26}$$

We also arrive to the following ordering of the discord measures:

$$D_1^A \leq D_2^A \leq D_3^A. \tag{27}$$

There are several important cases when the measures of discord coincide. Most importantly, they vanish simultaneously:

Property 3. $D_1 = 0 \Leftrightarrow D_2 = 0 \Leftrightarrow D_3 = 0$.
The proof follows from Eqs.(13) and (27). □.

For pure states the discord is equal to the degree of entanglement,

$$D_i^A(\phi_{AB}) = S(\phi_A) = E(\phi_{AB}), \qquad i = 1, 2, 3. \tag{28}$$

Discord is also independent of the basis of measurement if the state is invariant under local rotations [5]. Finally, if A is in a maximally mixed state, then $D_1^A = D_2^A$.

These coincidences make it interesting to check when the discords D_1 and D_2 are different. By returning to the measurement-dependent versions of the discords, we see that

$$D_1^{\Pi^A}(\rho_{AB}) = D_2^{\Pi^A}(\rho_{AB}) - [H(A_\rho^\Pi) - S(\rho_A)]. \tag{29}$$

Assume that $D_2^{\Pi^A}(\rho_{AB})$ reaches the minimum on the set of projectors $\check{\Pi}$, which are not the eigenprojectors of ρ_A. In this case $H(A_\rho^{\check{\Pi}}) - S(\rho_A) > 0$, so we can conclude that the strict inequality $D_1^A < D_2^A$ holds, because

$$\begin{aligned} D_1^A(\rho_{AB}) &\leq D_1^{\check{\Pi}}(\rho_{AB}) \\ &= D_2^A(\rho_{AB}) - [H(A_\rho^{\check{\Pi}}) - S(\rho_A)] < D_2^A(\rho_{AB}). \end{aligned} \tag{30}$$

For example, the state of Eq. (21), with $b = c = \frac{1}{2}$, satisfies $D_1^A \approx 0.05$, $D_2^A \approx 0.20$ and $D_3^A \approx 0.21$.

2.4. Some applications

The discord $D_1^A(\rho)$ has an operational meaning through quantum state merging [11, 12]. Different restrictions on the local Maxwell demons that operate on the subsystems, as compared to the power of a global demon, lead to the differences in maximal work extraction that are determined by the discords D_2^A and D_3^A [7, 8]. Further work on the discord in quantum open systems can be found in [13].

Since zero discord is thought to represent absence of quantum correlations, it is interesting to investigate the following question. Consider a set of pure orthogonal bipartite states, each of which may have a different prior probability, with the ensemble density matrix ρ_{AB}. Does the value of $D(\rho_{AB})$ tell us something about the ability to perfectly distinguish these states by local operations and classical communication (LOCC)? The answer is negative [8]: there is no relation between $D(\rho_{AB})$ and local distinguishability, as illustrated in Table 1. We use the nine teahouse states

$$|1\rangle \otimes |1\rangle, \quad |0\rangle \otimes |0 \pm 1\rangle/\sqrt{2}, \quad |2\rangle \otimes |1 \pm 2\rangle/\sqrt{2}, |1 \pm 2\rangle \otimes |0\rangle/\sqrt{2}, \quad |0 \pm 1\rangle|2\rangle/\sqrt{2}, \tag{31}$$

of [3], and the result [21] that any two orthogonal (entangled or not) states can be perfectly distinguished by LOCC to compile the table.

Table 1. Local measurability vs. discord

States	Discord	Locally Distiguishable
9 teahouse states, equal weights	$D^A = D^B = 0$	no
2 product bi-orthogonal states	$D^A = D^B = 0$	yes
2 entangled orthogonal states	$D_1^A > 0$	yes
9 teahouse states, unequal weights	$D_1^A > 0$	no

3. Restricted quantum gates

It is usually thought that quantum computers are potentially faster then their classical counterparts because of the ability to create and use entanglement during the computation [22]. Nevertheless, recent results indicate that it is discord rather than entanglement which is responsible for the speed-up [4, 23, 24, 25].

Using a simple example we show how changes in discord indicate the need for entanglement as a resource even if the processed states are unentangled. More general results may be found in [23]. The non-classicality of quantum gates can be examined by the amount of quantum resources required for the operation of the gate. Our method is to analyze the resources that are needed for a bi-local implementation of a gate.

A CNOT gate can be implemented bi-locally by Alice and Bob if they share one ebit of entanglement per gate use [26]. If they can perform only LOCC they cannot implement this gate even on a restricted set of unentagled inputs that are transformed into unentagled outputs. One such set $\mathfrak{L}$ is in Table 2.

Table 2. Four inputs/outputs for the CNOT gate

#	State	#	State
a	$\|1\rangle\|Y_+\rangle \to i\|1\rangle\|Y_-\rangle$	c	$\|Y_+\rangle\|X_-\rangle \to \|Y_-\rangle\|X_-\rangle$
b	$\|0\rangle\|Y_+\rangle \to \|0\rangle\|Y_+\rangle$	d	$\|Y_+\rangle\|X_+\rangle \to \|Y_+\rangle\|X_+\rangle$

Here $\sigma_y|Y_\pm\rangle = \pm|Y_\pm\rangle$, $\sigma_x|X_\pm\rangle = \pm|X_\pm\rangle$, where $\sigma_{x,y,z}$ are Pauli matrices.

We show that the ability to implement the CNOT gate on $\mathfrak{L}$ without shared entanglement allows one to unambiguously discriminate between these non-orthogonal states using just one input copy, which is impossible [17]. Without specifying the local operations of Alice and Bob we classify them according to their action on the state $|Y_+\rangle$. An operation Φ is *flipping* (F) if up to a phase $\Phi(|Y_+\rangle) = |Y_-\rangle$, *non-flipping* (N) if $\Phi(|Y_+\rangle) = |Y_+\rangle$, and undetermined otherwise.

Knowing the type of the operation allows Alice and Bob to narrow down the list of the possible inputs: e.g., Bob's F is incompatible with having input b, while for Alice's operation not to have a definite type excludes both c and d. If one of the operations done is neither F nor N, then the type of other operations allows to determine the input uniquely.

Any pair of outputs can be reset to the original inputs by local unitaries and resent through the gate. For example, if the overall operation was of FF type, the operation $\sigma_z^A \otimes \sigma_x^B$ transforms the outputs $\psi_a' = |1\rangle|Y_-\rangle$ and $\psi_c' = |Y_-\rangle|X_-\rangle$ into the inputs ψ_a and ψ_c.

The operations that implement the gate this time may be of the same type as before, or different. If the gate is such that there is a finite probability of having a different operation type, it will be realized after a finite number of trials. This other type (FN or NF in the above example) will uniquely specify the inputs. If the gate's design is such that a particular pair of inputs is always processed by the same type of operations, then the gate can be used to unambiguously distinguish between one state from this pair and at least one of the two remaining states in a single trial. □

The operation in the example changes the discord of the ensemble and has been extended in [23] to a more general setting involving a symmetrised type of discord defined as

$$D_2(\rho) := \min[D_2^A(\rho), D_2^B(\rho)] \neq 0. \tag{32}$$

It is however not yet clear if discord is indeed the best measure for this kind of non locality. It is noteworthy to mention that the algorithm described in [4] changes the discord of only one of the parties so that D_2 remains zero throughout the computation.

4. (Non) completely positive maps and discord

4.1. Completely positive maps and gate tomography

Environmental interactions are the major obstacle for practical quantum information processing. Together with the imperfect tailoring of gate Hamiltonians they are responsible for discrepancies between the ideal unitary gates and their experimental realizations. This is why characterization of quantum processes is an essential step in implementing quantum technology. The exposition below is based on [28]. Experimental realization of this discussion is currently in progress.

Usually the transformation of input to output states of an open system is described by completely positive (CP) maps. Any such map $\mathcal{E}(\rho)$ can be seen as a reduction of a unitary evolution of some initially uncorrelated system–environment state $\rho_{AB} = \rho_A \otimes \omega_B$, where the initial states of the system and the environment were ρ_A and ω_B, respectively. Under the joint evolution U_{AB} the state of the system is transformed to

$$\rho'_A \equiv \mathcal{E}(\rho_A) = \mathrm{tr}\,_B U \rho_{AB} U^\dagger = \sum_a M_a \rho_A M_a^\dagger, \tag{33}$$

where M_a are the Kraus representation matrices [1, 2]. Using the spectral decomposition $\omega = \sum_\nu p_\nu |\nu\rangle\langle\nu|$ one recovers the Kraus matrices from

$$\rho' = \sum_{\mu,\nu} \langle\mu|\sqrt{p_\nu}\, U|\nu\rangle\rho\langle\nu|\sqrt{p_\nu}\, U^\dagger|\mu\rangle. \tag{34}$$

This result is based on the absence of prior correlations between the system and the environment.

The action of a given gate is reconstructed using several methods of quantum process tomography. We focus on the standard process tomography, which consists in following. A set of input states $\{\rho_j\}$ is prepared and sent through the process $\mathcal{E}$. From the knowledge of the input and the reconstructed outputs $\mathcal{E}(\rho_k)$ states the process matrix χ [1, 27, 29] is reconstructed. If K_m, $m = 1, \ldots, d^2$ form the basis for operators acting on $\mathcal{H}_A$, $d = \dim \mathcal{H}_A$, then

$$\mathcal{E}(\rho) = \sum_{m,n} \chi_{mn} K_m \rho K_n^\dagger. \tag{35}$$

For a CP evolution all the eigenvalues of χ are non-negative, and the Kraus matrices are obtained as its (generalized) eigenvectors.

Various mathematical techniques are used to convert the relative frequencies of different measurement outcomes into the states $\mathcal{E}(\rho_k)$. Positivity of states ρ'_j and complete positivity of the process $\mathcal{E}$ are enforced, and its violations by the raw data are interpreted as a result of noise.

The combined state of a system (A) and its environment (B) can be represented in the Fano form

$$\rho_{AB} = \frac{1}{d_A d_B}\Big(\mathbb{1}_{AB} + \sum_i \alpha_i \sigma_i^A \otimes \mathbb{1}_B + \sum_j \beta_j \mathbb{1}_A \otimes \sigma_j^B + \sum_{ij} \gamma_{ij} \sigma_i^A \otimes \sigma_j^B\Big). \tag{36}$$

Here the σ_i^X, $i = 1, ..., d_X^2$ represent generators of SU(d_X), while the real vector $\vec{\alpha}$ (or $\vec{\alpha}$) of size $d_X^2 - 1$ is the generalized Bloch vector of the reduced density operator ρ_X. The correlations between subsystems A and B are characterized by

$$\Gamma_{ij} = (\gamma_{ij} - \alpha_i \beta_j)/d_A d_B. \tag{37}$$

Presence of correlations not only can lead to a non-CP evolution of the system, but blurs its boundary with the environment. Part of the controversy surrounding non-CP maps in literature [27] can be traced to this ambiguity as well as to the impossibility of unrestricted creation of arbitrary input states of the system (that are to form a tomographically complete basis in the standard CP paradigm) ρ_j *and* their correlations with the environment Γ_j.

This situation has an important bearing on the process tomography and action of quantum processing devices. In the following we show that while zero discord $D_A(\rho_{AB})$ ensures a complete positivity of the resulting evolution [14], it does not necessarily translate into a useful tomographic description of the ensuing evolution.

4.2. Role of preparations

State preparation is a vital aspect of quantum process tomography. When initial correlations are present in the input state, the preparation procedure used has considerable impact on the outcome. We discuss two such schemes: state preparation by a measurement and rotations and state preparation by projective measurements [30].

In the former case (Method 1) a single post-selected state that corresponds to a projection operator $\rho_H = \Pi_H$ is used. Once we have performed the projective measurement the required input states are obtained by applying the appropriate matrix from a set of unitary rotation matrices $\{R_a\}$ such that $R_a \rho_H R_a^\dagger = \rho_a$ where ρ_a is run over the tomographically complete set of input states.

Hence the preparation maps for Method 1 are given by

$$\mathcal{P}_a(\rho_{AB}^{\text{in}}) = \frac{1}{p_H} \left(R_a \Pi_H^A \otimes \mathbb{1}_B\right) \left(\rho_{AB}^{\text{in}}\right) \left(\Pi_H^A R_a^\dagger \otimes \mathbb{1}_B\right) = \rho_a \otimes \omega_H, \tag{38}$$

where $p_H = \text{tr}\,(\Pi_H^A \rho_A^{\text{in}})$ is the probability of detecting the outcome corresponding H, and R_j brings $|H\rangle\langle H|$ to one of the states $\rho_a \equiv |\psi_a\rangle\langle\psi_a|$. The post-preparation state of the environment is dependent on both the initial state ρ_{AB}^{in} *and* the measurement operator Π_H.

State preparation using *only* measurements (Method 2) utilizes the set of projectors $\{\Pi_a\}$ where $\Pi_a \equiv \rho_a$. Hence the preparation procedure for projective measurements is given by the collection of maps $\{\mathcal{P}_a\}$ where

$$\mathcal{P}_a(\rho_{AB}^{\text{in}}) = \frac{1}{p_a} \left(\Pi_a^A \otimes \mathbb{1}_B\right) \rho_{AB}^{\text{in}} \left(\Pi_a^A \otimes \mathbb{1}_B\right) = \rho_a \otimes \omega_a. \tag{39}$$

where $p_a = \text{tr}\,(\Pi_a^A \rho_A^{\text{in}})$ is the probability of detecting the outcome corresponding to Π_j. In this preparation scheme the state of the environment, ω_j depends on the probe state of the system, ρ_k, hence the process matrix will not necessarily be CP.

Lemma 1: Zero discord of the initial state does not guarantee a CP evolution in a tomographic procedure.

Consider the state

$$\rho_{AB}^{\text{in}} = \frac{1}{2} \left(|0\rangle\langle 0| \otimes |0\rangle\langle 0| + |1\rangle\langle 1| \otimes |1\rangle\langle 1|\right), \tag{40}$$

which satisfies even a stronger requirement $D^A(\rho_{AB}^{\text{in}}) = D^B(\rho_{AB}^{\text{in}}) = 0$. If the tomographic procedure is based on preparations by measurements, the resulting process matrix χ has the eigenvalues $(1 \pm \sqrt{3}/2, \pm\sqrt{3}/2)$ and hence the evolution is not CP. □

Preparation by measurement and rotations is guarantied to lead to a CP evolution. However, we should be concerned that the actual state on which the gate will operate is different from the ones prepared in this procedure, and hence the predictive power of the resulting process matrix χ will be low. Having a zero discord does not alleviate this difficulty, as shown below. We consider different reconstruction of process matrices for a generic fixed overall unitary U_{AB}.

Lemma 2: Reconstructed process matrices $\chi_{\mathcal{P}_a}$ that follow from the spanning set of measurement preparations $\{\mathcal{P}_a^A\}$ that are given by projectors Π_a^A, $a = 1, \ldots d_A^2 - 1$, applied to the same initial system–environment state $(\rho_{AB}^{\rm in})$, are equal if and only if this initial state is simply separable, $(\rho_{AB}^{\rm in}) = (\rho_A^{\rm in}) \otimes (\omega_B^{\rm in})$.

The sufficient part is obvious. To prove the necessary condition we recall that Kraus matrices form an eigendecomposition of the process matrix, hence from Eq. (34) it follows that all post-selected environmental states ω_k have to have the same spectrum. Hence up to a local reshuffling after the preparation by measurement Π_a

$$\rho_{AB}^{\rm in} \mapsto \mathcal{P}_a^A(\rho_{AB}^{\rm in}) = \Pi_a \otimes \omega_B^0. \tag{41}$$

Using Fano decomposition of Eq. (36) we find that for all a the post-measurement state of B is

$$\omega_B^0 = {\rm tr}\,_A \mathcal{P}_a^A(\rho_{AB}^{\rm in}) = \frac{1}{d_B}\big(\mathbb{1}_B + \sum_j \beta_j^{\rm in}\sigma_j^B\big) + \frac{1}{p_a}\sum_j \left(\sum_i \Gamma_{ij}^{\rm in}{\rm tr}\,(\Pi_a\sigma_i^A)\right)\sigma_j^B, \tag{42}$$

hence for any j

$$\sum_i \Gamma_{ij}^{\rm in}\langle\psi_a|\sigma_i|\psi_a\rangle = 0, \tag{43}$$

for all a in all sets of tomographically complete projectors. As a result, $\Gamma_{ij}^{\rm in} = 0$ so the initial state $\rho_{AB}^{\rm in}$ is indeed a direct product $\rho_A^{\rm in} \otimes \omega_B^0$. □

4.3. Examples

We illustrate the differences between the outcomes of Methods 1 and 2 by considering qubits as the system and environment and a CNOT gate with system A as the target being the overall unitary. Let the initial state of joint system AB to be the maximally entangled state $\rho_{\Phi^+} = |\Phi^+\rangle\langle\Phi^+|$ where $|\Phi^+\rangle = (|00\rangle + |11\rangle)/\sqrt{2}$.

The reconstructed process matrices are

$$\chi_1 = \begin{pmatrix} 1 & 0 & 0 & 1 \\ 0 & 0 & 0 & 0 \\ 0 & 0 & 0 & 0 \\ 1 & 0 & 0 & 1 \end{pmatrix}, \quad \chi_2 = \frac{1}{2}\begin{pmatrix} 2 & 0 & -1-i & 1 \\ 0 & 0 & 1 & 1+i \\ -1+i & 1 & 2 & 0 \\ 1 & 1-i & 0 & 0 \end{pmatrix}, \tag{44}$$

for Methods 1 and 2, respectively. As follows from Eq. (38), in the former case we have a CP evolution (actually it is an identity $\mathcal{E}_1 = \mathcal{I}$), while χ_2 on the other hand has eigenvalues $(1 + \frac{\sqrt{3}}{2}, -\frac{\sqrt{3}}{2}, \frac{\sqrt{3}}{2}, 1 - \frac{\sqrt{3}}{2})$ and hence the reconstructed process is not completely positive. To see why this happens consider the state of the environment after each preparation procedure. We have that all states are different,

$$\omega_H = \begin{pmatrix} 1 & 0 \\ 0 & 0 \end{pmatrix}, \ \omega_V = \begin{pmatrix} 0 & 0 \\ 0 & 1 \end{pmatrix}, \ \omega_D = \frac{1}{2}\begin{pmatrix} 1 & 1 \\ 1 & 1 \end{pmatrix}, \ \omega_R = \frac{1}{2}\begin{pmatrix} 1 & i \\ -i & 1 \end{pmatrix}. \tag{45}$$

While we used a maximally entangled input state, this result holds even for separable inputs. For example, for the input state $\rho_{AB}^{\rm in} = \frac{1}{2}(\rho_H \otimes \rho_A + \rho_D \otimes \rho_V)$, the process matrix χ_2 has the eigenvalues $\{1.642, 0.507, -0.253, 0.105\}$, so the evolution is still non-CP.

We would like to thank C. Rodriguez-Rosario and K. Modi for their helpful comments and suggestions.

References

[1] Nielsen M A and Chuang I L 2000 *Quantum Computation and Quantum Information* (Cambridge University Press)
[2] Bruß D and Leuchs G 2007 *Lectures on Quantum Information* (Wiley-VCH, Weinheim)
[3] Bennett C H, DiVincenzo D P, Fuchs C A, Mor T, Rains E, Shor P W, Smolin J A and Wootters W K 1999 *Phys. Rev. A* **59** 1070
[4] Datta A, Shaji A and Caves C M 2008 *Phys Rev Lett* **100** 050502
[5] Ollivier H and Zurek W H 2002 *Phys Rev Lett* **88** 017901
[6] Henderson L and Vedral V 2001 *J Phys A: Math Gen* **34** 6899–6905
[7] Zurek W H 2003 *Phys Rev A* **67** 012320
[8] Brodutch A and Terno D R 2010 *Phys Rev A* **81** 062103
[9] Dillenschneider R 2008 *Phys Rev B* **78** 224413
[10] Maziero J, Guzman H C, Céleri L C, Sarandy M S and Serra R M 2010 *Phys. Rev. A* **82** 012106
[11] Madhok V and Datta A 2010 *Preprint* arXive:1008.4135
[12] Cavalcanti D, Aolita L, Boixo S, Modi K, Piani M and Winter A 2010 *Preprint* arXive:1008.3205
[13] Cesar A. Rodriguez-Rosario, Gen Kimura, Hideki Imai, Alan Aspuru-Guzik 2010 *preprint* arXiv:1004.5405 ; Cesar A. Rodriguez-Rosario, E.C.G. Sudarshan 2008 *preprint* arXiv:0803.1183
[14] Rodríguez-Rosario C A, Modi K, Kuah A-m, Shaji A and Sudarshan E C G 2008 *J Phys A: Math Gen* **41** 205301
[15] Shabani A and Lidar D A 2009 Phys Rev Lett **102**, 100402
[16] Cover T M and Thomas J A 1991 *Elements of Information Theory* (John Wiley & Sons, New York)
[17] Peres A 1995 *Quantum Theory: Concepts and Methods* (Kluwer, Dordrecht)
[18] Horodecki M, Horodecki P, Horodecki R, Oppenheim J, Sen(De) A, Sen U and Synak-Radtke B 2005 *Phys. Rev. A* **71** 062307
[19] Ferraro A, Aolita L, Cavalcanti D, Cucchietti F M and Acín A 2010 *Phys. Rev. A* **81** 052318
[20] Wehrl A 1978 *Rev Mod Phys* **50** 221
[21] Walgate J, Short A J, Hardy L and Vedral V 2000 *Phys Rev Lett* **85** 4972
[22] Jozsa R and Linden N 2003 *Proc Roy Soc Lon A* **459** 2011
[23] Brodutch A and Terno D R 2011 *Phys Rev A* **83** 010301(R)
[24] Eastin B 2010 *Preprint* arXive:1006.4402
[25] Fanchini F F, Cornelio M F, de Oliveira M C and Caldeira A O 2010 *Preprint* arXive:1006.2460
[26] Eisert J, Jacobs K, Papadopoulos P and Plenio M B 2000 *Phys Rev A* **62** 052317
[27] Carteret H A, Terno D R and Życzkowski K 2008 *Phys Rev A* **77** 042113
[28] Wood C J 2009 *Non-Completely Positive Maps: Properties and Applications* (B Sc (Hon) thesis, Macquarie University) *Preprint* arXive:0911.3199
[29] Gilchrist A, Terno D R, Wood C J 2009 *Preprint* arXiv:0911.2539
[30] A. Kuah, K. Modi, C. Rodriguez-Rosario, and E. C. G. Sudarshan, 2007 *Phys Rev A* **76** , 042113

Experimental investigation of tunneling times using Bose-Einstein condensates

Donatella Ciampini[1,2], **Oliver Morsch**[2] **and Ennio Arimondo**[1,2]
[1] CNISM UdR Pisa, Dipartimento di Fisica 'E. Fermi', Lgo Pontecorvo 3, I-56127 Pisa, Italy
[2] INO-CNR, Dipartimento di Fisica 'E. Fermi', Lgo Pontecorvo 3, I-56127 Pisa, Italy

E-mail: ciampini@df.unipi.it

Abstract. The time it takes a quantum system to complete a tunneling event (which in the case of cross-barrier tunneling can be viewed as the time spent in a classically forbidden area) is related to the time required for a state to evolve to an orthogonal state, and an observation, i.e., a quantum mechanical projection on a particular basis, is required to distinguish one state from another. We have performed time-resolved measurements of Landau-Zener tunneling of Bose-Einstein condensates in accelerated optical lattices, clearly resolving the steplike time dependence of the band populations. The use of different protocols enabled us to access the tunneling probability, in two different bases, namely, the adiabatic basis and the diabatic basis. The adiabatic basis corresponds to the eigenstates of the lattice, and the diabatic one to the free-particle momentum eigenstates. Our findings pave the way towards more quantitative studies of the tunneling time for LZ transitions, which are of current interest in the context of optimal quantum control and the quantum speed limit.

1. Introduction

Tunneling is one of the hallmarks of quantum systems, and physical effects associated with quantum tunneling are important in many branches of science [1]. While the probability for quantum tunneling can be readily calculated for a variety of systems and has been measured experimentally to great accuracy, the time it takes a quantum system to complete a tunneling event is a much less well-defined notion. It is often hard to measure in experiment and in many cases still the subject of intense debate [2]. In this paper we address the problem of measuring the timescale associated with one of the conceptually simplest tunneling phenomena, Landau-Zener (LZ) tunneling.

Landau-Zener tunneling arises when two energy levels of a quantum system cross as a function of some parameter that varies in time. There is a possibility of a transition if the degeneracy at the level crossing is lifted by a coupling and the system is forced across the resulting avoided crossing by varying the parameter that determines the level separation. LZ tunneling was first studied theoretically in the early 1930's in the context of atomic scattering processes and spin dynamics in time-dependent fields [3, 4, 5, 6]

In its basic form the LZ problem can be described by a simple two-state model with a Hamiltonian given by

$$H_{\rm LZ} = \begin{pmatrix} \alpha t & \Delta E/2 \\ \Delta E/2 & -\alpha t \end{pmatrix} \tag{1}$$

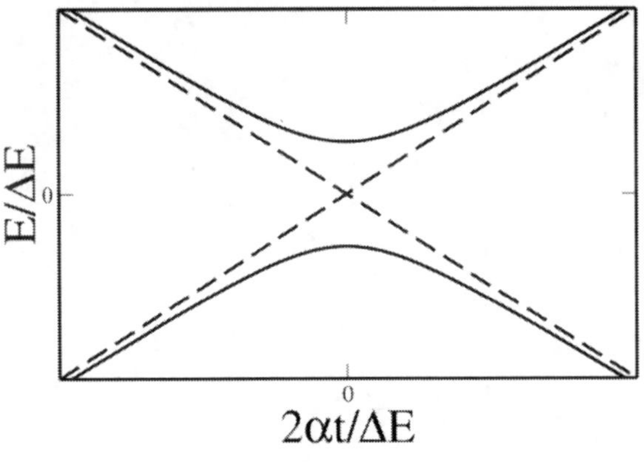

Figure 1. Energy levels as a function of time. The dashed lines show the so-called diabatic levels i.e., the energy position of states in the absence of the interaction. The solid lines demonstrate the so-called adiabatic levels, i.e., the eigenstates of the system corresponding to the instantaneous Hamiltonian.

where the off-diagonal term, $\Delta E/2$, is the coupling between the two states and α is the rate of change of the energy levels in time. The dynamics of the system can be described either in the *diabatic* or in the *adiabatic* basis. The diabatic basis is the basis of the bare states of Eq. (1) when there is coupling (i.e., no off-diagonal entries in the matrix). The adiabatic basis, on the other hand, is the basis of a system with a finite coupling $\Delta E/2$ between the two states. The Hamiltonian has two adiabatic energy levels $E_\pm = \pm\frac{1}{2}\sqrt{(2\alpha t)^2 + \Delta E^2}$.

Assuming that the system is initially, at $t \to -\infty$, in the ground energy level E_- and if the sweeping rate is small enough, it will be exponentially likely that the system remains in its adiabatic ground state E_- at $t \to +\infty$. The limiting value of the adiabatic LZ survival probabilities (for t going from $-\infty$ to $-\infty$) is [7],

$$P_{\mathrm{a}}(\infty) = 1 - \exp\left(-\frac{\pi}{\gamma}\right), \tag{2}$$

where we have introduced the dimensionless adiabaticity parameter $\gamma = 4\hbar\alpha/\Delta E^2$.

While the above analysis can predict the probability for LZ tunneling very accurately, in contains no reference to the dynamics around the avoided crossing and hence to the time it takes the system to complete the tunneling event which eventually results in the populations measured in the ground and excited bands (or in the corresponding diabatic states). Also, in the case of LZ tunneling, which occurs in an abstract space spanned by the energy levels of the system as a function of a parameter, the concept of tunneling time is less intuitive than in the case of cross-barrier tunneling in real space, to name just one example. Nevertheless, the tunneling time (or transition time or jump time, as it is called in certain contexts) associated with LZ is a meaningful concept referring to the timescale on which the system evolves around the avoided crossing. Analytical estimates for the LZ transition times have been derived in [8, 9] using the two-state model of Eq. 1. In a given basis, e.g., adiabatic or diabatic, different transition times are obtained. Vitanov [9] calculated the time-dependent diabatic/adiabatic survival probability at finite times. Analytical estimates for the LZ transition times were derived in [8] using some exact and approximate results for the transition probability.

In general, the LZ jump time in a given basis can be defined as the time after which the transition probability reaches its asymptotic value. From this definition one can expect to observe a step-like structure, with a finite width, in the time-resolved tunneling probability.

Because the step is not very sharp, it is not straightforward to define the initial and final times for the transition. It is even less obvious how to define the jump time for both small and for large coupling. Some possible choices have been used by Lim and Berry [10] and Vitanov [8, 9]. The problem is even more complicated when the survival probability shows an oscillatory behavior on top of the step structure. The oscillations give rise to an additional time scale in the system, namely an oscillation time and a damping time of the oscillations appearing in the transition probability after the crossing. Therefore, a measurement of the tunneling time depends very much on how these times are defined and also which basis is considered.

In [9] the jump time in the diabatic/adiabatic bases is defined as

$$\tau_{\mathrm{d/a}}^{jump} = \frac{P_{\mathrm{d/a}}(\infty)}{P'_{\mathrm{d/a}}(0)}. \tag{3}$$

where $P_{\mathrm{d/a}}$ is the transition probability between the two diabatic/adiabatic states, respectively. $P'_{\mathrm{d/a}}(0)$ denotes the time derivative of the tunneling probability evaluated at the crossing point. From this definition, the *diabatic* jump time is calculated as $\tau_{\mathrm{d}}^{jump} \approx \sqrt{2\pi\hbar/\alpha}$ is almost constant for large values of the adiabaticity parameter γ. For $\gamma \ll 1$, on the other hand, it decreases with γ , $\tau_{\mathrm{d}}^{jump} \approx 2\sqrt{\hbar}(\gamma\alpha)^{-1}$. In the *adiabatic* basis, when γ is large, the transition probability is similar to that in the diabatic basis. For a small adiabaticity parameter, because of the oscillations appearing on top of the transition probability step structure, it is not straightforward to define the initial and the final time for the transition. Vitanov defines the initial jump time as the time $t < 0$ at which the transition probability is very small (i.e., $P_{\mathrm{a}}(\tau) = \varepsilon P_{\mathrm{a}}(\infty)$, where ε is a proper small number). The final time of the transition $t > 0$ is defined as the time at which the non-oscillatory part of $P_{\mathrm{a}}(\tau)$ is equal to $(1+\varepsilon)P_{\mathrm{a}}(\infty)$. Using these definitions, Vitanov derived that the transition time in the adiabatic basis depends exponentially on the adiabaticity parameter, $\tau_{\mathrm{a}}^{jump} \approx (4/\epsilon)^{1/6}\gamma^{-1/3}\exp\left(\pi/(6\gamma)\right)\sqrt{\hbar/\alpha}$.

2. Experimental results

The Landau-Zener model is realized in our experiments using Bose-condensed rubidium atoms inside an optical lattice [11, 12]. Initially, we created Bose-Einstein condensates of 5×10^4 rubidium-87 atoms inside an optical dipole trap (mean trap frequency around 80 Hz). A one-dimensional optical lattice created by two counter-propagating, linearly polarized gaussian beams was then superposed on the BEC by ramping up the power in the lattice beams in 100 ms. The wavelength of the lattice beams was $\lambda = 842$ nm, leading to a sinusoidal potential with lattice constant $d_{\mathrm{L}} = \lambda/2 = 421$ nm. A small frequency offset $\Delta\nu(t)$ between the two beams could be introduced through the acousto-optic modulators in the setup, which allowed us to accelerate the lattice in a controlled fashion and hence, in the rest-frame of the lattice, to subject the atoms to a force $F_{\mathrm{LZ}} = Ma_{\mathrm{LZ}}$ with $a_{\mathrm{LZ}} = d_{\mathrm{L}}\frac{d\Delta\nu(t)}{dt}$.

The energy level structure of Bose condensates in optical lattices can be represented by energy bands in the Brillouin zone picture. At the edge of the Brillouin zone successive bands are separated by gaps, and in the vicinity of the zone edge our system approximates the LZ model very well. We can make time-resolved measurements of a single tunneling event in the following way: First, the Bose condensate is loaded adiabaticallly into a lattice, after which the lattice is accelerated to some finite quasimomentum. Thereafter, the instantaneous populations of the eigenstates of the system are measured, the exact protocol depending on the basis chosen. In detail, the protocols are as follows:

- For measurements in the *adiabatic* basis, after loading the BEC into the optical lattice the lattice was accelerated with acceleration a_{LZ} for a time t_{LZ}. The lattice thus acquired a final velocity $v = a_{\mathrm{LZ}}t_{\mathrm{LZ}}$. At time $t = t_{\mathrm{LZ}}$ the acceleration was abruptly reduced to a

smaller value $a_{\rm sep}$ and the lattice depth was increased to $V_{\rm sep}$ in a time $t_{\rm ramp} \ll T_{\rm B}$. These values were chosen in such a way that at time $t = t_{\rm LZ}$ the probability for LZ tunneling from the lowest to the first excited energy band dropped from between $\approx 0.1 - 0.9$ (depending on the initial parameters chosen) to less than ≈ 0.01, while the tunneling probability from the first excited to the second excited band remained high at about 0.95. This meant that at $t = t_{\rm LZ}$ the tunneling process was effectively interrupted and for $t > t_{\rm LZ}$ the measured survival probability $P(t) = N_0/N_{\rm tot}$ (calculated from the number of atoms N_0 in the lowest band and the total number of atoms in the condensate $N_{\rm tot}$) reflected the instantaneous value $P(t = t_{\rm LZ})$.
The lattice was then further accelerated for a time $t_{\rm sep}$ such that $a_{\rm sep}t_{\rm sep} \approx 2np_{\rm rec}/M$ (where typically $n = 2$ or 3). In this way, atoms in the lowest band were accelerated to a final velocity $v \approx 2np_{\rm rec}/M$, while atoms that had undergone tunneling to the first excited band before $t = t_{\rm LZ}$ underwent further tunneling to higher bands with a probability > 0.95 and were, therefore, no longer accelerated. At time $t_{\rm sep}$ the lattice and dipole trap beams were suddenly switched off and the expanded atomic cloud was imaged after 23 ms. In these time-of-flight images the two velocity classes 0 and $2np_{\rm rec}/M$ were well separated and the atom numbers N_0 and $N_{\rm tot}$ could be measured directly. Since the populations were effectively "frozen" inside the energy bands of the lattice, which represent the adiabatic eigenstates of the total Hamiltonian of the system, this experiment measured the time dependence of the LZ survival probability $P_{\rm a}$ in the *adiabatic* basis.

- For measurements in the *diabatic* basis, after the initial loading phase the lattice was accelerated with acceleration $a_{\rm LZ}$ for a time $t_{\rm LZ}$ as in the adiabatic case. At that point, however, the atomic sample was projected onto the free-particle diabatic basis by instantaneously (within less than $1\,\mu s$) switching off the optical lattice. After a time-of-flight the number of atoms in the $v = 0$ and $v = 2p_{\rm rec}/M$ momentum classes are measured and from these the survival probability (corresponding to the atoms remaining in the $v = 0$ velocity class relative to the total atom number) is calculated.

The results of typical measurements in the adiabatic and diabatic bases are shown in Fig. 2. The step-like behaviour of the survival probability around $t = 0.5T_B$ is clearly visible, which demonstrates that our experimental protocol does, indeed, allow us to access the timescale of the LZ transition. It is also obvious from the figure that while in the adiabatic basis the transition is smooth and can be well fitted with a sigmoid function, in the diabatic basis there are strong oscillations for times $t > 0.5\,T_B$. As a consequence, the tunneling time in the adiabatic basis can be easily identified as the width of the transition curve (indicated in the figure), while in the diabatic basis it is less obvious when the tunneling event is completed.

3. Conclusions

We have demonstrated that experiments with Bose-condensates in accelerated optical lattices allow access to the full dynamics of LZ tunneling and hence to the timescales involved in the tunneling process, both in the adiabatic and diabatic bases of the problem. Our experiments pave the way towards a thorough investigation of tunneling times and the quantum speed limit. The latter has recently been discussed theoretically in the context of optimal quantum control [13]QSL and more generalized Landau-Zener protocols involving non-linear sweep functions that are predicted to lead to shorter minimum times for completing a tunneling event. As our experimental setup allows us to realize arbitrary protocols for the lattice acceleration, such experiments should be relatively straightforward to realize.

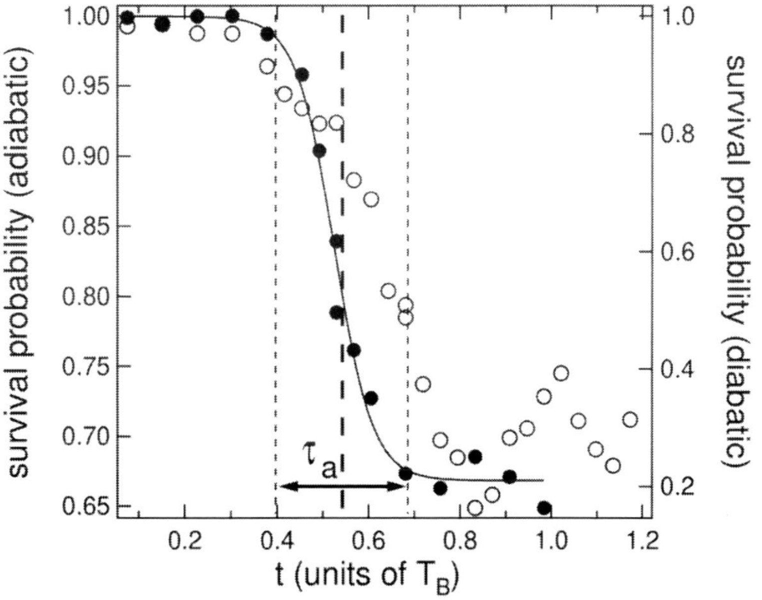

Figure 2. Time-resolved measurements of LZ tunneling in the adiabatic (filled circles) and diabatic bases (open circles)for $F_0 = 1.197$ and $V/E_{\rm rec} = 1.8$. The solid line is a sigmoid fit to the adiabatic survival probability, and the vertical dashed lines indicate the position of the zone edge at $t = 0.5\,T_B$ and the tunneling time τ_a.

Acknowledgments

The authors would like to thank the PhD students and post-docs participating in the experiments described in this article: A. Zenesini, H. Lignier and J. Radogostowicz. We acknowledge funding by the EU Project "NAMEQUAM" (EC FP7-225187) and the CNISM "Progetto Innesco 2007".

References

[1] Razavy M 2003 *Quantum Theory of Tunneling* (Singapore: World Scientific)
[2] L.S. Schulman L S 2008 *Lect. Notes Phys.* **734** 107
[3] Landau L D 1932 *Phys. Z. Sowjetunion* **2** 46
[4] Zener C 1932 *Proc. R. Soc.* A **137** 696
[5] Stückelberg E C G 1932 *Helv. Phys. Acta* **5** 369
[6] Majorana E 1932 *Nuovo Cimento* **9** 43
[7] Holthaus M 2000 *J. Opt.* B **2** 589
[8] Vitanov N V and Garraway B M 1996 *Phys. Rev.* A **53** 4288
[9] Vitanov N V 1999 *Phys. Rev.* A **59** 988
[10] Berry M V 1990 *Proc. R. Soc.* A **429** 61; Lim R and Berry M V 1991 *J. Phys.* A **24** 3255
[11] Jona-Lasinio M, Morsch O, Cristiani M, Malossi N, Müller J H, Courtade E, Anderlini M and Arimondo E 2003 *Phys. Rev. Lett.* **91** 230406
[12] Zenesini A, Lignier H, Tayebirad G, Radogostowicz J, Ciampini D, Mannella R, Wimberger S, Morsch O and Arimondo E 2009 *Phys. Rev. Lett.* **103** 090403
[13] Giovannetti V, Lloyd S and Maccone L 2003 *Phys. Rev.* A **67** 052109; Sillanpää M, Lehtinen T, Paila A, Makhlin Y and Hakonen P 2006 *Phys. Rev. Lett.* **96** 187002; Caneva T, Murphy M, Calarco T, Fazio R, Montangero S, Giovannetti V and Santoro G E 2009 *Phys. Rev. Lett.* **103** 240501

5th International Workshop DICE2010 IOP Publishing
Journal of Physics: Conference Series **306** (2011) 012032 doi:10.1088/1742-6596/306/1/012032

Some remarks on dissipative Ermakov systems and damping in Bose–Einstein condensates

Dieter Schuch
Institut für Theoretische Physik, Goethe-Universität Frankfurt am Main, Max-von-Laue-Str. 1, D-60438 Frankfurt am Main, Germany
E-mail: schuch@em.uni-frankfurt.de

Radhey Shyam Kaushal
Department for Physics and Astrophysics, University of Dehli, Dehli, 110 007, India
E-mail: rkaushal@physics.du.ac.in

Abstract. For physical systems where the Hamiltonian is no longer a constant of motion, like oscillators with time-dependent frequency or dissipative systems, a dynamical invariant of the Ermakov type might still exist. This invariant depends on the classical position and velocity and an auxiliary variable that is proportional to position uncertainty. It is possible to construct this invariant via an algebraic method based on the classical Poisson bracket. A modified version for dissipative systems that also employs anti-Poisson brackets and is related to a description in an expanding coordinate system will be presented. The resulting invariant is identical to that obtained from a logarithmic nonlinear Schrödinger equation. The uncertainties of position and momentum, expressed in terms of the auxiliary variable, can be compared with the so-called moment-method for the description of the time-evolution of Bose–Einstein condensates, corresponding to a description in terms of the cubic nonlinear Gross–Pitaevskii equation. It will be shown how this method can be extended to also include dissipative damping effects into the Bose–Einstein dynamics.

1. Introduction

For physical systems where the Hamiltonian is no longer a constant of motion, like the harmonic oscillator (HO) with time-dependent (TD) frequency or dissipative systems with energy loss due to interaction with an environment, there might still be another constant of motion that does not have the dimension of an energy, like the Hamiltonian, but that of an action.

In a first step it will be shown that such an invariant, first established by Vassili Petrovich Ermakov in 1880(!) for the HO with TD frequency [1] can also be obtained via an algebraic method based on the description of the time-evolution in terms of Poisson brackets. In section 3, it will be shown that the same invariant and the equations of motion leading to it, can also be received from the time-evolution of a Gaussian wave packet (WP) that is a solution of the corresponding TD Schrödinger equation (SE). In this context, the dynamics of quantum uncertainties will play an important role. The coupled closed set of differential equations describing this dynamics is closely related to the algebraic method mentioned above and will be crucial for the further analysis.

Section 4 will show how the interaction with a dissipative environment, expressed in the form of a linear velocity dependent friction force can be included into a formal Hamiltonian description, involving exponentially expanding coordinates and a non-canonical transformation, but supplying a Hamiltonian that is a constant of motion. It will be shown that also for this system a dynamical Ermakov invariant with the dimension of an action can be obtained via a modified algebraic method, now also including anti-commutator terms.

In section 5, it will be shown that the same invariant can also be obtained from a quantum mechanical description of the problem, taking into account the environmental effect via a complex logarithmic nonlinear term, turning the SE into a nonlinear (NL) SE.

The time-evolution of a Bose–Einstein condensate (BEC) can be described in the mean field approximation in terms of the cubic-nonlinear Gross–Pitaevskii equation for the corresponding WP. This equation cannot usually be solved analytically but, in certain cases, it is possible to describe the dynamcis of this system entirely in terms of so-called moments. Their definition and the closed set of coupled evolution equations will be given in section 6 and the relation to the equations of motion for the quantum uncertainties will be established.

Finally, it will be shown in section 7 how the results from the NLSE can be used to include dissipative effects into the dynamics of the BEC, based on the dynamical equations leading to the corresponding Ermakov invariant.

The last Section will summarize the results.

2. Classical Ermakov systems and dynamical algebra

In conservative classical Hamiltonian mechanics, the Hamiltonian function

- determines the dynamics of the system, and
- is a constant of motion that represents the conserved energy of the system.

In cases where the Hamiltonian is no longer a constant of motion, e.g., for the HO with TD frequency $\omega = \omega(t)$, there still may exist another constant of motion, the Ermakov invariant, that does not have the dimension of an energy, but essentially that of an action.

This invariant can be obtained from the corresponding classical Newtonian equation of motion for the (one-dimensional) position coordinate q,

$$\ddot{q} + \omega^2(t)\, q = 0 \tag{1}$$

and the equation for an auxiliary variable $\alpha(t)$,

$$\ddot{\alpha} + \omega^2(t)\, \alpha = \frac{1}{\alpha^3} \tag{2}$$

by eliminating $\omega^2(t)$ from these two equations (see Ermakov [1]), leading to the constant of motion

$$I = \frac{1}{2}\left[(\dot{q}\,\alpha - q\,\dot{\alpha})^2 + \left(\frac{q}{\alpha}\right)^2 \right] = \text{const.} \tag{3}$$

However, the same invariant can also be obtained in a different way via an **algebraic** approach (see, e.g., [2]). In this case, the Hamiltonian is rewritten in terms of (not explicitly TD) phase space functions Γ_n as

$$H = \sum_n h_n(t)\, \Gamma_n \tag{4}$$

where the dynamical algebra is the Lie algebra of the functions Γ_n, which is closed with relation to the Poisson brackets $\{\ ,\ \}_-$,

$$\{\Gamma_n, \Gamma_m\}_- = \sum_r C^r_{nm}\, \Gamma_r \tag{5}$$

with the C^r_{nm} being the structure constants of the algebra.

The time-evolution of any phase space function $F(q,p,t)$ is given by

$$\frac{d}{dt}F = \{F,H\}_- + \frac{\partial}{\partial t}F \; . \tag{6}$$

In particular, a **dynamical invariant** is characterized by $\frac{d}{dt}I = 0$, i.e.,

$$\frac{\partial}{\partial t}I = \{H,I\}_- \; . \tag{7}$$

Looking for an invariant that is also a member of the dynamical algerbra, i.e.,

$$I = \sum_n \kappa_n(t)\; \Gamma_n \; , \tag{8}$$

Eq. (7) leads to a coupled set of evolution equations for the expansion coefficients κ_n,

$$\dot{\kappa}_r + \sum_n \left(\sum_m C^r_{nm}\; h_m(t) \right) \kappa_n \; = \; 0 \; . \tag{9}$$

In the following, we consider the dynamical algebra of

$$\Gamma_1 = \frac{p^2}{2m} \; , \quad \Gamma_2 = p\; q \; , \quad \Gamma_3 = \frac{mq^2}{2} \tag{10}$$

with the Poisson brackets $\{\Gamma_1, \Gamma_3\}_- = -\,\Gamma_2$, $\{\Gamma_1, \Gamma_2\}_- = -2\,\Gamma_1$, $\{\Gamma_3, \Gamma_2\}_- = 2\,\Gamma_3$.

For the Hamiltonian of the HO with TD frequency,

$$H = \frac{1}{2m}\, p^2 \; + \; \frac{m}{2}\, \omega^2(t)\; q^2 \tag{11}$$

with $h_1 = 1$, $h_2 = 0$ and $h_3 = \omega^2(t)$, the set of equations for the κ_n can be written as

$$\begin{aligned} \dot{\kappa}_1 &= -\,2\,\kappa_2 \\ \dot{\kappa}_2 &= \omega^2\kappa_1 \; - \; \kappa_3 \\ \dot{\kappa}_3 &= 2\,\omega^2\; \kappa_2 \; . \end{aligned} \tag{12}$$

This coupled set of equations can be reduced to a single second order differential equation by introducing a new variable $\alpha(t)$ via $\kappa_1 = \frac{1}{m}\alpha^2$, leading to

$$\ddot{\alpha} + \omega^2(t)\; \alpha = \frac{k}{\alpha^3} \tag{13}$$

with constant k.

With the κ_n, expressed in terms of α and $\dot{\alpha}$, the invariant (8) can be written as

$$I \; = \; \frac{1}{2}\left[\alpha^2\frac{p^2}{m^2} \; - \; 2\dot{\alpha}\alpha\frac{p}{m}\, q \; + \; \left(\dot{\alpha}^2 + \frac{k^2}{\alpha^2}\right) q^2 \right] \; , \tag{14}$$

which is identical to (3) for $k = 1$ and $p = m\dot{q}$.

3. Quantum mechanical wave packets and Ermakov invariant

Now the (one-dimensional) TDSE for the HO with possibly TD frequency $\omega = \omega(t)$ shall be considered,

$$i\hbar \frac{\partial}{\partial t}\Psi(x,t) = \left\{ -\frac{\hbar^2}{2m}\frac{\partial^2}{\partial x^2} + \frac{m}{2}\omega^2 x^2 \right\} \Psi(x,t), \tag{15}$$

with exact analytic Gaussian WP solutions, written as [3]

$$\Psi_{WP}(x,t) = N(t) \exp\left\{ i\left[y(t)\tilde{x}^2 + \frac{1}{\hbar} < p > \tilde{x} + K(t) \right] \right\}, \tag{16}$$

where $\tilde{x} = x - < x >$ with $< x > = \int_{-\infty}^{+\infty} \Psi^* x \Psi dx = q(t) =$ classical trajectory. The **complex** TD quantity $y(t) = y_R(t) + i y_I(t)$ is connected with the WP width, or the position uncertainty, via $y_I = \frac{1}{4<\tilde{x}^2>}$ with $<\tilde{x}^2> = < x^2 > - < x >^2$.

Inserting WP (16) into the TDSE (15) yields the Newtonian equation

$$\ddot{q} + \omega^2(t)\, q = 0 \tag{17}$$

for the WP maximum and the **complex Riccati equation**

$$\frac{2\hbar}{m}\dot{y} + \left(\frac{2\hbar}{m} y \right)^2 + \omega^2 = 0 \tag{18}$$

for the WP width.

Introducing a new variable $\alpha_L(t)$ via $\frac{2\hbar}{m} y_I = \frac{1}{\alpha_L^2(t)}$ with $\alpha_L = \sqrt{\frac{2m<\tilde{x}^2>}{\hbar}}$, the Ricatti equation (18) leads to an Ermakov equation for α_L which is identical to the auxiliary equation (2) for α, so $\alpha_L \equiv \alpha$. Together with Eq. (17), this again yields the invariant $I_L = I = (3)$.

It can be shown that position and momentum uncertainties are related to α_L via

$$\begin{aligned}
< \tilde{x}^2 >_L &= \frac{\hbar}{2m}\, \alpha_L^2 , \\
< \tilde{p}^2 >_L &= \frac{\hbar m}{2} \left(\dot{\alpha}_L^2 + \frac{1}{\alpha_L^2} \right) , \\
< [\tilde{x}, \tilde{p}]_+ >_L &= \hbar \dot{\alpha}_L\, \alpha_L , \\
U_L &= < \tilde{x}^2 >_L < \tilde{p}^2 >_L = \frac{\hbar^2}{4} \left[1 + (\dot{\alpha}_L \alpha_L)^2 \right]
\end{aligned} \tag{19}$$

where $[\ ,\ ]_+$ denotes the anti-commutator.

This allows one to write the invariant I_L also in the form

$$\begin{aligned}
I_L &= \frac{1}{2}\left[\alpha_L^2 \frac{p^2}{m^2} - 2\dot{\alpha}_L \alpha_L \frac{p}{m} q + \left(\dot{\alpha}_L{}^2 + \frac{k^2}{\alpha_L^2} \right) q^2 \right] \\
&= \frac{1}{\hbar m} \left[< \tilde{x}^2 >_L p^2 - < [\tilde{x}, \tilde{p}]_+ >_L\, q\, p + < \tilde{p}^2 >_L q^2 \right] .
\end{aligned} \tag{20}$$

Comparison with I defined via the expansion coefficients κ_n, $I = \sum_n \kappa_n(t)\, \Gamma_n$, shows how these coefficients are related to the quantum mechanical uncertainties,

$$\begin{aligned}
\kappa_1 &= \frac{1}{m}\alpha_L^2 = \frac{2}{\hbar} < \tilde{x}^2 >_L , \\
\kappa_2 &= -\frac{1}{m}\dot{\alpha}_L \alpha_L = -\frac{1}{\hbar m} < [\tilde{x}, \tilde{p}]_+ >_L , \\
\kappa_3 &= \frac{1}{m}\left(\dot{\alpha}_L^2 + \frac{1}{\alpha_L^2} \right) = \frac{2}{\hbar m^2} < \tilde{p}^2 >_L .
\end{aligned} \tag{21}$$

Therefore, the coupled set of equations for the κ_n, Eqs. (12), represents a set of equations for the time-dependence of the quantum uncertainties.

4. Classical dissipative Hamiltonian in expanding coordinates

In the following a classical dissipative system with (linear) velocity dependent friction force shall be considered. It will be described in an exponentionally-expanding coordinate system [4] with the variables

$$Q \;=\; e^{\gamma t/2}\, q \quad , \; P \;=\; m\dot{Q} \;=\; m e^{\gamma t/2}(\dot{q} \;+\; \frac{\gamma}{2}\, q)\;, \tag{22}$$

where the transition $(q,p) \to (Q,P)$ is a **non-canonical** transformation.

The corresponding Hamiltonian function

$$\hat{H}_{exp} \;=\; \frac{1}{2m} P^2 \;+\; \frac{m}{2}\Omega^2 Q^2 \;=\; \text{const.} \tag{23}$$

with $\Omega^2 = \left(\omega^2 - \frac{\gamma^2}{4}\right)$ can be expressed in terms of the physical variables q and p as

$$\hat{H}_{exp} \;\hat{=}\; \left[\frac{1}{2m}p^2 \;+\; \frac{\gamma}{2}\; p\, q \;+\; \frac{m}{2}\omega^2 q^2\right]\; e^{\gamma t} \;=\; \frac{1}{2m}p_0^2 \;+\; \frac{m}{2}\omega^2 q_0^2 \;=\; E_0\;. \tag{24}$$

The corresponding Hamiltonian equations of motion are

$$\begin{aligned} \frac{\partial \hat{H}_{exp}}{\partial P} \;&=\; \frac{1}{m}\; P \;=\; \dot{Q} \;=\; e^{\gamma t/2}\left(\dot{q} + \frac{\gamma}{2}\; q\right) \\ -\frac{\partial \hat{H}_{exp}}{\partial Q} \;&=\; -\,\Omega^2 Q \;=\; -\left(\omega^2 - \frac{\gamma^2}{4}\right)\; e^{\gamma t/2}\; q \;=\; \dot{P} \;=\; e^{\gamma t/2}\left(\ddot{q} + \gamma\, \dot{q} + \frac{\gamma^2}{4}\; q\right)\;, \end{aligned} \tag{25}$$

equivalent to the Newtonian equation

$$\ddot{Q} + \left(\omega^2 - \frac{\gamma^2}{4}\right) Q = 0 \tag{26}$$

in the expanding coordinates, or, to a Newtonian equation including a friction term $\gamma\dot{q}$ with friction coefficient γ,

$$\ddot{q} + +\gamma\dot{q} + \omega^2(t)\; q = 0 \tag{27}$$

in the physical coordinate system.

Although the dissipative system in the expanding **canonical variables** $(\mathbf{Q}, \mathbf{P})$ can be described via the usual **canonical formalism**, this is obviously not possible in terms of the **physical variables** $(\mathbf{q}, \mathbf{p})$.

However, since the relations between the physical and the canonical variables are known, it is possible to transform the canonical results into the ones on the physical level.

In particular, if the **time-evolution** of any phase space function $\hat{F}(Q,P,t)$ on the **canonical level** is given by

$$\frac{d}{dt}\hat{F} = \{\hat{F}, \hat{H}_{exp}\}_{(Q,P)-} + \frac{\partial}{\partial t}\hat{F}\;, \tag{28}$$

on the **physical level**, additional Poisson brackets $\{\;,\;\}_-$ and anti-Poisson brackets $\{\;,\;\}_+$ occur,

$$\frac{d}{dt}F(q,p,t) \;=\; \{F,H\}_- + \{F, \frac{\gamma}{2}\; p\, q\}_- - \{F, \frac{\gamma}{2}\; p\, q\}_+ + \frac{\partial}{\partial t}F\;, \tag{29}$$

where H is given by Eq. (11) (for further details see [4]).

In this dissipative case, the dynamical invariant

$$I_{exp} = \sum_n \hat{\kappa}_{n,exp}(t)\; \Gamma_n \tag{30}$$

must now fulfil the relation

$$\begin{aligned}\frac{d}{dt} I_{exp} &= \{I_{exp}, H\}_- + \{I_{exp}, \frac{\gamma}{2}\, p\, q\}_- - \{I_{exp}, \frac{\gamma}{2}\, p\, q\}_+ + \frac{\partial}{\partial t} I_{exp} \\ &= \{I_{exp}, H\}_- + \frac{\gamma}{2}\, \{I_{exp}, \Gamma_2\}_- - \frac{\gamma}{2}\, \{I_{exp}, \Gamma_2\}_+ + \frac{\partial}{\partial t} I_{exp} \;=\; 0\;. \end{aligned} \tag{31}$$

In comparison with the conservative case, there are **additional terms** from $\{\Gamma_n, \Gamma_2\}_- - \{\Gamma_n, \Gamma_2\}_+$. This leads to a modified set of coupled equations for $\kappa_{n,exp} = e^{-\gamma t}\, \hat{\kappa}_{n,exp}$,

$$\begin{aligned}\dot{\kappa}_{1,exp} &= \gamma\kappa_{1,exp} \;-\; 2\;\kappa_{2,exp} \\ \dot{\kappa}_{2,exp} &= \omega^2\kappa_{1,exp} \;-\; \kappa_{3,exp} \\ \dot{\kappa}_{3,exp} &= \gamma\kappa_{3,exp} \;+ 2\;\omega^2\;\kappa_{2,exp}\;. \end{aligned} \tag{32}$$

With $\kappa_{1,exp} = \frac{1}{m}\alpha^2_{NL}$, this set of coupled equations can be condensed into the **modified Ermakov equation**

$$\ddot{\alpha}_{NL} + \left(\omega^2(t) - \frac{\gamma^2}{4}\right)\alpha_{NL} = \frac{1}{\alpha^3_{NL}} \quad . \tag{33}$$

The corresponding **Ermakov invariant**, written in terms of q and α^2_{NL}, reads

$$I_{exp} \;=\; \frac{1}{2} e^{\gamma t}\left[\left(\dot{\eta}\;\alpha_{NL} - \left(\dot{\alpha}_{NL} - \frac{\gamma}{2}\alpha_{NL}\right)\eta\right)^2 \;+\; \left(\frac{\eta}{\alpha_{NL}}\right)^2\right] \;=\; \text{const.} \tag{34}$$

Expressing q and $\dot{q}$ in terms of the expanding coordinate Q and velocity $\dot{Q}$, this invariant can be rewritten as

$$I_{exp} = \frac{1}{2}\left[\left(\dot{Q}\;\alpha_{NL} - Q\;\dot{\alpha}_{NL}\right)^2 + \left(\frac{Q}{\alpha_{NL}}\right)^2\right] \;=\; \text{const.} \tag{35}$$

In this form, it becomes obvious that the invariant is not only independent of ω, i.e., also existing for $\omega = \omega(t)$, but also **independent of** γ, i.e., also existing for $\gamma = \gamma(t)$! This form is identical to the one for the conservative case, only q being replaced by Q and α by α_{NL}. The meaning of the replacement $q \rightarrow Q$ has been explained above; the meaning of the replacement $\alpha \rightarrow \alpha_{NL}$ will become clear from the following description of an **effective quantum mechanical approach** for **dissipative systems**.

5. Nonlinear Schrödinger equation for dissipative systems

It has been shown that quantum systems interacting with some kind of dissipative environment may not only be described in terms of a reduced density operator coupled to a large number of environmental degrees of freedom (drastically increasing computational effort and thus severely limiting the systems that can be treated), but also in terms of pure state wave functions (or WPs) obeying an effective NLSE [5] where the NL Hamiltonian can even be non-Hermitian and still provide normalizable wave functions [6].

Our approach in this direction starts from the reversible continuity equation for the probability density $\varrho\,(x,t) \;=\; \Psi^*\,(x,t)\,\Psi\,(x,t)$ by adding a time-reversal symmetry-breaking diffusion term, leading to an **irreversible** (Fokker–Planck-type) **Smoluchowski equation** [6]

$$\frac{\partial}{\partial t}\varrho \;+\; \frac{\partial}{\partial x}\,(\varrho\; v_-) \;-\; D\frac{\partial^2}{\partial x^2}\varrho \;=\; 0 \tag{36}$$

with the convection velocity field $v_- \;=\; \frac{\hbar}{2mi}\left(\frac{\frac{\partial}{\partial x}\Psi}{\Psi} \;-\; \frac{\frac{\partial}{\partial x}\Psi^*}{\Psi^*}\right)$ and the (possibly TD) diffusion coefficient D.

To obtain the corresponding SE, Eq. (36) must be separated into two equations for Ψ or Ψ^*, respectively. Due to the diffusion term, this is not possible in general, only in particular cases. So, separation can be achieved with the ansatz

$$-D\frac{\frac{\partial^2}{\partial x^2}\varrho}{\varrho} = \gamma\,(\ln\varrho - <\ln\varrho>)\ . \tag{37}$$

This leads to the NLSE

$$\begin{aligned} i\hbar\frac{\partial}{\partial t}\Psi_{NL}(x,t) &= \{-\frac{\hbar^2}{2m}\frac{\partial^2}{\partial x^2} + V(x) + \gamma\frac{\hbar}{i}(\ln\Psi_{NL} - <\ln\Psi_{NL}>)\}\Psi_{NL}(x,t) \\ &= \{H_L + \tilde{W}\}\Psi_{NL}(x,t) = H_{NL}\ \Psi_{NL}(x,t) \end{aligned} \tag{38}$$

with the **complex logarithmic nonlinearity** $\tilde{W} = \gamma\frac{\hbar}{i}(\ln\Psi_{NL} - <\ln\Psi_{NL}>)$.

For this log NLSE, also exact Gaussian WP solutions exist [6, 7]. In particular, the motion of the **WP maximum** is determined by the Newtonian equation

$$\ddot{q} + \gamma\dot{q} + \omega^2(t)\ q = 0\ , \tag{39}$$

whereas the dynamics of the **WP width**, also in this case, follows from a modified complex Riccati equation,

$$\frac{2\hbar}{m}\dot{y} + \gamma\left(\frac{2\hbar}{m}y\right) + \left(\frac{2\hbar}{m}y\right)^2 + \omega^2(t) = 0\ . \tag{40}$$

With the same definition for the imaginary part in terms of a new variable $\alpha_{NL}(t)$ as in the case of the linear TDSE, i.e., $\frac{2\hbar}{m}y_I = \frac{1}{\alpha^2_{NL}(t)}$, the Riccati equation (40) turns into the modified Ermakov equation (33) for the auxiliary variable leading to the Ermakov invariant (34) in the expanding coordinate system. Therefore, the Newtonian equation (39) for $q(t)$ and this equation for $\alpha_{NL}(t)$ provide the same Ermakov invariant, i.e., $I_{NL} \equiv I_{exp}$

As in the conservative case, also in this dissipative case the quantity α_{NL} is related to the **position uncertainty** via $\alpha_{NL} = \sqrt{\frac{2m<\tilde{x}^2>_{NL}}{\hbar}}$.

All the uncertainties for the dissipative system, expressed in terms of α_{NL} and $\dot{\alpha}_{NL}$, now have the modified form

$$\begin{aligned} <\tilde{x}^2>_{NL} &= \frac{\hbar}{2m}\alpha^2_{NL} = \frac{\hbar}{2}\,\kappa_{1,exp}\ , \\ <\tilde{p}^2>_{NL} &= \frac{\hbar m}{2}\left[\left(\dot{\alpha}_{NL} + \frac{\gamma}{2}\alpha_{NL}\right)^2 + \frac{1}{\alpha^2_{NL}}\right] = \frac{\hbar m^2}{2}\,\kappa_{3,exp}\ , \\ <[\tilde{x},\tilde{p}]_+>_{NL} &= \hbar\left(\dot{\alpha}_{NL}\,\alpha_{NL} - \frac{\gamma}{2}\alpha^2_{NL}\right) = -\,\hbar m\,\kappa_{2,exp}\ , \\ U_{NL} &= <\tilde{x}^2>_{NL}\ <\tilde{p}^2>_{NL} = \frac{\hbar}{4}\left[1 + \left(\dot{\alpha}_{NL}\,\alpha_{NL} - \frac{\gamma}{2}\alpha^2_{NL}\right)\right] \end{aligned} \tag{41}$$

where the relations between the uncertainties and the expansion coefficients $\kappa_{n,exp}$ correspond to the ones in the conservative case. Therefore, the set of equations (32) for the time-evolution of the expansion coefficients $\kappa_{n,exp}$ now corresponds to a set of equations for the time-evolution of the uncertainties of the dissipative system.

Important for the following is that:

in the **conservative case** the **quantum uncertainties**, given by $<\tilde{x}^2>_L$, $<\tilde{p}^2>_L$, etc. are directly related to

- the κ_n from the classical algebraic approach and

- the α_L, $\dot{\alpha}_L$ from the WP solution of the TDSE;

in the **dissipative case** the **quantum uncertainties** are given by $< \tilde{x}^2 >_{NL}$, $< \tilde{p}^2 >_{NL}$, etc., derived from the log NLSE and uniquely related to

- the $\kappa_{n,exp}$ from the classical expanding coordinate system and
- the α_{NL}, $\dot{\alpha}_{NL}$ from the WP solution of the log NLSE.

6. Bose–Einstein condensates and moment method

In the mean field approximation, a BEC can be described by a macroscopic WP for the condensate, Ψ, which obeys the **Gross–Pitaevskii equation**

$$i\hbar\frac{\partial}{\partial t}\Psi = \left\{-\frac{\hbar^2}{2m}\Delta + V(\mathbf{r},t) + g|\Psi|^2\right\}\Psi \,. \tag{42}$$

The trapping potential $V(\mathbf{r},t)$ shall be given by $V(\mathbf{r},t) = \frac{m}{2}\omega^2(t)r^2$ with TD frequency (e.g., like in a Paul trap).

Although Eq. (42) cannot be solved analytically (even in the case of constant frequency ω), the **dynamics** of the **BEC** characterized by this equation can be described by a set of coupled differential equations for the so-called **moments** M_n ($n = 1-4$) (for details see, e.g., [8]).

These moments are defined by the following integral parameters (with $\hbar = m = 1$)

$$\begin{aligned}
M_1 &= \int d^2x\ |u|^2 && \sim \text{norm} \\
M_2 &= \int d^2x\ r^2\ |u|^2 && \sim \text{width} \\
M_3 &= i\int d^2x \left(\frac{\frac{\partial}{\partial r}u^*}{u^*} - \frac{\frac{\partial}{\partial r}u}{u}\right)|u|^2 && \sim \text{radial momentum} \\
M_4 &= \int d^2x \left(|\nabla u|^2 + \frac{k^2}{r^2}|u|^2 + |u|^4\right) && \sim \text{energy of WP}\,.
\end{aligned} \tag{43}$$

It can be shown that these moments M_n satisfy a set of coupled first order differential equations (where $\frac{d}{dt}M_1 = 0$, corresponding to conservation of probability or particle number),

$$\begin{aligned}
\frac{d}{dt}M_2 &= M_3\,, \\
\frac{d}{dt}M_3 &= -\,\omega^2(t)\ M_2\ +\ 4M_4\,, \\
\frac{d}{dt}M_4 &= -\,\frac{1}{2}\omega^2(t)\ M_3\,.
\end{aligned} \tag{44}$$

With the help of the invariant $2M_4M_2 - M_3^2 = 1$, this set of equations can be reduced to a single equation for M_2,

$$\frac{d^2}{dt^2}M_2\ -\ \frac{1}{2M_2}\left(\frac{d}{dt}M_2\right)^2\ +\ \omega^2(t)M_2\ =\ \frac{2}{M_2}\,. \tag{45}$$

Introducing a new variable $X(t)$ via $X(t) = \sqrt{M_2}$ = WP width, changes this equation into

$$\ddot{X} + \omega^2(t)\ X = \frac{k}{X^3}\,. \tag{46}$$

In comparison with the algebraic method for the conservative case and the WP solution of the corresponding TDSE it immediately follows (with k = 1) that

$$\begin{aligned}
M_2 &\hat{=} \frac{2m}{\hbar} < \tilde{x}^2 >_L &= \alpha_L^2 &= m\,\kappa_1 \\
M_3 &\hat{=} \frac{2}{\hbar} < [\tilde{x},\tilde{p}]_+ >_L &= 2\dot{\alpha}_L\alpha_L &= -2\,m\,\kappa_2 \\
M_4 &\hat{=} \frac{1}{m\hbar} < \tilde{p}^2 >_L &= \frac{1}{2}\left(\dot{\alpha}_L^2 + \frac{1}{\alpha_L^2}\right) &= \frac{1}{2}\,m\,\kappa_3\,.
\end{aligned} \tag{47}$$

Since $X = M_2^{\frac{1}{2}}$ is identical to α_L, the equation for $X(t)$ is just the Ermakov equation from the TDSE.

7. Bose–Einstein condensates with dissipation

To include dissipative energy loss effects, attempts at different modifications of the Gross–Pitaevskii equation have been made, e.g., by using imaginary (non-Hermitian) loss terms of the kind $i\gamma(x)\Psi$, etc. [9].

Considering the quantum mechanical results discussed in Section 5 for the inclusion of dissipative friction effects, a modification of the Gross–Pitaevskii equation by adding the complex logarithmic term should be used, i.e.,

$$i\hbar\frac{\partial}{\partial t}\Psi = \left\{-\frac{\hbar^2}{2m}\Delta + V(\mathbf{r},t) + g|\Psi|^2 + \gamma\frac{\hbar}{i}\left(\ln\Psi - <\ln\Psi>\right)\right\}\Psi\,. \tag{48}$$

Although this equation also cannot be solved analytically, the moment method can still be applied.

Since the moments M_n can be directly identified with the expansion coefficients κ_n (or $\kappa_{n,exp}$, respectively) of the Ermakov invariant which are uniquely connected with the quantum uncertainties, the set of equations for $\kappa_{n,exp}$, including the dissipative effects from the log NLSE, can be directly translated into the corresponding equations for the moments M_n, including dissipation.

Therefore, all results concerning the log NLSE can be transferred to the dissipative BEC problem. The moments can be identified with the uncertainties of the NLSE in the same way as given in Eqs. (47), only replacing $<\ldots>_L$ by $<\ldots>_{NL}$.

The set of equations of motion for the moments changes into

$$\begin{aligned}
\frac{d}{dt}M_2 &= M_3 + \gamma\,M_2 \\
\frac{d}{dt}M_3 &= -\,\omega^2(t)\,M_2 + 4M_4 \\
\frac{d}{dt}M_4 &= -\frac{1}{2}\omega^2(t)\,M_3 - \gamma\,M_4\,,
\end{aligned} \tag{49}$$

corresponding to the set of evolution equations (32) for the $\kappa_{n,exp}$ or the uncertainties of the log NLSE, respectively, which again, can be condensed into the Ermakov equation for $X = \sqrt{M_2} = \sqrt{\frac{2m<\tilde{x}^2>_{NL}}{\hbar}}$,

$$\ddot{X} + \left(\omega^2(t) - \frac{\gamma^2}{4}\right)X = \frac{1}{X^3}\,. \tag{50}$$

8. Conclusions

For the HO with TD frequency, a model that can describe an ion in a Paul trap [10] or a confined BEC [8], the Hamiltonian (with the dimension of an energy) is no longer a constant of motion but there still exits the Ermakov invariant (with the dimension of an action) for this system. As an alternative to the original derivation [1], this invariant can also be obtained via a classical dynamical algebra. The (conservative) dynamics of quantum mechanical WPs fulfilling the corresponding TDSE shows the connection between this invariant, the classical position (q) and the quantum mechanical position uncertainty (α_L).

Another system where the energy is no longer a constant of motion and, in addition, the usual Hamiltonian formalism for the physical position and momentum cannot be applied to provide correct equations of motion, is the HO with linear velocity dependent friction force. However, a description in terms of the canonical formalism becomes feasible via a non-canonical transformation to a new set of exponentially-expanding variables. In this case, the formal canonical Hamiltonian function $\hat{H}_{exp}$ is a constant of motion and corresponds to the initial energy of the system. It can be shown that for this system also an Ermakov invariant exists. The (non-canonical) backtransformation shows which changes in the conservative Hamiltonian formalism must be made in order to include the friction effect also into the equations that are expressed in the physical variables. This also leads to a modification of the algebraic method for the determination of the Ermakov invariant, particularly the set of coupled equations for the determination of the expansion coefficients κ_n is affected.

It can be shown that the same dissipative system can also be treated analytically in a qunatum mechanical context if a description in terms of an effective NLSE with complex logarithmic nonlinearity is used. The coupled equations of motion for the quantum uncertainties correspond, like already in the conservative case, to the coupled set of equations for the $\kappa_{n,exp}$ in the dissipative case. The Ermakov invariant of the NLSE is identical to the one obtained in the expanding coordinate system.

The dynamics of a conservative BEC can also be described by a (different) NLSE, the Gross–Pitaevskii equation with cubic nonlinearity. Although no analytical solutions to this equation exist, the dynamics can still be obtained using the so-called moment method. The relation betwen these moments, the quantum uncertainties of the TDSE and the expansion coefficients for the conservative Ermakov invariant have been clarified. The structural relationship is still valid if the dissipative effect is taken into account, thus allowing one to obtain the dynamics of a dissipative BEC just by comparing it with the results from the log NLSE.

Acknowledgment

One of the authors (R.S.K.) would like to thank the Alexander von Humboldt Foundation for supporting his visit to Germany.

References

[1] Ermakov V P 1880 *Univ. Izv. Kiev* **20** 1
[2] Korsch H J 1979 *Phys. Lett.* A **74** 294
[3] Schuch D and Moshinsky M 2006 *Phys. Rev.* A **73** 062111
[4] Schuch D 1990 *Int. J. Quantum Chem.; Quantum Chem. Symp* **24** 767
[5] May V and Kühn O 2003 *Charge and Energy Transfer Dynamics in Molecular Systems* 2nd ed (Berlin: Wiley-VCH) chapter 3.12.3
[6] Schuch D, Chung K-M and Hartmann H 1983 *J. Math. Phys.* **24** 1652; Schuch D and Chung K-M 1986 *Int. J. Quantum Chem.* **29** 1561
[7] Schuch D 2007 *J. Math. Phys.* **48** 122701
[8] Garcia -Ripoll J J, Perez-Garcia V M and Torres P 1999 *Phys. Rev. Lett.* **83** 1715
[9] Centurion M, Porter M A, Pu Y, Kevrekidis P G, Frantzeskakis D J and Psaltis D 2007 *Phys. Rev.* A **75** 063804
[10] Schleich W P 2001 *Quantum Optics in Phase Space* (Berlin: Wiley-VCh) p. 480

Why is more different: the appearance of microscopic classical particles

G. Doyen[1] **and D. Drakova**[2]

[1] Ludwig-Maximilians Universtät, München, Germany
[2] University of Sofia, Sofia, Bulgaria

E-mail: gerold@gerold-doyen.de and nhdd@wmail.chem.uni-sofia.bg

Abstract. Particles (electrons, atoms, molecules) in interaction with large ensembles of atoms, e.g. solids and surfaces, often display properties, which are unexpected, based on experimental and theoretical investigations of small ensembles. A frequently cited phrase has been coined by P.W. Anderson to focus on this problem: *"More is different"*. Theoretical descriptions of the unexpected properties commonly use ad hoc assumptions, justified by referring to the vogue word *"emergence"*. In this contribution some striking examples of behaviour changes are outlined with the transition from *"few"* to *"more"*. It is suggested that entanglement between the observable local system and unobservable continua of the environment (phonons, electron-hole pairs, gravitons) can contribute much to the understanding of the relationship between the behaviour of *"few"* and *"more"*. Solving the time dependent Schrödinger equation for coherent model worlds, comprising both the local system and the entangled continua of the environment, adsorbate motion at surfaces at low temperature is examined. In the time-dependent formalism the entanglement of the core movement of a gas atom to the high dimensional continuum of substrate environs leads to a qualitative behaviour as it is observed experimentally. The focus in the present article is the emergence of classical appearance of a local quantum system entangled with the excitations of the environment. The dynamical evolution of the quantum system is coherent and according to the time dependent Schrödinger equation. The environmental excitations can be phonons, tomonagons, gravitons, either observable or not observable in experiment.

1. Introduction

It is frequently observed that microscopic particles - atoms and molecules - do not apparently obey the rules of standard quantum mechanics when they are in interaction with solid surfaces and, hence, coupling to the high dimensional phase space of the solid excitations. In this case atoms often behave like classical particles and we refer to this phenomenon as dimensionality effect and the appearance of microscopic classical particles.

This kind of observations has been addressed nearly four decades ago by Anderson in a very influencial paper published in 1972 [1]. Anderson tried to subsume these phenomena under the principle of spontaneous symmetry breaking, but obviously not all kinds of classical behaviour can be classified as broken symmetry. Therefore many scientists prefer now to talk about emergence, meaning perhaps that everyday intuition combined with some brilliant mathematical idea are the proper tools for attacking these phenomena.

This neglect of reductionism contains the danger of overlooking hidden aspects and influences in the environment which might be responsible for the observed non-quantal behaviour. There might be acting something behind the scene which is the causal origin - in a reductionist sense - of what is observed. But using the concept of emergence we will neither be interested nor able of discovering the causal origin.

The purpose of this contribution is to explore this possibility for some apparently classical behaviour of microscopic particles at solids. As the classical behaviour of microscopic particles at surfaces is perhaps not too widely known outside the area of surface science, we shall describe some prominent examples. For the last example - localization at surfaces - we shall discuss the hidden aspects which might be responsible for the experimental observations. Classical appearance is usually described by decoherence theory which in its application to surface science has been summarized in ref. [2].

2. Experimental evidence of classical behaviour at solid surfaces

2.1. Scattering of noble gas beams at surfaces

Experiment demonstrates that rare gas atoms behave as expected from standard quantum mechanics, if they are scattered from nano gratings [3]. The situation is quite different, if rare gas atoms are scattered from solid surfaces. In this case only He scattering displays sharp diffreaction peaks [4], whereas all other rare gases show a scattering distribution which can be described by classical mechanics [5]. This includes the classical rainbow. He and Ne scattering results in diffraction patterns imposed on a so called inelastic background, whereas for Kr and heavier rare gas atoms scattering the structureless background dominates to such an extend that diffraction patterns are no longer discernible.

2.2. Accommodation at surfaces

This classical behaviour of scattering rare gas atoms is not only observed for elastic diffraction, but also for inelastic scattering at low temperatures. In this case even He behaves classically as can be seen from the analysis of the so called accommodation coefficient α which measures the deviation of the energy of the scattered atom from the average energy of the substrate atom. If α equals unity, the scattered atom is fully accommodated with the substrate motion. At zero temperature and low scattering energy quantum theory predicts $\alpha = 0$ [6], while classical theory predicts $\alpha = 1$ [8, 9]. Obviously experiment obeys classical theory [6, 7]. Stationary state 3D scattering theory and the generalized Ehrenfest theorem have been used earlier to explain the classical behaviour within the deformation resonance approach, accounting for the different dimensionality of the gas atom and substrate movement [10].

2.3. Metastable survival probabilities

One might expect that classical behaviour should be restricted to core movement, but electronic transtions are affected as well. There is a spectacular example: a He atom is electronically excited by transferring an electron from the 1s-orbital to the 2s-orbital with the two spins being parallel. This is a metastable state with a lifetime of the order of seconds. If such an electronically excited He atom is scattered from a CO molecule in the gas phase, there is a probability of roughly 10% that the He atom gets de-excited by an Auger process into its electronic ground state, i.e., 90% of the atoms survive in the electronically excited state. This process can quantitatively be described and calculated by standard quantum mechanics.

If we now form a densely packed overlayer of CO molecules on a smooth solid surface (e.g. Pd(111)) and scatter electronically excited He atoms from this layer, we would expect similar de-excitation probabilities as in the gas phase. But surprisingly the survival probability decreases by a factor of 10^4. This cannot be understood by the standard quantum mechanical description of Auger processes [11]. Even more surprisingly the scattering distribution is purely classical,

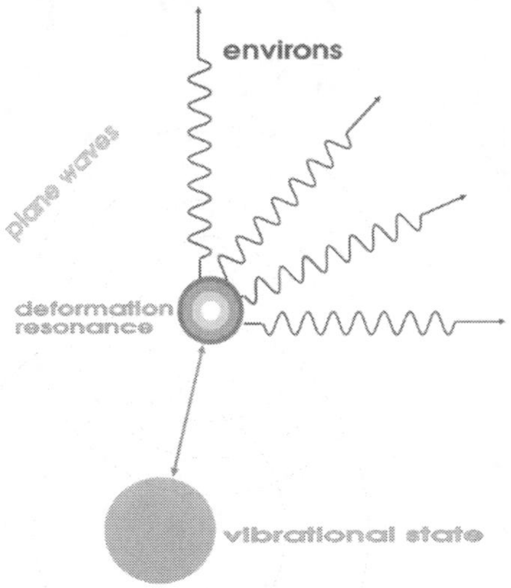

Figure 1. Illustration of the coherent world model

whereas the scattering of ground state helium shows the diffraction peaks predicted by quantum theory.

2.4. Adparticle localization

At cold solid surfaces adparticle movement does not follow the laws as predicted by Schrödinger's equation for static potential energy surfaces even if phonons are included in the calculations [12, 13]. The entanglement to gravitons is suggested as an ubiquitous decoherence mechanism and as the only relevant process for particle localization at low T. This mechanism is crucially dependent upon the high dimensionality of the graviton field and on the hidden dimensions where gravitons can propagate. The entanglement to the rest excitations of the sample (phonons, photons, tomonagons), though it is effective at high temperature, cannot be the mechanism for particle localization at low temperature. The typical wave lengths of the order of 180 Å of these excitations at low temperature are too long to allow to distinguish a H atom on two neighbouring adsorption sites, for instance.

3. The coherent world model

3.1. The model

We describe a particle (atom, electron) coupling to two continua of different dimensionality. The continuum of plane waves, if used to describe surface diffusion, is two-dimensional compared to the highly dimensional continuum of the environs. The different dimensionality should be retained in the calculations and still the method should be made possible on a computer. The model is illustrated in fig. 1. Via a localized state the particle couples to the environment. This localized state together with the perturbation it induces in the environment is called a deformation resonance. This can e.g. describe a gas particle incident from the vacuum onto a solid surface thereby deforming the solid surface locally. The environs will then be phonons. The deformation consists of a transient displacement of the substrate atoms near the point of impact.

Alternatively it could describe an adsorbed particle moving across the surface thereby hitting a substrate atom. If hidden dimensions are present, particle and substrate atom could experience a small but significant gravitational attraction. In this case the environs will be gravitons and the deformation resonance will be termed warp resonance.

3.2. Hamiltonian

The Hamiltonian includes terms describing all particles and their interactions:

$$\begin{aligned} H &= H_{gas\ atom} + H_{phonon} + H_{gas\ atom-phon} + H_{graviton} + H_{gas\ atom-graviton} \\ &= \sum_g E_g n_g + E_w n_w + \sum_p E_p n_p + \sum_g V_{loc}^g (c_g^+ c_w + c_w^+ c_g) \\ &+ \sum_p V_{gp}(c_g^+ c_p + c_p^+ c_g) + \sum_p V_{wp}(c_w^+ c_p + c_p^+ c_w) \\ &+ \omega_{phon} b_{phon}^+ b_{phon} + \sum_k E_k b_k^+ b_k + V_{phon} n_w (b_{phon} + b_{phon}^+) + \sum_k V_k b_{phon}^+ b_{phon} (b_k^+ + b_k) \\ &+ \varepsilon_{grav} a_{grav}^+ a_{grav} + \sum_\kappa \varepsilon_\kappa a_\kappa^+ a_\kappa + \sum_\kappa [V_{grav,w} n_w a_{grav}^+ a_\kappa + V_{w,grav} n_w a_\kappa^+ a_{grav}] \end{aligned} \quad (1)$$

The meaning of the symbols is: n_g, n_w: occupation number operators for the gas particle in the vibrationally ground and excited (parallel to the surface) core movement states and in the deformation resonance; c_i^+, c_j: creation and annihilation operators for the gas particle in the respective core movement states; V_{loc}^g: interaction potential between a gas particle state and the deformation resonance; V_{gp}, V_{wp}: interaction potentials between the gas particle states and plane waves; E_g, E_w, E_p: energies of the respective gas particle states and plane waves. ω_{phon}, E_k: energy of the local and continuum phonon modes; $b_{phon}^+, b_{phon}, b_k^+, b_k$: creation and annihilation operators for the local and continuum phonon modes; $\varepsilon_{grav}, \varepsilon_\kappa$: energy of local and continuum environs (gravitons); $a_{grav}^+, a_{grav}, a_\kappa^+, a_\kappa$: creation and annihilation operators for environs (gravitons); $V_{grav,w}$: interaction between the gas particle and the gravitons within the warp resonance (for the sake of a simpler notation the polarization degrees of freedom are not explicitly displayed); V_{phon}, V_k: interaction between the gas particle and the phonon modes within the deformation resonance (cf. [2]).

3.3. Formalism

The basis for the representation of the Hamiltonian is similar to CI in quantum chemistry. We have many-particle states, which are product states from basis states. The basis states involve states for the local particle (ground $\mid g\rangle$ and local deformation state (warp resonance) $\mid w\rangle$) and for its delocalized movement (plane waves) $\{\mid p\rangle\}$. The environ field consists of local environs and continuum environs: $\mid grav\rangle$, $\{\mid \kappa\rangle\}$ and $\mid phon\rangle$, $\{\mid k\rangle\}$. In a many-particle picture the dynamics of the system is described with the help of both localized and delocalized basis states for all particles in the model: adsorbate, solid excitations, gravitons. The interactions are between many-body states, which are product states with components referring to the gas particle core movement parallel to the surface, its vibrational state and environmental excitations.

The initial state of the particle depends upon the experiment to be described. For instance it is a plane wave if we describe electrons in a metal, it is a state with a local component if we describe adparticle diffusion on a solid surface. The time evolution of the initial states is visualized by following the time evolution of the density matrix. The density matrices are defined in the initial basis and in the basis of the 4D world.

The different dimensionality of the continua, the local region is entangled to, means different time and energy scales. However, it is not possible to have continua states at all energies. This difficulty is surmounted by an adequate discretization procedure.

In a coherent world model we follow the time evolution of an adsorbed particle localized on a substrate surface with 2D periodicity at its equilibrium position in the adsorption well. The localized wave packet used to represent the particle is not an eigenstate in a 2D periodic potential, hence it changes its shape with time as the time-dependent Schrödinger equation requires. The entanglement with different elementary excitations in the environment will affect the evolution of the local adsorbate state, leading eventually to particle localization, hence to decoherence. The competition between the environmental decoherence and the coherent time evolution depends upon their relative rates. As we showed in a recent paper [2], the entanglement with the lattice phonons and tomonagons in the solid surface is not fast enough to decohere a single adsorbed atom as it is seen in the low temperature STM images. A more efficient decoherence mechanism was suggested due to entanglement with the gravitons within the spacetime deformation (warp) resonance.

We use the philosophy and formalism of the Quantum Nano Dynamics (QND) theory, a field theoretical approach developed to treat correlated many-particle systems [2, 14, 15]. In the present paper we are concerned with the coherent time evolution of the core motion of a single adsorbed atom entangled with the continuum of delocalized plane waves and the environmental field provided by the surface atoms. Although in the presented numerical calculations the delocalized phonons, in which the local phonons decay, are not explicitly treated together with other environmental degress of freedom, we present the formalism including bosonized solid excitations (e.g. phonons) in order to demonstrate how bosonized solid excitations can be handled.

The coupling of the local gas particle states to the plane wave continuum leads to energy broadening and finite lifetime in the local gas particle states, hence to delocalization. We take into account the two lowest vibrational states of an adsorbate parallel to the solid surface $\mid g\rangle$ and $\mid w\rangle$ which is contracted (of the order of 1 bohr). Tunnelling out of the adsorption site into plane waves, hence delocalization and diffusion, can only occur via the state $\mid g\rangle$. The coupling of the atom core movement to the environ (graviton) continuum occurs within the deformed space which is represented with the more contracted basis function $\mid w\rangle$.

4. Coherent time evolution of the localized adsorbate entangled with the plane wave and environ continua

The time evolution of the adsorbate core movement state follows Schrödinger's time-dependent equation:

$$-i\hbar\frac{\mathrm{d}\Psi(t)}{\mathrm{d}t} = H\Psi(t) \qquad (2)$$

with Ψ an eigenfunction of the total Hamiltonian eq. 1. We assume that at $t_0 = 0$ the adparticle is in the local state $\mid g\rangle$ and an environ of high energy is excited: $\mid \kappa\rangle$. The approximate eigenfunctions Ψ_I (determined by a limited CI-expansion) of the QND hamiltonian eq. (1) are used to determine the time dependence according to the unitary time evolution:

$$\Psi_I(t) = \Psi_I(t_0)e^{iE_I t/\hbar} \qquad (3)$$

The expansion of the local adsorbate core movement state in the eigenfunctions $\{\mid I\rangle\}$ at time $t_0 = 0$ is then:

$$\mid g0_{t_0}\rangle = c^{-1}_{1,g0} \mid 1\rangle + c^{-1}_{2,g0} \mid 2\rangle + ... \qquad (4)$$

The time development of $\mid g0\rangle$ is obtained as:

$$\mid g0_{t_1}\rangle = c^{-1}_{1,g0}e^{iE_1 t_1/\hbar} \mid 1\rangle + c^{-1}_{2,g0}e^{iE_2 t_1/\hbar} \mid 2\rangle + ... \qquad (5)$$

We need the representation of the density operator at t_1 in the input basis, for instance:

$$\rho(t_1) = | g0_{t_1}\rangle\langle g0_{t_1} | \tag{6}$$

The procedure is repeated for each next time step to provide the coherent time evolution of the density operator (in the input basis) which can be used to illustrate the delocalization and diffusion of a localized adsorbate on a solid surface in interaction with the continua of plane waves and environs.

5. String theory of gravitational induced decoherence

In this section we discuss the background which leads to considering gravitational interaction as relevant for decoherence phenomena [16].

5.1. Large hidden dimensions: The Arkani-Hamed-Dimopoulos-Dvali theory

Arkani-Hamed, Dimopoulos and Dvali (ADD) [17] proposed that large extra dimensions could be underlying the Standard Model and solve the hierarchy problem. Rolled-up extra dimensions can be big, if the particles in our four dimensional world are confined to a brane and are not free to travel in the bulk of a higher dimensional space. This would not contradict experimental results.

In the ADD model all existing particles are confined to a single brane which does not bound space. It simply sits inside the extra curled-up dimensions, like a line along a cylinder. The ADD model may have up to seven extra dimensions. The reason that the extra dimensions are allowed to be large is that gravity is very feeble. The gravitational force between two electrons is 10^{41} times weaker than the electromagnetic force. It is therefore extremely difficult to investigate gravity at short distances experimentally. For light objects at close distances, gravity is so weak that its effects are readily overwhelmed by other forces. Based on experimental results, it is known that extra dimensions must be smaller than about a tenth of a millimeter. If gravity is the only force in the bulk of higher dimensional space, the existence of large extra dimensions would not contradict any experimental results.

In the absence of branes, millimeter-size extra dimensions would have to be ruled out. However, branes make large extra dimensions conceivable. Branes can trap all known particles so that only gravity experiences the full higher dimensionality of space. In the ADD scenario everything that does not involve gravity would look exactly the same as it would without the extra dimensions.

One can hope to see evidence of sub-millimeter-size dimensions in the ADD scenario only with an extremely sensitive gravity probe. We argue that in fact adsorbed microscopic particles on surfaces would provide such a sensitive probe.

The gravitational potential provided by a pointlike mass M in 4-dimensional space-time is:

$$V^{(4)}_{grav}(r) = -\frac{GM}{r} \tag{7}$$

In 11-dimensional space-time (10 space dimensions) the gravitational potential is:

$$V^{(11)}_{grav}(r) = -\frac{G^{(11)}M}{\pi^7 r^8} \tag{8}$$

$G^{(11)}$ is the gravitational constant in 11-dimensional space-time and has to be determined by comparison with experiment. The gravitational law eq. (8) cannot be valid for large separations r as this would violate the experimentally verified classical law eq. (7). Therefore the hidden dimensions are rolled-up (compactified) to a diameter $2a$ so that at large distances the separation in the hidden dimensions never exceeds $2a$. Equating the classical and the 11-dimensional gravitational law at large distances yields:

$$-\frac{GM}{r} = -\frac{G^{(11)}M}{(2a\pi)^7 r} \rightarrow G^{(11)} = (2a\pi)^7 G \tag{9}$$

This choice makes the two laws eq. (7) and eq. (8) agree at separations larger than $2a$. Inserting eq. (9) in eq. (8) yields at $r = 1$bohr:

$$V^{(11)}_{grav}(r = 1) = GM(2a)^7 \tag{10}$$

This means that at $r = 1$bohr the gravitational interaction is $(2a)^7$ times as strong as predicted by the classical law eq. (7). Assuming that at $r = 1$bohr the gravitational interaction strength between two electrons is 0.000027meV $= 10^{-9}$ Hartree one finds that the compactification radius has to be of order 10000 bohr = 500 nm.

If this is correct then experimenters will find a deviation from Newton's classical gravitational law at distances smaller than 10^{-3} mm. The best experiment we know of up to now concluded that at 100 nm the deviations from Newton's classical gravitational law are smaller than a factor of 10^9 [18]. We predict at 100 nm a deviation by a factor 10^7. So everything is still in agreement with experiment but it is getting close!

5.2. Gravitons, density of states (DOS) and relation to the DOS of the 2D plane waves

As in the electromagnetic case, we expand the gravitational radiation field in the eigenmodes of resonators with fixed frequency, polarization and field distribution. Although the selection of modes is in principle arbitrary, the most popular expansion uses the eigenmodes of rectangular boxes with perfectly reflecting walls. The mirrors at position 0 and a impose the boundary conditions that the field vanishes at those positions. The eigemodes have field distributions that vary with $\sin(n\pi x/a)$, where n is the mode number, x the spatial variable and a the separation of the two mirrors. The spatiotemporal variation of the field is

$$F_n = \sin(n\pi x/a)e^{i\omega_n t} \tag{11}$$

where the angular frequency is $\omega_n = \frac{n\pi c}{a}$. This defines the wave number as $k = (n\pi/a)$ and consequently the one dimensional density of modes in k-space as

$$\rho_k = \frac{a}{\pi} \tag{12}$$

Interestingly this density of k-points is the same as for a massive particle (mass M) in a perfectly reflecting box of the same size as can be seen by equating the energy levels for a particle in a box with the kinetic energy:

$$E_n = \frac{\pi^2 n^2}{2a^2 M} = \frac{k^2}{2M} \tag{13}$$

yielding as well eq. (12). Using the dispersion relation for gravitons $\epsilon_\kappa = \kappa c$ with c the velocity of light, $\kappa = \sqrt{\kappa_1^2 + \kappa_2^2 + \kappa_3^2 + \kappa_1^2 + ... + \kappa_d^2}$ and d the dimension of gravitational k-space we write for the graviton density of states:

$$\rho_{grav} = \sum_\kappa \delta(E - \epsilon_\kappa) = \int d^d\kappa \rho_\kappa^d \delta(E - \epsilon_\kappa). \tag{14}$$

Transforming to spherical coordinates [19], one obtains:

$$\rho_{grav} = \int d\eta \eta^{d-1} \delta(E - c\eta) \rho_\kappa^d \frac{\pi^{d/2}}{\Gamma(1+\frac{d}{2})}. \tag{15}$$

Specializing to 10 spatial dimensions and substituting $d\eta = d\epsilon/c$ one obtains

$$\rho_{grav} = \frac{1}{c} \frac{\pi^5}{\Gamma(1+5)} \rho_\kappa^d \int d\epsilon \frac{\epsilon^{d-1}}{c^{d-1}} \delta(E - \epsilon) \tag{16}$$

$$= \frac{E^{d-1}}{c^d} \frac{\pi^5}{5!} (\frac{L}{\pi})^3 (\frac{a}{\pi})^{d-3} \tag{17}$$

where L is the normalization length of the macroscopic 3-dim space and a is the normalization length of the hidden dimensions. The density of states for the two dimensional movement of an adparticle of mass M is $\rho_{2D} = \frac{2ML^2}{\pi}$ Therefore the relationship between the density of states of gravitons at energy E and the states for two dimensional adparticle motion is:

$$\frac{\rho_{grav}}{\rho_{2D}} = \frac{\frac{E^{d-1}}{c^d}}{2M} \frac{\pi^5}{5!} \frac{La^{d-3}}{\pi^{d-1}} \approx 10^{34}. \tag{18}$$

For the wave vector of the decohering graviton we assumed $\kappa \approx 10$, which is valid for a deformation resonance width of 1 bohr ($\lambda = 2\pi/\kappa \approx 1$ bohr), $L \approx 10^7$ bohr, $M = 2000$ and $a \approx 10^4$ bohr and hence a graviton energy of order $\epsilon_\kappa \approx 10^3$ Hartree.

In conclusion the graviton continuum appears infinitely dense compared to the continua describing any particle motion in our 3 dimensional space.

6. Numerical results

The model and its result scale with the energy unit, the time unit being related via the relationship $\Delta E \Delta t = 2\pi\hbar$, if the time is given in atomic units. What determines the behaviour of the model system is the coupling strength between gas particle and environment in relation to the coupling strength to the degrees of freedom of the local system. We do not present here results for a specific experimental situation, but rather give a general consideration without ascribing concrete values to the parameters of the model.
There are three critical couplings in the model:

(i) coupling between the vibrational state and the deformation resonance described by the parameter γ_{penetr}.

(ii) coupling between the deformation resonance and the plane wave continuum described by the parameter γ_{escape}

(iii) coupling between the deformation resonance and the environment described by the parameter $\gamma_{environ}$.

In our example the ratio of the three coupling parameters has been chosen as $\gamma_{penetr} : \gamma_{escape} : \gamma_{environ} = 1 : 10 : 14$. The separation of the energy levels of the local continuum (plane waves) will be 2.2×10^{-3} times smaller than γ_{penetr}. The absolute values will depend on the particular physical situation. The energy zero is chosen to agree with the energy of the non-interacting vibrational state ($E_g = 0$). Fig. 2 illustrates the energetic situation without including the coupling to the environment. The spectral weights are plotted only for the energy eigenvalues of the local system, where the eigenvalue energy separation corresponds to the number of states in the local continuum.

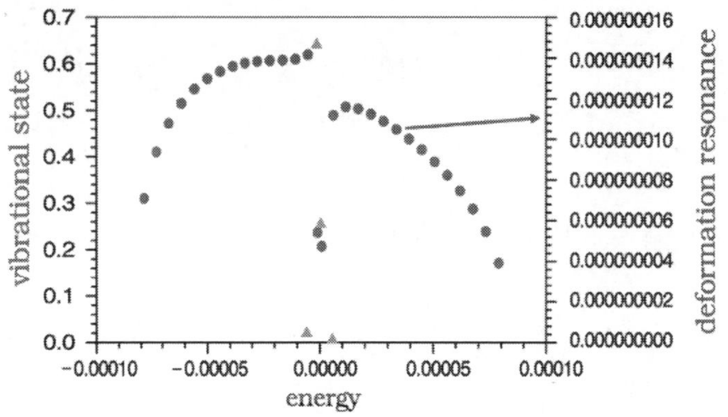

Figure 2. Spectral energy distribution of the local states without including coupling to the environment. The blue dots represent the spectral distribution of the deformation resonance in the eigenstates of the local system. The red triangles indicate the values for the vibrational state. Note the different scales.

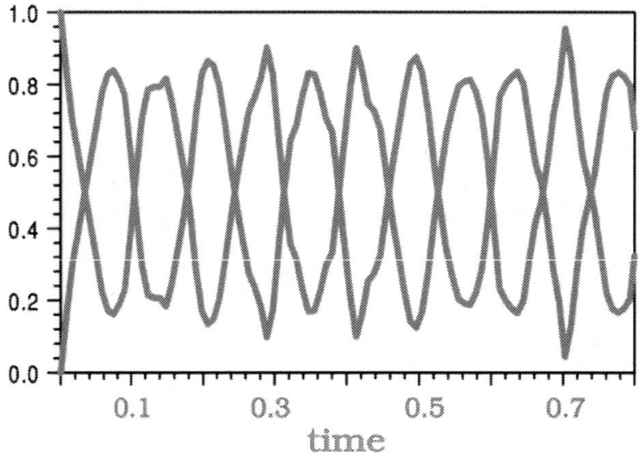

Figure 3. Time evolution of the occupation of the local states (vibrational state plus deformation resonance, red curve) and the occupation of the local continuum (blue curve) without including the coupling to the environment.

The time development of a wave packet consisting at time zero of the vibrational state only is displayed in fig. 3. After 0.06 relative time units the particle has nearly completely vanished from the vibrational state and delocalized into the local continuum (plane wave). From thereon the particle returns back to the vibrational state because of the discreteness of our model calculations. In the following there is an ongoing oscillation between the vibrational state and the states in the local continuum.

If the coupling to the environment is included we obtain the results displayed in fig. 4. Each discrete value of fig. 2 has turned into a narrow resonance due to coupling to the environment. The spectral distribution of the vibrational state became an on-shell resonance at energy zero comprising many levels of the environmental continuum.

The time development of the wave packet changes dramatically. Figure 5 demonstrates that up to 0.06 relative time units there is a tendency of the particle to escape into the local continuum. At later times, however, the interaction with the environmental continuum has been fully developed

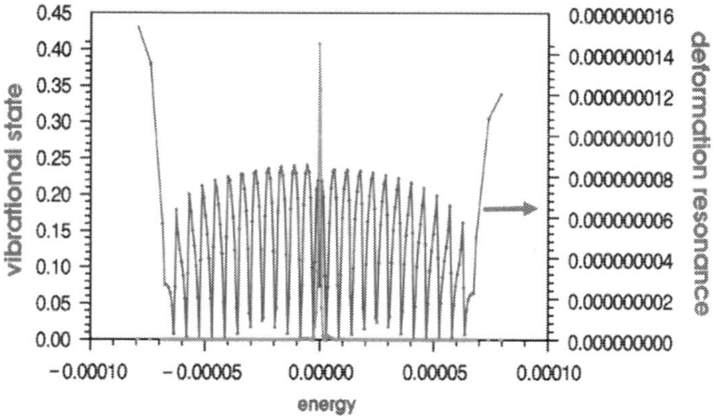

Figure 4. Spectral energy distribution of the local states including the coupling to the environment. Note the different scales indicated on the left and right hand sides of the figure.

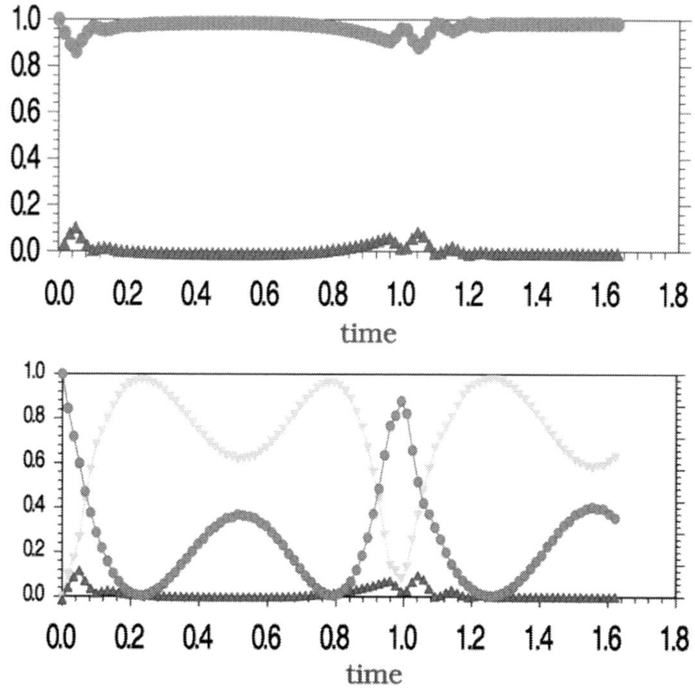

Figure 5. Time evolution of the occupation of the local states including coupling to the environment. In the upper panel the red and blue curve have the same meaning as in fig. 3. In the lower panel the green curve describes the particle in the deformation resonance and the red curve in the vibrational state, respectively.

and the particle is locally captured because of the entanglement with the many degrees of freedom of the environment. Around the time of one relative unit there is a recurrence to be observed which again is solely due to the discreteness of the environmental continuum in our model calculation. For infinitely fine discretization there would be no recurrence.

The energy of the non-interacting deformation resonance is in these calculations chosen as

$E_w = 10^3$ in the relative units indicated. Physically this means a relatively strong potential corrugation: the deformation (warp) resonance lies energetically high and is off-shell relative to the scattering energy. The on-shell coupling between the vibrational state and the environ continuum is therefore indirect and weak, which is the condition for the dimensionality effect to be demonstrated here. If the gas atom vanishes from the vibrational state, it localizes in the deformation resonance and cannot scatter into the Bloch states of the local system. The gas atom localizes in the deformation resonance as a result of the dimensionality effect. The dimensionality effect is recognized by its proportionality to the number of gravitons in an energy interval when starting from a rough discretization. For finer and finer discretizations the results will rather quickly converge to definite limiting values.

The dimensionality effect leads to particle stabilization in the state which results from the *weak* coupling between the initial particle state and a continuum of states of large dimensions. For instance a particle in a plane wave state, weakly entangled to the graviton continuum, will be localized because the dimensions of the localizing graviton continuum are much higher compared to those of the plane wave continuum. The weak coupling to the environs does not allow indirect mixing (via the deformation resonance) of other plane wave states, hence no scattering in different delocalized states is possible. The dominating scattering channel is in the states with graviton components, hence localization.

Apparent classical behaviour of atoms and molecules near surfaces can qualitatively be rationalized by the theory outlined in this contribution:

- The classical behaviour of inelastic scattering of atoms from surfaces at low temperature can be traced back to the different dimensions of the Hilbert spaces of the system and the environment and the weak interaction limit (dimensionality effect).
- The dimensionality effect explains also the small survival probabilities for metastable rare gas atoms scattering from surfaces.
- The classical scattering angular distributions of metastable rare gas atoms indicate decoherence due to strong system - phonon environment interaction leading to localized pointer states.

String theory's hidden dimensions provide a possibility of explaining localization experiments where the conventional environmental continua do not serve for this purpose.

7. References

[1] Anderson P W 1972, More is different *Science* **177** 393
[2] Doyen G and Drakova D 2009 *Surface Science* **603** 1579
[3] Bruch L W, Schöllkopf W and Toennies J P 2001 in: Atomic and Molecular Beams, Springer p. 165
[4] Boato G and Cantini P 1974 in: Proc. Int. Sch. Phys. Enrico Fermi, Course LVIII p. 707; Boato G, Cantini P and Mattera L 1976 *Surface Sci.* **55** 141
[5] O'Keefe D R, Palmer R L and Smith J N jr. 1971 *J. Chem. Phys.* **55** 4572
[6] Manson J R 1972 *J. Chem. Phys.* **56** 3451
[7] Finzel H U, Frank H, Hoinkes H, Luschka M, Nahr H, Wilsch H and Wonka M 1975 *Surface Sci.* **49** 577
[8] Goodman F O and Wachman H Y 1967 *J. Chem. Phys.* **46** 2376; Goodman F O 1971 *Surface Sci.* **24**, 667; Goodman F O 1974 *Prog. Surf. Sci.* **5** 261
[9] Cabrera N 1959 *Faraday Disc. Chem. Soc.* **28** 16; Zwanzig R W 1960 *J. Chem. Phys.* **32** 1173; McCarrol B and Ehrlich G 1963 *J. Chem. Phys.* **38** 523; Goodman F O and Wachman H Y 1967 *J. Chem. Phys.* **46** 2376
[10] Doyen G 1980 *Phys. Rev.* B **22** 497
[11] Conrad H, Doyen G, Ertl G, Küppers J and Sesselmann W 1982 *Chem. Phys. Lett.* **88** 281
[12] Mattsson Th, Engberg U and Wahnström G 1993 *Phys. Rev. Lett.* **71** 2615; Mattsson Th and Wahnstroem G 1997 *Phys. Rev.* B **56** 14944
[13] Lauhon L J and Ho W 2000 *Phys. Rev. Lett.* **85** 4566
[14] Drakova D and Doyen G 2007 *J. Physics: Conference ser.* **61** 262

[15] Drakova D and Doyen G 2005 *Physica Status Solidi* (b), **242** 103
[16] Randall L 2005 Warped passages: unraveling the mysteries of the Universe's hidden dimensions, New York, Ecco
[17] Arkani-Hamed N, Dimopoulos S and Dvali G 1998 The hierarchy problem and new dimensions at a millimeter *Physics Letters* **429** 263-272
[18] *Physics News Update* 2005, Number 734 # 2, June 22
[19] Zwiebach B 2004 A first course in string theory, Cambridge

Relativistic state reduction model

Daniel Bedingham
Blackett Laboratory, Imperial College, London, SW7 2BZ
E-mail: d.bedingham@imperial.ac.uk

Abstract. In order to address the measurement problem of quantum theory we make the assumption that quantum state reduction should be regarded as a genuine physical process deserving of a dynamical description. Generalizing the nonrelativistic spontaneous localization models of Ghirardi, Rimini, Weber, and Pearle, a relativistic state reduction mechanism is proposed. The mechanism involves nonlinear stochastic modifications to the standard description of unitary state evolution and the introduction of a mediating field to facilitate smearing of quantum field interactions.

1. Introduction

Here we shall address the question of how to understand quantum states and their evolution. The conventional view is that the quantum state is a complete description of a system. The state at some time can be determined from the state at some earlier time by using the Schrödinger equation. *That is*, unless there is a measurement, at which point a quantum state reduction occurs: the state changes in a discontinuous and stochastic way consistent with the observed measurement outcome. There is no underlying theory to decide precisely when a measurement occurs in this picture and the fact that such a fuzzy concept plays such a key role is the quantum measurement problem.

In an attempt to address this issue we shall make two assumptions: (i) the quantum state is a real physical structure (in principle it has some definite value at any given stage despite our possible inability to determine it accurately), and (ii) both the Schrödinger equation and quantum state reduction describe real physical processes which continually affect the quantum state (each to a greater or lesser extent at any given stage).

The task is now to find a set of dynamical equations, more general than the Schrödinger equation, which can be approximated by either the Schrödinger equation or quantum state reduction in situations where each of those descriptions is appropriate. In this way we avoid having to place the concept of measurement at the centre of our theory.

In particular we shall address the question of how to do all this whilst satisfying the constraints of relativity. After a brief review of some existing nonrelativistic models we shall outline our proposal for a relativistic state reduction mechanism. (For a general review of dynamical state reduction models see Ref. [1].)

2. Nonrelativistic state reduction models

2.1. Spontaneous localization of particles

We begin by reviewing the spontaneous localization model of Ghirardi, Rimini, and Weber [2]. The model concerns the quantum mechanics of a system of distinguishable particles.

Consider this system described at time t by the state $|\psi(t)\rangle$. For each particle we consider a sequence of Poisson distributed random points in time (with frequency of order $10^{-17}\mathrm{s}^{-1}$ making these events extremely rare). At each of these random times the state changes discontinuously under the action of a localization operator (e.g., for particle i)

$$|\psi(t)\rangle \to |\psi(t+)\rangle = L_i(\mathbf{z})|\psi(t)\rangle. \tag{1}$$

At all other times the state evolution satisfies the Schrödinger equation. The localization operator is defined by

$$L_i(\mathbf{z}) = \frac{1}{(\pi r^2)^{3/4}} e^{-\frac{(\mathbf{q}_i - \mathbf{z})^2}{2r^2}}, \tag{2}$$

where $\mathbf{q}_i$ is the (3D) position operator for particle i and $\mathbf{z}$ is a real random variable representing a random point in position space. The operator $L_i(\mathbf{z})$ has the effect of focusing the quantum amplitude (in position space) for particle i about the point $\mathbf{z}$. The parameter r is of order 10^{-7}m and determines the length scale at which localizations occur.

The localization centre $\mathbf{z}$ for particle i is chosen from a probability distribution defined by

$$\mathbb{P}_i(\mathbf{z} \in D) = \int_D d\mathbf{z}\, \frac{\langle\psi(t)|L_i^2(\mathbf{z})|\psi(t)\rangle}{\langle\psi(t)|\psi(t)\rangle}, \tag{3}$$

for some spatial region D. This probability rule ensures that the statistical connection between the quantum state, pre and post localization is maintained (i.e. the Born rule is satisfied).

Now consider a lump of matter composed of vast numbers of particles. Although the chance of a localization occurring in a small amount of time for any one particle is very small, for a sufficiently large number of particles, the chance for at least one particle to be localized can be large. Entanglements between particle positions then entail that bulk superpositions are suppressed. This is the mechanism by which state reductions occur and it explains why they can be neglected at the scale of individual particles. We must be careful to understand here that a measurement is caused by the occurrence of a localization event, not the other way around.

2.2. Spontaneous localization with fields

The major drawback for the spontaneous localization model outlined above is that all particles must be distinguishable. If we allow particles to be indistinguishable then the state must satisfy certain symmetries under particle interchange. If we continue to consider localization events occurring for individual particles then they will break these symmetries.

The resolution of this dilemma, as suggested by Ghirardi, Pearle, and Rimini [3], is to treat the system of particles in the manner of a quantum field. By this we mean that we consider annihilation and creation operators $a(\mathbf{x})$ and $a^\dagger(\mathbf{x})$ which respectively describe the annihilation and creation of particles at point $\mathbf{x}$ (the field operator at $\mathbf{x}$ is $a(\mathbf{x}) + a^\dagger(\mathbf{x})$ and $[a(\mathbf{x}), a^\dagger(\mathbf{y})] = \delta^3(\mathbf{x} - \mathbf{y})$). This enables us to define a smeared number density operator by

$$N(\mathbf{x}) = \int d\mathbf{y}\; g(\mathbf{x} - \mathbf{y}) a^\dagger(\mathbf{y}) a(\mathbf{y}), \tag{4}$$

where $g(\mathbf{x})$ is a spherically symmetric, positive real function peaked around $\mathbf{x} = \mathbf{0}$.

Now consider points Poisson distributed in *spacetime volume* with a certain fixed frequency density independent of the system. The state evolution satisfies the Schrödinger equation except for those times when the timeslice (in spacetime) crosses one of the random points. At these times the state is acted upon by the localization operator

$$L(Z_x) = \frac{1}{(\pi r^2)^{3/4}} e^{-\frac{(N(\mathbf{x}) - Z_x)^2}{2r^2}}, \tag{5}$$

where $x = (\mathbf{x}, t)$ is the random spacetime point and Z_x is a real positive random variable representing a preferred particle number density at x. The effect of the operator $L(Z_x)$ is to resolve different number density states at $\mathbf{x}$ and focus the quantum amplitude about $N(\mathbf{x}) = Z_x$. The probability rule for Z_x is

$$\mathbb{P}(Z_x \in D_x) = \int_{D_x} dZ_x \, \frac{\langle\psi(t)|L^2(Z_x)|\psi(t)\rangle}{\langle\psi(t)|\psi(t)\rangle}, \tag{6}$$

where D_x is some range of possible Z_x values. The localizations are now labelled by spacetime points x rather than particle index i and time.

This picture is quite different from before. Since the random distribution of points in spacetime has nothing to do with the quantum system itself, the points must be quite dense (to account for the possibility of state reductions occurring at any time). But in order to ensure that individual particles are not continuously being localized, the effect of the localization operator must be tuned such that it is only effective for states involving a large spread in $N(\mathbf{x})$ values (r must be sufficiently large). Such states correspond to bulk superpositions.

The smearing function $g(\mathbf{x})$ is necessary since without it the action of the localization operator would result in spatial discontinuities in the number density of particles. Such states have infinite energy. The length scale associated with g is in fact the length scale associated with the localizing behaviour of this model.

Note that in the model presented in Ref. [3], a limit is taken in which the localization operator acts at every single point (giving a continuous localization process). Here for simplicity we consider only discrete localization events.

The developments outlined here pave the way toward a relativistic formulation. Although the state is still described only at a sequence of times (in some frame of reference), the localization events (or hits) have an invariant distribution in spacetime (a Poisson distribution in spacetime volume is invariant since spacetime volumes are invariant). The challenge of formulating a fully relativistic model is to understand how the state moves through the invariant minefield of hits in a way which does not refer to any preferred timeslice.

3. Relativistic state reduction

3.1. Tomonaga picture

In the Tomonaga picture [4] a quantum state is assigned, not just to a sequence of timeslices in some preferred frame, but to every spacelike hypersurface σ. In this way we do not discriminate between different frames of reference, neither locally nor globally. Given an interaction Hamiltonian density $H_{\text{int}}(x)$, the dynamics of the state is given by the Tomonaga equation

$$i\frac{\delta}{\delta\sigma(x)}|\Psi(\sigma)\rangle = H_{\text{int}}(x)|\Psi(\sigma)\rangle, \tag{7}$$

where

$$\frac{\delta|\Psi(\sigma)\rangle}{\delta\sigma(x)} = \lim_{\sigma'\to\sigma} \frac{|\Psi(\sigma')\rangle - |\Psi(\sigma)\rangle}{\Delta\omega}, \tag{8}$$

and where $\Delta\omega$ is the spacetime volume enclosed by σ and σ' about point x (with no points in σ' to the past of σ). In order that the Tomonaga equation has covariant form the Hamiltonian density $H_{\text{int}}(x)$ must be a Lorentz scalar. Furthermore, it must be the case that $[H_{\text{int}}(x), H_{\text{int}}(y)] = 0$ for spacelike separated x and y. This rule means that there is no preferred ordering of spacelike separated points and ensures a consistent and frame-independent assignment of states to hypersurfaces.

3.2. Relativistic hits

The solution to the Tomonaga equation is given by

$$|\Psi(\sigma_1)\rangle \rightarrow |\Psi(\sigma_2)\rangle = Te^{-i\int_{\sigma_1}^{\sigma_2} d^4x\, H_{\rm int}(x)}|\Psi(\sigma_1)\rangle, \tag{9}$$

with no points in σ_2 to the past of σ_1 and where T describes path ordering of timelike separated events. This solution represents the unitary evolution of the quantum state in spacetime. Our goal is to supplement this background with random localization events or hits. As in section 2.2 we consider an invariant Poisson distribution of points in spacetime with fixed frequency density independent of the system.

Consider advancing the hypersurface σ in a timelike direction. We assume that when σ crosses one of the random spacetime points (at x say), in addition to the unitary evolution, the state changes by

$$|\Psi(\sigma)\rangle \rightarrow |\Psi(\sigma_+)\rangle = L(Z_x)|\Psi(\sigma)\rangle. \tag{10}$$

We suppose that $L(Z_x)$ is given by

$$L(Z_x) = \frac{1}{(\pi r^2)^{3/4}} e^{-\frac{(N(x)-Z_x)^2}{2r^2}}, \tag{11}$$

and base the following discussion around determining a possible form for relativistic version of the operator $N(x)$.

First of all $N(x)$ must be a Lorentz scalar in order that the (stochastic) equations of motion have covariant form (just as for $H_{\rm int}$). Next we must have

$$[N(x), N(y)] = 0 \quad \text{and} \quad [N(x), H_{\rm int}(y)] = 0, \tag{12}$$

for any spacelike separated x and y (meaning no preferred ordering of spacelike separated points).

The first obvious model to consider is a scalar field ϕ where we take $N = \phi$ (with $H_{\rm int}$ possibly equal to ϕ^4 or some interaction between ϕ and other fields) [5, 6]. This choice satisfies (12), but if we attempt to calculate the expected change in energy as a hit point is traversed by σ we find an infinite result. The reason is because the pointlike hits lead to spatial discontinuities in the ϕ state of the field. It is for this same reason that a smearing function g was employed in section 2.2. We must try to find a invariant *smeared* operator N for which (12) still holds.

3.3. Mediating field

Here we suggest a solution to this problem making use of a mediating field [7]. The mediating field we use in our state reduction mechanism has some unconventional properties designed to enable us to fulfill the constraints outlined above. We define annihilation and creation operators $a(x)$ and $a^\dagger(x)$ with the property

$$[a(x), a^\dagger(y)] = \delta^4(x-y). \tag{13}$$

Field states are defined by repeated application of the particle creation operator on the vacuum state $|0\rangle$ (which the annihilation operator annihilates). For example, a first excited state is given by

$$|h\rangle = \int d^4x\; h(x) a^\dagger(x)|0\rangle, \tag{14}$$

for some square integrable function h. Note that whereas the modes of a conventional quantum field describe a field configuration on a timeslice (or spacelike hypersurface), the degrees of freedom of the mediating field describe a field configuration over the whole of spacetime.

The point of constructing a field like this is that it enables us to specify a smeared number density operator satisfying (12). We define the number density operator by $n(x) = a^\dagger(x)a(x)$. We then define our smeared number density as

$$N(x) = \int d^4y\, f(x,y)n(y). \tag{15}$$

We also define a smeared field operator by

$$A(x) = \int d^4y\, g(x,y)\left[a(y) + a^\dagger(y)\right]. \tag{16}$$

We do not specify f and g as yet but assume for now that they are invariantly specified and sufficiently well behaved functions.

On its own the mediating field is static; it has no intrinsic dynamics. Its dynamics must result from interaction with other quantum fields. For this we propose an interaction Hamiltonian density of the form $H_{\rm int}(x) = J(x)A(x)$ where $J(x)$ is some (Lorentz scalar) matter density operator for a conventional quantum field (e.g., $J = \phi^2$). This interaction has the effect of exciting the mediating field in proportion to $J(x)$. To clarify the picture we will assume that the mediating field is initially in its vacuum state and that its only excitations are those caused by this interaction.

For all x and x' the smeared operators have the commutation properties $[N(x), N(x')] = 0$, $[A(x), A(x')] = 0$, and

$$[N(x), A(x')] = \int d^4y\, f(x,y)g(x',y)\left[a^\dagger(y) - a(y)\right]. \tag{17}$$

This means that (12) is satisfied if (i) $f(x,y)$ is only nonzero for y the causal past of x, and (ii) $g(x,y)$ is only nonzero for y the causal future of x. In this case $[N(x), A(x')] = 0$ for spacelike separated x and x' (and consequently $[N(x), H_{\rm int}(x')] = 0$ for spacelike separated x and x').

Before summarizing our relativistic state reduction mechanism we remark on the functions f and g. These functions may depend on any Lorentz invariant local properties of the model (e.g, local operator expectations invariantly specified by the state assigned to the past light cone hypersurface; realized values of random variables Z_x, see [7]). This enables us to find locally preferred frames invariantly determined by the state (e.g., its local rest frame), with respect to which the localizations occur. Without this feature we would not be able to talk about localization.

3.4. State reduction mechanism outline

With the mediating field initially in the ground state consider a quantum field in a superposition of two different matter density states (J states). This might correspond to a superposition of two displaced lumps of matter. The subsequent interaction between $J(x)$ and $A(x)$ (described by the Tomonaga equation) leads to an excitation of the mediating field. The result is an entangled superposition in which each matter density state is associated with a mediating field state representing a smeared image of the path of matter through spacetime. This can be thought of as similar to the effect of a particle passing through a cloud chamber where a record of the track is formed and left behind.

Next suppose that a random hit occurs at some spacetime point where the mediating field has been excited. If the mediating field is in a superposition of different $N(x)$ states at this point (each corresponding to different parts of the entangled superposition) then the localization operator will choose between the different possibilities by suppressing the quantum amplitude of states with $N(x)$ different from Z_x. Consequently, one of the matter density states in the superposition will be suppressed.

The absence of any point interactions in this mechanism ensures that the result is finite [7].

3.5. Probabilities
Suppose that a sprinkling of Poisson distributed points in spacetime is given and suppose that we are interested in the probabilities for $\{Z_x|\sigma_1 < x < \sigma_2\}$ (i.e. random variables Z_x at all random points enclosed by the surfaces σ_1 and σ_2). The probability that each Z_x is in the range D_x is given by

$$\mathbb{P}\left(\{Z_x \in D_x|\sigma_1 < x < \sigma_2\}\right) = \left(\prod_{\sigma_1<x<\sigma_2} \int_{D_x} dZ_x\right) \frac{\langle\Psi(\sigma_2)|\Psi(\sigma_2)\rangle}{\langle\Psi(\sigma_1)|\Psi(\sigma_1)\rangle}, \tag{18}$$

where it is understood that $|\Psi(\sigma_2)\rangle$ depends on $\{Z_x|\sigma_1 < x < \sigma_2\}$ via Eq. (10).

3.6. State reduction timescale
Consider an initial equal superposition of two displaced $J(x)$-states, each describing some localized lump of matter. We assume that each $J(x)$-state has some finite spatial extent (outside of which we can take $J(x) = 0$). Working in some fixed frame, let the spatial volume of points belonging to one but not both of the spatial extents of the two $J(x)$-states be $V_\triangle$. The state reduction timescale is then given (in this frame) by [7]

$$\tau \sim r^2\mu^{-1}V_\triangle^{-1}J^{-4}, \tag{19}$$

where μ is the frequency density of hit points in spacetime and J is the typical value of $J(x)$ in $V_\triangle$. This is a result which can in principle be experimentally tested [8].

3.7. Nonlocality
Any proposed completion of quantum theory must violate the concept of Bell locality. Here we briefly remark on the way in which this model is nonlocal.

Consider a state which describes two spacelike separated subsystems of an entangled global system (such as in an EPR-type experiment). Suppose that each of these subsystems undergoes a state reduction (such as that involved in a spin measurement). The probability rule (18) ensures not only that the outcomes of the state reduction for each subsystem occur individually with the correct quantum probabilities but also that the joint probabilities for outcomes in the two subsystems satisfy quantum predictions. This implies that Z_x values are correlated over spacelike separation in the physical probability measure. We might even think of Z globally as a nonlocal hidden variable.

4. Summary
We have outlined a relativistic mechanism describing quantum state reduction in which, by construction, outcomes have the probabilities predicted by standard quantum theory. This has been achieved without the need to talk about measurements. Indeed within this framework, measurements are just special examples of localization phenomena.

The model offers the possibility of consistently and objectively explaining the behaviour of micro and macro objects. The mechanism can be applied to any quantum field for which we can form a Lorentz scalar operator $J(x)$ (both bosons and fermions), and is compatible with gauge field interactions.

The key to this model is the use of the mediating field which enables interactions to be smeared without reference to a preferred foliation of spacetime.

References

[1] Bassi A and Ghirardi G C 2003 *Phys. Rep.* **379** 257
[2] Ghirardi G C, Rimini A and Weber T 1986 *Phys. Rev.* D **34** 470
[3] Ghirardi G C, Pearle P and Rimini A 1990 *Phys. Rev.* A **42** 78
[4] Tomonaga S 1946 *Prog. Theor. Phys.* **1** 27
[5] Pearle P 1990 *Sixty-Two Years of Uncertainty: Historical, Philosophical, and Physical Inquiries into the Foundations of Quantum Physics* ed A I Miller (New York: Plenum Press)
[6] Ghirardi G C, Grassi R and Pearle P 1990 *Found. Phys.* **20** 1271
[7] Bedingham D 2011 *Found. Phys.* **41** 686 (*Preprint* arXiv:1003.2774v2)
[8] Marshall W, Simon C, Penrose R and Bouwmeester D 2003 *Phys. Rev. Lett.* **91** 130401

A new approach to quantum measurement problem: cluster separability

Petr Hájíček

Institute for Theoretical Physics, University of Bern, Sidlerstrasse 5, CH-3012 Bern, Switzerland

E-mail: hajicek@itp.unibe.ch

Abstract. The paper describes a solution to the problem of quantum measurement that has been proposed recently. The literal understanding of the basic rule of quantum mechanics on identical particles violates the cluster separation principle and so leads to difficulties. A proposal due to Peres of how such difficulties could be removed is reformulated and extended. Cluster separability leads to a locality requirement on observables and to the key notion of separation status. Separation status of a microsystem is shown to change in preparation and registration processes. The indispensability of detectors plays an important role. Changes of separation status are alterations of kinematic description rather than some parts of dynamical trajectories and so more radical than 'collapse of the wave function'. Textbook quantum mechanics does not provide any information of how separation status changes run, hence new rules must be formulated. This enables to satisfy the objectification requirement for registrations. To show how the ideas work, a simplified model of registration apparatus is constructed.

1. Introduction

It is well known that the state of quantum theory of measurement is unsatisfactory. Let us mention just the excellent book [1] and more recent clearly written critical review [2].

In a series of papers [3, 4, 5, 6], we looked for solutions of three interrelated problems of quantum mechanics, those of measurement, of classical properties and of realist interpretation. The three main ideas that we proposed were very different from what had ever been published and it seems that we have been successful.

All previous attempts to formulate a realist interpretation of quantum mechanics failed because everybody was looking for objective properties where they could not be found, namely among values of observables. Quantum observables are not objective, see e.g. [1]. Their values can be obtained only after an interaction of a quantum system with a registration apparatus and so they are properties of a composite: microsystem + macrosystem, and not of the microsystem alone.

Hence, we have rejected all these attempts and proposed (an extended discussion is given in [3, 6]) instead the following:

Basic ontological hypothesis of quantum mechanics A property is objective if its value is uniquely determined by a preparation according to the rules of standard quantum mechanics. The 'value' is the value of the mathematical expression that describes the property and it may be more general than just a real number. No registration is necessary to establish such a property but a correct registration cannot disprove its value; in many cases, registrations can confirm the value.

Our interpretation of quantum mechanics differs further from Copenhagen interpretation by assuming the universality of quantum mechanics: all physical systems are quantum systems. Thus, classical systems are only a special kind of quantum systems. Our theory of classical properties is given in [4, 6]. It is based on the principle of maximum entropy rather than on the popular but incorrect ideas about coherent states. For example, we have introduced quantum states called maximum-entropy packets and shown them to match trajectories of classical mechanics better than Gaussian wave packets.

The present paper contains only a simplified version of some selected points concerning the problem of measurement.

2. Identical particles

We have to start with the standard theory of identical particles. The basic rule is:

Rule 1 *Let system $\mathcal{S}$ consists of N particles of the same type, each particle having Hilbert space* $\mathbf{H}$. *Then, the Hilbert space of $\mathcal{S}$ is the symmetrised tensor product of N copies of* $\mathbf{H}$, $\mathbf{H}_s^N$, *for bosons and the anti-symmetrised one,* $\mathbf{H}_a^N$, *for fermions. States and observables of $\mathcal{S}$ are then described by operators on the respective Hilbert space* $\mathbf{H}_s^N$ *or* $\mathbf{H}_a^N$.

It is known that the literal understanding of Rule 1 leads to problems (see e.g. [7] p 128). To explain the problems, let us consider two experiments.

Experiment I: State $\psi(\vec{x}_1)$ of particle $\mathcal{S}_1$ is prepared in our laboratory.

Experiment II: State $\psi(\vec{x}_1)$ is prepared as in Experiment I and state $\phi(\vec{x}_2)$ of particle $\mathcal{S}_2$ of the same type is prepared simultaneously in a remote laboratory.

Now, our laboratory claims: the state of $\mathcal{S}_1$ is $\psi(\vec{x}_1)$ because this has been prepared according to all rules of experimental art. An adherent of Rule 1 who knows about both laboratories will, however, claim: the true preparation is a combination of the two sub-preparations, one in each laboratory, and the state is

$$2^{-1/2}\big(\psi(\vec{x}_1)\phi(\vec{x}_2) \pm \phi(\vec{x}_1)\psi(\vec{x}_2)\big) \ , \tag{1}$$

as Rule 1 requires.

Thus, it seems that the notions of preparation and of state are ambiguous. In fact, Rule 1 requires to work with the state of *all* particles of the same type in the universe and the uncertainty is much higher than that in the above example.

Has this ambiguity any observable consequences? To answer this question, let us first consider Experiment I supplemented by a registration corresponding to the observable of $\mathcal{S}_1$ with kernel $a(\vec{x}_1; \vec{x}'_1)$, and let the registration be made in our laboratory. Measurements of this kind lead to average value

$$\int d^3x_1 d^3x'_1\, a(\vec{x}_1; \vec{x}'_1)\psi^*(\vec{x}_1)\psi(\vec{x}'_1) \ . \tag{2}$$

Second, perform Experiment II supplemented by the same registration in our laboratory as above. The correct observable corresponding to this registration, as required by Rule 1, now is:

$$a(\vec{x}_1; \vec{x}'_1)\delta(\vec{x}_2 - \vec{x}'_2) + \delta(\vec{x}_1 - \vec{x}'_1)a(\vec{x}_2; \vec{x}'_2) \ . \tag{3}$$

Journal of Physics: Conference Series **306** (2011) 012035 doi:10.1088/1742-6596/306/1/012035

Such measurements lead to the average value defined by Eqs. (1) and (3):

$$\int d^3x_1 d^3x'_1\, a(\vec{x}_1;\vec{x}'_1)\psi^*(\vec{x}_1)\psi(\vec{x}'_1) + \int d^3x_1 d^3x'_1\, a(\vec{x}_1;\vec{x}'_1)\phi^*(\vec{x}_1)\phi(\vec{x}'_1) \ . \tag{4}$$

Expressions (2) and (4) differ by a term that is equal to the average of observable $a(\vec{x}_1;\vec{x}'_1)$ in state ϕ. Hence, measurements on $\mathcal{S}_1$ are in general disturbed by other identical systems.

3. Cluster separability

Cluster separability principle is a kind of locality assumption that has been fruitful in several branches of quantum physics, see e.g. [8] and Chap. 4 of [9]. An application of the principle to identical particles can be found for instance in [7] p 128. We formulate it as follows:

Cluster Separability Principle No quantum experiment with a system in a local laboratory is affected by the mere presence of an identical system in remote parts of the universe.

Peres also observed that the principle was in general violated by results similar to those of the previous section, and he suggested a solution. This solution will now be reformulated and extended.

We introduce an important locality property of observables (for generalisation to composite systems, see [6]):

Definition 1 *Let $D \subset \mathbb{R}^3$ be open. Operator with kernel $a(\vec{x}_1;\vec{x}'_1)$ is D-local if*

$$\int d^3x'_1\, a(\vec{x}_1;\vec{x}'_1)f(\vec{x}'_1) = \int d^3x_1\, a(\vec{x}_1;\vec{x}'_1)f(\vec{x}_1) = 0 \ ,$$

for any test function f that vanishes in D.

Now assume for Experiment II that

(i) our laboratory is inside open set $D \subset \mathbb{R}^3$,

(ii) $\mathrm{supp}\,\phi \cap D = \emptyset$.

Then, clearly the second term in (4) vanishes for D-local observables and Eqs. (2) and (4) agree in this case (for a more general theorem see [6]). This suggests our strategy. In fact, what we shall do will formalise and explain the common practice that serenely ignores all unknown identical systems and that is intuitive, irrational but successful.

(i) We introduce the key notion of our theory:

Definition 2 *Let $\mathcal{S}$ be a particle and $D \subset \mathbb{R}^3$ an open set satisfying the conditions:*

- *Registrations of any D-local observable A of $\mathcal{S}$ lead to average $\langle\psi(\vec{x})|\mathsf{A}\psi(\vec{x})\rangle$ for all states $\psi(\vec{x})$ of $\mathcal{S}$.*
- *$\mathcal{S}$ is prepared in state $\phi(\vec{x})$ such that $\mathrm{supp}\,\phi \cap D \neq \emptyset$.*

In such a case, we say that $\mathcal{S}$ has separation status D.

(For generalisation to composite systems and non-vector states, see [6].) Thus, the registration is not disturbed by other states of identical systems. This can be the case e.g. if wave functions of all other identical systems vanish in D.

(ii) We assume: Any preparation of $\mathcal{S}$ must give it a non-trivial separation status $D \neq \emptyset$. Then D-local observables are individually registrable on $\mathcal{S}$ but only these are if $\mathcal{S}$ has not separation status D' such that D is a proper subset of D'. Indeed, registration of observables that are not D-local are then disturbed by other identical systems. Now,

Peres did not warn that standard observables of quantum mechanics (position, momentum, energy, angular momentum, spin...) were not D-local with $D \neq \mathbb{R}^3$. This problem can be removed by a construction of D-local observables that are registered in real experiments and that resemble, in certain sense, the standard ones [6].

(iii) We list the cases in which Rule 1 does not hold:

Rule 2 *Let system $\mathcal{S}$ be prepared in a state described by state operator* T *so that it has separation status $D \neq \emptyset$. Then its state is* T *and its observables form algebra $\mathbf{A}[\mathcal{S}]_D$ of all D-local observables of $\mathcal{S}$.*

Composition of such states and observables satisfy

Rule 3 *Let systems $\mathcal{S}_1$ and $\mathcal{S}_2$ be prepared in states T_1 and T_2 with separation statuses D_1 and D_2, respectively. Then system $\mathcal{S}_1 + \mathcal{S}_2$ has state $\mathsf{T}_1 \otimes \mathsf{T}_2$ and the algebra of its observables is $\mathbf{A}[\mathcal{S}_1]_{D_1} \otimes \mathbf{A}[\mathcal{S}_2]_{D_2}$.*

Clearly, the preparation described in Rule 3 is a preparation of $\mathcal{S}_1 + \mathcal{S}_2$ in state $\mathsf{T}_1 \otimes \mathsf{T}_2$ with separation status $D_1 \cup D_2$ only if $D_1 = D_2$.

Now, cluster separability holds.

4. Preparations and registrations

As usual, we assume that any measurement on microsystems can be divided into a preparation and a registration procedure. The preparation determines a state and the registration gives a value of an observable. What do the above ideas imply for preparation and registration?

As we have seen in point 2 of our strategy, any preparation must transfer system $\mathcal{S}$ from a trivial into a non-trivial separation status. Thus, the separation status changes during a preparation. What is the relation of registrations to separation status change?

An important assumption of our theory of measurement is:

Rule 4 *Any registration apparatus for microsystems must contain at least one detector and every 'reading of a pointer value' (see e.g. [1]) is a signal from a detector.*

Hence, during registration, system must enter the sensitive matter of a detector. In this way, its non-trivial separation status changes into a trivial one. More discussion of this important point is given in [5, 6].

Let us give an example of separation status change. Let the states in Rule 3 be vector states of identical particles,

$$\mathsf{T}_1 = \psi(\vec{x})\psi^*(\vec{x}') \ , \quad \mathsf{T}_2 = \Psi(\vec{x}_1, \cdots, \vec{x}_N)\Psi^*(\vec{x}'_1, \cdots, \vec{x}'_N) \ .$$

Then the state of $\mathcal{S}_1 + \mathcal{S}_2$ is

$$\psi(\vec{x})\Psi(\vec{x}_1, \cdots, \vec{x}_N) \ . \tag{5}$$

Let the time evolution lead to $\mathcal{S}_1$ entering D_2 and $\mathcal{S}_1 + \mathcal{S}_2$ having separation status D_2. Then the state of $\mathcal{S}_1 + \mathcal{S}_2$ will be

$$\mathsf{P}_{\mathbf{S}} \circ \mathsf{P}^{(N+1)}_{s,a}\big(\psi(\vec{x})\Psi(\vec{x}_1, \cdots \vec{x}_N)\big) \ , \tag{6}$$

where $\mathsf{P}^{(N+1)}_{s,a}$ denotes symmetrisation or anti-symmetrisation of $N+1$ arguments and $\mathsf{P}_{\mathbf{S}}$ is the projection to the unit sphere. The map

$$\mathsf{P}_{\mathbf{S}} \circ \mathsf{P}^{(N+1)}_{s,a} : \mathbf{H} \otimes \mathbf{H}^N_{s,a} \mapsto \mathbf{H}^{N+1}_{s,a}$$

is a non-invertible and non-linear map between two different Hilbert spaces. Second, in state (5), the observables that are registrable on $\mathcal{S}$ form the algebra $\mathbf{A}[\mathcal{S}_1]_{D_1}$. In state (6), the algebra would be $\mathbf{A}[\mathcal{S}_1]_\emptyset = \emptyset$: there are no observables that would be registrable individually on $\mathcal{S}_1$. Thus, the set of observables changes from $\mathbf{A}[\mathcal{S}_1]_{D_1}$ to $\mathbf{A}[\mathcal{S}_1]_\emptyset$.

In classical mechanics, the possible states of system $\mathcal{S}$ are all positive normalised functions (distribution functions) on the phase space $\mathbf{P}$ and possible observables are all real function on $\mathbf{P}$ (at least, all such observables have definite averages on $\mathcal{S}$ independently of external circumstances). $\mathbf{P}$ is fixed and uniquely associated with the system alone and forms the basis of this kinematic description. Hence, transitions between different sets of observables similar to those described above would be impossible in classical mechanics. They are only enabled in quantum mechanics by the non-objective character of observables: not only their values cannot be ascribed to microsystem $\mathcal{S}$ alone but some of them are not even registrable in principle due to external conditions in which $\mathcal{S}$ is.

We assume that the quantum kinematics of a microsystem is defined mathematically by the possible states represented by all positive normalised (trace one) operators, and possible observables represented by some self-adjoint operators, on the Hilbert space associated with the system. Then the transitions of states and observables that go with changes of separation status cannot be viewed as a part of a dynamical trajectory due to some new version of the dynamics of $\mathcal{S}$, but as a change of its kinematic description. Thus, although the change of separation status is similar to the collapse of the wave function (the non-local character included), it is both more radical and better understood.

What has been said up to now shows that textbook quantum mechanics is incomplete in the following sense:

(i) It accepts and knows only <u>two</u> separation statuses:
  (a) that of isolated systems, $D = \mathbb{R}^3$, with the standard operators as observables, and
  (b) that of a member of a system of identical particles, $D = \emptyset$, with no registrable observables of its own.

(ii) It provides <u>no</u> rules for changes of separation status.

Our main idea is: Quantum mechanics must be supplemented by a theory of general separation status and by new rules that govern processes in which separation status changes. The new rules must not contradict the rest of quantum mechanics and ought to agree with and explain observational facts.

5. Extended understanding of separation status

Definition 2 leaves open the question of what the nature of disturbances is that might prevent registrations on $\mathcal{S}$. As yet, all examples of such a disturbance had to do with entanglement of identical particles. In the light of Rule 4, the absence of such entanglement need not be sufficient for $\mathcal{S}$ to allow undisturbed registrations. To explain that, it is helpful to distinguish two kinds of registration.

A *direct registration* first manipulates $\mathcal{S}$ by classical fields and shields so that the prepared beam is split into spatially separated beams, each of which associated with one eigenvalue of the registered observable. Then, there is a set of detectors each of which can be hit by only one beam. Hence, $\mathcal{S}$ must be separated from other microsystems, not only of the same type, so that it is $\mathcal{S}$ that is available to those manipulation by fields and shields and has sufficient kinetic energy to excite a detector.

An *indirect registration*, such as scattering or QND measurement (see e.g. [10] and [7] p 400), lets $\mathcal{S}$ interact with another microsystem $\mathcal{S}'_1$ and it is only $\mathcal{S}'_1$ that is then subject to a direct registration. For the measurement to be QND, several further conditions must be satisfied, but this does not concern us here. After a QND procedure, $\mathcal{S}$ remains available to another one: another system $\mathcal{S}'_2$ of the same type as $\mathcal{S}'_1$ interacts with $\mathcal{S}$ and is then directly registered etc. Information given by the detectors of the direct registrations reveals also something about $\mathcal{S}$.

Thus, detectors are necessary for indirect registrations, too, and $\mathcal{S}$ must be suitably separated from other microsystems, not only of the same type, so that we can be sure that it was $\mathcal{S}$ that interacted with $\mathcal{S}'_1$, $\mathcal{S}'_2$ etc.

There are interesting consequences for macroscopic systems. In general, a macroscopic system $\mathcal{A}$ contains very many different particles. Consider the observable with kernel $a(\vec{x}_k;\vec{x}'_k)$ that concerns particle $\mathcal{S}$ of $\mathcal{A}$. Suppose that there would be an apparatus $\mathcal{B}$ associated with registration of $a(\vec{x}_k;\vec{x}'_k)$ on $\mathcal{S}$, if $\mathcal{S}$ were prepared individually. Then, the apparatus $\mathcal{B}$ cannot be applied to $\mathcal{A}$ to register $a(\vec{x}_k;\vec{x}'_k)$ because $\mathcal{S}$ is not suitably separated.

For example, in a direct registration, readings of $\mathcal{B}$ are signals of a detector that registers $\mathcal{S}$ and $\mathcal{S}$ must be isolated to be manipulable, have sufficient kinetic energy, etc. Hence, to register $a(\vec{x}_k;\vec{x}'_k)$, we need a method that makes measurements directly on $\mathcal{A}$ and which is, therefore, different from apparatus $\mathcal{B}$.

Consider a scattering as an example of indirect registration. Let $\mathcal{A}$ be a crystal. By scattering X-rays off it, relative positions of its ions can be recognised. But rather than a position of an individual ion it is a space dependence of the average density due to all ions. In general, scattering of a microsystem $\mathcal{S}'$ off a macrosystem $\mathcal{A}$ can be determined in terms of potential $V_k(\vec{x},\vec{x}_k)$ that describes the interaction between $\mathcal{S}'$ and k-th microscopic subsystem $\mathcal{S}_k$ of $\mathcal{A}$ so that the whole interaction Hamiltonian is the sum

$$\sum_k V_k(\vec{x},\vec{x}_k) \ .$$

Hence, the scattering yields only information on a sum of over all subsystems $\mathcal{S}_k$ of $\mathcal{A}$ that can interact with $\mathcal{S}'$ and $\mathcal{S}_k$'s need not be identical with $\mathcal{S}'$. Although ingenious potentials can be invented, nature provides only a small number of potentials.

Another example is the kinetic energy of $\mathcal{S}$. Again, this observable cannot be measured by the method kinetic energy is measured on individual systems of type $\mathcal{S}$ (proportional counter, scattering off a crystal, etc.). In the case that $\mathcal{A}$ is in a state of thermodynamic equilibrium, the average kinetic energy can be calculated from the temperature. Hence, a viable method to measure the average is to measure the temperature of $\mathcal{A}$. Again, this is a special case that works only under specific conditions. Further examples could be other additive quantities, such as momentum and angular momentum. Average total values of these quantities can be measured directly on $\mathcal{A}$. We notice that only the averages of some observables with rather large variances can be observed in these cases. It is impossible to obtain the single eigenvalues of these observables as results of registration (for an example, see Ref. [7] p 181).

Thus, the qualitative difference between macroscopic and microscopic systems is due to quantum properties of macroscopic systems rather than to some kind of limits on the validity of quantum mechanics for them.

6. Beltrametti-Cassinelli-Lahti model of quantum measurement

We shall proceed from account [1] (p 38) of model [11]. Let a discrete observable O of microsystem $\mathcal{S}$ with Hilbert space $\mathbf{H}$ is registered. Let o_k be the eigenvalues and $\{\phi_{kj}\}$ be the complete orthonormal set of eigenvectors of O,

$$\mathsf{O}\phi_{kj} = o_k\phi_{kj} \ .$$

We assume that $k = 1, \cdots, N$ so that there is only a finite number of different eigenvalues o_k. This is justified by the fact that no real registration apparatus can distinguish all elements of an infinite set from each other. It can therefore measure only a function of an observable that maps its spectrum onto a finite set of real numbers. Our observable O is such a function.

Let the registration apparatus be a quantum macrosystem $\mathcal{A}$ with Hilbert space $\mathbf{H}_{\mathcal{A}}$. Let $\mathcal{S}$ and $\mathcal{A}$ be prepared in some independent initial states and then interact for a finite time by coupling U, where U is a unitary transformation on $\mathcal{H}_{\mathcal{S}} \otimes \mathcal{H}_{\mathcal{A}}$. Then, a theorem has been shown in [11]:

Theorem 1 *For any initial vector state ψ of $\mathcal{A}$, there is a set $\{\varphi_{kl}\}$ of unit vectors in $\mathcal{H}_{\mathcal{S}}$ satisfying the orthogonality conditions*

$$\langle \varphi_{kl} | \varphi_{kj} \rangle = \delta_{lj}$$

such that U is a unitary extension of the map

$$\phi_{kl} \otimes \psi \mapsto \varphi_{kl} \otimes \psi_k \ , \tag{7}$$

where $\{\psi_k\}$ is a set of N orthonormal vectors in $\mathbf{H}_{\mathcal{A}}$.

One assumes that states ψ_k are uniquely associated with what will be read on the apparatus after the measurement. Then, an important requirement is that the apparatus is in one of the states $|\psi_k\rangle\langle\psi_k|$ after each individual registration. This is called *objectification requirement.*

Suppose that the initial state of $\mathcal{S}$ is an eigenstate, $\mathsf{T} = |\phi_{kl}\rangle\langle\phi_{kl}|$, with the eigenvalue o_k. Then, Eq. (7) implies that the final state of apparatus $\mathcal{A}$ is $|\psi_k\rangle\langle\psi_k|$, and the registration does lead to a definite result. However, suppose next that the initial state is an arbitrary vector state, $\mathsf{T} = |\phi\rangle\langle\phi|$. Decomposing ϕ into the eigenstates,

$$\phi = \sum_{kl} c_{kl} \phi_{kl} \ ,$$

we obtain from Eq. (7)

$$\mathsf{U}(\phi \otimes \psi) = \sum_k \sqrt{p_k} \Phi_k \otimes \psi_k \ , \tag{8}$$

where

$$\Phi_k = \frac{\sum_l c_{kl} \varphi_{kl}}{\sqrt{\langle \sum_l c_{kl} \varphi_{kl} | \sum_j c_{kj} \varphi_{kj} \rangle}} \tag{9}$$

and

$$p_k = \left\langle \sum_l c_{kl} \varphi_{kl} \middle| \sum_j c_{kj} \varphi_{kj} \right\rangle$$

is the probability that a registration of O performed on vector state ϕ gives the value o_k.

The final state of apparatus $\mathcal{A}$ is the partial trace over $\mathcal{S}$:

$$tr_{\mathcal{S}}[\mathsf{U}(\mathsf{T} \otimes \mathsf{T}_{\mathcal{A}})\mathsf{U}^\dagger] = \sum_{kl} \sqrt{p_k} \sqrt{p_l} \langle \Phi_k | \Phi_l \rangle |\psi_k\rangle\langle\psi_l| \ . \tag{10}$$

If the objectification requirement is to be satisfied, two condition must be met:

(A) The final state of the apparatus must the convex combination of the form

$$tr_{\mathcal{S}}[\mathsf{U}(\mathsf{T} \otimes \mathsf{T}_{\mathcal{A}})\mathsf{U}^\dagger] = \sum_j p_j |\psi_j\rangle\langle\psi_j| \ . \tag{11}$$

(B) The right-hand side of Eq. (11) must be the *gemenge structure* of the state.

The notion of gemenge will play an important role. The term has been introduced in Ref. [1], some authors use the term 'proper mixture' or 'direct mixture'. The crucial point is that the convex decomposition

$$\mathsf{T} = \sum_{k=1}^{n} w_k \mathsf{T}_k \tag{12}$$

of any state T can be a gemenge only if its preparation procedure $\mathbf{P}(\mathsf{T})$ is a random mixture with rates (frequencies) w_k of preparations $\mathbf{P}(\mathsf{T}_k)$, where each $\mathbf{P}(\mathsf{T}_k)$ is some preparation procedure for T_k, $k = 1, \cdots, n$. The preparation mixture can be done by humans or result from some process in nature.

Thus, gemenge concerns a physical property of preparation rather than any mathematical characteristic of the right-hand side of Eq. (12) (such as T_k being vector states or being mutually orthogonal, etc). From the mathematical point of view, many different convex decompositions of a general state T may exist. A state is 'extremal' if it cannot be written as a non-trivial convex combination. Extremal states are described by projections onto one-dimensional subspaces of the Hilbert space. A preparation of non-extreme state T selects only some of the mathematically possible convex decompositions of T.

A random mixture of preparations is not uniquely determined by the preparation process. It can be coarsened or refined i.e. some of $\mathbf{P}(\mathsf{T}_k)$ can be combined into one preparation procedure or $\mathbf{P}(\mathsf{T}_k)$ for some k can itself be a random mixture of other preparations.

Definition 3 *The finest convex decomposition of state* T *defined by its preparation as gemenge is called* gemenge structure *of* T.

Thus, gemenge structure of T is uniquely determined by its preparation.

It may be advantageous to distinguish the mathematical convex combination of states from its gemenge structure by writing the sum in Eq. (12) as follows

$$\mathsf{T} = \left(\sum_{k=1}^{n}\right)_{\text{gs}} w_k \mathsf{T}_k \tag{13}$$

in the case that the right-hand side is a gemenge structure of T.

The properties that follow directly from the definition of gemenge structure and that will be needed later are described by the following theorem.

Theorem 2 *(i) Gemenge structure is preserved by unitary dynamics,*

$$\mathsf{U}\left[\left(\sum_k\right)_{gs} w_k \mathsf{T}_k\right]\mathsf{U}^\dagger = \left(\sum_k\right)_{gs} w_k \mathsf{U}\mathsf{T}_k\mathsf{U}^\dagger \ :$$

if the sum on the left-hand side describes a gemenge structure of T*, then the gemenge structure of its evolution is described by the sum on the right-hand side.*

(ii) In the following sense, gemenge structure is also preserved by composition of systems. Let T *be a state of a composite system* $\mathcal{S} + \mathcal{S}'$*. The necessary and sufficient condition for the partial trace over* $\mathcal{S}'$ *to have the gemenge structure described by*

$$tr_{\mathcal{S}'}[\mathsf{T}] = \left(\sum_k\right)_{gs} w_k \mathsf{T}_k$$

is that T *itself has gemenge structure described by*

$$\mathsf{T} = \left(\sum_k\right)_{gs} w_k \mathsf{T}_k \otimes \mathsf{T}'_k \ ,$$

where T'_k *are some states of* $\mathcal{S}'$.

All these ideas on gemenges seem to be well known. Now, an important new point will be added: the Basic Ontological Hypothesis of Quantum Mechanics (see Introduction). It leads to a new meaning of gemenge structure: any individual system prepared in the state (13) is objectively in one of the states T_k, because each of the systems has been prepared by one of the preparations $\mathbf{P}(\mathsf{T}_k)$, and the probability that $\mathbf{P}(\mathsf{T}_k)$ has been used is w_k.

It follows that the contents of both points **(A)** and **(B)** can be concisely written as

$$tr_{\mathcal{S}}[\mathsf{U}(\mathsf{T} \otimes \mathsf{T}_{\mathcal{A}})\mathsf{U}^\dagger] = \left(\sum_j\right)_{\rm gs} p_j |\psi_j\rangle\langle\psi_j| \ .$$

Beltrametti-Cassinelli-Lahti model does not satisfy the objectification criterion because the end state $\mathsf{T} \otimes \mathsf{T}_{\mathcal{A}}$ of the system is $\mathsf{U}(\phi \otimes \psi)$ (Eq. (8)), which is a vector state and can therefore have only a trivial gemenge structure. Then, Point 2 of Theorem 2 implies that this is not compatible with state $tr_{\mathcal{S}}[\mathsf{U}(\mathsf{T} \otimes \mathsf{T}_{\mathcal{A}})\mathsf{U}^\dagger]$ being a non-trivial gemenge.

7. Model of registration apparatus

To show, how our ideas on measurement work, we have constructed a model of registration apparatus [5, 6]. It starts by extending and modifying Beltrametti-Cassinelli-Lahti model by additional assumptions:

(i) Particle $\mathcal{S}$ is prepared in state ϕ and separation status D.

(ii) $\mathcal{A}$ is an array of N detectors $\mathcal{A}_k$ and each detector has separation status D_k where all D_k are mutually disjoint and disjoint from D.

(iii) The support of state φ_{kl} is D_k for all k and l. It follows then that,

$$\langle\varphi_{kl}|\varphi_{mn}\rangle = \delta_{km}\delta_{ln} \quad k \leq N, m \leq N \ . \tag{14}$$

(iv) Each detector is composite, $\mathcal{A}_k = \mathcal{A}'_k + \mathcal{S}'_k$, where $\mathcal{S}'_k$ consists of all particles of $\mathcal{A}_k$ identical with $\mathcal{S}$. Let the number of particles in $\mathcal{S}'_k$ be M_k and let the state of $\mathcal{S}'_k$ be T_k. Systems $\mathcal{S}'_k$ either are 'natural' parts of $\mathcal{A}_k$'s or they are created quickly after the start of the measurement by pollution.

(v) $\mathcal{A}'_k$ is the part of the sensitive matter interacting with $\mathcal{S}$ so that ψ_k is a state of $\mathcal{A}'_k$.

(vi) State (8) is not of an end state of $\mathcal{S} + \sum_k \mathcal{A}'_k$ but a choice of an intermediate state before an amplification procedure.

.

The existence of systems $\mathcal{S}'_k$ leads to a change of the separation status of $\mathcal{S}$ after $\mathcal{S}$ enters a detector. There are several mathematical possibilities for the choice of the intermediate state. One of the possibilities is:

Rule 5 *The intermediate state of* $\mathcal{S} + \sum_k \mathcal{S}'_k + \sum_k \mathcal{A}'_k$ *(before amplification) is*

$$\mathsf{T}_{int} = \left(\sum_k\right)_{gs} p_k \nu_k^2 \mathsf{T}_1 \otimes \cdots \otimes \mathsf{T}_{k-1} \otimes \mathsf{W}_{kk} \otimes \mathsf{T}_{k+1} \otimes \cdots \otimes \mathsf{T}_N \otimes |\psi_k\rangle\langle\psi_k| \ . \tag{15}$$

Choices:

(i) The term

$$W_{kk} = P_{s,a}^{(M_k+1)}(|\varphi_k^1\rangle\langle\varphi_k^1| \otimes T_k)P_{s,a}^{(M_k+1)}$$

expresses our choice of symmetrisation or anti-symmetrisation and ν_k^2 is a normalisation factor that makes W_{kk} to a state (action of P_S).

(ii) Some correlations have been erased so that the state operator T_{int} is diagonal.

(iii) The convex decomposition (15) is postulated to be the gemenge structure of T_{int}.

The choices 2. and 3. are dictated by the objectification requirement which can be regarded as an experimental fact.

Finally, we can show that the new rule cannot be disproved by measuring the 'erased' correlations if standard quantum mechanical rules on jointly measurable observables hold [5, 6].

8. Conclusion

We have improved understanding of the theory of identical particles and removed a disorder in textbook quantum mechanics. This has lead to the notion of separation status. We have discovered the crucial role that detectors have in the theory. Then, both preparations and registrations include changes of separation status, which are changes of kinematic description and so even more radical than 'collapse of the wave function'. The theoretical freedom in changes of separation status has been used to satisfy objectification requirement. Thus, a deep revision of quantum theory of measurement results that has been derived with the help of standard principles of quantum mechanics from analysis of real measurement processes.

As yet, our results are limited to non-relativistic quantum mechanics and to measurements performed on microsystems. Moreover, only special kind of measurements have been considered, registrations of discrete observables with the help of unitary measurement couplings, the definition assumptions of Beltrametti-Cassinelli-Lahti model. Much work remains to be done.

Acknowledgments

The author is indebted to Štefan Jánoš and Jiří Tolar for important suggestions.

References

[1] Busch P, Lahti P J and Mittelstaedt P 1996 *The Quantum Theory of Measurement* (Berlin: Springer)
[2] Ghirardi G C 2009 The interpretation of quantum mechanics: where do we stand *J. Phys. Conf. Ser.* **174** 912013
[3] Hájíček P and Tolar J 2009 Intrinsic properties of quantum systems *Found. Phys.* **39** 411
[4] Hájíček P 2009 Quantum models of classical mechanics: maximum entropy principle *Found. Phys.* **39** 1072
[5] Hájíček P 2011 Quantum measurement problem and cluster separability *Found. Phys.* **41** 640
[6] Hájíček P and Tolar J 2010 Survey of an approach to quantum measurement, classical properties and realist interpretation problems *Acta Phys. Slovaca* **60** 613
[7] Peres A 1995 *Quantum Theory: Concepts and Methods* (Dordrecht: Kluwer)
[8] Keister B D and Polyzou W N 2002 Relativistic Hamiltonian dynamics in nuclear and particle physics *Advances in Nuclear Physics* vol 20, ed J W Negele and E Vogt (New York: Plenum) chapter 4 p 225
[9] Weinberg S 1995 *The Quantum Theory of Fields* vol I (Cambridge: Cambridge University Press)
[10] Braginsky V B and Khalili F Ya 1996 *Rev. Mod. Phys.* **68** 1
[11] Beltrametti E G, Cassinelli G and Lahti P J 1990 Unitary measurement of discrete quantities in quantum mechanics *J. Math. Phys.* **31** 91

5th International Workshop DICE2010 IOP Publishing
Journal of Physics: Conference Series **306** (2011) 012036 doi:10.1088/1742-6596/306/1/012036

Experimental tests of quantum mechanics – Pauli exclusion principle violation (the VIP experiment) and future perspective

C Curceanu (Petrascu)[1,2], S Bartalucci[1], S Bertolucci[3], M Bragadireanu[1,2], M Cargnelli[4], S Di Matteo[5], J-P Egger[6], C Guaraldo[1], M Iliescu[1], T Ishiwatari[4], M Laubenstein[7], J Marton [4], E Milotti[8], D Pietreanu[1], T Ponta[1,2], A Rizzo[1], A Romero Vidal[1], A Scordo[1], D L Sirghi[1,2], F Sirghi[1,2], L Sperandio[1], O Vazquez Doce[1], E Widmann[4] and J Zmeskal[4]

[1] INFN, Laboratori Nazionali di Frascati, CP 13, Via E. Fermi 40, I-00044, Frascati (Roma), Italy
[2] "Horia Hulubei" National Institute of Physics and Nuclear Engineering, Str. Atomistilor no. 407, P.O. Box MG-6, Bucharest - Magurele, Romania
[3] CERN, CH-1211, Geneva 23, Switzerland
[4] The Stefan Meyer Institute for Subatomic Physics, Boltzmanngasse 3, A-1090 Vienna, Austria
[5] Institut de Physique UMR CNRS-UR1 6251, Université de Rennes1, F-35042 Rennes, France
[6] Institut de Physique, Université de Neuchâtel, 1 rue A.-L. Breguet, CH-2000 Neuchâtel, Switzerland
[7] Laboratori Nazionali del Gran Sasso, S.S. 17/bis, I-67010 Assergi (AQ), Italy
[8] Dipartimento di Fisica, Università di Trieste and INFN– Sezione di Trieste, Via Valerio, 2, I-34127 Trieste, Italy

E-mail: catalina.curceanu@lnf.infn.it

Abstract. The Pauli exclusion principle (PEP), consequence of the spin-statistics connection, is one of the basic principles of the modern physics. Being at the very basis of our understanding of matter, as many other fundamental principles it spurs a lively debate on its possible limits, deeply rooted in the very foundations of Quantum Field Theory. Therefore, it is extremely important to test the limits of its validity. Quon theory provides a suitable mathematical framework of possible violation of PEP, where the violation parameter q translates into a probability of violating PEP. The VIP (VIolation of the Pauli exclusion principle) experiment established a new limit on the probability that PEP is violated by electrons, using the method of searching for PEP forbidden atomic transitions in copper. We describe the experimental method, the obtained results, both in terms of the q-parameter from quon theory and as probability of PEP violation, we briefly discuss them and present future plans to go beyond the actual limit by upgrading the experimental setup. We also shortly mention the possibility of using a similar experimental technique to search for eventual X-rays, generated in the spontaneous collapse models in quantum mechanics.

1. Introduction

The Pauli Exclusion Principle (PEP), which plays a fundamental role in our understanding of many physical and chemical phenomena, it is intimately connected to the basic axioms of quantum field theory [1]. Although the principle has been spectacularly confirmed by the number and accuracy of its predictions, its foundation lies deep in the structure of quantum theory and has defied all attempts to produce a simple proof, as for example stressed by Feynman [2]. Pauli himself in his Nobel lecture declared: "...Already in my original paper I stressed the circumstance that I was unable to give a logical reason for the exclusion principle or to deduce it from more general assumptions.....The impression that the shadow of some incompleteness (falls) here on the bright light of success of the new quantum mechanics seems to me unavoidable". Given its basic standing in quantum theory, it seems appropriate to carry out precise tests of the PEP validity and, indeed, mainly in the last 20 years, several experiments have been performed to search for possible small violations [3, 4, 5, 6, 7, 8]. Often, these experiments were born as by-products of experiments with a different objective (e.g., dark matter searches, proton decay, etc.), and most of the recent limits on the validity of PEP have been obtained for nuclei or nucleons.

In 1988 Ramberg and Snow [9] performed a dedicated experiment, searching for anomalous X-ray transitions, that would point to a small violation of PEP in a copper conductor. The result of the experiment was a probability [10] $\frac{\beta^2}{2} < 1.7 \times 10^{-26}$ that the PEP is violated by electrons. The VIP Collaboration set up a much improved version of the Ramberg and Snow experiment, with a higher sensitivity apparatus [11]. Our final aim is to improve the PEP violation limit for electrons by about 3 orders of magnitude by using high resolution Charge-Coupled Devices (CCDs) as soft X-rays detectors [12, 13, 14, 15, 16], and decreasing the effect of background by a careful choice of the materials and sheltering the apparatus in the LNGS underground laboratory of the Italian Institute for Nuclear Physics (INFN).

In the next sections we describe the experimental setup, the results of a first measurement performed in the Frascati National Laboratories (LNF) of INFN, along with a preliminary result obtained by VIP running at the underground Gran Sasso National Laboratory (LNGS) of INFN. We then briefly discuss the results and present future plans to go beyond the existing limit by using fast Silicon Drift Detectors (SDD) and a veto system, which opens, as well, new possibilities for more refined checks on PEP violation.

We will then conclude the paper by presenting some ideas to use a related experimental technique to perform measurements on effects predicted by spontaneous collapse models.

2. The VIP experiment

VIP is a dedicated experiment for the measurement of the probability of the Pauli Exclusion Principle violation for electrons. The experiment uses the same method of the Ramberg and Snow experiment, but with a much better soft X-ray detector and a low-background experimental area - the INFN Gran Sasso underground laboratory. The detector is an array of Charge-Coupled Devices (CCDs), characterized by the excellent background rejection capability, based on pattern recognition and good energy resolution (320 eV FWHM at 8 keV in the present measurement).

2.1. The VIP experiment

The experimental method, originally described in Ref. [9], consists in the introduction of "fresh" electrons into a copper strip, by circulating a current, and in the search for the X-rays resulting from the PEP forbidden radiative transitions that occur if one of these electrons is captured by a copper atom and cascades down to the 1S state which is already filled by two electrons. In particular we are looking for the $2P \rightarrow 1S$ (K_α transition). The energy of this non-Paulian transition would differ from the normal K_α transition energy by about 300 eV (7.729 keV instead of 8.040 keV) [17], providing an unambiguous signal of PEP violation. The measurement

Figure 1. The VIP copper target.

alternates periods without current in the copper strip, in order to evaluate the X-ray background in conditions where no PEP violating transitions are expected to occur, with periods in which current flows in the conductor, when we expect that the "fresh" electrons may lead to Pauli-forbidden transitions.

2.2. The VIP setup

The VIP setup consists of a copper cylinder, 45 mm radius, 50 μm thickness, and 88 mm height (see fig. 1), surrounded by 16 equally spaced "type 55" CCDs made by EEV [18]. The CCDs are at a distance of 23 mm from the copper cylinder, and paired one above the other. The setup is enclosed in a vacuum chamber, and the CCDs are cooled to about 168 K by a cryogenic system. The current flows in the thin cylinder made of ultrapure (99.995%) copper foil from the bottom of the vacuum chamber. The CCDs surround the cylinder and are supported by cooling fingers which protrude from the cooling heads in the upper part of the chamber. The readout electronics is just behind the cooling fingers; the signals are sent to amplifiers on top of the chamber and the amplified signals are read out by ADC boards in the data acquisition computer. More details on CCD-55 performance, as well on the analysis method used to reject background events, can be found in reference [19, 20]. An overall schematic view of the setup is shown in fig. 2.

VIP improves very significantly on the Ramberg and Snow measurement, thanks to the following features:

- use of CCD detectors instead of gaseous detectors, having much better energy resolution (4-5 times better) and higher stability;
- experimental setup located in the clean, low-background, environment of the underground LNGS Laboratory;
- collection of much higher statistics (longer DAQ periods, thanks to the stability of CCDs)

We make full use of these features to obtain an improvement of several orders of magnitude on previous limits.

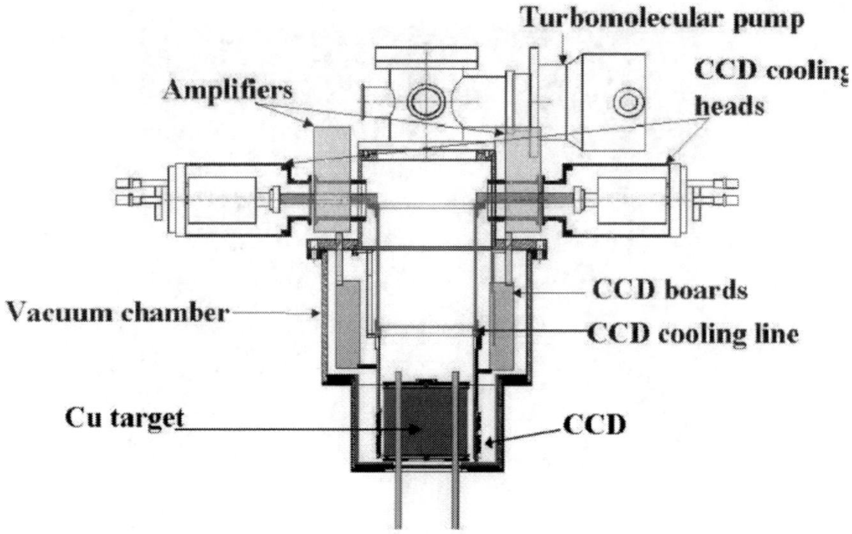

Figure 2. The VIP setup - schematic view

3. The VIP experimental results

3.1. Results obtained at LNF-INFN

The VIP setup took data in the low-background LNGS underground laboratory of INFN. Before installation in the Gran Sasso laboratory, it was prepared and tested at the LNF-INFN laboratory, where measurements were performed in the period 21 November - 13 December 2005. Two types of measurements were performed:

- 14510 minutes (about 10 days) of measurements with a 40 A current circulating in the copper target;
- 14510 minutes of measurements without current.

CCDs were read-out every 10 minutes. The resulting energy calibrated X-ray spectra are shown in figure 3.

These spectra include data from 14 CCD's out of 16, because of noise problems in the remaining 2. Both spectra, apart from the continuous background component, display clear Cu K_α and K_β lines due to X-ray fluorescence caused by the cosmic ray background and natural radioactivity. No other lines are present and this reflects the careful choice of the materials used in the setup, as for example the high purity copper and high purity aluminium, the last one with K-complex transition energies below 2 keV. The subtracted spectrum is shown in Figure 4 a) (whole energy scale) and b) (a zoom on the region of interest). Notice that the subtracted spectrum fluctuates around zero within the statistical error, and is structureless. This not only yields an upper bound for a violation of the Pauli Exclusion Principle for electrons, but also confirms the correctness of the energy calibration procedure and points to the absence of systematic effects.

To extract the experimental limit on the probability that PEP is violated for electrons, $\beta^2/2$, from our data, we used the same arguments of Ramberg and Snow: see references [9] and [21] for details of the analysis. The obtained value is:

$$\frac{\beta^2}{2} < 4.5 \times 10^{-28} \quad (1)$$

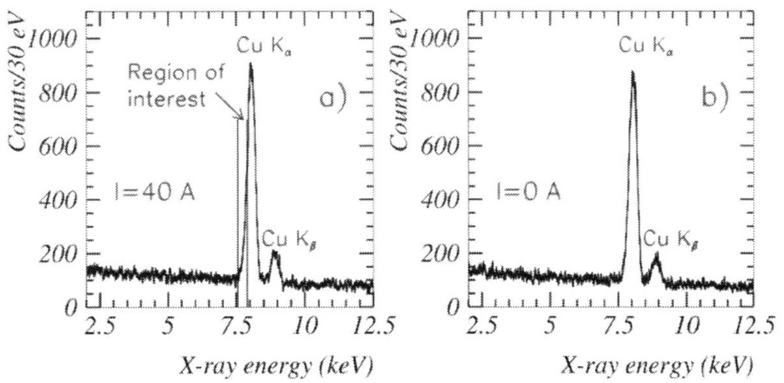

Figure 3. Energy spectra with the VIP setup at LNF-INFN: (a) with current (I = 40 A); (b) without current (I = 0).

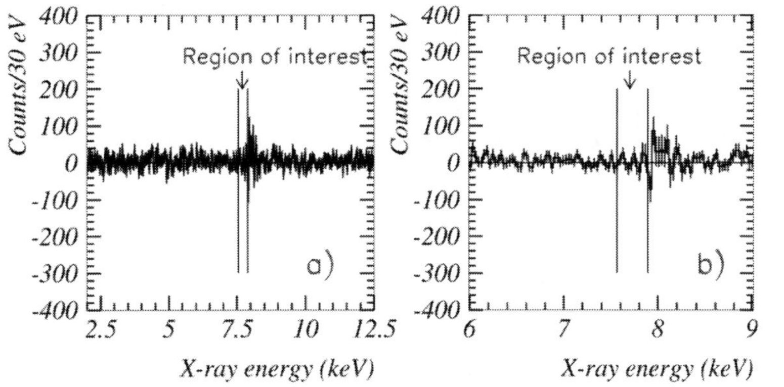

Figure 4. Subtracted energy spectra in the Frascati measurement, *current-on* minus *current-off*, giving the limit on PEP violation for electrons: a) whole energy range; b) expanded view in the region of interest (7.564 - 7.894 keV). No evidence for a peak in the region of interest is found.

Thus with this first measurement in an unshielded enviroment, we have improved the limit obtained by Ramberg and Snow by a factor $\sim$ 40.

3.2. Experimental results from LNGS

The experiment was installed at LNGS-INFN in Spring 2006 and took data till Spring 2010, alternating period with current on (signal) to periods with current off (background).

We have already established a new preliminary limit on PEP violation by electrons from the data taken at LNGS [22]:

$$\frac{\beta^2}{2} < 4.7 \times 10^{-29} \tag{2}$$

We are presently working on an improved version of the setup.

4. Discussion of the results

After the introduction of the straightforward quantum model of Ignatiev and Kuzmin in 1987 [10], Govorkov [23] showed that the model could not work in the wider framework of quantum field theory, because it led to many-particle states with negative norm. The situation changed with the introduction of quon theory [24], which turned out to be a consistent theory of *small* violations of PEP. The basic idea of quon theory is that (anti)commutators, are replaced by weighted sums

$$\frac{1-q}{2}\left[a_i, a_j^+\right]_+ + \frac{1+q}{2}\left[a_i, a_j^+\right]_- = a_i a_j^+ - q a_j^+ a_i = \delta_{i,j} \tag{3}$$

where $q = -1$ ($q = 1$) gives back the usual fermion (boson) commutators. The statistical mixture in equation (3) also shows that the PEP violation probability is just $(1+q)/2$ and thus our best experimental bound on q is

$$\frac{1+q}{2} < 4.7 \times 10^{-29} \tag{4}$$

A consistent interpretation of the VIP results can thus be based on quon theory; however here we note that is not easy to devise tests of PEP, because of many conceptual difficulties (see, e.g. [25]), and VIP (and shares some problems of its own with its precursor, the Ramberg and Snow experiment). The main problem lies in the definition of "fresh" electrons: in fact it is unclear how an electron originally injected by the current source into the copper strip can be set apart from the other electrons already present in the strip. One possibility is that some of these "new" electrons have a "wrong wavefunction" in the sense suggested by Rahal and Campa [26]. Yet another possibility is that some localization effect really allows at least a partial identification of electrons: this localization was intuitively rather obvious in the experiment of Goldhaber and Scharff-Goldhaber [27], originally devised to test the identity of β-rays and electrons, later reinterpreted by Reines and Sobel as a test of PEP [28]. In that experiment electrons were injected by a radioactive source, rather than a power supply, and thus the "fresh" electrons were simply those electrons impinging on the target from the radioactive source. This concept of novelty is related to electron localizability outside the target, and if an analogous process could be pinpointed for the power supply – which is required to achieve a large statistics, much larger than it is possible with a laboratory radioactive source – then this conceptual problem of VIP (and of the Ramberg and Snow experiment) would fade away. The required localization might be provided by some form of quantum decoherence: Yu and Eberly [29] have shown that in an idealized situation quantum coherence dies off in a finite time just because of quantum noise. Similarly we can conjecture that entanglement of the electron wavefunction in the copper strip could be limited in space and this could let us set apart "old" and "fresh" electrons.

At the moment these are just speculations, we do not yet have a final answer to these conceptual problems, however we do strongly feel that the test is meaningful and we are now planning an improved version.

5. Future perspectives: VIP upgrade and measurement of X rays from the collapse of wave function models

The present VIP setup uses CCD detectors which are excellent X-ray detectors (good energy resolution, background rejection based on pixel-size) but they are integrating detectors (no timing capability). We plan to switch to a new type of detectors for precision X-rays measurements, the triggerable Silicon Drift Detectors (SSD) which have a fast readout time ($\simeq 1\mu$s) and large collection area (100 mm^2). These detectors were successfully used in the SIDDHARTA experiment [30] for measurements of the kaonic atoms transitions at the DAΦNE

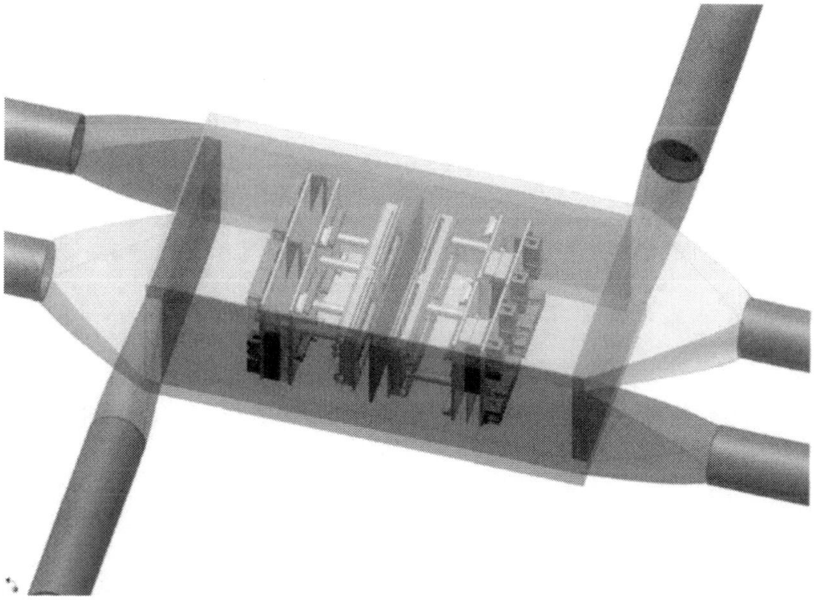

Figure 5. The possible implementation of the upgrade of the VIP experiment using SDD detectors and an external veto-system

accelerator of LNF-INFN; using a proper trigger system a background rejection factor of the order of 10^{-4} was achieved in SIDDARTHA.

With these new detectors it might then be possible to further reduce the background by using an external veto-system which would eliminate a large part allow to eliminate big part of the background produced by charged particles coming from the outside the setup. A schematic layout of the new setup is shown in fig. 5.

Presently, experimental tests are under way to define the new experimental setup, which will be more compact than the present VIP setup and, as such, more manageable.

Apart of the measurements of X rays related to the violation of PEP, we are presently considering the possibility to perform in the future measurements of X rays (having us such excellent X-ray detectors, as the CCDs and SDDs) generated as spontaneous radiation predicted by (some) collapse models. The collapse models are deal with the "measurement problem" in quantum mechanics by introducing a new physical dynamics that naturally collapses the state vector.

In the nonrelativistic collapse model developed by Ghirardi, Rimini, Weber [31] and Pearle [32] (see also ref. [33] for a review), namely the continuous spontaneous localization (CSL) model, the state vector undergoes a nonunitary evolution in which particles interact with a fluctuating scalar field. This interaction has not only the effect of collapsing the state vector towards the particle number density eigenstates in position space, but it increases the expectation value of particle's energy as well. This means, for a free charged particle (as the electron) electromagnetic radiation. This type of phenomenon is predicted by the CSL and is totally absent in the standard quantum mechanics.

Moreover, as shown in the [34], the measurement of X rays spontaneouslly emitted in the collapse models is the most sensitive method to the parameters of the collapse, see table 1 and [34] for related explanations.

In the paper [35] a pioneering work on this spontaneous emission of radiation was performed - the author analysed X-ray data measured in an underground experiment and interpreted them

Table 1. Upper bounds on the parameter λ from [34]

Laboratory experiments	Distance (decades) from the enhanced CSL value	Cosmological data	Distance (decades) from the enhanced CSL value
Fullerene difraction experiments	3-4	Dissociation of cosmic hydrogen	9
Decay of supercurrents (SQUIDS)	6	Heating of Intergalactic medium (IGM)	0
Spontaneous X-ray emmision from Ge	-2	Heating of protons in the universe	4
Proton decay	10	Heating of Interstellar dust grains	7

as a limit for the combination of the CSL parameter λ/a^2. It was shown that the highest sensitivity is at few keV X-rays, exactly in the range where our detectors are ideal. We plan to perform a feasibility study for a dedicated experiment to measure X rays coming from the spontaneous collapse models.

Acknowledgments
The VIP Collaboration wishes to thank all the LNGS laboratory staff for the precious help and assistance during all phases of preparation, installation and data taking. The support from the HadronPhysics FP6 (506078), HadronPhysics2 FP7 (227431) and MIUR PRIN2008 2008LH2X28-004 projects is acknowledged.

References

[1] Lüders G and Zumino B 1958 *Phys. Rev.* **110** 1450
[2] Feynman R P, Leighton R B and Sands M 1963 *The Feynman Lectures on Physics*, Addison-Wesley, Reading, MA
[3] Bernabei R *et al.* 1997 *Phys. Lett. B* **408** 439
[4] Borexino Colaboration, Back H O 2005 *et al. Eur. Phys. J.* **C37** 421
[5] Hilborn R C and Yuca C L 1996 *Phys. Rev. Lett.* **76** 2844
[6] Nemo Colaboration 2000 *Nucl. Phys. B (Proc. Suppl.)* **87** 510
[7] Nolte E *et al.* 1991 *J. Phys. G: Nucl. Part. Phys.* **17** S355
[8] Tsipenyuk Y, Barabash A, Kornoukhov V and Chapyzhnikov B, 1998 *Radiat. Phys. Chem.* **51** 507
[9] Ramberg E and Snow G A 1990 *Phys. Lett. B* **238** 438
[10] Ignatiev A Yu and Kuzmin V A 1987 *Yad. Fiz.* **46** 786; *Sov. J. Nucl. Phys.* **47** 6; see also the recent review paper Ignatiev A Yu 2006 *Rad. Phys. Chem.* **75** 2090
[11] The VIP proposal, LNF-LNGS Proposal, september, 2004, http://www.lnf.infn.it/esperimenti/vip.
[12] Culhane J L 1990 *Nucl. Instrum. Methods A* **310** 1
[13] Egger J P, Chatellard D and Jeannet E 1993 *Particle World* **3** 139
[14] Fiorucci G *et al.* 1990 *Nucl. Instrum. Methods A* **292** 141
[15] Varidel D *et al.* 1990 *Nucl. Instrum. Methods A* **292** 147
[16] Kraft R P *et al.* 1995 *Nucl. Instrum. Methods A* **372** 372
[17] Di Matteo S, Sperandio L, 2006 *VIP Note, IR-04, 26 April 2006; The energy shift has been computed by P. Indelicato, private communication*
[18] CCD-55 from EEV (English Electric Valve), Waterhouse Lane, Chelmsford, Essex CM1 2QU, UK.
[19] Ishiwatari T *et al.* 2004 *Phys. Lett. B* **593** 48; Beer G *et al.* 2005 *Phys. Rev. Lett.* **94** 212302
[20] Ishiwatari T *et al.* 2006 *Nucl. Instrum. Methods Phys. Res. A* **556** 509
[21] Bartalucci S *et al.* (VIP Collaboration) 2006 *Phys. Lett. B* **641** 18

[22] Sperandio L, 2008 *Ph D thesis "New experimental limit on the Pauli Exclusion Principle violation by electrons from the VIP experiment" at University "Tor Vergata", Roma, 5 March 2008*
[23] Govorkov A B 1989 *Phys. Lett. A* **137** 7
[24] Greenberg O W 1990 *Phys. Rev. Lett.* **64** 705
[25] Amado R D and Primakoff H 1980 *Phys. Rev.* **C 64** 1338
[26] Rahal V and Campa A 1988 *Phys. Rev.* **A 38** 3728
[27] Goldhaber M and Scharff-Goldhaber G 1948 *Phys. Rev.* **73** 1472
[28] Reines F and Sobel H W 1974 *Phys. Rev. Lett.* **32**
[29] Ting Yu and Eberly J H 2009 *Science* **323** 598
[30] Curceanu C *et al.* 2007 *Eur Phys* **J A31** 537-539; Bazzi M *et al.* 2009 *Phys. Lett.* **B681** 310
[31] Ghirardi G C, Rimini A and Weber T 1986 *Phys. Rev.* **D34** 470; ibid. 1987 3287; *Found. Phys.* **18** (1988) 1
[32] Pearle P 1989 *Phys. Rev* **A39** 2277; Ghirardi G C, Pearle P and Rimini A 1990 *Phys. Rev.* **A42** 78
[33] Bassi A 2007 *Journal of Physics* **67** 012013
[34] Adler S L and Bassi A 275 *Science* **325** 275
[35] Fu Q 1997 *Phys. Rev* **A56** 1806.

Physical flavor neutrino states

Massimo Blasone[† 1,2]

[1] Dipartimento di Matematica e Informatica, Università degli Studi di Salerno, Via Ponte don Melillo, I-84084 Fisciano (SA), Italy
[2] INFN Sezione di Napoli, Gruppo collegato di Salerno, Baronissi (SA), Italy

E-mail: [†]`blasone@sa.infn.it`

Abstract. The problem of representation for flavor states of mixed neutrinos is discussed. By resorting to recent results, it is shown that a specific representation exists in which a number of conceptual problems are resolved. Phenomenological consequences of our analysis are explored.

1. Introduction

The detailed study of neutrino mixing in the context of quantum field theory has led to the discovery of unexpected features associated to such a phenomenon. Indeed, it was found [1] that the Hilbert space where the mixed (flavor) field operators are defined is unitarily inequivalent, in the infinite volume limit, to the Hilbert space for the original (unmixed) field operators. This is due to the condensate structure of the vacuum state for the flavor fields, the flavor vacuum, which turns out to be a coherent state.

These results have been then found to have general validity independently of the type or number of field involved [2, 3, 4, 5, 6]. Flavor oscillation formulas were derived [7, 8, 9] exhibiting corrections with respect to the usual ones derived in quantum mechanics [10]. Effects of flavor vacuum structure have been discussed also in connection with cosmological constant [11] and Lorentz invariance [12]. Flavor states have also been studied as (relativistic) examples of single-particle entanglement [13]. More recently, a non-abelian gauge structure has been recognized to be associated to flavor mixing [14].

In this paper, we discuss a novel representation of flavor vacuum and flavor states which presents some advantages with respect to the one previously adopted [1]. Phenomenological consequences of this choice are explored.

2. Neutrino mixing in QFT

Let us denote by ν_e, ν_μ the neutrino fields with definite flavors and by ν_1, ν_2 the neutrino fields with definite masses m_1, m_2, respectively. We consider the lagrangian

$$\mathcal{L} = \bar{\nu}_e \left(i \not\partial - m_e\right) \nu_e + \bar{\nu}_\mu \left(i \not\partial - m_\mu\right) \nu_\mu - \ m_{e\mu} \ \left(\bar{\nu}_e \nu_\mu + \bar{\nu}_\mu \nu_e\right) \ . \qquad (1)$$

which is diagonalized by the transformation

$$\nu_e(x) \ = \ \nu_1(x) \, \cos\theta + \nu_2(x) \, \sin\theta \, , \qquad (2)$$

$$\nu_\mu(x) \ = \ -\nu_1(x) \, \sin\theta + \nu_2(x) \, \cos\theta \, , \qquad (3)$$

where θ is the mixing angle and $m_e = m_1 \cos^2\theta + m_2 \sin^2\theta$, $m_\mu = m_1 \sin^2\theta + m_2 \cos^2\theta$, $m_{e\mu} = (m_2 - m_1)\sin\theta\cos\theta$. The result is the sum of two free Dirac Lagrangians:

$$\mathcal{L} \;=\; \bar{\nu}_1 \,(i\,\not{\partial} - m_1)\,\nu_1 + \bar{\nu}_2 \,(i\,\not{\partial} - m_2)\,\nu_2 \,. \tag{4}$$

The expansions for ν_1 and ν_2 are:

$$\nu_i(x) = \sum_{r=1,2} \int \frac{d^3k}{(2\pi)^{\frac{3}{2}}} \left[u^r_{\mathbf{k},i}\alpha^r_{\mathbf{k},i}(t) + v^r_{-\mathbf{k},i}\beta^{r\dagger}_{-\mathbf{k},i}(t)\right] e^{i\mathbf{k}\cdot\mathbf{x}}, \qquad i = 1, 2\,, \tag{5}$$

with $\alpha^r_{\mathbf{k},i}(t) = e^{-i\omega_i t}\alpha^r_{\mathbf{k},i}(0)$, $\beta^r_{\mathbf{k},i}(t) = e^{-i\omega_i t}\beta^r_{\mathbf{k},i}(0)$ and $\omega_i = \sqrt{\mathbf{k}^2 + m_i^2}$. Here and in the following we use $t \equiv x_0$, when no misunderstanding arises. The vacuum for the α_i and β_i operators is denoted by $|0\rangle_{1,2}$: $\alpha^r_{\mathbf{k},i}|0\rangle_{12} = \beta^r_{\mathbf{k},i}|0\rangle_{12} = 0$. The anticommutation relations are the usual ones (see Ref. [1]). The orthonormality and completeness relations are:

$$u^{r\dagger}_{\mathbf{k},i}u^s_{\mathbf{k},i} = v^{r\dagger}_{\mathbf{k},i}v^s_{\mathbf{k},i} = \delta_{rs}\,, \quad u^{r\dagger}_{\mathbf{k},i}v^s_{-\mathbf{k},i} = v^{r\dagger}_{-\mathbf{k},i}u^s_{\mathbf{k},i} = 0\,, \quad \sum_r (u^r_{\mathbf{k},i}u^{r\dagger}_{\mathbf{k},i} + v^r_{-\mathbf{k},i}v^{r\dagger}_{-\mathbf{k},i}) = 1\!\!1\,. \tag{6}$$

The fields ν_e and ν_μ are thus completely determined through Eq. (3), which can be rewritten in the following form (we use $(\sigma, j) = (e, 1), (\mu, 2)$):

$$\nu_\sigma(x) \;=\; G^{-1}_\theta(t)\,\nu_j(x)\,G_\theta(t) = \sum_{r=1,2}\int \frac{d^3k}{(2\pi)^{\frac{3}{2}}} \left[u^r_{\mathbf{k},j}\alpha^r_{\mathbf{k},\sigma}(t) + v^r_{-\mathbf{k},j}\beta^{r\dagger}_{-\mathbf{k},\sigma}(t)\right] e^{i\mathbf{k}\cdot\mathbf{x}}, \tag{7}$$

$$G_\theta(t) \;=\; \exp\left[\theta \int d^3\mathbf{x} \left(\nu_1^\dagger(x)\nu_2(x) - \nu_2^\dagger(x)\nu_1(x)\right)\right], \tag{8}$$

where $G_\theta(t)$ is the generator of the mixing transformations (3) (see Ref. [1] for a discussion of its properties).

Eq.(7) gives an expansion of the flavor fields ν_e and ν_μ in the same basis of ν_1 and ν_2. The flavor annihilation operators are then identified with

$$\left(\begin{array}{c} \alpha^r_{\mathbf{k},\sigma}(t) \\ \beta^{r\dagger}_{-\mathbf{k},\sigma}(t) \end{array}\right) = G^{-1}_\theta(t) \left(\begin{array}{c} \alpha^r_{\mathbf{k},j}(t) \\ \beta^{r\dagger}_{-\mathbf{k},j}(t) \end{array}\right) G_\theta(t)\,. \tag{9}$$

The action of the mixing generator on the vacuum $|0\rangle_{1,2}$ is non-trivial and we have:

$$|0(t)\rangle_{e,\mu} \equiv G^{-1}_\theta(t)\;|0\rangle_{1,2}\,. \tag{10}$$

$|0(t)\rangle_{e,\mu}$ is the *flavor vacuum*, i.e., the vacuum for the flavor fields[1].

The explicit expression of the flavor annihilation operators is (we choose $\mathbf{k} = (0, 0, |\mathbf{k}|)$):

$$\left(\begin{array}{c} \alpha^r_{\mathbf{k},e}(t) \\ \alpha^r_{\mathbf{k},\mu}(t) \\ \beta^{r\dagger}_{-\mathbf{k},e}(t) \\ \beta^{r\dagger}_{-\mathbf{k},\mu}(t) \end{array}\right) = \left(\begin{array}{cccc} c_\theta & s_\theta\,|U_\mathbf{k}| & 0 & s_\theta\,\epsilon^r\,|V_\mathbf{k}| \\ -s_\theta\,|U_\mathbf{k}| & c_\theta & s_\theta\,\epsilon^r\,|V_\mathbf{k}| & 0 \\ 0 & -s_\theta\,\epsilon^r\,|V_\mathbf{k}| & c_\theta & s_\theta\,|U_\mathbf{k}| \\ -s_\theta\,\epsilon^r\,|V_\mathbf{k}| & 0 & -s_\theta\,|U_\mathbf{k}| & c_\theta \end{array}\right) \left(\begin{array}{c} \alpha^r_{\mathbf{k},1}(t) \\ \alpha^r_{\mathbf{k},2}(t) \\ \beta^{r\dagger}_{-\mathbf{k},1}(t) \\ \beta^{r\dagger}_{-\mathbf{k},2}(t) \end{array}\right), \tag{11}$$

where $c_\theta \equiv \cos\theta$, $s_\theta \equiv \sin\theta$, $\epsilon^r \equiv (-1)^r$, and

$$|U_\mathbf{k}| \equiv u^{r\dagger}_{\mathbf{k},2}u^r_{\mathbf{k},1} = v^{r\dagger}_{-\mathbf{k},1}v^r_{-\mathbf{k},2}\,, \tag{12}$$

$$|V_\mathbf{k}| \equiv \epsilon^r\; u^{r\dagger}_{\mathbf{k},1}v^r_{-\mathbf{k},2} = -\epsilon^r\; u^{r\dagger}_{\mathbf{k},2}v^r_{-\mathbf{k},1}\,. \tag{13}$$

The number of particles with definite mass condensed in the flavor vacuum is given by

$$ {}_{e,\mu}\langle 0(t)|\alpha^{r\dagger}_{\mathbf{k},i}\alpha^{r}_{\mathbf{k},i}|0(t)\rangle_{e,\mu} = {}_{e,\mu}\langle 0(t)|\beta^{r\dagger}_{\mathbf{k},i}\beta^{r}_{\mathbf{k},i}|0(t)\rangle_{e,\mu} = \sin^2\theta\,|V_{\mathbf{k}}|^2, \qquad i=1,2\,. \tag{14} $$

Following the usual procedure, we define the flavor charges for the flavor neutrino fields [8]:

$$ Q_\sigma(t) = \int d^3\mathbf{x}\,\nu^\dagger_\sigma(x)\,\nu_\sigma(x)\,, \qquad \sigma = e, \mu, \tag{15} $$

with $Q_e(t) + Q_\mu(t) = Q$, where Q is the total (conserved) lepton charge.

The flavor charges are diagonal when expressed in terms of the flavor ladder operators Eq (11):

$$ Q_\sigma(t) = \sum_r \int d^3\mathbf{k}\left(\alpha^{r\dagger}_{\mathbf{k}\sigma}(t)\alpha^{r}_{\mathbf{k}\sigma}(t) - \beta^{r\dagger}_{-\mathbf{k}\sigma}(t)\beta^{r}_{-\mathbf{k}\sigma}(t)\right). \tag{16} $$

The eigenstates of such charge operators (flavor neutrino states) are consistently defined as:

$$ |\nu^r_{\mathbf{k}\sigma}(t)\rangle \equiv \alpha^{r\dagger}_{\mathbf{k}\sigma}(t)|0(t)\rangle_{e,\mu}. \tag{17} $$

3. General expansion

It was noted [2], that in the expansion Eq. (7) one could use eigenfunctions with arbitrary masses μ_σ, and therefore not necessarily the same as the masses which appear in the (diagonalized) Lagrangian. On this basis, the flavor fields can be also written as [2, 15]

$$ \nu_\sigma(x) \;=\; \sum_{r=1,2}\int \frac{d^3k}{(2\pi)^{\frac{3}{2}}}\left[u^r_{\mathbf{k},\sigma}\widetilde{\alpha}^r_{\mathbf{k},\sigma}(t) + v^r_{-\mathbf{k},\sigma}\widetilde{\beta}^{r\dagger}_{-\mathbf{k},\sigma}(t)\right]e^{i\mathbf{k}\cdot\mathbf{x}}, \tag{18} $$

where u_σ and v_σ are the helicity eigenfunctions with mass μ_σ. We denote by a tilde the generalized flavor operators introduced in Ref. [2] in order to distinguish them from the ones defined in Eq. (9). The expansion Eq. (18) is more general than the one in Eq. (7) since the latter corresponds to the particular choice $\mu_e \equiv m_1$, $\mu_\mu \equiv m_2$.

The relation between the flavor and the mass operators is now:

$$ \left(\begin{array}{c}\widetilde{\alpha}^r_{\mathbf{k},\sigma}(t)\\ \widetilde{\beta}^{r\dagger}_{-\mathbf{k},\sigma}(t)\end{array}\right) = K^{-1}_{\theta,\mu}(t)\left(\begin{array}{c}\alpha^r_{\mathbf{k},j}(t)\\ \beta^{r\dagger}_{-\mathbf{k},j}(t)\end{array}\right)K_{\theta,\mu}(t)\,, \tag{19} $$

with $(\sigma,j) = (e,1),(\mu,2)$ and where $K_{\theta,\mu}(t)$ is the generator of the transformation (3) and can be written as

$$ K_{\theta,\mu}(t) \;=\; I_\mu(t)\,G_\theta(t)\,, \tag{20} $$

$$ I_\mu(t) \;=\; \prod_{\mathbf{k},r}\exp\left\{i\sum_{(\sigma,j)}\xi^{\mathbf{k}}_{\sigma,j}\left[\alpha^{r\dagger}_{\mathbf{k},j}(t)\beta^{r\dagger}_{-\mathbf{k},j}(t) + \beta^{r}_{-\mathbf{k},j}(t)\alpha^{r}_{\mathbf{k},j}(t)\right]\right\}, \tag{21} $$

with $\xi^{\mathbf{k}}_{\sigma,j} \equiv (\chi_\sigma - \chi_j)/2$ and $\cot\chi_\sigma = |\mathbf{k}|/\mu_\sigma$, $\cot\chi_j = |\mathbf{k}|/m_j$. For $\mu_e \equiv m_1$, $\mu_\mu \equiv m_2$ one has $I_\mu(t) = 1$. The explicit matrix form of the flavor operators is [2]:

$$ \left(\begin{array}{c}\widetilde{\alpha}^r_{\mathbf{k},e}(t)\\ \widetilde{\alpha}^r_{\mathbf{k},\mu}(t)\\ \widetilde{\beta}^{r\dagger}_{-\mathbf{k},e}(t)\\ \widetilde{\beta}^{r\dagger}_{-\mathbf{k},\mu}(t)\end{array}\right) = \left(\begin{array}{cccc} c_\theta\,\rho^{\mathbf{k}}_{e1} & s_\theta\,\rho^{\mathbf{k}}_{e2} & ic_\theta\,\lambda^{\mathbf{k}}_{e1} & is_\theta\,\lambda^{\mathbf{k}}_{e2}\\ -s_\theta\,\rho^{\mathbf{k}}_{\mu1} & c_\theta\,\rho^{\mathbf{k}}_{\mu2} & -is_\theta\,\lambda^{\mathbf{k}}_{\mu1} & ic_\theta\,\lambda^{\mathbf{k}}_{\mu2}\\ ic_\theta\,\lambda^{\mathbf{k}}_{e1} & is_\theta\,\lambda^{\mathbf{k}}_{e2} & c_\theta\,\rho^{\mathbf{k}}_{e1} & s_\theta\,\rho^{\mathbf{k}}_{e2}\\ -is_\theta\,\lambda^{\mathbf{k}}_{\mu1} & ic_\theta\,\lambda^{\mathbf{k}}_{\mu2} & -s_\theta\,\rho^{\mathbf{k}}_{\mu1} & c_\theta\,\rho^{\mathbf{k}}_{\mu2}\end{array}\right)\left(\begin{array}{c}\alpha^r_{\mathbf{k},1}(t)\\ \alpha^r_{\mathbf{k},2}(t)\\ \beta^{r\dagger}_{-\mathbf{k},1}(t)\\ \beta^{r\dagger}_{-\mathbf{k},2}(t)\end{array}\right), \tag{22} $$

where $c_\theta \equiv \cos\theta$, $s_\theta \equiv \sin\theta$ and

$$\rho^{\mathbf{k}}_{ab}\delta_{rs} \equiv \cos\frac{\chi_a - \chi_b}{2}\delta_{rs} = u^{r\dagger}_{\mathbf{k},a}u^{s}_{\mathbf{k},b} = v^{r\dagger}_{-\mathbf{k},a}v^{s}_{-\mathbf{k},b}\,, \tag{23}$$

$$i\lambda^{\mathbf{k}}_{ab}\delta_{rs} \equiv i\sin\frac{\chi_a - \chi_b}{2}\delta_{rs} = u^{r\dagger}_{\mathbf{k},a}v^{s}_{-\mathbf{k},b} = v^{r\dagger}_{-\mathbf{k},a}u^{s}_{\mathbf{k},b}\,, \tag{24}$$

with $a, b = 1, 2, e, \mu$. Since $\rho^{\mathbf{k}}_{12} = |U_{\mathbf{k}}|$ and $i\lambda^{\mathbf{k}}_{12} = \epsilon^r|V_{\mathbf{k}}|$, etc., the operators (22) reduce[1] to the ones in Eqs. (11) when $\mu_e \equiv m_1$ and $\mu_\mu \equiv m_2$.

The generalized flavor vacuum, which is annihilated by the flavor operators given by Eq. (22), is now written as [2]:

$$|\widetilde{0}(t)\rangle_{e,\mu} \equiv K^{-1}_{\theta,\mu}(t)|0\rangle_{1,2}\,. \tag{25}$$

For $\mu_e \equiv m_1$ and $\mu_\mu \equiv m_2$, this state reduces to the standard flavor vacuum $|0(t)\rangle_{e,\mu}$ of Eq. (10).

4. Gauge structure and choice of the representation

We have seen how it is possible to define exact flavor charge eigenstates for mixed neutrinos. We have also seen that an apparent arbitrariness exists in the choice of the basis of free fields respect to which the flavor fields are expanded. Such a choice cannot be arbitrary since the structure of the flavor vacuum and thus the physical results depend on it. A remarkable result was presented in Ref. [15], where it was shown that the oscillation formulas are insensitive to the choice of such a basis.

However, there are other aspects which need to be considered. One is that Lorentz invariance is broken, since the flavor vacuum is explicitly time-dependent. As a consequence, flavor states cannot be interpreted in terms of irreducible representations of the Poincaré group. A possible way to recover Lorentz invariance for mixed fields has been explored in Ref. [12] where non-standard dispersion relations for the mixed particles have been related to non-linear realizations of the Poincaré group [16].

A different way has been explored in Ref. [14], where it has been shown that a non-abelian gauge structure appears naturally in connection with flavor mixing. In this framework, it is then possible to account for the above-mentioned violation of Lorentz invariance due to the flavor vacuum having, at the same time, standard dispersion relations for flavor neutrino states.

To see how this is possible, let us note that the Lagrangian Eq. (1) can be rewritten as describing a doublet of Dirac fields in interaction with an external Yang–Mills field:

$$\mathcal{L} = \bar{\nu}_f(i\gamma^\mu D_\mu - M_d)\nu_f, \tag{26}$$

where $\nu_f = (\nu_e, \nu_\mu)^T$ is the flavor doublet and $M_d = \mathrm{diag}(m_e, m_\mu)$ is a diagonal mass matrix. We have introduced a covariant derivative and a gauge field as

$$D_\mu = \partial_\mu + i\,g\,\beta\,A_\mu, \tag{27}$$

$$A_\mu \equiv \frac{1}{2}A^a_\mu\sigma_a = n_\mu\delta m\,\frac{\sigma_1}{2} \in su(2), \qquad n^\mu \equiv (1,0,0,0)^T, \tag{28}$$

where $m_{e\mu} = \frac{1}{2}\tan 2\theta\,\delta m$, and $\delta m \equiv m_\mu - m_e$. We also define $g \equiv \tan 2\theta$ as the coupling constant for the mixing interaction. Flavor mixing can thus be seen as an interaction of the flavor fields with an $SU(2)$ constant gauge field.

Here α_i, $i = 1, 2, 3$ and β are the usual Dirac matrices in a given representation. For definiteness, we choose the following representation:

$$\alpha_i = \begin{pmatrix} 0 & \sigma_i \\ \sigma_i & 0 \end{pmatrix}, \qquad \beta = \begin{pmatrix} \mathbb{I} & 0 \\ 0 & -\mathbb{I} \end{pmatrix}, \tag{29}$$

[1] Up to a minus sign, see Ref. [15].

where σ_i are the Pauli matrices and $\mathbb{1}$ is the 2×2 identity matrix.

We now consider the energy momentum tensor associated with the flavor neutrino fields in interaction with the external gauge field [14]:

$$\widetilde{T}_{\rho\sigma} = \bar{\nu}_f i\gamma_\rho D_\sigma \nu_f - \eta_{\rho\sigma}\bar{\nu}_f(i\gamma^\lambda D_\lambda - M_d)\nu_f , \tag{30}$$

which is to be compared with the canonical energy momentum tensor associated with the Lagrangian (1):

$$T_{\rho\sigma} = \bar{\nu}_f i\gamma_\rho D_\sigma \nu_f - \eta_{\rho\sigma}\bar{\nu}_f(i\gamma^\lambda D_\lambda - M_d)\nu_f + \eta_{\rho\sigma} m_{e\mu}\bar{\nu}_f \sigma_1 \nu_f . \tag{31}$$

We then define a 4-momentum operator as $\widetilde{P}^\mu \equiv \int d^3\mathbf{x}\widetilde{T}^{0\mu}$ and obtain a conserved 3−momentum operator:

$$\begin{aligned}\widetilde{P}^i &= i\int d^3\mathbf{x}\,\nu_f^\dagger \partial^i \nu_f \\ &= i\int d^3\mathbf{x}\,\nu_e^\dagger \partial^i \nu_e + i\int d^3\mathbf{x}\,\nu_\mu^\dagger \partial^i \nu_\mu \\ &\equiv \widetilde{P}_e^i(t) + \widetilde{P}_\mu^i(t), \qquad\qquad i = 1,2,3\end{aligned} \tag{32}$$

and a non-conserved Hamiltonian operator:

$$\begin{aligned}\widetilde{P}^0(t) &\equiv \widetilde{H}(t) = \int d^3\mathbf{x}\,\bar{\nu}_f\left(i\gamma_0 D_0 - i\gamma^\mu D_\mu + M_d\right)\nu_f \\ &= \int d^3\mathbf{x}\,\nu_e^\dagger\left(-i\boldsymbol{\alpha}\cdot\boldsymbol{\nabla} + \beta m_e\right)\nu_e + \int d^3\mathbf{x}\,\nu_\mu^\dagger\left(-i\boldsymbol{\alpha}\cdot\boldsymbol{\nabla} + \beta m_\mu\right)\nu_\mu \\ &\equiv \widetilde{H}_e(t) + \widetilde{H}_\mu(t).\end{aligned} \tag{33}$$

Note that both the Hamiltonian and the momentum operators split in a contribution involving only the electron neutrino field and in another where only the muon neutrino field appears.

We remark that the tilde Hamiltonian is *not* the generator of time translations. This role competes to the complete Hamiltonian $H = \int d^3\mathbf{x}\, T^{00}$, obtained from the energy-momentum tensor Eq. (31).

We now show that it is possible to define flavor neutrino states which are simultaneous eigenstates of the 4-momentum operators above constructed and of the flavor charges. Such a non-trivial request requires a redefinition of the flavor vacuum. Indeed, one can show [14] that this is achieved by means of the expansion Eq. (7), provided one sets $\mu_e = m_e$ and $\mu_\mu = m_\mu$. From now on we use the tilde to denote the flavor operators so defined.

The tilde flavor operators are connected to those of Section 2 by a Bogoliubov transformation:

$$\begin{pmatrix} \widetilde{\alpha}^r_{\mathbf{k},\sigma}(t) \\ \widetilde{\beta}^{r\dagger}_{-\mathbf{k},\sigma}(t) \end{pmatrix} = J^{-1}(t)\begin{pmatrix} \alpha^r_{\mathbf{k},\sigma}(t) \\ \beta^{r\dagger}_{-\mathbf{k},\sigma}(t) \end{pmatrix} J(t), \tag{34}$$

$$J(t) = \prod_{\mathbf{k},r}\exp\left\{ i\sum_{(\sigma,j)}\xi^{\mathbf{k}}_{\sigma,j}\left[\alpha^{r\dagger}_{\mathbf{k},\sigma}(t)\beta^{r\dagger}_{-\mathbf{k},\sigma}(t) + \beta^r_{-\mathbf{k},\sigma}(t)\alpha^r_{\mathbf{k},\sigma}(t)\right]\right\}, \tag{35}$$

with $(\sigma, j) = (e,1), (\mu,2)$, and with $\xi^{\mathbf{k}}_{\sigma,j} \equiv (\chi_\sigma - \chi_j)/2$ and $\cot\chi_\sigma = |\mathbf{k}|/m_\sigma$, $\cot\chi_j = |\mathbf{k}|/m_j$. The new (physical) flavor vacuum is given by

$$|\widetilde{0}(t)\rangle_{e\mu} = J^{-1}(t)|0(t)\rangle_{e\mu}. \tag{36}$$

Notice that the flavor charges are invariant under the above Bogoliubov transformations [15], i.e., $\widetilde{Q}_\sigma = Q_\sigma$, with:

$$\widetilde{Q}_\sigma(t) = \sum_r \int d^3\mathbf{k} \left(\widetilde{\alpha}^{r\dagger}_{\mathbf{k}\sigma}(t)\widetilde{\alpha}^{r}_{\mathbf{k}\sigma}(t) - \widetilde{\beta}^{r\dagger}_{-\mathbf{k}\sigma}(t)\widetilde{\beta}^{r}_{-\mathbf{k}\sigma}(t)\right). \tag{37}$$

In terms of the tilde flavor ladder operators, the Hamiltonian and momentum operators Eqs. (32) and (33) read:

$$\widetilde{\mathbf{P}}_\sigma(t) = \sum_r \int d^3\mathbf{k}\; \mathbf{k} \left(\widetilde{\alpha}^{r\dagger}_{\mathbf{k},\sigma}(t)\widetilde{\alpha}^{r}_{\mathbf{k},\sigma}(t) + \widetilde{\beta}^{r\dagger}_{\mathbf{k},\sigma}(t)\widetilde{\beta}^{r}_{\mathbf{k},\sigma}(t)\right), \tag{38}$$

$$\widetilde{H}_\sigma(t) = \sum_r \int d^3\mathbf{k}\, \omega_{\mathbf{k},\sigma} \left(\widetilde{\alpha}^{r\dagger}_{\mathbf{k},\sigma}(t)\, \widetilde{\alpha}^{r}_{\mathbf{k},\sigma}(t) - \widetilde{\beta}^{r}_{\mathbf{k},\sigma}(t)\, \widetilde{\beta}^{r\dagger}_{\mathbf{k},\sigma}(t)\right). \tag{39}$$

Since all the above operators are diagonal, we can define common eigenstates as follows:

$$|\widetilde{\nu}^r_{\mathbf{k},\sigma}(t)\rangle = \widetilde{\alpha}^{r\dagger}_{\mathbf{k},\sigma}(t)|\widetilde{0}(t)\rangle_{e\mu}, \tag{40}$$

and similar ones for the antiparticles. We have

$$\begin{pmatrix} \widetilde{H}_\sigma(t) \\ \widetilde{\mathbf{P}}_\sigma(t) \end{pmatrix} |\widetilde{\nu}^r_{\mathbf{k},\sigma}(t)\rangle = \begin{pmatrix} \omega_{\mathbf{k},\sigma} \\ \mathbf{k} \end{pmatrix} |\widetilde{\nu}^r_{\mathbf{k},\sigma}(t)\rangle, \tag{41}$$

making explicit the 4−vector structure.

Note that the above construction and the consequent Poincaré invariance holds at a given time t. Thus, for each different time, we have a different Poincaré structure. Flavor neutrino fields behave (locally in time) as ordinary on-shell fields with definite masses m_e and m_μ, rather than those of the mass eigenstates of the standard approach, m_1 and m_2. Flavor oscillations then arise as a consequence of the interaction with the (constant) gauge field, which acts as a birefringent medium and can be seen as a *neutrino aether*.

The operator $\widetilde{H}$ can be viewed as the sum of the kinetic energies of the flavor neutrinos, or equivalently as the energy which can be extracted from flavor neutrinos by scattering processes, the mixing energy being "frozen". A thermodynamical picture is given in Ref. [14].

Finally we consider the condensation densities for the physical flavor vacuum Eq. (36). They are given by

$$_{e,\mu}\langle\widetilde{0}(t)|\alpha^{r\dagger}_{\mathbf{k},1}\alpha^{r}_{\mathbf{k},1}|\widetilde{0}(t)\rangle_{e,\mu} = {}_{e,\mu}\langle\widetilde{0}(t)|\beta^{r\dagger}_{\mathbf{k},1}\beta^{r}_{\mathbf{k},1}|\widetilde{0}(t)\rangle_{e,\mu} = \cos^2\theta\sin^2\xi^{\mathbf{k}}_{e,1} + \sin^2\theta\sin^2\xi^{\mathbf{k}}_{e,2}, \tag{42}$$

$$_{e,\mu}\langle\widetilde{0}(t)|\alpha^{r\dagger}_{\mathbf{k},2}\alpha^{r}_{\mathbf{k},2}|\widetilde{0}(t)\rangle_{e,\mu} = {}_{e,\mu}\langle\widetilde{0}(t)|\beta^{r\dagger}_{\mathbf{k},2}\beta^{r}_{\mathbf{k},2}|\widetilde{0}(t)\rangle_{e,\mu} = \cos^2\theta\sin^2\xi^{\mathbf{k}}_{\mu,2} + \sin^2\theta\sin^2\xi^{\mathbf{k}}_{\mu,1}, \tag{43}$$

which have to be compared with the result Eq.(14).

5. Phenomenology

A number of (possible) physical effects can be discussed starting from the results of last Section. Here we briefly consider some of them.

5.1. Neutrino oscillations

The flavor oscillation formulas are derived by computing, in the Heisenberg representation, the expectation value of the flavor charge operators on the flavor state [7].

In the physical representation above defined, we obtain[2]

$$\langle \widetilde{\nu}^r_{\mathbf{k},\rho}|\widetilde{Q}_\sigma(t)|\widetilde{\nu}^r_{\mathbf{k},\rho}\rangle = \left|\left\{\widetilde{\alpha}^r_{\mathbf{k},\sigma}(t), \widetilde{\alpha}^{r\dagger}_{\mathbf{k},\rho}(0)\right\}\right|^2 + \left|\left\{\widetilde{\beta}^{r\dagger}_{-\mathbf{k},\sigma}(t), \widetilde{\alpha}^{r\dagger}_{\mathbf{k},\rho}(0)\right\}\right|^2, \qquad \sigma,\rho = e,\mu. \tag{44}$$

A similar result holds if we consider expectation values on antiparticle states.

For the case of two flavors, the above formula turns out to be identical [15] to the one defined by means of the representation of Section 2. However, corrections arise for the case of three flavors, when CP violation is present [9].

5.2. Beta decay

In the above picture, flavor neutrinos states have standard dispersion relations and the oscillations are due to the interaction with the external gauge field (neutrino aether). Consequently, in experiments for the direct measurement of the neutrino masses, as the ones based on tritium (beta) decay, flavor neutrinos are predicted to exhibit the behavior of ordinary free particles with masses m_e and m_μ rather than superpositions of massive states with masses m_1 and m_2. This is discussed in detail in Ref. [14].

5.3. Cosmological constant

The energy of the flavor vacuum can provide a non-standard contribution to the cosmological constant. In Ref. [11] such a contribution was evaluated by means of the representation of Section 2. In the light of the above results, it needs to be recalculated by use of the correct flavor vacuum.

In Ref. [11] the contribution $\langle \rho^{mix}_{vac}\rangle$ of the neutrino mixing to the vacuum energy density is shown to be:

$$\langle \rho^{mix}_{vac}\rangle \eta_{00} = {}_{e,\mu}\langle \widetilde{0}|\sum_i T^{(i)}_{00}|\widetilde{0}\rangle_{e,\mu}\,, \tag{45}$$

where

$$T^{(i)}_{00} = \sum_r \int d^3\mathbf{k}\,\omega_{k,i}\left(\alpha^{r\dagger}_{\mathbf{k},i}\alpha^r_{\mathbf{k},i} + \beta^{r\dagger}_{-\mathbf{k},i}\beta^r_{-\mathbf{k},i}\right), \quad i = 1,2. \tag{46}$$

By using the result Eqs.(42), (43), we obtain

$$\langle \rho^{mix}_{vac}\rangle \eta_{00} = 4\int d^3\mathbf{k}\,\omega_{k,1}\left(\cos^2\theta\sin^2\xi^{\mathbf{k}}_{e,1} + \sin^2\theta\sin^2\xi^{\mathbf{k}}_{e,2}\right) \tag{47}$$

$$+ \; 4\int d^3\mathbf{k}\,\omega_{k,2}\left(\cos^2\theta\sin^2\xi^{\mathbf{k}}_{\mu,2} + \sin^2\theta\sin^2\xi^{\mathbf{k}}_{\mu,1}\right). \tag{48}$$

which is to be compared with the result of Ref. [11]. Notice that the above contribution is zero in the no-mixing limit when the mixing angle $\theta = 0$ and/or $m_1 = m_2$.

When a cutoff K is introduced, the above integral can be evaluated and the result turns out to be proportional to K^2, whereas the usual free-field zero-point energy contribution would be going like K^4.

[2] We use the notation $|\widetilde{\nu}^r_{\mathbf{k},\rho}\rangle \equiv \widetilde{\alpha}^{r\dagger}_{\mathbf{k},\rho}(0)|\widetilde{0}(0)\rangle_{e,\mu}$ and $|\widetilde{0}\rangle_{e,\mu} \equiv |\widetilde{0}(0)\rangle_{e,\mu}$.

5.4. Supersymmetry

It has been argued [17, 18] that the flavor vacuum structure can provide a mechanism for supersymmetry breaking. This can be seen [17] by considering a Lagrangian of the form[3]

$$
\begin{aligned}
\mathcal{L} &= -\frac{i}{2}\bar{\psi}_f\,(\partial\!\!\!/ + M)\,\psi_f - \frac{1}{2}\partial_\mu S_f\,\partial^\mu S_f - \frac{1}{2}S_f^T\,M^2\,S_f - \frac{1}{2}\partial_\mu P_f\,\partial^\mu P_f - \frac{1}{2}P_f^T\,M^2\,P_f, \\
&= -\frac{i}{2}\bar{\psi}\,(\partial\!\!\!/ + M_d)\,\psi - \frac{1}{2}\partial_\mu S\,\partial^\mu S - \frac{1}{2}S^T\,M_d^2\,S - \frac{1}{2}\partial_\mu P\,\partial^\mu P - \frac{1}{2}P^T\,M_d^2\,P\,, \qquad (49)
\end{aligned}
$$

with $M = \begin{pmatrix} m_a & m_{ab} \\ m_{ab} & m_b \end{pmatrix}$ and $M_d = \mathrm{diag}(m_1, m_2)$. The fields are two free Majorana fermions ψ_i, two free real scalars S_i, two free real pseudoscalars P_i: $\psi = (\psi_1, \psi_2)^T, S = (S_1, S_2)^T, P = (P_1, P_2)^T$. The flavor fields are defined as:

$$
\psi_f = U\psi, \qquad S_f = US, \qquad P_f = UP, \qquad (50)
$$

where $\psi_f = (\psi_a, \psi_b)^T$, $S = (S_a, S_b)^T$, $P = (P_a, P_b)^T$, and $U = \begin{pmatrix} \cos\theta & \sin\theta \\ -\sin\theta & \cos\theta \end{pmatrix}$. It is also $m_a = m_1\cos^2\theta + m_2\sin^2\theta$, $m_b = m_1\sin^2\theta + m_2\cos^2\theta$, and $m_{ab} = (m_2 - m_1)\sin\theta\cos\theta$.

The Fourier expansion of the fields ($i = 1, 2$) are:

$$
\begin{aligned}
\psi_i(x) &= \sum_{r=1}^{2}\int \frac{d^3\mathbf{k}}{(2\pi)^{\frac{3}{2}}}\, e^{i\mathbf{k}\mathbf{x}}\left[u^r_{\mathbf{k},i}\alpha^r_{\mathbf{k},i}e^{-i\omega_{k,i}t} + v^r_{-\mathbf{k},i}\alpha^{\dagger r}_{-\mathbf{k},i}e^{i\omega_{k,i}t}\right], \qquad (51)\\
S_i(x) &= \int \frac{d^3\mathbf{k}}{(2\pi)^{\frac{3}{2}}}\,\frac{1}{\sqrt{2\omega_{k,i}}}e^{i\mathbf{k}\mathbf{x}}\left[b_{\mathbf{k},i}e^{-i\omega_{k,i}t} + b^\dagger_{-\mathbf{k},i}e^{i\omega_{k,i}t}\right], \qquad (52)\\
P_i(x) &= \int \frac{d^3\mathbf{k}}{(2\pi)^{\frac{3}{2}}}\,\frac{1}{\sqrt{2\omega_{k,i}}}e^{i\mathbf{k}\mathbf{x}}\left[c_{\mathbf{k},i}e^{-i\omega_{k,i}t} + c^\dagger_{-\mathbf{k},i}e^{i\omega_{k,i}t}\right], \qquad (53)
\end{aligned}
$$

where $v^r_{\mathbf{k},i} = \gamma_0 C(u^r_{\mathbf{k},i})^*$ and $u^r_{\mathbf{k},i} = \gamma_0 C(v^r_{\mathbf{k},i})^*$ by the Majorana condition and the operators $\alpha^r_{\mathbf{k},i}$, $b_{\mathbf{k},i}$ and $c_{\mathbf{k},i}$ annihilate the vacuum $|0\rangle = |0\rangle^\psi \otimes |0\rangle^S \otimes |0\rangle^P$. The expectation value of the Hamiltonian on this vacuum is zero

$$
\langle 0|(H_\psi + H_B)|0\rangle = 0, \qquad (54)
$$

where $H_B = H_S + H_P$. The generators of the mixing transformations (50) are given by [6]:

$$
\begin{aligned}
G_\psi(\theta) &= \exp\left[\frac{\theta}{2}\int d^3\mathbf{x}\left(\psi_1^\dagger(x)\psi_2(x) - \psi_2^\dagger(x)\psi_1(x)\right)\right], \qquad (55)\\
G_S(\theta) &= \exp\left[-i\theta\int d^3\mathbf{x}(\pi_1^S(x)S_2(x) - \pi_2^S(x)S_1(x))\right], \qquad (56)\\
G_P(\theta) &= \exp\left[-i\theta\int d^3\mathbf{x}(\pi_1^P(x)P_2(x) - \pi_2^P(x)P_1(x))\right], \qquad (57)
\end{aligned}
$$

where $\pi_i^S(x)$ and $\pi_i^P(x)$ are the conjugate momenta of the fields $S_i(x)$ and $P_i(x)$, respectively. The flavor vacuum (at $t = 0$) is: $|0\rangle_f \equiv |0\rangle_f^\psi \otimes |0\rangle_f^S \otimes |0\rangle_f^P$, where

$$
|0\rangle_f^\psi \equiv G_\psi^{-1}(\theta)\,|0\rangle^\psi, \qquad |0\rangle_f^S \equiv G_S^{-1}(\theta)\,|0\rangle^S, \qquad |0\rangle_f^P \equiv G_P^{-1}(\theta)\,|0\rangle^P, \qquad (58)
$$

[3] The symbol M_d introduced here should not be confused with the one of Section 4.

are the flavor vacua of the fields $\psi_\sigma(x)$, $S_\sigma(x)$, $P_\sigma(x)$, respectively.

In a similar way as done above, we now introduce the physical flavor vacuum:

$$|\widetilde{0}\rangle_f \equiv |\widetilde{0}\rangle_f^\psi \otimes |\widetilde{0}\rangle_f^S \otimes |\widetilde{0}\rangle_f^P , \tag{59}$$

The expectation value of the fermionic part of H on $|\widetilde{0}\rangle_f$ is given by:

$$\begin{aligned} {}_f\langle\widetilde{0}|H_\psi|\widetilde{0}\rangle_f &= -\int d^3\mathbf{k}\,(\omega_{k,1}+\omega_{k,2}) \\ &+ 2\int d^3\mathbf{k}\,\omega_{k,1}\left(\cos^2\theta\sin^2\xi^{\mathbf{k}}_{a,1} + \sin^2\theta\sin^2\xi^{\mathbf{k}}_{a,2}\right) \\ &+ 2\int d^3\mathbf{k}\,\omega_{k,2}\left(\cos^2\theta\sin^2\xi^{\mathbf{k}}_{b,2} + \sin^2\theta\sin^2\xi^{\mathbf{k}}_{b,1}\right), \end{aligned} \tag{60}$$

while for the bosonic part we obtain:

$$\begin{aligned} {}_f\langle\widetilde{0}|H_B|\widetilde{0}\rangle_f &= \int d^3\mathbf{k}\,(\omega_{k,1}+\omega_{k,2}) \\ &+ 2\int d^3\mathbf{k}\,\omega_{k,1}\left(\cos^2\theta\sin^2\hat{\xi}^{\mathbf{k}}_{a,1} + \sin^2\theta\sin^2\hat{\xi}^{\mathbf{k}}_{a,2}\right) \\ &+ 2\int d^3\mathbf{k}\,\omega_{k,2}\left(\cos^2\theta\sin^2\hat{\xi}^{\mathbf{k}}_{b,2} + \sin^2\theta\sin^2\hat{\xi}^{\mathbf{k}}_{b,1}\right), \end{aligned} \tag{61}$$

with $\hat{\xi}^{\mathbf{k}}_{\sigma,i} \equiv \frac{1}{2}\ln\frac{\omega_{k,\sigma}}{\omega_{k,i}}$ (see Ref. [3]).

Combining Eqs.(60) and (61) we finally have:

$$\begin{aligned} {}_f\langle\widetilde{0}|(H_\psi + H_B)|\widetilde{0}\rangle_f &= 2\,\cos^2\theta\int d^3\mathbf{k}\,\omega_{k,1}\left(\sin^2\xi^{\mathbf{k}}_{a,1} + \sin^2\hat{\xi}^{\mathbf{k}}_{a,1}\right) + \omega_{k,2}\left(\sin^2\xi^{\mathbf{k}}_{b,2} + \sin^2\hat{\xi}^{\mathbf{k}}_{b,2}\right) \\ &+ 2\,\sin^2\theta\int d^3\mathbf{k}\,\omega_{k,1}\left(\sin^2\xi^{\mathbf{k}}_{a,2} + \sin^2\hat{\xi}^{\mathbf{k}}_{a,2}\right) + \omega_{k,2}\left(\sin^2\xi^{\mathbf{k}}_{b,1} + \sin^2\hat{\xi}^{\mathbf{k}}_{b,1}\right), \end{aligned} \tag{62}$$

which exhibits supersymmetry breaking associated to flavor mixing. The above result differs from that of Ref. [17].

6. Conclusions

In the framework of the quantum field theory treatment of particle mixing, we have discussed flavor states for two flavor neutrino mixing. We have given arguments for selecting a physically relevant representation for the flavor states. Phenomenological consequences of our discussion have been explored, also in connection to previous results.

Acknowledgments

We acknowledge partial financial support from MIUR and INFN.

References

[1] Blasone M and Vitiello G 1995 *Ann. Phys. (N.Y.)* **244** 283
[2] Fujii K, Habe C and Yabuki T 1999 *Phys.Rev.* D **59** 113003
[3] Blasone M, Capolupo A, Romei O and Vitiello G 2001 *Phys. Rev.* D **63** 125015
[4] Ji C R and Mishchenko Y 2002 *Phys. Rev.* D **65** 096015
[5] Hannabuss K C and Latimer D C 2000 *J. Phys.* A **33** 1369
[6] Blasone M and Palmer J 2004 *Phys. Rev.* D **69** 057301
[7] Blasone M, Henning P A, and Vitiello G 1999 *Phys. Lett.* B **451** 140
[8] Blasone M, Jizba P, and Vitiello G 2001 *Phys. Lett.* B **517** 471
[9] Blasone M, Capolupo A and Vitiello G 2002 *Phys. Rev.* D **66** 025033

[10] Bilenky S M and Pontecorvo B 1978 *Phys. Rep.* **41** 225
[11] Blasone M, Capolupo A, Capozziello S, Carloni S and Vitiello G 2004 *Phys. Lett.* A **323** 182
[12] Blasone M, Magueijo J and Pires-Pacheco P 2005 *Europhys. Lett.* **70** 600
[13] Blasone M, Dell'Anno F, De Siena S, Di Mauro M, and Illuminati F 2008 *Phys. Rev.* D **77** 096002; Blasone M, Dell'Anno F, De Siena S, and Illuminati F 2009 *EPL*, **85** 50002; 2010 *J. Phys.: Conf. Series* **237** 012007
[14] Blasone M, Di Mauro M and Vitiello G 2010 Non-abelian gauge structure in neutrino mixing *Phys. Lett.* B in print (*Preprint* arXiv:1003.5812 [hep-ph])
[15] Blasone M and Vitiello G 1999 *Phys. Rev.* D **60** 111302
[16] Magueijo J and Smolin L 2003 *Phys. Rev.* D **67** 044017
[17] Capolupo A, Di Mauro M and Iorio A 2010 Mixing-induced Spontaneous Supersymmetry Breaking *Preprint* arXiv:1009.5041 [hep-th]
[18] Mavromatos N E, Sarkar S and Tarantino W 2010 Flavour-Condensate-induced Breaking of Supersymmetry in Free Wess-Zumino Fluids *Preprint* arXiv:1010.0345 [hep-th]

Leptons, quarks, and their antiparticles from a phase-space perspective

Piotr Żenczykowski

Division of Theoretical Physics, Institute of Nuclear Physics, Polish Academy of Sciences, Radzikowskiego 152, 31-342 Kraków, Poland

E-mail: `piotr.zenczykowski@ifj.edu.pl`

Abstract. It is argued that antiparticles may be interpreted in macroscopic terms without explicitly using the concept of time and its reversal. The appropriate framework is that of nonrelativistic phase space. It is recalled that a quantum version of this approach leads also, alongside the appearance of antiparticles, to the emergence of 'internal' quantum numbers identifiable with weak isospin, weak hypercharge and colour, and to the derivation of the Gell-Mann-Nishijima relation, while simultaneously offering a preonless interpretation of the Harari-Shupe rishon model. Furthermore, it is shown that - under the assumption of the additivity of canonical momenta - the approach entails the emergence of string-like structures resembling mesons and baryons, thus providing a different starting point for the discussion of quark unobservability.

'It is utterly beyond our power to measure the changes of things by time (...) time is an abstraction at which we arrive by means of the changes of things; made because we are not restricted to any one definite measure, all being interconnected.'

Ernst Mach - [1]

1. Introduction

1.1. Charge conjugation and time

In the Standard Model (SM), elementary particles are grouped into multiplets of various symmetry groups such as $SU(2)$, $SU(3)$, etc. Particles and antiparticles belong then to complex conjugate representations (e.g. when coloured quarks are assigned to representation $\mathbf{3}$ of $SU(3)_C$, the antiquarks are assigned to $\mathbf{3}^*$). With the standard particle theory formulated on the background of classical space and time, the concepts of complex conjugation and time reversal are closely related. Accordingly, Stückelberg and Feynman interpret the antiparticles as particles 'moving backwards in time'.

The spacetime-based description of Reality provided by the Standard Model is very successful. Yet, there are many questions that go beyond what SM was devised to answer. These include: why the physical quantities that enter into the SM as parameters have their specific values (masses of elementary particles or mixing angles between particle generations), what is the origin of various internal quantum numbers, or how to describe both elementary particles and gravity in a single framework.

The latter issue suggests that the standard particle theory should give way to an approach in which spacetime serves no longer as a background but becomes a dynamical structure, in line with general relativity ideas. In fact, many physicists have argued that we need an approach in which macroscopic classical time and space emerge - in the limit of a large interconnected structure - out of a simple timeless and alocal quantum level [2],[3]. The description of this level is still expected to be complex, as quantum descriptions are supposed to be [4].

Can one interpret then classically the expected presence of particles and antiparticles without explicitly using the concept of macroscopic time? In order to deal with this question, we find it appropriate to recall first how the classical concept of background time was originally introduced.

1.2. Time and change

In classical approaches with background time, position is viewed as a function of time $\mathbf{x}(t)$, while its change $\delta\mathbf{x}$ is thought of as occurring in time, *'due'* to its increase: $\delta\mathbf{x} = \frac{d\mathbf{x}}{dt}\delta t$. However, a deeper insight undermines this concept of background time. Rather, it views time as an effective parameter that somehow parametrizes *change*, as stated by Ernst Mach (see motto).

In [5] Barbour described how energy conservation (in an isolated system) serves as a key ingredient that leads to a definition of an increment of ephemeris time through observed changes:

$$\delta t = \sqrt{\frac{\sum_i m_i(\delta\mathbf{x}_i)^2}{2(E-V)}}, \tag{1}$$

where $\delta\mathbf{x}_i$ are measured changes in the positions of astronomical bodies, E is total (and fixed) energy and V is the gravitational potential of all interacting bodies of the system. Thus, the (astronomical) time is defined by change, not vice versa. A more neutral expression of the relation between an increment of time δt and changes of position $\delta\mathbf{x}_i$ is that the two are correlated. It is then up to us to decide which of the two alternative formulations to choose: change in (background) time, or time defined from (observed) change.

Now, in full analogy with Eq. (1), it is natural to consider

$$\delta t\ \mathbf{P} = \sum_i m_i\ \delta\mathbf{x}_i, \tag{2}$$

where $\mathbf{P}$ is the total (and conserved) momentum of the system. For the sake of our discussion, we restrict the above equation to the one-particle case

$$\mathbf{p}\ \delta t = m\ \delta\mathbf{x}, \tag{3}$$

and observe that, again, one may view this relation in two ways:
1) either with momentum $\mathbf{p}$ calculated from the change of position $\delta\mathbf{x}$ in a given increment of (background) time δt, or
2) with time increment δt calculated from given $\mathbf{p}$ and $\delta\mathbf{x}$.

The latter standpoint - in which momentum is not calculated from a change in position but assumed as independent of position - is familiar from the Hamiltonian formalism, in which momenta and positions are treated as independent variables. Thus, the idea of time induced by change should be expressible in the language of phase space in which the macroscopic arena is not 3- but 6- dimensional. Furthermore, with independent $\mathbf{p}$ and $\mathbf{x}$ one may consider various independent transformations of $\mathbf{p}$ and $\mathbf{x}$, for example, study the symmetries of the time-defining equation Eq.(3). Then, standard 3D reflection corresponds to $(\mathbf{p},\mathbf{x}) \to (-\mathbf{p},-\mathbf{x})$ (and leaves time untouched), while the operation $(\mathbf{p}, \mathbf{x}) \to (\mathbf{p}, -\mathbf{x})$ leads to time reflection: $t \to -t$.

Moving now to the quantum description we observe that the canonical commutation relations naturally involve the imaginary unit (we use units in which $\hbar = 1$):

$$[x_k, p_l] = i\delta_{kl}. \tag{4}$$

In spite of containing i, Eq. (4) does not involve time explicitly. Consequently, since charge and complex conjugations are related, it should be possible to give an interpretation to the particle-antiparticle degree of freedom using the phase-space concepts of positions and momenta alone, i.e. without referring to the concept of explicit time. Charge conjugation should then be seen not only as connected to time reversal, but also as just one of several possible transformations in phase space.

2. Noncommuting phase space

In the standard picture with background spacetime there is a connection between the properties of the background (e.g. under 3D rotation) and the existence of the 'spatial' quantum numbers (e.g. spin). Therefore, if one is willing to enlarge the arena to that of phase space, with independent position and momentum coordinates, one might expect the appearance of additional quantum numbers [6]. From the point of view of the standard (3D space + time) formalism, such quantum numbers would necessarily appear 'internal'.

2.1. Born reciprocity

The issue of a possible relation between particle properties (such as quantum numbers or masses), and the concept of phase space was of high concern already to Max Born. In his 1949 paper [7] he discussed the difference between the concepts of position and momentum for elementary particles and noted that the notion of mass appears in the relation $p^2 = m^2$, while x^2, the corresponding invariant in coordinate space (with x^2 of atomic dimensions), does not seem to enter in a similar relation. At the same time, Born stressed that various laws of nature such as

$$\begin{aligned} &\dot{x}_k = \frac{\partial H}{\partial p_k}, \quad \dot{p}_k = -\frac{\partial H}{\partial x_k}, \\ &[x_k, p_l] = i\delta_{kl}, \\ &L_{kl} = x_k p_l - x_l p_k, \end{aligned} \tag{5}$$

are invariant under the 'reciprocity' transformation:

$$x_k \to p_k, \qquad p_k \to -x_k. \tag{6}$$

Noting that relation $p^2 = m^2$ is not invariant under this transformation, he concluded: *'This lack of symmetry seems to me very strange and rather improbable'.*

The simplest phase-space generalization of the 3D (nonrelativistic) concepts of rotation and reflection is obtained with a symmetric treatment of the two O(3) invariants, $\mathbf{x}^2$ and $\mathbf{p}^2$, via their addition (obviously, this procedure requires an introduction of a new fundamental constant of Nature of dimension [momentum/distance]):

$$\mathbf{x}^2 + \mathbf{p}^2. \tag{7}$$

Eq.(7) is invariant under $O(6)$ transformations and Born reciprocity in particular.

We may now treat $\mathbf{x}$ and $\mathbf{p}$ as operators, and require their commutators to be form invariant. The original $O(6)$ symmetry is then reduced to $U(1) \otimes SU(3)$. The appearance of the symmetry group present in the Standard Model leads us to ask the question whether phase-space symmetries could possibly lie at the roots of SM symmetries. As shown in [6] and also argued below, this seems quite possible.

Table 1. Decomposition of eigenvalues of Y into eigenvalues of its components.

colour	0	1	2	3
Y	-1	$+\frac{1}{3}$	$+\frac{1}{3}$	$+\frac{1}{3}$
Y_1	$-\frac{1}{3}$	$-\frac{1}{3}$	$+\frac{1}{3}$	$+\frac{1}{3}$
Y_2	$-\frac{1}{3}$	$+\frac{1}{3}$	$-\frac{1}{3}$	$+\frac{1}{3}$
Y_3	$-\frac{1}{3}$	$+\frac{1}{3}$	$+\frac{1}{3}$	$-\frac{1}{3}$

2.2. Dirac linearization

Our present theories are standardly divided into classical and quantum ones. Yet, as stressed by Finkelstein, the latter should be more appropriately regarded as belonging to a mixed, classical-quantum type. In Finkelstein's view, the classical (c), and classical-quantum (cq) theories should give way to a purely quantum (q) approach in which infinities would not appear and the concept of infinite and infinitely divisible background space would no longer exist [3]. This view is clearly in line with the arguments for a background-independent approach to quantum gravity.

The Dirac linearization prescription may be thought of as a procedure that has led us from the classical description of Nature to part of its quantum description. Indeed, the linearization of $\mathbf{p}^2$ leads to the appearance of Pauli matrices, which describe spin at the quantum level. It is therefore of great interest to apply Dirac's idea to the phase-space invariant of Eq.(7). Using anticommuting matrices A_k and B_k ($k = 1, 2, 3$) (for more details, see [6]) one finds

$$(\mathbf{A}\cdot\mathbf{p}+\mathbf{B}\cdot\mathbf{x})(\mathbf{A}\cdot\mathbf{p}+\mathbf{B}\cdot\mathbf{x}) = (\mathbf{p}^2+\mathbf{x}^2)+\sum_1^3 \sigma_k\otimes\sigma_k\otimes\sigma_3 \equiv R+R^\sigma. \tag{8}$$

The first term on the r.h.s., $R = \mathbf{p}^2+\mathbf{x}^2$, appears thanks to the anticommutation properties of A_k and B_l. The other term, R^σ, appears because x_k and p_k do not commute. These two terms sum up to a total $R^{tot} = R + R^\sigma$.

When viewed from Finkelstein's perspective, the invariant

$$\mathbf{A}\cdot\mathbf{p}+\mathbf{B}\cdot\mathbf{x} \tag{9}$$

connects then the cq-level of phase space (noncommuting positions $\mathbf{x}$ and momenta $\mathbf{p}$) with the presumably purely q-level structure: the Clifford algebra built from matrices $\mathbf{A}$ and $\mathbf{B}$.

3. Quantization

3.1. Gell-Mann-Nishijima relation

Just as R is quantized, so is R^σ. Thus, we have to find the eigenvalues of R^σ. For better correspondence with the standard definitions of internal quantum numbers, we introduce operator Y:

$$Y \equiv \frac{1}{3}R^\sigma B_7 = \frac{1}{3}\sum_1^3 \sigma_k\otimes\sigma_k\otimes\sigma_0 \equiv \sum_1^3 Y_k. \tag{10}$$

where $B_7 = iA_1A_2A_3B_1B_2B_3$ is the 7-th anticommuting element of the Clifford algebra.

Since Y_k commute among themselves, they may be simultaneously diagonalized. The eigenvalues of Y_k ($k = 1, 2, 3$) are $\pm 1/3$. The resulting pattern of possible eigenvalues of Y is shown in Table 1.

Table 2. Rishon structure of leptons and quarks with $I_3 = +1/2$.

	ν_e	u_R	u_G	u_B	e^+	$\bar{d}_R$	$\bar{d}_G$	$\bar{d}_B$
	VVV	VTT	TVT	TTV	TTT	TVV	VTV	VVT
Q	0	$+\frac{2}{3}$	$+\frac{2}{3}$	$+\frac{2}{3}$	$+1$	$+\frac{1}{3}$	$+\frac{1}{3}$	$+\frac{1}{3}$
Y	-1	$+\frac{1}{3}$	$+\frac{1}{3}$	$+\frac{1}{3}$	$+1$	$-\frac{1}{3}$	$-\frac{1}{3}$	$-\frac{1}{3}$

In [6] a conjecture was put forward that the electric charge Q is proportional to operator $R^{tot}B_7$, evaluated for the lowest level of R, i.e.:

$$Q = \frac{1}{6}(R_{lowest} + R^{\sigma})B_7 = I_3 + \frac{Y}{2}, \qquad (11)$$

where $R_{lowest} = (\mathbf{p}^2 + \mathbf{x}^2)_{lowest} = 3$, and $I_3 = B_7/2$.

The above equation is known under the name of the Gell-Mann-Nishijima relation (with I_3 of eigenvalues $\pm 1/2$ known as weak isospin and Y of eigenvalues $-1, +1/3$ known as weak hypercharge) and is considered to be a law of nature. It summarizes the pattern of charges of all eight leptons and quarks from a single SM generation. In the phase-space approach it is *derived* as a consequence of phase-space symmetries.

3.2. Harari-Shupe rishons

As shown in [6], the pattern in which the weak hypercharge Y is built out of 'partial hypercharges' Y_k corresponds *exactly* to the pattern in which electric charges are built in the Harari-Shupe (HS) model of quarks and leptons [8]. The HS approach describes the structure of a SM generation with the help of a composite model: it builds all eight fermions of a single generation from two spin-1/2 'rishons' V and T of charges 0 and $+1/3$. The proposed structure of leptons and quarks is shown in Table 2.

Our phase-space approach not only reproduces exactly the successful part of the rishon structure, but it also removes all the main shortcomings of the HS model. In particular, the approach is *preonless*, i.e. the phase-space 'rishons' are components of charge (hypercharge) only, with *no interpretation in terms of spin 1/2 subparticles*. Thus, there is no problem of rishon confinement. Consequently, our leptons and quarks are viewed as pointlike, in perfect agreement with the experimental knowledge. In summary, the phase-space approach explains the origin of the observed symmetries without introducing any subparticles.

One might object that the nontrivial combination of spatial and internal symmetries is forbidden by the Coleman-Mandula no-go theorem [9]. Yet, this theorem is neatly evaded by our construction: the theorem works at the S-matix level, while quarks are to be confined, as we expect and as will be argued later on. Furthermore, no additional dimensions (standardly understood) have actually been added in our framework. The only change was a shift in the conceptual point of view: instead of a picture based on 3D space and time, we decided to view the world in terms of a picture based on the 6D arena of canonically conjugated positions and momenta.

4. Compositeness and additivity

The linearized phase-space approach suggests that the Clifford algebra of nonrelativistic phase space occupies an important place in our description of leptons and quarks (see also [10]). Thus, one would like to use it also in a description of composite systems, in particular in the description of hadrons as composite systems of quarks. Unfortunately, it is not clear how to achieve this goal. Yet, the q-level construction in question must be related to the c-level description of Nature, and, consequently, should be interpretable at the purely classical level. Indeed, as Niels Bohr said: *'However far the phenomena transcend the scope of classical physical explanation, the account of all evidence must be expressed in classical terms'*. In fact, Eq.(9) provides the required connection through which transformations between leptons, quarks, and their antiparticles are related to those in phase space.

Now, in any description of composite systems, be it at the classical or the quantum level, an important ingredient is provided by the tacitly assumed concept of additivity. Indeed, additivity is assumed both at the quark level (e.g. additivity of spins or flavour quantum numbers), and at the classical level (e.g. additivity of momenta). The question then appears: can this property of the additivity of momenta at the classical level be somehow used to infer about the properties of such q-level objects as our quarks when these are viewed from the macroscopic classical perspective.

The relevant basic macroscopic observation is that for any direction in 3D space, and completely regardless of what happens in the remaining two directions, we have the *additivity of physical momenta of any number of ordinary particles or antiparticles, quite irrespectively of their internal quantum numbers* (e.g. $P_z = \sum_i p_{iz}$). (By ordinary particles we mean those that can be oberved *individually*, such as leptons and hadrons, but not quarks.) Note that the positions of ordinary particles are not additive in such a simple way, since a composite object is best described in terms of its center-of-mass coordinates: $X_z = \sum_i m_i x_{iz} / \sum_i m_i$. In the following, we shall present the connections between lepton-quark and phase-space transformations and discuss their implications for the concept of the additivity of momenta.

4.1. Charge conjugation

In a quantum description the transition from particles to antiparticles is effected by complex conjugation. Consider now a system of particles in which some are being transformed into antiparticles. We want to find the related transformation in phase space. We note that in order to preserve the principle of the additivity of physical momenta, one is not allowed to change the momenta of any of the particles being transformed. The invariance of $[x_k, p_l] = i\delta_{kl}$ requires then that

$$p_k \rightarrow p_k, \qquad x_k \rightarrow -x_k, \qquad i \rightarrow -i. \tag{12}$$

Using the invariance of Eq.(9), the ensuing transformations of A_k and B_k, and the definition of Q, I_3, and Y, one can check [11] that in this way we are indeed led from particles to antiparticles. Thus, in the phase-space picture, with time regarded as secondary, the antiparticles are related to particles via $i \rightarrow -i$ combined with the reflection of position space (with the time reflection $t \rightarrow -t$ induced via the invariance of Eq.(3)).

4.2. Isospin reversal

Similarly, one checks [11] that isospin reversal $I_3 \rightarrow -I_3$ corresponds to the transformation $A_k \rightarrow A_k, B_k \rightarrow -B_k$, and $i \rightarrow i$, and, consequently, to $p_k \rightarrow p_k$ and $x_k \rightarrow -x_k$. Yet, this is not the same as charge conjugation since in this case, with $i \rightarrow +i$, the momentum-position commutation relations do not stay invariant.

In summary, the phase-space representations of both the particle-antiparticle transformation and of the isospin reversal may be (and have been) chosen in a way that does not affect the momenta of any of the particles, thus preserving their additivity.

4.3. Lepton-to-quark transformations

Transformations from the lepton to the quark sector (in particular, the change $Y = -1 \rightarrow Y = +1/3$) require the use of SO(6) rotations going outside those generated by the familiar $1 + 8$ generators of $U(1) \times SU(3)$. The remaining six of the fifteen SO(6) generators form two SU(3) triplets, of which only one actually leads to transformations of some A_k into B_m, while keeping i and I_3 fixed [6].

Under these transformations, $\mathbf{A}$ is changed into some $\mathbf{A}^{Qn}$, and $\mathbf{B}$ into some $\mathbf{B}^{Qn}$, with $n = 1, 2, 3$ being the colour index [11]. For the transformed elements $\mathbf{A}^{Qn}$ (and similarly for $\mathbf{B}^{Qn}$) one can then choose the following (not unique) representation :

$$A^Q \equiv \begin{bmatrix} \mathbf{A}^{Q1} \\ \mathbf{A}^{Q2} \\ \mathbf{A}^{Q3} \end{bmatrix} = \begin{bmatrix} A_1 & -B_3 & +B_2 \\ +B_3 & A_2 & -B_1 \\ -B_2 & +B_1 & A_3 \end{bmatrix}, \tag{13}$$

with the quark *canonical momenta* $\mathbf{P}^{Qn}$ (and likewise, for quark canonical positions $\mathbf{X}^{Qn}$), obtained from the condition of the invariance of Eq.(9):

$$P^Q \equiv \begin{bmatrix} \mathbf{p}^{Q1} \\ \mathbf{p}^{Q2} \\ \mathbf{p}^{Q3} \end{bmatrix} = \begin{bmatrix} p_1^1 & -x_3^1 & +x_2^1 \\ +x_3^2 & p_2^2 & -x_1^2 \\ -x_2^3 & +x_1^3 & p_3^3 \end{bmatrix}. \tag{14}$$

The above forms of A^Q and P^Q remain unchanged for quarks of opposite isospin.

For the antiquarks, the relevant forms are

$$A^{\overline{Q}} = \begin{bmatrix} A_1 & +B_3 & -B_2 \\ -B_3 & A_2 & +B_1 \\ +B_2 & -B_1 & A_3 \end{bmatrix}, \tag{15}$$

and

$$P^{\overline{Q}} = \begin{bmatrix} p_1^1 & +x_3^1 & -x_2^1 \\ -x_3^2 & p_2^2 & +x_1^2 \\ +x_2^3 & -x_1^3 & p_3^3 \end{bmatrix}. \tag{16}$$

Again, these forms are independent of isospin.

As expected, the difference between quarks and antiquarks is represented by a change in sign in front of physical positions entering into the definitions of quark (antiquark) canonical momenta. *If* additivity of canonical momenta of quarks is a proper generalization of the additivity of physical momenta for leptons and other individually observable particles, then - on account of the relative (positive and negative) signs between position components in (14) and (16) - this additivity (separately in each of the relevant phase-space directions) leads to translationally invariant string-like expressions for quark-antiquark and three-quark systems (i.e. $x_2^1(q) - x_2^1(\overline{q})$ for $q\overline{q}$ and $(x_1^3(q_3) - x_1^2(q_2), x_2^1(q_1) - x_2^3(q_3), x_3^2(q_2) - x_3^1(q_1))$ for $q_1q_2q_3$), but not for qq or $qqqq$ systems. Additivity of canonical momenta leads therefore to the formation of 'mesons' and 'baryons' only.

In principle, the chain of arguments leading to Eq. (13,14) could involve ordinary reflections, e.g $(A_1, -B_3, +B_2) \rightarrow (A_1, +B_3, +B_2)$, *before* putting the latter expression and its cyclic counterparts together into a matrix form similar to (13) ('grouping'). The corresponding phase-space counterparts would then look somewhat different, i.e. $(p_1, -x_3, +x_2) \rightarrow (p_1, +x_3, +x_2)$ etc. The translational invariance of three-quark systems could then not be achieved by simply adding the appropriate canonical momenta of different quarks, because all physical position coordinates would enter an analog of (14) with positive signs. The point, however, is that one *may* choose the ordering of the operations of grouping and reflection in such a way that - by the simple procedure of addition - the translational invariance can be achieved at all. Furthermore, it seems nontrivial that this requires collaboration of phase-space representatives of quarks of three different colours. After all, the latter were originally defined - *in a way seemingly independent of the concept of additivity* - via a diagonalization procedure performed at the Clifford algebra level.

In other words, the q-level structure of coloured quark charges corresponds to a specific picture in the macroscopic arena. According to this picture, individual quarks are not observable at the classical level since their individual canonical momenta are not translationally invariant. On the other hand, translational invariance may be restored via the collaboration of quarks of different colours. When viewed from our classical point of view, the resulting composite systems possess standard particle-like properties, while at the same time exhibiting internal string-like features. Thus, quark unobservability is supposed to be connected to the very emergence and nature of space and time. Speaking more precisely, quarks are supposed to be unobservable because space and phase space are most probably just convenient classical abstractions, into the descriptive corset of which we try to force various pieces of Reality.

The often-used argument that 'space is standard' at the distances a few orders of magnitude smaller than proton's size is not sufficiently sound. After all, the existence of long-distance nonlocal quantum correlations indicates that - at least for some purposes - our classical spacetime concepts (into which we try to force our descriptions of elementary particles) are inadequate at much larger scales. In fact, statements about an 'unchanged nature of space' at a distance of 10^{-18} m or so follow from the success of the Standard Model (a cq-level theory) and are strictly valid only *within* the description it provides, not outside of it (e.g. not in a q-level theory in which space is to be an emergent concept only).

The above discussion suggests that the phase-space approach has the capability of describing the phenomenon of quark unobservability in a way seemingly different from the SM flux-tube picture of confinement. In fact, however, the phase-space approach does not have to be in conflict with the latter picture, just as the Faraday picture of 'real' fundamental field lines is not in conflict with the Maxwell concept of fields. Rather, we regard the idea of the linearized phase space as offering a possible q-level starting point. Our discussion is based on the invariance of Eq. (9) which provides a link between q- and cq- (c-) levels of description. Obviously, however, Eq. (9) does not specify how the macroscopic background phase space actually emerges from the underlying quantum level. Hence, at present there is no way to compare our ideas directly with the standard, background-dependent, QCD-based picture of confinement.

5. Conclusions

The linearized phase-space approach differs markedly from the standard frameworks, in particular from the $SU(5)$-based unifications. A brief comparison of some differences between the two schemes is given in Table 3. In the author's opinion, the SU(5) approach lacks a solid philosophical background. On the other hand, the phase-space approach, although overly simplistic, satisfies an important philosophical condition: the necessity to connect the q-level description of elementary particles to the c-level description of the macroscopic world. One can hear the echoes of this condition in the words of Roger Penrose, who stated in [12]: *'I do not believe that a real understanding of the nature of elementary particles can ever be achieved without a simultaneous deeper understanding of the nature of spacetime.'*

The phase-space approach provides a possible theoretical explanation of the structure of a single generation of the Standard Model. It gives us a tentative (pre)geometric interpretation of the origin of the Gell-Mann-Nishijima relation. It reproduces the structure of the Harari-Shupe preon model without actually introducing any preons at all, in line with the standard pointlike

Table 3. Comparison with $SU(5)$.

Simple group – $SU(5)$	Linearized phase space
$5^* + 10\ (+1)$ representations	two replicas of 4 and 4^* of $SU(4)$
fair prediction for Weinberg angle	–
anomaly free	–
no connection to macroscopic arena	connection to macroscopic phase space
proton decay	protons are forever
–	Harari – Shupe reproduced

description of fundamental fermions. It touches on the question of the origin of time. It suggests that the phenomenon of quark unobservability is related to the very nature of space and time, which are regarded as emergent (or abstract) classical concepts only.

The picture offered by the Clifford algebra of nonrelativistic phase space need not be regarded as 'the' q-level approach. Rather, it should be thought of as a possible deeper layer of description only. However, it has encouraging features, which - as I believe - will show up at the classical level if derived 'in an emergent way' from any suitable q-level description. Let me therefore end by paraphrasing Penrose's opinion: I do not believe that deeper understanding of elementary particles can be achieved without further studies of the proposed link between the elementary particles themselves and the properties and symmetries of nonrelativistic phase space.

This work has been partially supported by the Polish Ministry of Science and Higher Education research project No N N202 248135.

References

[1] Mach E 1960 (English translation) *The Science of Mechanics* (Chicago: Open Court)
[2] Penrose R 1971 Angular Momentum: an Approach to Combinatorial Space-Time *Quantum Theory and Beyond* ed T Bastin (Cambridge: Cambridge Univ. Press) p 151
Wheeler J A 1980 Pregeometry: Motivations and Prospects *Proceedings of a Symposium on Quantum Theory and Gravitation* ed A R Marlow (New York: Academic Press) p 1
Rovelli C 1991 *Phys. Rev.* D **43** 442
[3] Finkelstein D 1972 *Phys. Rev.* D **5** 320
[4] Barbour J 1993 *Phys. Rev.* D **47** 5422
[5] Barbour J 2003 *Lect. Notes Phys.* **633** 15
Barbour J 2009 The nature of time. First prize of the FQXi *The Nature of Time* contest (arXiv:0903.3489)
[6] Żenczykowski P 2007 *Acta Phys. Pol.* B **8** 2053, 2631
Żenczykowski P 2008 *Phys. Lett.* B **660** 567
Żenczykowski P 2009 *J. Phys.* A **42** 045204
Żenczykowski P 2009 *J. Phys. Conf. Ser.* **174** 012032
[7] Born M 1949 *Rev. Mod. Phys.* **21**, 463
[8] Harari H 1979 *Phys. Lett.* B **6** 83
Shupe M A 1979 *Phys. Lett.* B **86** 87
[9] Coleman S, Mandula J 1967 *Phys. Rev.* **159** 1251
[10] Hestenes D, Sobczyk G 1984 *Clifford Algebra to Geometric Calculus* (Dordrecht: D Reidel)
Pavšič M 2003 *Found. Phys.* **33** 1277
[11] Żenczykowski P 2010 *J. Theor. Phys.* **49** 2246
[12] Penrose R 1968 Structure of Spacetime *Batelle Rencontres: 1967 Lectures in Mathematics and Physics* ed C M DeWitt and J A Wheeler (New York: Benjamin) p 121

Journal of Physics: Conference Series **306** (2011) 012039 doi:10.1088/1742-6596/306/1/012039

Relativistic entanglement from relativistic quantum mechanics in the rest-frame instant form of dynamics

Luca Lusanna

Sezione INFN di Firenze, Polo Scientifico, Via Sansone 1, 50019 Sesto Fiorentino (FI), Italy

E-mail: lusanna@fi.infn.it

Abstract. After a review of the problems induced by the Lorentz signature of Minkowski space-time, like the need of a clock synchronization convention for the definition of 3-space and the complexity of the notion of relativistic center of mass, there is the introduction of a new formulation of relativistic quantum mechanics compatible with the theory of relativistic bound states. In it the zeroth postulate of non-relativistic quantum mechanics is not valid and the physics is described in the rest frame by a Hilbert space containing only relative variables. The non-locality of the Poincare' generators imply a kinematical non-locality and non-separability influencing the theory of relativistic entanglement and not connected with the standard quantum non-locality.

Atomic physics is an approximation to QED, in which the atoms are described as non-relativistic particles in quantum mechanics (QM) with a coupling to the electro-magnetic field of order $1/c$. For all the applications in which the energies involved do not cross the threshold of pair production, this description with a fixed number of particles is enough. Therefore atomic physics and the theory of entanglement are formulated in the absolute Euclidean 3-space and use Newton absolute time, namely they are formulated in Galilei space-time. The main drawback is that, due to the coupling to the electro-magnetic field there is not a realization of the kinematical Galilei group connecting non-relativistic inertial frames. On the other hand, if we want to arrive at an understanding of relativistic entanglement, we must reformulate the theory in Minkowski space-time with a well defined realization of the kinematical Poincare' group connecting relativistic inertial frames. This would lead to *relativistic atomic physics* as the quantization of a fixed number of classical relativistic charged scalar (or spinning) particles interacting with the classical electro-magnetic field.

In this framework it is possible to take into account the non-local aspects of the Lorentz signature of space-time and of the Poincare' group. Some of the implications are: i) the non-existence of a unique notion of relativistic center of mass; ii) the propagation of rays of light along the light-cone as implied by Maxwell equations. Instead in the non-relativistic theory used for experiments testing entanglement the world-lines of photons do not exist and only their polarization is described by means of a two-dimensional Hilbert space. While for experiments on Earth this is enough, teleportation between the Earth and the Space Station [1] will require a control on the world-lines of light rays. See Refs.[2, 3] for the existing attempt to formulate a relativistic theory of entanglement.

Till now all the attempts to define relativistic QM employ the so-called *zeroth postulate of QM* (see Zurek in Ref.[4]). According to it a composite system of two spatially separated subsystems is described by the *tensor product* of the Hilbert spaces of the subsystems. The notation $\mathcal{H} = \mathcal{H}_1 \otimes \mathcal{H}_2 = \mathcal{H}_{com} \otimes \mathcal{H}_{rel}$ means that the quantum 2-body isolated system can be imagined to be constituted either by the two single particle subsystems with masses m_1 and m_2 or as the tensor product of a decoupled center-of-mass particle of mass $m = m_1 + m_2$ carrying an internal space with an internal relative motion of reduced mass $\mu = m_1 m_2/m$ [1]. The two descriptions are connected by a unitary transformation and correspond to different choices of bases in $\mathcal{H}$. We will see that the zeroth postulate does not hold at the relativistic level, where the basic conceptual problem is the absence of an intrinsic notion of 3-space without which we cannot formulate a well posed Cauchy problem for Maxwell equations and we loose predictability. This is the problem of clock synchronization, whose clarification came from the attempts to get a consistent description of relativistic bound states.

Moreover the solution must be such that the transition from a simultaneity convention to another one has to be formulated as a gauge transformation (so that physical results are not influenced by the convention, which only modifies the appearances of phenomena).

The standard way out from the problem of 3-space is to choose the Euclidean 3-space of an inertial frame centered on an inertial observer and then use the kinematical Poincare' group to connect different inertial frames. In the absolute Minkowski space-time (replacing the Newtonian absolute time and absolute Euclidean 3-space) the Euclidean 3-spaces of the inertial frames centered on an inertial observer A are identified by means of Einstein convention for the synchronization of clocks: the inertial observer A sends a ray of light at x^o_i towards the (in general accelerated) observer B; the ray is reflected towards A at a point P of B world-line and then reabsorbed by A at x^o_f; by convention P is synchronous with the mid-point between emission and absorption on A's world-line, i.e. $x^o_P = x^o_i + \frac{1}{2}(x^o_f - x^o_i) = \frac{1}{2}(x^o_i + x^o_f)$.

However, this is not possible in general relativity (GR), where there is no absolute notion since also space-time becomes dynamical (with the metric structure satisfying Einstein's equations). The equivalence principle implies the absence of global inertial frames: in the restricted class of globally hyperbolic, asymptotically Minkowskian at spatial infinity, space-times the best we can have are global non-inertial frames connected by 4-diffeomorphisms (the gauge group of GR). As a consequence, also in SR we have to face the problem of reformulating physics in non-inertial frames centered on accelerated observers [6] as a first step before facing GR [2].

A metrology-oriented description of non-inertial frames in SR has been done with the *3+1 point of view* and the use of observer-dependent Lorentz scalar radar 4-coordinates [6, 7]. Let us give the world-line $x^\mu(\tau)$ of an arbitrary time-like observer carrying a standard atomic clock: τ is an arbitrary monotonically increasing function of the proper time of this clock. Then we give an admissible 3+1 splitting of Minkowski space-time, namely a nice foliation

[1] The zeroth postulate, i.e. $\mathcal{H} = \mathcal{H}_1 \otimes \mathcal{H}_2$, is based on a notion of *separability* (Einstein's one) independent from the Galilei group, which instead is at the basis of the decomposition $\mathcal{H} = \mathcal{H}_{com} \otimes \mathcal{H}_{rel}$ emphasizing that the center-of-mass momentum is a constant of motion for an isolated system so that one can do the separation of variables in the Schroedinger equation.

[2] Regarding the problem of clock synchronization in presence of gravity near the Earth let us underline the relevance of the ACES mission of ESA [5], programmed for 2013. It will make possible a measurement of the gravitational redshift of the Earth from the two-way link among a microwave clock (PHARAO) on the Space Station and similar clocks on the ground: the proposed microwave link should make possible the control of effects on the scale of 5 picoseconds. This will be a test of post-Newtonian gravity in the framework of Einstein's geometrical view of gravitation: the redshift is a measure of the $1/c^2$ deviation of post-Newtonian null geodesics from Minkowski ones. This is a non-perturbative effect (requiring a re-summation of the whole perturbative expansion) for every quantum field theory, which has to fix the background (and therefore the associated light-cones) to be able to define the quantum Fock space of the theory. As a consequence soon we will also need a formulation of relativistic entanglement in non-inertial frames in presence of post-Newtonian gravity.

with space-like instantaneous 3-spaces Σ_τ: it is the mathematical idealization of a protocol for clock synchronization (all the clocks in the points of Σ_τ sign the same time of the atomic clock of the observer). On each 3-space Σ_τ we choose curvilinear 3-coordinates σ^r having the observer as origin. These are the radar 4-coordinates $\sigma^A = (\tau; \sigma^r)$. If $x^\mu \mapsto \sigma^A(x)$ is the coordinate transformation from the Cartesian 4-coordinates x^μ of a reference inertial observer to radar coordinates, its inverse $\sigma^A \mapsto x^\mu = z^\mu(\tau, \sigma^r)$ defines the *embedding* functions $z^\mu(\tau, \sigma^r)$ describing the 3-spaces Σ_τ as embedded 3-manifold into Minkowski space-time. The induced 4-metric on Σ_τ is the following functional of the embedding ${}^4g_{AB}(\tau, \sigma^r) = [z^\mu_A\, \eta_{\mu\nu}\, z^\nu_B](\tau, \sigma^r)$, where $z^\mu_A = \partial\, z^\mu / \partial\, \sigma^A$ and ${}^4\eta_{\mu\nu} = \epsilon\, (+ - - -)$ is the flat metric ($\epsilon = \pm 1$ according to either the particle physics $\epsilon = 1$ or the general relativity $\epsilon = -1$ convention). While the 4-vectors $z^\mu_r(\tau, \sigma^u)$ are tangent to Σ_τ, so that the unit normal $l^\mu(\tau, \sigma^u)$ is proportional to $\epsilon^\mu{}_{\alpha\beta\gamma}\, [z^\alpha_1\, z^\beta_2\, z^\gamma_3](\tau, \sigma^u)$, we have $z^\mu_\tau(\tau, \sigma^r) = [N\, l^\mu + N^r\, z^\mu_r](\tau, \sigma^r)$ ($N(\tau, \sigma^r) = \epsilon\, [z^\mu_\tau\, l_\mu](\tau, \sigma^r)$ and $N_r(\tau, \sigma^r) = -\epsilon\, g_{\tau r}(\tau, \sigma^r)$ are the lapse and shift functions).

The foliation is nice and admissible if it satisfies the conditions [3]:
1) $N(\tau, \sigma^r) > 0$ in every point of Σ_τ (the 3-spaces never intersect, avoiding the coordinate singularity of Fermi coordinates);
2) $\epsilon\, {}^4g_{\tau\tau}(\tau, \sigma^r) > 0$, so to avoid the coordinate singularity of the rotating disk, and with the positive-definite 3-metric ${}^3g_{rs}(\tau, \sigma^u) = -\epsilon\, {}^4g_{rs}(\tau, \sigma^u)$ having three positive eigenvalues (these are the Møller conditions [6, 8]);
3) all the 3-spaces Σ_τ must tend to the same space-like hyper-plane at spatial infinity (so that there are always asymptotic inertial observers to be identified with the fixed stars).

In the 3+1 point of view the 4-metric ${}^4g_{AB}(\tau, \vec{\sigma})$ on Σ_τ has the components $\epsilon\, {}^4g_{\tau\tau} = N^2 - N_r\, N^r$, $-\epsilon\, {}^4g_{\tau r} = N_r = {}^3g_{rs}\, N^s$, ${}^3g_{rs} = -\epsilon\, {}^4g_{rs} = \sum_{a=1}^3 {}^3e_{(a)r}\, {}^3e_{(a)s} = \tilde{\phi}^{2/3} \sum_{a=1}^3 e^{2 \sum_{\bar{b}=1}^2 \gamma_{\bar{b}a}\, R_{\bar{b}}}\, V_{ra}(\theta^i)\, V_{sa}(\theta^i))$, where ${}^3e_{(a)r}(\tau, \sigma^u)$ are cotriads on Σ_τ, $\tilde{\phi}(\tau, \sigma^r) = \sqrt{det\, {}^3g_{rs}(\tau, \sigma^r)}$ is the 3-volume element on Σ_τ, $\lambda_a(\tau, \sigma^r) = [\tilde{\phi}^{1/3}\, e^{\sum_{\bar{b}=1}^2 \gamma_{\bar{b}a}\, R_{\bar{b}}}](\tau, \sigma^r)$ are the positive eigenvalues of the 3-metric ($\gamma_{\bar{a}a}$ are suitable numerical constants) and $V(\theta^i(\tau, \sigma^r))$ are diagonalizing rotation matrices depending on three Euler angles. The components ${}^4g_{AB}$ or the quantities N, N_r, γ, $R_{\bar{a}}$, θ^i, play the role of the *inertial potentials* generating the relativistic apparent forces in the non-inertial frame. It can be shown [6, 8] that the Newtonian inertial potentials are hidden in the functions N, N_r and θ^i. The extrinsic curvature ${}^3K_{rs}(\tau, \sigma^u) = [\frac{1}{2N}\, (N_{r|s} + N_{s|r} - \partial_\tau\, {}^3g_{rs})](\tau, \sigma^u)$, describing the *shape* of the instantaneous 3-spaces of the non-inertial frame as embedded 3-manifolds of Minkowski space-time, is a secondary inertial potential functional of the independent inertial potentials ${}^4g_{AB}$.

The description of isolated systems (particles, strings, fields, fluids) admitting a Lagrangian formulation in the non-inertial frames of SR is done by means of *parametrized Minkowski theories* [6, 7]. The matter variables are replaced with new ones knowing the 3-spaces Σ_τ. For instance a Klein-Gordon field $\tilde{\phi}(x)$ will be replaced with $\phi(\tau, \sigma^r) = \tilde{\phi}(z(\tau, \sigma^r))$; the same for every other field. Instead for a relativistic particle with world-line $x^\mu(\tau)$ we must make a choice of its energy sign: then it will be described by 3-coordinates $\eta^r(\tau)$ defined by the intersection of the world-line with Σ_τ: $x^\mu(\tau) = z^\mu(\tau, \eta^r(\tau))$. Differently from all the previous approaches to relativistic mechanics, the dynamical configuration variables are the 3-coordinates $\eta^r_i(\tau)$ and not the world-lines $x^\mu_i(\tau)$ (to rebuild them in an arbitrary frame we need the embedding defining that frame!). Then the matter Lagrangian is coupled to an external gravitational

[3] These conditions imply that global *rigid* rotations are forbidden in relativistic theories. In Ref.[6, 8] there is the expression of the admissible embedding corresponding to a 3+1 splitting of Minkowski space-time with parallel space-like hyper-planes (not equally spaced due to a linear acceleration) carrying differentially rotating 3-coordinates without the coordinate singularity of the rotating disk. It is the first consistent global non-inertial frame of this type.

field and the external 4-metric is replaced with the 4-metric $g_{AB}(\tau, \sigma^r)$ of an admissible 3+1 splitting of Minkowski space-time. With this procedure we get a Lagrangian depending on the given matter and on the embedding $z^\mu(\tau, \sigma^r)$, which is invariant under *frame-preserving diffeomorphisms* [9]. As a consequence, there are four first-class constraints (an analogue of the super-Hamiltonian and super-momentum constraints of canonical gravity) implying that the embeddings $z^\mu(\tau, \sigma^r)$ are *gauge variables*, so that all the admissible non-inertial or inertial frames are gauge equivalent, namely physics does *not* depend on the clock synchronization convention and on the choice of the 3-coordinates σ^r: only the appearances of phenomena change by changing the notion of instantaneous 3-space. Even if the gauge group is formed by the frame-preserving diffeomorphisms, the matter energy-momentum tensor allows the determination of the ten conserved Poincare' generators P^μ and $J^{\mu\nu}$ (assumed finite) of every configuration of the system; they are non-local quantities knowing the whole 3-space!.

If we restrict ourselves to inertial frames, we can define the *inertial rest-frame instant form of dynamics for isolated systems* by choosing the 3+1 splitting corresponding to the intrinsic inertial rest frame of the isolated system centered on an inertial observer: the instantaneous 3-spaces, named Wigner 3-space due to the fact that the 3-vectors inside them are Wigner spin-1 3-vectors [7], are orthogonal to the conserved 4-momentum P^μ of the configuration. In Ref.[6] there is the extension to admissible non-inertial rest frames, where P^μ is orthogonal to the asymptotic space-like hyper-planes to which the instantaneous 3-spaces tend at spatial infinity. This non-inertial family of 3+1 splittings is the only one admitted by the asymptotically Minkowskian space-times covered by the canonical gravity formulation of Refs.[10, 11].

In the inertial rest frames we can get the explicit form of the Poincare' generators (in particular of the Lorentz boosts, which, differently from the Galilei ones, are interaction dependent). We can also give the final solution to the old problem of the relativistic extension of the Newtonian center of mass of an isolated system. In its rest frame there are *only* three notions of collective variables, which can be built by using *only* the Poincare' generators (they are *non-local* quantities knowing the whole Σ_τ) [12]: the canonical non-covariant Newton-Wigner center of mass (or center of spin), the non-canonical covariant Fokker-Pryce center of inertia and the non-canonical non-covariant Møller center of energy. All of them tend to the Newtonian center of mass in the non-relativistic limit. See Ref.[7] for the Møller non-covariance world-tube around the Fokker-Pryce 4-vector identified by these collective variables. As shown in Refs.[12, 13, 14] these three variables can be expressed as known functions of the rest time τ, of the canonically conjugate Jacobi data (frozen Cauchy data) $\vec{z} = Mc\,\vec{x}_{NW}(0)$ ($\vec{x}_{NW}(\tau)$ is the standard Newton-Wigner 3-position) and $\vec{h} = \vec{P}/Mc$, of the invariant mass $Mc = \sqrt{\epsilon\, P^2}$ of the system and of its rest spin $\vec{\bar{S}}$. As a consequence, every isolated system (i.e. a closed universe) can be visualized as a decoupled non-covariant collective (non-local) pseudo-particle described by the frozen Jacobi data $\vec{z}$, $\vec{h}$ carrying a *pole-dipole structure* (see also Ref.[15]), namely the invariant mass and the rest spin of the system, and with an associated *external* realization of the Poincare' group. The universal breaking of Lorentz covariance is connected to this decoupled non-local collective variable and is irrelevant because all the dynamics of the isolated system leaves inside the Wigner 3-spaces and is Wigner-covariant. In each Wigner 3-space Σ_τ there is a *unfaithful internal* realization of the Poincare' algebra, whose generators are built by using the energy-momentum tensor of the isolated system. While the internal energy and angular momentum are Mc and $\vec{\bar{S}}$ respectively, the internal 3-momentum vanishes: it is the rest frame condition. Also the internal Lorentz boost (whose expression in presence of interactions is given for the first time) vanishes: this condition identifies the covariant non-canonical Fokker-Pryce center of inertia as the natural inertial observer origin of the 3-coordinates σ^r in each Wigner 3-space. As a consequence [16] there are three pairs of second class (interaction-dependent) constraints eliminating the internal 3-center of mass and its conjugate momentum inside the Wigner 3-spaces [16]: this avoids a double counting of the collective variables and allows to re-express the dynamics only in terms of internal Wigner-covariant relative variables. As a consequence, we find that disregarding the unobservable center of mass all the dynamics is described only by relative variables: this is a form of *weak relationism* without the heavy foundational problem of approaches like the one in Ref.[17].

In the case of relativistic particles the reconstruction of their world-lines requires a complex interaction-dependent procedure delineated in Ref.[14]. The final derived world-lines $x_i^\mu(\tau)$ turn out to be non-canonical predictive coordinates, i.e. already at the classical level there is a non-commutative structure implied by Lorentz signature. See Ref.[16] for the comparison with the other formulations of relativistic mechanics developed for the study of the problem of *relativistic bound states*. In Refs.[14, 18] there is the explicit form of the Lorentz boosts for some interacting systems.

In this framework it has been possible to obtain a relativistic formulation of the classical background of atomic physics, considered as an effective theory of positive-energy scalar (or spinning) particles with mutual Coulomb interaction plus the transverse electro-magnetic field of the radiation gauge valid for energies below the threshold of pair production. As shown in Refs.[13] and [16], this has been possible by considering Grassmann-valued electric charges for the particles ($Q_i^2 = 0$, $Q_i\, Q_j = Q_j\, Q_i \neq 0$ for $i \neq j$). It allows a) to make an ultraviolet regularization of Coulomb self-energies; b) to make an infrared regularization eliminating the photon emission; c) to express the Lienard-Wiechert potentials only in terms of the 3-coordinates $\eta_i^r(\tau)$ and the conjugate 3-momenta $\kappa_{ir}(\tau)$ in a way independent from the used (retarded, advanced,..) Green function. All this amount to reformulate the dynamics of the one-photon exchange as a Cauchy problem with well defined potentials. Moreover there is a canonical transformation [16] sending the above system in a transverse radiation field (in- or out-fields) decoupled, in the global rest frame, from Coulomb-dressed particles with a mutual interaction described by the sum of the Coulomb potential plus the Darwin potential. Therefore for the first time we are able to obtain results, previously derived from instantaneous approximations to the Bethe-Salpeter equation for the description of relativistic bound states (see the bibliography of Ref.[13]), starting from the classical theory. Moreover, for the first time, at least at the classical level, we have been able to avoid the Haag theorem according to which the interaction picture does not exist in QFT.

In refs.[21] there is the *multi-temporal quantization* of positive-energy free scalar and spinning particles in a family of global non-inertial frames of Minkowski space-time with space-like hyper-planes as 3-spaces and differentially rotating 3-coordinates defined in Ref.[6]. We take the point of view *not to quantize the inertial effects* (the appearances of phenomena): the embedding $z^\mu(\tau, \sigma^r)$ remains a c-number and we get results compatible with atomic spectra. Instead the problem of the reformulation of particle physics in non-inertial frames is unsolved due to the no-go theorem of Ref.[22] showing the existence of obstructions to the unitary evolution of a massive Klein-Gordon field between two space-like surfaces of Minkowski space-time. This problem has to be reformulated as the search of the class of admissible 3+1 splittings of Minkowski space-time admitting unitary evolution after quantization: this would allow to check whether the hypothesis of non-quantized inertial effects is valid also in field theory (it will be a crucial point for quantum gravity!).

The previous framework allowed to find a new formulation of *relativistic quantum mechanics and entanglement*, which is developed in Ref. [19].

As already said, in Galilei space-time non-relativistic QM, where all the main results about entanglement are formulated, describes a composite system with two (or more) subsystems with a Hilbert space which is the tensor product of the Hilbert spaces of the subsystems: $H = H_1 \otimes H_2$. This type of spatial separability is named *the zeroth postulate* of quantum mechanics. However, when the two subsystems are mutually interacting, one makes a unitary transformation to the tensor product of the Hilbert space H_{com} describing the decoupled Newtonian center of mass of the two subsystems and of the Hilbert space H_{rel} of relative variables: $H = H_1 \otimes H_2 = H_{com} \otimes H_{rel}$. This allows to use the method of separation of variables to split the Schroedinger equation in two equations: one for the free motion of the center of mass and another, containing the interactions, for the relative variables (this equation describes both the bound and scattering states). A final unitary transformation of the Hamilton-Jacobi type allows to replace H_{com} with $H_{com,HJ}$, the Hilbert space in which the decoupled center of mass is frozen and described by non-evolving Jacobi data. Therefore we have $H = H_1 \otimes H_2 = H_{com} \otimes H_{rel} = H_{com,HJ} \otimes H_{rel}$.

While at the non-relativistic level these three descriptions are unitary equivalent, this no more true in relativistic quantum mechanics, due to the previously described problems arising from Lorentz signature like the need of clock synchronization and the non-covariance of the (non-local) canonical center of mass.

In the approach of Ref.[19] we quantize the frozen Jacobi data of the canonical non-covariant decoupled center of mass and the Wigner-covariant relative variables on the Wigner hyper-plane. Since the center of mass is decoupled, its non-covariance is irrelevant: like for the wave function of the universe, who will observe it? Moreover its non-local nature implies that it is not a locally measurable quantity. Finally the use of the static Jacobi data for the external center of mass avoids the causality problems connected with the instantaneous spreading of wave packets (the Hegerfeldt theorem [20]). This viewpoint is in accord with relativistic bound states and relativistic atomic physics

The need of clock synchronization for the definition of the instantaneous 3-spaces and the non-local and non-covariant properties of the decoupled relativistic center of mass, described by the frozen Jacobi data $\vec{z}$ and $\vec{h}$, imply that the only consistent relativistic quantization is based on the Hilbert space $H = H_{com,HJ} \otimes H_{rel}$ ($H_{com,HJ}$ is the Hilbert space of the external center of mass in the Hamilton-Jacobi formulation, while H_{rel} is the Hilbert space of the internal relative variables). The Hilbert space H is not unitarily equivalent to $H_1 \otimes H_2 \otimes ...$, where H_i are the Hilbert spaces of the individual particles. This is due to the fact that already in the non-interacting two-particle case, in the tensor product of two quantum Klein-Gordon fields, $\phi_1(x_1)$ and $\phi_2(x_2)$, most of the states correspond to configurations in Minkowski space-time in which one particle may be present in the absolute future of the other particle, because the two times x_1^o and x_2^o are totally uncorrelated, or in other words there is no notion of instantaneous 3-space (clock synchronization convention). Also the scalar products in the two formulations are completely different as shown in Ref.[23]. In S-matrix theory this problem is eliminated by avoiding the interpolating states at finite (the problem of the Haag theorem) and going the the asymptotic (in the times x_i^o) limit of the free in- and out- states. However in atomic physics we need interpolating states, and not S-matrix, to describe a laser beam resonating in a cavity and intersected by a beam of atoms!

We have also $H \neq H_{com} \otimes H_{rel}$, because if instead of $\vec{z} = Mc\vec{x}_{NW}(0)$ we use the evolving (non-local and non-covariant) Newton-Wigner position operator $\vec{x}_{NW}(\tau)$, then we get a violation of relativistic causality because the center-of-mass wave packets spread instantaneously as shown by the Hegerfeldt theorem [20].

Therefore the only consistent Hilbert space is $H = H_{com,HJ} \otimes H_{rel}$, whose non-relativistic limit is the corresponding Newtonian Hilbert space. The main complication is the definition of H_{rel}, because we must take into account the three pairs of (interaction-dependent) second-class constraints eliminating the internal 3-center of mass inside the Wigner 3-spaces. When we are not able to make the elimination at the classical level and formulate the dynamics only in terms of Wigner-covariant relative variables, we have to quantize the particle Wigner-covariant 3-variables η_i^r, κ_{ir} and then to define the physical Hilbert space by adding the quantum version of the constraints a la Gupta-Bleuler.

The quantization defined in Ref.[19] leads to a first formulation of a theory for *relativistic entanglement*. The non validity of the zeroth postulate and the *non-locality* of Poincare' generators imply a *kinematical non-locality* and a *kinematical spatial non-separability* introduced by special relativity, which reduce the relevance of *quantum non-locality* in the study of the foundational problems of quantum mechanics which have to be rephrased in terms of relative variables. Einstein's notion of separability is not valid since in $H = H_{com,HJ} \otimes H_{rel}$ the composite system must be described by means of relative variables in a Wigner 3-space (this is a type of weak form of relationism different from the formulations induced by the Mach principle like in Ref.[17]).

The relativistic formulation of problems like the relevance of decoherence [24] for the selection of preferred robust pointer bases and the emergence of quasi-classical macroscopic objects from quantum constituents will have to be done in terms of relative variables. Moreover, the control of Poincare' kinematics will force to reformulate the experiments connected with Bell inequalities and teleportation in terms of isolated systems containing: a) the observers with their measuring apparatus (Alice and Bob as macroscopic quasi-classical objects); b) the particles of the protocol (but now the ray of light, the "photons" carrying the polarization, move along null geodesics); c) the environment (macroscopic either quantum or quasi-classical object). All these problems disappear as $1/c$ corrections in experiments on Earth, but will be relevant for space physics. A preliminary step for understanding this framework would be to reformulate the non-relativistic theory of entanglement in the Hilbert space $H = H_{com,HJ} \otimes H_{rel}$ and to forget about the decoupled center-of-mass Hilbert space $H_{com,HJ}$.

The final challenge will be a consistent inclusion of the gravitational field, at least at the post-Newtonian level! See Refs.[11] for the use of 3+1 splittings and of radar 4-coordinates in ADM canonical tetrad gravity in globally hyperbolic, asymptotically Minkowskian at spatial infinity, space-times. As shown in Ref.[25], the dynamical nature of space-time implies that each solution of Einstein's equations dynamically selects a preferred 3+1 splitting of the space-time, namely in GR the instantaneous 3-spaces are dynamically determined except for the trace of the extrinsic curvature (the York time ${}^3K(\tau, \sigma^r)$): this inertial gauge variable is the general relativistic remnant of the special relativistic gauge freedom in clock synchronization.

References

[1] QUEST - *Quantum Entanglement in Space Experiments*, http://www.quantum.at/quest
[2] Harshman N L, 2005 Basis States for Relativistic Dinamically Entangled Particles, *Phys.Rev.* A **71** 022312; 2005 Dynamical Entanglement in Particle Scattering, *Int.J.Mod.Phys.* A **20** 6220 (*Preprint* quant-ph/0506212); 2005 Poincare' Semigroup Symmetry as an Emergent Property of Unstable Systems *Preprint* hep-ph/0511298; 2005 Limits on Entaglement in Rotationally-Invariant Scattering of Spin Systems *Preprint* quant-ph/0509013.
[3] Peres A and Terno D R 2004 Quantum Information and Relativity Theory, *Rev.Mod.Phys.* **76** 93 (*Preprint* quant-ph/0212023).
Terno D R 2005 Introduction to Relativistic Quantum Information *Preprint* quant-ph/0508049.
Hay O and Peres A 1998 Quantum and Classical Descriptions of a Measuring Apparatus, *Phys.Rev.* A **58** 116 (*Preprint* quant-ph/9712044).
[4] Zurek W H 2003 Decoherence, Einselection, and the Quantum Origins of the Classical, *Rev.Mod.Phys.* **75** 715-775.
[5] Cacciapuoti L and Salomon C 2007 *ACES: Mission Concept and Scientific Objective*, ESA document, Estec (ACES_Science_v1_printout.doc).

Blanchet L, Salomon C, Teyssandier P and Wolf P 2000 Relativistic Theory for Time and Frequency Transfer to Order $1/c^3$, *Astron.Astrophys.* **370** 320.

[6] Alba D and Lusanna L 2010 Charged Particles and the Electro-Magnetic Field in Non-Inertial Frames: I. Admissible 3+1 Splittings of Minkowski Spacetime and the Non-Inertial Rest Frames, *Int.J.Geom.Methods in Physics* **7** 33 (*Preprint* arXiv 0908.0213) and 2010 II. Applications: Rotating Frames, Sagnac Effect, Faraday Rotation, Wrap-up Effect, *Int.J.Geom.Methods in Physics* **7** 185 (*Preprint* arXiv 0908.0215).

[7] Lusanna L 1997 The N- and 1-Time Classical Descriptions of N-Body Relativistic Kinematics and the Electromagnetic Interaction, *Int. J. Mod. Phys.* A **12** 645; 2007 The Chrono-Geometrical Structure of Special and General Relativity: A Re-Visitation of Canonical Geometrodynamics, lectures at *42nd Karpacz Winter School of Theoretical Physics: Current Mathematical Topics in Gravitation and Cosmology*, Ladek, Poland, 6-11 Feb 2006, *Int.J.Geom.Methods in Mod.Phys.* **4** 79 (*Preprint* gr-qc/0604120); 2000 Towards a Unified Description of the Four Interactions in Terms of Dirac-Bergmann Observables, invited contribution to the book *Quantum Field Theory: a 20th Century Profile*, of the Indian National Science Academy, ed.A.N.Mitra, forewards by F.J.Dyson (Hindustan Book Agency, New Delhi) (*Preprint* hep-th/9907081).

[8] Alba D and Lusanna L 2007 Generalized Radar 4-Coordinates and Equal-Time Cauchy Surfaces for Arbitrary Accelerated Observers, *Int.J.Mod.Phys.* D **16** 1149 (*Preprint* gr-qc/0501090).

[9] Schmutzer E and Plebanski J 1978 Quantum Mechanics in Noninertial Frames of Reference, *Fortschr.Phys.* **25** 37.

[10] Lusanna L 2001 The Rest-Frame Instant Form of Metric Gravity, *Gen.Rel.Grav.* **33** 1579 (*Preprint* gr-qc/0101048).
Lusanna L and Russo S 2002 A New Parametrization for Tetrad Gravity, *Gen.Rel.Grav.* **34** 189 (*Preprint* gr-qc/0102074).
DePietri R, Lusanna L, Martucci L and Russo S 2002 Dirac's Observables for the Rest-Frame Instant Form of Tetrad Gravity in a Completely Fixed 3-Orthogonal Gauge, *Gen.Rel.Grav.* **34** 877 (*Preprint* gr-qc/0105084).

[11] Alba D and Lusanna L 2007 The York Map as a Shanmugadhasan Canonical Transformationn in Tetrad Gravity and the Role of Non-Inertial Frames in the Geometrical View of the Gravitational Field, *Gen.Rel.Grav.* **39** 2149 (*Preprint* gr-qc/0604086, v2); 2009 The Einstein-Maxwell-Particle System in the York Canonical Basis of ADM Tetrad Gravity: I) The Equations of Motion in Arbitrary Schwinger Time Gauges, *Preprint* arXiv 0907.4087; 2010 II) The Weak Field Approximation in the 3-Orthogonal Gauges and Hamiltonian Post-Minkowskian Gravity: the N-Body Problem and Gravitational Waves with Asymptotic Background, *Preprint* arXiv 1003.5143; 2010 III) The Post-Minkowskian N-Body Problem, its Post-Newtonian Limit in Non-Harmonic 3-Orthogonal Gauges and Dark Matter as an Inertial Effect, *Preprint* arXiv 1009.1794).
Lusanna L 2010 Dark Matter as a Relativistic Inertial Effect in Einstein Canonical Gravity?, talk at *Group 28: The XXVIII International Colloquium on Group-Theoretical Methods in Physics*, Northumbria University, Newcastle, 26-30 July 2010 (*Preprint* arXiv 1011.2908).

[12] Alba D, Lusanna L and Pauri M 2006 New Directions in Non-Relativistic and Relativistic Rotational and Multipole Kinematics for N-Body and Continuous Systems, in *Atomic and Molecular Clusters: New Research*, ed.Y.L.Ping (Nova Science, New York) (*Preprint* hep-th/0505005); 2002 Centers of Mass and Rotational Kinematics for the Relativistic N-Body Problem in the Rest-Frame Instant Form, *J.Math.Phys.* **43** 1677-1727 (*Preprint* hep-th/0102087); 2004 Multipolar Expansions for Closed and Open Systems of Relativistic Particles , *J. Math.Phys.* **46** 062505 (*Preprint* hep-th/0402181).

[13] Crater H W and Lusanna L 2001 The Rest-Frame Darwin Potential from the Lienard-Wiechert Solution in the Radiation Gauge, *Ann.Phys.(N.Y.)* **289** 87 (*Preprint* hep-th/0001046).
Alba D, Crater H W and Lusanna, L 2001 The Semiclassical Relativistic Darwin Potential for Spinning Particles in the Rest-Frame Instant Form: Two-Body Bound States with Spin 1/2 Constituents, *Int.J.Mod.Phys.* A **16** 3365 (*Preprint* hep-th/0103109).

[14] Alba D, Crater H W and Lusanna l 2007 Hamiltonian Relativistic Two-Body Problem: Center of Mass and Orbit Reconstruction, *J.Phys.* A **40** 9585 (*Preprint* gr-qc/0610200).

[15] Leutwyler H and Stern J 1978 Relativistic Dynamics on a Null Plane, *Ann.Phys. (N.Y.)* **112** 94.

[16] Alba D, Crater H W and Lusanna L 2010 Towards Relativistic Atom Physics. I. The Rest-Frame Instant Form of Dynamics and a Canonical Transformation for a system of Charged Particles plus the Electro-Magnetic Field, *Canad.J.Phys.* **88** 379 (*Preprint* arXiv 0806.2383); 2010 Towards Relativistic Atom Physics. II. Collective and Relative Relativistic Variables for a System of Charged Particles plus the Electro-Magnetic Field, *Canad.J.Phys.* **88** 425 (*Preprint* arXiv 0811.0715).

[17] Rovelli C 1996 Relational Quantum Mechanics, *Int.J.Theor.Phys.* **35** 1637 (*Preprint* quant-ph/9609002).
Rovelli C and Smerlak M 2007 Relational EPR, *Found.Phys.* **37** 427 (*Preprint* quant-ph/0604064).

[18] Alba D, Crater H W and Lusanna L 2010 The Rest-Frame Instant Form and Dirac Observables for the Open

Nambu String, *Preprint* arXiv 1005.3653; 2010 Massless Particles plus Matter in the Rest-Frame Instant Form of Dynamics, *J.Phys.* A **43** 405203 (*Preprint* arXiv 1005.5521).

[19] Alba D, Crater H W and Lusanna L 2009 Relativistic Quantum Mechanics and Relativistic Entanglement in the Rest-Frame Instant Form of Dynamics, *Preprint* arXiv 0907.1816.

[20] Hegerfeldt G C 1974 Remark on Causality and Particle Localization, *Phys.Rev.* D **10** 3320; 1985 Violation of Causality in Relativistic Quantum Theory?, *Phys.Rev.Lett.* **54** 2395; 1998 Instantaneous Spreading and Einstein Causality in Quantum Theory, *Ann.Phys.Lpz.* **7** 716 (*Preprint* quant-ph/9809030); 1998 Causality, Particle Localization and Positivity of the Energy, in *Irreversibility and Causality in Quantum Theory - Semigroups and Rigged Hilbert Spaces*, eds. A.Bohm, H.D.Doebner and P.Kielanowski, Lecture Notes in Physics 504, p.238 (Springer , NewYork) (*Preprint* quant-ph/9806036).

[21] Alba D and Lusanna L 2006 Quantum Mechanics in Noninertial Frames with a Multitemporal Quantization Scheme: I. Relativistic Particles, *Int.J.Mod.Phys.* A **A21** 2781 (*Preprint* hep-th/0502194).
Alba D 2006 Quantum Mechanics in Non-Inertial Frames with a Multi-Temporal Quantization Scheme: II) Non-Relativistic Particles, *Int.J.Mod.Phys.* A **21** 3917 (*Preprint* hep-th/0504060).

[22] Torre C G and Varadarajan M 1999 Functional Evolution of Free Quantum Fields, *Clas. Quantum Grav.* **16** 2651-2668.
Helfer A D 1999 The Hamiltonian of Linear Quantum Fields *Preprint* hep-th/990811.
Arageorgis A, Earman J and Ruetsche L 2002 Weyling the Time Away: the Non-Unitary Implementability of Quantum Field Dynamics o n Curved Spacetimes, *Studies in History and Philosophy of Modern Physics* **33** 151.

[23] Longhi G and Lusanna L 1986 Bound-State Solutions, Invariant Scalar Products and Conserved Currents for a Class of Two-Body Relativistic Systems, *Phys.Rev.* D **34** 3707.

[24] Schlosshauer M 2007 *Decoherence and the Quantum-to-Classical Transition* (Springer, Berlin); 2004 Decoherence, the Measurement Problem and Interpretations of Quantum Mechanics, *Rev.Mod.Phys.* **76** 1267.

[25] Lusanna L and Pauri M 2006 The Physical Role of Gravitational and Gauge Degrees of Freedom in General Relativity. I: Dynamical Synchronization and Generalized Inertial Effects, *Gen.Rel.Grav.* **38** 187 (*Preprint* gr-qc/0403081); 2006 II: Dirac versus Bergmann Observables and the Objectivity of Space-Time, *Gen.Rel.Grav.* **38** 229 (*Preprint* gr-qc/0407007); 2007 Dynamical Emergence of Instantaneous 3-Spaces in a Class of Models of General Relativity, in *Relativity and the Dimensionality of the World*, ed. V.Petkov (Springer Series Fundamental Theories of Physics, Berlin) (*Preprint* gr-qc/0611045).

5th International Workshop DICE2010 IOP Publishing
Journal of Physics: Conference Series **306** (2011) 012040 doi:10.1088/1742-6596/306/1/012040

Is there a relation between the 2D Causal Set action and the Lorentzian Gauss-Bonnet theorem?

Dionigi M T Benincasa

Theoretical Physics Group, Blackett Laboratory, Imperial College, Prince Consort Rd., London SW7 2AZ, U.K.

E-mail: db1808@ic.ac.uk

Abstract. We investigate the relation between the two dimensional Causal Set action, $\mathcal{S}$, and the Lorentzian Gauss-Bonnet theorem (LGBT). We give compelling reasons why the answer to the title's question is no. In support of this point of view we calculate the causal set inspired action of causal intervals in some two dimensional spacetimes: Minkowski, the flat cylinder and the flat trousers.

1. The Gauss-Bonnet theorem

The Gauss-Bonnet (GB) theorem is a remarkable theorem first proved by Gauss, and later by Bonnet in a special case, that relates the geometry of a two-dimensional (2D) Riemannian manifold to its topology. In one of its variations it states that for any 2D compact Riemannian manifold $\mathcal{M}$ with boundary $\partial\mathcal{M}$

$$\frac{1}{2}\int_{\mathcal{M}} R dV + \int_{\partial\mathcal{M}} k_g\, ds + \sum \theta_{exterior} = 2\pi\chi(\mathcal{M}). \tag{1.1}$$

Here R is the scalar curvature of $\mathcal{M}$, k_g is the geodesic curvature of $\partial\mathcal{M}$, $\theta_{exterior}$ is the extrior angle at non-smooth points of $\partial\mathcal{M}$ and $\chi(\mathcal{M})$ is the Euler character of $\mathcal{M}$.

This theorem was first extended to pseudo-Riemannian 2D manifolds by Avez [1] and Chern [2]. The Lorentzian [1] version of the theorem can be viewed as a natural extension of (1.1) where the geometric quantities on the left-hand side are carried over to the Lorentzian context. The essential difference between the Riemannian and Lorentzian theorems then comes from the definition of $\theta_{exterior}$ which depends on the orientation preserving isometry group of their respective metrics: $SO(2)$ and $SO(1,1)$ for the Euclidean and Lorentzian signatures respectively. There are various versions of the LGBT, here we will be using that due to Law [3] (see also [4, 5])

Lorentzian GB Theorem. Let D be a domain in a 2D Lorentzian manifold $(\mathcal{M}, g)$ with piecewise smooth boundary Γ consisting of a finite number of nonnull smooth curves. Then

$$\frac{1}{2}\int_{D} R dV + \int_{\Gamma} k_g\, ds + \sum \theta_{exterior,\pm} = \pm 2\pi i\chi(D). \tag{1.2}$$

[1] Note that in 2D the only pseudo-Riemannian metric is Lorentzian

Where now $\theta_{exterior,\pm}$ is defined with respect to the group $SO(1,1)$ and the choice of sign depends on whether the boundary is traversed clockwise or anti-clockwise.

It is important to note the following things about this theorem: First, as we mentioned earlier, the right hand side is imaginary. Second, we required Γ to be composed of *nonnull* curves only. This is because the notion of angle is ill-defined at points where a null segment meets with timelike, spacelike or nullike segments. As we will see shortly both of these properties of the LGBT are in stark contrast with the causal set action which is real, and whose topological properties only become apparent when the boundary of the domain D is everywhere non-timelike.

2. The 2D causal set action

Recall that a *causal set* [7] (*causet* for short) is a locally finite partial order, *i.e.* it is a pair $(\mathcal{C}, \preceq)$ where $\mathcal{C}$ is a set and $\preceq$ is a relation on $\mathcal{C}$ which is reflexive ($x \preceq x$), acyclic ($y \preceq x \preceq y \Rightarrow y = x$) and transitive ($z \preceq y \preceq x \Rightarrow z \preceq x$). Local finiteness is the condition that the cardinality of any *order interval* is finite, where the (inclusive) order interval between a pair of elements $y \preceq x$ is defined to be $I(x,y) := \{z \in \mathcal{C} \,|\, y \preceq z \preceq x\}$. We call x the *top element* and y the *bottom element* of $I(x,y)$. We write $y \prec x$ when $y \preceq x$ and $y \neq x$. We define $n(x,y) := |I(x,y)|$ and call a relation $y \prec x$ a *link* if $n(x,y) = 2$.

We define the 2D action, $\mathcal{S}$, of a finite causal set $\mathcal{C}$ to be [8]

$$\mathcal{S}[\mathcal{C}] = N - 2N_1 + 4N_2 - 2N_3 \tag{2.3}$$

where N is the cardinality of $\mathcal{C}$, and N_m is the number of inclusive order intervals in $\mathcal{C}$ of cardinality $m+1$. N_1 therefore is the number of links in $\mathcal{C}$, N_2 is the number of order intervals that are 3-chains (3 element chains) and N_4 is the number of order intervals that are 4-chains plus the number that are "diamonds" (with two mutually unrelated elements between the top and bottom elements). The form of $\mathcal{S}$ as an alternating sum of (weighted) numbers of things is intriguingly reminiscent of certain topological indices.

In order to make a connection with the continuum we need to define the notion of sprinkling: *Sprinkling* is a random process that produces a causet which is a discretisation of a d-dimensional, causal, Lorentzian manifold $(\mathcal{M}, g)$. It is a Poisson process of selecting points in $(\mathcal{M}, g)$, independently at random, with density ρ so that the expected number of points sprinkled in a region of spacetime volume V is ρV, where $\rho = l^{-d}$ and l is of order the Planck length. This process generates a causet whose elements are (identified with) the sprinkled points and whose order is that induced by the manifold's causal order restricted to the sprinkled points. We say that a causet $\mathcal{C}$ is well approximated by a Lorentzian manifold $\mathcal{M}$ if it could have been generated, with relatively high probability, by sprinkling into $\mathcal{M}$.

The Poisson distribution gives for the mean of (2.3)

$$\langle \mathcal{S}_{\mathcal{M}} \rangle = \rho V - 2\rho^2 \int_{\mathcal{M}} d^d y \sqrt{-g(y)} \int_{\mathcal{M} \cap J^+(y)} d^d x \sqrt{-g(x)} \left(1 - 2\rho V_{xy} + \frac{1}{2}(\rho V_{xy})^2\right) e^{-\rho V_{xy}} \tag{2.4}$$

where V_{xy} is the volume of the spacetime causal interval, $[x,y] := J^+(y) \cap J^-(x)$, between x and y and d is the dimension of $\mathcal{M}$.

It can be shown that when $\mathcal{C}$ is a 2D sprinkling then the mean of $\mathcal{S}$ tends to something that contains a term $\frac{1}{4}\int_{\mathcal{M}} d^2x \sqrt{-g}R$ plus terms arising from boundary effects. But we know that in 2D gravity the Einstein-Hilbert action plus (the 2D analogue of) the Gibbons-Hawking boundary term are a topological invariant (Eqn. (1.2)), so it is natural to ask to what extent the 2D causet action $\mathcal{S}$ is a topological invariant? To answer this question we study the action of flat causal intervals in different topologies.

3. Causal intervals in $\mathbb{M}^2$

Consider first a causal interval in 2D Minkowski spacetime, $\mathcal{I} := [p,q] \subset \mathbb{M}^2$. For definiteness consider the interval to have fixed volume (area), V. Let us coordinates in which p and q lie on the time axis and q is at the origin. We consider null coordinates $u_x = \frac{1}{\sqrt{2}}(x^0 - x^1)$, $v_x = \frac{1}{\sqrt{2}}(x^0 + x^1)$ and similarly for u_y, v_y. Then the interval is defined by $u, v \in [0,a]$ for $a = \sqrt{V}$. The action of $\mathcal{I}$ is given by [9]

$$\langle \mathcal{S}_\mathcal{I} \rangle = \rho V - 2\int_0^a du_x \int_0^a dv_x \int_0^{u_x} du_y \int_0^{v_x} dv_y \, \rho^2 \, p(\rho \, \Delta u \, \Delta v) = 1 - e^{-\rho V} \tag{3.5}$$

where $\Delta u = u_x - u_y$, $\Delta v = v_x - v_y$. As $\rho \to \infty$, $\langle \mathcal{S}_\mathcal{I} \rangle \to 1$ and we write $\langle \mathcal{S}_\mathcal{I} \rangle \approx 1$ to denote this. This result is a first hint that $\mathcal{S}$ might be topological. Consider now splitting up the interval $\mathcal{I}$ into four smaller intervals $\mathcal{I}_i$, $i = 1, \ldots 4$, as shown in Fig. 1a. When computing the expected

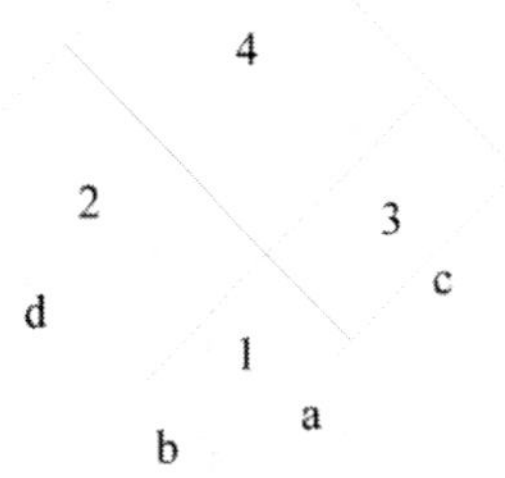

(a) a and c (b and d) are the v-coordinate (u-coordinate) lengths of the sides of the subintervals

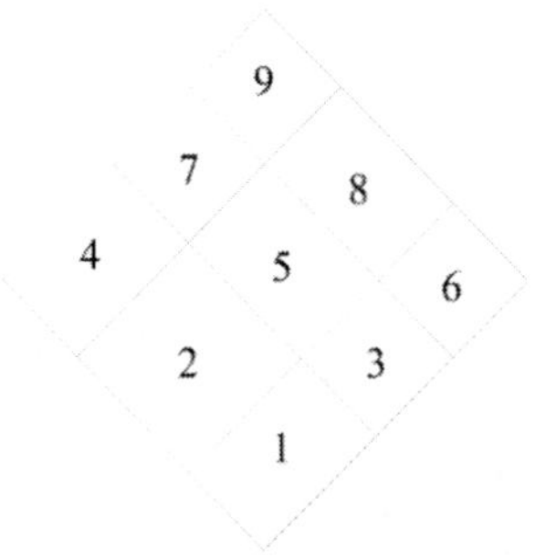

(b)

Figure 1: Splitting up a causal interval in 2D Minkowski to compute the action

value of $\mathcal{S}_\mathcal{I}$ one can split the integral up into the means of the actions of the four subintervals plus the "bilocal" contributions when x and y lie in two different subintervals. More concretely, given any subcausets, A and B of a causal set $\mathcal{C}$, we define the bilocal action,

$$\mathcal{S}[\mathcal{C}; A, B] = N(A,B) - 2N_1(A,B) + 4N_2(A,B) - 2N_3(A,B) \tag{3.6}$$

where $N(A,B)$ is the number of elements in $A \cap B$ and $N_m(A,B)$ is the number of inclusive order intervals in $\mathcal{C}$ of cardinality $m+1$ with top element in A and bottom element in B. The mean of $\mathcal{S}[\mathcal{C}; A, B]$ we will denote by $\langle \mathcal{S}_{\mathcal{M};X,Y} \rangle$ where X and Y are submanifolds of $\mathcal{M}$ into which $\mathcal{C}$ has been sprinkled, with subcausets A, B sprinkled into X, Y respectively. Note that $\mathcal{S}_{\mathcal{M};X,X} = \mathcal{S}_X$ if X is a *causally convex* subset of $\mathcal{M}$.[2]

Now, consider $\mathcal{I}$ and its subintervals. If we adopt $\mathcal{S}_{ij}$ as simplified notation for the bilocal action $\mathcal{S}_{\mathcal{I};\mathcal{I}_i,\mathcal{I}_j}$, then we have

$$\langle \mathcal{S}_\mathcal{I} \rangle = \sum_{i=1}^{4} \langle \mathcal{S}_{\mathcal{I}_i} \rangle + \sum_{\substack{i,j=1 \\ j<i}}^{4} \langle \mathcal{S}_{ij} \rangle \,. \tag{3.7}$$

[2] A causally convex region, X, of $\mathcal{M}$ is one such that $x, y \in X$ implies that the causal interval in $\mathcal{M}$ between x and y is a subset of X.

The bilocal summands can be computed using the integral in Eq. (3.5) and adjusting the boundaries:

$$\langle \mathcal{S}_{21} \rangle = -2 \int_0^a dv_x \int_b^{b+d} du_x \int_0^{v_x} dv_y \int_0^b du_y \, \rho^2 \, p(\rho \, \Delta u \, \Delta v) \approx -1$$

and

$$\langle \mathcal{S}_{41} \rangle = -2 \int_a^{a+c} dv_x \int_b^{b+d} du_x \int_0^a dv_y \int_0^b du_y \, \rho^2 \, p(\rho \, \Delta u \, \Delta v) \approx 1 \, .$$

The three other bilocal contributions $\langle \mathbf{S}_{ij} \rangle$ can be obtained from $\langle \mathbf{S}_{21} \rangle$ by changing the parameters appropriately. Putting together all parts of Eq. (3.7) one exactly recovers Eq. (3.1).

Now, one can continue this game and split up the interval even further as in Fig. 1b. To compute the mean of the action one must again calculate

$$\langle \mathcal{S}_{\mathcal{I}} \rangle = \sum_{i=1}^{9} \langle \mathcal{S}_{\mathcal{I}_i} \rangle + \sum_{\substack{i,j=1 \\ j<i}}^{9} \langle \mathcal{S}_{ij} \rangle \, . \tag{3.8}$$

We already know the contributions $\langle \mathcal{S}_{I_i} \rangle \approx 1$ and the bilocal contributions from two intervals that either share an edge or lie above and below a shared vertex (*e.g.* $\langle \mathcal{S}_{21} \rangle$ and $\langle \mathcal{S}_{51} \rangle$ in Fig. 1b). It remains to compute the bilocal contributions from pairs of intervals such as (4,1),(7,1) and (9,1) in Fig. 1b. It turns out they consist only of exponential terms that are small when intervening intervals are large on the discreteness scale. In the limit of large density, we are left with a contribution of 1 for every subinterval, -1 for every edge and 1 for every vertex. One could write

$$\langle \mathcal{S} \rangle \approx F - E + V \tag{3.9}$$

where F denotes the number of faces *i.e.* intervals, E the number of edges and V the number of vertices. This expression is the formula for the Euler character of a polyhedron. This result suggests that $\langle \mathcal{S} \rangle$ might be topological in nature. But if it is related to the LGBT then where is the factor of 2π ? And i? Also, how does the result depend on the form of the boundary of the region? It can be shown that for any causally convex region, $R \subset \mathbb{M}^2$, and for any "triangulation" of R such that no edge is timelike Eqn. (3.5) holds. The first two questions however remain unanswered; unless we take the view that they are proof that there is no connection between $\mathcal{S}$ and the LGBT. Let us not give up yet however and study the mean action of a causal interval on a cylinder.

4. The flat cylinder

Consider a causal interval, $\mathcal{I}_c$, of height T on a cylinder with circumference L with $L \leq T \leq 2L$. To calculate $\langle \mathcal{S}_{\mathcal{I}_c} \rangle$ consider tiling $\mathcal{I}_c$ into subintervals, $\mathcal{I}_i$, $i = 1, \ldots 8$, as shown in Fig. 2. Taking into account the topological identification, we have $F = 8, E = 12, V = 4$ thus yielding a predicted high-density expectation value of $\langle \mathbf{S}_{\mathcal{I}_c} \rangle \approx 0$. Indeed one can calculate $\langle \mathbf{S}_{\mathcal{I}_c} \rangle$ explicitly by using (2.2) and finds the same result (one needs to be very careful with intervals whose tips have more than one homotopically non-trivial causal curve connecting them, since in these cases the causal volume between them has to be modified appropriately, see [9, 10]). Once again the result is consistent with the Euler character of a cylider.

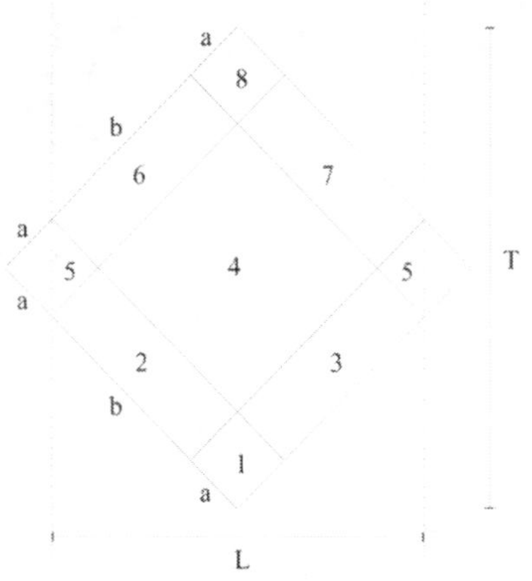

Figure 2: Tiling of the interval $\mathcal{I}_c$ with $L \leq T \leq 2L$. a and b are the u and v coordinate lengths of the sides of the subintervals shown.

5. The flat trousers

As our final example let us look at the action of a causally convex neighbourhood, $\mathcal{N}$, of the flat (1+1)D trousers. The trousers spacetime is a piece of $\mathbb{M}^2$ with cuts and identifications as shown in Fig. 3. Although the singularity, P, at which the topology changes is by some definitions not strictly in the spacetime since the metric degenerates there, nevertheless the causal order is well defined at the singularity: it is clear what the causal past and causal future of P are. Therefore we will consider P as a point of the manifold. Note that in any sprinkling into the trousers almost surely no element will be sprinkled at P. Using the same techniques used in the previous section we can calculate $\langle S \rangle$ for this neighbourhood and we find

$$\langle \mathbf{S}_{\mathcal{N}} \rangle = 4\ln(\rho a^2) + 4(\gamma - 1) + \mathrm{O}\left(\frac{1}{\rho a^2}\right) \tag{5.10}$$

where γ is Euler's constant. We see that the expected DA of the neighbourhood of the singularity does not tend to 1 or any constant but grows logarithmically with the density.

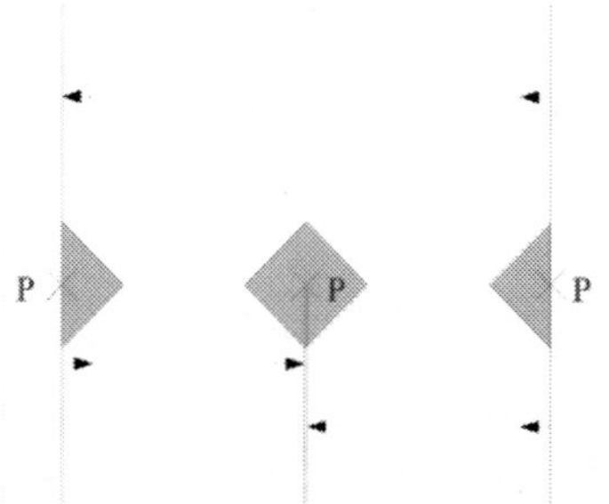

Figure 3: The trousers spacetime. P is the singularity – all three instances of P are identified – and the shaded region is a neighbourhood of P. There is a vertical cut down from the central copy of P with the two legs identified as shown.

6. Discussion

The result of the last section, together with the missing factor of 2π, and more crucially i, leads us to conclude that the answer to the title's question is no. But then a few natural questions arise: Why does $\mathcal{S}$ display topological properties for some regions of flat spacetimes and how are the boundaries contributing? How do these results carry over to curved spacetimes?

The first question we can answer as follows: Let D be some causally convex region of $\mathbb{M}^2$ whose boundary is composed of null segments. We know that in general "$\langle\mathcal{S}\rangle \approx$ bulk + boundary". So for D, $\langle\mathcal{S}\rangle \approx$ boundary. But since the geodesic curvature of the boundary is zero we do not expect these boundary contributions to come from the "edges". When the boundary contains spacelike segments we expect there to be boundary contributions as these segments can have nonzero geodesic curvature. However it is possible to argue that even for such boundaries there will be no contribution to $\langle\mathcal{S}\rangle$, see [9]. So the only points where some boundary contribution can come in, is from the points which are close to both past and future boundaries i.e. from the spacelike co-dimension 2 "corners" where the past and future boundaries intersect (There is no reason, from this argument, that timelike boundaries could not contribute and in fact there is evidence that they do, but not in a way which helps establish a connection with the LGBT [9]). Now the GB theorem can hold because as the geometry of the bulk is varied, the extrinsic curvature of the boundary changes in precisely the right way so that their sum remains constant. But if for the causet action the only boundary contribution comes from the S^0 corners , then we do not expect these to compensate for a changing bulk term and hence we wouldn't expect the LGBT to hold. This putative lack of boundary terms could explain why the expected causet action for any causally convex region of $\mathbb{M}^2$ is the same. The continuum bulk term is zero. If the mean of $\mathcal{S}$ is indeed close to the continuum bulk term plus only a contribution from the S^0 corners then that should be the same for all causally convex regions. Presumably, the difference for the neighbourhood of the singularity of the trousers comes from a boundary effect of the non-standard causal structure around the singularity which has a double lobed past and future. These issues all remain to be investigated.

The second question is currently being addressed. As things stand we can only express the hope that similar results will hold in curved 2D spacetimes. Here the bulk term would not be zero in general, so a result of the form "$\langle\mathcal{S}\rangle \approx$ constant" would suggest that the boundary term is getting contributions from the edges as well as the corners. If true, this would possibly revive hope in establishing a connection between the causet action and some modified form of the LGBT — which allows for null boundaries and is real. Of course one would also need the constant appearing on the right hand side to be the same as in flat space for topologically isomorphic causally convex regions, e.g. we would need $\langle\mathcal{S}\rangle \approx 1$ for all causally convex domains which are topologically disks. Further work is needed to illuminate these issues.

Acknowledgments

The author would like to thank Fay Dowker with whom much of this work was done, and Rafael Sorkin and Sumati Surya for useful and stimulating discussions. DMTB is supported by an EPSRC studentship.

References

[1] Avez A 1963 Formule de gauss-bonnet-chern en métrique de signature quelconque *Rev. Un. Mat. Argentina* **21** 191–197
[2] Chern S S 1963 Pseudo-riemannian geometry and gauss-bonnet formula *An. Acad. Brasil* **35** 17–26
[3] Law P R 1992 Neutral geometry and the Gauss-Bonnet theorem for two-dimensional pseudo-Riemannian manifolds *Rocky Mountain Journal of Mathematics* **22** 1365–1383
[4] Birman G S and Nomizu K 1992 The Gauss-Bonnet theorem for 2-dimensional spacetimes *Michigan Math. J.* **31** 77–81
[5] Jee D J 1984 Gauss-Bonnet formula for general lorentzian surfaces *Geometriae Dedicata* **15** 215–231

[6] Louko J and Sorkin R 1997 Complex actions in two-dimensional topology change *Class. Quant. Grav.* **14** 179–204

[7] Bombelli L, Lee J, Meyer D, and Sorkin R 1987 Space-time as a causal set *Phys. Rev. Lett* **59** 521

[8] Benincasa D M T and Dowker F 2010 The Scalar Curvature of a Causal Set *Phys. Rev. Lett.* **104** 181301

[9] Benincasa D M T and Dowker F 2010 The Random Discrete Action for 2-Dimensional Spacetime *Preprint* arXiv:1011.5191v2

[10] Schmitzer B 2010 Topology and curvature on causal sets MSc dissertation Imperial College London

Journal of Physics: Conference Series **306** (2011) 012041 doi:10.1088/1742-6596/306/1/012041

Causal set actions in various dimensions

Lisa Glaser
RWTH Aachen University, D-52056 Aachen, Germany
E-mail: `Lisa.Glaser@physik.rwth-aachen.de`

Abstract. Causal set theory is a discrete approach to quantum gravity. The idea is that space-time is not a smooth manifold but a discrete causal set. This means that differential structure can not be assumed but has to be replaced by discrete alternatives. In this paper a pattern for a discrete version of the scalar D'Alembertian operator in various dimensions is proposed and tested. Corrections to this operator in curved space-time are proportional to the curvature scalar, and the proportionality constant is calculated for various dimensions. These corrections can be used to define a Causal Set action which is then calculated for various dimensions.

1. Introduction

In Causal set theory the smooth space-time manifold of ordinary gravity is replaced by a fundamentally discrete description. The only information needed are the causality relation between the discrete points and a fundamental volume associated with each point. These information suffice to reconstruct the whole manifold. For an introduction to Causal sets the papers [1] and [2] can be recommended. If a fundamentally discrete description is to keep Lorentz symmetry we have to give up on locality, as Sorkin showed in his paper [3]. Although this seems a grave problem, there would be no contradiction between this nonlocality and the continuum physics that has been tested so far so long as the nonlocality can be confined to unobserved scales.

2. Discrete Operators

In his paper [3] Sorkin pointed out that in discrete space-times either locality or Lorentz invariance have to be sacrificed. The obvious discrete replacement for differential operators is some sum over nearest neighbours of every point. But in causal sets only a causal definition of nearest neighbour can make sense. So the nearest neighbours of a point are those which are linked to it[1]. In a causal set that arises from the Poisson process of sprinkling into a continuum space-time [3], most of the points linked and to the past to any given point lie very close to its past light cone and are in a sense quite far away from it. In a causal set sprinkled into infinite Minkowski space-time there is an infinite number of links for any point. For a causal set which is a sprinkling into Minkowski space-time of dimension d, and a real scalar field ϕ on it Sorkin proposed that the discrete operator to replace the d'Alembertian should have the form:

$$B_l\phi(x) = \frac{1}{l^2}\left(\alpha\phi(x) + \beta\sum_i^j C_i \sum_{y\in L_i}\phi(y)\right) .$$

[1] Linked meaning that one is in the past of the other and there exists no point causally between these two.

Journal of Physics: Conference Series **306** (2011) 012041 doi:10.1088/1742-6596/306/1/012041

where L_0 is the "zeroth layer," the set of elements linked to and in the past of x, L_1 is the set of past elements where one other point lies causally between them and x and so forth. C_i are some coefficients to be determined, C_0 is set to be 1 and j is some finite rather small number. This operator is nonlocal on the fundamental discreteness scale. As Sorkin elaborates in his paper [3] these sums actually fluctuate more strongly with growing sprinkling density $\rho = \frac{1}{l^d}$. Because infinite density is the continuum limit, this makes it necessary to introduce the nonlocality scale $k = \frac{1}{l_k}$ where $m_{EW} \ll k \ll \frac{1}{l}$ with m_{EW} being the scale at which current experiments probe space-time[2]. Introducing this intermediate scale k corresponds to smearing out the sum over layers of the causal set to a sum over a region around a layer, which suppresses the fluctuations. The nonlocal discrete operator in d dimensions is

$$B_k\phi(x) = \frac{\epsilon^2}{l^2}\left(\alpha\phi(x) + \beta\epsilon^d \sum_{y \prec x} f_d(n(x,y),\epsilon)\phi(y)\right),$$

where $n(x,y)$ is the number of elements of the causal set that lie between y and x. The parameter ϵ is a measure of the nonlocality and is given by $\epsilon = \frac{l}{l_k}$, and $f_d(n,\epsilon)$ is

$$f_d(n,\epsilon) = \sum_{i=0}^{j} C_i \left(\frac{\epsilon^d}{1-\epsilon^d}\right)^i \frac{n!}{(n-i)!i!}$$

with C_i dimension dependent constants and $C_0 = 1$.

Given a continuum space-time, a point x in it and a scalar field, ϕ on it of compact support, a continuum version of the nonlocal operator can be derived as the expected value, in the sprinkling process, of the discrete operator applied to the induced scalar field on the sprinkled causal set:

$$\Box_l\phi(x) := \frac{\alpha}{l^2}\phi(x) + \frac{\beta}{l^{d+2}} \int_{J^-(x)} \mathrm{d}^d y \sqrt{-g(y)}\phi(y) \sum_i^j C_i \left(\frac{V_{\diamond d}(x,y)}{l^d}\right)^i \frac{e^{-\frac{V_{\diamond d}(x,y)}{l^d}}}{i!} \tag{1}$$

where $V_{\diamond d}(x,y)$ is the space-time volume of the causal interval between x and y. We will first concentrate on the case where the space-time is Minkowski space-time in which case $\sqrt{-g} = 1$ and the volume of the causal interval depends only on $x - y$. We will choose coordinates with x at the origin.

Then, when $l_k \neq l$, the nonlocal analog to equation (1), *i.e.* the expected value of discrete operator (2), can be shown to be

$$\Box_k\phi(x) = \alpha k^2\phi(x) + \beta\, k^{d+2} \int_{J^-(x)} \mathrm{d}^d y \phi(x-y) \sum_i^j C_i \left(k^d V_{\diamond d}(x-y)\right)^i \frac{e^{-k^d V_{\diamond d}(x-y)}}{i!}$$

Looking at the sum under the integral we observe that

$$k^d V_{\diamond d}(x-y) e^{-k^d V_{\diamond d}} \propto k \frac{\partial}{\partial k} e^{-k^d V_{\diamond d}}$$

So we can use this derivative to generate the sum over layers. For that we simply introduce an Operator $\hat{\mathcal{O}}$ such that

$$\hat{\mathcal{O}} e^{-k^d V_{\diamond d}(x-y)} = \sum_i^j C_i \left(k^d V_{\diamond d}(x-y)\right)^i \frac{e^{-k^d V_{\diamond d}(x-y)}}{i!}$$

[2] The relation $m_{EW} \ll k$ is necessary for this theory not to contradict experimental data. But as $\frac{1}{l}$ is of the same order of magnitude as the Planck scale there is enough space between these constraints for k

The continuum nonlocal operator then can be written as

$$\Box_k\phi(0) = \alpha k^2\phi(0) + \beta k^{d+2}\hat{\mathcal{O}} \int\limits_{J^-(0)} \mathrm{d}^d x\ \phi(x)\ e^{-k^d V_{\diamond d}(x)}\ . \tag{2}$$

$\hat{\mathcal{O}}$ is a polynomial in $k\frac{\partial}{\partial k} = H$, the homogeneity operator, generating the layer structure. We now turn to finding the correct coefficients that will make $\Box_k\phi(0)$ approximately equal to the continuum D'Alembertian.

To actually perform the integrals we use light cone coordinate $v := \frac{1}{\sqrt{2}}(t+r)$ and $u := \frac{1}{\sqrt{2}}(t-r)$ and do a Taylor expansion of the scalar field to second order in x . This can be interpreted as a restriction on the test-functions to be slowly varying on the length-scales examined here.

2.1. 3d operator explicit treatment

To make the approach applied clearer we will show how this is done in 3 dimensions. The volume of a 3 dimensional Alexandrov neighbourhood is calculated in [4] as $V_{\diamond 3} := \frac{\pi}{12}\tau^3 = \frac{\pi}{\sqrt{2}\cdot 3}(u\cdot v)^{\frac{3}{2}}$. The operator in 3d is $\hat{\mathcal{O}} = \frac{1}{8}(H+2)(H+4)$ [5]. This gives a structure with three layers:

$$\Box_k\phi(0) :=\alpha k^2\phi(0) + \beta\, k^5 \int\limits_{-\infty}^{0} \mathrm{d}v \int\limits_{v}^{0} \mathrm{d}u \int\limits_{0}^{2\pi} \mathrm{d}\varphi \frac{(v-u)}{\sqrt{2}}\phi(x)$$
$$\times \left(1 - \frac{27}{8}k^3\frac{\pi}{\sqrt{2}\cdot 3}(u\cdot v)^{\frac{3}{2}} + \frac{9}{8}k^6\left(\frac{\pi}{\sqrt{2}\cdot 3}\right)^2(u\cdot v)^3\right) e^{-k^3\cdot\frac{\pi}{\sqrt{2}\cdot 3}(u\cdot v)^{\frac{3}{2}}} \tag{3}$$

After Taylor expanding, integrating over φ and plugging in u and v the field $\phi(x)$ is:

$$\phi(x) = 2\pi\phi + \sqrt{2}\pi(u+v)\partial_t\phi + \frac{\pi}{2}(u+v)^2\partial_t^2\phi + \frac{\pi}{4}(v-u)^2(\partial_x^2 + \partial_y^2)\phi$$

With this replacing $\phi(x)$ the other integrations in (3) can be solved. Leading to:

$$\Box_k\phi(0) = \alpha k^2\phi(0) + \beta\left(-k^2 - \left(\frac{3\sqrt{2}}{\pi}\right)^{\frac{2}{3}}\Gamma\left(\frac{5}{3}\right)(-\partial_t^2 + \partial_x^2 + \partial_y^2)\right)$$

There are no corrections to this if the integration is performed over the whole light cone. However for simulations on small finite causal sets we need to consider them(see 3.1 below for the corrections in the 3d case). Fixing $\beta = -\frac{\left(\frac{\pi}{3\sqrt{2}}\right)^{2/3}}{\Gamma\left(\frac{5}{3}\right)}$ and $\alpha = -\frac{\left(\frac{\pi}{3\sqrt{2}}\right)^{2/3}}{\Gamma\left(\frac{5}{3}\right)}$ the expression approximates the continuum d'Alembertian. The right operator is then

$$B_k\phi(0) = \frac{\epsilon^2}{l^2}\left(\alpha\phi(0) + \beta\epsilon^3\sum_{y\prec 0} f(n(0,y),\epsilon)\phi(y)\right)\ ,$$

where α and β are to be taken as above and

$$f(n,\epsilon) = (1-\epsilon^3)^n\left(1 - \frac{27\epsilon^3 n}{8(1-\epsilon^3)} + \frac{9\epsilon^6 n(n-1)}{8(1-\epsilon^3)^2}\right)$$

where $n(x,y)$ is the number of points lying causally between x and y.

2.2. Other dimensions

The principal course of action in other dimensions stays the same. First of all one needs to find the operator $\hat{\mathcal{O}}$. The form of the operator follows a simple pattern, being

$$\hat{\mathcal{O}} = \frac{(H+2)(H+4)\ldots(H+d+2))}{2^{\frac{d}{2}+1}(\frac{d}{2}+1)!} \tag{4}$$

for d even, and $\hat{\mathcal{O}}$ for $d = 2n+1$ is the same as for $d = 2n$. Up to the normalisation factor $\frac{1}{2^{\frac{d}{2}+1}(\frac{d}{2}+1)!}$ this is the so called Double Pochhammer triangle [6], which is well known in mathematics.

For the dimensions examined in this paper the resulting coefficients C_i are given in table (1). The case of one dimension -time only- does not follow the pattern and is just given for completeness, it is the usual discretization of the second derivative. The 2d case was examined in an earlier paper[7]. There is no simple closed form for the Coefficients as they are given as a

Table 1. Homogeneity operators and the coefficients arising from them.

	Operator $\hat{\mathcal{O}}$	Coefficients for discrete expression				
		C_1	C_2	C_3	C_4	C_5
1D		1	-2	1		
2D	$\frac{1}{8}(H+2)(H+4)$	1	-2	1		
3D	$\frac{1}{8}(H+2)(H+4)$	1	$-\frac{27}{8}$	$+\frac{9}{4}$		
4D	$\frac{1}{48}(H+2)(H+4)(H+6)$	1	-9	16	-8	
5D	$\frac{1}{48}(H+2)(H+4)(H+6)$	1	$-\frac{215}{16}$	$\frac{225}{8}$	$-\frac{125}{8}$	
6D	$\frac{1}{384}(H+2)(H+4)(H+6)(H+8)$	1	-34	141	-189	81
7D	$\frac{1}{384}(H+2)(H+4)(H+6)(H+8)$	1	$-\frac{6307}{128}$	$\frac{14749}{64}$	$-\frac{10633}{32}$	$\frac{2401}{16}$

super-position of the double Pochhammer triangle with numbers which arise from the derivatives action on the exponential, namely

$$H^n e^{-k^d V} = \left(k\frac{\partial}{\partial k}\right)^n e^{-k^d V} = d^n \sum_i^n (-1)^i i^{n-i} (k^d V)^i \tag{5}$$

The volume of the Alexandrov neighbourhood is dimension dependent, for those dimensions examined the volumes are given in table 2, taken from [4].

The pre-factors α and β are also given in table 2. They follow a clear pattern. For odd dimensions α is $-1/\Gamma\left(\frac{d+2}{d}\right) \cdot f_d^{2/d}$ where f_d is the factor from the Alexandrov volume in light cone coordinates. In even dimensions there is an additional factor of 2 so[3] $\alpha = -2/\Gamma\left(\frac{d+2}{d}\right) \cdot f_d^{2/d}$. Additionally β can be calculated from α as

$$\beta = -\frac{\alpha}{\hat{\mathcal{O}} \int \mathrm{d}V e^{-k^d V}} \,. \tag{6}$$

Here we will briefly list the complete expressions for the discrete operators in 4,5,6 and 7 dimensions.

[3] In 4-d by coincidence the factor comes out to be $\frac{-2}{\Gamma(\frac{6}{4})}\left(\frac{\pi}{6}\right)^{2/4} = -\frac{4}{\sqrt{6}}$

Table 2. Volumes of d- dimensional Alexandrov neighbourhoods and the prefactors alpha and beta

	2d	3d	4d
Volume	$u \cdot v$	$\frac{\pi}{3\sqrt{2}}(u \cdot v)^{3/2}$	$\frac{\pi}{6}(u \cdot v)^2$
α	$-2 = -\frac{2}{\Gamma(\frac{4}{2})}$	$-\frac{1}{\Gamma(\frac{5}{3})}\left(\frac{\pi}{3\sqrt{2}}\right)^{2/3}$	$-\frac{4}{\sqrt{6}} = -\frac{2}{\Gamma(\frac{6}{4})}\left(\frac{\pi}{6}\right)^{1/2}$
β	$4 = \frac{4}{\Gamma(\frac{4}{2})}$	$-\frac{1}{\Gamma(\frac{5}{3})}\left(\frac{\pi}{3\sqrt{2}}\right)^{2/3}$	$\frac{4}{\sqrt{6}} = \frac{2}{\Gamma(\frac{6}{4})}\left(\frac{\pi}{6}\right)^{1/2}$

	5d	6d	7d
Volume	$\frac{\pi^2}{20\sqrt{2}}(u \cdot v)^{5/2}$	$\frac{\pi^2}{45}(u \cdot v)^3$	$\frac{\pi^2}{168\sqrt{2}}(u \cdot v)^{7/2}$
α	$-\frac{1}{\Gamma(\frac{7}{5})}\left(\frac{\pi^2}{20\sqrt{2}}\right)^{2/5}$	$-\frac{2}{\Gamma(\frac{8}{6})}\left(\frac{\pi^2}{45}\right)^{2/6}$	$-\frac{1}{\Gamma(\frac{9}{7})}\left(\frac{\pi^2}{168\sqrt{2}}\right)^{2/7}$
β	$-\frac{3}{8\,\Gamma(\frac{7}{5})}\left(\frac{\pi^2}{20\sqrt{2}}\right)^{2/5}$	$\frac{4}{5\,\Gamma(\frac{8}{6})}\left(\frac{\pi^2}{45}\right)^{2/6}$	$-\frac{1}{8\,\Gamma(\frac{9}{7})}\left(\frac{\pi^2}{168\sqrt{2}}\right)^{2/7}$

2.2.1. 4 d operator

$$B_k\phi(0) = \frac{\epsilon^2}{l^2}\left(\alpha\phi(0) + \beta\epsilon^4 \sum_{y \prec x} f_4(n(x,y),\epsilon)\phi(y)\right),$$

where

$$\alpha = \frac{4}{\sqrt{6}} \quad \text{and} \quad \beta = \frac{4}{\sqrt{6}}$$

and

$$f_4(n,\epsilon) = (1-\epsilon^4)^n\left(1 - \frac{9\epsilon^4 n}{(1-\epsilon^4)} + \frac{8\epsilon^8 n(n-1)}{(1-\epsilon^4)^2} - \frac{4\epsilon^{12} n(n-1)(n-2)}{3(1-\epsilon^4)^3}\right)$$

2.2.2. 5 d operator

$$B_k\phi(0) = \frac{\epsilon^2}{l^2}\left(\alpha\phi(0) + \beta\epsilon^5 \sum_{y \prec x} f_5(n(x,y),\epsilon)\phi(y)\right),$$

where

$$\alpha = \frac{\pi^{4/5}}{2\,5^{2/5}\Gamma\left(\frac{7}{5}\right)} \quad \text{and} \quad \beta = -\frac{3\pi^{4/5}}{16\,5^{2/5}\Gamma\left(\frac{7}{5}\right)}$$

and

$$f_5(n,\epsilon) = (1-\epsilon^5)^n\left(1 - \frac{215\epsilon^5 n}{16(1-\epsilon^5)} + \frac{225\epsilon^{10} n(n-1)}{16(1-\epsilon^5)^2} - \frac{125\epsilon^{15} n(n-1)(n-2)}{24(1-\epsilon^5)^3}\right)$$

2.2.3. 6 d operator

$$B_k\phi(0) = \frac{\epsilon^2}{l^2}\left(\alpha\phi(0) + \beta\epsilon^6 \sum_{y \prec x} f_6(n(x,y),\epsilon)\phi(y)\right),$$

where

$$\alpha = \frac{2\left(\frac{\pi}{3}\right)^{2/3}}{5^{1/3}\Gamma\left(\frac{4}{3}\right)} \quad \text{and} \quad \beta = \frac{4\left(\frac{\pi}{3}\right)^{2/3}}{5\, 5^{1/3}\Gamma\left(\frac{4}{3}\right)}$$

and

$$f_6(n,\epsilon) = (1-\epsilon^6)^n\left(1 - \frac{34\epsilon^6 n}{(1-\epsilon^6)} + \frac{141\epsilon^{12}n(n-1)}{2(1-\epsilon^6)^2} - \frac{63\epsilon^{18}n(n-1)(n-2)}{2(1-\epsilon^6)^3} + \frac{27\epsilon^{24}n(n-1)(n-2)(n-3)}{8(1-\epsilon^6)^4}\right)$$

2.2.4. 7 d operator

$$B_k\phi(0) = \frac{\epsilon^2}{l^2}\left(\alpha\phi(0) + \beta\epsilon^7 \sum_{y \prec x} f_7(n(x,y),\epsilon)\phi(y)\right),$$

where

$$\alpha = \frac{\pi^{6/7}}{2\,(21)^{2/7}\Gamma\left(\frac{9}{7}\right)} \quad \text{and} \quad \beta = -\frac{\pi^{6/7}}{16\,(21)^{2/7}\Gamma\left(\frac{9}{7}\right)}$$

and

$$f_7(n,\epsilon) = (1-\epsilon^7)^n\left(1 - \frac{6307\epsilon^7 n}{128(1-\epsilon^7)} + \frac{14749\epsilon^{14}n(n-1)}{256(1-\epsilon^7)^2} - \frac{10633\epsilon^{21}n(n-1)(n-2)}{192(1-\epsilon^7)^3} + \frac{2401\epsilon^{28}n(n-1)(n-2)(n-3)}{384(1-\epsilon^7)^4}\right)$$

3. Simulations

To test if these discrete operators give the correct results they have to be applied in simulated causal sets. The best way to simulate a causal set which is in some sense manifold like is to sprinkle the Causal set into a preexisting manifold. For more on this see [8]. As one can only simulate finite Causal Sets the corrections resulting from the finite size L will have significant effects.

3.1. Corrections

The size of the Causal Set grows with L^d, as does the number of sprinkled points necessary to get sensible results. So for this paper only simulations in 3 d were done. Corrections appear for a finite cut-off L in the integral (3).

$$I_{gen} = \int_{-L}^{0} \mathrm{d}v \int_{v}^{0} \mathrm{d}u \int_{0}^{2\pi} \mathrm{d}\varphi\, \frac{1}{\sqrt{2}}(v-u)\phi(x)e^{-k^3 V_{\diamond 3}(x)}.$$

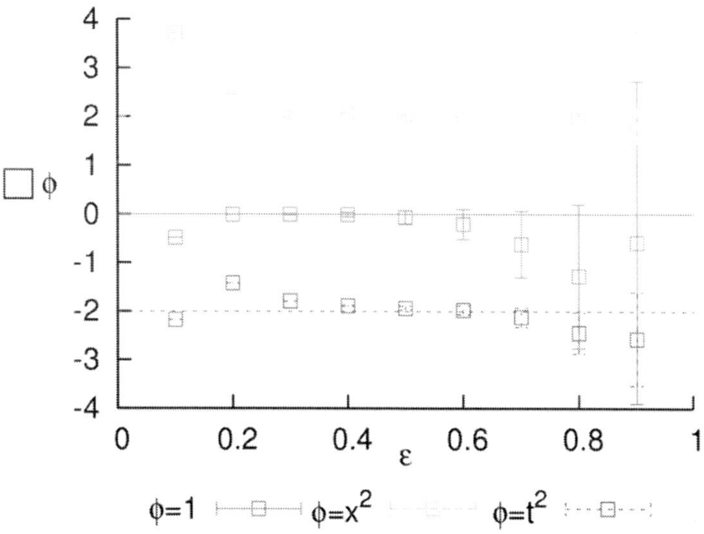

Figure 1. Simulation results

Solving this and using the values for α and β from 2 leads to

$$\Box_{k_3 d}\phi(0) = \Box\phi(0) - \left(\frac{3\sqrt{2}}{\pi}\right)^{\frac{1}{3}} \frac{\sqrt{2}\,\partial_t\phi(0)}{kL^2\,\Gamma(\frac{5}{3})} + \left(\frac{3\sqrt{2}}{\pi}\right)^{\frac{1}{3}} \frac{\partial_t^2\phi(0)}{kL\,\Gamma(\frac{5}{3})}$$
$$- \left(\frac{3\sqrt{2}}{\pi}\right)^{\frac{1}{3}} \frac{3\,(\partial_x^2 + \partial_y^2)}{2\,kL\,\Gamma(\frac{5}{3})}\phi(0) + \mathcal{O}\left(\frac{1}{k^2L^3}\right)$$

neglecting terms in $\Gamma\left(\frac{2}{3}, \frac{k^3L^3\pi}{3\sqrt{2}}\right)$ and $e^{-\frac{k^3L^3\pi}{3\sqrt{2}}}$ as these fall off very fast with L.

3.2. Simulation results

To test if the operator and the corrections give the right result simulations were done. The functions which where tested were $\phi = 1$ to test that a constant gives the right result, $\phi = x^2$ to test that functions of space work correctly and $\phi = t^2$ to test functions of time. The functions were chosen because the results if the d'Alembertian is applied to them are constants, so that the value does not depend on the causal set. Every point in the plot 1 corresponds to 100 000 causal sets with 1000 points. For these sprinkled sets L grows like $l \cdot N^{1/3}$. The corrections due to the finite size of the causal set are of order one. In the plot of the simulated data the variable used is $\epsilon = k \cdot l$. It is obvious that the best results are achieved with $\epsilon \approx 0.5$. This is to be expected, as for epsilon close to one the nonlocality is small so the fluctuations in the results are still large. For very small ϵ the nonlocality becomes very large. The errors here are due to the relatively small size of the sprinkled sets as the layers now are thick and would reach points that are not in the small sprinkled set. These simulations show that the discrete operator in three dimensions works.

4. Curvature Scalar R

As Dionigi Benincasa shows in his paper [9] in curved space-time there is a connection between these discrete operators and the curvature scalar. The relevant relation is:

$$\Box_k\phi(0) = \Box\phi(0) + aR\phi(0) + b\frac{1}{k}R^2\phi(0) + \dots \tag{7}$$

In his paper he finds that the constant a is $-\frac{1}{2}$ in 2 and 4 dimensions. The next step is then to calculate these coefficients for 3,5,6 and 7 dimensions. They are found to be $\frac{1}{2}$, $\frac{1}{2}$, $-\frac{1}{2}$ and $\frac{1}{2}$ so odd and even dimensions only differ by a sign.

4.1. Calculations

The continuum version of the discrete operator is defined in equation (2) to be

$$\Box_k\phi(0) = \alpha k^2\phi(0) + \beta k^{d+2}\int\limits_{J^-(0)} \mathrm{d}^d x\sqrt{-g}\phi(x)\, f(V)e^{-k^d V_{\diamond d}(x)} \,.$$

This expression is valid in flat space, so $\sqrt{-g} = 1$. To simplify the calculations we now take the frame to be Riemannian normal coordinates centred at the origin, as in [10]. The function $f(V)$ is given by $\hat{\mathcal{O}}e^{-k^d V_{\diamond d}(x)} = f(V)e^{-k^d V_{\diamond d}(x)}$. So it could be rewritten as function $f(k^d V)$ which will be used later. If space is curved there are corrections to $\sqrt{-g}$ and to V, to lowest order in the curvature these expressions are:

$$\sqrt{-g} = 1 - \frac{1}{6}R_{\mu\nu}x^\mu x^\nu \tag{8}$$

and

$$V_{\diamond d} = V_{\diamond 0 d}\left(1 - \frac{d}{24(d+1)(d+2)}R\tau^2 + \frac{d}{24(d+1)}R_{\mu\nu}T^\mu T^\nu + \dots\right) . \tag{9}$$

The last expression is taken from Solodukhin and Gibbons [4]. T^μ is a time-like vector, pointing along the unique geodesic that connects the origin and x. Plugging this into equation (7) and doing a little algebra we can see that actually the corrections are just proportional to the variation of $\Box_k$ and with an Operator approach similar to the one used before we can get that in d dimensions

$$\delta\Box_k\phi(0) = \frac{(-1)^{d+1}}{2}R \,. \tag{10}$$

For more details on this calculations please see [5], [10].

4.2. Causal Set Actions

Now using (10) we can generate the causal set actions by just applying the operator $\Box_k$ on constant fields ϕ with the value $(-1)^{d+1}2$. (For more detail compare to [9] This leads to the actions:

$$\frac{1}{\hbar\alpha}S^{(d)}[\mathcal{C}] = N + \frac{\beta}{\alpha}\sum_i^j C_i N_i \qquad \text{in d dimension}$$

leading to:

$$\frac{1}{\hbar}S^{(2)}[\mathcal{C}] = N - 2N_1 + 4N_2 - 2N_3 \qquad \text{in 2-d}$$
$$\frac{1}{\hbar}S^{(3)}[\mathcal{C}] = N + N_1 - \frac{27}{8}N_2 + \frac{9}{4}N_3 \qquad \text{in 3-d}$$
$$\frac{1}{\hbar}S^{(4)}[\mathcal{C}] = N - N_1 + 9N_2 - 16N_3 + 8N_4 \qquad \text{in 4-d}$$
$$\frac{1}{\hbar}S^{(5)}[\mathcal{C}] = N + \frac{3}{8}N_1 - \frac{645}{128}N_2 + \frac{675}{64}N_3 - \frac{375}{64}N_4 \qquad \text{in 5-d}$$
$$\frac{1}{\hbar}S^{(6)}[\mathcal{C}] = N - \frac{3}{5}N_1 + \frac{68}{5}N_2 - \frac{282}{5}N_3 + \frac{378}{5}N_4 - \frac{243}{5}N_5 \qquad \text{in 6-d}$$
$$\frac{1}{\hbar}S^{(7)}[\mathcal{C}] = N + \frac{1}{8}N_1 - \frac{6307}{1024}N_2 + \frac{14749}{512}N_3 - \frac{10633}{256}N_4 + \frac{2401}{128} \qquad \text{in 7-d}$$

with $N_i = \sum_{L_i}$ the number of causal set points in the i-th layer.

5. Conclusion

This shows that the method of finding a discrete operator devised in [7] works for dimensions higher and lower than 4. Thus it is necessary to be extremely careful not to break Lorentz invariance. This means we have to apply the generated operator $\hat{O}$ before solving the integral. Ordering things in this way makes the introduction of a cut off unnecessary, and therefore keeps Lorentz invariance.

The rule to find the operator $\hat{\mathcal{O}}$ is

$$\hat{\mathcal{O}} = \frac{(H+2)(H+4)\ldots(H+d+2))}{2^{\frac{d}{2}+1}(\frac{d}{2}+1)!} \tag{11}$$

for d even, and $\hat{\mathcal{O}}$ for $d = 2n+1$ is the same as for $d = 2n$.

It was possible to evaluate the proportionality constant for the Ricci Scalar in all dimensions evaluated here and it seems fair to guess that the so found pattern continues in all other dimensions. The coefficient to the Ricci Scalar is $-\frac{1}{2}$ in even and $\frac{1}{2}$ in odd dimensions. This gives a possibility to define gravity actions on the causal set.

Another interesting question about this operator pattern is how it is connected to the Greens functions on 2 and 4 dimensional Causal sets found by Steven Johnston [11], and what kind of Greens functions in other dimensions.

Acknowledgments

Acknowledgments I thank Fay Dowker for telling me about Causal Sets and taking me to the conference and Bernhard Schmitzer, Dionigi Benincasa and Steven Johnston for helpful discussions. Special thanks go to Noel Hustler who did the simulations.

References

[1] Dowker F, Henson J and Sorkin R D 2004 *Mod. Phys. Lett.* **A19** 1829–1840 (*Preprint* `gr-qc/0311055`)
[2] Sorkin R D 1990 *Relativity and Gravity, Classical and Quantum* (World Scientific, Singapore)
[3] Sorkin R D 2007 *Towards Quantum Gravity* ed Oriti D (Cambridge University Press)
[4] Gibbons G W and Solodukhin S N 2007 *Phys. Lett.* **B649** 317–324 (*Preprint* `hep-th/0703098`)
[5] Dowker F and Glaser L 2011 *publication in preparations*
[6] The on-line encyclopedia of integer sequences (*Preprint* `http://oeis.org/A039683`)
[7] Benincasa D M T 2009 *A Lorentz Invariant Discrete d'Alembertian* Master's thesis Imperial College London
[8] Reid D D 2003 *Phys. Rev. D* **67** 024034

[9] Benincasa D M T and Dowker F 2010 *Phys. Rev. Lett.* **104** 181301 (*Preprint* `gr-qc/1001.2725v3`)
[10] Benincasa D M T 2011 *publication in preparation*
[11] Johnston S 2008 *Class.Quant.Grav.25:202001* (*Preprint* `hep-th/0806.3083`)

Journal of Physics: Conference Series **306** (2011) 012042 doi:10.1088/1742-6596/306/1/012042

Algorithmic causets

Tommaso Bolognesi
CNR-ISTI, Via Moruzzi 1, 56124, Pisa, Italy
E-mail: t.bolognesi@isti.cnr.it

Abstract. In the context of quantum gravity theories, several researchers have proposed causal sets as appropriate discrete models of spacetime. We investigate families of causal sets obtained from two simple models of computation – 2D Turing machines and network mobile automata – that operate on 'high-dimensional' supports, namely 2D arrays of cells and planar graphs, respectively. We study a number of quantitative and qualitative emergent properties of these causal sets, including dimension, curvature and localized structures, or 'particles'. We show how the possibility to detect and separate particles from background space depends on the choice between a global or local view at the causal set. Finally, we spot very rare cases of pseudo-randomness, or deterministic chaos; these exhibit a spontaneous phenomenon of 'causal compartmentation' that appears as a prerequisite for the occurrence of *anything* of physical interest in the evolution of spacetime.

1. Introduction

The purpose of this paper is to progress investigation in an area of fundamental physics that we locate at the intersection between the *Causal Set Programme* and the *Computational Universe Conjecture*. Causality among events is widely recognized as a most fundamental aspect of spacetime, and the concept of *causal set* has been formulated in the homonymous research programme [1, 2, 3] in "an attempt to combine the twin ideas of discreteness and order to produce a structure on which a theory of quantum gravity can be based" [4]. The fundamental idea at the root of the Causal Set Programme is summarized by the conceptual equation:

"spacetime geometry = order + number".

A *causal set* (or '*causet*') is a finitary partially ordered set, that is, a set of events provided with a binary relation '$\prec$' which is reflexive, antisymmetric and transitive, and such that the number of elements between any two events is finite. A causet is meant to represent an instance of discrete spacetime in which causality is reflected by the '$\prec$' relation ('order'), and the volume of a given region is measured by the count of nodes in it ('number'). Causets are important because the order and number information that they encode is sufficient for completely determining the metric tensors of General Relativity (see e.g. [2]).

The idea of a discrete universe is also adopted in the Computational Universe Conjecture, which, in its most extreme form, suggests that all the complexity we observe in the physical universe should emerge from the iteration of a few simple transition rules, and could be fully reproduced by running a short computer program. In this case, the conceptual equation is:

"complexity in physics = emergence in computation".

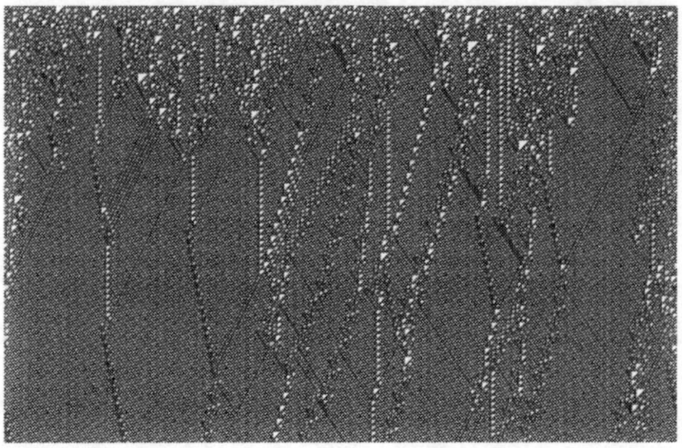

Figure 1. 500-step computation of Wolfram's Elementary Cellular Automaton 110. Interacting 'particles' always emerge, from any random initial configuration.

The fact that complex patterns resembling those found in nature can originate from simple computations is widely recognized today, and has been investigated and popularized in particular by S. Wolfram, with his extensive analysis of the 'spontaneous' behavior of cellular automata and other simple models [5]. An example of a structure emerging from the computation of an elementary cellular automaton is illustrated in Figure 1. In the depicted bit array, the first row is random and each bit in the subsequent rows is computed by a boolean function of three variables:

$$b_{i+1,j} = (\neg b_{i,j-1} \wedge b_{i,j} \wedge b_{i,j+1}) \oplus b_{i,j} \oplus b_{i,j+1},$$

where $\oplus$ is exclusive OR and column indices rotate, the last on the right being followed by the first on the left. Cellular automata diagrams are indeed commonly interpreted as space-time structures, with space expanding horizontally and time flowing downward. However, unlike in causets, neither events nor causal links are explicitly represented in this type of diagram.

We believe that causets represent an excellent opportunity for progressing research on the computational universe conjecture, due to the fact that they come with a clear, explicit physical interpretation, and due to their relevance in current quantum gravity research. Thus, our main objectives in this paper are: (i) to explore, by computer simulations, the causets that can be obtained from the 'spontaneous' computations of some simple *deterministic* algorithms, (ii) to analyze some of their quantitative and qualitative properties, with the idea to compare them with those considered in the Causal Set Programme. Our hope is to promote some cross-fertilization between the two approaches in terms of techniques and specific research items.

2. Causets from simple models of computation

The idea of describing computations as networks of causally related events has been first introduced, to our knowledge, by Levin: "A causal net can be interpreted as the time-space history of all elementary operations accomplished in the computing process, with their mutual dependencies indicated" [6]. The purpose of these nets, however, was to characterize computable functions; it is only by the more recent work of Wolfram that these directed graphs are viewed as possible instances of physical spacetime.

In [7] (see also [8]) we have described a general criterion for deriving causets from virtually any model of computation. In summary, the idea is as follows. A computation is defined as a sequence of composite states, and a computation step as a pair of adjacent states. Each computation step becomes an event in the causet. Then, a causal link is created from event e_i to event e_j, with $i < j$, whenever e_j is the first event occurring after e_i to read a global state component that was affected by e_i: that state component has acted as a *causality mediator* between the two events. [1] Finally, letting R denote the relation defined by the above links, the full causality relation is obtained by taking the *transitive closure* of R. Note that the above relation is antisymmetric by construction. It may be sometimes useful to also consider the *transitive reduction* of R, which is the minimal relation whose transitive closure is the same as the transitive closure of R. When R is antisymmetric and finite, the transitive reduction is unique.

In [7] we have explored causet families associated with a number of models, including various classes of one-dimensional Turing machines, mobile automata on tape, string rewrite systems and tag systems. All those models operate on a *one-dimensional* support, such as a tape or a string of symbols. In this paper we extend our exploration by considering two models of computation that operate on *higher-dimensional* supports, namely 2-D arrays of cells and planar trivalent graphs. As with our previous investigation, our attitude is one of avoiding predetermined expectations, while being open to anything interesting that might *emerge*. Two obvious items of interest, however, are the emergent dimensionality (or 'dimension') of the causet, and its curvature. Thus, before illustrating our algorithmic causet families, we briefly discuss how we have addressed these quantitative features.

2.1. Causet dimension and curvature by shortest-path node-shell growth

A simple way to assign a dimension to an *undirected* graph consists in considering the size $S(d)$ of the node-shells at progressive distance d from a generic node, where distance is measured by the edge count of the *shortest-path* between nodes. If $S(d) = O(d^n)$, we assign dimension $n + 1$ to the graph. In the case of undirected graphs this concept is defined, for example, in [9].

We apply the same idea to causets, that are *directed*, acyclic graphs. One can define a quantity:

$$D_x(d) := log(S_x(d))/log(d) + 1$$

relative to some node x of the causet, where $S_x(d)$ is the size of the node-shell at distance d from x. Then the causet has *node-shell dimension* $D_x := lim_{d->\infty} D_x(d)$, *starting from node* x, when such limit exists, and it has *node-shell dimension* D whenever $D_x = D$ for all x.

What if node-shell growth is not polynomial? In our experience, if we exclude cases in which large fluctuations prevent the identification of a definite growth rate, the only alternative is exponential growth. We interpret it as a sign of *negative curvature*, while suspending judgement about dimension. This choice is justified by an analogy with the continuous case: a circumference of radius r in a hyperbolic surface with negative Gaussian curvature K has length $c(r) = 2\pi r sinh(r/R)$, where $R = 1/\sqrt{-K}$, and in the expansion of the hyperbolic sine function $sinh(x) = 1/2 * (e^x - e^{-x})$ the predominant term is indeed a positive exponential.

2.2. Causets from two-dimensional Turing machines

A *two-dimensional Turing machine* is a Turing machine in which the control head operates on a 2-D array of cells, containing symbols from a given alphabet. For each value p of the current control head state, and each symbol a read in the current cell c, the state transition table describing the control head behavior provides a triple (p', a', d), indicating the new control head state p', the new symbol a' to be written in c, and the move of the control head, expressed as a vertical or horizontal, unit-length displacement d from c. By convention, we start computations with the array filled uniformly, say with 0's. No termination conditions are imposed.

[1] The total order of computation steps does not represent physical time; the latter, as well as space, can only emerge (if ever) from the growing structure of the causet, which does not keep explicit track (e.g. by node labels) of that total order.

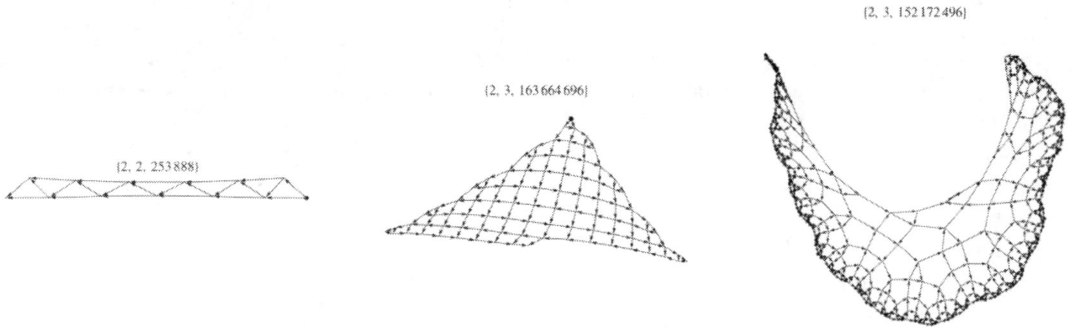

Figure 2. Three causets from simple 2-D Turing machines in class $(s = 2, k = 2)$ and class $(s = 2, k = 3)$. 1-D (left), 2-D flat (center), negatively curved (right).

The instantiation of the general causet construction method for this model is simple. Causal links are mediated by two global state components: tape cells and control state. Thus:

- Every step of the machine corresponds to an event of the causet.
- A directed edge connects event e_i to event e_j iff e_j reads a cell that was written by e_i, and by no other event in between.
- A directed edge connects each event e_i to its successor e_{i+1}, since the latter reads the control head state that is set by the former.

A consequence of the last bullet is that these causets will always be *totally ordered.* [2]

Using shortest-path node-shell growth analysis, we have examined families of causets from classes of 2-D Turing machines characterized by various settings of parameters s (number of control states) and k (tape alphabet size). The number of different 2-D Turing machines for given s and k is $(4sk)^{sk}$. The simplest class we consider is $(s = 2, k = 2)$, consisting of 65,536 elements. We find that all of the causets in this class are 1-D; an example is shown in Figure 2 (left). By considering class $(s = 2, k = 3)$, we still find that the vast majority of causets are 1-D, with exceptionally rare cases of two different types: regular, flat 2-D grids and negatively curved causets (see Figure 2). [3]

In search for more interesting behaviors, we have considered *turmites* [10]. They are a subclass of 2-D Turing machines for which the motion of the control head is described in terms of left and right turns relative to the current head orientation. The most famous turmite is Langton's ant [11]. In [10], 44 turmites with particularly interesting behavior have been identified; we have derived causets for all of them, thus discovering interesting quantitative and qualitative features. Two examples are shown in Figure 3.

The highest causet dimensions we have found in this family fall around 3.5, which is higher than any value found in [7]. For establishing these values we have computed the sizes of the node-shells at progressive distance d from the root, and then found a best fit for a prefix of this numeric sequence, for excluding boundary effects. The fitting is based on the non-integer monomial function $a * d^b$, where parameters a and b are positive real numbers: the estimated dimension is then $b + 1$. We also detect a case of exponential node-shell growth; this is not shown, for space reasons.

[2] We discuss this potential limitation in the next subsection, in connection with particles.

[3] These results closely resemble those found in [7] for 1-D Turing machines.

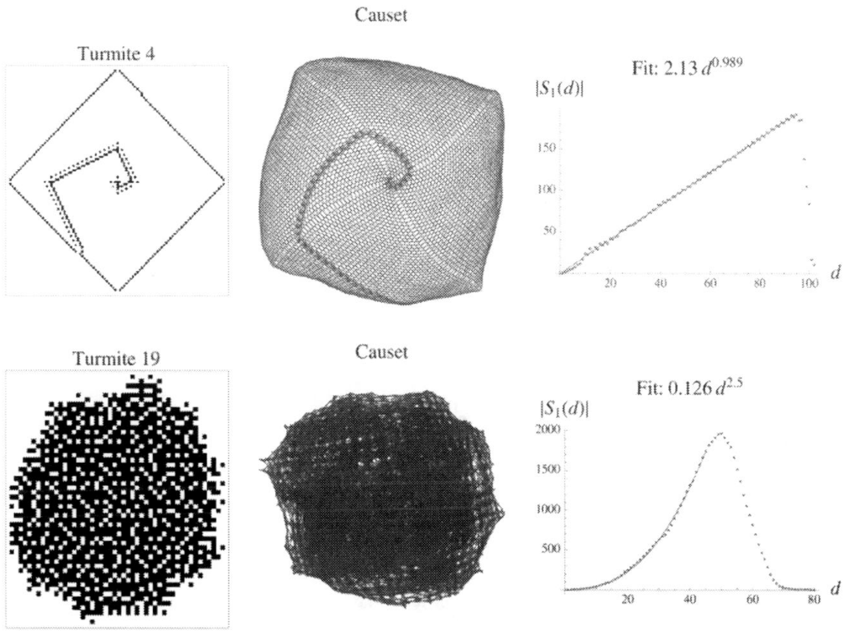

Figure 3. Two turmites: final configuration (left), causet (center), and best fit approximation of node-shell sizes at progressive distance d from the causet root, using non-integer monomial $a * d^b$. Turmite code numbers as in [10]. Computation lengths: 10000 and 50000 steps.

2.3. Particle detection: global vs. local view

Let us turn to qualitative aspects of turmite causets. One the most attractive features of cellular automata is the emergence of localized structures, or 'particles' (see Figure 1). In our previous experiments [7] our search for particles was successful only in part: causets exhibited at most one particle each, in the form of a spacetime trajectory at the border between two periodic 2-D regions. We have now the more elaborate case of turmite 4 in Figure 3. The causet for this machine consists of four triangular sectors formed by regular hexagonal grids, separated by four 'radii' formed by octagons. On top of this structure, a localized perturbation creates a spiral. By looking at both the spiral and the four radii as particles, the causet appears as a sort particle collision diagram. Note that the graph is not planar; the non-planarity is introduced precisely by the spiraling particle. We can however determine the *combinatorial curvature* for each node in the planar region. Recall that, for a planar graph, the combinatorial curvature of node x is defined as

$$cc(x) := 1 - degree(x)/2 + \sum_{f \sim x} (1/size(f))$$

where summation is over all faces f incident with x. In our graph, the four grid sectors are flat, their nodes having null combinatorial curvature, but the nodes of the octagons have three different non-null curvature values, namely -1/24, -1/12, +1/4, suggesting a possible relation between the motion of the spiraling particle and the curvature of its 'background'.

So far we have used the term 'particle' in a rather abstract way, for referring to emergent localized structures that we can visually spot in causet diagrams. In doing so we have sloppily ignored edge orientation, and, more importantly, we have considered causets exactly as obtained by the described construction criterion, while ignoring the fact that the transitive reduction of these graphs collapses into a trivial chain of spacetime events. We are not addressing here the problem of retrieving a notion of (dynamic) space from a generic causet; but it is clear that a necessary condition for appearance of space is the presence of pairs of events in a *spacelike* relation, or anti-chains, and these cannot be found in a totally ordered set.

Should we conclude that Turing machine computations are unable to produce causets of any interest for physics? Let c be the causet directly obtained from a Turing machine computation, and let $TC(c)$ and $TR(c)$ denote, respectively, the transitive closure and transitive reduction of c; clearly $TR(c) \subseteq c \subseteq TC(c)$. Then, the question is whether the 'redundant' edges in $c - TR(c)$ can be assigned any physical role. We believe they can: their meaning becomes apparent when switching from a global to a local view at the causet structure. When we compute the transitive reduction of the a totally ordered causet with redundant edges, these are washed away. But when the reduction is applied locally, to small regions of the causet, then (some of) these edges might survive, and reveal some spatial structure.

In the subsequent figures we illustrate this idea by focusing on small circular regions of causets from turmite computations. A disk with center node c and radius r shall include all nodes found at a distance at most r from c, regardless of edge orientation. When these disks are centered at the causet root, the total order of the included spacetime events is (almost) completely represented inside the disk, and the reduction process basically destroys all structure. But when the disk is moved off the root, information on causal relations is only partial; this affects the reduction process and may expose local space and/or particles.

In Figure 4 we show the causet for Langton's ant. The ant (the TM control head) settles to

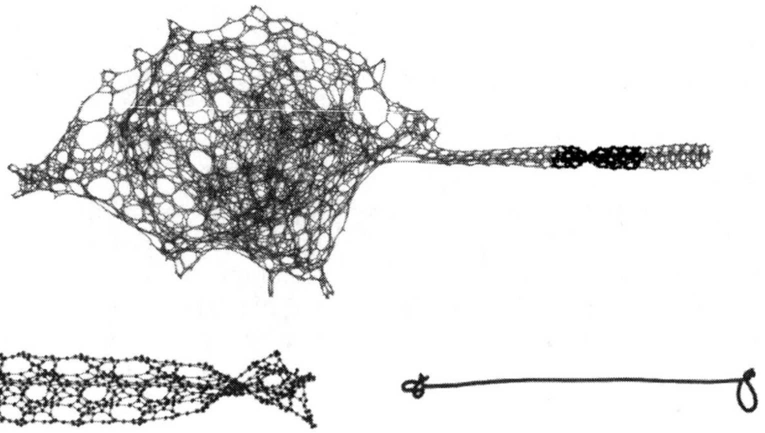

Figure 4. Causet from 12000-step computation of turmite 1 - 'Langton's ant' (upper), disk from the causet arm (left), and transitive reduction of the disk (right).

a periodic motion – a 'highway' – after about 11000 steps. This is reflected in the arm growing from the main body of the causet. The picture shows the transitive reduction of a disk centered at some event in the arm: this is essentially a plain chain. We conclude that, at the considered scale, the causet arm does not extend spatially.

Figure 5 provides an example in which space and particle are preserved by the local transitive reduction, unless the disk is centered to the causet root. Note that the particle pattern is simplified by the reduction.

Figure 6 provides a surprising example. The grid-like spatial structure is preserved by the transitive reduction, unless a particle is present: in this case the grid basically disappears,

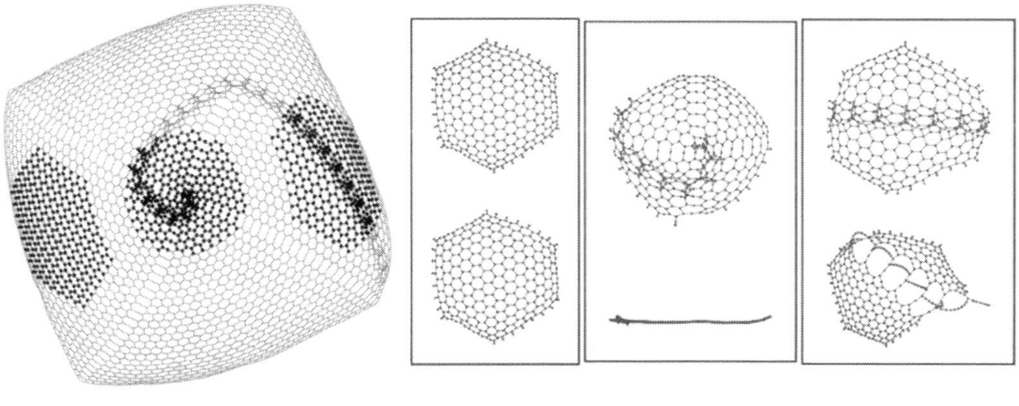

Figure 5. Causet from 4000-step computation of turmite 4, with three highlighted disks (left); the same disks and their transitive reductions (right).

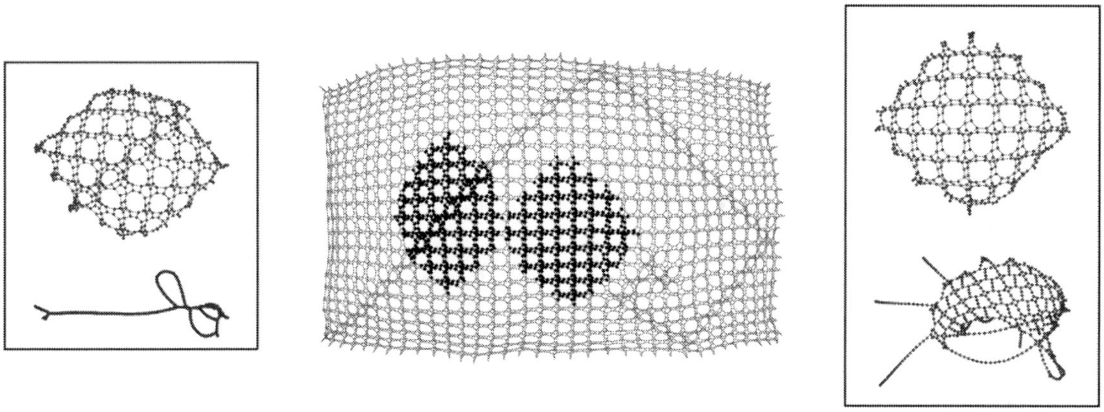

Figure 6. Causet from 10000-step computation of turmite 18, with two highlighted disks (center); the same disks and their transitive reductions (sides).

except for a portion corresponding to the 'shadow' of the original particle trajectory (this is barely visible in the lower-left diagram).

Figure 7 is the only case in which we initialize the cell array with a non-uniform 5×5 region. This computation, when observed on the cell array, exhibits the emergence of a 'two-way' highway (lower-left diagram) that is more interesting than the 'one-way' highway of Langton's ant: the ant now keeps oscillating between the two opposite endpoints of the linear structure, while progressively extending it in both directions, thus implementing a sort of wave behavior.

When taking local views at this causet, we find that basically all the structure is washed away by the transitive reduction, as with Langton's ant; the only space-like cells left (see lower-right diagram) are organized in two sequences that correspond to the ant entering and leaving the central, random-like region of the cell array.

We believe that the detection of particles in causets from computations is an interesting, original result of this paper; localized structures may appear also in turmite diagrams, but the two phenomena are not always coupled. The causet, not the 2D-grid of a Turing machine, or any other computational support, is the right place where to look for particles, or their worldlines.

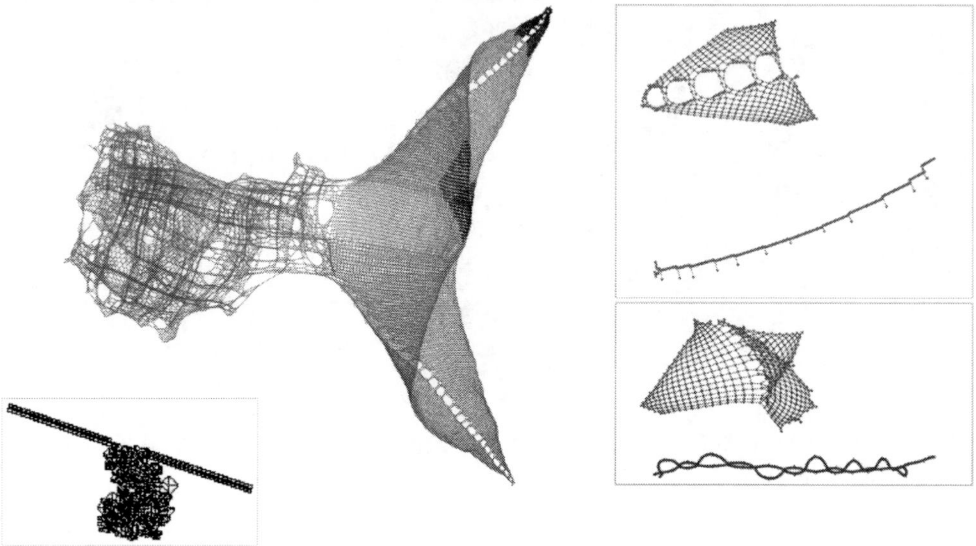

Figure 7. Causet from 26000-step computation of turmite 21, with two highlighted disks (center), and the final configuration of the cell array (lower left); the two disks and their transitive reductions (right).

2.4. Causets from network mobile automata on trivalent planar graphs

Network mobile automata have been introduced and shortly discussed in [5], and further studied in [12, 13]. A *trinet* is a planar, undirected, trivalent graph. A *trinet mobile automaton* is similar to a 2-D Turing machine, except that the stateless control head moves on a trinet rather than on an array of cells. At each computation step the control head performs a local graph rewriting and moves to some nearby location. Both the rewrite rule and the next location are chosen by some deterministic criterion. In our investigations we have always used 2-D Pachner moves, sometimes called *Expand*, *Contract* and *Exchange*, that find application also in Loop Quantum Gravity (see, e.g., [14]). Each of these rules affects three or four faces of the planar graph.

The instantiation of the general causet construction method is simple also in this case.

- Every step of the machine corresponds to an event of the causet.
- A directed edge connects event e_i to event e_j iff the rewriting performed by e_j affects a planar face of the trinet that has been previously affected by e_i (without further modifications in between).

In [12, 13] we have explored two variants of the model, in both of which rule *Contract* is not used. A first variant, qualified as *three-connectivity preserving*, is defined as follows:

(i) Start with a trinet consisting of two nodes connected by three parallel arcs. (This is the smallest possible three-connected graph. A connected graph is n-connected when n is the smallest number of edges one has to remove in order to disconnect it.)

(ii) Choose rule *Exchange* whenever it does not violate three-connectivity, otherwise choose *Expand*.

(iii) Move the control head to a new nearby location which depends on the applied rule and on a fixed parameter (see [13] for details).

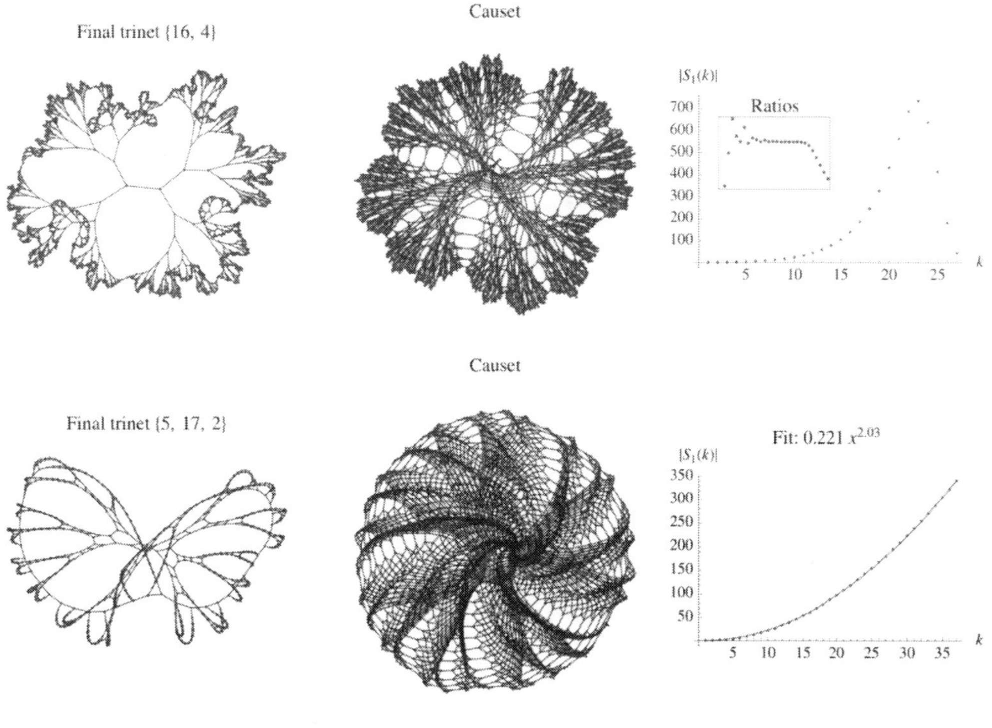

Figure 8. Two network mobile automata: final configuration (left), causet (center), and analysis of node-shell sizes at progressive distance d from the causet root. Numeric codes as in [12, 13].

In the second variant, qualified as *threshold-based*, point (ii) is modified as follows: choose *Exchange* whenever it does not create trinet faces with less than k sides, otherwise choose *Expand*. The two-parameter policy for control head moves is the same in the two variants, and always involves *short* steps, in analogy with Turing machines. As a consequence, the faces affected by a step always partially overlap with those affected by the next step, so that events e_i and e_{i+1} are always causally related, and the causet is, again, totally ordered. However, due to the construction rule, each node has 3 or 4 incoming, and 3 or 4 outgoing arcs, except for the first node and those at the growth boundary, so that causet structures can still be far from trivial, and they would remain so even after *local* transitive reduction.

Even with computations from this model, the large majority of causets are 1-D graphs of no interest. However, several more interesting cases are found; two of them are illustrated in Figure 8. In the first case the causet is negatively curved: the exponential node-shell growth is revealed by the flat region of the small inset diagram, showing the ratios of consecutive shell size values. In the second case the causet exhibits an approximate node-shell dimension 3 from the root, as revealed by the exponent of the non-integer monomial fit.

The causets considered above manifest a considerable degree of regularity, or predictability; this becomes even more apparent when tracing the motion of the control head (not shown here). With the objective of modeling spacetime in mind, we are certainly interested in more chaotic behaviors. In [12, 13] only three computations, out of a few thousand, could be singled out for the surprising pseudo-random character of control head motion and resulting trinet graphs. Their causets turn out to be rather similar, at least at small scales, and quite different from those in Figure 8. Their dimensionality analysis based on node-shell growth is made problematic

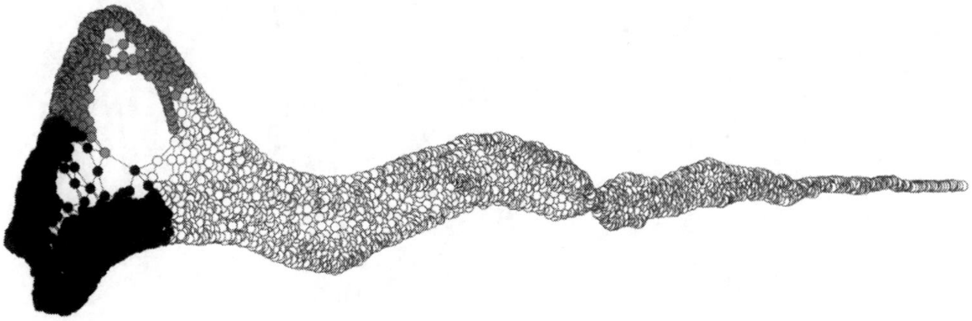

Figure 9. Pseudo-random causet from a computation of a trinet mobile automaton (code 4, 17, 8 in [12]). Nodes are white up to event 4020, then gray up to event 4649, then black up to event 6000.

by the wide oscillations of shell sizes. However, one of these pseudo-random computations lends itself to the illustration of a remarkable, emergent *qualitative* property, as discussed below.

2.5. Emergent compartmentation
What type of feature might we expect to see emerge in our computational spacetime models, and in particular in those with pseudorandom character, on larger and larger scales, beside particles and curvature? We suggest that one of the key features would be the self-organization into components that achieve some form of *independence* from one another, as a necessary basis for building further complexity and obtaining a multiplicity of phenomena (in principle, up to chemistry and biology). In Figure 9 we show a remarkable, pseudo-random causet. The directed graph is partitioned into three sub-segments, colored in white, gray, and black, in a progression that reflects the sequence of computation steps. All grey events have occurred between white and black events, and yet the white and black portions are in direct causal contact. Thus, the 'hole' appearing in the graph is not a pure accident of the specific layout algorithm. This phenomenon reflects the confinement of the 'ant' inside some region of the growing trinet for a relatively large number of steps.

As one would expect, in experiments we have conducted with 'genuinely random' causets, no spontaneous formation of similar macroscopic compartments has been observed.

3. Conclusions
In this paper we have provided some experimental evidence that causets created by simple deterministic algorithms exhibit a variety of interesting emergent properties of physical significance, that can not be obtained by probabilistic approaches.

Several items require further investigation. The node-shell growth analysis based on graph-theoretic, shortest-path distance is weak in two aspects: it does not separate dimension and curvature, and it does not match spacetime distance, which corresponds to the longest-path between points. We have started using longest-path distance in our most recent experiments; but we still need estimators that decouple dimension and curvature. We have identified localized structures in some algorithmic causets, and have called them 'particles', following terminology commonly used for cellular automata; but we still have to discover whether and how these structures exhibit wave/particle duality. We plan to experiment with ant-based models of computation that operate directly on the causet structure, rather than on an underlying memory support. We expect this approach to possibly account for important phenomena from quantum mechanics, e. g. Bell inequalities, that are out of the reach of our current models (this idea has been suggested by Alex Lamb.)

Acknowledgments
This work has been partially supported by CNR Project RSTL-XXL. I express my gratitude to Alex Lamb for many discussions on the presented material, and to Tommaso Toffoli for drawing my attention to Levin's causal nets.

References

[1] Bombelli L, Lee J, Meyer D and Sorkin R D 1987 *Phys. Rev. Lett.* **59** 521–524
[2] Reid D D 1999 Introduction to causal sets: an alternate view of spacetime structure Can.J.Phys. 79 (2001) 1-16 (*Preprint* `http://arxiv.org/abs/gr-qc/9909075 [gr-qc]`)
[3] Henson J 2010 Discovering the discrete universe (*Preprint* `gr-qc/1003.5890v1`)
[4] Sorkin R D 2003 Causal sets: Discrete gravity, Notes for the Valdivia Summer School (*Preprint* `gr-qc/0309009`)
[5] Wolfram S 2002 *A New Kind of Science* (Wolfram Media, Inc.) ISBN 1579550088
[6] Gacs P and Levin L A 1981 *Information and Control* **51** 1–19
[7] Bolognesi T 2010 Causal sets from simple models of computation - To appear in Int. Journ. of Unconventional Computing
[8] Bolognesi T 2010 Causal sets from simple models of computation (*Preprint* `comp-ph/1004.3128`)
[9] Nowotny T and Requardt M 1997 Dimension theory of graphs and networks (*Preprint* `hep-th/9707082`)
[10] Pegg E Turmites - The Wolfram Demo Project (*Preprint* `http://demonstrations.wolfram.com/Turmites/`)
[11] Langton C 1986 *Physica D* **22** 120–149
[12] Bolognesi T 2008 *CompleSystems* **18** 1–41 ISSN 0891-2513
[13] Bolognesi T 2009 *Inf. Process. Lett.* **109** 668–674 ISSN 0020-0190
[14] Markopoulou F 1997 Dual formulation of spin network evolution (*Preprint* `gr-qc/9704013`)

Entropy for theories with indefinite causal structure

Sonia Markes

E-mail: smarkes@perimeterinstitute.ca
Department of Applied Mathematics, University of Waterloo, Waterloo, Ontario, Canada, N2L 3G1
Perimeter Institute for Theoretical Physics, Waterloo, Ontario, Canada, N2L 2Y5

Lucien Hardy

E-mail: lhardy@perimeterinstitute.ca
Perimeter Institute for Theoretical Physics, Waterloo, Ontario, Canada, N2L 2Y5

Abstract. Any theory with definite causal structure has a defined past and future, be it defined by light cones or an absolute time scale. Entropy is a concept that has traditionally been reliant on a definite notion of causality. However, without a definite notion of causality, the concept of entropy is not all lost. Indefinite causal structure results from combining probabilistic predictions and dynamical space-time. The causaloid framework lays the mathematical groundwork to be able to treat indefinite causal structure. In this paper, we build on the causaloid mathematics and define a causally-unbiased entropy for an indefinite causal structure. In defining a causally-unbiased entropy, there comes about an emergent idea of causality in the form of a measure of causal connectedness, termed the Q factor.

1. Introduction

In Newtonian physics, physical processes are understood with respect to a fixed spatial coordinate system and a time parameter, which is absolute and ever increasing. Predictions are entirely deterministic. Quantum theory and general relativity depart from this classical picture in opposing manners. Quantum theory gives probabilistic predictions as to the outcomes of measurements, but retains fixed space and time coordinates. On the other hand, general relativity is deterministic, but shows that space and time form a dynamical structure. Reconciling these fundamental philosophical differences is one of the many challenges one is faced with in trying to construct a theory of quantum gravity. There have been many different approaches to this problem with many different results [1, 2, 3, 4, 5, 6]. One way of moving forward is to dismiss classical assumptions and create a probabilistic theory that has a dynamic causal structure. However, what results is indefinite casual structure. This is more radical than either probabilistic predictions or dynamical space-time structure. In general relativity, a separation between space-time locations is either space-like or time-like. An indefinite causal structure would allow for a separation between space-time locations to be something like a quantum superposition of a space-like and a time-like separation. While we may be uncertain of the causal structure of the path between measurements, we know where in space-time we make measurements, what measurements we have made, and what outcomes we get. With this data, we can examine probabilistic correlations for information. The causaloid framework ([7],[8],[9]) provides us with the necessary structure. We will outline the essentials of this framework in Section 2.

It is natural in discussions of causal structure to raise the question of entropy. The second law of thermodynamics tells us that in an isolated system, entropy can increase or remain the same, but it can never decrease [10]. In information theory, entropy is viewed as being a measure of uncertainty before we measure a state or equivalently, the amount of information gained by upon learning the state of a system [11]. As is the basis for other work [12, 13, 14] we shall take an equivalence of these two types of entropy. Inherent in this concept of entropy is an assumed causal structure, specifically that there exists a background time[15]. The standard definition of entropy is in the context of a definite causal structure with reference to absolute time. In order to make sense of entropy in an indefinite causal structure, a clear definition must be established. To do so requires consideration of the following questions:

What are the concepts from the usual picture of entropy in a definite causal structure that are necessary to define entropy? What are the analogues to these concepts in a picture with indefinite causal structure?

Using the formalism introduced in the causaloid framework, we are able to provide answers to these questions and then, define a causally-unbiased entropy.

In Section 2, we will review the relevant aspects of the causaloid framework. We then proceed with the new developments. In Section 3, we define a new type of product that is utilized in the work on entropy. The definition of causally-unbiased entropy and resulting features are developed in Section 4.

2. Causaloid framework

2.1. The Picture

Every experiment results in a set of data from making measurements on a system. Each piece of data could be thought of as a card with three indications on it; where the measurement is made in space-time, what is measured, and what the result of the measurement is. We will represent each card (or piece of data) as (x, f_x, y_x) where x denotes the space-time location, f_x represents the apparatus configuration, and y_x denotes the outcome of the measurement. The set of all possible cards (i.e. all possible measurements with all possible outcomes with every space-time configuration) is denoted V. We can imagine running an experiment an infinite number of times so as to be able to obtain relative frequencies. (Note that the concept of relative frequencies is utilized here only has a illustrative tool and is not meant to specify an ontology.) In order for the cards to tell us the relative frequencies, we must systematically sort them.

Each distinct x is defined as an *elementary region* of space-time. A *composite region*, denoted $\mathcal{O}_1$, is a set of elementary regions. (Note: These definitions of "elementary region" and "composite region" differ from those in [7, 8, 9].) Therefore, these cards can be sorted according to their associated space-time region. The set of all possible cards with the same space-time location x on them is the *elementary information set*. We denote this set as R_x. The *composite information set*, R_1, is the union of all elementary information regions for which the associated elementary regions are contained within the composite region $\mathcal{O}_1$. More concisely,

$$R_1 \equiv \bigcup_{x \in \mathcal{O}_1} R_x \tag{1}$$

Note that a composite information set has no more or less structure than an elementary region of space-time. Therefore, without adding structure or losing generality, we can treat the sets R_x as elementary regions. From this point forward, the term *region* will be used interchangeably to refer to objects of type x or $\mathcal{O}_1$ and type R_x or R_1.

If we chose a particular measurement to perform in the region x (indexed by α_x), the set of cards consistent with $f_x^{\alpha_x}$ is called the *procedure information set* and is denoted $F_x^{\alpha_x}$.

$$F_x^{\alpha_x} \equiv \bigcup_{all\, y_x} \{(x, f_x^{\alpha_x}, y_x)\} \tag{2}$$

For composite regions, we can then define the following:

$$F_1^{\alpha_1} \equiv \bigcup_{x\in\mathcal{O}_1} F_x^{\alpha_x} \tag{3}$$

For the measurement $f_x^{\alpha_x}$, the set of cards consistent with the outcome $y_x^{\alpha_x}$ is called the *outcome information set* and is denoted $Y_x^{\alpha_x}$

$$Y_x^{\alpha_x} \equiv \{(x, f_x^{\alpha_x}, y_x^{\alpha_x})\} \tag{4}$$

For composite regions, we have

$$Y_1 \equiv \bigcup_{x\in\mathcal{O}_1} Y_x^{\alpha_x} \tag{5}$$

The set of all measurement and outcome pairs (i.e. all α = (Y, F)) is denoted Υ.

Notice that the set of all cards V can be viewed as all the cards from all (elementary) regions.

$$V = \bigcup_{all\, x} R_x \tag{6}$$

So V is the largest of all regions that can be considered.

These definitions provide a firm foundation on which the causaloid framework rests both mathematically and conceptually.

2.2. First level physical compression

The most basic quantity that we would want to be able to calculate is the probability that a certain (set of) outcome(s) is observed given that a certain (set of) measurement(s) has been performed at a (set of) location(s) in space and time. Suppose that the set of locations we are interested in is $\mathcal{O}_1$. Then the information set of interest is R_1. The set comprised of all the cards not in R_1 is $V - R_1$. We call (Y_{V-R_1}, F_{V-R_1}) the *generalized preparation* because it is the information that surrounds R_1 not only from the immediate past, but from the future and the rest of space-time as well. By the choices we make in setting up the experiment, we can put conditions on the generalized preparation such that $\text{Prob}(Y_V|F_V)$ is well-defined. (See Ref.[7] for details.) Then we can write

$$\text{Prob}(Y_V|F_V) = \text{Prob}(Y_1, Y_{V-R_1}|F_1, F_{V-R_1}) \tag{7}$$

For a specific pair $\alpha_1 \Leftrightarrow (Y_1^{\alpha_1}, F_1^{\alpha_1})$, we can write this probability as

$$\text{Prob}(Y_1^{\alpha_1}, Y_{V-R_1}|F_1^{\alpha_1}, F_{V-R_1}) \tag{8}$$

We use the short-hand p_{α_1} to denote the probability defined in Eq.(8). One way to specify the state of a system is to list all the possible p_{α_1} for elements of R_1.

$$\begin{pmatrix} \vdots \\ p_{\alpha_1} \\ \vdots \end{pmatrix} \qquad \alpha_1 \in \Upsilon_1 \tag{9}$$

However, this over-specifies the state. We do not usually need to know the probability of every outcome of every measurement in order to determine what the complete state of the system is. Physical theories tell us what relationships exist between variables and what constraints those relationships place on the variables of the system. These relationships and constraints can be used to determine a reduced set of probabilities from which all other probabilities can be represented. The reduced set of probabilities is defined such that any probability can be written as a linear combination of the probabilities in the reduced set. Let us denote the reduced or fiducial set in R_1 as $\Omega_1 \subseteq \Upsilon_1$. This process of going from the set of all the probabilities to the smallest essential set we call *first level physical compression*. This can be expressed as

$$\mathbf{p} = \begin{pmatrix} \vdots \\ p_{l_1} \\ \vdots \end{pmatrix} \qquad l_1 \in \Omega_1 \subseteq \Upsilon_1 \tag{10}$$

such that

$$p_{\alpha_1} = \mathbf{r}_{\alpha_1} \cdot \mathbf{p} \tag{11}$$

where $\mathbf{r}_{\alpha_1}$ encodes the physical compression and therefore, is determined by the details of the physical theory. Of course, the compression is not unique. We can define a *decompression matrix*, $\Lambda^{l_1}_{\alpha_1}$ such that

$$\Lambda^{l_1}_{\alpha_1} \equiv \mathbf{r}_{\alpha_1}|_{l_1} \tag{12}$$

where $\mathbf{r}_{\alpha_1}|_{l_1}$ means the l_1 component of $\mathbf{r}_{\alpha_1}$. While the decompression matrix may seem superfluous at the first level, it becomes a useful notation for higher level compression.

2.3. Second level physical compression and the causaloid product

Let us consider two distinct regions $R_1, R_2 \subset V$. In a similar fashion to the single region case,

$$p_{\alpha_1\alpha_2} = \text{Prob}(Y_1^{\alpha_1}, Y_2^{\alpha_2}, Y_{V-R_1-R_2}|F_1^{\alpha_1}, F_2^{\alpha_2}, F_{V-R_1-R_2}) \tag{13}$$

We specify the state of the system by listing all $p_{\alpha_1\alpha_2}$.

$$\begin{pmatrix} \vdots \\ p_{\alpha_1\alpha_2} \\ \vdots \end{pmatrix} \qquad \alpha_1\alpha_2 \in \Upsilon_1 \times \Upsilon_2 \tag{14}$$

where $\times$ is the cartesian product. It can be shown that

$$p_{\alpha_1\alpha_2} = \sum_{l_1 l_2 \in \Omega_1 \times \Omega_2} \Lambda^{l_1}_{\alpha_1} \Lambda^{l_2}_{\alpha_2} p_{l_1 l_2} \tag{15}$$

which implies that the following list of probabilities is sufficient.

$$\begin{pmatrix} \vdots \\ p_{l_1 l_2} \\ \vdots \end{pmatrix} \qquad l_1 l_2 \in \Omega_1 \times \Omega_2 \subseteq \Upsilon_1 \times \Upsilon_2 \tag{16}$$

This is effectively first level compression on each index. But if a physical theory has some connection between the two regions, $\Omega_1 \times \Omega_2$ may no longer be the smallest set that is sufficient to represent all possible states. Then *second level physical compression* is possible. It is defined to be

$$\mathbf{p} = \begin{pmatrix} \vdots \\ p_{k_1k_2} \\ \vdots \end{pmatrix} \qquad k_1k_2 \in \Omega_{12} \subseteq \Omega_1 \times \Omega_2 \tag{17}$$

such that

$$p_{\alpha_1\alpha_2} = \mathbf{r}_{\alpha_1\alpha_2} \cdot \mathbf{p} = \sum_{k_1k_2 \in \Omega_{12}} \mathbf{r}_{\alpha_1\alpha_2}|_{k_1k_2} p_{k_1k_2} \tag{18}$$

When $\Omega_{12} = \Omega_1 \times \Omega_2$, second level compression is trivial. But it is proven in [8] that it is possible that $\Omega_{12} \subset \Omega_1 \times \Omega_2$.

Now we can define a second level decompression matrix. By comparing Eq.(15) and Eq.(18), we infer that

$$\mathbf{r}_{\alpha_1\alpha_2}|_{k_1k_2} = \sum_{l_1l_2 \in \Omega_1 \times \Omega_2} \Lambda^{l_1}_{\alpha_1} \Lambda^{l_2}_{\alpha_2} \Lambda^{k_1k_2}_{l_1l_2} \tag{19}$$

where

$$\Lambda^{k_1k_2}_{l_1l_2} = \mathbf{r}_{l_1l_2}|_{k_1k_2} \tag{20}$$

which is the desired second level decompression matrix. This matrix encodes how we move from $p_{l_1l_2}$'s to $p_{k_1k_2}$'s. Using the definition of the first level decompression matrix, Eq.(19) becomes

$$\mathbf{r}_{\alpha_1\alpha_2}|_{k_1k_2} = \sum_{l_1l_2 \in \Omega_1 \times \Omega_2} \Lambda^{k_1k_2}_{l_1l_2} \mathbf{r}_{\alpha_1}|_{l_1} \mathbf{r}_{\alpha_2}|_{l_2} \tag{21}$$

This defines the *causaloid product*, denoted $\mathbf{r}_{\alpha_1} \otimes^{\Lambda} \mathbf{r}_{\alpha_2}$ which unifies the different causal structure-specific products. Explicitly,

$$\mathbf{r}_{\alpha_1} \otimes^{\Lambda} \mathbf{r}_{\alpha_2} = \mathbf{r}_{\alpha_1\alpha_2} \tag{22}$$

It is this product that allows us to look at the probabilistic correlations between arbitrary locations in space-time without specifying the causal relationship.

We have shown second level compression for the case where we have two regions. This is easily generalized for any number of regions. The object that would encode the compression for three regions would be $\Lambda^{k_1k_2k_3}_{l_1l_2l_3}$, for four regions would be $\Lambda^{k_1k_2k_3k_4}_{l_1l_2l_3l_4}$, etc. After second level compression over multiple regions, we have

$$\begin{pmatrix} \Lambda^{k_1}_{\alpha_1} \\ \Lambda^{k_1k_2}_{l_1l_2} \\ \Lambda^{k_1k_2k_3}_{l_1l_2l_3} \\ \vdots \end{pmatrix} \tag{23}$$

There is a third level of physical compression that compresses these multi-region Λ-matrices to give the *Causaloid*, $\mathbf{\Lambda}$, which is defined as

$$\mathbf{\Lambda} \equiv \Big(\{\Lambda\} \mid \{\Lambda\} \subseteq \{\Lambda^{k_1}_{\alpha_1}, \Lambda^{k_1k_2}_{l_1l_2}, \ldots\}\Big) \tag{24}$$

where $\{\Lambda\}$ is determined by the rules of the physical theory (for detailed discussion of how this works see [2]). By decompressing the set $\{\Lambda\}$, we can obtain the Λ-matrix for any set of regions. This means that the Causaloid gives us the ability to perform any calculation that the physical theory allows for.

2.4. Well-defined probabilities

Up to this point we have exclusively dealt with probabilities conditioned on procedures. It is more useful to also be able to condition on outcomes. Specifically, we'd like an expression for the following:

$$\text{Prob}(Y_2^{\alpha_2}|Y_1^{\alpha_1}, F_1^{\alpha_1}, F_2^{\alpha_2}) \tag{25}$$

Using Bayes' Theorem, this becomes

$$\text{Prob}(Y_2^{\alpha_2}|Y_1^{\alpha_1}, F_1^{\alpha_1}, F_2^{\alpha_2}) = \frac{\text{Prob}(Y_1^{\alpha_1}, Y_2^{\alpha_2}|F_1^{\alpha_1}, F_2^{\alpha_2})}{\sum_{Y_2^{\beta_2}\sim F_2^{\alpha_2}} \text{Prob}(Y_1^{\alpha_1}, Y_2^{\beta_2}|F_1^{\alpha_1}, F_2^{\alpha_2})} \tag{26}$$

where $Y_2^{\beta_2} \sim F_2^{\alpha_2}$ denotes that the sum is over all possible outcomes corresponding to the measurement $F_2^{\alpha_2}$ (in R_2). (For simplicity, we have suppressed the part of the notation denoting the generalized preparation.) In the causaloid framework, this becomes

$$\text{Prob}(Y_2^{\alpha_2}|Y_1^{\alpha_1}, F_1^{\alpha_1}, F_2^{\alpha_2}) = \frac{\mathbf{r}_{\alpha_1\alpha_2} \cdot \mathbf{p}}{\mathbf{r}_{\alpha_1\frown 2} \cdot \mathbf{p}} \tag{27}$$

where $\mathbf{r}_{\alpha_1\frown 2} = \sum_{\beta_2} \mathbf{r}_{\alpha_1\beta_2}$. (The sum being over β_2 in this notation has the same meaning as the sum being over all outcomes consistent with F_2.) In order for this probability to be considered well-defined, the right hand side cannot depend on $V - R_1 - R_2$. Since $\mathbf{r}_{\alpha_1\alpha_2}$ and $\mathbf{r}_{\alpha_1\frown 2}$ are determined exclusively by the physical theory, neither has any dependence on $V - R_1 - R_2$. However, $\mathbf{p}$ does depend on $V - R_1 - R_2$. This implies that in order for the probability Eq.(27) to be well defined (i.e. not depend on $V - R_1 - R_2$), it must vary with $\mathbf{p}$. The dependence on $\mathbf{p}$ can be removed altogether by requiring that $\mathbf{r}_{\alpha_1\alpha_2}$ be parallel to $\mathbf{r}_{\alpha_1\frown 2}$. Therefore, the above probability is well defined if and only if

$$\mathbf{r}_{\alpha_1\alpha_2} \parallel \mathbf{r}_{\alpha_1\frown 2} \tag{28}$$

With this condition, we get

$$\text{Prob}(Y_2^{\alpha_2}|Y_1^{\alpha_1}, F_1^{\alpha_1}, F_2^{\alpha_2}) = \frac{|\mathbf{r}_{\alpha_1\alpha_2}|}{|\mathbf{r}_{\alpha_1\frown 2}|} \tag{29}$$

3. $\odot^{\Gamma}$ product

Consider two distinct regions; R_A and R_P. By definition

$$\mathbf{r}_{\alpha_A\alpha_P} = \mathbf{r}_{\alpha_A} \otimes^{\Lambda} \mathbf{r}_{\alpha_P}$$
$$\mathbf{r}_{\beta_A\alpha_P} = \mathbf{r}_{\beta_A} \otimes^{\Lambda} \mathbf{r}_{\alpha_P}$$

Suppose we wanted to take the dot product between two vectors of the above form. Using decompression matrices, we can write

$$\begin{aligned}\mathbf{r}_{\alpha_A\alpha_P} \cdot \mathbf{r}_{\beta_A\alpha_P} &= \left(\mathbf{r}_{\alpha_A} \otimes^{\Lambda} \mathbf{r}_{\alpha_P}\right) \cdot \left(\mathbf{r}_{\beta_A} \otimes^{\Lambda} \mathbf{r}_{\alpha_P}\right) \\ &= \sum_{k_A k_P} \left(\sum_{l_A l_P} \Lambda_{l_A l_P}^{k_A k_P} \mathbf{r}_{\alpha_A}|_{l_A} \mathbf{r}_{\alpha_P}|_{l_P}\right) \left(\sum_{l'_A l'_P} \Lambda_{l'_A l'_P}^{k_A k_P} \mathbf{r}_{\beta_A}|_{l'_A} \mathbf{r}_{\alpha_P}|_{l'_P}\right)\end{aligned} \tag{30}$$

where $k_A k_P \in \Omega_{AP}$, $l_A l_P \in \Omega_A \times \Omega_P$, and $l'_A l'_P \in \Omega_A \times \Omega_P$. Notice that we can write

$$\sum_{l_A l_P} \Lambda_{l_A l_P}^{k_A k_P} \mathbf{r}_{\alpha_A}|_{l_A} \mathbf{r}_{\alpha_P}|_{l_P}$$

as

$$\sum_{l_A \in \Omega_A} \left[\mathbf{r}_{\alpha_A}|_{l_A} \left(\sum_{l_P \in \Omega_P} \Lambda^{k_A k_P}_{l_A l_P} \mathbf{r}_{\alpha_P}|_{l_P} \right) \right]$$

Similarly,

$$\sum_{l'_A l'_P} \Lambda^{k_A k_P}_{l'_A l'_P} \mathbf{r}_{\beta_A}|_{l'_A} \mathbf{r}_{\alpha_P}|_{l'_P} = \sum_{l'_A \in \Omega_A} \left[\mathbf{r}_{\beta_A}|_{l'_A} \left(\sum_{l'_P \in \Omega_P} \Lambda^{k_A k_P}_{l'_A l'_P} \mathbf{r}_{\alpha_P}|_{l'_P} \right) \right]$$

Define

$$\Gamma^{k_A k_P}_{l_A}(\mathbf{r}_{\alpha_P}) \equiv \sum_{l_P \in \Omega_P} \Lambda^{k_A k_P}_{l_A l_P} \mathbf{r}_{\alpha_P}|_{l_P}$$

and, similarly,

$$\Gamma^{k_A k_P}_{l'_A}(\mathbf{r}_{\alpha_P}) \equiv \sum_{l'_P \in \Omega_P} \Lambda^{k_A k_P}_{l'_A l'_P} \mathbf{r}_{\alpha_P}|_{l'_P}$$

Using this, Eq.(30) becomes

$$\mathbf{r}_{\alpha_A \alpha_P} \cdot \mathbf{r}_{\beta_A \alpha_P} = \sum_{k_A k_P} \left(\sum_{l_A l'_A} \Gamma^{k_A k_P}_{l_A}(\mathbf{r}_{\alpha_P}) \Gamma^{k_A k_P}_{l'_A}(\mathbf{r}_{\alpha_P}) \mathbf{r}_{\alpha_A}|_{l_A} \mathbf{r}_{\beta_A}|_{l'_A} \right) \tag{31}$$

where $k_A k_P \in \Omega_{AP}$ and $l_A l'_A \in \Omega_A \times \Omega_A$. This suggests that the essence of $\mathbf{r}_{\alpha_A \alpha_P} \cdot \mathbf{r}_{\beta_A \alpha_P}$ is a relationship between $\mathbf{r}_{\alpha_A}$ and $\mathbf{r}_{\beta_A}$ mediated by matrices that depend on $\mathbf{r}_{\alpha_P}$. Therefore, we can view Eq.(30) as kind of product of $\mathbf{r}_{\alpha_A}$ and $\mathbf{r}_{\beta_A}$. Dot products of this form come up frequently enough that we will define this as the Γ-dot product and denote it as

$$\mathbf{r}_{\alpha_A} \odot^{\Gamma(\mathbf{r}_{\alpha_P})} \mathbf{r}_{\beta_A} \equiv \mathbf{r}_{\alpha_A \alpha_P} \cdot \mathbf{r}_{\beta_A \alpha_P} = \left(\mathbf{r}_{\alpha_A} \otimes^{\Lambda} \mathbf{r}_{\alpha_P} \right) \cdot \left(\mathbf{r}_{\beta_A} \otimes^{\Lambda} \mathbf{r}_{\alpha_P} \right) \tag{32}$$

We will make use of this product later in the paper.

4. Causally-unbiased entropy

Standard definitions of entropy assume fixed causal structure. Here we develop a causally-unbiased definition of entropy in the causaloid formalism.

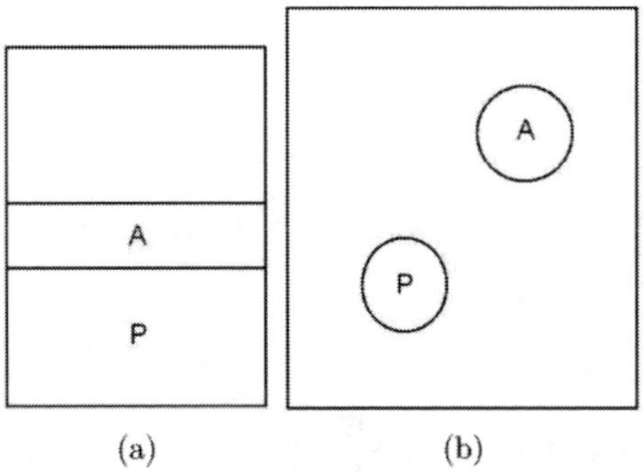

Figure 1: (a) Fixed causal structure (b) Indefinite causal structure

4.1. The picture

Shannon entropy [11] for a classical state is defined as

$$S = -\sum_i p_i log_2 p_i \tag{33}$$

The definition of p_i used in this equation requires that the structure of space-time be organized with the following features:

- a region of interest, A
- an immediate past space-time region, P
- sufficient data about what happened in P
- a measurement F_A
- a set of outcomes, $\{Y_A^i\}$, corresponding to F_A

This allows us to write

$$p_i = \text{Prob}(Y_A^i | F_A, \text{data}_P) \tag{34}$$

Removing all time bias from these features of space-time structure, we get

- a region of interest, A
- a reference region P
- an outcome/measurement pair in P, $\{\alpha_P\} = \{(F_P, Y_P)\}$
- a measurement F_A
- a set of outcomes, $\{Y_A^i\}$, corresponding to F_A

The reference region can be thought of as a kind of preparation region that is not limited to being in the causal past. In fact, the choice of reference region is arbitrary as illustrated in Fig. 1b.

The definition of p_i in a causally-unbiased structure is

$$p_i = \text{Prob}(Y_A^i | Y_P, F_P, F_A) \tag{35}$$

(Since P is arbitrary, we should technically say 'p_i with respect to the reference region P'. However, for the sake of brevity, we will assume that 'with respect to P' is implied much as 'with respect to the past' is taken as implied in the causally-biased situation.)

Using the above definition of p_i, we define the entropy relative to the reference data (F_P, Y_P) as

$$S = -\sum_i \text{Prob}(Y_A^i | Y_P, F_P, F_A) \log_2 \left(\text{Prob}(Y_A^i | Y_P, F_P, F_A) \right) \tag{36}$$

Notice that this reduces to the causally-biased definition of entropy when P is the past; F_A measures the microstate in the classical case or measures in the basis where $\hat{\rho}$ is diagonal in the quantum case.

4.2. In the causaloid framework

Taking the probability to be well-defined, Eq.(29) and Eq.(36) give the following definition of entropy:

$$S = -\sum_{\alpha_A} \frac{|\mathbf{r}_{\alpha_A \alpha_P}|}{|\mathbf{r}_{\frown_A \alpha_P}|} \log_2 \left(\frac{|\mathbf{r}_{\alpha_A \alpha_P}|}{|\mathbf{r}_{\frown_A \alpha_P}|} \right) \tag{37}$$

Of course, this equation requires that $\mathbf{r}_{\alpha_A\alpha_P} \parallel \mathbf{r}_{\frown_A\alpha_P}$. We can also consider what happens when $\mathbf{r}_{\alpha_A\alpha_P}$ is nearly parallel to $\mathbf{r}_{\frown_A\alpha_P}$, using the definition of the probability from Eq.(27). The entropy associated with this is

$$S^A = -\sum_{\alpha_A}\left(\frac{\mathbf{r}_{\alpha_A\alpha_P}\cdot\mathbf{p}}{\mathbf{r}_{\frown_A\alpha_P}\cdot\mathbf{p}}\right)\log_2\left(\frac{\mathbf{r}_{\alpha_A\alpha_P}\cdot\mathbf{p}}{\mathbf{r}_{\frown_A\alpha_P}\cdot\mathbf{p}}\right) \tag{38}$$

It becomes necessary to shorten the notation for the following work so $\mathbf{r}_{\alpha_A\alpha_P}$ will be denoted as $\mathbf{v}_i$ (where the index α_A is represented by i) and $\mathbf{r}_{\frown_A\alpha_P}$ will be denoted as $\mathbf{u}$. As with any vector, $\mathbf{v}_i$ can be decomposed into a component parallel to $\mathbf{u}$ and a component perpendicular to $\mathbf{u}$ (i.e. components in $\hat{\mathbf{u}}^{\parallel}$ and $\hat{\mathbf{u}}^{\perp}$, respectively). That is,

$$\mathbf{v}_i = v_i^{\parallel}\hat{\mathbf{u}}^{\parallel} + v_i^{\perp}\hat{\mathbf{u}}^{\perp} \tag{39}$$

Using the unit vectors as defined, $\mathbf{p}$ can be decomposed as

$$\mathbf{p} = p_x\hat{\mathbf{u}}^{\parallel} + p_y\hat{\mathbf{u}}^{\perp} + \mathbf{p}^{\perp} \tag{40}$$

where $\mathbf{p}^{\perp}$ is the component of $\mathbf{p}$ that is perpendicular to the plane defined by $\mathbf{u}$ and $\mathbf{v}_i$. The probability of interest, p_i, then becomes

$$\begin{aligned} p_i &= \frac{\mathbf{v}_i\cdot\mathbf{p}}{\mathbf{u}\cdot\mathbf{p}} \\ &= \frac{v_i^{\parallel}}{u} + k\frac{v_i^{\perp}}{u} \end{aligned} \tag{41}$$

where $k = \frac{p_y}{p_x}$. Notice that the first term is equivalent to a well-defined probability (Eq. 29). We require the second term to be small since the deviation from well-defined should be small. Since we have already required that $\mathbf{v}_i^{\perp}$ be small, we need only place restrictions on k.

4.3. Bounds on k

For the purposes of this subsection, we will work in the plane defined by $\mathbf{u}$ and $\mathbf{v}_i$. Define the angle between $\mathbf{u}$ and the projection of $\mathbf{p}$ into the plane to be θ. Define the length of the projection of $\mathbf{p}$ into the plane to be p_{xy}. Using basic trigonometry, we get

$$p_y = p_{xy}sin\theta \tag{42}$$
$$p_x = p_{xy}cos\theta \tag{43}$$

Therefore, k can be written in a form that is dependent on only one variable, as follows:

$$k = tan\theta \tag{44}$$

As θ tends towards $\pm\frac{\pi}{2}$, k tends to infinity. Therefore, to ensure that the second term of (41) is small, we require that k be finite. Assume it to be a property of the state space for $\mathbf{p}$ that there exists some $0 < \theta_{max}$. Clearly, $|\theta_{max}| < \frac{\pi}{2}$ in order for k to be finite. So θ is bounded as follows:

$$-\frac{\pi}{2} < -\theta_{max} \leq \theta \leq \theta_{max} < \frac{\pi}{2} \tag{45}$$

The k corresponding to θ_{max} will be denoted as k_{max}. Further bounds can be placed on k by the state space of the physical theory. For our purposes, it is sufficient that k is finite.

4.4. Q factor

In light of (41), entropy, as defined in (38), becomes

$$\begin{aligned} S^{\Lambda} &= -\sum_i \left(\frac{v_i^{\parallel}}{u} + k\frac{v_i^{\perp}}{u}\right) \log_2\left(\frac{v_i^{\parallel}}{u} + k\frac{v_i^{\perp}}{u}\right) \\ &= -\sum_i \left(\frac{v_i^{\parallel}}{u} + k\frac{v_i^{\perp}}{u}\right)\left[\log_2\left(1 + k\frac{v_i^{\perp}}{v_i^{\parallel}}\right) + \log_2\left(\frac{v_i^{\parallel}}{u}\right)\right] \end{aligned} \quad (46)$$

Since $v_i^{\perp}$ is very small (as is implied by the fact that $\mathbf{v_i}$ and $\mathbf{u}$ are nearly parallel) and k is finite, we can take a Taylor expansion (to leading order) of the first $\log_2$ term. Doing this gives

$$\begin{aligned} S^{\Lambda} &= -\sum_i \left(\frac{v_i^{\parallel}}{u} + k\left(\frac{v_i^{\perp}}{u}\right)\right)\left[\frac{k}{\ln 2}\left(\frac{v_i^{\perp}}{v_i^{\parallel}}\right) + \mathcal{O}\left({v_i^{\perp}}^2\right) + \log_2\left(\frac{v_i^{\parallel}}{u}\right)\right] \\ &= -\sum_i \left(\frac{v_i^{\parallel}}{u}\right)\log_2\left(\frac{v_i^{\parallel}}{u}\right) + k\left(\frac{v_i^{\perp}}{u}\right)\log_2\left(e\frac{v_i^{\parallel}}{u}\right) + \mathcal{O}\left({v_i^{\perp}}^2\right) \end{aligned} \quad (47)$$

Notice that the first term is equivalent to the definition of entropy where $\mathbf{u} \parallel \mathbf{v_i}$ and that S^{Λ} reduces to this definition when $v_i^{\perp} = 0$. That is, when $\mathbf{u} \parallel \mathbf{v_i}$ (or equivalently, $v_i^{\perp} = 0$)

$$S^{\Lambda} = S \equiv -\sum_i \left(\frac{v_i^{\parallel}}{u}\right)\log_2\left(\frac{v_i^{\parallel}}{u}\right) \quad (48)$$

For $v_i^{\perp} \neq 0$, we will define

$$Q = -\sum_i \left(\frac{v_i^{\perp}}{u}\right)\log_2\left(e\frac{v_i^{\parallel}}{u}\right) \quad (49)$$

Using k_{max} as defined in the previous section, we can regard $k_{max}Q$ as a kind of correction to the causally-biased entropy. Then, to leading order

$$S - k_{max}Q \leq S^{\Lambda} \leq S + k_{max}Q \quad (50)$$

4.5. Understanding Q

Q is an entirely new quantity with no direct classical analogue so understanding its physical interpretation is a non-trivial matter. If we consider entropy as a measure of uncertainty, then S is the measure of our uncertainty that the measurement F_A in region A will yield the specific outcome Y_A^i, given the data we have from the reference region P. Since our reference region P is arbitrary, one way to view Q is that it measures how completely the region P influences region A. In a definite causal structure, an immediate past region would be the exclusive influence on our region of interest and Q would be zero. However, in the causally-indefinite picture, we cannot require a priori that the reference region we have chosen will be the exclusive influence on our region of interest. If there are no influences on region A from outside region P, then the probability will be well-defined and Q will be zero. But if there are influences on region A from outside region P, then the magnitude of Q will reflect that.

4.6. Using the $\odot^{\Gamma}$ product

For the sake of completeness the $\mathbf{u}$'s and $\mathbf{v}_i$'s must be translated into $\mathbf{r}_{\frown_{A}\alpha_P}$'s and $\mathbf{r}_{\alpha_A\alpha_P}$'s. Notice that

$$\frac{v_i^{\parallel}}{u} = \frac{\mathbf{v_i}\cdot\mathbf{u}}{\mathbf{u}\cdot\mathbf{u}} \quad (51)$$

$$\frac{v_i^{\perp}}{u} = \sqrt{\frac{v_i^2 - {v_i^{\parallel}}^2}{u^2}} = \sqrt{\frac{\mathbf{v_i} \cdot \mathbf{v_i}}{\mathbf{u} \cdot \mathbf{u}} - \frac{(\mathbf{v_i} \cdot \mathbf{u})^2}{(\mathbf{u} \cdot \mathbf{u})^2}} \tag{52}$$

Substituting $\mathbf{r}_{\alpha_A \alpha_P}$ for $\mathbf{v}_i$ and $\mathbf{r}_{\frown_A \alpha_P}$ for $\mathbf{u}$ gives

$$\frac{v_i^{\parallel}}{u} = \frac{\mathbf{r}_{\alpha_A \alpha_P} \cdot \mathbf{r}_{\frown_A \alpha_P}}{\mathbf{r}_{\frown_A \alpha_P} \cdot \mathbf{r}_{\frown_A \alpha_P}} \tag{53}$$

$$\frac{v_i^{\perp}}{u} = \sqrt{\frac{\mathbf{r}_{\alpha_A \alpha_P} \cdot \mathbf{r}_{\alpha_A \alpha_P}}{\mathbf{r}_{\frown_A \alpha_P} \cdot \mathbf{r}_{\frown_A \alpha_P}} - \frac{(\mathbf{r}_{\alpha_A \alpha_P} \cdot \mathbf{r}_{\frown_A \alpha_P})^2}{(\mathbf{r}_{\frown_A \alpha_P} \cdot \mathbf{r}_{\frown_A \alpha_P})^2}} \tag{54}$$

Using the Γ-dot product the above equations simplify to

$$\frac{v_i^{\parallel}}{u} = \frac{\mathbf{r}_{\alpha_A} \odot^{\Gamma(\mathbf{r}_{\alpha_P})} \mathbf{r}_{\frown_A}}{\mathbf{r}_{\frown_A} \odot^{\Gamma(\mathbf{r}_{\alpha_P})} \mathbf{r}_{\frown_A}} \tag{55}$$

$$\frac{v_i^{\perp}}{u} = \sqrt{\frac{\mathbf{r}_{\alpha_A} \odot^{\Gamma(\mathbf{r}_{\alpha_P})} \mathbf{r}_{\alpha_A}}{\mathbf{r}_{\frown_A} \odot^{\Gamma(\mathbf{r}_{\alpha_P})} \mathbf{r}_{\frown_A}} - \left(\frac{\mathbf{r}_{\alpha_A} \odot^{\Gamma(\mathbf{r}_{\alpha_P})} \mathbf{r}_{\frown_A}}{\mathbf{r}_{\frown_A} \odot^{\Gamma(\mathbf{r}_{\alpha_P})} \mathbf{r}_{\frown_A}}\right)^2} \tag{56}$$

This allows us to completely specify the entropy of R_A relative to a preparation R_P in the causaloid framework. It is straightforward to generalize this to define the joint entropy of R_A and R_B with reference to a "preparation" R_P. Simply redefine $\mathbf{u}$ and $\mathbf{v_{ij}}$ as

$$\mathbf{v}_{ij} = \mathbf{r}_{\alpha_A \alpha_B \alpha_P} \tag{57}$$

$$\mathbf{u} = \mathbf{r}_{\frown_A \frown_B \alpha_P} \tag{58}$$

where

$$\mathbf{r}_{\alpha_A \alpha_B \alpha_P} = \mathbf{r}_{\alpha_A} \otimes^{\Lambda} \mathbf{r}_{\alpha_B} \otimes^{\Lambda} \mathbf{r}_{\alpha_P}$$

and

$$\mathbf{r}_{\frown_A \frown_B \alpha_P} = \sum_{\beta_A} \mathbf{r}_{\beta_A} \otimes^{\Lambda} \sum_{\beta_B} \mathbf{r}_{\beta_B} \otimes^{\Lambda} \mathbf{r}_{\alpha_P}$$

Using the same procedure as for one region, we get

$$\frac{v_{ij}^{\parallel}}{u} = \frac{\mathbf{r}_{\alpha_A \alpha_B} \odot^{\Gamma(\mathbf{r}_{\alpha_P})} \mathbf{r}_{\frown_A \frown_B}}{\mathbf{r}_{\frown_A \frown_B} \odot^{\Gamma(\mathbf{r}_{\alpha_P})} \mathbf{r}_{\frown_A \frown_B}} \tag{59}$$

$$\frac{v_{ij}^{\perp}}{u} = \sqrt{\frac{\mathbf{r}_{\alpha_A \alpha_B} \odot^{\Gamma(\mathbf{r}_{\alpha_P})} \mathbf{r}_{\alpha_A \alpha_B}}{\mathbf{r}_{\frown_A \frown_B} \odot^{\Gamma(\mathbf{r}_{\alpha_P})} \mathbf{r}_{\frown_A \frown_B}} - \left(\frac{\mathbf{r}_{\alpha_A \alpha_B} \odot^{\Gamma(\mathbf{r}_{\alpha_P})} \mathbf{r}_{\frown_A \frown_B}}{\mathbf{r}_{\frown_A \frown_B} \odot^{\Gamma(\mathbf{r}_{\alpha_P})} \mathbf{r}_{\frown_A \frown_B}}\right)^2} \tag{60}$$

In this manner, we can define causally-unbiased entropy in the causaloid framework for any number of regions.

5. Conclusions

In a definite causal structure, the only thing required for a definition of entropy that is not in an indefinite causal structure is an immediate past region. Since there is no reason in an indefinite causal structure to choose any reference region over any other, we simply choose an arbitrary region. This ensures that we do not hold on to any pre-conceived notions of space-time and its connection to causality. The definition of the causally-unbiased entropy resulted in a correction to the causally-biased definition of entropy. In a sense, the Q factor gives us an emergent idea of causality. It is a measure of the extent to which our region of interest is causally connected to our reference (or "preparation") region. If it is zero, the traditional ideas of causality are recovered. The next step would be determining how the Q factor could potentially be physically observed. To do so may require us to know more of the theoretical and mathematical properties of Q. Which mathematical properties of Shannon entropy hold for causally-unbiased entropy? What is the status of the Second Law of Thermodynamics in an indefinite causal structure? To go about answering this, we could consider how S^{Λ} "evolves" along tubes through indefinite space-times. These questions will be the subjects of continuing work in the near future.

Acknowledgments

SM was supported by a grant through an OGS from the Government of Ontario. Research at Perimeter Institute for Theoretical Physics is supported in part by the Government of Canada through NSERC and by the Province of Ontario through MRI.

References

[1] Penrose R and MacCallum M A H 1972 Twistor theory: an approach to the quantization of fields and space-time *Phys. Rept.* **6**, 241
[2] Hawking S W 1978 Quantum gravity and path integrals *Phys. Rev.* D **18**, 1747
[3] Rovelli C and Smolin L 1990 Loop space representation of quantum general relativity *Nucl. Phys.* B **331**, 80
[4] Thiemann T 2007 *Modern Canonical Quantum General Relativity* (Cambridge, UK: Cambridge Univ. Pr.)
[5] Sorkin R D 1994 On the role of time in the sum over histories framework for gravity *Int. J. Theor. Phys.* **33**, 523
[6] Ambjorn J, Jurkiewicz J and Loll R 2006 Quantum gravity, or the art of building spacetime *Preprint* hep-th/0604212
[7] Hardy L 2005 Probability theories with dynamic causal structure: a new framework for quantum gravity *Preprint* gr-qc/0509120
[8] Hardy L 2006 Towards quantum gravity: a framework for probabilistic theories with non-fixed causal structure *J. Phys.* A **40**, 3081 (*Preprint* gr-qc/0608043)
[9] Hardy L 2008 Formalism locality in quantum theory and quantum gravity *Preprint* gr-qc/0804.0054
[10] Landau L D and Lifshitz E M 1980 *Course of Theoretical Physics Vol. 5: Statistal Physics Pt. 1* (Oxford: Butterworth Heinemann)
[11] Nielsen M A and Chuang I L 2000*Quantum Computation and Quantum Information* (Cambridge: Cambridge University Press)
[12] Maroney O J E 2004 Information and entropy in quantum theory *Preprint* quant-ph/0411172
[13] Jacobson T 1994 Black hole entropy and induced gravity *Preprint* gr-qc/9404039
[14] Jaynes E T 1957 Information Theory and Statistical Mechanics *Phys. Rev.* **106** 620
[15] Rovelli C 2004 *Quantum Gravity* (Cambridge: Cambridge University Press)

Reasoning in Quantum Theory: Modus Ponens and the co-event interpretation

Petros Wallden

1. University of Athens, Physics Department, Nuclear & Particle Physics Section, Panepistimiopolis 157-71, Ilissia Athens, Greece
2. Technological Educational Institute of Chalkida, Department of Technological Applications, Psahna-34400, Greece

E-mail: petros.wallden@gmail.com, pwallden@phys.uoa.gr

Abstract.
Classical logic does not apply in quantum theory. However one would like to be able to reason according to some standard rules, for quantum systems as well (whether they are an electron or the universe itself). We examine what exactly is needed in order to be able to construct deductive arguments, and in particular to maintain the "Modus Ponens", which is the basic deductive rule of inference. It turns out that this requirement restricts the possible theories and in the context of the coevent interpretation, it results uniquely to the "multiplicative-scheme".

1. Introductory Remarks

Applying logic to (classical) physics, is something that is done so frequently and easily, that sometimes, one does not even notice it. Every time we wish to speak about the properties that a system has, we assign to those properties the truth value "True", and we expect the truth values we assign to behave in a Boolean way, and allow us to make deductions, using rules of inference such as the "Modus Ponens" (see below) and proofs by contradiction. In this contribution we will examine, what changes if one wishes to do the same in quantum theory, and what restrictions to a "quantum logic" we need to impose, in order to maintain the arguably vital ability to reason. In Section 2 we will see the general structure of applying logic to physics, in Section 3 we will apply this to classical physics and reason with Boolean logic. In Section 4 we see why this picture fails in quantum theory, and in Section 5 we deduce that requiring to have deductive reasoning requires, multiplicative maps which is central result of this contribution and was first proven by Fay Dowker and the author in [1]. In Section 6 we see what is implied by all this, in the co-event interpretation of quantum theory, which also serves as an extra motivation, for adapting this viewpoint. Finally we summarize and conclude in Section 7.

Journal of Physics: Conference Series **306** (2011) 012044 doi:10.1088/1742-6596/306/1/012044

2. Logic and Physics - General Structure

There are three basic ingredients, when one wishes to use logic to assign truth values corresponding to physical questions.

(1) The first important structure, is the space of all finest grained descriptions Ω. In classical physics it is also the space of potential realities (as we will se below), however this picture does not carry over in quantum theory[1]. In probability theory, it is called sample space, while in histories formulation, is called *history space*. Each element of this space $h \in \Omega$, corresponds to a full description of the system, specifying every detail and property (for example, a fine grained history gives the exact position of the system along with the specification of any internal degree of freedom, for every moment of time). Along with Ω, we define the space $\mathcal{U}$, which consists of the collection of subsets $A \subseteq \Omega$ and the Boolean algebra associated with them (called events algebra), where the addition is defined as the symmetric difference between subsets (see Figure 1) $A+B := A \triangle B$ and the multiplication given by the intersection of subsets $A \cdot B := A \cap B$. Each subset $A \subseteq \Omega$ is called *event*, since all physical questions that one can ask, correspond to one of those subsets. If, for example, one wants to ask "was the system at the region Δ at time t?", it simply corresponds to the subset A defined as $\{A : h_i \in A \text{ iff } h_i(t) \in \Delta\}$, i.e. all histories that at time t are at the region Δ.

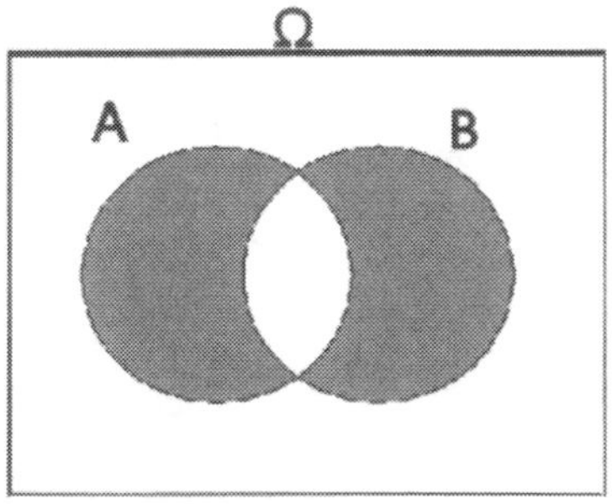

Figure 1. Symmetric Difference $A \triangle B$

(2) The second structure needed, is the space of truth values (for which we will use the notation $\mathcal{T}$), and the algebra associated with them. In the case of classical physics, the truth values are simply the two elements set $\mathcal{T} := \{\text{True, False}\}$ (or simply $\{1, 0\}$), and the Boolean algebra[2] associated (where $1 + 1 := 0, 0 + 0 := 0, 1 + 0 := 1, 0 \cdot 1 := 0, 1 \cdot 1 := 1, 0 \cdot 0 := 0$). As we will mention later, one could use as truth values a more general algebra, as for example a Heyting algebra.

(3) The third and final structure, is the space of possible *truth valuations maps* ($\phi : \mathcal{U} \to \mathcal{T}$). This corresponds, to an assignment of a truth value, to each of the possible questions (events). We shall use the notation $\mathcal{M}$ for this space.

However, the above is too general, to allow us to reason. In this description, there is no relation between the structure of $\mathcal{U}$ and that of the truth values assigned by the map ϕ. In particular, we would wish that the truth valuation map, respects the structure of the events algebra $\mathcal{U}$. For example, we wish to make arguments like: Statement 1: "All human, have two hands" , Statement 2: "Plato was human", therefore we deduce that "Deduction: Plato had

[1] We will see that reality can be thought of as a suitable co-event!

[2] called the truth values algebra

Journal of Physics: Conference Series **306** (2011) 012044 doi:10.1088/1742-6596/306/1/012044

two hands". This follows from the fact that the event "all human, have two hands = A", is subset of the event "Plato having two hands = B", and if $\phi(A) = 1$ then to respect the events algebra, the valuation map ϕ needs to give truth value for the event B ($\phi(B) = 1$). In the subsequent section, we will formalize the above.

3. Classical Physics & Reasoning

To fully respect the events algebra, we restrict the allowed truth valuation maps ϕ to maps that are homomorphisms between the events algebra $\mathcal{U}$ and the truth values algebra $\mathcal{T}$ in other words we require the map ϕ to obey:

(a) Multiplicativity

$$\phi(A \cdot B) = \phi(A \cap B) = \phi(A)\phi(B) \tag{1}$$

(b) Additivity (Linearity)

$$\phi(A + B) = \phi(A \triangle B) = \phi(A) + \phi(B) \tag{2}$$

These requirements, are fulfilled in classical physics. An important observation, is that maps ϕ that are homomorphisms, are in a one-to-one correspondence with single elements h of the space of potential realities Ω. One defines maps that are the characteristic function for a particular history h, i.e.

$$\begin{aligned} \phi_h(A) &= 1 \text{ if } h \in A \\ \phi_h(A) &= 0 \text{ otherwise} \end{aligned} \tag{3}$$

One can see easily that such maps are indeed homomorphisms between the Boolean event algebra and the Boolean truth values algebra. Moreover, *all* homomorphisms are of this type one for each of the (single) elements h of Ω. In classical physics, what is realized (in other words what is real) is simply one single element h of the space Ω (e.g. a particular history of the system). Due to the one-to-one correspondence of homomorphic maps and single elements h, we could equally well, assume that reality is a homomorphic map ϕ_h between the event algebra $\mathcal{U}$ and the truth values algebra $\mathcal{T}$.

However, in order to reason, one needs rules of inference. We define $A \to B$ to be the *material implication* (or simply implication). It means "if event A then also event B". We notice, that if A is False, implication ($A \to B$) is trivially True. In the case that A is True, then the implication is also True, only when the reality lies at the part where $A \cap B$, as it can be seen at the Venn diagram (see Figure 2).

The most important rule of inference, that is needed in order to construct deductive proofs, is the "Modus Ponens" (MP) which is the short version of the Latin "Modus Ponendo Ponens" or in English "the way that affirms by affirming". Mathematically it is:

- The sentence (event), "if A then B" ($A \to B$) is True.
- The sentence (event) "A" is also True.
- Therefore the sentence (event) "B" is also True.

Its validity, in classical physics, is guaranteed by the fact that ϕ's allowed, are homomorphisms, however as we will see later, this is a sufficient but not a necessary requirement. MP is the basis for any *deductive* proof. For example: Event A "today is Monday", event B "I will go to

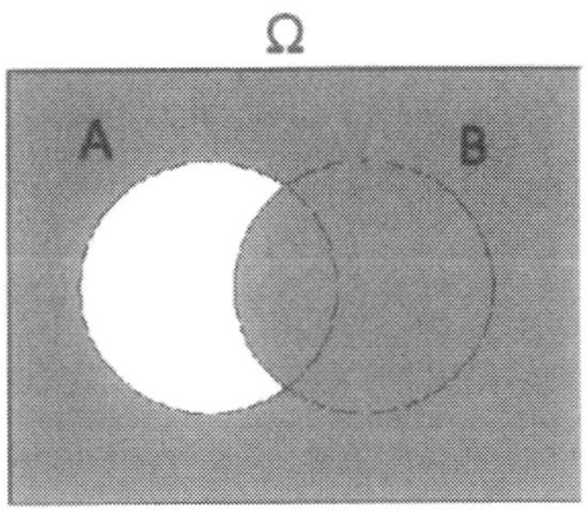

Figure 2. Material Implication $A \rightarrow B$

work". Event $A \rightarrow B$ ("if it is Monday I will go to work") is true. Event A is also true (it is Monday). Then we deduce, that I will go to work (event B going to work, is true).

Other than deductive proofs, one can construct proofs *by contradiction*. There are two types of such proofs, that both hold in classical physics (where allowed ϕ's are homomorphisms) and in that case, they are essentially the same:

(a) Event "A" is True. Its negation, event "$\neg A$", is False.

$$\phi(A) = 1 \Rightarrow \phi(\neg A) = 0 \tag{4}$$

(b) Event "A" is False. Its negation, event "$\neg A$" is True[3].

$$\phi(A) = 0 \Rightarrow \phi(\neg A) = 1 \tag{5}$$

While all this is obvious in classical physics, the above picture and consequently the straightforward way to reason, cannot be carried over to the case of quantum theory and quantum systems.

4. The trouble with Quantum Theory

Until this section we examined the potential realities (histories h or homomorphic maps ϕ), however nothing has been said about the *dynamics* or the *initial conditions* of the system. These are taken into account, by the existence of a measure (quantum measure for quantum systems[4]) μ defined on the space Ω, that gives the probability (or probability amplitude), that a particular event $A \subseteq \Omega$ occur. The least requirement that a map physical ϕ should satisfy is that any event with zero measure, gets the truth value "False", i.e.

$$\mu(A) = 0 \Rightarrow \phi(A) = 0 \tag{6}$$

and the map is called *preclusive*. This very simple requirement, along with requiring that ϕ is homomorphism, cannot be fulfilled in quantum theory. As a consequence of the Kochen

[3] This rule (the second type of contradiction), we shall refer to, as the NNR (No Name Rule) because it has no name and is important in the discussion we will do later.

[4] A quantum measure is a generalization of measure, where we weaken the Kolmogorov additivity condition. The reader is referred to the original paper [2].

and Specker theorem (original in [3] and in view of co-events and histories theories [4, 5]), we know that there exist quantum systems that the full history space Ω can be covered by (overlapping) zero quantum measure sets. A simple example of this arises if one considers a three-slits interference experiment. Consider a point at the screen, where crossing slit A destructively interferes with crossing through slit B and slit B destructively interferes with slit C, i.e. $\mu(\{h_1, h_2\}) = 0, \mu(\{h_2, h_3\}) = 0$ but $\mu(\{h_1, h_3\}) \neq 0$ (see Figure 3).

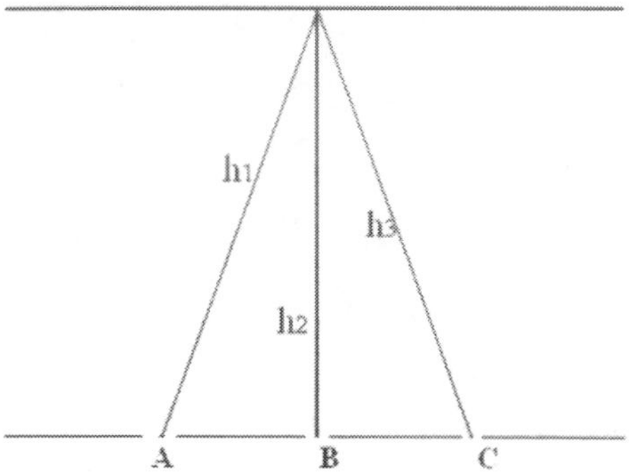

Figure 3. Three slits, h_1 is the history that passes from slit A and hits the central point on the screen, etc

If one requires a map ϕ that is homomorphic, then the above example leads to contradiction.

$$\phi(\{h_1, h_2\} \cap \{h_2, h_3\}) = \phi(\{h_1, h_2\}) \cdot \phi((\{h_2, h_3\}) \text{ implies } \phi(\{h_2\}) = 0 \tag{7}$$

$$\phi(\{h_1, h_2\} \triangle \{h_2, h_3\}) = \phi(\{h_1, h_2\}) + \phi((\{h_2, h_3\}) \text{ implies } \phi(\{h_1, h_3\}) = 0 \tag{8}$$

From Eqs 7 and 8 we get that $\phi(\{h_1, h_2, h_3\}) = 0$. However, since $\mu(\{h_1, h_2, h_3\}) \neq 0$ there are cases that the particle does hit the screen at the point we consider, while the above analysis would rule that out.

We therefore see that it is impossible to maintain the picture of classical physics and Boolean logic. There are three attempts to solve this problem, by changing one of the three structures analyzed in Section 2.

(i) Instead of considering the full set of possible questions the event algebra $\mathcal{U}$ (subsets of Ω), to somehow select a preferred set of classical questions, in other words a subalgebra of the event algebra (consistent/decoherent histories approach [6]). However, this approach (consistent histories) allows an arbitrary choice of which "Boolean subalgebra" to consider (see the critic in [7]) and in this sense, fails to give a satisfactory account of what actually occurs, unless some physical principle is discovered that selects a preferred classical domain.

(ii) Alter the space of truth values (and the algebra associated) $\mathcal{T}$. Instead of using a two values Boolean logic ({True, False}) one could use truth valuation maps that take truth values on a subobject classifier of a category, and the associated algebra is a Heyting algebras (see for example Isham, Butterfield and Doering in [8]). The resulting logic is intuitionistic logic which is deductive logic[5] but most importantly, contextual (the truth value, depends on the context/question asked).

(iii) Finally, one could alter the allowed maps ϕ from $\mathcal{U}$ to $\mathcal{T}$, while keeping both the event algebra and the truth values algebra, some as in classical physics. To do so we weaken the requirement that ϕ is homomorphism. In this line, is the co-event interpretation (see below). This cannot be done arbitrarily, but in a controlled way (keeping some of the structure) in order to maintain the ability to reason in the resulting logic. In the rest paper, we will see what restrictions are required and what they imply.

5. Modus Ponens and Multiplicativity

We therefore wish to weaken the requirement that the map is homomorphism, but in such a way that we are still able to reason. We therefore require that Modus Ponens holds and moreover, we require that the maps ϕ are unital, and from those two we derive restrictions. A map being unital means that $\phi(\Omega) = 1$ or in other words that the event "something happened" gets the answer "True".

It turns out that the following Theorem 1 holds, which will be proven in the rest part of this Section. This result is proven first in [1] by Fay Dowker and the author:

Theorem 1. : *If a valuation map ϕ, respects Modus Ponens, and it is unital, then necessarily it will be a multiplicative map, i.e. $\phi(A \cap B) = \phi(A)\phi(B)$.*

Theorem 2. : *(Filter property) Modus Ponens and unitality imply that "$\phi(A) = 1$ and $A \subseteq B \Rightarrow \phi(B) = 1$, i.e. in words, if event A is true, then any event containing A as subset, is also true.*

Proof. : The proof goes in four steps

(i) We know that: $A \to B = \neg(A \wedge \neg B)$

(ii) When $A \subseteq B$ from step (i) we get $A \to B = \Omega$.

(iii) Due to unitality, $\phi(\Omega) = 1$, and along with step (ii) we get $\phi(A \to B) = 1$

(iv) We can now apply MP (which by assumption holds) and using the fact that $\phi(A) = 1$ and the result of step (iii) deducts that that $\phi(B) = 1$.

□

Lemma 1. : *If $\phi(A) = 0$ for some event A, (and MP and unitality holds), then for* every *event $B \subseteq A$ it holds that $\phi(B) = 0$.*

Proof. If $\exists \quad B \subseteq A$ that $\phi(B) = 1$, then by Theorem 2 $\phi(A) = 1$, but by assumption $\phi(A) = 0$ and thus $\forall \quad B \subseteq A$, $\phi(B) = 0$. □

Theorem 3. : *If $\phi(A) = \phi(B) = 1$ and MP and unitality holds, then $\phi(A \cap B) = 1$.*

Proof. : The proof goes in three steps

(i) $B \subseteq (A \to B) = (A \to A \cap B)$

(ii) $\phi(B) = 1$ and thus by Theorem 2 and step (i): $\phi(A \to A \cap B) = 1$

(iii) $\phi(A) = 1$ and $\phi(A \to A \cap B) = 1$ so by MP (which by assumption holds) we get $\phi(A \cap B) = 1$

[5] Here again, as in the co-event interpretation, proofs by contradiction are not allowed, and only constructive/deductive proofs are acceptable.

□

From Lemma 1 and Theorem 3, Theorem 1 follows:

- If $\phi(A) = 0$ then by Lemma 1, $\phi(A \cap B) = 0$, and thus $\phi(A \cap B) = \phi(A)\phi(B)$. The same holds if $\phi(B) = 0$
- If $\phi(A) = \phi(B) = 1$ then by Theorem 3, $\phi(A \cap B) = 1$ and therefore in all cases $\phi(A \cap B) = \phi(A)\phi(B)$, which proves Theorem 1.

Theorem 1 is the key result, and we note that the additivity condition, Eq. (2), was not needed at all. Additivity is not necessary requirement for maps ϕ in order to have deductive proofs. However, to make proofs by contradiction something extra is needed, either additivity of the maps or the rule of Eq. (5). We do not expect to have proofs by contradiction in quantum theory. Even more interestingly, it seems that MP and multiplicativity are closely linked, since the (essentially) converse of Theorem 1 holds as well (see [1]):

Theorem 4. *: A valuation map ϕ that* is *multiplicative, obeys necessarily the Modus Ponens.*

Proof. We have $\phi(A)\phi(B) = \phi(A \cap B)$ and thus if any two of the three $\phi(A), \phi(B), \phi(A \cap B)$ are equal to 1, so is the third. To prove MP, let us assume that $\phi(A) = \phi(A \to B) = 1$. We get $\phi(A \cap (A \to B) \equiv A \cap B) = 1$. But now we have $\phi(A) = \phi(A \cap B) = 1$ and we are lead to the conclusion that $\phi(B)$ is also equal to 1, which completes the proof. □

6. The co-event Interpretation

It is an alternative interpretation of quantum theory, that was initiated and mainly developed by R. Sorkin [9, 10] , but also in [4, 5, 11] where one can find, other than developments of the interpretation, particular examples worked out. It is also called "Piombino Interpretation" because it was first presented in a talk by Sorkin at a D.I.C.E. meeting in Piombino 2006 (also known as anhomomorphic logic due to the anhomomorphic maps allowed). It is a histories theory and can be viewed as a development of the decoherent histories approach, though having many differences. The space Ω, is the space of possible histories h_i of the system, and a quantum measure μ is defined on it [2]. The quantum measure encodes the full dynamics and initial condition, and is related to standard quantum theory via the use of the decoherence functional. Potential realities are *co-events.* A co-event is defined to be a map ϕ from $\mathcal{U}$ to $\mathcal{T} = \{T, F\}$ obeying certain conditions. As analyzed earlier, maps cannot be homomorphic in general, if one wishes to describe quantum systems, and is unwilling to either restrict the possible questions or alter the space of truth values. Since, reasoning in the resulting theory is important, the requirement that the co-events are multiplicative maps, seems natural in light of the result of the previous section[6]. A multiplicative map (for finite Ω), corresponds to a characteristic function of some event $A \subseteq \Omega$ i.e.

$$\phi_A(B) = 1 \text{ iff } A \subseteq B \tag{9}$$

where A is called the support of the co-event ϕ_A. If A is a singleton (a single history event), then we are back to classical physics and the map is necessarily homomorphism. If it is not

[6] However, this was not the way it developed historically. Different relaxations of homomorphism, such as the additive-scheme was first considered. The multiplicative scheme, emerged as the main option, when other schemes contradicted some thought experiments with multiple-slits. Properties such as the unique classical domain and filter property, gave more arguments, while the present contribution, serves as one more very important reason to adopt the multiplicative scheme.

a singleton, then we are lead to some "paradoxes" stemming from anhomomorphisms. They appear when one asks a question B that intersects non-trivially both A and $\neg A$ and we get that both $\phi_A(B) = 0$ and $\phi_A(\neg B) = 0$. Due to this, it is apparent that we cannot use proofs by contradiction (at least no those that depend on Eq. (5))[7]. In principle, every event $A \in \mathcal{U}$ defines a multiplicative map, corresponding to Eq. (9). However, we impose further conditions for the physical co-events.

First, we require the possible maps ϕ to be preclusive, i.e. they have to obey Eq. (6). It is this point, that we (first) use the quantum measure, and thus dynamics enter the picture. Further, we would like to have co-events that are as close as possible (but not more) to classical co-events (singletons). The mathematical requirement for this, is that co-events are *primitive*. We say that a multiplicative co-event ψ dominates a multiplicative co-event ϕ if

$$\phi(A) = 1 \Rightarrow \psi(A) = 1 \quad \forall \quad A \in \mathcal{U} \tag{10}$$

A preclusive multiplicative co-event is *primitive* if it is not dominated by any other preclusive multiplicative co-event.

To sum up, possible realities, in the co-event interpretation, are maps ϕ from $\mathcal{U}$ to $\mathcal{T}$ called co-events, that are (i) multiplicative, (ii) preclusive and (iii) primitive. These are the conceivable realities. Remember, that in classical physics, where maps are homomorphisms, reality can be viewed as one single history h or in a dual manner, as the valuation (homomorphic) map ϕ_h. In the quantum case, we can still maintain this dual picture, where reality is either a multiplicative co-event ϕ_A (preclusive and primitive) or the event (coarse grained history) A. We now make few observations:

- In order to deal with probabilistic predictions, (where one assigns some chance of one of the co-events to be realized), we need to use once again the quantum measure μ and resort to the application of the " (weak) Cournot principle". The reader is referred to [11] for further details.
- An other very pleasing feature of the co-event interpretation, is that following the realist view point, that one co-event is actually realized, we can recover a unique, unambiguous, classical domain[8], which is given by the "Principle Classical Partition" (see Appendix A of [11]).
- Finally, we note that the resulting logic, which contains anhomomorphisms, depends on the set of preclusive sets, and thus on dynamics. It is justified, therefore, to say (in a bit provocative way) that we have a *dynamical logic*.

In the previous Sections of this contribution, we developed a full argument, why one would require for the valuation maps ϕ to be multiplicative if she/he wishes to be able to reason (Theorem 1). In this light, we see that choosing the physical co-events to be multiplicative maps, is not only well motivated, but we see that *any* other choice (different scheme), would not be suitable for applying deductive arguments (and use Modus Ponens).

7. Summary and Conclusions

We reviewed the use of logic in physics, which uses three structures: $\mathcal{U}$ the events algebra, $\mathcal{T}$ the truth values algebra and ϕ the maps between these algebras. In classical physics, the truth

[7] In this logic, we still have Eq.(4) holding and thus we can make this type of proofs by contradiction.

[8] c.f. in decoherent histories approach, the existence of many incompatible decoherent sets, introduced ambiguity in the interpretation of the approach and possibly, contextuallity.

valuation maps ϕ are homomorphisms between the algebras, and are into direct correspondence with fine grained histories h of Ω.

In quantum theory the above picture fails essentially due to interference, with more striking examples coming from three-slit example and Kochen Specker theorem. To deal with this, one has to change (at least) one of the above three structures described in Section 2. The present contribution, examined the case where one abandons homomorphisms as the only allowed maps.

In the main result of this contribution, we discovered the close link of Modus Ponens, and therefore of deductive logic, with the requirement that the maps are multiplicative (and unital). In particular, the additivity condition, was shown to be not necessary for having deductive logic (Theroem 1). This is an interesting result, on its own right, for logicians, and not directly tied to quantum theory.

Finally, we briefly introduced the co-event interpretation of quantum theory, and in the light of our previous result of Theorem 1 along with its converse, Theorem 4, justified why one selects as potential realities, maps/co-events that are multiplicative.

Acknowledgments

The author thanks Fay Dowker, for collaboration in [1] where Theorem 1 (and Theorem 4) are proven and would like to thank Yousef Ghazi-Tabatabai and Rafael Sorkin for discussions in earlier stages of the work. The author acknowledges I.K.Y. (State Scholarship Foundation) for postdoctoral scholarship during which this work was carried over.

References

[1] Dowker F and Wallden P *Modus Ponens and the Interpretation of Quantum Mechanics*, in preparation.
[2] Sorkin R D 1994 *Mod. Phys. Lett. A* **9**, 3119
[3] Kochen S and Specker E 1967 *J. Math. Mech.* **17**, 59
[4] Dowker F and Ghazi-Tabatabai Y 2008 *J. Phys. A* **41**, 105301
[5] Surya S and Wallden P 2010*Found. Phys.* **40**, 585
[6] Griffiths R B 1984 *J. Stat. Phys.* **36** 219; R. Omnès. 1988 *J. Stat. Phys.* **53**, 893; M. Gell-Mann and J. Hartle 1990, in *Complexity, Entropy and the Physics of Information, SFI Studies in the Science of Complexity, Vol. VIII*, edited by W. Zurek, (Addison-Wesley, Reading, 1990); M. Gell-Mann and J. Hartle. 1993 *Phys. Rev. D* **47**, 3345
[7] Dowker F and Kent K 1995 *Phys. Rev. Lett.* **75**, 3038; F. Dowker and A. Kent 1996 *J. Statist. Phys.* **82**, 1575
[8] Isham C J and Butterfield J 2002 *Int. J. Theor. Phys.* **41**, 613 and previous papers; Doering A and Isham C J 2008 *J. Math. Phys.* **49**, 053518 and previous papers.
[9] Sorkin R D 2007 *J. Phys. A* **40**, 3207; Sorkin R D 2007 *J. Phys. Conf. Ser.* **67**, 012018
[10] Sorkin R D, to appear in Ellis G F R, Murugan J and Weltman A (eds) *Foundations of Space and Time*, Cambridge University Press, e-preprint ArXiv:1004.1226 (quant-ph).
[11] Ghazi-Tabatabai Y and Wallden P 2009 *J. Phys. A: Math. Theor.* **42**, 235303; Ghazi-Tabatabai Y and Wallden P 2009 *J. Phys.: Conf. Ser.* **174**, 012054

Quantum measurement and the quantum to classical transition in a non-linear quantum oscillator

M. J. Everitt[1,2], W. J. Munro[3] and T. P. Spiller[4]

[1] Department of Physics, Loughborough University, Loughborough, Leics LE11 3TU, United Kingdom
[2] Centre for Theoretical Physics, The British University in Egypt, Postal No. 11837, P.O. Box 43, Egypt
[3] NTT Basic Research Laboratories, NTT Corporation, 3-1 Morinosato-Wakamiya, Atsugi, Kanagawa 243-0198, Japan
[4] Quantum Information Science, School of Physics and Astronomy, University of Leeds, Leeds LS2 9JT, UK

E-mail: m.j.everitt@physics.org

Abstract. We study a non-linear quantum mechanical oscillator, acting as a measurement device. Candidate systems for realising such apparatus range from superconducting devices through to nano-mechanical resonators. The measurement device comprises an oscillator circuit where the dynamics of expectation values, in its correspondence limit, are either chaotic-like or periodic depending on the measured state of the quantum object – in this case a qubit. In a previous work we showed how the classical like trajectories of such a quantum system can act as a model of a projective measurement process. Here we investigate the quantum to classical transition of the measurement device and postulate criteria for realisation of an effective implementation of such a device.

1. Introduction

The measurement problem has existed in quantum mechanics since Born's paper in 1926 [1] until the present day (see, for example, [2, 3, 4, 5, 6, 7, 8, 9, 10, 11, 12, 13, 14, 15]). Even now discussions, that can become quite lively, frequently occur on the metaphysical aspects of measurement. With the emergence of quantum technologies and specific applications such as feedback and control (see, for example, [16]) a more pragmatic approach is perhaps timely. As a detailed exploration of what might be achieved through modelling correspondence limit measurement devices simply in terms of Schrödinger evolution has yet to be undertaken, many of these metaphysical discussions could be considered somewhat pre-emptive. In other words, it is not yet clear if the process of measurement could be fully represented on a subset of the current axioms of quantum mechanics - i.e. without postulates relating directly to measurement such as the projection hypothesis.

Within the foundations theme of the DICE conference series, it is worth reflecting on the view that any attempt to identify a canonical set of axioms for quantum theory from a top down approach is confused by the very many different ways quantum mechanics can be motivated - from Schrödinger wave functions to Feynman path integrals. The method followed in this paper, as in [17], could be considered as tentative first steps taking the opposite, bottom up approach.

In other words, we seek to understand if it is possible to reproduce the effects associated with a subset of the standard axioms of quantum mechanics (in this case those relating to measurement) from other axioms. If this turns out to be possible, then we may reclassify once considered axioms as corollaries and better understand their origin and applicability in terms of a reduced and therefore simplified framework. In this way it may be possible to reduce the number of axioms in quantum theory and therefore remove some restrictions on those, such as the quantum gravity community, seeking to extend the theory.

In this work we illustrate how a classical record of a quantum measurement can arise and consider a mechanism for the realisation of effective projective measurements simply from such a perspective. The system dynamics are modified to allow for, and indeed operationally rely on, the effects of environmental degrees of freedom. Specifically, our model comprises an open quantum mechanical measurement device, which is arranged to operate in its classical-correspondence limit, coupled to a quantum system – in this case a qubit – whose state it measures through the natural emergence of a classical dynamical record.

In [17] we demonstrated that our proposed system could be considered to fully quantum mechanically model a projective measurement process. We studied both the ensemble and individual trajectories of the system using the master equation and quantum state diffusion [18] respectively. We showed that the oscillator, operating in its correspondence limit, develops dynamics in terms of its expectation values that are either chaotic-like or periodic depending on the measured value and projected state of a qubit. In our model no preferred basis is assumed to exist *a priori* – this simply emerges from the form of the coupling mechanism. Furthermore, our results were consistent with the Born rule and the Zeno effect. However, we did not demonstrate that our model necessarily requires the oscillators classical limit - an omission that we address in this paper.

2. The Model

In our model the (quantum) measurement device comprises a quantum oscillator that has a well defined correspondence limit and whose classical like dynamics is manifestly different depending on the quantum state of the other component of the system – in this case a qubit. The Hamiltonian takes the form [17]:

$$H = \frac{3}{4}p^2 + \frac{\beta^2}{4}q^4 - \frac{1}{4}q^2 + \frac{g}{\beta}\cos(t)q - H_{\text{int}}\ . \tag{1}$$

Here the interaction term

$$H_{\text{int}} = \frac{1}{4}\left(p^2 + q^2\right)\sigma_z = \frac{1}{2}\left(a^\dagger a + \frac{1}{2}\right)\sigma_z \tag{2}$$

represents a dispersive Jaymes Cummings coupling to a degenerate qubit. Here q and p are dimensionless position and conjugate momentum for the oscillator mode (i.e. defined such that the annihilation operator $a = (q+ip)/\sqrt{2}$) and σ_z is the usual Pauli operator for the qubit. The dimensionless parameter β represents a scaling of the classical action of the oscillator with respect to a Planck cell where, subject to appropriate decoherence effects, $\beta = 1.0$ is the quantum limit and the classical limit is achieved as β becomes vanishingly small (see [18, 19] for a comprehensive explanation of this quantum to classical transition). Finally, we note $g = 0.3$ is the strength of the applied driving term. We further note that the Hamiltonian is normalised by a characteristic oscillator energy $\hbar\omega$. Furthermore, time t is similarly rendered dimensionless with respect to the same characteristic frequency ω.

In order to gain an intuitive idea of the *modus operandi* of our model measurement device we present in Fig. 1 the (non-linear) potential of Eq. (1) as a dotted line. Invoking a hand-waving

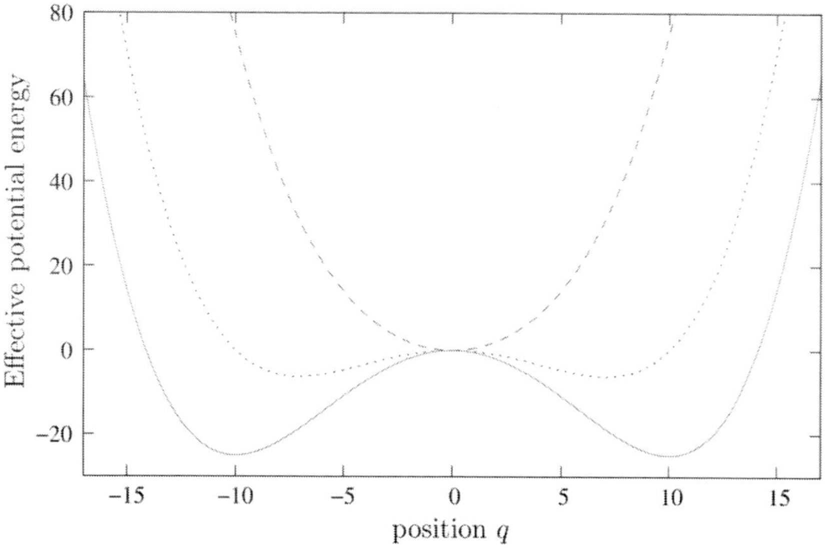

Figure 1. (colour online) Effective potential for the Hamiltonian of Eq. (1) where $\beta = 0.1$ and (a) $\langle\sigma_z\rangle = 1$ solid (red) line (b) $\langle\sigma_z\rangle = -1$ dashed (blue) line and (c) $\langle\sigma_z\rangle = 0$ dotted line (Figure and caption reproduced from [17]).

Born-Oppenheimer type approximation one may associate with the dashed and solid lines of Fig. 1 an effective oscillator potential associated with the ground state $|g\rangle$ ($\langle\sigma_z\rangle = -1$) or the excited state $|e\rangle$ ($\langle\sigma_z\rangle = 1$) respectively. That is, with the ground state of the qubit we associate a quartic potential in the oscillator and with the excited state we associate the well known potential of the Duffing or anharmonic oscillator. Here we have chosen $\beta = 0.1$ (see Eq. (1) which, for the double well potential, has been shown to produce dynamics in sufficiently good agreement the system's classical counterpart [18, 19].

In [17] we considered both the evolution of the ensemble average as well as individual quantum trajectories. For the purpose of this work the former will be sufficient. Hence, we model the effects associated with coupling the measurement device to an environment using a Lindblad [20] type master equation in the Markovian limit at zero temperature. The dynamics of this (reduced) density operator is given by:

$$\dot{\rho} = -i\,[H,\rho] + \sum_m \left[L_m \rho L_m^\dagger - \frac{1}{2} L_m^\dagger L_m \rho - \frac{1}{2} \rho L_m^\dagger L_m \right]. \tag{3}$$

In this work we assume only Ohmic type damping of the measurement device at zero temperature and hence have only one Lindblad operator $L = \sqrt{2\Gamma}a$ where a is the oscillator annihilation operator and $\Gamma = 0.125$.

In order to understand the dynamical evolution of the measurement device we will make use of its Wigner function representation. For a detailed discussion of this and other phase space methods in quantum systems see, for example, [21]. The Wigner function takes the form of a pseudo probability density function in the (q,p) phase space and can be defined, for the field, by:

$$W(q,p) = \frac{1}{2\pi} \int d\zeta \left\langle q + \frac{\zeta}{2} \middle| \mathrm{Tr}_{\mathrm{qubit}}\left[\rho(t)\right] \middle| q - \frac{\zeta}{2} \right\rangle \exp\left(-i\zeta p\right).$$

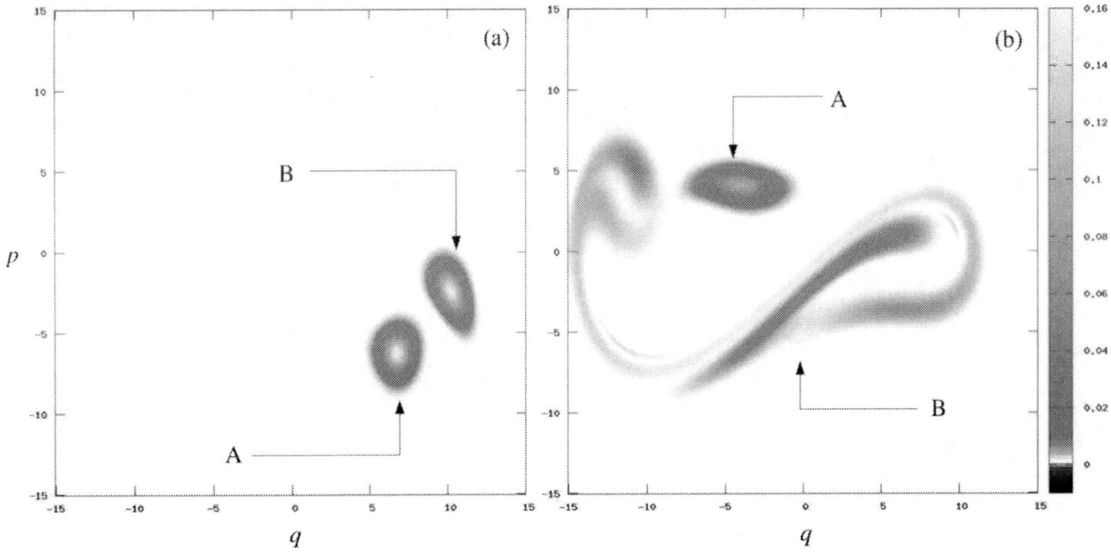

Figure 2. (colour online) Wigner function of the field for solutions to the master equation where the initial density operator corresponds to the pure state $\frac{1}{\sqrt{2}}(|g\rangle + |e\rangle) \otimes |\alpha \approx 6.8\rangle$. (a) at $t/2\pi = 0.1$ and (b) at $t/2\pi = 2.25$. Here region A and B corresponds to measuring the qubit in state $|g\rangle$ and $|e\rangle$ respectively (Figure and caption reproduced from [17]).

Where $\text{Tr}_{\text{qubit}}\rho(t)$ indicates a partial trace over the qubit space yielding a reduced density operator for the oscillator only. In Fig. 2 we show two snapshots of this Wigner function where, as stated in [17], "(a) is taken very early on in the system evolution at $t/2\pi = 0.1$. Here we see that the effect of the qubit on the measurement device is to split the coherent state up into two coherent state-like lumps. This is not surprising, as a very similar effect is seen in collapse and revival phenomena of the Jaynes-Cummings model of a qubit interacting with a harmonic oscillator. However, as the system evolves we see in Fig. 2(b) ($t/2\pi = 2.25$) the distribution becomes somewhat more interesting. One lump is still quite small and is associated with the qubit's ground state. The second is much larger, arising from the chaotic-like behaviour of the Duffing oscillator and is associated with the qubit's excited state.". In [17] we went on, amongst other things, to consider the qubits dynamics and demonstrate through an analysis of various entropic quantities that each of these lumps does indeed correspond to a measurement of one or the other state of the qubit.

3. Results

The quantum to classical transition of the Duffing oscillator has been studied in a great deal of detail. This transition was first clearly presented by Brun, Percival and Schack in [19]. In this work the authors demonstrated, using a quantum state diffusion approach, that the oscillator's Poincaré section begins to be clearly resolved for values of $\beta = 0.1$ (and becoming excellent by $\beta = 0.01$). It is for this reason, together with computational constrains, that we chose $\beta = 0.1$ in [17]. We now demonstrate that our model fails to achieve a measurement in the oscillator's quantum limit – i.e. its dynamics are no longer useful in clearly distinguishing if the qubit is in $|g\rangle$ or $|e\rangle$. Our model is therefore valid in, and only in, the device's classical limit. We will show this, as in Fig. 2, through examining the oscillator's Wigner function for different values of β. In Fig. 3 we show Wigner functions for the reduced density operator of the oscillator for $\beta = 0.1, 0.25, 0.5$ and 1.0. We note that, unlike in Fig. 2, in Fig. 3 we have allowed the dynamics

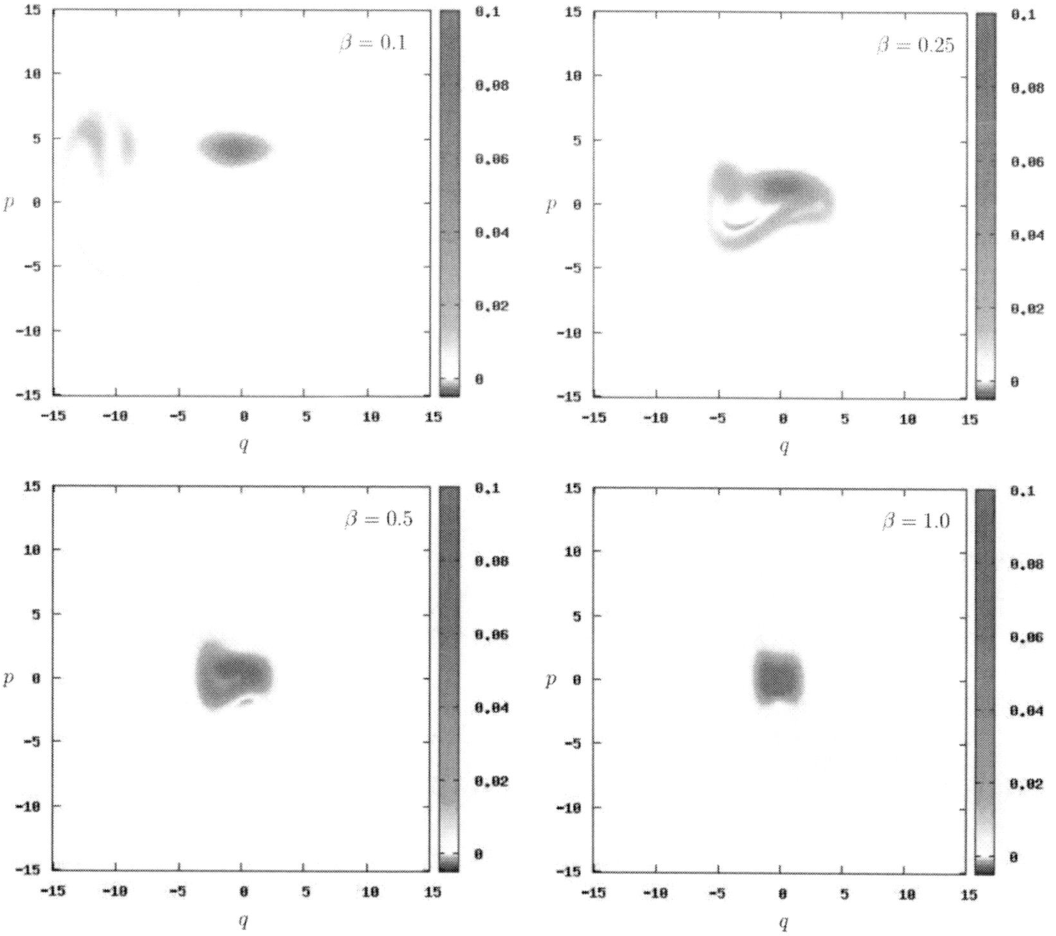

Figure 3. Wigner function of the field for solutions to the master equation where the initial density operator corresponds to the pure state $\frac{1}{\sqrt{2}}(|g\rangle + |e\rangle) \otimes |\alpha \approx 6.8\rangle$ at $t/2\pi = 3.4$ for four differt values of the scaling parameter β. Here we see that the regions corresponding to measuring the qubit in state $|g\rangle$ and $|e\rangle$ merge as $\beta \to 1$.

to continue until $t/2\pi = 3.4$ so as to enable the effects of decoherence to become more clearly manifest. We note that we have plotted each of these Wigner functions on the same scale to make clear the effect of β on the action.

In Fig. 3 for $\beta = 0.1$ we see that the Wigner function has evolved into two disjoint lumps. This would imply that, except possibly on a set of measure zero, it should be possible even in individual experiments to uniquely determine the state of the qubit. Indeed, such behaviour was observed in [17]. For $\beta = 0.25$ we see that, while these lumps are still identifiable, they have begun to merge. In terms of individual measurements, taken by examining the oscillator's phase space at a single point in time for $\beta = 0.25$, occasional errors would be therefore made in identifying the qubit's state. Nevertheless, it would still be possible to distinguish periodic from chaotic motion over an interval of time and therefore build up a measurement record. For $\beta = 0.5$ it appears that not even this would be possible as the structure becomes increasingly obscured until at $\beta = 1.0$ no discrimination whatsoever can be made.

Our above observations lead us to postulate criteria for the realisation of an instantaneous measurement for a device of this kind. That is, in order to perform a reliable measurement at a specific time the Wigner function representing the ensemble average of the measurement device must have separated into two disjoint and clearly identifiable regions in phase space. We note that due to environmental decoherence the state associated with individual trajectories will be localised to approximately one Planck cell. Hence, our condition could be made more strict by also requiring that the space between the disjoint lumps should be at least of the order of a Planck cell or more.

In constructing the above criteria we have based our observations on a somewhat idealised setup. In general it will not always be possible to realise dynamics of this kind. For example, if we consider a non-degenerate qubit being measured in an incompatible basis, this will lead to more complicated dynamics - which will be reflected in the device's Wigner function and our above criteria is unlikely to be effective. Nevertheless, while it may not be possible to produce an instantaneous measurement under such conditions, it may be possible to build up a measurement record by examining the oscillator's dynamics over an interval of time. In [17] we showed that our model measurement apparatus in its correspondence limit could indeed track the state of a qubit under such circumstances. However, a detailed investigation of the quantum to classical transition in this more complex situation is beyond the scope of this paper.

4. Conclusion

In [17] we demonstrated (i) a fully quantum mechanical model of a projective measurement process where the dynamics of expectation values, in the correspondence limit, are either chaotic-like or periodic depending on the measured value and projected state of a qubit. (ii) The preferred basis emerged from the coupling between the measurement device and the quantum object and was not assumed to exist *a priori*. (iii) Individual classical-like trajectories of an open quantum system can act as a record of the measurement of an individual qubit. (iv) Both the Born rule and Zeno effect arise as natural consequences of the systems dynamics.

In this paper we have further extended our analysis and have proposed, based on the Wigner function phase space representation of the measurement device, criteria for the effective realisation of pointer states for instantaneous measurement.

Acknowledgments

We thank the British University in Egypt's High Performance Computing Laboratory in conjunction with the Centre of Theoretical Physics for the use of their facilities.

References

[1] Born M 1926 *Z. Physik* **37** 863–867
[2] Wheeler J A and Zurek W H (eds) 1983 *Quantum Theory and Measurement* (Princeton, NJ: Princeton University Press)
[3] Joos E and Zeh H D 1985 *Z. Phys.* B **59** 223
[4] Zurek W H 1991 *Phys. Today* **44** 36
[5] Bell J S 1990 *Phys. World* **3** 33
[6] Schlosshauer M 2004 *Rev. Mod. Phys.* **76** 1267
[7] Bhattacharya T, Habib S and Jacobs K 2003 *Phys. Rev.* A **67** 042103
[8] Jacobs K and Steck D 2006 *Contemporary Physics* **47** 279–303
[9] Leggett A J 2008 *Rep. Progr. Phys.* **71** 022001
[10] Wiseman H M 1996 *Quant. Semiclass. Opt.* **8** 205–222
[11] Bhattacharya T, Habib S and Jacobs K 2000 *Phys. Rev.* Lett **85** 4852–4855
[12] Greenbaum B D, Habib S, Shizume K and Sundaram B 2007 *Phys. Rev.* E **76** 046215
[13] Penrose R and Marcer P 1998 *Phil. Trans. Royal. Soc.* **356** 1927–1939
[14] Everitt M J, Clark T D, Stiffell P B, Ralph J F, Bulsara A R and Harland C J 2005 *Phys. Rev.* E **72** 066209
[15] Everitt M J 2009 *New. J. Phys.* **11** 013014
[16] Ralph J F, Griffith E J, Clark T D and Everitt M J 2004 *Phys. Rev.* B **70** 214521

[17] Everitt M J, Munro W J and Spiller T P 2010 *Phys. Lett.* A **374** 2809–2815
[18] Percival I C 1998 *Quantum State Diffusion* (Cambridge: Cambridge University Press)
[19] Brun T A, Percival I C and Schack R 1996 *J. Phys.* A **29** 2077
[20] Lindblad G 1976 *Commun. Math. Phys.* **48** 119–130
[21] Schleich W P 2001 *Quantum Optics in Phase Space* (New York: Wiley)

Elements of sub-quantum thermodynamics: quantum motion as ballistic diffusion

G. Grössing, S. Fussy, J. Mesa Pascasio and H. Schwabl

Austrian Institute for Nonlinear Studies, Akademiehof, Friedrichstr. 10, 1010 Vienna, Austria

E-mail: ains@chello.at

Abstract. By modelling quantum systems as emerging from a (classical) sub-quantum thermodynamics, the quantum mechanical "decay of the wave packet" is shown to simply result from sub-quantum diffusion with a specific diffusion coefficient varying in time due to a particle's changing thermal environment. It is thereby proven that free quantum motion strictly equals ballistic diffusion. The exact quantum mechanical trajectory distributions and the velocity field of the Gaussian wave packet are thus derived solely from classical physics. Moreover, also quantum motion in a linear (e.g., gravitational) potential is shown to equal said ballistic diffusion. Quantitative statements on the trajectories' characteristic behaviours are obtained which provide a detailed "micro-causal" explanation in full accordance with momentum conservation.

1. Quantum mechanical dispersion of a free Gaussian wave packet exactly modelled by sub-quantum ballistic diffusion

Considering quantum theory as *emergent* [1], we propose that it may result from a deeper, more exact theory on a sub-quantum level. In our approach, one assumes that the latter can be described with the aid of non-equilibrium thermodynamics. We ask ourselves how quantum theory would have evolved, had the "tool" of modern non-equilibrium thermodynamics existed, say, a century ago. As one of us (G. G.) has recently shown, one can derive the exact Schrödinger equation with said tool, where the relation between energy E and frequency ω is used as the only empirical input, $E = \hbar\omega$ [2, 3], with the additional option that even the appearance of Planck's constant $\hbar$ may have its origin in classical physics [4]. For an extensive review of Refs. [2] and [3], for connections to similar work, and, in particular, to Fisher information techniques, see [1]. As to approaches in a similar spirit, see, for example, [5, 6, 7, 8, 9, 10, 11, 12].

In the thermodynamic approach to quantum behaviour [2, 3, 4], a particle of energy $E = \hbar\omega$ is characterized by an oscillator of angular frequency ω, which itself is a dissipative system maintained in a non-equilibrium steady-state by a permanent troughput of energy, or heat flow, respectively. The latter is a form of kinetic energy different from the "ordinary" kinetic energy of the particle, as it represents an additional, stochastic contribution to it, i.e., from the presence of zero point fluctuations. Therefore, one deals with a context-dependent total energy of the whole system (i.e., the particle as the "system of interest" in a narrower sense and the heat flow constituting the particle's noisy thermal embedding), which is assumed as

$$E_{\text{tot}} = \hbar\omega + \frac{(\delta p)^2}{2m} , \tag{1}$$

where $\delta p := mu$ is said additional, environment-related fluctuating momentum of the particle of mass m.

We let ourselves be inspired by the experiments with the classical "walkers" of Yves Couder's group (see, for example, [13, 14, 15, 16]). One has to do with a rapidly "bouncing" (oscillating) object, which itself is guided by an environment that contributes some fluctuating momentum to the walker's propagation: a bouncing droplet phase-locked with an oscillating fluid surface. However, the walker is also the cause of waves emanating from the particle "hitting" the surface, and it is the detailed structure of the emerging wave configurations, in turn, that stochastically influences the walker's path. Thus, if we imagine the bouncing of a walker in its fluid environment, the latter will become "excited" or "heated up" wherever the momentum fluctuations direct the particle to. After some time span, a whole area of the particle's environment will be coherently "heated up" in this way. We shall here try to implement a similar "particle-wave coupling" with respect to a quantum system.

To start with a first major aspect of our modeling, we consider a particle of initial velocity v in the midst of "well ordered" diffusion waves, which themselves emerge out of the noisy, Brownian-type diffusions of myriads of single sub-quantum particles. Being swept along with a diffusion wave, with initial ($t = 0$) location $x(0)$ and diffusion velocity u, a thus described "quantum" particle's distance to some heat accumulation's center x_0 at time t will be

$$x(t) = x(0) + ut \, , \tag{2}$$

such that one obtains the r.m.s. of (2) as

$$\overline{x^2(t)} = \overline{x^2(0)} + 2\, \overline{x(0) \cdot u(x,t)}\, t + \overline{u^2(x,t)}\, t^2 \, . \tag{3}$$

Now we introduce an essential point: we assume, as an emerging result out of the statistics of a vast number of diffusion processes, the complete statistical independence *at any time* t of the velocities u and the positions x, and thus also of u and v:

$$\overline{xu} = 0 \quad \text{and} \quad \overline{vu} = 0 \, . \tag{4}$$

This is justified considering the statistics of huge numbers, millions of millions of diffusive sub-quantum Brownian motions, which are supposed to bring forth the emergence of said larger-scale collective phenomenon, i.e., the diffusion wave fields as solutions to the heat equation [3]. Therefore, with the *average* orthogonality of classical (convective) momentum mv on one hand, and its associated diffusive momentum mu on the other, one gets rid of the term linear in t in Eq. (3), and thus of irreversibility, and one obtains

$$\overline{x^2(t)} = \overline{x^2(0)} + \overline{u^2}\, t^2 \, . \tag{5}$$

Eq. (5) is the result obtained for the "pure" emergent diffusive motion as given by (2).

Now, let us assume we have a source of identical particles, which are prepared in such a way that each one ideally has an initial (classical) velocity v. Even if we let them emerge one at a time only, say, from an aperture with unsharp edges (thus avoiding diffraction effects to good approximation), the probability density P will be a Gaussian one. This comes along with a heat distribution generated by the oscillating ("bouncing") particle(s), with a maximum at the center of the aperture $x_0 = vt$. As the classical diffusion equation, in one dimension for simplicity, is given by

$$\frac{\partial}{\partial t} P(x,t) = D \frac{\partial^2}{\partial x^2} P(x,t) \, , \tag{6}$$

where D is the "diffusion constant", we have the corresponding solution

$$P(x,t) = \frac{1}{\sqrt{2\pi}\sigma} \mathrm{e}^{-\frac{(x-x_0)^2}{2\sigma^2}} \tag{7}$$

with the usual variance $\sigma^2 = \overline{x^2(t)}$. Note that from Eq. (1) one has for the *averages* over particle positions and fluctuations (as represented via the probability density P)

$$\overline{E_{\mathrm{tot}}} = \overline{\hbar\omega} + \frac{\overline{(\delta p)^2}}{2m} = \text{const.}, \tag{8}$$

with the mean values (generally defined in $n-$dimensional configuration space)

$$\overline{(\delta p)^2} := \int P (\delta p)^2 \, \mathrm{d}^n x \,. \tag{9}$$

Equation (8) is a statement of total average energy conservation. This means that a variation in δp implies a varying "particle energy" $\hbar\omega$, and vice versa, such that each of the summands on the right hand side for itself is not conserved. In fact, as shall be detailed below, there will generally be an exchange of momentum between the two terms providing a net balance

$$m\delta v = m\delta u \,, \tag{10}$$

where δv describes a change in the "convective" velocity v paralleled by the "diffusive" momentum fluctuation $\delta(\delta p) := m\delta u$ in the thermal environment.

As elaborated in references [2] and [3], once Eq. (1) is assumed, considerations based on Boltzmann's relation between action and angular frequency of an oscillator provide, without any further reference to quantum theory, that

$$\delta p = mu := \hbar k_{\mathrm{u}} = -\frac{\hbar}{2}\nabla \ln P \,. \tag{11}$$

Further, as can easily be checked by integration, it holds that

$$\overline{(\nabla \ln P)^2} = -\overline{\nabla^2 \ln P} \,. \tag{12}$$

As in Eq. (8) only the kinetic energy varies, one has $\delta\overline{E_{\mathrm{kin}}}(t) = \delta\overline{E_{\mathrm{kin}}}(0) =$ const. Thus, with the Gaussian (7), this provides an expression for the averaged fluctuating kinetic energy, or heat, of a particle and its surroundings,

$$\begin{aligned} \delta\overline{E_{\mathrm{kin}}(t)} &= \frac{m}{2}\overline{(\delta v)^2} + \frac{m}{2}\overline{u^2} = \frac{m}{2}\overline{(\delta v)^2} + \frac{\hbar^2}{8m\sigma^2} = \\ &= \delta\overline{E_{\mathrm{kin}}(0)} = 0 + \frac{m}{2}\left.\overline{u^2}\right|_{t=0} = \frac{\hbar^2}{8m\sigma_0^2} =: \frac{m}{2}u_0^2 \,. \end{aligned} \tag{13}$$

Moreover, with the diffusion constant

$$D := \frac{\hbar}{2m} \tag{14}$$

Eq. (13) provides an expression for the initial velocity fluctuation

$$u_0 = \frac{D}{\sigma_0} \,. \tag{15}$$

However, in a next step we now take into account the small momentum fluctuations $m\delta u$, providing an altered convective velocity $v \to v + \delta v(t)$, and thus an additional displacement $\delta x = |\delta u|\, t = |\delta v|\, t$, i.e., as soon as $t > 0$. Therefore, returning to the simple rule of Eq. (2), one now must decompose $u(t)$ into its initial value u_0 and a fluctuating contribution $\delta u(t)$, respectively. Unless some thermal equilibrium were reached, the latter is typically given off from the "heated" thermal bath to the particle of velocity v,

$$u(t) = u_0 - \delta u(t) \; . \tag{16}$$

As is detailed in Ref. [17], this leads ultimately to the substitution of u by u_0 in Eq. (5), i.e.,

$$\overline{x^2(t)} = \overline{x^2(0)} + u_0^2 t^2 \; . \tag{17}$$

Inserting (15) into (17) for the particular case that $\overline{x^2(t)} \equiv \sigma^2$ (i.e., $\overline{x^2(0)} \equiv \sigma_0^2$) provides for the time evolution of the wave packet's variance

$$\sigma^2 = \sigma_0^2 \left(1 + \frac{D^2 t^2}{\sigma_0^4}\right) \; . \tag{18}$$

The quadratic time-dependence of the variance σ^2 is remarkable insofar as in ordinary diffusion processes the scenario is different. There, with the Gaussian distribution being a solution of the heat equation, for purely Brownian motion the variance grows only linearly with time, i.e., as described by the familiar relation

$$\overline{x^2(t)} = \overline{x^2(0)} + 2Dt \; . \tag{19}$$

However, as we have seen, the momentum exchange between the particle and its environment is characterized by both a changing velocity and by a changing thermal environment of the particle, i.e., also by a changing diffusivity. Therefore, Eq. (19) must be modified to allow for a time-dependent diffusivity. In other words, we shall have to deal with the field of *anomalous diffusion*. (For a short introduction, see, for example, Ref. [18].) This means that instead of the diffusion constant D in Eq. (6), we now introduce a time-dependent diffusion coefficient $D(t) = kt^\alpha$, where k is a constant factor. The exponent α was derived [19] to be $\alpha = 1$. Therefore, $k = D^2/\sigma_0^2 = u_0^2$, and the time-dependent diffusion coefficient becomes

$$D(t) = u_0^2\, t = \frac{D^2}{\sigma_0^2}\, t = \frac{\hbar^2}{4m^2\sigma_0^2}\, t \; . \tag{20}$$

Note that with the exponent of t being $\alpha = 1$, or the t^2-dependence of σ^2 in (18), respectively, one deals with the special case of anomalous diffusion usually named *ballistic diffusion*. We shall review some general properties of ballistic diffusion at the end of the paper. At this point, however, it is useful to recall that throughout our modelling of sub-quantum processes, we deal with various processes at different time scales. On the shortest scales, we have assumed Brownian-type motions (not detailed here), which, on the next higher level of (spatial and) temporal scales lead collectively to the emergence of a regular diffusion wave. The latter is characterized by a velocity u according to (2), and it is orthogonal on average to the particle's velocity v, thus providing the r.m.s. displacement (5) depending on $u(t)$. As a next step, we have introduced the noisy thermal bath of the particle's environment, i.e., essentially the effect of other diffusion wave configurations, which disturbs the relation (5) by introducing a fluctuating term δu. The net effect of the latter, however, is the r.m.s. displacement (17) with a dependence solely on the initial diffusive velocity u_0. This manifests itself also in the expression for $D(t)$ of the ultimately emerging ballistic diffusion, which is also dependent only on u_0. However, even on this level of ballistic diffusion one can recover the signature of Brownian motion. In fact, if one considers the time-average of $D(t)$ for large enough times $t \gg 1/\omega$, i.e.,

$$\langle D(t) \rangle := \frac{1}{t} \int_0^t D(t')\,\mathrm{d}t' = \frac{u_0^2}{2} t = \frac{D(t)}{2} \,, \tag{21}$$

one immediately obtains the linear-in-time Brownian relation

$$\overline{x^2(t)} = \overline{x^2(0)} + 2\,\langle D(t) \rangle\, t \quad \text{and} \quad \sigma^2 = \sigma_0^2 + 2\,\langle D(t) \rangle\, t \,, \tag{22}$$

which is, however, also in accordance with the t^2–dependence of Eq. (17).

Note that the diffusivity's rate of change is a constant,

$$\frac{\mathrm{d}D(t)}{\mathrm{d}t} = \frac{D^2}{\sigma_0^2} = u_0^2 = \text{const.}, \tag{23}$$

such that it is determined only by the initial r.m.s. distribution σ_0, providing also a re-formulation of Eq. (17) as

$$\overline{x^2(t)} = \overline{x^2(0)} + \frac{\mathrm{d}D(t)}{\mathrm{d}t} t^2 \,. \tag{24}$$

With the square root of (18),

$$\sigma = \sigma_0 \sqrt{1 + \frac{D^2 t^2}{\sigma_0^4}} \,, \tag{25}$$

we note that σ/σ_0 is a spreading ratio for the wave packet independent of x. This functional relationship is thus not only valid for the particular point $x(t) = \sigma(t)$, but for all x of the Gaussian. Therefore, one can generalize (25) for all x, i.e.,

$$x(t) = x(0)\frac{\sigma}{\sigma_0}, \quad \text{where} \quad \frac{\sigma}{\sigma_0} = \sqrt{1 + \frac{D^2 t^2}{\sigma_0^4}} \,. \tag{26}$$

Now we remind ourselves that we deal with a particle of velocity $v = p/m$ immersed in a wave-like thermal bath that permanently provides some momentum fluctuations δp. The latter are reflected in Eq. (25) via the r.m.s. deviation $\sigma(t)$ from the usual classical path. In other words, one has to do with a wave packet with an overall uniform motion given by v, where the position $x_0 = vt$ moves like a free classical particle. As the packet spreads according to Eq. (25), $x(t) = \sigma(t)$ describes the motion of a point of this packet that was initially at $x(0) = \sigma_0$. Depending on whether initially $x(0) > \sigma_0$ or $x(0) < \sigma_0$, then, respectively, said spreading happens faster or slower than that for $x(0) = \sigma_0$. In our picture, this is easy to understand. For a particle exactly at the center x_0 of the packet [$x(0) = 0$], the momentum contributions from the "heated up" environment on average cancel each other for symmetry reasons.

However, the further off a particle is from that center, the stronger this symmetry will be broken, i.e., leading to a position-dependent net acceleration or deceleration, respectively, or, in effect, to the "decay of the wave packet". Moreover, also the appearance of the time-dependent diffusivity $D(t)$ is straightforward in our model. Essentially, the "decay of the wave packet" simply results from sub-quantum diffusion with a diffusivity varying in time due to the particle's changing thermal environment: as the heat initially concentrated in a narrow spatial domain gets gradually dispersed, so must the diffusivity of the medium change accordingly.

Finally, one obtains with Eqs. (26) and (14) for the "smoothed out" *trajectories* (i.e., those averaged over a very large number of Brownian motions)

$$x_{\mathrm{tot}}(t) = vt + x(t) = vt + x(0)\frac{\sigma}{\sigma_0} = vt + x(0)\sqrt{1 + \frac{\hbar^2 t^2}{4m^2\sigma_0^4}} \,. \tag{27}$$

Also, one can now calculate the *average total velocity of a Gaussian wave packet*,

$$v_{\mathrm{tot}}(t) = \frac{\mathrm{d}x_{\mathrm{tot}}(t)}{\mathrm{d}t} = v(t) + \frac{\mathrm{d}x(t)}{\mathrm{d}t} \,, \tag{28}$$

providing [17]

$$v_{\mathrm{tot}}(t) = v(t) + [x_{\mathrm{tot}}(t) - vt]\,\frac{\hbar^2}{4m^2}\,\frac{t}{\sigma^2\sigma_0^2} \,. \tag{29}$$

2. Simulation of quantum phenomena with purely classical diffusion using coupled map lattices

It is straightforward to simulate the diffusion process of Eq. (22) in a simple computer model. Using coupled map lattices (CML), one approximates the heat equation as usual by

$$P[i, k+1] = P[i,k] + \frac{D[i,k]\Delta t}{\Delta x^2}\left\{P[i+1,k] - 2P[i,k] + P[i-1,k]\right\} \,, \tag{30}$$

with space and time grid indices i and k, respectively. For our anomalous ("ballistic") diffusion one simply inserts (20) into (30).

The result is depicted in Fig. 1, where the (macroscopic, classical) velocity is chosen as $v = 0$. (For examples with $v \neq 0$ and different σ_0, see [17].) Moreover, nine exemplary averaged Bohmian trajectories are shown in Fig. 1, and it must be stressed that *the Figure shows the emerging behaviour of the Gaussian packet following solely from the CML simulation of Eq.* (30). In addition, the emerging trajectories from the simulation (dark lines) are shown together with the calculated ones from Eq. (27) (bright, mostly superposed by the dark lines), providing *exactly the same trajectories* (i.e., up to resolution limits due to discretization).

Note that the trajectories are not the "real" ones, but only represent the averaged behaviour of a statistical ensemble. The results are in full concordance with quantum theory, and in particular with Bohmian trajectories. (For a comparison with the latter, see, for example, [20], or the Figures for the Gaussian wave packet example in [21], which are in excellent agreement with our Fig. 1.) This is so despite the fact that no quantum mechanics has been used yet, i.e., neither a quantum mechanical wave function, or the Schrödinger equation, respectively, nor a guiding wave equation, nor a quantum potential. Moreover, we want to stress that our model offers possible insights into the sub-quantum domain which must escape (Bohmian or orthodox) quantum theory because the latter simply does not employ the "language" necessary to express them. Note, for example, that the existence of the hyperbolic trajectories depicted in Fig. 1, which are given by the formula for the scale invariant wave packet spread (18), has a simple physical explanation in terms of sub-quantum processes. As the inflection points of the hyperbolas are, according to (18), characterized by the relation $D^2t^2/\sigma_0^4 \cong 1$, i.e., by the length scales $u_0^2t^2 \cong \sigma_0^2$, the trajectories' evolution is easily understood: as long as the main bulk of the heat "stored" in the initial Gaussian spreads well "inside" the distribution, $u_0^2t^2 < \sigma_0^2$, the average particle velocity v is not affected much. However, if said main bulk approximately reaches the distance σ_0, or spreads to regions $u_0^2t^2 > \sigma_0^2$, respectively, the particles will "feel" the

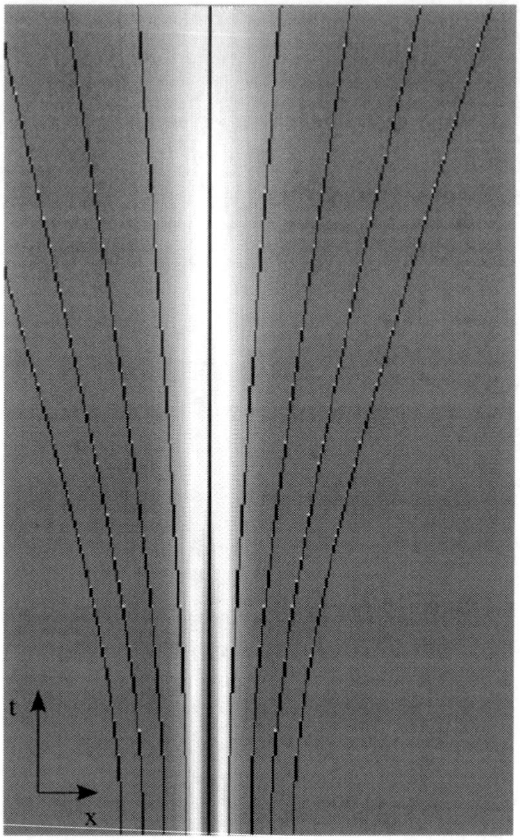

Figure 1. Classical CML simulation of the dispersion of a free Gaussian wave packet.

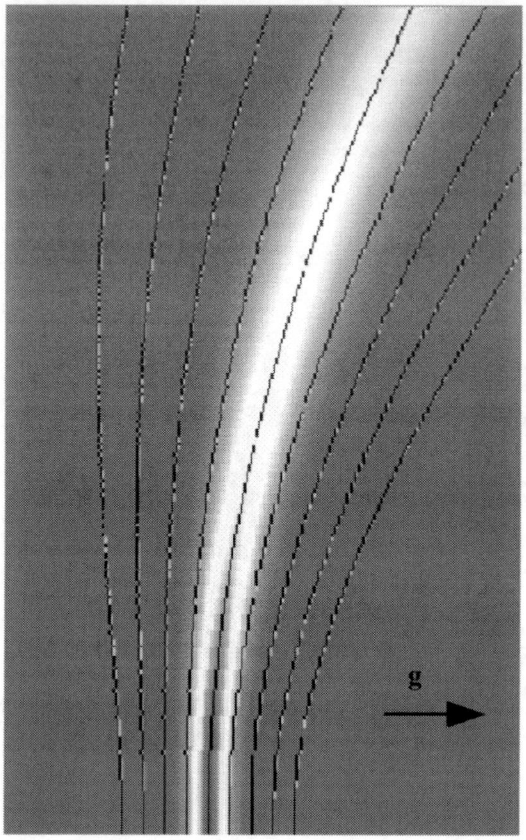

Figure 2. Dispersion of a Gaussian wave packet in a linear potential (e.g., a gravitational field).

full heat and get propagated into new directions. For $t \to \infty$, then, u_0 becomes the spreading rate of the whole Gaussian packet:

$$\frac{\mathrm{d}\sigma}{\mathrm{d}t} = \frac{\hbar^2 t}{4m^2\sigma_0^2\sigma} \stackrel{t\to\infty}{\longrightarrow} \frac{\hbar}{2m\sigma_0} = u_0 \; . \tag{31}$$

In other words, the "spreading" already begins at $t = 0$, but becomes "visible" in terms of deflected trajectories only when $t \cong \sigma_0/u_0$.

So far we have shown that free one-particle quantum motion is identical to sub-quantum ballistic diffusion. This is the basis of a research program that would eventually cover more and more complex situations beyond the case of free motion. As a first simple example, we extend the present scheme to include a linear potential. That is, we place the initial Gaussian packet (7) in a uniform potential $V = K \cdot x$, which may be an electric or a gravitational field, for example. For illustration, but without loss of generality, we substitute in the following K by mg, i.e., we deal with a Gaussian packet freely falling due to the potential $V = mg \cdot x$.

After a short calculation [19], one obtains the trajectories of particles in a gravitational field,

$$x_{\mathrm{tot}}(t) = vt - \frac{g}{2}t^2 + x(0)\sqrt{1 + \frac{\hbar^2 t^2}{4m^2\sigma_0^4}} \, , \tag{32}$$

and the particle acceleration

$$\ddot{x}_{\mathrm{tot}} = -g + x(0)\frac{u_0^2}{\sigma_0^2}\left[1 + \frac{u_0^2 t^2}{\sigma_0^2}\right]^{-3/2} . \tag{33}$$

In Fig. 2, exemplary trajectories of a Gaussian in a gravitational field are shown as obtained by the CML simulation of ballistic diffusion, modified by the substitution $v \to v - gt$. The trajectories (dark lines) superpose those (bright lines) derived directly from Eq. (32), and again exhibit excellent agreement.

Note that some trajectories of the dispersing Gaussian even overcome "gravity" for a well-defined period of time, as can also be seen in Fig. 2. In fact, our sub-quantum model provides a detailed explanation of why, and within which time limits, this "anti-gravity" effect becomes possible. A look at the last term of Eq. (33) provides the answer. Similarly to the discussion of the hyperbolas' inflection points in the free case, one deals also here with an extremum at the scale $u_0^2 t^2 \cong \sigma_0^2$. However, this time the corresponding expression (in rectangular brackets) is antagonistic to g. In other words, said scale describes the maximum of the "anti-gravity" effect, because it is there where the heat of the main bulk of the packet is consumed, which has via the kinetic energy counter-acted the effect of gravity. For larger times, then, the remaining heat gets gradually less, and therefore gravitational acceleration begins to dominate the trajectories' curvature.

Finally, instead of following just one Gaussian, we extend our simulation scheme to include two possible paths of a particle which eventually cross each other. For this, we use two Gaussians approaching each other. In the space-time diagram of Fig. 3, they are characterized by the same half-widths σ, whereas in Fig. 4 we choose Gaussians whose half-widths' ratio is 2:1.

Note that in both Figs. 3 and 4 one can observe a basic characteristic of the (averaged) particle trajectories, which, just because of the averaging, are identical with the Bohmian trajectories. To fully appreciate this characteristic, we remind the reader of the severe criticism of Bohmian trajectories as put forward by Scully and others (see [22], and references therein). The critics claimed that Bohmian trajectories would have to be described as "surreal" ones because of their apparent violation of momentum conservation. In fact, due to the "no crossing rule" for Bohmian trajectories in Young's double slit experiment, for example, the particles coming from, say, the right slit (and expected at the left part of the screen if "classical" momentum conservation should hold) actually arrive at the right part of the screen (and vice versa for the other slit). In Bohmian theory, this "no crossing rule" is due to the action of the non-classical quantum potential, such that, once the existence of a quantum potential is accepted, no contradiction arises and the trajectories may be considered "real" instead of "surreal".

Here we can note that in our sub-quantum approach an explanation of the "no crossing rule" is even more straightforward and actually a consequence of a detailed microscopic momentum conservation. As can be seen particularly clearly in Fig. 3, the (Bohmian) trajectories are repelled from the central symmetry line. However, in our case this is only implicitly due to a "quantum potential", but actually due to the identification of the latter with a kinetic (rather than a potential) energy: it is the "heat of the compressed vacuum" that accumulates along said symmetry line (i.e., as reservoir of "outward" oriented kinetic energy) and therefore repels the trajectories. In Fig. 4, moreover, one sees the resulting trajectories for the case that the symmetry of Fig. 3 is broken: by using different half-widths for the two paths, one obtains the skew-symmetric distribution of particle intensities and trajectories, respectively. Both Figs. 3 and 4 are in full concordance with Bohmian trajectories (cf., for example, [23] for comparison). However, as mentioned, in our case also a micro-causal explanation is provided, which brings the whole process into perfect agreement with momentum conservation on a more "microscopic" level.

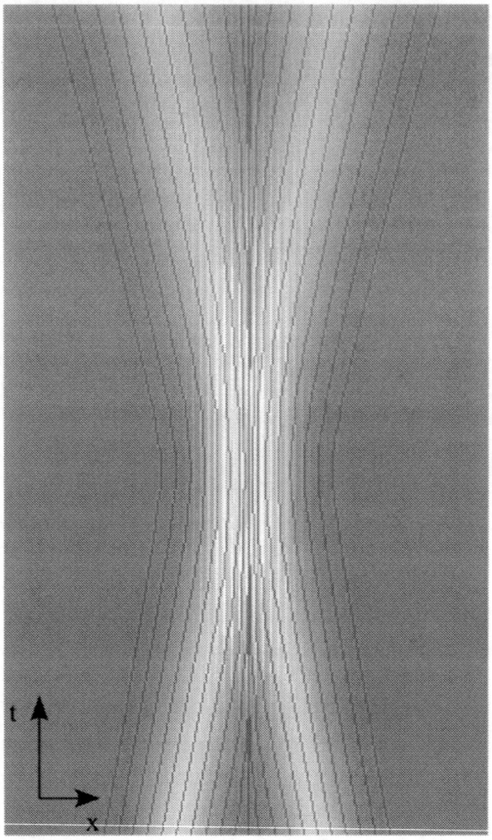

Figure 3. Trajectories of two Gaussians with the same half widths.

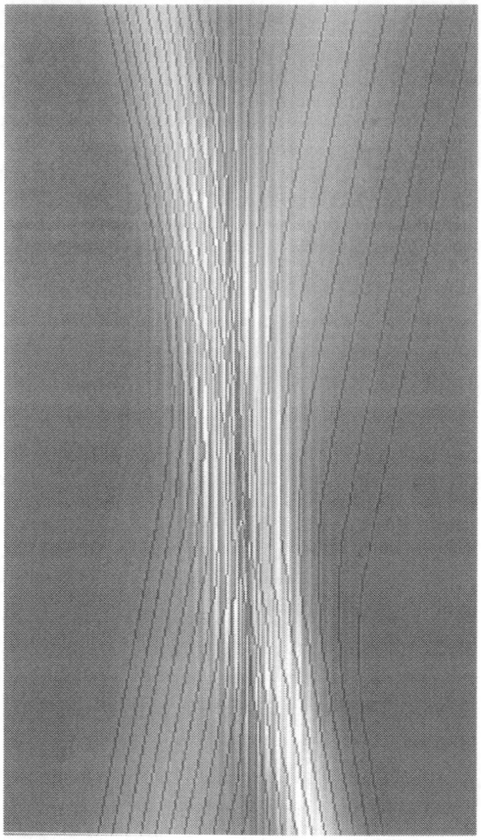

Figure 4. Trajectories of two Gaussians with half width ratio of 2:1.

To sum up, we have shown for the cases of free motion and motion in linear potentials, respectively, that the time evolution of a one-particle quantum system in the noisy heat bath of the surrounding "vacuum" exactly equals that of (classical) ballistic sub-quantum diffusion. Note that there are some well-known general characteristics of ballistic diffusion [24], and the results presented in this paper agree perfectly with them. For one thing, ballistic diffusion is the only type of diffusion that exhibits reversibility, and because of this it violates ergodicity (i.e., as in our cases). Also, if the ballistic system is not in equilibrium initially, it will never reach equilibrium (which is true here as well). Finally, the result of any measurement depends on the initial conditions. This can be clearly seen also from our results for the time evolution of the Gaussians and the corresponding averaged trajectories, which all depend on the initial values of u_0, or σ_0, respectively.

References

[1] Carroll R 2010 *On the Emergence Theme of Physics* (Singapore: World Scientific)
[2] Grössing G 2008 *Phys. Lett.* A **372**(25) 4556–4563 (*Preprint* quant-ph/0711.4954)
[3] Grössing G 2009 *Physica* A **388** 811–823 (*Preprint* quant-ph/0808.3539)
[4] Grössing G, Mesa Pascasio J and Schwabl H 2010 A classical explanation of quantization *Preprint* quant-ph/0812.3561
[5] Adler S L 2004 *Quantum Theory as an Emergent Phenomenon: The Statistical Mechanics of Matrix Models as the Precursor of Quantum Field Theory* (Cambridge: Cambridge University Press)
[6] Nelson E 1985 *Quantum Fluctuations* Princeton series in physics (Princeton, NJ: Princeton University Press)

[7] El Naschie M S 1995 *Chaos, Solitons & Fractals* **5**(5) 881–884
[8] Nottale L 1993 *Fractal Space-Time and Microphysics: Towards a Theory of Scale Relativity* (Singapore: World Scientific)
[9] Fritsche L and Haugk M 2003 *Ann. Phys. (Leipzig)* **12**(6) 371–403
[10] Elze H 2009 *Int. J. Qu. Info.* **7**(1) 83–96 (*Preprint* quant-ph/0806.3408)
[11] Rusov V D, Vlasenko D S and Mavrodiev S C 2009 *Ukr. J. Phys.* **54**(11) 1131–1138 (*Preprint* quant-ph/0906.1723)
[12] 't Hooft G 2001 *Quantum Mechanics and Determinism* in *Proc. 8th Int. Conf. on Particles, Strings and Cosmology* ed P Frampton and J Ng (Rinton Press) pp 275–285 (*Preprint* hep-th/0105105)
[13] Protière S, Boudaoud A and Couder Y 2006 *J. Fluid Mech.* **554** 85–108
[14] Couder Y and Fort E 2006 *Phys. Rev. Lett.* **97** 154101
[15] Eddi A, Fort E, Moisy F and Couder Y 2009 *Phys. Rev. Lett.* **102** 240401
[16] Couder Y, Protière S, Fort E and Boudaoud A 2005 *Nature* **437** 208–208
[17] Grössing G, Fussy S, Mesa Pascasio J and Schwabl H 2010 *Physica* A **389**(21) 4473–4484 (*Preprint* quant-ph/1004.4596)
[18] Vainstein M H, Lapas L C and Oliveira F A 2008 *Act. Phys. Pol.* B **39**(5) 1273–1281
[19] Grössing G 2010 *Entropy* **12**(9) 1975–2044 `http://www.mdpi.com/1099-4300/12/9/1975/`
[20] Holland P R 1993 *the quantum theory of motion: An Account of the de Broglie-Bohm Causal Interpretation of Quantum Mechanics* (Cambridge: Cambridge University Press)
[21] von Bloh K 2010 Causal interpretation of the free quantum particle `http://demonstrations.wolfram.com/CausalInterpretationOfTheFreeQuantumParticle/`
[22] Scully M O 1998 *Physica Scripta* **T76**(1) 41–46
[23] Sanz A S and Miret-Artes S 2008 *J. Phys. A: Math. Gen.* **41** 435303 (*Preprint* quant-ph/0806.2105)
[24] Vlahos L, Isliker H, Kominis Y and Hizandis K 2008 *Order and Chaos* vol 10 ed T Bountis (*Preprint* nlin/0805.0419)

Pilot-wave approaches to quantum field theory

Ward Struyve

Institute of Theoretical Physics, K.U.Leuven, Celestijnenlaan 200D, B–3001 Leuven, Belgium and Institute of Philosophy, K.U.Leuven, Kardinaal Mercierplein 2, B–3000 Leuven, Belgium

E-mail: Ward.Struyve@fys.kuleuven.be

Abstract. The purpose of this paper is to present an overview of recent work on pilot-wave approaches to quantum field theory. In such approaches, systems are not only described by their wave function, as in standard quantum theory, but also by some additional variables. In the non-relativistic pilot-wave theory of de Broglie and Bohm those variables are particle positions. In the context of quantum field theory, there are two natural choices, namely particle positions and fields. The incorporation of those variables makes it possible to provide an objective description of nature in which rather ambiguous notions such as 'measurement' and 'observer' play no fundamental role. As such, the theory is free of the conceptual difficulties, such as the measurement problem, that plague standard quantum theory.

1. Introduction

The pilot-wave theory of de Broglie and Bohm for non-relativistic system describes point particles in physical 3-space, which move under the guidance of the wave function. This theory reproduces the predictions of standard quantum theory (at least in so far those are unambiguous), given a state of equilibrium. Here we are concerned with generalizations of this theory to quantum fields. The purpose is to give an overview of our recent work and to put it into context. Much of this overview is based on [1], which contains a review and further development of field based approaches, on [2, 3] (with H. Westman) which deals with a minimalist approach and on [4] (with S. Colin) which deals with a Dirac sea approach.

In section 2, we first recall the pilot-wave theory of de Broglie and Bohm for non-relativistic systems and its extension to spin by Bell. Then we turn to quantum field theory. First, in section 3, we consider field based approaches. We present Bohm's approach to bosonic fields and discuss the problems with similar approaches to fermionic fields. We also present the minimalist approach, which avoids these problems. Then, in section 4, we consider particle based approaches: Bell's lattice approach, a continuum approach in terms of particles and anti-particles by Dürr *et al.* and the Dirac sea approach.

2. Non-relativistic pilot-wave theory

2.1. Pilot-wave theory of de Broglie and Bohm for spinless particles

Non-relativistic pilot-wave theory, which was first proposed by de Broglie in the 1920s [5], and rediscovered by Bohm in the 1950s [6, 7], is about point particles in physical 3-space, with positions $\mathbf{x}_1(t), \dots, \mathbf{x}_n(t)$, which are guided in their motion by the wave function $\psi(x,t) = \psi(\mathbf{x}_1, \dots, \mathbf{x}_n, t)$ (for detailed expositions, see [8–10]). The wave function satisfies the non-relativistic Schrödinger equation

$$\mathrm{i}\hbar\frac{\partial\psi(x,t)}{\partial t} = \left(-\sum_{k=1}^{n}\frac{\hbar^2}{2m_k}\nabla_k^2 + V(x)\right)\psi(x,t)\,. \tag{1}$$

The possible particle trajectories are solutions of the guidance equations

$$\frac{d\mathbf{x}_k(t)}{dt} = \frac{\hbar}{m_k}\mathrm{Im}\frac{\boldsymbol{\nabla}_k\psi(x,t)}{\psi(x,t)}\bigg|_{x=x(t)} = \frac{1}{m_k}\boldsymbol{\nabla}_k S(x,t)\bigg|_{x=x(t)}, \tag{2}$$

where S is the phase of the wave function, that is, $\psi = |\psi|\mathrm{e}^{\mathrm{i}S/\hbar}$.

The particle dynamics has the important property that it preserves the density $|\psi|^2$. That is, if a distribution of particles, all guided by the same wave function ψ, is given by $|\psi(x,t_0)|^2$ at a certain time t_0, then the distribution is given by $|\psi(x,t)|^2$ at any other time t. This property is called equivariance. It essentially follows from the continuity equation for $|\psi|^2$,

$$\frac{\partial|\psi|^2}{\partial t} + \sum_{k=1}^{n}\boldsymbol{\nabla}_k\cdot\left(\frac{\boldsymbol{\nabla}_k S}{m_k}|\psi|^2\right) = 0\,, \tag{3}$$

which itself follows from the Schrödinger equation.

The distribution $|\psi|^2$ plays the role of an equilibrium distribution and is called the quantum equilibrium distribution. Given the quantum equilibrium distribution and the fact that measurement results are ultimately recorded in positions of things, like instrument needles, computer outprint, etc., it almost follows immediately that the de Broglie-Bohm theory reproduces the standard quantum mechanical predictions.

Note that the equivariance of the equilibrium distribution $|\psi|^2$, which plays an important role in showing that the theory reproduces the standard quantum predictions, results from the fact that the velocity field is actually given by $\mathbf{j}_k^{\psi}/|\psi|^2$, where $\mathbf{j}_k^{\psi} = \hbar\mathrm{Im}(\psi^*\boldsymbol{\nabla}_k\psi)/m_k$ is the usual quantum current which appears in the continuity equation for $|\psi|^2$. For other quantum theories, in particular quantum field theories, one can develop a pilot-wave theory by introducing a velocity field of the same form (see [11] for the construction of such velocity fields for arbitrary Hamiltonians).

2.2. Treatment of spin

As shown by Bell, the theory can straightforwardly be extended to cover particles with spin [12, 13]. For example, a non-relativistic particle with spin-1/2 is then described by a spinor $\psi_a(\mathbf{x})$ $(a = 1,2)$, which satisfies the non-relativistic Pauli equation, and a position $\mathbf{x}$, which evolves according to the guidance equation

$$\frac{d\mathbf{x}(t)}{dt} = \frac{\hbar}{m}\mathrm{Im}\frac{\sum_a\psi_a^*(\mathbf{x},t)\boldsymbol{\nabla}\psi_a(\mathbf{x},t)}{\sum_a|\psi_a(\mathbf{x},t)|^2}\bigg|_{\mathbf{x}=\mathbf{x}(t)}. \tag{4}$$

The quantum equilibrium distribution is given by $\rho(\mathbf{x},t) = \sum_a|\psi_a(\mathbf{x},t)|^2$.

No extra variables are introduced to represent spin. There exist alternative approaches that do introduce such variables. For example, in one such approach [14], the spin-vector $\mathbf{s}$ is introduced, which is defined as

$$\mathbf{s}(t) = \frac{\sum_{a,a'}\psi_a^*(\mathbf{x},t)\frac{\hbar}{2}\boldsymbol{\sigma}_{aa'}\psi_{a'}(\mathbf{x},t)}{\sum_a|\psi_a(\mathbf{x},t)|^2}\bigg|_{\mathbf{x}=\mathbf{x}(t)}, \tag{5}$$

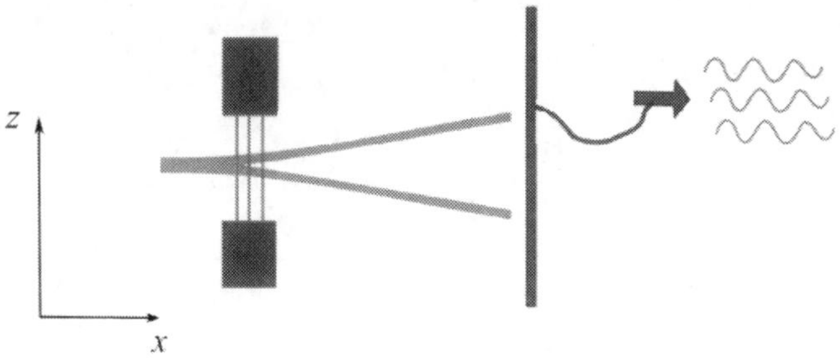

Figure 1. Setup of the Stern-Gerlach experiment. One can identify the Stern-Gerlach magnets, the corresponding magnetic field (represented by the three straight lines), the detecting screen and the pointer that is connected to the screen. The wavy lines represent the electromagnetic radiation surrounding the pointer.

where $\boldsymbol{\sigma}$ are the Pauli matrices and where the function on the right hand side is evaluated for the actual particle position $\mathbf{x}(t)$. As such, at each time, the spin vector is completely determined by the wave function and the actual particle position.

However, such extra variables, which seem rather artificial, are not necessary. Because results of measurements are generally recorded in positions of things, the theory reproduces the standard quantum predictions (in quantum equilibrium). For example, in a spin measurement with a Stern-Gerlach device, the measured spin of the particle will be up or down depending on whether the particle is detected in the upper or lower half of the detecting screen.

To conclude this section, we consider the Stern-Gerlach experiment in a bit more detail. Throughout the paper we will give an impression of the various ontologies by sketching the situation for the Stern-Gerlach experiment. The details of the setup are as follows (see Figure 1). We consider a source emitting electrons with spin up in the x-direction (meaning that their wave function is an eigenstate of the spin operator in the x-direction with eigenvalue 1). The particles pass through an inhomogeneous magnetic field with decreasing strength in the positive z-direction and end up at a detecting screen. The screen is connected to a pointer. In its steady state, the pointer points to the right. After the measurement, it will point up or down, depending on whether the particle was detected in the upper or lower half of the detecting screen. This pointer, which is absent in the usual setup, is introduced merely for convenience. In the usual setup, the incoming particle leaves an imprint on the detecting screen as a result of a chemical reaction, but this is much harder to display in our sketches of the ontology than the orientation of a pointer. For later purposes, we have also drawn the radiation near the pointer. This comprises the thermal radiation emitted by it and the radiation, such as visible light, that is scattered off it. The rest of the environment is ignored. In Figure 2, we sketched the situation of the Stern-Gerlach spin measurement according to Bell's model.

3. Field ontologies for quantum field theory

For quantum field theory, there are approaches in terms of particles and in terms of fields. We first consider field approaches. The particle approaches will be discussed in the next section.

The theories that we will consider are not Lorentz invariant at the fundamental level. First of all, there might be regulators such as ultra-violet momentum cut-offs (sometimes implicitly assumed), which violate the Lorentz invariance already at the level of the wave equation. Second, the non-locality as implied by the violation of the Bell inequalities, is hard to combine with Lorentz invariance. In quantum equilibrium, the theories reproduce the predictions of the regularized quantum theory. Hence, insofar the latter are compatible with Lorentz invariance,

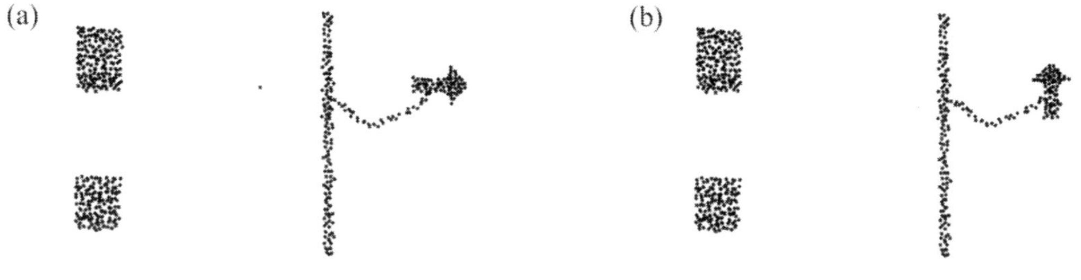

Figure 2. The Stern-Gerlach experiment according to Bell's model for non-relativistic particles with spin. (a) displays the situation just before the particle hits the detecting screen. (b) displays the situation just after the measurement.

so will be the equilibrium predictions of the corresponding pilot-wave theories. For an overview of possible ways to achieve Lorentz invariance also at the fundamental level, see [15].

In the following units are assumed for which $\hbar = c = 1$.

3.1. Field configurations for bosonic quantum fields

Already in his 1952 paper [7], Bohm presented a pilot-wave approach to quantum field theory. In this approach, which he considered for the free quantized electromagnetic field, there is an actual field configuration $\mathbf{A}^T(\mathbf{x})$, corresponding to the transverse part of the vector potential, which satisfies the guidance equation

$$\frac{\partial \mathbf{A}^T(\mathbf{x},t)}{\partial t} = \left.\frac{\delta S(\mathbf{A}^T,t)}{\delta \mathbf{A}^T(\mathbf{x})}\right|_{\mathbf{A}^T(\mathbf{x})=\mathbf{A}^T(\mathbf{x},t)}, \tag{6}$$

where S is phase of the wave functional $\Psi(\mathbf{A}^T,t)$, that is, $\Psi = |\Psi|e^{iS}$. The wave functional satisfies the functional Schrödinger equation

$$i\frac{\partial \Psi(\mathbf{A}^T,t)}{\partial t} = \frac{1}{2}\int d^3x \left(-\frac{\delta}{\delta \mathbf{A}^T(\mathbf{x})}\cdot\frac{\delta}{\delta \mathbf{A}^T(\mathbf{x})} - \mathbf{A}^T(\mathbf{x})\cdot\nabla^2\mathbf{A}^T(\mathbf{x})\right)\Psi(\mathbf{A}^T,t). \tag{7}$$

The quantum equilibrium distribution is given by $|\Psi(\mathbf{A}^T,t)|^2\mathcal{D}\mathbf{A}^T$. (A suitable regularization, which makes the equations of motion and the quantum equilibrium distribution well defined, is implicitly assumed.)

One can take a similar approach to other bosonic fields (see [1] for a review). But so far this approach has not been successfully applied to fermionic fields. There exist two attempts, one by Holland and one by Valentini, to which we now turn.

3.2. Euler angles for fermionic quantum fields

Holland considered a functional Schrödinger representation for the fermionically quantized non-relativistic Schrödinger field in which the wave functionals are defined on a space of fields $\boldsymbol{\alpha}(\mathbf{k}) = (\alpha(\mathbf{k}),\beta(\mathbf{k}),\gamma(\mathbf{k}))$ in a discretized momentum space [9,16]. For each momentum $\mathbf{k}$, $\boldsymbol{\alpha}(\mathbf{k})$ represents the Euler angles that parametrize the group $SU(2)$. Holland introduced an actual field configuration $\boldsymbol{\alpha}(\mathbf{k})$ which is guided by the wave functional $\Psi(\boldsymbol{\alpha},t)$. However, this configuration lives in momentum space and not in physical space, and Holland did not specify an ontology in physical space. Such an ontology could be constructed from the configuration $\boldsymbol{\alpha}(\mathbf{k})$ and the wave functional (in a way similar to that of the construction of the spin vector for non-relativistic particles in section 2.2), see [1]. However, the empirical adequacy of those theories is very hard to check.

In [1], we considered an alternative approach, which employs the sames techniques, but which is easier to analyse. In this approach, the actual configuration is given by a set of Euler angles $\boldsymbol{\alpha}(\mathbf{x}) = (\alpha(\mathbf{x}), \beta(\mathbf{x}), \gamma(\mathbf{x}))$ at each point $\mathbf{x}$ of a spatial lattice. The configuration is guided by a wave functional $\Psi(\boldsymbol{\alpha}, t)$. It was found that this model seems inadequate because it does not seem to give an adequate image of general macroscopic matter distributions. In order to see this, we considered the wave functional representing a region filled with ordinary matter and the wave functional representing an empty region. We compared the typical field configurations for those wave functionals. They should typically be distinct in order to be able to distinguish matter from empty space. For a cubic region with edge of length L, a lattice spacing a and a matter density ρ, we found that they are typically distinct if $La\rho^{2/3} \gg 1$. For a lattice spacing given by the Planck length, that is $a = 10^{-35}$m, and a density $\rho = 10^{30}/\text{m}^3$ (which corresponds to one particle per atomic distance cubed), L should be much bigger than 10^{15}m, which is clearly unacceptable. However, since we are dealing with a non-relativistic field theory, a much bigger lattice spacing could be considered. But even for a lattice spacing of the order of 10^{-15}m, the length L should be much bigger then 10^{-5}m, which is still too large. There also seems to be very little hope that applying this approach to relativistic quantum field theory will yield a better result. For example, for the quantized Dirac field, the configuration will be given by four sets of Euler angles at each point of the spatial lattice (one for each component of the Dirac spinor), but the lattice spacing should be taken much smaller.

3.3. Grassmann fields for fermionic quantum fields

Valentini explored a different approach to fermionic fields [17, 18]. His approach starts from a functional Schrödinger representation in which the wave functionals are defined on a space of Grassmann fields (that is, anti-commuting fields). The actual configurations are Grassmann fields, which are guided by the wave functional. However, there is a difficulty with this approach. While this functional Schrödinger picture bears some resemblance to the functional Schrödinger picture for bosonic fields, there are important differences. One is that the wave functionals are not complex valued, but take values in a Grassmann algebra. As a result of these differences, Bohm's approach to bosonic fields does not immediately carry over. The main difficulty is that there does not seem to be a natural measure on the space of Grassmann fields that could serve as the quantum equilibrium measure (also in the standard quantum field theoretical picture, integrals are defined algebraically, instead of in terms of measure theoretical notions). As such, it is not clear what possible guidance equations to adopt, as those are usually found by requiring that they preserve the quantum equilibrium distribution.

3.4. A minimalist approach: only configurations for the bosonic fields

In [2, 3], Westman and the present author considered a radically minimalist approach in which field configurations are introduced only for the bosonic degrees of freedom and no variables for the fermionic ones. This was illustrated for the case of scalar quantum electrodynamics. The wave functional, written as $\Psi_f(\mathbf{A}^T, t)$, is then a functional on the space of fields $\mathbf{A}^T(\mathbf{x})$, just as in the case of the free electromagnetic field, but it carries a label f representing the fermionic degrees of freedom. (We treat f as discrete, but it could also represent a continuous label.) The time evolution of the wave functional is governed by the appropriate Schrödinger equation which involves the Hamiltonian for scalar quantum electrodynamics.

An actual field configuration $\mathbf{A}^T(\mathbf{x})$ is introduced, just as in Bohm's approach to the free electromagnetic field, but no variables corresponding to the fermionic degrees of freedom. The

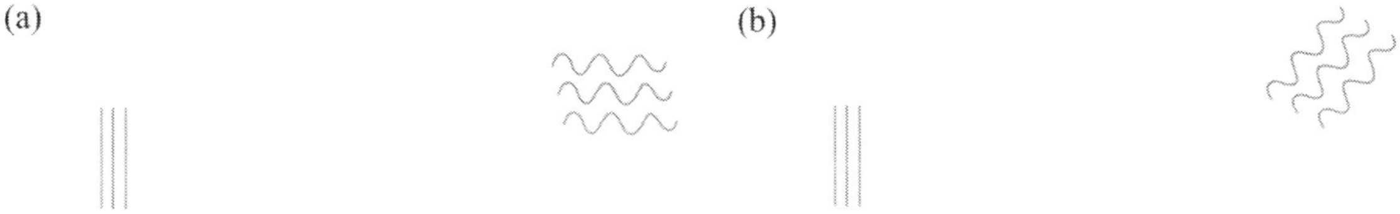

Figure 3. The Stern-Gerlach experiment according to the minimalist pilot-wave model. There is only the transverse vector potential for the electromagnetic field. The field produced by the magnets and the field surrounding the pointer are displayed. (a) and (b) respectively display the situation just before and just after the measurement.

guidance equation for this field is

$$\frac{\partial \mathbf{A}^T(\mathbf{x},t)}{\partial t} = \mathrm{Im}\frac{\sum_f \Psi_f^*(\mathbf{A}^T,t)\frac{\delta}{\delta \mathbf{A}^T(\mathbf{x})}\Psi_f(\mathbf{A}^T,t)}{\sum_f |\Psi_f(\mathbf{A}^T,t)|^2}\Bigg|_{\mathbf{A}^T(\mathbf{x})=\mathbf{A}^T(\mathbf{x},t)} . \tag{8}$$

The quantum equilibrium density is given by $\rho(\mathbf{A}^T,t) = \sum_f |\Psi_f(\mathbf{A}^T,t)|^2$.

While there are no variables representing matter, the wave functional which guides the field $\mathbf{A}^T(\mathbf{x})$ still contains the fermionic degrees of freedom. As such, the field configuration $\mathbf{A}^T(\mathbf{x})$ will in certain cases behave as if there was matter present. For example, it might look like radiation that has been scattered off some matter distribution, or like thermal radiation emitted by such a distribution. In this way, it was argued that the model is empirically adequate, because there will be an image of macroscopic matter distributions in the radiation field. Nevertheless such a model seems rather far removed from our everyday experience of the world and probably takes minimalism too far.

In Figure 3, the Stern-Gerlach experiment is displayed for this approach. Only the main contributions to the field are drawn. There is the magnetic field induced by the magnets and the radiation field near the pointer. The latter will encode and display the orientation of the pointer.

3.5. The minimalist approach with additional configurations for the fermionic fields

Starting from the minimalist model, one can actually introduce extra variables representing the fermionic degrees of freedom [3]. They could be constructed from the wave functional and the actual field configuration $\mathbf{A}^T(\mathbf{x},t)$ in a similar way as the spin vector in the case of non-relativistic particles with spin (cf. section 2.2). For example, one could introduce an energy density for the matter field, defined by

$$E(\mathbf{x},t) = \frac{\sum_{f,f'} \Psi_f^*(\mathbf{A}^T,t)\widehat{E}_{ff'}(\mathbf{x})\Psi_{f'}(\mathbf{A}^T,t)}{\sum_f |\Psi_f(\mathbf{A}^T,t)|^2}\Bigg|_{\mathbf{A}^T(\mathbf{x})=\mathbf{A}^T(\mathbf{x},t)} , \tag{9}$$

where $\widehat{E}_{ff'}(\mathbf{x}) = \langle f|\widehat{E}(\mathbf{x})|f'\rangle$, with $\widehat{E}(\mathbf{x})$ is the energy density operator for the fermionic fields. The functional on the right hand side is evaluated for the actual field configuration $\mathbf{A}^T(\mathbf{x},t)$. So the energy density $E(\mathbf{x},t)$ is defined completely in terms of the wave functional $\Psi_f(\mathbf{A}^T,t)$ and the actual configuration.

One could introduce other configurations, or extra configurations, such as for example a mass or charge density, in a similar way.

In Figure 4, the Stern-Gerlach experiment is displayed for this ontology. We have thereby ignored possible vacuum contributions to the energy density. (Whether there are such

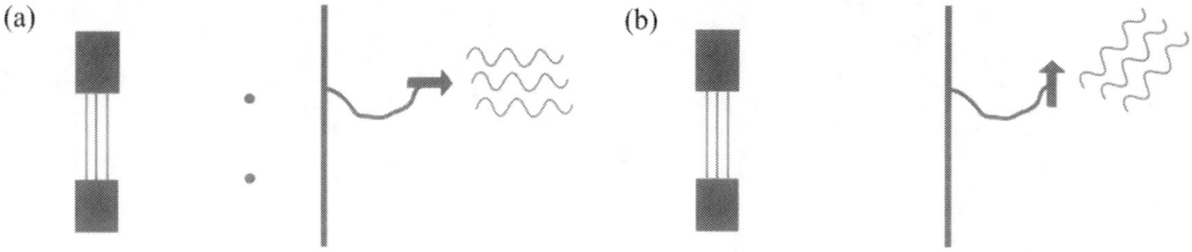

Figure 4. The Stern-Gerlach experiment according to the minimalist pilot-wave model with additional configurations for the fermionic fields. Apart from the vector potential, there is also the energy density of the fermionic fields. (a) and (b) respectively display the situation just before and just after the measurement.

contributions probably depends on the exact definition of the energy density operator. This has not been investigated yet.) After the spin-1/2 particle has been emitted from the source its energy density is approximately localized. When the density passes the magnets, its splits into two localized packets, one going up and one going down. During the measurement, the energy density will collapse to either packet, depending on the actual value of the radiation field. Similarly, the energy density corresponding to the pointer will display a definite orientation.

4. Particle ontologies for quantum field theory

4.1. Lattice approach in terms of fermion numbers

Bell was the first to present a pilot-wave approach to quantum field theory in terms of a particle ontology [19]. He considered a spatial lattice and introduced an actual configuration given by the fermion numbers at the lattice sites. No configurations were introduced for the bosonic degrees of freedom. The fermion numbers change in time in a stochastic way, according to a generalized guidance law.

4.2. Positions for particles and anti-particles

Dürr *et al.* developed a continuum generalization of Bell's model [20–23] (whereby Bell's lattice regularization is replaced by a suitable regularization compatible with a spatial continuum). They introduced positions for both particles and anti-particles. The trajectories are deterministic, interrupted by stochastic jumps, which usually correspond to creation and annihilation events. This approach has been applied to non-relativistic bosons and fermions, and to Dirac fermions, but not yet to relativistic bosons (the main challenge seems to be to find an appropriate position operator or position POVM which could yield the quantum equilibrium distribution).

4.3. Dirac sea approach

Instead of considering the usual particle–anti-particle picture of standard quantum field theory, as in the approach of Dürr *et al.*, one can also consider the Dirac sea picture. In this picture an anti-particle corresponds to a hole in a sea of negative energy particles. In the corresponding pilot-wave approach, positions are introduced for positive energies, as well as for the negative energies of the Dirac sea. No variables are introduced for the bosons.

This approach was considered for quantum electrodynamics by Bohm *et al.* [8, 24] and by Colin [25, 26], who introduced it as a continuum generalization of Bell's model,[1] and was later extended to the other interactions in the standard model by Colin and the present author [4].

[1] The reason that Colin obtained a different continuum generalization than Dürr *et al.* originates in a different

This approach is deterministic, at least as long as the fermion number is conserved (with fermion number being the number of particles minus the number of anti-particles). The standard model actually predicts a violation of fermion number, but only for very high energies (estimated to be accessible only in accelerators with a performance significantly higher than that of the Large Hadron Collider [4]). The implications for the Dirac sea approach have not yet been studied.

We also considered the empirical adequacy of this approach [4]. Since the particles that make up ordinary matter move against a background of particles corresponding to the sea, there might be the worry that the positions do not yield an image of macroscopic matter distributions. Therefore, we calculated how large a region in space should be in order to be able to distinguish a region with ordinary matter from a region without such matter. To simplify the calculation, we ignored the interactions. That is, we assumed that the wave function representing an empty region is given by that of the free vacuum (that is, the state corresponding to no particles and no anti-particles) and that the wave function of the matter is found by applying the particle creation operators on the free vacuum. Since we ignored interactions, we also assumed the physical masses for the particles, instead of the bare masses. A more rigorous analysis, which takes into account the interactions, should deal with the true vacuum and the bare masses of the particles.

With our simplified analysis, we found that the volume V of the region should satisfy $V \gg (\Lambda/\rho^2)^{3/5}$, where Λ is the ultra-violet momentum cut-off and ρ the particle density of the ordinary matter. Taking a cut-off of the order of the Planck scale, that is, $\Lambda \sim 10^{35}/\mathrm{m}$, a particle density $\rho \sim 10^{30}/\mathrm{m}^3$, and a spherical region with radius b, we found that the bound is given by $b \gg 10^{-6}\mathrm{m}$. This lower bound is rather large. However, it should be noted that the cut-off was taken at the Planck scale. In principle a lower cut-off could be considered. In addition, and more importantly, it is expected that this result will improve by taking into account the interactions. It is expected that the interactions will suppress the fluctuations of the density of background particles (so that they tend to be distributed more uniformly). So far, the only source for suppression of the fluctuations stems from the mutual repulsion of the particles due to the Pauli exclusion principle.

In Figure 5, the Stern-Gerlach experiment is displayed for this approach. Note that the ratio of the particle number density corresponding to a region filled with ordinary matter and the density corresponding to an empty region is not drawn to scale. The ratio is given by $1+8\pi^2\rho/\Lambda^3$, which is extremely close to one. Note also that, in these pictures, the electron that was fired to the screen can not be distinguished from the background particles. Maybe it could be distinguished by considering its time evolution. Nevertheless, as we mentioned above, there is good reason to expect that the approach will yield an image of macroscopic matter distributions (like that of the pointer), which is sufficient to account for the standard quantum predictions.

The approach of Dürr *et al.* actually yields a very similar picture. According to this approach, there are no particles corresponding to the Dirac sea, but instead the particle–anti-particle pairs that are associated to the vacuum fluctuations. While the expected number density of such particles is less than that of Dirac sea particles, it is presumably of the same order (that is, of the order of Λ^3).

Finally, note that in this approach and in the other approaches considered in this section, one could always introduce field configurations for the bosonic degrees of freedom.

5. Conclusions

We described a number of pilot-wave approaches to quantum field theory. Two types of approaches were distinguished: those based on a field ontology and those based on a particle

reading of Bell's notion of fermion number. Dürr *et al.* understood it as the number of particles plus the number of anti-particles, while Colin understood it as the number of particles minus the number of anti-particles [4].

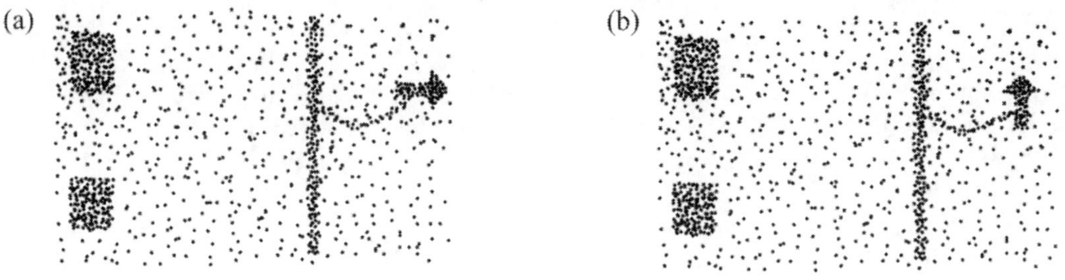

Figure 5. The Stern-Gerlach experiment according to the Dirac sea pilot-wave approach. (a) and (b) respectively display the situation just before and just after the measurement.

ontology. While the field approach can be applied to bosonic fields, there are problems to extend it to fermionic fields. These problems could be avoided by adopting a radically minimalist approach according to which configurations are introduced only for the bosonic degrees of freedom and nothing for the fermionic degrees of freedom. However, such an approach probably takes minimalism too far. Starting from the minimalist model, there actually is a simple way to introduce field configurations for the fermions, by constructing them out of the bosonic field configurations and the wave function. However, this approach has a rather arbitrary and artificial flavour to it.

On the other hand, the particle approach seems to work better for fermionic fields. There is the lattice approach by Bell, which is stochastic, and two continuum approaches, namely the particle–anti-particle approach by Dürr *et al.*, which is also stochastic, and the Dirac sea approach, which is deterministic.

Acknowledgments

It is a pleasure to thank Samuel Colin for valuable discussions. The author acknowledges support as Postdoctoral Fellow of the FWO-Flanders.

References

[1] Struyve W 2010 *Rep. Prog. Phys.* **73** 106001
[2] Struyve W and Westman H 2006 *AIP Conf. Proc.* **844** 321
[3] Struyve W and Westman H 2007 *Proc. R. Soc. A* **463** 3115
[4] Colin S and Struyve W 2007 *J. Phys. A* **40** 7309
[5] de Broglie L 1928 *Électrons et Photons: Rapports et Discussions du Cinquième Conseil de Physique* (Paris: Gauthier-Villars) p 105; English translation: Bacciagaluppi G and Valentini A 2009 *Quantum Theory at the Crossroads: Reconsidering the 1927 Solvay Conference* (Cambridge: Cambridge University Press)
[6] Bohm D 1952 *Phys. Rev.* **85** 166
[7] Bohm D 1952 *Phys. Rev.* **85** 180
[8] Bohm D and Hiley B J 1993 *The Undivided Universe* (New York: Routledge)
[9] Holland P R 1993 *The Quantum Theory of Motion* (Cambridge: Cambridge University Press)
[10] Dürr D and Teufel S 2009 *Bohmian Mechanics* (Berlin: Springer-Verlag)
[11] Struyve W and Valentini A 2009 *J. Phys. A* **42** 035301
[12] Bell J S 1966 *Rev. Mod. Phys.* **38** 447; reprinted in Bell J S 1987 *Speakable and Unspeakable in Quantum Mechanics* (Cambridge: Cambridge University Press) p 1
[13] Bell J S 1971 *Foundations of Quantum Mechanics. Proc. of the Int. School of Physics 'Enrico Fermi'*, course IL (New York: Academic) p 171; reprinted in Bell J S 1987 *Speakable and Unspeakable in Quantum Mechanics* (Cambridge: Cambridge University Press) p 29
[14] Dewdney C, Holland P R and Kyprianidis A 1986 *Phys. Lett. A* **119** 259
[15] Tumulka R 2007 *J. Phys. A* **40** 3245
[16] Holland P R 1988 *Phys. Lett. A* **128** 9

[17] Valentini A 1992 *On the Pilot-Wave Theory of Classical, Auantum and Subquantum Physics*, PhD Thesis, International School for Advanced Studies, Trieste, online `http://www.sissa.it/ap/PhD/Theses/valentini.pdf`
[18] Valentini A 1996 *Bohmian Mechanics and Quantum Theory: an Appraisal*, ed J T Cushing *et al* (Dordrecht: Kluwer) p 45
[19] Bell J S 1984 CERN preprint CERN-TH. 4035/84; reprinted in Bell J S 1987 *Speakable and Unspeakable in Quantum Mechanics* (Cambridge: Cambridge University Press) p 173
[20] Dürr D, Goldstein S, Tumulka R and Zanghì N 2003 *J. Phys. A* **36** 4143
[21] Dürr D, Goldstein S, Tumulka R and Zanghì N 2004 *Phys. Rev. Lett.* **93** 090402
[22] Dürr D, Goldstein S, Tumulka R and Zanghì N 2005 *Commun. Math. Phys.* **254** 129
[23] Dürr D, Goldstein S, Tumulka R and Zanghì N 2005 *J. Phys. A* **38** R1
[24] Bohm D, Hiley B J and Kaloyerou P N 1987 *Phys. Rep.* **144** 349
[25] Colin S 2003 *Phys. Lett. A* **317** 349
[26] Colin S 2004 *Ann. Fond. Louis de Broglie* **29** 273

Journal of Physics: Conference Series **306** (2011) 012048 doi:10.1088/1742-6596/306/1/012048

Incorporation particle creation and annihilation into Bohm's Pilot Wave model

Roman Sverdlov

[1] Raman Research Institute,
C.V. Raman Avenue, Sadashiva Nagar, Bangalore, Karnataka, 560080, India

Abstract.
The purpose of this paper is to come up with a Pilot Wave model of quantum field theory that incorporates particle creation and annihilation without sacrificing determinism; this theory is subsequently coupled with gravity.

1. Introduction

The purpose of this paper is to redo Pilot Wave model for the case of creation and annihilation of particles. It is easy to see, however, that creation and annihilation of particles, being discrete can not be part of a differential equation that defines determinism. Dürr and others have proposed that deterministic evolution of quantum states is being "interrupted" with discrete jumps; the timing of the latter is determined by probabilistic laws (see [1]).

In this paper, however, I will propose a competing view which allows the determinism to hold *at all times*. I will accomplish this by claiming that there is no true creation/annihilation of particles. The number of particles is always fixed, but some of the particles hide from our view. In particular, there is extra compactified coordinate $x_4 \in \Gamma$, $x_4 + 2\pi = x_4$. Whenever $0 \leq x_4 < \pi$ the particle is *visible* and when $\pi \leq x_4 < 2\pi$ it is not. A particle undergoes *continuous* trajectory in all four coordinates, and whenever it *happens* to intersect either $x_4 = 0$ or $x_4 = \pi$ an *appearance* of creation/annihilation takes place.

Noteably, this means that the number of *visible* particles can be smaller than the actual number of particles, but it can not be larger. Therefore, we are working within realm of effective field theory with an upper bound on the total number of particles. This, in particular, means that lack of unitarity is one price we have to pay. For the purposes of this paper, we will pretend that we already have such effective theory, and our goal is to come up with Pilot Wave model within that theory.

Sections 2 and 3 will be devoted to do what we set out above in the context of flat space time. Then, in Section 4 we will couple our model to gravity. It has to be understood, however, that the prescription of *adding* gravity in Section 4 is meant to be generic for many different Pilot Wave models; therefore, it is independent of details of Chapters 2 and 3.

2. Creation and annihilation of particles

Let us now formulate more precisely the model of creation and annihilation of particles mentioned in the previous section. We will introduce a notation: $\vec{x}^{(4)} = (x_1, x_2, x_3, x_4)$ (where $x_4 + 2\pi = x_4$), and $\vec{x}^{(3)} = (x_1, x_2, x_3)$ (thus, $\vec{x}^{(4)} = (\vec{x}^{(3)}, x_4)$). In this notation, the phenomenon of particles hiding from our view can be expressed in definition of field operator, $\phi(\vec{x}^{(4)}) = \phi(\vec{x}^{(3)}, x_4)$ as

$$\phi(\vec{x}^{(3)}, x_4) = \phi(\vec{x}^{(3)}),\ 0 \leq x_4 < \pi; \phi(\vec{x}^{(3)}, x_4) = 1,\ \pi \leq x_4 < 2\pi \tag{2.1}$$

This means that if we act with this operator on an empty state, we get

$$\phi(\vec{x}^{(3)}, x_4)|0\rangle = |\vec{x}^{(3)})\rangle,\ 0 \leq x_4 < \pi; \phi(\vec{x}^{(3)}, x_4)|0\rangle = |0\rangle,\ \pi \leq x_4 < 2\pi \tag{2.2}$$

Now, in terms of $\vec{x}^{(4)}$, the left hand side of the above equations always amounts to ONE particle being present, regardless of the vaues of x_4. The right hand side, on the other hand, can give us *either* one particle state (if $0 \leq x_4 < \pi$) *or* empty state (if $\pi \leq x_4 < 2\pi$). This essentially defines what we meant when we say particles *hide*.

For simplicity we will assume from now on that there is only one kind of particles, and it happens to be spin 0. Then the configuration space is $(\mathbb{R}^3 \times \Gamma)^N$, where Γ denotes the extra compactified coordinate and N is the total number of particles (or, in other words, the *largest* possible number of visible ones). Then we can define the probability amplitude in $(\mathbb{R}^3 \times \Gamma)^N$ as

$$\psi(x_1^{(4)}, ..., x_N^{(4)}) = \langle 0|\phi(\vec{x}_1^{(4)})...\phi(\vec{x}_N^{(4)})e^{iH(t-t_0)}|\psi(t_0)\rangle \tag{2.3}$$

It shoud be emphasized that on the right hand side of the above equation, both $|\psi(t)\rangle = e^{iH(t-t_0)}|\psi(t_0)\rangle$, as well as the Hamiltonian H are defined in terms of $\vec{x}^{(3)}$, *not* $\vec{x}^{(4)}$. The only part where $\vec{x}^{(4)}$ comes in is the definition of field operators $\vec{x}_k^{(4)}$. The reason we can mix $\vec{x}^{(3)}$ with $\vec{x}^{(4)}$ is that we have an expression of $\phi(\vec{x}^{(4)})$ in terms of $\phi(\vec{x}^{(3)})$ that we mentioned earlier.

Now, it is clear that the total number of visible particles can be smaller than the actual number of particles, but it can not be larger. Thus, we are dealing with effective field theory with upper bound on total number of particles. On the other hand, e^{iHt} can generate states with larger number of particles than allowed. Therefore, we will throw away these states, and, subsequently, include extra coefficient in transition from ψ to the probability density ρ in order to make sure that the latter is normalized to 1 *among allowed states*:

$$\rho(\vec{x}_1^{(4)}, ..., x_N^{(4)}) = \frac{|\psi(\vec{x}_1^{(4)}, ..., x_N^{(4)})|^2}{\int d^4x_1...d^4x_N\ |\psi(\vec{x}_1^{(4)}, ..., x_N^{(4)})|^2} \tag{2.4}$$

It is important to point out that ρ is *not* a probability that we woud postulate in Copenhaggen interpretation. We are seeking a deterministic model. The sole source of probability is the ignorance of the observer regarding initial conditions.

As far as the deterministic law goes, this is something we are *about* to write (see the rest of the chapter). Thus, strictly speaking, the above probabilistic equation should be a *consequence* of the latter. The reason we have written probability *first* is that we are "cheating": we have *first* decided what probability we *want* to get and *then* we will write the deterministic law that accomplishes this.

However, it would be easier to cheat if we will assume that ρ has some physical meaning. Thus, we will view ρ as a physical field on configuration space, and our goal is to come up with dynamics such that the *subjective* probability *happens* to coincide with *objective* ρ. This will allow us to use ρ as one of the parameters in our deterministic equation.

It should be noticed that, for any given $\vec{x}^{(3)}$, ρ is constant throughout $x_4 \in [0, \pi)$ and $x_4 \in [\pi, 2\pi)$, respectively, while there is a discontinuity at $x_4 = 0$ and $x_4 = \pi$. Thus, a particle should move *towards* the boundary of one semicircle *if* it thinks that *eventually* crossing that boundary might be favorable. Apart from that, *once* the particle has reached a boundary, its trajectory has to remain continuous. Surprisingly enough, both of the above features are common place in electrostatics: step function charge distribution can produce continuous electric field, and, of course, the Coulumb's law is non local. Therefore, we follow idea mentioned in [3][1] and introduces electrostatics into our theory, by defining electric field $\vec{E}$ and charge density σ as

$$\vec{E} = \rho\vec{v}\ ;\ \sigma = -\frac{\partial\rho}{\partial t} \tag{2.5}$$

Of course, this is *not* literal electric field or charge: the theory holds both for charged *and* neutral fields. We are only using electrostatic notation for mathematical analogy. On any event, it is easy to see that if the "Coulumb's law" holds, then the continuity equation holds as well, which guarantees that ρ will be the "equilibrium probabiilty density" of our deterministic Pilot Wave model.

We now use the method of images in order to compute $\vec{E}$ in compactified geometry. We consider an imaginary situation where coordinates x_{4k} extend to infinity, and charge density is periodic in these coordinates,

$$\sigma(\vec{x}) = \sigma(\vec{x} + 2\pi\hat{x}_{4k}) \tag{2.6}$$

By translational symmetry, we notice that

$$\vec{E}(\vec{x}) = \vec{E}(\vec{x} + 2\pi\hat{x}_{4k}) \tag{2.7}$$

from which it is easy to see that the identical copy of $\vec{E}$ satisfies the same differential equation in the compactified geometry, without having any discontinuities at $x_{4k} = 2\pi n$. Thus, this is a solution we are seeking. Now, from the non-compact case, we can show that the area of the sphere is given by

$$A = \frac{2\pi^{2N}}{(2N-1)!}r^{4N-1} \tag{2.8}$$

This tells us that the electric field is

$$\vec{E}(\vec{x}) = \frac{(2N-1)!}{2\pi^{2N}} \sum_{a_1,\ldots,a_N} \int d^{4N}x' \frac{\sigma(\vec{x}')(\vec{x}' - \vec{x} + \vec{R}_{a_1,\ldots,a_N})}{|\vec{x}' - \vec{x} + \vec{R}_{a_1,\ldots,a_N}|^{4N}} \tag{2.9}$$

where

$$\vec{R}_{a_1,\ldots,a_N} = \sum 2\pi a_i \hat{x}_{4i} \tag{2.10}$$

is a displacement of an image charge. Now by substituting back

$$\vec{E} \to \rho^{(4)}\vec{v}\ ;\ \sigma \to \frac{\partial|\psi|^2}{\partial t} \tag{2.11}$$

we obtain a guidance equation

$$\vec{v}(\vec{x}) = \frac{(2N-1)!}{2\pi^{2N}\rho^{(4)}(\vec{x})}\frac{\partial}{\partial t} \sum_{a_1,\ldots,a_N} \int d^{4N}x' \frac{\rho^{(4)}(\vec{x}')(\vec{x}' - \vec{x} + \vec{R}_{a_1,\ldots,a_N})}{|\vec{x}' - \vec{x} + \vec{R}_{a_1,\ldots,a_N}|^{4N}} \tag{2.12}$$

[1] While in [3] they mentioned *electrostatic idea*, the discussion seemed to be brief, so I am not sure whether or not they had the same motivation. On any event, as far as I can tell after brief look at the paper, they were not explicitly talking about particle creation/annihilation, the extra compactified coordinates or other things I am focusing on

3. Decoherence

So far we have successfully came up with a model that reproduces desired probability density. However, there is one more ingredient to Pilot Wave model: effective collapse. It is well known that sometimes wave function in a configuration space splits into several non-overlappnig branches that correspond to different outcomes of the measurement (such as, for example, arrow of measuring apparatus pointing in different directions). If a particle happened to be in one of these branches, it can not go to the other branch, since it is prevented from being in a region between branches due to low value of $|\psi|^2$ in that region.

This phenomenon, however, is strictly a consequence of locality. In particular, a particle can not enter the region between the branches because of the low value of $|\psi|^2$ in the *immediate* vicinity; it does not care that $|\psi|^2$ will become high again *if* it travels far enough to get into another branch. Our electrostatic model, however, is non-local. Thus, a possibility of unwanted travel between parallel universes is a natural concern that arises in the cocntext of our model.

More specific source of the problem is that the electric charge density, $\sigma = -\partial\rho/\partial t$ is nearly zero between the branches, so nothing keeps the particle from that region. The only reason the probability of finding a particle between the branches is small is that $\vec{v}$ is inversely proportional to ρ, which simply means that *once* it entered the region in question it will fly away very fast. But nothing prevents it from flying away into a *different* branch! In fact, since the velocity of the particle is parallel to electric field, we *know* it would be the case if electric field lines of one branch intersect the other one (this issue was also mentioned at the end of chapter 3 in [3])!

At the same time, it should be understood that the non-locality in our model ws introduced on purpose. After all, in light of the fact that ρ is constant along each of the semicircles of Γ, a particle needs to have non-local ways of seeing whether or not it should move towards the edge of Γ in order to cross on the other side. Thus, we would like to have it both ways: we want a particle to see the other side of Γ, but, at the same time, we do *not* want it to see the other branches of splitted wave function.

One idea that comes to mind is replacing Coulumb's interaction with Yukawa, since this would give us the short-range interaction that we desire. This, however, will create non-zero divergence in the region of zero charge, which would violate continuity equation. Therefore, this proposal is not appropriate. The failure of Yukawa's approach to adress the issue also has deeper physical meaning. A sequence of short range non-localities will, inevitably, lead to long range one. Thus, we have a choice of either unwanted long-range non-locality (Coulumb's case) or a *failure* of desired short-range ones (Yukawa's case) in a form of violatoin of continuity equation.

In this work we will stick with the first option: long range Coulumb's law and undesired long range non-locality. In order to avoid undesired effects from other branches, we will set up a dynamics in such a way that these branches disappear! In other words, yes, a particle *can* see other branches *if* they are out there. But, they are not! Of course, we now have to modify deterministic dynamics of a wave function in such a way that the latter, in fact, diminishes in the regions where we want it to. This can be formally done by adding to the standard quantum field theory Hamiltonian (H_Q) an extra one (H_B):

$$H = H_Q + H_B \tag{3.1}$$

where the H_B is newly introduced influence *of* a beable particle *on* a wave function. It is designed in such a way that the wave function within any of the branches *not* occupied by a beable dies out, while the wave function within the one that *is* occupied is left unchanged.

Technically, H_B diminishes the value of wave function in any given branch as long as a particle is *not* in that branch. Thus, if it is in the gap between the branches, then *all* of them die off; but, in light of the normalization, the uniform impact on ψ has no bearing on ρ. Now, suppose there are only two branches: 1 and 2. If branch number 1 happened to be bigger than branch number 2, then the particle spends more time within the former than it does within the latter.

Thus, branch 1 dies out at a slower rate. In light of normalization of ρ this means that branch 1 will actually *in*crease, until its total probability becomes 1, while branch 2 will decrease to 0.

Now, the above paragraph, by itself, is not what we want: it might imply that if the total probability of branch 1 is 0.7 while the one of branch 2 is 0.3, then the branch 1 will end up a survivor with probability much *higher* than 0.7. This is where more familiar effective collapse pheonomenon comes into picture to help us. If, at the *first moment* the particle *happened* to be within branch 2, then it would take some small portion of time for a particle to leave the branch 2. If the rate of disraction of a branch is high enough, then during this small time portion the size of the branch 1 decreased from 0.7 to 0.07.

The unbiased picture starts when the particle is in the region between branches. However, since the value of ρ is very small in that region, and velocity of a particle is inversely proportional to ρ, the particle spends very little time there. In particular, the travel from the boundary of branch 2 to the neutral point takes much shorter time than the time spent within branch 2. During this short travel time, the probabilities within either of the two branches only underwent very small change: the probability within branch 2 changed from 0.3 to 0.2 and within branch 1 it changed from 0.07 to 0.06. Then, the neutral competition starts between 0.06 and 0.2; and, predictably, 0.2 wins.

Thus, just like more standard models tell us, the deciding factor is which branch did a particle happen to occupy immediately after the split, *not* which branch is bigger. Just like in more standard Pilot Wave models, locality is a key element of the argument (here, as always, by locality we mean the one in multidimensional configuration space, *not* the one in ordinary space). This time, however, for the reasons stated earlier we could not use the locality of the influence of wave on a particle; instead, we have used the locality in the trajectory of the particle, itself (namely, the continuity of the latter). Thus, while the particle can non-locally see the other branch, it is not allowed to jump there. During the short time it is forced to take in order to leave the original branch, the other branch dies off which is the ultimate reason of collapse of a wave funciton.

Let us now go ahead and come up with a precise definition of H_B. In order to do this, we have to define the notion of being in the same branch as our particle. We will do that by postulating the field χ that obeys the following equation:

$$\frac{\partial \chi}{\partial t} = a\nabla^2 \chi + b e^{k|\vec{x}-\vec{x}_B|^2} - \frac{\lambda \chi(\vec{x})}{\rho(\vec{x})}, \tag{3.2}$$

where λ is assumed to be very small. From similarity with diffusion equation, $\rho \ll \lambda$ serves as a barrier of penetration of χ, while in the $\rho \gg \lambda$ region χ fills out the whole connected region, regardless of specifics of ρ. Thus, if a particle is inside one of the branches, χ will be nearly constant inside that one branch and nearly zero everywhere else.

Now, the simplest thing we can now do is define H_B as an *imaginary* expression $-i\delta/\chi$, where δ is some small constant. From the similarity with diffusion equation, we know that $\chi > 0$. Thus, if χ is reasonably large, $e^{-iH_B t} \approx 1$. At the same time, if $\chi \ll \delta$, we get $e^{-iH_B t} \approx 0$ which, of course, would diminish the value of ψ within the wrong branches.

We have to be a bit careful, since H was defined on ($\vec{x}^{(3)}$-based) Fock space, and *not* on $(\mathbb{R}^3 \times \Gamma)^N$ we are working on. In order to draw a link between the two spaces, for every state $|s\rangle$ in our Fock space we will define a *set* of points $S(|s\rangle)$ in our configuration space that correspond to $|s\rangle$. It is easy to see from our construction that the above set is an uncountable union of semicircles in x_4, each corresponding to different value of $\vec{x}^{(3)}$. We will now define H_B on the Fock space as follows:

$$H_B|s\rangle = -\Big(\int_{S(|s\rangle)} d^{4N}x \, \frac{i\delta}{\chi(\vec{x})}\Big)|s\rangle \tag{3.3}$$

In light of the structure of diffusion equation, it is possible that χ might have different values on the same semicircle which might lead to some ambiguity. It is, however, reasonable to expect that χ will either be consistenly large or consistently small throughout each semicircle; but that should be investigated more closely in further research.

4. Adding gravity to Pilot Wave model

In order to avoid the issue of quantum gravity altogether, I propose to introduce to view gravity as a classical field, outside of the quantum fields (such as ϕ) which are subjects to above model. Furthermore, in order to be able to use the results from previous sections, all of which are based on *flat* space, I will *not* view gravity as geometry either. Our space is fundamentally flat, and the presence of curvature is merely an illusion produced due to the coincidence that gravity *happened* to couple to all other fields (including itself) equally. This implies that we can unambiguously define Hilbert space, and also we can continue to use flat-space-based Coulumb's Law results from previous sections.

The coupling of gravity to other fields is expressed in explicit dependence of Hamiltonian on the metric. Thus, the evolution in $\vec{x}^{(3)}$ will change from $|\psi(t)\rangle = e^{iH(t-t_0)}|\psi(t_0)\rangle$ to

$$\frac{d}{dt}|\psi(t)\rangle = H(g_{\mu\nu}(\mathbb{R}^3))|\psi(t)\rangle \tag{4.1}$$

where $H(g_{\mu\nu}(\mathbb{R}^3))$ tells us that H depends on the value of metric *on* $t = const$ *hypersurface* $\mathbb{R}^3$; but, at the same time, it depends on *all* components of a metric, including the timelike ones. In light of flat geometry, after the above modification is made, the rest of the equations from previous section remain unchanged.

The hard part, however, is writing down how gravitational field is produced. In particular, since gravitational field is classical, we would like it to be an *exact* solution either to Einstein's equation or to some modification of it. Now, from Noether's Theorem, we know that the exact solution to Einstein's equation only exists if the energy-momentum tensor is conserved. Now, we have three options of sources (all of which, as we will see, will have some conceptual issues to adress, and the lack of conservation of energy-momentum is one of the problems):

1) Bohmian particle. In order to avoid black hole configuration, we can simply replace delta function with Gaussian. This will produce a continuous energy-momentum distribution:

$$T_{\mu\nu}(\vec{x}^{(3)}) = \sqrt{\frac{\pi}{a}} \sum_{k=1}^{N} \sum_{0\leq\mu,\nu\leq 3} \frac{dx_B^{5k+\mu}}{d\tau_k}\frac{dx_B^{5k+\nu}}{d\tau_k} m_k f(x_{5k+4}) e^{-a|\vec{x}^{(3)}-\vec{r}_k(\vec{x}_B)|^2} \tag{4.2}$$

where

$$\vec{r}_k(\vec{x}_B) = (x_B^{5k+1}, x_B^{5k+2}, x_B^{5k+3}) \; , \; d\tau_k = \sum_{0\leq\mu,\nu\leq 3} g_{\mu\nu} dx^\mu dx^\nu, \tag{4.3}$$

m_k is a mass of particle number k (which is introduced for a more general case where more than one kind of field is considered), and

$$f(x_4) = \frac{1}{2}(1 + \tan^{-1}(n \; sin \; x_4)) \tag{4.4}$$

is a differential approximation to step function that is needed in order to assure that the hidden particles do not contribute to gravitational field. The differential approximation to step function is needed in place of actual step function in order to avoid singular effects on gravity produced by the latter. Finally, in the above expressions the Einstein's convention was ignored in order to avoid confusion between N-term sums and 4-term ones.

2) The source of energy momentum tensor is a wave function. Thus, the energy momentum tensor is given by

$$T_{\mu\nu}(\vec{x}^{(3)}) = \sum_{k=1}^{N} \frac{\partial \psi}{\partial x^{5k+\mu}} \frac{\partial \psi}{\partial x^{5k+\nu}} \tag{4.5}$$

Furthermore, here are two options of defining ψ:
a) By using the evolution based on H_Q
b) By using the evolution based on $H_Q + H_B$

It is easy to see that option "2a" predicts gravitational interaction between different branches of wave function, which means that the gravitational field produced by objects in our universe will be lost among gravitational fields produced in all of the other parallel ones. On the other hand, in cases "1" and "2b", the influence of acceleraton of particle on a wave leads to energy momentum non-conservation. By the way, even in 2a, non-conservation occurs in a much milder form, namely it is a consequence of non-unitarity of the theory (due to the upper bound on particle number). Thus, in order to avoid *additional* problem, namely the interaction between universes, we will stick to options 1 and 2*b*, even though it means we have to face non-conservation of energy in a more severe form. (For this reason, it is important to be able to modify gravity in such a way that it has exact solutions for non-conserved sources. We can attempt to find a solution by evolving our metric along timelike curve. However, if we select timelike geodesics, they might intersect which would lead to possible contradiction between the results produced through the evolution along each of the two geodesics*unless* the sources meet some condition. That condition would be of a form of "curl equals zero" which is just as bad as energy conservation that we were trying to avoid.

Therefore, we instead introduce a flat set of coordinates that are guaranteed not to intersect, while gravity is viewed as merely a field in these coordinates (which, by the way, gives us a good excuse not to generalize our electrostatics from previous chapter to curved space). Since Einstein's equation for arbitrary metric is too complicated to analyze, we will consider a linear approximation. Within this approximation we will first convince ourselves that the equation is not invertible, and then find a way to modify it to make it such. The linear approximation to Einstein's equation is

$$R_{\mu\nu} - \frac{1}{2} R g_{\mu\nu} = -\frac{1}{2}\partial^\alpha \partial_\alpha h_{\mu\nu} + \frac{1}{2}\partial_\mu \partial_\alpha h^\alpha_\nu + \frac{1}{2}\partial_\nu \partial_\alpha h^\alpha_\mu - \frac{1}{2}\eta_{\mu\nu}\partial_\alpha \partial_\beta h^{\alpha\beta} - \frac{1}{2}\partial_\mu \partial_\nu h + \frac{1}{2}\eta_{\mu\nu}\partial^\alpha \partial_\alpha h \tag{4.6}$$

We will now compute the above expression for all choices of $(\mu\nu)$. However, we will leave out all terms except for the second time derivatives. We will denote the expressions *not* containing second time derivatives by dots. Thus, we get

$$R_{00} - \frac{1}{2} R g_{00} = \cdots \; ; \; R_{11} - \frac{1}{2} R g_{11} = \frac{1}{2}(\partial_0 \partial_0 h^{22} + \partial_0 \partial_0 h^{33}) + \cdots$$

$$R_{10} - \frac{1}{2} R g_{10} = \cdots \; ; \; R_{12} - \frac{1}{2} R g_{22} = -\frac{1}{2} \partial_0 \partial_0 h^{12} + \cdots \tag{4.7}$$

The rest of the components are obvious from the permutting indeces. Now, in order to find a dynamics, we have to solve the above equation for $\partial_0 \partial_0 h^{\mu\nu}$. Thus, in order to see whether or not the equation is solvable, we have to compute a determinant of corresponding 10×10 tensor. In order to write that tensor in the most transparent form, we will define coordinates of vectors it acts upon as follows:

$$(v_1, \cdots, v_{10}) = \partial_0 \partial_0 (h^{00}, h^{11}, h^{22}, h^{33}, h^{01}, h^{02}, h^{03}, h^{12}, h^{13}, h^{23}) \tag{4.8}$$

In these coordinates, our equation takes the form $A\vec{v} = \vec{w}$, where

$$A = \frac{1}{2}\,\mathrm{diag}\,(0_{1\times 1}, B_{3\times 3}, 0_{3\times 3}, -I_{3\times 3}) \tag{4.9}$$

and

$$B = \begin{pmatrix} 0 & 1 & 1 \\ 1 & 0 & 1 \\ 1 & 1 & 0 \end{pmatrix} \tag{4.10}$$

As a result of $0_{1\times 1}$ and $0_{3\times 3}$ components of a diagonal, the determinant is zero. This confirms our earlier prediction that Einstein's equation, as it stands, is not invertible. However, if we add a diagonal $\epsilon I_{10\times 10}$, then the 0-s will become ϵI-s (where I is an identity matrix). This, together with non-zero determinant of B will prove the non-zero determinant of $A + \epsilon I$.

Now, adding this extra term is equivalent to adding $\epsilon \partial_0 \partial_0 h^{\mu\nu}$ to our equations. Since we would like to have a covariant theory, we instead add a term $\epsilon \partial^\alpha \partial_\alpha h_{\mu\nu}$, which for a general metric generalizes to $\epsilon \partial^\alpha \partial_\alpha h_{\mu\nu}$:

$$R_{\mu\nu} - \frac{1}{2} R g_{\mu\nu} + \epsilon \partial^\alpha \partial_\alpha g_{\mu\nu} = T_{\mu\nu} \tag{4.11}$$

which corresponds to a Lagrangian

$$S = \int d^4x\,\sqrt{-g}\,\left(R + kT + \frac{\epsilon}{2}\,\partial^\alpha g^{\mu\nu} \partial_\alpha g_{\mu\nu}\right) \tag{4.12}$$

where ϵ-term likewise breaks the general relativistic covariance.

It is important to notice that, in light of ∂_α being used in place of ∇_α, the ϵ-term in the above equation is *not* covariant. In fact, in light of the fact that $\nabla_\alpha g_{\mu\nu} = 0$, we couldn't have possibly made it covariant even if we wanted to. However, it can be argued based on Noether's theorem that violation of covariance is necessary if we desire to violate a conservation law.

At the same time, however, not all violations of covariance would have served our purpose. For example, adding mass term would have had no impact on above matrix and, therefore, determinant would have continued to be zero. This can be explained by saying that, while mass term *does* violate the covariance we are familiar with, the Lagrangian might still have been symmetric with respect to something else that we don't easilly recognize. This would also lead to a conservation law that we would not recognize either, but the latter would have been just as restrictive as the energy momentum one mass term would have gotten rid of. On the other hand, the kinetic term above does *not* produce any unrecognized symmetries/conservation laws, since the determinant of coefficient is non-zero.

It is important to admit, however, that in the above analysis we were working exclusively in the linear approximation. For the general metric, it is quite possibe for determinant to accidentally become zero. In fact, if the determinant changes sign between two points, then its continuity implies a presence of hypersurface in which it is 0. The surface, itself, is infinitely smaller than any three dimensional region and, therefore, insignificant. We do, however, have to avoid the *impact* of that surface onto the future region.

In order to do this, we have to tame the behavior of time derivatives somehow. For now, we will settle on violating relativity (our excuse is that we have already done that, anyway, with Pilot Wave model), although for a future it might be worthwhile trying to find more covariant ways of doing that. As we said before, we can write the second time derivatives in a form of a single vector $\vec{v} \in \mathbb{R}^{10}$, which allows us to write the dynamics as $\vec{v} = A^{-1}\vec{w}$, for the tensor A specified earlier. We will now replace it with

$$\vec{v} = \left(\frac{\tan^{-1}(\eta\; det\; A^{-1})}{\eta\; det\; A^{-1}}\right)^{1/10} A^{-1}\vec{w} \tag{4.13}$$

where η is a very small number. It is easy to see that the determinant of modified version of A^{-1} coincides with the determinant of A^{-1} itself as long as the latter does not blow up. When it does, however, the former approaches $\pi/2\eta$.

To satisfy more mathematically minded people, it might be an interesting project for future to modify principle of least action in such a way that the above modification of dynamics is dictated to us by substitutting un-modified Lagrangian into modified Lagrange's equation (that would probably require us to single out the preferred variables in order for modification to be applied to $\partial_t \partial_t g_{\mu\nu}$ instead of, say, $(\partial_t \partial_t g_{\mu\nu})^2$, since the covariant problem of finding extreme point of an action does not give us the dynamics we want). Either that, or else we can search for the modifications of Lagrangian itself in order for principle of least action to dictate us the above modified dynamics. I will leave both of these questions for the future research.

As far as the above equation of motion, itself, is concerned, we can do even better by removing $1/10$ from the power of the above coefficient. In this case, near the singular region the second time derivatives will approach zero, which is just *the opposite* of the issue we were trying to avoid! We have to be careful if we want to do that, because this might potentially freeze the singularity once it is reached. What comes to our rescue is that the *first* time derivatives continue to be non-zero. The space variation of the latter will eventually allow us to get out of the singular region. But, of course, this needs to be investigated more closely.

While modification of gravity might look artificial, it is important to point out that it has some benefits beyond the Pilot Wave model. For example, it might adress the problem with singularities that arise in theories involving massive gravitons. As can be seen from the equation 2.151 of [4], the singularities are linked to the equation

$$-3m_g^2 h = \frac{\tilde{k}}{2} T \tag{4.14}$$

where m_g is a graviton mass. Now, by looking at the above equation, one can immediately see that it does not involve second time derivatives! In our language this means that the above equation puts a restrictions on initial conditions on $t = \text{const}$ hypersurface, and if these initial conditions are not met the equation *for second time derivatives* has no solution. In other words, the equation is not invertible.

Thus, by making it invertible through the kinetic ϵ-term, we automatically get rid of the above static equation, and thus allow for non-singular theories involving massive graviton. In fact, if the mass of the graviton is singificantly smaller than some function of ϵ then the the existence of solution around $m_g = 0$, toghether with continuity in m_g, implies a nice behavior of the solution. If, on the other hand, m_g is much larger than the above function of ϵ, then the $\epsilon \approx 0$ approximation will imply nearly-sigular behavior. This means that there might be some correlation between ϵ and the upper bound on graviton mass.

At the same time, we can use Pilot Wave model at hand to obtain a *lower* bound on ϵ (after all, if the one didn't exist, nothing would have stopped us from setting ϵ to 0). This would then imply a *lower* bound on the *upper* bound on m_g. It would be interesting to compare this to the graviton masses currently proposed by other arguments.

5. Conclusions

In this paper we have accomplished two things. First of all, we have introduced creation and annihilation of particles into Pilot Wave model. We have done that through introducing extra compactified dimension and postulating that the particles on the wrong side of the semicircle are invisible. This allowed us to view a discrete creation and annihilation events as merely manifestations of *continuous* trajectories in our extended space. We then introduced Pilot Wave model in order to describe the latter deterministically.

Then, in the second part of the paper, we have added gravity to the theory by associating energy momentum tensor with our beable particles (which is smeared out in space in order to avoid black hole configuration). The latter serves as a source of *classical* gravitational field. However, as a result of acceleration of beable particles, energy-momentum tensor is not conserved, which means that Einstein's equation, in its original form, has no exact solution.

Thus, we were forced to add extra kinetic term to the gravitational action with very small coefficient. That coefficient breaks general relativistic covariance and, at the same time, makes equation invertible and, therefore, guarantees the existence of a solution within a region bounded by some hypersufrace. Nevertheless, we still had to introduce *further* modification in order to tame the behavior of a solution in vicinity of that surface, which ultimately allows it to extend beynd that surface. Doing so required us to violate relativity.

Finally, we have pointed out that the modification of gravity had a side benefit: massive graviton can be introduced without resulting singularities! More precisely, the standard prediction that the graviton mass should be *exactly* zero (see [4]) is transformed into a conclusion that the mass is much smaller than some function of ϵ, where ϵ is a coefficient of our newly introduced kinetic term.

While massive gravitons are *not* needed for our Pilot Wave model, I believe that the prediction of the upper bound to their mass is important since it refutes a potential aesthetic objection that I went through the hoops of modifying gravity only for one purpose (that is, Pilot Wave model) which might, in some people's eyes, indicate dishonesty of the theory. For that same reason, it is important to explore other potential uses of modification of gravity.

One thing that naturally comes to mind is dark matter problem. There have been a lot of claims that dark matter does not exist and the phenomena that are typically interpretted as the effects of dark matter are, in fact, due to the modifications of gravity that are not felt on our everyday scales but become increasingly important cosmologically. It might be interesting to replace the modifications of gravity that has been proposed before with the one proposed in the current work and compare the differences in cosmological predictions.

Interestingly enough, while the above cosmological problem does not have a direct link to Pilot Wave model, it *will* have indirect one. In particular, we will use Pilot Wave model in order to judge the value of the constant ϵ to accomodate it. We will then use *this same* ϵ for the dark matter problem! This means that Pilot Wave model will begin to have predictive power, which will hopefully bring it into a mainstream interest.

References

[1] Dürr D, Goldstein S, Tumulka R and Zanghì N 2003 *J. Phys. A: Math. Gen.* **36** 4143-4149, and `arXiv:quant-ph/0208072`.
[2] Sverdlov R `arXiv:gen-ph:1010.0580`.
[3] Struyve W 2010 *Foundation of Physics* **40**, 1700-1711
[4] Maggiore M 2008 *Gravitational waves. Volume 1: Theory and Experiments*, Oxford University Press

de Broglie Deterministic Dice and emerging Relativistic Quantum Mechanics

Donatello Dolce

The University of Melbourne, School of Physics, Parkville VIC 3010, Australia.

E-mail: ddolce@unimelb.edu.au

Abstract. Generalizing the de Broglie hypothesis of intrinsically periodic matter fields, it can be shown that the basic quantum behavior of ordinary field theory can be retrieved in a semi-classical and geometrical way. The geometrodynamical description of interactions that arises in this theory provides an intuitive interpretation of the Maldacena's conjecture and it turns out to be of the same type of the one prescribed by general relativity.

1. Introduction

Similarly to a particle in a box, relativistic fields can be quantized by imposing their characteristic de Broglie periodicities as constraints [1, 2, 3, 4]. In this way elementary systems can be thought of as "de Broglie internal clocks", that is as relativistic fields with intrinsic dynamical de Broglie time periodicities T_t. Even for a light particle such as the electron, the intrinsic time periodicities are extremely fast ($T_t \lesssim 10^{-20} s$), many orders of magnitude away from the present experimental resolution in time. As in a dice rolling too fast with respect to a given resolution in time, these de Broglie internal clocks can be only described statistically. It can be shown that the effective statistical description emerging from such periodic dynamics matches (without fine tunings) ordinary relativistic Quantum Mechanics (QM) [1]. The idea is similar to the "stroboscopic quantization" [5] or to the 't Hooft determinism [6]. At the same time the underling classical-relativistic physics seems to solve many conceptual difficulties of the canonical quantum theory. In particular they do not involve any local-hidden-variable so that we can actually speak about deterministic quantization. Our assumption of dynamical periodic fields can be regarded as a combination of the Newton's law of inertia and de Broglie-Planck hypothesis of periodic matter waves: *elementary isolated systems must be supposed to have persistent periodicities as long as they do not interact.* To the energy of an elementary system $\bar{E} = \hbar\bar{\omega}$ we associate an intrinsic time periodicity is $T_t = 2\pi/\bar{\omega}$. Furthermore T_t must be considered to be dynamical, since it is related to the energy through the de Broglie like relation $T_t = h/\bar{E}$. This dynamical behavior of the periodicities guarantees full consistence with relativistic causality and allows time ordering. Interactions can be introduced in the theory as variations of the de Broglie periodicities of the fields. They can be equivalently formulated in terms of geometric deformations of the compact space-time dimensions of fields with Periodic Boundary Conditions (PBCs), in a way that mimics very closely the geometrodynamics of General Relativity (GR). The good behaviors of the theory arise from the fact that it adds a cyclic property to the usual notion of relativistic time. After all time can only be defined by counting the number of cycles of isolated phenomena *supposed* to be periodic[1] such as the Cs atom ($T_{Cs} \sim 10^{-10}s$). This approach turns out to be of particularly interest to address the problem of the arrow of time.

2. Cyclic fields

The relativistic generalization of the Newton's law of inertia states that every isolated elementary system has persistent four-momentum $\bar{p}_\mu = \{\bar{E}/c, \bar{\mathbf{p}}\}$. On the other hand, the de Broglie-Planck formulation of QM prescribes that the four-momentum must be associated to the four-angular-frequency $\bar{\omega}_\mu = \bar{p}_\mu c/\hbar$ of a corresponding field. Here we will assume that every elementary system is described in terms of intrinsically periodic fields whose periodicities are the usual de Broglie-Planck periodicities $T^\mu = \{T_t, \vec{\lambda}_x/c\}$. As the Newton's law of inertia doesn't imply that every point particle moves on a straight line, our assumption of intrinsic periodicities does not mean that the physical world should appear to be periodic. In fact, the four-periodicity T^μ is fixed dynamically by the four-momentum throught the de Broglie-Planck relation

$$T^\mu = \frac{2\pi}{\bar{\omega}_\mu} = \frac{h}{\bar{p}_\mu c}. \quad (1)$$

Thus, the variation of four-momentum that occurs during interactions implies a variation of the intrinsic periodicities of the fields. This will guarantee time ordering and relativistic causality.

Considering for the sake of simplicity only the case of periodicity $T_t(\mathbf{p})$ along the time component, the cyclic field turns out to be decomposed as a tower of frequency eigenmodes

$$\Phi(\mathbf{x}, t) = \sum_n A_n \phi_n(\mathbf{x}) u_n(t) , \qquad \text{where} \qquad u_n(t) = e^{-i\omega_n(\bar{\mathbf{p}})t} . \quad (2)$$

Its quantized angular frequency spectrum is

$$\omega_n(\mathbf{p}) = n\bar{\omega}(\bar{\mathbf{p}}) = n\frac{2\pi}{T_t(\bar{\mathbf{p}})} . \quad (3)$$

The eigenmodes correspond to the harmonics modes of a string vibrating in a compact dimension with compactification length $T_t(\mathbf{p})$ and PBCs, *i.e.* in a cyclic dimension. Such a system can be imagined as a so called "de broglie internal clock", that is a periodic phenomenon whose time periodicity is fixed dynamically by the inverse of its energy $T_t(\mathbf{p}) = h/\bar{E}(\mathbf{p})$. Thus, according to the de Broglie-Planck relation eq.(1), we have a quantized energy spectrum

$$E_n(\mathbf{p}) = n\bar{E}(\mathbf{p}) = n\frac{h}{T_t(\mathbf{p})} . \quad (4)$$

The dependence on the spatial momentum $\bar{\mathbf{p}}$ means that the time periodicity depends on the reference frame. Indeed $T_t(\mathbf{p})$ is a time interval. Therefore it transforms dynamically according to the Lorentz transformations.

In a relativistic framework it is easy to figure out that time periodicity induces periodicities on the spatial dimensions as well. They must be necessarily considered to have a consistent and Lorentz invariant theory. Since in this case the whole physical information of the system is contained in a single four-period T^μ, our intrinsically four-periodic free field can be described by the following action in compact four dimensions with PBCs

$$\mathcal{S}_{\lambda_s} = \int_0^{T^\mu} d^4x \mathcal{L}_{\lambda_s}(\partial_\mu \Phi, \Phi) . \quad (5)$$

[1] *"By a clock we understand anything characterized by a phenomenon passing periodically through identical phases so that we must assume, by the principle of sufficient reason, that all that happens in a given period is identical with all that happens in an arbitrary period."* A. Einstein [7]

It is important to note that PBCs minimize the action at the boundaries, in particular of the compact time dimension. Therefore they have the same formal validity of the usual (Synchronous) BCs assumed in ordinary field theory (fixed variation at the time boundaries, *i.e.* $\delta\Phi|_\Sigma = 0$). This is an essential feature because it guarantees that all the symmetries of the relativistic theory are preserved (as in usual field theory), even though we are assuming PBCs. In particular it guarantees that the theory is Lorentz invariant. For instance we can assume a generic Lorentz transformation

$$dx^\mu \to dx'^\mu = \Lambda^\mu_\nu \, dx^\nu \,, \qquad \bar{p}_\mu \to \bar{p}'_\mu = \Lambda^\nu_\mu \, \bar{p}_\nu \,. \tag{6}$$

By definition, the four-periodicity T^μ is such that the phase of the field is invariant under four-periodic translations $\exp\left[-ix^\mu \bar{p}_\mu\right] = \exp\left[-i(x^\mu + T^\mu)\bar{p}_\mu\right]$. In this way we see that the four-periodicity is actually a contravariant four-vector. It transforms under Lorentz transformations in the following way

$$T^\mu \to T'^\mu = \Lambda^\mu_\nu \, T^\nu \tag{7}$$

and the phase of the field is a scalar quantity under Lorentz transformations - de Broglie phase harmony. This can be also inferred by noticing that after the transformation of variables eq.(6), the free action eq.(5) turns out to be

$$\mathcal{S}_{\lambda_s} = \int_0^{T'^\mu} d^4x' \mathcal{L}_{\lambda_s}(\partial'_\mu \Phi, \Phi) \,. \tag{8}$$

Therefore, in the new reference system, the new four-periodicity T'^μ of the field is actually given by eq.(7), that is eq.(8) describes a system with four-momentum $\bar{p}'_\mu$ given by eq.(6).

Naming the proper-time periodicity of massive particles (known as the "deBroglie periodic phenomenon") as T_τ, it is easy to figure out that the four-periodicity transforms in the relativistic way eq.(7) according to the following constraint

$$\frac{1}{T_\tau^2} = \frac{1}{T_t^2(\mathbf{p})} - \frac{c^2}{\vec{\lambda}_x^2(\mathbf{p})} = \frac{1}{T_\mu}\frac{1}{T^\mu} \,. \tag{9}$$

It can be derived for instance by dividing the relativistic dispersion relation $\bar{M}^2c^4 = \bar{E}^2(\bar{\mathbf{p}}) - \bar{\mathbf{p}}^2c^2 = \bar{p}^\mu \bar{p}_\mu c^2$ by the Planck constant, that is by using eq.(1).

If we give energy to a field through interaction or we move away from the rest frame we have a relativistic deformation of its intrinsic time periodicity, as shown in fig.(1.b) and eq.(7). Therefore every level of the quantized energy spectrum transforms according to a relativistic dispersion relation. The resulting quantized energy spectrum of the field is

$$E_n(\bar{\mathbf{p}}) = nE(\bar{\mathbf{p}}) = n\sqrt{\bar{M}^2c^4 - \bar{\mathbf{p}}^2c^2} \,. \tag{10}$$

Thus, we find that such periodic fields have the same energy spectrum of the ordinary second quantized fields after normal ordering, as shown in fig.(1.a). In this way we see that this quantization prescription represents a generalization to relativistic fields of the semi-classical quantization of a particle in a box. This approach has also interesting analogies with the Matsubara [8] and with the Kaluza-Klein (KK) [9, 10] theories.

It must be noticed from eq.(9) that the proper-time periodicity T_τ is related to the mass of the field $T_\tau = T_t(0) = h/\bar{M}c^2$. Moreover, since the proper-time τ is proportional to the world-line parameter $s = c\tau$, we find that in this theory the world-line parameter is a cyclic variable with compactification length

$$\lambda_s = cT_\tau = \frac{h}{\bar{M}c} \,. \tag{11}$$

It actually corresponds to the Compton wavelength of matter fields.

The proper-time periodicity fixes the upper bond of the time periodicity, $T_\tau \geq T_t(\bar{\mathbf{p}})$, since the mass is the lower bond of the energy, $\bar{M}c^2 \leq \bar{E}(\bar{\mathbf{p}})$. The heavier the mass the faster the proper-time periodicity. For instance, even a light particle such as the electron has (in a generic reference frame) intrinsic time periodicity equal or faster than $10^{-20}s$, *i.e.* faster than its proper-time periodicity. It should be noted that it is many orders of magnitude away from the Cs atomic clock whose periodicity is by definition of the order of $10^{-10}s$ and it is extremely fast even if compared with the present experimental resolution in time ($\sim 10^{-17}s$).

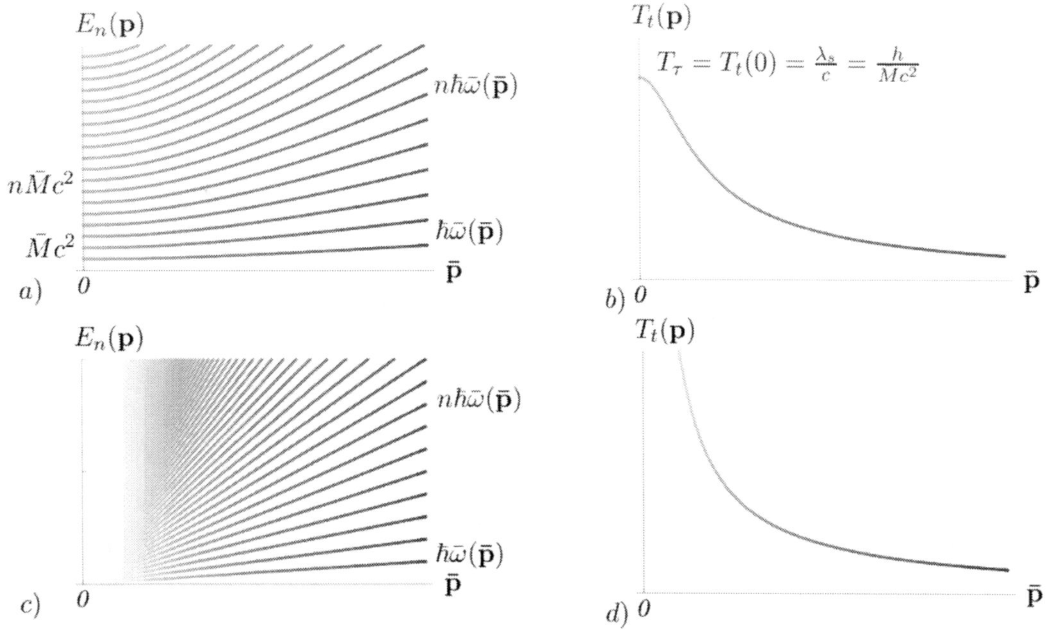

Figure 1. Dispersion behaviour *a*) - *c*) - of the quantized energy spectrum $E_n(\mathbf{p})$ and *b*) - *d*) - the corresponding time periodicity $T_t(\mathbf{p})$ of a massive - massless - field in compact dimensions.

It is interesting to point out that since the world-line parameter s is compact, we have an deep analogy with ordinary string theory where one of the two world-sheet parameters is supposed to be compact or periodic. Thus our field theory can be classified as a particular kind of string theory - the periodic field eq.(2) is a vibrating string. We will also see that our field theory turns out to be "dual" to a extra-dimensional field theory whose compactification length is λ_s.

3. Correspondence with Quantum Field Theory

Now we briefly show that this description provides a remarkable matching with the canonical formulation of QM as well as with the Feynman Path Integral (FPI) formulation. The evolution along the compact time dimension is described by the so called bulk equation of motions $(\partial_t^2 + \omega_n^2)\phi_n(x,t) = 0$ - for the sake of simplicity in this section we assume a single spatial dimension x. Thus the time evolution of the energy eigenmodes can be written as first order differential equations $i\hbar\partial_t\phi_n(x,t) = E_n\phi_n(x,t)$. The periodic field eq.(2) is a sum of on-shell standing waves. Actually this is the typical case where a Hilbert space can be defined. In fact, the energy eigenmodes form a complete set with respect to the inner product

$$\langle\phi|\chi\rangle \equiv \int_0^{\lambda_x} \frac{dx}{\lambda_x}\phi^*(x)\chi(x)\,. \tag{12}$$

Therefore we can define the Hilbert eigenstates as $\langle x|\phi_n\rangle \equiv \phi_n(x)/\sqrt{\lambda_x}$. On this base we can formally build a Hamiltonian operator $\mathcal{H}|\phi_n\rangle \equiv \hbar\omega_n|\phi_n\rangle$ and a momentum operator $\mathcal{P}|\phi_n\rangle \equiv -\hbar k_n|\phi_n\rangle$, where $k_n = n\bar{k} = nh/\lambda_x$. Thus the time evolution of an generic state $|\phi(0)\rangle \equiv \sum_n a_n|\phi_n\rangle$ is actually described by the familiar Schrödinger equation

$$i\hbar\partial_t|\phi(t)\rangle = \mathcal{H}|\phi(t)\rangle. \tag{13}$$

Moreover the time evolution is given by the usual time evolution operator $\mathcal{U}(t';t) = \exp[-\frac{i}{\hbar}\mathcal{H}(t-t')]$ which turns out to be a Marcovian operator: $\mathcal{U}(t'';t') = \prod_{m=0}^{N-1}\mathcal{U}(t'+t_{m+1};t'+t_m-\epsilon)$ where $N\epsilon = t''-t'$.

From the fact that the spatial coordinate is in this theory a cyclic variable, by using the definition of the expectation value of an observable $\hbar\partial_x F(x)$ between to generic initial and final states $|\phi_i\rangle$ and $|\phi_f\rangle$ of this Hilbert space, and after integrating by parts eq.(12), we find

$$\langle\phi_f|\hbar\partial_x F(x)|\phi_i\rangle = \langle\phi_f|\mathcal{P}F(x) - F(x)\mathcal{P}|\phi_i\rangle\ . \tag{14}$$

Assuming now that the observable is such that $F(x) = x$ [11] we obtain the usual commutation relation of ordinary QM: $[x,\mathcal{P}] = i\hbar$. With this result we have checked the correspondence with canonical QM.

Furthermore, it is possible to prove the correspondence with the FPI formulation. In fact, it is sufficient to plug the completeness relation of the energy eigenmodes in between the elementary time evolutions of the Marcovian operator. With this elements at hand and proceeding in a complete standard way we find the usual FPI which, in phase space, can be written in this way

$$\mathcal{Z} = \lim_{N\to\infty}\int_0^{\lambda_x}\left(\prod_{m=1}^{N-1}dx_m\right)\prod_{m=0}^{N-1}\langle\phi|\, e^{-\frac{i}{\hbar}(\mathcal{H}\Delta\epsilon_m - \mathcal{P}\Delta x_m)}\,|\phi\rangle\ . \tag{15}$$

This important result has been obtained without any further assumption than PBCs. It can be understood intuitively even in a graphical way. For the sake of simplicity we again consider only time periodicity. We have to imagine a cylinder where there is an infinite set of possible classical - straight - paths with different winding numbers that link every given initial and final points. If we imagine to open this cylinder, the paths with different winding number are lines, but the initial and final points are no more the same - they differ by periods - fig.(2.a). Since these evolutions are Marcovian and invariant under periodic translations, we can ideally cut these paths in elementary parts, translate the resulting elementary paths by periods and combine them, fig.(2.b). In the resulting combination of paths the degeneracy under periodic translations has been removed, that is all the resulting paths obtained from this combination of elementary periodic paths have the same initial and final points. Repeating this procedure several times, fig.(2.c), it is easy to see that it reproduces - see [2] for a mathematical proof - the variation around a give classical paths of the usual Feynman formulation, see [1, 2]. In this way we can figure out that with cyclic dimensions there are many possible evolutions of a field from an initial configuration to a final configuration, that can self-interfere, similarly to the non-classical path of the FPI. However there is a fundamental conceptual difference with respect to the usual Feynman formulation, in fact now all these possible paths are classical paths, that is they are equivalent to classical periodic paths. This means that in this path integral formulation it is not necessary to relax the classical variational principle in order to have self-interference.

Since we have essentially standing waves, it is easy to see that there is an underlying Heisenberg uncertain relation [1, 2]. Moreover, since periodicity means that the only possible energy eigenmodes are those with an integer number of cycles, we obtain the Bohr-Sommerfeld quantization condition (for instance the periodicity condition $E_n T_t = nh$ can be more in general written as $\oint E_n dt = nh$). This allows to solve many non-relativistic quantum problems [1, 2].

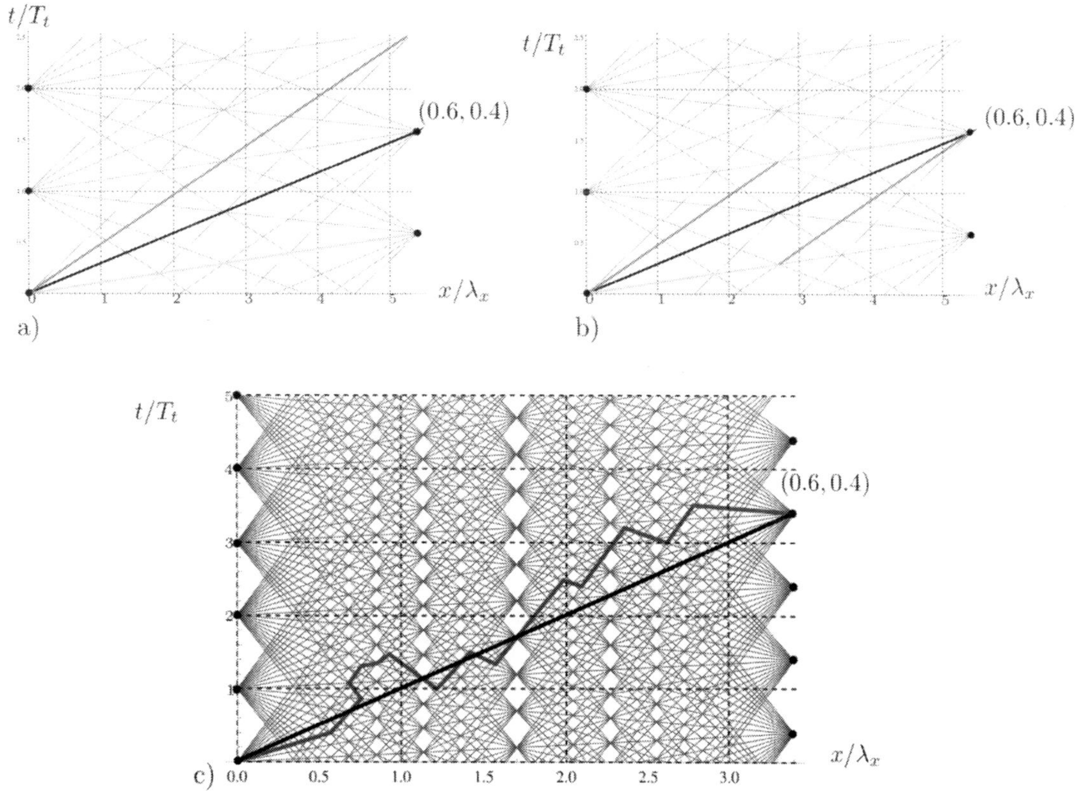

Figure 2. Periodic paths a) can be combined b) in order to reproduce Feynman paths c).

The non-quantum limit of a massive field, that is the non-relativistic single particle description, is obtained by putting the mass to infinity so that, as shown in [1, 2], in an effective classical limit, only the first level of the energy spectrum must be considered. This leads to a consistent interpretation of the wave/particle duality and of the double slit experiment. The quantities describing only the first energy level are addressed by the bar sign. For instance, the Lagrangian of the fundamental mode $\bar{\Phi}(x)$ is $\bar{\mathcal{L}}_{\lambda_s}(\partial_\mu\bar{\Phi}(x), \bar{\Phi}(x))$.

On the other hand a massless field has infinite Compton wavelength and thus an infinite proper-time periodicity, fig.(1.d). Its quantum limit is at high frequency where the PBCs are important. In this limit we have discretized energy spectrum, in agreement with the ordinary description of the black-body radiation (no UV catastrophe), fig.(1.c). The opposite limit is when the time periodicity tends to infinity and we get back a continuous energy spectrum.

A remarkable property of this theory is that QM emerges from PBCs without involving any hidden-variable, so that the theory can in principle violates the Bell's inequality and we can actually speak about determinism. The idea is similarly to the stroboscopic quantization or to the 't Hooft determinism. In fact, the intrinsic time periodicities are typically extremely fast. Thus, we inevitably have a too low revolution in time, so that at every observation the system appears in an aleatoric phase of its evolution, just like a rolling clock or a dice observed under a stroboscopic light. Thus, the de Broglie intrinsic clock of elementary particles can be imagined as a "de Broglie deterministic dice", that is a dice rolling with time periodicity T_t. Since such a dice rolls too fast with respect our time resolution, we can only predict its outcomes statistically. The results presented in this section show us that the statistical description associated to intrinsically periodic phenomena actually matches ordinary QM.

4. Matching with relativistic geometrodynamics

To introduce interactions we must bear in mind that the four-periodicity T^μ is fixed by the inverse of the four-momentum $\bar{p}_\mu$ according to the de Broglie-Planck relation eq.(1).

As already said, an isolated elementary system (*i.e.* free field) has persistent four-momentum. On the other hand, an elementary system under a generic interaction scheme can be described in terms of corresponding variations of its four-momentum along the world-line evolution $s = s(x)$ with respect to the free case

$$\bar{p}_\mu \rightarrow \bar{p}'_\mu(s) = e^a_\mu(s)\bar{p}_a\,. \tag{16}$$

In other words we describe interaction in terms of the so call tetrad (or virebein) $e^a_\mu(s)$. Thus the interaction scheme eq.(16) turns out to be encoded in the corresponding variation of the space-time periodicities

$$T^\mu \rightarrow T'^\mu(s) = e^\mu_a(s)T^a\,, \tag{17}$$

that is the corresponding deformation of the compactification lengths of a periodic field. Roughly speaking, interactions can be thought of as stretching of the compact dimensions of the theory. Equivalently, the interaction eq.(16) turns out to be encoded in the corresponding curved space-time background

$$\eta_{\mu\nu} \rightarrow g_{\mu\nu}(s) = e^a_\mu(s)e^b_\nu(s)\eta_{ab}\,. \tag{18}$$

This result can be double checked by considering the transformation of space-time variables

$$dx_\mu \rightarrow dx'_\mu(s) = e^a_\mu(s)dx_a\,. \tag{19}$$

After this transformation of variables (diffeomorphism) with determinant of the Jacobian $\sqrt{-g(s)}$, the free action eq.(5) turns out to be

$$\mathcal{S}_{\lambda_s} = \int^{e^\mu_a(s)T^a} d^4x\sqrt{-g(s)}\mathcal{L}_{\lambda_s}(e^a_\mu(s)\partial_a\Phi(x), \Phi(x))\,. \tag{20}$$

Therefore, the periodic field which minimizes this action has four-periodicity T'^μ, eq.(17), or equivalently has four-momentum $\bar{p}_\mu$, eq.(16). We conclude that a field under the interaction scheme eq.(16) is described by the solutions of the bulk equations of motion with PBCs on the deformed compact background eq.(18) and compactification lengths eq.(17).

This geometrodynamical approach to interactions is interesting because it actually mimics very closely the usual geometrodynamical approach of GR. In fact, if we suppose a weak Newton potential $V(\mathbf{x}) = -GM_\odot/|\mathbf{x}| \ll 1$, we find that the energy on a gravitational well varies (with respect to the free case) as $\bar{E} \rightarrow \bar{E}' \sim (1 + GM_\odot/|\mathbf{x}|)\,\bar{E}$. According to eq.(17) or eq(1), this means that the de Broglie clocks in a gravitational well are slower with respect to the free clocks $T_t \rightarrow T'_t \sim (1 - GM_\odot/|\mathbf{x}|)\,T_t$. Thus we have a gravitational redshift $\bar{\omega} \rightarrow \bar{\omega}' \sim (1 + GM_\odot/|\mathbf{x}|)\,\bar{\omega}$. With this schematization of interactions we have retrieved two important predictions of GR.

Besides this, [12], we must also consider the analogous variation of spatial momentum $\bar{\mathbf{p}} \rightarrow \bar{\mathbf{p}}' \sim (1 + GM_\odot/|\mathbf{x}|)\,\bar{\mathbf{p}}$, and the corresponding variation of spatial periodicities $\vec{\lambda} \rightarrow \vec{\lambda}' \sim (1 - GM_\odot/|\mathbf{x}|)\,\vec{\lambda}$. According to the relation eq.(18), this means that the weak newtonian interaction turns out to be encoded in the usual Schwarzschild metric

$$ds^2 = \left(1 - \frac{GM_\odot}{|\mathbf{x}|}\right)dt^2 - \left(1 + \frac{GM_\odot}{|\mathbf{x}|}\right)d\mathbf{x}^2\,. \tag{21}$$

We have found that the geometrodynamical approach to interactions actually can be used to describes linearized gravity and that the geometrodynamics of the compact space-time dimensions correspond to the usual relativistic ones.

As well known, see for instance [12], it is possible to retrieve ordinary GR from a linear formulation by including self-interactions. More naively, as we will show in detail in a forthcoming paper, we can add "by hand" a kinetic term to the Lagrangian (with appropriate coupling), in order to describe the dynamics of the metric $g_{\mu\nu}$ which is the new *d.o.f.* of the theory eq.(22). Thus, neglecting quantum corrections, we can replace the classical limit $\sqrt{-g}\bar{\mathcal{L}}_{\lambda_s}$ of the Lagrangian in eq.(22) with the Hilbert-Einstein Lagrangian

$$\bar{\mathcal{L}}_{HE} = \sqrt{-g}\left[-\frac{g^{\mu\nu}\mathcal{R}_{\mu\nu}}{16\pi G_N} + \bar{\mathcal{L}}_{\lambda_s}(e^a_\mu \partial_a \bar{\Phi}(x), \bar{\Phi}(x))\right] . \quad (22)$$

It can be shown that this procedure is parallel to what we usually do in electromagnetism when we add the term $-F_{\mu\nu}F^{\mu\nu}/4e^2$ to describe the kinematics of the gauge field. Intuitively, because of its geometrical meaning, the Ricci tensor is the correct mathematical object to describes the variations of the space-time compactification lengths in different space-time points. Bearing in mind eq.(1), we note that actually such a kinetic term must encodes the content of four-momentum in different space-time points. By varying the metric (neglecting issues related to the variation of boundary terms of the Hilbert-Einstein action and related BCs [13]) eq.(22) yields to the usual Einstein equation $\mathcal{R}^{\mu\nu} = -8\pi G_N T^{\mu\nu}$. With these simple and heuristic arguments we have shown that field theory in compact space-time is in agreement not only with special relativity but also with GR. In forthcoming papers we will show that, by writing eq.(16) as a minimal substitution, such a geometrodynamical approach to interactions can be also used to describe ordinary gauge interactions. Gauge fields will turn out to "tune" the variation of periodicities, allowing a semi-classical interpretation of superconductivity [2]

5. "Virtual" Extra Dimension (VXD)

The very same geometrodynamical approach provides an intuitive interpretation of AdS/CFT whose essential meaning is summarized by the Witten's words: "In this description, quantum phenomena [...] are coded in classical geometry" [14]. To understand this point we should note that our field theory in compact space-time is "dual" to an extra dimensional field theory. For instance we may note that the energy spectrum eq.(10) in the rest frame ($\bar{\mathbf{p}} \equiv 0$) reproduces the usual KK mass tower $M_n = E_n(0)/c^2 = nM = nh/\lambda_s c$, see fig.(1.a). Practically, the receipt to obtain our periodic fields from a corresponding extra dimensional field on a flat five dimensional metric ($dS^2 = dx_\mu dx^\mu - ds^2$) and with zero five dimensional mass ($dS^2 = 0$), is to identify the compact extra dimension s with our compact world-line parameter $s = c\tau$ and integrating it out, see [1, 2] for more detail. In this way we get back the usual four-dimensional minkowskian metric $ds^2 = dx_\mu dx^\mu$. In particular we find that the compactification length of the eXtra Dimension (XD) corresponds to the compactification length λ_s of the compact world-line parameter s, *i.e.* to the Compton wavelength eq.(11). When we identify the XD with the world-line parameter we say that it is a "Virtual" eXtra Dimension (VXD) and the KK mode are the "virtual" modes of the field[2]. Thus, in our theory, the KK modes play the role of energy eigenmodes of the same periodic field. They are not different elementary fields as in the usual KK scenario. Finally, since periodic fields reproduce QM and, on the other hand, they turn out to be dual to extra dimensional fields, we infer that the classical geometrodynamics on a VXD could be used to describe quantum behaviors, similarly to AdS/CFT.

Shortly, to describe the interaction scheme eq.(16) through the VXD formalism we have to generalize the deformed metric eq.(18) to the following deformed metric in VXD (and related

[2] Note that in the original Kaluza's formulation [9] the XD was introduced not like an "real" XD but as a "mathematical trick", whereas in the original Klein's proposal [10] the assumption of cyclic (*i.e.* compact with PBCs) XD was used as a semi-classical quantization condition.

compactification lengths)

$$g_{\Gamma\Delta}(s) = \begin{pmatrix} g_{\mu\nu}(s) & 0 \\ 0 & 1 \end{pmatrix}, \tag{23}$$

where the capital letters label the 5D Lorentz indices. Thus we expect to find out that the classical geometrodynamics of the field in such a deformed metric with VXD reproduces the quantum behaviors of the corresponding interaction scheme eq.(16). Similarly to Maldacena conjecture, this correspondence can be summarized by the mnemonic relation

$$\mathcal{Z} = \int_0^{\lambda_x} \mathcal{D}x \; e^{-\frac{i}{\hbar}\mathcal{S}(s,s')[\dot{x},x]} \leftrightarrow \lambda_x \; e^{-\frac{i}{\hbar}\mathcal{S}^{VXD}(s,s')} . \tag{24}$$

For instance [2], we can consider a collider experiment where the Quark-Gluon-Plasma (QGP) can be approximated as a volume of massless fields at high four-momentum[3]. As predicted by a simple Bjorken hydrodynamical model [15], or by thinking to the analogy of QCD with a thermodynamical system [16], the four-momentum of the fields decays exponentially and conformally as the QGP radiates hadronically and electromagnetically[4]: $\bar{p}_\mu \to \bar{p}'_\mu(s) \simeq e^{-ks}\bar{p}_\mu$. The QGP freeze-out is, in first approximation, encoded in the tetrad $e^a_\mu(s) \simeq \delta^a_\mu e^{-ks}$. This means that the space-time periodicities of the fields have a conformal and exponential dilatation $T^\mu \to T'^\mu(s) \simeq e^{ks}T^\mu$. Thus the QGP freeze-out is effectively described by the virtual AdS metric

$$dS^2 \simeq e^{-2ks}dx_\mu dx^\mu - ds^2 \equiv 0 . \tag{25}$$

By using holography [17, 18], it is easy to prove that the propagation of classical massless fields in such a warped virtual metric[5] is given - in natural units - by the classical correlator

$$\Pi^{Holo}(q^2) \simeq -\frac{q^2}{2kg_5^2}\log\frac{q^2}{\Lambda^2} , \tag{26}$$

which, assuming $1/kg_{5^2} = N_c/12\pi^2$, can be matched to the quantum two point function of QCD. Thus we obtain semi-classically the quantum running of the strong coupling constant

$$\frac{1}{e^2_{eff}(q)} \simeq \frac{1}{e^2} - \frac{N_c}{12\pi^2}\log\frac{q}{\Lambda} . \tag{27}$$

This is a fundamental characteristic of the AdS/QCD models.

6. Conclusions

To understand the conceptual meaning of this theory we have to note that time can only be defined by counting the number of cycles of phenomena *supposed* to be periodic. Only by assuming periodicity we can ensure that the duration of a unit of time is always the same; in the past, in the present and in the future, [7]. In particular, the usual - non compact - time axis ($t \in \mathbb{R}$) is defined with reference to the Cs-133 atomic clock whose characteristic periodicity is thus of the order of $10^{-10}s$. Hence, for a consistent formulation of natural laws we believe that there must be an assumption of periodicity in physics. We explicitly introduce this assumption by imposing to every free field its de Broglie intrinsic periodicity as constraint. As shown in

[3] This example is of particular interest because of the recent data of LHC

[4] The most simple Bjorken kinetic model [15] describing the exponential gradient of temperature of the QGP freeze-out can be regarded as the analogous of the Newton's law of cooling in QCD thermodynamics.

[5] To evaluate the low energies effective propagator we assume Neumann BCs at the UV scales Λ and boundary field $A_\mu(q)$ at the IR scale μ, in the hypothesis of Euclidean momentum q such that $\Lambda \gg |q| \gg \mu$. [18, 2].

detail in [1], see also [2, 3, 4], and as we have summarize in this paper, under this assumption ordinary QM emerges as an effective description of the underlying deterministic field theory by a process of "information loss". It is important to note that in our case there are no local-hidden-variables being time a physical variable that can not be eliminated from the theory, and being the PBCs an element of non locality. Thus the present theory is not constrained by the Bell's or similar theorems. It can in principle violate the Bell's inequality and reproduce the predictions of QM. This allow us to speculate about a scenario where, by observing the de Broglie dice of an electron with resolution in time greater that the ZHz, it could be possible to resolve its underlying deterministic dynamics and in principle predict the outcomes of the quantum dice.

These cyclic fields can therefore be identified with the so call "de Broglie internal clocks" or "de Broglie periodic phenomena". Similarly to a calendar or to a stopwatch where every moment in time is determined by the combination of the phases of periodic cycles (typically: years, months, days, hours, minutes and seconds), every value of our external temporal axes (defined in terms of the "ticks" of the Cs-133 atomic clock) is characterized by a combination of the phases of the de Broglie internal clocks, *i.e.* by the "ticks" of all the elementary fields constituting the system under investigation. In this scenario the long time scales are provided by massless fields with low frequencies (low energies implies long time periodicities). This is a simplified picture since the clocks can vary periodicity through interactions (exchange of energy) and since their periodicities depend on the reference systems according to the relativistic laws. Moreover the combination of two or more clocks - that is to say a non elementary system - with irrational ratio of periodicities, gives ergodic - or even more chaotic - evolutions. Once that the de Broglie internal clocks are fixed to be clockwise or anticlockwise, the flow of time is determined by the combinations of their "ticks" and the variations of their periodicities through interactions. Hence, this means that the external time axis can be in principle dropped and the flow of time can be effectively described in terms of the "ticks" of these de Broglie internal clocks. In this scenario the flow of time doesn't depend on the assumption of clockwise or anticlockwise rotations of the de Broglie internal clocks. The resulting formulation of the flow of time is particularly interesting for the problem of the time arrow in physics.

References

[1] Dolce D 2011 *Found. Phys.* **41**(2) 178
[2] Dolce D 2009 Compact time and determinism for bosons *Preprint* 0903.3680v4
[3] Dolce D 2010 *AIP Conf. Proc.* **1232** 222
[4] Dolce D 2010 *AIP Conf. Proc.* **1246** 178
[5] Elze H T 2003 *Phys. Lett. A* **310** 110
[6] 't Hooft G 2003 *Int. J. Theor. Phys.* **42** 355
[7] Einstein A 1910 *Arch. Sci. Phys. Natur.* **29**
[8] Matsubara T 1955 *Prog. Theor. Phys.* **14** 351
[9] Kaluza T 1921 *Sitzungsber. Preuss. Akad. Wiss. Berlin* **1921** 966
[10] Klein O 1926 *Z. Phys.* **37** 895
[11] Feynman R P *Feynman's Thesis* (Northwestern University, USA: Laurie M Brown)
[12] Ohanian H and Ruffini R 1994 *Gravitation and Space-Time* (New York: Norton)
[13] York J 1986 *Found. Phys.* **16**(3) 249
[14] Witten E 1998 *Adv. Theor. Math. Phys.* **2** 505
[15] Magas V K, Anderlik A, Anderlik C and Csernai L P 2003 *Eur. Phys. J. C* **30** 255
[16] Satz H 2008 The thermodynamics of quarks and gluons *Preprint* 0803.1611
[17] Pomarol A 2000 *Phys. Rev. Lett.* **85** 4004
[18] Arkani-Hamed N, Porrati M and Randall L 2001 Holography and phenomenology *J. High Energy Phys.* JHEP08(2001)017

Chaos and the quantum: how nonlinear effects can explain certain quantum paradoxes

Wm. C. McHarris

Departments of Chemistry and Physics/Astronomy, Michigan State University, East Lansing, MI 48824, USA

E-mail: mcharris@chemistry.msu.edu

Abstract. In recent years we have suggested that many of the so-called paradoxes resulting from the Copenhagen interpretation of quantum mechanics could well have more logical parallels based in nonlinear dynamics and chaos theory. Perhaps quantum mechanics might not be strictly linear as has been commonly postulated, and indeed, during the past year experimentalists have discovered signatures of chaos in a definitely quantum system. As an illustration of what can go wrong when quantum effects are forced into a linear interpretation, I examine Bell-type inequalities. In conventional derivations of such inequalities, classical systems are found to impose upper limits on the statistical correlations between, say, the properties of a pair of separated but entangled particles, whereas quantum systems allow greater correlations. Numerous experiments have upheld the quantum predictions (greater statistical correlations than allowed classically), which has led to inferences such as the instantaneous transmission of information between effectively infinitely separated particles — Einstein's "spooky action-at-a-distance," incompatible with relativity. I argue that there is nothing wrong with the quantum mechanical side of such derivations (the usual point of attack by those attempting to debunk Bell-type arguments), but implicit in the derivations on the classical side is the assumption of independent, uncorrelated particles. As a result, one is comparing uncorrelated probabilities versus conditional probabilities rather than comparing classical versus quantum mechanics, making moot the experimental inferences. Further, nonlinear classical systems are known to exhibit correlations that can easily be as great as and overlap with quantum correlations — so-called nonextensive thermodynamics with its nonadditive entropy has verified this with numerous examples. Perhaps quantum mechanics does contain fundamental nonlinear elements. Nonlinear dynamics and chaos theory could well provide a bridge between the determinism so dear to Einstein and the statisical interpretation of the Copenhagen school. Einstein and Bohr both could have been right in their debates.

1. Introduction

Ever since the Copenhagen interpretation of quantum mechanics gained status as the orthodox interpretation, many so-called paradoxes have accrued. These have perplexed many of us, but on the other hand, many others have taken an almost perverse sense of pride in them, touting to all who will listen that the quantum world is indescribably different from the classical world — in a quantum world indeterminacy reigns, particles can be in several places or states at the same instant, time perhaps can flow backward with the future affecting the present, and, contrary to and incompatible with relativity, not merely superluminal but instantaneous signals can pass from particle to particle in an entangled system. Such quantum peculiarities have led to predictions such as quantum teleportation and, perhaps more realistically, quantum computation

— which rest firmly on the principle of superposition of states and the fact that quantum mechanics be a strictly linear science.

During the last several decades, however, the very glamour of some of these predictions has led to a reinvestigation of some of the foundations of quantum mechanics. In particular, the burgeoning field of quantum information has led to questioning of some of the fundamental postulates. During the same period chaos theory has finally become somewhat tractable, thanks to the advent of modern high-speed computers with their associated graphics. It should be noted that the founders of quantum mechanics did not have access to modern chaos theory — or even to anything but minimal nonlinear dynamics — so they were forced to cast quantum mechanics in a linear mold. There have been a few, relatively unsuccessful attempts to apply nonlinear dynamics to quantum mechanics, but these have been primarily along the line of nonlinear perturbations applied to basically linear systems — situations where chaos cannot develop. The resulting frustration is succinctly stated by Mielnik [1]: "I cannot help concluding that we do not know truly whether or not nonlinear QM generates superluminal signals — or perhaps it resists embedding into too narrow a scheme of tensor products. After all, if the scalar potentials were an obligatory tool to describe the vector fields, some surprising predictions could as well arise! ...the nonlinear theory would be in a peculiar situation of an Orwellian 'thoughtcrime' confined to a language in which it cannot even be expressed. ...A way out, perhaps, could be a careful revision of all traditional concepts..."

In a series of papers I have raised the question as to whether it is possible to explain some quantum paradoxes by nonlinear, even chaotic, parallels [2]. Thus far, progress has been more or less in a "botanical" or species-collecting mode, so I emphasize that here I am primarily raising the possibility of fundamental nonlinearities underlying quantum mechanics — certainly not stating this as a fact. However, certain quantum paradoxes do have nonlinear parallels, and these parallels are considerably more logical (hence, plausible?) than their quantum mechanical counterparts. Most scientists dealing with quantum mechanics have had relatively little exposure to modern nonlinear dynamics and its extreme manifestation, chaos theory. If they examine it closely, they will find that superficially it is every bit as perplexing and counterintuitive as quantum theory; yet, its peculiarities develop out of the strict logic of physics rather than more ephemeral philosophical arguments. In this it is closer in spirit to relativity than to quantum mechanics.

2. Quantum paradoxes having nonlinear parallels

The following so-called paradoxes found in quantum mechanics have parallel explanations in nonlinear dynamics, usually in a realm on the verge of chaos or actually in a chaotic region. Some of them are clearer and better understood than others, so here I list them in descending order of understanding or plausibility. Because of space limitations, I cover them rather briefly and suggest that the reader examine the extensive references found in Ref. [2] for greater details. The important fact to glean here is that *classical* systems exhibit these properties, properties that are commonly associated only with quantum systems.

• **Exponential decay laws.** Exponential decay laws for radioactive decay or for any other first-order quantum kinetics process were one of the earliest quantum mysteries encountered. It is impossible to determine ahead of time exactly when an individual radioactive nucleus will decay; yet, given a statistically significant number of such nuclei, one can be confident that they will decay according to a well-behaved exponential decay law, exhibiting a well-defined half-life. The statistical nature of their decay is often likened to actuarial statistics: The mortality of an individual can be difficult to predict; yet, again given a statistically significant number of people, insurance companies can be confident of well-behaved statistical economics. However, the analogy is completely false. Mortality rates depend on a myriad of complex factors — different people, different life styles, different diseases, etc. On the other hand, one of the most basic tenets of quantum mechanics is the indistinguishabilty of individual species — a radioactive ^{22}Na nucleus is a radioactive ^{22}Na nucleus is a radioactive ^{22}Na nucleus...! Why, then, should there be any differences in the time required for different nuclei to decay?

Chaos theory provides a relatively straightforward answer involving the "butterfly effect," the exponential sensitivity of chaotic systems to minute differences in initial conditions. It is quite well known that differences in initial conditions do exist — the Uncertainty Principle involving energy and time demonstrates one aspect of this, with Γ, the width in energy of a state varying inversely with its lifetime. One can "derive" an exponential decay law in various ways with chaos theory, but perhaps the simplest is to iterate a map such as the quadratic or logistic map. (This is carried out in considerable detail in Refs. [2] and [3].) With such a map in a well-defined chaotic region, one can iterate and follow points selected from a miniscule initial interval (corresponding to the initial nuclear state) until they "escape" into a second interval (the final state); the process of iteration corresponds to physical processes such as the oscillations of a dipole for γ decay or to the collisions of an α particle with a coulomb barrier. Iterating 10,000 randomly chosen points from an initial interval having a width of 10^{-11}, then plotting the survivors against time (iteration number) produces a well-defined exponential curve. And the slight variation in initial conditions is consistent with the Uncertainty Principle.

•Bell's inequality and conditional probabilities. Bell's theorem and the various inequalities built around it have been very active fields of research, both theoretically and experimentally, during recent decades, with extreme advocates both for and against the validity and interpretations of Bell's theorem. In particular, the series of Swedish conferences on "Quantum Theory: Reconsideration of Foundations" has attracted anti-Bell critics [4], whereas the Vienna school, centered around experimental tests of Bell-type inequalities, has strongly advocated pushing nonlocal interpretations to the limit [5,6]. As an example of "anti-Bell" reasoning, Hess and Philipp [7] make an excellent case for the irrelevance of nonrelativistic Bell-type inequalities. In much of the latter part of this paper, however, I take a different approach, examining the possible misuse (or misinterpretation) of conditional probabilities.

•Quantization itself and preferred eigenstates. The whole concept of emergent systems is based on self-organization. Even many simple nonlinear systems are "self-quantized," i.e., have preferred oscillation modes, and their dynamics are governed by eigenvalue equations. Their phase space is characterized by separate (sometimes complicated or "riddled") basins of attraction for these quantized modes [8]. In addition, most Hamiltonian nonlinear, chaotic systems are characterized by islands of stable (periodic or quasi-periodic) motion in a sea of otherwise chaotic motion. Thus, mesoscopic and macroscopic *strictly classical* systems can exhibit behavior strikingly similar to that of quantum systems.

In addition, during the past year Chadhury et al. [9] published the first experimental evidence for chaotic behavior in a clearly quantum system, further blurring the distinction between classical and quantum realms. Using an ensemble of supercooled ^{133}Cs atoms in the $F = 3$ hyperfine ground state, they performed kicked-top experiments. A magnetic field induced spin about the y-axis, while laser pulses produced kicked twists about the x-axis. Because spin magnitude is conserved, the phase space for this experiment is a spherical surface, in which islands of regularity are separated by a sea of chaos. Each initial state was followed for 40 periods of the kicked-top Hamiltonian. They found that initial states lying in an island of regularity were relatively stable to perturbations, although tunneling did occur between islands. Conversely, initial states lying in the sea of chaos were extremely sensitive to perturbations. Also, there was very little mixing between regular, periodic orbits and chaotic orbits. The implication is that here the quantum regime respects "classical" boundaries that were previously thought to apply only to mesoscopic or macroscopic systems.

•Spontaneous symmetry breaking — nonconservation of parity. Spontaneous symmetry breaking is a common feature of nonlinear systems. It has been thoroughly studied through the iteration of odd-order maps, including the cubic and sine maps. Shinbrot and Muzzio [10] give some vivid examples of this, showing the separation of mixed powders in a nonlinear tumbler into bands and/or their components.

•**Barrier penetration.** Although not emphasized in most elementary physics books, barrier penetration is a fairly common phenomenon in classical optics. For example, when light is beamed into a glass plate, striking its far surface at an angle greater than the critical angle, it is totally reflected. However, if a second plate is brought close to the first, some "extraordinary" light penetrates through the back surface into the second plate — barrier penetration. And it obeys the same type of exponential relation to barrier height and width as is common in quantum systems, such as α decay.

•**Duality and wave versus particle behavior.** This is currently more speculative than the previous parallels, but there are intriguing similarities between chaotic scattering and wave-like behavior. Bleher, Grebogi, and Ott [11] give an excellent overview of this in the initial article for an entire issue of *Physica D*, and an even clearer exposition is given by José and Saletan [12] in their book, *Classical Dynamics: A Contemporary Approach.* Chaotic scattering produces patterns that look remarkably like diffration patterns. This comes about partly because of the close association of windows of (periodic) stability within chaotic regions of phase space. Thus, it seems possible to obtain "constructive interference" within regions of stability as opposed to "destructive interference" within the chaotic regions. The double-slit experiment is often touted as the very epitome of quantum puzzlement; yet, the double-slit experiment is commonly presented in a much-oversimplified geometrical optics picture. Remember that even a single-slit produces some diffration if examined closely enough, so, if the double-slit experiment were re-examined on a small-enough (microscopic or atomic?) scale, it is quite conceivable that one might obtain chaotic scattering. However, at present this remains a speculative task for future work.

•**Decoherence and reduction of wave-functions.** Most nonlinear Hamiltonian systems not only contain islands of stability scattered about in a sea of chaos, but also these islands often have a knife-edged stability — the very slightest perturbation can and will drive them into dissipative behavior. An elaborate theory of this "collapse" has been developed, involving the orderly disintegration of KAM tori [13], with the "most irrational" becoming the last to decohere. It is worth noting that in 1917 Einstein touched upon the rather similar problem of how to quantize what are now known as nonintegrable (chaotic) systems [14].

In short, nonlinear classical systems are now known to behave in seemingly paradoxical manners in many ways analogous to how quantum systems behave. It is thus short-sighted to emphasize the "peculiar" behavior of quantum systems while treating all classical mechanics as the well-behaved linear systems depicted in textbooks. Of course, arguments by analogy can be dangerous; however, the fact that certain so-called quantum paradoxes do have analogs in nonlinear classical systems should raise the question as to whether quantum and classical systems are so diametrically opposite and incompatible as is so commonly touted.

3. Bell's inequality and conditional probabilities.

As mentioned above, in recent years there has been an enormous amount of interest in Bell's theorem and Bell-type inequalities. Several dozen elegant and intricate experiments have been performed to test Bell-type inequalities, and in essentially all cases (except for some stumbling in the earliest attempts), these experiments have evinced the greater statistical correlations allowed by quantum mechanics rather than the lesser correlations of "classical" mechanics. Interpretations have varied, but the most common inferences have centered around eliminating the *local reality* defined by Einstein, Podolsky, and Rosen in their 1935 EPR paper [15]. Any extension of quantum theory involving *hidden variables* [16] must be non-local; otherwise, nature must not have a well-defined, predictable existence independent of and existing before measurement. In other words, Schrödinger's cat arguments must make sense on the microscopic scale even though they were designed to be nonsense on the macroscopic scale! — and Einstein's "spooky action-at-a-distance" must be an actuality. Zeilinger's new book, *Dance of the Photons* [6], gives an excellent summary of this extreme point of view.

Most of the experimental tests have involved the CHSH inequality [17], which is a more specific, experimetally friendly version of Bell's original inequality. It involves generating a pair of entangled particles (e.g., a pair of electrons with their spins in the singlet state), then sending one each to Alice and Bob, the stereotypical quantum information experimentalists separated at an incommunicado (effectively infinite) distance. The entangled pair has an overall spin, $J = 0$, but the spin of each individual particle is undefined — the particles have not yet "determined" their individual spins, in the sense of the Copenhagen interpretation. Alice can measure her particle's spin with respect to, say, a vertical (measurement Q) or a horizontal (measurement R) axis, getting a result of up or down ($+1$ or -1) with respect to her axis of measurement. Similarly, Bob can make measurement S or T on his particle. After accruing a statistically meaningful number of measurements, Alice and Bob get together to compare notes. I show a straightforward derivation of the CHSH inequality in Refs. [2,3], with the result:

$$E(QS) + E(RS) + E(RT) - E(QT) \leq 2$$

This is the inequality they use for the comparison. Each of the measurement can be $+1$ or -1, and the efficiencies (E) lower the correlations. This is the so-called classical CHSH inequality. For the quantum case, we start with the particles in the Bell singlet state:

$$|\Psi> = \frac{|01> - |10>}{\sqrt{2}}$$

Working out the expectation values, using the Pauli spin matrices, we get a higher degree of correlation:

$$< QS > + < RS > + < RT > - < QT > = 2\sqrt{2}$$

Most of the experiments have involved polarized photons, but the basic math is the same. And essentially all of the experiments have vindicated quantum mechanics.

As mentioned earlier, I see nothing wrong with the quantum mechanical predictions, although that has most often been the point of attack for those who wish to discredit this sort of Bell-type arguments. Instead, I would point out that *implicit in the so-called classical side is the idea of independent — not entangled! — particles.* This can be seen most clearly by going back to the original Bell inequality [18], which, after all, was derived from a classical viewpoint.

Stated simply, if the elements of a system can have three properties, A, B, and C, we can derive an inequality concerning having or not having these properties. Here we say that $N(A)$, $N(B)$, and $N(C)$ are the numbers of elements in the set that do have the properties, while $N(\overline{A})$, $N(\overline{B})$, and $N(\overline{C})$ are the numbers of elements that do not have the properties. Bell's inequality thus is:

$$[N(A) + N(\overline{B})] + [N(B) + N(\overline{C})] \geq [N(A) + N(\overline{C})]$$

This can be seen most easily by using a diagram [19], as in Figure 1. Here the elements having each property are represented by the (overlapping) circles. For example, region 1 has property A but not B or C; region 2 has both A and C but not B, and region 3 has all three properties. It is rather clear from the diagram that the areas of the regions obey the inequality:

$$[1 + 2] + [7 + 4] \geq [1 + 4]$$

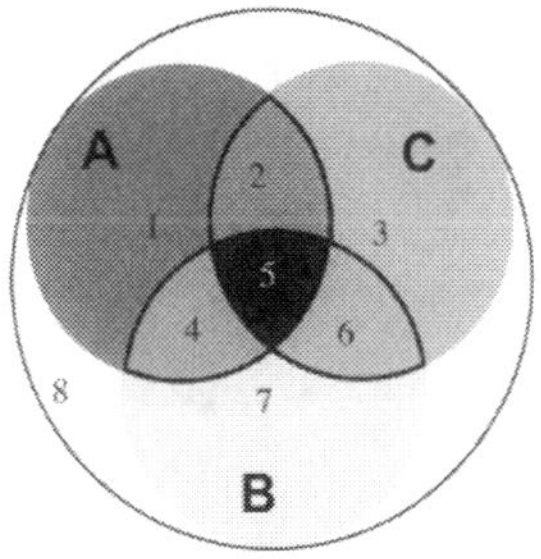

Figure 1. Diagram to aid in demonstrating Bell's inequality.

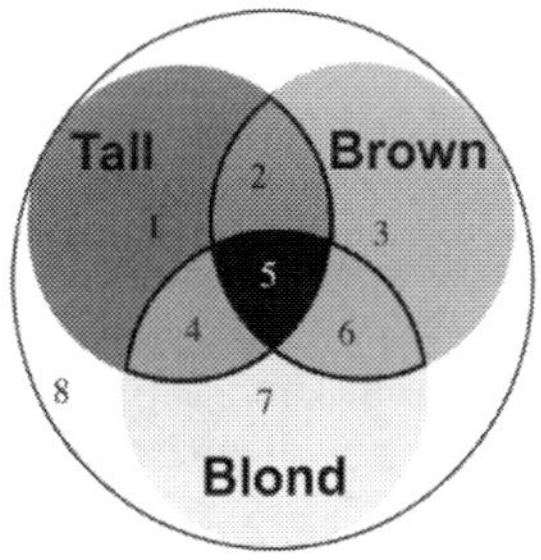

Figure 2. Diagram altered to demonstrate pairs of twins having different heights, hair colors, and eye colors.

To make the picture even more concrete, we can consider pairs of twins having different attributes. Figure 2 shows the same diagram for twins sorted according to height, hair color, and eye color [20]. (Note that short = not tall, brunet = not blond, and blue = not brown.) There are eight possible combinations:

•Tall, blue-eyed, blond
•Tall, blue-eyed,brunet
•Tall, brown-eyed, blond
•Tall, brown-eyed, brunet
•Short, blue-eyed, blond
•Short, blue-eyed, brunet
•Short, brown-eyed, blond
•Short, brown-eyed, brunet

It is obvious that the areas in Figure 2 add up in the same manner as those in Figure 1, resulting in a specific example for applying Bell's inequality:

$$[N(\text{tall}) + N(\text{brunet})] + [N(\text{blond}) + N(\text{blue})] \geq [N(\text{tall}) + N(\text{blue})]$$

Implicit in these diagrams is the fact that the properties are independent: Height, hair color, and eye color are independent variables. Note that the sizes of the sets and the degree of overlap do not affect the validity of the inequality, since it is the presence of the extra overlap areas on the left that force the inequality in the first place.

But what happens if the properties are not truly independent? Again, a concrete example perhaps can clarify matters. It is well known that many blue-eyed white cats are deaf. This

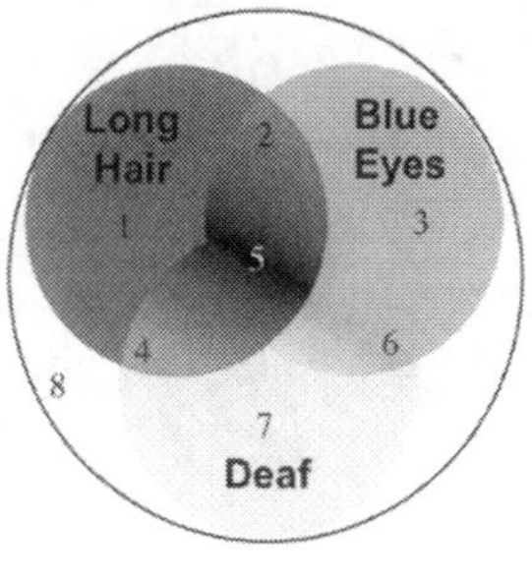

Figure 3. Diagram altered to show the conditional probabilities involved with blue-eyed white cats. There is a subtle link between blue-eyes and deafness, thought to involve a mutation on a specific gene. It is clear that the areas of overlap between blue eyes and deaf are not well defined, so one cannot apply the same straightforward reasoning here to obtain a variant of Bell's inequality.

link is thought to involve a single mutation on a specific gene. The diagram for this is shown in Figure 3.

We are not innately responsive to conditional probabilities. A rather frivolous example of this can be found in the so-called Monty Hall paradox [21], named for the host of a long-gone television game show. It goes like this: There are three closed doors, behind two of which are goats and behind the third is a new sports car, but who knows which is which? The contestant picks one of the doors but does not open it. The host then opens one of the two remaining doors to reveal a goat. The question then is — should the contestat stick to his or her original choice, or should she or he switch to the remaining unopened door? Almost everyone will say it does not make any difference – the chances are one in three that he/she will get the car. This is not true. If the choice of which door to open were independent or random, it would make no difference, However, the host may not choose which door to open or do so at random. The conditional probability involved is that he must open the door that has the goat behind it. Here are the final odds: If the contestant first chose the door with the car behind it, it would make no difference whether she/he switched or not — the odds of this happening are one in three. However, had the contestant chosen first a door with a goat behind it, that would have forced the host to open the remaining door with a goat behind it, and it pays the contestant to switch. The odds of this are two in three, so switching doubles the chance of winning. Think about it for a moment. This rather trivial example contains the essence of conditional probabilities.

On a more serious note, classical nonlinear systems are known to exhibit correlations, ranging from the directions of particles in tornadoes to the distribution of energies in cosmic rays — and at times these can be great enough to overlap with quantum correlations. A more quantitative understanding of these can be had with the use of "nonextensive" thermodynamics, including nonextensive entropy introduced by Tsallis in 1988 [22]. He defined a gerneralized entropy:

$$S_q = \frac{1 - \sum_{i=1}^{W} p_i^q}{q - 1}$$

Here the phase space has been divided into W cells of equal measure, with p_i being the probability of occupying cell i. When the exponent (the "entropic index"), $q = 1$, this reduces to the standard Boltzmann entropy:

$$S_1 = -\sum_{i=1}^{W} p_i \ln p_i$$

The greater the differene of q from 1, the greater the deviation from standard distributions, with apparent long-range correlations resulting in long tails in the distributions. When $q < 1$, the entropy of a combined system is greater (superextensive) than the sum of the individual entropies, and when $q > 1$, it is less. Thus, $q > 1$ indicates that a system has long-range correlations that interfere constructively, characteristic, e.g., of many emergent systems.

The concept of nonextensive entropy has found rather widespread applications in classical systems, especially in biological systems, and thre have been several international conferences related to its use — see, e.g., Ref. [23] for examples. And it has been found that non-ergodic distributions in phase space can easily masquerade as long-range interactions. The quantitative expression for adding the entropies of two nonlinear systems includes feedback or interference between the systems as a cross-term:

$$\frac{S_q(A+B)}{k} = \frac{S_q(A)}{k} + \frac{S_q(B)}{k} + (1-q)(\frac{S_q(A)}{k})(\frac{S_q(B)}{k})$$

In summary, one could almost as well use the term "entangled" for some interacting nonlinear classical systems as for quantum systems.

4. Conclusions.

Because of space limitations the arguments in this paper are necessarily somewhat sketchy. Also, I would emphasize once again that its intent is more to raise questions about the possibility of nonlinear correlations in quantum systems rather than to try to prove things quantitatively. Nevertheless, arguments about the fundamentals of quantum mechanics are far from over. The originators and early developers of quantum theory did not have access to modern nonlinear dynamics and chaos theory, so they were forced to deal with quantum ideosyncracies within a linear, if perturbative framework. The result has been beautiful insofar as current practical quantitative applications are concerned. However, the specter of nonlinear contributions could wreak havoc with some of the basic ideas involved with quantum computing, and even more so with more far-out ideas such as quantum teleportation.

Finally, perhaps both Einstein and Bohr were fundamentally correct, and their disagreement was only apparent. Chaos and nonlinear dynamics as applied to quantum mechanics has nothing to do with hidden variables, but it does provide the fundamental determinism so dear to Einstein's heart. On the other hand, it can be interpreted only statistically, in line with Bohr's Copenhagen school of thought. Perhaps it can provide a bridge over the chasm between the two sides.

References

[1] Mielnik B 2001 *Phys Lett* **A289** 1
[2] McHarris Wm C 2007 *Complexity* **12**(4) 12
[3] McHarris Wm C 2006 *arXiv* quant-ph/0610234
[4] Khrennikov A, Ed *Proceedings of the Conference on Quantum Theory: Reconsideration of Foundations, Växjö, Sweden* (Until 2003 published by the University of Växjö Press; after 2005 published as an American Institute of Physics Proceedings.)
[5] Bertlemann R A and Zeilinger A, Eds 2003 *Quantum [Un]Speakables* (Berlin: Springer)
[6] Zeilinger A 2010 *Dance of the Photons: From Einstein to Quantum Teleportation* (New York: Farrar, Strauss and Giroux)
[7] Hess K and Philipp W 2005 in Khrennikov A, Ed. *Proceedings of the Conference on Quantum Theory: Reconsideration of Foundations* (New York: AIP Proceedings)
[8] For a good introduction, see Hilborn R C 2000 *Chaos and Nonlinear Dynamics, 2nd Ed* (Oxford: Oxford University Press).

[9] Chadhury S, Smith A, Anderson B E, Ghose S and Jessen P S 2009 *Nature* **461**, 768; Merali 2009 *Nature News* (7 Oct 2009) 091007
[10] Shinbrot T and Muzzio F J 2000 *Phys Today* **53**(3) 25
[11] Bleher S, Grebogi C and Ott E 1990 *Physics D* **46** 87
[12] José J V and Saletan E J 2002 *Classical Dynamics: A Contemporary Approach* (Cambridge: Cambridge University Press)
[13] Gutzweller M C 1990 *Chaos in Classical and Quantum Mechanics* (New York: Springer)
[14] Einstein A 1997 *The Collected Papers of Albert Einstein*, Vol. 6 (Princeton: Princeton University Press)
[15] Einstein A, Podolsky B and Rosen N 1935 *Phys Rev* **47** 777
[16] Bohm D 1951 *Quantum Theory* (Englewood Cliffs, NJ: Prentice-Hall)
[17] Clauser J, Horne M A, Shimony A and Holt R 1969 *Phys Rev Lett* **23** 880
[18] Bell J S 1993 *Speakable and Unspeakable in Quantum Mechanics* (Cambridge: Cambridge University Press)
[19] Susskind L 2006 *Stanford on iTunes U*, Physics 24: Modern Theoretical Physics (Quantum Entanglement), Lecture 3
[20] Ref. [6], Appendix gives a somewhat different approach to the same idea.
[21] A very elaborate literature exists about the so-called Monty Hall paradox; Wikipedia is a good starting place.
[22] Tsallis C J 1988 *J Stat Phys* **52** 479
[23] Gell-Mann M and Tsallis C J, Eds 2004 *Nonextensive Entropy: Interdisciplinary Applications* (New York: Oxford University Press)

Gauge Theories with Fuzzy Extra Dimensions and Noncommutative Vortices and Fluxons

Seçkin Kürkçüoğlu

Middle East Technical University, Department of Physics, Inönü Boulevard, 06531, Ankara, Turkey

E-mail: kseckin@metu.edu.tr

Abstract. A $U(2)$ Yang-Mills theory on the space $\mathcal{M} \times S_F^2$ is considered, where it is assumed that $\mathcal{M}$ is an arbitrary noncommutative space and S_F^2 is a fuzzy sphere spontaneously generated from a noncommutative $U(\mathcal{N})$ Yang-Mills theory on $\mathcal{M}$ coupled to a triplet of scalars in the adjoint of $U(\mathcal{N})$. SU(2)-equivariant reduction of this theory leads to a noncommutative $U(1)$ gauge theory coupled adjointly to a set of scalar fields. The emergent model is studied on the Groenewald-Moyal plane $\mathbb{R}_\theta^2$ and it is found that, in certain limits, it admits noncommutative, non-BPS vortex as well as fluxon solutions.

1. Introduction

Dimensional reduction of Yang-Mills theories over coset spaces of the form G/H has long been an interesting theme in modern particle physics. It was first formulated in a systematic manner by Forgacs and Manton [1]. The essential idea in this context can be clearly illustrated by considering a Yang-Mills theory over $\mathcal{M} \times G/H$, where $\mathcal{M}$ is a given manifold. G has a natural action on its coset, and requiring the Yang-Mills gauge fields to be invariant under the G action up to a gauge transformation leads to the dimensional reduction of the theory after integrating over the coset space G/H. This technique is usually called "coset space dimensional reduction" (CSDR), and it has been widely used as a method in attempts to obtain the standard model on the Minkowski space M^4 starting from a Yang-Mills-Dirac theory on the higher dimensional space $M^4 \times G/H$; for a review on this topic see [2]. Recently, CSDR techniques have also been applied to Yang-Mills theories over $\mathbb{R}_\theta^{2d} \times S^2$ [3], where $\mathbb{R}_\theta^{2d}$ is the $2d$ dimensional Groenewald-Moyal space; a prime example of a noncommutative space. In this framework, Donaldson-Uhlenbeck-Yau (DUY) equations of a $U(2k)$ Yang-Mills theory have been reduced to a set of equations on $\mathbb{R}_\theta^{2d}$ whose solutions are given by BPS vortices on $\mathbb{R}_\theta^{2d}$ and the properties of the latter have been elaborated.

On another front, there have been significant advances in understanding the structure of gauge theories possessing fuzzy extra dimensions (for a review on fuzzy spaces see [4]). Gauge theories with fuzzy extra dimensions using CSDR scheme were first studied in [5]. Later on this was followed by [6], where it was shown that an $SU(\mathcal{N})$ Yang-Mills theory on the four dimensional Minkowski space M^4 coupled to an appropriate set of scalar fields develops fuzzy extra dimensions in the form of fuzzy spheres S_F^2 via spontaneous symmetry breaking. The vacuum expectation values (VEV) for the scalar fields form the fuzzy sphere, while the fluctuations around this vacuum are interpreted as gauge fields over S_F^2. The resulting theory can therefore be viewed as a gauge theory over $M^4 \times S_F^2$ with a smaller gauge group; which is further corroborated by the expansion of a tower of Kaluza-Klein modes of the gauge fields over $M^4 \times S_F^2$. Inclusion of the fermions into this theory was considered in [7]. Gauge theory on $M^4 \times S_F^2 \times S_F^2$ has recently been investigated in [8]. For a review on these and related results [9] can be consulted. It is also worthwhile to mention that, in [10] starting from a suitable matrix gauge theory, noncommutative gauge theories on $\mathbb{R}_\theta^4$ possessing extra dimensions have been proposed. Extra dimensions are interpreted as scalars on $\mathbb{R}_\theta^4$ coupled to the gauge fields, and it was shown that scalars could take vacuum expectation values leading to their identification as fuzzy spheres. Consequently, spontaneous symmetry breaking in a particular gauge theory has been investigated, and its content is compared with that of the standard model.

In a recent article together with D. Harland [11], the question of dimensional reduction of gauge theories over fuzzy coset spaces has been addressed. For this purpose, a $U(2)$ Yang-Mills theory over $\mathcal{M} \times S_F^2$ is considered, where $\mathcal{M}$ is taken as a Riemannian manifold and S_F^2 is as the fuzzy sphere. The equivariant reduction of this model over S_F^2 has been performed by finding the most general $SU(2)$-equivariant gauge field over $\mathcal{M} \times S_F^2$ and applying the well-known CSDR techniques. It was shown that for $\mathcal{M} = \mathbb{R}^2$ the emergent theory has non-BPS vortex solutions depending on the parameters in the model, which correspond to instantons in the original theory.

In this paper, we report on some of the results of a recent work in which we consider a $U(2)$ Yang-Mills theory $\mathcal{M} \times S_F^2$ where $\mathcal{M}$ is an arbitrary noncommutative manifold. Performing SU(2)-equivariant dimensional reduction of this theory leads to a noncommutative $U(1)$ gauge model coupled adjointly to a set of scalar fields [12]. We find that on the Groenewald-Moyal plane $\mathcal{M} = \mathbb{R}_\theta^2$ the emergent models admit noncommutative, non-BPS vortex as well as fluxon solutions. A broader discussion and further details can be found in [12].

2. Yang-Mills Theory on $\mathcal{M} \times S_F^2$

In this section, we collect the essential features of gauge theory on $\mathcal{M} \times S_F^2$. We start with considering the following $U(\mathcal{N})$ Yang-Mills theory over a suitable noncommutative space $\mathcal{M}$ which we leave unspecified for the time being:

$$S = \int_{\mathcal{M}} \mathrm{Tr}_{\mathcal{N}}\Big(\frac{1}{4g^2} F_{\mu\nu}^\dagger F_{\mu\nu} + (D_\mu \phi_a)^\dagger (D_\mu \phi_a)\Big) + \frac{1}{\tilde{g}^2} V_1(\phi) + a^2 V_2(\phi) \,. \tag{1}$$

Here, $\phi_a \, (a = 1,2,3)$ are anti-Hermitian scalars, transforming in the adjoint of SU($\mathcal{N}$), $D_\mu \phi_a = \partial_\mu \phi_a + [A_\mu \, , \phi_a]$ are the covariant derivatives and A_μ are the $u(\mathcal{N})$ valued anti-Hermitian gauge fields associated to the curvature $F_{\mu\nu}$. We take the potentials of the form

$$V_1(\phi) = \mathrm{Tr}_{\mathcal{N}}\big(F_{ab}^\dagger F_{ab}\big) \,, \quad V_2(\phi) = \mathrm{Tr}_{\mathcal{N}}\big((\phi_a \phi_a + \tilde{b})^2\big) \,. \tag{2}$$

where in $V_1(\phi)$ we have defined

$$F_{ab} := [\phi_a \, , \phi_b] - \varepsilon_{abc} \phi_c \,, \tag{3}$$

whose purpose will become evident shortly.

In the expressions above a, $\tilde{b}$, g and $\tilde{g}$ are constants and $\mathrm{Tr}_{\mathcal{N}} = \mathcal{N}^{-1}\mathrm{Tr}$ denotes a normalized trace. We further note that ϕ_a transform in the vector representation of an additional global $SO(3)$ symmetry and that V_1 and V_2 are invariant under this symmetry.

As its commutative counterpart [6], this theory spontaneously develops extra dimensions in the form of fuzzy spheres. Following [6], let us very briefly see how this actually comes about. We observe that the potential $\tilde{g}^{-2} V_1 + a^2 V_2$ is positive definite, and that solutions of

$$F_{ab} = [\phi_a \, , \phi_b] - \varepsilon_{abc} \phi_c = 0 \,, \quad -\phi_a \phi_a = \tilde{b} \tag{4}$$

are evidently a global minima. Most general solution to this equation is not known. However depending on the values taken by the parameter $\tilde{b}$, a large class of solutions has been found in [6]. Here we restrict ourselves to the simplest situation and refer the reader to [6] for a general discussion and its physical consequences.

Taking the value of $\tilde{b}$ as the quadratic Casimir of an irreducible representation of SU(2) labeled by ℓ, $\tilde{b} = \ell(\ell+1)$ with $2\ell \in \mathbb{Z}$ and assuming further that the dimension $\mathcal{N}$ of the matrices ϕ_a is $(2\ell+1)n$, (4) is solved by the configurations of the form

$$\phi_a = X_a^{(2\ell+1)} \otimes \mathbf{1}_n \,, \tag{5}$$

where $X_a^{(2\ell+1)}$ are the (anti-Hermitian) generators of SU(2) in the irreducible representation ℓ, which has dimension $2\ell+1$. We observe that this vacuum configuration spontaneously breaks the $\mathrm{U}(\mathcal{N})$ down to $\mathrm{U}(n)$ which is the commutant of ϕ_a in (5).

Fluctuations about the vacuum (5) may be written as

$$\phi_a = X_a + A_a \,, \tag{6}$$

where $A_a \in u(2\ell+1) \otimes u(n)$ and we have used the short-hand notation $X_a^{(2\ell+1)} \otimes \mathbf{1}_n =: X_a$. Then A_a $(a = 1, 2, 3)$ may be interpreted as three components of a $\mathrm{U}(n)$ gauge field on the fuzzy sphere S_F^2.

A short definition of the fuzzy sphere and some of its properties are in order here. The fuzzy sphere at level ℓ is defined to be the algebra of $(2\ell+1)\times(2\ell+1)$ matrices $\mathrm{Mat}(2\ell+1)$. The three Hermitian "coordinate functions"

$$\hat{x}_a := \frac{i}{\sqrt{\ell(\ell+1)}} X_a^{(2\ell+1)} \tag{7}$$

satisfy

$$[\hat{x}_a \,, \hat{x}_b] = \frac{i}{\sqrt{\ell(\ell+1)}} \varepsilon_{abc} \hat{x}_c \,, \quad \hat{x}_a \hat{x}_a = 1 \,, \tag{8}$$

and generate the full matrix algebra $\mathrm{Mat}(2\ell+1)$. Derivations of functions can be given by the adjoint action of $su(2)$ on S_F^2:

$$f \to ad X_a^{(2\ell+1)} f := [X_a^{(2\ell+1)} \,, f] \,, \quad f \in \mathrm{Mat}(2\ell+1) \,. \tag{9}$$

In the limit $\ell \to \infty$, the functions $\hat{x}_a$ are identified with the standard coordinates x_a on $\mathbb{R}^3$, restricted to the unit sphere, and the infinite-dimensional algebra $\mathcal{C}^\infty(S^2)$ of functions on the sphere is recovered. Also in this limit, the derivations $[X_a^{(2\ell+1)}, \cdot]$ become the vector fields $-i\mathcal{L}_a = \varepsilon_{abc} x_a \partial_b$, induced by the usual action of $SO(3)$.

Tracking back to our main discussion in this section, we observe that ϕ_a are the "covariant coordinates" on S_F^2 and (3) defines the associated curvature F_{ab}. The latter may be expressed in terms of the gauge fields A_a as:

$$F_{ab} = [X_a \,, A_b] - [X_b \,, A_a] + [A_a \,, A_b] - \varepsilon_{abc} A_c \,. \tag{10}$$

Obviously, the term V_1 corresponds to the Yang-Mills action on S_F^2. However, with this term alone, gauge theory on the sphere is not recovered in the commutative limit, since the fuzzy gauge field has three components rather than two. Rather, one obtains gauge theory with an additional scalar; the scalar is more precisely the component of the gauge field pointing in the radial direction when S^2 is embedded in $\mathbb{R}^3$. The purpose of the term V_2 in the action is to suppress this scalar.

To summarize, with (6) the action in (1) takes the form of a U(n) gauge theory on $\mathcal{M} \times S_F^2(2\ell+1)$ with the gauge field components $A_M(\hat{y}) = (A_\mu(\hat{y}), A_a(\hat{y})) \in u(n) \otimes u(2\ell+1)$ and field strength tensor ($\hat{y}$ are a set of coordinates for the noncommutative manifold $\mathcal{M}$)

$$\begin{aligned} F_{\mu\nu} &= \partial_\mu A_\nu - \partial_\nu A_\mu + [A_\mu, A_\nu] \\ F_{\mu a} &= D_\mu \phi_a = \partial_\mu \phi_a + [A_\mu, \phi_a] \\ F_{ab} &= [\phi_a, \phi_b] - \epsilon_{abc}\phi_c . \end{aligned} \tag{11}$$

3. The SU(2)-Equivariant Gauge Field

Let us focus on the case of a U(2) gauge theory on $\mathcal{M} \times S_F^2$. The construction of the most general SU(2)-equivariant gauge field on S_F^2 was given in a recent article by the present author with D. Harland [11]. This construction uses essentially the representation theory of SU(2). Here we give a brief account and refer the reader to [11] for further details.

We begin by selecting

$$\omega_a = X_a^{(2\ell+1)} \otimes \mathbf{1}_2 - \mathbf{1}_{2\ell+1} \otimes \frac{i\sigma^a}{2}, \quad \omega_a \in u(2) \otimes u(2\ell+1)\,, \text{for } a = 1,2,3 \tag{12}$$

These ω_a are the generators of the representation $\underline{1/2} \otimes \underline{\ell}$ of SU(2), where by $\underline{m}$ we denote the spin m representation of SU(2) of dimension $2m+1$. The two terms which make up ω_a generate rotations and gauge transformations, therefore imposing ω-equivariance amounts to requiring that rotations can be compensated by gauge transformations. There are certainly more possible choices for ω_a; for example, $\omega_a = X_a^{(2\ell+1)} \otimes \mathbf{1}_2$ was studied in [5].

SU(2)-equivariance of the theory requires the fulfillment of the symmetry constraints,

$$[\omega_a\,, A_\mu] = 0\,, \quad [\omega_a, \phi_b] = \epsilon_{abc}\phi_c, \tag{13}$$

on the gauge field. A consistency condition on these constraints is $[\omega_a, \omega_b] = \varepsilon_{abc}\omega_c$ and it is readily satisfied by our choice of ω_a.

The solutions to these constraints are obtained using the representation theory of $SU(2)$ and are presented in [11]. They are conveniently parametrized as

$$A_\mu = \frac{1}{2} Q a_\mu(\hat{y}) + \frac{1}{2} i b_\mu(\hat{y})\,, \tag{14}$$

$$A_a = \frac{1}{2}\varphi_1(\hat{y})[X_a, Q] + \frac{1}{2}(\varphi_2(\hat{y}) - 1)Q[X_a, Q] + i\frac{1}{2}\varphi_3(\hat{y})\frac{1}{2}\{\hat{X}_a, Q\} + \frac{1}{2}\varphi_4(\hat{y})\hat{\omega}_a,\,, \tag{15}$$

with $\phi_a = X_a + A_a$ and a_μ, b_μ are Hermitian U(1) gauge fields, φ_i are Hermitian scalar fields over $\mathcal{M}$, the curly brackets denote anti-commutators throughout, and we have used

$$\hat{X}_a := \frac{1}{\ell + 1/2} X_a\,, \quad \hat{\omega}_a := \frac{1}{\ell + 1/2}\omega_a. \tag{16}$$

We have further introduced the anti-Hermitian matrix

$$Q := \frac{X_a \otimes \sigma^a - i/2}{\ell + 1/2}\,, \quad Q^\dagger = -Q\,, \quad Q^2 = -\mathbf{1}_{2(2\ell+1)}\,. \tag{17}$$

Indeed, Q is the fuzzy version of $q := i\sigma \cdot \mathbf{x}$ and converges to it in the $\ell \to \infty$ limit.

4. Reduced Action

Using the SU(2)-equivariant gauge field in the noncommutative $U(2)$ Yang-Mills theory on $\mathcal{M} \otimes S_F^2$, we can explicitly trace it over the fuzzy sphere and reduce it to a theory on $\mathcal{M}$. The reduced action has the general form

$$S = \int_{\mathcal{M}} \mathcal{L}_F + \mathcal{L}_G + \frac{1}{\tilde{g}^2}\tilde{V}_1 + a^2\tilde{V}_2 \tag{18}$$

where $\mathcal{L}_F$ stands for the curvature term, $\mathcal{L}_G$ for the gradient term and $\tilde{V}_1, \tilde{V}_2$ for the reduced forms of $V_1\,, V_2$, respectively. The explicit form of these terms and the related results are omitted here, they can be found in [12].

It turns out that the presence of extra degrees of freedom, namely $\varphi_3\,, \varphi_4$, in the $SU(2)$-equivariant gauge field on S_F^2 leads to a further symmetry breaking in the reduced action and the reduced action is invariant only under a noncommutative $U(1)$ gauge group and it has the form [12],

$$\begin{aligned} S = \int_{\mathcal{M}} \frac{1}{4g^2}\,|F_{\mu\nu}|^2 + \frac{1}{2}\frac{\ell^2+\ell}{(\ell+1/2)^2} D_\mu\varphi D_\mu\varphi^\dagger + \frac{1}{8}\frac{\ell^2+\ell}{(\ell+1/2)^2}\left(\frac{(\ell+\frac{3}{2})(\ell-\frac{1}{2})}{(\ell+\frac{1}{2})^2}+1\right)(D_\mu\varphi_3)^2 \\ + \frac{\ell^2+\ell+\frac{3}{4}}{4(\ell+\frac{1}{2})^2}(D_\mu\varphi_4)^2 + \frac{\ell^2+\ell}{4(\ell+\frac{1}{2})^3}\{D_\mu\varphi_3\,, D_\mu\varphi_4\} + \frac{1}{\tilde{g}^2}\tilde{V}_1 + a^2\tilde{V}_2\,. \end{aligned} \tag{19}$$

where we have

$$D_\mu\cdot = \partial_\mu\cdot + [c_\mu\,,\cdot]\,, \quad F_{\mu\nu} = \partial_\mu c_\nu - \partial_\nu c_\mu + i[c_\mu\,, c_\nu]\,, \quad c_\mu = \frac{1}{2}b_\mu\,. \tag{20}$$

5. Solutions of the Reduced Theory on $\mathbb{R}^2_\theta$

We now wish to study the classical solutions of the system governed by the action given in (19) on the Groenewald-Moyal plane $\mathbb{R}^2_\theta$. As emphasized in [11] there is no canonical choice for the coefficient a^2 of the fuzzy constraint term; we will consider the two extreme cases of $a^2 = \infty$ and $a^2 = 0$ corresponding, respectively, to imposing the constraint $\phi_a\phi_a + \ell(\ell+1) = 0$ in full (i.e. "by hand") and imposing no constraint at all. In both cases, we consider the large ℓ limit; in the $a = \infty$ theory, we include only terms appearing at $O(\ell^{-2})$, whereas for the case $a = 0$, we assume $\ell = \infty$.

5.1. Definitions and Conventions for the Groenewald-Moyal Plane $\mathbb{R}^2_\theta$

Using the operator formalism, $\mathbb{R}^2_\theta$ is defined by two operators $\hat{y}_1\,, \hat{y}_2$ acting on the standard Harmonic oscillator Fock space $\mathcal{H}$. They fulfill the Heisenberg algebra commutation relation

$$[\hat{y}_1\,, \hat{y}_2] = i\theta\,, \tag{21}$$

where θ is the noncommutativity parameter.

It is often useful to switch to the complex basis which we take as

$$z = \frac{1}{\sqrt{2}}(y_1 + iy_2)\,, \quad \bar{z} = \frac{1}{\sqrt{2}}(y_1 - iy_2)\,, \tag{22}$$

fulfilling

$$[z\,, \bar{z}] = \theta\,. \tag{23}$$

The derivatives on $\mathbb{R}^2_\theta$ maybe expressed as

$$\partial_\mu \cdot = -\frac{i}{\theta}\varepsilon_{\mu\nu}[\hat{y}_\nu\,,\cdot]\,,\quad \partial_z \cdot = -\frac{1}{\theta}[\bar{z}\,,\cdot]\,,\quad \partial_{\bar{z}} \cdot = \frac{1}{\theta}[z\,,\cdot]\,. \tag{24}$$

The integration over $\mathbb{R}^2$ becomes a trace over the Fock space $\mathcal{H}$ on $\mathbb{R}^2_\theta$:

$$\int_{\mathbb{R}^2} d^2y \longrightarrow 2\pi\theta\,\mathrm{Tr}_{\mathcal{H}}\,. \tag{25}$$

5.2. *Case 1: The constraint fully imposed*

After performing the dimensional reduction, the fuzzy constraint $\phi_a\phi_a + \ell(\ell+1) = 0$ can be solved order by order in powers of the parameter $\frac{1}{\ell}$ to obtain φ_3 and φ_4 in terms of φ_1 and φ_2. Substituting back into the action yields an action involving only the scalar $\varphi = \varphi_1 + i\varphi_2$.

For large but finite ℓ, one can solve the constraint approximately by expanding it to leading order in powers of ℓ^{-1} around the $\ell = \infty$. Performing this to order $O\left(\ell^{-3}\right)$, we find [12]

$$\varphi_3 = -i\frac{4}{\ell}[\varphi_1\,,\varphi_2] + \frac{1}{2\ell^2}(\varphi_1^2 + \varphi_2^2 - 1) + O\left(\ell^{-3}\right)\,, \tag{26}$$

$$\varphi_4 = -\frac{1}{2\ell}(\varphi_1^2 + \varphi_2^2 - 1) + i\frac{3}{\ell^2}[\varphi_1\,,\varphi_2] + O\left(\ell^{-3}\right)\,. \tag{27}$$

Note that these indeed preserve the gauge symmetry since both sides transform covariantly under the action of the gauge group.

Using these in (19), we find

$$S = 2\pi\theta\mathrm{Tr}_{\mathcal{H}}\left[\frac{1}{4g^2}\left|F_{\mu\nu}\right|^2 + \frac{1}{2}\left(1 - \frac{1}{4\ell^2}\right)D_\mu\varphi D_\mu\varphi^\dagger + \frac{1}{\ell^2}\left(D_\mu[\varphi\,,\varphi^\dagger]\right)^2 + \frac{1}{32\ell^2}\left(D_\mu\{\varphi\,,\varphi^\dagger\}\right)^2\right.$$
$$\left. + \frac{1}{\tilde{g}^2}\left(\left(\frac{1}{2} + \frac{1}{4\ell^2}\right)\left(\frac{1}{2}\{\varphi\,,\varphi^\dagger\} - 1\right)^2 + \frac{1}{8}\left(1 - \frac{1}{\ell} - \frac{3}{4\ell^2}\right)[\varphi\,,\varphi^\dagger]^2\right) + O\left(\ell^{-3}\right)\right]. \tag{28}$$

To obtain this result we have also used the cyclicity property of the trace $\mathrm{Tr}_{\mathcal{H}}$, under which the terms proportional to $[\varphi\,,\varphi^\dagger]$ and $\left\{\{\varphi\,,\varphi^\dagger\}\,,[\varphi\,,\varphi^\dagger]\right\}$ vanish. The expression (28) is clearly invariant under the noncommutative $U(1)$ gauge symmetry, as it should be.

It is possible to employ the solution generating techniques introduced in [13, 14] to find noncommutative vortex type solutions of (28). To this end we proceed as follows. Let us first define the covariant coordinates

$$X = -\frac{1}{\theta}\bar{z} + ic_z\,,\quad X^\dagger = -\frac{1}{\theta}z - ic_{\bar{z}}\,, \tag{29}$$

where we have used the complex combinations $c_z = \frac{1}{\sqrt{2}}(c_1 - ic_2)$, $c_{\bar{z}} = \frac{1}{\sqrt{2}}(c_1 + ic_2)$. The covariant derivatives and the field strength may be expressed as

$$D_z\varphi = [X\,,\varphi]\,,\quad D_{\bar{z}}\varphi = -[X^\dagger\,,\varphi]\,. \tag{30}$$

$$\begin{aligned} F_{z\bar{z}} &= \partial_z c_{\bar{z}} - \partial_{\bar{z}} c_z + i[c_z\,,c_{\bar{z}}]\,, \\ &= i[X\,,X^\dagger] + \frac{i}{\theta}\,, \end{aligned} \tag{31}$$

All the basic constituents of the action (28) transform covarianty under the gauge symmetry

$$X \longrightarrow UXU^\dagger\,, \quad \varphi \longrightarrow U\varphi U^\dagger\,, \quad D_z\varphi \longrightarrow UD_z\varphi U^\dagger\,, \quad F_{z\bar{z}} \longrightarrow UF_{z\bar{z}}U^\dagger\,. \tag{32}$$

It follows that the equations of motion will transform covariantly, that is,

$$\frac{\delta S}{\delta X} \longrightarrow U\frac{\delta S}{\delta X}U^\dagger\,, \quad \frac{\delta S}{\delta \varphi} \longrightarrow U\frac{\delta S}{\delta \varphi}U^\dagger\,, \tag{33}$$

under a partial isometry U satisfying

$$U^\dagger U = 1\,, \quad UU^\dagger = P\,, \tag{34}$$

where P is a projection operator [13]. Thus, the partial isometries (34) generate solutions from a known solution.

A trivial solution to the equations of motion of (28) may easily found to be $X = -\frac{1}{\theta}\bar{z}\,, \varphi = 1$. Taking $U = S^m$, where S is the usual shift operator $S = \sum_{k=0}^{\infty} |k+1\rangle\langle k|$, we can write a set of non-trivial solutions for the theory governed by (28) as

$$\begin{aligned} \varphi &= S^m S^{\dagger m} = 1 - P_m\,, \\ X &= -\frac{1}{\theta} S^m \bar{z} S^{\dagger m}\,, \end{aligned} \tag{35}$$

where

$$P_n = \sum_{k=0}^{n-1} |k\rangle\langle k|\,, \tag{36}$$

is the projection operator of rank m. The corresponding field strength is $F_{12} = -iF_{z\bar{z}} = \frac{1}{\theta}P_m$. We can view these solutions as noncommutative vortices [13, 14, 15, 16, 17] carrying m units of flux:

$$2\pi\theta\,\mathrm{Tr}F_{12} = 2\pi m\,. \tag{37}$$

It is useful to evaluate the value of the action (28) on these solutions; we find

$$S = \pi\theta m\left(\frac{1}{g^2\theta^2} + \frac{1}{\tilde{g}^2}\left(1 + \frac{1}{2\ell^2}\right)\right) + O\left(\ell^{-3}\right)\,. \tag{38}$$

This corresponds to the energy of the static vortices in $2+1$ dimensions, $\mathbb{R}^2_\theta \times \mathbb{R}^1$ with $\mathbb{R}^1$ standing for time. We observe that to leading order in ℓ^{-1} there is a ℓ^{-2} contribution adding to the energy, which is a residue of the fact that the present model has descended from a model with a fuzzy sphere of order ℓ, $S^2_F(\ell)$ as extra dimensions.

Two limiting cases may also be easily recorded from (38). For $\tilde{g} \to \infty$, our solutions collapse to the fluxon solutions discussed in [18, 19]; whereas, for $\theta \to \infty$, the action gets a contribution only from the potential term, and our vortex solution collapses to a noncommutative soliton solution of the type first discussed in [20].

5.3. Case 2: No constraint

With $a = 0$ and $\ell = \infty$, the action reduces to

$$S = 2\pi\theta\mathrm{Tr}_{\mathcal{H}}\left(\frac{1}{4g^2}\,|F_{\mu\nu}|^2 + \frac{1}{2}D_\mu\varphi D_\mu\varphi^\dagger + \frac{1}{4}(D_\mu\varphi_3)^2 + \frac{1}{4}(D_\mu\varphi_4)^2 + \frac{1}{\tilde{g}^2}\tilde{V}_1\right)\,. \tag{39}$$

where $\tilde{V}_1$ is

$$\tilde{V}_1 \underset{\ell\to\infty}{=} \frac{1}{4}\Bigg((\varphi\varphi^\dagger)^2 + (\varphi^\dagger\varphi)^2 + \{\varphi\varphi_3\,,\varphi^\dagger\varphi_3\} + \{\varphi\,,\varphi^\dagger\}\left(\varphi_3^2 + 2(\varphi_3 - 1)\right) + 2(\varphi_3-1)^2$$
$$\left\{[\varphi\,,\varphi^\dagger]\,,\varphi_4\right\} + 2\varphi_4^2\Bigg)\,. \quad (40)$$

We see that there are linear terms in the potential (40) in φ_3, which will prevent us from applying the solution generating technique used in the previous subsection since these lead to terms in the equations of motion proportional to identity, and, therefore, they do not transform adjointly under the solution generating transformations [13, 14]. However, in the present model this situation can be remedied by defining a new field $\varphi_3' = \varphi_3 - 1$. In this manner, all the terms in the potential are quadratic or higher order or a constant, while the gradient term involving φ_3 is unaffected by this substitution. The equations of motion are

$$(D_zD_{\bar{z}} + D_{\bar{z}}D_z)\phi - \frac{\partial V_1'}{\partial\phi} = 0\,, \quad \text{for } \phi : \varphi\,, \varphi_3', \varphi_4\,,$$
$$\frac{1}{g^2}D_zF_{z\bar{z}} + i(\varphi D_{\bar{z}}\varphi^\dagger - D_{\bar{z}}\varphi\varphi^\dagger) + i[\varphi_3'\,, D_{\bar{z}}\varphi_3'] + i[\varphi_4\,, D_{\bar{z}}\varphi_4] = 0\,. \quad (41)$$

We observe that a trivial solution to these equations is given by $\varphi = 1\,, \varphi_3' = -1\,, \varphi_4 = 0\,, X = -\frac{1}{\theta}\bar{z}$. Applying the solution generating technique with $U = S^m$, we find

$$\varphi = S^m S^{\dagger m} = 1 - P_m\,, \quad \varphi_3' = -S^m S^{\dagger m} = P_m - 1\,, \quad \varphi_4 = 0\,, \quad X = -\frac{1}{\theta}S^m \bar{z} S^{\dagger m}\,, \quad (42)$$

where S and X are defined as in the previous subsection.

Evaluating the value of the action (39) on these solutions we find

$$S = \frac{\pi m}{g^2\theta}\,, \quad (43)$$

and the flux carried by these solutions is again

$$2\pi\theta\,\mathrm{Tr}F_{12} = 2\pi m\,. \quad (44)$$

As it turns out, there is in fact no contribution to (43) from the potential term. Thus, we can interpret (42) as flux-tube solutions carrying m units of flux [18]. It is easy to see that (42) satisfies the equations of motion (41), by noting that

$$D_z\phi = [X\,,\phi] = 0\,, \quad D_{\bar{z}}\phi = -[X^\dagger\,,\phi] = 0\,, \quad (45)$$

where ϕ are the solutions for $\varphi, \varphi_3', \varphi_4$ given in (42).

We also wish to remark that the field redefinition for φ_3 used above works only in the infinite ℓ limit. In fact, there does not appear to be a field redefinition at finite ℓ or at leading order around $\ell = \infty$ which will allow the use of solution generating transformations to construct non-trivial solutions.

Results of this section are generalized to $\mathbb{R}^{2d}_\theta$ in a rather straightforward manner, for details we refer the reader to [12].

Acknowledgments
This work is supported by the Middle East Technical University under Project No. BAP-08-11-2010-R-108.

References

[1] Forgacs P and Manton N S 1980 *Commun. Math. Phys.* **72** 15
[2] Kapetanakis D and Zoupanos G 1992 *Phys. Rept.* **219** 4
[3] Lechtenfeld O, Popov A D and Szabo R J 2003 Noncommutative instantons in higher dimensions, vortices and topological K-cycles *J. High Energy Phys.* JHEP0312(2003)022
[4] Balachandran A P, Kurkcuoglu S and Vaidya S 2007 *Lectures on Fuzzy and Fuzzy SUSY Physics* (Singapore: World Scientific)
[5] Aschieri P, Madore J, Manousselis P and Zoupanos G 2004 Dimensional reduction over fuzzy coset spaces *J. High Energy Phys.* JHEP0404(2004)034
[6] Aschieri P, Grammatikopoulos T, Steinacker H and Zoupanos G 2006 Dynamical generation of fuzzy extra dimensions, dimensional reduction and symmetry breaking *J. High Energy Phys.* JHEP0609(2006)026
[7] Steinacker H and Zoupanos G 2007 Fermions on spontaneously generated spherical extra dimensions *J. High Energy Phys.* JHEP0709(2007)017
[8] Chatzistavrakidis A, Steinacker H and Zoupanos G 2010 *Fortsch. Phys.* **58** 537
[9] Chatzistavrakidis A and Zoupanos G 2010 Higher-dimensional unified theories with fuzzy extra dimensions SIGMA **6** 063
[10] Grosse H, Lizzi F and Steinacker H 2010 *Phys. Rev.* D **81** 085034
[11] Harland D and Kurkcuoglu S 2009 *Nucl. Phys.* B **821** 380
[12] Kurkcuoglu S 2010 *Phys. Rev.* D **82** 105010
[13] Harvey J A, Kraus P and Larsen F 2000 Exact noncommutative solitons *J. High Energy Phys.* JHEP0012(2000)024
[14] Harvey J A 2001 Komaba lectures on noncommutative solitons and D-branes *Preprint* hep-th/0102076
[15] Jatkar D P, Mandal G and Wadia S R 2000 Nielsen-Olesen vortices in noncommutative Abelian Higgs model *J. High Energy Phys.* JHEP0009(2000)018
[16] Bak D, Lee K M and Park J H 2001 *Phys. Rev.* D **63** 125010
[17] Bak D 2000 *Phys. Lett.* B **495** 251
[18] Polychronakos A P 2000 *Phys. Lett.* B **495** 407
[19] Aganagic M, Gopakumar R, Minwalla S and Strominger A 2001 Unstable solitons in noncommutative gauge theory *J. High Energy Phys.* JHEP0104(2001)001
[20] Gopakumar R, Minwalla S and Strominger A 2000 Noncommutative solitons *J. High Energy Phys.* JHEP0005(2000)020

Fermionic cosmologies

L P Chimento[1], F P Devecchi[2], M Forte[1], G M Kremer[2], M O Ribas[2] and L L Samojeden[2]

[1] Physics Department, UBA, 1428 Buenos Aires, Argentina
[2] Physics Department, UFPR, 81531-990 Curitiba, Brazil

E-mail: kremer@fisica.ufpr.br, devecchi@fisica.ufpr.br, chimento@df.uba.ar

Abstract.
In this work we review if fermionic sources could be responsible for accelerated periods during the evolution of a FRW universe. In a first attempt, besides the fermionic source, a matter constituent would answer for the decelerated periods. The coupled differential equations that emerge from the field equations are integrated numerically. The self-interaction potential of the fermionic field is considered as a function of the scalar and pseudo-scalar invariants. It is shown that the fermionic field could behave like an inflaton field in the early universe, giving place to a transition to a matter dominated (decelerated) period. In a second formulation we turn our attention to analytical results, specifically using the idea of form-invariance transformations. These transformations can be used for obtaining accelerated cosmologies starting with conventional cosmological models. Here we reconsider the scalar field case and extend the discussion to fermionic fields. Finally we investigate the role of a Dirac field in a Brans-Dicke (BD) context. The results show that this source, in combination with the BD scalar, promote a final eternal accelerated era, after a matter dominated period.

1. Introduction

The search for constituents responsible for accelerated periods in the evolution of the universe is a fundamental topic in cosmology. The formulations include usually elements of general relativity, field theory and thermodynamics, analysing the evolution of space-time variables like the scale factor, its acceleration and energy densities. Several candidates has been tested for describing both the inflationary period and the present era: scalar fields, exotic equations of state and cosmological constants[1]. Another possibility is to consider fermionic fields as gravitational sources in an FWR expanding universe. Fermionic sources has been tested using several approaches, with results including exact solutions, anisotropy-to-isotropy scenarios and cyclic cosmologies (see, for example[2, 6]).

In our work the connection between general relativity and the Dirac equation is done via the tetrad formalism; the interactions between the constituents can be modeled through the presence of a non-equilibrium pressure term in the source's energy-momentum tensor. Besides, it can be considered a self-interaction term for the fermion, in the form of a potential that can assume several forms (included the Nambu-Jona-Lasinio case[7]). While testing fermionic sources as responsible of accelerated periods, different regimes are possible[3, 4]. In a young universe scenario the fermion produces a fast expansion where matter is created till it starts to predominate and the initial accelerated expansion gives place to a decelerated era, which ends when the Dirac field predominates again leading to an accelerated era[3]. In this case the fermionic field plays the role of the inflaton in the early universe and of dark energy for the old universe, without the need of a cosmological constant term or a scalar field. In the old universe case, an initially matter dominated period gradually turns into a dark (fermionic) energy era when an accelerated regime starts and remains for the rest of the systems evolution.

On the other hand, taking a analytical approach, the idea of form-invariant symmetries has been invoked to extract different evolution regimes from Friedmann-Robertson-Walker cosmologies. This is the case when we obtain the so-called phantom cosmologies [14] from ordinary regimes like universes filled with barotropic fluids. These models generate solutions that contain negative pressure eras, therefore promoting positive accelerated expansions [13]. The case were the source is a bosonic field was investigated in[13]. An obvious question is what happens when the field promoting the universe expansion obeys Dirac's equation[15]. Taking into account these ideas, the main point of discussion of the second part of this manuscript is to investigate the form-invariance transformations when the Klein-Gordon dynamics is replaced by the Dirac equations in a FRW space-time. What is verified, among other results, is that the process exibits a new class of possible regimes, when in comparison with the bosonic case [15]. As another approach, fermionic sources can also be tested in a Brans-Dicke context[5]. In this particular case the fermionic field (in combination with the Brans-Dicke scalar field $\varphi(t)$) can be responsible for a final eternally accelerated universe, after an initial matter dominated period.

The manuscript is structured as follows: In section II we make a brief review of the tetrad formalism used to include fermionic and matter fields in a dynamical curved space-time. The field equations for an FRW universe are derived in sections II and III and then we present the analysis of the scenarios in which the fermion answers for accelerated eras and transitions to decelerated periods. In section IV we review the basic ideas behind the form-invariance transformations in cosmology; in section V we focus on the fermionic field case. Finally in section VI we consider a fermionic source in the Brans-Dicke (BD) context, showing that this source combines with the BD scalar to promote a final accelerated regime in a FRW universe; to finish we display our conclusions.

2. Fermionic accelerated regimes

As it is well-known[8, 9, 10] the gauge group of GR does not admit a spinor representation and the tretad formalism solves the problem[8, 9, 10]. Following the general covariance principle, a connection between the tetrad and the metric tensor $g_{\mu\nu}$ is established, $g_{\mu\gamma} = e^a_\mu e^b_\gamma \eta_{ab}, a = 0, 1, 2, 3$, where e^a_μ denotes the tetrad or vierbein and η_{ab} is the Minkowski metric tensor; latin indices refer to the local inertial frame an greek indices to the general system.

Considering that we want to construct a fermionic cosmology we need to describe the behavior of fermions in the presence of a gravitational field, so the next step is to construct an action for this system. The Dirac lagrangian density in Minkowski space-time is

$$L_D = \frac{\imath}{2}[\overline{\psi}\gamma^a\partial_a\psi - (\partial_a\overline{\psi})\gamma^a\psi] - m\overline{\psi}\psi - V, \tag{1}$$

where the spinors are treated as classically commuting fields[9]. m is the fermionic mass, $\overline{\psi} = \psi^\dagger\gamma^0$ denotes the adjoint spinor field and the term V, which is a function of ψ and $\overline{\psi}$, describes the potential density of self-interaction between fermions. The general covariance principle imposes that the Dirac-Pauli matrices γ^a must be replaced by their generalized counterparts $\Gamma^\mu = e^\mu_a\gamma^a$, whereas the generalized Dirac-Pauli matrices satisfy the extended Clifford algebra, i.e., $\{\Gamma^\mu, \Gamma^\nu\} = 2g^{\mu\nu}$.

In a second step we need to substitute the ordinary derivatives by their covariant versions $\partial_\mu\psi \to D_\mu\psi = \partial_\mu\psi - \Omega_\mu\psi$, where the spin connection Ω_μ is given by

$$\Omega_\mu = -\frac{1}{4}g_{\mu\nu}[\Gamma^\nu_{\sigma\lambda} - e^\nu_b(\partial_\sigma e^b_\lambda)]\gamma^\sigma\gamma^\lambda, \tag{2}$$

with $\Gamma^{\nu}_{\sigma\lambda}$ denoting the Christoffel symbols. Hence, the generally covariant Dirac lagrangian becomes

$$L_D = \frac{\imath}{2}[\overline{\psi}\,\Gamma^{\mu}D_{\mu}\psi - (D_{\mu}\overline{\psi})\Gamma^{\mu}\psi] - m\overline{\psi}\psi - V. \tag{3}$$

The field equations are obtained from the total action $S(g,\psi,\overline{\psi}) = \int\sqrt{-g}\,L_t\,d^4x$, where $L_t = L_g + L_D + L_m$ is the total lagrangian density. $L_g = R/2$, with R denoting the curvature scalar is the free Einstein lagrangian. L_D is the Dirac lagrangian density (3) and L_m is the lagrangian density of the matter field. From the lagrangian density (3), through Euler-Lagrange equations, we obtain the Dirac equations for the spinor field and its adjoint coupled to the gravitational field

$$\imath\Gamma^{\mu}D_{\mu}\psi - m\psi - \frac{dV}{d\overline{\psi}} = 0, \quad \imath D_{\mu}\overline{\psi}\,\Gamma^{\mu} + m\overline{\psi} + \frac{dV}{d\psi} = 0. \tag{4}$$

The variation of the total action with respect to the metric tensor leads to Einstein field equations $R_{\mu\nu} - \frac{1}{2}g_{\mu\nu}R = -T_{\mu\nu}$, where $T_{\mu\nu}$ is the total energy-momentum tensor which is a sum of the energy-momentum tensor of the fermionic field $T_f^{\mu\nu}$ and of the matter field $T_m^{\mu\nu}$, i.e., $T^{\mu\nu} = T_m^{\mu\nu} + T_f^{\mu\nu}$. The symmetric form of the energy-momentum tensor of the fermionic field is given by

$$T_f^{\mu\nu} = \frac{\imath}{4}\left[\overline{\psi}\Gamma^{\mu}D^{\nu}\psi + \overline{\psi}\Gamma^{\nu}D^{\mu}\psi - D^{\nu}\overline{\psi}\Gamma^{\mu}\psi - D^{\mu}\overline{\psi}\Gamma^{\nu}\psi\right] - g^{\mu\nu}L_D. \tag{5}$$

The Robertson-Walker metric mirrors the homogeneity and isotropy properties of the universe, we have $ds^2 = dt^2 - a(t)^2(dx^2 + dy^2 + dz^2)$, where $a(t)$ is the cosmic scale factor. The total energy-momentum tensor for an isotropic and homogeneous universe is given by $(T^{\mu}{}_{\nu}) = \mathrm{diag}(\rho, -p-\varpi, -p-\varpi, -p-\varpi)$, where the total energy density ρ and the total pressure p are given as a sum of the corresponding terms of the fermionic and matter fields, i.e., $\rho = \rho_f + \rho_m$ and $p = p_f + p_m$. Moreover, the quantity ϖ refers to a non-equilibrium pressure which is related to dissipative processes during the evolution of the universe and represents an irreversible process of energy transfer between the matter and the gravitational field [12].

Thanks to the Bianchi identities, the covariant differentiation of Einstein field equations lead to the conservation law of the total energy-momentum tensor $T^{\mu\nu}{}_{;\nu} = 0$, hence it follows by using the representation (12) the evolution equation for the total energy density:

$$\dot{\rho} + 3H(\rho + p + \varpi) = 0, \tag{6}$$

where the dot refers to a differentiation to time and $H = \dot{a}(t)/a(t)$ denotes the Hubble parameter. Furthermore, from Einstein equations follow the Friedmann and acceleration equations

$$H^2 = \frac{1}{3}\rho, \qquad \frac{\ddot{a}}{a} = -\frac{1}{6}(\rho + 3p + 3\varpi), \tag{7}$$

respectively. Only two equations from (13) and (14) are linearly independent. For the FRW metric the tetrad reads $e_0^{\mu} = \delta_0^{\mu}, e_i^{\mu} = \frac{1}{a(t)}\delta_i^{\mu}$. Also, the Dirac matrices become

$$\Gamma^0 = \gamma^0, \quad \Gamma^i = \frac{1}{a(t)}\gamma^i, \quad \Gamma^5 = -\imath\sqrt{-g}\,\Gamma^0\Gamma^1\Gamma^2\Gamma^3 = \gamma^5, \tag{8}$$

from which the spin connection components are obtained, yielding $\Omega_0 = 0, \Omega_i = \frac{1}{2}\dot{a}(t)\gamma^i\gamma^0$. For a FRW universe the fermionic field is an exclusive function of time, so the Dirac equations (4) become

$$\dot{\psi} + \frac{3}{2}H\psi + \imath m\gamma^0\psi + \imath\gamma^0\frac{dV}{d\overline{\psi}} = 0. \tag{9}$$

Combining the equations above, the non-zero components of the energy-momentum tensor become

$$(T_f)^0{}_0 = m(\overline{\psi}\psi) + V, (T_f)^1{}_1 = (T_f)^2{}_2 = (T_f)^3{}_3 = V - \frac{dV}{d\psi}\frac{\psi}{2} - \frac{\overline{\psi}}{2}\frac{dV}{d\overline{\psi}}, \tag{10}$$

which are only functions of ψ and $\overline{\psi}$. By identifying the components of the energy-momentum tensor of the fermionic field as $(T_f)^\mu{}_\nu = \mathrm{diag}(\rho_f, -p_f, -p_f, -p_f)$, one can obtain from the equations above[3] the conservation law for the energy density of the fermionic field: $\dot{\rho}_f + 3H(\rho_f + p_f) = 0$. Hence, the evolution equation for the energy density of the fermionic field decouples from the energy density of the matter field, and we obtain[3]: $\dot{\rho}_m + 3H(\rho_m + p_m) = -3H\varpi$, where the term $-3H\varpi$ could be interpreted as the energy density production rate of the matter field (see e.g.[12]). According to the extended thermodynamic theory the non-equilibrium pressure $\varpi(t)$ obeys an evolution equation, whose linear form reads $\tau\dot{\varpi} + \varpi = -3\eta H$, where τ denotes a characteristic time and η is the so-called bulk viscosity.

3. Cosmological Solutions

In order to analyze cosmological solutions for the system of the previous section, we have to specify the fermionic potential V. According to the Pauli-Fierz theorem V is an exclusive function of the scalar invariant $(\overline{\psi}\psi)^2$ and of the pseudo-scalar invariant $(i\overline{\psi}\gamma^5\psi)^2$, i.e., $V = V((\overline{\psi}\psi)^2, (i\overline{\psi}\gamma^5\psi)^2)$. Here we are interested in self-interaction potentials of the form

$$V = \lambda\left[\beta_1(\overline{\psi}\psi)^2 + \beta_2(i\overline{\psi}\gamma^5\psi)^2\right]^n, \tag{11}$$

where the coupling constant λ is a non-negative quantity and n is a constant exponent. We shall analyze three cases, namely, (i) $\beta_1 = 1$ and $\beta_2 = 0$ where V is a function only of the scalar invariant; (ii) $\beta_1 = 0$ and $\beta_2 = 1$ where V depends only on the pseudo-scalar invariant and (iii) $\beta_1 = \beta_2 = 1$ where V is a combination of the scalar and pseudo-scalar invariants. The Nambu-Jona-Lasinio potential [7] is represented by the last case with $n = 1$.

The energy density and the pressure of the fermions for the potential V are given by

$$\rho_f = m(\overline{\psi}\psi) + \lambda\left[\beta_1(\overline{\psi}\psi)^2 + \beta_2(i\overline{\psi}\gamma^5\psi)^2\right]^n, p_f = (2n-1)[\rho_f - m(\overline{\psi}\psi)]. \tag{12}$$

The fermions could be classified according to the value of the exponent n. Indeed, for $n \geq 1/2$ the fermions represent a matter field with positive pressure ($n > 1/2$) or a pressureless fluid ($n = 1/2$), whereas for $n < 1/2$ the pressure of the fermions is negative and they could represent either the inflaton or the dark energy[3].

For massless fermions, the pressure is connected with the energy density by a simple barotropic equation of state $p_f = (2n-1)\rho_f$ and it follows from the conservation equation (20) for the fermions that $\rho_f \propto 1/a^{6n}$. We shall not analyze this case here, since the behavior of the fermionic field does not differ from that of a matter field when $n > 1/2$ or from that of a bosonic field when $n < 1/2$. Moreover, for the massive case, we shall deal with only the case where the fermionic field behaves as inflaton or dark energy, i.e., the case where $n < 1/2$.

For the pressure of the matter field we shall adopt a barotropic equation of state, i.e., $p_m = w_m\rho_m$ with $0 \leq w_m \leq 1$. Furthermore, the coefficient of bulk viscosity η and the characteristic time τ are assumed to be related with the energy density ρ by $\eta = \alpha\rho$ and $\tau = \eta/\rho$, where α is a constant[12].

The system of field equations we shall investigate in order to find the cosmological solutions are: (a) the acceleration equation $\frac{\ddot{a}}{a} = -\frac{1}{6}(\rho_f + \rho_m + 3p_f + 3p_m + 3\varpi)$; (b) the evolution equation for the energy density of the matter field and the non-eq. pressure $\dot{\rho}_m + 3H(\rho_m + p_m + \varpi) = 0$, $\tau\dot{\varpi} + \varpi = -3\eta H$, and c) the Dirac equation. This consists of a system of seven coupled ordinary differential equations for the fields $a(t)$, $\rho_m(t)$, $\varpi(t)$, $\psi_1(t)$, $\psi_2(t)$, $\psi_3(t)$ and $\psi_4(t)$ and in the following subsections we shall find solutions of this system of equations for given initial conditions.

Let us analyze the case that corresponds to the evolution of the early universe, where the fermionic field plays at the beginning the role of an inflaton and the matter is created by an irreversible process through the presence of a non-equilibrium pressure ϖ. The initial conditions we have chosen for $t = 0$ are $a(0) = 1, \psi_1(0) = 0.1, \psi_2(0) = 1, \psi_3(0) = 0.3, \psi_4(0) = \imath, \rho_m(0) = 0, \varpi(0) = 0$. The last two initial conditions correspond to a vanishing energy density of the matter field and a vanishing non-equilibrium pressure at $t = 0$. The conditions chosen here characterize qualitatively an initial proportion between the constituents in the corresponding era (i.e., in this case we have a predominating fermionic field over the matter, indicating the initial inflationary state). We need to specify now an initial condition for $\dot{a}(0)$. This condition follows from the Friedmann equation, i.e., $\dot{a}(0) = a(0)\sqrt{\frac{\rho_f(0)+\rho_m(0)}{3}}$. Besides, we also have to specify some parameters in order to obtain numerical solutions of the coupled system above, these parameters are: (a) λ, β_1, β_2 and n which define the self-interacting potential V; (b) m which is related to the mass of the fermionic field; (c) w_m which defines the matter field through its barotropic equation of state $p_m = w_m\rho_m$ and (d) α which is connected with the bulk viscosity term $\eta = \alpha\rho$. We have chosen the following values for these parameters: $\lambda = 0.1, \beta_1 = \beta_2 = 1, n = 0.3, m = 0.01, w_m = 1/3, \alpha = 1.0, \text{and}\alpha = 1.2,$. Which correspond to a fermionic field with a negative pressure, described by a self-interacting potential that depends on the scalar and pseudo-scalar invariants and a matter field of massless particles that could describe a radiation field. . The numerical integration of the system[3] indicates that there exists an accelerated period where the fermionic field dominates and the matter field is created at the expenses of an irreversible process of energy transfer between the matter and the gravitational field. The accelerated period is followed by a decelerated era which is dominated by the matter field. Due to fact that the self-interaction potential tends to a constant value for large values of time, the energy density of the femionic field overcomes the energy density of the matter field and the universe goes into another accelerated period. It is noteworthy that for the bosonic case where the potentials are exponentials or inverse power-laws, one has to add a constant value – which is similar to introduce a cosmological constant term – in order to return to an accelerated era after the accelerated-decelerated period (see, for example[12]). Here the same self-potential interaction which plays the role of an inflaton field at the beginning plays the role of a cosmological constant for later times and could be identify as dark energy. The numerical integration[3] also indicates that for large values of the coefficient α it follows that: (a) the energy density of fermionic field decays more rapidly causing a larger accelerated period and (b) the energy density of the matter field has a more significant growth and leads to a larger decelerated period. In this first part we have investigated the possibility that a fermionic field – with a self-interacting potential – could be the responsible for accelerated regimes in the evolution of the universe. We have shown that the fermionic field behaves like an inflaton field for the early universe, where the matter field was created by an irreversible process connected with a non-equilibrium pressure. As a final comment to this section we must mention that there is in fact another possible scenario in this model; the fermion playing the role of dark energy[3]. For an old decelerated universe dominated by ordinary matter the fermionic field drives the universe to a final eternal accelerated era[2, 3, 4].

4. Internal symmetry in FRW

When we turn our attention to analytical solutions in cosmological models one interesting idea is the use of symmetries. The issue of form-invariant symmetries has been invoked recently to extract different evolution regimes from Friedmann-Robertson-Walker cosmologies [14]. This is the case when we obtain the so-called phantom cosmologies [13] from ordinary regimes like universes filled with barotropic fluids [14]. These new solutions contain negative pressure regimes, therefore promoting positive accelerated expansions [13]. The case were the gravitational source is a bosonic field was investigated in [14]. In spatially flat perfect fluid FRW cosmologies, such symmetries can be viewed as a prescription relating the quantities a, H, ρ and p in a given initial scenario to quantities $\bar{a}$, $\bar{H}$, $\bar{\rho}$ and $\bar{p}$ corresponding to a new cosmological model. As a starting point we present a quick review of this internal symmetry with the purpose of investigating it in bosonic and fermionic cases [13, 19, 15]. As is well-known, the Einstein equations for a flat FRW cosmological model with scale factor a, and filled with a perfect fluid with energy density ρ and pressure p, are

$$3H^2 = \rho, \quad \dot{\rho} + 3H(\rho + p) = 0, \tag{13}$$

where $H = \dot{a}/a$. The correspondent form-invariance transformation is given by [13]

$$\bar{\rho} = \bar{\rho}(\rho), \; \bar{H} = \left(\frac{\bar{\rho}}{\rho}\right)^{1/2} H, \bar{p} = -\bar{\rho} + \left(\frac{\rho}{\bar{\rho}}\right)^{1/2} (\rho + p)\frac{d\bar{\rho}}{d\rho} \tag{14}$$

where $\bar{\rho} = \bar{\rho}(\rho)$ is an invertible arbitrary function. This set of transformations give rise to the form-invariance symmetry group (FISG)[14, 13, 15]. For the perfect fluid case, $p = (\gamma - 1)\rho$, the barotropic indexes of the original and transformed fluid are related by $\bar{\gamma} = \left(\frac{\rho}{\bar{\rho}}\right)^{3/2} \frac{d\bar{\rho}}{d\rho}\gamma$. For later application it will useful to investigate the transformation rule for the quantities which characterize the geometry and the fluid when the transformation is generated by [13] $\bar{\rho} = n^2\rho$, where n is the constant parameter of the FISG. Hence, for the geometrical quantities and equation of state we get

$$\bar{H} = nH, \rightarrow \bar{a} = a^n, \bar{w} = \bar{\gamma} - 1 = \frac{\gamma}{n} - 1, \quad \bar{p} = \bar{w}\bar{\rho} = \left(\frac{\gamma}{n} - 1\right)\bar{\rho}, \tag{15}$$

Let us analyze the equation of state (15) when the seed fluid represents a normal fluid with $\gamma > 0$. We can infer: (i) when $0 < n < \gamma$ it follows that $\bar{w} > 0$ and the cosmological fluid behaves as a normal fluid; (ii) when $n > \gamma$ it follows that $-1 < \bar{w} < 0$ and the cosmological fluid behaves as a quintessence constituent; (iii) when $n < 0$ it follows that $\bar{w} < -1$ and the cosmological fluid behaves as a phantom constituent[14].

5. The fermionic case

In this section we extend the form-invariance transformation to a fermionic field satisfying the Dirac equation in curved space-time. As required by the equations above we must calculate the energy density and the hydrostatic pressure of the fermionic field. We have [3]

$$\rho_\psi = m(\overline{\Psi}\Psi) + V, \quad p_\psi = \frac{dV}{d\Psi}\frac{\Psi}{2} + \frac{\overline{\Psi}}{2}\frac{dV}{d\overline{\Psi}} - V. \tag{16}$$

To obtain the transformation properties of ρ_ψ and p_ψ we express the potential in terms of the scalar invariant $X = (\overline{\Psi}\Psi)^2$ and the pseudo-scalar invariant $Y = (i\overline{\Psi}\gamma^5\Psi)^2$. To this end, we suppose a generic potential $V = V(X,Y)$, that include, among others, the Nambu-Jona-Lasinio potential [3]. For this kind of potential the hydrostatic pressure becomes

$$p = 2X\frac{\partial V}{\partial X} + 2Y\frac{\partial V}{\partial Y} - V. \tag{17}$$

Applying the transformation rules (14) to the energy density (16) and the pressure we get

$$\bar{m}\sqrt{\bar{X}} + \bar{V} = n^2(m\sqrt{X} + V), \tag{18}$$

$$2\bar{X}\frac{\partial \bar{V}}{\partial \bar{X}} + 2\bar{Y}\frac{\partial \bar{V}}{\partial \bar{Y}} + \bar{m}\sqrt{\bar{X}} = n\left[2X\frac{\partial V}{\partial X} + 2Y\frac{\partial V}{\partial Y} + m\sqrt{X}\right]. \tag{19}$$

We also obtain the following transformation equation[?]

$$\left(\frac{m}{2\sqrt{X}} + \frac{\partial V}{\partial X}\right)\left[\bar{X}\frac{\partial X}{\partial \bar{X}} + \bar{Y}\frac{\partial X}{\partial \bar{Y}} - \frac{X}{n}\right] + \frac{\partial V}{\partial Y}\left[\bar{X}\frac{\partial Y}{\partial \bar{X}} + \bar{Y}\frac{\partial Y}{\partial \bar{Y}} - \frac{Y}{n}\right] = 0.$$

All transformations satisfying these last conditions represent internal symmetries of the Einstein's equations with a fermionic source, provide that condition becomes an identity for any potential and particle mass. Then the two square brackets must vanish

$$\bar{X}\frac{\partial X}{\partial \bar{X}} + \bar{Y}\frac{\partial X}{\partial \bar{Y}} = \frac{X}{n}, \quad \bar{X}\frac{\partial Y}{\partial \bar{X}} + \bar{Y}\frac{\partial Y}{\partial \bar{Y}} = \frac{Y}{n}. \tag{20}$$

In addition, by imposing the condition $\bar{m}(n = 1, \bar{X} = X, \bar{V} = V) = m$, on the transformed fermionic mass, any solution of the coupled system of partial differential equations (20) is a form-invariance symmetry transformation of the Einstein-Dirac dynamics. For the general massive fermionic field whose potential depends only on X, the solutions lead to the general form-invariance symmetry transformation[?]

$$\bar{X} = X^n, \bar{V} = n^2V + n^2m(\sqrt{X} - X^{n/2}), \bar{m} = n^2m. \tag{21}$$

Following the same methodology of the bosonic scalar field, let us investigate two simple cases, (i) the free massive fermionic particle and (ii) the massless fermionic particle with a polynomial potential. In the free case, the energy density (16) and pressure (17) of the fermionic particle are given by $\rho = m\sqrt{X}, p = 0, V = 0$. Solving the Einstein equations we obtain

$$a = a_0t^{2/3}, \qquad X = \frac{16}{9m^2}\left(\frac{a_0}{a}\right)^6, \tag{22}$$

where a_0 is an integration constant. Suppose now that we want to in comparing this cosmological model with other one, this can be done by applying the transformations (14) together with Eq. (22) to $a(t)$ and the potential, which is dependent only on X. Inserting the seed solution into the generated equations we find the full set of power law solutions along with the potential, energy density and pressure characterizing the transformed fermionic configuration:

$$\bar{a} = \bar{a}_0(\pm t)^{2n/3}, \qquad \bar{a}_0 = a_0^n, \bar{X} = \left(\tfrac{16}{9m^2}\right)^n \left(\tfrac{\bar{a}_0}{\bar{a}}\right)^6, \tag{23}$$

$$\bar{V} = \bar{m}(\bar{X}^{1/2n} - \bar{X}^{1/2}), \qquad \bar{m} = n^2m, \bar{\rho} = \bar{m}\bar{X}^{1/2n}, \qquad \bar{p} = \left(\tfrac{1}{n} - 1\right)\bar{\rho}. \tag{24}$$

The Dirac's equations in a FRW background for the seed and transformed configurations read

$$\dot{\psi} + \frac{3}{2}H\psi + \imath m\gamma^0\psi + \imath\gamma^0\frac{dV}{d\bar{\psi}} = 0, \tag{25}$$

along with the corresponding equation for $\bar{\psi}$ and $\bar{\bar{\psi}}$. In order to the physics of the fermionic particles maintain their own characteristics we assume that the Dirac's matrices remain invariants under a form invariance symmetry transformation. For $n = 1$ the latter becomes the spinor field corresponding to the seed solution (22). The Einstein equations for the new configuration are $3\bar{H}^2 = \bar{\rho}, -2\dot{\bar{H}} = \bar{\rho} + \bar{p}$; so the new solution (23)-(24) can be associated to a fermionic barotropic fluid whose equation of state is $\bar{\gamma} = 1 + \bar{p}/\bar{\rho} = 1/n$. When $n = 1$, identical transformation, the solution of the Einstein's equations become the seed solution (22), which can be identified with some kind of fermionic dust matter[15]. Now, we consider the massless case and investigate a spinor field driven by a potential $V = V(X)$ depending on X. The imposition of the energy density transformation

$\bar{\rho} = n^2\rho$ allow us to apply the transformation rules (14) for X and the potential V:

$$\bar{X} = X^n, \qquad \bar{V} = n^2 V. \tag{26}$$

By considering the case where the potential is given by a power law $V = X^\alpha$, where α is a constant, we choose the seed potential $V = \sqrt{X}$, and from the Einstein equations it follows

$$a = a_0 t^{2/3}, \qquad X = \frac{16}{9}\left(\frac{a_0}{a}\right)^6. \tag{27}$$

This solution and the corresponding spinor field can be obtained from the above seed solution (22) by making $m = 1$. However, both cases, $m \neq 0$ and $m = 0$, are absolutely different because a form invariance transformation that relates them does not exist, as can be seen from Eq. (27). In fact, for the $m = 0$ case, the solution for the transformed fermionic configuration read

$$\bar{a} = \bar{a}_0 t^{2n/3}, \bar{a}_0 = a_0^n, \bar{X} = \left(\frac{16}{9}\right)^n \left(\frac{\bar{a}_0}{\bar{a}}\right)^6, \bar{V} = n^2 \bar{X}^{1/2n}, \bar{\rho} = n^2 \bar{X}^{\alpha/n}, \bar{p} = \left(\frac{1}{n} - 1\right)\bar{\rho}. \tag{28}$$

Thus, comparing the transformed configurations (23,24) with the original ones for $m = 1$, we see that both potentials are different and the dynamics too.

6. Brans-Dicke fermionic cosmology

Einstein's General Relativity (GR) is the usually invoked theory for describing the evolution of the fundamental space-time variables in cosmological models[8]. As an alternative to GR we have the Brans-Dicke theory of gravitation[8], a scalar-tensor formulation that rules the gravitational phenomena through the interplay between the metric tensor and a scalar field φ that controls the intensity of the gravitational constant G. On the other hand, cosmological models that include dark energy are strong candidates for explaining the present regime of positive acceleration of the universe[1]. Going in that direction other possibility, as discussed in the previous approaches, is to consider fermionic fields as gravitational sources for those accelerated universes. As was exposed in the previous sections, these authors[3] proposed a cosmological model (based on GR) in a dissipative Universe and showed that in a young universe scenario the fermionic field produces a fast expansion where matter (included via a barotropic equation of state) is created till it starts to predominate and the initial accelerated period gives place to a decelerated era. In this case the fermionic field plays the role of the inflaton in an early period and the role of dark energy for an old universe; without the need of a cosmological constant term or a scalar field[3]. In an exclusive old universe scenario an initially matter dominated period gradually turns into a dark (fermionic) energy period when an accelerated regime rules for the rest of the system evolution[3, 4]. Testing those properties in a universe ruled by Brans-Dicke gravitation would be of interest and this is the main focus of this section.

As we saw in the previous discussions, for a FRW universe the fermionic field becomes an exclusive function of time. Besides the Dirac field, we have considered as gravitational source a matter constituent that is ruled by a barotropic equation of state, namely $p_m = b_m p_m$, where b_m is the barotropic coefficient ($0 \leq b_m \leq 1$); then the total energy density ρ is given as a sum of these individual contributions, i.e., $\rho = \rho_D + \rho_m = T_{00}$, where $T_{\mu\nu}$ is the total energy-momentum tensor of sources. The energy density ρ satisfies the conservation law $\dot{\rho} + 3\dot{a}/a(\rho + p) = 0$. Combining the Dirac and Einstein equations one can obtain an independent conservation law for the energy density of the fermionic field[3]. This implies into a decoupled conservation equation for the energy density of the matter constituent, i.e., $\dot{\rho}_m + 3H(\rho_m + p_m) = 0$. In order to analyze the cosmological solutions of our model we have to define first some sources properties. As in section II we consider a fermionic self-interaction potential, $V = \left[(\bar{\psi}\psi)^2\right]^n$, where n is a constant real number[3]. On the other hand the matter field, is again following a barotropic equation of state[3, 5]. After some algebraic manipulation we can put the model dynamics in the following form:

$$2\frac{\ddot{a}}{a} + H^2 + \alpha^2\ddot{\varphi} + 2\alpha H\dot{\varphi} + \frac{\alpha^2\omega}{2}\dot{\varphi}^2 = e^{-\alpha\varphi}\left[(2n-1)(\bar{\psi}\psi)^{2n} + p_m\right], \tag{29}$$

$$\alpha^2\ddot{\varphi} + 3\alpha H\dot{\varphi} = \frac{e^{-\alpha\varphi}}{3+2\omega}\left[V + \rho_m - 3(2n-1)(\bar{\psi}\psi)^{2n} - 3p_m\right], \tag{30}$$

$$\dot{\psi} + \frac{3}{2}H\psi + \imath\gamma^0\frac{dV}{d\bar{\psi}} = 0, \quad \dot{\rho_m} + 3H(\rho_m + p_m) = 0 \tag{31}$$

We consider a pressureless matter field ($p_m = 0$), so that one can obtain from conservation that $\rho_m(t) = \rho_m(0)/a(t)^3$[5]. The equations (29-31) constitute a highly non-linear system of differential equations and we proceed to solve it numerically. We analyze the time evolution of our model choosing first the conditions for $t = 0$: $a(0) = 1, \dot{a}(0) = 1, [\bar{\psi}\psi](0) = 0.001, \quad \rho_m(0) = 1, \varphi(0) = 1, \dot{\varphi}(0) = 0.001$. These conditions characterize qualitatively an initial proportion between the constituents; an era when matter predominates over the fermionic density. Besides that, we have to specify the magnitude of the remaining parameters: we suppose initially that $\alpha = 1.0$, $\omega = 4 \times 10^4$ [5] and a value for the potential power n, $n = 0.2$ [3]. These choices are reference values that permit final adjustments to follow several cosmological constraints, like the spectrum of the Brans-Dicke coupling ω[11] and the present value of the Brans-Dicke scalar field φ. These parameters can be in fact adjusted due to invariance properties of the Brans-Dicke gravitational field equations, by using the following change of variables: $\alpha \to \bar{\alpha}, \quad \varphi \to \bar{\varphi}, \quad \bar{\alpha}\bar{\varphi} = \gamma/\varphi + \alpha$ where $\exp(-\gamma) = x_0$ is associated to the asymptotic value of $\bar{\varphi}_0 = (2\omega+4)/(2\omega+3)$. These transformations show how a value of the Brans-Dicke parameter ω is linked to the definition of a new Brans-Dicke scalar field $\bar{\varphi}$. In fact, after numerical integration, it is possible to verify that these features are included in the φ evolution that imply into the evolution of the gravitational "constant" as $G(t) = (2\omega + 4)/[(2\omega + 3)\bar{\varphi}(t)]$.

The numerical integration of the system shows[5] that initially the universe is expanding with negative acceleration, a period where the matter constituent predominates over the fermionic field. With the evolution of time we have increasing values in energy transference to the fermion (this is happening via the gravitational field $a(t)$ and the scalar field $\varphi(t)$, as the equations of motion show). The negative pressure of $\psi(t)$ helps in fact to promote a final accelerated period indicating that in this model the dark energy role would be played by a *combination of the fermionic constituent with the scalar* $\varphi(t)$. Other interesting results appear when we choose the power n to be in the neighborhood of values $n \approx 0.33$. In fact, for a fixed value of $\dot{\varphi}(0)$ and increasing values of n what emerges is a universe that fails to show a final accelerated period, even when the fermionic field still exhibits a negative pressure ($p_D = (2n-1)\left[(\bar{\psi}\psi)^2\right]^n$). On the other hand, for values $n \leq 0.33$ we will find a universe that is permanently in accelerated expansion. Another important remark here is that this general qualitative behavior depends strongly on the initial value of the time derivative of the scalar field $\varphi(t)$: what we verify is that increasing values of $\dot{\varphi}(0)$ promote an earlier entrance on the accelerated period[5].

For the energy densities of the fermionic field ρ_D and matter ρ_m the numerical results show[5] that eventually the energy density of the femionic field overcomes the energy density of the matter field, although this does not coincide with the instant when the universe goes into an accelerated period (opposed to what occurs in some Einstein gravity based models[12]). Another feature showed by the numerical results is that for larger values of n it follows that: (a) the energy density of fermionic field grows more slowly causing a larger decelerated period and (b) the energy density of the matter field has a less significant decay and also promotes a larger decelerated period. On the other hand, both densities have decreasing values in time due to the permanent expansion of the universe (this was verified with plots of the scale factor against time[5]). Again, all cases are strongly dependent on the exponent n of the self-interacting potential and one can obtain different behaviors, that are in tune with the acceleration patterns presented above. Finally we verify that the scalar field φ approaches a final constant value that validates the accord between Brans-Dicke and Einstein gravitation for large t.

7. Conclusion

In this work we have reviewed several scenarios where fermionic field is responsible for accelerated regimes in the evolution of a FRW universe. Through numerical integrations is is possible to verify that the fermion behaves like an inflaton field for the early universe and later on as a dark energy field[3], whereas ordinary matter is created by including a non-equilibrium pressure. On the analytic solutions analysis we have found the symmetry transformations under which the Einstein equations, for a FRW universe filled with bosons or fermions, preserve their form. This group of transformations has been used to obtain power-law solutions from a seed one. Starting from a contracting spatially flat FRW cosmological model we get, after using the dual transformation, a super-accelerated spatially flat FRW cosmological model, i.e., the phantomization of the model[15]. Finally, using the Brans-Dicke theory of gravitation we investigated a hypothetical FRW universe filled with a fermionic field (with a self interaction potential) and a matter constituent ruled by a barotropic equation of state. The results show that the fermionic field (in combination with the Brans-Dicke scalar field $\varphi(t)$) can be responsible for a final eternal accelerated era, after an initial matter dominated period.

[1] Peebles P J and Ratra S 1988 *Astrophys. J. Lett.* L**17** 325; Copeland E, Sami M and Tsujikawa S 2006 *Int. J. Mod. Phys.* D**15** 1573; Capozziello S, De Laurentis S, Nojiri S, Odintsov S 2009 *Phys.Rev.* D**79** 124007; Baccigalupi C, Liddle A R and Perrotta F 2005 *Phys Rev. D* **71** 104025.
[2] Obukhov Y N 1993 *Phys. Lett.* A **182** 214; Saha B and Shikin G N, 1997 *Gen. Relativ. Gravit.* **29** 1099; Armendáriz-Picón C and Greene P B 2003 *Gen. Relativ. Gravit.* **35** 1637; Saha B 2006 *Phys. Rev.* D **74** 124030.
[3] Ribas M O, Devecchi F P and Kremer G M 2005 *Phys. Rev.* D **72** 123502.
[4] Ribas M O, Devecchi F P and Kremer G M 2008 *Europhys. Lett.* **81** 14001.
[5] Samojeden L L, Devecchi F P and Kremer G M 2010 *Phys. Rev.* **81** 27301.
[6] Saha B and Boyadjiev T 2004 *Phys. Rev.* D **69** 124010.
[7] Nambu Y, Jona-Lasinio G 1961 *Phys. Rev.* **122** 345.
[8] Wald R M 1984 *General Relativity* (Chicago: UCP); Weinberg S 1972 *Gravitation and Cosmology* (NYC: Wiley).
[9] Ryder L 1996 *Quantum Field Theory* (Cambridge: Cambridge).
[10] Birrel N D and Davies P C 1982 *Quantum fields in curved space-time* (Cambridge: Cambridge).
[11] Will C M 1994 *Theory and experiment in gravitational physics* (Cambridge: Cambridge).
[12] Kremer G M and Devecchi F P 2003 *Phys. Rev.* D **67** 047301.
[13] Chimento L P 2002 *Phys. Rev.* D **65** 063517.
[14] Chimento L P and Lazkoz R 2003*Phys. Rev. Lett.* **91** 211301.
[15] Chimento L P, Devecchi F P, Forte M, Kremer G M and Ribas M O 2008 *Phys. Rev.* D **96** 211301.
[16] Hoffman M and Trodden M 2003 *Phys. Rev.* D **68** 023509.
[17] Caldwell R R 2002 *Phys. Lett.* B**545** 23.
[18] Mersini L, Bastero-Gil M and Kanti P 2001 *Phys. Rev.* D **64** 043508.
[19] Aguirregabiria J M, Chimento L P, Jakubi A and Lazkoz R 2003 *Phys. Rev.* D **67** 083518.
[20] Veneziano G 1991 *Phys. Lett.* B **265** 287.

Infinite degeneracy of states in quantum gravity

Jonathan Hackett and Yidun Wan

Perimeter Institute for Theoretical Physics,
31 Caroline st. N., Waterloo, Ontario N2L 2Y5, Canada, and
Department of Physics, University of Waterloo,
Waterloo, Ontario N2J 2W9, Canada

E-mail: jhackett@perimeterinstitute.ca and ywan@perimeterinstitute.ca

Abstract. The setting of Braided Ribbon Networks is used to present a general result in spin-networks embedded in manifolds: the existence of an infinite number of species of conserved quantities. Restricted to three-valent networks the number of such conserved quantities in a given network is shown to be determined by the number of nodes in the network. The implication of these conserved quantities is discussed in the context of Loop Quantum Gravity.

1. Introduction

Spin networks[1] have been a part of the search for quantum gravity for over three decades. More recently they have gained prominence in Loop Quantum Gravity as a basis for the kinematical Hilbert space[2, 3]. Though a great deal of the understanding of spin-networks is derived directly from graph theory, the spin-network states of quantum gravity possess a richer structure owing to their embedding. Recently a great deal of attention has been focused on understanding the possibility that we may have to work with an even further enriched structure of spin networks with width [4, 5, 6, 7, 8, 9, 10, 11, 12, 13, 14]. This work was motivated from the suggestion that in studying loop quantum gravity with a positive cosmological constant we may need to consider framed spin networks [15, 16, 17].

One of the major revelations of the work on framed spin networks has been the existence of conserved topological structures. This was joined in [18] with the existence of conserved quantities even in the absence of an embedding.

We shall present an overview of embedded three-valent framed spin networks (which we will call three-valent Braided Ribbon Networks), we will then describe the existence of conserved topological structures in these networks and demonstrate the existence of countably infinite species of these structures. We will then finally lift the result to embedded three-valent spin networks without framing. The results on the classification of the structures will be general in the valence of the spin-network though the results in the invariance of the number of conserved structures will only apply to the three-valent case.

2. Braided Ribbon Networks

Braided ribbon networks (BRNs) generalize spin networks by extending the edges into a higher dimensional structure and associating non-isotopic embeddings of the same spin network to different basis states in a quantum space of a theory of quantum gravity. Trivalent BRNs are constructed through taking the union of trinions - 2-surfaces of the form of Fig. 1a. By considering a trinion to be the combination of three legs and a node (the union of the three legs), we can allow the legs of the trinion to be twisted about and to cross over parts of other trinions before being joined together into a BRN. This gives us generalized structures with features such as those in Fig. 1b. We will use the word ribbon to refer to the extended edges of the BRN, and use the term edges to refer instead to the boundaries of the ribbons. The BRN is considered to evolve subject to the standard evolution algebra of trivalent spin networks $\mathcal{A}_{evol}$ whose generators consist of the $1-3$ move and the $2-2$ move (Fig. 2). Whether these correspond to evolution with respect to time, or some other concept is irrelevant to this work, and so will be left to the reader's taste. In the present work we shall use what is referred to as the Smolin-Wan rule in [18] which means that a $2-2$ move cannot be performed when the shared ribbon (labeled b in the exchange move of Fig. 2) has additional content from the embedding (for example a knot, or a twist). We shall also consider the simpler case of embedded spin networks, an analogous situation where we do not extend the graph edges, but still consider non-isotopic embeddings of the same abstract spin network to be distinct (these admit similar structures but without twisting). In discussing embedded spin networks, we will use the word edge in the sense of graph theory - the meaning of the word edge should be clear from its context.

Additionally before we proceed we will clarify what we mean by a 'knot' in an embedded spin network:

Definition 1. *We will call a* ***local knot*** *any locally knotted arc that we can place a closed compact surface around with the network only intersecting the surface precisely twice at the single edge making up the locally knotted arc.*

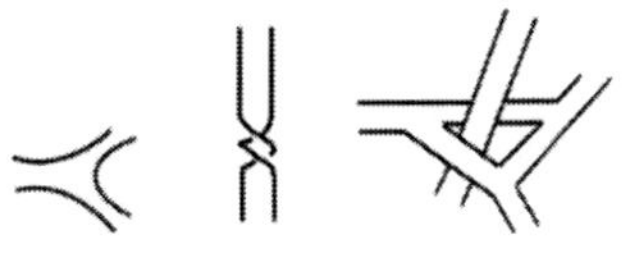

(a) (b)

Figure 1: Trinions, twists and braiding

In [5] the reduced link was introduced as a tool for understanding the non-trivial topological content of a braided network. The reduced link consists of removing the interior of the 2-surface - leaving only the edges of the ribbons - and then removing all un-linked un-knotted loops. An example of taking the reduced link is shown in Fig. 3 to demonstrate the process. It has been shown that the reduced link is an invariant of the generators of $\mathcal{A}_{evol}$ and therefore could be used to demonstrate invariants of the BRN.

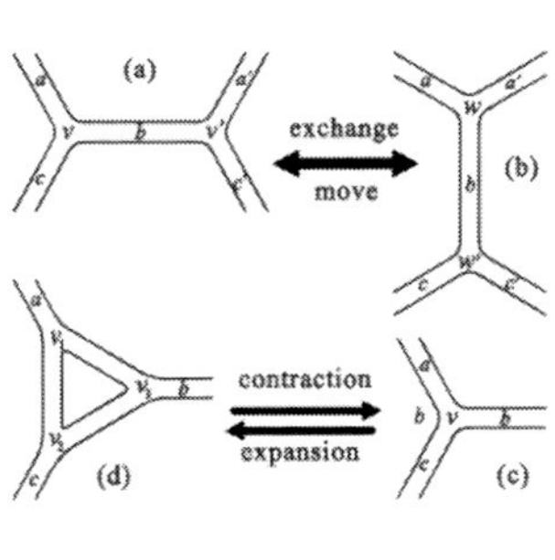

Figure 2: Generators of $\mathcal{A}_{evol}$

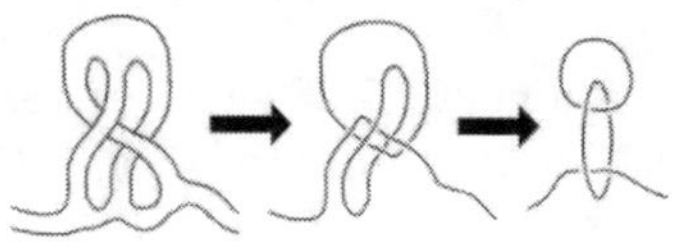

Figure 3: The Reduced Link

3. Conserved Structures in Braided Ribbon Networks

Given the existence of conserved structures in braided ribbon networks we shall consider specifically the more local structures (in the sense of micro-locality from [6]). This specifically shall ignore structures where ribbons are knotted or braided with ribbons that are distant under the standard distance function of graph theory. We can consider the local structures in order of reducing locality: the most local are those that involve only a single ribbon (ultra local), then those that involve ribbons sharing a node (1st degree local), then those that involve the ribbons of two adjacent nodes (2nd degree local), and so on. In particular, we know from [18] that a structure such as that in Fig. 4 is conserved as we are unable to perform the $2-2$ move, and is therefore an ultra local conserved structure. The only exception to this is if the ribbon we are concerned with connects two 'halves' of the network (i.e. that the ribbon divides the network between two parts that are not connected to one another through anything other than that single ribbon - in the language of the next section, it corresponds to both b and b' being tethers of isolated substructures). In this situation we are able to remove the knotting and twisting by isotopy. As this case is artificial in nature and uninteresting, we shall ignore it for our investigation.

3.1. Isolating the conserved structures

In [6] the concept of an isolated substructure was introduced as a means of understanding the ability to translate features through a braided ribbon graph. An isolated substructure is a subset of a graph which connects to the rest of the graph only through a single ribbon (called its tether) and is not part of any larger topological features. If a substructure can be evolved into an isolated substructure by a sequence of applications of elements of the evolution algebra, we will call them isolatable. Isolatable substructures are essentially propagating locally conserved quantities[6] able to move via the evolution algebra to any point edge-connected to its tether, and having conserved structure inside of it. We can also see via the form of the reduced link of an isolated substructure - and the invariance of the reduced link under the evolution moves - that an isolatable substructure corresponds to a 'piece' of the reduced link that is essentially cut and paste into the link of the edge its tether is on (see for example fig.3). That a general reduced link can be considered a direct product of these 'pieces' means that a structure being isolatable does not require the evolution to acquire its meaning as a part of an invariant of the network, and that this meaning is invariant under interpretation of the meaning of the evolution. We shall demonstrate that a specific class of ultra-local conserved quantities are isolatable. To do this we shall first introduce several definitions.

Definition 2. *Two edge segments are said to be* **edge connected** *if they are connected in the space consisting of the edge of the network. Equivalently two edge segments are said to be edge connected if they are part of the same link in the reduced link of the network. The path in the space of edges between the two edge segments is called the* **edge path**.

Definition 3. *Two ribbons or two nodes in a network have a* **path** *between them if there exists a sequence of ribbons and nodes that can be traversed between them. The sequence of ribbons and nodes taken is called the path.*

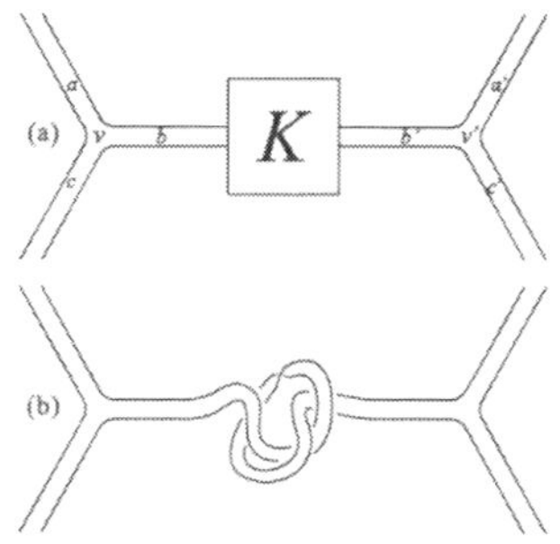

Figure 4: An example of an ultra-local structure

Definition 4. *A **free path** is a path which does not have any twists, knots or links along it. Specifically, each ribbon connecting the nodes of the path does not have any knotting or twisting on it, and there is no ribbon that crosses a ribbon in the path in such a way that cannot be undone by the Reidemeister moves applied to the ribbons.*

Definition 5. *A **free edge path** is an edge path which does not have any knots or links along it. This corresponds similarly to requiring that each ribbon the edges of the path belong to does not have any knotting or twisting on it, and there is no ribbon that crosses one of these ribbons in such a way that cannot be undone by the Reidemeister moves applied to the ribbons.*

We shall now prove that a general class of ultra-local structures (see Fig. 4) can be made isolated.

Theorem 1. *An ultra-local structure in a BRN which possesses a free edge path between one of the edges of each of the external ribbons can be isolated.*

Figure 5: 0 Node Case

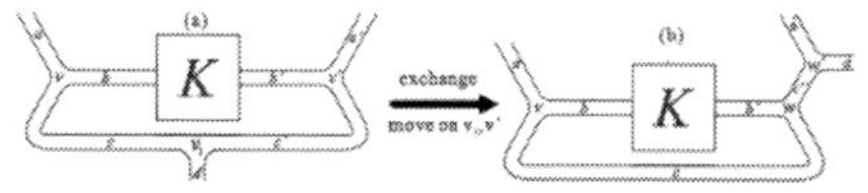

Figure 6: 1 Node Case

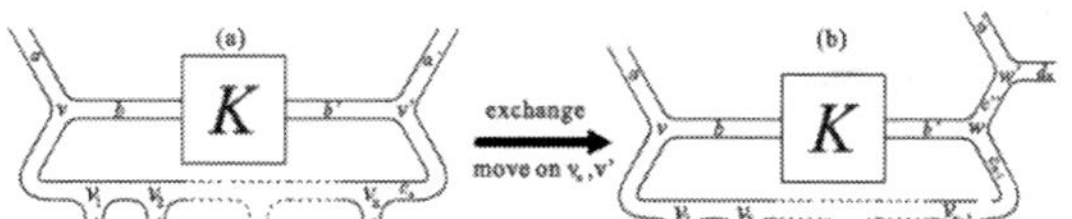

Figure 7: n Node Case

Proof. We shall prove this using induction on the number of intervening nodes. First we shall prove for one intervening nodes, then assume true for $n-1$ nodes and prove true for n nodes. Consider the situation depicted in Fig. 6 (where the apparent orientation of the nodes is for simplicity, and is in fact general), the application of the exchange move between nodes v_1 and v' reduces the situation to that of Fig. 5 which can then be isolated by using the exchange move on the two nodes involved. Now, we examine the situation in Fig. 7 to demonstrate that the n

node situation can be reduced to $n-1$ nodes by applying the exchange move on v' and v_n. We can then use the assumption of truth on the $n-1$ case to isolate the knot. □

This result lets us examine a peculiar situation: that where a knot on a ribbon is *not* a conserved quantity. Examining Fig. 8 (where the unattached ribbons connect to a larger network) we can see that it is possible in certain situations to reduce the number of ultra local structures, in the sense that two knots, e.g. the K_1 and K_2 in the figure, merge with each other. From this we see that we can always obtain a graph with only conserved local structures remaining.

Theorem 2. *A knot in an embedded graph which possesses a free path between its two external edges can be isolated.*

Proof. The proof of this follows inherently from the above proof and the fact that in an embedded graph - instead of a BRN - one can rotate an edge without introducing a twist. □

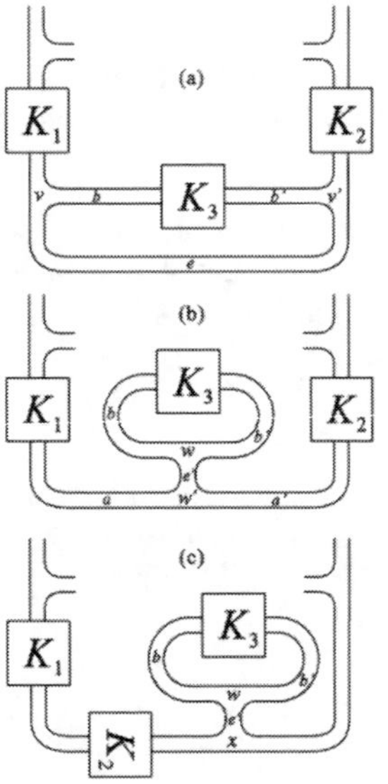

Figure 8: Reduction of the number of ultra local structures

We can apply the above theorems to reduce less local structures to more local situations. Consider for example the situation in Fig. 9a, we can apply the results of the above theorems to transform it to Fig. 9b if ribbons a and b, and a' and b' are connected by a free edge path.

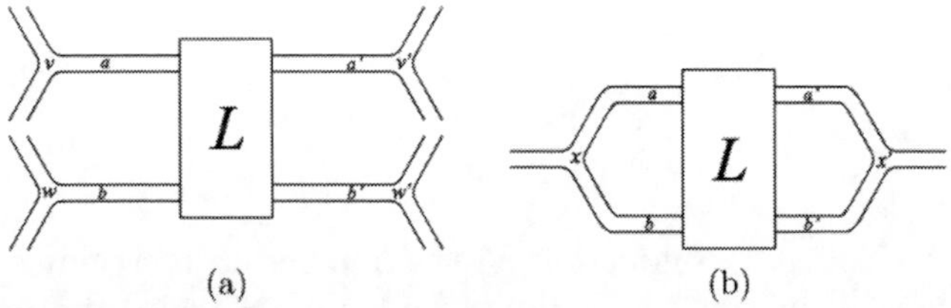

Figure 9: Less Local knottings

We can also see from these results that if there is no (edge) free path between the two edges of any of the knots in a network, there is no means to combine the knots onto single edges. This leads us to conclude that if one isolates all isolatable knots on a graph the remaining knots are completely invariant.

3.2. Immediate Results of Ultra-Local Structures

Considering the idea of ultra local structures, we find that there is only a single type of conserved structure. A structure formed by the topological deformation of a single ribbon can only possess two features: knots and twists. As twists can be passed through the knotting of a ribbon by an isotopy, we can consider any ultra local structure to be exactly characterized by a half integer (corresponding to the number of rotations) and a knot or a connected sum of knots.

This leads us to our first results:

> ***The existence of countably infinite many species of local conserved structures***
> There exist infinitely many species of local conserved structures.
>
> Any edge can be replaced by an edge with an isolated edge with some half-integer twist and a knot or a connected sum of knots. As there are infinitely many half-integers and knots, there are therefore infinitely many such species of structures.
>
> ***Maximal number of local conserved quanities***
> For a closed 3-valent BRN with N nodes, the maximum number of ultra-local conserved quantities is $\frac{3N}{2}$.

We can immediately lift these results to the scenario of un-framed spin networks: excepting the twists, all the results follow immediately. Additionally the first result does not depend on the valence of the spin-network involved in any way and is therefore a general result for embedded spin-networks.

4. Conclusions and Discussion

We have demonstrated the existence of a countable infinity of species of local conserved structures within Braided Ribbon Networks (and embedded spin-networks in general). We have also provided several results of use for isolating these structures and understanding when they are actually preserved.

In a theory of Quantum Gravity where the states are given by spin-networks embedded in a 3-manifold all of these states will be part of the Hilbert space. The difficulty this poses comes from the fact that these locally conserved structures correspond to an infinite number of conserved quantities that don't correspond with anything that commutes with the constraints of general relativity. This poses a significant problem in any attempt to recover the classical limit from a generic embedded spin-network - there is no reason to believe that these conserved quantities will simply cease to exist in the classical limit. This leaves a significant dilemma: we must change something in the theory for general relativity to be the classical limit.

There are two immediately obvious alternatives for resolving this, the first being to modify the hamiltonian constraint in such a way that we introduce new generators or the evolution algebra. The alternative to this is that we should reduce the physical Hilbert space of a theory of quantum gravity to require that there do not exist any knots or linking. Our ability to consider this super selection rule and still do certain things (including considering embedded spin networks) is questionable and requires investigation before this can be adopted as an 'easy' solution.

Acknowledgments

The authors are indebted to their Advisor, Lee Smolin, for his discussion and critical comments. We thank Laurent Freidel, Sudance Bilson-Thompson, Louis Kauffman and Isabeau Premont-Schwarz for helpful discussions. We are further indebted to Sundance Bilson-Thompson for the use of Fig. 3 with permission. Similarly Fig. 1a is taken from [6] with permission. Research at Perimeter Institute for Theoretical Physics is supported in part by the Government of Canada through NSERC and by the Province of Ontario through MRI.

References

[1] Penrose R 1971 Angular momentum: an approach to combinatorial space-time *Quantum Theory and Beyond*, ed T Bastin (Cambridge: Cambridge University Press)
[2] Rovelli C 1998 Loop quantum gravity *Living Rev. Rel.* **1** (*Preprint*: arXiv:gr-qc/9710008)
[3] Rovelli C 2004 *Quantum Gravity* (Cambridge: Cambridge University Press)
[4] Bilson-Thompson S 2005 A topological model of composite preons *Preprint* hep-ph/0503213
[5] Bilson-Thompson S, Markopoulou F and Smolin L 2007Quantum gravity and the standard model *Class. Quant. Grav.* **24** 3975
[6] Hackett J 2007 Locality and translations in braided ribbon networks *Class. Quant. Grav.* **24** 5757
[7] Wan Y 2007 On braid excitations in quantum gravity *Preprint* hep-th/0710.1312
[8] Bilson-Thompson S, Hackett J, Kauffman L and Smolin L 2008 Particle identifications from symmetries of braided ribbon network invariants *Preprint* hep-th/0804.0037
[9] Wan Y 2008 Effective theory of braid excitations of quantum geometry in terms of Feynman diagrams *Nucl. Phys.* B **805** 1
[10] Bilson-Thompson S Hackett J and Kauffman L 2009 Particle topology, braids, and braided belts *Preprint* math.at/0903.1376
[11] Hackett J and Wan Y 2008 Conserved quantities for interacting four valent braids in quantum gravity *Preprint* hep-th/0803.3203
[12] Smolin L and Wan Y 2007 Propagation and interaction of chiral states in quantum gravity *Nucl. Phys.* B **796**, 331
[13] He S and Wan Y 2008 Conserved quantities and the algebra of braid excitations in quantum gravity *Nucl. Phys.* B **804**, 286
[14] He S and Wan Y 2008 C, P, and T of braid excitations in quantum gravity *Nucl. Phys.* B **805**, 1
[15] Major S and Smolin L 1996 Quantum deformation of quantum gravity *Nucl. Phys.* B **473**, 267
[16] Borissov R, Major S and Smolin L 1996 The geometry of quantum spin networks *Class. Quant. Grav.* **13**, 3183
[17] Smolin L 2002 Quantum gravity with a positive cosmological constant *Preprint* hep-th/0209079
[18] Markopoulou F and Premont-Schwarz I 2008 Conserved topological defects in non-embedded graphs in quantum gravity *Class. Quant. Grav.* **25**, 205015
[19] Kribs D W and Markopoulou F 2005 Geometry from quantum particles *Preprint* gr-qc/0510052
[20] Smolin L 1996 The classical limit and the form of the Hamiltonian constraint in non-perturbative quantum general relativity *Preprint* gr-qc/9609034

Is it Possible to Find a Probable Solution for the Problem of Divergence in QFT?

Farrin Payandeh

Department of Physics, Payame Noor University, Tabriz 19395-4697 Iran

E-mail: f.payandeh@pnu.ac.ir

Abstract. There are some fundamental questions in quantum physics and quantum field theory which can not be answered considering the positive energy states, alone. Negative energy states show up in many places in physics and could help us to provide an answer to some 'awkward' questions of physics [1, 2]. In recent years much research has been done on extending classical mechanics, quantum mechanics and quantum field theory into the complex domain [3, 4, 5]. On the other hand, applying negative energy states in QFT and performing the field quantization results in a theory without ultraviolet and infrared divergences and it seems as if evaluating divergent integrals of QED space leads to convergent values.

Due to Dirac, "negative energies and probabilities should not be considered as nonsense...", so that comparison of different approaches and reviving quantum theories that were thought to be dead may be helpful in leading us toward a general formalism and finding an appropriate answer to the problem of divergence in QFT.

1. Introduction

Quantum field theory is plagued by infinities, and it is therefore considered ugly [6]. The singular behavior of Green function at short/large relative distances leads to main divergences in quantum field theory. Entering the negative frequency states, which are also solutions of the field equation, leads to Krein space quantization. As a result, the vacuum energy becomes automatically zero and all ultraviolet divergences of QFT except the light-cone singularity are removed. The latter is removed with the consideration of quantum metric fluctuations in the linear approximation [7, 8]. Krein space method of quantization could be extended to free boson and spinor fields in Minkowski spacetime [9, 10], Wick's theorem and Feynman rules could be rewritten and the basic structure of QFT could be reformulated [11]. It is remarkable that the auxiliary negative frequency states play the role of regularization of the theory.

2. Origion of Divergence in QFT and a Probable Solution

The origin of divergences in standard quantum field theory lies in the singularity of the Green's function. The divergence appears in the imaginary part of the Feynman propagator, and the real part is convergent [12]:

$$G_F^P(x,x') = -\frac{1}{8\pi}\delta(\sigma_0)+\frac{m^2}{8\pi}\,\theta(\sigma_0)[\frac{J_1(\sqrt{2m^2\sigma_0})-iN_1(\sqrt{2m^2\sigma_0})}{\sqrt{2m^2\sigma_0}}]-\frac{im^2}{4\pi^2}\,\theta(-\sigma_0)\frac{K_1(\sqrt{2m^2(-\sigma_0)})}{\sqrt{2m^2(-\sigma_0)}}$$

Consideration of negative frequency states removes singularity of the Green function with exception of delta function singularity:

$$G_T(x,x') = -\frac{1}{8\pi}\delta(\sigma_0) + \frac{m^2}{8\pi}\theta(\sigma_0)\frac{J_1(\sqrt{2m^2\sigma_0})}{\sqrt{2m^2\sigma_0}}, \quad \sigma_0 \geq 0 .$$

However, considering the quantum metric fluctuations removes the latter singularity:

$$< G_T(x,x') >= -\frac{1}{8\pi}\sqrt{\frac{\pi}{2<\sigma_1^2>}}\exp(-\frac{\sigma_0^2}{2<\sigma_1^2>}) + \frac{m^2}{8\pi}\theta(\sigma_0)\frac{J_1(\sqrt{2m^2\sigma_0})}{\sqrt{2m^2\sigma_0}}$$

where $<\sigma_1^2>$ is related to the density of gravitons. When $\sigma_0 = 0$, due to the metric quantum fluctuation $<\sigma_1^2>\neq 0$, and we have

$$< G_T(0) >= -\frac{1}{8\pi}\sqrt{\frac{\pi}{2<\sigma_1^2>}} + \frac{m^2}{16\pi} .$$

By using the Fourier transformation, we obtain [13]

$$< \tilde{G}_T(p) >= \tilde{G}_T(p) + PP\frac{m^2}{p^2(p^2-m^2)} .$$

However, in the one-loop approximation, the contribution of delta function is negligible and the Green function in Krein space quantization appearing in the transition amplitude is

$$< \tilde{G}_T(p) >|_{one-loop} \equiv \tilde{G}_T(p)\,|_{one-loop} \equiv PP\frac{m^2}{p^2(p^2-m^2)} .$$

In the standard QED the divergent quantities are found in the self-energy graph, the vacuum polarization graph and the vertex graph. Applying the above resulted Green function can be regarded as a new method for regularization and considering the renormalization point at $p^2 = m^2$, the previous results for physical quantities in standard QED are obtained.

Since applying Krein method to QFT leads to the standard results, so it is safer to regard this procedure as a convenient way or another viewpoint of performing standard procedure of introducing counter-terms, specially the wave function renormalization term. The Krein space method takes care of divergence on the propagator level, while the standard method takes care of it on the interaction level by introducing counter-terms.

The role of 'unphysical' states appears in the method as a natural tool for removing the divergences of the theory. The crucial point is that while imposing the physical boundary conditions on the field operator, the positive frequency states are only affected. The negative frequency states do not interact with the physical states or real physical world. So that, they can not be affected by the physical boundary conditions as well indicating that the former are technical artifacts rather than real objects of deeper meaning. The situation is quite similar to the 'method of images' for finding an electromagnetic field in electrostatics. As the simplest example, suppose one tries to find out the electromagnetic field produced by a positively charged point-particle, placed in front of an infinite flat conductor plate. One can find out the field configuration either by a system of a point charge and an infinite flat conductor or a virtual system of a point charge and a negatively charged virtual point-particle. In this case, the 'positive charge', the conductor plate and the virtual charge correspond to, respectively, positive frequency modes, counter-terms and negative frequency modes. One can replace one object (the conductor plate or counter-terms) with another object (the virtual charge or negative frequency modes) to get the same result. However, no new aspects are added to the situation, so that nobody claims that the virtual negative charge is a real object. In the same manner, we may introduce unphysical modes if convenient as far as they do not contradict with the correct results. It does not mean, however at a first glance, that they are real objects of physical meaning. A question arises here: Can this method provide an answer when asked for what he had won the Nobel prize, Feynman replied, "For sweeping the infinities under the rug"?

However, at a second glance it is not bad to mention that negative energy states show up in other places and help to explain some fundamental behaviors and provide an answer to some 'awkward' questions of physics, such as Klein paradox, original version of EPR paradox [2], the miracle of creation, etc. As stated by Dirac, "negative energies and probabilities should not be considered as nonsense. They are well-defined concepts mathematically, like a negative sum of money, since the equations which express the important properties of energies and probabalities can still be used when they are negative. Thus, negative energies and probabilities should be considered simply as things which do not appear in experimental results. The physical interpretation of relativistic theory involves these things and is thus in contradiction with experiment. We, therefore, have to consider ways of modifying or supplementing this interpretation."[14, 15].

3. Conclusion

In standard quantum field theory the procedures of normal ordering and renormalization have been adopted for removing the divergences that appear in the physical quantities for free and interacting fields. Relaxation of the requirement for the positivity of norm and energy, results in the disappearance of the divergences. Addition of the new unphysical states leads us to the Krein space quantization. It is remarkable that they play no role in the physical world but an element for removing the divergences of the theory. It could be shown that the divergent integrals of QED in Krein space quantization leads to convergent values.

Negative energy states are entered in different approaches to quantum mechanics and quantum field theory, so that the comparison of them will be helpful, as stated by Dirac, "negative energies and probabilities should not be considered as nonsense...".

Reviving quantum theories that were thought to be dead, is it possible to find an appropriate answer to the problem of divergence in QFT?

Acknowledgments

This work was supported by Payame Noor University Research Council. Also, the author would like to thank Prof. T. Elze chief organizer of DICE2010.

4. References

[1] Hotson D L 2002 *Infinite Energy* **43** 43
[2] Ni G J and Guan H 1999 *Preprint* quant-ph/9901046
[3] Bender C M 2010 *Preprint* hep-th/0912.4659
[4] Bender C M 2010 *Preprint* hep-th/1003.3881
[5] Yang C D 2008 *Chaos, Solitons and Fractals* **38** 316
[6] 't Hooft G 2001 *Theory and Experiment Heading for New Physics* ed A Zichichi (Singapore: World Scientific) p 656
[7] Ford H L 1997 *Preprint* gr-qc/9707062
[8] Rouhani S and Takook M V 2009 *Int. J. Theor. Phys.* **48** 2740
[9] Payandeh F, Mehrafarin M and Rouhani S, Takook M V 2008 *Ukr. J. Phys.* **53** 1203
[10] Payandeh F, Mehrafarin M and Takook M V 2007 *AIP Conf. Proc.* **957** 249
[11] Payandeh F, Mehrafarin M and Takook M V 2009 *Science in China Series G: Physics, Mechanics and Astronomy* **52** 212
[12] Payandeh F 2009 *J. Phys. Conf. Ser.* **174** 012056
[13] Refaei A and Takook M V 2011 *Mod. Phys. Lett.* A **26** 31
[14] Dirac P A M 1942 Bakerian Lecture *Proc. Roy. Soc. Lond.* A **180** 1
[15] Dirac P A M 1933 Nobel Lecture, December 12

Relativistic quantum equation in nonstationary space-time geometry

L Gaffour

Department of Physics, Laboratory of Elaboration and Caracterisation of Materials, Faculty of Sciences, Sidi-Bel-Abbes university, 22000, Algeria

E-mail: `lgaffour.usba@yahoo.fr`

Abstract. An analytical solution for the 1-1 D relativistic quantum equation in dependent space-time geometry is given when the imaginary part of the particle mass is negligible. The solution is accomplished through covariant transformations of wave equation. The solutions have a modal structure and the modes are dynamics. The relativistic redshift turns out to be a manifestation of the space-time geometry motion. As illustrations, two space-time geometry motions corresponding to nonlinear mappings and the linear Lorentz transformations are discussed.

1. Introduction

This study follows up on previous works [1]-[2] proposing a new formulation of quantum theory in nonstationary space-time geometry paralleling the usual theory. In these previous papers the mass of particle turns out to be complex and affected by the space-time geometry motion. In view of their importance in the study of various quantum systems under the influence of nonstationary space-time geometry, exact solution of relativistic quantum equation is of a great interest. However, because of the interdependence of the space and time coordinates, the classical analytical methods cannot, in general, be applied. Therefore, the present work provides, for the 1-1D relativistic quantum equation, an analytical solution when the imaginary part of the mass is assumed negligible in relation to the real part. The solution is accomplished through covariant transformations of wave equation. We show that the solutions are dynamics and turn out to be the generalization of the classical ones. As illustrations, two space-time geometry motions corresponding to nonlinear mappings and the linear Lorentz transformations are discussed. The paper is organized as follows: Section 2 briefly summaries the analytical method used in this work. Section 3 outlines the position of the problemsec. Section 4 gives the analytical solution of the problem. Section 5 examines two cases of space-time geometry motion. Section 6 yields a conclusion.

2. Brief Summary of the used method

For more clarity of the work, we provide a brief summary of a covariant transformation of the wave equation [1]-[3]. The method consists in transforming the wave equation

$$\frac{\partial^2 \psi}{\partial x^2} - \frac{\partial^2 \psi}{\partial \tau^2} = \theta(\tau, x) \tag{1}$$

for the field $\Psi(\tau, x)$ in the nonstationary space-time domain $R(\tau, x)$ into a wave equation of the covariant form, namely

$$[\frac{\partial^2\psi}{\partial\eta^2} - \frac{\partial^2\psi}{\partial\xi^2}]\varphi\prime(\tau - x)\,\varphi\prime(\tau + x) = \theta(\xi, \eta) \tag{2}$$

for the field $\Psi(\xi, \eta)$ in the fixed domain $S(\xi, \eta)$. $\varphi\prime$ is the first derivative of a transformation function φ defined by the relationships between the coordinates in the two domains $S(\xi, \eta)$ and $R(\tau, x)$:

$$\xi + \eta = \varphi(\tau + x) \tag{3}$$

$$\xi - \eta = \varphi(\tau - x) \tag{4}$$

or

$$\eta = \frac{1}{2}[\varphi(\tau + x) - \varphi(\tau - x)] \tag{5}$$

$$\xi = \frac{1}{2}[\varphi(\tau + \mathrm{x}) + \varphi(\tau - \mathrm{x})] \tag{6}$$

It should be noted that the real transformation functions

$$\xi = \xi(\tau, x) \tag{7}$$

$$\eta = \eta(\tau, \tau) \tag{8}$$

satisfy the following fundamental conditions

$$\frac{\partial\xi}{\partial\tau} = \frac{\partial\eta}{\partial x}, \frac{\partial\xi}{\partial x} = \frac{\partial\eta}{\partial\tau} \tag{9}$$

In addition to the relations 9, the transformation functions 7 and 8 themselves satisfy wave equation

$$\frac{\partial^2\xi}{\partial\tau^2} = \frac{\partial^2\xi}{\partial x^2}\frac{\partial^2\eta}{\partial\tau^2} = \frac{\partial^2\eta}{\partial x^2} \tag{10}$$

Finally, we can say that the real function transformations, which leave the wave equation of the covariant form and map every point of a given nonstationary space-time geometry into a point of a static space-time geometry, are of the form 7-8 and satisfy the fundamental conditions 9.

It should be noted that $\varphi(\tau, x)$ and its inverse $F(\xi, \eta)$ are functions of a real variable not of a complex variable, since the conditions 9 play a similar role as the Cauchy-Riemann conditions in the transformation of the Laplace equation but are different.

3. Position of the problem

The relativistic quantum equation of a particle of a mass m evolving in fixed space-time geometry is given by the well known Klein-Gordon equation. However, when considering nonstationary space-time geometry, the evolution of the same particle is described by the following equation[1]:

$$\frac{\partial^2\psi}{\partial x^2}-\frac{\partial^2\psi}{\partial\tau^2}-\frac{m^2c^2}{\hbar^2 k}[k\varphi\prime^2-i\varphi\prime\prime]\Psi=0 \tag{11}$$

or

$$\frac{\partial^2\psi}{\partial x^2}-\frac{\partial^2\psi}{\partial\tau^2}-\frac{M^2c^2}{\hbar^2}\Psi=0 \tag{12}$$

with

$$M^2=x^2\frac{k\varphi\prime^2-i\varphi\prime\prime)}{k} \tag{13}$$

where $\varphi\prime$ and $\varphi\prime\prime$ are respectively the first and second derivative of $\varphi(\tau-x)$ and $\tau=ct$. It should be noted that φ is an identity when the space-time geometry becomes static, so $\varphi\prime=1$ and $\varphi\prime\prime=0$.

Equation 11 is identical in appearance with the classical Klein-Gordon equation(pseudo-covariance). The mass M is complex and depends on the space-time geometry motion. We are interested in this study by the case of a stable particle that is when the imaginary part of the mass is negligible. Thus, equation 11 can be expressed as follows:

$$\frac{\partial^2\psi}{\partial x^2}-\frac{\partial^2\psi}{\partial\tau^2}-\frac{m^2c^2\varphi\prime^2}{\hbar^2}\Psi=0 \tag{14}$$

with the relativistic energy-momentum relation:

$$E_{ns}^2=P_{ns}^2+m^2\varphi\prime^2c^4 \tag{15}$$

E_{ns} and P_{ns} are respectively the energy and momentum in nonstationary space-time geometry.

We note that energy at the rest ($P_{ns}=0$) is $E_{ns}=m\varphi\prime c^2$ and when the space-time geometry becomes static ($\varphi\prime=1$), we find again the Einsteins relation. Relation 15 means that equation 14 allows positive and negative energy as solutions like in classical case. However, because of the interdependence of the space and time coordinates, the classical analytical methods cannot be applied for determining these solutions. Actually, more general methods [4] - [10] are used, which are mathematically rather involved. In our case, we use the covariant transformation of wave equation cited in section 2. Therefore, the relativistic quantum equation 14 in nonstationary space-time geometry $R(\tau,x)$ is transformed into modified Klein-Gordon equation in corresponding static space-time geometry $S(\xi,\eta)$ and writes

$$\frac{\partial^2\psi}{\partial\eta^2}-\frac{\partial^2\psi}{\partial\tau^2}-\frac{m^2c^2}{\hbar^2}\frac{F\prime(\xi+\eta)}{F\prime(\xi-\eta)}\Psi=0 \tag{16}$$

where $F\prime(\xi\pm\eta)$ is the first derivative of the inverse transformation function $F(\xi\pm\eta)$ of $\varphi(\tau\pm x)$.

Next, let be

$$\frac{F\prime\,(\xi+\eta)}{F\prime\,(\xi-\eta)} = G\,(\xi,\eta) \tag{17}$$

Finally, equation 14 is transformed in $S(\xi,\eta)$ into the following equation of the covariant form:

$$\frac{\partial^2\psi}{\partial\eta^2} - \frac{\partial^2\psi}{\partial\tau^2} - \frac{m^2c^2}{\hbar^2}G\,(\xi,\eta)\,\Psi = 0 \tag{18}$$

4. Analytical solution

For determining the solutions of equation 14 in $R(\tau,x)$, first of all, we have to solve eq. 18. However, equation 18 will not separate unless $G(\xi,\eta)$ turns out to be the sum of a function of ξ alone and a function of η alone. In other words, G has to be in the following form:

$$G\,(\xi,\eta) = G_1\,(\xi) + G_2\,(\eta) \tag{19}$$

In this condition, equation 18 becomes of separable form and writes:

$$\frac{\partial^2\psi}{\partial\eta^2} - \frac{\partial^2\psi}{\partial\tau^2} - \frac{m^2c^2}{\hbar^2}[G_1\,(\xi) + G_2\,(\eta)]\Psi = 0 \tag{20}$$

So, we seek Ψ of the form:

$$\Psi\,(\xi,\eta) = L\,(\xi)\,K\,(\eta) \tag{21}$$

Therefore, we obtain two differential equations which are:

$$\frac{\partial^2 L}{\partial\xi^2} + [\mu^2 + \frac{m^2c^2}{\hbar^2}G_1]L = 0 \tag{22}$$

$$\frac{\partial^2 K}{\partial\eta^2} + [\mu^2 - \frac{m^2c^2}{\hbar^2}G_2]K = 0 \tag{23}$$

The solutions of equations 22 and 23 need the expression of $G(\xi,\eta)$ which itself depends on the function transformation. However, since the 2th derivative $\varphi\prime\prime$ is assumed negligible, so the first derivative $\varphi\prime$ can be assumed as a constant. As a consequence, according to 1 and 12, equation 14 is transformed in fixed space-time geometry $S(\xi,\eta)$ into Klein-Gordon equation and writes:

$$\frac{\partial^2\psi}{\partial\eta^2} - \frac{\partial^2\psi}{\partial\xi^2} - \frac{m^2c^2}{\hbar^2}\Psi = 0 \tag{24}$$

The solutions of 24 for positive energy are the plane wave type of the form:

$$\psi(\xi,\eta)=\psi_0\exp -ik(\xi-\eta) \tag{25}$$

Transforming back to the original variables τ, x in $R(\tau,x)$ by means of 4, we obtain the solution of equation 14 in non stationary space-time geometry. This solution writes as follows:

$$\psi\left(\tau,\eta\right)=\psi_0\exp -ik\varphi\left(\tau-x\right) \tag{26}$$

We ascertain that solutions are, in general, dynamics and of functional plane wave type. In practical, we are interesting by particular solutions corresponding to real situations. This case will be treated at the next subsection.

4.1. Particular solutions

Consider the 1-1D relativistic quantum equation 14 defined on the following nonstationary space-time geometry such as $x=a(\tau)$ expanding or shrinking the x domain:

$$0\leq x\leq a\left(\tau\right) \tag{27}$$

according to 5, the corresponding fixed domain is given by

$$0\leq\eta\leq\eta_0 \tag{28}$$

with

$$\eta_0=\frac{1}{2}[\varphi\left(\tau+a\left(\tau\right)\right)-\varphi\left(\tau-a\left(\tau\right)\right)] \tag{29}$$

The boundary conditions are needed and they vary with the respective applications. For instance, for the following boundary conditions in $R(\tau,x)$

$$\Psi\left(\tau,0\right)=0,\Psi\left(\tau,a\left(\tau\right)\right)=0, \tag{30}$$

the corresponding boundary conditions in $S(\xi,\eta)$ are

$$\Psi\left(\xi,0\right)=0,\Psi\left(\xi,\eta_0\right)=0, \tag{31}$$

Thus, the general solution of equation 24 is given by:

$$\Psi\left(\xi,\eta\right)=A_n\sum_{1}^{\infty}\sin\left(k_n\eta\right)\cos\left(\varpi_n\xi+\alpha_n\right) \tag{32}$$

with

$$k_n=\frac{n\pi}{\eta_0} \tag{33}$$

$$\varpi_n = \left(\frac{m^2c^2}{\hbar^2} + \frac{n^2\pi^2}{\eta_0^2}\right)^{\frac{1}{2}} \tag{34}$$

Finally, transforming back to the variables τ, x in $R(\tau, x)$ by means of the function transformation φ, we deduce the general solution of equation 14 in $R(\tau, x)$. Indeed, the substitution of η, ξ and η_0 in equations 32-34 by their respective expression 5, 6 and 29, we obtain

$$\Psi(\tau, x) = A_n \sum_1^{\infty} \left\{ \begin{array}{c} \sin k_n \frac{1}{2}[\varphi(\tau + x) - \varphi(\tau - x)] \times \\ \cos\left(\varpi_n \frac{1}{2}[\varphi(\tau + x) + \varphi(\tau - x)] + \alpha_n\right) \end{array} \right\} \tag{35}$$

Remarks

- 'Expression 35 shows that the modes are expressed in terms of functional series and are dynamics.'
- 'The boundary conditions 30 are satisfied by 35.'
- 'When the space-time geometry is fixed, the function transformation φ becomes identity, so that we find again the classical solutions'
- 'We remind that the pulsation '

$$\omega_n = \varpi_n C \tag{36}$$

5. Illustrations

5.1. Nonlinear transformations

For a linear motion $a(\tau) = a_0 + \beta\tau$, expanding or shrinking the x domain, the corresponding real function transformations are [3], [11]:

$$\eta = \frac{1}{2} \ln\left[a_0 + \beta(\tau + x)\right]/\left[a_0 + \beta(\tau - x)\right] \tag{37}$$

$$\xi = \frac{1}{2} \ln\left[a_0 + \beta(\tau + x)\right]\left[a_0 + \beta(\tau - x)\right]/a_0^2 \tag{38}$$

Thus, according to 3 and 4, the transformation φ and its first derivative are respectively given by:

$$\varphi(\tau \pm x) = \ln[1 + \frac{\beta}{a_0}(\tau \pm x)] \tag{39}$$

$$\varphi\prime(\tau \pm x) = \frac{\beta}{\beta(\tau \pm x) + a_0} \tag{40}$$

and 29 gives

$$\eta_0 = \frac{1}{2}\ln\frac{1+\beta}{1-\beta} \tag{41}$$

with $\tau = ct$ and $\beta = c/v$, v is the velocity motion increasing or decreasing the given domai n. Next, since the first derivative is not a constant, we have to solve equations 22 and 23. However, since the velocity v is assumed to be very small in relation to c ($v \ll co$ or $\beta \ll 1$), the following approximations hold extremely well:

$$\varphi(\tau \pm x) = \frac{\beta}{a_0}(\tau \pm x), \varphi\prime(\tau \pm x) = \frac{\beta}{a_0}, \eta = \frac{\beta}{a_0}x, \xi = \frac{\beta}{a_0}\tau, \tag{42}$$

Hence,

$$k_n = \frac{n\pi}{\beta}, \varpi_n = \left(\frac{m^2c^2}{\hbar^2} + \frac{n^2\pi^2}{\beta^2}\right)^{\frac{1}{2}} \tag{43}$$

Therefore, the general solution 35 writes as follows:

$$\Psi(\tau, x) = A_n \sum_1^\infty \sin\left(\frac{n\pi}{a_0}x\right)\cos\left(\frac{\beta}{a_0}\varpi_n\tau + \alpha_n\right) \tag{44}$$

or

$$\Psi(\tau, x) = A_n \sum_1^\infty \sin(K_n x)\cos(\bar{\Omega}_n\tau + \alpha_n) \tag{45}$$

with

$$K_n = \frac{n\pi}{a_0} = k_n\varphi\prime \tag{46}$$

$$\Omega_n = \frac{\beta}{a_0}\omega_n = \omega_n\varphi\prime \tag{47}$$

We note a general Doppler shift which, according to the dispersion relation, affects also the mass m.

5.2. Linear Lorentz transformations

Consider a 1-1 D. confined region moving with the same velocity v expanding or shrinking the space-time geometry. This means that locally the geometry of the considered region is assumed static. This region constitutes, with its corresponding transformed region in the fixed space-time geometry, inertial frames. Therefore, the two regions are related by the linear Lorentz transformations. The linear Lorentz transformations are:

$$\xi = \gamma\,(\tau - \beta x) \tag{48}$$

$$\eta = \gamma\,(x - \beta\tau) \tag{49}$$

With $\beta = v/c$ and $\gamma = (1-\beta)^{-\frac{1}{2}}$ The Lorentz transformations satisfy the fundamental conditions 9 and are solutions of wave equation 10. From 48 and 49, we deduce the transformation f and its first derivative φ:

$$\xi \pm \eta = \varphi\,(\tau \pm x) = (\frac{1 \mp \beta}{1 \pm \beta})^{\frac{1}{2}}\,(\tau \pm x) \tag{50}$$

$$\varphi\prime\,(\tau \pm x) = \left(\frac{1 \mp \beta}{1 \pm \beta}\right)^{\frac{1}{2}}(\tau \pm x) \tag{51}$$

As we have seen, Klein- Gordon equation admits, in fixed space-time geometry, solutions of the type 25. So by transforming back to the (τ, x) variables, we find the solution of equation 14. It writes:

$$\Psi\,(\tau, x) = \Psi_0 \exp[-ik\left(\frac{1+\beta}{1-\beta}\right)^{\frac{1}{2}}(\tau - x)] \tag{52}$$

or

$$\Psi\,(\tau, x) = \Psi_0 \exp[-iK_{ns}\,(\tau - x)] \tag{53}$$

with

$$K_{ns} = k\left(\frac{1+\beta}{1-\beta}\right)^{\frac{1}{2}} = k\varphi\prime \tag{54}$$

$$\Omega_{ns} = \omega\left(\frac{1+\beta}{1-\beta}\right)^{\frac{1}{2}} = \omega\varphi\prime \tag{55}$$

We ascertain that we obtain the usual solution with the relativistic redshift which affects all the observables even the mass.

6. Conclusion

The relativistic quantum equation in nonstationary space-time geometry has been solved analytically when the imaginary part of the particle mass is assumed negligible. This study leads to the following results:

- The solutions expresses in terms of functional modes.
- The modes are, in general, dynamics.
- All the physical characteristics of the solutions are affected by a factor representing the redshift which is a manifestation of the space-time geometry rather than an effect due to object motions in space-time. The factor depends on the motion expanding or shrinking the space-time geometry.
- We ascertain that the mass of the particle is also affected by the redshift.
- The usual relativistic Dopplers effect turns out to be only a particular case corresponding specifically to Lorentz transformations.

References

[1] Gaffour L 2008 *Int. J. of Theo. Phys.* **47** 4 859-864.
[2] Gaffour L 2009 *Phys. essays.* **22** 288.
[3] Gaffour L 1998 *Prog. In Elect. Reseach.*, PIER **20** 63-73.
[4] Balazs W 1961 *J. Math. Anal. App.* **3**(9) 472.
[5] Gonzalez N 1998 *J. Math. Anal. App.* **228**(1) 51-59.
[6] Chen G, Chen Z, and Lou Z 2004 *Phys. Lett.* A **331** 374.
[7] Chen G 2005 *Phys. Lett.* A **339** 300.
[8] de Castro A 2005 *Phys. Lett.* A **338** 81.
[9] Alhaidari A, Bahlouli H and Al-Hassani A 2006 *Phys. Lett.* A **349** 47.
[10] Gaffour L 1999 *Prog. In Elect.. Reseach.*, PIER **23** 263-274.
[11] Morse P, Feshbach H 1953 *Theoretical Physics, Part I* (McGraw-Hill: New York) pp 1223.

Journal of Physics: Conference Series **306** (2011) 012056 doi:10.1088/1742-6596/306/1/012056

A review of the decoherent histories approach to the arrival time problem in quantum theory

James M Yearsley

Theoretical Physics Group, Blackett Laboratory, Imperial College, London SW7 2BZ, UK

E-mail: james.yearsley@imperial.ac.uk

Abstract. We review recent progress in understanding the arrival time problem in quantum mechanics, from the point of view of the decoherent histories approach to quantum theory. We begin by discussing the arrival time problem, focussing in particular on the role of the probability current in the expected classical solution. After a brief introduction to decoherent histories we review the use of complex potentials in the construction of appropriate class operators. We then discuss the arrival time problem for a particle coupled to an environment, and review how the arrival time probability can be expressed in terms of a POVM in this case. We turn finally to the question of decoherence of the corresponding histories, and we show that this can be achieved for simple states in the case of a free particle, and for general states for a particle coupled to an environment.

1. Introduction

Questions involving time in quantum theory have a long and controversial history [1]. Quantities such as arrival and dwell times, despite being measureable [2], still lack concrete grounding within "standard" quantum theory. Arrival times, in particular, have attracted much interest, as the natural procedure of quantising the appropriate classical quantity gives rise to an operator which is not self-adjoint and thus, in standard quantum theory at least, cannot easily be considered as an observable.

Despite these difficulties, if one considers a free particle in an initial state $\rho = |\psi\rangle\langle\psi|$ localized in $x > 0$ consisting entirely of negative momenta, and asks for the probability $p(t_1, t_2)$ that the particle crosses the origin during the time interval $[t_1, t_2]$ there *is* a semi-classical answer, given by the difference between the probability of being in $x > 0$ at the initial and final time [1]. Defining $P(t) = \theta(\hat{x}(t))$,

$$\begin{aligned} p(t_1,t_2) &= \mathrm{Tr}(P(t_1)\rho) - \mathrm{Tr}(P(t_2)\rho) = -\int_{t_1}^{t_2} dt \mathrm{Tr}(P\dot{\rho}_t) \\ &= \int_{t_1}^{t_2} dt J(t), \end{aligned} \tag{1}$$

where

$$J(t) = \frac{i}{2m}\left(\psi^*(0,t)\frac{\partial\psi(0,t)}{\partial x} - \frac{\partial\psi^*(0,t)}{\partial x}\psi(0,t)\right)$$

is the standard quantum mechanical probability current. (We denote the state at time t by ρ_t, and set $\hbar = 1$ throughout.) This can also be written in the following two forms,

$$p(t_1, t_2) = \int_{t_1}^{t_2} dt \int dpdq \frac{(-p)\delta(q)}{m} W_t(p, q), \tag{2}$$

where $W_t(p, q)$ is the Wigner function, defined later, and

$$\begin{aligned} p(t_1, t_2) &= \mathrm{Tr}(C\rho) \\ C &= \int_{t_1}^{t_2} dt \frac{(-1)}{2m} \left(\hat{p}\delta(\hat{x}(t)) + \delta(\hat{x}(t))\hat{p}\right) \\ &= P(t_1) - P(t_2). \end{aligned} \tag{3}$$

These expressions agree with the classical result, with the Wigner function W replaced by the classical distribution function w, provided that the classical trajectories are straight lines. They are also correctly normalized to 1 as $t_2 \to \infty$ with $t_1 = 0$, but they are not generally positive, even for states consisting entirely of negative momenta. This genuinely quantum phenomenon is called backflow and arises because the operator C, positive classically, has negative eigenvalues [3, 4, 5]. This means we cannot generally regard Eq.(1) as defining an arrival time distribution. There is, in addition, a more fundamental problem with Eq.(1), which is that probabilities in quantum theory should be expressible in the form [6]

$$p(\alpha) = \mathrm{Tr}(P_\alpha \rho). \tag{4}$$

Here P_α is a projection operator, or more generally a POVM, associated with the outcome α. Eq.(1) cannot be expressed in this form, and we therefore conclude that it is not a fundamental expression in quantum theory, so must be the result of some approximation.

One interesting question is whether heuristic expressions for the arrival time probabilities of the form Eqs.(1, 2) can be derived in some limit from a more axiomatic approach. In this contribution we will show how this may be done in the decoherent histories approach to quantum theory. This contribution is based on a series recent papers [7, 8, 9, 10], but also draws on earlier work by Hartle, Halliwell and others [11].

Another issue is the relationship between the various time scales present in the problem. How long does it take a wavepacket to cross the origin? Is there a minimum time scale on which we can meaningfully talk about probabilities for crossing? One of the aims of this contribution is to clarify some of these issues.

The rest of this paper is structured as follows. In section 2 we briefly review the decoherent histories approach to quantum theory in general, and the arrival time problem in particular. In section 3 we examine how complex potentials may be used to define class operators for the arrival time problem. In section 4 we then discuss the arrival time problem for open quantum systems, reviewing how environmentally induced decoherence allows the arrival probabilities to be written in terms of a POVM. In section 5 we then turn to the question of whether the corresponding histories decohere. We summarise and discuss some open questions on section 6.

2. The decoherent histories approach to the arrival time problem

2.1. The Decoherent Histories approach to Quantum Theory

We begin by briefly reviewing the decoherent histories approach to quantum theory. More extensive discussions can be found in the references [12].

Alternatives at fixed moments of time in quantum theory are represented by a set of projection operators $\{P_a\}$, satisfying the conditions

$$\sum_a P_a = 1, \quad P_a P_b = \delta_{ab} P_a, \tag{5}$$

where we take a to run over some finite range. In the decoherent histories approach to quantum theory histories are represented by class operators C_α which are time-ordered strings of projections

$$C_\alpha = P_{a_n}(t_n) \cdots P_{a_1}(t_1) \tag{6}$$

or sums of such strings [13]. Here the projections are in the Heisenberg picture and α denotes the string $(a_1, \cdots a_n)$. All class operators satisfy the condition

$$\sum_\alpha C_\alpha = 1. \tag{7}$$

Probabilities are assigned to histories via the formula

$$p(\alpha) = \mathrm{Tr}\left(C_\alpha \rho C_\alpha^\dagger\right). \tag{8}$$

However probabilities assigned in this way do not necessarily obey the probability sum rules, because of quantum interference. We therefore introduce the decoherence functional

$$D(\alpha, \beta) = \mathrm{Tr}\left(C_\alpha \rho C_\beta^\dagger\right), \tag{9}$$

which may be thought of as a measure of interference between pairs of histories. We require that sets of histories satisfy the condition of decoherence, which is

$$D(\alpha, \beta) = 0, \quad \alpha \neq \beta. \tag{10}$$

This ensures that all probability sum rules are satisfied.

We note that when there is decoherence Eq.(7) and Eq.(10) together imply that the probabilities $p(\alpha)$ are given by the simpler expressions

$$q(\alpha) = \mathrm{Tr}\left(C_\alpha \rho\right). \tag{11}$$

Decoherence ensures that $q(\alpha)$ is real and positive, even though it is not in general. In this way decoherent histories may reproduce probabilities of the form Eq.(3).

2.2. The Decoherent Histories aproach to the arrival time problem

We turn now to the definition of the arrival time problem in decoherent histories. It is natural to start by considering the class operator for not crossing the origin in a given time interval $[0, \tau]$. Splitting this time into smaller intervals of size ϵ and letting $\tau = n\epsilon$, the natural class operator for not crossing the origin during this time would be

$$C_{nc} = P(n\epsilon)...P(2\epsilon)P(\epsilon). \tag{12}$$

This object represents the history where the state is in $x > 0$ at the times $\epsilon, 2\epsilon...$ but can be anywhere at intermediate times. An obvious thing to try is to take the limit as the spacing between projections approaches zero, with the hope of obtaining a precise specification of the behaviour of the state. However that this limit yields the restricted propagator.

$$\lim_{\epsilon \to 0} P(n\epsilon)...P(2\epsilon)P(\epsilon) = g_r(\tau, 0). \tag{13}$$

$g_r(\tau, 0)$ generates unitary evolution in the subspace of states with support only in $x > 0$, so one find that the probability of crossing the origin is zero. This is a manifestation of the quantum Zeno effect [14]. The resolution of this problem is simple in principle, one simply declines to take the limit $\epsilon \to 0$ in Eq.(12), and instead works with histories which have some finite "resolution" ϵ. In practice, however, the situation is complicated for two reasons. The first is that the object Eq.(12) is difficult to work with analyticaly, although semi-classical approximations are avaliable [8]. The second is that it is not clear what the limits are on the "resolution" ϵ.

The solution to both of these problems is to show that the evolution represented by Eq.(12) is equivalent to evolution in the presence of a simple complex potential $V(x) = -iV_0\theta(-x)$ [9]. One can then use this equivalence either to derive the class operators for crossing and not crossing directly [7], or one can instead use it to deduce the minimum spacing allowed between the projections in Eq.(12) [9].

3. The relationship between complex potentials and strings of projection operators

Probably the most significant development in the histories approach to time in quantum theory to have occured in the past few years has been the understanding of the relationship between class operators defined in the manner of Eq.(12), and those defined via complex potentials. This development has been significant for three reasons,

- It has made the calculation of arrival time probabilities within decoherent histories relatively straightforward.
- It has led to a much better understanding of the relationship between the different timescales that occur in the problem, in particular the Zeno time.
- It has suggested new connections between the histories approach and some of the various detector based models for the computation of arrival time probabilities.

The proof of this equivalence can be found in [9], building on earlier work by Echonabe et al [15]. Here we will simply state the equivalence, and discuss some of the consequences for the arrival time problem.

The equivalence is expressed as follows,

$$e^{-iH\tau}P(n\epsilon)...P(2\epsilon)P(\epsilon)\left|\psi\right\rangle \approx e^{-iH\tau - V_0\bar{P}\tau}\left|\psi\right\rangle \tag{14}$$

where

$$P = \theta(\hat{x}), \quad \bar{P} = \theta(-\hat{x}), \quad \text{and } n\epsilon = \tau \tag{15}$$

The equivalence holds provided

$$V_0\epsilon \sim 1, \quad \text{and } \epsilon << \frac{1}{\Delta H} \tag{16}$$

Since we know from studies of the complex potential that we begin to have reflection when $V_0 \sim E$, where E is the energy of the initial state, it follows that the time scale on which the projections cause significant reflection is given by $\epsilon \sim 1/E$. There is therefore a regime,

$$\frac{1}{E} << \epsilon << \frac{1}{\Delta H} \tag{17}$$

in which evolution under strings of projection operators is equivalent to that under the complex potential *and* there is negligible reflection.

In Refs.[7, 8] the class operators corresponding to a first crossing of the origin in the time interval $[t_{k-1}, t_k]$ were computed to be

$$C_k = \bar{P}(t_k)P(t_{k-1})\cdots P(t_2)P(t_1), \tag{18}$$

for $k \geq 2$, with $C_1 = \bar{P}(t_1)$. These clearly describe histories which are in $x > 0$ at times $t_1, t_2, \cdots t_{k-1}$ and in $x < 0$ at time t_k, so, approximately, describe a first crossing between t_{k-1}

and t_k. These class operators can also be expressed in terms of evolution under the complex potential [7],

$$C_k = \int_{t_k}^{t_{k+1}} dt C(t) \tag{19}$$

$$C(t) = e^{-iH(\tau-t)} V e^{-iHt-Vt} \tag{20}$$

The object $C(t)$ is the class operator for crossing at a time t. However this designation is somewhat misleading, because any state that enters within a time $\sim 1/V_0$ of t will also contribute to this crossing probability, because of the finite time it takes the complex potential to absorb the state. Differently put, the "detector" modeled by the complex potential has a time resolution of order $\sim 1/V_0$, and thus we cannot use it to define arrival times more accurately than this.

Given this, it is natural to use class operators like Eq.(19) to define arrival time. If the time interval is much greater than $1/V_0$ the dependence on V_0 drops out and we can show that [7]

$$\begin{aligned} e^{iH\tau} C_k &= \int_{t_k}^{t_{k+1}} dt e^{iHt} V e^{-iHt-Vt} \\ &\approx P(t_k) - P(t_{k+1}), \quad \text{providing } t_{k+1} - t_k >> 1/V_0 \end{aligned} \tag{21}$$

In Ref [8] Eq.(21) was also derived from Eq.(18) with the help of a simple semiclassical approximation [1]. In fact, in [10] it was shown that this semi-classical approximation is closely related to the decoherence conditions discussed below.

In the case of a free particle without an environment, this class operator reproduces Eq.(1) under the conditions of decoherence. This is because, assuming decoherence, the arrival time probability is given by

$$p(t_2, t_1) = \mathrm{Tr}(C\rho C^\dagger) = \mathrm{Tr}(C\rho), \tag{22}$$

using Eq.(11). This is an important result, we have succeeded in deriving probabilities of the form Eq.(1) from decoherent histories, in the limit that the corresponding histories decohere.

4. Arrival time for open systems

We will see below that decoherence of arrival time histories can only be achieved for a generic initial state in the presence of an environment. It is therefore important to look more generally at the arrival time problem for the case of a state coupled to an environment. There are two aspects to consider. The first is, given that we expect the arrival time probabilities to be given in terms of the current, what are the relavent properties of the current for a system coupled to an environment? The second issue is to examine how the inclusion of an environment helps to produce decoherence of the corresponding arrival time histories. We will deal with the first issue in this section, and put off the discussion of decoherence till later.

The arrival time probability $p(t_1, t_2)$ can be written in terms of the Wigner function, as

$$p(t_1, t_2) = \int_{t_1}^{t_2} dt \int dp dq \frac{(-p)\delta(q)}{m} W_t(p, q), \tag{23}$$

where the Wigner function is defined as

$$W(p, q) = \frac{1}{2\pi} \int d\xi \rho(q + \xi/2, q - \xi/2) e^{-ip\xi} \tag{24}$$

[1] The situation is more subtle than implied in the derivation in [7, 8]. The class operators Eq.(12, 19) are the quantum analogue of *first* crossing operators, they are non-zero only for states incoming from the right, and they are non-negative for any initial state. By contrast, the class operators Eq.(21) are non-zero for states incoming from the left, and can be negative for states incoming from the right. The first of these difficulties can be handled straightforwardly [16], but the second, a manifestation of backflow, requires conditions on the state. Fortunately it can be shown that states exhibiting decoherence of arrival time probabilities will not display backflow [7].

The difficulty with Eq.(23) is that for a general state $W(p, q)$ need not be positive, it is therefore possible that this "probability" could be negative [2]. An obvious solution to this problem is to consider a state coupled to an environment. This is because it is known that such evolution causes the Wigner function of the system to become positive on a very short timescale. After this time, an expression like Eq.(23) may be an acceptable, if still heuristic, arrival time distribution for a system coupled to an environment.

In [10] this scheme was examined, using as an environment the model of quantum brownian motion [17]. The model can be expressed in terms of a master equation, in general quite complicated, but which simplifies in the limit of high temperatures and negligible dissipation to the following [10],

$$\frac{\partial \rho_t}{\partial t}(x, y) = \frac{i}{2m}\left(\frac{\partial^2}{\partial x^2} - \frac{\partial^2}{\partial y^2}\right)\rho_t(x, y) - D(x-y)^2\rho_t(x, y) \tag{25}$$

A further interesting limit of this evolution is the so called "near-deterministic limit" [10] of $t << \tau_s = p_0^2/D$. It is easy to show that τ_s is the time scale on which an initial state sharply peaked around momentum $p_0 < 0$ begins to have significant probability to be found with positive momentum. For times much less than this, although the evolving state will have a finite momentum width it will still approximately follow the deterministic classical trajectory. This is obviously an interesting limit for the arrival time problem. It was found that, in this limit, Eq.(23) defines an acceptable arrival time distribution for a generic initial state, after a time

$$t = \left(\frac{3}{16}\right)^{1/4}\left(\frac{2m}{D}\right)^{\frac{1}{2}} = \left(\frac{3}{16}\right)^{1/4}\tau_l \tag{26}$$

Introducing the notation,

$$\mathbf{z} = \begin{pmatrix} p \\ q \end{pmatrix} = \begin{pmatrix} z_0 \\ z_1 \end{pmatrix} \tag{27}$$

and the class of Gaussian phase space functions,

$$g(\mathbf{z}; A) = \frac{1}{2\pi|A|^{\frac{1}{2}}}\exp\left(-\frac{1}{2}\mathbf{z}^T A^{-1}\mathbf{z}\right) \tag{28}$$

After this time we may write the current as

$$J(t) = \mathrm{Tr}(F\rho) \tag{29}$$

where

$$F = \int d\mathbf{z} P_{\mathbf{z}}\frac{(-z_0)}{m}\delta(z_0 + z_1 t/m) \tag{30}$$

and we have defined the POVM

$$P_{\mathbf{z}} = \frac{1}{\pi}\int d\mathbf{z}' \left|\mathbf{z}'\right\rangle\left\langle\mathbf{z}'\right| g(\mathbf{z}-\mathbf{z}'; B), \quad \text{with } B = \begin{pmatrix} 2Dt - s^2 & Dt^2/m \\ Dt^2/m & 2Dt^3/3m^2 - 1/4s^2 \end{pmatrix} \tag{31}$$

We may now go further, and write the arrival time probability as,

$$p(t_1, t_2) = \int_{t_1}^{t_2} dt\mathrm{Tr}(F\rho) \approx \mathrm{Tr}(E\rho) \tag{32}$$

[2] Note that this negativity of the current had nothing to do with the possibility of having momentum components with both signs in the initial state. We have chosen our initial state to have only negative momenta. Clearly Eq.(23) will also have to be modified for the case of an initial state with momenta of both signs, but this is a separate issue that occurs equally in the classical case, and it can be dealt with in a reasonably straightforward manner, [1, 16]

where

$$E = \int d\mathbf{z} P_{\mathbf{z}} [\theta(z_1 + z_0 t_1/m) - \theta(z_1 + z_0 t_2/m)] \tag{33}$$

is a POVM representing arrival at $x = 0$ between t_1 and t_2. This operator is just a smeared version of the classical phase space operator, but crucially it is positive for $p < 0$. Since the decoherent histories approach to the arrival time problem gives the current as the correct arrival time distribution when there is decoherence, Eq.(32) gives the arrival time probabilities computed via decoherent histories.

The important point here is that we have managed to write the arrival time probabilities Eq.(23), in terms of the expectation value of a POVM, thus bringing them into line with other expressions for probabilities in quantum theory.

5. Decoherence of arrival time probabilities

We turn now to discussing the conditions under which we have decoherence of crossing probabilities. The general picture we have in mind is an initial wavepacket defined at $t = 0$, evolved until time t_1, possibly in the presence of an environment , and then we wish to compute the probability of crossing between t_1 and t_2.

Consider first the decoherence functional for the histories correspnding to crossing in the intervals $[t_k, t_{k+1}]$ and $[t_j, t_{j+1}]$, where without loss of generality we take $t_{j+1} \leq t_k$

$$\begin{aligned} D_{kj} &= \langle\psi| (P(t_k) - P(t_{k+1}))(P(t_j) - P(t_{j+1})) |\psi\rangle \\ &= \langle\psi| (P(t_k) - P(t_{k+1}))(\bar{P}(t_{j+1}) - \bar{P}(t_j)) |\psi\rangle \end{aligned}$$

It is a sum of terms of the form

$$d_{kj} = \langle\psi| P(t_k)\bar{P}(t_j) |\psi\rangle \tag{34}$$

with $t_j \leq t_k$. Note that

$$|d_{kj}|^2 \leq \Delta_{kj} = \langle\psi| P(t_k)\bar{P}(t_j)P(t_k) |\psi\rangle \tag{35}$$

The key point is that Δ has the form of a probability, it is essentially the probability to find the state in $x < 0$ at t_j and then in $x > 0$ at t_k. Since we will be considering left moving wavepackets, this probability should be small. To show we have decoherence it therefore suffices to compute the quantity Δ_{kj} defined in Eq.(35). We will do this both for a free particle and a particle coupled to an environment. If we include an environment we anticipate that this will be small simply from the form of Eq.(35). This is because the effect of the environment is to cause the density matrix to become tightly peaked around the classical path and classically for $p < 0$ the probability given by Δ_{kj} is zero. We will see how this works in a specific example below.

We take $k = 1, m = 2$ without loss of generality, and we drop the subscript from now on

$$\Delta = \mathrm{Tr}(P(t_2)\bar{P}(t_1)\rho\bar{P}(t_1)) = \int_\alpha \mathcal{D}x \int_\alpha \mathcal{D}y \exp\left(iS[x] - iS[y] - D\int dt(x-y)^2\right)\rho_{t_1}(x_1, y_1), \tag{36}$$

where the histories α are those that start at $x_1, y_1 < 0$ and finish at $x_2 > 0$. In terms of the density matrix propagator [18]

$$\begin{aligned} \Delta &= \int_0^\infty dx_2 \int_{-\infty}^0 dx_1 \int_{-\infty}^0 dy_1 J(x_2, x_2, t_2|, x_1, y_1, t_1)\rho_{t_1}(x_1, y_1) \\ &= \int_0^\infty dx_2 \int_{-\infty}^0 dx_1 \int_{-\infty}^0 dy_1 \left(\frac{m}{\pi t}\right) \\ &\quad \exp\left(\frac{im}{2(t_2 - t_1)}((x_2 - x_1)^2 - (x_2 - y_1)^2) - \frac{D(t_2 - t_1)}{3}(x_1 - y_1)^2\right)\rho_{t_1}(x_1, y_1). \end{aligned}$$

Transforming to new variables

$$X = \frac{x_1 + y_1}{2}, \quad \xi = x_1 - y_1, \tag{37}$$

and writing the density matrix in terms of the Wigner function via

$$\rho(x, y) = \int_{-\infty}^{\infty} dp e^{ip(x-y)} W(p, \frac{x+y}{2}), \tag{38}$$

we obtain

$$\begin{aligned} \Delta &= -\int_0^{\infty} dx_2 \int_{-\infty}^{0} dX \int_{-X}^{X} d\xi \int_{-\infty}^{\infty} dp \left(\frac{m}{2\pi(t_2 - t_1)}\right) \\ &\exp\left(\frac{im}{(t_2 - t_1)}\xi(X - x_2) + ip\xi - \frac{D(t_2 - t_1)}{3}\xi^2\right) W_{t_1}(X, p). \end{aligned} \tag{39}$$

We see from this expression that there is a time scale, $\tau_l = \sqrt{2m/D}$, set by the environment on which localisation effect are important. This timescale is the same as that on which the current becomes positive, as we saw earier.

There are three cases to explore here, the first is where there is no environment, $D = 0$. This is the case covered in Ref.[7]. The second case is the intermediate one, $t_1/\tau_l >> 1$ but $(t_2 - t_1)/\tau_l << 1$. This is the most general case in which we can expect to have environmentally induced decoherence. The final case is where $t_1/\tau_l, (t_2 - t_1)/\tau_l >> 1$. This is the case of very strong environmental coupling.

The details of the calculation of Δ in each case are given in [10], and also for the free particle case in [7], so we simply quote the result in each case.

5.1. *Free particle case*

We assume that our state at t_1 is of the form

$$W_{t_1}(X, p) = \frac{1}{\pi} \exp\left(-\frac{(X - X_0 - p_0 t_1/m)^2}{2\sigma^2} - 2\sigma^2(p - p_0)^2\right), \tag{40}$$

and that σ^2 is large, so the state is tightly peaked in momentum. After some calculation, and assuming $p_0^2(t_2 - t_1)/2m >> 1$, we obtain

$$\Delta < \frac{1}{\sigma|p_0|} << 1, \tag{41}$$

there will therefore be decoherence for gaussian wavepackets tightly peaked in momentum, ie $|p_0|\sigma >> 1$, provided $E_0(t_2 - t_1) >> 1$. In [7] it was argued that this conclusion also holds for orthogonal superpositions of gaussians.

5.2. *Weak environment*

Turning to the intermediate case, although we can still neglect the effects of the environment for the time it takes to cross the origin, the key point is that $W_{t_1}(X, P)$ is the initial state evolved with an environment, and since $t_1/\tau_l >> 1$ this should be significant. Because the Wigner function propagator is a gaussian the analysis is similar to the first case. We find, after some calculation,

$$\Delta < \sqrt{\frac{1}{E_0 t_1}\left(\frac{\tau_l}{t_1}\right)} << 1. \tag{42}$$

5.3. Strong environment

Finally we have the case of strong system-environment coupling. In this case we can show that Δ will be very small, provided $(t_2 - t_1) << \tau_s$. This gives us an *upper* bound on the time interval, $[t_1, t_2]$, rather than a lower one. The lower time scale is provided by the condition $t_2 - t_1 >> \tau_l = \sqrt{2m/D}$. This lower time scale is compatible with the condition that the current be positive.

Note however that this lower limit is state independent. There will be states for which arrival time probabilities decohere on much shorter time scales than this, for example the simple cases which decohere in the absence of any environment will continue to do so in the presence of an environment, at least until a time $\sim \tau_s$.

5.4. Discussion

What one draws from studying these cases is that whether or not one can assign arrival time probabilities in decoherent histories depends on the form of the state at the time it crosses the origin. Environmentally induced decoherence produces mixtures of gaussian states from generic initial ones, and thus after a short time arrival time probabilities can be defined whatever the initial state. However it is not necessary for the system to be monitored whilst it is crossing the origin. The smallest time interval, δt, for which we can define an arrival time probability is therefore set by the energy of the system and not the details of the environment and we must have $E\delta t >> 1$. This is in agreement with Ref.[7], and also with the results of earlier works, concerning the accuracy with which a quantum system may be used as a clock [19].

In conclusion, for a general state, decoherence of histories requires that we evolve for a time much greater than $\tau_l = \sqrt{2m/D}$ before the first crossing time. This is because τ_l is the time scale on which quantum correlations disappear and our state begins to resemble a mixture of gaussian states. After this time, we may define arrival time probabilities to an accuracy δt, provided only that $E\delta t >> 1$. States which start as gaussians may be assigned arrival time probabilities without this initial period of evolution. This is in line with the general result that some coarse-graining is always required to achieve a decoherent set of histories in quantum theory.

6. Summary and open questions

In this contribution we have given an overview of the way in which decoherent histories gives insight into the arrival time problem in quantum theory. The key steps are as follows, in section 2 we discussed how probabilities are defined in decoherent histories, and how they can reduce to the form Eq.(11) when there is decoherence. Next in section 3 we saw how the class operators can be derived using complex potentials, and how they reduce to the form Eq.(21) under certain conditions. Then in section 4 we showed that the same is true for a particle coupled to an environment, under certian conditions. Finally in section 5 we proved that we have decoherence of histories for an interesting class of states for a free particle, and for a generic state for a particle coupled to an environment. Although space has prevented us from giving all the details, we hope that the general picture is clear.

Although the arrival time probabilities computed from decoherent histories agree with the heuristic ones when we have decoherence, their derivation in this way represents a significant advance in our understanding. The difficulty with regarding the current as the arrival time distribution is the arbitrary way in which one accepts these "probabilities" when they are positive, but declines to do so when they are not. Because decoherence is an essential part of the histories formalism, this arbitrariness is replaced with a consistent set of rules governing when probabilities may or may not be assigned. Whilst this may be of no consequence in the setting of a laboratory, it may prove hugely important in the analysis of closed systems, in particular the study of quantum cosmology [20].

Although the arrival time problem is now well understood from this perspective, there are a number of other areas, such as dwell or tunneling times, where the methods outlined here could be usefully applied. Finally, we mentioned right at the beginning of this contribution that there are a multitiude of different approaches to tackling the arrival time problem. An interesting question is how the decoherent histories approach is related to other approaches, particularly to those operational approaches that rely on coupling to model clocks. Further study of this issue may shed light on the reasons why there seem to be so many divergent approaches to the arrival time problem.

Acknowledgments

I would like to thank the organisers of DICE 2010, especially Hans-Thomas Elze, for giving me the opportunity to take part in this conference. The majority of the work reported here was carried out in collaboration with Jonathan J Halliwell, to whom I am also indebted for many years of support and stimulating discussion.

References

[1] See for example; Muga J G, Mayato R S and Egusquiza I L ed 2002 *Time in Quantum Mechanics* (Berlin: Springer); Muga J G, Ruschhaupt A and del Campo A ed 2010, *Time in Quantum Mechanics - Vol. 2* (Berlin: Springer); Muga J G and Leavens C R 2000 , *Phys.Rep.* **338** 353
[2] Damborenea J A, Egusquiza I L, Hegerfeldt G C, and Muga J G 2002, *Phys. Rev.* A **66** 052104; Hegerfeldt G C, Seidel D, Muga J G, 2003 *Phys. Rev.* A **68**
[3] Muga J G, Palao J P and Leavens C R 1999 *Phys. Lett.* A **253** 21
[4] Bracken A J and Melloy G F 1994 *J.Phys.* A **27** 2197
[5] Eveson S P, Fewster C J and Verch R 2005 *Ann.Inst. H.Poincaré* **6** 1; Penz M, Grübl G, Kreidl S and Wagner P 2006 *J.Phys.* A **39** 423
[6] See for example; Nielsen M A and Chuang I L 2000 *Quantum computation and quantum information* (Cambridge: CUP); Busch P, Grabowski M and Lahti P 1997 *Operational Quantum Physics* (Berlin: Springer-Verlag)
[7] Halliwell J J and Yearsley J M 2009 *Phys. Rev.* A **79** 062101
[8] Halliwell J J and Yearsley J M 2009 *Phys. Lett.* A **374** 154
[9] Halliwell J J and Yearsley J M 2010 *J.Phys.* A **43** 445303
[10] Yearsley J M 2010 *Phys. Rev.* A **82** 012116
[11] See for example, Hartle J B 1991 *Phys. Rev.* D **44** 3173; Halliwell J J and Zafiris E 1998 *Phys.Rev.* D **57** 3351; Micanek R J and Hartle J B 1996 *Phys.Rev.* A **54** 3795; Yamada N and Takagi S 1991 *Prog.Theor.Phys.* **85** 985; 1991 **86** 599; 1992 **87** 77; Yamada N 1992 *Sci. Rep. Tôhoku Uni., Series 8* **12** 177
[12] Gell-Mann M and Hartle J B 1990 *Complexity, Entropy and the Physics of Information, SFI Studies in the Sciences of Complexity, Vol. VIII* ed Zurek W (Reading: Addison Wesley); 1993 *Phys.Rev.* D **47** 3345; Griffiths R B 1984 *J.Stat.Phys.* **36** 219; 1993 *Phys.Rev.Lett.* **70** 2201; 1987 *Am.J.Phys.* **55** 11; Omnès R 1988 *J.Stat.Phys.* **53** 893; 1988 **53** 933; 1988 **53** 957; 1989 **57** 357; 1991 **62** 841; 1990 *Ann.Phys.* **201** 354; 1992 *Rev.Mod.Phys.* **64** 339; 1994 *The Interpretation of Quantum Mechanics* (Princeton: Princeton University Press); Halliwell J J 1994 *Fundamental Problems in Quantum Theory* ed Greenberger G and Zeilinger A *Annals of the New York Academy of Sciences, Vol 775* **726**; Dowker H F and Halliwell J J 1992 *Phys. Rev.* D **46** 1580
[13] Isham C 1994 *J.Math.Phys.* **35** 2157
[14] Misra B and Sudarshan E C G 1977 *J.Math.Phys.* **18** 756; Peres A 1980 *Am. J. Phys.* **48** 93
[15] Echanobe J, del Campo A and Muga J G 2008 *Phys. Rev.* A **77** 032112
[16] Yearsley J M (unpublished).
[17] Caldeira A O and Leggett A J 1983 *Physica* A **121** 587
[18] Halliwell J J and Zoupas A 1997 *Phys. Rev.* D **55** 4697
[19] Peres A 1993 *Quantum Theory: Concepts and Methods* (Dordrecht: Kluwer Academic, Dordrecht); Aharonov Y, Oppenheim J, Popescu S, Reznik B and Unruh W G 1998 *Phys. Rev.* A **57** 4130
[20] Halliwell J J 2009 *Phys. Rev.* D **80** 124032

5th International Workshop DICE2010 IOP Publishing
Journal of Physics: Conference Series **306** (2011) 012057 doi:10.1088/1742-6596/306/1/012057

Unidirectionality of time induced by T violation

Joan A. Vaccaro
Centre for Quantum Dynamics, Griffith University, Nathan 4111, Brisbane, Australia
E-mail: `J.A.Vaccaro@griffith.edu.au`

Abstract. The physical nature of time has been an enigma for centuries. Despite the progress made in linking various arrows of time together, the underlying reason for the observed direction of time has remained a deep mystery. However, an increasing number of experiments are confirming a violation of time reversal invariance (T) in meson decay processes. T violation represents a fundamental break in the symmetry between the past and future. I show here that such processes induce destructive interference between different paths that the universe can take through time. The interference eliminates all paths except for two that represent continuously forwards and continuously backwards time evolution. These results illustrate that there are previously-unknown large-scale physical effects arising from the T violation processes in meson decay. Moreover these effects offer a solution to the age-old problem of the cause of the unidirectionality of time.

1. Introduction

Nature has been shown to violate the time reversal (T) and charge-parity conjugation (CP) invariances [1]. For T violating processes

$$H \neq THT^{-1}$$

where H is the Hamiltonian for one specific direction of time, THT^{-1} is the version of the Hamiltonian in the reversed direction of time and T is Wigner's time inversion operation [2]. In principle, one can write down the Schrödinger equation for T violating processes provided the direction of time evolution and the corresponding version of the Hamiltonian are known. Then by applying the time reversal operation we obtain the Schrödinger equation involving the THT^{-1} version of the Hamiltonian for evolution in the opposite direction of time. But we do not have a dynamical equation of motion for the situation where the *direction of time evolution cannot be specified* and for which there is no argument for favouring one version of the Hamiltonian over the other. This problem becomes critical when we attempt to describe the universe as a closed system because being closed precludes any external clock-like device for use as a reference for the direction of time. There is no argument for favouring one direction of time over the other, and so there is no reason to favour H over THT^{-1}. This means that if H is included in the dynamical equation of motion of the universe then so must THT^{-1}. Current quantum theory is unable to provide a description in this case because of its inability to deal with two different Hamiltonians for a single system in one dynamical equation. There is, therefore, no satisfactory quantum formalism for describing a universe that exhibits T violation processes. The resolution of this problem calls for *a major shift in the way we think about time and dynamical equations of motion.*

This hurdle has thwarted previous attempts to understand the implications of T violation for the nature of time [3, 4, 5]. I will show here, however, that it can be overcome by using Feynman's sum over histories method [6] to construct the set of all possible paths that the universe can take through time. This set includes paths that zigzag forwards and backwards through time, for which the universe evolves according to H in one direction of time and THT^{-1} for the reverse direction. T violation induces destructive interference that restricts the possible paths through time the universe can take. The interference eliminates paths which meander forwards and backwards in time and leaves only two main paths corresponding to evolution continuously forwards and evolution continuously backwards. This analysis resolves the problem of modelling the dynamics of T violation processes in the absence of a preferred direction of time by incorporating two Hamiltonians, one for forwards and one for backwards evolution in a single dynamical equation. Remarkably it demonstrates that there are *large-scale* physical effects of T violations and it offers a physical explanation of the unidirectionality of time. It stands to overhaul conventional notions of time.

In what follows I will describe how the *paths through time* are constructed, how T violation processes induce destructive *interference between paths* and how this gives rise to the *unidirectional nature of time*. Fuller details can be found in Ref. [7].

2. Paths through time

Let the universe be modelled as a closed system. This immediately excludes any external clock devices that could be used as references for the direction of time. As a consequence, the evolution of the universe needs to be unbiased with respect to the direction of time. For convenience, let the two directions of time be labelled as "forward" and "backward" although these are arbitrary labels. We begin with the universe in a state $|\psi_0\rangle$ which we shall call the "origin state" without reference to the direction of time. Let the evolution of the universe in the forwards direction over the time interval τ be

$$|\psi_F(\tau)\rangle = U_F(\tau)|\psi_0\rangle$$

where $U_F(\tau) = \exp(-i\tau H_F)$ is the forwards time evolution operator, and H_F is the Hamiltonian for forwards time evolution. (Throughout this paper we shall use units in which $\hbar = 1$.) Then the backwards time evolution operator is given by

$$U_B(\tau) = TU_F(\tau)T^{-1} = \exp(i\tau H_B)$$

where $H_B = TH_FT^{-1}$ is the Hamiltonian for backwards time evolution. This gives the state of the universe after backwards evolution over a time interval of the same magnitude as

$$|\psi_B(\tau)\rangle = U_B(\tau)|\psi_0\rangle\ .$$

The matrix elements $\langle\phi|U_F(\tau)|\psi_0\rangle$ and $\langle\phi|U_B(\tau)|\psi_0\rangle$ represent the probability amplitudes for the universe in state $|\psi_0\rangle$ to evolve over the time interval τ to $|\phi\rangle$ via two paths in time corresponding to the forwards and backwards directions, respectively. Given that we have no basis for favouring one path over the other, we follow Feynman [6] and attribute an equal statistical weighting to each. Thus, *the total probability amplitude for the universe to evolve from one given state to another is proportional to the sum of the probability amplitudes for all possible paths through time between the two states.* In the current situation we have only two possible paths and so the total amplitude is proportional to

$$\langle\phi|U_F(\tau)|\psi_0\rangle + \langle\phi|U_B(\tau)|\psi_0\rangle = \langle\phi|\,[U_F(\tau) + U_B(\tau)]\,|\psi_0\rangle\ . \tag{1}$$

This result holds for all states $|\phi\rangle$ of the universe and so the evolution of $|\psi_0\rangle$ via both paths can be written as

$$|\Psi(\tau)\rangle = [U_F(\tau) + U_B(\tau)]\,|\psi_0\rangle \tag{2}$$

which is called the *symmetric time evolution* of the universe. To avoid unnecessary detail unnormalised vectors will be used to represent states of the universe. It follows that the symmetric time evolution of the universe in state $|\Psi(\tau)\rangle$ over an additional time interval of τ is given by

$$|\Psi(2\tau)\rangle = [U_F(\tau) + U_B(\tau)]\,|\Psi(\tau)\rangle = [U_F(\tau) + U_B(\tau)]^2\,|\psi_0\rangle\,.$$

Repeating this for N such time intervals yields the general expression for symmetric time evolution

$$|\Psi(N\tau)\rangle = [U_F(\tau) + U_B(\tau)]^N\,|\psi_0\rangle\,. \tag{3}$$

Figure 1 gives a graphical interpretation of this result in terms of a binary tree. It is useful to write the expansion of the product on the right side of Eq. (3) as

$$|\Psi(N\tau)\rangle = \sum_{n=0}^{N} S_{N-n,n}\,|\psi_0\rangle \tag{4}$$

where $S_{m,n}$ represents a sum containing $\binom{n+m}{n}$ different terms each comprising n factors of $U_F(\tau)$ and m factors of $U_B(\tau)$, and where

$$\binom{k}{j} = \frac{k!}{(k-j)!j!}$$

is the binomial coefficient. A more manageable expression for $S_{m,n}$, with all the U_B factors to the left of the U_F factors, is [7]

$$S_{m,n} \;=\; U_B(m\tau)U_F(n\tau)\sum_{v=0}^{m}\cdots\sum_{\ell=0}^{s}\sum_{k=0}^{\ell}\exp\Big[(v+\cdots+\ell+k)\tau^2[H_F,H_B]\Big] + \exp[O(\tau^3)] \tag{5}$$

where $[A, B]$ is the commutator of A and B, and there are n summations on the right side.

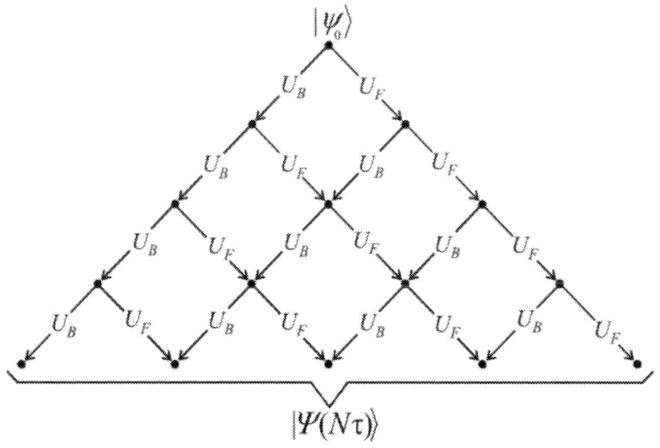

Figure 1. Binary tree representation of the generation of the state $|\Psi(N\tau)\rangle$ from the origin state $|\psi_0\rangle$ according to Eq. (3) with $N = 4$. States are represented as nodes (solid discs) and unitary evolution by links (arrows) between them. The root node (at the top) represents the origin state $|\psi_0\rangle$ and the leaf nodes (on the bottom row) represent components of the state $|\Psi(4\tau)\rangle$.

The expression $\langle\phi|S_{N-n,n}|\psi_0\rangle$ represents the evolution of the universe from $|\psi_0\rangle$ to $|\phi\rangle$ over a set of $\binom{N}{n}$ paths through time, where each path comprises a total of n steps in the forwards direction and $N-n$ steps in the backwards direction. The set of paths includes all possible orderings of the forwards and backwards steps. This set of paths is the focus of the remaining analysis.

Journal of Physics: Conference Series **306** (2011) 012057 doi:10.1088/1742-6596/306/1/012057

There is an important point to be made about the size of the time steps τ. To explore the consequences of the limit $\tau \to 0$ of infinitely-small time steps for a fixed total time $t_{\rm tot}$, we set $\tau = t_{\rm tot}/N$ and consider Eq. (3) for increasing values of N. We can write

$$2^{-N}[U_F(\tau)+U_B(\tau)]^N = \left\{\exp[-i\tfrac{1}{2}(H_F - H_B)\tau] + \mathcal{O}(\tau^2)\right\}^N$$

which becomes $\exp[-i\frac{1}{2}(H_F - H_B)t_{\rm tot}]$ as $N \to \infty$. In this limit, the universe evolves according to the Hamiltonian $\frac{1}{2}(H_F - H_B)$. This means that isolated subsystems of the universe that obey T invariance, such as a conventional model of a clock, would not exhibit any evolution. As clocks are taken to measure time intervals, the universe as a whole would not exhibit dynamics in the conventional sense in this $\tau \to 0$ limit. The lack of dynamics is an unphysical result if our system is to model the visible universe. This suggests that the time interval τ should be a small non-zero number for a universe-like system. Accordingly, in the following we set the value of τ to be a small time interval, for which we pick the Planck time for definiteness, i.e. $\tau \approx 5 \times 10^{-44} s$. The need for a finite time interval should not be alarming. Indeed the discreteness of space at the Planck scale is regarded as one of the triumphs of Loop Quantum Gravity [8, 9, 10].

3. Interference between paths

Being anti-Hermitian, the commutator $[H_F, H_B]$ has imaginary eigenvalues. Its presence in the nested summations in Eq. (5) gives rise to interference terms in the basis of eigenstates of $[H_F, H_B]$ of the form [7]

$$\sum_{v=0}^{m} \cdots \sum_{\ell=0}^{s} \sum_{k=0}^{\ell} \exp\left[-(v + \cdots + \ell + k)\tau^2 i\lambda\right]$$

where λ is an eigenvalue of $i[H_F, H_B]$. This leads to interference between the set of paths represented by $\langle\phi|S_{N-n,n}|\psi_0\rangle$.

An estimate of typical values for λ can be made using neutral K meson (kaon) evolution as a prototypical T-violating process. The phenomenological model of Lee and Wolfenstein [11] and the empirical values of Yao *et al.* [12] gives a spectrum of values of λ which has a mean of zero and standard deviation of $\lambda_{\rm SD} \approx \sqrt{f}10^{57}\,s^{-2}$. Here f is a fraction given by dividing the number of particles associated with T-violating processes by the total number of particles in the universe (which is assumed to be 10^{80}).

It is shown in Ref. [7] that destructive interference leads to Eq. (4) being replaced with $|\Psi(N\tau)\rangle = [\sum_{n\approx 0} S_{N-n,n} + \sum_{n\approx N} S_{N-n,n}]|\psi_0\rangle$ for total times $N\tau > 10^{-17}s$. Ignoring terms of order τ gives

$$\begin{aligned} |\Psi(N\tau)\rangle &= \left\{[U_B(\tau)]^N + [U_F(\tau)]^N\right\}|\psi_0\rangle \\ &= [U_B(N\tau) + U_F(N\tau)]\,|\psi_0\rangle\ , \end{aligned} \tag{6}$$

and then on writing the time as $t = N\tau$ we arrive at a key result, the *bievolution equation of motion*:

$$|\Psi(t)\rangle = [U_B(t) + U_F(t)]\,|\psi_0\rangle\ , \tag{7}$$

which is illustrated in Fig. 2. The term "bievolution" here refers to the dual evolution generated by the two different Hamiltonians. The approximations made in deriving this equation become exact in the limit $t \gg f^{-1/2}10^{-13}s$.

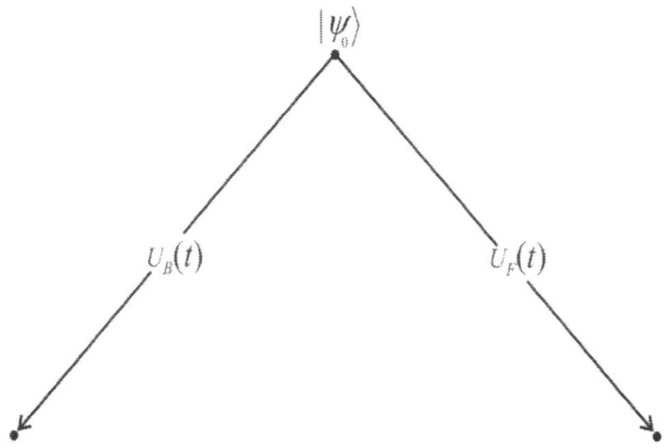

Figure 2. Binary tree representation in the presence of sufficient T violation processes. Destructive interference has eliminated all paths except for two corresponding to either continuous evolution in the forwards or continuous evolution in the backwards directions of time.

4. Unidirectional nature of time

Accelerator experiments are able to discern which Hamiltonian H_F or H_B is responsible in neutral K and B meson decay experiments. Such experiments yield evidence of just *one* of the Hamiltonians H_F or H_B. Our experience, therefore, is consistent with just one of the terms on the right side of the bievolution equation Eq. (7). The opposite signs in the exponents of the definitions of U_F and U_B imply that they represent time evolution in opposing directions of time. We conclude that in Eq. (7) the expressions $U_F(t)|\psi\rangle$ and $U_B(t)|\psi\rangle$ represent the universe evolving in opposite directions of time and in each case the evolution leaves evidence of the associated Hamiltonian in the state of the universe.

To expand on this point a little, the bievolution equation Eq. (7) describes our observation of the universe as evolving according to either

$$|\Psi(t)\rangle = U_F(t)\,|\psi_0\rangle$$

or

$$|\Psi(t)\rangle = U_B(t)\,|\psi_0\rangle$$

depending on whether we observe the version H_F or H_B, respectively, of the corresponding Hamiltonian in meson decay experiments. Each of these equations is the solution of the conventional Schrödinger equation for the corresponding version of the Hamiltonian and its associated direction of time. This demonstrates how the *unidirectional nature of time emerges phenomenologically from our physical observations.* It also demonstrates the *consistency between the bievolution equation and the conventional quantum theory.*

5. New standing of the matter-antimatter arrow of time

These results significantly elevate the importance of the matter-antimatter arrow in the study of the nature of time. The underlying T violation is based on the empirical evidence that different versions of the Hamiltonian operate for different directions of time evolution. Moreover, as discussed in in the previous section, each direction of time can be uniquely identified by physical evidence left by the Hamiltonian. This is quite different to other arrows of time. In particular, in a universe in which all physical laws are T invariant, the big bang expansion occurs equally in both directions of time and, by symmetry, the entropy of the universe also increases equally in both directions of time. The thermodynamic arrow, therefore, does not distinguish between the different directions of time evolution because entropy increases in the direction of time evolution *regardless of the direction.* This suggests that T violation and the matter-antimatter arrow play a far more significant role in determining the nature of time than previously imagined [3, 4, 5].

6. Absence of T violation processes

In many cosmological models, the early universe would not have had significant numbers of T violation processes [13]. In such models, T violation processes become significant only towards the end or even after the inflation period. It is instructive, therefore, to compare the above analysis with a universe that obeys T invariance. As $H_F = H_B = H$ in this case we can write $U_F + U_B = 2\cos(H\tau)$ and so

$$|\Psi(N\tau)\rangle = 2^N \cos^N(H\tau)|\psi_0\rangle \ . \tag{8}$$

According to the power method of determining the largest eigenvalue of a matrix, as N increases the right side approaches the eigenstate of $\cos^N(H\tau)$ that has the largest absolute eigenvalue, which in this case is the eigenstate of H with zero eigenvalue. Hence in the limit as $N \to \infty$

$$H|\Psi(N\tau)\rangle = 0 \ . \tag{9}$$

According to Eq. (3), $|\Psi(N\tau)\rangle$ is a superposition of states each of which represents a net time ranging from $-N\tau$ to $+N\tau$. It therefore represents a *history* of the universe. The zero energy eigenstate implies that the positive energy of matter fields is balanced by the negative energy of the gravitational potential energy [14, 15, 16].

Eq. (9) corresponds to the Hamiltonian constraint of the Wheeler-DeWitt equation [17, 18]. The solution of the Wheeler-DeWitt equation represents the entire history of the universe with no bias towards either direction of time. Time symmetric evolution is unbiased in the same way in the sense that it represents evolution without a fixed direction of time and the state in Eq. (9) represent the history of the universe before T violation processes were significant. This suggests that there is a close connection between the Wheeler-DeWitt equation and time symmetric evolution.

7. Conclusion

Our general understanding of T violating processes has previously been hindered by the lack of a formalism for accommodating their time asymmetry in a single dynamical law. We have removed this obstacle by deriving a general method for incorporating both versions of the Hamiltonian for a T violating process, one for forwards and the other for backwards evolution, in a single dynamical equation of motion. Moreover we have shown that T violation in meson decay provides, at least in principle, a physical mechanism for the phenomenological unidirectionality of time. This analysis, therefore, offers a new picture of the physical nature of time itself.

Acknowledgments

The author thanks Prof. D.T. Pegg for sharing his wealth of knowledge on the nature of time and for the generous advice he offered in the development of this article, Prof. S.M. Barnett for his encouragement and help with K meson dynamics and Prof. H.M. Wiseman for helpful discussions.

References

[1] Angelopoulos A *et al* 1998 First direct observation of time-reversal non-invariance in the neutral-kaon system *Phys. Lett.* B **444** 43-51
[2] Wigner E P 1959 *Group Theory and its Application to the Quantum Mechanics of Atomic Spectra* (New York: Academic Press)
[3] Price H 1996 *Time's Arrow and Archimedes' Point* (New York: Oxford University Press)
[4] Novikov I D 1998 *The River of Time* translated into English by Vitaly Kisin (Cambridge: Cambridge University Press)
[5] Zeh H D 2007 *The Physical Basis for the Direction of Time* (Berlin: Springer)

[6] Feynman R P 1948 Space-Time Approach to Non-Relativistic Quantum Mechanics *Rev. Mod. Phys.* **20** 367-387
[7] Vaccaro J A 2009 Large scale physical effects of T violation in mesons *Preprint* arXiv:0911.4528v2
[8] Loll R 1995 Volume Operator in Discretized Quantum Gravity *Phys. Rev. Lett.* **75** 3048-3051
[9] Rovelli C 2008 Loop Quantum Gravity *Living Rev. Relativity* **11** 5
[10] Smolin L 2004 An invitation to loop quantum gravity *Preprint* arXiv:hep-th/0408048v3
[11] Lee T D and Wolfenstein L 1964 Analysis of CP-Noninvariant Interactions and the K_1^0, K_2^0 System *Phys. Rev.* **138** B1490-B1496
[12] Yao W-M *et al* 2006 Review of Particle Physics *J. Phys. G: Nucl. Part. Phys.* **33** 666-684
[13] Liddle A R and Lyth D H 2000 *Cosmological Inflation and Large-Scale Structure* (Cambridge: Cambridge University Press)
[14] Tolman R C 1930 On the Use of the Energy-Momentum Principle in General Relativity *Phys. Rev.* **35** 875-895
[15] Tryon E P 1973 Is the Universe a Vacuum Fluctuation? *Nature* **246** 396
[16] Hartle J B and Hawking S W 1983 Wave function of the Universe *Phys. Rev.* D **28** 2960
[17] DeWitt B S 1967 Quantum Theory of Gravity. I. The Canonical Theory *Phys. Rev.* **160** 1113
[18] Alvarez E 1989 Quantum gravity: an introduction to some recent results *Rev. Mod. Phys.* **61** 561

A New Perspective on Relativistic Quantum Mechanics

Otto C.W. Kong

Department of Physics and Center for Mathematics and Theoretical Physics, National Central University, Chung-li, TAIWAN 32054

E-mail: otto@phy.ncu.edu.tw

Abstract. Based on a linear realization formulation of a quantum relativity, — proposed relativity for 'quantum space-time', we introduce the new Poincaré-Snyder relativity and Snyder relativity as relativities in between the latter and the well known Galilean and Einstein cases. While there is supposed to be not separate notion of classical and quantum mechanics at the level of the very unconventional quantum relativity, the Poincaré-Snyder relativity is more like a mathematically extended form of Einstein relativity on which we can write down a formal canonical classical and quantum mechanics. We discuss how the Poincaré-Snyder relativity may provide a stronger framework for the description of the usual (Einstein) relativistic quantum mechanics and present a first look of the interesting picture from the new perspective.

1. Introduction

Our title may give the impression that our topic is just about (Einstein) relativistic quantum mechanics. We are, however, presenting the new perspective from the consideration of relativities beyond the usual (Einstein) relativity. To put it in another way, Einstein relativity symmetry maybe considered a part of a bigger symmetry, the full consideration of which gives a boarder perspective on the usual relativistic quantum mechanics. The full symmetry is a symmetry for what we called the Poincaré-Snyder relativity [1]. The the new Poincaré-Snyder relativity, and Snyder relativity, are introduced as relativities in between the well known Galilean and Einstein cases and a proposed Quantum Relativity [2]. The latter is supposed to be the relativity for the 'quantum space-time'.

Our Quantum Relativity idea is based on the following aspects of our new approach to think about a plausible formulation of the the physics of 'quantum space-time' [2, 3] and eventually 'quantum gravity'. We consider there being a true microscopic/quantum structure of space-time itself and seek a direct description of that. Such a perspective of quantum gravity being beyond a quantized version of a classical theory of gravity has been advocated by many theorists, including particularly speakers at this Workshop [4]. It may then be just natural consider the 'quantum space-time' has its own relativity symmetry — the Quantum Relativity. The latter idea maybe considered as first introduced by Snyder in 1947, with the perspective that the Einstein Relativity has to be modified or deformed to accommodate a reference frame independent quantum scale [5]. Such an issue of a deformed special relativity theory rekindled quite some attention at the beginning of the century [6]. We added two important idea to the basic approach towards the theory of Quantum Relativity. Firstly, the 'quantum space-time' maybe described as something beyond 'space-time' [3]. We introduced the u-coordinates to supplement the s(pace) and t(time) coordinates for the classical geometric arena for formulating the Quantum Relativity [3, 7]. As space is the arena for Galilean Relativity and space-time the one for Einstein Relativity, 'space-time-u..' will be the one for the Quantum Relativity. In ref.[2], we introduced further the idea that a single 'quantum scale' is not enough to get to the Quantum Relativity. We have the well known quantum scale as one scale through assuming $\hbar$. To build also $\hbar$ itself into the framework, we need both a quantum momentum and a quantum length parameters. We are only talking about the special theory of Quantum Relativity though, and we will need the general theory for the 'quantum gravity'.

The picture we sketch is more than a simple speculation. It is given by the mathematics of Lie algebra deformation/stabilization [8] and the simple linear realization [3]. From a pure mathematical point of view, stabilization gives a new symmetry that has the old one as a contraction limit. The Lorentz symmetry is exactly a stabilization/deformation of the Galilean counterpart, with the deformation parameter given by $1/c^2$. Philosophically, one can argue that only stable symmetries can be scientifically verified to be correct. The contracted unstable limit is a singular point on the 'parameter space', requiring infinite experimental precision to be confirmed against the stable symmetry which admits other parameter values. And typically, the symmetry is independent of the exactly value of the parameter, so long as its nonzero. Following the line of thinking, we implemented in Ref.[3] a linear realization perspective. The linear realization scheme, applied to the Einstein Relativity as the deformation of Galilean Relativity on 3D space, is nothing other than the 4D Minkowski space-time picture. Such a mathematically conservative approach, however, leads to a very radical physics perspective, that space-time is to be described as part of something bigger [3], what we called the quantum world in Ref.[2] with the 'space-time-u..' coordinates. In the latter paper we identify the Quantum Relativity symmetry as $SO(2,4)$, with a linear realization on a 6D classical geometry beyond space-time providing a description of a 4D noncommutative (quantum) space-time. The quantum world is really the coset space $SO(2,4)/SO(2,3)$, *i.e.* the hypersurface $\eta_{\mathcal{MN}}X^{\mathcal{M}}X^{\mathcal{N}} = 1$ $[\eta_{\mathcal{MN}} = (-1,1,1,1,1,-1)]$. For the lack of a better terminology, we denote it by 'AdS$_5$'. ‡ With the $SO(2,4)$ relativity, fundamental constants $\hbar$ and G are essentially putting on exactly the same footing as c. §

2. Getting to the Quantum Relativity Symmetry $SO(2,4)$

In summary, we have the Lie algebra sequence of stabilizations and extensions by translations :

$$\begin{aligned} &\boldsymbol{ISO(3)} \rightarrow \boldsymbol{SO(1,3)} \hookrightarrow \boldsymbol{ISO(1,3)} \\ &\rightarrow \boldsymbol{SO(1,4)} \hookrightarrow \boldsymbol{ISO(1,4)} \rightarrow \boldsymbol{SO(2,4)} \end{aligned}$$

The $\boldsymbol{ISO(3)}$ algebra corresponds to the symmetry for 3D space. It is unstable, with $\boldsymbol{SO(1,3)}$ a possible stabilization result. The only other mathematical option is $\boldsymbol{SO(4)}$. It is the physics picture of the deformation parameter being identified as invariant speed that fix the right choice. A linear realization of the $SO(1,3)$ group adopts the 4D Minkowski space-time M^4 over that of 3D space $I\!R^3$ as the arena for the relativity theory; $SO(1,3)$ is the isometry of M^4. With the linear realization comes the natural extension of the Lorentz symmetry to Poincaré symmetry $ISO(1,3)$, by adding the coordinate translations. The latter algebra is again an unstable symmetry, to be stabilized to $\boldsymbol{SO(1,4)}$, with the Planck momentum κc as an energy-momentum invariant. The physics behind the stabilization step as well as the last one is summarized in Table 1.

‡ Note that we used a different metric sign convention in our earlier papers, refs.[2, 3], and partly during this talk presentation. We apologize for the confusion.

§ We do believe the generic noncommutative geometry, as the natural mathematical generalization of the classical/commutative geometry [9], should be the right mathematical setting for quantum gravity. In more exact terms, we believe, or postulate that, *Non-Commutative Geometry is to Quantum Gravity as Non-Euclidean Geometry is to Classical (Einstein) Gravity.* We see our 'AdS$_5$' quantum world as providing an alternative description of the asymptotic flat limit of the 4D noncommutative geometry, somewhat similar to the embedding of a non-Euclidean geometry into a higher dimensional Euclidean geometry.

Table 1. Physics of the Relativity Deformations Summarized:- The first column shows the familiar Galilean to Einstein case. Having an invariant speed c change the velocity from a unconstrained three-vector v^i to a constrained four-vector u^μ, matched to a 4D arena with an extra coordinate ct. Mathematically, the zero commutator between Galilean boosts is deformed to a rotation. The latter two columns similarly summarized the physics of the two further steps towards the quantum relativity of Ref.[2]. For example (2nd column), having an invariant magnitude κc for $p^\mu = \frac{dx^\mu}{d\sigma}$ change the unconstrained four-vector to a constrained five-vector π^A, to be matched to a 5D arena with an extra coordinate $\kappa c\sigma$. The originally zero commutator between two momentum boosts, to be described before imposing the constraint as $\Delta x^\mu = p^\mu \sigma$ on 4D space-time, is deformed to a Lorentz transformation. The introduction of the σ parameter/coordinate and the momentum boosts before the deformation are the key features behind the Poincaré-Snyder relativity introduced in this letter.

$\Delta x^i(t) = v^i \cdot t$	$\Delta x^\mu(\sigma) = p^\mu \cdot \sigma$	$\Delta x^A(\rho) = z^A \cdot \rho$
$\|v^i\| \leq c$ $\eta_{ij} v^i v^j = c^2 \left(1 - \frac{1}{\gamma^2}\right)$	$\sqrt{-\eta_{\mu\nu} p^\mu p^\nu} \leq \kappa c$ $\eta_{\mu\nu} p^\mu p^\nu = -\kappa^2 c^2 \left(1 - \frac{1}{\Gamma^2}\right)$	$\|z^A\| \leq \ell$ $\eta_{AB} z^A z^B = \ell^2 \left(1 + \frac{1}{G^2}\right)$
$M_{0i} \equiv N_i \sim P_i$ $[N_i, N_j] \longrightarrow M_{ij}$	$J_{\mu 4} \equiv O_\mu \sim P_\mu$ $[O_\mu, O_\nu] \longrightarrow M_{\mu\nu}$	$J_{A5} \equiv O'_A \sim P_A$ $[O'_A, O'_B] \longrightarrow J_{AB}$
$\vec{u}^4 = \frac{\gamma}{c}(c, v^i)$ $-\eta_{\mu\nu} u^\mu u^\nu = 1$ $I\!R^3 \rightarrow SO(1,3)/SO(3)$	$\vec{\pi}^5 = \frac{\Gamma}{\kappa c}(p^\mu, \kappa c)$ $\eta_{AB} \pi^A \pi^B = 1$ $M^4 \rightarrow SO(1,4)/SO(1,3)$	$\vec{X}^6 = \frac{G}{\ell}(z^A, \ell)$ $\eta_{\mathcal{MN}} X^{\mathcal{M}} X^{\mathcal{N}} = 1$ $M^5 \rightarrow SO(2,4)/SO(2,3)$

As in the case of ct in going from $I\!R^3$ to M^4, a linear realization asks to have a fifth coordinate $\kappa c\sigma$ be incorporated to form the arena for the $SO(1,4)$ relativity. The σ-coordinate is quite peculiar. It has a space-like geometric signature and a physics dimension of $(\frac{time}{mass})$. It is neither space or time. After all, whether the extra coordinate possibly has the physics meaning of a space-time one is a question of how one formulate the physics theory. For example, taking the metric of the 5D geometry as a gravitational field will be giving σ a space-time (indeed space rather than time) interpretation. The latter kind of ready-to-be-formulated theories will be in conflict with the role of σ as seen from the relativity symmetry stabilization physics, or equivalently for having the right Einstein limit. The linear realization does suggest a radical departure of our physics thinking. It provokes even the question of if we could still formulate dynamics in the way we used to. Anyway, we have again a natural extension of the symmetry algebra to $\boldsymbol{ISO(1,4)}$, and again a stabilization awaits. It may look like such a sequence of stabilization followed by extension will go on forever. However, as illustrated in Table 1, we take the last stabilization as corresponding a 'length' invariant, as terminate the sequence. The idea is that the quantum scale as usually described equally effectively by Planck mass κ and Planck length ℓ assumes $\hbar$. To retrieve $\hbar$ from the symmetry stabilization picture similar to c, one should rather use both Planck momentum and Planck length for the two deformations following the familiar one with c, getting $\hbar$ from the identity $\hbar = \kappa c \ell$. The deformation gives an $\boldsymbol{SO(2,4)}$ algebra. The c constraint enforces the velocity space to be the coset $SO(1,3)/SO(3)$. Similarly, the second deformation curves the momentum space and the last curves the 'quantum space-time' arena itself. Hence, though the Quantum Relativity symmetry $SO(2,4)$ is to be linearly realized as an isometry of a 6D (beyond space-time) geometry, the relevant quantum world is only the hypersurface given by $\eta_{MN} X^M X^N = 1$. Coordinate translations are simply not admissible symmetries. ||

We have the Quantum Relativity symmetry obtained through the the Lie algebra stabilization scheme with quite limited physics inputs. To really construct a theory to be tested experimentally, we need to take it beyond kinematics. The radical beyond space-time picture posts a daunting challenge, as we have to figure out the role of the new σ and ρ coordinates (the u-coordinates) in any picture of dynamics. At the most conventional level of classical physics, dynamics is a study of motion and motion is characterized by change of spatial position with respect to time. The σ and ρ coordinates would have apparently nothing to do with motion then. Obviously, understanding the physics role of σ and ρ will be a key to confront the theoretical challenge ahead.

3. The Poincaré-Snyder Relativity

In order to develop thinking about a plausible formulation of the challenging dynamics of the Quantum Relativity, we first take a step background. We want to work on intermediate limiting theories that are more similar to the familiar one, yet allow up to probe the the new conceptual structures.

In ref.[1], we introduced explicitly the intermediate relativity theories, which are summarized in Table 2. In particular, we focus on the Poincaré-Snyder Relativity the classical and quantum mechanics of which are studied in refs.[1, 10]. The symmetry algebra $\boldsymbol{G(1,3)}$ for the relativity is the Inönü-Wigner contraction limit from $\boldsymbol{ISO(1,4)}$ and $\boldsymbol{SO(2,4)}$. It has, besides the $\boldsymbol{ISO(1,3)}$ generators, generators for the σ-dependent momentum boosts in the commuting limit. It is a Galilean-like symmetry on the 4D Minkowski space-time with an external absolute σ-parameter taking the role of Galilean time. As $dt = \frac{dx}{v}$, we can start thinking about σ through $p^\mu = \frac{dx^\mu}{d\sigma}$. While this change the conventional definition of momentum, it actually allow a standard canonical formulation

|| For the readers who are not familiar with the idea of symmetry stabilization, we present here a simply argument for an easy appreciation of the $\boldsymbol{ISO(3)}$ to $\boldsymbol{SO(1,3)}$ story. Firstly, note that the only difference in the two algebras is the commutator of two velocity boosts. Commutators of Lorentz boost generators are rotation generators, as given by $[N_i, N_j] = -i\,\frac{1}{c^2}\,M_{ij}$, where $\frac{1}{c^2}$ has been explicitly shown. The commuting algebra of Galilean boosts is retrieved at the $\frac{1}{c^2}$ goes to zero limit. The latter is unstable, as taking any small change in value of the zero structural constant changes the algebra. The Lorentz algebra is stable; different values of c give isomorphic algebra connected by a simple scaling. A more direct way is see it is to realize that the mathematics of symmetry algebra sees no units in physics. The value of c depends, though, on our choice of units, we can make it 3×10^{-7} (km ps^{-1}) or 10^{28} (A yr^{-1}). The symmetry of space-time is of course independent of what units physicists cooked up to measure things inside. The value of c tells only when we would be in a regime where the Galilean symmetry, as an approximation of the Lorentz symmetry, is good enough to describe physics.

Table 2. The various relativities – matching the generators : The table matches out the generators for the various relativity symmetries from a pure mathematical point of view. Note as algebras, the mathematical structures of translations (denoted by $P_.$) or the boosts (denoted by $K_.$ and $K'_.$ – the so-called Lorentz boosts not included as they are really space-time rotations) in relation to rotations $J_{..}$ are the same. Algebraically, translation and boost generators are distinguished only by the commutation with the Hamiltonian (H and H'). Successive contractions retrieve $G(1,3)$ and $ISO(1,4)$ from $SO(2,4)$, similar to the more familiar $G(3)$ and $ISO(1,3)$ from $SO(1,4)$. In the physics picture under discussion, however, $SO(1,4)$ part of our so-called Snyder relativity $ISO(1,4)$ is *different* from the usual de-Sitter $SO(1,4)$ contracting to $ISO(1,3)$. We consider simply keeping only the P_μ and $J_{\mu\nu}$ generators to reduce from our Poincaré-Snyder $G(1,3)$ to the Einstein $ISO(1,3)$.

Relativity Symmetry	Quantum $SO(2,4)$	Snyder $ISO(1,4)$	Poincaré-Snyder $G(1,3)$	Einstein $ISO(1,3)$	Galilean $G(3)$
Arena	'AdS$_5$'	M^5	M^4 (with σ)	M^4	$I\!R^3$ (with t)
SO(1,4) part	J_{ij}	J_{ij}	J_{ij}	J_{ij}	J_{ij}
	J_{i0}	J_{i0}	J_{i0}	J_{i0}	K_i
	J_{40}	J_{40}	K'_0	P_0	H
	J_{4i}	J_{4i}	K'_i	P_i	P_i
	J_{50}	P_0	P_0		
	J_{5i}	P_i	P_i		
	J_{54}	P_4	H'		

of Hamiltonian dynamics as σ-evolution. The single particle (σ-)Hamiltonian has the form

$$H' = \frac{1}{2} p_a^\mu p_{a\mu} + V(x^\mu) \ . \tag{1}$$

The new definition for momentum is admissible so long as the physics theory guarantees $p^\mu = m\,c\,u^\mu$, the Einstein energy-momentum at the proper limit. From the latter requirement, one retrieves for such cases $\sigma = \frac{\tau}{m}$, the proper time over the rest mass of an Einstein particle. That is very interesting for a coordinate with a space-like geometric signature indeed. As a solid complement, the straightforward canonical formulation of the classical σ-Hamiltonian dynamics for the free particle case can be reduced to the dynamics of an Einstein particle with $p^\mu = m\,c\,u^\mu$. The classical Poincaré-Snyder dynamics looks hence like a generalization of Einstein particle dynamics which in fact gets around the undesirable situation with the no-interaction theorem [15].

It is interesting to note that parameter having a similar role ro that of σ has been used quite lot in various approaches to (Einstein) relativistic quantum mechanics in somewhat ambiguous ways [11]. A major difference in our new perspective as given from the Poincaré-Snyder Relativity is the idea that σ should not be taken as a parameter providing some measure of time at all. If we want a time-evolution picture, we have to go to a relational framework. That is to say, the σ-dynamics gives definite relation between time, as a σ-dynamics variable, and the other space variables. some examples have been given in ref.[10].

4. Quantum Mechanics as σ-Hamiltonian Dynamics

Aldaya and de Azcárraga [12] presented a particularly nice approach to geometric quantization in which the quantum dynamical description of the system can be obtained with the symmetry group as the basic starting point. In fact, while the approach gives an elegant presentation for the quantization of the Galilean system, its application to the case of Einstein Relativity is less than equally appealing. For the former case, the group to be considered is a $U(1)$ central extension of the symmetry for the corresponding classical system — $G(3)$. The essentially unique nontrivial central extension is depicted by the modified commutator $[K_i, P_j] = m\delta_{ij}\Xi$, where Ξ is a central charge (m the particle mass). The $ISO(1,3)$ symmetry, however, admits no such nontrivial central extension. However, representation for a $U(1)$ central extension is simply equivalent to projective representation as required in quantum mechanics. So, the above looks like a short-coming for Einstein Relativity itself. It is easy to see from Table 2 that the $G(1,3)$ symmetry for the Poincaré-Snyder case has a mathematical structure mostly parallel to that of the Galilean $G(3)$. Hence, we should have the same kind of nontrivial $U(1)$ central extension available for the implementation of such a quantization scheme.

The standard action of $G(1,3)$ on the Minkowski space-time (x^μ) with the extra, external, parameter σ is given by

$$x'^\mu = \Lambda^\mu{}_\nu x^\nu + p^\mu \sigma + A^\mu \,,$$
$$\sigma' = \sigma + b \,. \tag{2}$$

An element of our extended $G(1,3)$ group may be written as $g = (b, A^\mu, p^\mu, \Lambda^\mu{}_\nu, e^{i\theta})$, with group product rule given by

$$A''^\mu = \Lambda'^\mu{}_\nu A^\nu + p'^\mu b + A'^\mu \,, \qquad b'' = b' + b \,,$$
$$p''^\mu = \Lambda'^\mu{}_\nu p^\nu + p'^\mu \,, \qquad \Lambda''^\mu{}_\nu = \Lambda'^\mu{}_\rho \Lambda^\rho{}_\nu \,, \tag{3}$$

and the nontrivial $U(1)$ extension of given by

$$\theta'' = \theta' + \theta + z\left[A'_\mu \Lambda'^\mu{}_\nu p^\nu + b(p'^\mu \Lambda'_\mu{}^\nu p_\nu + \frac{1}{2} p'^\mu p'_\mu)\right] \,. \tag{4}$$

The last term is the cocycle the exact choice of which is arbitrary up to a coboundary [14]; z corresponds to a value of the central charge is taken as an arbitrary constant at this point.

The right-invariant vector fields are given by

$$\mathbf{X}^R_b = \frac{\partial}{\partial b} \,, \qquad \mathbf{X}^R_{A^\mu} = \frac{\partial}{\partial A^\mu} + z p_\mu \frac{i\zeta}{\hbar}\frac{\partial}{\partial \zeta} \,,$$
$$\mathbf{X}^R_{p^\mu} = b\frac{\partial}{\partial A^\mu} + \frac{\partial}{\partial p^\mu} + z b p_\mu \frac{i\zeta}{\hbar}\frac{\partial}{\partial \zeta} \,,$$
$$\mathbf{X}^R_{\omega^{\mu\nu}} = \widetilde{\mathbf{X}}^R_{\omega^{\mu\nu}} + A_\nu \frac{\partial}{\partial A^\mu} - A_\mu \frac{\partial}{\partial A^\nu} + p_\nu \frac{\partial}{\partial p^\mu} - p_\mu \frac{\partial}{\partial p^\nu} \,,$$

$$\mathbf{X}^R_\zeta = \frac{i\zeta}{\hbar}\frac{\partial}{\partial\zeta} , \tag{5}$$

where we skip the details of $\widetilde{\mathbf{X}}^R_{\omega^{\mu\nu}}$, the invariant vector field for the $SO(1,3)$ subgroup [with $\Lambda(\omega) = e^{\frac{-i}{2}\omega^{\mu\nu}J_{\mu\nu}}$] [1]. Note that $\zeta = \exp(\frac{i}{\hbar}\theta)$ with $\frac{i\zeta}{\hbar}\frac{\partial}{\partial\zeta} = \frac{\partial}{\partial\theta}$ locally. The left-invariant vector fields are given by

$$\begin{aligned}
&\mathbf{X}^L_b = \frac{\partial}{\partial b} + p^\mu \frac{\partial}{\partial A^\mu} + \frac{z}{2}p^\mu p_\mu \frac{i\zeta}{\hbar}\frac{\partial}{\partial\zeta} , \\
&\mathbf{X}^L_{A^\mu} = \frac{\partial}{\partial A^\nu}\Lambda^\nu{}_\mu , \qquad \mathbf{X}^L_{p^\mu} = \frac{\partial}{\partial p^\nu}\Lambda^\nu{}_\mu + zA_\nu\Lambda^\nu{}_\mu \frac{i\zeta}{\hbar}\frac{\partial}{\partial\zeta} , \\
&\mathbf{X}^L_{\omega^{\mu\nu}} = \widetilde{\mathbf{X}}^L_{\omega^{\mu\nu}} , \qquad \mathbf{X}^L_\zeta = \frac{i\zeta}{\hbar}\frac{\partial}{\partial\zeta} ;
\end{aligned} \tag{6}$$

again we skip explicit expression for the $SO(1,3)$ vector field $\widetilde{\mathbf{X}}^L_{\omega^{\mu\nu}}$. We have the quantization form given by the left-invariant 1-forms conjugate to $\mathbf{X}^L_\zeta$, the vertical 1-form; explicitly

$$\Theta = -zA^\nu\Lambda_\nu{}^\mu dp_\mu - \frac{z}{2}p^\mu p_\mu db + \frac{\hbar d\zeta}{i\zeta} . \tag{7}$$

The characteristic module is defined through the conditions $i_{\mathbf{X}}\Theta = 0$ and $i_{\mathbf{X}}d\Theta = 0$, characterizing the differential system given by the vector field $\mathbf{X}^L_b$. We have the equations of motion given by

$$\begin{aligned}
&\frac{db}{d\sigma} = 1 , \qquad \frac{dA^\mu}{d\sigma} = p^\mu , \\
&\frac{dp^\mu}{d\sigma} = 0 , \qquad \frac{d\Lambda^\mu{}_\nu}{d\sigma} = 0 \quad \left(\frac{d\omega^{\mu\nu}}{d\sigma} = 0\right) , \\
&\frac{d\zeta}{d\sigma} = \frac{z}{2}p^\mu p_\mu \frac{i\zeta}{\hbar} .
\end{aligned} \tag{8}$$

Identifying the integration parameter as σ gives

$$\begin{aligned}
&b = \sigma , \qquad A^\mu = p^\mu\sigma + K'^\nu\Lambda_\nu{}^\mu , \qquad p^\mu = P^\mu , \\
&\zeta = \zeta_o \exp(\frac{iz}{2\hbar}p^\mu p_\mu \sigma) .
\end{aligned} \tag{9}$$

Naturally, A^μ is to be identified as x^μ giving p^μ as $\frac{dx^\mu}{d\sigma}$, showing consistence with our original introduction of the momentum boosts (see table 1) as the extra symmetry transformations to supplement $SO(1,3)$ and hence getting to $G(1,3)$. We have constants of motion, $K'^\nu\Lambda_\nu{}^\mu$, P^μ, and ζ_o which parameterize the quantum manifold. Passing to the latter, Θ takes the form

$$\Theta_P = -zK^\mu dP_\mu + \frac{\hbar d\zeta_o}{i\zeta_o} . \tag{10}$$

The symplectic 2-form is given by $\omega = d\Theta$. Taking $z = 1$, $H' = \frac{1}{2}p^\mu p_\mu$, and $K'^\mu = A^\nu\Lambda_\nu{}^\mu$ we have

$$\omega = d\Theta_P = -dK'^\mu \wedge dp_\mu , \tag{11}$$

where we have taken $z = 1$ corresponding to $H' = \frac{1}{2}p^\mu p_\mu$, which gives the right form for the classical σ-Hamiltonian [10]. The expression suggested the identification of

(K'^{μ}, b) as particle 'position' variables (x^{μ}, σ) and H' as the σ-Hamiltonian generating 'evolution' in the absolute parameter σ. Note that we also need $z = 1$ for the extraction of the Heisenberg commutation relation (see also ref.[2, 3]). The prequantum operator associated with a real function f on the classical phase space acting on wavefunction ψ is given by

$$\hat{f}\psi \equiv -i\hbar\tilde{X}f \cdot \psi = -i\hbar X_f \cdot \psi + [f - \Theta(X_f)]\psi \,, \tag{12}$$

where $i_{X_f}\omega = -df$. In particular,

$$\begin{aligned} \hat{K'}^{\mu} &= i\hbar\frac{\partial}{\partial P_{\mu}} \,, & \hat{P}_{\mu} &= -i\hbar\frac{\partial}{\partial K'^{\mu}} + P_{\mu} \,, \\ \hat{\sigma} &= i\hbar\frac{\partial}{\partial H'}\sigma \,, & \hat{H}' &= i\hbar\frac{\partial}{\partial \sigma} \,, \end{aligned} \tag{13}$$

where the an extra negative sign is adopted in $\hat{H}'$. The operators K'^{μ} and P_{μ} can also be obtained from $\mathbf{X}^{R}_{A^{\mu}}$ and $\mathbf{X}^{R}_{p^{\mu}}$. The full polarization subalgebra can be taken as spanned by $\{\mathbf{X}^{L}_{b}, \mathbf{X}^{L}_{A^{\mu}}, \mathbf{X}^{L}_{\omega^{\mu\nu}}\}$, giving the momentum space wavefunction $\phi(P_{\mu})$, *i.e.* the wavefunction as dependent only on P_{μ} but not K'^{μ}. We have then simply $\hat{P}_{\mu}\phi(P_{\mu}) = P_{\mu}\phi(P_{\mu})$. $\mathbf{X}^{L}_{b}$ generates the Euler-Lagrange equation

$$i\hbar\partial_{\sigma}\psi + \frac{\hbar^2}{2}\partial_{\mu}\partial^{\mu}\psi = 0 \tag{14}$$

for the Fourier transform ψ of 'momentum' space wavefunction ϕ. Note that $\hat{H}'$ and $\hat{P}_{\mu}$ constitute a complete set of commuting observables for the configuration space wavefunction. The equation expresses an operator form of $H' = \frac{1}{2}p^{\mu}p_{\mu}$ with $\hat{P}_{\mu}$ reduced to just $-i\hbar\frac{\partial}{\partial K'^{\mu}}$, *i.e.* $-i\hbar\frac{\partial}{\partial x'^{\mu}}$. Eq.(14) is of the same form as the so-called (Lorentz) covariant Schrödinger equation studied in the literature[11], except with the parameter σ replacing the proper time τ (or rather τ/m). The equation, with again essentially the proper time as evolution parameter, is also what is obtained in Ref.[13]. One can see that the rest mass, or $m^2/2$ to be exact, of an Einstein particle is just the $\hat{H}'$ eigenvalue. Without considering the spin degree of freedom, the usual (Einstein) relativistic quantum mechanics corresponds to the σ independent eigenvalue equation for $\hat{H}'$, obtainable from Eq.(14) separation of the σ variable from the x^{μ} variables. The eigenvalue equation is the Klein-Gordon equation. Recall that for an Einstein particle, *i.e.* taking the Poincaré-Snyder free particle to the Einstein relativistic limit, we have $\sigma \to \tau/m$.

5. Final remarks

We have outlined a scheme to think about the physics of the Quantum Relativity. We consider the subject a sensible and very interesting theoretical endeavor, though conceptually quite radical in comparison to other approaches in the literature. It demands creative but careful thinking about physics beyond the usual framework. The subject is in a very primitive stage, and the challenge ahead is formidable. We introduced the Poincaré-Snyder Relativity as a limit of the Quantum Relativity at the level of the familiar Einstein Relativity. It gives the perspective that the latter maybe just a part of a bigger symmetry, the full usage of which allows a formulation of a richer machines naturally bypassing the no-interaction theorem. We focus here more of the quantum mechanics. The latter gives a new perspective onto relativistic quantum mechanics in a Lorentz covariant form, with a formal evolution parameter beyond space-time. Further theoretical and maybe experimental investigations of the mechanics will help to shed light on our physics understanding of the evolution parameter σ, which will be very useful in our final task for the dynamics of the Quantum Relativity.

Acknowledgments

Research work of the author is partially supported by the research Grant No. 99-2112-M-008-003-MY3 from the NSC of Taiwan.

References

[1] Kong O C W and Lee H-Y 2009 *Preprint* NCU-HEP-k036, arXiv:0909.4676 [gr-qc]
[2] Kong O C W 2008 *Phys. Lett.* B**665** 58; see also *Preprint* arxiv:0705.0845 [gr-qc] for an earlier version with some different background discussions.
[3] Das A and Kong O C W 2006 *Phys. Rev.* D**73** 124029
[4] See, in particular, discussions in the talk by B.-L. Hu in this conference.
[5] Snyder H S 1947 *Phys. Rev.* **71** 38; see also Yang C N 1947 *Phys. Rev.* **72** 874
[6] Amelino-Camelia G 2001 *Phys. Lett.* B**510** 255; 2002 *Int. J. Mod. Phys.* D**11** 35; Magueijo J and Smolin L 2002 *Phys. Rev. Lett.* **88** 190403; 2003 *Phys. Rev.* D**67** 044017; Kowalski-Glikman J and Smolin L 2004 *Phys. Rev.* D**70** 065020
[7] Kong O C W 2006 *Preprint* NCU-HEP-k027, talk presented at the *Miami 2006 Conf.* (Fort Lauderdale, Florida, USA)
[8] Mendes R V 1994 *J. Phys.* A **27** 8091; Chryssomalakos C and Okon E 2004 *Int. J. Mod. Phys.* D **13** 1817; *ibid.* D **13** 2003
[9] Connes A 1994 *Noncommutative Geometry* (Academic Press)
[10] Kong O C W and Lee H-Y 2010 *Preprint* NCU-HEP-k037, arXiv:1010.3515 [gr-qc]
[11] Kyprianidis A 1987 *Phys. Rep.* **155** 1, and references therein.
[12] Aldaya V and J.A. de Azcárraga J A 1982 *J. Math. Phys.* **23** 1297
[13] Aldaya V and J.A. de Azcárraga J A 1985 *J. Phys.* A **18** 2639
[14] de Azcárraga J A and Izquierdo J M 1995, *Lie Groups, Lie Algebras, Cohomology and Some Applications in Physics* (Cambridge University Press)
[15] See for example, Esposito G, Marmo G, and Sudarshan G 2004 *From Classical to Quantum Mechanics* (Cambridge Universtiy Press)

5th International Workshop DICE2010 IOP Publishing
Journal of Physics: Conference Series **306** (2011) 012059 doi:10.1088/1742-6596/306/1/012059

The Poincaré-Einstein synchronization: historical aspects and new developments

E. Minguzzi

Dipartimento di Matematica Applicata, Università degli Studi di Firenze, Via S. Marta 3, I-50139 Firenze, Italy

E-mail: ettore.minguzzi@unifi.it

Abstract. An historical introduction to the Poincaré-Einstein synchronization and the problem of its transitivity is given. Inspired by cohomology theory I show that it is possible to improve the Poincaré-Einstein synchronization convention so as to obtain a consistent time foliation even in rotating frames for every assigned measure on space. This fact implies that there is, contrary to a widespread belief, a privileged time foliation of spacetime in several interesting cases.

1. Introduction

In this work I am going to investigate how to consistently synchronize clocks at rest in some frame and distributed in a space S. This is an old problem which has several motivations. The presence of a global time is often necessary for legal reasons, for instance if we want to make sense of some contract. It is essential to clarify priorities over some discovery or in order to avoid the mess that would arise from the proliferation of overlapping local times [1]. It a key ingredient to improve the computational efficiency of computer networks, or to time-order the writing-reading access rights of different processes inside one single computer. Other motivations come from quantum mechanics, quantum cosmology and quantum gravity. Here canonical quantization schemes use time dependent Hamiltonians H(t) which are generically not foliation independent. As a consequence, these methods might run into problem in absence of a privileged time foliation. This is one aspect of the so called problem of time of quantum gravity which would be clarified if we could show that some privileged foliation does indeed exist. This is indeed what I will prove in this work although I will leave some details for a different article. I will also use this opportunity to clarify the historical development of several ideas connected with synchronization and its consistency.

The synchronization problem we are considering will make sense in rather general settings. We wish to synchronize distant clocks by making use of signal exchanges (thus we do not consider slow clock transport synchronization). Nevertheless, we do not need to specify the nature of the signal; it could be sound in air, light in vacuum, or an electric signal propagating along copper wires.

For our mathematical framework we need just a set S, whose points are called *clocks*, and a function $a : S \times S \to \mathbb{R}$ which gives the time $a(s_1, s_2)$ needed by the signal to go from point s_1 to point s_2. To *resynchronize* the clocks means to replace the function a with

$$a'(s_1, s_2) = a(s_1, s_2) + c(s_2) - c(s_1)$$

where $c : S \to \mathbb{R}$, and $c(s)$ is how much we have moved forward the clock s in the resynchronization. In this mathematical description is already embodied a condition which we denote $\mathbf{z} = \mathbf{0}$, namely that the time it takes the signal to move from s_1 to s_2 does not depend on the instant of departure but only on the initial and final space points. This condition implies that if two signals are emitted at times t_1 and $t_1 + \Delta t$ from s_1, then they are received at times t_2 and $t_2 + \Delta t$ at s_2 (thus $a(s_1, s_2) = t_2 - t_1$). The condition $\mathbf{z} = \mathbf{0}$ essentially states that clocks s_1 and s_2 go at the same rate, i.e. they are syntonized. Without this condition the problem of synchronization would not make sense as two clock initially synchronized would run out of synch.

We shall impose some additional conditions on the map a. They are, for every $s, s_1, s_2 \in S$,

(i) $a(s_1, s_3) \le a(s_1, s_2) + a(s_2, s_3)$,

(ii) $a(s, s) = 0$,

(iii) $a(s_1, s_2) + a(s_2, s_1) = 0$ only if $s_1 = s_2$.

Condition (i) states that the time it takes the signal to reach s_3 from s_1 is equal or smaller than the time it would take by passing through s_2. This is not really a condition on the propagation of the signal but on our way of defining the time of arrival. Since the events which happen at the clock s_3 are totally ordered by the time of s_3, we use the prenscription that the time of arrival is that of the first signal (according to the order at s_3) that arrives at s_3 after emission at s_1, independently of the path followed from s_1. With this Fermat's type prescription condition (i) is automatically fulfilled. Condition (ii) states that the signal cannot arrive at s before it is emitted, thus it is a causality condition. By the way this condition was already included in the assumption that a has finite values because, if (ii) was violated then by (i) the only possibility could be $a(s, s) = -\infty$. Finally, condition (iii) states that the speed of the signal is finite in the sense that the two-way time between two distinct point is greater than zero (use (i) and (ii)). This condition could have been omitted by passing to a quotient space, but for simplicity we include it.

Let us now define (radar distance)

$$d_r(s_1, s_2) = \frac{1}{2}[a(s_1, s_2) + a(s_2, s_1)].$$

By summing the two equations

$$\begin{aligned} a(s_1, s_3) &\le a(s_1, s_2) + a(s_2, s_3), \\ a(s_3, s_1) &\le a(s_3, s_2) + a(s_2, s_1). \end{aligned}$$

we find easily that d_r is a distance on S, that is it satisfies the triangle inequality

$$d_r(s_1, s_3) \le d_r(s_1, s_3) + d_r(s_1, s_3),$$

and $d_r(s_1, s_2) \ge 0$ with equality if and only if $s_1 = s_2$. Furthermore, the equality holds in the triangle inequality if and only if

$$a(s_1, s_3) = a(s_1, s_2) + a(s_2, s_3), \tag{1}$$

$$a(s_3, s_1) = a(s_3, s_2) + a(s_2, s_1). \tag{2}$$

Before we go on with the description of our model, and our proposed synchronization we make a short historical detour in the most known synchronization method: the Poincaré-Einstein synchronization.

1.1. The Poincaré-Einstein synchronization

The Poincaré-Einstein synchronization establishes that clock s_1 is synchronized with clock s_2 if the time taken by the signal to go from s_1 to s_2 equals the time taken by the signal to go from s_2 to s_1, in formulas

$$a(s_1, s_2) = a(s_2, s_1). \tag{3}$$

Two events are said to be simultaneous with respect to the Poincaré-Einstein convention if the readings of the Poincaré-Einstein synchronized clocks at those events coincide.

In practice in order to synchronize the clocks an observer at s_2 sends a signal towards s_2 where it is reflected back to s_1. Calling t_1, s_1-time of emission; t_2, s_2-time of reflection; and t_1', s_1-time of reception then the observer at s_2 regulates its clock in such a way that

$$t_2 = \frac{1}{2}(t_1' + t_1). \tag{4}$$

Of course this regulation can be made only at posteriori when the observer s_2 receives a further signal from s_1 with the information on the data t_1 and t_1' (the follow up signal).

I am going to prove some equivalent ways of stating the same convention. Suppose to define on S the radar metric, and let s_2 be a point in the middle of s_1 and s_3 which means $d_r(s_1, s_3) = 2d_r(s_1, s_2) = 2d_r(s_2, s_3)$. This implies in particular $d_r(s_1, s_3) = d_r(s_1, s_2) + d_r(s_2, s_3)$ thus Eqs. (1)-(2). Their difference can be rewritten

$$\begin{aligned} a(s_1, s_3) - a(s_3, s_1) &= a(s_1, s_2) + a(s_2, s_3) - [a(s_3, s_2) + a(s_2, s_1)] \\ &= 2[d_r(s_1, s_2) - d_r(s_2, s_3)] - 2[a(s_2, s_1) - a(s_2, s_3)] \\ &= -2[a(s_2, s_1) - a(s_2, s_3)]. \end{aligned}$$

This equation can be read as follows:

> *Suppose that s_1 and s_3 admit a middle point s_2 according to the radar distance. Clocks s_1 and s_3 are synchronized according to Poincaré-Einstein, i.e. $a(s_1, s_3) - a(s_3, s_1) = 0$, if and only if a signal sent from s_2 reaches s_1 and s_3 at the 'same' time, i.e. $a(s_2, s_1) - a(s_2, s_3) = 0$.*

The difference between Eqs. (1)-(2) can also be rewritten

$$\begin{aligned} a(s_1, s_3) - a(s_3, s_1) &= a(s_1, s_2) + a(s_2, s_3) - [a(s_3, s_2) + a(s_2, s_1)] \\ &= -2[d_r(s_1, s_2) - d_r(s_2, s_3)] + 2[a(s_1, s_2) - a(s_3, s_2)] \\ &= 2[a(s_1, s_2) - a(s_3, s_2)]. \end{aligned}$$

which can be read as follows:

> *Suppose that s_1 and s_3 admit a middle point s_2 according to the radar distance. Clocks s_1 and s_3 are synchronized according to Poincaré-Einstein, i.e. $a(s_1, s_3) - a(s_3, s_1) = 0$, if and only if two signals sent from s_3 and s_1 at the 'same' time reach s_2 simultaneously, i.e. $a(s_1, s_2) - a(s_3, s_2) = 0$.*

As pointed out by Jammer [2] the first historical mention of the Poincaré-Einstein synchronization method is contained in Augustine of Hippo's Confessions [3, Book 7, Chp. 6]. Augustine relates a story which would prove the unreliability of astrological divinations. According to it two families established that their babies were born at the same time, although they later had very different fortunes in life. The families established the simultaneity of the births by sending messengers to the other family at the time of birth. According to the story the messengers met middle-way and so they had the proof that the babies were born at the 'same' time.

Of course, in this story our signal is represented by the messengers. The correspondence with the Poincaré-Einstein synchronization is clear once one considers the last of the equivalent definitions given above.

Even before its formal formulation the Poincaré-Einstein synchronization convention was currently used by telegraphers of the middle of the XIX century to take into account delays in telegraphic signals running on cables laid under the Atlantic ocean [1, 4]. At the time accurate time measurements were needed to get longitudes and draw accurate maps. The first formal statement of the convention was given by Poincaré in the 1900 Leiden lecture [5], where he considers a reference frame moving uniformly with respect to the ether and synchronizing its clocks demanding isotropy. In his words

> Let us suppose that there are some observers placed at various points, and they synchronize their clocks using light signals. They attempt to adjust the measured transmission time of the signals, but they are not aware of their common motion, and consequently believe that the signals travel equally fast in both directions. They perform observations of crossing signals, one traveling from A to B, followed by another traveling from B to A. The local time t is the time indicated by the clocks which are so adjusted.

It is clear that for Poincaré this time, called by him *local time*, was fictitious. Indeed, he assumed the existence of the ether [6]. In his talk at the St. Louis conference in 1904 he takes up again this idea [7]

> A sends its signal when its clock marks the hour 0, and the station B perceives it when its clock marks the hour t. The clocks are adjusted if the slowness equal to t represents the duration of the transmission, and to verify it the station B sends in its turn a signal when its clock marks 0; then the station A should perceive it when its clock marks t. The time-pieces are then adjusted.

Finally, Einstein in his 1905 paper [8] gives the same stipulation. Here the obtained time is no more fictitious because he has dismissed the existence of the ether

> We have not defined a common "time" for A and B, for the latter cannot be defined at all unless we establish by definition that the "time" required by light to travel from A to B equals the "time" it requires to travel from B to A. Let a ray of light start at the "A time" t_A from A towards B, let it at the "B time" t_B be reflected at B in the direction of A, and arrive again at A at the "A time" t'_A. In accordance with definition the two clocks synchronize if $t_B - t_A = t'_A - t_B$.

Both recognized the theoretical importance of this synchronization when coming to light signals. As Peter Galison points out [4], Poincaré, as a permanent member of the French Bureau of Longitude, was concerned about accurate maps and delay times in telegraphic transoceanic signals. Einstein, while working at the Swiss patent office in Bern, reviewed patents for synchronizing clocks. We might say that they were at the right place at the right time.

Once clocks have been synchronized one can measure the one-way speed of light. It must be noted that its isotropy is imposed by the way we have synchronized clocks. This observation has lead some authors to a common misconception which can be summarized by the following statement

> The one-way speed of light with its isotropy depends on the synchronization convention adopted, so to say that the speed of light is constant on the frame is devoid of physical content.

Poincaré was a conventionalist and would have perhaps shared this view which, however, is incorrect. In order to clarify this point we need first to to learn more on the consistency of synchronization, because as a matter of fact the Poincaré-Einstein synchronization could not work.

2. The transitivity of synchronization

In this section we wish to answer the following question: Is the Poincaré-Einstein synchronization consistent? Does it really lead to a time foliation? In order to answer this question we need to prove

(a) Clocks once synchronized remain synchronized,

(b1) Reflexivity: clock s_1 is synchronized with itself,

(b2) Symmetry: if clock s_1 is synchronized with clock s_2 then s_2 is synchronized with s_1,

(b3) Transitivity: if clock s_1 is synchronized with clock s_2 and s_2 is synchronized with s_3 then clock s_1 is synchronized with clock s_3.

Condition (a) is essentially the condition $\mathbf{z} = \mathbf{0}$ and thus it is already embodied in our formalism based on the Fermat's a map. Condition (b1) follows from (ii), namely $a(s, s) = 0$. Condition (b2) is a consequence of our definition of synchronization, namely $a(s_1, s_2) = a(s_2, s_1)$. It remains to establish under which conditions (b3) holds.

Unfortunately, in most recent relativity textbooks the approach to this consistency problem is the same as Einstein's (1905) who wrote

> We assume that this definition of synchronism is free from contradictions, and possible for any number of points; and that the following relations are universally valid ... [he writes (b2) and (b3)].

This position ignores completely further developments by Von Laue [9], Silberstein [10], Weyl [11], Reichenbach [12, 13, 14], Macdonald [15] and the author [16, 17].

The fundamental idea is that while the one-way time of propagation relies on the synchronization adopted, the time taken by the light signal over a round-trip does not. Some round-trip condition can therefore guarantee that properties (a),(b1)-(b3) hold. The final result can be expressed with the following theorem known to Reichenbach [12, 13] (see also [14, Chap. 4]) (who however did not introduce in a neat way the condition $\mathbf{z} = \mathbf{0}$) but clearly stated for the first time by Alan Macdonald [15]

Theorem 2.1. *The conditions (a),(b1),(b2) and (b3) hold if and only if the following two experimentally verifiable (as synchronization independent) conditions hold*

$\mathbf{z} = \mathbf{0}$: *Clocks are syntonized, that is, if two signals are sent from s_1 at times t_1 and t'_1 and they are respectively received at times t_2 and t'_2 at s_2 then $t'_1 - t_1 = t'_2 - t_2$.*

$\triangle$: *(Reichenbach's round-trip condition) The time it takes light to cover a triangular path (through reflections over suitable mirrors) does not depend on the direction taken around the triangle.*

Under the condition $\mathbf{z} = \mathbf{0}$ we can apply our initial formalism under which it is possible to restate Reichenbach's condition by introducing the following Sagnac effect function

$$w(s_1, s_2, s_3) = [a(s_1, s_2) + a(s_2, s_3) + a(s_3, s_1)] - [a(s_1, s_3) + a(s_3, s_2) + a(s_2, s_1)]. \tag{5}$$

Then Reichenbach's condition holds if and only if $w = 0$.

Another even older approach based on a different round-trip condition was started by Max Von Laue [9, p.36], and then taken up again by Silberstein [10, Chap. 4], Weyl [11, Sect. 23] and Macdonald and the author [16, 17]. Basically one imposes that

Theorem 2.2. *Let (S, d) be a metric space. The conditions (a),(b1),(b2) and (b3) hold and the one-way speed of light of the so synchronized clocks is a constant (denote it c) throughout the frame if and only if the following condition holds*

L/c *law: the average speed of light over a closed (polygonal) path is a constant (c) independent of the path.*

The condition **L/c** is stronger than Reichenbach's and so are the conclusions that follow from it. Max Von Laue [9], Silberstein [10] and Weyl [11] proved this theorem under the additional tacit assumption that the clocks are syntonized ($\mathbf{z} = \mathbf{0}$), however the condition $\mathbf{z} = \mathbf{0}$ follows already from **L/c** (see [17]).

We can prove that actually the Reichenbach's round-trip condition and the **L/c** law are essentially the same if as distance is used the radar distance.

Theorem 2.3. *Under condition $\mathbf{z} = \mathbf{0}$, let us define on S the radar distance d_r and let us measure the distances with it. Then the Reichenbach's round trip condition holds if and only if the* **L/c** *law holds.*

Proof. It is clear that the **L/c** law implies $\triangle$. Let us prove the converse. Let $s_1 s_2 s_3 \cdots s_k$, $s_k = s_1$ be a polygonal path and let us consider a signal which starts at s_1-time t_0 from s_1 and covers the path finally returning at s_1 at time t_1. Let us consider a second signal which starts at the same instant but covers the path in the opposite direction to finally reach s_1 at time t_2. Moreover, let us consider the similar propagations but this time over the triangulated path $s_1 s_2 s_3 s_1 s_3 s_4 s_1 s_4 s_5 s_1 \cdots s_k$. Let us denote with t_1' and t_2' the corresponding arrival times. We have the identities

$$t_1' - t_0 = t_1 - t_0 + 2[d_r(s_1, s_3) + d_r(s_1, s_4) + d_r(s_1, s_5) + \cdots + d_r(s_1, s_{k-2})] \tag{6}$$

$$t_2' - t_0 = t_2 - t_0 + 2[d_r(s_1, s_3) + d_r(s_1, s_4) + d_r(s_1, s_5) + \cdots + d_r(s_1, s_{k-2})] \tag{7}$$

from which we obtain $t_1' - t_2' = t_1 - t_2$. However, since the path $s_1 s_2 s_3 s_1 s_3 s_4 s_1 s_4 s_5 s_1 \cdots s_k$ is the succession of triangles $s_1 s_2 s_3 s_1$, $s_1 s_3 s_4 s_1$, $s_1 s_4 s_5 s_1 \cdots$ we can also write

$$t_1' - t_2' = w(s_1, s_2, s_3) + w(s_1, s_3, s_4) + w(s_1, s_4, s_5) + \cdots$$

which vanishes because of Reichenbach's condition $w = 0$. We have $t_2 = t_1$ and

$$\begin{aligned} t_1 - t_0 &= \frac{1}{2}[(t_1 - t_0) + (t_2 - t_0)] = \frac{1}{2}[(a(s_1, s_2) + a(s_2, s_3) + \cdots) + (a(s_k, s_{k-1}) + a(s_{k-1}, s_{k-2}) + \cdots)] \\ &= d_r(s_1, s_2) + d_r(s_2, s_3) + d_r(s_3, s_4) + \cdots = L \end{aligned}$$

because the last expression is the length of the polygonal path. Since $t_1 - t_0$ is the time it takes the signal to cover the path, the last equality proves the **L/c** law.

□

Although these theorems are formulated for light signals, it must be remarked that they do not depend on the nature of the signal.

It seems that in the historical and philosophical literature, as far as these developments are concerned, the contribution of Reichenbach has been overestimated (the very informative book by Jammer [2], which has the merit of pointing out several references, is no exception). While it is certainly true that he put more effort than others in the clarification of the many subtle issues raised by special relativity, his explanations were often lengthy and obscure. This criticism was indeed first moved to him by H. Weyl [18]. Rynasiewicz [19] clarifies the relationship between the two scientists as follows

> their common interest in the foundations of relativity theory and geometry kept them in communications during the 1920's. By mid-decade, though, they had had it with one another, exchanging unkind words in print.

The very round-trip condition that goes under Reichenbach's name is probably not his own, indeed in the work "Bericht über eine Axiomatik der Einsteinschen Raum-Zeit-Lehre," Physikalische Zeitschrift, vol. 22 (1921), pp. 6837, reprinted in [14, Chap. 4], after the introduction of the round trip condition $\triangle$, in a footnote he states that

> Einstein has already pointed out the significance of these axioms for synchronization in his lectures.

Reichenbach had attended Einstein's lectures in Berlin and it is known that Einstein gave them between 1914 and 1917. However, as Jammer points out [2, Chap. 7], Einstein had probably already read Max Von Laue book [9] back in 1912 when in the manuscript called "Einstein's 1912 manuscript on the special theory of relativity" he adopts some of his terminologies. Therefore, as far as we know, the credit for the introduction of the round-trip conditions should be given to Max Von Laue.

Coming to our earlier issue, we can say that the content of the naive statement "the speed of light is constant over the frame" is actually "the speed of light is constant over closed paths", and this is indeed all we need for synchronization and what is ultimately checked in experiments. Of course, the condition $\mathbf{L}/\mathbf{c}$ is not satisfied in all frames, this is the case of the rotating platform over which Poincaré-Einstein's synchronization cannot be consistently applied (this lack of transitivity is related to the Sagnac effect). Indeed, a good definition of inertial reference frame could be

Definition 2.4. An *inertial reference frame* is a space of clocks S over which the radar distance is Euclidean and the propagation of light satisfies the $\mathbf{L}/\mathbf{c}$ law.

The idea that the concept of inertial frame should be related to the geometry of light propagation has also been proposed by Martin Strauss [2, Chap. 7].

In summary we have learned in this section that the consistency of the Poincaré-Einstein synchronization method depends on condition $\mathbf{z} = \mathbf{0}$ and on some round-trip condition. While the former is necessary, for otherwise synchronized clocks would run out of synch, the latter does not seems strictly necessary. Of course it is necessary to get transitivity of Poincaré-Einstein synchronization but can we generalize the synchronization method so as to remove this additional assumption? Is there a method that works also when $w \neq 0$, hence for rotating frames? The answer is affirmative and will be given in the next section.

3. The new synchronization method

In the previous section we have defined the function $w(s_1, s_2, s_3)$ as the difference between the time it takes the signal to cover the path $s_1 s_2 s_3 s_1$ and the opposite path $s_1 s_3 s_2 s_1$. The function w is (twice) the Sagnac holonomy and $w = 0$ if and only if Reichenbach's round-trip condition holds. From the given expression for w, Eq. (5), it follows that w is skew-symmetric and satisfies the cocycle condition

$$w(s_2, s_3, s_4) - w(s_3, s_4, s_1) + w(s_4, s_1, s_2) - w(s_1, s_2, s_3) = 0.$$

Our method works if on the space of clocks S it is defined a natural unit measure μ (e.g. S is a Riemannian compact space). In this case, integrating the previous equation we obtain,

$$\delta(s_1, s_2) = \int_S w(s_1, s_2, s) \mathrm{d}\mu(s),$$

is a skew-symmetric function such that

$$w(s_1, s_2, s_3) = \delta(s_1, s_2) + \delta(s_2, s_3) + \delta(s_3, s_1). \tag{8}$$

A skew-symmetric function δ which satisfies the latter equality leads to a consistent synchronization (i.e. reflexive, symmetric and transitive) by setting (with the same interpretation of Eq. (4))

$$t_2 = \frac{1}{2}(t'_1 + t_1) + \frac{1}{2}\delta(s_1, s_2).$$

Indeed, this equation can be rewritten

$$a(s_1, s_2) - a(s_2, s_1) = \delta(s_1, s_2), \tag{9}$$

which replaces Eq. (3). From this expression reflexivity and symmetry follow from the antisymmetry of δ. As for transitivity, let s_1 be synchronized with s_2 and let s_2 be synchronized with s_3 then

$$a(s_1, s_2) - a(s_2, s_1) = \delta(s_1, s_2),$$
$$a(s_2, s_3) - a(s_3, s_2) = \delta(s_2, s_3).$$

Summing the two equations and using Eq. (5)

$$w(s_1, s_2, s_3) - [a(s_3, s_1) - a(s_1, s_3)] = \delta(s_1, s_2) + \delta(s_2, s_3)$$

and using Eq. (8)

$$a(s_1, s_3) - a(s_3, s_1) = \delta(s_1, s_3), \tag{10}$$

that is, clocks s_1 and s_3 are synchronized.

4. Conclusions

I have given an historical account of the origins of the Poincaré-Einstein's synchronization method, and in particular, of the issue of its transitivity. In the last section I have shown that it can be improved so as to work in more general circumstances. The new synchronization method adds a correction to Poincaré-Einstein's which amounts to an average of the Sagnac effect (which is observable). Contrary to Poincaré-Einstein's it is consistent even in absence of round-trip conditions. The new synchronization method reduces to Poincaré-Einstein's for vanishing Sagnac effect (which is the case in which Poincaré-Einstein's convention works). Even more it can be proved that with this choice of synchronization the map $a(s_1, s_2)$ becomes non-negative which means that the so obtained time is non-decreasing over successive causally related events.

As for applications it is not difficult to prove that once applied to the rim of a rotating platform the method gives as simultaneity slices those of the inertial (non-rotating) reference frame. I am presently studying the connection between this time and the one which is usually used in slowly rotating planets (we might call it Kerr time). In general it could be complicated to establish the connection between our proposed time and that displayed in some metrics. This fact does not mean that the method is not useful in practice, on the contrary it is often more practical than a metric approach. For instance it applies readily at the surface of a rotating planet because this surface is equipotential and hence the condition $\mathbf{z} = \mathbf{0}$ holds there. This surface need not be spherical or axisymmetric, the fact that it is equipotential guarantees that the synchronization method can be applied consistently and lead to a time foliation. The method can be readily applied for instance by modifying the synchronization protocols which are used in computational networks. The existence of such reference total order simplifies the problems mentioned in the introduction. In particular this development proves that in several interesting cases, contrary to a widespread belief, the reference frame does indeed posses a privileged time foliation.

Acknowledgments
This work has been partially supported by GNFM of INDAM.

References

[1] Kern S 1983 *The Culture of Time and Space 1880-1918* (Cambridge Massachusetts: Harvard University Press)
[2] Jammer M 2006 *Concepts of simultaneity* (Baltimore: The Johns Hopkins University Press)
[3] St Augustine 1992 *Confessions* World's Classics (Oxford: Oxford University Press) translated by H. Chadwick
[4] Galison P 2003 *Einstein's Clocks, Poincaré's Maps: Empires of Time* (New York: Norton and Company)
[5] Poincaré H 1900 *Archives Néerlandaises de Sciences Exactes et Naturelles, 2iéme série* **5** 252–278 the "Leyden lecture"
[6] Reignier J 2004 *Solvay Workshops and Symposia, Symposium Henri Poincaré* vol 2 ed Gaspard P, Henneaux M and Lambert F pp 175–189
[7] Poincaré H 1904 *Bull. des Sci. Math.* **28** 302–324
[8] Einstein A 1905 *Annalen der Physik* **17** 891–921 reprinted in *The Principle of Relativity*, trans. W. Perrett and G. B. Jeffrey, Dover Publications, New York, 1923.
[9] Laue M 1911 *Das Relativitätsprinzip* (Braunschweig: Friedr. Vieweg & Sohn)
[10] Silberstein L 1914 *The theory of relativity* (London: Macmillan)
[11] Weyl H 1988 *Raum Zeit Materie* (New York: Springer-Verlag) seventh edition based on the fifth German edition (1923)
[12] Reichenbach H 1924 *Axiomatik der relativistichen Raum-Zeit-Lehre* (Braunschweig: F. Vieweg)
[13] Reichenbach H 1969 *Axiomatization of the Theory of Relativity* (Berkeley: University of California Press)
[14] Reichenbach H 2006 *Defending Einstein: Hans Reichenbach writings on space, time, and motion* (Cambridge: Cambridge University Press)
[15] Macdonald A L 1983 *Am. J. Phys.* **51** 795–797
[16] Minguzzi E 2002 *Found. Phys. Lett.* **15** 153–169
[17] Minguzzi E and Macdonald A 2003 *Found. Phys. Lett.* **16** 593–604
[18] Weyl H 1924 *Deutsche Literaturzeitung* **30** 2122–2128
[19] Rynasiewicz R 2005 *Weyl vs. Reichenbach on Lichtgeometrie* (Boston: The Center for Einstein Studies) chap 9, pp 137–156

Journal of Physics: Conference Series **306** (2011) 012060 doi:10.1088/1742-6596/306/1/012060

Dynamics of Unruh-DeWitt detectors in a relativistic quantum field

Shih-Yuin Lin

Department of Physics, National Changhua University of Educatoin, Changhua 50007, Taiwan

E-mail: sylin@cc.ncue.edu.tw

Abstract. Based on our results for the detector-field system in Gaussian states, I am addressing the following issues about the quantum dynamics of atoms or detectors moving in a relativistic quantum field: First, the dynamics of the detectors as an open quantum quantum system, hence the entanglement creation and sudden death processes between the detectors, are non-Markovian in general. Second, the excitations of the accelerated detectors in the Unruh effect depend only on the kinematics of the detectors. The event horizon for the detector is not essential in this effect, and the particle notion of the quantum field is not needed here. Third, entanglement dynamics of the detectors are independent of time-slicing scheme. Local projective measurements on point-like detectors are also consistent in different time-slicing schemes even in the presence of relativistic quantum fields.

1. Introduction

Recent years have seen growing interest in relativistic quantum informaion (RQI) science, in which people are applying new knowledge from quantum information science to relativistic systems, while using relativity to examine the consistency of quantum information. The most intriguing notion of the RQI is the tension between relativistic locality and quantum nonlocality. Such a tension manifests when performing a quantum measurement on two spacelike seperated, spatially localized but quantum entangled objects. The sudden transition of quantum entangled state at a distance ("wave-function collapse") before and after measurement appears to violate the principles of relativity.

To study this notion, one of the simplest tools is the Unruh-DeWitt detector model [1, 2, 3]. In (3+1)D Minkowski space, it is described by the action

$$S = -\frac{1}{2}\int d^4x\sqrt{-g}\partial_\alpha\Phi\partial^\alpha\Phi + \frac{1}{2}\sum_\Gamma\left\{\int d\tau_\Gamma\left[(\partial_\Gamma Q_\Gamma)^2 - \omega_\Gamma^2 Q_\Gamma^2\right] + \lambda\int d\tau_\Gamma Q_\Gamma(\tau_\Gamma)\int d^4x\Phi(x)\delta^4(x^\alpha - z_\Gamma^\alpha(\tau_\Gamma))\right\}, \quad (1)$$

which consists of harmonic oscillators (HO) Q_Γ as internal degrees of freedom of point-like objects in space, called the detectors, moving along prescribed trajectories $z_\Gamma^\alpha(\tau_\Gamma)$ in a massless scalar field Φ. Here $\alpha = 0, 1, 2, 3$, Γ is labelling the detectors: $\Gamma = A$ for one-detector case, $\Gamma = A, B$ for two-detector case, and $\partial_\Gamma \equiv d/d\tau_\Gamma$. This simple atom-field interacting system is explicitly covariant. It is a linear system, so the solutions once obtained are crystal clear. Indeed, in simple cases analytic results have been obtained in the whole parameter range without any approximation. Still it is complicated enough to give nontrivial results and insights. Note that, with the point-like detectors and the retarded field sourced from them present in spacetime, the field state does not respect the full Lorentz symmetry.

Below I am discussing three issues in the quantum dynamics of atoms or detectors moving in a relativistic quantum field: (i) Non-Markovian dynamics of the detector as an open quantum system, (ii) relations between the Unruh effect and the event horizon, and (iii) the scheme independence of time-slicing for the reduced dynamics of the detectors and the projective measurement on the detectors. Although most of our insights are coming from specific models and the calculations are restricted for Gaussian states, I expect the following discussions would be valid for more general cases.

2. Non-Markovian dynamics in open system

For simplicity, the initial state in our model is usually taken as a factorized state,

$$| \psi(0) \rangle = | q \rangle \otimes | 0_M \rangle , \tag{2}$$

which is a direct product of a Gaussian state of the detector $| q \rangle$ and the Minkowski vacuum of the field $| 0_M \rangle$, which is also a Gaussian state in Schrödinger picture and is the no-particle (field quanta) state for an inertial observer in Minkowski space. The quantum state (2) is not a stationary state of the detectors -field interacting system. Thus the solution of the Schrödinger equation with this initial state will be time-varying, and there are interesting dynamics after the combined system started to evolve.

One can express a Gaussian state of the combined system in the (K, Δ)-representation of the Wigner functional [11] as

$$\begin{aligned} \rho(K, \Delta) &= \int \mathcal{D}\Sigma e^{iK\cdot\Sigma/\hbar} \psi\left[\Sigma - (\Delta/2)\right] \psi^*\left[\Sigma + (\Delta/2)\right] \\ &= N \exp -\frac{1}{2\hbar}\left[K_\mu \mathcal{Q}^{\mu\nu} K_\nu - \Delta_\mu \mathcal{R}^{\mu\nu} K_\nu + \Delta_\mu \mathcal{P}^{\mu\nu} \Delta_\nu\right] \end{aligned} \tag{3}$$

where the indices μ, ν are running over $\{\Gamma\}$ and $\{\mathbf{x}\}$, N is the normalization factor, and the time-dependent factors $\mathcal{Q}^{\mu\nu}$, $\mathcal{P}^{\mu\nu}$, and $\mathcal{R}^{\mu\nu}$ are completely determined by the symmetric two-point correlators $\langle A, B \rangle \equiv \langle AB + BA \rangle /2$ as

$$\langle \Phi_\mu, \Phi_\nu \rangle = \left[\frac{\hbar\delta}{i\delta K_\mu}\frac{\hbar\delta}{i\delta K_\nu}\rho(K,\Delta)\right]_{\Delta=K=0} = \hbar\mathcal{Q}^{\mu\nu}, \tag{4}$$

$$\langle \Pi_\mu, \Pi_\nu \rangle = \left[\frac{i\hbar\delta}{\delta\Delta_\mu}\frac{i\hbar\delta}{\delta\Delta_\nu}\rho(K,\Delta)\right]_{\Delta=K=0} = \hbar\mathcal{P}^{\mu\nu}, \tag{5}$$

$$\langle \Pi_\mu, \Phi_\nu \rangle = \left[\frac{i\hbar\delta}{\delta\Delta_\mu}\frac{\hbar\delta}{i\delta K_\nu}\rho(K,\Delta)\right]_{\Delta=K=0} = \frac{\hbar}{2}\mathcal{R}^{\mu\nu}, \tag{6}$$

where Π_μ are conjugate momenta to Φ_μ, and we have written $\Phi_\Gamma \equiv Q_\Gamma$ and $\Pi_\Gamma \equiv P_\Gamma$ for convenience. Because of the linearity of the Hamiltonian derived from the action (1), a Gaussian state of the combined system will always be a Gaussian state during the evolution, which can always be fully determined by the two-point correlators. So it is sufficient to calculate the evolution of the the two-point correlators, either in Schödinger picture or in Heisenberg picture, to see how the quantum state evolves.

In Fig. 1 we illustrate the time-evolution of the self correlator $\langle Q^2 \rangle$ of a single uniformly accelerated detector with the initial state $| q \rangle$ being its ground state. One can see that the value

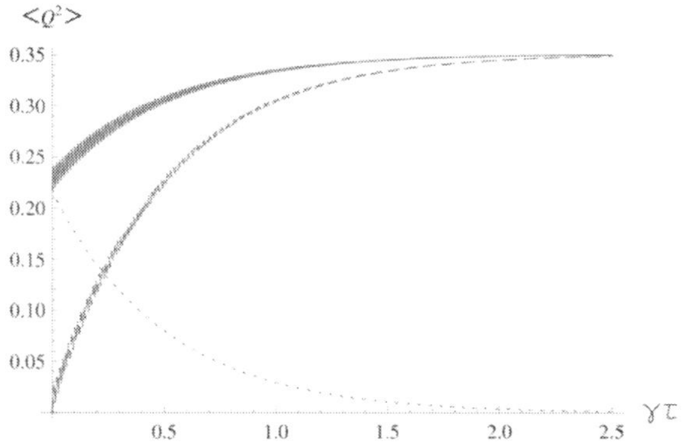

Figure 1. Evolution of the two-point correlator $\langle\, Q^2(\tau)\,\rangle$ of a single uniformly accelerated UD detector (solid curve). The initial state (2) at $\tau = 0$ is a direct product of the ground state of the detector and the Minkowski vacuum of the field. The dotted curve represents how the zero-point fluctuations of the detector dissipate to the field, the dashed curve represents the response of the detector driven by vacuum fluctuations of the field, and the solid curve is the sum of these two. The small oscillation on top of the growing solid curve is due to the sudden switching-on of the coupling. Here $\gamma = 0.01$, $\Omega = 2.3$, $\Lambda_0 = 20$, $a = 10$, and τ is the proper time of the detector.

of $\langle\, Q^2\,\rangle$ evolves from a value for the ground state of a HO, $\hbar/2\Omega m$ (≈ 0.2174 in Fig. 1) with the mass $m = 1$ and the Planck constant $\hbar = 1$, to a value for a HO in equilibrium with the heat bath at the Unruh temperature $T_U = a/2\pi$. In the weak-coupling limit $\gamma\Lambda_0 \ll a, \Omega$, where $\gamma \equiv \lambda^2/8\pi m$ is the strength of coupling and Λ_0 is a constant corresponding to the time-scale of switching-on the coupling [4], the latter value is approximately $\hbar \coth(\pi\Omega/a)/2\Omega m$ (≈ 0.3515 in Fig. 1). At early times ($\gamma(\tau - \tau_0) \ll 1$ but $\tau - \tau_0 \gg a^{-1}$ with τ_0 the initial moment, roughly in the interval $0.01 < \gamma\tau < 0.2$ in Fig. 1, where $\gamma = 0.01$ is not quite weak to see its effect in the plot, though), the value of $\langle\, Q^2\,\rangle$ grows almost linearly in time in the weak-coupling limit. This is the regime where the conventional result, the transition probability

$$P_{0\to 1} = \frac{\hbar}{4\pi m}\frac{\tau - \tau_0}{e^{2\pi\Omega_r/a} - 1}\,, \tag{7}$$

from the ground state of the detector to its first excited state by the Unruh effect, is valid [5]. Indeed, this transition probability is obtained by time-dependent perturbation theory (TDPT), which describes the transient behavior only at early times of the evolution and only in perturbative regime.

Dynamics of the correlators of two detectors, hence the dynamics of quantum entanglement between them, are non-Markovian in general. One origin of the non-Markovian property is the correlations in the field state at zero or low temperature. If the field is not initially at high temperature and the coupling of the field and the detectors are not weak enough, then such long-range correlations will manifest in the dynamics. The most obvious case is observed in the Rob-AntiRob problem [8], where the cross correlators started to evolve at initial moment $-t_0$ will have a sudden change of behavior at $+t_0$, just like they have the memory of the initial moment (see Fig. 2).

The other origin of the non-Markovian or memory effect is the mutual influences mediated by the retarded field sourced from one point-like detector then hitting the other detector [6, 7]. The influenced detectors then generate the secondary or higher-order back reactions to the field, which propagate back and forth between the detectors. Note that these retarded mutual influences are of quantum origin and explicitly causal.

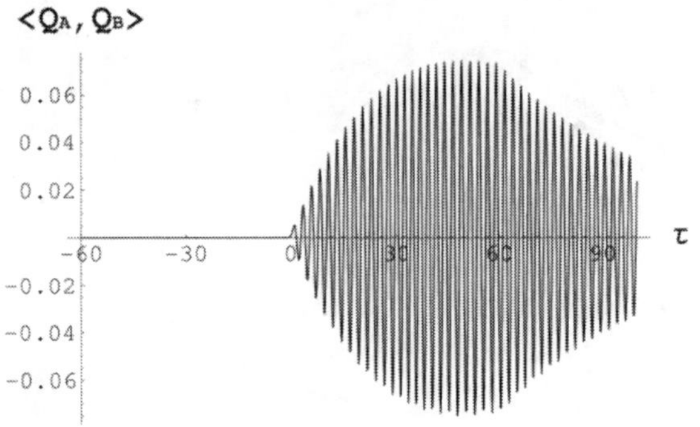

Figure 2. Evolution of the cross correlator $\langle Q_A, Q_B \rangle$ for a pair of back-to-back uniformly accelerated detectors A and B initially in their ground state (For more details, see Ref. [8].) Here $\gamma = 0.01$, $\Omega = 1.3$, $a = 2$, $\hbar = 1$, and the coupling was started at $\tau_0 = -60$. One can see that when $\tau \approx -\tau_0 = 60$, the bahavior of $\langle Q_A, Q_B \rangle$ shows a transition.

Non-Markovian property from both origins can decrease, increase, eliminate or create entanglement, which is measured by the logarithmic negativity E_N or the quantity Σ [7]. For example, quantum entanglement will have a sudden death very quickly after the coupling of the field and the detectors is turned on in strong coupling regime [6]. In other examples, entanglement between two spacelike separated detectors initially in a separable state can be created either by the field correlations they experienced in specific motion (the Rob-AntiRob problem in [8]), or by the mutual influences between two detectors at rest and sitting very close to each other [7].

Of course, such entanglement generation between two detectors may not improve the fidelity of quantum teleportation using these detectors. Suppose Alice and Bob share an entangled pair of detectors, and Alice want to teleport the quantum state of Charlie's detector to Bob. Recall that to perform a successful quantum teleportation, Bob needs the knowledge of the initial state to do the correct operation (post-selection) on his detector (or qubit) according to the outcome of the Gaussian (or Bell) measurement on Alice and Charlie's detectors. Once the quantum state of the Alice-Bob pair deflects from the initial state, Bob's ability of reproducing Charlie's quantum state (i.e. the fidelity of quantum teleportation) will decrease, no matter quantum entanglement between Alice and Bob's detectors (or qubits) right before Alice's measurement was increased or decreased compared with the initial state.

3. Event Horizon and Unruh Effect

There are mainly two approaches to the Unruh effect in literature [1, 2, 3]. In the first approach the Minkowski vacuum of a quantum field is re-expressed as a two-mode squeezed state in Rindler space. When the degrees of freedom behind the event horizon for an observer living in one of the left and right Rindler wedges are traced out, the field state in the observer's wedge will appear to be a thermal state of Rindler particle at certain Unruh temperature. In the other approach one applies the TDPT to see a uniformly accelerated Unruh-DeWitt detector initially in the ground state can be excited by the Minkowski vacuum.

Event horizon is essential in the former approach, but not in the latter. Indeed, in [9], we calculated the deviation from the self correlator $\langle Q^2(\tau) \rangle_{\rm v} |_{T_U=0}$ for inertial UD detectors in the Minkowski vacuum,

$$\delta \langle Q^2(\tau) \rangle \equiv \lim_{\tau' \to \tau} \left[\langle Q(\tau)Q(\tau') \rangle - \langle Q(\tau)Q(\tau') \rangle_{T_U=0} \right] \tag{8}$$

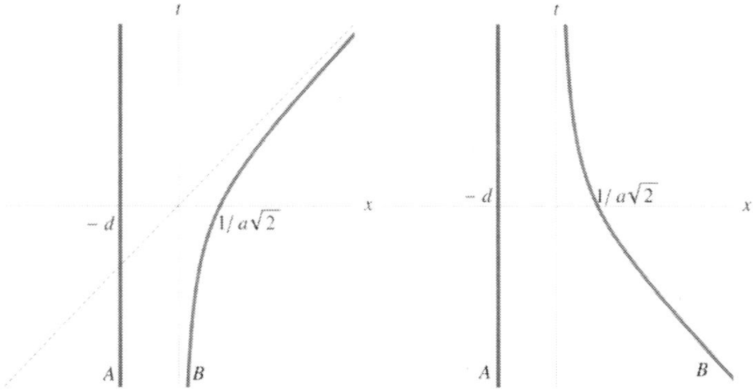

Figure 3. Two setups of the trajectories of the detectors A and B in non-uniform acceleration in [9]. The detector A is at rest in both setups. In the left plot, the detector B is almost at rest when $t < 0$, and almost uniformly accelerated when $t > 0$. The dashed line is the event horizon for the detector B. In the right plot, the detector B is linearly accelerated when $t < 0$ and becomes inertial when $t > 0$, so there is no event horizon for the detector B.

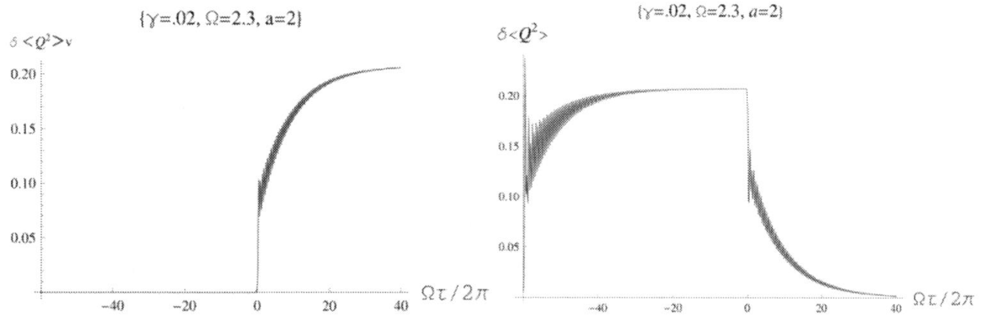

Figure 4. Evolutions of $\delta \langle Q^2(\tau) \rangle$ defined in (8) for the detector B in the left and the right plots in Fig. 3, respectively.

by numerical method. We found that during each session of almost-uniform acceleration with some constant instantaneous proper acceleration, the self correlators of the accelerated detector behave the same as those uniformly accelerated in the same proper acceleration, as illustrated in Fig. 4. As we showed in [5], the correlators of a UD detector can be related to the transition probability of excitation in TDPT. So our results imply that the excitations in the detector moving in the Minkowski vacuum at some proper acceleration will look the same as those due to the Unruh effect.

If the acceleration of the detector eventually vanishes, there will be no event horizon for the detector, but those excitations in the detector driven by vacuum fluctuations of the field will still occur at early times when the detector is almost-uniformly accelerated. Therefore it is the kinematics of the detector rather than the global structure of spacetime relevant to the excitations in the detector.

One may wonder, without the event horizon, how can one incorporate the notion of particles such as the Rindler particles of quantum fields in a non-inertial frame? The answer is simple. Recall that the concept of particle of a quantum field is not fundamental and sometimes not well-defined in curved space [3], and even in inertial frame, the particle notion of the field is not clear in the non-perturbative regime. In both cases one may consider the transitions between the neighboring energy levels of an atom or a detector as an operational definition of the particle (field quanta), but it works only in the TDPT regime. Off the TDPT regime and the asymptotic region, the particle notion of the field will not be well defined and one should look at the overall response of the detector rather than the transitions between the neighboring energy levels.

4. Time-slicing scheme independence

Since the UD detectors are point-like objects in space, quantum entanglement of the detectors at each single moment is independent of the time-slicing scheme we are using in calculation to specify that moment. Those quantum fields $\Phi_{\mathbf{x}}$ living on different choices of time slices $(t, \mathbf{x})$ but passing through the same two spacelike separated events of the detectors will be traced out to give the same reduced density matrix (RDM) of the detectors at the time associated with that time slice, namely,

$$\rho_R(Q, Q'; t) = \mathrm{Tr}_\Phi \rho(Q, \Phi_{\mathbf{x}}, Q', \Phi'_{\mathbf{x}}; t) \equiv \int \mathcal{D}\Phi_{\mathbf{x}} \rho(Q, \Phi_{\mathbf{x}}, Q', \Phi_{\mathbf{x}}; t) \ , \tag{9}$$

where ρ is the density matrix of the combined system. Here the RDM ρ_R of the detector can be completely determined by the two-point correlators of the detectors by virtue of the Gaussianity. Now the RDM is not explicitly dependent on the data over the time slice $(t, \mathbf{x})$ except those right at the position of the detectors. From these RDM one obtains the same correlators of the detectors at the moment t from different time-slicing schemes, which give the same measure of quantum entanglment at that moment.

Note that, as we have observed in [6], the evolution of the two-point correlators of the detectors in one coordinate (e.g. $\langle\, Q_A(\tau_A(t)) Q_B(\tau_B(t))\,\rangle$ in t) could be quite different from the ones in a different coordinate (e.g. $\langle\, Q_A(\tau_A(t')) Q_B(\tau_B(t'))\,\rangle$ in t'), since τ_A and τ_B evolve in different ways in different coordinates. So entanglement dynamics of the detectors in a time interval do depend on the coordinate of the observer.

A more interesting question is, whether the projective measurement (or wave function collapse) is consistent in different time-slicing scheme. For example, if we choose an alternative coordinate in Minkowski space,

$$\eta = t - A \sin t \cos x \ , \tag{10}$$

$$\xi = x - A \sin x \cos t \ , \tag{11}$$

with a constant $A < 1$. Then

$$ds^2 = -dt^2 + dx^2 = \gamma(\eta, \xi) \left(-d\eta^2 + d\xi^2\right) \ , \tag{12}$$

where

$$\gamma(\eta, \xi) = \left[1 - 2A \cos t \cos x + A^2 \cos(t + x) \cos(t - x)\right]^{-1} \tag{13}$$

with $t = t(\eta, \xi)$ and $x = x(\eta, \xi)$ according to (10) and (11). η-slices and t-slices will overlap whenever $t = n\pi$ with integer n. Off these time-slices, η-slices are the wavy ones in Fig. 5, where t-slices would be the horizontal straightlines. Now, the wave functional of a quantum field in Minkowski space is defined on the whole time-slice at each moment. Suppose the projective measurement works in both frames, and a local measurement is done on a point-like detector coupled with the field at some moment $t_1 \neq n\pi$, then which time-slice, t_1-slice or η_1-slice, would the wave functional of the combined system collapse onto? If both happen in their coordinates, will they produce different post-measurement states?

An explicit calculation in a Raine-Sciama-Grove model [12], which is an Unruh-DeWitt like detector theory in (1+1) dimensional Minkowski space, has been done in [10] where the case with a detector A at rest at $x = \xi = 0$ was considered. Suppose the combined system is initially

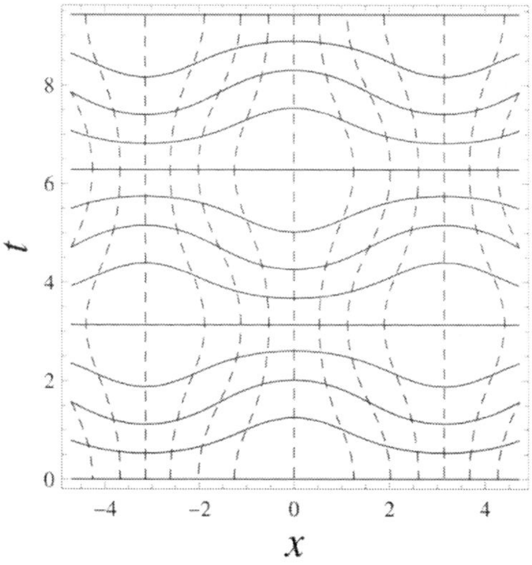

Figure 5. An alternative coordinate (η, ξ) of Minkowski space given by (10) and (11) with $A = 1/2$. The solid curves are constant η slices while the dashed curves are constant ξ hypersurfaces.

in a Gaussian state at $t = \eta = 0$, and at some moment $t = t_1$, $\eta_1 \equiv \eta(t_1) \in (0, \pi)$, a projective Gaussian measurement collapses the wave functional of the combined system to a new Gaussian state,

$$\tilde{\rho} = \rho_A \otimes \rho_\Phi, \tag{14}$$

instantaneously on the whole t_1-slice or η_1-slice, depending on which coordinate the observer is using in Minkowski space. Here the detector state ρ_A after measurement is supposed to be a squeezed Gaussian state and the field is also in a Gaussian state ρ_Φ which is derived according to ρ_A. After t_1 or η_1, the post-measurement state (14) keeps evolving to, say, $t = t_2 = \pi$ and $\eta = \eta_2 = \pi$, when t and η-slices overlap so we can make a direct comparison between these two quantum states of the combined system collapsed over two different time-slices by the same local Gaussian measurement on the detector.

In the conventional coordinate (t, x) of Minkowski space, the two-point correlators at $t = t_2$ determining the wave functional (3) can be expressed as combinations of the mode-functions evolving from t_1 to t_2, together with the initial data on the t_1-slice in the form of the correlators of the field at space points on the slice. For instance,

$$\begin{aligned} \langle\, \hat{\Phi}_x, \hat{\Phi}_y \,\rangle_\pi &= \mathrm{Tr}\hat{\Phi}_x(t_2 - t_1)\hat{\Phi}_y(t_2 - t_1)\tilde{\rho}(\Phi, \Pi; t_1) \\ &\sim \int dx' dy' \phi_x^{x'}(t_2 - t_1)\phi_y^{y'}(t_2 - t_1)\,\langle\, \hat{\Phi}_{x'}, \hat{\Phi}_{y'} \,\rangle_{t_1} + \cdots \end{aligned} \tag{15}$$

where x' and y' are points on the t_1-slice, and $\phi_x^{x'}(t_2 - t_1)$ is a real mode-function in the expansion

$$\hat{\Phi}_x(t) = \sum_{\Gamma'} \left[\phi_x^{\Gamma'}(t)\hat{Q}_{\Gamma'} + \Phi_x^{\Gamma'}(t)\hat{P}_{\Gamma'}\right] + \int dx' \left[\phi_x^{x'}(t)\hat{\Phi}_{x'} + \Phi_x^{x'}(t)\hat{\Pi}_{x'}\right], \tag{16}$$

$$\hat{Q}_\Gamma(\tau_\Gamma) = \sum_{\Gamma'} \left[\chi_\Gamma^{\Gamma'}(\tau_\Gamma)\hat{Q}_{\Gamma'} + X_\Gamma^{\Gamma'}(\tau_\Gamma)\hat{P}_\Gamma\right] + \int dx' \left[\chi_\Gamma^{x'}(\tau_\Gamma)\hat{\Phi}_{x'} + X_\Gamma^{x'}(\tau_\Gamma)\hat{\Pi}_{x'}\right]. \tag{17}$$

According to the equations of motion for $\phi_x^{x'}(t_2 - t_1)$, it can be interpreted as a retarded field $\phi(t_2, x)$ sourced from the point-like detector driven by the vacuum fluctuations at (t_1, x'), which is a point on the t_1-slice. In the alternative coordinate (η, ξ), the form of the correlator is similar, except that there η_1-slice is different from t_1-slice. Now the two-point correlators, or equivalently, the wave functional at $t_2 = \eta_2$, appear to depend on the time-slicing scheme.

Nevertheless, the dependence of the data on the t_1-slice or η_1-slice can be removed. If the measurement at t_1 or η_1 was absent, one would have

$$\langle\, \hat{\Phi}_x, \hat{\Phi}_y \,\rangle_{t_2} = \langle\, \hat{\Phi}_x(t_{20}), \hat{\Phi}_y(t_{20}) \,\rangle_{t_0} = \phi_x^A(t_{20})\phi_y^A(t_{20}) \,\langle\, \hat{Q}_A^2 \,\rangle_{t_0} + \cdots, \tag{18}$$

with $t_{mn} \equiv t_m - t_n$, while

$$\begin{aligned} \langle\, \hat{\Phi}_x, \hat{\Phi}_y \,\rangle_{t_2} &= \langle\, \hat{\Phi}_x(t_{21}), \hat{\Phi}_y(t_{21}) \,\rangle_{t_1} = \phi_x^A(t_{21})\phi_y^A(t_{21}) \,\langle\, \hat{Q}_A^2 \,\rangle_{t_1} + \cdots \\ &= \phi_x^A(t_{21})\phi_y^A(t_{21}) \,\langle\, \hat{Q}_A^2(t_{10}) \,\rangle_{t_0} + \cdots \\ &= \phi_x^A(t_{21})\phi_y^A(t_{21})\chi_A^A(t_{10})\chi_A^A(t_{10}) \,\langle\, \hat{Q}_A^2 \,\rangle_{t_0} + \cdots \end{aligned} \tag{19}$$

is also true. A comparison between the above two expressions implies a set of mathematical identities such as

$$\begin{aligned} \phi_x^\mu(t_{20}) &= \sum_{\Gamma'} \left[\phi_x^{\Gamma'}(t_{21})\chi_{\Gamma'}^\mu(t_{10}) + \Phi_x^{\Gamma'}(t_{21}) p_{\Gamma'}^\mu(t_{10}) \right] + \\ &\quad \int dx' \left[\phi_x^{x'}(t_{21})\phi_{x'}^\mu(t_{10}) + \Phi_x^{x'}(t_{21})\pi_{x'}^\mu(t_{10}) \right], \end{aligned} \tag{20}$$

in this initial value problem, where $p_\Gamma^\mu(\tau_\Gamma(t))$, $P_\Gamma^\mu(\tau_\Gamma(t))$, $\pi_x^\mu(t)$, and $\Pi_x^\mu(t)$ are conjugate variables to the mode-functions χ_Γ^μ, X_Γ^μ, ϕ_x^μ, and Φ_x^μ, respectively. These identities can be interpreted as the Huygens principle of the mode functions, and can be verified explicitly by inserting the solutions for those mode-functions into the identities.

It turns out that all the dependence in the wave functional at $t = \eta = \pi$ on the data living on the t_1-slice or η_1-slice, namely, those $\int dx'$ terms, can be replaced using these identities, e.g., one has $\int dx'[\cdots] = \phi_x^\mu(t_{20}) - \sum_{\Gamma'}[\cdots]$ in (20). Then the wave functional at $t = \eta = \pi$ can be expressed as a functional of the functions representing the point-like detector as well as the retarded field sourced from the detector. All these objects are explicitly covariant, so that the two wave functionals in different coordinates and collapsed on different time-slices can be identified with each other by a coordinate transformation at $t_2 = \eta_2 = \pi$. This means that these two quantum states after measurement are actually the same, and the projective measurement is consistent in the detector theory even in the presence of relativistic quantum fields.

This consistency still holds if the initial Gaussian state is a mixed state, and if one has two or more successive Gaussian measurements on the detector before comparison. It will be interesting to see if this consistency still hold for non-Gaussian states, non-Gaussian measurements, or in nonlinear systems.

Acknowledgments
I would like to thank Bei-Lok Hu for his consistent support and illuminated discussions. This work is supported by the National Science Council of Taiwan under the grant NSC 99-2112-M-018-001-MY3 and in part by the National Center for Theoretical Sciences, Taiwan.

References

[1] Unruh W G 1976 *Phys. Rev.* D **14** 870
[2] DeWitt B S 1979 *General Relativity: an Einstein Centenary Survey* ed S W Hawking and W Israel (Cambridge: Cambridge University Press)

[3] Birrell N D and Davies P C W 1982 *Quantum Fields in Curved Space* (Cambridge: Cambridge University Press)
[4] Lin S-Y and Hu B L 2006 *Phys. Rev.* D **73** 124018
[5] Lin S-Y and Hu B L 2007 *Phys. Rev.* D **76** 064008
[6] Lin S-Y, Chou C-H and Hu B L 2008 *Phys. Rev.* D **78** 125025
[7] Lin S-Y and Hu B L 2009 *Phys. Rev.* D **79** 085020
[8] Lin S-Y and Hu B L 2010 *Phys. Rev.* D **81** 045019
[9] Ostapchuk D C M, Lin S-Y, Mann R B and Hu B L 2011 Entanglement Dynamics with Non-Uniformly Accelerated Detectors *in preparation*
[10] Lin S-Y 2011 Is Projective Measurement Consistent in a Relativistic Quantum Field? *Preprint* arXiv:1104.0772
[11] Unruh W G and Zurek W H 1989 *Phys. Rev.* D **40** 1071
[12] Raine D J, Sciama D W and Grove P G *Proc. R. Soc. Lond.* A **435** 205

Journal of Physics: Conference Series **306** (2011) 012061 doi:10.1088/1742-6596/306/1/012061

Representations of the Dirac wave function in a curved spacetime

Mayeul Arminjon [1,2] **and Frank Reifler**[3]

[1] CNRS (Section of Theoretical Physics), France

[2] Laboratory "Soils, Solids, Structures, Risks" (CNRS, UJF, and G-INP), Grenoble, France

[3] Lockheed Martin Corporation, MS2 Division, Moorestown, New Jersey, USA

E-mail: `arminjon@hmg.inpg.fr`

E-mail: `frank.j.reifler@lmco.com`

Abstract. The Dirac wave function in a curved spacetime is usually defined as a quadruplet of scalar fields. It can alternatively be defined as a four-vector field. We describe these two representations in a common geometrical framework and we prove theorems that relate together the different representations and the different choices of connections. In particular, the standard Dirac equation in a curved spacetime, with any choice of the tetrad field, is equivalent to a particular realization of the Dirac equation for a vector wave function, in the same spacetime.

1. Introduction

To our knowledge, the observations of quantum-mechanical properties of matter in the classical gravitational field are the only direct experimental proof of the interplay between gravity and the quantum. Such observations are obtained e.g. by cold neutron interferometry in a plane whose inclination can be varied [1], or by measuring the transmission of ultra-cold neutrons through a thin horizontal slit [2]. Since neutrons are spin $\frac{1}{2}$ particles, this motivates work on the curved spacetime Dirac equation.

Weyl [3] and Fock [4] recognized that, on changing the coordinates, transforming the Dirac wave function under the spin group is not an option in a curved spacetime, or already in a flat spacetime with affine coordinates. They proposed what has become the standard version of the Dirac equation in a curved spacetime [3, 4], hereafter referred to as the Dirac-Fock-Weyl equation. As is well known [5, 6, 7], the wave function ψ transforms as a quadruplet of complex scalars under a coordinate transformation of the Dirac-Fock-Weyl equation. In contrast, in two recently discovered alternative extensions of the Dirac equation to a curved spacetime, the wave function ψ is a complex four-vector, while the set of the components of the Dirac matrices γ^μ builds a $\binom{2}{1}$ tensor [8]. Thus, in a curved spacetime, the wave function ψ can be either a quadruplet of complex scalars—this is the Quadruplet Representation of the Dirac field (QRD) to which the Dirac-Fock-Weyl equation belongs. Or, it can be a complex four-vector—this is the Tensor Representation of the Dirac field (TRD) to which the two alternative equations of Ref. [8] belong. With constant Dirac gamma matrices in a flat spacetime in Cartesian coordinates, all Dirac equations (QRD and TRD) are equivalent to the original Dirac equation with spinor transformation, because: i) The explicit expression of the Dirac equation in a Cartesian coordinate system is the same as with the spinor transformation, and ii) There is no influence of the possible set of constant Dirac matrices [9].

However, in a general Lorentzian spacetime $(\mathrm{V}, g_{\mu\nu})$ in general coordinates, there is a whole continuum of different possible choices for the field of Dirac matrices, with no possibility of a unique choice [10, 11, 12]. In our previous studies of the basic quantum mechanics for the standard (Dirac-Fock-Weyl) equation and for the two alternative equations based on TRD, we found that those three equations behave similarly: e.g. the same hermiticity condition of the Hamiltonian applies to the three [10], and similar non-uniqueness problems of the Hamiltonian theory occur [11, 12]. The similar behaviour which we found for the Dirac-Fock-Weyl equation (with ψ being a 4-scalar) and our alternative equations based on TRD led us to study the *relations between the two representations in a curved spacetime:* ψ 4-scalar vs. ψ 4-vector.

In the work [13] summarized in the present conference paper, the two representations are formulated in a common geometrical framework. We study and compare the two possible representations of the Dirac wave function in a curved spacetime. We prove several theorems that relate together the QRD and TRD representations in a general spacetime.

2. A common geometrical framework

The Dirac-Fock-Weyl equation belongs to the more general "quadruplet representation of the Dirac field" (QRD). For both QRD and the tensor representation (TRD), the wave function is living in some complex vector bundle with base V (the spacetime manifold), and with dimension 4, denoted E. In spite of being complex, that vector bundle is *very simple* in either case:

- For QRD, the wave function is a complex four-scalar field, that is a section of the trivial bundle $\mathrm{V} \times \mathsf{C}^4$, thus
$$\mathrm{E} = \mathrm{V} \times \mathsf{C}^4 \qquad \text{for QRD.} \tag{1}$$
- Instead, for TRD, the wave function is a complex vector field, that is a section of the complexification of the tangent bundle TV, thus
$$\mathrm{E} = \mathrm{T}_{\mathsf{C}}\mathrm{V} \qquad \text{for TRD.} \tag{2}$$

Selecting a local frame field $X \mapsto (e_a(X))$ of the relevant vector bundle E, the wave function ψ has the local expression (valid in some open subset U of the spacetime V)

$$\psi = \Psi^a \, e_a. \tag{3}$$

Then, to define the Dirac matrices γ^μ, we may consider a section γ of the following tensor product bundle:

$$\mathrm{F} = \mathrm{TV} \otimes \mathrm{E} \otimes \mathrm{E}^\circ, \tag{4}$$

where E° is the dual vector bundle of E. To any such section, and depending on the local chart $\chi : X \mapsto (x^\mu)$ and on the local frame field (e_a) on E, both being assumed defined on some open subset U of the spacetime V, we associate a unique local expression, valid in U:

$$\gamma = \gamma^{\mu a}_{b} \; \partial_\mu \otimes e_a \otimes \theta^b, \tag{5}$$

with (∂_μ) the coordinate basis associated with the chart χ and (θ^b) the dual frame of the selected frame field (e_a). The Dirac matrices γ^μ are defined to be the matrices with components

$$(\gamma^\mu)^a{}_b \equiv \gamma^{\mu a}_{b}, \tag{6}$$

thus they are defined locally and do depend on the local chart and the local frame field. They are of course assigned to satisfy the anticommutation relation

$$\gamma^\mu \gamma^\nu + \gamma^\nu \gamma^\mu = 2 g^{\mu\nu} \, \mathbf{1}_4, \quad \mu, \nu \in \{0, ..., 3\} \quad (\mathbf{1}_4 \equiv \mathrm{diag}(1,1,1,1)). \tag{7}$$

A global section γ of F, such that (7) be thus satisfied in any local coordinates (x^μ) and in any local frame field (e_a), will be called "an intrinsic field of Dirac matrices". The existence of at least one intrinsic field of Dirac matrices is guaranteed if the spacetime is four-dimensional, noncompact, and admits a spinor structure. In that case, indeed, there exists an orthonormal global tetrad field [14]. Using this fact, it is easy to show that one may define a global section of F, such that the associated Dirac matrices (6) satisfy Eq. (7) [13]. This applies to both representations, i.e., whether the wave function lives in the trivial bundle $V \times \mathrm{C}^4$ or in the complex tangent bundle $\mathrm{T_C V}$.

Select a representation, and consider a γ field for which the anticommmutation relation (7) is satisfied, and a connection, say D, on the relevant bundle E, Eq. (1) or Eq. (2). By the definition of a connection on a vector bundle with base V [15], it associates to any section ψ of the bundle E, a section $D\psi$ of the tensor product bundle $\mathrm{TV}^\circ \otimes \mathrm{E}$. (Here TV° is the dual bundle of TV, hence is the cotangent bundle.) The corresponding Dirac equation in the curved spacetime $(\mathrm{V}, g_{\mu\nu})$ may be written as:

$$\gamma : D\psi = -im\psi, \tag{8}$$

where $\gamma : D\psi$ is the doubly-contracted product of these sections of $\mathrm{TV} \otimes \mathrm{E} \otimes \mathrm{E}^\circ$ and $\mathrm{TV}^\circ \otimes \mathrm{E}$, corresponding to the two dual pairs $(\mathrm{TV}, \mathrm{TV}^\circ)$ and $(\mathrm{E}^\circ, \mathrm{E})$ appearing in their respective component spaces. Using a local coordinate system (x^μ) and a local frame field (e_a), $D\psi$ decomposes as

$$D\psi = D_\mu \Psi^b \, \mathrm{d}x^\mu \otimes e_b, \tag{9}$$

with $(\mathrm{d}x^\mu)$ the dual frame of the coordinate frame (∂_μ). The local expression of the Dirac equation (8) is accordingly:

$$\gamma^{\mu a}_{b} D_\mu \Psi^b = -im\Psi^a. \tag{10}$$

This local expression is manifestly covariant under a change of either the coordinate system or the frame field. The connection D on E can be given explicitly by providing the expression of the connection matrices $\Gamma_\mu = \left((\Gamma_\mu)^b{}_a\right)$, defined by [cf. Eq. (9)]:

$$De_a = (\Gamma_\mu)^b{}_a \, \mathrm{d}x^\mu \otimes e_b. \tag{11}$$

The explicit expression of $D\psi$ is then from (3):

$$D_\mu \Psi^b \equiv \frac{\partial \Psi^b}{\partial x^\mu} + (\Gamma_\mu)^b{}_a \Psi^a. \tag{12}$$

The link with the more classical writing of the Dirac equation:

$$\gamma^\mu D_\mu \Psi = -im\Psi \tag{13}$$

is got by defining Ψ and $D_\mu \Psi$ for each $\mu = 0, ..., 3$ as column vectors with the following components:

$$\Psi \equiv \left(\Psi^b\right) \tag{14}$$

and

$$D_\mu \Psi \equiv \left(D_\mu \Psi^b\right). \tag{15}$$

By (6) and (15), the classical writing (13) is equivalent to the manifestly-covariant local expression (10). Hence, Eq. (13) also is covariant under changes of the coordinate chart and the frame field as well.

The definitions (3) and (6) give rise to two different transformation behaviours, depending on whether QRD and TRD is considered.

For QRD ($\mathrm{E} = \mathrm{V} \times \mathsf{C}^4$), thus including for the standard, Dirac-Fock-Weyl equation, the canonical basis of C^4 is a preferred frame field on E. The wave function ψ behaves as a scalar under a coordinate change, i.e., it remains *invariant*, while the Dirac matrices transform as a four-vector:

$$\gamma'^\mu = L^\mu_{\ \nu}\, \gamma^\nu, \qquad L^\mu_{\ \nu} \equiv \frac{\partial x'^\mu}{\partial x^\nu}. \tag{16}$$

For TRD ($\mathrm{E} = \mathrm{T}_{\mathsf{C}}\mathrm{V}$), the frame field on E can be taken to be the coordinate basis (∂_μ). In that case, on changing the coordinate chart, ψ behaves as an usual four-vector:

$$\Psi'^\mu = L^\mu_{\ \nu} \Psi^\nu, \tag{17}$$

and the Dirac matrices transform as an usual $\binom{2}{1}$ tensor [8, 16]:

$$\gamma'^{\mu\rho}_{\nu} = L^\mu_{\ \sigma}\, L^\rho_{\ \tau}\, \left(L^{-1}\right)^\chi_{\ \nu}\, \gamma^{\sigma\tau}_{\chi}. \tag{18}$$

The anticommutation relation (7) is covariant under a change of chart, for either of the two transformation modes (16) and (18), thus including for TRD [16]. It is also covariant under a change of the local frame field (e_a) on E.

3. Four classes of Dirac equations

1) The standard, Dirac-Fock-Weyl equation, which is a QRD equation, obtains when one assumes simultaneously [5, 7] that:

- The field γ is deduced from some real *tetrad field*:

$$\gamma^\mu(X) = a^\mu_{\ \alpha}(X)\, \gamma^{\natural\alpha}. \tag{19}$$

  Here $(\gamma^{\natural\alpha})$ is a fixed set of constant Dirac matrices, i.e., a constant solution of Eq. (7) with $g^{\mu\nu} = \eta^{\mu\nu} \equiv \mathrm{diag}(1, -1, -1, -1)$. The $a^\mu_{\ \alpha}(X)$'s are the components in the coordinate basis of a (real) "tetrad field" (u_α), i.e., a frame field on the (real) tangent bundle TV: $u_\alpha = a^\mu_{\ \alpha}\, \partial_\mu$. The tetrad field (u_α) is assigned to satisfy the orthonormality condition:

$$g_{\alpha\beta} \equiv g(u_\alpha, u_\beta) \equiv g_{\mu\nu}\, a^\mu_{\ \alpha}\, a^\nu_{\ \beta} = \eta_{\alpha\beta}, \tag{20}$$

  which ensures that the field of matrices (19) satisfies the anticommutation relation (7).
- The connection D on $\mathrm{V} \times \mathsf{C}^4$ depends on γ in such a way that $D\gamma = 0$.

The Dirac-Fock-Weyl equation can then be written as Eq. (13).

2) The QRD–0 equations fix the connection D by assuming that

$$D\, E_a = 0, \tag{21}$$

where (E_a) is the canonical basis of $\mathrm{V} \times \mathsf{C}^4$. Therefore, the connection matrices Γ_μ in Eq. (11) are zero when the frame field (e_a) is taken to be (E_a). It follows that the covariant derivatives (12) entering the Dirac equation (13) reduce to partial derivatives:

$$D_\mu \Psi = \partial_\mu \Psi \tag{22}$$

for the QRD–0 version, when the frame field is taken to be (E_a).

3) The TRD–0 equations fix the connection D by assuming that

$$D\,u_a = 0, \tag{23}$$

where (u_a) is some global orthonormal (and generally complex) frame field on $\mathrm{T_C V}$. [For the spacetime that we consider, there does exist at least one global orthonormal real tetrad field, say (u_α), on the real tangent bundle TV: see after Eq. (7). The induced frame field (u_a) on the complex tangent bundle $\mathrm{T_C V}$, i.e., $u_a \equiv \delta^\alpha_a u_\alpha$, is then a global orthonormal frame field on $\mathrm{T_C V}$.]
4) The TRD–1 equations assume the Levi-Civita connection, extended from TV to $\mathrm{T_C V}$.

For the three classes (2) to (4), the Dirac equation is written with an additional term [10, 11] as compared with Eq. (13):

$$\mathcal{D}(\gamma, \mathcal{A}, D)\Psi \equiv \gamma^\mu D_\mu \Psi + \frac{1}{2} A^{-1}(D_\mu B^\mu)\Psi = -im\Psi, \tag{24}$$

where

$$B^\mu \equiv A\gamma^\mu. \tag{25}$$

In Eq. (24), $\mathcal{A}$ is a Hermitian metric with respect to which the Dirac matrices γ^μ are Hermitian [9, 13, 17], and A is the matrix of $\mathcal{A}$: given a local frame field (e_a) on the bundle E, with dual frame field (θ^a) on E°, the local expression of $\mathcal{A}$ is [13]

$$\mathcal{A} = A_{ab}\,(\theta^a)^* \otimes \theta^b \tag{26}$$

(where $(\theta^a)^*$ is the complex conjugate of the one-form θ^a). This expression is determined by the data of the regular Hermitian 4×4 matrix field $A \equiv (A_{ab})$, called the field of the "hermitizing matrix" [9, 10, 18]. The matrix field B^μ, Eq. (25), is made of all the components at fixed μ of a tensor field $\mathcal{B}$, obtained by contraction:

$$\mathcal{B} \equiv \mathcal{A}.\gamma, \quad B^\mu_{ab} \equiv A_{ac}\,\gamma^{\mu c}_b. \tag{27}$$

Thus, the matrix $D_\mu B^\mu$ in Eq. (24) is made of all components of a second-order tensor field: $D_\mu B^\mu \equiv \left(D_\mu B^\mu_{ab}\right)$. Like Eq. (13), Eq. (24) is covariant under a change of either the coordinate system (x^μ) or the frame field (e_a).

A direct way of explaining the occurrence of this additional term on the l.h.s. of Eq. (24) is the following one [11]: For a general field of Dirac matrices γ^μ, the standard "Dirac Lagrangian" which applies to the Dirac-Fock-Weyl equation has to be extended to include the field A in the exact place of the constant Dirac matrix $\gamma^{\natural 0}$. For the Dirac-Fock-Weyl equation, we have always $D_\mu B^\mu = 0$, hence the additional term in Eq. (24) vanishes [10].

The QRD–0 equations, as well as the TRD–0 equations, are new. The TRD–1 equations (together with another class of equations, "TRD–2", that is based on a connection deduced from an assumed preferred reference frame) were introduced in Ref. [8].
Note that any two tetrad fields lead to two equivalent Dirac-Fock-Weyl equations [5], except for some non-trivial topologies [19]. Thus, the Dirac-Fock-Weyl equation (1) is normally a unique equation, not a class of equations.

In contrast, none of the three classes (2) to (4) does correspond to a unique equation. For each of the classes (2) to (4), the connection D is fixed, but the field γ is restricted only by the anticommutation relation (7). In general, two fields $\gamma \neq \gamma'$ give inequivalent Dirac equations. This is the price to pay for the simplicity of the definition of these equations and for their greater generality, as compared with the Dirac-Fock-Weyl equation. However, if we consider not the Dirac equation itself but the Hamiltonian operator which is associated to the Dirac equation in a given reference frame, then we find [11, 12] that the Dirac-Fock-Weyl Hamiltonian is non-unique, too—just as is the Hamiltonian operator associated with, for example, the TRD–1 equation [or with any class of Dirac equations among the classes (2) to (4)]. In other words, the fact that the Dirac-Fock-Weyl equation is unique in a "reasonable" spacetime does not prevent the corresponding Hamiltonian from being non-unique (in a given reference frame or even in a given coordinate system).

4. Equivalence theorems between classes

In the work [13], summarized here, we prove technically precise versions of three theorems, which we state below in more concise terms. (In addition, Theorem 1 in Ref. [13] is more general than is Theorem 1 below.)

Theorem 1. *The QRD–0 and TRD–0 equations are equivalent for given coefficient fields (γ^μ, A). Moreover, when the QRD–0 equation is a "normal" one, i.e., when $D_\mu B^\mu = 0$ in Eq. (24), then the corresponding TRD–0 equation is also a normal one.*

Here, "for given coefficient fields (γ^μ, A)" means: **i)** that the γ field used to write the QRD–0 equation and the γ' field used to write the TRD–0 equation correspond to the same Dirac matrices in any local coordinate system (x^μ) defined on an open subset U of the spacetime V — that is, $\gamma^\mu = \gamma'^\mu$ — when the frame fields on E are, respectively: the canonical basis (E_a) of $\mathrm{V} \times \mathsf{C}^4$, for the QRD–0 equation; and the global tetrad field (u_a) on $\mathrm{T}_\mathsf{C}\mathrm{V}$ that appears in Eq. (23), for the TRD–0 equation. **ii)** Moreover, in these respective frame fields, the respective hermitizing matrix fields coincide, $A = A'$. With any fields γ and $\mathcal{A}$ valid for QRD, one may indeed easily associate fields γ' and $\mathcal{A}'$ valid for TRD, and such that the two equalities $\gamma^\mu = \gamma'^\mu$ and $A = A'$ be verified in the conditions specified above [13].
The proof of Theorem 1 is not difficult: due to Eqs. (21) and (23) respectively, the covariant derivatives entering the local expression (24) of the Dirac equation reduce to partial derivatives, both for the QRD–0 or the TRD–0 equation, when the frame field on the relevant bundle E is (E_a) or (u_a), respectively. (See Eq. (22) for QRD–0.) And since the coefficient fields (γ^μ, A) are, by hypothesis, the same for both equations, it follows that the local expression (24) is the same for the QRD–0 equation and the TRD–0 equation. Thus, the local expressions of these two equations, in the arbitrary local coordinate system (x^μ) which is considered, and in the respective global frame fields (E_a) or (u_a), are indeed equivalent. Finally, to say that the Dirac equation is a "normal" one, i.e., that $D_\mu B^\mu = 0$ in Eq. (24), means for both the QRD–0 and the TRD–0 equation that $\partial_\mu B^\mu = 0$. Since the coefficient fields (γ^μ, A) are, by hypothesis, the same for the QRD–0 and the TRD–0 equation which are considered, it follows that the validity of the condition $D_\mu B^\mu = 0$ for the QRD–0 equation is equivalent to its validity for the TRD–0 equation.

□

Theorem 2. *Let γ be any "intrinsic field of Dirac matrices", let $\mathcal{A}$ be an associated hermitizing metric, and let D be any connection on* E*. Let D' be any other connection on* E.

Then, there is another "intrinsic field of Dirac matrices", $\tilde{\gamma}$, with an associated hermitizing metric $\tilde{\mathcal{A}}$, such that the Dirac equation based on γ, $\mathcal{A}$ and D is equivalent to that based on $\tilde{\gamma}$, $\tilde{\mathcal{A}}$ and D'. In particular, any form of the QRD (TRD) equation is equivalent to a QRD–0 (TRD–1) equation.

The proof of this theorem [13] is outlined in the next section.

Theorem 3. The Dirac-Fock-Weyl equation is equivalent to a TRD–1 equation (thus with vector wave function) in the same spacetime.

Theorem 3 is an easy consequence of Theorems 1 and 2, at least locally.

5. Theorem 2: outline of the proof

For given fields γ and $\mathcal{A}$, the difference between the Dirac operators $\mathcal{D}(\gamma, \mathcal{A}, D)$ [Eq. (24)] and $\mathcal{D}(\gamma, \mathcal{A}, D')$, corresponding with two different connections D and D' on E, is found after some algebra [13] to depend just on the matrix

$$K \equiv \gamma^\mu K_\mu, \tag{28}$$

so that $\mathcal{D}(\gamma, \mathcal{A}, D) = \mathcal{D}(\gamma, \mathcal{A}, D')$ if $K = 0$. Here, the γ^μ 's are the Dirac matrices associated by Eq. (6) with γ, in the local chart χ and in the local frame field (e_a) which are considered; and we define

$$K_\mu \equiv \Gamma_\mu - \Gamma'_\mu, \tag{29}$$

where Γ_μ and Γ'_μ are the connection matrices (11) of D and D' respectively, in that chart and in that frame field.

Consider a new field $\tilde{\gamma}$ in the same spacetime. In the local chart χ and in the local frame field (e_a), the new field $\tilde{\gamma}$ corresponds to a new field of Dirac matrices: $X \mapsto \tilde{\gamma}^\mu(X)$. Similarly, the hermitizing metric $\tilde{\mathcal{A}}$ associated with $\tilde{\gamma}$ corresponds in the local frame field (e_a) to a new field $\tilde{A}$ of the hermitizing matrix. The field $\tilde{\gamma}$, as well as the associated hermitizing metric $\tilde{\mathcal{A}}$, are necessarily deduced from γ and $\mathcal{A}$ by a local similarity transformation

$$S : \mathrm{V} \to \mathsf{GL}(4, \mathsf{C}). \tag{30}$$

In the local chart χ and in the local frame field (e_a), this writes in the following way [18, 9, 10]:

$$\tilde{\gamma}^\mu(X) = S(X)^{-1}\gamma^\mu(X)S(X) \tag{31}$$

and

$$\tilde{A}(X) = S(X)^\dagger A(X)S(X). \tag{32}$$

We know how to change D for a new connection $\tilde{D}$ so that the starting Dirac equation $\mathcal{D}(\gamma, \mathcal{A}, D)\Psi = 0$ is equivalent to the Dirac equation $\mathcal{D}(\tilde{\gamma}, \tilde{\mathcal{A}}, \tilde{D})\tilde{\Psi} = 0$, where we define $\tilde{\Psi} \equiv S^{-1}\Psi$. [1] As in Eqs. (28) and (29), set $\tilde{K}_\mu \equiv \tilde{\Gamma}_\mu - \Gamma'_\mu$ and $\tilde{K} \equiv \tilde{\gamma}^\mu \tilde{K}_\mu$. Thus, if $\tilde{K} = 0$, then the Dirac equation $\mathcal{D}(\tilde{\gamma}, \tilde{\mathcal{A}}, D')\tilde{\Psi} = 0$ is equivalent to $\mathcal{D}(\tilde{\gamma}, \tilde{\mathcal{A}}, \tilde{D})\tilde{\Psi} = 0$, hence to the starting Dirac equation $\mathcal{D}(\gamma, \mathcal{A}, D)\Psi = 0$.

The condition for $\tilde{K} \equiv \tilde{\gamma}^\mu \tilde{K}_\mu = 0$ can then be shown [13] to be [2]

$$\gamma^\mu D'_\mu S = -KS. \tag{33}$$

This is a system of sixteen first-order linear partial differential equations for the sixteen components of S. It turns out that this system can be rewritten as a symmetric hyperbolic system [13]. One consequence is that, by assuming analytic coefficients and initial data, the Cauchy-Kovalevskaya theorem ensures that it can always be solved locally. Another consequence [13] is that, by using a more advanced theorem, the system (33) can be solved in certain a priori given (large) domains.

□

[1] Namely, we have to define the new connection by the matrices $\tilde{\Gamma}_\mu = S^{-1}\Gamma_\mu S + S^{-1}(\partial_\mu S)$ [13].

[2] In Eq. (33), we have $D'_\mu S = \partial_\mu S + \Gamma'_\mu S - S\Gamma'_\mu$ [11]. This can be defined in intrinsic terms [13].

6. Conclusion

In the present work [13], summarized in this conference paper, we described in a common geometrical framework the Quadruplet Representation of the Dirac field (QRD) and the Tensor Representation of the Dirac field (TRD). We proposed a simple form ("QRD–0") of the Dirac equation in a curved spacetime, in which the covariant derivatives reduce to partial derivatives. We proved several theorems that relate together the QRD and TRD representations in a general spacetime. In particular, by Theorem 2, the Dirac-Fock-Weyl equation is equivalent to a particular case of the simple QRD–0 equation. By Theorem 3, the Dirac-Fock-Weyl equation is also equivalent to a particular case of either of the two alternative equations based on TRD. Thus, these alternative Dirac equations in a curved spacetime, in which the wave function is a complex *four-vector*, can be seen as *extensions* of the standard version of the Dirac equation in a curved spacetime, in which the wave function is a quadruplet of complex scalars.

References

[1] Colella R Overhauser A W and Werner S A 1975 *Phys. Rev. Lett.* **34** 1472
[2] Nesvizhevsky V V *et al.* 2002 *Nature* **415** 297
[3] Weyl H 1929 *Z. Phys.* **56** 330
[4] Fock V A 1929 *Z. Phys.* **57** 261
[5] Brill D R and Wheeler J A 1957 *Rev. Modern Phys.* **29** 465; Erratum 1961 *Rev. Modern Phys.* **33** 623
[6] de Oliveira C G and Tiomno J 1962 *Nuovo Cim.* **24** 672
[7] Chapman T C and Leiter D L 1977 *Am. J. Phys.* **44** 858
[8] Arminjon M 2008 Dirac-type equations in a gravitational field, with vector wave function *Found. Phys.* **38** 1020 (*Preprint* arXiv:gr-qc/0702048)
[9] Arminjon M and Reifler F 2008 Dirac equation: Representation independence and tensor transformation *Braz. J. Phys.* **38** 248 (*Preprint* arXiv:0707.1829 [quant-ph])
[10] Arminjon M and Reifler F 2010 Basic quantum mechanics for three Dirac equations in a curved spacetime *Braz. J. Phys.* **40** 242 (*Preprint* arXiv:0807.0570 [gr-qc])
[11] Arminjon M and Reifler F 2009 A non-uniqueness problem of the Dirac theory in a curved spacetime *Preprint* arXiv:0905.3686 [gr-qc]
[12] Arminjon M and Reifler F 2010 Non-uniqueness of the Dirac theory in a curved spacetime (*First Mediterranean Conference on Classical and Quantum Gravity, Kolymbari, Greece, Sept. 2009*) *J. Phys.: Conf. Series.* **222** 012042 (*Preprint* arXiv:1001.0460 [gr-qc])
[13] Arminjon M and Reifler F 2010 Four-vector vs. four-scalar representation of the Dirac wave function (in preparation)
[14] Geroch R 1968 *J. Math. Phys.* **9** 1739
[15] Chern S S Chen W H and Lam K S 1999 *Lectures on Differential Geometry* (Singapore: World Scientific) p. 113
[16] Arminjon M 2006 Dirac equation from the Hamiltonian and the case with a gravitational field *Found. Phys. Lett.* **19** 225 (*Preprint* arXiv:gr-qc/0512046)
[17] Lawson H B and Michelsohn M-L 1989 *Spin Geometry* (Princeton: University Press)
[18] Pauli W 1936 *Ann. Inst. Henri Poincaré* **6** 109
[19] Isham C J 1978 *Proc. Roy. Soc. London A* **364** 591

A field theoretic generalization of Hajicek and Kuchar's quantization scheme in 3+1 canonical quantum gravity

Evangelos Melas

National and Capodistrian University of Athens, 8 Pesmazoglou Street, 105 59 Athens

E-mail: evangelosmelas@yahoo.co.uk

Abstract. The 3+1 (canonical) decomposition of all geometries admitting two-dimensional space-like surfaces is exhibited as a generalization of a previous work. A proposal, consisting of a specific re-normalization **Assumption** and an accompanying **Requirement**, which has been put forward in the 2+1 case is now generalized to 3+1 dimensions. This enables the canonical quantization of these geometries through a generalization of Kuchař's quantization scheme in the case of infinite degrees of freedom. The resulting Wheeler-deWitt equation is based on a re-normalized manifold parameterized by three smooth scalar functionals. The entire space of solutions to this equation is analytically given, a fact that is entirely new to the present case. This is made possible by exploiting the freedom left by the imposition of the **Requirement** and contained in the third functional.

1. Introduction

In a previous work of ours [4], a method has been developed which enables a generalization of Kuchar's quantization scheme [2], [3] to be applied to a system of infinite degrees of freedom. The system consisted of a midi-superspace axially symmetric model in 2+1 dimensions plus a Λ term. In our present work [?], we apply the same method to the case of a 3+1 midi-superspace model under the assumption of the existence of maximally symmetric two dimensional subsurfaces, i.e. spheres (constant positive curvature), planes (zero curvature) and Gauss-Bolyai-Lobachevsky (henceforth GBL) spaces (constant negative curvature).

The structure of the paper is as follows: In the second section the reduced metrics, the space of classical solutions and the Hamiltonian formulation of the reduced Einstein-Hilbert action principle, resulting in one (quadratic) Hamiltonian and one (linear) momentum first class constraint, are presented. In the third section we consider the quantization of this constraint system by implementing the quantum operator constraints as conditions annihilating the wave-function [5], according to a conceptual generalization of the quantization scheme developed in [2], [3] for constraint systems with finite degrees of freedom. This generalization is possible through the use of a specific re-normalization **Assumption** and an associated **Requirement**. After the symmetry reduction, the system still represents an one-dimensional field theory since all remaining metric components depend on time and the radial coordinate. Nevertheless, we manage to extract a Wheeler-deWitt equation in terms of three smooth scalar functionals of the reduced geometries. The exploitation of a residual freedom left by the imposition of the

Requirement enables us to acquire the entire solution space to this equation. Finally, some concluding remarks are included in the Discussion.

2. Classical configuration space and Hamiltonian Formulation

Our starting point is the two-dimensional spaces of positive, zero and negative constant curvature. Their line elements are respectively:

$$ds^2 = d\theta^2 + \sin^2\theta\, d\phi^2, \qquad ds^2 = d\theta^2 + \theta^2\, d\phi^2, \qquad ds^2 = d\theta^2 + \sinh^2\theta\, d\phi^2 \tag{1}$$

with an obvious range of the coordinates for each case. The corresponding (maximal) symmetry groups are generated by the following KVF's:

$$\xi_1 = \frac{\partial}{\partial\phi}, \quad \xi_2 = -\cos\phi\tfrac{\partial}{\partial\theta} + \cot\theta\sin\phi\tfrac{\partial}{\partial\phi}, \quad \xi_3 = \sin\phi\frac{\partial}{\partial\theta} + \cot\theta\cos\phi\frac{\partial}{\partial\phi} \tag{2}$$

$$\xi_1 = \frac{\partial}{\partial\phi}, \quad \xi_2 = -\cos\phi\tfrac{\partial}{\partial\theta} + \tfrac{\sin\phi}{\theta}\tfrac{\partial}{\partial\phi}, \quad \xi_3 = \sin\phi\frac{\partial}{\partial\theta} + \frac{\cos\phi}{\theta}\frac{\partial}{\partial\phi} \tag{3}$$

$$\xi_1 = \frac{\partial}{\partial\phi}, \quad \xi_2 = -\cos\phi\tfrac{\partial}{\partial\theta} + \coth\theta\sin\phi\tfrac{\partial}{\partial\phi}, \quad \xi_3 = \sin\phi\frac{\partial}{\partial\theta} + \coth\theta\cos\phi\frac{\partial}{\partial\phi} \tag{4}$$

We next promote these KVF's to four-dimensional fields by adding to each and every of them the zero-sum $0\,\frac{\partial}{\partial t} + 0\,\frac{\partial}{\partial r}$. We then enforce these vector fields as symmetries of a generic space-time metric $g_{IJ}(t,r,\theta,\phi)$, i.e. we demand that its Lie derivative with respect to all three fields for each family vanishes. We thus arrive at three classes of metrics, collectively described by the following line element:

$$ds^2 = \left(-\alpha(t,r)^2 + \frac{\beta(t,r)^2}{\gamma(t,r)^2}\right) dt^2 + 2\,\beta(t,r)\,dt\,dr + \gamma(t,r)^2\,dr^2 + \psi(t,r)^2\,d\theta^2 + \psi(t,r)^2\,f(\theta)^2\,d\phi^2 \tag{5}$$

where $f(\theta) = \sin\theta$ in the case of spherical symmetry, $f(\theta) = \theta$ in the case of plane symmetry, and $f(\theta) = \sinh\theta$ in the case of the GBL symmetry.

For each type of symmetry, there exists one solution in which the metric components depend only on the radial coordinate ("point-like") and another with only time dependence ("cosmological") [?]. For example, in the case of spherical symmetry and $\Lambda = 0$, we have the well known Schwarzschild and Kantowski-Sachs metrics respectively.

Employing the Hamiltonian formulation (presented, e.g., in chapter 9 of [6]) for the system (5) we get the primary constraints $P_\alpha \equiv \delta L/\delta\dot\alpha \approx 0$, $P^\beta \equiv \delta L/\delta\dot\beta \approx 0$ and

$$\mathcal{H}_o = \frac{1}{2} G^{\alpha\beta}\,\pi_\alpha\,\pi_\beta + V, \tag{6}$$

$$\mathcal{H}_1 = -\gamma\,\pi_\gamma' + \psi'\,\pi_\psi, \qquad \mathcal{H}_2 = 0, \qquad \mathcal{H}_3 = 0, \tag{7}$$

where $' = \dfrac{d}{dr}$, the indices $\{\alpha,\beta\}$ take the values $\{\gamma,\psi\}$ and $H = \int \left(N^o\,\mathcal{H}_o + N^i\,\mathcal{H}_i\right) dr$ is the Hamiltonian of our system with $N^o = \alpha(t,r)$ and $N^i = (\frac{\beta(t,r)}{\gamma(t,r)^2}, 0, 0)$ the lapse and shift functions, respectively. The reduced Wheeler-deWitt super-metric $G^{\alpha\beta}$ appearing in (6) reads

$$G^{\alpha\beta} = \begin{pmatrix} \frac{\gamma}{4\psi^2} & -\frac{1}{4\psi} \\ -\frac{1}{4\psi} & 0 \end{pmatrix},$$

while the potential V is

$$V = -2\,\epsilon\,\gamma + 2\Lambda\gamma\,\psi^2 - 2\frac{\psi'^2}{\gamma} + 4\left(\frac{\psi\,\psi'}{\gamma}\right)' \tag{9}$$

with $\epsilon = \{1, 0, -1\}$ for the families (5) of two-dimensional subspaces with positive, zero or negative constant curvature, respectively. The open Poisson bracket algebra of the secondary constraints

$$\mathcal{H}_o \approx 0, \qquad \mathcal{H}_1 \approx 0 \tag{10}$$

produced by demanding the time preservation of the primary constraints is the following

$\{\mathrm{H}_o(r), \mathcal{H}_o(\tilde{r})\} = \left[\frac{1}{\gamma^2(r)}\mathcal{H}_1(r) + \frac{1}{\gamma^2(\tilde{r})}\mathcal{H}_1(\tilde{r})\right]\delta'(r,\tilde{r}),$

$\{\mathcal{H}_1(r), \mathcal{H}_o(\tilde{r})\} = \mathcal{H}_o(r)\delta'(r,\tilde{r}),$

$\{\mathcal{H}_1(r), \mathcal{H}_1(\tilde{r})\} = \mathcal{H}_1(r)\delta'(r,\tilde{r}) - \mathcal{H}_1(\tilde{r})\delta(r,\tilde{r})'.$

This means that (10) are first class and that Dirac's algorithm is terminated.

In order to unveil the nature of the action of $\mathcal{H}_1$ on $\gamma(t,r)$, $\psi(t,r)$ we have to investigate their behavior under r re-parameterizations of the form $r \to \tilde{r} = h(r)$. By observing (5) it can be readily deduced that ψ is a scalar and γ, ψ' are covariant rank 1 tensors (i.e. one-forms). A very important consequence of this feature is that the scalar derivative is given by $\gamma^{-1}d/dr$ and not by d/dr. By considering now an infinitesimal r re-parametrization $r \to \tilde{r} = r - \eta(r)$ and by using the aforementioned results we can easily infer that the corresponding changes on the fields $\gamma(t,r)$ and $\psi(t,r)$ are nothing but the one-dimensional analogues of the appropriate Lie derivatives, i.e. $\delta\,\gamma(r) = (\gamma(r)\,\eta(r))'$ and $\delta\,\psi(r) = \psi'(r)\,\eta(r)$, respectively. (Notice that the time dependence has been suppress for the sake of brevity.) One is in position now to specify the nature of the action of the linear constraint on our basic configuration variables as that of the generators of spatial diffeomorphisms:

$$\left\{\gamma(r), \int d\tilde{r}\,\eta(\tilde{r})\,\mathcal{H}_1(\tilde{r})\right\} = (\gamma(r)\,\eta(r))', \left\{\psi(r), \int d\tilde{r}\,\eta(\tilde{r})\,\mathcal{H}_1(\tilde{r})\right\} = \psi'(r)\,\eta(r). \tag{11}$$

With this knowledge in hand we are legitimated to consider $\mathcal{H}_1$ as the representative, in phase-space, of an arbitrary infinitesimal re-parametrization of the radial coordinate. In the same manner we can also see that the action of the quadratic constraint $\mathcal{H}_o$ on the basic configuration space variables is identical to an infinitesimal change of the time coordinate.

3. Quantization

In this section a quantization scheme for the Hamiltonian system described in the preceding chapter is proposed. The main motivation behind our approach closely can be traced in Dirac's desire to construct a quantum theory manifestly invariant under the "gauge" generated by the quadratic and linear (first class) constraints (10). In order to achieve this goal we must seek i) a proper realization of our first class constraints as quantum operators and ii) an appropriate wave function that will be annihilated by them. To begin with, let us note that, despite the simplification brought by the imposition of the symmetries (2), (3), (4), the system is still a field theory as all configuration variables and their canonical conjugate momenta depend not only on time, but also on the radial coordinate r. Thus, to canonically quantize the system in the Schrödinger representation, we have to realize the classical dynamical variables as operators in the following sense

$$\hat{\gamma}(r) = \gamma(r), \qquad \hat{\psi}(r) = \psi(r), \qquad \hat{\pi}_\gamma(r) = -i\,\hbar\,\frac{\delta}{\delta\,\gamma(r)}, \qquad \hat{\pi}_\psi(r) = -i\,\hbar\,\frac{\delta}{\delta\,\psi(r)}. \tag{12}$$

In the rest of the article we will assume that $\hbar = 1$. These operators obey the canonical commutation relations

$$[\hat{q}^k(r), \hat{\pi}_j(\tilde{r})] = i\,\delta^k{}_j\,\delta(r,\tilde{r}),$$

where $\hat{q}^k = \{\hat{\gamma}, \hat{\psi}\}$ and $\hat{\pi}_j = \{\hat{\pi}_\gamma, \hat{\pi}_\psi\}$. Our next step is to specify the initial space of state vectors in a way that the action of the operators (12) be well-defined. By this is meant that their action does not produce $\delta(0)$ terms. Consider, for example, the action of a momentum operator on some function of the configuration field variables, say

$$\hat{\pi}_\gamma(r) f(\gamma(\tilde{r})) = -i \frac{\partial f(\gamma(\tilde{r}))}{\partial \gamma(\tilde{r})} \delta(\tilde{r}, r).$$

Obviously, when the momentum operator acts at the same point as the function, i.e. $\tilde{r} = r$, the Dirac delta-function that appears on the r.h.s of the above equation produces a $\delta(0)$. In addition, the action of the momentum operator produced a distribution instead of a function. A way to bypass both of these unwanted features is to choose as our initial collection of states all *smooth functionals* (i.e., integrals over r) of the configuration variables $\gamma(r), \psi(r)$ and their derivatives of any order:

$$\Phi \equiv \int f(\gamma, \psi, + \text{ derivatives of any order})\, d\tilde{r}. \tag{13}$$

By applying this choice to the above example, i.e. by replacing $f(\gamma(\tilde{r}))$ with $\int f(\gamma(\tilde{r}))\, d\tilde{r}$ the action of the momentum operator becomes well-defined; no $\delta(0)$ terms and no distributions appear at the r.h.s. Unfortunately, the above choice of initial states can not prevent the occurrence of $\delta(0)$s when local expressions quadratic in momenta are considered, i.e. the action of $\hat{\pi}_\gamma(r)\hat{\pi}_\gamma(r)$ on $\int f(\gamma(\tilde{r}))\, d\tilde{r}$ apparently produces $\delta(0)$ terms.

An other problem of equal, if not greater, importance has to do with the number of derivatives (with respect to r) considered: A momentum operator acting on a smooth functional of degree n in derivatives of $\gamma(r), \psi(r)$ will, in general, produce a function of degree $2n$. Thus, clearly, more and more derivatives must be included if we desire the action of momentum operators to keep us inside the space of integrands corresponding to the initial collection of smooth functionals; eventually, we have to consider $n \to \infty$. This, in a sense, can be considered as the reflection to the canonical approach, of the non-re-normalizability results existing in the so-called covariant approach. Loosely speaking, the way to deal with these problems is to regularize (i.e., render finite) the infinite distribution limits, and re-normalize the theory by, somehow, enforcing n to terminate at some finite value.

Despite the fact that the choice (13) prevents the occurrence of $\delta(0)$ terms when expressions linear in momenta are considered, we still have to find a way to exclude their appearance when expressions quadratic in momenta are included. In addition, we have to take care of the unceasing occurrence of higher and higher derivatives emanating from the action of the momentum operators. With the motivation to dispose off these unwanted features, we are going, in the rest of this article, to present a quantization scheme of our system which manages to:

(a) avoid the occurrence of $\delta(0)$s,

(b) reveal the value $n = 1$ as the only *natural* (i.e. without *ad-hoc* cut-offs) possibility to obtain a self-contained space of state vectors,

(c) extract a finite-dimensional Wheeler-deWitt equation governing the quantum dynamics.

The scheme comprises a conceptual generalization of the quantization developed in [2], [3] for finite systems with one quadratic and a finite number of linear constraints. A detailed description of the method is given in [4] (pp. 8-9). Here, we are going to give only a brief outline of the basic ideas underlying it. The central idea is to Dirac-quantize our classical system while keeping the resulting quantum theory invariant under the transformations that leave the classical theory unaffected. This means that the action of the quantum constraints on the wave function has to be chosen in such a way as to secure that the whole scheme is independent under i) the "gauge" transformations generated by the quadratic and linear constraints, ii) the mixing of the linear constraints, iii) the gauging of the quadratic constraint with the linear constraints and iv) the scaling of the quadratic constraint. To construct a quantum theory that respects the above restrictions we have to i) realize the linear constraints with the momentum operators to the right and ii) define an induced metric and realize the quadratic in momenta part of the quadratic constraint as the conformal Laplace-Beltrami operator based on the induced metric. It is noteworthy that in our case the aforementioned procedure resolves two more problems, namely the occurrence of $\delta(0)$ terms emanating from the action of the quadratic constraint and the ordering of the operators inside the constraints. Let us now to apply the above described method to our system.

Firstly, the quantum analogue of the linear constraint $\mathcal{H}_1(r) \approx 0$ is defined with the momentum operators to the far right, i.e.

$$\hat{\mathcal{H}}_1(r)\Phi = 0 \leftrightarrow -\gamma(r)\,(\frac{\delta\,\Phi}{\delta\,\gamma(r)})' + \psi'(r)\,\frac{\delta\,\Phi}{\delta\,\psi(r)} = 0, \tag{14}$$

where Φ is given by 13. It can be proven (see [4] references therein and [7]) that the general solution to the previous equation is any scalar functional:

$$\Phi \;=\; \int \gamma(\tilde{r})\, L\left(\Psi^{(0)}, \Psi^{(1)}, \ldots, \Psi^{(n)}\right) d\tilde{r}, \tag{15}$$

$$\Psi^{(0)} \;\equiv\; \psi(\tilde{r}), \quad \Psi^{(1)} \equiv \frac{\psi'(\tilde{r})}{\gamma(\tilde{r})}, \ldots, \Psi^{(n)} \equiv \frac{1}{\gamma(\tilde{r})}\frac{d}{d\tilde{r}}\left(\underbrace{\cdots}_{n-1}\psi(\tilde{r})\right) \tag{16}$$

with L being any arbitrary function of its arguments. Thus, the initial collection of state vectors is, by the imposition of the quantum linear constraint $\hat{\mathcal{H}}_1(r)$, reduced to the above given smooth scalar functionals.

Secondly, we define the equivalent of Kuchař's induced metric on the so far space of "physical" states (15). It must be pointed out that in our case these "physical" states are the analogues of Kuchař's physical variables q^α. We start by considering **one** initial candidate of the form (15). Then, the induced metric is defined as:

$$g^{\Phi\Phi} = G^{\alpha\beta}\,\frac{\delta\Phi}{\delta\,x^\alpha}\,\frac{\delta\Phi}{\delta\,x^\beta}, \quad \text{where} \quad x^\alpha = \{\gamma, \psi\} \tag{17}$$

and $G^{\alpha\beta}$ is given by (8). A gratifying property of this metric is the non-appearance of $\delta(0)$'s, since only first functional derivatives are employed. However, the following problem arises: $g^{\Phi\Phi}$ *is* a local function while the "physical" states (15) *are* functionals. This means that the problem of whether $g^{\Phi\Phi}$ is composed of the so far physical states cannot even be addressed. It is thus clear that, in order for such a property of $g^{\Phi\Phi}$ to be imposed, a correspondence between local functions and smooth functionals must be established as a prerequisite. A way to achieve this is to adopt the following ansatz:

Assumption: *We assume that, as part of the re-normalization procedure, we are permitted to map local functions to their corresponding smeared expressions e.g.,* $\psi(r) \leftrightarrow \int d\tilde{r}\psi(\tilde{r})$.

For a detailed discussion about the nature of this Assumption the interested reader is referred to [4] (p. 12). The use of this anzatz allows us to start examining the validity of the above mentioned crucial property of $g^{\Phi\Phi}$. In the case of systems with finite degrees of freedom this feature is automatically satisfied (see [2], [3]). But, in the infinite case here discussed, things

become more complicated because of the dependence of γ and ψ on the radial coordinate r. This feature will cause more and more derivatives to appear as the action of a functional derivative, say $\frac{\delta}{\delta\psi(r)}$, on a functional whose integrand contains $\Psi^{(n)}$, will produce a maximum term proportional to $\Psi^{(2n)}$. Even with the use of the Assumption $g^{\Phi\Phi}$ is not automatically composed out of the initial physical states. Therefore, such a property must be *enforced.* This difference between the finite and infinite case can be traced to the fact that the first Poisson bracket in (2) does not vanish as in the finite case (see [4]). Thus, a non-trivial mixing of the dynamical evolution generator $\mathcal{H}_o$ and the linear generator $\mathcal{H}_1$ is brought forth in the infinite case.
According to the above reasoning, it clear now that if we want to proceed further with the generalization of Kuchař's method, the property of $g^{\Phi\Phi}$ being composed out of the physical states must be enforced. Consequently, we demand that:

Requirement: $L\left(\Psi^{(0)},\dots,\Psi^{(n)}\right)$ *must be such that* $g^{\Phi\Phi}$ *becomes a general function, say* $F\left(\gamma(r)\,L(\Psi^{(0)},\dots,\Psi^{(n)})\right)$ *of the integrand of* Φ*, so that it can be considered a function of this state:* $g^{\Phi\Phi} \overset{Assumption}{\equiv} F\left(\int\gamma(\tilde r)\,L(\Psi^{(0)},\dots,\Psi^{(n)})d\tilde r\right) = F(\Phi)$.

At this point, it must emphasized that the application of the **Requirement** in the subsequent development of our quantum theory will result in very severe restrictions on the form of (15). Essentially, all higher derivatives of $\psi(r)$ (i.e $\Psi^{(2)}\dots\Psi^{(n)}$)) are eliminated from Φ (see (18), (19) bellow). This might, at first sight, strike as odd; indeed, the common belief is that all the derivatives of the configuration variables should enter the physical states. However, before the imposition of both the linear *and* the quadratic constrains there are no truly physical states. Thus, no physical states are lost by the imposition of the **Requirement**; ultimately the only true physical states are the solutions to (26).

Let us now see how the **Assumption** and the **Requirement** further restrict our "physical" states (15). As we argued before, the functional derivatives $\frac{\delta}{\delta\psi(r)}$ and $\frac{\delta}{\delta\gamma(r)}$ acting on a functional containing in its integrand $\Psi^{(n)}$ will, upon partial integration of the nth derivative of the Dirac delta function, produce a term proportional to $\Psi^{(2n)}$ and $\Psi^{(2n-1)}$ respectively. Thus, by inserting (15) in (17) one can, after a somewhat tedious calculation, get

$$\begin{aligned} g^{\Phi\Phi} &= \dots + 2\,G^{12}\frac{\delta\Phi}{\delta\gamma(r)}\frac{\delta\Phi}{\delta\psi(r)} = \\ &= \dots + 2\,G^{12}(-1)^{n-1}\frac{\partial^2 L}{\partial\left(\Psi^{(n)}\right)^2}\,\Psi^{(2n-1)}\,\Psi^{(1)}(-1)^n\gamma\,\frac{\partial^2 L}{\partial\left(\Psi^{(n)}\right)^2}\,\Psi^{(2n)} = \\ &= \dots - \frac{\gamma}{2\psi}(-1)^{2n-1}\left(\frac{\partial^2 L}{\partial\left(\Psi^{(n)}\right)^2}\right)^2\Psi^{(1)}\,\Psi^{(2n-1)}\,\Psi^{(2n)}, \end{aligned}$$

where $\dots$ stand for all other terms, not involving $\Psi^{(2n)}$. According to the **Requirement** we demand $g^{\Phi\Phi} = F(\gamma L)$. For this to happen the coefficient of $\Psi^{(2n)}$ must vanish, i.e. $\partial^2 L/\partial\left(\Psi^{(n)}\right)^2 = 0$. Thus, by integrating this necessary condition, L is reduced to the form

$$L = L_1\left(\Psi^{(0)},\dots,\Psi^{(n-1)}\right)\Psi^{(n)} + L_2\left(\Psi^{(0)},\dots,\Psi^{(n-1)}\right).$$

Notice that the term of Φ corresponding to L_1 is, up to a surface term, equivalent to a general term depending on $\Psi^{(0)},\dots,\Psi^{(n-1)}$ only:

$$\Phi_1 = \int\gamma(\tilde r)L_1\frac{1}{\gamma(\tilde r)}\frac{d}{d\tilde r}\Psi^{(n-1)}d\tilde r,$$

which upon subtraction of the surface term

$$A=\int d\tilde{r}\frac{d}{d\tilde{r}}\left(\int d\Psi^{(n-1)}L_1\right)$$

produces a smooth functional with arguments up to $\Psi^{(n-1)}$ only. Since a surface term in Φ does not affect the outcome of the variational derivatives $\frac{\delta\Phi}{\delta\psi(r)}$ and $\frac{\delta\Phi}{\delta\gamma(r)}$, we conclude that only L_2 is important for the local part of Φ. By repeating the entire argument successively for $n-1,\ n-2,\ldots,\ 2$ we can show that all $\Psi^{(n)}$s are suppressed from L except when $n=1$. Thus, we have

$$\Phi\equiv\int\gamma(\tilde{r})L\left(\psi,\Psi^{(1)}\right)d\tilde{r}. \tag{18}$$

It can be shown [?], with the use of the re-normalization **Assumption** and the accompanying **Requirement**, that the reduced re-normalized manifold of the physical states Φ is finally parameterized by the following three smooth scalar functionals:

$$y^1=\int\gamma(\tilde{r})d\tilde{r},\qquad y^2=\int\gamma(\tilde{r})\psi(\tilde{r})^2d\tilde{r},\qquad y^3=\int\gamma(\tilde{r})L(\Psi^{(1)})d\tilde{r}. \tag{19}$$

The components of the induced metric g^{AB} based on the functionals (19) read

$$g^{AB}_{ren}(y^1,y^2,y^3)=-\frac{1}{4}\begin{pmatrix}-\frac{(y^1)^2}{y^2} & y^1 & -\frac{y^1y^3}{y^2}\\ y^1 & 3\,y^2 & y^3\\ -\frac{y^1y^3}{y^2} & y^3 & -\frac{(y^1)^2}{y^2}W(\frac{y^3}{y^1})\end{pmatrix}. \tag{20}$$

To demonstrate the complexity of these calculations we present the computation of the trickiest component, namely of g^{33}:

$$\begin{aligned}g^{33} &= \frac{\gamma}{4\psi^2}\left(L-\Psi^{(1)}L'\right)^2+\frac{\gamma}{2\psi}\left(L-\Psi^{(1)}L'\right)\Psi^{(2)}L'' \overset{Assumption}{\Longleftrightarrow}\\ g^{33}_{ren} &= \int dr\left[\frac{\gamma}{4\psi^2}\left(L-\Psi^{(1)}L'\right)^2+\frac{\gamma}{2\psi}\left(L-\Psi^{(1)}L'\right)\Psi^{(2)}L''\right]-\\ &- \int dr\frac{d}{dr}\int d\Psi^{(1)}\left(\frac{\left(L-\Psi^{(1)}L'\right)L''}{2\psi}\right)\Leftrightarrow\end{aligned}$$

$$g^{33}_{ren}=\int\frac{\gamma}{4\psi^2}\left[\left(L-\Psi^{(1)}L'\right)^2-\Psi^{(1)}\int\frac{d\Psi^{(1)}}{\Psi^{(1)}}\frac{\partial}{\partial\Psi^{(1)}}\left(L-\Psi^{(1)}L'\right)^2\right]dr. \tag{21}$$

The expression inside the square brackets of g^{33}_{ren} above, being a generic function of $\Psi^{(1)}$, can also be considered as a function of L, say $W(L(\Psi^{(1)}))$. It is thus clear that the **Requirement** is satisfied for any $L(\Psi^{(1)})$:

$$\overset{Assumption}{\Longleftrightarrow}\ g^{33}_{ren}=\frac{(\int\gamma dr)^2}{4\int\gamma\psi^2dr}W(\frac{\int\gamma Ldr}{\int\gamma dr})=\frac{(y^1)^2}{4y^2}W(\frac{y^3}{y^1}).$$

We are legitimated to parameterize the expression $W(L(\Psi^{(1)}))$ in the following way

$$L\left(\Psi^{(1)}\right)^2-\frac{4\,F[L\left(\Psi^{(1)}\right)]^2}{3\,F'[F[L\left(\Psi^{(1)}\right)]]^2}. \tag{22}$$

This "peculiar" parametrization of the arbitrariness in $L\left(\Psi^{(1)}\right)$ has been chosen in order to facilitate the subsequent proof that the freedom in the choice of L (left by the imposition of the **Requirement**) is a pure general coordinate transformation (gct) of the induced re-normalized metric. Taking into account (22) the components of the induced metric (20) read

$$g^{AB}_{ren}(y^1,y^2,y^3) = -\frac{1}{4}\begin{pmatrix} -\frac{(y^1)^2}{y^2} & y^1 & -\frac{y^1 y^3}{y^2} \\ y^1 & 3\,y^2 & y^3 \\ -\frac{y^1 y^3}{y^2} & y^3 & -\frac{(y^3)^2}{y^2} + \frac{4\,(y^1)^2 F\left(\frac{y^3}{y^1}\right)^2}{3\,y^2 F'\left[F\left(\frac{y^3}{y^1}\right)\right]^2} \end{pmatrix}, \text{ and}$$

$$g_{AB_{ren}}(y^1,y^2,y^3) = \begin{pmatrix} \frac{3\,y^2}{(y^1)^4}\left((y^1)^2 - \frac{(y^3)^2 F'\left[F\left(\frac{y^3}{y^1}\right)\right]^2}{F\left(\frac{y^3}{y^1}\right)^2}\right) & -\frac{1}{y^1} & \frac{3\,y^2 y^3 F'\left[F\left(\frac{y^3}{y^1}\right)\right]^2}{(y^1)^3 F\left(\frac{y^3}{y^1}\right)^2} \\ -\frac{1}{y^1} & -\frac{1}{y^2} & 0 \\ \frac{3\,y^2 y^3 F'\left[F\left(\frac{y^3}{y^1}\right)\right]^2}{(y^1)^3 F\left(\frac{y^3}{y^1}\right)^2} & 0 & -\frac{3\,y^2 F'\left[F\left(\frac{y^3}{y^1}\right)\right]^2}{(y^1)^2 F\left(\frac{y^3}{y^1}\right)^2} \end{pmatrix}. \tag{23}$$

One can readily show that any function $\Psi(y^1,y^2,y^3)$ on this manifold is annihilated by the quantum linear constraint, i.e.

$$\hat{\mathcal{H}}_1\Psi(y^1,y^2,y^3) = \frac{\partial\Psi(y^1,y^2,y^3)}{\partial y^1}\,\hat{\mathcal{H}}_1\,y^1 + \frac{\partial\Psi(y^1,y^2,y^3)}{\partial y^2}\,\hat{\mathcal{H}}_1\,y^2 + \frac{\partial\Psi(y^1,y^2,y^3)}{\partial y^3}\,\hat{\mathcal{H}}_1\,y^3 = 0$$

since the derivatives with respect to r are transparent to the partial derivatives of Ψ (which are, just like the y^A's, r-numbers).

The covariant metric (23) describes a three dimensional conformally flat geometry, since the corresponding Cotton-York tensor vanishes. The Ricci scalar is $R = \frac{3}{8\,y^2}$, indicating that the arbitrariness in F (and thus also in L) is a *pure gauge*. Indeed, the change of coordinates

$$(y^1,y^2,y^3) = (e^{-\frac{1}{8}(5\,Y^1+3\,Y^3)}, e^{Y^1+Y^2+Y^3}, e^{-\frac{1}{8}(5\,Y^1+3\,Y^3)}F^{-1}(e^{\frac{1}{24}(-9\,Y^1+8\,Y^2-15\,Y^3)})) \tag{24}$$

(where F^{-1} denotes the function inverse to F, i.e $F^{-1}(F(x)) = x$) brings the metric to the manifestly conformally flat form:

$$g_{AB_{ren}}(Y^1,Y^2,Y^3) = \begin{pmatrix} e^{Y^1+Y^2+Y^3} & 0 & 0 \\ 0 & -\frac{4}{3}\,e^{Y^1+Y^2+Y^3} & 0 \\ 0 & 0 & -e^{Y^1+Y^2+Y^3} \end{pmatrix}, \tag{25}$$

in which all the F dependence has indeed disappeared.

Journal of Physics: Conference Series **306** (2011) 012062 doi:10.1088/1742-6596/306/1/012062

We are now finally in a position to impose the final restriction on the form of the wave function Ψ by the quantum analog of the quadratic constraint $\mathcal{H}_o$: We define that the quantum gravity of the geometries given by (5) be described by the following partial differential equation (in terms of the Y^A's)

$$\hat{\mathcal{H}}_o \Psi \equiv [-\frac{1}{2}\,{}_c + V_{ren}]\, \Psi(Y^1, Y^2, Y^3) = 0 \tag{26}$$

with

$${}_c = + \frac{d-2}{4\,(d-1)}\, R \tag{27}$$

being the conformal Laplacian based on $g_{AB\,ren}(Y^1, Y^2, Y^3)$, R the Ricci scalar, and d the dimensions of $g_{AB\,ren}$. The metric (25) is conformally flat with Ricci scalar $R = \frac{3}{8}\, e^{-Y^1 - Y^2 - Y^3}$, and its dimension is $d = 3$. The re-normalized form of the potential (9) offers us the possibility to introduce, in a dynamical way, topological effects into our wave functional: Indeed, under our **Assumption**, the first two terms become $-2\,\epsilon\, y^1$ and $2\,\Lambda\, y^2$, respectively, while the last, being a total derivative, becomes $A_T \equiv 4\, \frac{\psi \psi'}{\gamma} \,|_\alpha^\beta\ (if\ \alpha < r < \beta)$. In the spirit previously explained we should drop this term, however one could also keep it. The re-normalized form of the remaining, third, term of the potential can be obtained as follows

$$y^3 \;=\; \gamma\, L(\Psi^{(1)}) \Leftrightarrow L(\Psi^{(1)}) = \frac{y^3}{\gamma} \overset{Assumption}{\Longleftrightarrow} L(\Psi^{(1)}) = \frac{y^3}{y^1} \Leftrightarrow \Psi^{(1)} = L^{-1}\left(\frac{y^3}{y^1}\right),$$

thus finally

$$\frac{\psi'}{\gamma} = L^{-1}\left(\frac{y^3}{y^1}\right)$$

and the third term becomes $-2\, y^1 \left[L^{-1}\left(\frac{y^3}{y^1}\right)\right]^2$. Finally, effecting the transformation (24) the form of the re-normalized potential is

$$\begin{aligned} V_{ren} \;=\; & -2\,\epsilon\, e^{-\frac{1}{8}(5\,Y^1 + 3\,Y^3)} - 2\, e^{-\frac{1}{8}(5\,Y^1 + 3\,Y^3)} \left[L^{-1}\left(F^{-1}(e^{\frac{1}{24}(-9\,Y^1 + 8\,Y^2 - 15\,Y^3)})\right)\right]^2 + \\ & 2\,\Lambda\, e^{Y^1 + Y^2 + Y^3} + A_T \end{aligned} \tag{28}$$

and the Wheeler-deWitt equation is given as

$$\begin{aligned} & -2\,\epsilon\, e^{\frac{1}{8}(3\,Y^1 + 5\,Y^3) + Y^2}\, \Psi(Y^1, Y^2, Y^3) + 2\,\Lambda\, e^{2(Y^1 + Y^2 + Y^3)}\, \Psi(Y^1, Y^2, Y^3) - \\ & 2\, e^{\frac{1}{8}(3\,Y^1 + 5\,Y^3) + Y^2} \left[L^{-1}\left(F^{-1}(e^{\frac{1}{24}(-9\,Y^1 + 8\,Y^2 - 15\,Y^3)})\right)\right]^2 \Psi(Y^1, Y^2, Y^3) + \\ & A_T\, e^{Y^1 + Y^2 + Y^3}\, \Psi(Y^1, Y^2, Y^3) - \frac{3}{128} \Psi(Y^1, Y^2, Y^3) - \frac{1}{4} \frac{\partial \Psi(Y^1, Y^2, Y^3)}{\partial Y^1} + \\ & \frac{3}{16} \frac{\partial \Psi(Y^1, Y^2, Y^3)}{\partial Y^2} + \frac{1}{4} \frac{\partial \Psi(Y^1, Y^2, Y^3)}{\partial Y^3} - \frac{1}{2} \frac{\partial^2 \Psi(Y^1, Y^2, Y^3)}{\partial\, (Y^1)^2} + \frac{3}{8} \frac{\partial^2 \Psi(Y^1, Y^2, Y^3)}{\partial\, (Y^2)^2} + \\ & \frac{1}{2} \frac{\partial^2 \Psi(Y^1, Y^2, Y^3)}{\partial\, (Y^3)^2} = 0. \end{aligned}$$

Since F^{-1} is an arbitrary function of its arguments, we may contemplate the choice:

$$F^{-1}\left(e^{\frac{1}{24}(-9\,Y^1 + 8\,Y^2 - 15\,Y^3)}\right) = L\left(\sqrt{e^{\frac{1}{24}(-9\,Y^1 + 8\,Y^2 - 15\,Y^3)} - \epsilon}\right). \tag{29}$$

Of course there is a question of existence for such a choice: since F which appears in (22) is nothing but a convenient parametrization of the expression inside the square brackets of (21), any demand that F has a specified form (much more in terms of L) constitutes an implicit restriction on the form of L itself. Subsequently, at least the existence of such an L must be secured. Indeed, it can be shown [?] that an appropriate L exists; its form is given by :

$$L(\Psi^{(1)}) = m + \int \frac{(\Psi^{(1)})^{3/2}}{((\Psi^{(1)})^2 - \epsilon)^{13/16}} e^{k - \frac{3\epsilon}{16((\Psi^{(1)})^2-\epsilon)}} d\Psi^{(1)} \quad \text{where} \quad c_1\, m + c_2 + c_3\, e^k = 0.$$

This choice for F reduces the Wheeler-deWitt equation to the final almost-separable form

$$\begin{aligned} &2\,\Lambda\, e^{2(Y^1+Y^2+Y^3)}\Psi(Y^1,Y^2,Y^3) - 2\,\epsilon\, e^{\frac{4}{3}Y^2}\Psi(Y^1,Y^2,Y^3) + A_T\, e^{Y^1+Y^2+Y^3}\Psi(Y^1,Y^2,Y^3) - \\ &\frac{3}{128}\Psi(Y^1,Y^2,Y^3) - \frac{1}{4}\frac{\partial\Psi(Y^1,Y^2,Y^3)}{\partial Y^1} + \frac{3}{16}\frac{\partial\Psi(Y^1,Y^2,Y^3)}{\partial Y^2} + \frac{1}{4}\frac{\partial\Psi(Y^1,Y^2,Y^3)}{\partial Y^3} - \\ &\frac{1}{2}\frac{\partial^2\Psi(Y^1,Y^2,Y^3)}{\partial\,(Y^1)^2} + \frac{3}{8}\frac{\partial^2\Psi(Y^1,Y^2,Y^3)}{\partial\,(Y^2)^2} + \frac{1}{2}\frac{\partial^2\Psi(Y^1,Y^2,Y^3)}{\partial\,(Y^3)^2} = 0. \end{aligned} \tag{30}$$

This equation is separable for $\Lambda = 0$ and $A_T = 0$. In this case it can readily be solved: assuming $\Psi(Y^1,Y^2,Y^3) = \Psi^1(Y^1)\,\Psi^2(Y^2)\,\Psi^3(Y^3)$ and dividing (30) by Ψ we get the three ordinary differential equations:

$$\begin{aligned} \frac{1}{4\,\Psi^1(Y^1)}\frac{d\Psi^1(Y^1)}{dY^1} + \frac{1}{2\,\Psi^1(Y^1)}\frac{d^2\Psi^1(Y^1)}{d\,(Y^1)^2} &= m, \\ \frac{3}{16\,\Psi^2(Y^2)}\frac{d\Psi^2(Y^2)}{dY^2} + \frac{3}{8\,\Psi^2(Y^2)}\frac{d^2\Psi^2(Y^2)}{d\,(Y^2)^2} - 2\,\epsilon\, e^{\frac{4}{3}Y^2} &= n, \\ \frac{1}{4\,\Psi^3(Y^3)}\frac{d\Psi^3(Y^3)}{dY^3} + \frac{1}{2\,\Psi^3(Y^3)}\frac{d^2\Psi^3(Y^3)}{d\,(Y^3)^2} - \frac{3}{128} &= m - n, \end{aligned}$$

where m and n are separation constants. Their solutions are:

$$\begin{aligned} \Psi^1(Y^1) &= c_1\, e^{\frac{1}{4}\left(-1-\sqrt{1+32\,m}\right)Y^1} + c_2\, e^{\frac{1}{4}\left(-1+\sqrt{1+32\,m}\right)Y^1}, \\ \Psi^2(Y^2) &= c_3\, e^{-Y^2/4}\, I_{-\frac{\sqrt{3}}{8}\sqrt{3+128\,n}}\left(2\,\epsilon\,\sqrt{3}\,e^{2\,Y^2/3}\right) + c_4\, e^{-Y^2/4}\, I_{\frac{\sqrt{3}}{8}\sqrt{3+128\,n}}\left(2\,\epsilon\,\sqrt{3}\,e^{2\,Y^2/3}\right) \\ \Psi^3(Y^3) &= c_5\, e^{\frac{1}{8}\left(-2-\sqrt{7+128\,m-128\,n}\right)Y^3} + c_6\, e^{\frac{1}{8}\left(-2+\sqrt{7+128\,m-128\,n}\right)Y^3}, \end{aligned}$$

where $I_{\pm\frac{\sqrt{3}}{8}\sqrt{3+128\,n}}\left(2\sqrt{3}\,e^{2\,Y^2/3}\right)$ are modified Bessel functions of the first kind and non-integer order.

References

[1] Christodoulakis T, Doulis G, Terzis P A, Melas E, Grammenos T, Papadopoulos G O and Spanou A 2008 *Class. Quantum Grav.* **25** 235014
[2] Hajicek P and Kuchař K V 1990 *Phys. Rev. D* **41** 1091
[3] Hajicek P and Kuchař K V 1990 *JMP* **31** 1723
[4] Christodoulakis T, Doulis G, Terzis P A, Melas E, Grammenos T, Papadopoulos G O and Spanou A 2010 *Class. Quantum Grav.* **27** 1455018
[5] Dirac P A M 1964 *Lectures on Quantum Mechanics* (New York: Academic Press)
[6] Sundermeyer K 1982 *Constrained Dynamics* (Lecture Notes in Physics vol 169) (Heidelberg: Springer-Verlag)
[7] Thomas T T 1991 *The Differential Invariants of Generalized Spaces* (New York: Chelsea Publishing Company) p 140

An Information Geometric Analysis of Entangled Continuous Variable Quantum Systems

D.-H. Kim
Institute for the Early Universe, Ewha Womans University, Seoul 120-750, South Korea

S. A. Ali
Department of Physics, State University of New York at Albany, 1400 Washington Avenue, Albany, NY 12222, USA
Department of Arts and Sciences, Albany College of Pharmacy and Health Sciences, 106 New Scotland Avenue, Albany, NY 12208, USA

C. Cafaro and S. Mancini
School of Science and Technology, Physics Division, University of Camerino, I-62032 Camerino, Italy

Abstract. In this work, using information geometric (IG) techniques, we investigate the effects of micro-correlations on the evolution of maximal probability paths on statistical manifolds induced by systems whose microscopic degrees of freedom are Gaussian distributed. Analytical estimates of the information geometric entropy (IGE) as well as the IG analogue of the Lyapunov exponents are presented. It is shown that the entanglement duration is related to the scattering potential and incident particle energies. Finally, the degree of entanglement generated by an s-wave scattering event between minimum uncertainty wave-packets is computed in terms of the purity of the system.

1. Introduction

One of the most important, interesting and often debated features of (composite) quantum mechanical systems is their ability to become entangled [1, 2]. By quantum entanglement we mean quantum correlations among the distinct subsystems of the entire composite quantum system. For such correlated quantum systems, it is not possible to specify the quantum state of any subsystem independently of the remaining subsystems [1]. Quantum entanglement is an indispensable resource for quantum information processes. Continuous variable quantum systems (CVQS) are also an interesting topic in quantum information theory [3]. By CVQS we refer to quantum mechanical systems on which one can - in principle - perform measurements of certain observables whose eigenvalue spectrum is continuous. The most realistic approach to the generation of entangled continuous variable systems is via dynamical interaction, of which local scattering events (collisions) are a natural, ubiquitous type [4]. The primary focus of the present article is to explore the potential utility of the information geometric approach to chaos (IGAC) for analyzing quantum mechanical systems. The IGAC [5, 6, 7, 8, 9, 10] is a theoretical framework developed to study chaos in informational geodesic flows on statistical manifolds associated with probabilistic descriptions of physical, biological or chemical systems. For a system of two spin-0, non-relativistic particles in one dimension with no internal degrees of freedom, a complete set of commuting observables is furnished by the momentum operators of each particle [11]. The continuous variables in our case are therefore taken to be the momentum of each particle. In the present work we consider two CVQS with Gaussian continuous degrees of freedom that are prepared independently, interact via a scattering processes mediated by an interaction (scattering) potential and separate again. We investigate the entanglement of the two-particle wave-function of the system generated by such a scattering event.

The layout of this article is as follows. In Section II, we briefly review the phenomenon of quantum entanglement for two colliding wave packets. In Section III, we describe the general steps through which the statistical manifold $\mathcal{M}$ in presence of micro-correlations is constructed. In Section IV, we apply IG techniques in conjunction with standard s-wave scattering theory to study our specific continuous two-variable micro-correlated Gaussian model. In Section V, the IGAC is used to investigate the chaoticity arising in the information constrained dynamics of our quantum entangled system. Our final remarks are presented in Section VI.

2. Entanglement and Wave-packet Scattering Processes

For the purpose of modeling a head-on collision we consider two identical spin-0 particles in momentum space, each represented by minimum uncertainty Gaussian wave-packets. Before collision, particles 1 and 2 are initially located far from each other - a linear distance R_{o} - each having the initial average momentum $\langle p_1\rangle_{\text{o}} = p_{\text{o}} = \hbar k_{\text{o}}$ and $\langle p_2\rangle_{\text{o}} = -p_{\text{o}} = -\hbar k_{\text{o}}$, respectively, with equal momentum dispersion $\sigma_{\text{o}} = \hbar\sigma_{k\text{o}}$. The normalized, separable (i.e., non-entangled) two-particle Gaussian wave function representing the situation before collision is then given by [12]

$$\begin{aligned}\psi_{\text{before}}(k_1,k_2) &= \left(\frac{1}{2\pi\sigma_{k\text{o}}^2}\right)^{1/2}\exp\left[-\frac{(k_1-k_{\text{o}})^2}{4\sigma_{k\text{o}}^2}-\frac{(k_2+k_{\text{o}})^2}{4\sigma_{k\text{o}}^2}\right]\\ &\times\; e^{i\left[-\frac{1}{2}(k_1-k_{\text{o}})R_{\text{o}}+\frac{1}{2}(k_2+k_{\text{o}})R_{\text{o}}\right]}.\end{aligned}\tag{1}$$

After collision, the wave function for the two-particle system in the long time limit takes the form [12]:

$$\begin{aligned}\psi_{\text{after}}(k_1,k_2,t) = (N)^{-1/2}\Big[&\;\psi_{\text{before}}(k_1,k_2)\, e^{-i\left(k_1^2+k_2^2\right)t/(2\hbar m)}\\ &+\varepsilon\psi_{\text{scat}}(k_1,k_2,t)\Big],\end{aligned}\tag{2}$$

where N and ε are normalization constants such that ψ_{after} and ψ_{scat} are both normalized to unity, and m denotes the mass of each particle. We treat $|\varepsilon| \ll 1$ as a small number. Following [12], one can specify

$$\begin{aligned}\varepsilon\psi_{\text{scat}}(k_1,k_2,t) &= \left(\frac{1}{2\pi\sigma_{k\text{o}}^2}\right)^{1/2}\exp\left[-\frac{K^2+4(k-k_{\text{o}})^2}{4\sigma_{k\text{o}}^2}\right]\\ &\times\frac{if(k)\,\sigma_{k\text{o}}^2}{2k_{\text{o}}-i\sigma_{k\text{o}}^2R_{\text{o}}}e^{-\hbar K^2t/(2M)+i\left[(k-k_{\text{o}})R_{\text{o}}-\hbar k^2t/(2\mu)\right]},\end{aligned}\tag{3}$$

where we have adopted the center of mass and relative coordinates such that the conjugate momenta, $K \equiv k_1 + k_2$ and $k \equiv \frac{1}{2}(k_1 - k_2)$ are used along with the total mass, $M = 2m$ and the reduced mass, $\mu = m/2$, and $f(k) \equiv \frac{e^{i2\theta(k)}-1}{2ik}$ is the s-wave scattering amplitude due to the s-wave scattering phase shift, $\theta(k)$. In describing a head-on collision between two identical particles we may choose (for simplicity) the total momentum of the system to be vanishing, i.e. we choose the momenta of each particle to be equal in magnitude but opposite in sign such that $K=0$. Then using (2) and (3), and assuming that our wave-packet is well-localized around $k=k_{\mathrm{o}}$, we find

$$|\psi(k_1,k_2,t)|^2 \sim \exp\left(-\frac{\tilde{k}^2}{\sigma_{k\mathrm{o}}^2}\right)\left[1+2\Re(\varrho_{\mathrm{o}})\exp\left(-\frac{\tilde{k}^2}{2\sigma_{k\mathrm{o}}^2}\right) + |\varrho_{\mathrm{o}}|^2\exp\left(-\frac{\tilde{k}^2}{\sigma_{k\mathrm{o}}^2}\right)\right], \tag{4}$$

where $\tilde{k}\equiv\frac{1}{2}[(k_1-k_{\mathrm{o}})-(k_2+k_{\mathrm{o}})]$ and $\varrho_{\mathrm{o}}\equiv\frac{if(k_{\mathrm{o}})\sigma_{k\mathrm{o}}^2}{2k_{\mathrm{o}}-i\sigma_{k\mathrm{o}}^2R_{\mathrm{o}}}$.

It is worth noting that by choosing $K=0$ in computing $|\psi(k_1,k_2,t)|^2$ we render $|\psi(k_1,k_2,t)|^2$ time-independent. This implies that with the appropriate choice of reference frame, in which $K=0$ (experiencing an exact head-on collision), the probability associated with the quantum scattering process can be interpreted as static.

2.1. On the Two-variable Micro-correlated Gaussian Statistical Model

We want to make reliable macroscopic predictions when only partial knowledge about the micro-structure of a system is available. The complexity of such predictions is quantified in terms of the IGE and the Lyapunov exponents. Consider micro-correlated Gaussian statistical models with probability distribution $P(x,y|\mu_x,\mu_y,\sigma)$ [13],

$$P(x,y|\mu_x,\mu_y,\sigma)=\frac{\exp\left\{-\frac{1}{2(1-r^2)}\left[\frac{(x-\mu_x)^2}{\sigma^2}-2r\frac{(x-\mu_x)(y-\mu_y)}{\sigma^2}+\frac{(y-\mu_y)^2}{\sigma^2}\right]\right\}}{2\pi\sigma^2\sqrt{1-r^2}}, \tag{5}$$

which is a bivariate normal distribution, where $\mu_x\equiv\langle x\rangle$, $\mu_y\equiv\langle y\rangle$ and $\sigma=\sqrt{\left\langle(x-\langle x\rangle)^2\right\rangle}=\sqrt{\left\langle(y-\langle y\rangle)^2\right\rangle}>0$, and the micro-correlation coefficient $r=r(x,y)\overset{\mathrm{def}}{=}\frac{\langle xy\rangle-\langle x\rangle\langle y\rangle}{\sigma^2}$ assumes values in the domain $(-1,1)$.

We introduce the Fisher-Rao information metric [20]

$$\begin{aligned} g_{ab}(\Theta) &= \int dX P(X|\Theta)\,\partial_a\ln P(X|\Theta)\,\partial_b\ln P(X|\Theta) \\ &= 4\int dX\partial_a\sqrt{P(X|\Theta)}\partial_b\sqrt{P(X|\Theta)}, \end{aligned} \tag{6}$$

where $a,b=1,2,3$ and $X=(x,y)$, $\Theta=(\mu_x,\mu_y,\sigma)$, $\partial_a=\frac{\partial}{\partial\Theta^a}$, which assigns an information geometry to the space of macrostates. Upon substituting (5) in (6), the Fisher-Rao information metric $g_{ab}(\mu_x,\mu_y,\sigma;r)$ becomes

$$g_{ab}(\mu_x,\mu_y,\sigma;r)=\frac{1}{\sigma^2}\begin{pmatrix} -\frac{1}{r^2-1} & \frac{r}{r^2-1} & 0 \\ \frac{r}{r^2-1} & -\frac{1}{r^2-1} & 0 \\ 0 & 0 & 4\end{pmatrix}. \tag{7}$$

The metric for the non-correlated case, $g_{ab}(\mu_x, \mu_y, \sigma; 0)$ can be obtained from (7) by setting $r = 0$. With $r \ll 1$, we may split the micro-correlated geometry into two pieces, i.e. $g_{ab}(\mu_x, \mu_y, \sigma; r) = g_{ab}(\mu_x, \mu_y, \sigma; 0) + h_{ab}(\mu_x, \mu_y, \sigma; r)$. We comment on this metric decomposition in our Final Remarks.

In what follows we limit our analysis to the study of non-negative micro-correlations, that is we will consider $r \in [0, 1)$.

2.2. Information Dynamics on $\mathcal{M}^{3D}_{\text{corr.}}$ (three-dim statistical manifold in presence of micro-correlations)

The information dynamics can be derived from a standard principle of least action of Jacobi type [14]. The geodesic equations for the macro-variables of our Gaussian model are given by

$$\frac{d^2\Theta^a}{d\tau^2} + \Gamma^a_{bc}\frac{d\Theta^b}{d\tau}\frac{d\Theta^c}{d\tau} = 0, \tag{8}$$

where $a, b, c = 1, 2, 3$ and we denote $\Theta^1 = \mu_1 = \mu_x$, $\Theta^2 = \mu_2 = \mu_y$, $\Theta^3 = \sigma$. The connection coefficients Γ^a_{bc} appearing in (8) are defined as $\Gamma^a_{bc} = \frac{1}{2}g^{ad}(\partial_b g_{dc} + \partial_c g_{bd} - \partial_d g_{bc})$ [15], and computed by means of 7. After integration of (8), the geodesic trajectories become: for the correlated Gaussian system,

$$\begin{aligned} \mu_1(\tau; r) &= -\sqrt{(1-r)\left(p_o^2 + 2\sigma_o^2\right)}\tanh\left(A_o\tau\right), & (9)\\ \mu_2(\tau; r) &= \sqrt{(1-r)\left(p_o^2 + 2\sigma_o^2\right)}\tanh\left(A_o\tau\right), & (10)\\ \sigma(\tau; r) &= \sqrt{\frac{1}{2}p_o^2 + \sigma_o^2}\frac{1}{\cosh\left(A_o\tau\right)}, & (11)\end{aligned}$$

where $p_o \equiv \mu_1(-\tau_o; 0)$, $\sigma_o \equiv \sigma(-\tau_o; 0)$ and the quantity A_o is defined as

$$\begin{aligned} A_o &\equiv \frac{1}{\tau_o}\sinh^{-1}\left(\frac{p_o}{\sqrt{2}\sigma_o}\right) \\ &\overset{\frac{\sigma_o}{p_o}\ll 1}{=} \frac{1}{\tau_o}\left\{\ln\left(\frac{\sqrt{2}p_o}{\sigma_o}\right) + \frac{1}{2}\left(\frac{\sigma_o}{p_o}\right)^2 - \frac{3}{8}\left(\frac{\sigma_o}{p_o}\right)^4 + \mathcal{O}\left[\left(\frac{\sigma_o}{p_o}\right)^6\right]\right\} > 0. \end{aligned} \tag{12}$$

The geodesic trajectories on $\mathcal{M}^{3D}_{\text{non-corr.}}$ for the non-correlated Gaussian system may be obtained from (9), (10) and (11) by setting $r = 0$.

3. Application to Quantum Physics

We employ IG techniques in conjunction with standard partial wave quantum scattering theory to investigate quantum entanglement produced by a head-on collision between two Gaussian wave-packets in momentum space.

3.1. Momentum-space Gaussian Statistical Models

By letting $x \to p_1 = \hbar k_1$, $y \to p_2 = \hbar k_2$ and $\mu_x \to \langle p_1\rangle_o = p_o = \hbar k_o$, $\mu_y \to \langle p_2\rangle_o = -p_o = -\hbar k_o$ in (5), we can identify the two particle probability density before collision, given by (1), with the non-correlated probability distribution obtained from (5) with $r = 0$. In a similar manner, the probability density $|\psi(k_1, k_2, t)|^2$ in (4) is approximated with the Gaussian probability distribution (5) with $r \neq 0$. Comparison of (4) and (5) suggests that when both $|\varrho_o|$ and r

are very small: $P_{\rm QM} \simeq P_{\rm Corr-Gaussian}$, where

$$P_{\rm QM} = \frac{1}{\sqrt{\pi}\sigma_{ko}\left[1+\frac{2\sqrt{2}}{\sqrt{3}}\Re\left(\varrho_o\right)+\frac{1}{\sqrt{2}}\left|\varrho_o\right|^2\right]}\exp\left(-\frac{\tilde{k}^2}{\sigma_{ko}^2}\right) \times \left[1+2\Re\left(\varrho_o\right)\exp\left(-\frac{\tilde{k}^2}{2\sigma_{ko}^2}\right)+\left|\varrho_o\right|^2\exp\left(-\frac{\tilde{k}^2}{\sigma_{ko}^2}\right)\right], \tag{13}$$

and

$$P_{\rm Corr-Gaussian} = \frac{1}{\sqrt{\pi}\sigma_{ko}\sqrt{1-r}}\exp\left[-\frac{\tilde{k}^2}{\sigma_{ko}^2\left(1-r\right)}\right], \tag{14}$$

respectively. Here we have reduced the expression via $\tilde{K} \equiv (k_1 - k_o) + (k_2 + k_o) = K = 0$ and $\tilde{k} \equiv \frac{1}{2}\left[(k_1 - k_o) - (k_2 + k_o)\right]$, and the normalization factor for each distribution has been weighted.

We are thus able to model the system of pair of particles, which describes the pre-collision and post-collision scenarios, by the normalized, bipartite Gaussian statistical system. In this system we join two different charts of Gaussian statistical manifolds, one without correlation (before collision) and the other with correlation (after collision). To examine how the correlation affects the geodesic curves, we compute the relative conjugate-momentum curves, $\langle p\left(\tau;0\right)\rangle \equiv \frac{1}{2}\left[\mu_2\left(\tau;0\right)-\mu_1\left(\tau;0\right)\right]$ and $\langle p\left(\tau;r\right)\rangle \equiv \frac{1}{2}\left[\mu_2(\tau;r)-\mu_1(\tau;r)\right]$ and find that at any arbitrary time $\tau \geq 0$

$$\langle p\left(\tau;0\right)\rangle \geq \langle p\left(\tau;r\right)\rangle = \sqrt{1-r}\,\langle p\left(\tau;0\right)\rangle. \tag{15}$$

This implies that the correlation causes the momentum to reduce for any $\tau \geq 0$ (relative to the non-correlated situation). This situation is analogous to the change in momentum caused by a repulsive scattering potential. It is then reasonable to assume there exists some connection between the scattering potential and the correlation. That is, with the correlation, the wave-packets experience the effect of a repulsive potential; the magnitude of the wave vectors (momenta) decreases relative to the corresponding non-correlated value. Given the potential $V(x) = V$ for $0 \leq x \leq L$ and $V(x) = 0$ for $x > L$, one may write [21]

$$k_r \cot\left(k_r L\right) = k_o \cot\left(k_o L + \theta_o\right), \tag{16}$$

$$k_r = \frac{\sqrt{2\mu\left(\mathcal{E}-V\right)}}{\hbar}, \quad 0 < x < L, \quad k_o = \frac{\sqrt{2\mu\mathcal{E}}}{\hbar}, \quad x > L. \tag{17}$$

The quantities V and L denote the height ($V > 0$; repulsive potential) and range of the potential, respectively, and $\theta_o \equiv \theta\left(k_o\right)$ the s-wave scattering phase shift, and k_r and k_o the wave vectors with and without the correlation, respectively, and μ and $\mathcal{E}$ being the reduced mass and kinetic energy of the two-particle system in the relative coordinates, respectively. From (15) one finds that the correlation renders

$$k_o \longrightarrow k_r \equiv \sqrt{1-r}k_o. \tag{18}$$

Then using (17) and (18), we determine the scattering potential,

$$V = r\mathcal{E} = r\frac{p_o^2}{2\mu}, \tag{19}$$

where $p_o = \hbar k_o$. With the potential determined, one can determine the scattering phase shift for low energy s-wave scattering $k_o L \ll 1$ by combining (16), (17) and (19), the result being

$$\theta_o \approx -\frac{r k_o^3 L^3}{3} = -\frac{r p_o^3 L^3}{3\hbar^3} = -\frac{2\mu V p_o L^3}{3\hbar^3}, \tag{20}$$

which is in agreement with [16].

3.2. Correlation vs. Entanglement: Connection Established via Scattering and Purity

Thepurity of a system of two particles engaged in a head-on collision was calculated in [12] by deriving the two-particle wave function modified by s-wave scattering amplitudes. They utilized the purity function $\mathcal{P}$ as a measure of entanglement. For pure two-particle states, the smaller the value of $\mathcal{P}$ the higher the entanglement. That is, the loss of purity provides an indicator of the degree of entanglement. Hence, a disentangled product state corresponds to $\mathcal{P} = 1$. For the system being considered, it was found in [12] that the purity of final state is evaluated approximately as

$$\mathcal{P} \approx 1 - 2\left|\varepsilon\right|^2; \ \left|\varepsilon\right| \approx \sqrt{2}\left|f\left(p_{\text{o}}\right)\right| \frac{\sigma_{\text{o}}}{\hbar\chi} \tag{21}$$

for the wave-packets whose widths are very narrow in momentum space: the subscript, "$_{\text{o}}$" denotes the initial state, and $f\left(p_{\text{o}}\right) \equiv \frac{\hbar\left(e^{i2\theta_{\text{o}}}-1\right)}{2ip_{\text{o}}}$ is the s-wave scattering amplitude with θ_{o} representing the s-wave scattering phase shift, and $\chi \equiv \sqrt{1+\left(\frac{\sigma_{\text{o}}^2 R_{\text{o}}}{2\hbar p_{\text{o}}}\right)^2}$ corresponding to the spreading factor of the spatial width of the wave-packets. Using (21), and employing the scattering cross section $\Sigma\left(p_{\text{o}}\right) = 4\pi\left|f\left(p_{\text{o}}\right)\right|^2$, and requiring that if the particles are not correlated (i.e. if $r = 0$) after collision then no entanglement should be present (i.e. $\mathcal{P} = 1$ implying $\theta_{\text{o}} = f\left(p_{\text{o}}\right) = \Sigma\left(p_{\text{o}}\right) = 0$), we may express the purity in an alternative manner as

$$\mathcal{P} = 1 - 4\left(\frac{\sigma_{\text{o}}}{\hbar\chi}\right)^2\left|f\left(p_{\text{o}}\right)\right|^2 = 1 - \frac{\left(\frac{\sigma_{\text{o}}}{\hbar\chi}\right)^2\Sigma\left(p_{\text{o}}\right)}{\pi} = 1 - \frac{16\mu^2\sigma_{\text{o}}^2V^2L^6}{9\hbar^6\left[1+\left(\frac{\sigma_{\text{o}}^2R_{\text{o}}}{2\hbar p_{\text{o}}}\right)^2\right]}. \tag{22}$$

The correlation coefficient r can now be expressed in terms of the physical quantities such as the scattering potential, the scattering cross section and purity as follows

$$r = \frac{2\mu V}{p_{\text{o}}^2} = \frac{3\hbar^2\sqrt{\Sigma\left(p_{\text{o}}\right)}}{2\sqrt{\pi}p_{\text{o}}^2L^3} = \frac{3\hbar^3\chi\sqrt{1-\mathcal{P}}}{2p_{\text{o}}^2\sigma_{\text{o}}L^3}. \tag{23}$$

From (23) we observe that the micro-correlation coefficient r is directly associated with the quantum scattering process, and thus with the quantum entanglement. For example, the cross term $\langle p_1p_2\rangle$ in $r = r\left(p_1, p_2\right) \overset{\text{def}}{=} \frac{\langle p_1p_2\rangle - \langle p_1\rangle\langle p_2\rangle}{\sigma^2}$ may represent the average interference between transmitted/reflected modes in the momentum degrees of freedom of particle 1 and 2, respectively.

From (15) it is observed that for the micro-correlated Gaussian system considered here, more time is required to attain the same momentum value compared with the non-correlated Gaussian system. For example, to attain the same value of initial momentum p_{o}, the non-correlated system and the micro-correlated system would require time intervals τ_{o} and τ_*, respectively, where

$$p_{\text{o}} = \sqrt{p_{\text{o}}^2 + 2\sigma_{\text{o}}^2}\tanh\left(A_{\text{o}}\tau_{\text{o}}\right), \tag{24}$$

$$p_{\text{o}} = \sqrt{\left(1-r\right)\left(p_{\text{o}}^2 + 2\sigma_{\text{o}}^2\right)}\tanh\left(A_{\text{o}}\tau_*\right). \tag{25}$$

Combining (24) and (25) and using (12), we find

$$\Delta \propto \left|\ln\left\{1 - \left[\left(1-r\right)^{-1/2} - 1\right]\cdot\eta_\Delta\right\}\right|, \tag{26}$$

where $\Delta \equiv \tau_* - \tau_{\text{o}}$ represents the prolongation and

$$\eta_\Delta \equiv \frac{1}{2}e^{2A_{\text{o}}\tau_{\text{o}}} = \left(\frac{p_0}{\sigma_0}\right)^2\exp\left[\left(\frac{\sigma_0}{p_0}\right)^2 - \frac{3}{4}\left(\frac{\sigma_0}{p_0}\right)^4 + \mathcal{O}\left[\left(\frac{\sigma_0}{p_0}\right)^6\right]\right]$$

for $\frac{\sigma_0}{p_0} \ll 1$. See the Final Remarks for further discussion of the prolongation.

4. Complexity

In this Section, we investigate the complexity of the dynamical system being considered.

4.1. Information Geometric Indicators of Chaoticity

4.1.1. Jacobi Fields and Lyapunov Exponents The relative geodesic spread on $\mathcal{M}^{3D}_{\text{corr.}}$ is formally characterized by the Jacobi-Levi-Civita (JLC) equation [22, 23],

$$\frac{D^2 J^a}{D\tau^2} + \mathcal{R}^a_{bcd} \frac{\partial \Theta^b}{\partial \tau} J^c \frac{\partial \Theta^d}{\partial \tau} = 0, \tag{27}$$

where $a, b, c, d = 1, 2, 3$ and $\frac{D^2}{D\tau^2}$ denotes the second order covariant derivatives and the Riemann curvature tensor $\mathcal{R}_{abcd}$ is defined as $\mathcal{R}^a_{bcd} = \partial_c \Gamma^a_{bd} - \partial_d \Gamma^a_{bc} + \Gamma^a_{fc}\Gamma^f_{bd} - \Gamma^a_{fd}\Gamma^f_{bc}$, and the Jacobi vector field components J^a are given by $J^a = \delta_{\varsigma}\Theta^a \equiv \left.\frac{\partial \Theta^a(\tau;\vec{\varsigma})}{\partial \varsigma^b}\right|_\tau \delta\varsigma^b$ with τ and $\vec{\varsigma} = (\varsigma^1, \varsigma^2, \varsigma^3)$ denoting affine parameters. Through some involved analysis by means of (7) and (27), we reduce the JLC equation to

$$\frac{D^2 \mathcal{J}_{\mathcal{M}}}{D\tau^2} - A_{\text{o}}^2 \mathcal{J}_{\mathcal{M}} = 0. \tag{28}$$

where the Jacobi vector field intensity is defined as $\mathcal{J}_{\mathcal{M}} \equiv \left(g_{ab}J^a J^b\right)^{\frac{1}{2}} = \left(J^a J_a\right)^{\frac{1}{2}}$ and A_{o} is given by (12). Solutions of equation (28) assume the form $\mathcal{J}_{\mathcal{M}}(\tau) = \text{const} \times \sinh(A_{\text{o}}\tau)$. In our approach the quantity $\lambda_{\mathcal{M}}$,

$$\lambda_{\mathcal{M}} \overset{\text{def}}{=} \lim_{\tau\to\infty} \frac{1}{\tau} \ln \left[\frac{\left\|\mathcal{J}_{\mathcal{M}}(\tau)\right\|^2 + \left\|\frac{d\mathcal{J}_{\mathcal{M}}(\tau)}{d\tau}\right\|^2}{\left\|\mathcal{J}_{\mathcal{M}}(0)\right\|^2 + \left\|\left.\frac{d\mathcal{J}_{\mathcal{M}}(\tau)}{d\tau}\right|_{\tau=0}\right\|^2} \right] = 2A_{\text{o}} > 0 \tag{29}$$

would play the role of the conventional Lyapunov exponents. It is evident from (29) that the Lyapunov exponents can be determined solely from the initial conditions.

4.1.2. Information Geometric Entropy It is known [18, 19] that a suitable indicator of temporal complexity within the IGAC framework is provided by the *information geometric entropy* (IGE) $\mathcal{S}_{\mathcal{M}^{3D}_{\text{corr.}}}(\tau)$,

$$\mathcal{S}_{\mathcal{M}^{3D}_{\text{corr.}}}(\tau) \overset{\text{def}}{=} \lim_{\tau\to\infty} \ln \widetilde{vol}\left[\mathcal{D}^{(\text{geodesic})}_{\Theta}(\tau)\right], \tag{30}$$

where the quantity $\widetilde{vol}\left[\mathcal{D}^{(\text{geodesic})}_{\Theta}(\tau)\right]$ is termed the *information geometric complexity* (IGC). The IGC is defined as the temporal average of the dynamical statistical volume, and for the system under investigation, via (7), (9), (10) and (11) it is given by

$$\begin{aligned} \widetilde{vol}\left[\mathcal{D}^{(\text{geodesic})}_{\Theta}(\tau)\right] &\overset{\text{def}}{=} \lim_{\tau\to\infty}\left(\frac{1}{\tau}\int_0^\tau d\tau' \int_{\mu_1(0)}^{\mu_1(\tau')}\int_{\mu_2(0)}^{\mu_2(\tau')}\int_{\sigma(0)}^{\sigma(\tau')} \sqrt{g}\, d\mu_1 d\mu_2 d\sigma\right) \\ &= \frac{4\sqrt{\frac{1-r}{1+r}}}{A_{\text{o}}}\left[-\frac{3}{2}A_{\text{o}} + \frac{1}{4}\frac{\sinh(2A_{\text{o}}\tau)}{\tau} + \frac{\tanh(A_{\text{o}}\tau)}{\tau}\right]. \end{aligned} \tag{31}$$

By means of (30) and (31), the IGE for the correlated Gaussian statistical model is given by

$$\mathcal{S}_{\text{corr.}}(\tau; r) \overset{\tau\to\infty}{=} \lambda_{\mathcal{M}}\tau + \frac{1}{2}\ln\left(\frac{1-r}{1+r}\right). \tag{32}$$

5. Final Remarks

In this article, micro-correlated and non-correlated Gaussian statistical models were used to model scattering-induced quantum entanglement. The manifolds $\mathcal{M}^{3D}_{\text{corr.}}$ and $\mathcal{M}^{3D}_{\text{non-corr.}}$ were used to model the quantum entanglement induced by head-on scattering of two spin-0 particles, each represented by minimum uncertainty wave-packets. When $r = 0$, the quantities θ_o, Σ and V are each zero while $\mathcal{P} = 1$ (indicating that the system is not entangled). When $r \neq 0$, the wave-packets experience the effect of a repulsive potential and the magnitude of the wave vectors (momenta) decreases relative to their corresponding non-correlated value. It was found that the micro-correlation coefficient r depends inversely on p_o. The role played by r in the quantities $\mathcal{P}$, Σ, θ_o, and V suggests that information about quantum scattering and therefore about quantum entanglement is encoded in the statistical micro-correlation, specifically in the covariance term appearing in the definition of r.

Furthermore, from the notion $g_{ab}(\mu_x, \mu_y, \sigma; r) = g_{ab}(\mu_x, \mu_y, \sigma; 0) + h_{ab}(\mu_x, \mu_y, \sigma; r)$ for $r \ll 1$, we observe that the quantum entanglement manifests as a geometric perturbation of the statistical space in analogy to the interpretation of a static gravitational field as a perturbation of flat space. After scattering the two particles maintain a correlation among their microscopic momentum degrees of freedom regardless of the extent of their seperation in statistical space. This fact, together with the time-independence of the statistical geometry (i.e. the information metric is Riemannian since its signature is positive definite) leads to a notion of statistical non-locality. The perturbation of statistical geometry is associated with the scattering phase shift in the statistical momentum space.

The prolongation of the system entanglement, denoted Δ, was defined as the time required for the observed momentum difference between the correlated and non-correlated momenta to vanish. From (26), it was found that for r values close to its upper bound, the prolongation Δ becomes infinitely large. On the other hand, with r vanishing (i.e., no micro-correlation) Δ is identically zero. With r fixed however, the prolongation Δ depends on p_o and σ_o. Thus, the prolongation Δ may be taken to represent the duration of quantum entanglement for a given correlated system where the entanglement duration can be controlled by the initial conditions p_o and σ_o as well as r. Maximal prolongation occurs when r is greatest and the ratio σ_o/p_o is smallest. For small initial r and p_o, Δ would be correspondingly small, suggesting that for such scenarios quantum entanglement is transient.

The Lyapunov exponents on both manifolds were found to be constant and positive definite, i.e. $\lambda_{\mathcal{M}} = 2A_o > 0$. In the presence of micro-correlations the IGE assumes a smaller initial value relative to the non-correlated case while the growth characteristics of both correlated and non-correlated IGEs were found to be similar. Furthermore, in the presence of micro-correlations, the IGE is attenuated in a correlation-dependent manner such that $S_{\text{corr.}}(\tau; r)$ decreases as the magnitude of the correlation increases. It is important to observe that this is not an asymptotic (long-time limit) feature of the IGE.

Acknowledgments

This work was partially supported by WCU (World Class University) program of NRF/MEST (R32-2009-000-10130-0) and by the European Community's Seventh Framework Program FP7/2007-2013 under grant agreement 213681 (CORNER Project).

References

[1] Einstein A, Podolsky B, Rosen N 1935 *Phys. Rev.* **47** 777
[2] Schrödinger E 1935 *Naturwiss.* **48** 807; **49** 823; **50** 844
[3] Law C K 2004 *Phys. Rev. A* **70** 062311
[4] Bubhardt M and Freyberger M 2007 *Phys. Rev. A* **75** 052101
[5] Cafaro C and Ali S A 2008 *EJTP* **5** 139
[6] Cafaro C and Ali S A 2007 *Physica D* **234** 70

[7] Cafaro C and Ali S A 2008 *Physica A* **387** 6876
[8] Ali S A, Cafaro C, Kim D H, Mancini S 2010 *Physica A* **389** 3117
[9] Cafaro C and Mancini S 2010 *Phys. Scr.* **82**, 035007
[10] Cafaro C, Giffin A, Ali S A and Kim D H 2010 *Appl. Math. Comput.* **217** 2944
[11] Harshman N L and Hutton G 2007 *Preprint* arXiv:quant-ph/0710.5776
[12] Wang J, Law C K and Chu M C 2006 *Phys. Rev. A* **73** 034302
[13] Rozanov Y A 1997 *Probability Theory: A Concise Course* (New York, NY: Dover Publications)
[14] Caticha A 2002 *Bayesian Inference and Maximum Entropy Methods in Science and Engineering* ed Fry R L (*AIP Conf. Proc.* **617** 302)
[15] De Felice F and Clarke J S 1990 *Relativity on Curved Manifolds* (Cambridge, UK: Cambridge University Press)
[16] Mishima K, Hayashi M and Lin S H 2004 *Phys. Lett. A* **333** 371
[17] Caticha A 2001 *Maximum Entropy and Bayesian Methods in Science and Engineering* ed. Mohammad-Djafari A (*AIP Conf. Proc.* **568** 72)
[18] Cafaro C 2008 *The Information Geometry of Chaos* (SUNY at Albany, NY: PhD Thesis)
[19] Cafaro C 2009 *Chaos, Solitons & Fractals* **41** 886
[20] Amari S and Nagaoka H 2000 *Methods of Information Geometry* (Oxford, UK: Oxford University Press)
[21] Hunt M *Lecture notes on Nuclear Physics II* (available online at http://www.physics.gla.ac.uk)
[22] Misner C W, Thorne K S and Wheeler J A 1973 *Gravitation* (San Francisco, CA: Freeman & Co.)
[23] Do Carmo M P 1992 *Riemannian Geometry* (Boston, MA: Birkhauser)

How to use the cosmological Schwinger principle for energy flux, entropy, and "atoms of space–time" to create a thermodynamic space–time and multiverse

Andrew Beckwith
71 Lakewood court, apt 7, Moriches, New York, 11955, USA
E-mail: beckwith@iibep.org

Abstract. We make explicit an idea by Padmanabhan in DICE 2010 [1], as to finding "atoms of space–time" permitting a thermodynamic treatment of emergent structure similar to Gibbs treatment of statistical physics. That is, an ensemble of gravitons is used to give an "atom" of space–time congruent with relic GW. The idea is to reduce the number of independent variables to get a simple emergent space–time structure of entropy. An electric field, based upon the cosmological Schwinger principle, is linked to relic heat flux, with entropy production tied in with candidates as to inflaton potentials. The effective electric field links with the Schwinger 1951s result of an E field leading to pairs of e^+e^- charges nucleated in space–time volume $V \cdot t$. Note that in most inflationary models, the assumption is for a magnetic field, not an electric field. An electric field permits a kink–anti-kink construction of an emergent structure, which includes Glinka's recent pioneering approach to a Multiverse. Also an E field allows for an emergent relic particle frequency range between one and 100 GHz. The novel contribution is a relic E field, instead of a B field, in relic space–time "atom" formation and vacuum nucleation of the same.

1. Introduction. What is to be analyzed?

T. Padmanabhan at DICE 2010 introduced the theme of this document [1]. That is, to reverse engineer GR emergent structure into initial component space–time "atoms," to permit a "Gibbs" style treatment of the thermodynamics of space–time physics [2, 3, 4]. To make a link with graviton physics, we emphasize entropy generation of initial structure via a relic electric field. Note that Padmanabhan [5] has, for weak gravitational fields emphasized generation of electric field, but the treatment of the author as given is for relic E fields. This, if one uses a Lenz's law transformation has been discussed in relic space–time conditions for the early formation of electromagnetic fields in between early galaxies [6]. The present document, while assuming an eventual magnetic field via a Lenz's law transformation of a relic E field brings up electromagnetics in very early universe conditions via conditions similar to a cosmological Swinger effect [7]. The formation of a relic electric field would be to initiate formation of relic particles from the initial inflaton structure.

This paper answers the question as to what would be optimal conditions for initial entropy production. To begin this inquiry, we start by examining candidates for the initial configuration of the normalized energy density. The normalized energy density of gravitational waves, as given by Maggiore [8] is

$$\Omega_{gw} \equiv \frac{\rho_{gw}}{\rho_c} \equiv \int_{\nu=0}^{\nu=\infty} d(\log \nu) \cdot \Omega_{gw}(\nu) \Rightarrow h_0^2 \Omega_{gw}(\nu) \cong 3.6 \cdot \left[\frac{n_\nu}{10^{37}}\right] \cdot \left(\frac{\nu}{1\ \text{kHz}}\right)^4. \quad (1)$$

n_ν is a frequency-based count of gravitons per unit cell of phase space. Eq. (1) leads to, as given in Figure 1, candidates as to early universe models to be investigated experimentally. The author, Beckwith, wishes to determine inputs into n_ν above, in terms of frequency, and also initial temperature. What is in the brackets of the exponential is a way of counting the number of space–time e^+e^- charges nucleated in a space–time volume $V \cdot t$. Beckwith used a very similar constructions with Density wave physics [9] and also has extended this idea to use in graviton physics, in a kink–anti-kink construction [9]. The idea, after one knows how to obtain a counting algorithm, with additional refinements will be to use what can be understood by the above analogy, assuming a minimal mass, $m \cong m_{\text{eff}}$, as Beckwith brought up [10]. This will be discussed as inputs into the models represented by Figure 1.

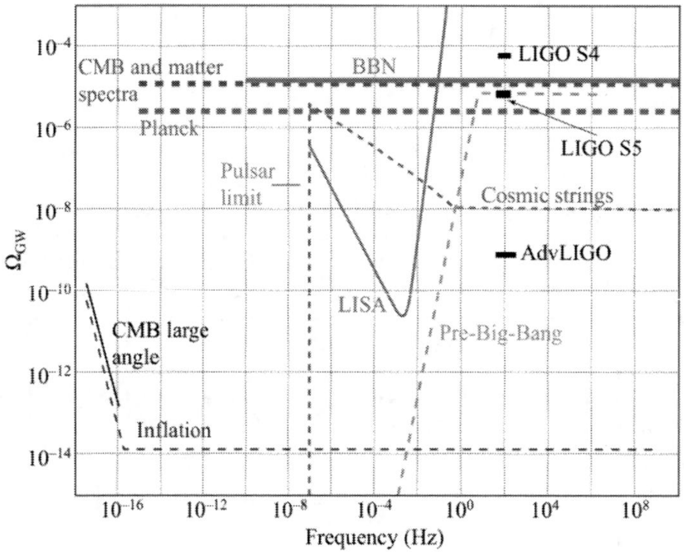

Figure 1. From the LIGO Scientific Collaboration [11] shows the relation between Ω_g and frequency.

Beckwith has derived a way to use two Friedman equations for estimating Entropy [12]. Also, by applying a Gaussian bifurcation mapping, Beckwith defined the initial degrees of space–time freedom up to over 1000 per unit of phase space during inflation [12]. We define entropy in terms of slow roll parameters and an effective electric field. The effective E field is a way to join vacuum nucleation with inflaton physics, and also nucleation of counted "pairs" of vacuum nucleated structures [9], allowing the use of Ng's "infinite quantum statistics" [13] for inflaton/inflationary physics. Finally, the electric field is a way to present creation of relic particles up to the point of chaotic dynamics becoming dominant, in the permitted frequency range. The route to chaos has a well-defined sequence of evolution, and formulation of the emergent structure electric field permits identification of key bandwidth frequency ranges for relic particles, before the creation of a stochastic background of signal-to-noise chaos, permitting identification of key frequency patches to identify as high-frequency signals of the relic particles created at the start of inflation. What Beckwith intends to do is to use the emergent structure set by the cosmological

Schwinger method, as outlined by Martin [7] to obtain the number of emergent particles in initial phase-space counting, and tie in the phase space numerical counting and entropy with different candidates for the inflaton potential. Beckwith, via dimensional analysis [10] made the following identifications. $Force = qE = [T\Delta S/dist] = \hbar[[\omega_{\text{final}} - \omega_{\text{initial}}]/dist]$ which in terms of inflaton physics leads to (if $V' = \mathrm{d}V/\mathrm{d}\phi$, $V'' = \mathrm{d}^2V/\mathrm{d}\phi^2$ where V is an inflaton potential, and $dist$ = distance of Planck length, or more)

$$\left[\frac{T\Delta S}{dist}\right] = \left[\frac{\hbar}{dist}\right]\left[2k^2 - \frac{1}{\eta^2}\left[M_P^2 \cdot \left[\frac{6}{16\pi} - \frac{3}{4\pi}\right] \cdot \left[\frac{V'}{V}\right]^2 - \frac{3}{4\pi} \cdot M_P^2 \cdot \left[\frac{V''}{V}\right]\right]\right]^{\frac{1}{2}}. \tag{2}$$

Eq. (2) divided by a charge, q, gives a relic electric field. While the existence of a charge, q, as an independent entity, at the onset of inflation is open to question, what the author, Beckwith [10], is paying attention to is using inputs into free energy, as can be specified by the following: If one identified the evolution of temperature, with energy, and made the following identification, $\breve{T}$ for time, and Ω_0 for a special frequency range, as inputs into [10],

$$E_{\text{thermal}} = \frac{5}{2}k_B T_{\text{temp}} \propto [\Omega_0 \breve{T}] \sim \widetilde{\beta} \tag{3}$$

Here, the thermal energy, as given by temperature ranging as $T_{\text{temp}} \in (0, 10^{19}\ \text{GeV})$, up to the Planck interval of time $t_P \sim 10^{-44}$ sec, so that one is looking at $\widetilde{\beta} \approx |F| \equiv \frac{5}{2}k_B T \cdot \overline{N}$, as a free energy. For this parameter, if $\overline{N}$, as an initial entropy, arrow of time configuration, were fixed, then the change in temperature would lead to change in "free energy," so that work, is here change in energy, and $\mathrm{d}E = T\mathrm{d}S - p\mathrm{d}V$. In basic physics, this would lead to force being work (change in energy) divided by distance. Then $\Delta\widetilde{\beta} \cong (5/2)k_B \Delta T_{\text{temp}} \cdot \overline{N} \sim Force \times dist$. We assume that there would be an initial fixed entropy arising, with $\overline{N}$ a nucleated structure arising in a short time interval as a temperature $T_{\text{temp}} \in (0, 10^{19}\ \text{GeV})$ arrives. So [10] leads to a force value

$$\frac{\Delta\widetilde{\beta}}{dist} = \frac{5}{2}k_B \Delta T_{\text{temp}} \cdot \frac{\overline{N}}{dist} \sim qE_{\text{net electric field}} \sim [T\Delta S/dist]. \tag{4}$$

The parameter, as given by $\Delta\widetilde{\beta}$ will be one of the parameters used to define chaotic Gaussian mappings, whereas the right-hand side of Eq. (4) is Eq. (2), which is force, which is linked to inflaton physics in Eq. (1). Next will be the identification of inflation physics, as dimensionally argued in both Eq. (1) and Eq. (3), to choices in the inflaton potential. To see that, consider the following, as given by Eq. (6) below [10]. Candidates as to the inflaton potential would be in powers of the inflaton—i.e., in terms of $V \sim \phi^N$, with perhaps $N = 2$ an admissible candidate (chaotic inflation). For $N = 2$, one gets the result [10]

$$[\Delta S] = [\hbar/T]\left[2k^2 - \frac{1}{\eta^2}\left[M_P^2 \cdot \left[\left[\frac{6}{4\pi} - \frac{12}{4\pi}\right] \cdot \left[\frac{1}{\phi}\right]^2 - \frac{6}{4\pi}\left[\frac{1}{\phi^2}\right]\right]\right]\right]^{1/2} \sim n_{\text{particle count}}, \tag{5}$$

which in full dimensional correctness takes finally the form

$$[\Delta S] = \frac{\hbar c}{T}\left\{2k^2 + 8\left(\sqrt{\frac{3}{8\pi}}\frac{1}{\ell_P}\frac{M_P c/\hbar}{\eta\phi}\right)^2\right\}^{\frac{1}{2}}. \tag{6}$$

Making a comparison with a weighted average of $\Delta S \sim 10^5$ and varying values of a scalar field of $0 < \phi < 2\pi$, when one has $\eta \in (10^{-44}\ \text{sec}, 0)$, and $0 < T \leq T_P \sim 10^{19}$ GeV leads to a rich

Journal of Physics: Conference Series **306** (2011) 012064 doi:10.1088/1742-6596/306/1/012064

phenomenology, where one could see variations as of a time parameter, and how the wave vector value, $k = 2\pi/\lambda$ where λ is wavelength, evolved, especially if $\Delta S \sim 10^5$ remains constant. That is, why did the value of the wave vector value, k, vary so much, in a short period of time—less than Planck time? As mentioned before in [10], this question asks how the initial wave vector, k, forms and to what degree variation in the inflaton $0 < \phi < 2\pi$ occurs. That is, it gives a way to vary the inflaton, and understand relic entropy generation.

2. First principle evaluation of initial bits of information, as opposed to numerical counting, and entropy

A consequence of Verlinde's [14] generalization of entropy as also discussed by Beckwith [10] and the number of "bits" yields the following consideration, which will be put here for startling effect. Namely, if a net acceleration is such that $a_{\text{accel}} = 2\pi k_B cT/\hbar$ as mentioned by Verlinde [15] and Beckwith [12] as the Unruh result, and that the number of "bits" is

$$n_{\text{Bit}} = \frac{\Delta S}{\Delta x} \cdot \frac{c^2}{\pi \cdot k_B^2 T} \approx \frac{3 \cdot (1.66)^2 g^\star}{[\Delta x \cong \ell_P]} \cdot \frac{c^2 \cdot T^2}{\pi \cdot k_B^2}, \tag{7}$$

Eq. (7) has a T^2 temperature dependence for information bits, as opposed to [10]

$$S \sim 3 \cdot \left[1.66 \cdot \sqrt{\widetilde{g}_\star}\right]^2 T^3 \sim n_f. \tag{8}$$

Should the $\Delta x \cong \ell_P$ order of magnitude minimum grid size hold, then when $T \sim 10^{19}$ GeV [10],

$$n_{\text{Bit}} \approx \frac{c^2}{\pi \cdot k_B^2 T} \approx \frac{3 \cdot (1.66)^2 g^\star}{[\Delta x \cong \ell_P]} \cdot \frac{c^2 \cdot T^2}{\pi \cdot k_B^2} \sim 3 \cdot \left[1.66 \cdot \sqrt{\widetilde{g}_\star}\right]^2 T^3. \tag{9}$$

The situation for which one has [10, 14] $\Delta x \cong \ell^{1/3}\ell_P^{2/3}$ with $\ell \sim \ell_P$ corresponds to $n_{\text{Bit}} \propto T^3$ whereas $n_{\text{Bit}} \propto T^2$ if $\Delta x \cong \ell^{1/3}\ell_P^{2/3} \gg \ell_P$. This issue of either $n_{\text{Bit}} \propto T^3$ or $n_{\text{Bit}} \propto T^2$ will be analyzed in future publications. If the bits of information can be related to a numerical count, the next step will be to make a linkage between thermal heat flux, due to the initial start of inflation, with degrees of freedom rising from a point, almost zero to over 1000 in a Planck time interval.

2.1. How to set up a bifurcation diagram for creation of $N(T) \sim 10^3$ degrees of freedom at the start of inflation

In a word, the way to introduce the expansion of the degrees of freedom from nearly zero, at the maximum point of contraction to having $\mathbf{N(T) \sim 10^3}$ is to first of all define the classical and quantum regimes of gravity in such a way as to minimize the point of the bifurcation diagram affected by quantum processes [10]. That is, classical physics, with smoothness of space–time structure down to a gird size of $\ell_P \sim 10^{-33}$ centimeters at the start of inflationary expansion. When doing this construction, what would be needed would be to look at the maximum point of contraction, set at $\ell_P \sim 10^{-33}$ centimeters as the quantum "dot," as a de facto measure zero set, as the bounce point, with classical physics behavior before and after the bounce "through" the quantum dot.

Dynamic-systems modeling could be directly employed right "after" evolution through the "quantum dot" regime, with a transfer of crunched in energy to Helmholtz free energy, as the driver "force" for a Gauss map type chaotic diagram right after the transition to the quantum "dot" point of maximum contraction. The diagram would look like an application of the Gauss mapping of [10, 15]:

$$X_{i+1} = \exp\{-\alpha \cdot x_i^2\} + \widetilde{\beta}. \tag{10}$$

In dynamic-systems type parlance, one would achieve a diagram, with tree structure looking like what was given by Binous [16], using material written up by Lynch [15, 10]—by looking at his bifurcation diagram for the Gauss map. Binous's demonstration plots the bifurcation diagram for user-set values of the parameter x_{i+1}. For the author's purposes, the parameter x_{i+1} and x_I^2 as put in Eq. (10) would represent the evolution of a number of degrees of freedom, with ironically, the near zero behavior, plus a Helmholtz degree of freedom parameter set fed into $\widetilde{\beta}$. The quantum "dot" contribution would be a measure set zero glitch in the mapping given by Eq. (10), with the understanding that where the parameter $\widetilde{\beta}$ "turns on" would be right AFTER the "bounce" through the infinitesimally small quantum "dot" regime. Far from being trivial, there would be a specific interactive chaotic behavior initiated by the turning on of parameter $\widetilde{\beta}$, corresponding, as brought up by Dickau [17], with a connection between octo-octonionic space and the degrees of freedom available at the beginning of inflation. That is, turning on the parameter $\widetilde{\beta}$ would be a way to have Lisi's E8 structure [18] be nucleated at the beginning of space–time. As the author sees it, $\widetilde{\beta}$ would be proportional to the Helmholtz free energy, F, where as Mandl [19, p. 272] relates, the usual definition of $F = E - TS$, becomes, instead, here, using partition function, Z, with $\overline{N}$ a "numerical count factor," so that [10, 19]

$$F = -k_B T \cdot \ln Z(T, V, \overline{N}). \tag{11}$$

Note that Ng [13] sets a modification of $Z_N \sim \frac{1}{N!}\left(\frac{V}{\lambda^3}\right)^N$ as in the use of his infinite quantum statistics, with the outcome that one will be setting a temperature dependent free energy [5] $F = -k_B T(\ln(V/\lambda^3) + 5/2)$ with $V \sim \ell_P^3$ and the Entropy obeying [10, 13]

$$S \approx N \cdot (\ln \left\lfloor V/N\lambda^3 \right\rfloor + 5/2) \xrightarrow[\text{Ng infinite Quantum Statistics}]{} N \cdot (\ln \left\lfloor V/\lambda^3 \right\rfloor + 5/2) \approx N. \tag{12}$$

2.2. Linking driving "frequency" as a result of change in temperature to relic particles

After presenting this initial argument, we will bring up similar, more refined attempts by Donnelly and Jacobson [20], and also Rosenblum, Pikovsky, and Kurths [21] via their phased synchronization of chaotic oscillators to suggest necessary refinements ss to the linkage of the chaotic Gaussian mapping with structure formation. The first part of the following discussion is meant to be motivational, with references from the other papers being refinements that will be worked on in additional publications. The main idea is that $\widetilde{\beta}$ increasing up to a maximum temperature T would enable the evolution and spontaneous construction of the Lisi E8 structure as given by [18]. As Beckwith wrote up [10], including in additional energy due to an increase of $\widetilde{\beta}$ due to increasing temperature T, would have striking similarities to the following. Observe the following argument as given by Mukhanov and Winitzki [22, 10], as to additional particles being "created" due to what is an infusion of energy in an oscillator, obeying two equations of motion:

$$\begin{aligned} &\ddot{q}(t) + \omega_0^2 q(t) = 0, \quad \text{for} \quad t < 0 \text{ and } t > \breve{T}; \\ &\ddot{q}(t) - \Omega_0^2 q(t) = 0, \quad \text{for} \quad 0 < t < \breve{T}. \end{aligned} \tag{13}$$

Given $\Omega_0 \breve{T} \gg 1$ with a starting solution of $q(t) = q_1 \sin \omega_0 t$ if $t < 0$, Mukhanov and Winitzki state that for $t > T$

$$q_2 \approx \frac{1}{2}\sqrt{1 + \frac{\omega_0^2}{\Omega_0^2}} \cdot \exp\left\{\Omega_0 \breve{T}\right\}. \tag{14}$$

The Mukhanov and Winitzki argument leads to an exercise which Mukhanov and Winitzki [22] claim is a solution. The exercise yields an increase in number count, as can be given by first setting the oscillator in the ground state with $q_1 = \omega_0^{-1/2}$, with the number of particles linked to amplitude by $\breve{n} = (1/2)(q_0^2\omega_0 - 1)$, leading to

$$\breve{n} = (1/2)\left(1 + \frac{\omega_0^2}{\Omega_0^2}\right) \cdot \sinh^2\left\{\Omega_0 \breve{T}\right\} \tag{15}$$

That is, for nonzero $\Omega_0 \breve{T}$, Eq. (15) leads to exponential expansion of the numerical state. For sufficiently large $\Omega_0 \breve{T}$, Eq. (13) and Eq. (14) are equivalent to placing of energy into a system, leading to vacuum nucleation. A further step in this direction is given on p. 82 of Mukhanov and Winitzki, leading to a Bogoliubov particle number density becoming exponentially large:

$$n \sim \sinh^2[m_0 \eta_1]. \tag{16}$$

Interestingly, Glinka et al [23] using quantum field theory methods showed that number of universes generated from stabilized vacuum is

$$n = \sinh^2 r(\phi), \tag{17}$$

where $r(\phi)$ is one of the Bogoliubov angles, and $\phi = \phi_0 a(x^0)$ is the cosmological scale factor with $\phi_0 = \sqrt{\frac{3}{8\pi}} M_P$.

Eq. (15) and (16) are, for sufficiently large $\Omega_0 \breve{T}$, a way to quantify what happens if initial thermal energy is placed in a harmonic system, leading to vacuum particle "creation." Eq. (16) is the formal Bogoliubov coefficient limit of particle creation. Note that $\ddot{q}(t) - \Omega_0^2 q(t) = 0$ for $0 < t < \breve{T}$ corresponds to a thermal flux of energy into a time interval $0 < t < \breve{T}$. If $\breve{T} \approx t_P \propto 10^{-44}$ sec or some multiple of t_P and if $\Omega_0 \propto 10^{10}$ Hz, then Eq. (13), and Eq. (15) plus its generalization as given in Eq. (16) may be a way to imply either vacuum nucleation, or transport of gravitons from a prior to the present universe. To generalize, what is done from Eq. (13) to Eq. (16), as brought up by Rosenblum, Pikovsky, and Kurths [21], would be to seek an explicit coupling of the coupled oscillations used to set up Eq. (10) with Eq. (13) to Eq. (16) explicitly. In order to take such a coupling, of chaotic oscillators, one would need to move beyond the idea given in this document's section E to look at $Force = qE = [T\Delta S/dist] = \hbar[\omega_{\text{final}} - \omega_{\text{initial}}]/dist$, possibly setting the initial $\omega_{\text{final}} = 0$. That is, look at candidates for ω_{final} from the perspective of frequencies about and around $\Omega_0 \propto 10^{10}$ Hz. The author's initial results seem to indicate that in doing so, there is a range of permitted Ω_0 values from 1 GHz up to 100 GHz, and this range of permitted Ω_0 values due to the choices in grid size from a minimum value $\ell_P \propto 10^{-33}$ centimeters, to those several thousand times larger, for reasons brought up earlier. Secondly, Donnelly and Jacobson's linearized perturbations [20] and dispersions relations sections, in terms of an aether background have in its Eq. (28) and Eq. (30) potentially useful limits as to additional constraints for the frequency $[T\Delta S/dist] = \hbar[\omega_{\text{final}} - \omega_{\text{initial}}]/dist$ to obey, in the limits that $k \to 0$ as in Eq. (3) above. If, as an example, Donnelly and Jacobson's dispersion relationship [20] as of their Eq. (30) can be reconciled with the limits of $k \to 0$ as in our Eq. (3), and $[T\Delta S/dist] = \hbar[\omega_{\text{final}} - \omega_{\text{initial}}]/dist$, this may entail a rethinking concerning if Donnelly and Jacobson's damped harmonic oscillator for the inflaton, with a driving term—such as their Eq. (85)—is applicable, i.e., his [21],

$$\ddot{\phi} + \theta\dot{\phi} + m^2\phi + \mu N\phi = 0. \tag{18}$$

This Eq. (18) is a driven damped harmonic oscillator, with a "source" term. That is, the idea is that with an aether, or some similar background, possibly with the aether having the same function as DE, up to a point, with damping put in, as seen above, as a further modification as to inflaton potentials which inputs as to our Eq. (3) to Eq. (7) should be considered. All this will be tried in future extensions of this project. The author's own work, at least in low red shift regimes up to a billion years ago [9], emphasizes the role of massive gravitons with having, in higher than four dimensions qualitative overlap in dynamic behavior as to the speed of cosmological expansion a billion years ago as attributed to DE. Beckwith's own work [9], as also Alves et al. [24], involves what is known as the deceleration parameter to show a resumption of a speed up of cosmological acceleration one billion years ago. Making use of Eq. (18) would be a way, to perhaps link between massive gravitons, as given by Beckwith [9] as well as Alves et al. [24] and the implications of Donnelly and Jacobson's damped driven harmonic oscillator [20] as represented by Eq. (18) above, which would make our investigation more comprehensive.

The remainder of this document will be in setting the "atoms" of space–time brought up in [1], via a comparison of energy for GW, for relic gravitons, with an eye toward thermal inputs into the inflation field geometry.

3. Effective "electric field" as proportional to temperature, to the first power. Its interpretation.

$$n_f = [1/4] \cdot \left[\sqrt{\frac{\nu(a_{\text{initial}})}{\nu(a)}} - \sqrt{\frac{\nu(a)}{\nu(a_{\text{final}})}} \right] \tag{19}$$

Eq. (19) above, factually derived and used by Glinka [25, 26] in his pioneering approach to Multiverse, could be investigated as being part of the bridge between phenomenology of what inflaton potentials should be used. That is, the inputs into Eq. (4) lead to a number of permissible inputs into the inflaton potential which should be looked at: the values of the inflaton field which are acceptable, for $\phi \in (0, 2\pi)$. What remains to be seen would be if the Schwinger numerical counting formula [27] as given in the argument of the exponential of Eq. (2) leads to a way to represent a numerical counting procedure giving reality to Ng's infinite quantum statistics [13], via $S \sim N$ with N representing a numerical counting of some assembly of effective mass of gravitons [9, 13]. The goal would be to make a linkage between the instanton–anti-instanton construction as Beckwith used in [10, 28] below, for electric fields, and an emergent graviton, along the lines of the false vacuum nucleation procedure seen below [29].

In order to connect with GR, one needs to have a higher dimensional analog of this pop up, as discussed in [9]. With comparisons between Eq. (4) and Figure 2, we hope to find appropriate inflaton potentials and their scaling behavior so Eq. (4) and the space–time nucleation, with an effective electric field, are clearly understood by an extension of the Schwinger result given in Eq. (2) and its Ng. counterpart $S \sim N$. Also one of the most exciting parts of this inquiry would be to find if there is a way to make a linkage with Eq. (19) and DE with the work done by the author with massive gravitons, to mimic DE one billion years ago. The author's motivation was in his work as reported in Dark Side of the Universe 2010 [29], to obtain a working DM/DE joint model, in a way better than current work done with Chaplygin Gas models [30, 31]. Having a linkage given between Eq. (18) of a damped driven inflaton field, with an "aether" in space–time, possibly linked to DE would be a way to show to what degree inflaton physics can be linked to a synthesis of DM/DE models [9]. If such a linkage can be made, with the author's view of space–time inflation being driven by an E field inflaton field, as a source of entropy, it would give insight as to what role the formation of relic particles plays in entropy production. Last, but not least. A model using a Gaussian bifurcation map to generate chaotic dynamics as to creation of many degrees of freedom—up to 1000—in place of usual maximum between 100 and 120, with a turn on of Helmholtz free energy right after the start of inflation [10]. The author is convinced that the generation of Helmholtz free energy, as portrayed, is a classical phenomenon, but that the quantum gravity congruent grid size, possibly

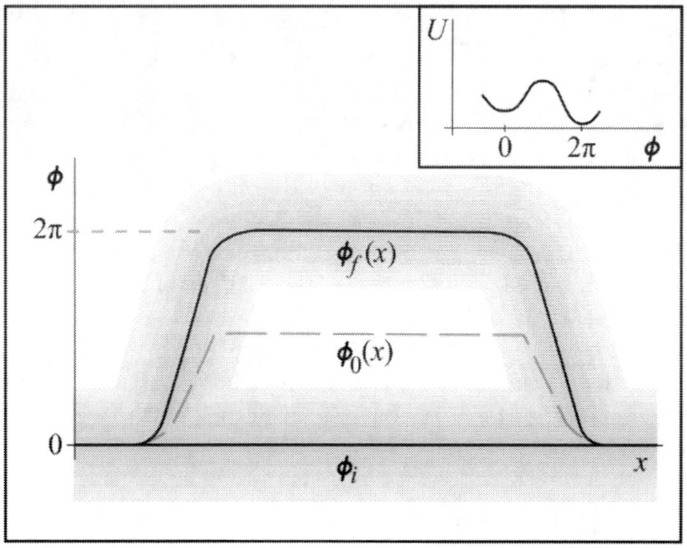

Figure 2. The pop-up effects of an instanton–anti-instanton in Euclidian space from references [9, 28].

as small as a Planck length is very important. A great deal of analytical work has been done to isolate the regime where Classical and Quantum Gravity intersect. It is the author's firm conviction that this debate can be settled by ascribing the regime of space–time as of the order of, or slightly larger than the length of a Planck size grid size, $\ell_P \sim 10^{-33}$ centimeters (at best it would be about two to three orders of magnitude larger) as where gravity is a purely quantum phenomenon, whereas when the Helmholtz free energy term [10] is turned on, right AFTER the growth of space–time past $\ell_P \sim 10^{-33}$ centimeters as where chaotic, classical dynamics plays a dominant role. Proving this latter conjecture would allow for, among other things helping to isolate the frequency ranges for production of relic particles, whose faint traces show up in an otherwise almost perfect "white noise" of relic GW/relic graviton frequencies from the big bang. Numerous authors, including Buonanno [32] have stated that this white noise is not analyzable. If one takes into account that the approach to chaos has well defined steps up to its initiation, physically, the author asserts that this last assumption may be falsifiable experimentally.

4. Conclusion: Analyzing the problem of graviton "counting," atoms of space–time, GW, and Khrennikov's signal theory treatment of QM with regards to the representation of gravitons and inflaton fields

According to Khrennikov [33], the classical and quantum probabilities can be delineated, in CM, by

$$\langle f \rangle_\mu = \int_M f(\mu) \mathrm{d}\mu(\phi) \tag{20}$$

Here, M is the state space, and f is a functional in classical probability to be estimated, while μ is the measure. Eq. (20) should be contrasted with a QM presentation of the probability to be estimated with

$$\langle \hat{A} \rangle_\rho = \mathrm{Tr} \rho \hat{A} \tag{21}$$

Khrennikov's main claim is that randomness is the same in classical mechanics as in QM, and furthermore delineates a way to make a linkage between Eq. (20) and Eq. (21) via use of, if

t is the time scale of fluctuation, and T is the time of measurement that one can write, to order $\vartheta(t/T)$

$$\int_M f_A(\mu)\mathrm{d}\mu(\phi) = \langle \hat{A} \rangle_\rho = \mathrm{Tr}\rho\hat{A} + \vartheta(t/T) \tag{22}$$

Note that our supposition, via its cartoon sketch in Figure 2 above which we delineate as a vacuum nucleation of a graviton as a kink–anti-kink super position in space–time, with the overlap of order $\vartheta(t/T)$ being the source for the tiny four-dimensional graviton rest mass given by Padmanabhan [4, 5]. That is, this is our way of stating that QM has an embedding, via Eq. (21) above in a more complex nonlinear theory to be built up. How that is built up will enable a fuller creation of "space–time atoms" as stated by Padmanabhan [1, 4] for a thermodynamic treatment of the evolution of the inflaton, in terms of gravitons and GW physics as mentioned by the author [10]. We seek to do this by, among other things using Elze's negative probability [34] as a way to delineate the dynamics of evolution of the pop up represented in Figure 2 above in future work. The hope is to establish the details of the embedding of QM as part of a larger nonlinear theory and also to answer L. Lusanna's [35] statement of the problem as he sees it. That is, a Swinger space–time kink–anti-kink pair would have a different embedding group structure for its representation than that of the Poincaré group commonly used in GR. The group structure is different, so reconciling the different structures would be part and parcel of resolving issues as of Eq. (22) above.

Acknowledgments
The author benefitted from numerous valuable discussions from Lukasz Glinka and Jonathan Dickau.

References
[1] Padmanabhan T 2010 Lessons for quantum structure of spacetime from classical gravity, *J. Phys.: Conf. Series* this issue
[2] Padmanabhan T 2010 Surface density of spacetime degrees of freedom from equipartition law in theories of gravity *Phys. Rev.* **D81** 124040 (*Preprint* arXiv:1003.5665[gr-qc])
[3] Padmanabhan T 2010 Thermodynamical aspects of gravity: New insights *Rep. Prog. Phys.* **73** 046901 (*Preprint* arXiv:0911.5004[gr-qc])
[4] Padmanabhan T 2009 A dialogue on the nature of gravity *Preprint* arXiv:0910.0839[gr-qc]
[5] Padmanabhan H and Padmanabhan T 2010 Aspects of electrostatics in a weak gravitational field *Gen. Rel. Grav.* **42**, 1153 (*Preprint* arXiv:0910.0926[gr-qc])
[6] Ando S and Kusenko A 2010 Evidence for gamma-ray halos around active galactic nuclei and the first measurement of intergalactic magnetic fields *Astrophys. J.* **722** L39 (*Preprint* arXiv:1005.1924[astro-ph.HE])
[7] Martin J 2008 *Inflationary Cosmology* vol. 738, ed. M Lemione, J Martin and P Peter (New York: Springer) p. 193
[8] Maggiore M 2008 *Gravitational Waves* vol. 1 *Theory and Experiment* (Oxford: University Press)
[9] Beckwith A W 2010 Applications of euclidian snyder geometry to the foundations of space time physics *Electron. J. Theor. Phys.* **7** 1 (*Preprint* viXra:0912.0012)
[10] Beckwith A W 2010 Supplying conditions for having up to 1000 degrees of freedom in the onset of inflation, instead of 2 to 3 degrees of freedom, today, in space–time *Preprint* viXra:1009.0001
[11] The LIGO scientific collaboration and the Virgo collaboration 2009 *Nature* **460** 990
[12] Beckwith A W 2009 *Electron. J. Theor. Phys.* **6** 257; 2006 *Mod. Phys. Lett.* **B20** 849
[13] Ng Y J 2008 *Entropy* **10** 441
[14] Verlinde E P 2010 On the origin of gravity and the laws of Newton *Preprint* arXiv:1001.0785[hep-th]
[15] Lynch S 2007 *Dynamical Systems with Applications using Mathematica* (Boston: Birkhäuser)
[16] Binous H 2010 *Bifurcation Diagram for the Gauss Map* from The Wolfram Demonstrations Project http://demonstrations.wolfram.com/BifurcationDiagramForTheGaussMap
[17] Dickau J J, Priv. communications
[18] Lisi A 2007 An exceptionally simple theory of everything *Preprint* arXiv:0711.0770
[19] Mandl F 1971 *Statistical Physics* (London: Wiley)

[20] Donnelly W and Jacobson T 2010 Coupling the inflaton to an expanding aether *Preprint* arXiv:1007.2594[gr-qc]
[21] Rosenblum M, Pikovsky A and Kurths J 1996 *Phys. Rev. Lett.* **76** 1804
[22] Mukhanov V F and Winitzki S 2007 *Introduction to Quantum Effects in Gravity* (Cambridge: Cambridge University Press)
[23] Glinka L A and Pervushin V N 2008 *Concepts Phys.* **5** 31
[24] Alves M E S, Miranda O D and de Araujo J C N 2010 Can massive gravitons be an alternative to dark energy? *Preprint* arXiv:0907.5190[astro-ph.CO]
[25] Glinka L A 2007 Quantum information from graviton-matter gas *SIGMA* **3** 87 (*Preprint* arXiv:0707.3341[gr-qc])
[26] Glinka L A 2008 Multiparticle quantum cosmology *AIP Conf. Proc.* **1018** 94 (*Preprint* arXiv:0801.4157[gr-qc]); 2007 Many-particle quantum cosmology *Preprint* arXiv:0712.2769[hep-th]; 2007 On quantum cosmology as field theory of bosonic string mass groundstate *Preprint* arXiv:0712.1674[gr-qc]; 2007 Preliminaries in many-particle quantum gravity. Einstein–Friedmann spacetime *Preprint* arXiv:0711.1380[gr-qc]; 2009 Thermodynamics of space quanta models quantum gravity *Preprint* arXiv:0906.3827[gr-qc]
[27] Schwinger J 1951 *Phys. Rev.* **82** 664
[28] Beckwith A W 2001 *Classical and Quantum Models of Density Wave Transport, a Comparative Study*, PhD dissertation, University of Houston, Physics Department
[29] Beckwith A W 2010 Inquiry as to if higher dimensions can be used to unify DM and DE, if massive gravitons are stable *Preprint* viXra:1006.0027 to appear in DSU 2010 Proceedings
[30] Debnath U and Chakrabory S 2008 *Int. J. Theor. Phys.* **47** 2663
[31] Roos M 2003 *Introduction to Cosmology* 3rd ed. (Chichester: Wiley)
[32] Buonanno A 2006 *Particle Physics and Cosmology: The Fabric of Space–Time* vol. 86, ed. F Bernardeau, C Grojean and J Dalibard p. 10 (Amsterdam: Elsevier)
[33] Khrennikov A Y 2009 Bell's inequality: Physics meets probability *Information Science* **179** 492 (*Preprint* arXiv:0709.3909); 2010 Quantum mechanics as classical signal theory on noisy background, *J. Phys.: Conf. Series* this issue
[34] Elze H-T 2010 General linear dynamics: quantum, classical or hybrid *J. Phys.: Conf. Series* this issue
[35] Lusanna L 2010 Priv. Communications

Journal of Physics: Conference Series **306** (2011) 012065 doi:10.1088/1742-6596/306/1/012065

Geometric phase and gauge theory structure in quantum computing

A. Bruno[1,3], **A. Capolupo**[2,3], **S. Kak**[4], **G. Raimondo**[1,3] **and G. Vitiello**[2,3]

[1]Dipartimento di Fisica dell'Università di Salerno, 84100 Salerno, Italy
[2]Dipartimento di Matematica e Informatica dell'Università di Salerno, 84100 Salerno, Italy
[3] Istituto Nazionale di Fisica Nucleare, Gruppo Collegato di Salerno, 84100 Salerno, Italy
[4]Department of Computer Science, Oklahoma State University, Stillwater, OK 74078, USA

E-mail: bruno@sa.infn.it

Abstract. We discuss the presence of a geometrical phase in the evolution of a qubit state and its gauge structure. The time evolution operator is found to be the free energy operator, rather than the Hamiltonian operator.

As well known, gauge theories provide important tools in the study of the dynamics of elementary particles and condensed matter systems. The physics of many systems is indeed described by Lagrangians which are invariant under local gauge transformation groups. The relevance of global geometric properties and gauge invariance has been displayed also in the quantum computing framework [1, 2, 3]. In this paper, we report recent results [4] according to which the time evolution of a two level system, such as a qubit state, is governed by a gauge field structure and geometric phases appear. We construct the covariant derivative operator and show that it is related to free energy.

Let us start by considering the standard orthonormal basis of two vectors $|0\rangle$ and $|1\rangle$, eigenvectors of the operator $H = \omega_1|0\rangle\langle 0| + \omega_2|1\rangle\langle 1|$ with eigenvalues ω_1 and ω_2, respectively. By use of a rotation operator in the plane $\{|0\rangle, |1\rangle\}$ one may prepare the states

$$|\phi\rangle = \alpha\,|0\rangle + \beta\,|1\rangle\,, \tag{1}$$

$$|\psi\rangle = -\beta\,|0\rangle + \alpha\,|1\rangle\,. \tag{2}$$

The coefficients α and β fulfill the relation $|\alpha|^2 + |\beta|^2 = 1$. Then, in full generality, we may put $\alpha = e^{i\gamma_1}\cos\theta$ and $\beta = e^{i\gamma_2}\sin\theta$. Here we assume $\gamma_1 = \gamma_2 = 0$ (in this paper, we do not consider the initialization problem [5, 6, 7]).

Let t denote the time parameter. H plays the role of the Hamiltonian operator and the time evolution of the states (1) and (2) is

$$|\phi(t)\rangle = e^{-iHt}|\phi(0)\rangle = e^{-i\omega_1 t}(\cos\theta\,|0\rangle + e^{-i(\omega_2-\omega_1)t}\sin\theta\,|1\rangle)\,, \tag{3}$$

$$|\psi(t)\rangle = e^{-iHt}|\psi(0)\rangle = e^{-i\omega_1 t}(-\sin\theta\,|0\rangle + e^{-i(\omega_2-\omega_1)t}\cos\theta\,|1\rangle)\,. \tag{4}$$

with $\langle\phi(t)|\psi(t)\rangle = 0$ and $\langle\psi(t)|\psi(t)\rangle = 1$, $\langle\phi(t)|\phi(t)\rangle = 1$, for all t. We note that the states $|\phi(t)\rangle$ and $|\psi(t)\rangle$, for all t and $\omega_1 \neq \omega_2$, are not eigenstates of H. From these states we have

$$\langle\phi(t)|\; H \;|\phi(t)\rangle = \omega_1\; \cos^2\theta + \omega_2\; \sin^2\theta \;=\; \omega_{\phi\phi}\;, \tag{5}$$

$$\langle\psi(t)|\; H \;|\psi(t)\rangle = \omega_1\; \sin^2\theta + \omega_2\; \cos^2\theta \;=\; \omega_{\psi\psi}\;, \tag{6}$$

$$\langle\psi(t)|\; H \;|\phi(t)\rangle = \frac{1}{2}(\omega_2 - \omega_1)\; \sin 2\theta \;=\; \omega_{\phi\psi}\;, \tag{7}$$

and $\omega_{\phi\psi} = \omega_{\psi\phi}$. Notice that the matrix elements of H in these equations are time-independent. We also have

$$tg2\theta \;=\; \frac{2\omega_{\phi\psi}}{\delta\omega}\;, \tag{8}$$

where $\delta\omega \equiv \omega_{\psi\psi} - \omega_{\phi\phi}$. By using Eqs. (3) and (4), the operator H can be expressed for any t as

$$H = \omega_{\phi\phi}|\phi(t)\rangle\langle\phi(t)| + \omega_{\psi\psi}|\psi(t)\rangle\langle\psi(t)| + \omega_{\phi\psi}(|\phi(t)\rangle\langle\psi(t)| + |\psi(t)\rangle\langle\phi(t)|)\;. \tag{9}$$

We come now to the discussion of the Berry-like phase involved in the qubit time evolution. The state $|\phi(t)\rangle$, apart from a phase factor, reproduces the original state $|\phi(0)\rangle$ after a period $T = \frac{2\pi}{\omega_2 - \omega_1}$:

$$|\phi(T)\rangle = e^{i\varphi}|\phi(0)\rangle\;, \qquad \varphi = -\frac{2\pi\omega_1}{\omega_2 - \omega_1}\;, \tag{10}$$

where it is essential that $\omega_1 \neq \omega_2$. One thus recognizes that such a time evolution does contain a purely geometric part, i.e. the Berry-like phase. Indeed, we separate the geometric and "dynamical" phases following the standard procedure [8]:

$$\beta_\phi \;=\; \varphi + \int_0^T \langle\phi(t)|\; i\partial_t \;|\phi(t)\rangle\, dt \;=\; 2\pi \sin^2\theta\,. \tag{11}$$

β_ϕ is independent from ω_1 and ω_2 and depends only on the "mixing angle" θ. Similarly, for the state $|\psi(t)\rangle$ one finds $\beta_\psi = 2\pi\cos^2\theta$ and thus $\beta_\psi + \beta_\phi = 2\pi$ for any θ. We can thus rewrite (10) as

$$|\phi(T)\rangle = e^{i2\pi\sin^2\theta}e^{-i\omega_{\phi\phi}T}|\phi(0)\rangle\,. \tag{12}$$

The meaning of Eqs. (11)-(12) can be better understood by noticing that, for any t,

$$\langle\phi(0)|\phi(t)\rangle \;=\; e^{-i\omega_1 t}\cos^2\theta + e^{-i\omega_2 t}\sin^2\theta\,. \tag{13}$$

Thus, as an effect of the non vanishing difference $\omega_- = \omega_2 - \omega_1$ of the phases, the components $|0\rangle$ and $|1\rangle$ evolve with different "weights" and the state $|\phi(t)\rangle$ "rotates" as shown by Eq. (13). At $t = T$,

$$\langle\phi(0)|\phi(T)\rangle \;=\; e^{i\varphi} \;=\; e^{i\beta_\phi}e^{-i\omega_{\phi\phi}T}\,, \tag{14}$$

i.e. $|\phi(T)\rangle$ differs from $|\phi(0)\rangle$ by a phase φ, part of which is a geometric "tilt" (the Berry-like phase β_ϕ) and the other part is of dynamical origin. This is similar to what happens in the context of particle mixing (see ref. [9], which we closely follow in our presentation below). In general, for $t = T + \tau$, we have

$$\begin{aligned}\langle\phi(0)|\phi(t)\rangle \;&=\; e^{i\varphi}\,\langle\phi(0)|\phi(\tau)\rangle \\ &=\; e^{i2\pi\sin^2\theta}e^{-i\omega_{\phi\phi}T}\left(e^{-i\omega_1\tau}\cos^2\theta + e^{-i\omega_2\tau}\sin^2\theta\right)\,.\end{aligned} \tag{15}$$

Also notice that

$$\langle\psi(0)|\phi(t)\rangle = \frac{1}{2}e^{i\varphi}e^{-i\omega_1\tau}\sin 2\theta\left(e^{-i(\omega_2-\omega_1)\tau}-1\right), \qquad \text{for } t=T+\tau\,, \tag{16}$$

which is zero only at $t = T$. Eq. (16) expresses the fact that $|\phi(t)\rangle$ "oscillates", getting a component of the $|\psi(0)\rangle$ state, besides getting the Berry-like phase. At $t = T$, $|\phi(t)\rangle$ and $|\psi(0)\rangle$ are again each other orthogonal states.

Generalization to $n-$cycles, $n = 1, 2, 3, ...,$ is also interesting. Eq. (11) can be rewritten for the $n-$cycle case as

$$\beta_\phi^{(n)} = \int_0^{nT}\langle\phi(t)|\,i\partial_t - \omega_1\,|\phi(t)\rangle\,dt = 2\pi\,n\,\sin^2\theta\,, \tag{17}$$

and Eq. (15) becomes

$$\langle\phi(0)|\phi(t)\rangle = e^{in\varphi}\,\langle\phi(0)|\phi(\tau)\rangle\,, \qquad \text{for} \quad t=nT+\tau\,. \tag{18}$$

Similarly, in Eq. (16) one obtains the phase $e^{in\varphi}$ instead of $e^{i\varphi}$. Eq. (17) shows that the Berry-like phase acts as a "counter" of $|\phi(t)\rangle$ oscillations, adding up $2\pi\,\sin^2\theta$ to the phase of the $|\phi(t)\rangle$ state after each complete oscillation. We observe that $\beta_\phi^{(n)}$ in Eq. (17) can be rewritten as

$$\beta_\phi^{(n)} = \int_0^{nT}\langle\phi(t)|\,U^{-1}(t)\,i\partial_t\left(U(t)\,|\phi(t)\rangle\right)dt = \int_0^{nT}\langle\tilde{\phi}(t)|\,i\partial_t|\tilde{\phi}(t)\rangle\,dt = 2\pi\,n\,\sin^2\theta\,, \tag{19}$$

with $U(t) = e^{-if(t)}$, where $f(t) = f(0) - \omega_1 t$, with $f(0)$ an arbitrary constant, and

$$|\tilde{\phi}(t)\rangle \equiv U(t)|\phi(t)\rangle = e^{-if(0)}\left(\cos\theta\,|0\rangle + e^{-i(\omega_2-\omega_1)t}\sin\theta\,|1\rangle\right). \tag{20}$$

We can regard $|\phi(t)\rangle \to U(t)|\phi(t)\rangle = |\tilde{\phi}(t)\rangle$ as a local (in time) gauge transformation of the state $|\phi(t)\rangle$. In contrast with the state $|\phi(t)\rangle$, the gauge transformed state $|\tilde{\phi}(t)\rangle$ is not "tilted" in its time evolution:

$$\langle\tilde{\phi}(0)|\tilde{\phi}(t)\rangle = \langle\tilde{\phi}(0)|\tilde{\phi}(\tau)\rangle\,, \qquad \text{for} \quad t=nT+\tau\,, \tag{21}$$

which has to be compared with Eq. (18). From Eq. (20) we see that time evolution only affects the $|1\rangle$ component of the state $|\tilde{\phi}(t)\rangle$. The gauge transformation acts as a "filter" freezing out time evolution of the $|0\rangle$ state component, so that we have

$$\begin{aligned} i\partial_t|\tilde{\phi}(t)\rangle &= (\omega_2-\omega_1)e^{-if(0)}e^{-i(\omega_2-\omega_1)t}\sin\theta|1\rangle \\ &= (H-\omega_1)e^{-if(0)}\left(\cos\theta\,|0\rangle + e^{-i(\omega_2-\omega_1)t}\sin\theta\,|1\rangle\right) \\ &= (H-\omega_1)|\tilde{\phi}(t)\rangle\,, \end{aligned} \tag{22}$$

namely

$$-i(\partial_t + iH)|\tilde{\phi}(t)\rangle = \omega_1|\tilde{\phi}(t)\rangle\,. \tag{23}$$

Eq. (19) actually provides an alternative way for defining the Berry-like phase [8], which makes use of the state $|\tilde{\phi}(t)\rangle$ given in Eq. (20). Eq. (19) directly gives us the geometric phase because the quantity $i\langle\tilde{\phi}(t)|(i\partial_t|\tilde{\phi}(t)\rangle\,dt)$ is the overlap of $|\tilde{\phi}(t)\rangle$ with its "parallel transported" $(i\partial_t|\tilde{\phi}(t)\rangle\,dt)$ at $t + dt$.

Another geometric invariant is the Anandan–Aharonov phase [10], defined as $s = 2\int \Delta\omega(t)dt$. It has the advantage to be well defined also for systems with non-cyclic evolution. Since $\Delta\omega = \omega_{\psi\phi}$, we have

$$s_n = 2\int_0^{nT} \omega_{\psi\phi}\, dt = 2\,\pi\, n \sin 2\theta\,. \tag{24}$$

Such an invariant represents the distance between qubit evolution states, as measured by the Fubini–Study metric, in the projective Hilbert space $\mathcal{P}$ [4] (see also ref. [11]).

We now show the gauge structure underlying the time evolution of two level systems. The motion equations for the $|\phi(t)\rangle$ and $|\psi(t)\rangle$ are computed by using the operator H in Eq.(9) written in terms of time dependent states. We have $H|\phi(t)\rangle = \omega_{\phi\phi}\,|\phi(t)\rangle + \omega_{\phi\psi}\,|\psi(t)\rangle$ and $H|\psi(t)\rangle = \omega_{\psi\psi}\,|\psi(t)\rangle + \omega_{\phi\psi}\,|\phi(t)\rangle$. Thus, in compact form, we have

$$i\,\partial_t\,|\zeta(t)\rangle = \omega_d\,|\zeta(t)\rangle + \omega_{\phi\psi}\,\sigma_1\,|\zeta(t)\rangle\,, \tag{25}$$

where $|\zeta(t)\rangle = (|\phi(t)\rangle\,, |\psi(t)\rangle)^T$ and $\omega_d = diag(\omega_{\phi\phi}, \omega_{\psi\psi})$. We use the notation $g \equiv \tan 2\theta$ and $\omega_{\phi\psi} = \frac{1}{2} g\,\delta\omega$ from Eq. (8) and we set $A_0 = A_0^{(1)}\sigma_1 \equiv \frac{1}{2}\,\delta\omega\,\sigma_1$. The covariant derivative is defined by the following relation

$$D_t = \partial_t + i\,\omega_{\phi\psi}\,\sigma_1 = \partial_t + i\,g\,A_0^{(1)}\,\sigma_1 \tag{26}$$

where g behaves as a coupling constant and A_0 represents the gauge operator field. The motion equation (25) can be then expressed as

$$iD_t\,|\zeta(t)\rangle = \omega_d\,|\zeta(t)\rangle\,. \tag{27}$$

We point out that under the following transformations

$$D'_t = \partial_t + i\,g\,(A_0^{(1)}\,\sigma_1 + \partial_t\,\lambda(t)\,\sigma_1)\,, \tag{28}$$

$$|\zeta'(t)\rangle = e^{-ig\,\lambda(t)\,\sigma_1}|\zeta(t)\rangle\,, \tag{29}$$

Eq.(27) becomes

$$iD'_t\,|\zeta'(t)\rangle = \omega_d\,|\zeta'(t)\rangle\,. \tag{30}$$

Thus defining $U(t) \equiv e^{-ig\,\lambda(t)\,\sigma_1}$, we have

$$U(t)\,(iD_t\,|\zeta(t)\rangle) = iD'_t\,U(t)\,|\zeta(t)\rangle \tag{31}$$

$$g\,A_0^{(1)'}\,\sigma_1 = U(t)\,g\,A_0^{(1)}\,\sigma_1\,U^{-1}(t) + i\,(\partial_t\,U(t))\,U^{-1}(t)\,, \tag{32}$$

as it should be indeed for a gauge field transformation (see also ref. [12] where neutrino mixing is studied). The constant A_0 is the only non-vanishing component of A_μ, this implies that the strength field $F_{\mu\nu}$ is identically zero. This is a feature which, for example, occurs when the gauge potential is a pure gauge (with non-singular gauge functions). A typical case (with non-integrable phase conditions) is the one of the Aharonov–Bohm effect [13].

We now express Eqs. (25) as

$$(H - \omega_{\phi\psi}\,\sigma_1)\,|\zeta(t)\rangle = \omega_d\,|\zeta(t)\rangle\,, \tag{33}$$

and introduce the operator F so that

$$F = H - \omega_{\phi\psi}\,\sigma_1\,. \tag{34}$$

Such an operator may be considered as the free energy operator provided that the term $\omega_{\phi\psi}\,\sigma_1$ is identified with the entropy term TS in the traditional free energy expression. In such a case we may put the "temperature" $T = g$ and the entropy $S = A_0 = \frac{1}{2}\,\delta\omega\,\sigma_1$. The entropy term TS, in terms of the time dependent states, is written as

$$TS = \omega_{\phi\psi}(|\phi(t)\rangle\langle\psi(t)| + |\psi(t)\rangle\langle\phi(t)|)\ . \qquad (35)$$

Moreover, we note that the geometric invariant s defined in Eq. (24) is related to the entropy term; indeed we have

$$\int_0^{nT} \langle\zeta(t)|TS\sigma_1|\zeta(t)\rangle\, dt = \int_0^{nT} \langle\zeta(t)|g\,A_0^{(1)}|\zeta(t)\rangle\, dt = 2\int_0^{nT} \omega_{\phi\psi}\, dt = s_n. \qquad (36)$$

Thus, the relation between TS and the variance of the energy $\Delta\,\omega = \omega_{\phi\psi}$ is induced by the non diagonal elements of H.

In conclusion, we have shown the presence of geometric phases in the evolution of a two level system and studied its gauge structure. We have computed the covariant derivative and pointed out that it acts as the free energy with the gauge field acting as the entropy. In such a picture time evolution is thus controlled by the free energy. When applied to a qubit state, these results may be of interest in quantum computing studies.

Acknowledgements

Partial financial support from INFN is acknowledged.

References

[1] Pachos J, Zanardi P and Rasetti M 1999 *Phys. Rev.* A **61** 010305
[2] Zanardi P and Rasetti M 1999 *Phys. Lett.* A **264** 94
[3] F. Wilczek and A. Zee 1984 *Phys. Rev. Lett.* **52** 2111
[4] Bruno A, Capolupo A, Raimondo G, Kak S and Vitiello G 2010 Gauge theory and two level systems *Preprint*
[5] Kak S 1999 *Foundations of Physics* **29** 267
[6] Kak S 2006 *ACM Ubiquity* **7** 1
[7] Kak S 2003 *Information Sciences* **152** 195
[8] Berry M V 1984 *Proc. Roy. Soc. London* A **392** 45
[9] Blasone M, Henning P A and Vitiello G 1999 *Phys.Lett.* B **466** 262
[10] Anandan J and Aharonov Y 1990 *Phys. Rev. Lett.* **65** 1697
[11] Blasone M, Capolupo A, Celeghini E and Vitiello G 2009 *Phys. Lett.* B **674** 73
[12] Blasone M, Di Mauro M and Vitiello G 2011 Non-abelian gauge structure in neutrino mixing *Phys. Lett.* B in print (*Preprint* arXiv:1003.5812v1 [hep-ph])
[13] Aharonov Y and Bohm D 1959 *Phys. Rev.* **115** 485

Journal of Physics: Conference Series **306** (2011) 012066 doi:10.1088/1742-6596/306/1/012066

CP and CPT violation and non-cyclic phase

Antonio Capolupo

Dipartimento di Matematica e Informatica, Università di Salerno and
Istituto Nazionale di Fisica Nucleare, Gruppo Collegato di Salerno, 84100 Salerno, Italy.

Abstract. The effective Hamiltonian describing the time evolution of neutral mixed meson systems like $K^0 - \bar{K}^0$ is non-Hermitian and non-normal. It is then necessary diagonalize it by a complete biorthonormal set of eigenvectors. Moreover the presence of non-cyclic phases for oscillating mesons is shown and such phases turn out to be related to CP and CPT violating parameters.

Meson mixing and oscillations have been studied extensively in the frameworks of quantum mechanics (QM) [1], [2] and of quantum field theory (QFT) [3]-[5]. In particular, modifications to the QM oscillation formulas have been presented in Refs.[3]-[5]. It has been found that the field-theoretical corrections, due to the nonperturbative vacuum structure associated with particle mixing, may be as large as $5 - 20\%$ for strongly mixed systems, such as $\omega - \phi$ or for $\eta - \eta'$, [4]. On the contrary in K^0, B_d^0, B_s^0 and D^0 these corrections are negligible. Then, although the QFT analysis discloses features which cannot be ignored (see for example Refs.[6]-[8]), nevertheless a correct phenomenological description of systems as $K^0 - \bar{K}^0$ can be done in the context of QM, neglecting the nonperturbative field-theoretical effects.

In this report I present a preliminary result [9], obtained in the QM framework, according to which the CPT invariance in neutral meson systems as $K^0 - \bar{K}^0$ can be tested by measuring the non-cyclic phases generated in the time evolution of mixed meson systems. Indeed I show that the Aharonov–Anandan invariants [10] and the Mukunda–Simon phases [11] for mixed mesons are connected to CP and CPT violating parameters. In order to do that, I analyze some of the features of the formalism depicting the evolution of mixed neutral mesons in QM.

In the Wigner-Weisskopf approximation [12], the effective Hamiltonian H for such systems, in the presence of CP violation, is non-Hermitian $H \neq H^\dagger$ and non-normal $[H, H^\dagger] \neq 0$. This implies that one cannot diagonalize H with its eigenstates since they do not satisfy orthogonality and completeness relations. The diagonalization of H is obtained by utilizing the rules of non-Hermitian quantum mechanics. In particular, the biorthonormal basis formalism [13]-[16] is used to describe the time evolution of mixed mesons in the presence of CP violation.

The time evolution of a beam of neutral boson system and of its decay products can be described as

$$|\psi(t)\rangle = \psi_{M^0}(t)|M^0\rangle + \psi_{\bar{M}^0}(t)|\bar{M}^0\rangle + \sum_n d_n(t)|n\rangle\,, \tag{1}$$

where M^0 denotes K^0, B_d^0, B_s^0 or D^0, $\bar{M}^0$ the corresponding antiparticles, $|n\rangle$ are the decay products, t is the proper time and $\psi_{M^0}(t)$, $\psi_{\bar{M}^0}(t)$ and $d_n(t)$ are time dependent functions. Since the decay products are absent at the instant of the production of M^0 and $\bar{M}^0$, the state vector at initial time $t = 0$ is then given by

$$|\psi(0)\rangle = \psi_{M^0}(0)|M^0\rangle + \psi_{\bar{M}^0}(0)|\bar{M}^0\rangle\,. \tag{2}$$

In the Wigner-Weisskopf approximations [12], the wave functions $\psi_{M^0}(t)$ and $\psi_{\bar{M}^0}(t)$, describing the evolution of $|\psi(t)\rangle$ in the space formed by $|M^0\rangle$ and $|\bar{M}^0\rangle$, satisfy the Schrodinger equation

$$i\frac{d}{dt}\begin{pmatrix} \psi_{M^0}(t) \\ \psi_{\bar{M}^0}(t) \end{pmatrix} = \begin{pmatrix} H_{11} & H_{12} \\ H_{21} & H_{22} \end{pmatrix}\begin{pmatrix} \psi_{M^0}(t) \\ \psi_{M^0}(t) \end{pmatrix}, \tag{3}$$

where the effective Hamiltonian $H = \begin{pmatrix} H_{11} & H_{12} \\ H_{21} & H_{22} \end{pmatrix}$ governing the time evolution of the system is written as $H = M - i\frac{\Gamma}{2}$ with M and Γ Hermitian. The matrix elements of H are constrained by the conservation of discrete symmetries [2], indeed, CPT conservation implies $H_{11} = H_{22}$, T conservation entails $|H_{12}| = |H_{21}|$ and CP conservation requires $H_{11} = H_{22}$ and $|H_{12}| = |H_{21}|$. As pointed out above, in the presence of CP violation H is non-Hermitian and non-normal. Then H is diagonalized by means a bi-orthonormal basis [13]-[16].

Let $\lambda_j = m_j - i\Gamma_j/2$, with $j = a, b$, be the eigenvalues of H and with $|M_j\rangle$ and $\langle\widetilde{M}_j|$ the corresponding right and left eigenvectors:

$$H|M_j\rangle = \lambda_j|M_j\rangle\,, \qquad \langle\widetilde{M}_j|H = \langle\widetilde{M}_j|\lambda_j\,. \tag{4}$$

Since H is non-Hermitian and non-normal, the conjugate states $\langle\widetilde{M}_j|^\dagger \equiv |\widetilde{M}_j\rangle$ are not isomorphic to $|M_j\rangle$, i.e. $|\widetilde{M}_j\rangle \neq |M_j\rangle$ and similar for $|M_j\rangle^\dagger \equiv \langle M_j|$: $\langle M_j| \neq \langle\widetilde{M}_j|$. However, it is easy to show that the set of states $\{|M_j\rangle, \langle\widetilde{M}_j|\}$, with $j = a, b$, ($a = S$ =short lived state, $b = L$ =long lived state in the kaon system case) forms a complete bi-orthonormal basis [9], [16]:

$$\langle\widetilde{M}_j|M_i\rangle = \langle M_i|\widetilde{M}_j\rangle = \delta_{ij}\,, \qquad \sum_j |M_j\rangle\langle\widetilde{M}_j| = \sum_j |\widetilde{M}_j\rangle\langle M_j| = 1\,. \tag{5}$$

The existence of a complete biorthonormal set of eigenvector of H implies that H is diagonalizable. Indeed the matrix V exists such that $V^{-1}HV = diag(\lambda_a, \lambda_b)$ with

$$V = \begin{pmatrix} p_a & p_b \\ q_a & -q_b \end{pmatrix}, \qquad V^{-1} = \frac{1}{q_a p_b + q_b p_a}\begin{pmatrix} q_b & p_b \\ q_a & -p_a \end{pmatrix}. \tag{6}$$

where

$$\frac{q_a}{p_a} = \frac{H_{21}}{\lambda_a - H_{22}} = \frac{\lambda_a - H_{11}}{H_{12}}\,, \qquad \frac{q_b}{p_b} = \frac{H_{21}}{H_{22} - \lambda_b} = \frac{H_{11} - \lambda_b}{H_{12}}\,. \tag{7}$$

The mass eigenstates $|M_a\rangle$ and $|M_b\rangle$ are then written in terms of $|M^0\rangle$, $|\bar{M}^0\rangle$ as

$$|M_a\rangle = p_a\,|M^0\rangle + q_a\,|\bar{M}^0\rangle\,, \tag{8}$$
$$|M_b\rangle = p_b\,|M^0\rangle - q_b\,|\bar{M}^0\rangle\,, \tag{9}$$

and, in a similar way, $\langle\widetilde{M}_a|$ and $\langle\widetilde{M}_b|$ are expressed as

$$\langle\widetilde{M}_a| = \frac{1}{q_a p_b + q_b p_a}\left[q_b\,\langle\widetilde{M^0}| + p_b\,\langle\widetilde{\bar{M}^0}|\right], \tag{10}$$
$$\langle\widetilde{M}_b| = \frac{1}{q_a p_b + q_b p_a}\left[q_a\,\langle\widetilde{M^0}| - p_a\,\langle\widetilde{\bar{M}^0}|\right]. \tag{11}$$

Since the time evolution operator associated with H: $U(t) = e^{-iHt}$ is not unitary, we introduce the time evolution operator of $H^\dagger$: $\widetilde{U}(t) = e^{-iH^\dagger t}$ such that $U\widetilde{U}^\dagger = \widetilde{U}^\dagger U = 1$. The spectral form of the Hamiltonian and of the operators $U(t)$ and $\widetilde{U}(t)$ are given respectively by

$$H = \sum_j \lambda_j |M_j\rangle\langle\widetilde{M}_j|, \qquad U(t) = \sum_j e^{-i\lambda_j t}|M_j\rangle\langle\widetilde{M}_j|, \qquad \widetilde{U}(t) = \sum_j e^{-i\lambda_j^* t}|\widetilde{M}_j\rangle\langle M_j|. \quad (12)$$

Thus, by using Eqs.(8)-(12), the states $|M^0(t)\rangle$ and $|\bar{M}^0(t)\rangle$ are expressed in terms of $|M_a\rangle$ and $|M_b\rangle$, for any time t, as [9]:

$$|M^0(t)\rangle = \frac{1}{q_a p_b + q_b p_a}\left[q_b\,|M_a\rangle\, e^{-i\lambda_a t} + q_a\,|M_b\rangle\, e^{-i\lambda_b t}\right], \quad (13)$$

$$|\bar{M}^0(t)\rangle = \frac{1}{q_a p_b + q_b p_a}\left[p_b\,|M_a\rangle\, e^{-i\lambda_a t} - p_a\,|M_b\rangle\, e^{-i\lambda_b t}\right], \quad (14)$$

$$\langle\widetilde{M^0(t)}| = p_a\,\langle\widetilde{M}_a|\, e^{i\lambda_a t} + p_b\,\langle\widetilde{M}_b|\, e^{i\lambda_b t}, \quad (15)$$

$$\langle\widetilde{\bar{M}^0(t)}| = q_a\,\langle\widetilde{M}_a|\, e^{i\lambda_a t} - q_b\,\langle\widetilde{M}_b|\, e^{i\lambda_b t}. \quad (16)$$

The states in Eqs.(13)–(16) have to be used to derive the expressions of the asymmetries describing the violations of time-reversal invariance and of CPT invariance [9] and to obtain the Aharonov–Anandan and the Mukunda–Simon phases in the $M^0 - \bar{M}^0$ systems.

Before to study the non-cyclic phases, I introduce the parameters describing the CP and CPT violations. The constrains on H imposed by CP and T invariance suggest to adopt the following CP and T violation parameter:

$$\varepsilon = \frac{|p_b/q_b| - |q_a/p_a|}{|p_b/q_b| + |q_a/p_a|} = \frac{|p/q| - |q/p|}{|p/q| + |q/p|} = \frac{|H_{12}| - |H_{21}|}{|H_{12}| + |H_{21}|}, \quad (17)$$

where we define

$$\frac{q}{p} = \sqrt{\frac{q_a q_b}{p_a p_b}} = \sqrt{\frac{H_{21}}{H_{12}}}. \quad (18)$$

Moreover, the CPT invariance imposes the equality of the diagonal elements of the Hamiltonian H in Eq.(3): $H_{11} = H_{22}$, thus in order to test the CPT invariance the difference $H_{22} - H_{11}$ has to be parametryzed. Such a parametrization have to be invariant under phase transformation as:

$$|M^{0\prime}\rangle = e^{i\alpha}|M^0\rangle, \qquad |\bar{M}^{0\prime}\rangle = e^{-i\alpha}|\bar{M}^0\rangle. \quad (19)$$

Under the transformation (19) we have $H_{11} \to H_{11}$, $H_{22} \to H_{22}$, $H_{12} \to e^{-2i\alpha}H_{12}$, and $H_{21} \to e^{2i\alpha}H_{21}$, i.e. the diagonal element of H and the product $H_{12}H_{21}$ are invariant, then a possible parameter denoting the CPT violation is

$$\delta = \frac{H_{22} - H_{11}}{\sqrt{H_{12}H_{21}}}. \quad (20)$$

Note that in the case of CPT invariance: $p/q = p_a/q_a = p_b/q_b$ and $\delta = 0$.

Let us begin by studying the Aharonov–Anandan invariant. For an Hermitian Hamiltonian it is defined as [10]

$$s = 2\int_0^t \Delta E(t')\,dt' \quad (21)$$

where ΔE is the variance of the energy E. In the case of a non-Hermitian system with a complete bi-orthonormal basis $\{|\psi(t)\rangle, \langle\widetilde{\psi}(t)|\}$, the variance is given by

$$\Delta E^2(t) = \langle\widetilde{\psi}(t)|H^2|\psi(t)\rangle - (\langle\widetilde{\psi}(t)|H|\psi(t)\rangle)^2, \tag{22}$$

and the generalization of Eq.(21) to such a system is [14]

$$s = 2\int_0^t |\Delta E(t')|\, dt' . \tag{23}$$

In particular, in the meson mixing systems, we have following variances:

$$\Delta E_{M^0M^0}(t) = \Delta E_{\bar{M}^0\bar{M}^0}(t) = \frac{(\lambda_a - \lambda_b)}{\sqrt{4+\delta^2}} = (\lambda_a - \lambda_b)\left(\frac{1}{2} - \frac{\delta^2}{16} + O(\delta^3)\right), \tag{24}$$

$$\Delta E_{M^0\bar{M}^0}(t) = \langle\widetilde{M^0}(t)|H|\bar{M}^0(t)\rangle = \frac{p}{q}\frac{(\lambda_a - \lambda_b)}{\sqrt{4+\delta^2}} = \frac{p}{q}(\lambda_a - \lambda_b)\left(\frac{1}{2} - \frac{\delta^2}{16} + O(\delta^3)\right), \tag{25}$$

$$\Delta E_{\bar{M}^0M^0}(t) = \langle\widetilde{\bar{M}^0}(t)|H|M^0(t)\rangle = \frac{q}{p}\frac{(\lambda_a - \lambda_b)}{\sqrt{4+\delta^2}} = \frac{q}{p}(\lambda_a - \lambda_b)\left(\frac{1}{2} - \frac{\delta^2}{16} + O(\delta^3)\right). \tag{26}$$

The corresponding Aharonov–Anandan invariants are then:

$$s_{M^0M^0}(t) = s_{\bar{M}^0\bar{M}^0}(t) = 2\int_0^t |\Delta E_{M^0M^0}(t')|\, dt' = \frac{|\lambda_a - \lambda_b|}{|\sqrt{4+\delta^2}|} t, \tag{27}$$

$$s_{M^0\bar{M}^0}(t) = 2\int_0^t |\Delta E_{M^0\bar{M}^0}(t')|\, dt' = \left|\frac{p}{q}\right| \frac{|\lambda_a - \lambda_b|}{|\sqrt{4+\delta^2}|} t, \tag{28}$$

$$s_{\bar{M}^0M^0}(t) = 2\int_0^t |\Delta E_{\bar{M}^0M^0}(t')|\, dt' = \left|\frac{q}{p}\right| \frac{|\lambda_a - \lambda_b|}{|\sqrt{4+\delta^2}|} t. \tag{29}$$

respectively. Eqs.(27)-(29) show that the phases s are related to the CPT violating parameter δ and that $s_{M^0\bar{M}^0}(t)$ and $s_{\bar{M}^0M^0}(t)$ are different because of CP violation, independently from CPT violation. In the case of CPT invariance, Eqs.(27)-(29) become

$$s_{M^0M^0}(t) = s_{\bar{M}^0\bar{M}^0}(t) = \frac{|\lambda_a - \lambda_b|}{2} t, \tag{30}$$

$$s_{M^0\bar{M}^0}(t) = \left|\frac{p}{q}\right| \frac{|\lambda_a - \lambda_b|}{2} t, \tag{31}$$

$$s_{\bar{M}^0M^0}(t) = \left|\frac{q}{p}\right| \frac{|\lambda_a - \lambda_b|}{2|} t, \tag{32}$$

respectively. Moreover, by using Eqs.(28) and (29) we can consider the quantity:

$$\frac{s_{M^0\bar{M}^0} - s_{\bar{M}^0M^0}}{s_{M^0\bar{M}^0} + s_{\bar{M}^0M^0}} = \frac{|p/q| - |q/p|}{|p/q| + |q/p|} = \frac{|H_{12}| - |H_{21}|}{|H_{12}| + |H_{21}|}. \tag{33}$$

It coincides with the CP and T violating parameter ε given in Eq.(17).

Let us now analyze the Mukunda–Simon phase for the mixed meson systems $M^0 - \bar{M}^0$. In general, it is defined as [11]

$$\Phi(t) = \Upsilon(t) + \frac{1}{\hbar}\int_0^t \langle\psi(t')|H|\psi(t')\rangle dt', \tag{34}$$

where

$$\Upsilon(t) = \arg\langle\psi(0)|\psi(t)\rangle\,. \tag{35}$$

In the case of a system with a diagonalizable non-Hermitian Hamiltonian, Eqs.(34) and (35) become respectively:

$$\Phi_{NH}(t) = \Upsilon_{NH}(t) + \frac{1}{\hbar}\int_0^t |\langle\widetilde{\psi}(t')|H|\psi(t')\rangle|dt'\,, \tag{36}$$

and

$$\Upsilon_{NH}(t) = \arg\langle\widetilde{\psi}(0)|\psi(t)\rangle\,. \tag{37}$$

In particular, in the meson system, we have the following quantities

$$\Phi_{M^0M^0}(t) = \arg\langle\widetilde{M^0}(0)|M^0(t)\rangle + \frac{1}{\hbar}\int_0^t |\langle\widetilde{M^0}(t')|H|M^0(t')\rangle|dt'\,, \tag{38}$$

$$\Phi_{\bar{M}^0\bar{M}^0}(t) = \arg\langle\widetilde{\bar{M}^0}(0)|\bar{M}^0(t)\rangle + \frac{1}{\hbar}\int_0^t |\langle\widetilde{\bar{M}^0}(t')|H|\bar{M}^0(t')\rangle|dt'\,. \tag{39}$$

Denote with $\Delta m = m_b - m_a$, $m = m_a + m_b$, $\Delta\Gamma = \Gamma_a - \Gamma_b$ and $\Gamma = \Gamma_a + \Gamma_b$ and assume $\frac{\Delta\Gamma t}{2}$ small, that is $t \ll 2/\Delta\Gamma$. A straightforward computation leads to the expressions of $\Phi_{M^0M^0}(t)$ and $\Phi_{\bar{M}^0\bar{M}^0}(t)$ in terms of δ:

$$\begin{aligned}\Phi_{M^0M^0}(t) \simeq\ & \arctan\left[\frac{\Re\delta\cos(\frac{mt}{2})\sin(\frac{\Delta mt}{2}) - \sin(\frac{mt}{2})\left[2\,(1+\Re\delta)\cos(\frac{\Delta mt}{2}) + \Im\delta\sin(\frac{\Delta mt}{2})\right]}{\Re\delta\sin(\frac{mt}{2})\sin(\frac{\Delta mt}{2}) + \cos(\frac{mt}{2})\left[2\,(1+\Re\delta)\cos(\frac{\Delta mt}{2}) + \Im\delta\sin(\frac{\Delta mt}{2})\right]}\right] \\ & + \frac{t}{2}\sqrt{\frac{m^2 + \frac{\Gamma^2}{4} + \Re\delta\left[m^2 + m\Delta m + \frac{\Gamma^2 - \Gamma\Delta\Gamma}{4}\right] - \Im\delta(m\Delta\Gamma + \Delta m\Gamma)}{1+\Re\delta}}\,.\end{aligned} \tag{40}$$

and

$$\begin{aligned}\Phi_{\bar{M}^0\bar{M}^0}(t) \simeq\ & \arctan\left[\frac{-\Re\delta\cos(\frac{mt}{2})\sin(\frac{\Delta mt}{2}) - \sin(\frac{mt}{2})\left[2\,(1+\Re\delta)\cos(\frac{\Delta mt}{2}) - \Im\delta\sin(\frac{\Delta mt}{2})\right]}{-\Re\delta\sin(\frac{mt}{2})\sin(\frac{\Delta mt}{2}) + \cos(\frac{mt}{2})\left[2\,(1+\Re\delta)\cos(\frac{\Delta mt}{2}) - \Im\delta\sin(\frac{\Delta mt}{2})\right]}\right] \\ & + \frac{t}{2}\sqrt{\frac{m^2 + \frac{\Gamma^2}{4} + \Re\delta\left[m^2 - m\Delta m + \frac{\Gamma^2 + \Gamma\Delta\Gamma}{4}\right] + \Im\delta(m\Delta\Gamma + \Delta m\Gamma)}{1+\Re\delta}}\,.\end{aligned} \tag{41}$$

In the case of CPT invariance, we have:

$$\Phi_{M^0M^0}(\Delta t) = \Phi_{\bar{M}^0\bar{M}^0}(\Delta t) = \frac{mt}{2}\left(\sqrt{1 + \frac{\Gamma^2}{4m^2}} - 1\right)\,, \tag{42}$$

This implies that deviation from the equality (42) indicates CPT violation.

In conclusion, a complete biorthonormal set of vectors has been used to diagonalize the effective non-Hermitian and non-normal Hamiltonian of mixed neutral meson systems. By using such states, the non-cyclic phases for oscillating mesons have been shown to be related to CP and CPT violating parameters. This fact suggests that new interferometric methods applied to mixed meson systems capable of measuring the non-cyclic phases studied in this work could represent a new way to test the CPT invariance in mixed meson systems.

Acknowledgements
Partial financial support from INFN and Miur is acknowledged.

References

[1] Kabir P K 1968 *The CP Puzzle* (London: Academic Press); Nachtmann O 1990 *Elementary Particle Physics: Concepts and Phenomena* (Berlin: Springer)
[2] Fidecaro M and Gerber H J 2006 *Rept. Prog. Phys.* **69** 1713; 2006 Erratum-ibid. **69** 2841 and references therein
[3] Blasone M, Capolupo A, Romei O and Vitiello G 2001 *Phys. Rev.* D **63**, 125015
[4] Capolupo A, Ji C R, Mishchenko Y and Vitiello G 2004 *Phys. Lett.* B **594** 135
[5] Capolupo A 2004 *Ph.D. Thesis* arXiv:hep-th/0408228
[6] Blasone M, Capolupo A, Capozziello S, Carloni S and Vitiello G 2004 *Phys. Lett.* A **323** 182
[7] Capolupo A, Capozziello S and Vitiello G 2007 *Phys. Lett.* A **363** 53
[8] Capolupo A, Capozziello S and Vitiello G 2009 *Phys. Lett.* A **373** 601
[9] Capolupo A Meson mixing, CP and CPT violation and non-cyclic phase, work in progress
[10] Anandan J and Aharonov Y 1990 *Phys. Rev. Lett.* **65** 1697
[11] Mukunda N and Simon R 1993 *Ann.Phys.(N.Y)* **228** 205
[12] Weisskopf V, Wigner E P 1930 *Z. Phys.* **63** 54
[13] Bender C M and Boettcher S 1998 *Phys. Rev. Lett.* **80** 5243; Tanaka T 2006 *J. Phys. A: Math. Gen.* **39** 7757
[14] Nesterov A I 2009 *Phys. Lett.* A **373** 3629
[15] Baker H C and Singleton Jr. R L 1990 *Phys. Rev. A* **42** 10
[16] Dattoli G, Torre A and Mignani R 1990 *Phys. Rev. A* **42** 1467

Extreme Physics and Informational/Computational Limits

Paolo Di Sia

Department of Computer Science, Faculty of Science, Verona University, Strada Le Grazie 15, I-37134 Verona, Italy

Faculty of Computer Science, Free University of Bozen, Piazza Domenicani 3, I-39100 Bozen-Bolzano, Italy

E-mail: paolo.disia@univr.it; 10alla33@virgilio.it

Abstract. A sector of the current theoretical physics, even called "extreme physics", deals with topics concerning superstring theories, multiverse, quantum teleportation, negative energy, and more, that only few years ago were considered scientific imaginations or purely speculative physics. Present experimental lines of evidence and implications of cosmological observations seem on the contrary support such theories. These new physical developments lead to informational limits, as the quantity of information, that a physical system can record, and computational limits, resulting from considerations regarding black holes and space-time fluctuations. In this paper I consider important limits for information and computation resulting in particular from string theories and its foundations.

1. Introduction

Many of the fundamental limits on information processing derive from physics, in particular from thermodynamics, relativity, quantum mechanics and the following unified theories, like string theories. Recent physically motivated computing paradigms like reversible computing and quantum computing may help for same aspects, but in every case they remain subject to some basic limits. Computers are physical systems, therefore the physical laws establish their operativity. In particular, the speed with which a physical device can process information is limited by its energy and the amount of information which it can process by the number of degrees of freedom possessed. These physical limits of computation appear depending from important physical constants, the speed of light c, the quantum scale $\hbar$, and the gravitational constant G [1]. The so called "Extreme Physical Information", seen as an extension of the information theory, includes much theoretical physics and it has been used to derive some fundamental laws not only in the sectors of physics, but also for biology, chemistry, economics. The holographic principle is a property of string theories and quantum gravity declaring that the description of a volume of space can be thought as encoded on the region boundary, of preference a light-like boundary. Through this theory the entire universe can be seen as a 2-D information structure "painted" on the cosmological horizon, such that the 3 observed dimensions result to be a description at macroscopic scales and low energies. Therefore the 3-D universe could be totally equivalent to alternative quantum fields and physical laws "painted" on a big and distant surface [2]. The physics of black holes suggests credibility to this theory; studies about black holes have showed that the maximum entropy or information content of a space region is defined by its surface area and not by its volume. In particular, a hole with a horizon area A has $A/4$ units of entropy, with the Planck area, approximately $10^{-66}\,cm^2$, the fundamental quantum unit of area. The holographic principle shows the complete equivalence between two universes with different dimensions and obeying different physical laws. From a theoretical viewpoint this property has been demonstrated mathematically for 5-D anti-de Sitter spacetimes and the corresponding 4-D boundary, considering the 5-D universe recorded like a hologram on the 4-D surface. As example, a black hole in the 5-D spacetime appears equivalent to hot radiation on the hologram; the hole and the radiation have the same entropy even if the physical origin of the entropy is completely different. Theoretical studies suggested also that a black hole with appropriate properties can serve as a basis for a description of string theory [3].

2. Informational limits

Entropy is measurable in bits, if considered as information. The total quantity of bits is related to the total degrees of freedom of matter/energy. Considering a given volume, there is an upper limit to the density of information in relation to all the particles costituting the matter in that volume. The fundamental difference between the thermodynamic entropy of physics and the Shannon's information entropy regards the units of measure; the former is expressed in units of energy divided by temperature, the latter in bits of information.
By information theory, the maximal information I_{MAX} of a system, obtainable knowing the details of its state, is given by:

$$I_{MAX} = S_{MAX}/K_B \ln 2 \tag{1}$$

with K_B the Boltzmann constant. If this information collapses into a black hole, the resulting black hole has an entropy S_{BH} related to S_{MAX} as:

$$I_{MAX}\,K_B \ln 2 = S_{MAX} < S_{BH} < \frac{K_B\,A}{4\,L_{PL}^2} \tag{2}$$

with L_{PL} the Planck length. From relation (2) it follows:

$$I_{MAX} < \frac{A}{4\,L_{PL}^2\,\ln 2}\,. \tag{3}$$

The important bound (3) was deduced by 't Hooft (1993) and Susskind (1995). Previously, in 1981, Bekenstein had reasoned in this way: putting an information content of overall radius R and total mass-energy E into a black hole, it is possible to obtain the bound:

$$I_{MAX} < \frac{2\,\pi\,R\,E}{\hbar\,c\,\ln 2}\,. \tag{4}$$

The bound (4) does not contain the gravitational constant G and for simple closed systems it can also be derived from quantum statistical considerations without considering black holes [4]. Numerically, it requires that 1 cm^3 of ordinary material holds no more than $\approx 10^{38}$ bits. The statistical mechanics allows the computation of the entropy of its molecules, apparently ignoring the atomic and subatomic deeper degrees of freedom of matter as contributions to the total entropy; but "de facto" all atomic and subatomic particles gravitate. This is one of the interesting peculiarities of black hole thermodynamics with respect to detailed aspects of physical systems. The thermodynamics of black holes allows therefore to deduce important limits on the density of entropy or information in a lot of circumstances.
Of interest for the contest is important to remember two important bound: the holographic bound and the universal entropy bound. The holographic bound explains how much information can be contained in a specific space region. Let's consider an approximately spherical matter

distribution contained within a surface of area A; the matter is induced to collapse to form a black hole. The black hole's area must be smaller than A, so its entropy must be less than $A/4$. For the fact that entropy cannot decrease, it follows that the original distribution of matter also must carry less than $A/4$ units of entropy or information.
The universal entropy bound establishes how much information can be carried by a mass m of diameter D. For its derivation we can imagine that a capsule of matter falls in a black hole of about the same dimension. The size of the black hole increases and this gives a limit on how much entropy is contained in the capsule. This limit is tighter than the holographic bound, except the case in which the density of the capsule is about the same with respect to the black hole (in this case the two bounds become equivalent).
The information content of a big quantity of computer chips increases in proportion with the occuped volume. Continuing to add chips, we must stop when the information exceeds the holographic bound, which depends on the surface area. The "breakdown" will occur when the huge quantity of chips collapses to form a black hole [5].

3. Computational limits and quantum fluctuations of space-time

The laws of quantum mechanics and gravitation lead to physical bounds on computation and on precision of clocks; the same laws control also the quantum fluctuations of space-time. The limits regard the speed of computation v and the memory space I of a simple computer. These evaluations hold also for the so-called "*simple* clocks", intending for "simple" that a separation of components of a clock is not involved (and itself in relation to "simple computer"). These bounds are saturated for black holes, which, for some aspects, may be regarded as the ultimate simple computers and ultimate simple clocks [6,7]. It is possible to use quantum mechanics to find fundamental limits on any system with mass m that serves as a time-registering device. For a given T and t, with T the total running time over which a simple clock can remain accurate and t the smallest time interval with the clock capable of resolving, the lower bound on m will be [8]:

$$m \geq \frac{\hbar}{c^2\, t}\left(\frac{T}{t}\right) . \tag{5}$$

It is possible to obtain also an upper bound considering the general relativity. Suppose a simple light-clock consisting of two parallel mirrors, everyone with mass $m/2$, and with a beam of light bouncing between them. The separation d between the mirrors must satisfy the relation $d \leq c\,t$ for resolving a time interval of the order of t, and d is necessarily larger than the Schwarzschild radius $G\,m/c^2$. These two requirements permit to find the upper bound on m:

$$t \geq \frac{G\,m}{c^3} . \tag{6}$$

A bound on the speed of computation v of any information processor [9] is obtainable considering the mean input power $P = m\,c^2/T$. Eq. (5) and the fastest possible processing frequency $v = 1/t$ imply:

$$v^2 \leq \frac{m\,c^2}{\hbar\, T} = \frac{P}{\hbar} . \tag{7}$$

Eqs. (5) and (6) lead to a relation between T and t:

$$T \leq t\left(\frac{t}{t_{PL}}\right)^2 \tag{8}$$

with t_{PL} the Planck time. If we consider now that the ratio T/t (maximum step number of information processing), is (except constant factors) the amount of information I recordable by the computer, it is possible to write a limit for the memory space I of a simple computer:

$$I \approx \frac{T}{t} \leq \frac{1}{(v\,t_{PL})^2} \approx \frac{\hbar}{P\,t_{PL}^2} \,. \tag{9}$$

The combination of Eqs. (7) and (9) is very interesting:

$$I\,v^2 \leq \frac{1}{t_{PL}^2} = \frac{c^5}{\hbar\,G} \approx 10^{86}\big/s^2 \,, \tag{10}$$

because of its independence from mass, size, and details of the simple computer. This expression (valid for simple computers) unites together the concepts of information, gravity, and quantum uncertainty. This bound is realized in particular for black holes [10]. Considering that at short distance scales the space-time is foamy, the same physics underlies the foaminess of space-time, the limits to computation, and clock precision. The bound given from Eq. (10) is saturated for black holes, with both v and I bounds saturated, confirming the conceptual importance of black holes as the simplest and most fundamental constructs of space-time.
The experimental research, in particular in the gravitational-wave interferometer field, can sure help these theoretical aspects providing confirmations or denials.

4. Computational capacity of the universe

It is possible to quantify the amount of information processing performed by the Universe, considered as a whole, since the big bang. The Universe has the capacity to perform a maximum of $I_{UNIV} \leq 10^{122}$ (10^{123} considering all the constant factors like 2π, etc...) elementary quantum logic operations on $\approx 10^{90}$ bits registered in quantum fields, considering $t \approx 10^{10} yr$ the age of the Universe. If we consider a closed Universe, the previous numbers represent the amount of elementary logic operations (ops) and bits available in the entire Universe; if on the contrary the Universe is open, and of infinite extension, then the numbers give the amount of computation performed within the event horizon [2].
The previous information limit can be derived looking at the link between physics and information, discovered by Bekenstein (1973) and Hawking (1975), when they applied quantum mechanics to black holes. They have found that an uncharged and not rotating black hole has entropy S given by:

$$S = \frac{4\,\pi\,K_B\,G\,M^2}{\hbar\,c} = \frac{K_B}{4L_{PL}^2}\,A \tag{11}$$

with M and A respectively mass and area of the black hole horizon. Seeing the entropy as a measure of the information I (or lost information) through the relation $S = K_B \ln 2\,I$, Eq. (11) binds the total information content of a space region to the surface area including such volume. The idea to associate entropy and information with the area of the horizon has been soon enlarged to include all the event horizons, not only those that surround a black hole.

5. Black hole computation

Supposing to perform a high speed serial computation, a way can be to compress the computer dimensions to reduce the time of sending signals. "Smallering" the computer, for fixed values of energy, the density to its inside increases, requesting the "exploration" of different regimes of the high energy physics. At first we find the weak unification scale, therefore that of great unification; when the linear dimensions of the computer are near to its Schwarzschild radius, the Planck scale is reached. To such scale both gravitational and quantum effects become

important; the Compton wavelength of a particle of mass m, $\lambda_C = 2\pi\hbar/mc$, is of order of its Schwarzschild radius $R_S = 2mG/c^2$. A theory of quantum gravity unification or a superstring theory is required.
Although it is not possible to know the exact bit number recordable from a computer of the mass of 1 Kg confined in the volume of 1 litre, it is possible to know the exact bit number recordable from a computer of 1 Kg which has been compressed to the dimension of a black hole [11]. Using Page's results in relation to the mechanism of black holes evaporation [12], the time needed for a black hole to evaporate is given by:

$$t_{EVAP} = \frac{m^3 G^2}{3\hbar c^4 C}, \tag{12}$$

where C is a constant depending on the number of particles kinds with mass smaller than $K_B T$, with T the black hole temperature. In the interval $10 - 100$ of such kinds, C is of order of $10^{-3} - 10^{-2}$ and this implies a lifetime around $\approx 10^{-19} s$ for a black hole of 1 Kg of, during which the black hole can treat around 10^{32} operations on its 10^{16} bits. The effective massless particle number at Hawking temperature for a black hole of 1 Kg is greater than 100. The Schwarzschild radius of a computer of 1 Kg results to be $R_S = 1.485 \cdot 10^{-27} m$. Therefore, it is possible to evaluate the information quantity stored in a black hole through the relation:

$$I = \frac{4\pi m^2 G}{\hbar c \ln 2}. \tag{13}$$

For a computer of 1 Kg in the black hole limit, relation (13) gives $I = 3.827 \cdot 10^{16}\, bits$, from which follows $N = 5.426 \cdot 10^{50}\, ops/s$.

6. Velocity, memory space, relativistic effects

An interesting evaluation concerns the calculation of the computational capacity of a computer that we could call "extreme" or "ultimate". The actual netbooks have a mass of order of 1 Kg. We consider therefore a mass of 1 Kg and look for a limit regarding the speed of such a device. An elementary logic operation in a time Δt requires an average energy given by:

$$E \geq \frac{\pi\hbar}{2\Delta t} \;\Rightarrow\; N = \frac{1}{\Delta t} \approx \frac{2E}{\pi\hbar}. \tag{14}$$

For a mass of 1 Kg, $E = mc^2 \cong 8.987 \cdot 10^{16} J$ and consequently $N = 5.426 \cdot 10^{50}\, ops/s$.
A real computer operates much slower, because a lot of energy constitutes the mass of particles of which it is composed, leaving only a very small fraction for logical execution; mereover a conventional computer employs many degrees of freedom (billions of electrons) to record a single bit.
Thermodynamics and statistical mechanics provide also a fundamental limit to the quantity of performing information bits through the use of an energy quantity confined in a given volume. A system with N accessible states can record $\log_2 N$ bits of information. The accessible state number W of a physical system is also related to its thermodynamic entropy $S = K_B \ln W$. The quantity of information recordable from a physical system, i.e the total number of available memory space bits is given by:

$$I = \frac{S(E,V)}{K_B \ln 2}, \tag{15}$$

with $S(E,V)$ the thermodynamic entropy of a system with energy E confined in a volume V. It is possible to estimate this entropy considering the volume occupied by the computer as set of elementary particles modes with average total energy E. Let's assume as conserved quantities the angular momentum and the electric charge, considered equal to zero. Encapsulating the number of kinds of particles/antiparticles, the polarization number and the particle statistics into a single quantity r, it is possible to write the entropy as function of energy:

$$S = \frac{4}{3} K_B \left(\frac{\pi^2 r V}{30 \hbar^3 c^3} \right)^{1/4} E^{3/4} . \tag{16}$$

Assuming entropy and energy dominated by black hole radiation consisting of photons, we have $r = 2$. For a computer of 1 Kg mass confined in a 1 litre volume, we find: $S \cong 2.04 \cdot 10^8 \, J/K$; $I \cong 2.13 \cdot 10^{31} \, bits$ [13].
With the increase of clock speed, also relativistic effects appear. The combination of relativistic and quantum effects in the logical circuits needs the use of relativistic quantum mechanics and quantum field theory. An important relativistic effect in quantum computation concerns the impossibility to locate an electron in a volume less than that characterized by the Compton wavelength of the electron. This implies that the time required by the light to cross such small distance, $\approx 10^{-21} s$, represents a minimum time for a logic process involving electrons, and therefore a working maximum frequency of order of $10^{21} \, Hz$. At such frequency, the energy-time Heisenberg relation involves voltage values going beyond the threshold of production of $e^+ - e^-$ pairs.
Estimating the power consumption of a quantum relativistic computer, we consider a device working on a clock cycle of $10^{-21} \, s$ and representing every bit through a single electron. From Heisenberg relation we deduce:

$$\Delta E \cong f \hbar = 1.06 \cdot 10^{-13} \, J . \tag{17}$$

Supposing that the elementary circuit process involves 10^3 electrons, at the uncertainty limit the power is given by the product of the number of electrons operating in parallel, times the energy of every electron, times the frequency. The required power in this case results in:

$$P = N_{e^-} \cdot \Delta E \cdot f = 1.06 \cdot 10^{11} \, W . \tag{18}$$

Since pratically the total of such power would go out of the computer in the form of gamma rays, a lot of problems would sure rise. Among the fastest fundamental processes known today in physics there is the creation or disintegration of the Z^0 boson. Its average timelife is of order of $10^{-26} \, s$, corresponding to a frequency of $10^{26} \, Hz$. Therefore if we assume that a circuit cannot operate in faster way, an ultimate limit to the computational speed is found. For the relativistic effects, the electrons cannot be measured in times less than $10^{-21} \, s$. If electrons are used, the maximum clock speed for an electronic circuit will be found in the range of $10^{21} \, Hz$ [14].

7. Closing remarks

The dimensions of the elementary circuit constituents are becoming smaller and smaller. The clock speed of computers is notably grown in the last years and it can well fit to an exponential growth. A huge progress has been done in the simulation of the material properties on computers and these applications accelerate manufacturing new, better devices. Quantum mechanics, together with the elementary properties of matter and radiation, has become fundamental for the future of computers and computation. This implies that the "ultimate computer" might necessarily incorporate the principles of relativistic quantum physics. A quantum computer allows a best understanding not only of subatomic world, but it also deeply penetrates the meaning of computation. Perhaps a final theory might be concerned not with fields and spacetime, but rather with information exchanged among physical processes. If so, the vision of information as stuff of the world would find a complete embodiment.

8. References

[1] Lloyd S 2000 *Nature* **406** 1047
[2] Lloyd S 2002 *Phys. Rev. Lett.* **88** 23
[3] Bekenstein J D 2003 *Sci. Am.* **17** 59
[4] Bekenstein J D 2001 *Stud. Hist. Phil. Mod. Phys.* **32** 511
[5] Bekenstein J D 1999 *Proceedings of the Eighth Marcel Grossman Meeting on General Relativity*, ed T Piran and R Ruffini (Singapore: WorldScientific) p 92
[6] Ng Y J and van Dam H 1994 *Mod. Phys. Lett.* A **9** 335
[7] Sasakura N 1999 *Prog. Theor. Phys.* **102** 169
[8] Salecker H and Wigner E P 1958 *Phys. Rev.* **109** 571
[9] Barrow J D 1996 *Phys. Rev.* D **54** 6563
[10] Ng Y J 2001 *Phys. Rev. Lett.* **86** 14
[11] Novikov I D and Frolov V P 1986 *Black Holes* (Berlin: Springer-Verlag)
[12] Page D N 1976 *Phys. Rev.* D **13** 198
[13] Lloyd S 2000 Ultimate physical limits to computation *Preprint* quant-ph/9908043v3
[14] Pakvasa S, Simmons W and Tata X 1999 Relativistic Quantum Computing *Preprint* quant-ph/9911091v1

Adsorbate localization versus diffusion: a coherent time dependent theory of coupling to gravitons in hidden large dimensions

D. Drakova[1] **and G. Doyen**[2]

[1] University of Sofia, Sofia, Bulgaria
[2] Ludwig-Maximilians Universtät, München, Germany

E-mail: nhdd@wmail.chem.uni-sofia.bg and gerold@gerold-doyen.de

Abstract. A long standing unsolved puzzle is the very small rate of surface diffusion of adsorbates in the low temperature STM in the regime of quantum diffusion, differing by orders of magnitude from the results of modern quantum theories of surface diffusion (QTSD) based on ab initio potential energy surfaces (refs. [1]-[5]). We investigate whether decoherence mechanisms, neglected in QTSD, can shed some light on these discrepancies. A quantum-mechanical theory is presented treating the entanglement of adsorbates with their environment (phonons, electron-hole pairs, gravitons) [6]. Time dependent wave packet evolution including the environmental modes indicates possible paths towards an explanation of the observed discrepancies and gives an insight into phenomenological descriptions like adsorbate localization, quantum Zeno effect (permenant "measurement" by specific environmental excitations) and wave function collapse.

1. Introduction

Within the same experiment, using the scanning tunnelling microscope (STM) at 9K, Lauhon and Ho observed the transition between the classical and quantum behaviour of a single hydrogen atom adsorbed on the Cu(001) surface [7]. The adsorbate displays classical behaviour, being imaged as a localized particle on the same adsorption site for thousands of seconds, which competes with the quantum tunnelling diffusion from one adsorption site to the next one. This is not a singular observation in surface science and it is not limited to light atoms like hydrogen. We mention that even a heavy fullerene C_{70} and a fullerene derivative $C_{60}F_{48}$ are displayed as stationary wave packets in the STM, whereas beams of these molecules show interference patterns from nanolattices in a Talbot-Lau interferometer [8].

Decoherence theory aims at explaining the transiton to classical appearance and localization of microscopic particles. The state-of-the-art decoherence theory, based on collapse and wave function reduction, yields results for the time for particle localization due to "measurement" by the environment [9], which is orders of magnitude longer than the localization time needed to reproduce the experimentally measured rate of hydrogen quantum diffusion above Cu(001) at the low temperature of 9K [7].

In the present article we approach the problem of the localization of an adsorbate on a solid surface. The two-dimensional (2D) periodicity of the surface implies that the adsorbate should be delocalized in a 2D Bloch wave on a time scale of 10^{-10} seconds [6]. This is, however, not observed in STM. The adsorbates in the low temperature STM are imaged as localized wave packets, which are no eigenstates. The localization of adsorbates at low temperature can be responsible for the extremely low diffusion rates, measured experimentally. The entanglement to gravitons is suggested as an ubiquitous decoherence mechanism and as the only relevant decoherence mechanism for particle localization at low temperature. This mechanism is crucially dependent upon the high dimensinality of the graviton field, as suggested by string theory, and on the hidden space dimensions where gravitons can propagate. The entanglement to the rest excitations of the sample (phonons, photons, tomonagons, plasmons), though it is effective at high temperature, cannot be the mechanism for particle localization at low temperature. The typical wave lengths of the order of 180 Å of phonons and tomonagons at low temperature are too long to allow to distinguish a hydrogen atom on two neighbouring adsorption sites, for instance.

In the next section a brief summary is presented of the main ideas suggested by 11D spacetime string theory, made use of in our theory of particle localization. The major features of our theory are described in previous papers [6] and in the contribution by Doyen and Drakova in the present volume [10]. In a separate section we present the adaptation of the formalism to the geometric model of adsorbate diffusion, followed by the results and their interpretation.

2. The hidden large dimensions of string theory and gravitation in 11 dimensional spacetime

The gravitational interaction $V_{grav}^{(4D)}(r) = -\frac{G^{(4D)}M}{r}$ between two electrons in four dimensional spacetime (4D) is 41 orders of magnitude weaker than their electrostatic interaction, though the distance dependence of the two interactions is the same. Increasing the dimensionality in 11D M-theory and supergravity yields the distance law:

$$V_{grav}^{(11D)}(r) = -\frac{G^{(11D)}M}{\pi^7 r^8}, \tag{1}$$

which means a much stronger gravitational interaction at small distances. The additional 7 spacial dimensions must be rolled up to radii of upto millimeter [11] to avoid discrepancy with Newton's gravitational law at distances larger than distances of the order of 10^6 bohr, at which the classical law is experimentally verified. Equating eq. 1 with Newton's law at $r = 10^6$ bohr yields an enhancement of the gravitational constant $G^{(11D)}$ in 11D of 39 orders of magnitude compared to Newton's gravitational constant $G^{(4D)}$ in 4D at distances smaller than the rolled up hidden dimensions. This result justifies the use of the gravitational interaction as an environmental interaction, contributing to the localization of quantum particles.

3. A world model for adsorbate diffusion and localization

We reformulate the formalism of decoherence in a closed quantum world, as it is described in detail in ref. [6], in view of the geometric model for adsorbate diffusion, which is represented on fig. 1. It consists of two adsorption sites coupling to a warp resonance, which mediates the coupling to the gravitons. The basis states are many-particle product states with components describing the local adsorption states and the state of the graviton field: $\mid g1\rangle \mid 0\rangle$ and $\mid g2\rangle \mid 0\rangle$ with $\mid 0\rangle$ denoting the vacuum state of the graviton field. $\mid g1\rangle$ and $\mid g2\rangle$ are the basis functions used to describe the adsorbate on site 1 and site 2, respectively. With the adsorbate in the warp resonance $\mid w\rangle \mid \gamma\rangle$ the state of the graviton field $\mid \gamma\rangle$ deviates from the vacuum state $\mid 0\rangle$ because

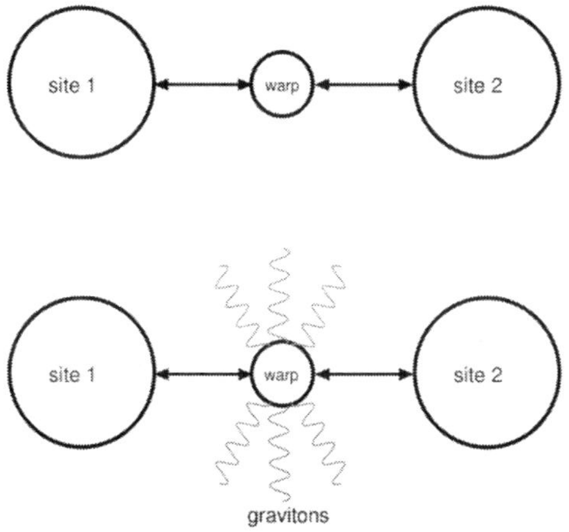

Figure 1. The geometrical model for adsorbate diffusion and localization includes two adsorption sites 1 and 2 described by the basis states $\mid g1\rangle \mid 0\rangle$, $\mid g2\rangle \mid 0\rangle$ and a contracted warp resonance $\mid w\rangle \mid \gamma\rangle$, mediating the diffusion from site 1 to site 2. Upper part: no interaction with gravitons, lower part: interaction with gravitons within the warp resonance.

gravitons are excited, emitted and absorbed, into and from the hidden dimensions. Coupling of the adsorbate to the gravitational field within the warp resonance means deformation of 11D spacetime, which is warranted by graviton excitations. Continuum gravitons, projected on the warp resonance, belong to the local system and their excitation state is labelled by γ or by $grav$. The rest gravitons, belonging to the graviton continuum, are labelled by κ.

The Hamiltonian is:

$$\begin{aligned} H &= \sum_{g=1}^{2} E_g n_g + E_w n_w + \sum_{g=1}^{2} V_{loc}^{g}(c_g^+ c_w + c_w^+ c_g) \\ &+ \varepsilon_{grav} a_{grav}^+ a_{grav} + \sum_{\kappa} \varepsilon_{\kappa} a_{\kappa}^+ a_{\kappa} + \sum_{\kappa} [V_{grav,w} n_w a_{grav}^+ a_{\kappa} + V_{w,grav} n_w a_{\kappa}^+ a_{grav}] \end{aligned} \quad (2)$$

The meaning of the symbols is: n_g, n_w: occupation number operators for the gas particle in the core movement states parallel to the surface and in the warp resonance; c_i^+, c_j: creation and annihilation operators for the gas particle in the respective core movement states; V_{loc}^g: interaction potential between a gas particle state and the warp resonance; $\varepsilon_{grav}, \varepsilon_{\kappa}$: energy of local and continuum gravitons; $a_{grav}^+, a_{grav}, a_{\kappa}^+, a_{\kappa}$: creation and annihilation operators for local and continuum gravitons; $V_{grav,w}$: interaction between the gas particle and the gravitons within the warp resonance (for the sake of simpler notation the polarization degrees of freedom are not explicitly displayed (cf. [6]).

The procedure for the time dependent solution of this hamiltonian is described in ref. [10].

4. Apparent classical behaviour in a time dependent quantum mechanical theory, taking into account the entanglement to gravitons: results and interpretation

The model for adsorbate diffusion depicted in the upper part of fig. 1 represents two degenerate adsorption states coupling to the warp resonance. The particle is adsorbed initially on site 1. The variation with time of the occupations of the two adsorption sites is plotted in fig. 2. The time for adsorbate diffusion from site 1 towards site 2 can be defined as twice the time needed

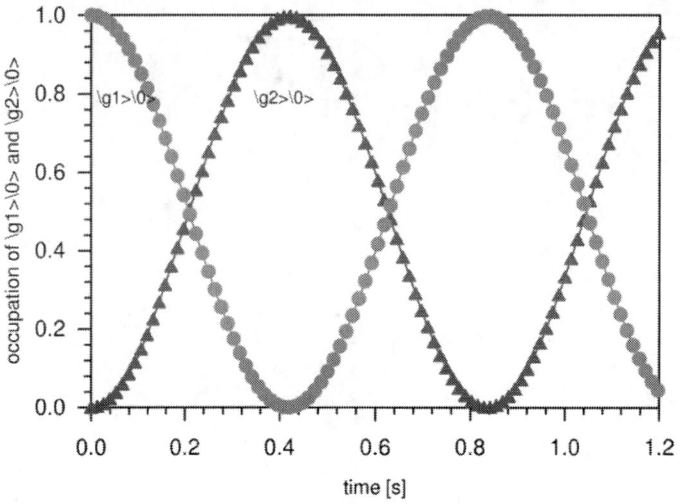

Figure 2. Variation of the occupation of adsorption site 1 and site 2 by an adsorbate with time via coupling to the warp resonance $\mid w\rangle \mid 0\rangle$. The geometric model for adsorbate diffusion is the one displayed in the upper panel of fig. 1.

for the probability of occupying site 1 to reduce by 0.5 and the probability for occupying site 2 to rise by 0.5. The diffusion rate is therefore approximately 1 particle per 0.4 seconds, which compares nicely to the ab initio result of Wahnström et al. for hydrogen diffusion on Ni(100) [5]. In the system of two degenerate adsorption states, coupling to the warp resonance $\mid w\rangle \mid 0\rangle$ (fig. 1 upper panel), the particle, initially adsorbed on the first site 1, is scattered by the interaction with the warp resonance into the second adsorption site 2. Two scattering channels are available: backscattering in the initial state $\mid g1\rangle \mid 0\rangle$ and forward scattering in $\mid g2\rangle \mid 0\rangle$. The two adsorbate states $\mid g1\rangle \mid 0\rangle$ and $\mid g2\rangle \mid 0\rangle$ give rise to two degenerate linear conbinations $\psi_s = N_s(\mid g1\rangle \mid 0\rangle + \mid g2\rangle \mid 0\rangle)$ and $\psi_{as} = N_{as}(\mid g1\rangle \mid 0\rangle - \mid g2\rangle \mid 0\rangle)$. The symmetric combination alone can couple to the warp resonance. Their coupling lifts the degeneracy between ψ_s and ψ_{as}. The adsorbate is involved in Rabi oscillations between ψ_{as} and the ψ_s-warp derived states, displayed in fig. 2, with a frequency corresponding to diffusion forth and back of the adsorbate between site 1 and site 2 with the rate of 2.5 particles per second. The scattering potential in the warp resonance is chosen such, that the diffusion time from the first adsorption site to the second site is of the same order as resulting from the ab initio theory of Wahnström et al. [5].

The entanglement of the adsorbate dynamics to the gravitons within the warp resonance (fig. 1 lower panel) leads to nearly four orders of magnitude slow down of diffusion, compared to the case when the entanglement to the gravitons is neglected, as the plots in figs. 3 and 2 show. The slow down of particle diffusion between the adsorption sites is defined as localization of the adsorbate. The diffusion rate equals 8×10^{-4} particles per second, which is in nice agreement with the experimental measurement on the rate of hydrogen quantum diffusion on Cu(001) in the low temperature STM [7]. Localization of the hydrogen atom due to entanglement to gravitons is associated with this effect.

The strong warp-graviton coupling splits the warp resonance into bonding- $\mid w\rangle \mid \gamma_1\rangle$ and anti-bonding $\mid w\rangle \mid \gamma_2\rangle$ parts, straddling widely around energy zero (cf. eqs. 9). In fig. 4 the spectral

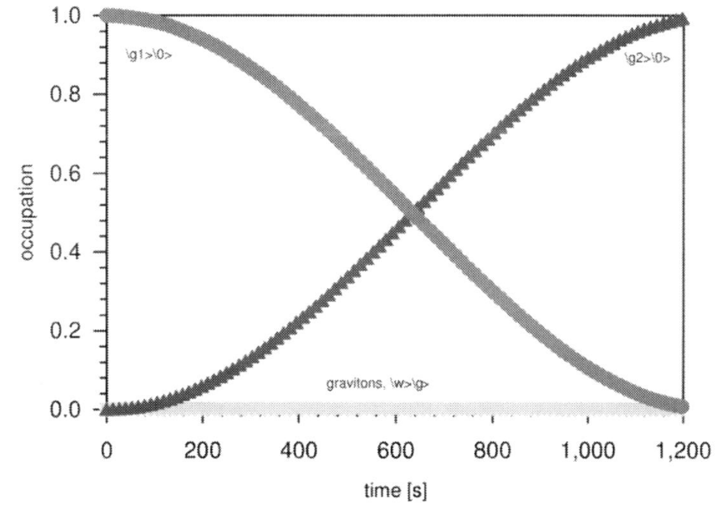

Figure 3. Entanglement to gravitons is included: variation of the occupation of adsorption site 1 and site 2 by the adsorbate with time. The warp resonance $| w\rangle \, | \gamma\rangle$ mediates the entanglement of the adsorbed particle dynamics to the gravitons (cf. the lower panel of fig. 1).

energy distribution of the eigensates on the basis states in the 4D world (fig. 4a) and in the world model, including the gravitons (fig. 4b), is displayed for the static case. The lower right panel in fig. 4b shows the splitting between $| w\rangle \, | \gamma_1\rangle$ and $| w\rangle \, | \gamma_2\rangle$, whereas in the lower left panel a detailed spectral distribution of the eigenstates around energy zero is plotted. If $E_{bonding} + E_{anti-bonding}$ were equal to zero, the states, resulting from the interaction of ψ_s with $| w\rangle \, | \gamma_1\rangle$ and $| w\rangle \, | \gamma_2\rangle$ would lie exactly at energy zero. Since $E_{warp} \neq 0$ and $E_{warp} \ll V_{g-warp}$, the resulting ψ_s-derived state is not exactly at energy zero, but significantly nearer to zero than without graviton coupling (cf. fig. 4a and 4b, left panel). This reduced energy difference $\Delta E_{\pm} = E_{\psi_s} - E_{\psi_{as}}$ between the states ψ_s and ψ_{as}, involved in the diffusion of the particle, is responsible for the slow down of diffusion, when the entanglement to the gravitons is accounted for. In the dynamic situation the straddling warp energies result in fast vibrations of the gas atom between ψ_{as} and the state derived from entanglement of the adsorbate ψ_s-state to gravitons in the warp resonance. One can say that the fast vibrations of the adsorbate back and forth towards the warp resonance are the stamp mark of the coupling, which forces $\Delta E_{\pm}$ to small values. In this way the fast $g-$warp vibrations are responsible for the slow down. This effect appears similar to the quantum Zeno effect [21] and will be discussed later in this section.

The interpretation of these results is in the spirit of the dynamics of a many-body system consisting of both localized adsorbate states as well as delocalized graviton states belonging to a high dimensional continuum of states. The warp resonance provides an additional channel for scattering of the adsorbate, i.e. scattering into states entangled with the gravitons. The interaction with the gravitons leads to splitting into bonding $| w\rangle \, | \gamma_1\rangle$ and antibonding $| w\rangle \, | \gamma_2\rangle$ many-particle states with dominating component due to the warp resonance. The remaining graviton continuum states are decoupled due to orthogonalization in the diagonalization of the Hamiltonian. Due to state splitting the Rabi oscillations between the split states, dominated by the warp resonance, are very fast. The fast oscillations of the occupation of the warp resonance on a very short time scale can be seen on the right panel of fig. 5. Within each oscillation of the warp resonance the gravitons entangle to the local particle and gain weight (the red curve in fig.

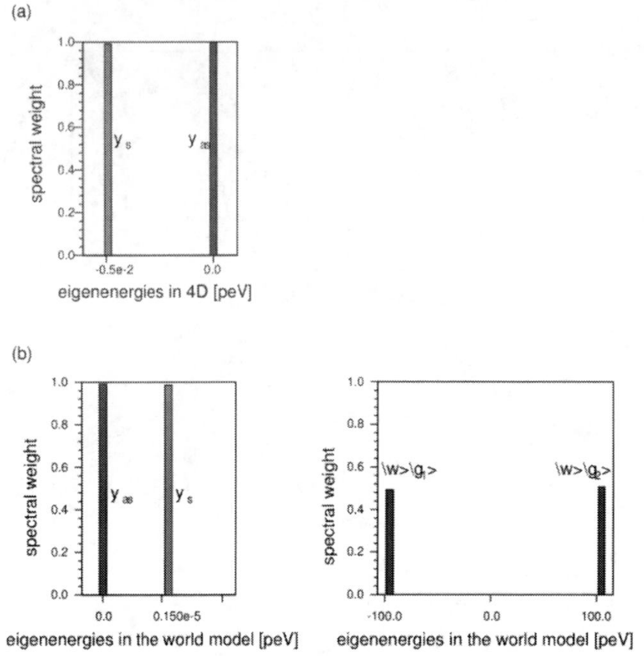

Figure 4. Spectral energy distribution of the eigenstates of the 4D world (a), derived from ψ_s (red) and ψ_{as} (blue) and of the eigenstates of the world model (b) in the static case. The energy positions of the split warp-derived bonding- and anti-bonding states due to the strong warp-graviton coupling are shown in the lower right panel (gray). The lower left panel shows a more detailed view of the splitting between ψ_{as} and the ψ_s-derived state around energy zero. As a consequence of the entanglement to gravitons within the warp resonance their splitting is significantly reduced, compared to the case, when the entanglement to the gravitons is neglected (panel a).

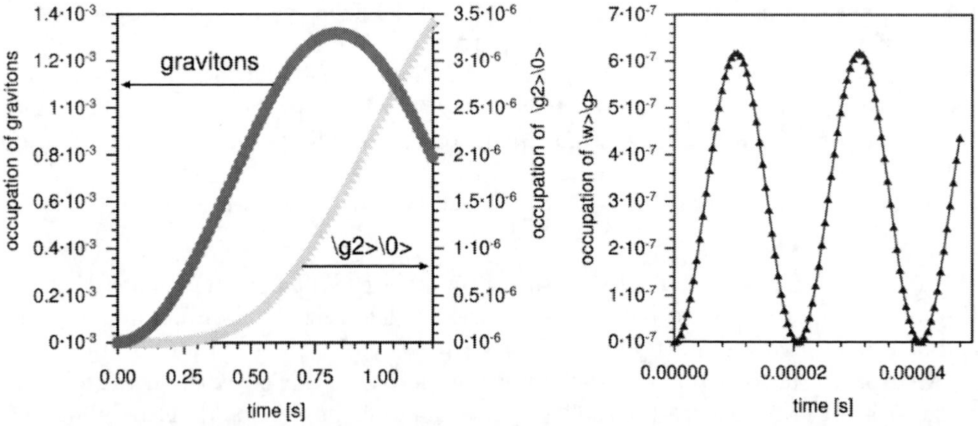

Figure 5. Entanglement to gravitons is included. Left panel: variation of the occupation of the gravitons (red curve) and the second adsorption site (green curve) at short time. The right panel shows the oscillations of the occupation of the warp resonance on a very short time scale.

5). This happens because the gravitons are slower due to degeneracy with the particle states $\mid g1\rangle \mid 0\rangle$ and $\mid g2\rangle \mid 0\rangle$. Furthermore, they need time to span the hidden dimensions, the so called recurrence time. Therefore, in a single recurrence cycle of the gravitons the particle can oscillate many times into the warp resonance, leading to enhancement of the graviton occupation. The gravitons are just a component of the total wave function, which involves the rest local state components in a product with the graviton part. Therefore, before the recurrence of the gravitons from the hidden dimensions the particle cannot leave the local state $\mid g1\rangle \mid 0\rangle$, because otherwise energy conservation will be violated. Hence the particle is localized on site 1 in $\mid g1\rangle \mid 0\rangle$ via the entanglement to the gravitons. This entangled state evolves coherently with time and with each next (fast) oscillation of the particle into the warp resonance the occupation of the gravitons enhances. Slow diffusion and partial occupation of site 2 $\mid g2\rangle \mid 0\rangle$ occur even before the gravitons recur back from the hidden dimensions because the graviton continuum entangles to the other localized state as well.

5. Discussion

5.1. Collapse models

To make contact with other theories we remind that particle localization comes out of the collapse model of decoherence theory as well [9, 12]. Wave function reduction induced by gravity in three-dimensional space is treated by Ghirardi et al. [13], Diósi [14] and Frenkel [15] within a stochastic modification of the coherent Schrödinger quantum dynamics. However, localization within the collapse model is the result of the collapse of the total wave function on a local state due to "permanent measurement" of the particle by the environmental excitations. Actually in this model collapse is measurement. Hence, collapse means interrupted coherent time evolution and it is introduced for the sake of interpreting measurement within quantum mechanics. In terms of the many-worlds interpretation [16]-[20], though the world wave function is a superposition of all alternative states of the local system+environment and evolves unitarily with time, "favoured" classical states are singled out via the interaction with the environment and the dynamical reduction, i.e. collapse, of the density matrix is the result. It should be emphasized that in our theory collapse does not occur and it is not the mechanism for particle localization. Localization is a consequence of the coherent time evolution of the total wave function of the local system entangled with the gravitons. The coherent quantum behaviour results in the appearance of the adsorbates as localized particles, because of entanglement to the gravitons in 11-dimensional spacetime.

5.2. Quantum Zeno effect

We try to reformulate the features and the results of our localization theory in terms of the traditional decoherence theory, pointing out the differences. Due to the entanglement to gravitons, in the short time the particle is within the warp resonance, it appears exactly like being "permanently measured". However this entanglement does not lead to collapse, as it is assumed within decoherence theory. It appears as a non-destructive von Neumann type of measurement, in which the local system changes negligibly or not at all (in the ideal von Neumann measurement), whereas the environment is changed. The gravitons are propagated in the hidden dimensions, but their entanglement to the local system is not broken, the coherent time development of the system is not interrupted by the entanglement. The entanglement of the adsorbed particle and the gravitons via the warp resonance to the two adsorption sites warrants the diffusion of the particle from site 1 towards site 2 and backwards, similar to the case, when the entanglement to the gravitons is ignored. However, the occupation of site 2 grows with time on a much slower time scale, as it is displayed by the green curve in fig. 5. The slow down of adsorbate diffusion is due to particle localization, induced by the entanglement to the gravitons within the warp resonance. The fast vibrations of the adsorbed particle between the adsorption site and the warp resonance, induced by the coupling to gravitons, appears similar to the quantum Zeno effect [21]. The latter implies no change in the particle state, if it is measured very fast by the environmental excitations, due to collapse of the particle on the initial state after each "measurement". Two issues make the difference between our theory of particle localization and the quantum Zeno effect [21]. The first issue refers to what determines how fast the environmental measurement is. In our theory, since the entanglement to the gravitons occurs within the warp resonance, the rate of localization is an indication of how often the particle is within the warp resonance. And this depends, of course, on the gravitational interaction, which influences the rate of particle Rabi oscillations between the local states. The second difference is that in our theory coherence is not lost at each "measurement" by the environment, as it is the case with the quantum Zeno effect. The total system continues its coherent time development, taking into account the entanglement to the gravitons. The localization mechanism we suggest may be regarded as the coherent version of the quantum Zeno effect. Each "measurement" by the environment ensues in a slight change of the total system. Both the local part and the environment are slightly changed, but no collapse on the initial state of the adsorbate has occurred. The amplitude squared of the component $\mid g2\rangle \mid 0\rangle$ enhances coherently with time, but only very slowly, because the particle is localized on the initial site 1 $\mid g1\rangle \mid 0\rangle$ due to the entanglement to the gravitons within the warp resonance.

5.3. Why gravitons?

Why do we need the gravitons? Why do the other excitations of the solid, e.g. phonons, electron-hole pairs, magnons, etc., not do the job of localizing the adsorbate and slowing down adsorbate diffusion? The reason is: we need environmental excitations with short wave length. In the hidden dimensions we have very many gravitons with small wave length and high energy, which constitute our vacuum state of the 11-dimensional gravitational field. We need deformation with energy of the order of peV which extends only over a region ot the order of 1 bohr. This is not possible with solid excitations which at small energies are necesseraly of very long wave length. On the other hand we can assume that the hidden dimensions contain a huge energy reservoir with gravitons of short wave length. This energy reservoir constitutes the "vacuum state" and therefore we have short wavelength deformations of 11-dimensional spacetime. Only this kind of short wavelength deformations are capable of including fast warp-g vibrations and slow down of diffusion.

All solid state excitations have very long wave lengths at $T \to 0$K and very low energy. They would excite a joint vibration of both the adsorbate and the warp resonance, however no vibration of the adsorbate and the warp resonance relative to each other. The interaction with the warp resonance is therefore weak and not local, hence no split off states (of the order of some hundred peV) with the warp resonance as component are generated. The result is very slow oscillations of the occupation of the warp resonance and very slow entanglement of the adsorbate to excitations of this kind, i.e. the entanglement and adsorbate localization by these environmental excitations is very slow. The scattering of the adsorbate from one adsorption site to the neighbouring one via the warp resonance will not be hindered, diffusion will not be slowed down. Split off states with the warp resonance as component, which are responsible for the fast localization of the adsorbate via entanglenment to gravitons, are generated only by the local interaction with gravitons. The strong and very local coupling of the warp resonance to gravitons is the only entanglement to environmental excitations which can warrant this effect.

5.4. Pointer states and emergence

As it was argued in ref. [6], the use of stationary state scattering theory and the generalized Ehrenfest theorem (GET) in the 4D world plus environment is equivalent to the reduced density matrix approach used to evaluate the expectation value of an operator A, which acts in the 4D world and corresponds to an observable in the 4D world:

$$A \otimes I_{env} = A \otimes (\mid 0\rangle\langle 0 \mid + \mid grav\rangle\langle grav \mid), \tag{3}$$

where I_{env} is the identity operator acting on the environment. $\mid grav\rangle$ is a superposition of excited graviton states, which is defined later in the context of the present model (eq. 10). Moreover, in ref. [6] the pointer states, introduced by Zurek [22] in conjunction with decoherence, are mathematically exactly defined and it was proven that, in the presence of decoherence, scattering can occur out of pointer states or into pointer states. The pointer states are many-particle product states, involving the scattering state of the particle due to the gravitational potential it experiences within the warp resonance, and the state of the environmental gravitons, which are mutually orthogonal. Having this in mind, in the present two-state world model we define the scattering state $\mid g1+\rangle$, generated from $\mid g1\rangle$, approximately as:

$$\mid g1+\rangle \approx \mid g1\rangle + a \mid w\rangle \tag{4}$$

$\mid w\rangle$ is the adsorbate component in 4D in the warp resonance, which is mixed in with a small coefficient a. The pointer states in the GET cannot be the asymptotic incoming and outgoing scattering states of three dimensional scattering theory, because they are not directly entangled to the environment of the measured local system. The pointer states can be only states, which entangle the environment directly to the measured local system. In the present model for adsorbate localization the pointer state is just the warp resonance entangled to the gravitons $\mid w\rangle \mid \gamma\rangle$, with γ denoting the graviton continuum deformed by the adsorbate, when it is in $\mid w\rangle$, and $\mid w\rangle$ denoting the 4D part of the pointer state.

The operator A for the localization rate is equivalent to the operator for the rate of change of the population of the initial adsorbate state $\mid g1+\rangle$. Localization via entanglement to gravitons in the warp resonance, being associated with the gravitational interaction potential $V_{grav}^{(11D)}$ in 11D string theory, allows to express the operator A in the form:

$$A = \frac{2\pi}{\hbar} V_{grav}^{(11D)} \mid w\rangle\langle w \mid V_{grav}^{(11D)} \delta(E) \tag{5}$$

In the last equation the 4D part of the warp resonance $\mid w\rangle$ serves as an intermediate state, mediating the coupling of the 4D states to the gravitons in the hidden dimensions. When and only when the adsorbed particle is in $\mid w\rangle$, it can be strongly entangled to the gravitons. Otherwise, if the diffusion process does not involve the intermediate $\mid w\rangle$ state, the result for the expectation value of A would be identically equal to zero. The effect of the entanglement of the adsorbate in the intermediate state to gravitons is as if the pointer of a measuring apparatus has indicated its presence in the gravitons. The strong interaction with the gravitons means fast entanglement and adsorbate localization in the 4D state $\mid g1+\rangle$.

In the stationary case the world wave function $\mid \Psi\rangle$ is an expansion over all many-particle basis states:

$$\mid \Psi\rangle = \sum_i \alpha_i \mid g1\rangle \mid \gamma_i\rangle + \sum_i \beta_i \mid w\rangle \mid \gamma_i\rangle \tag{6}$$

and it can be written as:

$$\mid \Psi\rangle = \sum_i d_i \mid g1+\rangle \mid \gamma_i\rangle \tag{7}$$

using eq. 4 and a correct choice of the coefficients d_i, so that both expressions for $\mid \Psi\rangle$ eq. 6 and 7 are identical. $\mid \Psi\rangle$ resulting from the solution of the Hamiltonian is the superposition of the dominating bonding and antibonding states, generated by the strong coupling of the warp resonance to the gravitons:

$$\begin{aligned} \mid \Psi\rangle &\approx \mid g1\rangle \mid 0\rangle + c1 \mid w\rangle \mid \gamma_1\rangle + c2 \mid w\rangle \mid \gamma_2\rangle \\ &= \mid g1\rangle \mid 0\rangle + \mid w\rangle \mid grav\rangle \end{aligned} \tag{8}$$

where $\mid \gamma_1\rangle$ and $\mid \gamma_2\rangle$ are the bonding and antibonding superpositons of the unperturbed gravitons and the gravitons, perturbed by the adsorbate, when it is in the warp resonance:

$$\begin{aligned} \mid \gamma_1\rangle &= \frac{1}{\sqrt{2}}(\mid 0\rangle + \mid \gamma\rangle) \\ \mid \gamma_2\rangle &= \frac{1}{\sqrt{2}}(\mid 0\rangle - \mid \gamma\rangle) \end{aligned} \tag{9}$$

and

$$\mid grav\rangle = c1 \mid \gamma_1\rangle + c2 \mid \gamma_2\rangle. \tag{10}$$

The expectation value of the operator A (eq. 3) in the 4D world, using $\mid \Psi\rangle$ eq. 7, eqs. 5 and 10 and the orthogonality of the many-particle states, equals:

$$\begin{aligned} \langle \Psi \mid A \otimes I_{env} \mid \Psi\rangle &= \sum_{i,j} d_i d_j \langle g1+ \mid A \mid g1+\rangle \{\langle \gamma_i \mid 0\rangle\langle 0 \mid \gamma_j\rangle + \langle \gamma_i \mid grav\rangle\langle grav \mid \gamma_j\rangle\} \\ &= \frac{2\pi}{\hbar} \mid a \mid^2 \mid V_{grav}^{(11D)} \mid^2 \sum_{i,j} d_i d_j \langle \gamma_i \mid grav\rangle\langle grav \mid \gamma_j\rangle \delta(E) \\ &= \frac{2\pi}{\hbar} \mid V_{grav}^{(11D)} \mid^2 \mid a \mid^2 (d_1 c1 + d_2 c2)^2 \delta(E) = \mid a_1 \mid^2 \mid V_{grav}^{(11D)} \mid^2 \delta(E) \end{aligned} \tag{11}$$

It is obvious that this result plays a role in the 4D world alone. Apparently the gravitons have disappeared from the last equation. But only apparently. Their effect is hidden in the values of a and $V_{grav}^{(11D)}$. Evaluating a and the perturbing potential in the 4D world, we recover the results of the ab initio theory [5] and the missing agreement with experiment. Hence, reducing the theory to 4D, we might claim that quantum mechanics fails. However, including the gravitons in the hidden dimensions of an 11D world, yields different values for a, $V_{grav}^{(11D)}$ and diffusion rate, which is in agreement with experiment.

Using the generalized Ehrenfest theorem of stationary state scattering theory leads to a single term for the expectation value of the localization rate, weighted by $\mid a \mid^2$, the modulus squared of the coefficient of the initial adsorbate state in the warp resonance (eq. 4). This result gives the impression as if in our coherent theory collapse has occurred and as if the coupling to gravitons destroys coherence in the 4D world, if coherence is defined as the result of Schrödinger's time-dependent theory in 4D. The appearance of collapse in the 4D world is interpreted as decoherence. In this sense the notion of pointer states, introduced by Zurek [22], is meaningful. However, in the 11D world model there is no collapse. We emphasize that coherence in 11D spacetime is not lost. In the present world model the pointer state is $\mid w\rangle \mid \gamma\rangle$, being a superposition of $\mid w\rangle \mid \gamma_1\rangle$ and $\mid w\rangle \mid \gamma_2\rangle$, it is not a world eigenstate. It has an energy expectation value of 0 and, since the world eigenstates are not on the energy shell, no selection can occur, because otherwise energy conservation would be violated. If one likes, in the present theory the effect of the graviton continuum can be reduced to the renormalization of the coupling in the 4D world. In this sense, a new 4D theory has "emerged". But in our case the emergence of this new 4D theory can be traced back to its causal origin: the coupling to gravitons in hidden dimensions. This reduction can lead to deeper insight and new predictions, whereas a renormalized 4D description would just be a phenomenological theory.

6. Conclusion

The localization of an adsorbed particle, which leads to slow down of surface diffusion in experiment, is, according to the present theory, the result of the coherent time evolution of the local quantum system in entanglement with the gravitons. The total quantum system involves the gravitons, which propagate in the hidden dimensions of 11D string theory and are not available for experimental measurement in our 4D world. At certain time intervals, however, the local system, which can be experimentally observed, appears as if it behaves classically, e.g. adsorbates appearing as localized wave packets. This seemingly classical appearance of the local system emerges from the coherent quantum mechanical behaviour of the total system, the environmental gravitons included. The hidden dimensions of 11D string theory and supergravity provide presently the only possibility of explaining, within the framework of conventional quantum field theory, the experimentally observed slow adsorbate diffusion in the quantum diffusion regime at low temperature in the STM.

References

[1] Auerbach A, Freed K F and Gomer R 1987 *J. Chem. Phys.* **86** 2356
[2] Kua J, Lauhon L J, Ho W and Goddard W A III 2001 *J. Chem. Phys.* **115** 5620
[3] Cao G X, Nabighian E and Zhu X D 1997 *Phys. Rev. Lett.* **79** 3696
[4] Heinrich A J, Lutz C P, Gupta J A and Eigler D M 2002 *Science* **298** 1381
[5] Mattsson Th, Engberg U and Wahnström G 1993 *Phys. Rev. Lett.* **71** 2615; Mattsson Th and Wahnström G 1997 *Phys. Rev.* B **56** 14944
[6] Doyen G and Drakova D 2009 *Surface Science* **603** 1579; Drakova D and Doyen G 2010 *e-J. Surf. Sci. Nanotech.* **8** 6
[7] Lauhon L J and Ho W 2000 *Phys. Rev. Lett.* **85** 4566
[8] Arndt M, Hackermüller L, Hornberger K and Zeilinger A 2005 in Decoherence, entanglement and information protection in complex quantum systems, Springer; Hackermüller L, Hornberger K, Brezger B, Zeilinger A and Arndt M 2004 *Nature* **427** 711; Hackermüller L, Uttenthaler S, Hornberger K, Reiger E, Brezger B, Zeilinger A and Arndt M. 2003 *Phys. Rev. Lett.* **91** 090408
[9] Joos E and Zeh H D 1985 *J. Phys. Cond. Matter* **59** 223; Joos E and Zeh H D 1985 *Z. Phys. B: Cond. Matter* **59** 223-243
[10] Doyen G and Drakova D present volume p....
[11] Arkani-Hamed N, Dimopoulos S and Dvali G 1998 *Phys. Lett.* B **429** 263; Antoniadis I, Arkani-Hamed N, Dimopoulos S and Dvali G 1998 *Phys. Lett.* B **436** 257
[12] Ghirardi G C, Rimini A and Weber T 1985 in Quantum probability and applications, eds. L. Accardi et al. Springer, Berlin; 1986 *Phys. Rev.* D **34** 470
[13] Ghirardi G C, Grassi R and Rimini A 1990 *Phys. Rev.* A **42** 1057
[14] Diósi L 1989 *Phys. Rev.* A **40** 1165
[15] Frenkel A 1990 *Found. Phys.* **20** 159
[16] Everett H 1957 *Rev. Mod. Phys.* **29** 454
[17] Wheeler J A 1957 *Rev. Mod. Phys.* **29** 463
[18] DeWitt B S 1970 *Phys. Today* **23** 30
[19] Deutsch D 1985 *Int. J. Theor. Phys.* **24** 1
[20] Tegmark M and Wheeler J A 2001 *Sci. Am.* **284** 68
[21] Misra B and Sudarshan E C G 1977 *J. Math Phys.* **18(4)** 756
[22] Zurek W H 2003 *Rev. Mod. Phys.* **75** 715

Geometric quantum phase in the spacetime of topological defects

K. Bakke

Department of Physics, University of Oxford, Clarendon Laboratory, Parks Road, Oxford, OX1 3PU, United Kingdom.

C. Furtado and J. R. Nascimento

Departamento de Física, Universidade Federal da Paraíba, Caixa Postal 5008, 58051-970, João Pessoa, Pb, Brazil.

Abstract. In this contribution, we study the quantum dynamics of a neutral particle in the presence of a topological defect. We investigate the appearance of a geometric phase in the relativistic quantum dynamics of a neutral particle which possesses permanent magnetic and electric dipole moments in the presence of an electromagnetic field in this curved background. We also study the influence of noninertial effects of a rotating frame and and obtain several contributions to the relativistic geometric phase due to the noninertial effects and the topology of spacetime. The analogous Aharonov-Casher and He-Mckellar-Wilkens effects are investigated in the nonrelativistic dynamics with the presence of a topological defect and under the influence of noninertial effects. We also obtain effects analogous to the Sagnac effect and Mashhoon effect due to the presence of the topological defect.

1. Introduction

After the discovery of the Aharonov-Bohm effect [1] fifty years ago and the publication of Berry's work [2] twenty-five years ago, the study of geometric phases is still a subject of a great interest in several branches of Physics. In the classic paper of phase shifts in non-quantal polarized light, Pancharatnam [3] anticipated the quantal geometric phases. It is worth mentioning the related effects to the Aharonov-Bohm effect have been anticipated by Franz [4] and by Ehrenberg and Siday [5]. The Aharonov-Bohm effect is a topological effect given by a phase shift on the wave function of a charged particle moving around a infinite thin solenoid that contains a magnetic flux inside it. Berry [2] showed that this phase shift constitutes a particular class of quantum phases that it is termed geometric phases. These quantum phases have been generalized to the case of a non-adiabatic evolution of a quantum system by Aharonov and Anandan [6]. In any case, the phase depends on the geometric nature of the path where the system evolves dynamically. Aharonov and Casher [7] studied the quantum behavior of a neutral particle with a permanent magnetic dipole moment interacting with an external electric field and obtained a topological quantum phase due to this interaction. This topological effect is called the Aharonov-Casher effect. The effect dual to the Aharonov-Casher effect has been proposed by He and MacKellar [8] and, independently, by Wilkens [9] by investigating the quantum dynamics of a neutral particle with a permanent electric dipole moment in the presence of an external magnetic field.

The phase shift acquired by the wave function of the neutral particle due to the interaction between the electric dipole moment and the magnetic field is called the He-McKellar-Wilkens effect. Following this way, Dowling *et al* [10] and Furtado *et al* [11] studied the dual effect of the Aharonov-Bohm effect by using the Maxwell duality transformation and the quantum dynamics of a magnetic monopole in the presence of a electric solenoid, respectively. A general study of the quantum phases for neutral particles with permanent dipoles has been done by Anandan [12] and Silenko [13]. Ribeiro and Furtado [14] demonstrated the Anandan quantum phase is a case of the Berry quantum phase. Recently, the dual Aharonov-Bohm effect in the dynamics of a composite particles has been discussed by Horsley *et al* [15].

The gravitational analogue of the Aharonov-Bohm effect has been investigated and discussed by several authors. Cai and Papini [16] obtained the covariant generalization of the Berry phase and applied their results by using the approach of weak gravitational fields. Corichi and Pierri [17] investigated the behavior of a scalar particle in a class of stationary spacetime backgrounds and studied the emergence of geometric phases in the dynamics of a particle in the presence of a rotating cosmic string. Mazur [18] obtained the gravitational analogue of the Aharonov-Bohm effect in the spinning cosmic string spacetime background. Mostafazadeh [19] investigated the relativistic geometric phase for a scalar particle in the background of a rotating cosmic string by employing the two-component formalism. Recently, one of us [22, 23] has investigated the Berry quantum phase for a scalar particle in the presence of a magnetic chiral cosmic string background. Many authors have investigated the geometric phase in a series of gravitational backgrounds [24, 25]. The gravitational analogue of the Aharonov-Casher effect has been obtained by Anandan [20] and Resnik [21].

Topological defects is a topic widely studied in various areas of the modern Physics. An excellent review of the applications and the problems involving topological defects in condensed matter physics, cosmology and gravitation can be found in the books of Kleinert [26]. The presence of topological defects changes the physical properties of the medium in which it is present. The study of the influence of topological defects in various physical systems have been analyzed, for instance, in superfluids [27], superconductors [26], graphene [28, 29, 30], liquid crystals [31], elastic solids [32, 33], effects on the electromagnetic field [34], the Casimir effect [35], the Landau quantization [36], etc. The study of the Berry quantum phase in media with dislocations has been done in Ref. [37]. Ribeiro *et al* [38] investigated the geometric phase for an induced electric dipole by using the the field configuration proposed by He, McKellar and Wilkens [8, 9]. Recently, geometric quantum phases for neutral particles with permanent magnetic dipole moment have been investigated in curved spacetime backgrounds [39], and in the presence of noninertial references frames [40, 41]. In this paper, we briefly review the results of the studies of geometric quantum phases in the presence of topological defects obtained by the authors in the articles mentioned above.

2. The cosmic string spacetime and the field configuration in a rotating frame

In this section, our focus is to present the curved spacetime background and show how we build a rotating frame in such a way that we can obtain a field configuration that produces no torque on the dipole moment of the neutral particle. The chosen curved spacetime background in this work is called the cosmic string spacetime. The cosmic string is a topological defect described by the line element:

$$ds^2 = -dT^2 + dR^2 + \eta^2 R^2 d\Phi^2 + dZ^2. \tag{1}$$

where $\eta = 1 - 4\nu$ is a parameter associated with the deficit angle of the cosmic string spacetime. it is defined in the range $0 < \eta < 1$, and ν is the linear mass density. In this work, we consider the units where $\hbar = c = 1$. The azimuthal angle varies in the interval: $0 \leq \varphi < 2\pi$. The parameter η in the cosmic string spacetime can assume only values $\eta < 1$. Values for η greater than 1 correspond to a spacetime with negative curvature, and it does make sense only in the description of linear topological defects in solids [32, 42]. The geometry of the cosmic string possesses a conical singularity [43] which gives rise to the curvature concentrated on the cosmic string axis. This conical singularity is represented by the curvature tensor $R^{\rho,\varphi}_{\rho,\varphi} = \frac{1-\eta}{4\eta}\,\delta_2(\vec{r})$, where $\delta_2(\vec{r})$ is the two-dimensional delta function. Since we are interested in discussing the influence of the noninertial effects and the topology of the cosmic string spacetime on the geometrical quantum phases of a neutral particle with a permanent dipole moment, is it possible to build a noninertial frame in such a way that the field configuration induced by the noninertial effects does not produce any torque on the dipole moment? The answer for this question is yes. By using the approach of the quantum field theory in the curved spacetime background [44], we can build local reference frames for the observers and study the influence of the noninertial effects on the geometric quantum phases. This study starts when we carry out the coordinate transformation $T = t,\quad R = \rho,\quad \Phi = \varphi + \omega t,\quad Z = z$, where ω is the constant angular velocity of the rotating frame and we consider $\omega\rho << 1$. With this transformation, the line element (1) becomes

$$ds^2 \;=\; -\left(1-\omega^2\eta^2\rho^2\right)\,dt^2 + 2\omega\eta^2\rho^2 d\varphi\,dt + d\rho^2 + \eta^2\rho^2 d\varphi^2 + dz^2. \tag{2}$$

Noninertial effects have been discussed in the literature for a long time. The Sagnac effect [45] is the best famous effect related to the appearance of a phase shift in the wave function of a quantum particle. Another well-known noninertial effect is the Mashhoon effect [46], a quantum phase given by the coupling of the spin of the particles with the angular velocity of the rotating frame. More discussions about the noninertial effects above can be found in [47, 48]. Recently, noninertial effects have been investigated in different contexts, for instance, in the studies of holonomies [49], Landau quantization for neutral particles [51, 50] and for bound states for a neutral particle analogous to a quantum dot [52]. How exactly can we build a rotating frame based on the line element (2) and no torque is produced on the magnetic dipole moment of a neutral particle? Since the spinors are defined locally in the curved spacetime background [44], we can build a local reference for the observer through a noncoordinate basis $\hat{\theta}^a = e^a_{\ \mu}(x)\,dx^\mu$, where the components $e^a_{\ \mu}(x)$ satisfy the relation $g_{\mu\nu}(x) = e^a_{\ \mu}(x)\,e^b_{\ \nu}(x)\,\eta_{ab}$. The components $e^a_{\ \mu}(x)$ are called *tetrads* or *vierbein*, $g_{\mu\nu}(x)$ is the metric tensor and $\eta_{ab} = \text{diag}(-+++)$ is the Minkowsky tensor [53, 54]. The tetrads have an inverse defined as $dx^\mu = e^\mu_{\ a}(x)\,\hat{\theta}^a$, where $e^a_{\ \mu}(x)\,e^\mu_{\ b}(x) = \delta^a_{\ b}$ and $e^\mu_{\ a}(x)\,e^a_{\ \nu}(x) = \delta^\mu_{\ \nu}$. In that way, we choose the tetrads in the form [40]:

$$\hat{\theta}^0 = \sqrt{1-\beta^2}\,dt - \frac{\omega\eta^2\rho^2}{\sqrt{1-\beta^2}}\,d\varphi;\quad \hat{\theta}^1 = d\rho;\quad \hat{\theta}^2 = \frac{\eta\rho}{\sqrt{1-\beta^2}}\,d\varphi;\quad \hat{\theta}^3 = dz, \tag{3}$$

where we have called $\beta = \omega\eta\rho$ in (3). Considering an electric charge density λ concentrated on the z axis in the inertial frame, we have that this distribution of charges creates a cylindrically symmetric electric field in the rest frame of the observer $E^{\rho}_{(rf)} = \frac{\lambda}{\sqrt{-g}}$, where $g = \det(g_{\mu\nu})$. Since we have chosen the 1-axis of the local reference frame parallel to the ρ axis of the spacetime, we can write $E^1 = e^1_{\ \rho}\,E^{\rho}_{(rf)} = \lambda/\sqrt{-g}$. Thus, in the corotating frame of the observers (3), we have only one non-null components of the electric field [40]:

$$E^\rho = \frac{1}{\sqrt{1-\beta^2}}\,\frac{\lambda}{\eta\rho}, \tag{4}$$

that is, the noninertial effects of the rotating frame (3) induce only one component of the electric field and no components of the magnetic field are induced in the rotating frame (3). It has been shown in [50, 52] that non-null components of the magnetic field are induced by the noninertial effects of a nonrotating frame called the Fermi-Walker reference. Our next step in this work is to discuss the dynamics of a neutral particle in this background.

3. Relativistic dynamics of a neutral particle

In this section, we discuss the appearance of relativistic geometric phases in the quantum dynamics of a neutral particle with a permanent magnetic dipole moment. The Dirac equation in the curved spacetime background with the interaction of the magnetic dipole moment of the neutral particle with an external electric field is given by the following expression:

$$i\,\gamma^{a}\,e^{\mu}_{\ a}\,(x)\;\partial_{\mu}\psi+i\,\gamma^{\mu}\,\Gamma_{\mu}\,(x)\,\psi+\frac{\mu}{2}\,F_{\mu\nu}\,\Sigma^{\mu\nu}\,\psi=m\psi, \tag{5}$$

where $\Gamma_{\mu}\,(x)=\frac{i}{4}\,\omega_{\mu ab}\,(x)\;\Sigma^{ab}$ is the spinorial connection, $F_{\mu\nu}$ is the electromagnetic tensor ($F_{0i}=E_{i}$ and $F_{ij}=-\epsilon_{ijk}\,B^{k}$) and $\Sigma^{ab}=\frac{i}{2}\left[\gamma^{a},\gamma^{b}\right]$. The indices $(a,b,c=0,1,2,3)$ indicate the local reference frame, while the greek indices indicate the spacetime coordinates. The γ^{a} matrices are defined in the local reference frame and they are identical to the Dirac matrices in the flat spacetime, *i.e.*,

$$\gamma^{0}=\begin{pmatrix}1&0\\0&-1\end{pmatrix};\;\gamma^{i}=\begin{pmatrix}0&\sigma^{i}\\-\sigma^{i}&0\end{pmatrix};\;\Sigma^{i}=\begin{pmatrix}\sigma^{i}&0\\0&\sigma^{i}\end{pmatrix}, \tag{6}$$

with $\vec{\Sigma}$ being the spin vector and σ^{i} are the Pauli matrices satisfying the relation $(\sigma^{i}\,\sigma^{j}+\sigma^{j}\,\sigma^{i})=2\,\eta^{ij}$. The spin connections $\omega_{\mu ab}\,(x)$ can be obtained by solving the Maurer-Cartan structure equations $d\hat{\theta}^{a}+\omega_{\mu ab}\,(x)\;dx^{\mu}\wedge\hat{\theta}^{b}=0$ [54], and they provide us the following spinorial connections [40]

$$\begin{aligned}\Gamma_{t}&=-\frac{1}{2}\,\frac{\omega^{2}\eta^{2}\rho}{\sqrt{1-\beta^{2}}}\,\hat{\alpha}^{1}-\frac{i}{2}\,\frac{\omega\eta}{\sqrt{1-\beta^{2}}}\,\Sigma^{3};\quad\Gamma_{\rho}=\frac{1}{2}\,\frac{\omega\eta}{(1-\beta^{2})}\,\hat{\alpha}^{2}; &(7)\\ \Gamma_{\varphi}&=-\frac{1}{2}\,\frac{\omega\eta^{2}\rho}{\sqrt{1-\beta^{2}}}\,\hat{\alpha}^{1}-\frac{i}{2}\,\frac{\eta}{\sqrt{1-\beta^{2}}}\,\Sigma^{3}. &(8)\end{aligned}$$

Taking the induced field configuration (4), it is easy to check that there is no torque $\vec{\tau}=\vec{\mu}\times\vec{B}$ on the magnetic dipole moment of the neutral particle. In that way, we are able to study the influence of the noninertial effects of the rotating frames (3) on the geometric quantum phases for neutral particles with permanent magnetic dipole moment. The Dirac equation in the cosmic string spacetime background with rotating frames has the form

$$\begin{aligned}m\psi&=\frac{i}{\sqrt{1-\beta^{2}}}\,\gamma^{0}\,\frac{\partial\psi}{\partial t}+i\gamma^{2}\frac{\omega\eta\rho}{\sqrt{1-\beta^{2}}}\,\frac{\partial\psi}{\partial t}+i\,\gamma^{1}\left(\partial_{\rho}-\mu\,\hat{\beta}\,E_{\rho}+\frac{1}{2\rho}\right)\psi\\&+\frac{i}{\sqrt{1-\beta^{2}}}\,\frac{\gamma^{2}}{\eta\rho}\,\frac{\partial\psi}{\partial\varphi}-i\gamma^{2}\frac{\omega^{2}\eta\rho}{\sqrt{1-\beta^{2}}}\,\frac{\partial\psi}{\partial\varphi}+i\,\gamma^{3}\,\frac{\partial\psi}{\partial z}-\frac{\gamma^{0}}{2}\,\frac{\eta\,\vec{\omega}\cdot\vec{\Sigma}}{(1-\beta^{2})}\,\psi.\end{aligned} \tag{9}$$

The relativistic geometric quantum phase acquired by the wave function of the particle is given by using the Dirac phase factor method [55]. The Dirac phase factor method consists in writing the wave function in the form $\psi=e^{i\phi}\psi_{0}$, where ϕ is the phase shift and ψ_{0} is the solution of the Dirac equation in the absence of fields. Substituting $\psi=e^{i\phi}\psi_{0}$ into (9), we find that the wave function acquires a phase shift given by

$$\begin{aligned}\phi&=-\mu\gamma^{0}\oint\frac{\left(\vec{\Sigma}\times\vec{E}\right)_{\varphi}}{\sqrt{1-\beta^{2}}}\,d\varphi+\oint\frac{1}{2}\,\frac{\eta}{\sqrt{1-\beta^{2}}}\,\Sigma^{3}\,d\varphi+\frac{1}{2}\oint\frac{\left(\vec{\Sigma}\times\vec{\mathcal{E}}\right)_{\varphi}}{(1-\beta^{2})^{3/2}}\,d\varphi\\&-\frac{\eta}{2}\oint\frac{\vec{\omega}\cdot\vec{\Sigma}}{\sqrt{1-\beta^{2}}}\,dt,\end{aligned} \tag{10}$$

where we have defined the vector $\vec{\mathcal{E}} = (\omega^2\eta^2\rho)\,\hat{\rho}$, with $\hat{\rho}$ being a unit vector in the ρ-direction. The expression for the relativistic quantum phase (10) is the relativistic analog of the Anandan quantum phase [12] given in a curved spacetime background and in a noninertial frame. This relativistic quantum phase has four independents contributions [40]: the first contribution in (10) is given by the interaction between the permanent magnetic moment and the external electric field; the second contribution is given by the topology of the cosmic string spacetime; the third contribution is given by the noninertial effects of the rotating frame (3), where the vector $\vec{\mathcal{E}}$ plays the role analogous to the electric field [48]; the last contribution to the relativistic phase is given by spin-rotation coupling. We can see that every term in (10) has the presence of the term $\left(1-\beta^2\right)^{1/2}$. This term corresponds to a correction term due to rotation of the local reference frames of the observers. In particular, integrating the last contribution for the relativistic quantum phase (10), we have $\phi' = \frac{\omega^2\eta^2}{\left(1-\beta^2\right)^{3/2}}\,\vec{A}\cdot\vec{\Sigma}$, where $\vec{A} = A\,\hat{n}$, with A being the area perpendicular to the symmetry axis of the cosmic string and $\hat{n}$ the unitary vector perpendicular to the area A. Taking the limit $\eta \rightarrow 1$ in ϕ', we obtain the relativistic phase shift induced by the noninertial effects analogous to that one obtained by Anandan and Suzuki in [48] without using the weak field approximation, up to the correction term $\left(1-\beta^2\right)^{3/2}$ due to the rotation of the local reference frame of the observers [40].

Another point of interest is when we take $\omega = 0$. In this way, we recover the relativistic Anandan quantum phase in the cosmic string spacetime background obtained in [39] in an inertial frame. It is interesting to note that, taking $\omega = 0$ in the second term of (10), this second term corresponds to the relativistic Berry phase proposed by Cai and Papini [16] for a spin-half particle without using the weak field approximation [39]. Moreover, substituting the value of the defect parameter η into the second term of (10) (with $\omega = 0$), we obtain the phase shift $\phi'' \propto (8\pi G)\,\nu\,\Sigma^3$. This phase shift ϕ'' corresponds to the phase shift obtained by Resnik [21] for the gravitational effect analogue of the Aharonov-Casher effect [39].

4. The nonrelativistic dynamics of the neutral particle

In this section, we discuss the appearance of geometric quantum phases in the nonrelativistic dynamics of the neutral particle with a permanent dipole moment interacting with external fields. Taking the nonrelativistic of the Dirac equation (9), we obtain the following the Schrödinger-Pauli equation [40]

$$\begin{aligned} i\frac{\partial\psi}{\partial t} &= \frac{1}{2m}\left(\vec{p}+\vec{\Xi}\right)^2\psi + \frac{\mu}{2m}\vec{\nabla}\cdot\vec{E}\,\psi - \frac{\mu^2E^2}{2m}\,\psi - m\,\mathcal{A}_0\,\psi \\ &- \frac{\eta}{2}\,\vec{\omega}\cdot\vec{\sigma}\,\psi + \frac{\mu\lambda\eta\omega^2}{4m}\,\psi + \mathcal{O}\left(\frac{\beta^2}{2m}\right), \end{aligned} \tag{11}$$

where we have considered $\omega\eta\rho \ll 1$ to obtain the equation (11). The vector $\vec{\Xi}$ has the following components

$$\Xi_k = \mu\left(\vec{\sigma}\times\vec{E}\right)_k - \frac{1}{2\rho}\sigma^3\,\delta_{k2} - m\mathcal{A}_k - \frac{1}{2}\left(\vec{\sigma}\times\vec{\mathcal{E}}\right)_k. \tag{12}$$

We have defined in the expressions (11) and (12) the vector components: $\mathcal{A}_0 = \frac{1}{2}\eta^2\left(\vec{\omega}\times\vec{r}\right)^2 = \frac{1}{2}\omega^2\eta^2\rho^2$, $\mathcal{A}_\varphi = \eta\left(\vec{\omega}\times\vec{r}\right)_\varphi$ and $\mathcal{E}^\rho = \omega^2\eta^2\rho$. The nonrelativistic geometric phases acquires by the wave function of the neutral particle can be obtained by using again the Dirac phase factor method [55], where $\psi = e^{i\phi}\psi_0$. Since the terms proportional to $\vec{\nabla}\cdot\vec{E}$, E^2 and $\mathcal{A}_0$ are local terms, they do not contribute to the geometric phase [39, 40]. Hence, the terms which contribute to the nonrelativistic geometric phase are the vector $\vec{\Xi}$ and the spin-rotation coupling $\vec{\omega}\cdot\vec{\sigma}$. Here, we have that ψ_0 is the solution of the equation

$$-\frac{1}{2m}\nabla^2\psi_0 - \frac{\mu^2 E^2}{2m} + m\mathcal{A}^0\psi_0 = 0. \tag{13}$$

Taking the radial electric field given in (4) (with $\omega\eta\rho \ll 1$), we have that the nonrelativistic Anandan quantum phase corresponds to the effect analogous to the Aharonov-Casher effect [7] given in a noninertial frame in the presence of a topological defect. The expression for the nonrelativistic geometric phase is [40]:

$$\phi_{AC} = 2\pi\,\mu\,\lambda\,\sigma^3 - \eta\,\pi\,\sigma^3 - 2m\eta\,\vec{\omega}\cdot\vec{A} - \omega^2\eta^2\vec{A}\cdot\vec{\sigma} + \frac{\eta}{2}\vec{\omega}\cdot\vec{\sigma}\,T. \tag{14}$$

Thus, we can see that the nonrelativistic geometric phase is given by five independents contributions. In particular, the third contribution for the geometric phase (14) is given by the gauge field $m\vec{\mathcal{A}}$. This result comes from $\phi'_{NR} = -m\oint \mathcal{A}_k\, e^k_{\ \varphi}\, d\varphi = -2m\eta\vec{\omega}\cdot\vec{A}$, where $A = \pi\eta\rho^2$ is the area enclosed by the path of the neutral particle, perpendicular to the angular velocity ω. This phase shift appears due to the rotation frame and give us an effect analogous to the Sagnac effect [45, 47, 48] in the presence of the topological defect.

The fourth contribution to the geometric phase (14) is given by the noninertial effects of the rotating frames, where we have the interaction between the effective electric field $\vec{\mathcal{E}}$ and the spin of the neutral particle, that is, $\phi''_{NR} = -\frac{1}{2}\int\left(\vec{\sigma}\times\vec{\mathcal{E}}\right)_{\varphi}d\varphi = -\omega^2\eta^2\vec{A}\cdot\vec{\sigma}$. Note that, taking the limit $\eta\rightarrow 1$ in the phase shift ϕ''_{NR}, we recover the results obtained in [48] without making the weak field approximation. The last contribution to the nonrelativistic geometric phase (14) is given by the rotation-spin coupling, that is, $\phi_{NR'''} = \frac{\eta}{2}\int\vec{\omega}\cdot\vec{\sigma}\,dt = \frac{\eta T}{2}\vec{\omega}\cdot\vec{\sigma}$, where T is the time which the particle spends moving along a closed path around the symmetry axis of the topological defect.

We can observe that taking $\omega = 0$ in (14), we recover the effect analogous to the Aharonov-casher effect obtained by the authors in [39], that is, we have that $\phi'_{AC} = 2\pi\,\mu\,\lambda\,\sigma^3 - \eta\,\pi\,\sigma^3$. The term $\eta\,\pi\,\sigma^3$ corresponds to the contribution given by the presence of the topological defect. In the limit $\eta\rightarrow 1$, that is, the absence of the topological defect, we recover the original Aharonov-Casher effect [7], up to a constant phase factor. In the same way of the relativistic case, the contribution given by the presence of the topological defect in the nonrelativistic geometric phase provide us the gravitational effect analogous to the Aharonov-Casher effect $\phi'' \propto (8\pi G)\,\nu\,\sigma^3$ obtained by Resnik in [21].

Our last comment is related to the He-McKellar-Wilkens effect [8, 9]. In the same way of the effect analogue of the Aharonov-Casher effect that we have obtained in (14), the effect analogous to the He-McKellar-Wilkens effect can be achieve by making the transformation: $\mu\rightarrow d$, $\vec{E}\rightarrow -\vec{B}$ and $\lambda\rightarrow\lambda_m$, where d is the permanent electric dipole moment of the neutral particle and λ_m is a linear distribution of magnetic charges. In that way, the geometric phase (14) becomes ϕ_{HMW} and corresponds to the effect analogue of the He-McKellar-Wilkens effect.

5. Conclusions

We have discussed the appearance of geometric quantum phases for neutral particles with a permanent magnetic and electric dipole moments in the relativistic and in the nonrelativistic regime. We have seen that we can build a rotating frame in the cosmic string spacetime background in such a way that the field configuration induced by the noninertial effects produces no torque on the magnetic dipole moment of the neutral particle. In the relativistic regime, we have obtained that the Anandan quantum phase was given by four independent contributions: one from the interaction between the electric field and the magnetic dipole moment of the neutral particle, one from the topology of the cosmic string spacetime background and two from the noninertial effects of the rotating frames of the observers. We have also seen that in the limit $\omega = 0$, we can recover the relativistic Anandan quantum phase in the cosmic string background obtained in [39], where the contribution given by the topology of the spacetime provide us a phase shift analogous to the gravitational Aharonov-Casher effect [21], without making the weak field approximation [39]. In the nonrelativistic dynamics of the neutral particle, we have obtained the effect analogous to the Aharonov-Casher effect in the presence of a topological defect in a rotating frame. This geometric phase was given by five independent contributions, where we could observe effects analogous to the Sagnac effect and the Mashhoon effect. We have also seen that the topology of the defect provided a flux equivalent to the gravitational Aharonov-Casher effect [21]. At the end, we have seen that, taking $\omega = 0$, the analogous Aharonov-Casher effect given by the presence of a topological defect [39] has been recovered.

Acknowledgments

This work was partially supported by NANOBIOTEC/CAPEs, INCT-FCx, FINEP, CNPq and CAPES/PROCAD, C. F. thanks the organizing committe of the DICE2010 for hospitality)

References

[1] Aharonov Y and Bohm D 1959 *Phys. Rev.* **115** 485
[2] Berry M V 1984 *Proc. R. Soc. London A* **392** 45
[3] Pancharatnam S 1956 *Proc. Indian Acad. Sci. A* **44** 247
[4] Franz W 1939 *Verh. Dtsch. Phys. Ges.* **20** 65
[5] Ehrenberg W and Siday R E 1949 *Proc. Phys. Soc. London, Sect. B* **62** 8
[6] Aharonov Y and Anandan J 1987 *Phys. Rev. Lett.* **58** 1593
[7] Aharonov Y and Casher A 1984 *Phys. Rev. Lett.* **53** 319
[8] He X-G and McKellar B H J 1993 *Phys. Rev. A* **47** 3424
[9] Wilkens M. 1994 *Phys. Rev. Lett.* **72** 5
[10] Dowling J P, Williams C and Franson J D 1999 *Phys. Rev. Lett.* **83** 2486
[11] Furtado C and Duarte G 2005 *Phys. Scr.* **71** 7
[12] Anandan J 1989 *Phys. Lett. A* **138** 347
Anandan J 2000 *Phys. Rev. Lett.* **85** 1354
[13] A. Ya. Silenko 2005 *Russ. Phys. J.* **48** 788
[14] Furtado C and de Lima Ribeiro C A 2004 *Phys. Rev. A* **69** 064104
[15] Horsley S A R and Babiker M 2007 *Phys. Rev. Lett.* **99** 090401
Horsley S A R and Babiker M 2008 *Phys. Rev. A* **78** 012107
[16] Cai Y Q and Papini G 1989 *Mod. Phys Lett A* **4** 1143
Cai Y Q and Papini G 1990 *Class. Quant. Grav.* **7** 269
[17] Corichi A and Pierri M 1995 *Phys. Rev. D* **51** 5870
[18] Mazur P O 1986 *Phys. Rev. Lett.* **57** 929
[19] Mostafazadeh A 1998 *J. Phys. A* **31** 7829
[20] Anandan J. 1992 *Quantum Coherence and Reality, Proc. Conf. on Fundamental Aspects of Quantum Theory* (*Preprint* arXiv:gr-qc/9504002).
[21] Resnik B 1995 *Phys. Rev. D* **51** 3108
[22] de Assis J G, Furtado C and Bezerra V B 2000 *Phys. Rev D* **62** 045003
[23] Carvalho J de S, Passos E, Furtado C and Moraes F 2008 *Eur. Phys. J. C* **57** 817
[24] Feng L L and Lee W 2001 *Int. J. Mod. Phys. D* **10** 961
[25] Ji P, Bai Y and Wang L 2007 *Phys. Rev. D* **75** 024010
[26] Kleinert H 1989 *Gauge fields in condensed matter, vol 2* (Sigapure: World Scientific).
[27] Volovik G E 2003 *The Universe in a Helium Droplet* (Oxford: Oxford University Press).
[28] Vozmediano M A H, Katsnelson M I and Guinea F 2010 *Physics Reports* **496** 109
[29] Furtado C, Moraes F and Carvalho A. M. de M. 2008 *Phys. Lett. A* **372** 5368
[30] Bakke K, Furtado C and Sergeenkov S 2009 *EPL* **87** 30002
[31] Sátiro C and Moraes F 2007 *Europhys. Lett.* **80** 46002
[32] Katanaev M O and Volovich I V 1992 *Ann. Phys.* (NY) **216** 1
[33] Moraes F 1996 *Phys. Lett. A* **214** 189
de Padua A, Parisio-Filho F and Moraes F 1998 *Phys. Lett. A* **238** 153

[34] Dias L and Moraes F 2005 *Braz. J. Phys.* **35** 636
[35] Moraes F 1995 *Phys. Lett. A* **204** 399
[36] Furtado C and Moraes F 1999 *Europhys. Lett.* **45** 279
Silva Neto A L and Furtado C 2008 *J. Phys.: Cond. Mat.* **20** 125209
Bakke K, Ribeiro L R, Furtado C and Nascimento J R 2009 *Phys. Rev. D* **79** 024008
Bakke K, Ribeiro L R and Furtado C 2010 *Cent. Eur. J. Phys.* **8** 893
Bakke K and Furtado C 2010 *Int. J. Mod. Phys. D* **19** 1
[37] Furtado C, Bezerra V B and Moraes F 2001 *Phys. Lett. A* **289** 160
[38] de Lima Ribeiro C A, Furtado C and Moraes F 2003 *EPL* **62** 306
[39] Bakke K, Nascimento J R and Furtado C 2008, *Phys. Rev. D* **78** 064012
Bakke K, Furtado C and Nascimento J R 2009 *Eur. Phys. J. C* **60** 501
Bakke K, Furtado C and Nascimento J R 2009 *Eur. Phys. J. C* **64** 169
Bakke K and Furtado C 2010 *Ann. Phys. (Berlin)* **522** 447
[40] Bakke K and Furtado C 2009 *Phys. Rev D* **80** 024033
[41] Bakke K and Furtado C 2010 *Eur. J. Phys. C* **69** 531
[42] Furtado C and Moraes F 1994 *Phys. Lett. A* **188** 392
[43] Sokolov D D and Starobinskii A A 1977 *Sov. Phys. Dokl.* **22** 312
[44] Birrell N D and Davies P C W 1982 *Quantum Fields in Curved Space* (Cambridge: Cambridge University Press).
[45] Sagnac M G 1913 *C. R. Acad. Sci. (Paris)* **157** 708
Sagnac M G 1913 *C. R. Acad. Sci. (Paris)* **157** 1410
[46] Mashhoon B 1988 *Phys. Rev. Lett.* **61** 2639
[47] Hehl F W and Ni W-T 1980 *Phys. Rev. D* **42** 2045
Iyer B R 1982 *Phys. Rev. D* **26** 1900
Post E J 1967 *Rev. Mod. Phys.* **39** 475
Anandan J 1981 *Phys. Rev. D* **24** 338
[48] Anandan J and Suzuki J 2004 *Relativity in Rotating Frames, Relativistic Physics in Rotating Reference Frame* ed G. Rizzi and M. L. Ruggiero (Dordrecht: Kluwer Academic Publishers) p 361 (*Preprint* arXiv:quant-ph/0305081).
[49] Bakke K, Carvalho A M de M and Furtado C 2009 *Eur. Phys. J. C* **63** 149
[50] Bakke K 2010 *Phys. Rev. A* **81** 052117
[51] Bakke K and Furtado C 2010 *Phys. Rev. D* **82** 084025
[52] Bakke K 2010 *Phys. Lett. A* **374** 3143
Bakke K 2010 *Phys. Lett. A* **374** 4642
[53] Misner C W, Thorne K S and Wheeler J A 1973 *Gravitation* (San Francisco: W. H. Freeman).
[54] Nakahara M 1998 *Geometry, Topology and Physics*, (Bristol: Institute of Physics Publishing).
[55] Dirac P M A 1931 *Proc. R. Soc. London, Ser. A* **133** 60
Berry M V 1980 *Eur. J. Phys.* **1** 240

Journal of Physics: Conference Series **306** (2011) 012070 doi:10.1088/1742-6596/306/1/012070

Nonlinear statistical effects in relativistic mean field theory

G Gervino[1,3], **A Lavagno**[2,3], **D Pigato**[2,3]

[1]Dipartimento di Fisica, Università di Torino, I-10126 Torino, Italy
[2]Dipartimento di Fisica, Politecnico di Torino, C.so Duca degli Abruzzi 24, Italy
[3]Istituto Nazionale di Fisica Nucleare (INFN), Sezione di Torino, Italy

Abstract. We investigate the relativistic mean field theory of nuclear matter at finite temperature and baryon density taking into account of nonlinear statistical effects, characterized by power-law quantum distributions. The analysis is performed by requiring the Gibbs conditions on the global conservation of baryon number and electric charge fraction. We show that such nonlinear statistical effects play a crucial role in the equation of state and in the formation of mixed phase also for small deviations from the standard Boltzmann-Gibbs statistics.

1. Introduction

Several experimental observations and theoretical calculations clearly indicate that hadrons dissociate into a plasma of their elementary constituents, quarks and gluons (QGP), at density several times the nuclear matter density and/or at temperature above few hundreds MeV. Such a QGP is expected to have occurred in the early stages of the Universe and can be found in dense and hot stars, neutron stars, nucleus-nucleus high energy collisions where heavy ions are accelerated to relativistic energies [1]. After collision, a fireball is created which may realize the conditions of the QGP. The plasma then expands, cools, freezes-out into hadrons, photons, leptons that are detected and analyzed [2].

It is a rather common opinion that, because of the extreme conditions of density and temperature in ultrarelativistic heavy ion collisions, memory effects and long–range color interactions give rise to the presence of non–Markovian processes in the kinetic equation affecting the thermalization process toward equilibrium as well as the standard equilibrium distribution [3, 4, 5]. A rigorous determination of the conditions that produce a nonextensive behavior, due to memory effects and/or long–range interactions, should be based on microscopic calculations relative to the parton plasma originated during the high energy collisions. At this stage we limit ourselves to consider the problem from a qualitative point of view on the basis of the existing theoretical calculations and experimental evidences.

On the other hand, over the last years, there has been an increasing evidence that the generalized non-extensive statistical mechanics, proposed by Tsallis [6, 7, 8, 9] and characterized by a power-law stationary particle distribution, can be considered as a basis for a theoretical framework appropriate to incorporate, at least to some extent and without going into microscopic dynamical description, long-range interactions, long-range microscopic memories and/or fractal space-time constraints. A considerable variety of physical issues show a quantitative agreement

Journal of Physics: Conference Series **306** (2011) 012070 doi:10.1088/1742-6596/306/1/012070

between experimental data and theoretical analysis based on Tsallis' thermostatistics. In particular, there is a growing interest in high energy physics applications of non-extensive statistics [10, 11, 12, 13, 14, 15, 16]. Several authors outline the possibility that experimental observations in relativistic heavy-ion collisions can reflect non-extensive statistical mechanics effects during the early stage of the collisions and the thermalization evolution of the system [17, 18, 19, 20, 21]. In this context, it is relevant to observe that the statistical origin of the nonextensive statistics lies in the deformation of the Boltzmann entropy. From the above considerations, it appears reasonable that in regime of high density and temperature both hadron and quark-gluon Equation of State (EOS) can be sensibly affected by nonextensive statistical effects [22, 23]. Furthermore, in this context it is very remarkable to observe that the relevance of these effects on the relativistic hadronic equation of state has also been recently investigated in Ref. [24].

The aim of this paper is to study the behavior of the nuclear equation of state at finite temperature and baryon density and to explore the existence of a hadron-quark mixed phase at a fixed value of the proton fraction Z/A.

2. Nonextensive hadronic and quark-gluon equation of state

In this Section we study the nonextensive hadronic EOS in the framework of a relativistic mean field theory in which nucleons interact through the nuclear force mediated by the exchange of virtual isoscalar-scalar (σ), isoscalar-vector (ω) and isovector-vector (ρ) meson fields [25, 26, 27].

The nonlinear Lagrangian density describing hadronic matter can be written as

$$\mathcal{L} = \mathcal{L}_{QHD} + \mathcal{L}_{\rm qfm} \,, \tag{1}$$

where [27]

$$\begin{aligned} \mathcal{L}_{QHD} &= \bar{\psi}[i\gamma_\mu\partial^\mu - (M - g_\sigma\sigma) - g_\omega\gamma_\mu\omega^\mu - g_\rho\gamma^\mu\vec{\tau}\cdot\vec{\rho}_\mu]\psi + \frac{1}{2}(\partial_\mu\sigma\partial^\mu\sigma - m_\sigma^2\sigma^2) \\ &\quad -U(\sigma) + \frac{1}{2}m_\omega^2\omega_\mu\omega^\mu + \frac{1}{2}m_\rho^2\vec{\rho}_\mu\cdot\vec{\rho}^{\,\mu} - \frac{1}{4}F_{\mu\nu}F^{\mu\nu} - \frac{1}{4}\vec{G}_{\mu\nu}\vec{G}^{\mu\nu} \,, \end{aligned} \tag{2}$$

and $M = 939$ MeV is the vacuum baryon mass. The field strength tensors for the vector mesons are given by the usual expressions $F_{\mu\nu} \equiv \partial_\mu\omega_\nu - \partial_\nu\omega_\mu$, $\vec{G}_{\mu\nu} \equiv \partial_\mu\vec{\rho}_\nu - \partial_\nu\vec{\rho}_\mu$, and $U(\sigma)$ is a nonlinear potential of σ meson

$$U(\sigma) = \frac{1}{3}a\sigma^3 + \frac{1}{4}b\sigma^4 \,, \tag{3}$$

usually introduced to achieve a reasonable compression modulus for equilibrium nuclear matter.

Following Ref.s [28, 29], $\mathcal{L}_{\rm qfm}$ in Eq.(1) is related to a (quasi) free gas of pions with an effective chemical potential (see below for details).

The field equations in a mean field approximation are

$$(i\gamma_\mu\partial^\mu - (M - g_\sigma\sigma) - g_\omega\gamma^0\omega - g_\rho\gamma^0\tau_3\rho)\psi = 0 \,, \tag{4}$$

$$m_\sigma^2\sigma + a\sigma^2 + b\sigma^3 = g_\sigma < \bar{\psi}\psi >= g_\sigma\rho_S \,, \tag{5}$$

$$m_\omega^2\omega = g_\omega < \bar{\psi}\gamma^0\psi >= g_\omega\rho_B \,, \tag{6}$$

$$m_\rho^2\rho = g_\rho < \bar{\psi}\gamma^0\tau_3\psi >= g_\rho\rho_I \,, \tag{7}$$

where $\sigma = \langle\sigma\rangle$, $\omega = \langle\omega^0\rangle$ and $\rho = \langle\rho_3^0\rangle$ are the nonvanishing expectation values of meson fields, ρ_I is the total isospin density, ρ_B and ρ_S are the baryon density and the baryon scalar density, respectively. They are given by

$$\rho_B = 2\sum_{i=n,p}\int \frac{d^3k}{(2\pi)^3}[n_i(k) - \overline{n}_i(k)]\,, \tag{8}$$

$$\rho_S = 2\sum_{i=n,p}\int \frac{d^3k}{(2\pi)^3}\frac{M_i^*}{E_i^*}\,[n_i^q(k) + \overline{n}_i^{\,q}(k)]\,, \tag{9}$$

where $n_i(k)$ and $\overline{n}_i(k)$ are the q-deformed fermion particle and antiparticle distributions:

$$n_i(k) = \frac{1}{[1+(q-1)\,\beta(E_i^*(k)-\mu_i^*)]^{1/(q-1)}+1}\,, \tag{10}$$

$$\overline{n}_i(k) = \frac{1}{[1+(q-1)\,\beta(E_i^*(k)+\mu_i^*)]^{1/(q-1)}+1}\,. \tag{11}$$

The nucleon effective energy is defined as $E_i{}^*(k) = \sqrt{k^2 + M_i{}^{*2}}$, where $M_i{}^* = M_i - g_\sigma\sigma$. The effective chemical potentials μ_i^* are given in terms of the meson fields as follows

$$\mu_i^* = \mu_i - g_\omega\omega - \tau_{3i}g_\rho\rho\,, \tag{12}$$

where μ_i are the thermodynamical chemical potentials $\mu_i = \partial\epsilon/\partial\rho_i$. At zero temperature they reduce to the Fermi energies $E_{Fi} \equiv \sqrt{k_{Fi}^2 + M_i^{*2}}$ and the nonextensive statistical effects disappear. The meson fields are obtained as a solution of the field equations in mean field approximation and the related meson-nucleon couplings (g_σ, g_ω and g_ρ) are the free parameters of the model. In the following, they will be fixed to the parameters set marked as GM2 of Ref. [27].

The thermodynamical quantities can be obtained from the thermodynamic potential in the standard way. More explicitly, the baryon pressure P_B and the energy density ϵ_B can be written as

$$P_B = \frac{2}{3}\sum_{i=n,p}\int \frac{d^3k}{(2\pi)^3}\frac{k^2}{E_i^*(k)}[n_i^q(k) + \overline{n}_i^q(k)] - \frac{1}{2}m_\sigma^2\sigma^2 - U(\sigma) + \frac{1}{2}m_\omega^2\omega^2 + \frac{1}{2}m_\rho^2\rho^2\,, \tag{13}$$

$$\epsilon_B = 2\sum_{i=n,p}\int \frac{d^3k}{(2\pi)^3}E_i^*(k)[n_i^q(k) + \overline{n}_i^q(k)] + \frac{1}{2}m_\sigma^2\sigma^2 + U(\sigma) + \frac{1}{2}m_\omega^2\omega^2 + \frac{1}{2}m_\rho^2\rho^2\,. \tag{14}$$

It is important to observe that Eq.s(9), (13) and (14) apply to $n_i^q \equiv (n_i)^q$ rather than n_i itself, this is a direct consequence of the basic prescription related to the q-mean expectation value in nonextensive statistics [8, 15]. In addition, since all equations must be solved in a self-consistent way, the presence of nonextensive statistical effects influences the many-body interaction mediated by the meson fields.

Especially in regime of low density and high temperature the contribution of the lightest mesons to the thermodynamical potential (and, consequently, to the other thermodynamical quantities) becomes relevant. As quoted before, we have included the contribution of pions considering them as a (quasi) ideal gas of nonextensive bosons with effective chemical potentials expressed in terms of the corresponding effective baryon chemical potentials [29].

In Fig. 1, the total pressure P and energy density ϵ are plotted as a function of μ_B for different values of temperature and q. The different behavior from P and ϵ reflects essentially the nonlinear combinations of the meson fields and the different functions under integration in Eq.s (13) and (14). Concerning the pressure, we have that becomes stiffer by increasing the q

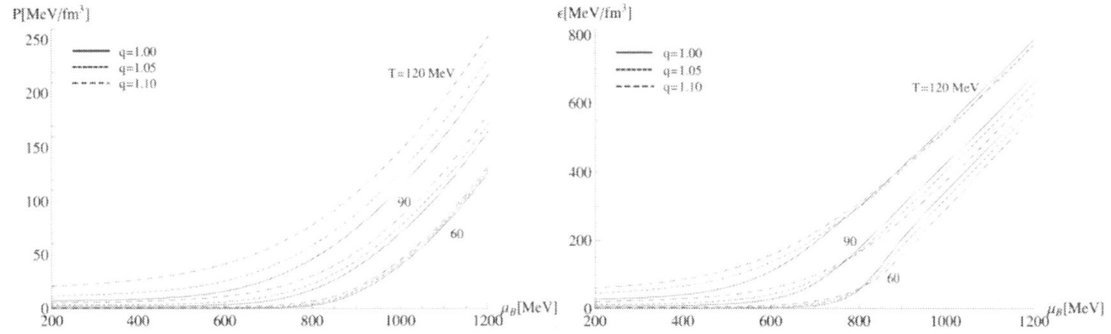

Figure 1. Pressure (left panel) and energy density (right panel) versus baryon chemical potential for different values of temperature and q.

parameter. On the other hand, the behavior of the energy density presents features very similar to the σ field one. At low μ_B, nonextensive effects make the energy density greater with respect to the standard case. At medium-high μ_B, the standard ($q = 1$) component of the energy density becomes dominant, this effect is essentially due to the reduction of the σ field for $q > 1$. The intersection point depends, naturally, on the physical parameters of the system.

Concerning the quark-gluon EOS, we use the MIT bag model [30]. In this model, quark matter is described as a gas of free quarks with massless up and down quarks. All the non-perturbative effects are simulated by the bag constant B which represents the pressure of the vacuum. Following this line, the pressure, energy density and baryon number density for a relativistic Fermi gas of quarks in the framework of nonextensive statistics can be written, respectively, as

$$P_q = \frac{\gamma_f}{3} \sum_{f=u,d} \int_0^\infty \frac{\mathrm{d}^3k}{(2\pi)^3} \frac{k^2}{e_f} [n_f^q(k) + \overline{n}_f^q(k)] - B\,, \tag{15}$$

$$\epsilon_q = \gamma_f \sum_{f=u,d} \int_0^\infty \frac{\mathrm{d}^3k}{(2\pi)^3} e_f\, [n_f^q(k) + \overline{n}_f^q(k)] + B\,, \tag{16}$$

$$\rho_q = \frac{\gamma_f}{3} \sum_{f=u,d} \int_0^\infty \frac{\mathrm{d}^3k}{(2\pi)^3} [n_f(k) - \overline{n}_f(k)]\,, \tag{17}$$

where the quark degeneracy for each flavor is $\gamma_f = 6$, $e_f = (k^2 + m_f^2)^{1/2}$, $n_f(k)$ and $\overline{n}_f(k)$ are the q-deformed particle and antiparticle quark distributions

$$n_f(k) = \frac{1}{[1 + (q-1)(e_f(k) - \mu_f)/T]^{1/(q-1)} + 1}\,, \tag{18}$$

$$\overline{n}_f(k) = \frac{1}{[1 + (q-1)(e_f(k) + \mu_f)/T]^{1/(q-1)} + 1}\,. \tag{19}$$

Similar expressions for the pressure and the energy density can be written for gluons treating them as a massless q-deformed Bose gas with zero chemical potential. Explicitly, we can calculate the nonextensive pressure P_g and energy density ϵ_g for gluons as

$$P_g = \frac{\gamma_g}{3} \int_0^\infty \frac{\mathrm{d}^3k}{(2\pi)^3} \frac{k}{[1 + (q-1)\,k/T]^{q/(q-1)} - 1}\,, \tag{20}$$

$$\epsilon_g = 3\,P_g\,, \tag{21}$$

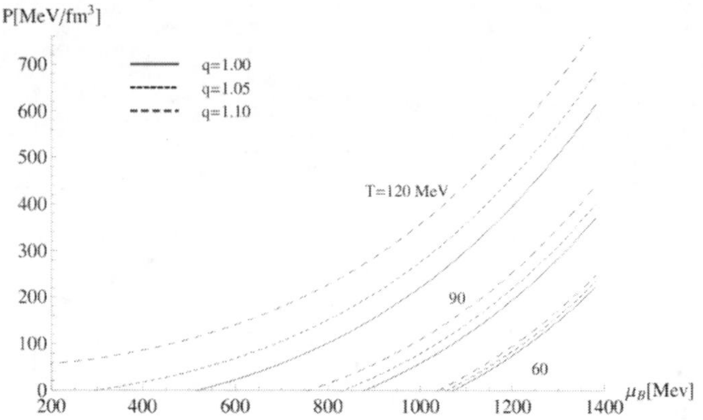

Figure 2. Pressure of the quark-gluon phase as a function of baryon chemical potential for different values of temperature and q.

with the gluon degeneracy factor $\gamma_g = 16$. In the limit $q \to 1$, one recovers the usual analytical expression: $P_g = 8\pi^2/45\, T^4$.

Let us note that, since one has to employ the fermion (boson) nonextensive distributions, the results are not analytical, even in the massless quark approximation. Hence a numerical evaluations of the integrals in Eq.s (15)–(17) and (20) must be performed.

In Fig. 2, we report the total pressure as a function of the baryon chemical potential for massless quarks and gluons, for different values of q and at fixed value of $Z/A = 0.4$. The bag constant is set equal to $B^{1/4}$=190 MeV. In presence of nonextensive effects, as in the case of hadronic phase, the pressure is significantly increased even for small deviations from standard statistics.

3. The hadron to quark-gluon phase transition

In this Section we investigate the hadron-quark phase transition at finite temperature and baryon chemical potential by means of the previous relativistic EOSs. Lattice calculations predict a critical phase transition temperature T_c of about 170 MeV, corresponding to a critical energy density $\epsilon_c \approx 1$ GeV/fm^3 [1]. In a theory with only gluons and no quarks, the transition turns out to be of first order. In nature, since the u and d quarks have a small mass, while the strange quark has a somewhat larger mass, the phase transition is predicted to be a smooth cross over. However, since it occurs over a very narrow range of temperatures, the transition, for several practical purposes, can still be considered of first order. Indeed the lattice data with 2 or 3 dynamical flavours are not precise enough to unambigously control the difference between the two situations. Thus, by considering the deconfinement transition at finite density as a the first order one, a mixed phase can be formed, which is typically described using the two separate equations of state, one for the hadronic and one for the quark phase.

The phase transition is described by using the Gibbs formalism applied to systems where more than one conserved charge is present [31]. In fact, because we are going to describe the nuclear EOS, we have to require the global conservation of two "charges": baryon number and electric charge. Each conserved charge has a conjugated chemical potential and the systems is described by two independent chemical potentials: μ_B and μ_C. The structure of the mixed phase is obtained by imposing the following Gibbs conditions for chemical potentials and pressure [32, 33, 34]

$$\mu_B^{(H)} = \mu_B^{(Q)}\,, \quad \mu_C^{(H)} = \mu_C^{(Q)}\,, \tag{22}$$

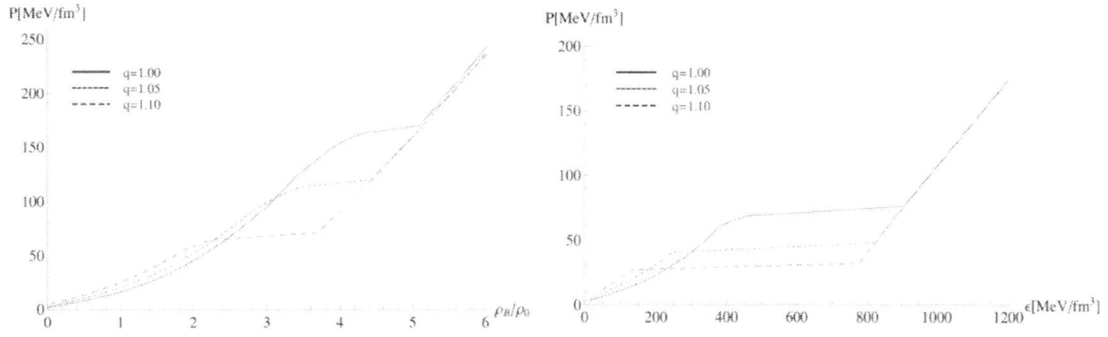

Figure 3. Pressure as a function of baryon density (left panel) and energy density (right panel) in the mixed phase for different values of q. The temperature is fixed at $T = 90$ MeV.

$$P^H(T, \mu_B, \mu_C) = P^Q(T, \mu_B, \mu_C)\,. \qquad (23)$$

Therefore, at a given baryon density ρ_B and at a given net electric charge density $\rho_C = Z/A\,\rho_B$, the chemical potentials μ_B are μ_C are univocally determined by the following equations

$$\rho_B = (1-\chi)\,\rho_B^H(T, \mu_B, \mu_C) + \chi\,\rho_B^Q(T, \mu_B, \mu_C)\,, \qquad (24)$$
$$\rho_C = (1-\chi)\,\rho_C^H(T, \mu_B, \mu_C) + \chi\,\rho_C^Q(T, \mu_B, \mu_C)\,, \qquad (25)$$

where $\rho_B^{H(Q)}$ and $\rho_C^{H(Q)}$ are, respectively, the net baryon and electric charge densities in the hadronic (H) and in the quark (Q) phase and χ is the fraction volume of quark-gluon matter in the mixed phase. In this way we can find out the phase coexistence region, for example, in the (T, μ_B) plane. We are particularly interested in the lower baryon density (baryon chemical potential) border, i.e. the first critical transition density $\rho_{\rm cr}^I$ $(\mu_{\rm cr}^I)$, in order to check the possibility of reaching such conditions in a transient state during a heavy-ion collision at relativistic energies.

In Fig. 3, we report the pressure at $T = 90$ MeV as a function of baryon density (in units of nuclear saturation density $\rho_0 = 0.153$ fm^{-3}) (left panel) and energy density (right panel). It is interesting to observe that pressure as a function of baryon density (or energy density) is stiffer in the pure hadronic phase for $q > 1$ but appears a strong softening in the mixed phase. This feature results in significant changes in the incompressibility and may be particularly important in identifying the presence of nonextensive effects in high energy heavy ion collisions experiments. Related to this aspect, let us observe that possible indirect indications of a significative softening of the EOS at the energies reached at AGS have been discussed several times in the literature [34].

In Fig. 4, it is reported the phase diagram in the plane $T - \rho_B$ for different values of q. The curves labelled with the index I and II represent, respectively, the beginning and the end of the mixed phase. For $q > 1$, both the first and the second critical densities are sensibly reduced, even if the shape of the mixed phase is approximately the same. Related to this aspect, let us mention that the simplest version of the MIT bag model, considered in this investigation, appears to be not fully appropriate to describe a large range of temperature and density. To overcome this shortcoming, a phenomenological approach can therefore be based on a density or temperature dependent bag constant [34, 35]. Moreover, in regime of high temperature and small baryon chemical potential the first order phase transition may end in a (second order) critical endpoint with a smooth crossover. These features cannot be incorporated in the considered

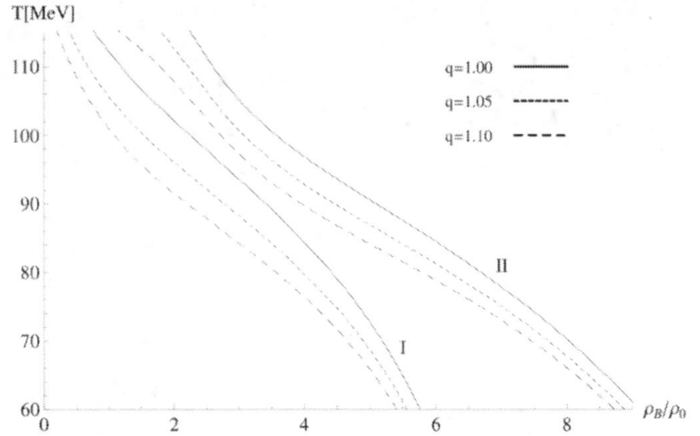

Figure 4. Phase diagram $T - \rho_B$ for different values of q. The curves with index I and II indicate, respectively, the beginning and the end of the mixed phase.

mean field approach. In our investigation, because we are focusing to nonextensive statistical effects on the nuclear EOS, instead of introducing additional parameterizations, we work with a fixed bag constant and limit our analysis to a restricted range of temperature and density, region of particular interest for high energy compressed nuclear matter experiments.

Let us now explore in more details the variation of the first transition baryon density $\rho_{\rm cr}^I$ as a function of different physical parameters. In Fig. 5, we report the dependence of $\rho_{\rm cr}^I$ as a function of Z/A for different values of q (y axis in logarithmic scale). It is interesting to note a significant reduction of $\rho_{\rm cr}^I$ in presence of nonextensive statistics; as in the previous cases, this effect increases with the temperature. The dependence of the first transition baryon density as a function of Z/A is essentially a consequence of the ρ meson field behavior in the hadronic phase because it is directly connected with the isospin density of the system (as appears from Eq.(7)). In this context, let us observe that, at fixed value of q, $\rho_{\rm cr}^I$ is significantly reduced by decreasing Z/A only at lower temperatures ($T = 60$ MeV) while, as expected, at higher temperatures ($T = 120$ MeV) the transition baryon density becomes very low and its isospin dependence becomes negligible, also in the framework of nonextensive statistics. This matter of fact is a consequence of fact that at low baryon chemical potentials (or baryon densities) the ρ meson field becomes almost constant and its absolute value significantly decreases.

4. Conclusions

Following the basic prescriptions of the Tsallis' nonlinear relativistic thermodynamics, we investigate the relevance of nonextensive statistical effects on the relativistic nuclear and subnuclear equation of state. We have focused our investigation in regime of finite temperature and baryon chemical potential, reachable in high-energy heavy-ion collisions, for which the deconfinement phase transition can be still considered of the first order.

In the first part of the work, we have investigated the hadronic equation of state and the role played by the meson fields in the framework of a relativistic mean field model which contains the basic prescriptions of nonextensive (nonlinear) statistical mechanics. We have shown that, also in presence of small deviations from standard Boltzmann-Gibbs statistics, the meson fields and, consequently, the EOS appear to be sensibly modified. In the second part, we have analyzed the QGP proprieties using the MIT Bag model and also in this case the EOS becomes stiffer in presence of nonextensive effects. Finally, we have studied the proprieties of the phase transition from hadronic matter to QGP and the formation of a relative mixed phase by requiring the

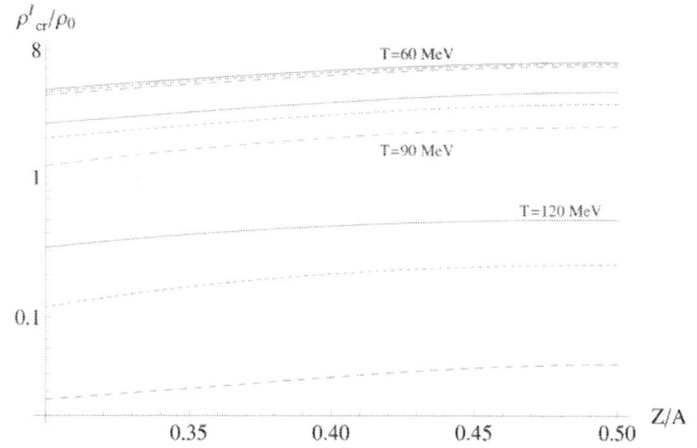

Figure 5. Variation of the first transition baryon density as a function of the net electric charge fraction Z/A for different temperatures and values of q ($q = 1$, solid lines; $q = 1.05$, short dashed lines; $q = 1.10$, long dashed lines).

Gibbs conditions on the global conservation of baryon number and electric charge fraction. We have seen that nonextensive effects play a crucial role in the deconfinement phase transition. Moreover, although pressure as a function of baryon density is stiffer in the hadronic phase, we have shown that a strong softening in the mixed phase takes place in presence of nonextensive statistics. Such a behavior implies an abruptly variation in the incompressibility and could be considered as a signal of nonextensive statistical effects in high energy heavy ion collisions.

From a phenomenological point of view, the nonextensive index q is considered here as a free parameter, even if, actually should not be treated as such because, in principle, it should depend on the physical conditions generated in the reaction, on the fluctuation of the temperature and be related to microscopic quantities (such as, for example, the mean interparticle interaction length, the screening length and the collision frequency into the parton plasma). Moreover, let us remember that, in the diffusional approximation, a value $q > 1$ implies the presence of a superdiffusion among the constituent particles (the mean square displacement obeys to a power law behavior $\langle x^2 \rangle \propto t^\alpha$, with $\alpha > 1$) [36].

References

[1] Hwa R C and Wang X N 2004 *Quark Gluon Plasma 3*, (Singapore: World Scientific) and references therein
[2] Biró T S 2008 *J. Phys. G: Nucl. Part. Phys.* **35** 044056
[3] Heiselberg H and Wang X N 1996 *Phys. Rev.* C **53** 1892
[4] Biró T S and Greiner C 1997 *Phys. Rev. Lett.* **79** 3138
[5] Alberico W M, Lavagno A and Quarati P 2000 *Eur. Phys. J.* C **12** 499
[6] Tsallis C 1998 *J. Stat. Phys.* **52** 479
Tsallis C, Mendes R S and Plastino A R 1998 *Physica* A **261** 534
[7] Gell-Mann M and Tsallis C 2004 *Nonextensive Entropy: Interdisciplinary Applications*, (New York: Oxford University Press)
[8] Tsallis C 2009 *Introduction to Nonextensive Statistical Mechanics* (New York: Springer-Verlag)
[9] Kodama T, Elze H -T, Aiguiar C E and Koide T 2005 *Europhys. Lett.* **70** 439
Elze H -T and Kodama T 2005 *Phys. Lett.* A **335** 363
Kodama T and Koide T 2009 *Eur. Phys. J.* A **40** 289
[10] Bediaga I, Curado E M F and de Miranda J M 2000 *Physica* A **286** 156
[11] Beck C 2000 *Physica* A **286** 164
Beck C 2009 *Eur. Phys. J.* A **40** 267
[12] Walton D B and Rafelski J 2000 *Phys. Rev. Lett.* **84** 31

[13] Wilk G and Włodarczyk Z 2000 *Phys. Rev. Lett.* **84** 2770
Wilk G and Włodarczyk Z 2009 *Eur. Phys. J.* A **40** 299
[14] Lavagno A and Quarati P 2001 *Phys. Lett.* B **498** 47
Lavagno A and Quarati P 2000 *Nucl. Phys.* B [PS] **87** 209
[15] Lavagno A 2002 *Phys. Lett.* A **301** 13
Lavagno A, Scarfone A M and Swamy P N 2007 J. Phys. A **40** 8635
Lavagno A, Scarfone A M and Swamy P N 2006 Eur. Phys. J. C **47** 253
[16] Biró T S and Purcsel G 2005 *Phys. Rev. Lett.* **95** 162302
Biró T S, Purcsel G and Ürmössy K 2009 *Eur. Phys. J.* A **40** 325
[17] Alberico W M and Lavagno A 2009 *Eur. Phys. J.* A **40** 313
[18] Biró T S and Peshier A 2006 *Phys. Lett.* B **632** 247
[19] Biyajima N *et al* 2006 *Eur. Phys. J.* C **48** 597
[20] Lavagno A 2002 *Physica* A **305** 238
Alberico W M, Czerski P, Lavagno A, Nardi M and Somá V 2008 *Physica* A **387** 467
[21] Cleymans J, Hamar G, Levai P, Wheaton S 2009 *J. Phys. G: Nucl. Part. Phys.* **36** 064018
[22] Lavagno A, Quarati P and Pigato D 2010 *J. Phys. G: Nucl. Part. Phys.* **37** 115102
Drago A, Lavagno A and Quarati P 2004 *Physica* A **344** 472
[23] Rożynek J and Wilk G 2009 *J. Phys. G: Nucl. Part. Phys.* **36** 125108
[24] Pereira F I M, Silva R and Alcaniz J S 2007 *Phys. Rev.* C **76** 015201
[25] Walecka J D 1974 *Ann. of Phys.* **83** 491
[26] Boguta J and Bodmer A R 1977 *Nucl. Phys.* A **292** 413
[27] Glendenning N K and Moszkowski S A 1991 *Phys. Rev. Lett.* **67** 2414
[28] Müller H 1997 *Nucl. Phys.* A **618** 349
[29] Lavagno A 2010 *Phys. Rev.* C **81** 044909
[30] Chodos A *et al* 1974 *Phys. Rev.* D **9** 3471
[31] Glendenning N K 1992 *Phys. Rev.* D **46** 1274
Müller H and Serot B D 1995 *Phys. Rev.* C **52** 2072
[32] Drago A and Lavagno A 2001 *Phys. Lett. B* **511** 229
Drago A, Lavagno A and Pagliara G 2005 *Phys. Rev. D* **71** 103004
[33] Di Toro M, Drago A, Gaitanos T, Greco V and Lavagno A 2006 *Nucl. Phys.* A **775** 102
Drago A, Lavagno A and Parenti I 2007 *ApJ.* **659** 1519
[34] Bonanno L, Drago A and Lavagno A 2007 *Phys. Rev. Lett.* **99** 242301
[35] Burgio G F *et al* 2002 *Phys. Rev.* C **66** 025802
[36] Tsallis C and Bukmann D J 1996 *Phys. Rev.* E **54** R2197

Journal of Physics: Conference Series **306** (2011) 012071 doi:10.1088/1742-6596/306/1/012071

Does the Schrödinger wave function describe a real physical wave?

Kurt Jung
Fachbereich Physik, Technische Universität Kaiserslautern,
Erwin-Schrödinger-Str. 56, D-67663 Kaiserslautern, Germany,
E-mail: jung@physik.uni-kl.de

Abstract. In 1924 Louis de Broglie presented the phase wave model [1]. He postulated that particle and wave coexist. Both objects are coupled to each other by the phase harmony principle, which always tends to synchronize the oscillations of wave and particle.

The escort wave model [2] replenishes the phase wave model by attributing a stochastic velocity component and a potential dependent rest mass to the particle. Moreover it introduces a novel method to deal with standing waves. In this new model a particle's trajectory is continuous but not deterministic because the particle is not only subjected to regular forces derived from external potentials but also to fluctuating forces. On this condition the space dependence of the wave escorting a particle agrees with the Schrödinger wave function.

By a measurement process the phase harmony coupling is broken off. Contrary to the standard interpretation of quantum mechanics the escort wave does not collapse but disappears in the ocean of fluctuating waves, because it is no longer stabilized by the particle. The particle immediately starts to build up a new escort wave by adjusting its phase to the phases of fluctuating waves, whose frequencies and rest systems nearly agree with the corresponding magnitudes of the particle.

1. Introduction

Since modern quantum mechanics has been established by Werner Heisenberg and Erwin Schrödinger numerous suggestions have been made how the abstract terms in the Hilbert space formalism have to be interpreted. Until now no satisfactory answer has been presented. Actually we do not need a new interpretation but we have to unify quantum mechanics with classical physics. If this is possible the interpretation will become evident.

Two approaches are carried out at the moment. Several physicists, e.g. [3] [4], study decoherence effects caused by the interaction of microscopic systems with the environment in order to explain why interference effects are not observed in macroscopic physics. These scientists assume that quantum mechanics is the universal theory and classical physics is a special case only valid for macrosystems.

The escort wave model [2] is based on the idea that classical physics is the groundwork of physics and that quantum phenomena arise from the interplay of particles with their associated waves. This statement cannot be proved within the quantum mechanical formalism but one has to start from conceptions used in classical physics. The uncertainty relation implies that position and momentum of a particle cannot be exactly determined at the same time. This does not necessarily means that both magnitudes cannot be always well defined. In the escort wave model it is taken for granted that particles always move on continuous trajectories. In order to account for the fact that quantum mechanics contains stochastic elements it will be assumed that particles are subjected to nonclassical forces exerted by fluctuating waves.

Many scientists have introduced stochastic elements into quantum mechanics. Most of these attempts have been performed in combination with the guidance wave concept. Especially Bohm [5], de Broglie [6, 7] and Vigier [8–11] have to be mentioned here. Vigier was able to reproduce quantum mechanical results on the assumption that massive particles run on continuous trajectories. Because Vigier only looked for diffraction experiments, he did not realize that continuous trajectories are in conflict with Born's probability density interpretation of excited states.

The influence of fluctuating waves will be the stronger the better energy and momentum of particle and fluctuating wave agree. Because of the limited coherence length of the waves, which will be discussed below, the exerted forces will smoothly vary in time causing unpredictable modifications of the trajectory's curvature. Thus the non-deterministic trajectories remain continuously differentiable. Even if a single particle is studied the physical system is not isolated from the rest of the universe.

2. The escort wave associated with a free particle

For free particles the escort wave model is equivalent to the phase wave model presented by Louis de Broglie 1924 in his doctor thesis [1]. Louis de Broglie ascribes to each particle with finite rest mass a periodic phenomenon, which will be called in the following internal oscillation. The angular frequency ω of the internal oscillation is attained by combining Planck's law $E = \hbar\omega$ and Einstein's relation $E = mc^2$ with the total energy E, Planck's constant $\hbar = h/(2\pi)$ and the velocity of light c. A resting particle with mass m_0 and eigen frequency

$$\omega_0 = m_0 c^2/\hbar \tag{1}$$

is always escorted by a wave of frequency ω_0 and infinite wave length. For a particle at rest the escort wave has the form

$$\psi(x,t) = \phi e^{-i\omega_0 t}. \tag{2}$$

The minus sign in the exponent has been chosen in order to obtain a positive energy, if the energy operator $i\hbar\partial/\partial t$ is applied to the wave. The plus sign in the exponent is then associated with the corresponding antiparticle. Particle and wave are coupled to each other by the phase harmony principle, which always tends to adjust the particle's internal oscillation to the oscillation of the escorting wave at the location of the particle.

In the near zone of scattering and diffraction experiments the oscillations of particle and wave are most often out of phase. Thus trajectory and wave will be steadily modified in order to adjust the phases. Although the total deflection of the trajectory due to the particle wave interaction will generally be small the diffraction pattern may be considerably modified.

Figure 1, which summarizes the first two diagrams of de Broglie's thesis, shows the trajectory of a freely moving particle and three phase lines of the escorting wave. The trajectory is subdivided by circles in the distance of the internal oscillation period $c\tau_0 = 2\pi c/\omega_0$. The axes of the particle's rest system (x', ct') and of the laboratory system (x, ct) are tilted with respect to each other. The time coordinate ct as well as the space coordinate x are subdivided into units of the wave period in the laboratory system $c\tau_E = 2\pi c/\omega_E$.

The trajectory coincides with the ct'-axis

$$ct = \frac{c}{v}x, \tag{3}$$

where v is the velocity of the particle. The lines of constant phase run parallel to the x'-axis

$$ct = \frac{v}{c}x. \tag{4}$$

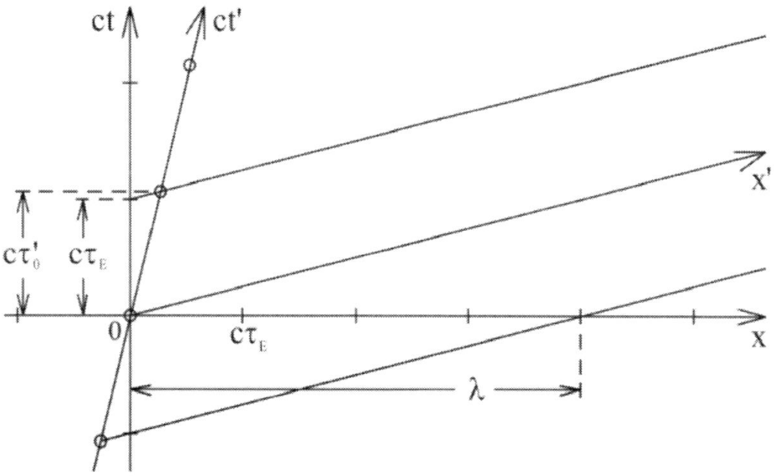

Figure 1. Minkowski diagram of the (steeply rising) trajectory of a free particle and three (slowly rising) phase lines of the escorting wave. The velocity v of the particle is arbitrarily chosen to be $v = c/4$. Details are given in the text.

Thus

$$ct = \frac{v}{c}x + c\tau_E. \tag{5}$$

and

$$ct = \frac{v}{c}x - c\tau_E. \tag{6}$$

denote the upper and lower phase lines in figure 1, respectively.

The period of the particle's internal oscillation measured in the laboratory system is given by

$$c\tau_0' = \frac{c\tau_0}{\sqrt{1-(v/c)^2}}. \tag{7}$$

On the other hand the intercept $c\tau_0'$ is the time component of the intersection of trajectory and upper phase line. Thus $c\tau_0'$ can also be expressed by

$$c\tau_0' = \frac{c\tau_E}{1-(v/c)^2}. \tag{8}$$

Combining (7), (8) and $\omega = 2\pi/\tau$ the frequency ω_E can be written in the form

$$\omega_E = \frac{\omega_0}{\sqrt{1-(v/c)^2}}. \tag{9}$$

Thus the frequency of the escort wave transforms equally like the associated mass

$$m_E = \frac{m_0}{\sqrt{1-(v/c)^2}}. \tag{10}$$

That means the relation $\hbar\omega = mc^2$ is Lorentz invariant.

The wave length of the escort wave in the laboratory system is given by the distance of the crossings of two consecutive phase lines with the x-axis. In figure 1 the middle phase line, which coincides with the x'-axis, crosses the x-axis at the origin. The lower phase line (6) intersects the x-axis at

$$\lambda = c\tau_E \frac{c}{v} = \frac{2\pi\hbar}{m_E v}. \tag{11}$$

The postulated wave length λ of matter waves has experimentally been verified by scattering electrons on nickel crystals [12]. It is traditionally called de Broglie wave length.

For non-relativistic movements the wave frequency $\hbar\omega_E$ may be written in the form

$$\hbar\omega_E = \frac{\hbar\omega_0}{\sqrt{1-(v/c)^2}} \approx \hbar\omega_0(1+\frac{1}{2}\frac{v^2}{c^2}) = m_0c^2 + E_{kin}. \tag{12}$$

Thus the escort wave for a free particle

$$\psi(x,t) = \phi e^{-i(\omega_E t - kx)} \tag{13}$$

with the wave number

$$k = \frac{2\pi}{\lambda} = \frac{m_E v}{\hbar} \tag{14}$$

is a solution of the time-dependent Schrödinger equation

$$(i\hbar\frac{\partial}{\partial t} - m_0c^2)\psi(x,t) = -\frac{\hbar^2}{2m_E}\frac{\partial^2}{\partial x^2}\psi(x,t) \approx -\frac{\hbar^2}{2m_0}\frac{\partial^2}{\partial x^2}\psi(x,t). \tag{15}$$

The mass m_E may be substituted by m_0 because second order terms in v/c may be neglected if the considerations are restricted to non-relativistic problems. Contrary to the traditional notation the energy $E = \hbar\omega_E$ does not only denote the kinetic energy but also comprises the rest energy m_0c^2 of the particle. Thus the rest energy term m_0c^2, which is already known from the Dirac and Klein-Gordon equations, has to be inserted into the time-dependent Schrödinger equation. The time-independent Schrödinger equation remains unchanged because the renormalization of the wave frequency has no influence on the space dependence of the escort wave.

The escort wave (13) only depends on the energy $\hbar\omega_E$ and on the momentum $\hbar k$ of the particle, but not upon the particle's position. The variable x is an argument of the escort wave and does not denote the position of the particle at time t. Therefore a wave is in phase harmony with a particle moving on infinitely many trajectories postponed in space and time.

The escort wave concept combines Edward Nelson's stochastic mechanics [13] with de Broglie's phase wave model. Caused by the fluctuating forces the probability distribution progresses to spread in space just like wave packets in quantum mechanics.

The stochastic velocity component makes sure that the wave escorting a free particle cannot be described by a single frequency but by a (narrow) frequency band. Thus the wave can approximately be written in the form of an eigen differential

$$\Phi_{\Delta\omega}(x,t;\omega) = \frac{1}{\sqrt{\Delta\omega}}\int_{\omega+\Delta\omega/2}^{\omega-\Delta\omega/2} \phi(x,t;\omega')d\omega' \tag{16}$$

with sufficient small $\Delta\omega$. Therewith the wave packed associated with a quantum state in the continuous spectrum has a finite norm. The smearing out of the unique frequency to a frequency band also leads to a finite coherence length of the escort wave.

3. The escort wave associated with a particle moving in stationary potentials

In his doctor thesis Louis de Broglie compared the Fermat principle with the principle of Maupertius. Both principles are variational methods with the aim to find least action rays of waves and least action trajectories of particles, respectively. De Broglie came to the conclusion that the trajectories of particles and the rays of the associated waves agree. That means in the laboratory frame trajectories and phase lines have reciprocal slopes at the crossing points. Thus trajectory and phase lines locally have the same geometry as for free particles.

If no photons are emitted or absorbed the total energy $E = m_0c^2 + E_{pot} + E_{kin}$ of a particle and the frequency ω_E of the escorting wave are constant in stationary potentials. That means the vertical distance $c\tau_E$ of subsequent phase lines is equal everywhere. As for free particles the wave length $\lambda = 2\pi\hbar/(m_E v) \approx 2\pi\hbar/(m_0 v)$ only depends on the momentum $m_E v \approx m_0 v$ of the particle and not on the potential. Relations (9) and (10) can only be valid if the internal frequency ω_0 and the rest mass m_0 are replaced by potential dependent terms

$$\omega_P = \omega_0 + E_{pot}/\hbar \tag{17}$$

and

$$m_P = m_0 + E_{pot}/c^2. \tag{18}$$

That means the rest mass of a particle changes with the potential.

This change of the rest mass with the potential has also been postulated by Louis de Broglie in 1971 [14], when he carefully analyzed the phase harmony principle for particles moving in stationary potentials. Although the change of the rest mass with the potential is not explicitely discussed in textbooks the effect is well known. The mass defect in chemical and nuclear reactions is a direct consequence of the rest mass change in potentials derived from electromagnetic and strong forces, respectively. The change of the rest mass in graviational potentials has already been postulated by Albert Einstein [15] in 1915 and has often been verified in the meantime.

If relation (17) is also valid for very high energies the rest masses of two antiparticles will practically go to zero if they come very close together in case of central collisions. This effect may erraneously simulate extremely high momentum transfers indicating that for electrons and positrons the Coulomb law is still valid in the femtometer region.

If the Schrödinger equation (15) for a free particle is extended to the case of a particle moving in stationary potentials m_0c^2 has to be replaced by $m_Pc^2 = m_0c^2 + E_{pot}$ leading to the equation

$$(i\hbar\frac{\partial}{\partial t} - m_0c^2)\psi(x,t) = \Big(-\frac{\hbar^2}{2m_0}\frac{\partial^2}{\partial x^2} + E_{pot}(x)\Big)\psi(x,t) \tag{19}$$

with $\psi(x,t) = \psi(x)e^{-i\omega_E t}$.

For stationary potentials the application of the energy operator $i\hbar\partial/\partial t$ to the time dependence factor $e^{-i\omega_E t}$ leads to the usual time-independent Schrödinger equation

$$(E - m_0c^2)\psi(x) = \Big(-\frac{\hbar^2}{2m_0}\frac{\partial^2}{\partial x^2} + E_{pot}(x)\Big)\psi(x). \tag{20}$$

or if masses and energies are expressed by frequencies

$$(\omega_E - \omega_0)\psi(x) = \Big(-\frac{c^2}{2\omega_0}\frac{\partial^2}{\partial x^2} + \omega_V(x)\Big)\psi(x) \tag{21}$$

with $\omega_V(x) = E_{pot}(x)/\hbar$.

That means the escort waves associated with particles moving on classical trajectories are solutions of the Schrödinger equation.

Not only the Schrödinger equation but all formulae in quantum mechanics can be written without $\hbar$ if energies and masses are consequently expressed in form of frequencies. For example the uncertainty relation $\Delta x \Delta p \geq \hbar/2$ can be expressed in the form $\Delta x \Delta k \geq 1/2$, which is already well known from classical physics for wave packets.

Obviously Planck's constant has the character of a conversion constant comparable with the Boltzmann constant k_B. When Ludwig Boltzmann and Max Planck succeeded to express the mean energy of particles in thermal equilibrium by $k_B t$ he embedded thermodynamics into classical physics. Textbooks of electrodynamics handle the movement of particles in external fields and the radiation of moving particles but not the interaction of particles with waves radiated by the particles. The new point of view may help to understand quantum mechanics as a completion of classical physics. The mutual interaction of accelerated particles and associated waves leads to the quantization of states and to the intensity redistribution observed in diffraction experiments.

The particle-wave phenomena in microphysics arise from the interplay of classical particles and classical waves, which are coupled together via the phase harmony principle. Thus if Planck's relation $E = \hbar\omega$ is considered as an universal relation of physics quantum mechanics can be understood as an extension of classical physics.

4. Escort waves associated with bound particles

Without fluctuating forces a particle in ground state would permanently stay at rest in the potential minimum and the zero point energy would be zero. Edward Nelson [13] derived the formalism of quantum mechanics from statistical mechanics of point particles without presuming the existence of waves. After having determined the density distribution of particles he formally derived a wave. For ground states stochastic mechanics provides the probability densities and zero point energies known from quantum mechanics.

The trajectories of ground states are continuous but mainly determined by fluctuating forces. For excited states with increasing quantum numbers the stochastic velocity component becomes less and less important. The trajectories approach classical trajectories in agreement with the correspondence principle formulated by Ehrenfest [16] after having compared classical trajectories with the movement of wave packets .

Escort waves of excited states generally enclose standing waves. For three-dimensional central force potentials the wave function can be expressed by a product of three one-dimensional functions, where two of the functions are standing waves.

In the following the consequent application of the phase harmony principle to standing waves will be demonstrated on the $n = 4$ state of a linear harmonic oscillator. The vibration frequency is chosen to be $\omega_{vib} = \omega_E/20$.

Figure 2 shows phase lines of the two counter-propagating travelling waves, which constitute the standing wave, and four of the infinitely many possible classical trajectories. For all these trajectories the particle is in phase harmony with the travelling wave, whose rest system agrees with that of the particle. Because of a special kind of relativistic Doppler effect the frequency of the counter-propagating wave does not agree with the internal oscillation of the particle. The frequency shift by a factor

$$G = \frac{1 + (v/c)^2}{1 - (v/c)^2} \approx 1 + 2\left(v/c\right)^2 \tag{22}$$

prevents the phase coupling of the counter-propagating wave with the particle.

The trajectories in figure 2 are assumed to be deterministic. If the fluctuating velocity component is taken into account particle and wave are most often out of phase. However, driven by the phase harmony principle the particle stays close to the trajectory shown in the figure. If the deviations are too strong the particle will undergo a spontaneous transition to a lower quantum state.

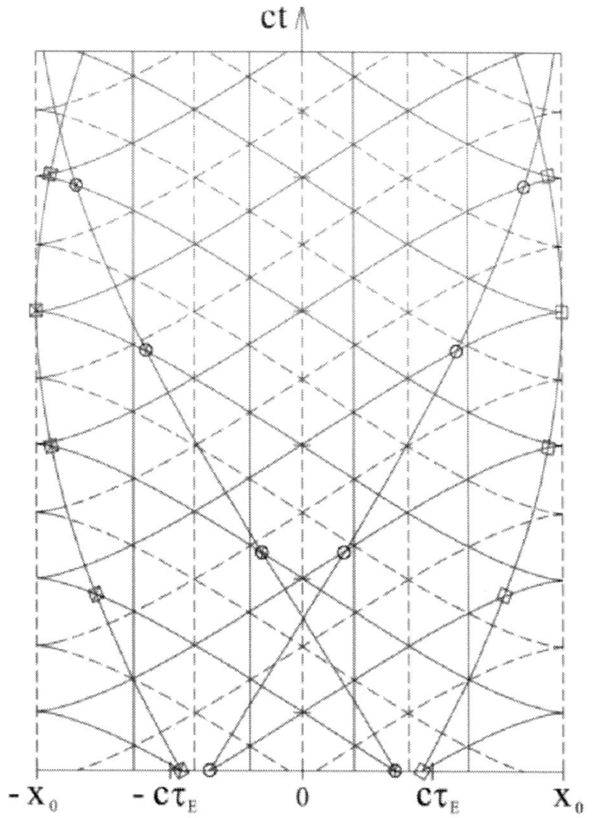

Figure 2. Four trajectories of a particle and the space-time structure of the two counter-propagating waves associated with the back and forth motion of the particle in the $n = 4$ state of a harmonic oscillator. The solid and dashed phase lines connect points with phase factors of 1 and -1, respectively. Both axes of the Minkowski-diagram are divided into units of $c\tau_E$. The distances between identical symbols indicate the particle's eigen period.

The velocities of the particle on its way back and forth are just opposite to each other. Thus their complex valued amplitudes have the same modulus at each location. Because the amplitudes are counter-rotating in the complex plane with opposite angular velocities the sum of both waves has everywhere the same phase. If the residual phase factor is arbitrarily set to one, the standing wave is real everywhere.

The intensity varies between $\psi^2(x)$ and zero, where $\psi(x)/2$ is the amplitude of the travelling waves at the location x. Without fluctuating forces the probability of finding the particle at a given location is identical with the classical probability. The fluctuations lead to a smearing out of the classical probability density, which is quite well approximated by the folding of the classical probability density with a normal distribution. The width of the normal distribution is chosen to be $\sigma = 0.48c/\sqrt{\omega_0\omega_{vib}} \approx 0.62c\tau_E$.

The occurrence of interferences is a characteristic property of waves, which has no counterpart in the probability density of particles. Whereas the standard interpretation of quantum mechanics postulates zeros in the probability density distributions the escort wave concept predicts smoothly varying probability densities.

Soon after the presentation of wave mechanics by Erwin Schrödinger Max Born [17] formulated the hypothesis, that in case of scattering experiments the wave intensity far away from

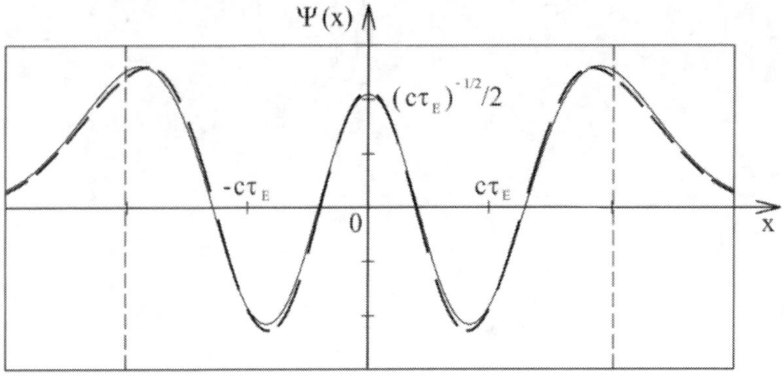

Figure 3. Comparison of the approximated space dependence of the escort wave $\phi_{approx}(x) \sin \varphi_{classic}(x)$ (dashed curve) with the Schrödinger wave function $\psi(x)$ (solid curve) for the $n = 4$ state of the harmonic oscillator. The dashed vertical lines indicate the classical turning points.

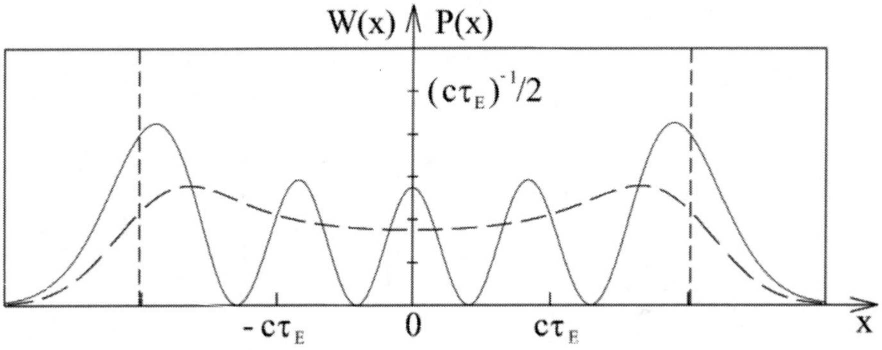

Figure 4. Comparison of the probability density of the particle derived from the approximated escort wave $P(x) = \phi^2_{approx}(x)$ (dashed curve) with the absolute square of the Schrödinger wave function $W(x) = \psi(x)\psi^*(x)$ (solid curve) for the $n = 4$ state of the harmonic oscillator. The dashed vertical lines indicate the classical turning points. Obviously the absolute square of the wave function does not give the probability of finding a particle at a given location.

the scattering zone is proportional to the probability of finding particles at a given scattering angle. This proportionality is one of the best verified laws in physics. Day by day it is confirmed by countless experiments.

Without any experimental evidence the scope of applicability of the ad-hoc assumption has been extended [18] to the probability density in bound states and in scattering zones. However, the application of the hypothesis to standing waves is not at all stringent.

A scattering experiment does not provide the probability density of a bound particle. The result is a matrix element, which contains besides the transition operator only the wave functions of the initial and final state. Thus one does not test the probability density but the amplitudes and the relative phases of the associated escort waves. Until now Born's rule cannot be verified experimentally. Recently there are even doubts on its universal validity [19]. Because Born's rule is an assumption, which has no influence on the formalism of quantum mechanics, it may be left off without any consequences. The rule must only be valid for the asymptotic behaviour of particles in order to properly describe scattering and diffraction experiments.

Experimental findings, which seem to support the probability density interpretation, are often misunderstood. So the intensity variation of the tunnel current on the surface of quantum corrals [20] can be traced back to the local variation of the exit energy. The experiments of Ketterle and his coworkers [21] showing the density modulation of particles going out from two correlated Bose-Einstein condensates can be understood as a modified double slit experiment.

Also other experiments are invoked in order to prove the probability density interpretation. In general they can be easily refused because experimental findings are misinterpreted. For example defocused high energy particle beams exhibit a characteristic density substructure. The observed beam intensity pattern comes from the diffraction of the high energy particles at the apertures arranged in the accelerator tube. In diffraction experiments one observes the intensity far from the diffraction device. In these cases the trajectories run practically parallel to the zero intensity planes. The particles, which are randomly tumbling under the influence of the fluctuating waves, are trapped in the regions of high wave intensity. This argument is no longer valid if particles just cross planes of zero intensity, which happens in the near zone of scattering processes and in excited states.

Sometimes the disturbance of the classical motion leads to a spontaneous quantum transition. But generally the particle will stay in its quantum state because the modified trajectory is in phase harmony with the original escort wave too. A single measurement process can never provide the full information about a quantum state because the particle is accidentally found in one of the infinitely many trajectories of the same class. Only by frequently repeating the measurement with the same initial conditions the structure of the wave can be fully revealed.

Experimentally only the probabilities for possible final states can be determined. The uncertainty principle prevents to find out the actual trajectory. Therefore it makes sense to subsume all trajectories belonging to the same wave to a quantum state $\psi(x)$. Quantum states of bound particles are generally found by solving eigenvalue problems in Hilbert space. In quantum mechanics many mathematically equivalent representations of the Hilbert space formalism are used.

However, the Schrödinger equation has a prominent status, because it is suited to determine the real waves escorting massive particles. Within the Schrödinger formalism the probability for a given final state can be derived from the Schrödinger wave function $\psi(x)$, which is identical with the space dependence of the escort wave. Thus the wave mechanics of Schrödinger is the appropriate method to derive quantum physics from classical physics.

That means, although the actual trajectory cannot be determined it is necessary to postulate well defined trajectories. The assumption that particle and wave coexist and are coupled to each other via the phase harmony principle reveals the proper interpretation of the Schrödinger wave function.

5. Final remarks

In the beginning of the 20th century Rutherford [22] refers in his publication to the atom model of Nagaoka [23] with electrons orbiting the nucleus. This model, which is most often attributed to Rutherford, has soon been abandoned, because in the framework of classical physics the electrons would continuously loose energy. In the (non-classical) escort wave model accelerated electrons do not loose energy, when they are in phase harmony with the associated wave. So the Rutherford model is quite adequate to descibe atomic shells, but one has to bear in mind that the orientations of the trajectories are permanently changing thus forming a rotational symmetric charge cloud around the nucleus.

The extension of the escort wave concept to many particle systems is not trivial. In case of interacting particles the ensemble of waves can no longer be described by a product of single particle waves because the movements of the particles and the associated waves are correlated. Thus the waves escorting massive particles depend on the locations and momenta of all particles involved. These coordinates span the configuration space.

References

[1] de Broglie L 1924 *Recherches sur la théorie des quanta* (Sorbonne Univ., Paris); de Broglie L 1925 The doctor thesis has completely been published in: *Annales de Physique* **3** 22. The doctor thesis translated into English by Kracklauer A F . Can be found on the web site of the foundation Louis de Broglie at the adress http://www.aflb.esnmp.fr/LDB-oeuvres/De_ Broglie _Kracklauer.htm.
[2] Jung K 2009 *Ann. Fond. Louis de Broglie* **34** 143
[3] Zeh D 1970 *Found. Phys.* **1** 69
[4] Zurek W H 2003 *Reviews of Modern Physics* **75** 715
[5] Bohm B 1952 *Phys. Rev.* **85** 166; Bohm B 1952 *Phys. Rev.* **85** 180
[6] de Broglie L 1956 *Une tentative d'interprétation causale et non-linéaire de la mécanique ondulatoire: la théorie de la double solution* (Gauthiers-Villars, Paris) English translation: Non-linear wave mechanics, (Elsevier, Amsterdam, 1960)
[7] de Broglie, L 1978 *Jalons pour une nouvelle microphysique* (Gauthier-Villars, Paris)154
[8] Bohm D and Vigier J P 1954 *Phys. Rev.* **96** 208
[9] Cufara-Petroni N and Vigier J P 1992 *Found. Phys.* **22** 1
[10] Vigier J P 1994 *Found. Phys.* **24**, 61
[11] Gueret P and Vigier J P 1982 *Found. Phys.* **12** 1057
[12] Davisson C and Germer L H 1927 *Phys. Rev.* **30** 705
[13] Nelson E 1966 *Phys. Rev.* **150** 1079
[14] de Broglie L 1971 *La réinterprétation de la mécanique ondulatoire* (Gauthier-Villars, Paris) 24
[15] Einstein A 1916 *Ann. Physik* **49** 769
[16] Ehrenfest P 1927 *Z. Phys.* **A 45** 455
[17] Born M 1926 *Z. Phys.* **37**, 863. English translation in Wheeler J A und Zurek W H eds., 1983 Quantum Theory and Measurement 52-55 , (Princeton University Press)
[18] Born M 1954 *Nobel lecture, Dec. 11.* Born M 1964 *in Nobel lectures Physics 1942-1962* (Elsevier Publ. Comp., Amsterdam)
[19] Khrennikov A 2009 *Found. Phys.* **39**, 997
[20] Crommie M F, Lutz C P and Eigler D M 1993 *Science* **262** 218
[21] Andrews M R, Townsend C G, Miesner H J, Durfee D S, Kurn D M and Ketterle W 1997 *Science* **275**, 637
[22] Rutherford E 1911 *Phil. Mag.* **21** 669
[23] Nagaoka H 1904 *Phil. Mag.* **7** 445

Emission of photons through cavity mirrors in the absence of external driving

A Beige[1], A Capolupo[2], A Kurcz[1], E Del Giudice[3], and G Vitiello[2]
[1]The School of Physics and Astronomy, University of Leeds, Leeds, LS2 9JT, United Kingdom
[2]Dipartimento di Matematica e Informatica and
I.N.F.N., Universitá di Salerno, Fisciano (SA)-84084, Italy
[3]I.N.F.N. - via Celoria 16 - Milano, Italy

Abstract. We present a mechanism of energy concentration in a system composed by an optical cavity and a large number of strongly confined atoms, which cannot be described in the rotating wave approximation. The mechanism consists in the emission of photons through the cavity mirrors even in the absence of external driving.

1. Introduction

Quantum electrodynamics phenomena in cavities are among the most intriguing topics of quantum optics and quantum information. In particular, the spontaneous emission of photons from a number of atoms trapped inside an optical cavity has been analyzed extensively. Experimental results are well described by models based on the rotating wave approximation (RWA). However, in some cases the counter-rotating term neglected in the RWA becomes important. It has been shown, for example, that the counter-rotating term relates to the entropy operator and generates an irreversible time evolution [1].

In this paper, we report on recent results [2] according to which, under proper conditions, quantum systems can leak energy into the environment even in the absence of external driving. This effect is due to the counter-rotating terms in the interaction Hamiltonian which are ignored in the RWA [1].

We analyze single and composite quantum systems consisting of many atoms inside an optical resonator. For a large number of atoms inside the cavity, we find that the stationary state photon emission rate can be experimentally detected. Similar phenomena might contribute to the heating in sonoluminescence experiments [3].

2. The model

We consider the Schrödinger picture and the Born and the dipole approximation. The Hamiltonian H describing the evolution of the system is $H = H_0 + H_{\text{int}}$ with

$$\begin{aligned} H_0 &= \hbar\omega \, s^+ s^- + \sum_{\mathbf{k},\lambda} \hbar\omega_k \, a^\dagger_{\mathbf{k}\lambda} a_{\mathbf{k}\lambda} \, , \\ H_{\text{int}} &= \sum_{\mathbf{k},\lambda} \hbar \left(g_{\mathbf{k}\lambda} \, a_{\mathbf{k}\lambda} + \tilde{g}_{\mathbf{k}\lambda} \, a^\dagger_{\mathbf{k}\lambda} \right) s^+ + \text{h.c.} \end{aligned} \tag{1}$$

and $|\tilde{g}_{\mathbf{k}\lambda}| = |g_{\mathbf{k}\lambda}|$. $a_{\mathbf{k}\lambda}$ are the annihilation operators of photons with polarizations λ and frequencies ω_k. In the case of an optical cavity, $\omega \equiv \omega_c$ is the frequency of its field mode and $s^\pm$ are the photon creation and annihilation boson operators $c^\dagger$ and c. In the case of a large number of tightly confined two-level atoms with states $|0\rangle$ and $|1\rangle$, $\hbar\omega \equiv \hbar\omega_0$ is the energy of the excited state $|1\rangle$ of a single atom and $s^\pm$ are related to the collective raising and lowering operators $S^\pm$, with $[S^-, S^+] = 1$ and $[S^+, S^+] = 0 = [S^-, S^-]$.

Let us consider the case of atoms confined in a region with linear dimensions much smaller than the wavelength of the emitted light: $|\mathbf{k}\cdot(\mathbf{r}_j - \mathbf{r}_i)| \ll 1$. Then all particles have approximately the same $g_{\mathbf{k}\lambda}$ and $\tilde{g}_{\mathbf{k}\lambda}$. The structure of the Hamiltonian H remains the same if we replace s^+s^- in H_0 by σ_3 and $s^\pm$ in $H_{\rm int}$ by $\sigma^\pm$ where $\sigma^\pm$ and σ_3 are given by

$$\sigma^\pm \equiv \sum_{i=1}^N \sigma_i^\pm\ , \quad \sigma_3 \equiv \sum_{i=1}^N \sigma_{3i} \tag{2}$$

with $\sigma_{3i} = \frac{1}{2}\left(|1\rangle_{ii}\langle 1| - |0\rangle_{ii}\langle 0|\right)$, $\sigma_i^+ = |1\rangle_{ii}\langle 0|$ and $\sigma_i^- = |0\rangle_{ii}\langle 1|$ the su(2) spin-like operators of atom i: $\left[\sigma_{3i}, \sigma_i^\pm\right] = \pm\sigma_i^\pm$, $\left[\sigma_i^-, \sigma_i^+\right] = -2\sigma_{3i}$. Assuming that the atoms are initially all in their ground state, under the action of $\sigma^\pm$ and σ_3, they evolve into the states:

$$|l\rangle_{\rm p} \equiv [|0_1 0_2 0_3 \ldots 0_{N-l} 1_{N-l+1} 1_{N-l+2} \ldots 1\rangle + \ldots + |1_1 1_2 \ldots 1_l 0_{l+1} 0_{l+2} \ldots 0\rangle] \Big/ \binom{N}{l}^{1/2} \tag{3}$$

which are the eigenstates of σ_3. The difference between excited and unexcited particles is counted by σ_3, since ${}_{\rm p}\langle l|\sigma_3|l\rangle_{\rm p} = l - \frac{1}{2}N$. For any l we have [4]:

$$\sigma^+\ |l\rangle_{\rm p} = \sqrt{l+1}\sqrt{N-l}\ |l+1\rangle_{\rm p}\ , \qquad \sigma^-\ |l\rangle_{\rm p} = \sqrt{N-(l-1)}\sqrt{l}\ |l-1\rangle_{\rm p}\ . \tag{4}$$

Then $\sigma^\pm$ and σ_3 are represented on $|l\rangle_{\rm p}$ by [5, 6] $\sigma^+ = \sqrt{N}S^+A_s$, $\sigma^- = \sqrt{N}A_sS^-$ with $\sigma_3 = S^+S^- - \frac{1}{2}N$, $A_s = \sqrt{1 - S^+S^-/N}$, $S^+|l\rangle_{\rm p} = \sqrt{l+1}|l+1\rangle_{\rm p}$, and $S^-|l\rangle_{\rm p} = \sqrt{l}|l-1\rangle_{\rm p}$ for any l. The σ's satisfy the su(2) algebra; however, for $N \gg l$, Eq.(4) becomes $\sigma^\pm\ |l\rangle_{\rm p} = \sqrt{N}S^\pm\ |l\rangle_{\rm p}$. This implies that for large N, the su(2) algebra, in terms of $S^\pm$ and $S_3 \equiv \sigma_3$, contracts to the projective algebra e(2) [6] $[S_3, S^\pm] = \pm S^\pm$, $[S^-, S^+] = 1$. The $s^\pm$ operators in the cavity case and in the many atom case are then formally the same.

We assume that the state of a single system $|\varphi\rangle$ at $t = 0$ is known. Moreover, we point out that spontaneously emitted photons leave at a very high speed and cannot be reabsorbed. Thus the free radiation field is initially in a state (denoted by $|\mathcal{O}\rangle$) with only a negligible photon population in the optical regime [7, 8, 9]. In the case of *no* photon emission in $(0, \Delta t)$, the (unnormalised) state vector of system and bath is

$$|\mathcal{O}\rangle|\varphi_{\rm I}^0\rangle = |\mathcal{O}\rangle\langle\mathcal{O}|\ U_{\rm I}(\Delta t, 0)\ |\mathcal{O}\rangle|\varphi\rangle\ . \tag{5}$$

By using second order perturbation theory in the interaction picture we find

$$|\varphi_{\rm I}^0\rangle = [1 - A\ s^+s^- - B\ s^-s^+ - C\ s^{+2} - D\ s^{-2}]\ |\varphi\rangle \tag{6}$$

with

$$\begin{aligned} A &= \int_0^{\Delta t} {\rm d}t \int_0^t {\rm d}t' \sum_{\mathbf{k},\lambda} g_{\mathbf{k}\lambda}\tilde{g}^*_{\mathbf{k}\lambda}\ {\rm e}^{{\rm i}(\omega-\omega_k)(t-t')}\ , \\ B &= \int_0^{\Delta t} {\rm d}t \int_0^t {\rm d}t' \sum_{\mathbf{k},\lambda} g^*_{\mathbf{k}\lambda}\tilde{g}_{\mathbf{k}\lambda}\ {\rm e}^{-{\rm i}(\omega+\omega_k)(t-t')}\ , \end{aligned}$$

$$C = \int_0^{\Delta t} \mathrm{d}t \int_0^t \mathrm{d}t' \sum_{\mathbf{k},\lambda} g_{\mathbf{k}\lambda} \tilde{g}_{\mathbf{k}\lambda}\, \mathrm{e}^{\mathrm{i}(\omega-\omega_k)t+\mathrm{i}(\omega+\omega_k)t'} ,$$
$$D = \int_0^{\Delta t} \mathrm{d}t \int_0^t \mathrm{d}t' \sum_{\mathbf{k},\lambda} g^*_{\mathbf{k}\lambda} \tilde{g}^*_{\mathbf{k}\lambda}\, \mathrm{e}^{-\mathrm{i}(\omega+\omega_k)t-\mathrm{i}(\omega-\omega_k)t'} . \tag{7}$$

In the case of an emission, the (unnormalised) density matrix of the system is

$$\rho_{\mathrm{I}}^{>} = \mathrm{Tr}_{\mathrm{R}} \left[\sum_{\mathbf{k},\lambda} a^\dagger_{\mathbf{k}\lambda} a_{\mathbf{k}\lambda}\, U_{\mathrm{I}}(\Delta t, 0)\, \tilde{\rho}\, U_{\mathrm{I}}^\dagger(\Delta t, 0) \right] \tag{8}$$

where $\tilde{\rho} = |\mathcal{O}\rangle\langle\mathcal{O}| \otimes \rho$ is the initial state of system and bath. Using second order perturbation theory we have

$$\rho_{\mathrm{I}}^{>} = \tilde{A}\, s^- \rho s^+ + \tilde{B}\, s^+ \rho s^- + \tilde{C}\, s^- \rho s^- + \tilde{D}\, s^+ \rho s^+ , \tag{9}$$

with [2]

$$\tilde{A} = 2\mathrm{Re}A\,, \quad \tilde{C} = C^* + \mathrm{e}^{-2\mathrm{i}\omega\Delta t}\, C\,, \quad \tilde{B} = 2\mathrm{Re}B\,, \quad \tilde{D} = D^* + \mathrm{e}^{2\mathrm{i}\omega\Delta t}\, D\,, \tag{10}$$
$$C = D^* = \frac{1}{2} f\, \gamma_{\mathrm{C}}\,, \quad \tilde{C} = \tilde{D}^* = f^*\, \gamma_{\mathrm{C}}\,, \quad f \equiv \mathrm{e}^{\mathrm{i}\omega\Delta t}\, \sin(\omega\Delta t)/\omega\,, \tag{11}$$

and γ_{C} real. The density matrix is obtained by averaging over the subensemble with and the subensemble without photon emission (cf. (6) and (9)) at Δt; we find

$$\begin{aligned} \rho_{\mathrm{I}}(\Delta t) &= \rho - \left[(A\, s^+ s^- + B\, s^- s^+)\, \rho + \mathrm{h.c.} \right] \\ &\quad - \frac{1}{2}\gamma_{\mathrm{C}} \left[(f\, s^{+2} + \mathrm{h.c.})\, \rho + \mathrm{h.c.} \right] + 2\mathrm{Re}A\, s^- \rho s^+ \\ &\quad + 2\mathrm{Re}B\, s^+ \rho s^- + \gamma_{\mathrm{C}} \left[f\, s^+ \rho s^+ + \mathrm{h.c.} \right] . \end{aligned} \tag{12}$$

By using second order perturbation theory we derive the following master equation:

$$\begin{aligned} \dot{\rho} &= -\frac{\mathrm{i}}{\hbar} \left[H_{\mathrm{cond}} \rho - \rho H^\dagger_{\mathrm{cond}} \right] + \mathcal{R}(\rho)\,, \\ \mathcal{R}(\rho) &= \gamma_{\mathrm{A}}\, s^- \rho s^+ + \gamma_{\mathrm{B}}\, s^+ \rho s^- + \gamma_{\mathrm{C}}\, (s^- \rho s^- + \mathrm{h.c.})\,, \\ H_{\mathrm{cond}} &= -\frac{\mathrm{i}}{2}\hbar [\gamma_{\mathrm{A}}\, s^+ s^- + \gamma_{\mathrm{B}}\, s^- s^+ + \gamma_{\mathrm{C}}\, (s^{+2} + \mathrm{h.c.})] + \hbar\tilde{\omega}\, s^+ s^- \end{aligned} \tag{13}$$

with $\gamma_{\mathrm{A}} = 2\mathrm{Re}A/\Delta t$, $\gamma_{\mathrm{B}} = 2\mathrm{Re}B/\Delta t$, and $\tilde{\omega}$ the shifted bare transition frequency.

3. Photon emission from a single quantum system

Let us study the possibility of photon emission from a single quantum system. The probability density $I_\gamma = \mathrm{Tr}(\mathcal{R}(\rho))$ for a photon emission of a system prepared in ρ is

$$I_\gamma = \left\langle \gamma_{\mathrm{A}}\, s^+ s^- + \gamma_{\mathrm{B}}\, s^- s^+ + \gamma_{\mathrm{C}} \left(s^{+2} + s^{-2} \right) \right\rangle . \tag{14}$$

Such a quantity can be computed by considering the time evolution of the expectation values $\mu_1 \equiv \langle s^+ s^- \rangle$, $\xi_1 \equiv \mathrm{i}\langle s^{-2} - s^{\dagger+2} \rangle$, and $\xi_2 \equiv \langle s^{-2} + s^{+2} \rangle$. We obtain a closed set of rate equations,

$$\begin{aligned} \dot{\mu}_1 &= -(\gamma_{\mathrm{A}} - \gamma_{\mathrm{B}})\, \mu_1 + \gamma_{\mathrm{B}}\,, \\ \dot{\xi}_1 &= -(\gamma_{\mathrm{A}} - \gamma_{\mathrm{B}})\, \xi_1 + 2\tilde{\omega}\, \xi_2\,, \\ \dot{\xi}_2 &= -(\gamma_{\mathrm{A}} - \gamma_{\mathrm{B}})\, \xi_2 - 2\tilde{\omega}\, \xi_1 - 2\gamma_{\mathrm{C}}\,. \end{aligned} \tag{15}$$

Setting these derivatives equal to zero, we find that the stationary photon emission rate of a single bosonic system is

$$I_\gamma = \frac{2\gamma_A\gamma_B}{\gamma_A - \gamma_B} - \frac{2\gamma_C^2(\gamma_A - \gamma_B)}{4\widetilde{\omega}^2 + (\gamma_A - \gamma_B)^2} \, . \tag{16}$$

I_γ is zero in the absence of external driving only when $\gamma_B = \gamma_C = 0$. However, this assumption is based on how the integrals in (7) are evaluated and whether relations like $\widetilde{D}(\omega) = \widetilde{C}(-\omega)$ are taken into account or not.

4. Photon emission from a composite quantum system

Let us now analyze the photon emission from a composite quantum system. We consider a large number N of tightly confined atoms inside an optical cavity. The energy of this composite system in the Born and dipole approximations is $H = H_0 + H_{int}$ where

$$\begin{aligned} H_0 &= \hbar\,\omega_c\, c^\dagger c + \hbar\omega_0\, S^+S^- + \sum_{\mathbf{k},\lambda} \hbar\omega_k\, a_{\mathbf{k}\lambda}^\dagger a_{\mathbf{k}\lambda} \, , \\ H_{\rm int} &= \sum_{\mathbf{k},\lambda} \hbar(g_{\mathbf{k}\lambda}\, a_{\mathbf{k}\lambda} + \tilde{g}_{\mathbf{k}\lambda}\, a_{\mathbf{k}\lambda}^\dagger)\, c^\dagger + \sqrt{N}\hbar\, (q_{\mathbf{k}\lambda}\, a_{\mathbf{k}\lambda} \\ &\quad + \tilde{q}_{\mathbf{k}\lambda}\, a_{\mathbf{k}\lambda}^\dagger)\, S^+ + \sqrt{N}\hbar g_c\, (c + c^\dagger)\, S^+ + \text{h.c.} \end{aligned} \tag{17}$$

with g_c, $g_{\mathbf{k}\lambda}$, $\tilde{g}_{\mathbf{k}\lambda}$, $q_{\mathbf{k}\lambda}$ and $\tilde{q}_{\mathbf{k}\lambda}$ being coupling constants. For simplicity, the cavity photon states should be chosen such that g_c becomes real. Proceeding as in the single system case, we have again the master equation (13) but with

$$\begin{aligned} \mathcal{R}(\rho) &= \kappa\; c\rho c^\dagger + N\Gamma\; S^+\rho S^- \, , \\ H_{\rm cond} &= \hbar\Big(\widetilde{\omega}_c - \frac{\rm i}{2}\kappa\Big)\, c^\dagger c + \hbar\Big(\widetilde{\omega}_0 - \frac{\rm i}{2}N\Gamma\Big)\; S^+S^- \\ &\quad + \sqrt{N}\hbar g_c\, (c + c^\dagger)(S^+ + S^-) \, , \end{aligned} \tag{18}$$

where $\widetilde{\omega}_c$ and $\widetilde{\omega}_0$ are the bare atom and cavity frequencies, κ is the cavity decay rate, and Γ is the decay rate of the excited state of a single atom. The crucial difference with respect to the usual Jaynes-Cummings model [10] is the presence of the cS^- and the $c^\dagger S^+$ term in (17) which are zero in the RWA. These operators generate a non-zero stationary state population in excited states and the continuous emission of photons, even without external driving.

To calculate this rate, we neglect γ_B and γ_C, since this assures that no emissions occur in the absence of external driving in the single system case. We obtain the following closed set of the rate equations:

$$\begin{aligned} &\dot{\mu}_1 = \sqrt{N}g_c\eta_1 - \kappa\mu_1 \, , \quad \dot{\mu}_2 = \sqrt{N}g_c\eta_2 - N\Gamma\mu_2 \, , \\ &\dot{\eta}_1 = 2\sqrt{N}g_c(1 + 2\mu_2 + \xi_4) + \widetilde{\omega}_0\eta_3 + \widetilde{\omega}_c\eta_4 - \tfrac{1}{2}\zeta\eta_1 \, , \\ &\dot{\eta}_2 = 2\sqrt{N}g_c(1 + 2\mu_1 + \xi_2) + \widetilde{\omega}_0\eta_4 + \widetilde{\omega}_c\eta_3 - \tfrac{1}{2}\zeta\eta_2 \, , \\ &\dot{\eta}_3 = -2\sqrt{N}g_c(\xi_1 + \xi_3) - \widetilde{\omega}_0\eta_1 - \widetilde{\omega}_c\eta_2 - \tfrac{1}{2}\zeta\eta_3 \, , \\ &\dot{\eta}_4 = -\widetilde{\omega}_0\eta_2 - \widetilde{\omega}_c\eta_1 - \tfrac{1}{2}\zeta\eta_4 \, , \\ &\dot{\xi}_1 = 2\sqrt{N}g_c\eta_4 + 2\widetilde{\omega}_c\xi_2 - \kappa\xi_1 \, , \\ &\dot{\xi}_2 = -2\sqrt{N}g_c\eta_1 - 2\widetilde{\omega}_c\xi_1 - \kappa\xi_2 \, , \\ &\dot{\xi}_3 = 2\sqrt{N}g_c\eta_4 + 2\widetilde{\omega}_0\xi_4 - N\Gamma\xi_3 \, , \\ &\dot{\xi}_4 = -2\sqrt{N}g_c\eta_2 - 2\omega_0\xi_3 - N\Gamma\xi_4 \end{aligned} \tag{19}$$

with $\mu_1 \equiv \langle c^\dagger c \rangle$, $\mu_2 \equiv \langle S^+ S^- \rangle$, $\eta_{1,2} \equiv \mathrm{i}\langle (S^- \pm S^+)(c \mp c^\dagger) \rangle$, $\eta_{3,4} \equiv \langle (S^- \mp S^+)(c \mp c^\dagger) \rangle$, $\xi_1 \equiv \mathrm{i}\langle c^2 - c^{\dagger 2} \rangle$, $\xi_2 \equiv \langle c^2 + c^{\dagger 2} \rangle$, $\xi_3 \equiv \mathrm{i}\langle S^{-2} - S^{+2} \rangle$, $\xi_4 \equiv \langle S^{-2} + S^{+2} \rangle$, and $\zeta \equiv \kappa + N\Gamma$. By using (14) and the stationary state of these equations we obtain the cavity photon emission rate

$$I_\kappa = \frac{N\zeta\kappa g_c^2 \left[8\zeta g_c^2 + \zeta^2 \Gamma + 4\Gamma \left(\widetilde{\omega}_0 - \widetilde{\omega}_c \right)^2 \right]}{16\zeta^2 g_c^2 \widetilde{\omega}_0 \widetilde{\omega}_c + 2\zeta^2 \kappa \Gamma \left(\widetilde{\omega}_0^2 + \widetilde{\omega}_c^2 \right) + 4\kappa\Gamma \left(\widetilde{\omega}_0^2 - \widetilde{\omega}_c^2 \right)^2} \tag{20}$$

which applies for $N\Gamma, \sqrt{N} g_c, \kappa \ll \widetilde{\omega}_0, \widetilde{\omega}_c$. Utilizing the parameters of the recent cavity experiment with ^{85}Rb [11] and $N = 10^4$ we have $I_\kappa = 301$ s^{-1} which can be detected experimentally.

5. Conclusions

In conclusion, by analyzing a system composed by an optical cavity and a large number of tightly confined particles without using RWA approximation, we have shown that even un-excited and un-driven quantum systems might constantly leak energy into their environment. For sufficiently many atoms, a detectable signal can be obtained. This effect is due to non-zero decay rates and to the counter-rotating terms in the interaction Hamiltonian which are neglected in the RWA.

Acknowledgements

This work has been partially supported by the UK Research Council EPSRC, the EU Research and Training Network EMALI, University of Salerno, and INFN.

References

[1] Kurcz A, Capolupo A, Beige A, Del Giudice E and Vitiello G 2010 *Phys. Lett.* A **374** 3726
[2] Kurcz A, Capolupo A, Beige A, Del Giudice E and Vitiello G 2010 *Phys. Rev.* A **81** 063821
[3] Kurcz A, Capolupo A and Beige A 2009 *New J. Phys.* **11** 053001
[4] Beige A, Knight P L and Vitiello G 2005 *New J. Phys.* **7** 96
[5] Holstein T and Primakoff H 1940 *Phys. Rev.* **58** 1098
[6] Shah M N, Umezawa H and Vitiello G 1974 *Phys. Rev.* B **10** 4724
[7] Hegerfeldt G C 1993 *Phys. Rev.* A **47** 449
[8] Dalibard J, Castin Y and Mølmer K 1992 *Phys. Rev. Lett.* **68** 580
[9] Carmichael H 1993 *An Open Systems Approach to Quantum Optics* Lecture Notes in Physics vol 18 (Berlin: Springer)
[10] Shore B W and Knight P L 1993 *J. Mod. Opt.* **40** 1195
[11] Trupke M, Goldwin J, Darquie B, Dutier G, Eriksson S, Ashmore J and Hinds E A 2007 *Phys. Rev. Lett.* **99** 063601

Journal of Physics: Conference Series **306** (2011) 012073 doi:10.1088/1742-6596/306/1/012073

Light speed and the expansion of the universe[1]

Shigeto Nagao[2]
Private, Fujizuka, Kohoku-ku, Yokohama, Kanagawa, 222-0012 Japan
E-mail: snagao@lilac.plala.or.jp

Abstract. Based on the formerly proposed model of the time and universe, a time-course of the light speed along with the expansion of the universe is discussed. Formulas of light propagated distance of a supernova are derived, whose graphs showed an excellent consistency to the observed supernovae data. This outcome verifies that the universe has been expanding at a constant speed by our observed time as the model predicts.

1. Introduction

There is a big mystery that the supernova data seem to imply the universe is expanding in an accelerating manner [1] - [5], for which successful explanation is missing. On the other hand, we unintentionally perceive the time without a clear definition. The author of this article proposed before a model for the development of the universe as well as the definition of a time dimension and what the time we observe passing at a constant speed is [6] [7].

Key aspects of the model are as follows:

- Energy is a vibration in multiple dimensions.
- In order to express a vibration, which is a movement, a dimension tracing values in other dimensions is essential.
- A vibration of energy in an additional space dimension vests an additional energy.
- If a vibration of energy in a certain dimension reaches a steady state for a minimum duration of observation, showing a stationary wave, it can be treated as a particle exhibiting the mass corresponding to the energy of the vibration. We can define that the vibration in the dimension is "*quantized*" and the dimension is "*compactified*" for the vibration.
- The gravitational force is given by the Newtonian equation provided that the mass is not the rest mass but is corresponding to the total energy.
- The space of universe is the area where energy is spread.
- At the Big Bang, the energy distribution transformed to be spread with expansion in 3-dimensional surface of a 4-dimensional sphere in tracing by the original time dimension. A reference image for the expansion is illustrated in Fig. 1.
- Vibrations in other hidden dimensions remain being quantized and the dimensions are compactified for the expansion of the universe.

[1] This paper is a private work independent from the company.
[2] Business Development and Licensing, Nippon Boehringer Ingelheim Co., Ltd.

- Vibrations in the hidden dimensions render intrinsic energy (referred to as the "*Spacia*") to the space.
- A vibration of the intrinsic energy (Spacia) in the 3-D space dimensions vests additional energy, which is our observable energy as light or other quantum particles.
- The time we observe passing constantly (referred to as the "*Time*") is the movement of the space energy in the radius dimension of the 4-D sphere.
- The gravitational attraction of an object from the space energy or the whole universe is equal to all directions and is offset each other. What we observe as a mass of an object is not for the whole energy of the space occupied by the object but is the additional energy vested by the vibration of the space energy in the 3-D space.
- The mass for gravitational interaction for the expansion of the universe is different from the mass for gravitational interaction between particles in the 3-D space.

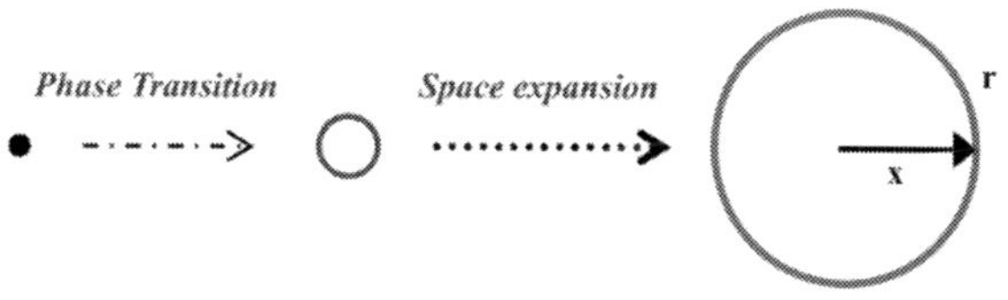

Figure 1. Expansion of the energy distribution of the space

According to this model, the space of the universe should expand at a constant speed with tracing by our observed Time (T) as shown by the equation

$$\frac{dr}{dT} = \frac{dr}{dx} = \theta \ , \tag{1}$$

where **r** denotes an arc vector in the 3-D surface, x denotes the radius of the 4-D sphere, and θ is the angle from the center of universe corresponding to **r** [7].

2. Light speed

This model implies the presence of the medium for light, which contradicts with the two propositions of the special relativity [8]; the principle of relativity and the principle of invariant light speed. The special relativity requires a constant speed of light even when the observer is moving. If there is a medium for light as I propose, light speed is independent from the light emitter's speed but is variable if the observer is moving toward the medium.

In this model, all quantum particles including photon are vibrations of the Spacia in the 3-D space. The propagating speed of the light should depend on the energy density or temperature of the medium, the Spacia. Along with the expansion of the universe, the energy density of the Spacia is getting lower. The light speed would be getting slower either by the time or the Time passing, I expect. A simulation of the light speed by analogy with the speed of sound in air [9] was given in my previous article [6]. Currently, the traveled time period of the light of supernovae is calculated to divide the luminosity distance (fainter) by the constant light speed throughout the whole time period. Therefore, the traveled time of the light of a distant supernova should be calculated longer than the real value. For getting Hubble Plots, the Time back from present when light was emitted should be adjusted by the change of the light speed by the Time. There is a possibility that the real expansion by the Time would be at a constant speed as expected from this model.

If the light is a wave of the Spacia, the Spacia should be equal to the light medium "luminiferous aether", which had been introduced before Einstein released the Special theory of relativity. It is well accepted that the Michelson-Morley experiment did not detect the relative speed of the observer to the aether but it became the most successful experiment having proved the absence of the aether and the constant speed of the light independent from observer's movement [10] - [12]. However, there is a critical mistake in the experimental design itself. Let's check key aspects subjected the presence of the aether (or Spacia).

As the light is a vibration of the medium, its propagation speed is a constant independent from the relative speed of the light emitter to the medium. However, if the observer moves to the medium, it detects the light speed depending on its own speed. In the Michelson-Morley experiment, a coherent light beam is split by a half-mirror into a straight beam and a reflected right angle beam. Both beams are reflected at the end of arms respectively, return to a half-mirror and are combined for detection.

The round traveled time t_a, from the half-mirror light splitter to the mirror and return, for the light beam parallel to the aether wind is given as

$$t_a = \frac{L}{c+v} + \frac{L}{c-v} = \frac{2cL}{c^2 - v^2} , \tag{2}$$

where c is the light speed to the aether, v is the speed of the apparatus toward the aether, and L is the distance from the splitter to the mirror. On the other hand, the round traveled time t_b for the light beam perpendicular to the aether wind is given as

$$t_b = \frac{2L}{\sqrt{c^2 - v^2}} . \tag{3}$$

The difference of the traveled time among the two beams, Δt , becomes as follows.

$$\Delta t = t_a - t_b = \frac{2L}{c^2 - v^2}\left(c - \sqrt{c^2 - v^2}\right) \tag{4}$$

Unless $v = 0$, Δt is greater than 0 and the two beams make an interference fringe. By rotating the apparatus, v varies and a change in the interference fringe should be observed. This was the expectation for the experiment.

In case two beams are split from the identical coherent light, the two beams have the same phase when they are split.

$$\varphi_a(t_0) = \varphi_b(t_0) \tag{5}$$

If there is a difference in time, Δt, to return to the detector, the pair of the respective phases of the beams, which separated at the same instant, can not be combined simultaneously to interfere. In a real experiment, continuous light beam is used. Because the fringe of the two split beams is detected, it is an essential requirement that the respective phases of the two beams should be combined for detector simultaneously. However, the phases, which are combined for interference, should have left the splitter at different time points. One should split Δt earlier than the other.

The parallel beam A takes Δt longer time than the right-angle beam B.

$$t_a = t_b + \Delta t \tag{6}$$

On the other hand, the phase of the beam B combined with the beam A should be released at the splitter Δt earlier than the phase of the beam A is released.

$$t_{0,b} = t_{0,a} + \Delta t \tag{7}$$

The phase of the beam A is equal to that of the beam B when they are combined for detection, independent from the value of v as shown below.

$$\varphi_a = \varphi(t_{0,a} + t_a) = \varphi(t_{0,a} + t_b + \Delta t) \quad (8)$$

$$\varphi_b = \varphi(t_{0,b} + t_b) = \varphi(t_{0,a} + \Delta t + t_b) = \varphi_a \quad (9)$$

As another possibility, if Δt is zero but the traveled distances are different among the two beams, such fringe would be expected. However, the light speed is constant in the medium independent from light emitter's speed. Therefore the traveled distance is identical for both beams because their traveled time periods are equal.

In conclusion, any change in the interference fringes can not be expected theoretically by the Michelson-Morley experiment.

The equation for the light propagation speed by the Time proposed in my previous article [6] is derived as follows. The speed of light should depend on the density of the medium, the Spacia. The sound speed in the air is given as

$$c_{air} = 331.3\ m \cdot s^{-1} \sqrt{\frac{T}{273.15}}, \quad (10)$$

in which T denotes the absolute temperature in kelvins [9, 13]. Speculated by analogy with the speed of sound, the light propagating velocity in the Spacia would be proportional to square root of the energy density or temperature of the Spacia. The energy density is inversely proportional to cube of the size of the space. The light speed by the original tracing time dimension would be given as follows.

$$\frac{dL}{dt} = \frac{K_1}{\sqrt{x^3}} \quad (11)$$

L is the light propagated distance, and x is the radius of the universe. The expansion speed of the radius by the time is shown as follows from energy preservation of potential and kinetic energies based on the Newtonian theory of gravity [6].

$$\frac{dx}{dt} = \pm\sqrt{2GM\,(1/x - K_2)} \quad (12)$$

M denotes the mass of the whole universe. The light speed by the Time is accordingly given as

$$\frac{dL}{dT} = \frac{dL}{dt} \cdot \frac{dt}{dT} = \frac{dL}{dt} \cdot \frac{dt}{dx} = \frac{K_1}{\sqrt{2GMx^3\,(1/x - K_2)}} = \frac{K}{x\sqrt{1 - K_2 x}}. \quad (13)$$

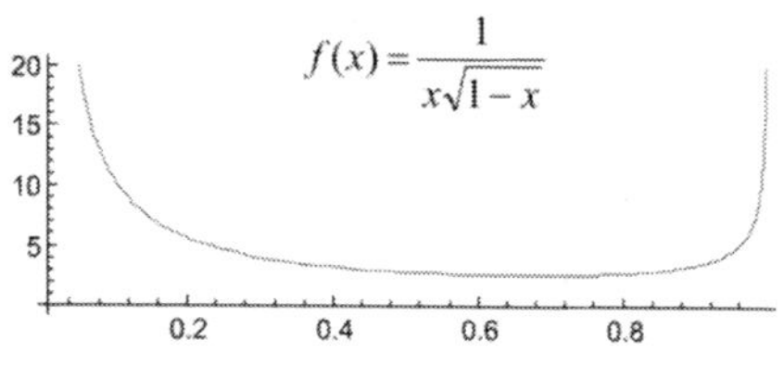

Figure 2. Time-course of light speed by the Time

If we use the unit of Time as 1 for the maximum radius of the universe, the space expansion speed by t at $T = 1$ is zero. Therefore $K_2 = 1$ from the equation (12), and the formula (13) becomes as follows.

$$\frac{dL}{dT} = \frac{K}{x\sqrt{1 - x}} \quad (14)$$

The graph of the Time-course of light speed by the Time is shown in Fig. 2. It has the minimum at $x = 0.75$.

3. Light propagated distance of a supernova

The light speed expected by this model is expressed as the equation (14). Let's see a light propagated distance of a supernova. Let the Time at present be T_P, and the back in Time from T_P when the light was emitted be T_B. The propagated distance L is given as follows from the equation (14).

$$L = \int_{T_P - T_B}^{T_P} \frac{K}{x\sqrt{1-x}} dT = \int_{T_P - T_B}^{T_P} \frac{K}{x\sqrt{1-x}} dx \tag{15}$$

$$\int \frac{1}{x\sqrt{1-x}} dx = \log\left|1 - \sqrt{1-x}\right| - \log\left|1 + \sqrt{1-x}\right| = \log\left|\frac{1-\sqrt{1-x}}{1+\sqrt{1-x}}\right| \tag{16}$$

$$\begin{aligned} L &= K\left(\log\left|\frac{1-\sqrt{1-T_P}}{1+\sqrt{1-T_P}}\right| - \log\left|\frac{1-\sqrt{1-T_P+T_B}}{1+\sqrt{1-T_P+T_B}}\right|\right) \\ &= K \cdot \log\left|\frac{1-\sqrt{1-T_P}}{1+\sqrt{1-T_P}} \cdot \frac{1+\sqrt{1-T_P+T_B}}{1-\sqrt{1-T_P+T_B}}\right| \end{aligned} \tag{17}$$

We don't know yet T_P or K. The graphs of L for various T_P values are shown in Fig. 3.

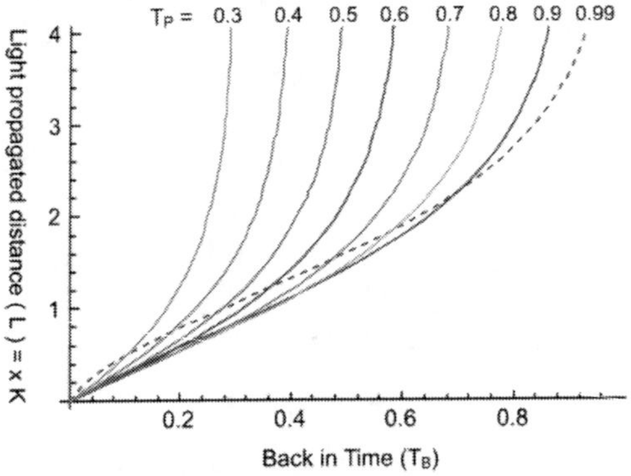

Figure 3. Light propagated distance (L) versus Back in Time (T_B)

Fig. 4 is the Hubble plots from the Supernova Cosmology Project [14]. In order to verify if there is a set of T_P and K values, which nicely matches the supernova observation in Fig. 4, let's modify the axes of Fig. 3. The value 1.0 of the redshift Z corresponds to a half of the present Time T_P as the back in Time T_B because the size of universe is expanded to two fold since the light was emitted. The equation (17) is alternatively expressed by the relative back in Time, which is defined as $T_{BR} \equiv T_B/T_P$, instead of T_B, as follows.

$$L(T_{BR}) = K \cdot \log\left|\frac{1-\sqrt{1-T_P}}{1+\sqrt{1-T_P}} \cdot \frac{1+\sqrt{1-T_P+T_P \cdot T_{BR}}}{1-\sqrt{1-T_P+T_P \cdot T_{BR}}}\right| \tag{18}$$

As for the vertical axis, $L(T_{BR})/L(0.1)$ is used to superimpose the lines for various T_P values in a way that its values fit the observed supernova data in the range of T_{BR} from 0 to 0.1. The relative light propagated distance to $L(0.1)$ versus the relative Back in Time is expressed by the formula (19) and shown in Fig. 5.

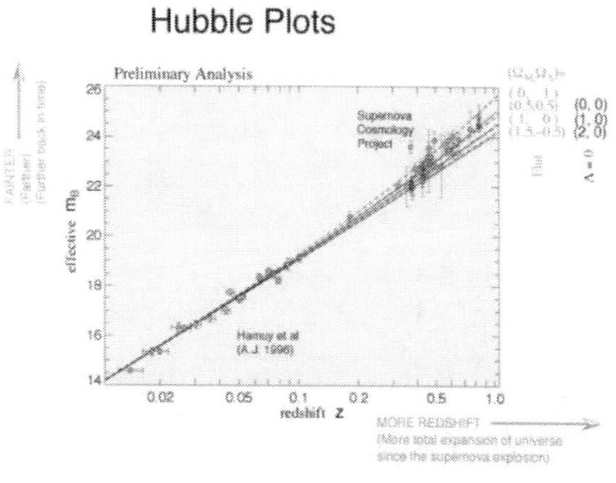

Figure 4. Hubble plots from the Supernova Cosmology Project

$$\frac{L(T_{BR})}{L(0.1)} = \log\left|\frac{1-\sqrt{1-T_P}}{1+\sqrt{1-T_P}}\cdot\frac{1+\sqrt{1-T_P+T_P\cdot T_{BR}}}{1-\sqrt{1-T_P+T_P\cdot T_{BR}}}\right| \times\left(\log\left|\frac{1-\sqrt{1-T_P}}{1+\sqrt{1-T_P}}\cdot\frac{1+\sqrt{1-0.9\cdot T_P}}{1-\sqrt{1-0.9\cdot T_P}}\right|\right)^{-1} \tag{19}$$

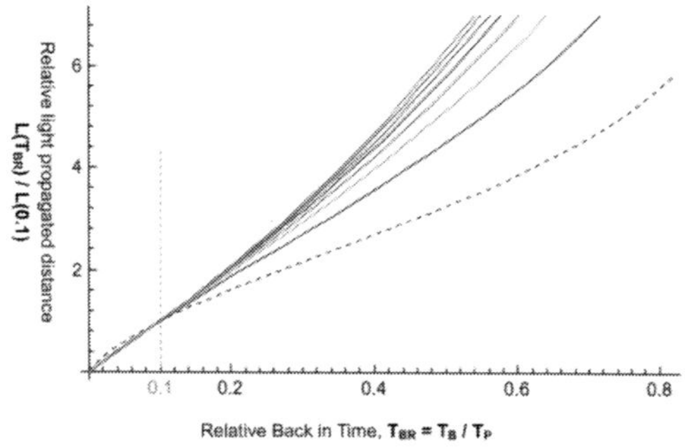

Figure 5. Relative light propagated distance, $L(T_{BR})/L(0.1)$, versus Relative Back in Time, $T_{BR} = T_B/T_P$

Fig. 6 is superimposed graphs of Fig. 5 to Fig. 4 for $T_P = 0.6$, 0.7 and 0.8, respectively. The superimposed line of the light propagated distance for $T_P = 0.7$ gives an excellent consistency to the real observation. If we can find a T_P value, whose graph in Fig. 5 meets the observed supernova data in Fig. 4 for the whole range of the back in Time, we can conclude that the back in Time T_B and the space expansion show a perfect linearity, that is, the universe has been expanding at a constant speed by our observed Time.

Provided that the current Time T_P is roughly 0.7, then the K value is given as follows from the equation (14), where c denotes the current light speed.

$$K \approx \left(0.7\times\sqrt{0.3}\right)\cdot c \approx 0.3834\times c \approx 1.15\times 10^8\ m\cdot s^{-1} \tag{20}$$

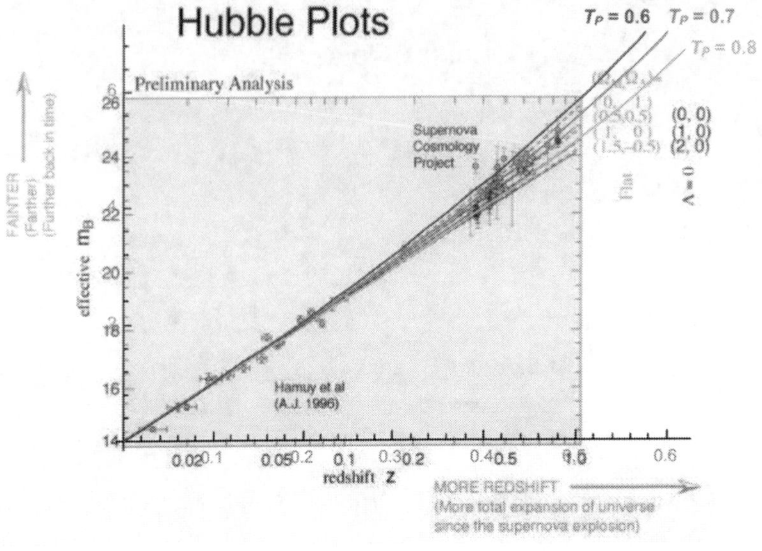

Figure 6. Adjustment lines of light propagated distance versus Time for $T_P = 0.6$, 0.7 and 0.8

Why I insist that the time we observe passing at a constant speed is not the original tracing dimension but the radius dimension of the universe is mainly based on this linearity of the space expansion.

4. Discussion

Many evidences for claiming the constant speed of light propagation are in fact supporting the independence from the speed of the light emitter. Although the special relativity is based on the invariance of the light speed even if the observer of light is moving [8], there might be no evidence that has proven it other than the Michelson-Morley type experiment. As I argued in this article, the experimental design of the Michelson-Morley experiment is theoretically impossible to detect the change of phases between the two split beams.

Most of evidences supporting the general relativity, e.g. the gravitational lensing, should be explained also by the Newton's gravity theory provided that the working mass is not the rest mass but the mass corresponding to the total energy including kinetic and potential energies.

The model of the universe discussed in this and former articles [6] [7] is based on classical physics on gravity with modified interpretation of working mass. It would be rather hypothetical. On the other hand, the definition of a time dimension by analyzing vectors, which is discussed in my preceding article [7], should be theoretical. The combination of the 4-D sphere model with the new interpretation of our observed Time led consequences that meet so beautifully our real observations of the space expansion from supernova data.

As shown in this article, I strongly insist that the universe has been expanding at a constant speed by our observed Time. Consequently, the dark energy should not exist but was introduced due to the lack of adjustment of the Time in back when light was emitted from supernovae, while the adjustment is required from the changing speed of light.

As a testable effect predicted from this model of the time and the universe, I wish professional physicists would investigate in more detail the adjustment line of the back in Time from the light propagated distance for supernovae.

Lastly, it is rather natural to expect that the Dark Matter would be the space energy referred to as the Spacia in this article. Verifying this possibility would be also highly appreciated.

References

[1] Perlmutter S *et al* 1997 *Astrophys. J.* **483** 565
[2] Perlmutter S *et al* 1998 *Nature* **391** 51
[3] Perlmutter S *et al* 1999 *Astrophys. J.* **517** 565
[4] Schmidt B *et al* 1998 *Astrophys. J.* **507** 46
[5] Perlmutter S and Schmidt B 2003 Measuring cosmology with supernovae *Astrophysics* arXiv:astro-ph/0303428
[6] Nagao S 2009 Nature of tracing dimension, imaginary order of freedom and our observed time passing at constant speed *J. Phys.: Conf. Ser.* **174** 012069 doi: 10.1088/1742-6596/174/1/012069
[7] Nagao S 2010 Definition of a time dimension and analysis of metrics for gravitational interactions *private website* http://www3.plala.or.jp/MiTiempo/metrics-MiTiempo.pdf (posted in August 2010)
[8] Einstein A 1905 *Annalen der Phys.* **17** 891
[9] Speed of sound *Wikipedia*
(http://en.wikipedia.org/wiki/Speed_of_sound)(accessed in Nov. 2008)
[10] Michelson A A 1881 *American Journal of Science* **22** 120
[11] Michelson A A and Morley E W 1887 *American Journal of Science* **34** 333
[12] Michelson-Morley experiment *Wikipedia*
http://en.wikipedia.org/wiki/Michelson-Morley_experiment (accessed in July 2009)
[13] Everest F 2001 *The Master Handbook of Acoustics* (New York: McGraw-Hill) pp 262-263
[14] High redshift supernova search *Supernova Cosmology Project*
http://www-supernova.lbl.gov/public/papers/aasposter198dir/wwwposter2b.jpg (accessed in Jan. 2008)

Quantum Description of Electromagnetic Waves in Time-Dependent Linear Media

I A Pedrosa

Departamento de Física, CCEN, Universidade Federal da Paraíba, Caixa Postal 5008, 58059-900, João Pessoa, PB, Brazil

E-mail: iapedrosa@fisica.ufpb.br

Abstract. In this work, we present a quantum description of electromagnetic waves propagating through time-dependent homogeneous nondispersive linear media without sources. By using the Coulomb gauge, we show that this description can be perfomed in terms of a time-dependent quantum harmonic oscillator. In addition, we construct conherent states for the quantized electromagnetic waves and evaluate the quantum fluctuations in coordinate and momentum as well as the uncertainty product for each mode of the electromagnetic field.

1. Introduction

The quantum description of the electromagnetic field is traditionally performed in empty cavities or in free space by associating a quantum-mechanical harmonic oscillator with each mode of the radiation field [1, 2].The behavior of electromagnetic waves is well understood in the case of electromgnetic fields in empty cavities or in free space.However,in the presence of external currents and others sources, the behavior of these waves is not so clear. Recently, the study of the quantum properties of electromagnetic waves propagating through material media has drawn special attention of physicists [3, 4, 5, 6, 7, 8, 9], motived partly by the advent of modern optical materials such as optical fibers and photonic crystal and partly by the growth of experiments on quantum optics processes taking place inside material [6, 8, 9]. Several approaches have been employed to tackle the problem of the quantization of the electromagnetic field inside material media [3, 9], which have mainly considered materials with electric permittivity [5, 6, 10, 11, 12, 13, 14]: (i) real (nondispersive) and either homogeneous or inhomogeneous media; (ii) real and dependent on time and position (time-dependent nonuniform media); or (iii) position and frequency dependent(dispersive inhomogeneous media). It is worth remarking here that, in dispersive media, the inclusion of losses into the system is, in general, introduced by a reservoir (a continuum of harmonic oscillators) which leads to an energy flow from the medium to the reservoir [8, 9, 15]. On the other hand, the problem of conducting media has also been considered [16, 17, 18, 19, 20].In this case,the losses were introduced phenomenologically as a set of time-dependent parameters.

In this work we present a simple and comprehensive quantum description of electromagnetic waves propagating through homogeneous nondispersive linear media, in the absence of sources, with time-dependent electric permittivity and conductivity. Similarly to the case of empty cavities, we show that this quantization can be performed by associating a damped or attenuated quantum-mechanical time-dependent harmonic oscillator with each mode of the electromagnetic field. Afterwards,we construct coherent states for the quantized electromagnetic waves and evaluate the quantum fluctuations in coordinate and momnetum as well as the uncertainty product for each mode of the electromagnetic field.

We organize this paper as follows. In section 2, we investigate classical and quantum properties of electromagnetic waves propagating through time-dependent conducting and nonconducting linear media without charge density.In section 3, we construct coherent states for the quantized electromagnetic waves and evaluate the quantum fluctuations in coordinate and momentum as well as the uncertainty product for each mode of the electromagnetic field.In section 4, we conclude the paper with a short summary.

2. Light propagation in time-dependent linear media

2.1. *Classical description*

In the absence of charge sources,the electromagnetic field dynamics in time-dependent homogeneous conducting linear media is governed by the phenomenological Maxwell's equations.Further,the relations between the fields and current are given by $\vec{D} = \varepsilon(t)\vec{E}$, $\vec{B} = \mu_0\vec{H}$ and $\vec{J} = \sigma(t)\vec{E}$. Here, $\varepsilon(t)$ and $\sigma(t)$ are heuristically introduced as the time-dependent electric permittivity and conductivity, respectively, while μ_0 is the magnetic permeability. In general, the electric permittivity and the magnetic permeability are complex. However, we will restrict our discussion to materials where they are real. This is the case [21, 22], for instance, of poor conductors and other materials for frenquency below the resonant frequency. Since Maxwell's equations are gauge invariant, we are free to choose the most appropriate gauge for our problem. A convenient gauge choice is the Coulomb gauge [1, 2] for which the divergence of the vector potential $\vec{A}$ is zero and the scalar potential is null in the absence of sources. Consequently, both the electric $\vec{E}$ and magnetic $\vec{B}$ fields are determined from the vector potential as

$$\vec{B} = \nabla \times \vec{A} \quad \text{and} \quad \vec{E} = -\frac{\partial \vec{A}}{\partial t}. \tag{1}$$

It is worth mentioning that in the Coulomb gauge the vector potential is purely transverse [1, 2]. Therefore, it is easy to verify that it obeys the damped wave equation

$$\nabla^2 \vec{A} - \mu_0(\dot{\varepsilon} + \sigma)\frac{\partial \vec{A}}{\partial t} - \mu_0\varepsilon\frac{\partial^2 \vec{A}}{\partial t^2} = 0. \tag{2}$$

where the dot represent a time-derivative. From Eq.(2) we notice the appearance of the unusual term $\dot{\varepsilon}$. Consequently, the time dependence of the electric permittivity, which, in principle, may be associated with the internal response of the localized charges to an external perturbation, causes an additional attenuation of the electromagnetic field (for $\dot{\varepsilon} > 0$). Furthermore, for nondispersive nonconducting dielectric media, $\sigma(t) = 0$, the media become absorbing because of the time dependence, just as if it were in contact with a reservoir (time-dependent background medium) or if it were a conductor.

Now, we move our attention to the solutions of Eq.(2). By using the well-known separation of variables method, we write the vector potential in terms of the mode $\vec{u}_l(\vec{r})$ and amplitude $q_l(t)$ functions of each cavity mode [1, 2, 16, 20]

$$\vec{A}(\vec{r}, t) = \sum_l \vec{u}_l(\vec{r})q_l(t). \tag{3}$$

If we substitute Eq.(3) into the damped wave Eq.(2) we get that

$$\nabla^2\vec{u}_l(\vec{r}) + \frac{\omega_l^2}{c_0^2}\vec{u}_l(\vec{r}) = 0, \tag{4}$$

$$\frac{\partial^2 q_l}{\partial t^2} + \frac{\dot{\varepsilon} + \sigma}{\varepsilon}\frac{\partial q_l}{\partial t} + \Omega_l^2(t)q_l = 0, \tag{5}$$

where ω_l is the natural frequency of the mode l, $c_0 = 1/\sqrt{\mu_0\varepsilon(0)}$ is the velocity of light inside the medium at $t = 0$ and $\Omega_l(t)$ is a modified frequency defined as

$$\Omega_l = c(t)\omega_l/c_0, \tag{6}$$

with $c(t) = 1/\sqrt{\mu_0\varepsilon(t)}$ being the velocity of the electromagnetic wave in the time-dependent medium. Here, we notice that the equations of motion for the amplitudes $q_l(t)$ given by Eq.(5) can be directly obtained from the classical Hamiltonian

$$H_l(t) = e^{-\wedge(t)}\frac{p_l^2}{2\varepsilon_0} + \frac{1}{2}e^{\wedge(t)}\varepsilon_0\Omega_l^2(t)q_l^2, \tag{7}$$

where q_l and p_l are canonical conjugated variables, and $\wedge(t)$ is given by

$$\wedge(t) = \int_0^t \frac{\dot{\varepsilon}(\tau) + \sigma(\tau)}{\varepsilon(\tau)} d\tau. \tag{8}$$

Hence, the total Hamiltonian of the electromagnetic field is a sum of individual Hamiltonians corresponding to each mode, that is, $\sum_l H_l$.

Let us now concentrate on Eq.(4). Considering the electromagnitc field to be contained in a certain cubic volme V of nonrefracting media, the mode functions are required to satisfy the transversality condition, $\nabla \cdot \vec{u}_l(\vec{r}) = 0$ and to form a complete orthonormal set [1, 2]. Further, assuming periodic boundary conditions on the surface, the mode function may be written in terms of plane waves as [1, 2, 16, 20]

$$\vec{u}_{l\nu}(\vec{r}) = L^{-3/2}e^{\pm i\vec{k}_l\cdot\vec{r}}\hat{e}_{l\nu}, \tag{9}$$

where $L = V^{1/3}$ is the size of the cube, $|\vec{k}_l| = \omega_l/c_0$ is the wave vector, and $\hat{e}_{l\nu}$ are unit vectors in the directions of polarization ($\nu = 1, 2$), which must be perpendicular to the wave vector because of the transvesality condition.With the spatial mode functions $\vec{u}_l$ completely determined, we only need the canonical variable $q_l(t)$ in order to obtain the vector potential and, consequently, a complete classical description of the electromagnetic field. Then, the electric field confined in the cubic volume of side L can be written as

$$\vec{E}(\vec{r}, t) = \frac{e^{-\wedge(t)}}{\varepsilon_0 L^{3/2}}\sum_l\sum_\nu \hat{e}_{l\nu}e^{\pm i\vec{k}_l\cdot\vec{r}}p_l(t). \tag{10}$$

Here we observe that we have quoted some results of this subsection in a recent paper [23].

2.2. Quantum description

In order to obtain a quantum description of electromagnetic waves propagating in a conducting linear media with time-dependent electric permittivity and conductivity we need to quantize the electromagnetic field. Now as the spatial mode functions $\vec{u}_l(\vec{r})$ are completely determined, the amplitude of each normal mode in Eq.(3) needed to specify a particular field configuration is $q_l(t)$ [1]. Further, for each canonical operator q_l the electromagnetic field is completely specified since $\vec{E}$ and $\vec{B}$ fields operators may be derived by inserting the operator $\vec{A}$ given by Eq.(3) into Eq.(1). So, we move our attention to the canonical operator $q_l(t)$ in order to obtain the vector potential. For this purpose, we must solve the Schrödinger equation associated with the Hamiltonian (7)

$$H_l(t)\Psi(q(t), t) = i\hbar\frac{\partial}{\partial t}\Psi(q(t), t), \tag{11}$$

where p_l is now the moment operator $p_l = -i\hbar\partial/\partial q$ with $[q_l, p_l] = i\hbar$. The solutions of Eq. (11) can be obtained with the aid of the dynamical invariant method devised by Lewis and Riesenfeld [24, 25]. Following this method, we look for a nontrivial Hermitian operator $I_l(t)$ which satisfies the equation

$$\frac{dI_l}{dt} = \frac{1}{i\hbar}[I_l, H_l] + \frac{\partial I_l}{\partial t} = 0. \tag{12}$$

If the exact invariant $I(t)$ (constant of motion), does not contain any time-dependent operator, the Schrödinger equation solutions are straightforwardly written in terms of the orthonormalized eigenfunctions $\phi_n(q_l, t)$ of $I_l(t)$,

$$I_l(t)\phi_n(q_l, t) = \lambda_n \phi_n(q_l, t), \tag{13}$$

and of the phase functions $\beta_n(t)$ as

$$\Psi_n(q_l, t) = e^{i\beta_n(t)}\phi_n(q_l, t). \tag{14}$$

Here, the λ_n are time-independent eigenvalues and the phase functions $\beta_n(t)$ are determined from the equation

$$\hbar\frac{d\beta_n(t)}{dt} = \left\langle \phi_n \middle| i\hbar\frac{\partial}{\partial t} - H_l \middle| \phi_n \right\rangle, \tag{15}$$

with the orthonormality condition $< \phi_{n'}|\phi_n >= \delta_{n'n}$.

Linear invariant operators satisfying Eq.(12) are innumerable [26, 27], but in this paper we are interested in dealing with a quadratic Hermitian invariant. Now,it is known that one such quadratic invariant is given by [28, 29]

$$I_l(t) = \frac{1}{2}\left[\left(\frac{q_l}{\rho_l}\right)^2 + \left(\rho_l p_l - \varepsilon_0 e^{\wedge(t)}\dot{\rho}_l q_l\right)^2\right], \tag{16}$$

where $\rho_l(t)$ is a time-dependent real function satisfying the Milne-Pinney equation [28, 30, 31]

$$\ddot{\rho}_l(t) + \frac{\dot{\varepsilon} + \sigma}{\varepsilon}\dot{\rho}_l(t) + \Omega_l^2(t)\rho_l = \frac{e^{-2\wedge(t)}}{\varepsilon_0^2\rho_l^3}. \tag{17}$$

Next, we look for the eigenstats $\phi_n(q, t)$ of $I_l(t)$. For this purpose, we consider the unitary transformation [28, 29]

$$\phi_n'(q_l, t) = U\phi_n(q_l, t), \tag{18}$$

with

$$U = exp\left(-\frac{i\varepsilon_0 e^{\wedge(t)}\dot{\rho}_l}{2\hbar\rho_l}q_l^2\right). \tag{19}$$

Making use of this transformation, we can rewrite the eigenvalue equation (13) as

$$I_l'\phi_n'(q_l, t) = \lambda_n\phi_n'(q_l, t), \tag{20}$$

where

$$I_l' = UI_lU^\dagger = -\frac{\hbar^2}{2}\rho^2\frac{\partial}{\partial q_l^2} + \frac{1}{2}\frac{q_l^2}{\rho_l^2}. \tag{21}$$

If we now define a new variable $z_l = q_l/\rho_l$, we can express Eq. (20) as

$$\left[-\frac{\hbar^2}{2}\frac{\partial^2}{\partial z_l^2}+\frac{q_l^2}{2}\right]\varphi_n(z_l)=\lambda_n\varphi_n(z_l), \tag{22}$$

where φ_n is related to ϕ'_n by

$$\varphi_n(z_l)=\varphi_n(q_l/\rho_l)=\rho_l^{1/2}\phi'_n(q_l,t). \tag{23}$$

The factor $\rho_l^{1/2}$ has been introduced to satisfy the normalization condition. Therefore, the solution $\varphi_n(z_l)$ of Eq. (22) are the eigenfunctions

$$\varphi_n(z_l)=\left[\frac{1}{\pi^{1/2}\hbar^{1/2}n!2^n}\right]^{1/2} exp\left(-\frac{z_l^2}{2\hbar}\right)H_n\left[\left(\frac{1}{\hbar}\right)^{1/2}z_l\right], \tag{24}$$

with the respective eigenvalues

$$\lambda_n=\hbar(n+\frac{1}{2}). \tag{25}$$

Here H_n is the Hermite polynomial of order n. So,using Eqs.(18), (19), (23) and (24) we find that

$$\begin{aligned}\phi_n(q_l,t) &= \left[\frac{1}{\pi^{1/2}\hbar^{1/2}n!2^n\rho_l}\right]^{1/2} exp\left[\frac{i\varepsilon_0 e^{\wedge(t)}}{2\hbar}\left(\frac{\dot{\rho}_l}{\rho_l}+\frac{ie^{-\wedge(t)}}{\varepsilon_0\rho_l^2}\right)q_l^2\right]\\ &\times\ H_n\left[\left(\frac{1}{\hbar}\right)^{1/2}\frac{q_l}{\rho_l}\right].\end{aligned} \tag{26}$$

The next step is to find the phase function given by Eq. (15). After some basic calculations, we get that

$$\beta_n(t)=-\left(n+\frac{1}{2}\right)\int_0^t\frac{e^{-\wedge(\tau)}}{\varepsilon_0\rho_l^2(\tau)}d\tau. \tag{27}$$

Therefore, we can write the solutions of the Schrödinger equation (11) as

$$\begin{aligned}\phi_n(q_l,t) &= exp[i\beta_n(t)]\left[\frac{1}{\pi^{1/2}\hbar^{1/2}n!2^n\rho_l}\right]^{1/2}\\ &\times exp\left[\frac{i\varepsilon_0 e^{\wedge(t)}}{2\hbar}\left(\frac{\dot{\rho}_l}{\rho_l}+\frac{ie^{-\wedge(t)}}{\varepsilon_0\rho_l^2}\right)q_l^2\right]H_n\left[\left(\frac{1}{\hbar}\right)^{1/2}\frac{q_l}{\rho_l}\right],\end{aligned} \tag{28}$$

with the phase finction $\beta_n(t)$ given by Eq. (27). The above expression represents the exact wave functions for each mode of the electromagnetic field. Finally, we observe that for ϵ and σ constants, the results of this section coincide with those of Ref. [29].

3. Coherent states for the quantized electromagnetic waves

In this section, we construct coherent states for the quantized electromagnetic waves propagating in time-dependent conducting media.In doing so, we introduce the annihilation and creation operators defined as

$$b'_l = \left(\frac{1}{2\hbar}\right)^{1/2}\left[\frac{q_l}{\rho_l} + i\rho_l p_l\right], \tag{29}$$

$$b'^{\dagger}_l = \left(\frac{1}{2\hbar}\right)^{1/2}\left[\frac{q_l}{\rho_l} - i\rho_l p_l\right], \tag{30}$$

with $[b'_l, b'^{\dagger}_l] = 1$. In terms of these operators, the invariant I'_l given by Eq.(21) can be rewritten as

$$I'_l = \hbar(b'^{\dagger}_l b'_l + \frac{1}{2}), \tag{31}$$

whose coherent states have the form [28, 32, 33]

$$\varphi_\alpha(z_l, t) = exp\left(-\frac{|\alpha|^2}{2}\right)\sum_n \frac{\alpha^n}{(n!)^{1/2}} exp\left[i\beta_n(t)\right]\varphi_n(z_l), \tag{32}$$

where α is an arbitrary complex number. Then, using Eqs.(18),(19),(23) and (32) we find that the coherent states for the system described by the Hamiltonian (7) are given by

$$\phi_\alpha(q_l, t) = \frac{1}{\rho_l^{1/2}} exp\left[\frac{i\varepsilon_0 e^{\wedge(t)}\dot{\rho}_l^2}{2\hbar\rho_l} q_l^2\right]\varphi_\alpha(z_l, t). \tag{33}$$

These states satisfy the eigenvalue equation

$$b_l\phi_\alpha(q_l, t) = \alpha_l(t)\phi_\alpha(q_l, t), \tag{34}$$

with b_l and b'_l related by

$$b_l = U^\dagger b'_l U = \left(\frac{1}{2\hbar}\right)^{1/2}\left[\frac{q_l}{\rho_l} + i(\rho_l p_l - \varepsilon_0 e^{\wedge(t)}\dot{\rho}_l q_l)\right], \tag{35}$$

and

$$\alpha_l(t) = \alpha_l exp[2i\beta_0(t)] , \tag{36}$$

$$\beta_0(t) = -\frac{1}{2}\int_0^t \frac{e^{-\wedge(t)}}{\varepsilon_0\rho_l^2(\tau)} d\tau. \tag{37}$$

Note that, in terms of the operator b_l, the invariant in Eq.(16) can be expressed as $I_l = \hbar\left(b_l^\dagger b_l + \frac{1}{2}\right)$. Hence, the expectation values of q_l and p_l in the coherent states $\phi_\alpha(q_l, t)$ are

$$< q_l > = (2\hbar|\alpha|^2\rho_l^2)^{1/2}\sin(-2\beta_0(t) + \delta_l), \tag{38}$$

$$< p_l > = (2\hbar|\alpha|^2)^{1/2}[\varepsilon_0 e^{\wedge(t)}\dot{\rho}_l \sin(-2\beta_0(t) + \delta_l) - \frac{1}{\rho_l}\cos(-2\beta_0(t) + \delta_l)], \tag{39}$$

where δ_l is the argument of the complex number α. Here we note that the expression (38) is a solution of Eq. (5). Hence, the center of the coherent state wave packet follows the motion of a classical particle . Therefore, the above result agrees with the original idea of Schrödinger about the coherent states, who was interested in finding quantum mechanical states which followed the motion of a classical particle in a given potential [34]. Also, we can use the result (39) to show that when the electric field (10) is in a state $(\dot{\phi})_\alpha(q_l, t)$, its expectation value looks like a classical field, as it should be. In what follows we evaluate the quantum fluctuations in q_l and p_l in the state $\phi_\alpha(q_l, t)$. After some algebra we find that

$$< \Delta q >^2 = < q_l^2 > - < q_l >^2 = \frac{\hbar}{2}\rho_l^2, \quad (40)$$

$$< \Delta p >^2 = < p_l^2 > - < p_l >^2 = \frac{\hbar}{2}\left[\frac{1}{\rho_l^2} + (\varepsilon_0 e^{\wedge(t)}\dot{\rho}_l)^2\right]. \quad (41)$$

Thus, the uncertainty product is expresed as

$$(\Delta q_l)(\Delta p_l) = \frac{\hbar}{2}\left[1 + (\varepsilon e^{\wedge(t)}\rho_l\dot{\rho}_l)^2\right]^{1/2}. \quad (42)$$

Here, we observe that the uncertainty relation (42), in general, does not attain its minimum value. This ocurrs because the states $\phi_\alpha(q_l, t)$ correspond to the well-known squeezed states[29, 35, 36, 37, 38, 39].It is also worth mentioning that if the medium is a dielectric material $[\sigma(t) = 0]$ with constant permitivity $[\varepsilon(t) = \varepsilon_0 = cte]$ and if we take the particular solution $\rho_l = (1/\varepsilon_0\omega_l)^{1/2}$ of the Milne-Pinney equation (17), the uncertainty product attains its minimum value. This ocurrs because, in this case, both the Hamiltonian (7) and the coherent states $\phi_\alpha(q_l, t)$ reduce to the ordinary harmonic oscillator model. Finally, we remark that we have quoted some of the results of this section and of the previous subsection 2.2 in [40].

4. Summary

In this work, we have presented a direct and simple quantum description of electromagnetic waves propagating in time-dependent conducting and nonconducting media. We have shown that this description can be performed by associating a damped quantum-mechanical oscillator with each mode of the electromagnetic field. As a consequence, we have established an unification of the procedure to obtain the quantum behavior of electromagnetic waves in empty cavities (or free space) and cavities filled with a (time-dependent) material medium. In the former case, it is usually performed by the association of an ordinary oscillator with each mode of the quantized field, and in the latter one it can be performed by associating a time-dependet harmonic oscillator. We have also constructed coherent states for the quantized electromagnetic waves and have calculated the quantum fluctuations in coordinate and momentum as well as the uncertainty product for each mode of the electromagnetic field. We have seen that this uncertainty product, in general, does not attain its minimum value.Finally, we expected that the approach developed in this work can be useful to investigate subjects related to the quantization of electromagnetic waves propagating in conducting and nonconducting media with material properties varying in time.

Acknowledgments

The author would like to thank CNPq and Capes/nanobiotec for financial support.

References

[1] Louisell W H 1993 *Quantum Statistical Properties of Radiation* (New York: Wiley)
[2] Walls D F and Milburn G J 1995 *Quantum Optics* (Berlin: Springer)
[3] Glauber R J 1991 *Quantum Theory of Particle by Oscillating Fields* (Cambridge, MA USA: Lyman Laboratory of Physics, Havard University)

[4] Hunter B and Barnett S M 1992 *Phys. Rev.* A **46** 4306
[5] Matloob R, Loudon R, Barnett S M and Jeffer J 1995 *Phys. Rev.* A **52** 4823
[6] Santos D J and Loudon R 1995 *Phys. Rev.* A **52** 1538
[7] Gruner T and Welsch D - G 1996 *Phys. Rev.* A **53** 1818
[8] Matloob R 2004 *Phys. Rev.* A **69** 052110
[9] Matloob R 2004 *Phys. Rev.* A **70** 022108
[10] Tip A, Moroz A and Combes J 2000 *J. Phys. Rev.* A **33** 6223
[11] Tip A, Knöll L, Scheel S and Welsch 2001 *Phys. Rev.* A **63** 043806
[12] Tip A 1998 *Phys. Rev.* A **57** 4818
[13] Dung H T, Knöll L and Welsch D 1998 *Phys. Rev.* A **57** 3931
[14] Scheel S, Knöll L and Welsch D 1998 *Phys. Rev.* A **58** 700
[15] Pedrosa I A and Baseia B 1984 *Phys. Rev.* D **30** 765
[16] Choi J R 2003 *J. Opt.* B **5** 409
[17] Crocce M, Dalvit D A R, Lombardo F C and Mazzitelli 2004 *Phys. Rev.* A **70** 033811
[18] Ying Z and Ben-Qing G 2005 *Chin. Phys. Lett.* **22** 446
[19] Choi J R 2006 *J. Phys.* B **39** 669
[20] de Lima A L, Rosas A and Pedrosa I A 2008 *J. Phys.* B **41** 2008
[21] Landau L D, Lifshitz E M and Pitaevskii 1998 *Electrodynamics of Continuos Media*, 2nd ed. (Oxford: Butterworth Heinemmann)
[22] Jackson J D 1975 *Classical Electrodynamics*, 2nd ed. (New York: Wiley)
[23] Pedrosa I A and Rosas A 2009 *Phys. Rev Lett.* **103** 010402
[24] Lewis H R Jr 1967 *Phys. Rev. Lett.* **18** 510
[25] Lewis H R Jr and Riesenfeld W B 1969 *J. Math. Phys.* **10** 1458
[26] Pedrosa I A, Rosas A and Guedes I 2005 *J. Phys. A: Math. Gen.* **38** 7757
[27] de Lima A L, Rosas A and Pedrosa I A 2008 *Ann. Phys.* **333** 2253
[28] Pedrosa I A 1997 *Phys. Rev.* A **55** 3219
[29] de Lima A L, Rosas A and Pedrosa J A 2009 *J. Mod. Opt.* **56** 41
[30] Milne W E 1930 *Phys. Rev.* D **35** 863
[31] Schief W K 1997 *Appl. Math. Lett.* **10** 13
[32] Hartley J G and Ray J R 1982 *Phys. Rev.* D **25** 382
[33] Ray J R 1982 *Phys. Rev.* D **25** 3417
[34] Nieto M M and Simmons L M Jr 1979 *Phys. Rev.* D **20** 1321
[35] Stoler D 1970 *Phys. Rev.* D **1** 3217
[36] Yuen H P 1976 *Phys. Rev.* A **13** 2226
[37] de M Carvalho MM, Furtado C and Pedrosa I A 2004 *Phys. Rev.* D **70** 123523
[38] Walls D F 1983 *Nature* **306** 141
[39] Pedrosa I A 1987 *Phys. Rev.* D 36 1279
[40] Pedrosa I A 2010 Quantum electromagnetic waves in nonstationary linear media, submited to publication

Journal of Physics: Conference Series **306** (2011) 012075 doi:10.1088/1742-6596/306/1/012075

Plausibility of quantum coherent states in biological systems

V Salari[1,2,3,6], J Tuszynski[4,5], M Rahnama[6,7] and G Bernroider[8]

[1] Institut de Mineralogie et de Physique des Milieux Condenses, Universite Pierre et Marie Curie-Paris 6, CNRS UMR7590, France

[2] Boite courrier 115, 4 place Jussieu, 75252 Paris cedex 05, France

[3] BPC Signal, 15 rue Vauquelin, 75005 Paris, France

[4] Department of Experimental Oncology, Cross Cancer Institute, 11560 University Avenue Edmonton, AB T6G 1Z2, Canada

[5] Department of Physics, University of Alberta, Edmonton, AB Canada

[6] Department of Physics, Shahid Bahonar University of Kerman, Kerman, Iran

[7] Afzal Research Institute, Kerman, Iran

[8] Department of Organismic Biology, University of Salzburg, Hellbrunnerstrasse 34, Salzburg, Austria

E-mail: `vahid.salari@impmc.upmc.fr`

Abstract. In this paper we briefly discuss the necessity of using quantum mechanics as a fundamental theory applicable to some key functional aspects of biological systems. This is especially relevant to three important parts of a neuron in the human brain, namely the cell membrane, microtubules (MT) and ion channels. We argue that the recently published papers criticizing the use of quantum theory in these systems are not convincing.

1. Quantum theory and biological systems

Biological systems operate within the framework of irreversible thermodynamics and nonlinear kinetic theory of open systems, both of which are based on the principles of non-equilibrium statistical mechanics. The search for physically-based fundamental models in biology that can provide a conceptual bridge between the chemical organization of living organisms and the phenomenal states of life and experience has generated a vigorous and so far unresolved debate [1,2]. Recently published experimental evidence has provided support for the hypothesis that biological systems use some type of quantum coherence in their functions. The nearly 100% efficient excitation energy transfer in photosynthesis is an excellent example [3]. Living systems are composed of molecules and atoms, and the most advanced physical theory describing interactions between atoms and molecules is quantum mechanics. For example, making and breaking of chemical bonds, absorbance of frequency specific radiation (e.g. in photosynthesis and vision), conversion of chemical energy into mechanical motion (e.g. ATP cleavage) and single electron transfer through biological polymers (e.g. in DNA or proteins) are all quantum effects. Regarding the efficient functioning of biological systems, the relevant question to ask is, how can a biological system with billions of semi-autonomous components function effectively and coherently? Quantum coherence is a plausible mechanism responsible for the efficiency and co-ordination exhibited by biological systems and may play an essential role in answering this question.

The hypothesis invoking long-range coherence in biological systems was proposed by H. Fröhlich [4–6] and followed by detailed investigations by Tuszynski *et al.* [7–21], Pokorný [22–24], Mesquita *et al.* [25–27] and others for over three decades. The possible role played by coherent states manifested outside low temperature physics has attracted considerable interest in both the physics and biology communities. Despite the potential power of quantum mechanics to explain coherent phenomena, there are serious challenges involved in applying it in the context of living systems. For instance, in order to have a very high degree of coherence between bio-molecules, Bose-Einstein condensation may be a viable mechanism but we note that the ambient temperature in a living system is likely to be too high for this phenomenon to occur. Also, the sizes of bio-molecules are very large by physical standards in order to be regarded as typical quantum systems. Moreover, because of the noisy environment according to decoherence theory, quantum states of these mesoscopic bio-molecules would collapse very rapidly. However there is no obvious limitation placed on the Schrodinger equation for its use only in atomic-scale systems. It is a universal equation and it can even be used for the entire universe (as is the case with quantum gravity applications). The boundary between quantum theory and classical physics is still largely unknown. Quantum theory obviously applies on length scales smaller than atomic radii but beyond that it is not clear how exactly at which time- or length-scale it should be superseded by Newtonian mechanics. Superconductors, lasers, superfluids, semiconductors etc., are examples of macroscopic-scale physical systems that behave quantum mechanically, so it is basically feasible that biological systems are effected by quantum principles at least in some of their functions. Here we argue that recent criticisms of the use of quantum mechanics in biology are not very convincing since they ignore the already existing evidence for quantum effects in biological systems. In this paper we investigate three particular systems of special importance: the cell membrane, microtubules (MTs), and ion channels which are some of the most important parts of a neuron in the human brain. We argue that these subsystems are the best candidates for possible sites of quantum effects.

2. A brief overview of the criticism of coherence and decoherence

2.1. Membrane

The original Fröhlich model was very general and did not limit the mechanism of biological coherence to any particular cellular structure. In his model, when the energy supply exceeds a critical level, the dipolar ensemble of biologically relevant molecules populates a steady state of non-linear vibrations characterized by a high degree of structural and functional order [28]. This (electrically polarized) ordered state expresses itself in terms of long-range phase correlations, which are physically similar to such phenomena as superconductivity and superfluidity, where the behaviour of particles is collective and inseparable. The existence of very strong static electric fields across the cell membrane led Fröhlich to consider cellular membranes to be the source of the postulated coherent vibrations.

Fröhlich considered a model for biological systems consisting of three parts:

1) A system of oscillators
2) A heat bath
3) An external source which pumps energy into the system incoherently

Fröhlich [5] derived a formula for the net rate of energy loss L_i^1 of the mode with frequency ω_i (containing n_i quanta) which can be written as

$$L_i^1 = \phi(T)\,(n_i e^{\frac{\hbar\omega_i}{kT}} - 1 - n_i) \tag{1}$$

where $\hbar$ is Planck's constant $(\frac{h}{2\pi})$, k is Boltzmann's constant, T is the temperature of the system and the heat bath, and $\phi(T)$ is a function of T. Recently, Reimers *et al.* [2] have shown that a very fragile Fröhlich coherent state may occur at sufficiently high temperatures and concluded that there is no possibility for the existence of Fröhlich coherent states in biological systems. However, it should be noted that in equation (1) they replaced the function $\phi(T)$ with a constant φ at 0 Kelvin. This replacement causes equation (1) to be plagued by serious problems in the limit

$$T \rightarrow 0,\ \frac{\hbar\omega_i}{kT} \rightarrow \infty,\ e^{\frac{\hbar\omega_i}{kT}} \rightarrow \infty \Rightarrow L_i^1 \rightarrow \infty \tag{2}$$

This in turn implies that the net rate of energy loss tends to infinity near the absolute zero, so it makes this assumption and its consequences unphysical. In the second order considered by these authors, in the case of a three body system, the same problem emerges. Also they have demonstrated several diagrams in terms of effective temperature which was defined by the authors as $T_{eff} = \frac{T_s}{T}$, where T_s is the temperature of system and T is the temperature of the thermal bath. Their effective temperature relation is

$$\frac{T_s}{T} \approx 1 + \frac{s}{\phi} e^{-\frac{\hbar\omega_0}{2kT}} \tag{3}$$

where s, ϕ and ω_0 are constants. Here T_{eff} is not a well-defined parameter because on both sides of the relation in (3) we have temperature of the heat bath T. Therefore, in the limit, both sides of the relation cannot be reconciled with each other:

$$T \rightarrow 0, -\frac{\hbar\omega_0}{2kT} \rightarrow -\infty,\ e^{-\frac{\hbar\omega_0}{2kT}} \rightarrow 0\ \ thus\ \ T_{eff} \rightarrow \infty\ \ while\ \ (1 + \frac{s}{\varphi} e^{-\frac{\hbar\omega_0}{2kT}}) \rightarrow 1 \tag{4}$$

$$T \rightarrow \infty, -\frac{\hbar\omega_0}{2kT} \rightarrow 0,\ e^{-\frac{\hbar\omega_0}{2kT}} \rightarrow 1\ \ thus\ \ T_{eff} \rightarrow 0\ \ while\ \ (1 + \frac{s}{\varphi} e^{-\frac{\hbar\omega_0}{2kT}}) \rightarrow 1 + \frac{s}{\varphi} \tag{5}$$

They have used the effective temperature parameter for the Wu-Austin Hamiltonian [29–31] and considered it in the high temperature limit. Their diagrams are mostly based on the effective temperature parameter and hence are, in our opinion, not acceptable due to the self-contradictory arguments used in their derivations.

2.2. Microtubules

The criticism raised by Reimers *et al.* [2, 36] is mainly directed against the so-called Orch OR model which was proposed by Penrose and Hameroff to introduce a physical basis for consciousness. In some formulations of the OrchOR model, a manifestation of quantum coherence involved Fröhlich coherent states in MTs [32–34]. MTs are highly ordered in the neurons of the brain and can indeed be regarded as good candidates for supporting Fröhlich coherent states. In this context, the conclusions of our discussion above also apply to MTs as well. Therefore, we believe that it is still hypothetically possible to generate Fröhlich coherent states in MTs. However, another issue that arises when considering quantum states for MTs is the rapid *decoherence* problem. The question is "how is it possible for MTs to be in a coherent state while the environment is relatively hot, wet and noisy?" According to decoherence theories any interactions with the environment cause decoherence, which destroys quantum effects [35]. For macroscopic particles there are two main 'natural' ways of experiencing this decoherenc: First, decoherence due to collisions with other particles and second the thermal emission of radiation due to the internal heat of an object [37, 40]. In the latter case, the decoherence time of the system is given by

$$T_{dec} = \frac{1}{10^{36}a^6T^9(\Delta x)^2} \qquad (6)$$

where a is the size of the object (diameter or length), T is the absolute temperature and Δx is the superposition distance of the object [37]. At room temperature $T = 300K$ and a given size of $a = 8nm$ for the tubulin dimer (i.e. the elementary constituent of an MT) and considering $\Delta x = 8nm$ for its superposition distance, the decoherence time would be on the order of $T_{dec} : 10^6 s$. This indicates that thermal photons cannot cause decoherence in this case. Hence we concentrate on the first case which is decoherence due to the scattering by environmental particles. Tegmark [38] has calculated decoherence times for MTs based on the scattering between MTs and environmental particles. Based on the collisions of ions with MTs, he has obtained the decoherence times in the order of:

$$\tau = \frac{D^2\sqrt{mkT}}{Ngq^2} \approx 10^{-13}s \qquad (7)$$

where D is the tubulin diameter, m is the mass of the ion, k is Boltzmanns constant, T is temperature, N is the number of elementary charges in the microtubule interacting system, $g = \frac{1}{4\pi\varepsilon_0}$ is the Coulomb constant and q is the charge of an electron. Hagan *et al.* [39] have shown that Tegmark used wrong assumptions for his investigation of MTs. Another main objection to the estimate in equation (7) is that Tegmarks formula yields decoherence times that increase with temperature contrary to well-established physical laws and the behavior of quantum coherent states. In view of these (and other) problems in Tegmarks estimates, Hagan *et al.* [39] assert that the values of quantities in Tegmarks relation are not correct and thus the decoherence time should be approximately 10^{10} times larger leading to a ms range of values for typical decoherence times. According to Hagan *et al.*, MTs in neurons could possibly avoid decoherence via several mechanisms for quantum processing to occur there.

2.3. Ion channel and selectivity filter (SF)

Ion channels are proteins in the membranes of excitable cells that cooperate for the onset and propagation of electrical signals across membranes by providing a highly selective conduction of charges bound to ions through a channel like structure. The SF is a part of the protein forming a narrow tunnel inside the ion channel which is responsible for the selection process and fast conduction of ions across the membrane. The determination of the atomic resolution structure of the ion channel and selectivity filter by MacKinnon led to the award of the Nobel prize for chemistry in 2003 [41]. The SF is a very narrow region of the protein. In the potassium selective KcsA bacterial channel, that frequently serves as a model structure, this regions extends to about 1.2 nm length and 0.3 nm in diameter. At the atomic scale the filter region exposes negative charges owing to the lone electron pairs from oxygen ions bound to 20 carbonyl groups arranged into 5 rings of the lining 'P-loop' peptide. Alltogether this structure provides a highly ordered atomic coordination pattern among oxygens and ions. If a positive charged ion, such as sodium or potassium enters the SF, the ion can be transiently trapped by Coulombic interactions with the negative charges provided by the oxygen ions. The physical action orders associated with the transient trapping of ions from the macroscopic scale down to the quantum scale have been analysed previously [42]. It has been concluded [43–45] that quantum theory is needed to explain the states of ions in the selectivity filter and hence the function of an ion channel. In contrast, Tegmark has calculated decoherence times for the superposition of ions crossing the entire membrane based on a simple ion-pore diffusion model [38]. The calculated decoherence times for crossing ions in this work are derived from the scattering with environmental particles based on the Coulomb interaction between ions and particles. He has assumed that ions are in a superposition state of inside and outside of the cell and are separated by a distance of 10 nm. In the view of atomic scaled resolution maps and recent molecular dynamics studies of the filter region in this protein, this type of interaction is oversimplified and the pore-diffusion-scattering model is not applicable to describe ion protein interactions. Further, Tegmark introduced a function for the decoherence rate [46] which is composed of two parts: one for short wavelengths and the other for long wavelengths. Every scattering calculation based on the Coulomb interaction and Tegmarks decoherence rate function leads to decoherence times that are proportional to temperature according to relations such as $\tau_{dec} = \sqrt{T}$, $\tau_{dec} = \sqrt{T^2}$, $\tau_{dec} = \sqrt{T^3}$, etc. Therefore, it can be expected that subsequent calculations based on these criteria are flawed in the high-temperature limit, i.e. as temperature approaches infinity, decoherence time increases too, and if temperature approaches absolute zero decoherence time approaches zero, a very unphysical situation. In contrast to Tegmarks conclusions, we suggest that his formulation of the problem cannot apply to the ion-protein coordinated states and his calculations do not address the temperature dependence of decoherence times correctly. Instead it is obvious that this problem remains unresolved and there is no simple and general relation between decoherence time and temperature in ion channels.

Based on a number of physical examples, we may expect biological coherence to occur under special circumstances at physiological temperatures. For example, it is well-known that lasers can maintain their coherence at high temperatures due to external pumping. Moreover, quantum spin transfer between quantum dots connected by benzene rings (the same structures that is found in aromatic hydrophobic amino acids) is more efficient at warm temperature than at absolute zero [47]. Further, a simple calculation of the translocation time of ions through an ion channel if based on a particle point of view, is quite different from what is obtained from experimental data based on X-ray crystallography and MD simulations [48–51]. The difference in the estimates of these translocation times can involve a factor up to 10^2 to 10^5 which indicates that the previous calculations using the particle point of view (i.e. classical physics) should be corrected when applied to the SF of channel proteins. Ions coordinated by the filter carbonyls should be considered as wave-packets rather than particles and their translocation as well as filter states must involve some interferences that account for the observed discrepancies between classical estimations and real observations. Recently, some of the expected consequences behind quantum interferences in the filter region of ion channels have been introduced into the classical equations of motion that lead to Hodgkin-Huxley type membrane potentials. As a result, these 'semi-classical' equations of ion motion are shown to carry the signatures of quantum-interferences [44] and can predict signal onset characteristics that are in accord with real recordings from central brain neurons [52].

3. Conclusion

We have argued above that the objections raised against the feasibility and role of quantum effects in biological systems are not tenable. It is still an open question as to how a macroscopic object is classical while its constituent atoms and molecules are quantum mechanical in nature. As atomic-scale quantum systems compose into large molecules they become classical objects. In biology however, the challenging question is whether and how the initial quantum properties extend into the 'functional domain' of the emerging classical systems. In the case of energy transfer in photosynthesis, magnetic compass sensing with receptor proteins, microtubules and ion conduction in channel proteins, transient quantum coherences may well play a decisive role to explain the observed functional states of classical molecular systems. However, it seems that decoherence theory has not solved this problem, and hence we pose an important question: *"What is the meaning of classicality when a large or complex system (as a quantum system) collapses to become a classical entity while the components (atoms or molecules) are still quantum mechanical?"*. At least in biological systems, we can expect that the emergence of classicality will involve some quantum signatures that cannot be ignored in functional explanations. Answers along this way will most probably play an increasing role for the understanding of organizational complexity and functions in living systems.

Acknowledgments
Research support from NSERC (Canada) awarded to J.A.T. is gratefully acknowledged. We thank Roman Fuchs from the Neurosignaling Unit, Univ. Salzburg, Austria for kindly assisting in preparing the manuscript.

References
[1] Abbott D, Gea-Banacloche J, Davies P C W, Hameroff S, Zeilinger A, Eisert J, Wiseman H, Bezrukov S M and Frauenfelder H 2008 Plenary debate: quantum effects in biology, trivial or not? *Fluct. Noise Lett.*, **8** c5-c26
[2] Reimers J R, McKemmish L K, McKenzie R H, Mark A E and Hush NS 2009 *Proc. Nat Acad. Sci.* **106** 4219-24
[3] Cheng Y C and Fleming G R 2009 Dynamics of light harvesting in photosynthesis *Annu. Rev. Phys. Chem.* **60** 24162
[4] Fröhlich H 1968 *Int. J. Quantum Chem.* **2** 641
[5] Fröhlich H 1970 *Nature* **228** 1093
[6] Fröhlich H 1975 *Proc. Nat. Acad. Sci.* **72** 4211
[7] Tuszynski J A and Dixon J M 2001 Quantitative analysis of the frequency spectrum of the radiation emitted by cytochrome oxidase enzymes *Phys. Rev.* E **64** 051915
[8] Tuszynski J A, Paul R, Chatterjee R and Sreenivasan S R 1984 Relationship between Fröhlich and Davydov models of biological order *Phys. Rev.* A **30** 2666e2675
[9] Tuszynski J A and Paul R 1991 Reply to "comment on 'relationship between Fröhlich and Davydov models of biological order' " *Phys. Rev.* A **43** 3179e3181
[10] Tuszynski J A, Bolterauer H and Satarić M V 1992 Self-organization in biological membranes and the relationship between Fröhlich and Davydov theories *Nanobiol.* **1** 177e190
[11] Tuszynski J A 1985 On the existence of Fröhlich's long-range interactions in dipolar chains *Phys. Lett.* A **107** 225e229
[12] Tuszynski J A 1985 Comments on the Davydov Hamiltonian for quasi-one-dimensionalmolecular chains *Int. J. Quantum Chem.* **29** 379e391
[13] Tuszynski J A 1985 Generalization of Fröhlich's thermodynamic model of the living state phase transition *Phys. Lett.* A **108** 177e182
[14] Tuszynski J A 1987 Biological membranes as dissipative structures *Nucl. Phys.* B *Proc. Suppl.* **2** 618
[15] Tuszynski J A 1988 On possible mechanisms of rouleau formation in human erythrocytes *J. Theor. Biol.* **132** 369e373
[16] Bolterauer H, Tuszynski J A 1989 Fröhlich's condensation in a biological membrane viewed as a Davydov soliton *J. Biol. Phys.* **17** 41e50
[17] Bolterauer H, Tuszynski J A and Satarić M V 1991 Fröhlich and Davydov regimes in the dynamics of dipolar oscillations of biological membranes *Phys. Rev.* A **44** 1366e1381
[18] Chatterjee R, Tuszynski J A and Paul R 1983 Comments on the Bose condensation of phonons in the biological systems *Int. J. Quantum Chem.* **23** 709e712
[19] Paul R, Chatterjee R, Tuszynski J A and Fritz O G 1983 Theory of long-range coherence in biological systems. I. The anomalous behaviour of human erythrocytes *J. Theor. Biol.* **104** 169e185
[20] Paul R, Tuszynski J A and Chatterjee R 1984 Dielectric constant of biological systems *Phys. Rev.* A **30** 2676e2685
[21] Portet S, Tuszynski J A, Hogue C W V and Dixon J M 2005 Elastic vibrations in seamless microtubules *Eur. Biophys. J.* **34** 912e920
[22] Pokorný J 1982 Multiple Fröhlich coherent states in biological systems: computer simulation. *J. Theor. Biol.* **98** 21e27
[23] Pokorný J 1999 Conditions for coherent vibrations in cytoskeleton *Bioelectrochemistry and Bioenergetics* **48** 267e271
[24] Pokorný J 2009 Biophysical cancer transformation pathway. *Electromagnetic Biology and Medicine* **28** 105e123

[25] Mesquita M V, Vasconcellos A R, Luzzi R and Mascarenhas S 2005 Large-scale quantum effects in biological systems *Int. J. Quantum Chem.* **102** 111630
[26] Mesquita M V, Vasconcellos A R and Luzzi R 1996 Near-dissipationless coherent excitations in biosystems *Int. J. Quantum Chem.* **60** 68997
[27] Mesquita M V, Vasconcellos A R, Luzzi R and Mascarenhas S 2005 Systems Biology: An Information-Theoretic-Based Thermo-Statistical Approach *Brazilian Journal of Physics* **34** 459-88
[28] Cifra M, Fields J Z and Farhadi A 2010 Electromagnetic cellular interactions, Progress in Biophysics and Molecular Biology, doi:10.1016/j.pbiomolbio.2010.07.003, (2010).
[29] Wu T M and Austin S 1977 *Phys. Lett.* A **64** 1512
[30] Wu T M and Austin S 1978 *Phys. Lett.* A **65** 746
[31] Wu T M and Austin S 1981 *J. Bio. Phys.* **9** 97107
[32] Penrose R and Hameroff S R 1995 *J. Conscious. Stud.* **2** 98
[33] Hameroff S R and Penrose R 1996 *J. Conscious. Stud.* **3** 36
[34] Hameroff S R 1998 *Philos. Trans. R. Soc.* A **356** 1869
[35] Rosa L P and Faber J 2004 Quantum models of the mind: are they compatible with environment decoherence? *Phys. Rev.* E **70** 031902.
[36] McKemmish L K, Reimers J R, McKenzie R H, Mark A E and Hush N S, 2009 Penrose-Hameroff orchestrated objective-reduction proposal for human consciousness is not biologically feasible *Phys. Rev.* E **80** 021912
[37] Schlosshauer M 2007 *Decoherence and Quantum-to-Classical Transition* (Berlin: Springer)
[38] Tegmark M 2000 *Phys. Rev.* E **61** 4194-206.
[39] Hagan S, Hameroff S R and Tuszynski J A 2002 Quantum computation in brain MTs: decoherence and biological feasibility *Phys. Rev.* E **65** 061901.
[40] Joos and Zeh, Z. Phys. B, 59, 223-243, (1985).
[41] MacKinnon R 1994 *Biophys. J.* **66** 1061-7
[42] Bernroider G and Roy S 2004 Quantum-classical correspondence in the brain: scaling, action distances and predictability behind neural signals *FORMA* **19** 55-68
[43] Bernroider G and Roy S 2005 Quantum entanglement of K ions, multiple channel states and the role of noise in the brain *Proc. SPIE* **5841** 205-14.
[44] Summhammer J and Bernroider G 2007 Quantum entanglement in the voltage dependent sodium channel can reproduce the salient features of neuronal action potential initiation *Preprint* arXiv:0712.1474v1
[45] Vaziri A and Plenio M B 2010 *New J. Phys.* **12** 085001
[46] Tegmark M 1993 *Found. Phys. Lett.* **6** 571
[47] Shrivastava I H, Tieleman D P, Biggina P C and Sansom M S P 2002 *Biophys. J.* **83** 633-45
[48] Shrivastava I H and Sansom M S P 2000 *Biophys. J.* **78** 557-70
[49] Shrivastava I H and Sansom M S P 2002 *Eur. Biophys. J.* **31** 207-16
[50] Guidoni L and Carloni P 2002 Potassium permeation through the KcsA channel: a density functional study *Biochimica et Biophysica Acta* **1563** 1-6
[51] Guidoni L, Torre V and Carloni P 1999 *Biochemistry* **38** 8599-604
[52] Naundorf B, Wolf F and Volgushev M 2006 *Nature* **440** 1060

AUTHOR INDEX

AUTHOR INDEX